# The Litmus Test for Flowers

*T*he beautiful colors that abound in the plant world are the result of the presence of organic compounds that absorb certain wavelengths (frequencies) of visible light and reflect the rest. The light that humans can perceive ranges from deep violet (short wavelengths) through blue, green, yellow, and orange to deep red (long wavelengths). In most cases, different compounds are responsible for different colors— for example, the reds and yellows of autumn leaves are due to several compounds. These compounds are always present in the leaves, but their colors are masked by the green of chlorophyll during the growing season. On the other hand, a single compound called cyanidine is responsible for both the red of a poppy and the blue of a cornflower. In the acidic sap of the poppy, cyanidine exists as the cationic (positively charged) part of a salt. In this form, it absorbs mainly blue and green light and reflects red light, which is perceived by our eyes. In the basic sap of the cornflower, cyanidine exists as a zwitterion (a neutral species with equal numbers of positive and negative charge centers). In this form, cyanidine absorbs red and green light; we see the cornflower as blue.

*Reproduction of a woodblock print, in the authors'
collection, from a book published in 1497 in Basel,
Switzerland, depicting an alchemist and his two
assistants, one working at the "fume hood" and the other
taking a sample from the cask. The alchemist is holding a
retort, an all-in-one distillation apparatus in which the
long snout serves as the condenser. Another retort is in use
in the fume hood, and a third one is on the floor.*

# Organic Chemistry

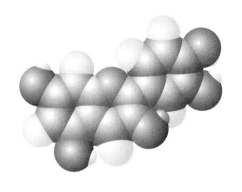

SECOND EDITION

Marye Anne Fox

James K. Whitesell

The University of Texas
*Austin, Texas*

Jones and Bartlett Publishers

*Sudbury, Massachusetts*

Boston   London   Singapore

*Editorial, Sales, and Customer Service Offices*
Jones and Bartlett Publishers
40 Tall Pine Drive
Sudbury, MA 01776
508-443-5000
info@jbpub.com
http://www.jbpub.com

Jones and Bartlett Publishers International
Barb House, Barb Mews
London W6 7PA UK

Acquisitions Editor: Christopher W. Hyde
Senior Production Administrator: Mary Sanr
Manufacturing Manager: Dana L. Cerrito
Design: Deborah Schneck
Editorial Production Service: Lifland et al., Bokmakers
Illustrations: JAK Graphics Ltd.
Typesetting: York Graphic Services
Cover Design: Hannus Design Associates
Printing and Binding: R.R. Donnelley & SonCompany
Cover Printing: Henry N. Sawyer Co., Inc.
Cover Photographs: © Jeff Marc, PHOTO/NATS, Inc.; © North Wind Picture Ahives



**Library of Congress Cataloging-in-Publication Data**

Fox, Marye Anne, 1947–
Organic chemistry / Marye Anne Fox, James K. Whitesell. —2nd ed.
p. cm.
Includes index.
ISBN 0-7637-0178-5
1. Chemistry, Organic. I. Whitesell, James K. II. Title.

QD251.2.F69 1997
547—dc21

96-46985
CIP

*Art and Photo Credits:* **p. 28:** © Ben Phillips, PHOTO/NATS, Inc. **p. 98:** © E. F. Anderson, Visuals Unlimited. **p. 135:** © George Calef, Masterfile. **Fig. 4.22, p. 188:** Courtesy of GE Medical Systems. **Fig. 4.23, p. 189:** Courtesy of Matthew C. Ohlson, Central Texas Imaging Center, Austin. **p. 190:** Spectra reprinted with permission from Garrett and Grisham, BIOCHEMISTRY, Harcourt Brace College Publishers, 1995. **Fig. 4.24, p. 191:** Reprinted with permission from S. L. Dieckman et al., 1992. *J. Am. Chem. Soc.,* 114:2717. Copyright 1992 American Chemical Society. **p. 202:** Courtesy Professor Frank N. Bash, Director, McDonald Observatory, University of Texas. **p. 253 (upper):** © Ann Reilly, PHOTO/NATS, Inc. **p. 253 (lower):** © Sidney Karp, PHOTO/NATS, Inc. **p. 258:** © S. S. Groet/Stock Boston. **p. 292:** © Carolina Biological Supply Company/Phototake–NYC. p. 472: Reprinted from *Neblette's Handbook of Photography and Reprography, Seventh Edition,* John M. Sturge, Ed., p. 310, Figure 12-67 (part a), New York: Van Nostrand Reinhold, 1977. **p. 550 (top):** © Sidney Karp, PHOTO/NATS, Inc. **p. 550 (middle):** © Coni Kaufman, Index Stock Photography. **p. 558:** © David M. Stone, PHOTO/NATS, Inc. **p. 800:** Photo © 1996 Zaleski; spectrum courtesy of Dr. Ben Shoulders. **p. 825:** Drawing of alpha helix by Linus Pauling and photo of Pauling both courtesy of Dorothy Munro, Linus Pauling Institute of Science and Medicine. **Fig. 16.12, p. 827:** Redrawn with permission from *Chemistry in Britain,* October 1989, p. 107. **p. 17-16:** © Mary Walker, PHOTO/NATS, Inc. **p. 17-39:** © Sidney Karp, PHOTO/NATS, Inc. **p. 18-26:** © James Beveridge, Visuals Unlimited. **p. 19-13:** © Bob DeGoursey, Visuals Unlimited. **p. 20-23:** © Runk/Schoenberger, Grant Heilman. **Fig. 20.13, p. 20-40:** Courtesy of Professor Brent Iverson, University of Texas. **p. 22-26:** © Andre Jenny, Unicorn Stock Photos. **Fig. 23.10, p. 23-23:** Courtesy of Victor Lorian, M.D., Bronx Lebanon Hospital Center. **p. 23-25:** © Runk/Schoenberger, Grant Heilman. **Endpaper (left):** © North Wind Picture Archives. **Endpaper (right):** © Hal Horwitz, PHOTO/NATS, Inc.

Printed in the United States of America
01 00 99 98 97 10 9 8 7 6 5 4 3 2 1

P 566 Acetanilide
P 554 → 557
diazotisation

# Brief Contents

# Contents

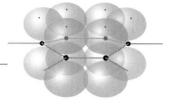

*Chapter* $2$ — Alkenes, Aromatic Hydrocarbons, and Alkynes  *43*

CHEM TV®

*Chapter* $3$ — Functional Groups Containing Heteroatoms  *87*

CHEM TV®

## Chapter 4

Chromatography and
Spectroscopy    155

## Chapter 5

Stereochemistry    *223*

CIM CD ▶

CIM CD ▶    CHEM TV ▶
CIM CD ▶

*Chapter* **6** Understanding Organic Reactions  *279*

**Chapter 7**  Mechanisms of Organic Reactions  *331*

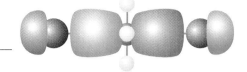

  CHEM TV®

*Chapter* **8**

Substitution by Nucleophiles at *sp*$^3$-Hybridized Carbon *385*

 CHEM TV®

*Chapter* **9**

Elimination
Reactions   435

# Chapter *10*  Addition to Carbon–Carbon Multiple Bonds    485

  ⊕CHEM TV®

**xiii**

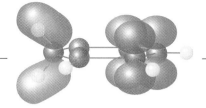

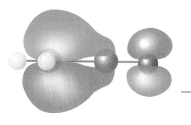

# Chapter *13* Substitution Alpha to Carbonyl Groups  *659*

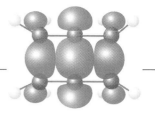

# Chapter *16*   Polymeric Materials   *789*

CHEMISTRY IN MOTION   CHEM TV®

*Chapter* **19** Noncovalent Interactions and Molecular Recognition *19-1*

 CHEM TV®

*Chapter* **20**    Catalyzed Reactions    *20-1*

 CHEM TV®

# Chapter *21* Cofactors for Biological Reactions    *21-1*

# Chapter 22  Energy Storage in Organic Molecules    22-1

# Chapter *23*  Molecular Basis for Drug Action    *23-1*

 CHEM TV®

CIM CD ▶

# Preface

Each year, most of the thousands of students who finish a first course in organic chemistry clearly express their dissatisfaction with what they have learned. They convey their displeasure both vocally and, even more persuasively, by "voting with their feet"—that is, by not enrolling in other advanced science courses. Ask a typical group of such students what was wrong with their course and you will hear the same answer that this query draws from deans of medical schools, from educational psychologists who specialize in the instruction of mathematics and science, from university administrators, and even from many instructors of the courses: all say that a typical organic chemistry textbook contains too much information, much of which is excruciatingly detailed, disconnected from "real life," irrelevant to other parts of a technical or liberal education, and just plain boring. Even the strongest students can emerge from a year of organic chemistry without a good picture of what a practicing organic chemist does.

## Concentrating on Fundamentals

Adopting a "less is more" philosophy, we have tried in this book to address each of these common criticisms in an intellectually demanding year-long introductory course.

- First, the course is developed as a "story," with each chapter containing only those topics and reactions that are needed to understand the intellectual roots of organic chemistry as it is currently practiced.

- Second, specific examples are included at each stage to illustrate familiar, concrete uses of the chemistry under discussion.

- And, third, the material is intended to enhance the student's appreciation of the significance of chemistry in other science and preprofessional courses, in undergraduate research in a modern organic chemistry laboratory, and in industrial and biomedical research.

In attempting to accomplish these objectives, we have had to take an approach that is substantially different from that in virtually all other currently available organic texts. Like most synthetic chemists, we began by working backward. We first asked ourselves what topics a well-informed student should understand after a one-year course in organic chemistry. We consulted extensively with health-profession faculty and with chemists of every stripe (industrial and academic, synthetic and mechanistic, material and biological), both in the United States and abroad. These conversations confirmed our initial supposition that an understanding of polymer chemistry, naturally occurring compounds, energy conversion and storage within organic molecules, molecular recognition and information transfer, modes of action of natural and artificial catalysts, and design criteria for new materials and biologically active molecules is of key importance if a student is to comprehend the contributions of organic chemistry to civilization. Most currently available texts, if they treat these topics at all, do so only as brief subsidiary applications rather than as intrinsic intellectual goals of the course.

Providing greater coverage of these topics, however, meant that something else would have to go, if we were to adhere to our first objective of concise presentation.

- We have tried to remove redundancy, believing that it is unnecessary, for example, to treat the complex metal hydride reductions of aldehydes, ketones, esters, and amides as four separate, seemingly unrelated reactions. This approach has required that we move away from the functional-group organization that has been widely used since the early 1960s as a means of tabulating reactions—an organization that has become unwieldy, owing to the ongoing development of large numbers of new reagents.

- We have tried to exercise restraint in choosing which chemical topics and reactions to include. Only those reactions that recur in the book's unfolding chemical story are retained, along with closely related ones that illustrate basic chemical principles and mechanisms for these essential reactions. We reasoned that good pedagogy does not oblige us to include every chemical topic and detail known to either of us. Rather, we sought to identify those topics absolutely required to reach our objective of giving the student sufficient information to understand the principles and practice of modern organic chemistry.

## Organic Chemistry, Second Edition: A Unique Organizational Structure

These goals led to an organizational structure that begins with seven chapters that deal primarily with the three-dimensional structures of various organic functional groups (Chapters 1 through 5) and the relation between structure and reactivity, from both a thermodynamic point of view and a kinetic one (Chapters 6 and 7). As soon as the student has been exposed to the range of organic functional groups, spectroscopy is introduced (Chapter 4) to facilitate work in the laboratory. The next seven chapters (Chapters 8 through 14) deal with specific reaction types, organized by common mechanism rather than by functional group. These chapters are followed by an integrative chapter (Chapter 15) that incorporates these reactions into strategies for planning the synthesis of new compounds. Finally, Chapters 16 through 23 illustrate how the structural features considered in the first part of the book, together with the specific reactions covered in the second part, can be sources of insight into the chemical structure and function of important naturally occurring and manufactured materials: polymers, proteins, and enzymes. We use examples to show how these materials accomplish specific chemical conversions in biological systems by molecular recognition, catalysis, and energetic coupling with cofactor conversions, and conclude by describing the function of pharmaceutical agents in the last chapter.

This textbook presupposes only the knowledge of chemistry typically attained in a high school course or in the first semester of standard college chemistry. If the curriculum requires it, the self-contained course presented in this book can be offered in the freshman year, without the quantitative development provided by a one-year general chemistry course. The topics covered here afford a solid basis for a description of common natural organic phenomena, which might effectively instill in students a greater enthusiasm for the more abstract topics of introductory physical chemistry.

## Tools for Student Success

Apart from organizing the text itself in a better way, we have included a number of learning aids and motivational stimulants.

- Each chapter contains exercises for testing immediate mastery of the concepts in a section, as well as end-of-chapter problems that help to integrate the concepts in the chapter as a whole. Both the exercises and the problems range in difficulty from those that provide basic reinforcement of a concept to those that require the student to apply the concept to a new situation. We have written detailed answers for the exercises and problems, preparing the *Study Guide and Solutions Manual* ourselves to ensure that the explanations given in the manual correspond with the presentation of concepts in the text.

- Each chapter contains boxed material—short stories relating the practical utility of the reactions and materials being considered.

- Each chapter includes a summary of the principal ideas of importance in the chapter. These summaries, together with lists of important topics in the *Study Guide and Solutions Manual,* are intended to help the student recognize and learn the main concepts presented in a chapter. Most chapters contain a list of reactions that are new to the chapter, and Chapters 8 through 14 also include tables that group the reactions considered according to what they accomplish as synthetic transformations.

- The book includes a comprehensive index that allows easy access to a given topic, if reinforcement is needed when it is discussed in a new context in a later chapter.

- Finally, a glossary of key terms is included in the text, supplying a definition and a citation to the chapter and section in which a term is introduced and developed. A chapter-by-chapter glossary is provided in the *Study Guide and Solutions Manual* to assist the student in preparing for examinations; the definitions constitute an additional means of reviewing the concepts developed in each chapter.

We hope that students will enjoy and benefit from the experience of learning modern organic chemistry as it is presented in this book. We will be grateful indeed to our readers for their evaluation of our work.

## What's New in This Edition?

In this second edition, we have incorporated significant revisions in response to the many positive comments we received from faculty who used the first edition. In making these changes, however, we have adhered resolutely to the intellectual objectives that originally motivated us to write an organic textbook: we maintain that the functional-group approach used in most organic texts no longer serves as an appropriate framework for teaching the fundamental concepts of organic chemistry. Instead, we believe that a thorough understanding of a small number of key principles intrinsic to the study of the structure and reactions of carbon-based compounds provides a much better basis for retaining this knowledge base and extending it to practical applications in other areas of science.

To help realize our objectives, we made the following changes:

■ Large portions of the text have been rewritten to make them more readable for the lower-level college student. Material has been added to motivate students and to emphasize the instructor's key role in the learning process.

■ Many new exercises and problems have been added to each chapter. In addition, an extensive set of supplementary problems now augments the problems at the end of each chapter. Solutions for these supplementary problems have been intentionally omitted from the *Study Guide and Solutions Manual,* so that instructors can assign them for take-home exams or graded homework sets. The added exercises, problems, and supplementary problems have a broad range of difficulties and call for skills ranging from simple algorithmic manipulations and lower-order responses through tests of higher-order cognitive skills.

■ Because Chapter 8 (on nucleophilic substitution) was deemed too long by many adopters of the first edition, it has been extensively revised, with enolate chemistry moved to an entirely new Chapter 13. This latter chapter is perhaps the best example of how a mechanism-based approach can bring together related subjects that are artificially separated and disconnected in a functional-group approach.

■ Chapter 23 (Molecular Basis of Drug Action) has been significantly expanded and now includes discussions of the chemistry underlying viral infections and cancer, as well as chemical treatments for these disease states.

■ The number of chemical highlights (now called Chemical Perspectives) has been significantly increased. Many of our students have commented that these chemical asides helped them correlate organic chemistry with their everyday lives and motivated them to stick with their study.

■ Ball-and-stick as well as space-filling models have been incorporated throughout the text to help the student appreciate the three-dimensional structures of molecules. These were created with Chem3D Pro® (Cambridge Scientific) from structures obtained by energy-minimized molecular mechanics calculations. Representations of molecular orbitals were derived from semi-empirical calculations using the AM-1 basis set with the Cache® suite of calculation programs. These models are thus state-of-the-art three-dimensional representations of the relevant structures and their molecular orbitals.

- CHEMISTRY IN MOTION™ icons throughout the book indicate a figure or illustration that comes to life in short animations created by Jim Whitesell and Mika Hase on the CHEMISTRY IN MOTION CD-ROM.

- The number of pages in the text has increased with the additional exercises and problems, the expanded Chemical Perspectives, and the extensive use of molecular representations. The chemical content, however, has remained essentially the same, consistent with what we believe can be covered realistically in a one-year course.

## Customized for You

Recognizing new trends in the curriculum and the desire of some faculty and students for a more manageable text, Jones and Bartlett, the publisher of this text, now offers you choices that allow you to customize a package to meet your specific needs.

- **Organic Chemistry, Second Edition** (ISBN 0-7637-0178-5). As outlined on pages xxvi and xxvii.

- **Core Organic Chemistry** (ISBN 0-7637-0367-2). Consists of Chapters 1 through 16 from *Organic Chemistry, Second Edition*. Instructors seeking a truly "less is more" approach will be well served by this intellectually demanding introduction to organic chemistry, which is quite suitable for use in full during a year-long course.

- **Chem Modules.** If you adopt Fox and Whitesell's *Core Organic Chemistry,* you can create a customized text that includes only the advanced topics in organic chemistry that *you* teach in *your* course. Creating a set of chem modules that match the content and sequence of your course is fast and easy. Simply choose one or more of the seven chapters (modules) that address advanced concepts in organic chemistry (Chapters 17 through 23 in *Organic Chemistry, Second Edition*). Jones and Bartlett will package these modules with the *Core Organic Chemistry* text. Your students assemble the modules into a single, easy-to-use *Chem Modules* book that serves as the perfect complement to *Core Organic Chemistry. Chem Modules* are the ideal solution for instructors who want to teach the core concepts of organic chemistry and only a few of the advanced topics. Available chapters cover naturally occurring compounds (Chapters 17 and 18), noncovalent interactions and molecular recognition (Chapter 19), catalyzed reactions (Chapter 20), cofactors and energy storage in biological systems (Chapters 21 and 22), and the chemical basis for drug action (Chapter 23). For details, ask your Jones and Bartlett representative, or visit the Fox and Whitesell home page at http://www.jbpub.com

## Supplementary Material

Various supplementary materials are available to assist instructors and aid students in mastering organic chemistry:

- **Study Guide and Solutions Manual.** Written entirely by the authors, Marye Anne Fox and James K. Whitesell, this manual contains key concepts, answers to questions, and solutions to problems. It includes the *Nucleophile/Electrophile Reaction Guide* by Dr. Donna Nelson of the University of Oklahoma, which facilitates students' recognition of patterns in these reactions. The *Study Guide and Solutions Manual* is available free to instructors; students can purchase a version of the manual for either *Organic Chemistry, Second Edition* (ISBN 0-7637-0413-X) or *Core Organic Chemistry* (ISBN 0-7637-0440-7).

- **Test Bank.** This evaluation tool, prepared by the authors, contains more than 600 questions, with at least twenty-five questions per chapter. Available free to instructors.

- **Electronic Test Bank.** An electronic version of the test bank that instructors can use to prepare customized tests is available for Windows and Macintosh operating systems.

- **Lecture Success CD-ROM.** The Lecture Success CD-ROM is an easy-to-use instructional device that contains many figures from the text, including their full captions. Images are arranged by chapter, topic, and figure number. It is designed as a lecture demonstration aid that replaces traditional transparency masters.

- **CHEMISTRY IN MOTION™ CD-ROM.** Included on the inside front cover of the text, this CD is an invaluable tool that helps students better visualize challenging concepts. CHEMISTRY IN MOTION icons throughout the text indicate figures and illustrations that come alive in short animations on this CD. In addition to the animations, more than 500 practice problems are provided.

- **CHEM TV®.** This visualization aid, by Dr. Betty Luceigh of the University of California at Los Angeles, and MECHANISMS IN MOTION, by Dr. Bruce Lipschitz of the University of California at Santa Barbara, may be available through your college or university's chemistry department. CHEM TV icons are placed throughout the text to match discussion with animation appearing on the CD that can be used effectively both in lecture and by the individual student. If your department does not already own copies of this visualization tool, consider acquiring a copy to aid students.

- **Reaction Flash Cards.** This set of preprinted flash cards has the reactants and reagents on the front of a card and the products on the back for all of the reactions covered in Chapters 8–14. A convenient way for students to learn reactions as they are encountered and to test their knowledge as they study.

## Acknowledgments

Preparing an organic chemistry text that departs so markedly from the traditional pedagogical approach of the past three decades has been a fascinating experience that has been significantly aided by the very useful and detailed criticisms of a number of reviewers, whose names are given below. We are indeed grateful to each of them. Their comments were universally helpful; any errors or deviations from their advice are our own responsibility.

Shelby R. Anderson, *Trinity College*
Eric Anslyn, *University of Texas, Austin*
Steven W. Baldwin, *Duke University*
Tadgh Begley, *Cornell University*
Eric Block, *State University of New York, Albany*
Erich C. Blossey, *Rollins College*
William T. Brady, *University of North Texas*
John I. Brauman, *Stanford University*
Keith Brown, *University of Saskatchewan*
Jared A. Butcher, Jr., *Ohio University*
William D. Closson, *State University of New York, Albany*
Imre Csizmadia, *University of Toronto*
Dennis P. Curran, *University of Pittsburgh*
William P. Dailey, *University of Pennsylvania*
Kurt Deshayes, *Bowling Green State University*
Vera Dragisich, *University of Chicago*
Graham W. L. Ellis, *Bellarmine College*
Jacqueline Gervay, *University of Arizona*
Warren Giering, *Boston University*
Rainer Glaser, *University of Missouri*
Joseph J. Grabowski, *University of Pittsburgh*
Roland Gustafsson, *Umeå Universitet, Sweden*
Martha A. Hass, *Albany College of Pharmacy*
Henk Hiemstra, *University of Amsterdam*
Robert Hoffman, *New Mexico State University*
Jack Isidor, *Montclair State University*
Eric N. Jacobsen, *Harvard University*
William P. Jencks, *Iowa State University*
George L. Kenyon, *University of California, San Francisco*
John L. Kice, *University of Denver*
Doris Kimbrough, *University of Colorado, Denver*
Jack F. Kirsch, *University of California, Berkeley*

G. J. Koomen, *University of Amsterdam*
Paul J. Kropp, *University of North Carolina, Chapel Hill*
David M. Lemal, *Dartmouth College*
Robert Lemieux, *Queens University, Canada*
Marshall W. Logue, *Michigan Technological University*
David Magee, *University of New Brunswick*
Kenneth L. Marsi, *California State University, Long Beach*
Daniell L. Mattern, *University of Mississippi*
Mark L. McLaughlin, *Louisiana State University*
Maher Mualla, *Adrian College*
Donna J. Nelson, *University of Oklahoma*
Ernest G. Nolen, *Colgate University*
Daniel J. O'Leary, *Pomona College*
Daniel J. Pasto, *University of Notre Dame, Indiana*
C. Dale Poulter, *University of Utah*
Suzanne Purrington, *North Carolina State University*
Harold R. Rogers, *California State University, Fullerton*
David A. Shultz, *North Carolina State University*
William M. Scovell, *Bowling Green State University*
Donald Slavin, *Community College of Philadelphia*
Richard T. Taylor, *Miami University, Ohio*
William P. Todd, *State University of New York, Brockport*
Guy Tourigny, *University of Saskatchewan*
Christopher T. Walsh, *Harvard Medical School*
Walter W. Zajac Jr., *Villanova University*
Charles K. Zercher, *University of New Hampshire*

We are also deeply grateful for the highly professional developmental editing of Philippa Solomon and copyediting of Leona Greenhill. Chris Hyde's tireless promotion of the first edition and efforts as editor for the second edition are also appreciated. The production team, directed by Mary Sanger at Jones and Bartlett and Jane Hoover at Lifland et al., Bookmakers, has been extraordinarily helpful in transforming crude copy into a visually appealing textbook. The moral support and direction of Dave Phanco will always be deeply appreciated.

Special recognition goes to Hal Rogers of the California State University at Fullerton for his important work as an accuracy reviewer of both the text and the *Study Guide and Solutions Manual*. Finally, we wish to thank all our colleagues who adopted the first edition and to acknowledge those who provided extensive reviews of the first and second editions, offering invaluable suggestions and comments.

## Tricks of the Trade—A Special Message to the Student

Mastering organic chemistry is likely to be among the most stimulating learning experiences you will have at an undergraduate level. Being able to understand the structures and functions of new synthetic molecules, as well as naturally occurring ones, will enable you to appreciate the excitement of this fascinating science. Yet, because of its reputation as a difficult course, organic chemistry is sometimes regarded with apprehension. You can take several steps, however, to help ensure success:

- Prepare adequately for lectures. This means reading the material in the text before it is presented in a lecture. This is crucial if you are to ask intelligent questions about a topic.

- Attend lectures regularly. You should set out to extract as much information as possible from your instructor. After difficult lectures, review your notes carefully; you may find it helpful to consult with classmates.

- Do the in-chapter exercises conscientiously while proceeding through each chapter. You should work the exercises on your own and then consult the *Study Guide and Solutions Manual* to confirm your answers.

- Work the end-of-chapter problems promptly after finishing each chapter. This activity, together with diligence in working the in-chapter exercises, will help assure that you integrate the concepts in the chapter as a whole.

- Use the Review of Reactions, Summary, and tables of synthetic reactions to review what you have learned in each chapter.

- Design your own learning aids. Everyone has a personal learning style and techniques. You should develop learning aids that suit your style—perhaps using molecular models to visualize structures or compiling a set of index cards to review important reactions.

- Seek additional assistance. You should take advantage of your instructor's or teaching assistant's office hours, participate in recitation or help sessions, seek supporting materials (such as handouts, sample tests, and computer programs), review audio or video tapes, and form regularly scheduled discussion groups.

- Make the most of laboratory experiences. Actually working with organic compounds when you have prepared sufficiently for the experiments reinforces the utility of the reactions learned.

*Marye Anne Fox*
*James K. Whitesell*

# Structure and Bonding in Alkanes

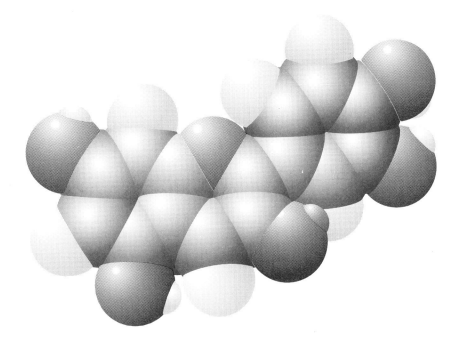

The compound called cyanidine is responsible for the colors of the flowers on the cover of this book—both the red of the poppy and the blue of the cornflowers. The difference in the colors results from a variation in the pH of the sap of the flowers. The model above represents a cyanidine molecule as it exists in the acidic sap of the poppy; in the basic sap of the cornflower, one proton (shown in green) is removed. The loss of this single proton causes a significant shift in the compound's absorption of visible light, producing a different color.

$O$rganic chemistry—what is it? And why is a full-year course devoted to the subject? Chemistry is the study of the properties and transformations of matter. Organic chemistry, a subset of chemistry, is concerned with compounds that contain the element carbon. The astounding number and complexity of these compounds is due to the bonding characteristics of carbon: carbon can form bonds to as many as four other atoms. It can bond with other carbons to form long chains composed of hundreds, even thousands of atoms. Carbon can form stable bonds with atoms of many different elements in the periodic table. It can form different types of bonds—single, double, and triple. The diversity of carbon-based chemistry is not so surprising in view of the differences in the forms of elemental carbon: diamond, graphite, and the newly discovered fullerenes. Diamond is hard and colorless; graphite is soft and black; and fullerenes are dark blue. The differences in the properties of these forms correspond to differences in structure (Figure 1.1).

The most fascinating aspect of organic chemistry is that it is the chemistry of life. Indeed, the very name *organic chemistry* reflects the old belief that certain substances could be produced only by living organisms. Chemists now know that what these substances produced by living things have in common is that they all contain the element carbon. The creatures that form the web of life are diverse, but both their structures and the energy that powers them are based on organic chemistry. And the fundamental chemistry operating in single-cell organisms is the same as that operating in human cells.

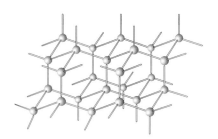 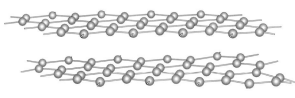 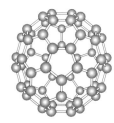

**FIGURE 1.1**

Three-dimensional representations of a subunit of diamond (left), a subunit of graphite (middle), and a fullerene (right).

## 1.1

## The Development and Study of Organic Chemistry

As a result of the explosive growth of scientific knowledge in the twentieth century, scientists now have a good understanding of the complex chemistry of living systems. But how did the science of organic chemistry get to this point?

The beginnings of human association with organic chemistry arose long before the time of Christ. Throughout the ancient world, people were familiar with organic materials, their uses and transformations. Soap was produced from animal fats and plant oils, and wood tar, a resin prepared from charcoal, was an important article of trade. The crystalline sugar sucrose, obtained from sugar cane, and plant extracts for flavorings and perfumes were also valued by the ancients. So was the dye Tyrian purple, used to color the clothes of rulers.

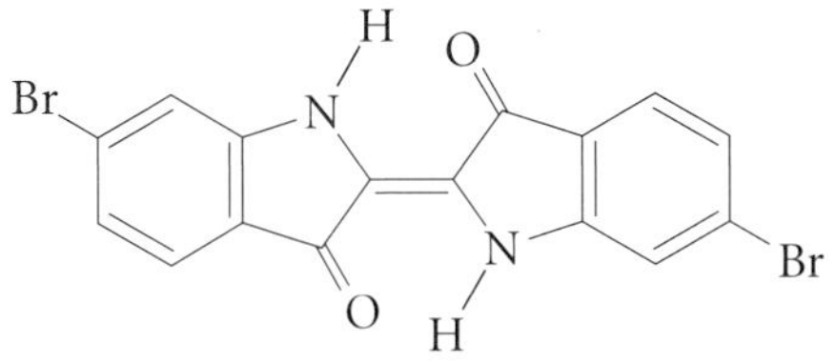

**Tyrian purple**

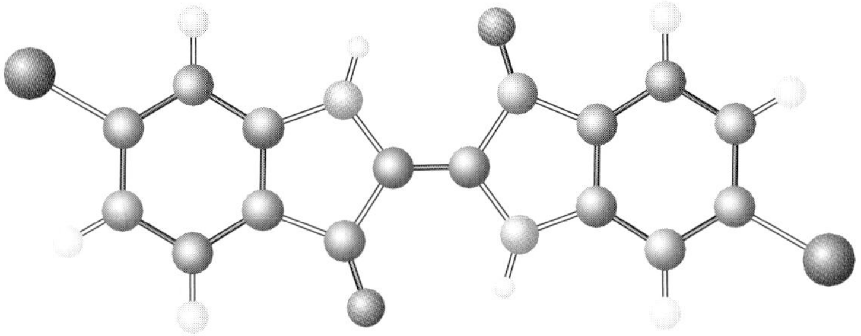

Red cloth was also greatly esteemed, and a small insect (*Coccus cacti* L.), when crushed, provided the deep red compound carminic acid.

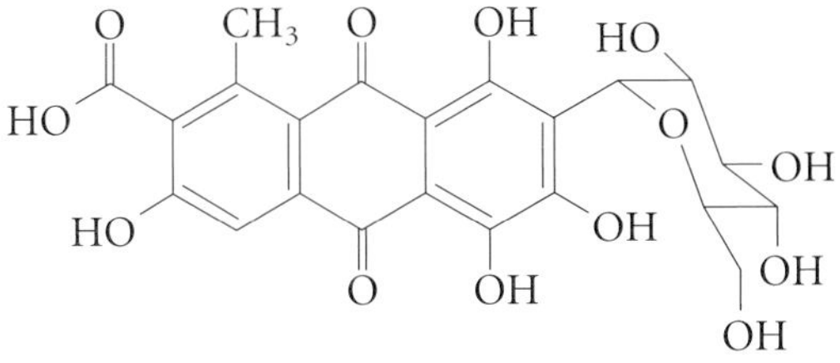

**Carminic acid**

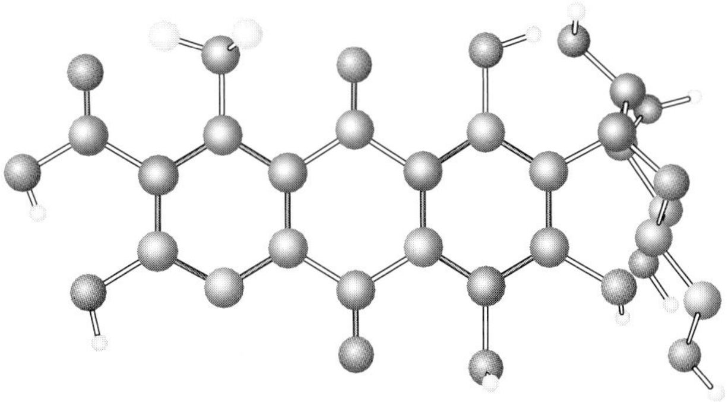

Poisons, too, were known—for example, coniine was derived from poison hemlock (*Conium maculatum*) and was taken by Socrates when he was sentenced to die.

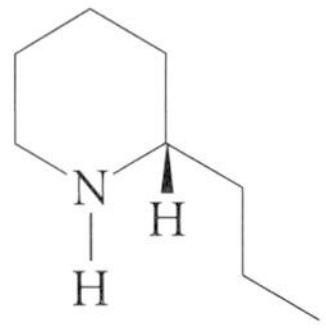

**Coniine**

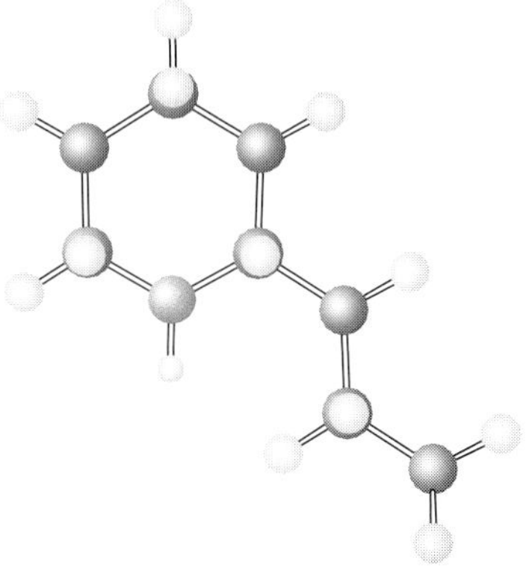

Other, more structurally complicated toxins are obtained from the plants nightshade and foxglove. Yet what is toxic in large doses may be beneficial in controlled amounts. The principal toxin from nightshade, atropine, is useful in small amounts as a mydriatic (to dilate the pupils) and as an anticholinergic (to block the action of the neurotransmitter acetylcholine).

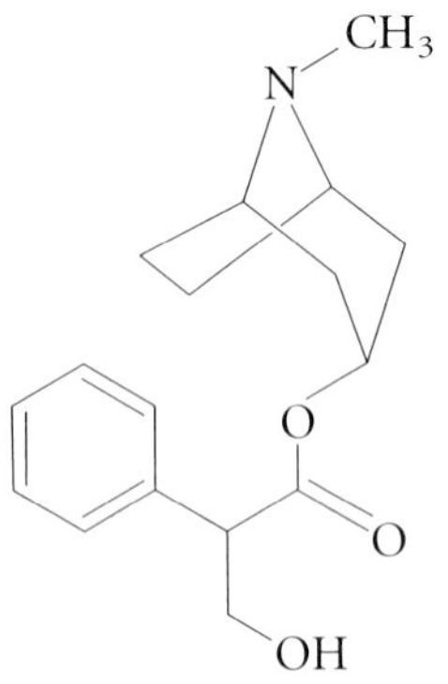

**Atropine**

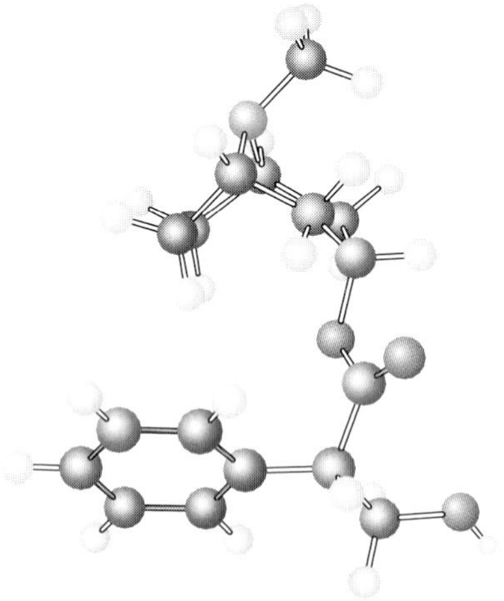

The extract from foxglove is known as digitalis, and both the crude mixture and some of the purified components such as digoxin are used to stimulate the heart.

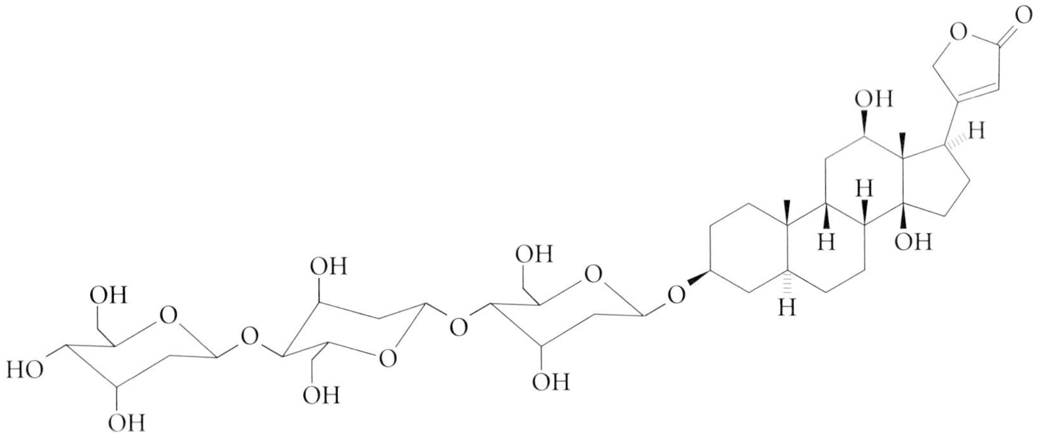

**Digoxin**

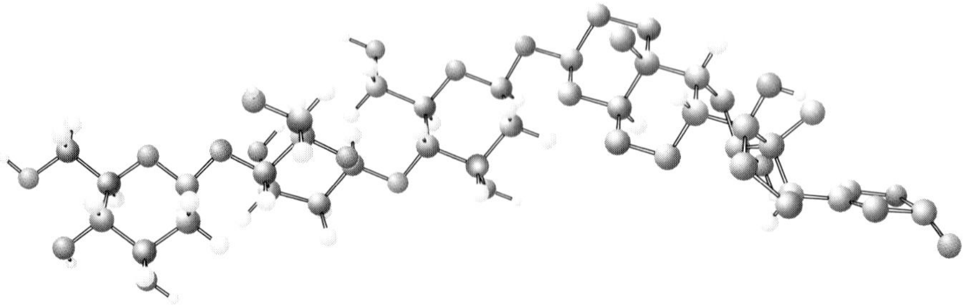

The struggles of the alchemists (like the fellows in the frontispiece of this book) to make gold from less valuable metals diverted attention from compounds of carbon until the sixteenth century, when scientists began to turn their attention to more practical (and less greedy) endeavors. In particular, Philippus Paracelsus, a German–Swiss holding a chair in medicine at the University of Basel, became convinced that drugs could be found that would relieve suffering due to pain. Indeed, he was the first to recognize that opium, an extract of the poppy plant, could serve as a pain reliever. Yet even though it was recognized that naturally occurring compounds could be useful, organic chemistry did not flourish. The structures of organic compounds were not understandable until the English scientist John Dalton (1766–1844) advanced his atomic theory in 1803. This revival and application of the atomic theory first proposed by the ancient Greek philosopher Democritus provided a foundation for understanding the nature of molecules and explaining the composition and reaction of chemical substances. At that time, many chemists began to focus their attention on organic compounds obtained from nature; morphine, the active constituent of opium, was isolated in 1804 by the French chemist Séguin. Nonetheless, it was not until 1847 that the empirical formula of morphine was determined and another three-quarters of a century before the correct structure was proposed in 1925.

**Morphine**

Morphine was first synthesized in the laboratory by Gates and Tschudi in 1952. The unfolding of the chemistry of morphine—starting with isolation of the pure substance, moving through determination of the empirical formula and the three-dimensional structure, and culminating in synthesis—typifies the development of classical organic chemistry.

The year 1828 was a milestone in the development of organic chemistry. In that year Friedrich Wohler (1800–1882) accidentally discovered that urea, previously isolated from mammalian urine, could be made by heating ammonium cyanate, an inorganic salt:

$$^{\oplus}NH_4 \, ^{\ominus}OCN \longrightarrow$$

**Ammonium cyanate**          **Urea**

Wohler's synthesis led to the realization that molecules found in nature can be described, handled, and synthesized in the same way as minerals and metals. What an astounding insight—that atoms and molecules move freely between the living and nonliving worlds, that the living and nonliving share fundamental attributes that can be studied. With this discovery, organic chemistry was born.

*CHEMICAL PERSPECTIVES*

FIRST TO PRESS, OR FIRST TO LECTURE?

In most historical accounts, the German apothecary F. W. A. Sertürner is credited with the first isolation of morphine, and, indeed, he was the first to publish his findings (*Trommsdorff's J. der Pharmazie, 13,* 234, 1806). Although Séguin had reported his findings in an oral presentation before the Institute of France in 1804, for some reason he did not publish the results until 1814 (*Ann. Chim., 92,* 225). Another figure emerges in the history of this alkaloid: C. L. Derosne published an account of the isolation of crystalline morphine in 1803 (*Ann. Chim., 45,* 257), but he failed to realize that morphine was basic—he thought that the green coloration produced from syrup of violets was the result of residual alkali from his extraction rather than the compound itself. Séguin used rhubarb paper for his pH test, correctly attributing the brown coloration to morphine, the first pure organic base isolated.

Who is the first to make a discovery in science (and who comes in second) can have a major impact on careers. Before the twentieth century, when communication was slow, credit was usually given to the first person who put results into print. Today, an appearance in the media (including radio and television) is often used as the criterion, although many still hold to the idea that science is not "official" until it appears on the printed page. The increasing encroachment of high-speed communication methods, especially the Internet, on both professional and personal lives will lead to significant questions of attribution in the future.

The lack of detailed structural information for organic compounds did not prevent the chemists of the nineteenth century from applying their knowledge to the preparation of new and much less costly dyes and to the isolation and purification of compounds from plants for medicinal purposes. For example, the toxic extract of the foxglove plant, previously used only as a poison, was purified and turned to beneficial use as a heart stimulant, and it is still in use today. And important structural features were recognized in naturally occurring large molecules, such as those in cotton, silk, and wool. As the number of useful compounds obtained from nature increased, so did interest in organic chemistry.

The curiosity and tireless drive of chemists in the twentieth century have yielded a detailed understanding of the inner workings of some cells and a fairly complete chemical picture of the organic chemistry behind the complex operations essential for multicelled animals. With this knowledge, chemists have synthesized sophisticated compounds with properties that can enhance the quality of life. The variety of useful organic compounds is truly amazing, ranging from the natural and synthetic polymers that are the basis of many materials used to clothe, house, equip, and transport people to the modern wonder drugs, such as the antibiotics used to treat many human diseases.

Your objective for this course is to develop sufficient knowledge of organic chemistry to be able to understand the structure and reactions of seemingly complicated molecules, such as organic polymers and penicillin antibiotics. Unique bonding states are available to carbon, and first we will explore the nature of the covalent bonds (and especially the multiple bonds) that are readily formed by carbon. We will then consider the structures of various types of organic molecules, how these structures are determined, and a variety of typical reactions for different classes of compounds. A good grasp of organic structure and reactivity will enable you to understand how modern chemistry is practiced: how syntheses of new compounds and materials are planned; how the properties of synthetic and naturally occurring polymers—for example, DNA—are explained, predicted, and manipulated; and how the structure and function of natural substances containing oxygen and nitrogen are investigated.

The basic principles underlying the relationship between structure and reactivity can be extended to explain the use of cofactors as biological reagents, the basis for molecular recognition, and the role of enzymes in controlling reactions. Even such complex processes as the storage and release of energy in fats and sugars and the transfer of information in the replication of genetic code reduce to relatively simple principles of structure and reactivity. A firm foundation in the basic concepts of organic chemistry provides sufficient background for understanding diverse living systems.

We begin the study of organic chemistry by reviewing some important topics (usually covered in high-school and general-chemistry courses) that are crucial to an understanding of molecular structure and reactivity.

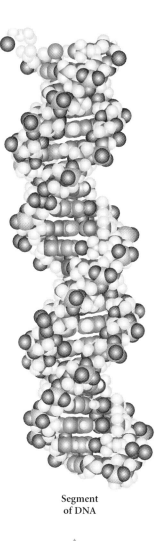

Segment
of DNA

## 1.2

## The Formation of Molecules

An understanding of the structure of organic molecules can be developed by considering how molecules can be formed from the combination of individual discrete atoms. This simplified approach enables us to describe and predict molecular structure by using what we know about atomic structure to explain how molecules might be formed from atoms. We then consider additional concepts needed to explain the properties of molecules. Despite the beguiling nature of this logical construction of molecules from atoms, you should not forget that, in reality, almost everything that we come into contact with exists in molecular form, and that in the real world, molecules are formed from other molecules. Molecules of oxygen ($O_2$) react with molecules of hydrogen ($H_2$) to form water ($H_2O$), rather than two individual atoms of hydrogen combining with one atom of oxygen.

### Atomic Structure

An understanding of the atom starts with the model of a massive, positively charged nucleus surrounded by a moving cloud of electrons, whose negative charge balances the positive nuclear charge. In 1926, Paul Dirac, Werner Heisenberg, and Erwin Schrödinger proposed independently that electrons could be considered as a type of wave and that the motion of electrons in the atom could be represented by mathematical wave equations. Solution of these equations for the hydrogen atom suggests that, within the atom, electrons are arranged in layers, or shells, and that each electron is found within a specific region of space, called an **orbital.** These orbitals, whose shapes and energies are calculated for the hydrogen atom, are assumed to be applicable to the atoms of heavier elements.

***Atomic Orbitals.*** An atomic orbital can be thought of as the picture that would be obtained if we could perform time-lapse photography of an electron within an atom—a sort of cloud of electrons about the nucleus. Of course, we can't really do this experiment, but if we solved the Schrödinger wave equations and plotted the probability of finding a given electron at a particular distance from the nucleus in three dimensions, we would have performed the theoretical equivalent of time-lapse electron photography. The picture thus obtained shows that each electron is localized within the atom, in regions whose shape and dimension are determined by quantum numbers.

***Quantum Numbers.*** Quantum numbers specify allowed energy states, each of which corresponds to a specific region within the atom, the atomic orbital. There are four quantum numbers:

**1.** The *principal quantum number, n,* has values 1, 2, 3, . . . , and is the major determinant of an electron's *energy* and its *distance* from the nucleus—that is, the orbital *size.* Thus, an electron with quantum number 2 is more energetic and farther from the nucleus than an electron with quantum number 1.

**2.** The *angular momentum quantum number, l,* has values 0, 1, 2, . . . , $n - 1$. If the principal quantum number is 2, $l$ can have the values 0 and 1. The angular momentum quantum number defines the *shape* of the orbital occupied by an electron. Orbitals are designated by the letters *s, p, d,* and *f,* which correspond to $l$ values of 0, 1, 2, and 3, respectively. The *s* orbitals are spherical; the *p* orbitals are dumbbell-shaped. The *d* and *f* orbitals have more complex shapes, which will not concern us here because they are unoccupied in the first- and second-row elements with which we will be mainly concerned (carbon, hydrogen, oxygen, and nitrogen).

**3.** The *magnetic quantum number, m,* has values $-l$, . . . , $-2$, $-1$, 0, $+1$, $+2$, . . . , $+l$. The magnetic quantum number defines the spatial orientation of an orbital and, as a corollary, the number of each type of orbital. For each principal quantum number, there is one *s* orbital. For $n \geq$ 2, there are three *p* orbitals oriented at right angles to one another. For $n \geq$ 3, there are five *d* orbitals. For $n \geq 4$, there are seven *f* orbitals. Orbitals with different magnetic quantum numbers are distinguished by subscripts: for example, $2p_x$, $2p_y$, $2p_z$.

**4.** The *spin quantum number,* which has the value $+\frac{1}{2}$ or $-\frac{1}{2}$, refers to the orientation (or spin) of the electron with respect to an external magnetic field. The spin quantum number is significant in determining the electron configuration (see below).

***Shapes and Dimensions of Orbitals.*** Atomic orbitals can be pictured as graphs of the probability surfaces within which the electrons are likely to be found. In considering the chemistry of carbon compounds containing hydrogen, oxygen, and nitrogen, we will focus on the spherical 1*s* and 2*s* orbitals and the three dumbbell-shaped 2*p* orbitals (Figure 1.2). The shapes of the *s* and *p* orbitals are similar for all elements of the periodic table. Keep in mind, however, that all of the third-level orbitals (3*s*, 3*p*, and 3*d*) are substantially larger than the second-level 2*s* and 2*p* orbitals.

Complete occupancy of any set of orbitals (for example, the 1*s* orbital, the 2*s* orbital, the three 2*p* orbitals, the 3*s* orbital, the three 3*p* orbitals, or the five 3*d* orbitals) leads to a spherical distribution of electron density

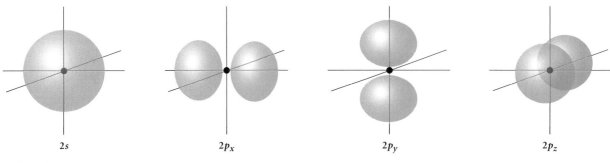

$2s$           $2p_x$           $2p_y$           $2p_z$

**FIGURE 1.2**

Shapes of the $2s$ and the three $2p$ atomic orbitals of hydrogen. There are many different ways to depict atomic orbitals. The surfaces used in this book uniformly contain 40% of the electron density—that is, 40% of the total electron density lies between the surface shown and the nucleus. Such a surface also roughly corresponds to the orbital's highest electron density.

about the central atom. This concept is easy to grasp for the $s$ orbitals, but also holds for the three equivalent $p$ orbitals with their lobes directed along three mutually orthogonal axes (Figure 1.3), as well as for the $d$ orbitals. The geometric sum of a completely filled set of $p_x$, $p_y$, and $p_z$ orbitals is a sphere. At the center of this sphere (at the nucleus), the probability of encountering an electron is negligible. The nucleus of the atom is said to be at a node of each of the $p$ orbitals, a position at which electron density is zero. (The same is true for the $d$ orbitals: the sum of electron density when all five $d$ orbitals are filled is a sphere, and each $d$ orbital has zero electron density at the center of this sphere—that is, the nucleus.)

*Pauli Exclusion Principle.* The *Pauli exclusion principle* states that each electron in an atom is uniquely defined by a distinct set of quantum numbers. The first three quantum numbers define the orbital—for example, the $2p_x$ orbital. The fourth quantum number defines the relative spin of the electron in the orbital. When two electrons occupy the same orbital, one has a spin of $+\frac{1}{2}$, and the other has a spin of $-\frac{1}{2}$. These two electrons are described as **spin-paired.** The electron spin is sometimes indicated by an arrow or by a plus or minus sign. However, because absolute spin is arbitrary, these labels are often omitted.

*Valence Shell.* Because only two spin quantum numbers are possible for an electron, an orbital is completely filled by two electrons of opposite spin. Thus, an $s$ orbital can accommodate exactly 2 electrons, and each of the three $p$ orbitals can accommodate two electrons for a total of 6. When all the orbitals with the same principal quantum number are filled, the atom is said to have a complete, or filled, valence shell. For each row of the periodic table, it is possible to determine the number of electrons needed to fill the valence shell for that row: 2 electrons fill the valence shell of a first-row element; an additional 8 electrons are needed to complete the shell for a second-row element; 18 more for a third-row element; and so forth. The electrons in an incomplete valence shell are referred to as **valence electrons.** Atoms react so as to achieve a filled valence shell, either by losing or gaining electrons (forming ions) or by sharing their unpaired electrons with other atoms (**covalent bonding**).

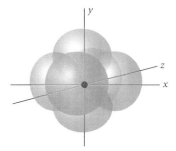

**FIGURE 1.3**

A three-dimensional representation of mutually orthogonal $p$ orbitals. The $p_x$ orbital is directed along the $x$ axis; the $p_y$, along the $y$ axis; and the $p_z$, along the $z$ axis.

***Van der Waals Radii.*** As the principal quantum number increases, the size of the orbital increases. Therefore, progressing down the periodic table, the valence electrons have higher principal quantum numbers and are found farther and farther from the nucleus. As a consequence, the effective size of the atom, described as its **van der Waals radius,** increases. On the other hand, progressing across the periodic table, the principal quantum number remains unchanged, and the orbitals become smaller as the increased nuclear charge pulls the electrons closer to the atom's center. The net effect is that the van der Waals radii of atoms *increase* going down a column of the periodic table and *decrease* going from left to right (Figure 1.4). Indeed, the radius of fluorine lies between that of boron and hydrogen! The van der Waals radii of atoms are important in determining the effective size of molecules and are a factor in reactivity.

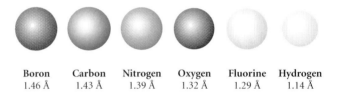

| Boron | Carbon | Nitrogen | Oxygen | Fluorine | Hydrogen |
| --- | --- | --- | --- | --- | --- |
| 1.46 Å | 1.43 Å | 1.39 Å | 1.32 Å | 1.29 Å | 1.14 Å |

**FIGURE 1.4**

The van der Waals radii (given here in angstroms) of atoms decrease across the periodic table because of the increasing number of protons in the nucleus. The greater the nuclear charge, the greater is the attraction of the nucleus for the surrounding electrons.

| | |
| --- | --- |
| H: | $1s^1$ |
| He: | $1s^2$ |
| Li: | $1s^2 2s^1$ |
| Be: | $1s^2 2s^2$ |
| B: | $1s^2 2s^2 2p^1$ |
| C: | $1s^2 2s^2 2p^2$ |
| N: | $1s^2 2s^2 2p^3$ |
| O: | $1s^2 2s^2 2p^4$ |
| F: | $1s^2 2s^2 2p^5$ |
| Ne: | $1s^2 2s^2 2p^6$ |

**FIGURE 1.5**

Electron configurations of first- and second-row elements. The number (1 or 2) preceding each letter is the principal quantum number that defines the valence shell, the letter (*s* or *p*) designates the orbital shape, and the superscript number (1 through 6) specifies the number of electrons in the orbital or suborbital.

***Electron Configuration.*** Filling the orbitals in an atom with electrons starting with the lowest-energy orbital and moving up to the highest-energy one (the *Aufbau principle*) yields the electron configuration for that atom. The electron configuration of hydrogen is denoted as $1s^1$. The initial number 1 indicates that hydrogen has a single spherically symmetric *s* orbital; the superscript 1 means that it is occupied by one electron. The two electrons of helium completely fill its valence shell; the electron configuration of helium is $1s^2$.

Second-row elements have a filled 1*s* orbital and additional electrons in 2*s* and 2*p* orbitals. Each *s* and *p* orbital can hold 2 electrons. The second-row valence shell is therefore filled when an atom has 10 electrons: 2 electrons in the 1*s* orbital and 8 electrons in the second shell (2 in the 2*s* orbital and 2 in each of the three 2*p* orbitals).

In organic chemistry, the primary focus is on the atomic structure of carbon. Its atomic number is 6. There are six protons in the nucleus, so a neutral carbon atom must have six electrons. Placing these electrons in the lowest-lying orbitals yields the electron configuration of carbon, which is $1s^2 2s^2 2p^2$.

The electron configurations of the first- and second-row elements are shown in Figure 1.5.

The electron configurations of ions are derived in the same way. For example, a lithium atom has the electron configuration $1s^2 2s^1$, but a lithium ion ($Li^\oplus$) has only two electrons (one fewer than a lithium atom) and thus has the electron configuration $1s^2$.

**EXERCISE 1.1**

Specify the atomic orbitals ($1s$, $2s$, $2p$, $3s$, $3p$, $3d$, etc.) and their occupancy to define the electron configuration of each of the following atoms or ions:

(a) atomic boron

(b) metallic magnesium

(c) $Mg^{2+}$

(d) elemental phosphorus

(e) $S^{2-}$

**EXERCISE 1.2**

How many electrons must be removed from or added to each of the following ions to achieve a filled valence shell?

(a) $H^-$      (b) $Ca^{2+}$      (c) $H^+$      (d) $Mg^+$      (e) $Cl^-$

Another important aspect of electron configuration is the distribution of electrons among orbitals of equal energy. In the electron configuration of carbon, there are two electrons in the $2p$ orbitals. There are three $p$ orbitals, all equal in energy, so more than one arrangement of the two electrons is possible. Both electrons can occupy the same orbital (it doesn't matter which one because they are all equivalent in energy), or the two electrons can occupy different orbitals. **Hund's rules** state that the preferred (lowest-energy) state is that in which as many orbitals as possible are occupied by single electrons, and that the spins of the electrons in these orbitals are parallel.

Following these rules, the detailed electron configuration for carbon is $1s^2 2s^2 2p_x^1 2p_y^1$ (Figure 1.6). This electron configuration shows that carbon has four valence electrons, two of which are unpaired. Note that a second electron is placed in the $2s$ orbital (the $2s$ orbital is filled) before electrons are placed in a $2p$ orbital because the $2s$ orbital is *lower in energy* than the $2p$ orbital, and this energy difference is enough to offset the energy disadvantage due to having two negatively charged electrons occupying the same $2s$ orbital. The electron configuration for carbon given above is consistent with experimentally determined spectroscopic data for carbon in the gas state.

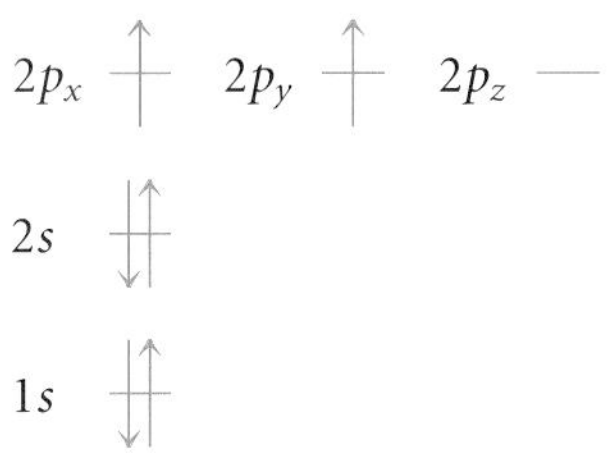

$$2p_x \uparrow \qquad 2p_y \uparrow \qquad 2p_z —$$

$$2s \uparrow\downarrow$$

$$1s \uparrow\downarrow$$

**FIGURE 1.6**

Electron configuration of carbon showing relative energy levels of the orbitals.

### Bonding

Bonding occurs when two or more atoms share electrons, thus forming a molecule. Many aspects of molecular structure can be understood in terms of two simple concepts—bond length and bond angle. However, a more detailed look at molecular structure requires the concepts of covalent and ionic bonding, hybridization, and molecular orbitals (introduced in Chapter 2). These concepts are valuable for explaining not only molecular structure, but also the reactivity of various types of molecules.

***Bond Length.*** One way of looking at bonding is to say that it occurs because energy is released when electrons are shared between atoms. When atoms are in close proximity in molecules, the attraction of the negatively charged electrons for the positively charged nuclei exceeds the electrostatic repulsions arising from the interactions of nucleus with nucleus and electrons with electrons. For the hydrogen molecule, $H_2$, this net attraction can be viewed as the energy released when two atoms of hydrogen combine to form the molecule. Experimentally, this energy, the energy of the H—H bond, has been found to be 104 kcal/mole.

We can plot energy versus interatomic distance for the hydrogen molecule, using a Schrödinger wave equation. (Schrödinger wave equations similar to those for atoms can be written for simple molecules such as $H_2$.) It turns out that there is a very narrow range of interatomic distances for which the energy is a minimum (Figure 1.7). In other words, the distance between the two hydrogen atoms in a hydrogen molecule, known as the *bond length,* is tightly controlled by energy requirements. This is also true for molecules other than hydrogen, and experimental evidence from a number of sources has allowed the determination of bond lengths in many compounds containing carbon, hydrogen, oxygen, nitrogen, and other elements. These bond lengths can be used to predict the bond lengths in new compounds containing these elements.

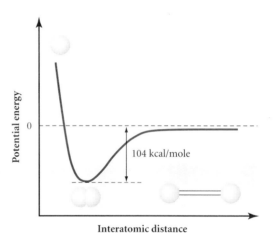

**FIGURE 1.7**

Plot of the relationship between calculated potential energy and interatomic distance for the hydrogen molecule.

*Bond Angles.* Interactions between atoms that are not bonded directly to one another in a molecule can be explained in terms of repulsive forces between electrons in the valence shells. To minimize repulsion, the atoms joined to the central atom assume positions as far away from one another as possible. Applying this principle reveals a molecule's *bond angles,* and thus its molecular shape.

To see how this works, let's look at the simple hydrocarbon methane, $CH_4$, in which a carbon atom is bonded to four hydrogen atoms. The molecule adopts an arrangement in which the four C—H bonds (or the four hydrogen nuclei, which amounts to the same thing) are as far from each other as possible. This produces a tetrahedron-shaped molecule (Figure 1.8) with an H—C—H bond angle of 109.5°.

When the central carbon atom has only three substituent atoms, as in formaldehyde, $CH_2O$, minimal electrostatic repulsion between the bonding electrons is achieved by the trigonal planar arrangement, in which all atoms lie in the same plane and the angle between substituent atoms (H—C—H

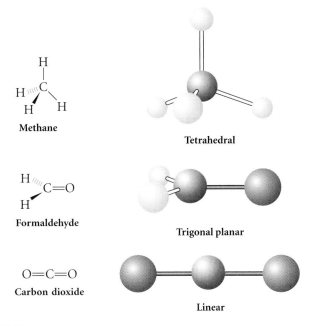

H
|
H∥C
/ \
H   H

**Methane**

**Tetrahedral**

H
∥
C=O
/
H

**Formaldehyde**

**Trigonal planar**

O=C=O

**Carbon dioxide**

**Linear**

**FIGURE 1.8**

Representative tetrahedral, trigonal planar, and linear molecules.

or H—C—O) is 120°. For a carbon atom with only two substituent atoms, such as in carbon dioxide, $CO_2$, the electrons localized between the carbon atom and the two oxygen atoms are farthest apart when all three atoms are colinear—that is, when the O—C—O bond angle is 180°.

**EXERCISE 1.3**

On the basis of maximum separation of electrons, predict all of the bond angles in the following structures:

(a) $H_3CCl$       (b) $H_2C=CH_2$       (c) $H_3C—C≡CH$

## Covalent Bonding

Bonding results from the sharing of electrons between atoms—but how does this sharing occur? For the simple diatomic molecule $H_2$, it is clear that if two hydrogen atoms approach one another closely, the unpaired electron from each can be shared most effectively when it is in the region between the two nuclei, as if the two individual atomic $1s$ orbitals overlapped. A similar picture results for the molecule $F_2$ from the overlap of the $2p$ orbitals of each fluorine atom. In these diatomic molecules, two electrons (one unpaired electron from each atom) are shared to complete the valence shell (two electrons for each hydrogen atom in $H_2$, eight electrons for each fluorine atom in $F_2$).

The situation becomes more complicated when we consider overlapping the atomic orbitals of carbon so that it can achieve a filled valence shell. To solve the difficulties that arise, chemists introduced the concepts of molecular orbitals and hybridization.

#01    Atomic Orbitals

***Orbital Overlap and Molecular Orbitals.*** The concept of orbitals can be extended from atoms to molecules; that is, the electrons in molecules, just like those in atoms, are constrained to certain energy states, or certain regions of space, called *orbitals*. Intuitively and, as it turns out, mathematically, a reasonable approximation of molecular orbitals can be arrived at by overlapping the orbitals of the valence electrons of the individual atoms. Thus, in the simplest example, $H_2$, the $1s$ orbitals of the two unpaired electrons (one from each atom) overlap in the region between the nuclei. The resulting **sigma bond** ($\sigma$ bond) is a region of increased electron density that is symmetric about the axis between two nuclei. A $\sigma$ bond can also be formed from the overlap of $p$ orbitals. For example, in the fluorine molecule, the $2p$ orbitals of the two unpaired electrons (one from each atom) overlap to form $F_2$. This approach works fine for the simplest diatomic molecules, but complications arise with more complex molecules.

***Hybridization.*** Let's look at the structure of $CH_4$ that would result if bonding were the result of overlapping the atomic orbitals of carbon and hydrogen. The electron configuration of carbon is $1s^2 2s^2 2p_x^1 2p_y^1$. One way for carbon to form four bonds and achieve a stable valence configuration would be to assume the more energetic configuration $1s^2 2s^1 2p_x^1 2p_y^1 2p_z^1$ and recoup the energy required to do this by forming bonds with hydrogen. However, if these four valence orbitals overlap with those of four hydrogen atoms, we would expect a molecule with one $\sigma$ bond formed by the overlap of a $2s$ and a $1s$ orbital and three $\sigma$ bonds formed by the overlap of three $2p$ orbitals from carbon with three $1s$ orbitals from three hydrogen atoms. This would be a molecule with two different kinds of $\sigma$ bonds, and we might reasonably expect that at least three of the bonds (those formed by overlap with $p$ orbitals) would be at right angles to one another. However, all the available experimental evidence indicates that when carbon forms bonds to four atoms of the same element, as in $CH_4$, all four bonds have the same energy and they are equidistant from one another—that is, directed toward the apices of a tetrahedron (with bond angles of 109.5°).

$$H \overset{\overset{\textstyle C}{\curvearrowright}}{\underset{109.5°}{}} H$$

To explain the observed bonding characteristics of carbon (and other second-row elements), it was proposed that when a carbon atom bonds to another atom, it undergoes hybridization. The carbon orbitals that overlap to form bonds are neither $s$ orbitals nor $p$ orbitals, but intermediate in character between the two. For example, the $2s$ and $2p$ orbitals can be mixed to form a new type of orbital referred to as an $sp^3$-hybrid orbital. The mixing of four atomic orbitals (one $s$ and three $p$) produces four $sp^3$-hybrid orbitals:

$$\text{C: } 1s^2 \; \underbrace{2s \; 2p_x \; 2p_y \; 2p_z}$$

$$\downarrow$$

Four $sp^3$-hybrid orbitals

These hybrid orbitals occupy separate regions of space directed as far as possible from one another. The resulting tetrahedral geometry (Figure 1.9) allows maximum overlap with orbitals of other atoms—for example, the $1s$

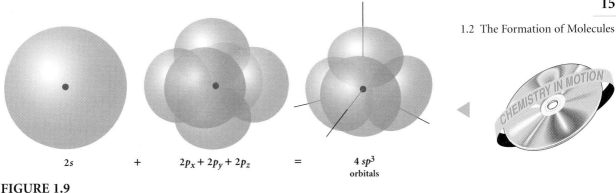

$$2s \quad + \quad 2p_x + 2p_y + 2p_z \quad = \quad 4\,sp^3$$
orbitals

**FIGURE 1.9**

Combination of the $2s$ and the three $2p$ orbitals yields four $sp^3$-hybrid orbitals.

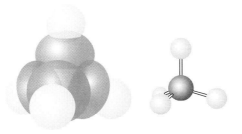

**FIGURE 1.10**

Three-dimensional (tetrahedral) representations of methane: an orbital picture (left) and a ball-and-stick model (right). Each of the four equivalent C—H bonds is formed by the overlap of a carbon $sp^3$-hybrid orbital with a hydrogen $1s$ orbital. These four bonds are directed to the corners of a tetrahedron, just like the $sp^3$-hybrid orbitals from which they are formed. They are thus as far from each other as possible, minimizing electron–electron repulsion. The carbon atom is shown in gray and the hydrogen atoms off-white, a conventional color code.

orbitals of hydrogen in methane, $CH_4$ (Figure 1.10). In methane, four bonding orbitals are formed by the overlap of four $sp^3$-hybrid orbitals of carbon with four $1s$ orbitals of four hydrogen atoms.

The four bonding orbitals in methane accommodate 8 electrons: 4 from carbon and 1 from each of the four hydrogen atoms. Because of the localization of the bonding electrons, the valence requirements of all the atoms are fulfilled (8 electrons for carbon, 2 in each of four bonds; 2 electrons for each hydrogen, both in one $\sigma$ bond).

***Characteristics of Hybrid Orbitals.*** The spatial characteristics of hybrid orbitals are based on those of the orbitals from which they were derived. For $sp^3$-hybrid orbitals, the substantial contribution of $p$ orbitals (three parts in four) means that the orbitals are elongated compared with $s$ orbitals, and the fractional $s$ contribution (one part in four) fattens them compared with $p$ orbitals. Each $sp^3$-hybrid orbital has two distinct parts, known as *lobes,* with a region of zero electron density, a **nodal surface,** in between. In the picture below, the lobes of these hybrid orbitals have been moved away from the nucleus to show their shapes better. Note that in reality the smaller back lobe of each orbital becomes buried underneath the larger lobes (at the right in Figure 1.9) and thus does not participate significantly in bonding to other atoms.

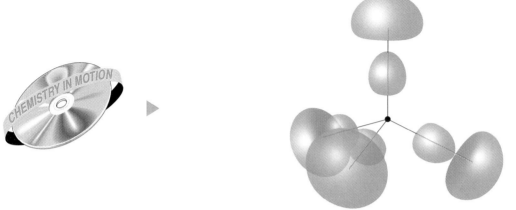

Lobes of $sp^3$ orbitals

### EXERCISE 1.4

The H—N—H bond angles in ammonia are 107°, and the H—O—H bond angle in water is 105°.

(a) Why is ammonia not planar, and why is water not linear?

(b) Why are their bond angles near the tetrahedral bond angle of 109.5°? (*Hint:* Consider the electronic configuration of the nitrogen atom in $NH_3$ and the oxygen atom in $H_2O$.)

(c) Explain why the bond angles in ammonia and water are *less* than 109.5°.

| H | C | N | O | F |
|---|---|---|---|---|
| 2.2 | 2.5 | 3.1 | 3.5 | 4.1 |
| | | P | S | Cl |
| | | 2.1 | 2.4 | 2.8 |
| | | | | Br |
| | | | | 2.7 |

**FIGURE 1.11**

The most electronegative elements.

*Electronegativity.* Electronegativity is a measure of the tendency of a particular atom to attract electrons. The most electronegative atoms are toward the top right of the periodic table, and these, along with hydrogen, are the most common bonding partners of carbon (Figure 1.11). (The electronegativities for all elements appear in the periodic table inside the back cover of this book.) Electronegativity increases from left to right across a row of the periodic table. For example, the electronegativity order for second-row elements is C < N < O < F. Electronegativity also increases from the bottom to the top of a column. Thus, among the halogens, the order is I < Br < Cl < F. For organic chemists, electronegativity is important primarily because it allows prediction of bond polarities.

*Polar and Nonpolar Bonds.* Up to this point we have assumed that, when bonding occurs between atoms, the electrons in the bond are shared equally. This is largely true for bonds between atoms like carbon and hydrogen that have similar electronegativities (2.5 and 2.2, respectively). The C—H bonds in methane and other hydrocarbons are described as nonpolar; there is little charge polarization associated with them. However, when a highly electronegative atom such as fluorine is attached to carbon, the electrons in the C—F bond are not shared equally. Instead, a shift of electrons occurs, placing a partial negative charge on fluorine and a partial positive charge on carbon. Knowledge of electronegativity trends in the periodic table can be used to predict the likelihood of **polar covalent bonding** (unequal sharing of the electrons in a covalent bond connecting two atoms), as well as the direction of polarization. We will consider the chemical and physical consequences of bond polarization in more detail in Chapter 3.

### EXERCISE 1.5

Based on the relative electronegativities of the atoms, choose the more polar bond in each pair of compounds:

(a) HO—H or $H_2N$—H

(b) $CH_3$—H or $CH_3$—F

(c) $H_3C$—OH or $H_3C$—$NH_2$

(d) $H_3C$—OH or $H_3C$—SH

(e) $H_3C$—OH or $H_3C$—Br

*Lewis Dot Structures.* Lewis dot structures are a useful way to summarize certain information about bonding and may be thought of as "electron bookkeeping." In Lewis dot structures, each dot represents an electron. A pair of dots between chemical symbols for atoms represents a bond. In the following Lewis dot structure of methane, four of the electrons are shown in blue to emphasize the fact that methane's covalent bonds are formed by the sharing of one of the four electrons of carbon with a valence electron of one of four hydrogen atoms. Note that no geometry is implied by a Lewis dot structure.

$$\begin{array}{c} H \\ \vdots \\ H \!:\! \overset{\displaystyle .}{\underset{\displaystyle .}{C}} \!:\! H \\ \vdots \\ H \end{array}$$

**Lewis dot structure of methane**

Here are Lewis structures for some other simple compounds:

$$H \!:\! \overset{\displaystyle ..}{\underset{\displaystyle ..}{O}} \!:\! H \qquad \overset{\displaystyle ..}{\underset{\displaystyle ..}{O}} \!::\! C \!::\! \overset{\displaystyle ..}{\underset{\displaystyle ..}{O}} \qquad H \!:\! C \!:::\! N \!:$$

**Water**     **Carbon dioxide**     **Hydrogen cyanide**

Note that each hydrogen atom has 2 electrons associated with it, and the atoms of the second-row elements each have 8 electrons. Nonbonding electrons (lone pairs) are easily identified in Lewis dot structures, as shown for oxygen and nitrogen atoms in the examples above. Usually bonding pairs of electrons are represented by a bond line, allowing an abbreviated form:

$$H\!-\!\overset{\displaystyle ..}{\underset{\displaystyle ..}{O}}\!-\!H \qquad H\!-\!\overset{\displaystyle H}{\underset{\displaystyle H}{\overset{\displaystyle |}{\underset{\displaystyle |}{C}}}}\!-\!H \qquad \overset{\displaystyle ..}{\underset{\displaystyle ..}{O}}\!=\!C\!=\!\overset{\displaystyle ..}{\underset{\displaystyle ..}{O}} \qquad H\!-\!C\!\equiv\!N\!:$$

**Water**     **Methane**     **Carbon dioxide**     **Hydrogen cyanide**

In a correct Lewis dot structure: (1) the total number of electrons should equal the sum of the valence electrons of all the atoms; (2) each atom should attain a filled valence shell (2 electrons for hydrogen, 8 electrons for second- and third-row elements).

### EXERCISE 1.6

Draw a Lewis dot structure for each of the following molecules. Be sure to include all valence electrons.

(a) $H_3CCH_3$    (b) $H_3COH$    (c) $H_2C{=}CH_2$    (d) $HC{\equiv}CH$

*Formal Charges.* The **formal charge (FC)** of an atom in a molecule is the charge calculated by assuming that electrons in covalent bonds are shared equally between the partners. Calculating the formal charge reveals where positive and negative charges end up in the molecule. The formal charge is calculated by noting the number of valence electrons in the free, or neutral, atom and then subtracting the number of unshared (nonbonding) electrons in the bonded atom and half the number of electrons shared by that atom. The numbers of shared and unshared electrons are derived from the Lewis dot structure.

$$FC = \text{number of valence electrons in free atom}$$
$$- \text{ number of unshared electrons in bonded atom}$$
$$- \tfrac{1}{2} \text{ number of shared electrons in bonded atom}$$

The formal charge of carbon in $CH_4$ is

$$FC = 4 \text{ valence electrons} - 0 \text{ unshared electrons} - \tfrac{1}{2}(8) \text{ shared electrons}$$
$$= 0$$

The formal charge of oxygen in $H{-}\ddot{\underset{..}{O}}\!:$ is

$$FC = 6 \text{ valence electrons} - 6 \text{ unshared electrons} - \tfrac{1}{2}(2) \text{ shared electrons}$$
$$= -1$$

### EXERCISE 1.7

Calculate the formal charge of each second-row atom in the following Lewis dot structures:

(a) $H\!:\!\overset{H}{\underset{H}{\overset{..}{C}}}\!:\!C\!:\!::N\!:$

(b) $:\!\ddot{\underset{..}{O}}\!:\!\overset{:\ddot{O}:}{\underset{..}{C}}\!:\!\ddot{\underset{..}{O}}\!:$

(c) $\ddot{\underset{..}{O}}\!::\!\ddot{\underset{..}{O}}\!:\!\ddot{\underset{..}{O}}\!:$

(d) $H\!:\!\overset{H}{\underset{H}{C}}\!:\!\overset{:\ddot{O}\cdot}{C}\!:\!\overset{H}{\underset{H}{C}}\!:\!H$

(e) $H\!:\!\overset{H\ \ H}{\underset{H\ \ H}{N\!:\!C}}\!:\!H$

### Ionic Bonding

So far we have considered covalent bonds, in which both atoms share the bonding electrons equally, and polar covalent bonds, in which the electrons are polarized toward the more electronegative atom. Advancing along this continuum, through compounds in which the atoms differ more and more in electronegativity, we reach a point at which the covalent bond is no longer a good representation of the atomic interaction, and molecular structure can be described more accurately by the ionic bonding model.

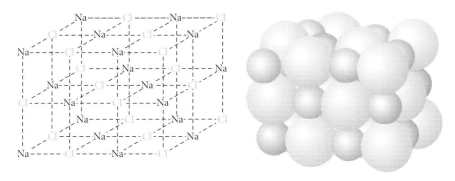

**FIGURE 1.12**

Crystal structure of sodium chloride.

An ionic compound such as sodium chloride is formed of sodium ions and chloride ions. In the solid state, sodium chloride consists of a crystal lattice held together by electrostatic attraction between the positive and negative ions. A sodium ion is attracted to its nearest neighbors (6 chloride ions) and to a lesser extent to all the chloride ions in the lattice. A similar situation exists for the chloride ions, each of which is surrounded by 6 sodium ions (Figure 1.12). In contrast, nonionic organic molecules are held together in molecular crystals by much weaker van der Waals attractions, which result from the attraction of the bonded electrons of one molecule for the nuclei of another. The amount of energy required to disrupt the strong electrostatic forces in an ionic crystal is much greater than that needed to interfere with van der Waals forces in an organic molecular solid. The melting points of (ionic) salts are thus often very high (for example, NaCl, mp 801 °C), whereas many organic compounds have such low melting points that they exist as liquids or even gases at room temperature.

Because of carbon's position near the center of its row in the periodic table and because of its relatively low electronegativity, ionic bonds to carbon are rare. Covalent bonding is the norm, and polar covalent bonds occur when hydrogen is replaced by a more (or sometimes less) electronegative atom, as we shall see later in more complex structures.

## Representing Molecules

You have already seen several ways in which chemists represent organic molecules. It is important that you understand that none of these representations actually looks like a molecule. Indeed, molecules are too small to ever be "seen" because the wavelength of visible light is too large to differentiate among molecular features. Despite the fact that some modern techniques (AFM, atomic force microscopy, and STM, scanning tunneling microscopy) are coming close to atomic scale resolution—meaning that images can be generated that show the relative position of atoms—molecules can not yet be seen in the usual sense.

This fact creates a dilemma. How should molecules be represented visually in a book? The answer depends on the information that needs to be conveyed. If all we are concerned with is the number of atoms present, we can use a molecular formula, $CH_4$ for methane, for example. From this formula and a knowledge of valency, you might deduce that the carbon is at

#03   Models for Visualization

CH$_4$

H
|
H—C—H
|
H

H
|
H$\sim$C
H   H

**FIGURE 1.13**

Three representations of a
methane molecule.

the center with four hydrogen atoms surrounding it. But you could arrive at this conclusion more quickly from a stick representation (Figure 1.13, center). (Note that this stick figure could be converted to the Lewis dot structure by replacing each "stick" with a pair of dots.) From this stick figure, you know that carbon is bonded to four hydrogen atoms because the lines represent pairs of bonding electrons between the atoms. From this simple stick figure and your knowledge that the hydrogens are arranged in a tetrahedral fashion about the carbon atom, you might be able to "picture" their three-dimensional arrangement. The representation with wedges and dashed lines (Figure 1.13, bottom) helps you arrive at an accurate three-dimensional image once you know that the hydrogen connected to the carbon with a **solid wedge** is intended to be in front of the plane containing the other atoms and the hydrogen connected with hatched lines is behind it.

Artists have known for centuries that two "tricks" help people perceive a three-dimensional image from a two-dimensional picture: Your mind thinks smaller objects are farther away than larger ones (so long as you think that the objects are in reality the same size), and objects partially covered by others are perceived as more remote. Compare the representations of methane in Figure 1.14. The first three are called **ball-and-stick models,** and your model kit can be used to construct similar representations. In the model on the far left, all four hydrogen atoms are the same size (and smaller, of course, than the carbon atom). In the middle ball-and-stick model, one hydrogen "ball" is larger and one is smaller than the other two. The size is varied in this picture so that the larger hydrogen atom appears closest to you and the smaller appears farthest away. This helps you visualize one hydrogen atom as pointing toward you and one away from you. The ball-and-stick representation on the right is slightly rotated so that the foremost hydrogen partially covers the one in the rear, giving a truer three-dimensional picture. The representation on the far right is called a **space-filling model,** and the relative size of each atom as shown is based on its van der Waals radius. Note how large the space-filling model appears relative to the other models (this is not a mistake!). This representation conveys an idea of the size and shape of a methane molecule, especially as it might be "perceived" when bumping into other molecules.

Which representation of methane in Figures 1.13 and 1.14 is most useful? In part, it depends on what information is to be conveyed. If the idea is to understand the ratio of hydrogen to carbon (as in a discussion of va-

**FIGURE 1.14**

Four "pictures" of methane, depicting the hydrogen atoms as off-white spheres and the carbon atom as a gray sphere.

lency), then either the top or middle representation in Figure 1.13 is sufficient. If the three-dimensional structure of methane is the issue, the representation at the bottom in Figure 1.13 or any of those in Figure 1.14 are
preferred. Note that as the information content of the representations of
methane increases, so does the complexity of the drawing, and this factor
often dictates which representation is chosen: the simplest one that conveys
the needed information.

## EXERCISE 1.8

Use your model set to construct methane and ethane. Hold one end of the ethane
molecule (a methyl group), and rotate the other end about the bond between the
carbon atoms. Note the changes that take place in the relative positions of the hydrogen atoms on the two methyl groups. Are similar changes possible for methane?

## Drawing Three-Dimensional Structures

The three-dimensional character of alkanes is easily depicted by using
single solid lines to represent $\sigma$ bonds lying in the plane of the page, solid
wedges to indicate those coming toward the observer, and hatched lines to
indicate those going away from the observer. One of many possible arrangements of the hydrogen atoms in ethane is shown in Figure 1.15.

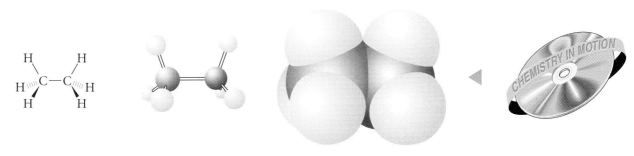

**FIGURE 1.15**

Representations of ethane: three-dimensional sawhorse (left), ball-and-stick (center), and space-filling (right).

The representation with wedges and hatched lines resembles that of the
legs of a sawhorse. This method for depicting three-dimensional structures
is therefore called a **sawhorse representation.** At this point, you should construct a three-dimensional model of the structure of ethane and correlate
it with the representations in Figure 1.15. This is also an opportune time
to use the model to assure yourself that a tetrahedral geometry can be maintained at both carbon atoms even though there is free rotation about the
C—C bond. You will often find that making a three-dimensional model
helps you to visualize a molecule's structure much more clearly than simply reading the text. You are encouraged to make such a model whenever a
new type of molecule is described.

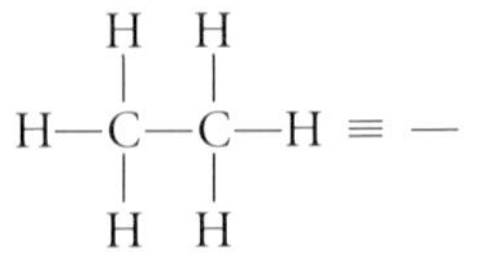

**FIGURE 1.16**

The line notation at the right is a shorthand method for representing the carbon skeleton of ethane.

Hydrocarbon skeletons can also be represented by a line notation in which each line segment represents a carbon–carbon bond, as shown for ethane in Figure 1.16. No C—H bonds are shown; their presence is inferred as needed to meet the valence requirement of carbon. Other atoms besides carbon and hydrogen in a molecule are drawn in specifically. This convention does not show three-dimensional structure; it is merely a useful way to depict structural isomers, the subject of the next section. Although this is a convenient shorthand, it conveys less structural information about ethane than do the three-dimensional representations of Figure 1.15. On the other hand, the line notation does clearly indicate the attachment of one atom to another in the molecule, which is called the **connectivity** of the molecule. As the number of atoms in a molecule increases, so does the complexity of bonding, and the simplicity of line notations becomes increasingly valuable.

### EXERCISE 1.9

With the three-dimensional structure of ethane in mind, replace each of the six hydrogen atoms in turn with a chlorine atom. Draw a sawhorse representation of the structure for each product. With a molecular model in hand, orient the molecule to correspond to the sawhorse representations you have drawn. Note that all six representations are pictures of the same molecule.

## 1.3

### Simple Hydrocarbons

Hydrocarbons are familiar in everyday life. The natural gas we use to cook our food, the liquid gasoline that powers our vehicles, and the bottles for soft drinks, cooking oils, and shampoo—all consist of hydrocarbons. Different hydrocarbons exist at room temperature as gases, liquids, or solids. Hydrocarbon molecules contain only carbon and hydrogen atoms; they are the most fundamental group of organic compounds. Hydrocarbons can be divided into several classes: alkanes, alkenes, alkynes, and arenes. We will begin by examining the simplest of these classes, the alkanes. Even this class, however, exhibits amazing structural diversity.

#### Properties of Hydrocarbons

Because carbon and hydrogen atoms have similar electronegativities, there is minimal charge polarization in the bonds of hydrocarbons, and polar interaction between these molecules is weak. Hydrocarbons are consequently described as nonpolar. The lack of polarity results in hydrocarbons being relatively *chemically inert*; they do not readily undergo chemical reactions. Also, hydrocarbons are relatively insoluble in polar liquids, such as water. Because highly polar molecules interact strongly with one another, the positive portion of one molecule being attracted to the negative portion of another, the molecules of polar liquids interact more strongly with one another than with hydrocarbon molecules.

TABLE 1.1

Physical Properties of Alkanes

| Name | Formula | Boiling Point (°C) | Melting Point (°C) |
|---|---|---|---|
| Methane | $CH_4$ | −164 | −182 |
| Ethane | $C_2H_6$ | −89 | −183 |
| Propane | $C_3H_8$ | −42 | −190 |
| Butane | $C_4H_{10}$ | −0.5 | −138 |
| 2-Methylpropane (isobutane) | $C_4H_{10}$ | −12 | −159 |
| Hexane | $C_6H_{14}$ | 69 | −95 |
| Cyclohexane | $C_6H_{12}$ | 81 | 6 |
| Octane | $C_8H_{18}$ | 126 | −57 |
| 2,2,4-Trimethylpentane (isooctane) | $C_8H_{18}$ | 99 | −107 |

The major molecular interaction of hydrocarbons is **van der Waals attraction,** in which the electrons of one molecule are attracted to the nuclei of another. These attractions are relatively weak and are easily disrupted. This accounts for the fact that the lower-molecular-weight hydrocarbons are gases at room temperature. The more atoms in a given molecule (the higher the molecular weight), the greater is the sum of the van der Waals attractions for another molecule of its kind. Thus, as molecular weight increases, van der Waals attractions generally increase, producing stronger molecular interactions and higher boiling and melting points. These trends are apparent in Table 1.1.

## Alkanes (Saturated Hydrocarbons)

The simplest member of the alkane family is methane, $CH_4$; the next member is ethane, $C_2H_6$. What happens when an alkane molecule has two carbon atoms? The two atoms can form a single bond between them by overlap of the $sp^3$-hybrid orbitals, and each carbon atom then has three additional bonding sites (Figure 1.17).

**FIGURE 1.17**

A C—C covalent bond formed by overlap of $sp^3$-hybridized orbitals. (For clarity, only the overlapping orbitals are shown. Note the position of the carbon nuclei, the two small black spheres, buried in the back lobe of the $sp^3$ orbitals.)

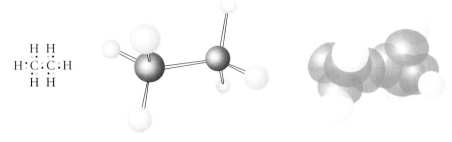

$$
\begin{array}{cc}
H & H \\
\cdot\cdot & \cdot\cdot \\
H \!:\! C \!:\! C \!:\! H \\
\cdot\cdot & \cdot\cdot \\
H & H
\end{array}
$$

**FIGURE 1.18**

Covalent bonding in ethane, shown by overlapping orbitals (right), ball-and-stick representation (center), and Lewis dot structure (left).

When hydrogen atoms are covalently bonded to carbon at these sites, the stable molecule **ethane** ($H_3CCH_3$) is formed (Figure 1.18). This structure satisfies the valence shell requirements of each carbon and hydrogen atom, as the Lewis dot structure shows. Extending the carbon chain forms larger members of the alkane family. Imagine removing a hydrogen atom from the end of the chain, say in ethane, and replacing it with a carbon atom bearing three hydrogen atoms, for the net addition of one carbon atom and two hydrogen atoms. This process can be extended indefinitely. The alkanes, also called *saturated hydrocarbons,* are therefore represented by the overall molecular formula $C_nH_{2n+2}$.

***Sigma Bonds in Alkanes.*** As you learned earlier, a $\sigma$ (sigma) bond can be formed from the overlap of two $1s$ orbitals, a $1s$ and an $sp^3$ orbital, or two $sp^3$ orbitals. In ethane, the second member of the alkane family, the two $sp^3$-hybridized carbon atoms form a $\sigma$ bond by overlap of $sp^3$ orbitals. Because the hybrid orbitals are directional and oriented toward the apices of a tetrahedron, only one orbital of each atom can point directly toward the other, and the region of $2sp^3$–$2sp^3$ orbital overlap is located along the line (axis) that connects the nuclei of the two atoms. The C—C bond length is 1.54 Å. Ethane also has six C—H bonds formed by $2sp^3$–$1s$ orbital overlap. The C—H bond length is 1.10 Å.

Because the electron density is symmetric about the internuclear axis, the extent of orbital overlap is not affected if the atoms rotate about this axis. Thus, free rotation can occur about a $\sigma$ bond without affecting bond strength. For example, holding one of the carbons of ethane fixed while rotating the other need not break the C—C or any C—H bond. Because all bonds in ethane are $\sigma$ bonds, the participating atoms can rotate freely about each of them, yielding an infinite number of three-dimensional structures for this single molecule. Molecules that differ only by rotation about $\sigma$ bonds are known as **conformational isomers.** We will discuss isomers and stereoisomerism in detail in Chapter 5.

## Structural Isomers

Now let's consider some larger alkanes. Replacing a hydrogen atom of ethane with a methyl group ($CH_3$) yields **propane,** $C_3H_8$ (Figure 1.19). (All six hydrogen atoms of ethane are identical, so it makes no difference which one is replaced.)

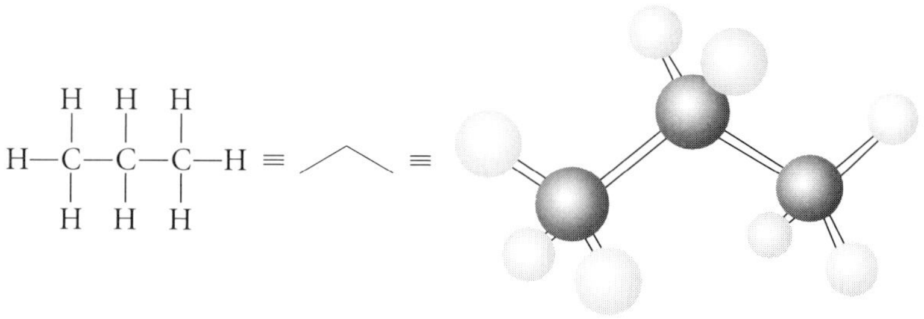

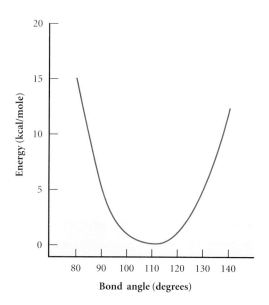

**FIGURE 1.19**

Three representations of the structure of propane.

The C—C—C bond angle in propane is slightly larger (111.7°) than the 109.5° tetrahedral angle. This is not surprising in light of the fact that the four substituents bonded to each carbon are not equivalent, so that there are slightly different repulsive interactions between the electrons of the various bonds. The amount of energy required to change the C—C—C bond angle from 109.5° to 111.7° is slight, only 0.08 kcal/mole. Indeed, even a 5° deviation from the tetrahedral angle raises the energy of propane only slightly. However, as the C—C—C bond angle deviates further from 109.5°, the energy increases more rapidly (Figure 1.20).

**FIGURE 1.20**

Plot of the calculated change in energy of propane as the C—C—C bond angle is varied. Increasing or decreasing the bond angle from the lowest-energy arrangement at 111.7° raises the molecule's energy. The curve is quite shallow near the minimum but becomes quite steep as the angle deviation becomes large.

The structure of **butane** ($C_4H_{10}$), the next member of the alkane family, results when a hydrogen atom in propane is replaced by a methyl group. Because there are two different types of hydrogen atoms in propane (the

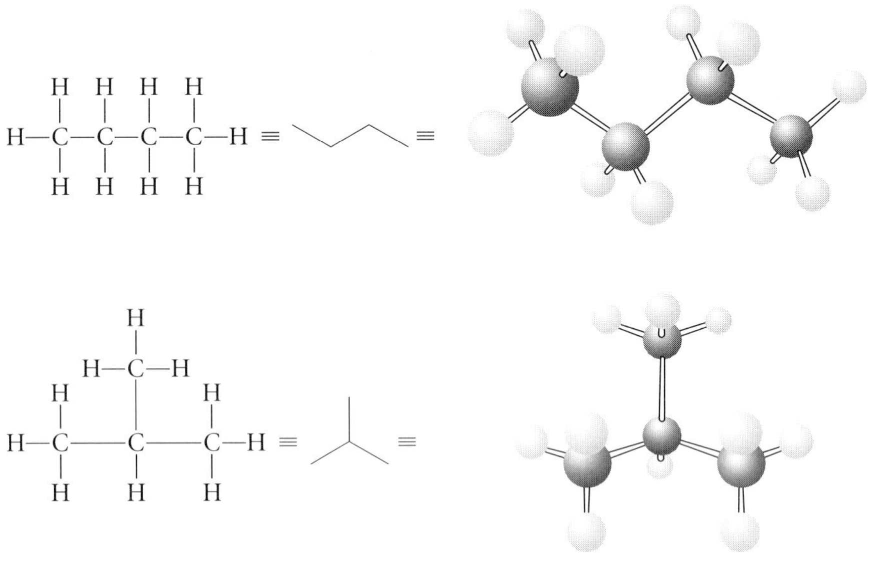

**FIGURE 1.21**

Two possible structures for butane ($C_4H_{10}$) obtained by replacing one of
propane's hydrogen atoms with a methyl group.

six hydrogens of the two methyl groups at the ends of the molecule and the
two hydrogens on the central carbon atom), two different structures for butane are obtained (Figure 1.21). The $C_4H_{10}$ isomer at the top (with all the
carbons in a row) is derived from propane by the addition of a methyl group
to one end of the skeleton, whereas the isomer at the bottom, also $C_4H_{10}$, is
obtained by the addition of a methyl group to the central carbon of propane.

These two butane molecules do not have the same sequence of chemical bonds no matter how they are oriented in space. Note that one of the
carbon atoms of the structure at the bottom in Figure 1.21 is attached to
three other carbon atoms, whereas each carbon atom in the structure at the
top is attached to no more than two other carbon atoms. The normal tetrahedral bond angles and bond lengths are maintained in both structures,
and they have the same molecular formula. Yet these are different molecules: they are **structural isomers** of the compound butane and differ in
their carbon backbones.

The number of possible structural isomers increases as the number of
carbon atoms in an alkane increases. In the next exercise you will determine
the number of structural isomers that exist for alkanes with five, six, and
seven carbons.

### EXERCISE 1.10

Draw all possible isomeric carbon skeletons with the following overall formulas:

(a) $C_5H_{12}$     (b) $C_6H_{14}$     (c) $C_7H_{16}$

(*Hint:* It will help greatly if you draw the skeletal structures first. Then add the
number of hydrogens necessary to complete the valence of each carbon. This is a

more important exercise than you might think. Its main purpose is to help you draw chemical structures and to recognize when different drawings represent the same structure, as below.)

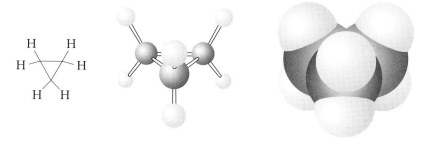

## 1.4

# Cycloalkanes

### Structures and Formulas

For alkanes with three or more carbon atoms, an alternative type of structure is possible. For example, three carbons can bond in such a way that each one is bonded to both of the others. This bonding produces the structure shown in Figure 1.22, a cyclic, three-carbon compound called **cyclopropane**.

#10   Cycloalkanes
(Monocyclic)

**FIGURE 1.22**

Three representations of the structure of cyclopropane: line drawing (left), ball-and-stick (center), and space-filling (right).

Because cyclopropane's three carbon atoms are constrained in a ring, its molecular formula is different from propane's. The formula for cyclopropane is $C_3H_6$; that for straight-chain propane is $C_3H_8$. Each time a ring of carbon atoms is formed, two fewer hydrogens are needed to satisfy the valence requirements of the carbon atoms. You can imagine that cyclopropane might be formed from propane by removing a hydrogen atom from each end carbon atom and then linking these end carbons together:

Thus, the formula for cyclopropane should have two hydrogen atoms fewer than that for propane. As mentioned in Section 1.3, the overall formula for isomeric hydrocarbons composed entirely of $sp^3$-hybridized atoms is $C_nH_{2n+2}$. A cycloalkane with one ring has the overall formula $C_nH_{2n}$. For molecules containing only $\sigma$ bonds, any deviation from the overall formula must be due to the introduction of rings: each time a ring is formed, two fewer hydrogens are needed. Thus, you can recognize from the formula of an alkane whether it includes one or more rings. To obtain the number of rings in an alkane with $n$ carbon atoms, subtract the number of hydrogen atoms ($m$) from $2n + 2$ and then divide by 2:

$$\text{Number of rings} = \frac{(2n + 2) - m}{2} \tag{1}$$

($m$ = number of hydrogen atoms present in
a hydrocarbon containing $n$ carbon atoms)

For example, for an alkane with the molecular formula $C_9H_{16}$, equation 1 gives

$$\text{Number of rings} = \frac{(2 \times \mathbf{9} + 2) - \mathbf{16}}{2} = 2$$

---

### CHEMICAL PERSPECTIVES

#### A CYCLOPROPANE-CONTAINING INSECTICIDE

Few naturally occurring compounds contain cyclopropane rings. Some compounds that do are the pyrethrins—for example, pyrethrin II.

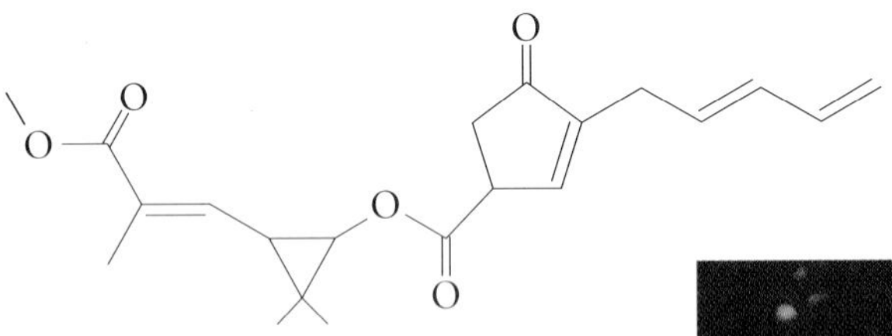

**Pyrethrin II**

Pyrethrins are potent insecticides. They got their name because they are found in chrysanthemums, flowers belonging to the genus *Pyrethrum*. Although these compounds are isolated from natural sources and thus are frequently assumed to be innocuous by many people, they cause severe allergic dermatitis and systemic allergic reactions in some individuals.

## EXERCISE 1.11

Assuming that each of the following formulas represents a hydrocarbon containing only $\sigma$ bonds, predict whether the compound has zero, one, or two rings, and draw at least three possible carbon skeletons corresponding to your prediction.

(a) $C_5H_{10}$   (b) $C_5H_8$   (c) $C_6H_{12}$   (d) $C_6H_{10}$   (e) $C_7H_{14}$

(*Hint:* It will be easiest to start with the largest ring possible and then make the ring smaller and smaller.)

## EXERCISE 1.12

Determine the molecular formula for each of the following cyclic compounds. Then use the formula and equation 1 to determine the number of rings. (*Warning:* You cannot always determine the number of rings by simply counting those you can see.)

(a)    (b) 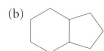   (c)    (d)

### Ring Strain

Close inspection of the structure of cyclopropane reveals a difficulty. Because three points determine a plane, the three carbon nuclei in cyclopropane must be coplanar. Simple geometry requires the sum of the C—C—C angles within this cyclic structure to be 180°, so each C—C—C angle must be 60°. However, all three carbon atoms are formally $sp^3$-hybridized and would form much stronger bonds if they could assume the normal tetrahedral bonding angle of about 109°. This deviation of 49° from the normal bonding angle for $sp^3$-hybridized atoms confers appreciable ring strain on this molecule and destabilizes it. The ring strain causes cyclopropane to have a higher potential energy content than it would otherwise have. The "extra" energy released when cyclopropane is burned, called its **strain energy,** is about 28 kcal/mole.

Let's now consider ring strain in the cycloalkane with four carbons. Planar and nonplanar representations of cyclobutane are shown in Figure 1.23. With all four carbon atoms of cyclobutane in a plane, the C—C—C bond angles have to be 90° (if all C—C bond lengths are equal). Moving

**FIGURE 1.23**

Two conformations of cyclobutane: planar (left) and puckered (right). The nonplanar conformation is lower in energy by a few kilocalories per mole.

one carbon atom out of the plane of the other three *decreases* the C—C—C bond angles, increasing the ring strain. However, other unfavorable interactions (discussed in Chapter 5) are reduced when cyclobutane is not planar. The overall result of these two factors is that the minimum energy arrangement of cyclobutane is the one in which the carbon skeleton of the ring is **puckered.** The planar and puckered forms of cyclobutane differ only in the relative orientation of atoms about $\sigma$ bonds and are thus conformational isomers, similar to those presented previously for ethane.

Cyclopentane ($C_5H_{10}$) can have five, four, or only three of its carbon atoms coplanar (Figure 1.24). As with cyclobutane, moving one or two of the carbon atoms out of the plane of the others makes the bond angles within the cyclopentane ring smaller than those of a regular pentagon (108°) but decreases other unfavorable interactions. The energy differences between the planar and nonplanar arrangements are small, but they favor the two nonplanar arrangements. All three conformations of cyclopentane shown in Figure 1.24 have significantly less strain energy than either of the conformations of cyclobutane shown in Figure 1.23 because the bond angles in cyclopentane are closer to the ideal tetrahedral angle of 109.5°.

**FIGURE 1.24**

Three conformations of cyclopentane. The two nonplanar conformations (center and right) are nearly equal in energy, and both are lower in energy than the planar conformation (left).

If the carbon skeleton of cyclohexane ($C_6H_{12}$) were planar, the C—C—C bond angles would be those of a regular hexagon, 120°. By moving two of the atoms out of the plane of the other four, the bond angles can be reduced to 109.5°. Since this nonplanar conformation of cyclohexane (Figure 1.25, right) is free of ring strain, cyclohexane—like other saturated, six-member cyclic carbon compounds—has a unique stability.

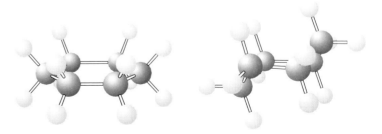

**FIGURE 1.25**

Planar (left) and nonplanar (right) cyclohexane conformations differ in energy; the nonplanar conformation is considerably more stable.

TABLE 1.2

Strain Energies of Cyclic Hydrocarbons

| Name | Structure | Strain Energy (kcal/mole) | Name | Structure | Strain Energy (kcal/mole) |
|------|-----------|---------------------------|------|-----------|---------------------------|
| Cyclopropane | △ | 27.6 | Cyclooctane | | 9.6 |
| Cyclobutane | □ | 26.4 | | | |
| Cyclopentane | ⬠ | 6.5 | Cyclononane | | 12.6 |
| Cyclohexane | ⬡ | 0 | Cyclodecane | | 12.0 |
| Cycloheptane | | 6.3 | Cyclododecane | | 2.4 |

To summarize, the smallest cycloalkanes have large strain energies. Increasing ring size is accompanied by a trend to lower strain energy, reaching a minimum for cyclohexane. Further increases in ring size result in increases in ring strain, but the increases do not follow a regular pattern. The strain energies of cycloalkanes are listed in Table 1.2. The three-dimensional structures of cyclic hydrocarbons are discussed in more detail in Chapter 5.

### EXERCISE 1.13

There is very little difference in strain energy between cyclopropane (27.6 kcal/mole) and cyclobutane (26.4 kcal/mole) even though the bond angles of cyclobutane are considerably closer to the tetrahedral angle (60° versus ~90°). Why are the strain energies not significantly different? (*Hint:* The number of carbon atoms differs in these two compounds.)

## 1.5

# Nomenclature

### IUPAC Rules

Because of the great number of compounds containing carbon and the range in complexity of their skeletons, each specific compound requires a unique name. The **International Union of Pure and Applied Chemistry**

## TABLE 1.3

IUPAC Nomenclature for Simple Hydrocarbons

| Alkanes | | Alkenes | | Alkynes | | Cycloalkanes | |
|---|---|---|---|---|---|---|---|
| Name | Structure | Name | Structure | Name | Structure | Name | Structure |
| Methane | $CH_4$ | | | | | Cyclopropane | |
| Ethane | $CH_3CH_3$ | Ethene | | Ethyne | | Cyclobutane | |
| Propane | | 1-Propene | | 1-Propyne | | Cyclopentane | |
| Butane | | 1-Butene | | 1-Butyne | | Cyclohexane | |
| Pentane | | 1-Pentene | | 1-Pentyne | | Cycloheptane | |
| Hexane | | 1-Hexene | | 1-Hexyne | | Cyclooctane | |
| Heptane | | 1-Heptene | | 1-Heptyne | | Cyclononane | |
| Octane | | 1-Octene | | 1-Octyne | | Cyclodecane | |
| Nonane | | 1-Nonene | | 1-Nonyne | | Cyclododecane | |
| Decane | | 1-Decene | | 1-Decyne | | | |

(IUPAC) has provided a set of rules for naming organic compounds in an exact way. In accord with the IUPAC rules, a hydrocarbon is specifically identified by a root, which indicates the number of carbon atoms in the longest continuous chain, and a suffix, which describes the kind of bonds present in the molecule. Prefixes indicate where side chains of carbons or other substituents are attached.

### Straight-Chain Hydrocarbons

A hydrocarbon in which all carbon atoms are $sp^3$-hybridized is a member of the class of **alkanes,** designated by the suffix -**ane.** When a double bond is present, the suffix is -**ene,** and the compound is an **alkene.** When a triple bond is present, the suffix is -**yne,** and the compound is an **alkyne.** The root names indicating the number of carbon atoms in the longest continuous chain are derived from Greek or Latin, except for the first four members of the series. For a cyclic structure, the prefix **cyclo-** is inserted before the root. Table 1.3 gives the skeletons and names of hydrocarbons containing up to ten carbon atoms. Comparable names apply to larger systems.

### Branched Hydrocarbons

For branched hydrocarbons, IUPAC rules dictate that the longest continuous carbon chain be identified as the root, with branching groups named as alkyl substituents. An **alkyl group** can be considered as an alkane from which one hydrogen has been removed; the name is derived by replacing the suffix -**ane** with -**yl.** Thus, $CH_3$ is a methyl group, $C_2H_5$ an ethyl group, and so forth. The position along the main carbon chain where an alkyl group is attached is designated by a number. The numbering of carbon atoms in the chain starts at the end closest to where the substituent is attached so that the lower of two possible numbers can be assigned to that position.

Figure 1.26 presents the names for some six-carbon hydrocarbons. Note that the same compound can often be drawn in more than one way, as shown for 2-methylpentane. However, whether you number from the right or the left along the carbon chain, as long as you assign the carbon to which the methyl group is attached the lowest possible number, you obtain the same unique name for either representation. The two structures shown for 2-methylpentane are really the same compound. (Use your molecular model set to convince yourself that they are indeed identical.) The presence of more than one alkyl group along a carbon chain is indicated by a Greek prefix (di-, tri-, tetra-, penta-, etc., for 2, 3, 4, 5 alkyl substituents). In this

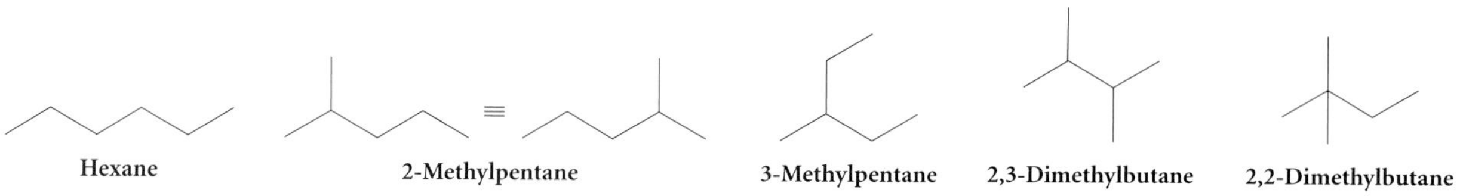

Hexane    2-Methylpentane    ≡    3-Methylpentane    2,3-Dimethylbutane    2,2-Dimethylbutane

**FIGURE 1.26**

The isomeric hexanes ($C_6H_{14}$).

case, each alkyl group must be assigned a number (as low as possible) to indicate its position. Figure 1.26 shows two isomeric dimethylbutanes, as well as 3-methylpentane. (The latter is not named 2-ethylbutane. Recall that the IUPAC rules stipulate that the longest continuous carbon chain is the root in the compound name.)

### Alkyl Groups

#05    Small Alkyl Groups

Alkyl groups with more than two carbons are often designated by common rather than IUPAC names. The common names designate not only the structure of the alkyl group but also the point on the group at which it is attached to the main chain. Table 1.4 shows the structures and names of some alkyl groups that are encountered frequently and should be memorized. The prefix *n-* (normal) refers to a straight-chain alkyl group, whose point of attachment is at a **primary carbon**—that is, one bonded to only one other carbon. The prefix *iso-* describes an alkyl group in which the point of attachment is at the end of a carbon chain that bears a methyl group at the second carbon from the opposite end. (This is easier to picture than to describe; look at the isopropyl, isobutyl, and isopentyl groups in Table 1.4.) The name *s-*butyl designates an alkyl group whose point of attachment is at the second carbon of a four-carbon straight chain. Here, *s-* is short for "secondary," indicating attachment at a **secondary carbon**—that is, one attached to two other carbons. The name *t-*butyl indicates attachment at the group's central carbon. Here, *t-* means "tertiary," indicating attachment at a **tertiary carbon**—that is, one bonded to three other carbons. The prefixes *s-* and *t-* are used only for butyl groups because longer alkyl groups often contain more than one type of secondary carbon, and so either of these designations would not be unique. The prefixes *n-* and *iso-* are used for longer chains. The prefix *neo-* is used almost exclusively for the neopentyl group, —$CH_2C(CH_3)_3$.

**TABLE 1.4**

Some Alkyl Groups and Their Common Names

| Name | Structure | Name | Structure | Name | Structure |
|---|---|---|---|---|---|
| methyl (Me) | —$CH_3$ | isobutyl (*i*-Bu) | $CH_3$ <br> $\vert$ <br> —$CH_2CH$ <br> $\vert$ <br> $CH_3$ | *n*-pentyl (*n*-Pent) | —$CH_2CH_2CH_2CH_2CH_3$ |
| ethyl (Et) | —$CH_2CH_3$ | | | isopentyl (*i*-Pent) | $CH_3$ <br> $\vert$ <br> —$CH_2CH_2CH$ <br> $\vert$ <br> $CH_3$ |
| *n*-propyl (*n*-Pr) | —$CH_2CH_2CH_3$ | *s*-butyl (*s*-Bu) | $CH_2CH_3$ <br> $\vert$ <br> —$CH$ <br> $\vert$ <br> $CH_3$ | | |
| isopropyl (*i*-Pr) | $CH_3$ <br> $\vert$ <br> —$CH$ <br> $\vert$ <br> $CH_3$ | *t*-butyl (*t*-Bu) | $CH_3$ <br> $\vert$ <br> —$C$—$CH_3$ <br> $\vert$ <br> $CH_3$ | neopentyl (*neo*-Pent) | $CH_3$ <br> $\vert$ <br> —$CH_2$—$C$—$CH_3$ <br> $\vert$ <br> $CH_3$ |
| *n*-butyl (*n*-Bu) | —$CH_2CH_2CH_2CH_3$ | | | | |

Write a correct name for each of the following hydrocarbons:

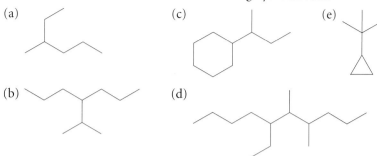

(a)

(c)

(e)

(b)

(d)

---

## *Cis* and *Trans* Isomers

Because of the bond angles that result from the $sp^3$-hybridization of carbon atoms, alkyl groups attached to the ring carbons in a cycloalkane are located either above or below the atoms that form the plane of the ring. If only one such group is present, the designations "above" and "below" are arbitrary because there is no point of reference. However, if two or more groups are present, the relative positions of the groups are fixed: Two groups on the same side of the ring are said to be in a ***cis*** **arrangement;** two groups on opposite sides of the ring are said to be in a ***trans*** **arrangement.**

*cis*                  *trans*

These two arrangements cannot be interconverted by rotation about a $\sigma$ bond and are referred to as ***cis*** and ***trans*** **isomers.** Names of compounds with two or more substituents on a cycloalkane skeleton must designate the substituents as either *cis* or *trans.* For example, the structures shown above are correctly named *cis-* and *trans*-1,2-dimethylcyclohexane.

---

## EXERCISE 1.15

To develop your skills in naming alkanes, assign names to the isomers you drew in Exercises 1.9 and 1.10.

## EXERCISE 1.16

Draw the structure that corresponds to each of the following IUPAC names:

(a) 3,3,4-trimethyloctane

(d) *cis*-1,2-dimethylcyclopentane

(b) *n*-propylcyclopentane

(e) *trans*-1,4-dimethylcyclohexane

(c) 3-ethyl-2-methylhexane

# Alkane Stability

In equilibrium reactions, the more stable compound predominates, and in reactions that can yield more than one product, it is generally (but not always) the more stable product that is formed in the higher yield. Thus, it is important to understand the relative stability of organic compounds.

## ▓ Heat of Combustion

Isomeric alkanes usually have slightly different energies (different free-energy contents). One method of determining the order of stability of isomers is to measure the **heat of combustion** ($\Delta H_c^\circ$), the amount of heat released when each isomer is converted into common products.

Alkanes burn in air, producing water and carbon dioxide. The reaction is as follows:

$$C_nH_{2n+2} + \left(\frac{3n+1}{2}\right)O_2 \longrightarrow n\,CO_2 + (n+1)\,H_2O + \text{heat}\,(\Delta H^\circ)$$

When an alkane is completely burned to carbon dioxide and water, heat is given off; the amount of heat is determined by the carbon and hydrogen content of the hydrocarbon. The greater the amount of heat given off (per mole of carbon dioxide released), the higher was the alkane's energy content and the less stable it was. A **calorimeter** is an instrument that is used to measure precisely the amount of heat released in a reaction. The device consists of a small closed vessel containing a measured quantity of the alkane to be burned (Figure 1.27). The vessel is immersed in a liquid, typically water, and the heat released in the chemical reaction warms the liquid. From the resulting change in temperature and the known heat capacity of the liquid, the heat released in the reaction is calculated. This heat is then converted to a molar basis to obtain the heat of combustion.

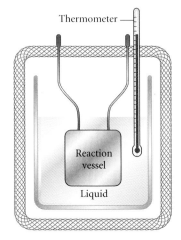

**FIGURE 1.27**

Schematic diagram of a calorimeter.

---

## CHEMICAL PERSPECTIVES

### HYDROCARBON BRANCHING AFFECTS GASOLINE QUALITY

The degree of branching of a hydrocarbon affects not only how much heat is released when the hydrocarbon undergoes combustion, but also how rapidly it reacts. Overly rapid burning of gasoline in an internal combustion engine leads to the sound referred to as "knocking"; the fuel burns so rapidly that it explodes. Isooctane (the "trivial," or non-IUPAC, name for 2,2,4-trimethylpentane) is used as a standard against which other hydrocarbons and mixtures of hydrocarbons are rated. Isooctane is arbitrarily assigned the value 100 on the "octane scale," with *n*-heptane representing zero octane. Measurements of the rate of combustion of gasoline samples are done both in the laboratory and in actual engines, and the average from the two methods serves as the octane rating.

**TABLE 1.5**

Heats of Combustion of Some Alkanes

| Alkane | $\Delta H_c^\circ$ (kcal/mole)[*] | Alkane | $\Delta H_c^\circ$ (kcal/mole)[*] |
|---|---|---|---|
| $CH_4$ | 212.9 | △ | 499.8 (166.6/C)[†] |
| $CH_3CH_3$ | 373.0 | ▢ | 656.3 (164.1/C)[†] |
| (n-propane) | 530.4 | ⬠ | 793.6 (158.7/C)[†] |
| (n-butane) | 687.8 | ⬡ | 944.7 (157.4/C)[†] |
| (n-pentane) | 845.0 | | |
| (isopentane) | 843.4 | ⬡(7) | 1108.3 (158.3/C)[†] |
| (neopentane) | 840.0 | | |

[*]$\Delta H_c^\circ$ is the heat released when 1 mole of compound is completely oxidized to carbon dioxide and water under standard conditions (1 atm $O_2$ at 0 °C).
[†]Values in parentheses are obtained by dividing $\Delta H_c^\circ$ by the number of carbon atoms.

Table 1.5 presents heats of combustion (on a molar basis) for various alkanes. Several trends are clear. The greater the number of carbon atoms in an alkane, the greater is its molar heat of combustion, because more molecules of carbon dioxide and water are produced, releasing energy as they are formed. In a series of isomeric hydrocarbons, linear alkanes (for example, *n*-pentane) have higher heats of combustion than do more highly branched isomeric alkanes (for example, neopentane, or 2,2-dimethyl-propane). A branched alkane is therefore more stable than its straight-chain, unbranched isomer. Cyclopropane and cyclobutane have higher heats of combustion, per carbon atom, than do larger cycloalkanes because of ring strain.

**EXERCISE 1.17**

Calculate the heat of combustion for each of the following compounds. (The heat capacity of water is 1.0 cal/g °C. Assume that no heat is lost during the measurement.)

(a) Combustion of 1.0 g $C_3H_6$ produces enough heat to warm 1000 g of water by 12 °C.

(b) Combustion of 1.0 g $C_6H_{12}$ warms 250 g of water by 45 °C.

### Heat of Formation

A second way to determine the relative stabilities of isomeric alkanes is to measure their heats of formation. The **heat of formation** ($\Delta H_f^\circ$) is a theoretical number that describes the energy that would be released if a molecule were formed from its component elemental atoms in their standard states.

$$nC + (n + 1) H_2 \longrightarrow C_nH_{2n+2} + \text{heat } (\Delta H_f^\circ)$$

**Graphite**          **(Gas)**

Thus, the heat of formation of an alkane represents a measure of the amount of heat that would be released if the alkane were formed from elemental carbon and hydrogen. Alkanes' heats of formation are often calculated from their heats of combustion by the following sequence. (The heats of formation of carbon dioxide and water are constant irrespective of the materials from which they are formed.)

$$C_nH_{2n+2} + \left(\frac{3n + 1}{2}\right) O_2 \quad \Delta H_c^\circ \text{ (alkane)}$$

$$\Delta H_f^\circ \text{ (alkane)} \qquad\qquad n\,CO_2 + (n+1)\,H_2O$$

$$nC + (n + 1) H_2 + \left(\frac{n + 1}{2}\right) O_2 \quad \Delta H_f^\circ (CO_2 + H_2O)$$

## Summary

**1.** Hydrocarbon structures are based on a tetravalent carbon atom: that is, the valence requirement of the Group IV element carbon is satisfied by forming four bonds with other atoms. That the valence requirement of carbon is met can be checked by drawing a Lewis dot structure, which specifies the position of shared electrons between atoms and accounts for non-bonded electrons.

**2.** In carbon, four equivalent $sp^3$-hybrid orbitals are directed toward the apices of a tetrahedron, to minimize electron–electron repulsion. Hydrocarbons containing only $sp^3$-hybridized atoms are called alkanes and are considered saturated. The bond angles at an $sp^3$-hybridized carbon are approximately 109°, and the bond lengths are about 1.54 Å for a carbon–carbon bond and 1.10 Å for a carbon–hydrogen bond. The carbon skeleton is held together by sigma ($\sigma$) bonds.

**3.** Hydrocarbons are nonpolar and, as a group, have relatively low melting points and boiling points. Hydrocarbons are most soluble in other nonpolar liquids and exhibit only weak intermolecular forces dominated by van der Waals attractions.

**4.** Alkanes without rings (acyclic alkanes) have the overall formula $C_nH_{2n+2}$.

**5.** Monocyclic alkanes have the general formula $C_nH_{2n}$. For each ring present, a cycloalkane requires two fewer hydrogen atoms than does a straight-chain alkane with the same number of carbons. Except for cyclohexane, cycloalkanes exhibit ring strain.

**6.** Alkanes are named by IUPAC rules. Each name combines a root, which specifies the number of carbons in the longest continuous chain of carbon atoms, with the suffix *-ane*, which specifies the chemical family of the compound as an alkane. Branches attached to the longest chain are named as alkyl groups, and their positions are indicated by a number that specifies the point of attachment along the chain. Alkanes containing rings are named by inserting the prefix *cyclo-* before the root descriptor.

7. Linear alkanes are less stable (that is, have higher heats of combustion) than their more highly branched isomers. The relative stability of an alkane is determined by measuring the amount of heat released when the alkane is completely burned to water and carbon dioxide (its heat of combustion) or the amount of heat that would be released if the alkane were formed from elemental carbon and hydrogen (its heat of formation).

## Review Problems

**1.1** Identify the atom or ion represented by each of the following electron configurations:

(a) a monocation, $1s^2 2s^2 2p^6$    (c) a dianion, $1s^2 2s^2 2p^4 3s^2 3p^6$

(b) a dication, $1s^2 2s^2 2p^6$    (d) a neutral atom, $1s^2 2s^2 2p^6$

**1.2** Classify the bond shown in red in each of the following structures as polar covalent, nonpolar covalent, or ionic.

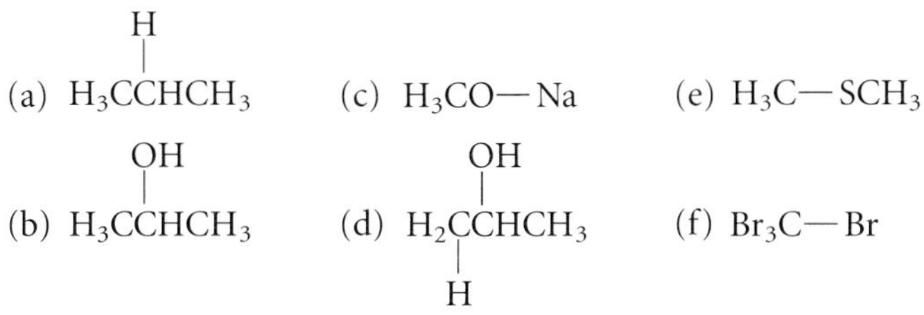

(a) $H_3CCHCH_3$ with H above the central C    (c) $H_3CO-Na$    (e) $H_3C-SCH_3$

(b) $H_3CCHCH_3$ with OH above the central C    (d) $H_2CCHCH_3$ with OH above and H below the central C    (f) $Br_3C-Br$

**1.3** Calculate the formal charges on each atom in the following compounds and ions.

(a) tetrafluoroborate, $BF_4^{\ominus}$    (d) hydronium, $H_3O^{\oplus}$

(b) ammonium, $NH_4^{\oplus}$    (e) molecular hydrogen, $H_2$

(c) methane, $CH_4$

**1.4** Draw a Lewis dot structure for each of the following:

(a) CO    (d) carbonate, $CO_3^{2\ominus}$

(b) $CO_2$    (e) formaldehyde, $H_2C{=}O$

(c) acetylene, $HC{\equiv}CH$

**1.5** Draw a Lewis dot structure for each of the following:

(a) cyclopropane    (c) $CH_3CHClCH_3$

(b) propane    (d) ammonium cation, $NH_4^{\oplus}$

**1.6** In each of the following compounds or ions, identify the atom that bears formal positive charge:

(a)

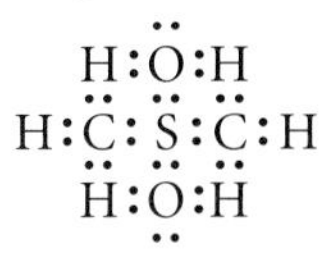

$$\text{H:}\overset{\cdot\cdot}{\text{O}}\text{:H}$$
$$\text{H:}\overset{\cdot\cdot}{\underset{\cdot\cdot}{\text{C}}}\text{:}\overset{\cdot\cdot}{\underset{\cdot\cdot}{\text{S}}}\text{:}\overset{\cdot\cdot}{\underset{\cdot\cdot}{\text{C}}}\text{:H}$$
$$\text{H:}\overset{\cdot\cdot}{\underset{\cdot\cdot}{\text{O}}}\text{:H}$$

(b)

$$\left[ \text{H:}\overset{\cdot\cdot}{\text{O}}\text{::}\overset{\cdot}{\underset{\cdot}{\text{C}}}\overset{\displaystyle\cdot\text{H}}{\underset{\displaystyle\text{H}}{}} \right]^{\oplus}$$

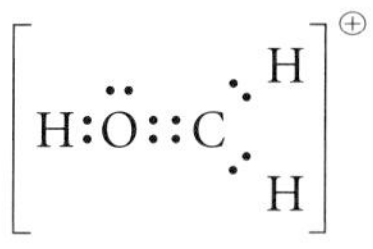

(c)

(d)

**1.7** Draw a structure that corresponds to each of the following names:

(a) 5-*s*-butylnonane

(c) 4-isopropyloctane

(b) *trans*-1,3-diethylcycloheptane

(d) *cis*-1-*t*-butyl-3-methylcyclopentane

**1.8** Draw the structure of each of the following alkyl groups:

(a) *t*-butyl      (b) isopropyl      (c) *s*-butyl      (d) ethyl

**1.9** Provide the IUPAC name for each of the following structures:

(a)

(f)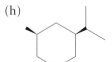

(b)

(g)

(c)

(h)

(d)

(i)

(e)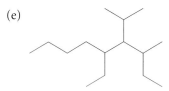

**1.10** It is possible to replace individual hydrogen atoms in an alkane with halogen atoms. If nonequivalent hydrogens are substituted in this way, isomers are formed. The equivalence or nonequivalence of various hydrogens in an alkane can be assessed by considering whether their substitution by halogen atoms results in the formation of isomers. Draw the three possible isomeric structures for $C_5H_{12}$, and determine which of the skeletons has exactly one monofluoro derivative, three different monofluoro derivatives, and four different monofluoro derivatives.

**1.11** Which isomer of the following pairs has the higher heat of combustion?

(a) 2-methylhexane or heptane

(b) 2,2-dimethylpropane or 2-methylbutane

(c) octane or *cis*-1,2-dimethylcyclohexane

**1.12** Compare the following structure with each of compounds A–D:

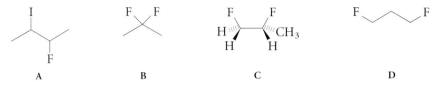

Are they isomers? The same compound (differing only by rotation about a single bond)? Or compositionally different compounds?

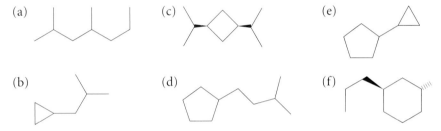

A                B                C                D

## Supplementary Problems

**1.13** Suggest acceptable names for each of the following compounds:

(a)        (c)        (e)

(b)        (d)        (f)

**1.14** Draw a structure correctly representing each of the following:

(a) 3-methylpentane

(b) 3-ethyloctane

(c) 1,1-dimethylcyclopropane

(d) isopropylcyclooctane

**1.15** Suggest an alkane that has:

(a) a lower heat of combustion than butane

(b) a higher heat of formation than butane

(c) a lower boiling point than butane

**1.16** Give the molecular formula for each of the following:

(a) an acyclic alkane with eight carbon atoms

(b) a cyclic alkane (one ring) with six carbon atoms

(c) an alkane with four carbons and no rings

(d) an alkane with twelve carbons arranged into three rings

**1.17** Each of the following incorrect names provides sufficient information to draw a unique structure. Draw each compound, and then determine why the name is incorrect according to IUPAC rules. Name the compound correctly.

(a) 1,1,1-trimethylbutane        (c) 3-*n*-propylpentane

(b) 3-dimethylbutane              (d) 2-isopropylheptane

**1.18** Is each of the following structures the same as 2,4-dimethyloctane or an isomer?

(a)

(b)

$$\underset{\underset{CH_3}{|}}{H_3C-CH-\underset{\overset{|}{CH_3}}{CH}-CH_2-CH_2-CH_2}$$

(c) $\underset{\overset{|}{CH_3}}{H_2C-CH_2-CH_2}-\underset{\overset{|}{CH_3}}{CH}-CH_2-\underset{\overset{|}{CH_3}}{CH}-CH_3$

**1.19** If gasoline is spilled into a lake, an oil slick floating on the surface forms rapidly. What properties of alkanes are responsible for such oil slicks?

**1.20** Assign the relative stabilities of the $C_5H_{12}$ isomers shown. The heat of combustion of pentane is 845.0 kcal/mole, that of 2-methylbutane is 843.4 kcal/mole, and that of 2,2-dimethylpropane is 840.0 kcal/mole.

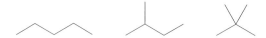

**1.21** All of the hydrogens of a $C_5H_{12}$ isomer are known to be equivalent. Which isomer is it?

**1.22** Differentiate a covalent from an ionic bond using an explanation that would be appropriate for someone who had some science background.

**1.23** Can the relative strengths of the covalent bonds in $O_2$ and $Cl_2$ be determined solely from the melting points? The boiling points?

**1.24** The structures of diamond, graphite, and $C_{60}$ (a fullerene) are shown in Figure 1.1. Predict which of these carbon allotropes has the lowest boiling point.

**1.25** Hydrogen forms a stable compound with each of the atoms in the second row of the periodic table except neon. Based on the positions of these elements in the periodic table, provide formulas for the hydrides of beryllium, boron, carbon, nitrogen, oxygen, and fluorine, and explain why no stable compound $H_xNe$ has yet been prepared.

**1.26** Write a line structure to represent each of the following alkanes, omitting any C—H bonds present:

(a) 2-methylpentane    (b) $CH_3CH_2CH(CH_3)CH_2CH_2CH(CH_3)_2$

**1.27** Define an alkyl group.

# Alkenes, Aromatic Hydrocarbons, and Alkynes

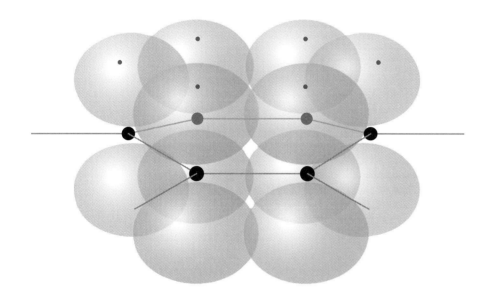

*A three-dimensional representation of benzene showing atoms, bonds, and orbital overlap.*

*A*lkanes, discussed in Chapter 1, are hydrocarbons in which all carbon atoms are $sp^3$-hybridized. Because the carbon and hydrogen atoms in an alkane are connected only by sigma ($\sigma$) bonds, the alkanes constitute the simplest class of organic compounds. To understand the chemical and physical properties of other classes of organic compounds, you must study those of carbon atoms that are not $sp^3$-hybridized and, later, those of elements other than carbon and hydrogen attached to the carbon backbone.

In this chapter, you will study the structures of hydrocarbons that have double bonds (alkenes) or triple bonds (alkynes) between carbon atoms, as well as those that have several such multiple bonds (dienes and aromatic compounds). Hydrocarbons with multiple bonds are said to be *unsaturated*. Double and triple bonds give rise to a characteristic geometry at the atoms involved. In addition, a multiple bond constitutes an area of special reactivity in a molecule (a functional group). When there is more than one multiple bond in a molecule, interactions between groups of multiple bonds can occur, giving rise to compounds with unusual stability (conjugated and aromatic compounds).

As is true of alkanes, unsaturated hydrocarbons (alkenes, dienes, aromatic hydrocarbons, and alkynes) are relatively nonpolar and interact intermolecularly primarily through van der Waals attractions. Compounds in these families thus have relatively low melting points and boiling points (Table 2.1) and are soluble in nonpolar solvents. As with alkanes, melting points and boiling points of unsaturated hydrocarbons increase with increasing molecular weight.

## 2.1

## Alkenes

In alkanes, the valence requirements of each carbon atom are satisfied by the formation of four $\sigma$ bonds. In **alkenes,** the valence requirements of at least two adjacent carbon atoms are satisfied by the formation of three $\sigma$ bonds and one $\pi$ bond between these neighboring atoms. Taken together, a $\sigma$ bond plus a $\pi$ bond constitutes a **double bond.** The high electron density associated with the double bond confers special reactivity on this part of the molecule; such a group of atoms that undergoes characteristic and selective reactions is called a **functional group.** The geometry and chemical characteristics of the double bond can be explained in terms of hybridization at carbon. Unlike the $sp^3$-hybridized carbon atoms found in alkanes, the carbon atoms of the double bond in alkenes are $sp^2$-hybridized and participate in $\pi$ bonding.

### ▓ Hybridization

In discussing alkanes, we considered hybridization in which the $2s$ orbital is mixed with all three $2p$ orbitals of carbon to form four $sp^3$-hybrid orbitals. For alkenes, the geometry and chemical characteristics can be explained on the basis of $sp^2$ hybridization, in which the $s$ orbital is mixed with only two $p$ orbitals. The doubly bonded carbon atoms thus have three equivalent $sp^2$-hybrid orbitals and one unchanged $p$ orbital.

**TABLE 2.1**

Physical Properties of Some Unsaturated Hydrocarbons

| Name | Formula | Boiling Point (°C) | Melting Point (°C) |
|---|---|---|---|
| Ethene | $C_2H_4$ | −104 | −169 |
| Ethyne | $C_2H_2$ | −84 | −81 |
| Propene | $C_3H_6$ | −48 | −185 |
| Propyne | $C_3H_4$ | −23 | −102 |
| 1-Butene | $C_4H_8$ | −6 | −185 |
| *trans*-2-Butene | $C_4H_8$ | 1 | −105 |
| *cis*-2-Butene | $C_4H_8$ | 4 | −139 |
| 1-Butyne | $C_4H_6$ | 8 | −126 |
| 2-Butyne | $C_4H_6$ | 27 | −32 |
| Benzene | $C_6H_6$ | 80 | 5.5 |

C:     1s    $\underbrace{2s\ 2p_x\ 2p_z}$    $\underbrace{2p_y}$

Three    One *p*
$sp^2$-hybrid   orbital
orbitals

Like $sp^3$-hybrid orbitals, the $sp^2$-hybrid orbitals are directed as far away as possible both from each other and from the remaining *p* orbital, which does not take part in the hybridization. In this arrangement, the three $sp^2$-hybrid orbitals lie in one plane and are directed toward the vertices of a regular triangle (at 120° angles). The remaining *p* orbital is perpendicular to this plane, as shown in Figure 2.1.

    The electron density distribution for an $sp^2$-hybridized carbon atom differs significantly from that for an $sp^3$-hybridized carbon. Because an $sp^2$-hybrid orbital has a larger fraction of *s* character, it has greater electron density near the nucleus. As a consequence, the electrons in an $sp^2$ orbital are held more tightly by the nucleus than are electrons near an $sp^3$ orbital. When forming a bond, an $sp^2$-hybridized carbon atom behaves as if it were more

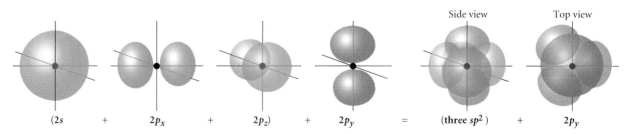

Side view     Top view

$(2s$   +   $2p_x$   +   $2p_z)$   +   $2p_y$   =   **(three $sp^2$ )**   +   $2p_y$

**FIGURE 2.1**

Combination of the 2s and the $2p_x$ and $2p_z$ orbitals produces three hybrid orbitals referred to as $sp^2$ orbitals. The $2p_y$ orbital not involved in hybridization is shown in gray.

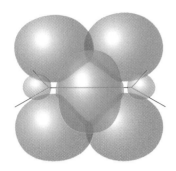

**FIGURE 2.2**

Sigma and pi overlap between $sp^2$-hybridized carbon atoms. For clarity, the $sp^2$-hybrid orbitals are shown in blue; the $p$ orbitals participating in $\pi$ bonding are gray. (Each carbon has two additional $sp^2$-hybrid orbitals involved in additional $\sigma$ bonds.)

#02　Bonding (Alkenes)

electronegative than an $sp^3$-hybridized carbon atom. We will see in later chapters how this difference influences the relative stability of anions formed at carbons of different hybridization.

### Sigma Bonding

A carbon–carbon $\sigma$ bond can be formed by the overlap of two $sp^2$-hybrid orbitals, one from each of two adjacent carbon atoms, pointed toward each other (Figure 2.2). Although an $sp^2$-hybrid orbital has less $p$ character than does the $sp^3$-hybrid orbital used for $\sigma$ bonding in alkanes, the $sp^2$–$sp^2$ $\sigma$ bond in alkenes exhibits many characteristics similar to those of $sp^3$–$sp^3$ $\sigma$ bonds.

### Pi Bonding

In the three-dimensional arrangement shown in Figure 2.2, the $p$ orbitals on adjacent $sp^2$-hybridized carbon atoms are coplanar. In this orientation, they interact above and below the molecular plane, thus forming a second bond in addition to the $\sigma$ bond formed by overlap of $sp^2$-hybrid orbitals. In this **pi ($\pi$) bond,** there are two regions of maximum density, one above and the other below a plane containing the two carbon atoms. (Recall that the electron density in $\sigma$ bonds appears as a cylinder along the axis connecting the two carbon atoms.) Overlap of two $p$ orbitals on adjacent carbon atoms in this fashion cannot be accomplished without simultaneous overlap of $sp^2$ orbitals. Thus, a $\pi$ bond between two atoms is always accompanied by a $\sigma$ bond.

### Molecular Orbitals

Bonds between atoms result from the overlap of atomic orbitals. These bonds may be $\sigma$ or $\pi$. Sigma bonds result from direct overlap of hybrid orbitals having some $s$ character, and $\pi$ bonds result from edge-to-edge overlap of $p$ orbitals. It is useful to view bonds as orbitals themselves, referred to as **molecular orbitals** to differentiate them from simple and hybrid atomic orbitals.

***Orbital Phasing:  The Wave Nature of the Electron.*** An electron has properties that are characteristic of both a particle and a wave. In the restricted environment of the molecular orbital, the electron behaves more like a wave than a particle. For any standing wave that has a node, the **phase** of the wave is opposite from one side of the node to the other. Waves with nodes are found in a vibrating string of fixed length, such as a vibrating guitar string. The string has its lowest energy of vibration (the *fundamental frequency*) when there are no nodes between the two fixed end points. The motion of all parts of the string at any point in time is uniform in direction (but not in magnitude). When the middle of the string is moving up, all other parts are also moving up, and all parts of the string are **in phase.**

The string can also vibrate at a higher frequency (and energy), known as the *first harmonic,* which has a single node in the middle. In this case, all elements on one side of the node move in unison, but in the direction opposite to that of all elements on the other side of the node. Thus, if the left side of the string is moving up, the right side is moving down. Note that we do not know which way a part of the string is moving at any particular

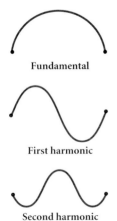

Fundamental

First harmonic

Second harmonic

time, only that the two ends are doing opposite things. The next higher energy of vibration, the *second harmonic,* has two nodes separating three individual segments of the string. As the middle moves in one direction, the two outside portions move in the opposite direction (both outside segments thus always move in the same direction).

When two strings are vibrating at the same frequency, what we hear depends on whether they are in phase or out of phase with each other. If both strings are moving up, for example, they are moving the air in the same direction, and the combined sound is louder than one string alone. On the other hand, if they are moving in opposite directions, and are thus **out of phase,** they tend to cancel each other and the sound produced is muted. (You may have experienced this yourself if you have ever connected the speakers of a stereo system "out of phase." The result is not total silence, because the signal from each speaker arrives at your ears not only directly but also by reflection from the walls and ceiling.) Similarly, when orbitals overlap in phase, electron density increases in the region of overlap. Conversely, when orbitals are combined out of phase, the result is diminished electron density in the region of overlap. Further, there are regions where the electron densities of the two combining orbitals are equal in magnitude but opposite in phase. The resulting molecular orbital will have no electron density in this region, known as a **node.**

***Sigma Molecular Orbitals for Hydrogen.*** As a simple example, let's consider the overlap of two hydrogen $1s$ orbitals, each containing one electron, to form the $\sigma$ bond of molecular hydrogen, $H_2$. Combining two hydrogen $1s$ orbitals that are in phase (Figure 2.3) results in an increase of electron density between the hydrogen atoms. When the orbitals are combined out of phase, as at the top of the figure, the points that are at equal distance from each hydrogen atom (a plane) have a density in each orbital that is equal but opposite in phase. The result is total cancellation, and no electron density exists in this plane, which is thus a node.

Combining atomic orbitals results in the formation of molecular orbitals, a process similar to the formation of hybrid atomic orbitals. Molecular orbitals formed from atomic $s$ orbitals or hybrid orbitals with $s$ character are referred to as **sigma ($\sigma$) molecular orbitals.**

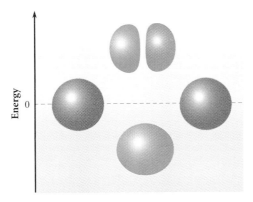

**FIGURE 2.3**

In-phase overlap of $1s$ atomic orbitals of hydrogen produces a bonding $\sigma$ molecular orbital. Out-of-phase overlap of the same orbitals produces an antibonding $\sigma$ molecular orbital. (Here and in other figures an orbital that is blue is in phase with another blue orbital and out of phase with one that is green.)

Recall that molecules are more stable than separated atoms only to the extent that the attraction between particles of opposite charge exceeds the repulsion between like-charged particles. This is best accomplished when the electrons are between the nuclei. In-phase overlap of atomic orbitals accomplishes this objective, whereas out-of-phase overlap results in a nodal surface just where electron density should be maximal for bonding to occur. In-phase overlap of atomic orbitals generates **bonding molecular orbitals;** out-of-phase overlap forms **antibonding molecular orbitals.** Bonding and antibonding $\sigma$ molecular orbitals are symbolized by $\sigma$ and $\sigma^*$, respectively. Electrons in bonding orbitals lower a molecule's energy, whereas electrons in antibonding orbitals increase the molecule's energy compared with the separated atoms. For the hydrogen molecule, the two electrons of the individual atoms can both be placed in the bonding $\sigma$ molecular orbital, resulting in a stable molecule.

The orbital picture in Figure 2.3 is often represented schematically as shown in Figure 2.4. In this diagram, the separate atomic $s$ orbitals are represented at the left and right, with the hydrogen atoms sufficiently far apart that there is no interaction. As the atoms come together, the in-phase overlap decreases in energy while the out-of-phase overlap increases.

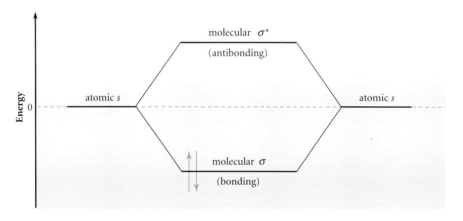

**FIGURE 2.4**

Schematic representation of the orbital picture shown in Figure 2.3. The bonding molecular orbital that results from the in-phase overlap of the 1$s$ atomic orbitals of the hydrogens contains the two spin-paired electrons (shown as blue arrows).

***Sigma Molecular Orbitals for a Carbon–Carbon Bond.*** Let's consider the formation of a $\sigma$ bond using $sp^2$-hybrid orbitals. At the start, the two carbon atoms are separated so that there is no interaction between them (Figure 2.5). Then these atoms move closer together until they are separated by the bonding distance. As these carbons and their orbitals approach each other, phasing becomes important. The in-phase combination decreases in energy, while the energy of the out-of-phase combination increases by an equal amount. The two available electrons, one from each carbon atom, naturally occupy the lower-lying, bonding $\sigma$ molecular orbital, resulting in a molecule that is more stable than the separated atoms. The higher-energy $\sigma^*$ molecular orbital is normally not populated because all of the valence electrons of the two atoms, which completely fulfill those atoms' valence requirement, are accommodated in the $\sigma$ molecular orbital.

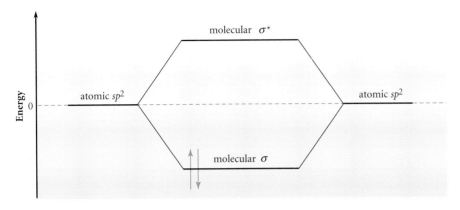

**FIGURE 2.5**

Overlap of two $sp^2$-hybrid atomic orbitals to form a $\sigma$ molecular orbital. Both in-phase and out-of-phase combinations are possible. The former results in a bonding molecular orbital, and the latter in an antibonding molecular orbital.

These molecular orbitals, like the atomic orbitals from which they are constructed, are mathematical surfaces that describe the likely positions of electron density. That two molecular orbitals ($\sigma$ and $\sigma^*$) result from the overlap of the two atomic hybrid orbitals should not surprise you: you have already seen that when atomic orbitals combine in the process of hybridization, the number of hybrid orbitals that results is exactly the same as the number of atomic orbitals from which they were formed.

***Pi Molecular Orbitals for a Carbon–Carbon Bond.*** Overlap of $p$ atomic orbitals of two carbon atoms leads to the formation of bonding and antibonding **pi ($\pi$) molecular orbitals** (Figure 2.6). Two electrons are avail-

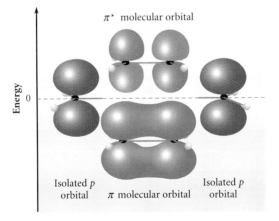

**FIGURE 2.6**

Interaction of $p$ orbitals to form $\pi$ bonding and $\pi^*$ antibonding molecular orbitals in ethylene. The two noninteracting $p$ orbitals are shown in gray because phasing is of no consequence in isolated orbitals. An in-phase combination of $p$ orbitals produces the $\pi$ bonding molecular orbital, and an out-of-phase combination produces the $\pi^*$ antibonding orbital. The bonding orbital has one node (the plane containing the carbon and hydrogen atoms). The antibonding orbital has an additional node, a plane between the carbon atoms and perpendicular to the C—C bond.

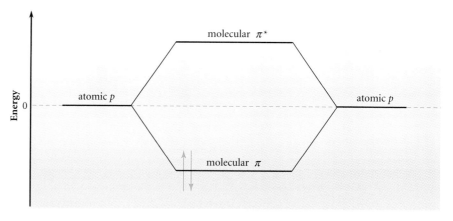

**FIGURE 2.7**

The $\pi$ bonding molecular orbital that results from the in-phase combination of $p$ atomic orbitals holds the two spin-paired electrons in the most stable situation.

able to fill these orbitals, one from each carbon atom. The molecule is most stable when both electrons occupy the bonding $\pi$ molecular orbital (Figure 2.7).

The electron density of hybrid orbitals with $s$ atomic orbital contribution is located primarily on one side of the nucleus, whereas only half of the electron density of a $p$ atomic orbital is found on one side of the nucleus (in any direction). Thus, overlap of hybrid atomic orbitals to form $\sigma$ bonds is greater than overlap of $p$ atomic orbitals to form $\pi$ bonds.

Bonding and antibonding orbitals are equidistant from the zero point of energy, where the atomic orbitals do not interact at all. Indeed, as we will explore in Chapter 3, placing one electron in the $\pi$ molecular orbital and one in the $\pi^*$ molecular orbital results in no net $\pi$ bond, and free rotation about the underlying $\sigma$ bond between the two carbons becomes possible.

## Structures of Alkenes

The structures of alkenes differ from those of alkanes in three important respects: bond angle, bond length, and hindered rotation about the double bond.

***Bond Angle.*** Let's consider a two-carbon molecule containing a double bond. Each $sp^2$-hybridized carbon atom participating in $\sigma$ and $\pi$ bonding with a neighboring carbon atom requires two additional $\sigma$ bonds to satisfy its valence electron requirement. For each carbon atom, two carbon–hydrogen bonds can be formed: two hydrogen $1s$ orbitals overlap with two carbon $sp^2$-hybrid orbitals. The molecule constructed in this way, **ethene** (also called **ethylene**), can be represented by the Lewis dot structure and line structure shown. The $\sigma$ bonds of a doubly bonded carbon atom form a plane with bond angles of 120°.

**Ethene (or ethylene)**

***Bond Length.*** A subtle but significant consequence of the presence of two bonds (rather than one) between atoms is that the atoms are held more closely to each other. The distance between carbons connected by a single bond (two $sp^3$-hybridized carbons) is typically 1.54 Å, whereas that between carbons joined by a double bond (two $sp^2$-hybridized carbons) is 1.34 Å. The length of the bond between an $sp^3$ carbon and an $sp^2$ carbon is intermediate between these two values (see the discussion of higher alkenes).

***Hindered Rotation.*** The existence of a $\pi$ bond between carbon atoms in ethene requires the overlapping $p$ orbitals to be coplanar. This makes rotation about the carbon–carbon $\sigma$ bond impossible without disruption of the $\pi$ bond. For example, if the aligned geometry shown in Figure 2.2 were altered by a 90° rotation about the carbon–carbon bond, to a geometry in which the $p$ orbitals were perpendicular, as in Figure 2.8, overlap between the $p$ orbitals would be reduced to zero. Because overlap is necessary for bonding, this rotation effectively breaks the $\pi$ bond and makes it impossible for the two $\pi$ electrons to be shared between the two atoms. The structure at the left in Figure 2.8 can be represented by a Lewis dot structure in which four electrons are shared between the carbon atoms; the structure at the right has only two electrons shared in the $\sigma$ bond between carbons, with a single electron localized on each carbon.

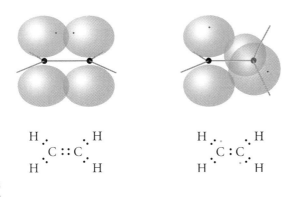

**FIGURE 2.8**

Twisting about a carbon–carbon $\pi$ bond in ethylene. At the left, the $p$ orbitals overlap and the two electrons are shared between the two carbon atoms. When the $p$ orbitals are perpendicular, as at the right, they can no longer overlap, and each electron in a $p$ orbital is held by only one nucleus. Thus, this rotation destroys the $\pi$ bond.

A structure bearing a single unpaired electron is called a **radical.** Because the structure on the right in Figure 2.8 has two noninteracting radical centers, it is called a **biradical.** In the biradical, the valence requirement of neither carbon atom is satisfied, and the orthogonal (perpendicular) geometry is expected to be unstable compared with that of the isomeric structure at the left. Breaking the $\pi$ bond through rotation costs energy, and thus free rotation (like that in molecules containing only $\sigma$ bonds) is not possible about a carbon–carbon $\pi$ bond.

***Higher Alkenes.*** The $sp^2$-hybridized carbon atoms of an alkene can be substituted with hydrogen atoms, as in ethene, or with additional carbon atoms. For example, replacing a hydrogen of ethene with a methyl group results in propene.

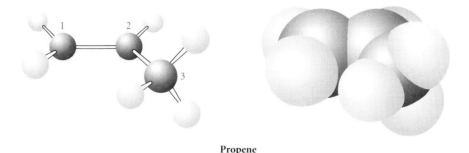

Propene

The angle defined by the two $\sigma$ bonds connecting the $sp^2$-hybridized central carbon atom to the other $sp^2$-hybridized carbon atom and to the $sp^3$-hybridized carbon (C-1—C-2—C-3) is 120°, as is consistent with the hybridization of the central atom. The bond angle at C-3 (C-2—C-3—H-3) is 109°, owing to the $sp^3$-hybridization of C-3. The length of the $\sigma$ bond between C-2 and C-3 (between $sp^2$- and $sp^3$-hybridized carbons) is only slightly shorter (1.50 Å) than that of a typical $sp^3$–$sp^3$ carbon–carbon $\sigma$ bond (1.54 Å) but longer than that of the carbon–carbon double bond (1.34 Å). Ball-and-stick and space-filling models of propene are shown.

Propene

***Degree of Unsaturation.*** A hydrocarbon with a single $\pi$ bond has two fewer hydrogen atoms than does one in which all carbon atoms are $sp^3$-hybridized. An alkene can be thought of as being formed from an alkane by the removal of two hydrogen atoms from adjacent carbon atoms and then linking the two carbon atoms together with the remaining electrons. (Indeed, it is possible to prepare an alkene from an alkane, although not by the formalism shown here.)

An alkane                                                 An alkene

By comparison with the corresponding alkane having the same number of carbon atoms, an alkene has two fewer hydrogen atoms and is said to be **unsaturated.** An alkane, which has no $\pi$ bonds, is referred to as **saturated.** For each double bond present in a hydrocarbon (as for each ring), the molecular formula will have two fewer hydrogen atoms than that required for a saturated, acyclic hydrocarbon, $C_nH_{2n+2}$; that is, simple alkenes have the overall formula $C_nH_{2n}$. The molecular formula of a hydrocarbon can be used to determine the total of double bonds and rings present, although the formula alone will not tell the number of each.

Earlier it was pointed out that the number of rings in a hydrocarbon can be determined from the formula $(2n + 2 - m)/2$, where $n$ is the number of carbon atoms and *m* is the number of hydrogen atoms. Because formation of a double bond and formation of a ring both require the formal loss of two hydrogen atoms, this formula enables us to determine, from the molecular formula, the number of double bonds plus the number of rings present. This value is called the **index of hydrogen deficiency,** or the **degree of unsaturation.**

## EXERCISE 2.1

For each molecule, determine the number of double bonds and/or rings that are present. In each case, draw three carbon skeletons that correspond with your prediction.

(a) $C_5H_{10}$     (b) $C_5H_8$     (c) $C_6H_{12}$

## Isomerism in Alkenes

In a four-carbon straight-chain hydrocarbon containing a double bond, there are two options for isomerism: constitutional and geometric. In **constitutional isomers** of alkenes, the position of the double bond differs. (The term *structural isomers* was formerly used for these isomers.) In **geometric isomers** of alkenes, the molecules differ in the relative disposition of one or more groups about the double bond.

For butene, two constitutional isomers are possible since the double bond can be located between the first and second atoms of the chain (1-butene) or between the second and third (2-butene). Furthermore, because there is restricted rotation about the carbon–carbon double bond, there are two possible geometric isomers for 2-butene: one in which the two hydrogen atoms are on the same side of the double bond, and another in which they are on opposite sides. If the same groups are on one side, the molecule is referred to as a *cis* isomer; if they are on opposite sides, it is called a *trans* isomer.

1-Butene          *cis*-2-Butene          *trans*-2-Butene

The barrier for interconversion of *cis*- and *trans*-2-butene, which requires a rotation that breaks the $\pi$ bond, can be used as a measure of the energy of the $\pi$ bond. An average value for the energy of a carbon–carbon $\pi$ bond is about 63 kcal/mole. This is well above the energy available to molecules at room temperature; therefore, at room temperature *cis*- and *trans*-2-butene exist as distinct chemical entities.

## Nomenclature for Alkenes

In the name of an alkene, the position of the functional group (the double bond) is indicated by a number immediately before the root name and its designated ending (*-ene*). Numbering of the longest carbon chain containing the double bond starts at the end closest to that functional group, allowing it to have the lowest possible number.

Here are some six-carbon isomeric alkenes and their IUPAC names:

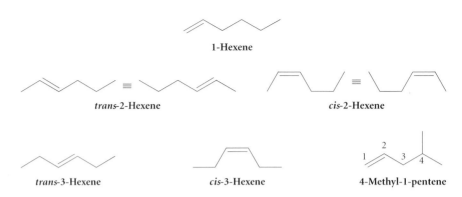

Note that for 4-methyl-1-pentene, the lower of the two possible numbers is assigned to the functional group (the double bond), not to the substituent methyl group. The numbering sequence along the longest chain starts at the end nearest the double bond; that is, the position of the functional group takes precedence over that of the alkyl group.

In summary, use the following steps to apply the IUPAC rules for naming simple alkenes:

**1.** Determine the longest continuous carbon chain that contains the double bond, and name it with the appropriate root and the suffix -*ene*.

**2.** Assign the first carbon atom of the double bond the lowest possible number.

**3.** Name substituent branches as alkyl groups with their positions indicated by numbers.

**4.** Indicate multiple substituents with the appropriate Greek prefix.

### EXERCISE 2.2

Write the IUPAC name for each of these alkenes:

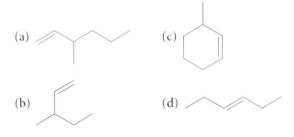

### EXERCISE 2.3

Draw a structure that corresponds to each of the following IUPAC names:

(a) 2-cyclopropyl-1-hexene      (c) 2-isobutyl-1-heptene

(b) 3-ethyl-1-octene

***Naming Geometric Isomers of Alkenes: E and Z Designations.*** The *cis* and *trans* designations for isomeric structures of alkenes are clear in simple cases such as 2-butene, in which there are two identical substituents at each end of the double bond. The *cis* and *trans* designations become ambiguous, however, if the substituents are different, as, for example, for these isomeric alkenes:

Which one is *cis* and which *trans*? To resolve such ambiguity, IUPAC has adopted a way to specify such isomers uniquely. The method consists of establishing group priorities at each end of the double bond and then specifying whether the groups of higher priority at each end are on the same or opposite sides of the double bond.

At each carbon participating in a double bond, assign priorities to the two attached groups according to the atomic number of the attached atom. For example, in 3-methyl-2-pentene, a hydrogen atom and a carbon atom are attached to C-2. Carbon, being of higher atomic number, has priority over hydrogen, and therefore the methyl group has priority at C-2. A methyl and an ethyl group are attached to C-3; there is no atomic number difference of the attached atom. Next, move out along each chain until the point at which an atomic number difference occurs. In both the methyl and the ethyl groups, a $CH_2$ (methylene) group is attached to C-3; however, the methyl then has a hydrogen, whereas the ethyl has a carbon substituent. Because carbon has a higher atomic number than hydrogen, ethyl takes priority over methyl at C-3.

Next, the spatial relation between the groups having priority is indicated by using the designations **E** (from *entgegen*, German for "opposite") and **Z** (from *zusammen*, "together"). In (*E*)-3-methyl-2-pentene, the ethyl group at C-3 is on the *opposite* side of the double bond from the methyl group at C-2. In the *Z* isomer, the ethyl and methyl groups are on the *same* side.

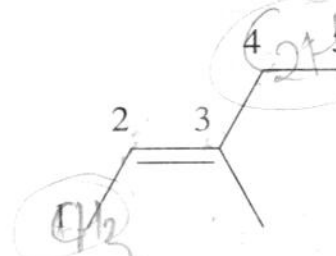

(*E*)-3-Methyl-2-pentene 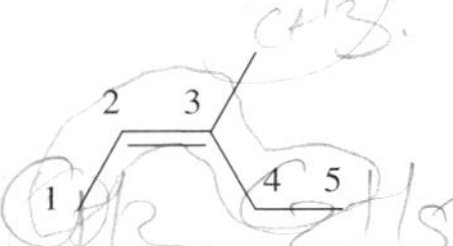          (*Z*)-3-Methyl-2-pentene

Some alkenes need no stereochemical designator: 2-methyl-2-butene has no *cis* or *trans* isomers because it has two identical groups on one of the doubly bonded carbon atoms.

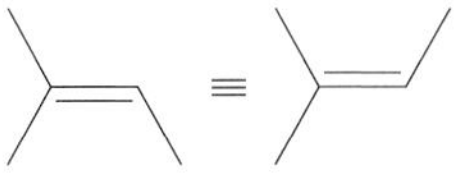

For more practice, you may return to Exercise 2.1 and name the isomeric hydrocarbons you drew.

2-Methyl-2-butene

## EXERCISE 2.4

Write a IUPAC name for each of the following hydrocarbons:

(a)

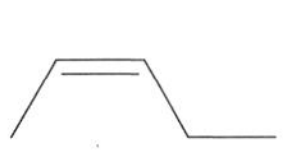

(b)

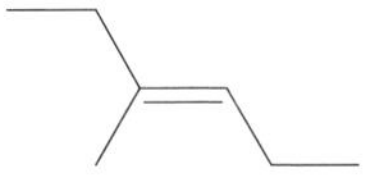

(c)

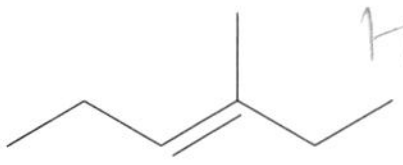

(d)

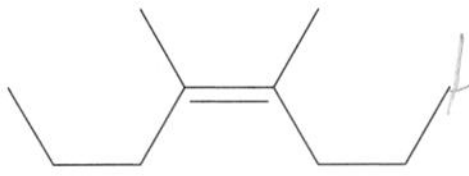

**EXERCISE 2.5**

Draw the structure that corresponds to each of the following IUPAC names:

(a)  (*E*)-2-octene        (c)  *cis*-2-octene

(b)  (*Z*)-3-octene        (d)  *trans*-3-octene

## Alkene Stability

Like isomeric alkanes, isomeric alkenes usually have slightly different energies. And as for the alkanes, the order of stability of alkene isomers can be determined by measuring the heat released upon conversion of each isomer into a common product.

*Heats of Combustion.* Both alkanes and alkenes burn in air. Only the reaction stoichiometry differs:

$$C_nH_{2n+2} + \left(\frac{3n+1}{2}\right) O_2 \longrightarrow n\, CO_2 + (n+1)\, H_2O + \text{heat} \ (\Delta H^{\circ}_c)$$

**Alkane**

$$C_nH_{2n} + \left(\frac{3n}{2}\right) O_2 \longrightarrow n\, CO_2 + \qquad n\, H_2O + \text{heat} \ (\Delta H^{\circ}_c)$$

**Alkene**

The heats of combustion of various alkenes have been measured (Table 2.2). As in alkanes, the greater the number of carbon atoms in an alkene, the greater is its molar heat of combustion. Within a series of alkenes, the *cis* isomer has a higher heat of combustion and is thus less stable than the *trans* isomer. For the butenes, for example, *cis*-2-butene has a higher heat of combustion than does *trans*-2-butene, and that of 1-butene is higher still. Thus, the order of stability of isomeric butenes is *trans*-2-butene > *cis*-2-butene > 1-butene. The *cis* isomer is generally less stable than the *trans* isomer because if the two substituents were on the same side of the double bond they would "bump" into each other if the bond angles were the ideal 120°:

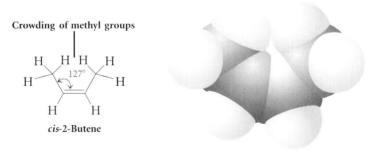

To avoid this interaction, the groups move away from each other, resulting in bond angle distortion that raises the alkene's energy.

# TABLE 2.2

Heats of Combustion* ($\Delta H_c^\circ$) and Hydrogenation† ($\Delta H_h^\circ$) of Several Alkenes

| Hydrocarbon | $\Delta H_c^\circ$ (kcal/mole) | $\Delta H_h^\circ$ (kcal/mole) | Hydrocarbon | $\Delta H_c^\circ$ (kcal/mole) | $\Delta H_h^\circ$ (kcal/mole) |
|---|---|---|---|---|---|
| $H_2C{=}CH_2$ | | $-32.8$ | (cyclohexene) | | $-28.6$ |
| (structure) | | $-30.1$ | (structure) | | $-20.7$ |
| (structure) | $-649.5$ | $-30.3$ | (structure) | | $-24.0$ |
| (structure) | $-805.2$ | $-30.3$ | (structure) | $-607.4$ | $-57.1$ |
| (structure) | | $-30.3$ | (structure) | $-761.7$ | $-54.1$ |
| (structure) | $-647.8$ | $-28.6$ | (structure) | $-768.8$ | $-60.8$ |
| (structure) | $-646.8$ | $-27.6$ | (structure) | | $-60.5$ |
| (structure) | $-645.4$ | $-28.4$ | | | |
| (structure) | $-803.4$ | $-28.5$ | | | |
| (structure) | $-801.8$ | $-26.9$ | | | |
| (structure) | | $-26.6$ | | | |

*$\Delta H_c^\circ$ is the heat released when 1 mole of the alkene is completely oxidized to carbon dioxide and water under standard conditions (1 atm $O_2$ at 0 °C).
†$\Delta H_h^\circ$ is the heat released when 1 mole of the alkene is completely hydrogenated under standard conditions (1 atm $H_2$ at 0 °C).

## EXERCISE 2.6

Based on the generalizations outlined in this section, predict which member of the following isomeric pairs has the higher heat of combustion.

(a) 1-hexene or (E)-2-hexene

(b) (E)-2-hexene or 2-methyl-2-pentene

(c) octane or 2,5-dimethylhexane

(d) (Z)-2-pentene or (E)-2-pentene

***Heats of Hydrogenation.*** Isomeric alkenes can also be ranked in order of stability by measuring the heat released upon the addition of hydrogen to generate a common alkane. For example, as shown in Figure 2.9, 1-butene and *cis-* and *trans-*2-butene all produce butane when hydrogen is added across the double bond. As was true of heats of combustion, a lower heat of hydrogenation indicates a more stable isomer.

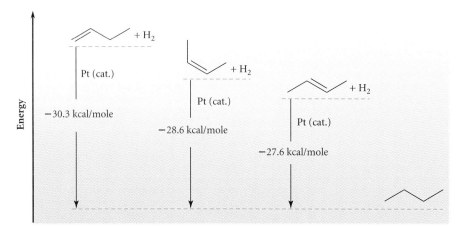

**FIGURE 2.9**

Heats of hydrogenation for the three isomeric butenes 1-butene, *cis*-2-butene, and *trans*-2-butene.

Addition of hydrogen to an alkene requires the presence of a **catalyst,** a compound that is not directly involved in the stoichiometry of the reaction but is necessary for the reaction to proceed at a reasonable rate. Platinum is often used as the catalyst for the addition of hydrogen, and the general reaction is referred to as **catalytic hydrogenation.**

**Catalytic Hydrogenation**

This type of reaction is also called an **addition reaction,** because two simple molecules combine to form a product of higher molecular weight. The product of hydrogenation of an alkene is an alkane lacking carbon–carbon $\pi$ bonds. The letter R is used to represent a hydrogen atom or a general carbon substituent of unspecified nature. When more than one R group is present in a structure, they may be the same or different.

Hydrogenation specifically affects the double bond. The heat of hydrogenation therefore describes not the overall stability of the molecule, but rather the relative stability of the reactive part of the molecule, the carbon–carbon double bond. (Recall that the $\pi$ bond of an alkene constitutes its functional group.) Thus, the heats of hydrogenation of the three isomeric butenes given in Figure 2.9 show that *trans*-2-butene is 1.0 kcal/mole more stable than *cis*-2-butene and 2.7 kcal/mole more stable than 1-butene.

ALKENES AS HIGH-QUALITY GASOLINE COMPONENTS

Alkenes have a higher octane rating than do the corresponding alkanes. It is common practice in petroleum refineries to treat petroleum mixtures with catalysts at high temperatures, at which the equilibrium between alkane and alkene plus hydrogen favors the latter mixture. The hydrogen produced is a valuable by-product that can be used in other processes.

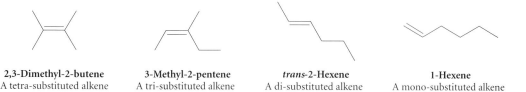

Heats of hydrogenation for several alkenes are given in Table 2.2. Many of the same trends that emerge from combustion analysis are also apparent from calorimetric measurements of hydrogenation, as is seen, for example, in the stability of the various butenes. Heats of hydrogenation of alkenes can be grouped according to the number of alkyl substituents at the double bond. In the isomeric hexenes ($C_6H_{12}$), for example, 2,3-dimethyl-2-butene has four alkyl groups attached to the doubly bonded carbons, 3-methyl-2-pentene has three such alkyl groups, 2-hexene has two, and 1-hexene has only one.

**2,3-Dimethyl-2-butene**
A tetra-substituted alkene

**3-Methyl-2-pentene**
A tri-substituted alkene

*trans*-**2-Hexene**
A di-substituted alkene

**1-Hexene**
A mono-substituted alkene

The more stable alkene has the more highly substituted double bond. This is revealed by a lower heat of hydrogenation.

The electron density of an $sp^2$-hybridized atom is held closer to the nucleus than that of an $sp^3$-hybridized atom. Because alkyl groups are more polarizable than hydrogen atoms, they more readily satisfy the electron demand of the $sp^2$-hybridized carbons of a $\pi$ bond. Therefore, the replacement of hydrogen atoms on a double bond by alkyl groups stabilizes the alkene and accounts for the observed order of stability.

**EXERCISE 2.7**

Rank the following groups of compounds in order of decreasing heats of hydrogenation (that is, by increasing stability):

(a) 1-heptene, 3-heptene, 2-methyl-2-hexene

(b) 1-methylcyclooctene, 3-methylcyclooctene, 1,2-dimethylcyclooctene

(c) 3-ethyl-1-octene, 2-ethyl-1-octene, 3-ethyl-2-octene

***Exhaustive Hydrogenation.*** Hydrogenation also allows chemists to determine the relative contribution of double bonds and rings to the index of hydrogen deficiency. Through exhaustive hydrogenation, hydrogen is added to each double bond until no further unsaturation remains. Rings are generally not affected by catalytic hydrogenation. (However, the strained cyclopropanes will undergo reduction with cleavage of a $\sigma$ bond, although conditions more vigorous than those required for reduction of an alkene are required.) Any hydrogen deficiency remaining after catalytic hydrogenation is taken to completion must be due to the presence of rings rather than multiple bonds.

***Hyperconjugation.*** The stabilizing effect of alkyl groups on adjacent $\pi$ bonds can be explained in terms of hyperconjugation. **Hyperconjugation** results from interaction of the $\pi$ molecular orbital system with adjacent $\sigma$ bonds (illustrated for propene in Figure 2.10). Thus, hyperconjugation occurs when there is net overlap between the $p$ molecular orbitals and the $sp^3$-hybrid orbitals on the adjacent carbon. Because a nodal plane separates the two lobes of the $p$ orbitals of the doubly bonded carbons, overlap with an adjacent hybrid orbital is maximal when the interacting $\sigma$ orbital is perpendicular to the nodal plane of the $\pi$ system and zero when it is in this plane. When there are three identical substituents on the adjacent carbon atom (the three hydrogen atoms of the methyl group in the case of 1-propene), the total hyperconjugative interaction of all three substituents with the $\pi$ system is the same regardless of rotation about the $sp^2$–$sp^3$ $\sigma$ bond.

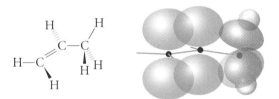

**FIGURE 2.10**

Overlap of C—H bonds of an adjacent carbon with the $\pi$ molecular orbital system results in net stabilization. The $sp^3$-hybrid orbitals are shown in gray, and the $p$ orbitals comprising the $\pi$ bond in blue. Such overlap requires that the hybrid orbital (and therefore the C—H bond) *not* lie in the nodal plane of the $\pi$ system.

Each time an alkyl group replaces a hydrogen atom on a carbon–carbon double bond, the number of possible hyperconjugative interactions increases. For example, in *trans*-2-butene (Figure 2.11), two methyl groups with a total of six $\sigma$ bonds can have hyperconjugative interactions with the $\pi$ system, whereas in the isomeric 1-butene, there are only three such bonds (one C—C and two C—H).

It is primarily through hyperconjugation that alkyl groups donate electron density to, and therefore stabilize, double bonds. The greater stability

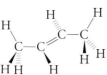

**FIGURE 2.11**

Hyperconjugative interactions in *trans*-2-butene.

of *trans*-2-butene relative to 1-butene (2.7 kcal/mole) can be attributed to the difference in hyperconjugative stabilization from two methyl groups in *trans*-2-butene versus that from one methyl group in 1-butene. The more alkyl groups on the double bond, the greater is the number of atoms that can hyperconjugate and the more stable is the double bond. This stability reveals itself in a lower heat of hydrogenation for the more highly substituted double bond (Table 2.2).

**EXERCISE 2.8**

Draw a three-dimensional representation of each of the following compounds that illustrates hyperconjugative stabilization of the double bond.

(a) 1-butene      (b) *trans*-2-butene      (c) 2,3-dimethyl-2-butene

*Heats of Formation.* The relative stabilities of isomeric alkenes can also be ranked according to their heats of formation. As defined in Chapter 1, the heat of formation is a theoretical number that describes the energy that would be released if a molecule were formed from its component elemental atoms in their standard states.

$$n \text{ C} \;\; + n \text{ H}_2 \longrightarrow \text{C}_n\text{H}_{2n} + \text{heat } (\Delta H_f^\circ)$$
$$\text{Graphite} \quad \text{(Gas)}$$

As a means of ordering alkene stability, heats of formation provide the same information as do heats of combustion. As we saw in Chapter 1, heats of formation can be obtained indirectly from measured heats of combustion and the heats of formation of water and carbon dioxide. In any case, chemists are most interested in the *differences* between the heats of combustion (or hydrogenation or formation) of isomers and not in the absolute values.

**EXERCISE 2.9**

Describe how a measured heat of hydrogenation of an alkene, together with the heat of combustion of hydrogen and the heat of combustion of the alkene's hydrogenation product, can give the heat of combustion of the alkene.

#01   Conjugated Pi Systems

## 2.2

# Dienes and Polyenes

Compounds having two double bonds are called **dienes.** As in simple alkenes, the position of each double bond in a diene is indicated by a number.

The reactivity of dienes varies with the positional relationship of the double bonds—conjugated, isolated, or cumulated. In 1,3-butadiene (Figure 2.12), there are $p$ orbitals on four adjacent atoms, so that the double bonds interact directly. Such a diene is referred to as **conjugated.** In 1,4-pentadiene the array of $p$ orbitals is interrupted by an $sp^3$-hybridized carbon atom at C-3, so that the two double bonds do not interact directly. The diene is referred to as **isolated.** In 2,3-pentadiene, the two double bonds are abutting, and the diene is referred to as **cumulated.** Note that the two double bonds in a cumulated diene are not aligned as in a conjugated diene, and, in fact, C-3 is *not sp²*-hybridized but *sp*-hybridized (see the discussion of *sp* hybridization later in this chapter).

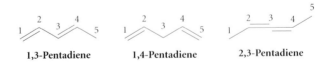

Butadiene (conjugated)                    1,4-Pentadiene (isolated)

**FIGURE 2.12**

Butadiene (left) is a conjugated diene, and 1,4-pentadiene (right) is an isolated diene. For clarity, these representations show only the underlying carbon atoms and $\sigma$ bonds (as lines).

The introduction of a second double bond into a molecule to form a diene (with the removal of two hydrogen atoms from adjacent carbons) changes the overall formula from $C_nH_{2n}$ (for alkenes) to $C_nH_{2n-2}$. Thus, the following five-carbon dienes all have the molecular formula $C_5H_8$:

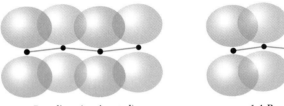

1,3-Pentadiene        1,4-Pentadiene        2,3-Pentadiene

The interaction of the $p$ orbitals differs strikingly in conjugated and isolated diene systems. In the conjugated diene system, $\pi$ orbital overlap between the aligned double bonds stabilizes the molecule. For example, 1,3-pentadiene is more stable than 1,4-pentadiene. The most stable geometry is shown in Figure 2.12. Here, all the atoms lie in one plane (and thus all the $p$ orbitals are perfectly aligned). This geometry is more stable than one in which rotation about the $\sigma$ bond between C-2 and C-3 puts the $\pi$ bonds in perpendicular planes.

The difference in energy between 1,3- and 1,4-pentadiene can be assessed by comparing their heats of hydrogenation (Figure 2.13). Not all of

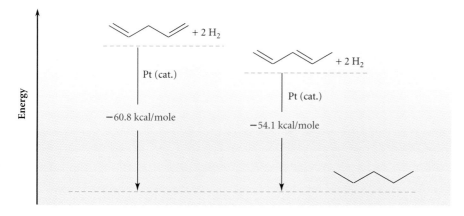

**FIGURE 2.13**

Hydrogenation of either 1,4- or 1,3-pentadiene produces pentane. The difference between the heats of hydrogenation reflects the difference in stability between the nonconjugated and conjugated diene.

the difference (6.7 kcal/mole) can be ascribed to the interaction of the two double bonds in 1,3-pentadiene, the conjugated isomer. Also having some effect is the fact that 1,4-pentadiene has two monosubstituted double bonds (as in 1-butene), whereas 1,3-pentadiene has one mono and one *trans* di-substituted double bond (as in *trans*-2-butene). The degree of substitution of the double bonds in these two isomeric alkenes by itself would stabilize the conjugated isomer by 2.7 kcal/mole compared with the isolated isomer. Thus, although the difference in the heats of hydrogenation for the two isomers is 6.7 kcal/mole, the stability conferred by the interaction of the *p* bonds is 4.0 (6.7 − 2.7) kcal/mole.

Many compounds have more extended conjugated systems. For example, β-carotene, the compound responsible for the yellow-orange color of carrots, and vitamin A, a compound needed for light sensitivity in human vision, contain long conjugated π systems that make them sensitive to light in the visible region (Figure 2.14). These compounds are called **polyenes,** because of the presence of several (*poly* is Greek for "many") double bonds. The dependence of light absorption on extended conjugation is discussed further in Chapter 4.

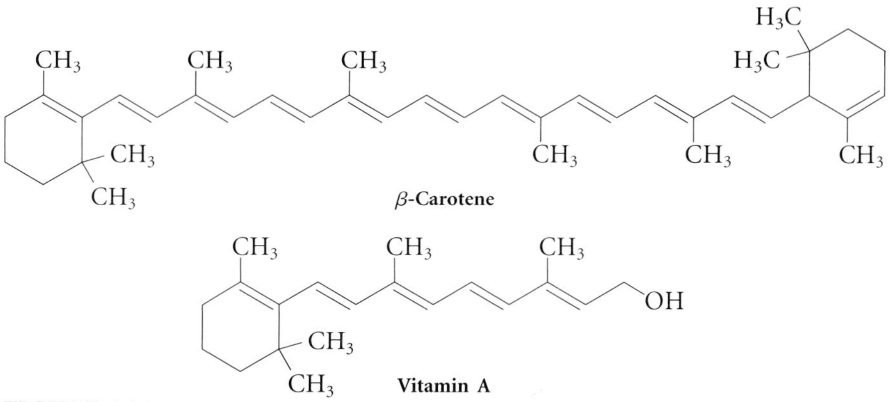

**FIGURE 2.14**

Two naturally occurring polyenes.

## EXERCISE 2.10

Write an IUPAC name for each of the following hydrocarbons:

(a)

(d)

(b)

(c)

(e)

## EXERCISE 2.11

Draw a structure that corresponds to each of the following IUPAC names:

(a) 2-(1-cyclobutenyl)-1-hexene

(c) 1,2-pentadiene

(b) *trans*-1,4-heptadiene

(d) (*E*)-3-methyl-1,3-pentadiene

## 2.3

# Aromatic Hydrocarbons

Planar, conjugated, cyclic, unsaturated molecules constitute another important class of hydrocarbons with $sp^2$-hybridized carbon atoms. Some of these compounds have unusual stability and are referred to as **aromatic hydrocarbons,** originally because of their characteristic odor. The parent compound of this family is benzene, which is a problematic molecule to depict using valence bond representations with pairs of electrons localized between adjacent atoms. Benzene, $C_6H_6$, consists of an array of six $sp^2$-hybridized carbons, each attached by a $\sigma$ bond to a hydrogen atom. From an orbital representation of benzene like that shown at the beginning of this chapter (and also in Figure 2.17), it can be seen that each carbon atom contributes a $p$ orbital with one electron to a $\pi$ system. But which carbons should be joined by double and which by single bonds?

The formulation of benzene's structure as a planar cyclic arrangement of CH units is one of the classic tales of organic chemistry, the structure having been imagined in a daydream in 1865 by the German chemist August Kekulé as a snake biting its tail.

### Resonance Structures

Benzene is known to have equivalent carbon–carbon bond lengths at each position around the ring; so a description of benzene as cyclohexatriene (with alternating single and double bonds) must be wrong. If benzene existed as cyclohexatriene, the double bonds would be shorter than the single bonds, with an *alternation* in bond length from one atom to the next around the ring.

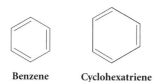

**Benzene**      **Cyclohexatriene**

The fact that the structure of benzene cannot be represented properly with a single structure using only single and double bonds reveals a fundamental problem with the system for drawing chemical structures. However, the bond alternation implied by the cyclohexatriene structure can be avoided by recognizing that there are two possible arrangements for the $\pi$ bonds in benzene. This is indicated in the two representations in Figure 2.15, in which a double bond can be placed either between C-1 and C-2 or between C-1 and C-6. There is no reason for an energetic preference for one or the other of these representations, which are called **Kekulé structures.** Benzene is better represented as a combination of both these structures.

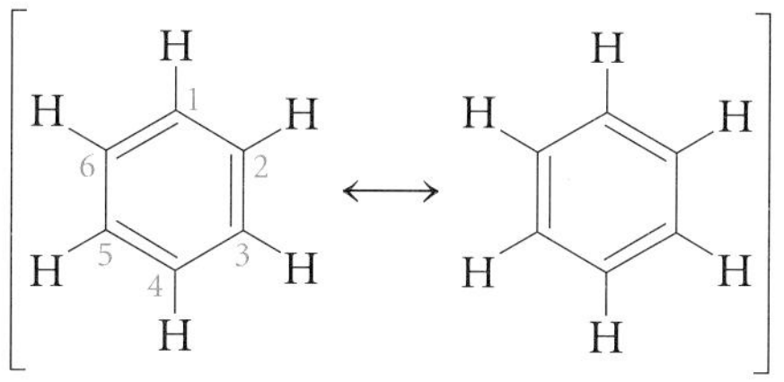

**FIGURE 2.15**

Kekulé structures (resonance contributors) for benzene.

Although the Kekulé structures, which depict benzene as having localized double bonds, are not an accurate representation of the benzene structure, they serve as a convenient shorthand for counting double bonds and electrons. By convention, chemists use a double-headed arrow between such structures to indicate that they are **resonance contributors,** or **resonance structures,** differing only with respect to the formal localization of electrons and *not* with respect to positions of atoms.

**Resonance Structures of Benzene**

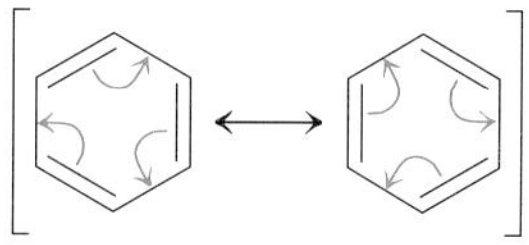

One resonance structure can therefore always be converted to another by moving only electrons. Curved arrows are used to show the motion of electrons.

Another notation in which a hexagon (representing the ring carbons of benzene) encloses a circle (representing the conjugated array of $\pi$ orbitals) is often used to indicate the equal contributions of benzene's resonance structures. This representation avoids the problems of trying to draw descriptive structures with conventional valence bond representations.

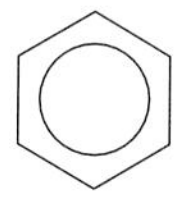

**Alternative notation for benzene**

**EXERCISE 2.12**

Draw an alternative resonance structure for each of the following:

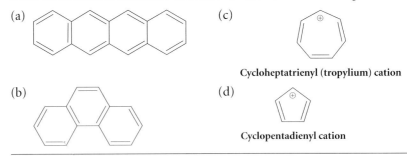

(a)

(c)

Cycloheptatrienyl (tropylium) cation

(b)

(d)

Cyclopentadienyl cation

Many other possible resonance contributors can be drawn to depict possible electron distributions in benzene. These resonance structures are drawn by keeping the position of each atom fixed while shifting electrons within the $\pi$ system. In the drawings shown here, two electrons from a $\pi$ bond are localized on a single carbon, making that atom negatively charged and some other atom of the conjugated system positively charged.

**Kekulé form**          **Zwitterions**

These structures, which are neutral overall but contain equal numbers of locally charged (plus and minus) centers, are called **zwitterions.**

These zwitterionic resonance contributors are of higher energy than the uncharged Kekulé form, shown at the left. Not only does it cost energy to create charged centers from a neutral species, but these structures also have one fewer covalent bond than do the Kekulé contributors. Thus, the importance of zwitterions in describing the electron distribution in benzene is minor, and chemists usually ignore such structures in their thinking. Much the same can be said about biradical contributors, in which single electrons are localized on two of the atoms of the conjugated systems.

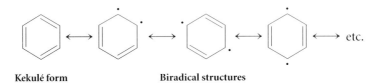

**Kekulé form**          **Biradical structures**

(Recall the twisted structure of ethylene discussed earlier.) Unlike the zwitterionic contributors, biradical structures do not have charge-separated states. However, a biradical contributor lacks one of the $\pi$ bonds that is intact in a Kekulé form.

***Electron Pushing.*** In converting one resonance structure of benzene to another, curved arrows show the flow of electrons as some bonds disappear and others appear. We will make extensive use of these curved arrows especially in depicting the conversion of one molecule to another, because the arrows help to focus attention on the parts that are changing. These arrows have specific meaning and must be drawn with care. An arrow begins at the source of electrons, and the arrowhead points at their destination.

As a simple example, consider the reaction of hydronium ion ($H_3O^{\oplus}$) with water. In the course of this reaction, a proton is transferred from one oxygen atom to another. The electrons originally localized between the hydrogen nucleus and the oxygen of the hydronium ion become a lone pair of electrons on oxygen, and a new O—H bond of the product hydronium ion is formed from a lone pair of electrons on the original $H_2O$ molecule.

**Reaction of Hydronium Ion with Water**

These changes in the location of electrons are clearly represented with two curved arrows. One begins at the middle of the O—H bond and points at the oxygen atom; the other begins at one of the lone pairs of electrons of the water molecule and points at the hydrogen atom that is being transferred.

## Stability

The unusual stability of benzene (and related structures) is seen in both its heat of hydrogenation and its chemical reactivity, which differ appreciably from those usually observed in conjugated alkenes, dienes, and trienes. The differences in reactivity will be discussed further in Chapter 11, but we consider the hydrogenation data here. Adding hydrogen catalytically to benzene to generate cyclohexane requires more severe conditions of temperature and pressure than does adding hydrogen to analogous alkenes.

$(\Delta H° -49.3 \text{ kcal/mole})$

Benzene     Cyclohexane

The stoichiometry of the reaction tells us that three moles of hydrogen are taken up per mole of benzene. If the double bonds were noninteractive (as in a hypothetical cyclohexatriene), the heat of hydrogenation of benzene would be approximately three times that of cyclohexene. The difference between the heat of hydrogenation of benzene (49.3 kcal/mole) and three times that of cyclohexene ($3 \times 28.4$ kcal/mole) is 35.9, or approximately 36 kcal/mole.

$(\Delta H° -28.4 \text{ kcal/mole})$

Cyclohexene     Cyclohexane

This difference cannot be due to a simple conjugation effect because the difference between the heat of hydrogenation of 1,3-cyclohexadiene (54.9 kcal/mole) and twice the heat of hydrogenation of cyclohexene (2 × 28.4 kcal/mole) is small (approximately 2 kcal/mole).

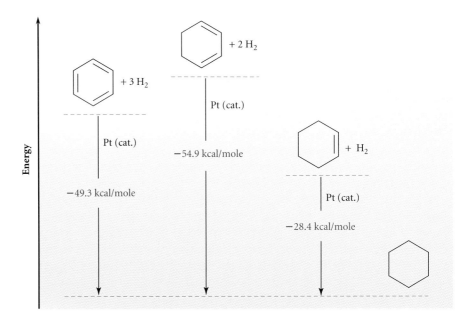

(The various heats of hydrogenation are summarized in Figure 2.16.) The larger energy difference seen with benzene must therefore have a different origin, and the special stability afforded by a planar cyclic array of *p* orbitals containing six electrons is known as **aromaticity.**

**FIGURE 2.16**

Heats of hydrogenation for the reductions of benzene, 1,3-cyclohexadiene, and cyclohexene to cyclohexane.

Other evidence that aromaticity is not a simple conjugation effect can be seen in the differing stabilities of benzene and cyclic hydrocarbons containing four and eight CH units, respectively. Despite the fact that cyclobutadiene can be written with two resonance contributors analogous to the Kekulé structures of benzene, it is an exceedingly unstable molecule that can be prepared and studied only at very low temperature under special conditions.

**Resonance Contributors
of Cyclobutadiene**

## CHEMICAL PERSPECTIVES

### CYCLOOCTATETRAENE: A NONAROMATIC, CONJUGATED, CYCLIC HYDROCARBON

Cyclooctatetraene was produced in large quantities during World War II by German chemists seeking to convert acetylene (ethyne) into benzene. Acetylene is a low-boiling hydrocarbon that can explode spontaneously when stored under pressure. It can be readily prepared from calcium carbide, which, in turn, is prepared by heating $CaCO_3$ and coal. Trimerization of acetylene yields benzene; the joining of four acetylene molecules yields cyclooctatetraene.

$$CaCO_3 + C_n \text{ (coal)} \xrightarrow{\Delta} CaC_2$$

Benzene                    Cyclooctatetraene

In fact, cyclobutadiene was not prepared until the late 1960s, whereas benzene was isolated early in the nineteenth century and can be stored easily at room temperature. For benzene to react, much more rigorous conditions are required than with simple unsaturated hydrocarbons; that is, reactions of aromatic compounds are induced only with some difficulty.

Cyclooctatetraene has been shown to exist not as a planar hydrocarbon but rather as a tub-shaped unsaturated molecule. In this geometry, the $p$ orbitals are not well aligned for interaction.

Cyclooctatetraene

### Aromaticity and Hückel's Rule

Both cyclobutadiene and cyclooctatetraene contain multiples of four electrons—that is, four and eight electrons, respectively. Benzene, on the other hand, contains two electrons more than a multiple of four (that is, $6 = 4n + 2$, in which $n$ is an integer—in the case of benzene, $n = 1$). This

#02   Benzene and Aromatic Compounds

distinction was recognized in 1938 by Erich Hückel, a German chemist who generalized this observation into what has come to be known as **Hückel's rule:** any planar, cyclic, conjugated system containing $(4n + 2)$ $\pi$ electrons (where $n$ is an integer) experiences unusual aromatic stabilization, whereas those containing $(4n)$ $\pi$ electrons do not. For aromatic molecules to be stabilized by orbital interaction, as predicted by Hückel's rule, the $p$ orbitals (Figure 2.17) must be aligned in a planar geometry.

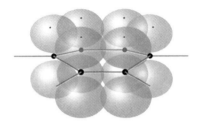

**FIGURE 2.17**

Benzene shown as a cyclic set of six $sp^2$-hybridized carbon atoms. The unhybridized $p$ orbitals of the six carbons overlap, with each contributing one electron for a total of six $\pi$ electrons. Benzene is thus a Hückel aromatic system.

Larger or smaller cyclic structures that maintain this alignment also are subject to Hückel's rule. For example, a cyclic, planar, conjugated array containing $4n + 2$ electrons in a smaller ring is found in the cyclopropenyl cation ($n = 0$) and in the cyclopentadienyl anion ($n = 1$):

**Cyclopropenyl cation**
A Hückel aromatic (2-electron) system
$4n + 2, n = 0$

**Cyclopentadienyl anion**
A Hückel aromatic (6-electron) system
$4n + 2, n = 1$

Note that, in the cyclopropenyl cation, one $p$ orbital is vacant and the ion bears a formal $(+1)$ charge. We can write three resonance structures for this cation (two of which are shown here) by shifting electrons in the $\pi$ bond from the position between C-1 and C-2 to the position between C-2 and C-3. All of the atoms remain in the same position in each resonance structure; only the position of the electrons is changed.

In the cyclopentadienyl anion, one $p$ orbital is doubly occupied, and the ion bears a formal $(-1)$ charge. By shifting electrons in the $\pi$ system as we did in the cyclopropenyl cation, we can write five equivalent resonance contributors (only two are shown above).

We can count the number of electrons in such systems by recognizing that each $p$ orbital in a formal $\pi$ bond is populated by a single electron. In the cyclopropenyl cation, two of the $p$ orbitals (those participating in the $\pi$ bond) contain one electron. The third $p$ orbital is vacant. In the cyclopentadienyl anion, four of the $p$ orbitals are singly occupied and one is doubly occupied. In these analyses, a double bond contributes two electrons to the $\pi$ system, a center with a vacant $p$ orbital (positively charged atom) contributes zero, and a center with a doubly occupied $p$ orbital (negatively charged atom) contributes two, as is consistent with our earlier discussion of formal charge calculation. Thus, the cyclopropenyl cation is a two-electron Hückel system $[(4 \times 0) + 2]$ and the cyclopentadienyl anion is a six-electron Hückel molecule $[(4 \times 1) + 2]$.

The cyclopropenyl anion and the cyclopentadienyl cation each contain four $\pi$ electrons ($4n$, where $n = 1$) and therefore lack the aromatic stabilization characteristic of a Hückel system. These structures are known to be so unstable that they have in fact been called **antiaromatic.**

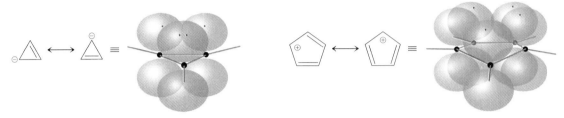

Cyclopropenyl anion
A Hückel antiaromatic (4-electron) system
$4n, n = 1$

Cyclopentadienyl cation
A Hückel antiaromatic (4-electron) system
$4n, n = 1$

### EXERCISE 2.13

Using Hückel's rule, predict which of the following hydrocarbons will exhibit aromatic stabilization. (One resonance contributor for each is shown.)

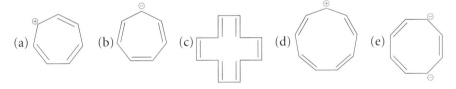

### EXERCISE 2.14

Use curved arrows to show the electron movement necessary to convert one resonance structure of cyclopentadienyl anion to a different one. Do the same for the cyclopentadienyl cation.

### Arenes

Derivatives of benzene obtained by the replacement of hydrogen by other groups or by the fusion of additional rings are called **arenes.** The delocalized $p$ orbitals of benzene shown in Figure 2.17 are perpendicular to the $\sigma$ bonds by which the six hydrogen atoms are attached to the carbon atoms. Thus, these $\sigma$ bonds are orthogonal to the $\pi$ system and do not significantly affect aromaticity. The hydrogen atoms of benzene can be

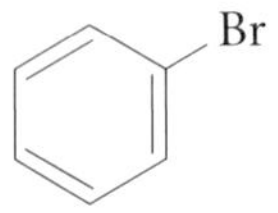

**Bromobenzene**

replaced by other substituents—for example, by bromine to form bromobenzene. The carbon–bromine bond is also in the plane of the six-membered ring. Substituted benzenes such as bromobenzene maintain a Hückel number of electrons in the $\pi$ system and, to a first approximation, exhibit the same aromaticity as the parent benzene.

*Polycyclic Aromatic Hydrocarbons.* The particular stability of the benzene ring is also found in fused cyclic aromatic hydrocarbons. These **polycyclic aromatic hydrocarbons** exhibit chemical stability similar to that of benzene. A representative sample, together with their common names, is shown in Figure 2.18.

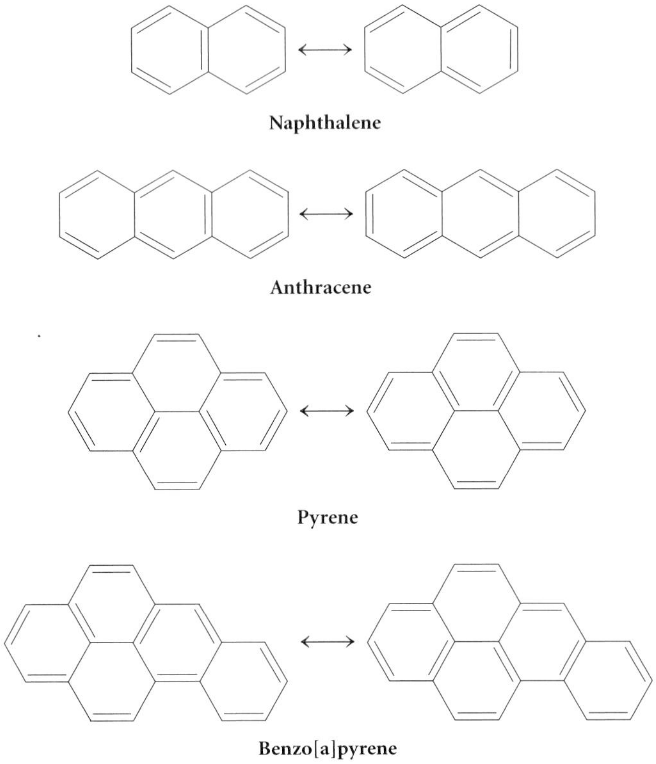

Naphthalene

Anthracene

Pyrene

Benzo[a]pyrene

**FIGURE 2.18**

Some fused-ring (polycyclic) aromatic hydrocarbons.

Like monocyclic aromatic compounds, polycyclic aromatic compounds have several important resonance contributors. Bear in mind that each such compound can have more than one important contributing resonance structure. Benzo[a]pyrene was one of the first clearly identified **carcinogens,** or cancer-inducing agents. This compound, found in soot from the partial combustion of wood, was shown in the nineteenth century to be responsible for inducing scrotal cancer in chimney sweeps in London.

## ▓ Nomenclature for Aromatic Hydrocarbons

The IUPAC system of nomenclature for aromatic hydrocarbons retains many of the common names that were in use long before the Union was formed. Thus, although each of the following compounds could be named

72

## CHEMICAL PERSPECTIVES

### CARCINOGENICITY OF BENZO[a]PYRENE

A derivative of benzo[a]pyrene, rather than the hydrocarbon itself, is the real culprit in inducing cancer. Because benzo[a]pyrene is a large hydrocarbon with a very low solubility in water, it collects in the liver, which is composed in part of fats—hydrocarbon-rich molecules that will be discussed in Chapter 17. There are many enzymes in the liver that carry out oxidation reactions on waste products of metabolism, as well as on unneeded materials consumed in the diet. These oxygenated materials have a higher solubility in water and can thus be excreted. Unfortunately, the oxidation product of benzo[a]pyrene interacts with DNA and results in abnormal cell growth (cancer).

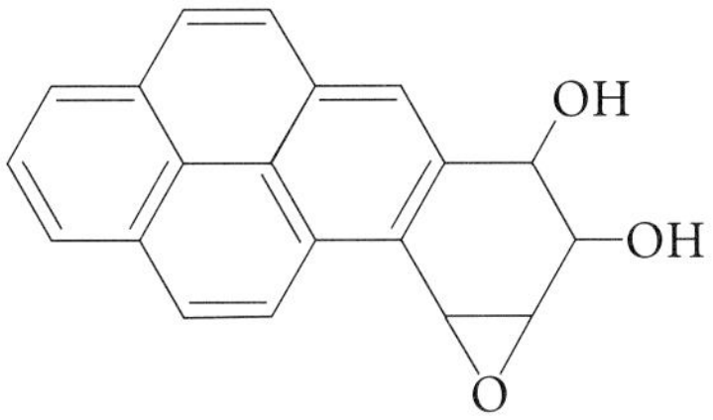

**Oxidation product of benzo[a]pyrene**

as a substituted benzene (for example, methylbenzene for toluene), the common names shown are used almost universally:

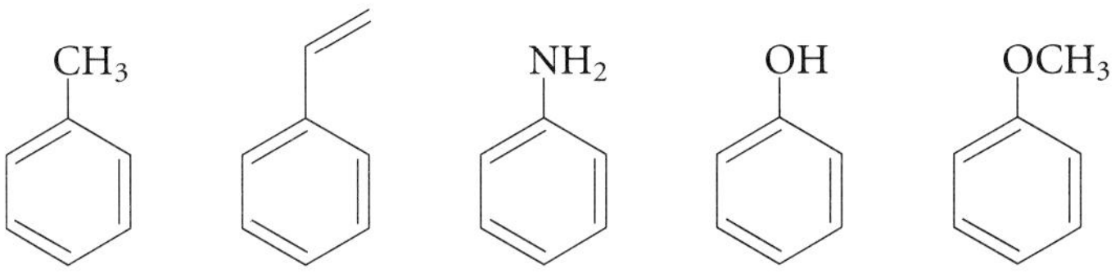

**Toluene**  **Styrene**  **Aniline**  **Phenol**  **Anisole**

There are three constitutional isomers of any disubstituted benzene, and there are two acceptable methods for describing the relative orientation of the substituents: using relative position numbers, and using the designations *ortho, meta,* and *para.* (*Ortho, meta,* and *para* are often abbreviated as *o, m,* and *p,* respectively.) The following three dimethylbenzenes are also referred to as xylenes:

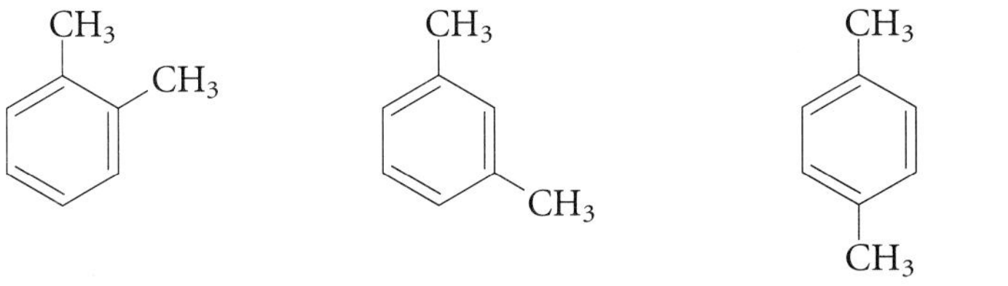

**1,2-Dimethylbenzene**
*ortho*-Dimethylbenzene
*ortho*-Xylene

**1,3-Dimethylbenzene**
*meta*-Dimethylbenzene
*meta*-Xylene

**1,4-Dimethylbenzene**
*para*-Dimethylbenzene
*para*-Xylene

(Note that a 1,5 isomer would be identical with a 1,3 isomer, and a 1,6 isomer would be identical with a 1,2 isomer. In all cases, the lowest possible number is used.)

Aromatic compounds with two different substituents are named, if possible, as a derivative of one of the common monosubstituted aromatics.

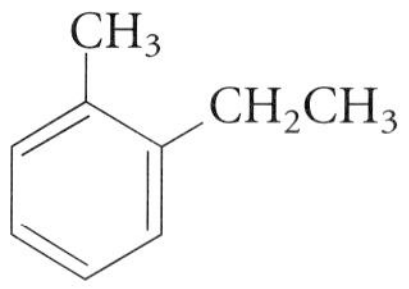

*ortho*-Ethyltoluene 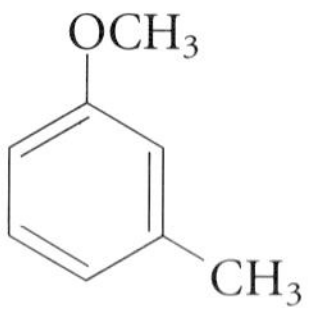        *meta*-Methylanisole 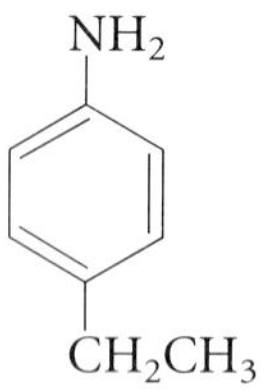        *para*-Ethylaniline

With three and more substituents on a benzene ring, numbers are always used to indicate the relative positions. The number scheme that gives the lowest numbers is always used.

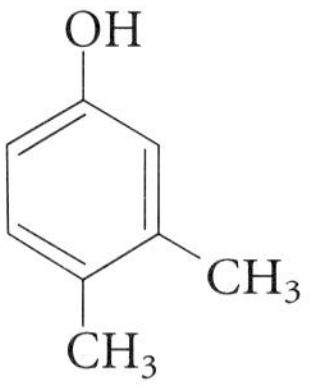

**3,4-Dimethylphenol** 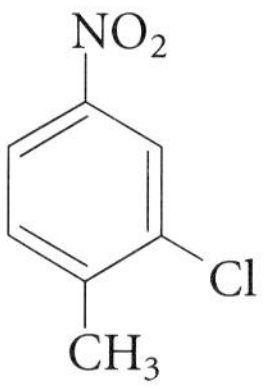        **3-Chloro-4-methylnitrobenzene**
(*not* 4-methyl-5-chloronitrobenzene)

## EXERCISE 2.15

Write an acceptable name for each of the following hydrocarbons:

(a)        (b) 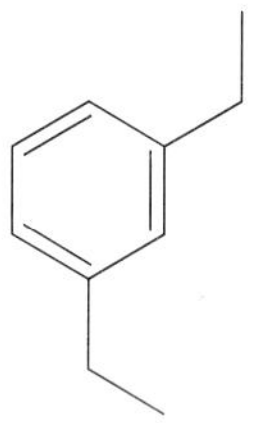        (c) 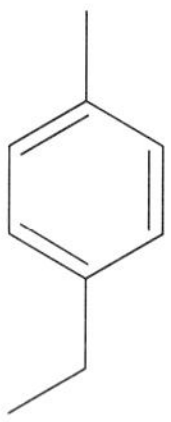        (d)

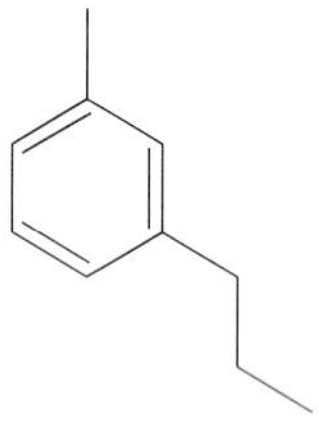

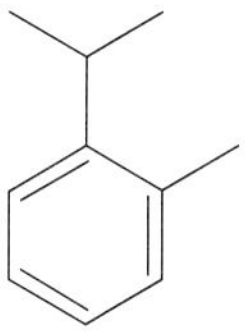

## EXERCISE 2.16

Draw a structure that corresponds to each of the following names:

(a)  *ortho*-diethylbenzene

(b)  *meta*-diethylbenzene

(c)  *para*-isopropyltoluene

## EXERCISE 2.17

Draw all of the possible isomers for a trisubstituted benzene that has one methyl, one ethyl, and one *n*-propyl substituent.

### ALLYL: A GROUP FOUND IN GARLIC

Many common names of organic compounds and groups derive from the botanical names of the plants from which they were first isolated. The allyl group is present in alliin (a key amino acid) and in allicin (responsible for the odor of garlic), both isolated from garlic. The term *allyl* derives from the botanical name for garlic (*Allium sativum*), which comes from the Latin word *allium,* from a Celtic word meaning "pungent."

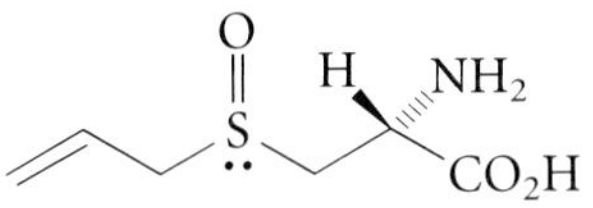

**Alliin**

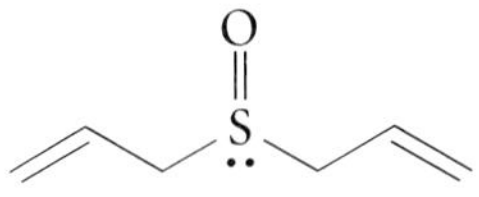

**Allicin**

Garlic supposedly protects against stroke, coronary thrombosis, and hardening of the arteries. Its extracts also have antibacterial and antifungal activity—and allegedly repel vampires.

**Unsaturated Substituent Groups.** A benzene ring as a substituent on a carbon chain is called a **phenyl group,** and a (generic) arene as a substituent is called an **aryl group.** An alkene substituent is called a **vinyl group** when the attachment is to one of the carbons in the double bond. When a three-carbon alkenyl chain is attached at the atom adjacent to the double bond, the substituent is called an **allyl group.**

In the aryl group, X is an undefined substituent for which the point of attachment can be at the *ortho, meta,* or *para* position. It is common for an aromatic ring to have multiple substituents. In each structure, the wavy line through the bond at the left indicates that this is the bond that links the substituent to some other molecular fragment.

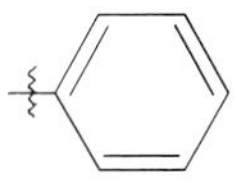

**Phenyl** 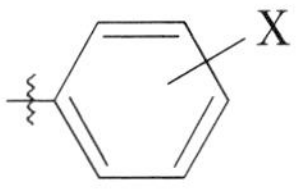  **Aryl**

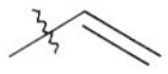

**Vinyl**   **Allyl**

### EXERCISE 2.18

Write an acceptable name for each of the following hydrocarbons:

(a)

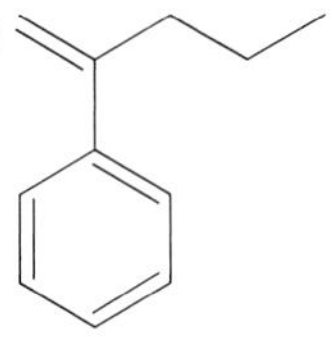

(b)

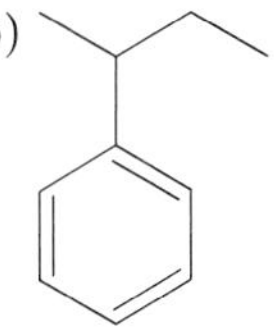

(c)

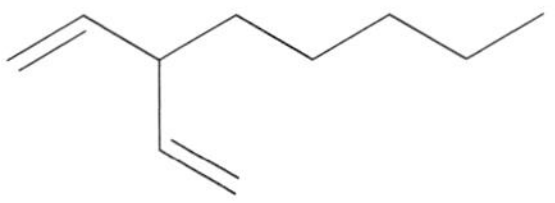

(d)

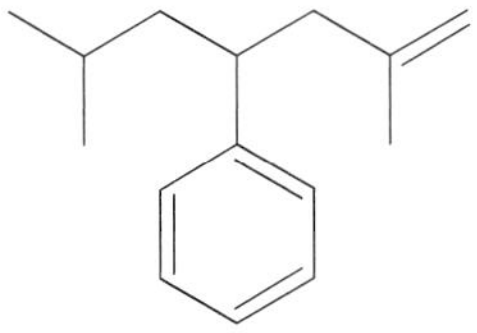

Use curved arrows to show how the electrons move in converting one resonance structure of naphthalene to the other (Figure 2.18). Do the same for pyrene.

#02   Bonding (Alkynes)

## 2.4

# Alkynes

When a carbon atom forms $\sigma$ bonds to two atoms, rather than three or four, its valence requirement is satisfied by the formation of two $\pi$ bonds. When both $\pi$ bonds are directed to the same atom, a triple bond (composed of one $\sigma$ and two $\pi$ bonds) is formed. In this section, we will discuss the geometry and chemical character of triple bonds and how $sp$ hybridization (rather than the $sp^3$ hybridization in alkanes or the $sp^2$ hybridization in alkenes) is required to form triple bonds.

### ■ *sp* Hybridization

In the third fundamental type of carbon hybridization, an $s$ orbital mixes with one of the $p$ orbitals to produce two hybrid orbitals.

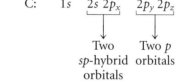

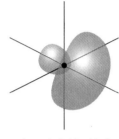

**An *sp*-hybrid orbital**

The two hybrid orbitals are directed as far from each other as possible, producing a 180° bond angle at the **sp-hybridized** atom, as shown in Figure 2.19. The remaining $2p$ orbitals (which do not participate in hybridization) are orthogonal to each other and to the hybrid orbitals. The two $sp$-hybrid orbitals lie along the $x$-axis, pointing in opposite directions. One of the $sp$-hybrid orbitals is shown by itself in the margin, revealing the "tail" lobe that is common to $sp$-, $sp^2$-, and $sp^3$-hybrid orbitals.

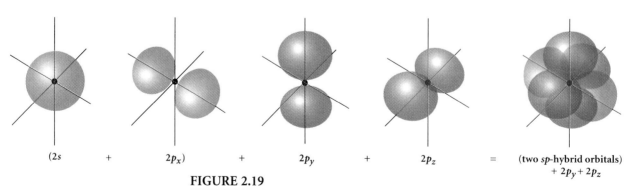

**FIGURE 2.19**

Combination of the $2s$ and $2p_x$ orbitals produces two $sp$-hybrid orbitals aligned along the $x$-axis. The $2p_y$ and $2p_z$ orbitals not involved in hybridization are shown in gray.

The *sp*-hybrid orbitals on adjacent carbon atoms can be used to form a carbon–carbon σ bond. At the same time, the $p_y$ and $p_z$ orbitals overlap to form two π bonds (Figure 2.20). The remaining *sp*-hybrid orbitals overlap with orbitals from substituents to form a second σ bond to each carbon atom. The two carbon atoms and the substituents at either end are all colinear. This geometry also allows for optimal overlap between the aligned $p_y$ and $p_z$ orbitals on adjacent carbons, above and below the σ bond and in front of and behind it. (The directional subscripts, $p_y$ and $p_z$, are arbitrary.) These overlaps thus form one σ and two π bonds between the carbons. The σ bond connecting triply bonded carbons is completely surrounded by a cloud of π-electron density; its bond length (1.20 Å) is less than that of single and double bonds. The Lewis dot structure and the line structure for the simplest alkyne **ethyne** (also called **acetylene**), $C_2H_2$, are shown.

H:C:::C:H     H—C≡C—H

**Ethyne, or acetylene**

Because the carbon–carbon σ bond at an *sp*-hybridized atom has less *p* character than those at *sp²*- or *sp³*-hybridized atoms, the *sp*-hybrid orbitals are less elongated and the σ bond formed from them is shorter, whether the *sp*-hybridized carbon is bonded to an *sp*-, *sp²*-, or *sp³*-hybridized atom. Table 2.3 (on page 78) summarizes the dependence of bond lengths and bond angles on hybridization for several hydrocarbons.

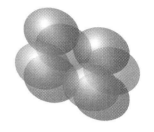

**FIGURE 2.20**

A three-dimensional view of carbon–carbon σ (top) and π bonding (bottom) in an alkyne.

### Higher Alkynes

A hydrocarbon containing a triple bond is called an **alkyne.** Having formally lost two hydrogen atoms in forming each π bond, an alkyne has the overall formula $C_nH_{2n-2}$, that is, four fewer hydrogen atoms than the corresponding alkane. Thus, the presence of one triple bond is equivalent to two units of unsaturation.

**An alkyne**

Like the carbon–carbon double bond, the carbon–carbon triple bond is a functional group. Similar to the structures for dienes (presented in Section 2.2), it is possible to have conjugated diynes, conjugated enynes (enynes are hydrocarbons with both a double and a triple carbon–carbon bond), isolated diynes, and isolated enynes.

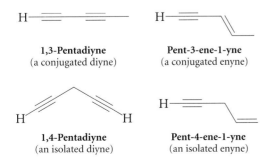

**1,3-Pentadiyne**
(a conjugated diyne)

**Pent-3-ene-1-yne**
(a conjugated enyne)

**1,4-Pentadiyne**
(an isolated diyne)

**Pent-4-ene-1-yne**
(an isolated enyne)

**TABLE 2.3**

Bond Lengths and Angles in Representative Hydrocarbons

| Bond Type | Hybridization | Bond Length | Examples |
|-----------|---------------|-------------|----------|
| C—H | $sp^3$—H | 1.10 Å | |
| | $sp^2$—H | 1.09 Å | |
| | $sp$—H | 1.06 Å | |
| C—C | $sp^3$—$sp^3$ | 1.54 Å | |
| | $sp^3$—$sp^2$ | 1.50 Å | |
| | $sp^2$—$sp^2$ | 1.47 Å | |
| | $sp^3$—$sp$ | 1.46 Å | |
| | $sp^2$—$sp$ | 1.43 Å | |
| | $sp$—$sp$ | 1.37 Å | |
| | $sp^2$⚌$sp^2$ | 1.40 Å | |
| C=C | $sp^2$—$sp^2$ | 1.34 Å | |
| C≡C | $sp$≡$sp$ | 1.20 Å | |

Like alkanes and alkenes, alkynes can be ranked in order of stability by determining their heats of combustion or heats of hydrogenation. As we observed for substituted alkenes earlier, alkyl substitution stabilizes the multiple bond, making, for example, 2-butyne more stable than 1-butyne. Because of the linearity imposed by the carbon–carbon triple bond, *E-Z* isomerism does not exist for alkynes.

PHYSIOLOGICALLY ACTIVE ALKYNES

Several naturally occurring compounds containing alkynes have been isolated from microbes. Among them are the dynemicins. Dynemicin A contains two triple bonds and one double bond in a conjugated system (shown
in red) as well as other functional groups. These compounds have potent antibacterial and anticancer activity but, unfortunately, are probably too toxic
to mammals to be used as effective pharmaceutical agents.

Dynemicin A

## Nomenclature for Alkynes

Alkynes are named according to the IUPAC rules presented in Section
2.1 for alkenes, except that the suffix **-yne** is used to indicate the presence
of a triple bond:

**1.** Find the longest chain that contains the triple bond.

**2.** Begin numbering the chain so as to assign the functional group (the
triple bond) the lowest possible number.

**3.** Name branches as alkyl groups.

**4.** Use Greek prefixes to indicate multiple substituents.

## EXERCISE 2.20

Write an IUPAC name for each of the following hydrocarbons:

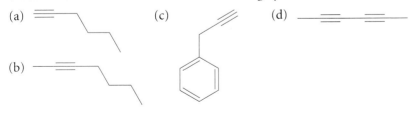

(a)

(b)

(c)

(d)

## EXERCISE 2.21

Draw a structure that corresponds to each of the following IUPAC names:

(a) 2-heptyne     (c) 2-methyl-3-hexyne

(b) 3-hexyne     (d) 1,5-octadiyne

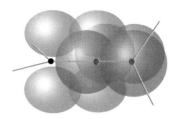

**FIGURE 2.21**

A three-dimensional view of π bonding in an allene (1,2-propadiene).

### Allenes

Although the most common functional group that incorporates *sp*-hybridized carbon atoms is the alkyne (a compound with a carbon–carbon triple bond), it is possible to have an *sp*-hybridized atom in a molecule that has two double bonds emanating in opposite directions from one carbon atom. These compounds are referred to as **allenes.** The π bonding in allenes is illustrated in Figure 2.21, where a pair of blue *p* orbitals form one π bond and a pair of gray *p* orbitals form the second π bond at right angles to the first. The central *sp*-hybridized carbon of an allene has a σ bond and a π bond to each of its two neighboring *sp²*-hybridized carbons. The two π bonds are orthogonal because the $p_y$ and $p_z$ orbitals of the *sp*-hybridized atom are at right angles. Thus, the plane containing the carbon atom and its two substituents at one end of the system is at right angles to the plane containing the carbon atom with its substituents at the other end.

An allene is also referred to as a *cumulated diene* because the double bonds share a common carbon atom. Because the two π bonds are orthogonal, they cannot interact as they do in a conjugated diene. Therefore, a cumulated diene is less stable than a conjugated diene.

### EXERCISE 2.22

Using sawhorse representations, draw the two possible three-dimensional structures of 2,3-pentadiene. Translate these two drawings into molecular models, and convince yourself that the two isomers are not identical.

## Summary

**1.** In alkenes (hydrocarbons having at least two *sp²*-hybridized carbons), maximum stability is achieved when the bond angles are about 120° and the *p* orbitals are aligned, resulting in a planar structure. The doubly bonded carbons are held together by a sigma (σ) and a pi (π) bond. A carbon–carbon double bond is shorter (1.34 Å) than a carbon–carbon single bond.

**2.** When a molecule contains more than one double bond, these bonds can be conjugated, isolated, or cumulated. In conjugated systems, delocalization of electron density stabilizes the molecule.

**3.** Molecules composed of *sp²*-hybridized atoms in planar, cyclic, conjugated arrays display aromatic properties if they contain $4n + 2$ electrons. The unusual stability of aromatic molecules can be predicted empirically by Hückel's rule. Aromatic systems usually have two or more dominant resonance contributors, which together describe the delocalized electron density of the molecule.

**4.** In hydrocarbons having adjacent *sp*-hybridized carbons (the alkynes), the hybrid orbitals are directed at 180° to minimize electron repulsion, with the atoms bonded to the *sp*-hybridized carbon atoms being colinear. The triple bond, composed of one σ bond and two π bonds orthogonal to one another, is approximately 1.20 Å in length, shorter than either a double or a single bond.

5. Since the bond energy of a $\sigma$ bond is greater than that of a $\pi$ bond, the latter are more reactive and form the basis for localized chemical reactivity. A multiple bond can thus be considered a functional group.

6. Isomers can be ranked as to their relative stability on the basis of heats of combustion, heats of hydrogenation, or heats of formation. A more highly substituted alkene is more stable than a less highly substituted isomer because of the higher electronegativity of an $sp^2$-hybridized atom and because of hyperconjugation.

7. The introduction of a double bond into a hydrocarbon backbone reduces the number of hydrogen atoms in an acyclic alkane $(C_nH_{2n+2})$ by two; so an acyclic alkene has the overall formula $C_nH_{2n}$. The introduction of each double bond reduces the number of hydrogen atoms by two. The presence of a triple bond reduces the number of hydrogen atoms in an alkyne by four from the alkane formula. Thus, an acyclic alkyne has the overall formula $C_nH_{2n-2}$.

8. All hydrocarbons can be named according to IUPAC rules.

## Review of Reactions

Catalytic Hydrogenation

Hydrocarbon Combustion

## Review Problems

**2.1** Draw structures for each of the following names:

(a) *cis*-1,3-pentadiene

(b) 4-methylcyclopentene

(c) 3-*t*-butyl-1-hexene

(d) *m*-bromotoluene

(e) oct-1-ene-4-yne

**2.2** For each of the following pairs, determine whether catalytic hydrogenation (with the molar equivalent of hydrogen uptake) can be used to distinguish between the compounds.

(a) cyclohexane and 1-hexene

(b) 1-hexene and (Z)-2-hexene

(c) cyclohexane and methylcyclopentane

(d) cyclohexane and cyclohexene

(e) 1-butene and 1-butyne

(f) 1-butene and 1-pentene

**2.3** Where possible, assign an *E* or *Z* designation to and provide a correct IUPAC name for the following alkenes:

(a)

(b)

(c)

(d) 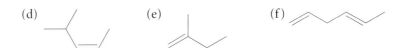 (e) (f)

**2.4** Calculate the index of hydrogen deficiency for each of the following naturally occurring hydrocarbons. From this value, calculate the number of double bonds, given the indicated number of rings for each compound.

(a) limonene (responsible for "citrus" odor), $C_{10}H_{16}$, one ring

(b) acenaphthene (in coal tar and sauna mud), $C_{12}H_{10}$, three rings

(c) benzo[a]pyrene (a carcinogen in soot), $C_{20}H_{12}$, five rings

(d) $\beta$-pinene (in pine needles and bark), $C_{10}H_{16}$, two rings

(e) caryophyllene (oil of cloves), $C_{15}H_{24}$, two rings

(f) $\beta$-cadinene (produces odor of cedar), $C_{15}H_{24}$, two rings

**2.5** Determine whether *cis–trans* isomerism is possible for each of the following compounds, and draw the structures of the geometric isomers if so.

(a) 1-hexene                    (e) 2-fluoro-2-butene

(b) 2-pentene                   (f) 1,2-dichlorocyclohexane

(c) 2-methyl-1-butene           (g) 1,2-dimethylcyclobutene

(d) 2-methyl-2-butene           (h) 3,4-dimethylcyclobutene

**2.6** (a) Arrange the following alkenes in order of relative stability: *trans*-3-heptene; 1-heptene; 2-methyl-2-hexene; *cis*-2-heptene; 2,3-dimethyl-2-pentene.

(b) For which pairs of compounds in part (a) can relative stabilities be determined by comparing heats of hydrogenation?

**2.7** Which compound would you expect to have the larger heat of hydrogenation? Explain.

(a) *cis*-cyclooctene or *trans*-cyclooctene          (b) *cis*-2-hexene or *trans*-2-hexene

**2.8** 1-Methylcyclohexene and methylenecyclohexane exist in equilibrium when dissolved in strong aqueous acid. Assuming that the stability of the alkene controls the equilibrium, which alkene is present at the higher concentration?

**2.9** $\alpha$-Phellandrene, $C_{10}H_{16}$, a naturally occurring product found in wormwood, is responsible for the odor of bitter fennel. Upon treatment of $\alpha$-phellandrene with an excess of hydrogen in the presence of a platinum catalyst, a compound with the formula $C_{10}H_{20}$ is produced.

(a) What is the index of hydrogen deficiency of $\alpha$-phellandrene?

(b) How many rings does $\alpha$-phellandrene have?

**2.10** Draw a structure and identify the hybridization of each carbon in each of the following compounds:

(a) 1,2,6-heptatriene           (c) vinylcyclopropane

(b) 3-phenyl-1-propyne          (d) *m*-xylene

**2.11** Which of the following pairs represent resonance contributors?

(a)          (b)          (c)

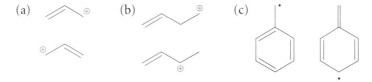

(d)   (e)

**2.12** Draw a significant resonance contributor that describes an electron distribution different from that shown in each of the following cases:

(a) allyl radical       (b) a cyclic pentadienyl cation       (c) pentadienyl anion

 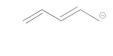

## Supplementary Problems

**2.13** Predict whether pentalene ($C_8H_6$) is stable as a planar hydrocarbon. Explain your reasoning.

Pentalene

**2.14** Unlike most hydrocarbons, azulene ($C_{10}H_8$) is highly colored (deep blue). Although its isomer naphthalene does not have significant zwitterionic character, azulene does. For example, azulene dissolves in aqueous acid; naphthalene does not.

Azulene

(a) Draw a resonance structure of azulene in which the five-member ring is anionic and the seven-member ring is cationic. (*Hint:* Consider resonance contributors in which the two electrons of a $\pi$ bond are moved to a single $p$ orbital, producing a formal negative charge in that orbital and a formal positive charge in another vacant $p$ orbital.)

(b) Can azulene be considered aromatic?

(c) The azulene molecule has an appreciable dipole moment. What does this observation imply about the relative importance of the resonance structure you drew in part (a)?

(d) Can a similar charge-separation argument explain the properties of pentalene (see Problem 2.13)? Why or why not?

**2.15** Although cyclopentene is a stable compound with chemical reactivity similar to that of a typical alkene, cyclopentyne is much less stable than a typical acyclic alkyne and cannot be stored at room temperature. Explain this difference in stability on the basis of what you know about hybridization and the preferred geometries for alkenes and alkynes.

**2.16** Calculate from the formula for each of the following compounds the sum of the number of double bonds and rings in each:

(a) $C_7H_{14}$       (b) $C_8H_{12}$       (c) $C_{10}H_{10}$       (d) $C_4H_{10}$

**2.17** Draw a structure that corresponds to each of the following names:

(a) 4-ethyl-(*E*)-2-octene

(b) 2-octyne

(c) 3-methyl-1-heptyne

(d) *trans*-3-hexene

(e) *n*-propylbenzene

(f) (*Z*)-2-hexene

(g) 1,4-hexadiene

(h) 3-methyl-*cis*-cyclododecane

(i) 2-*n*-propyl-1,4-pentadiene

(j) 1-*t*-butyl-1,3-cyclohexadiene

(k) 4-methyl-1,2-hexadiene

(l) (*E*)-1-cyclopentylpropene

**2.18** Provide a correct name for each of the following structures:

(a)    (c)    (e)

(b)    (d)

**2.19** Choose the isomer reasonably expected to have the higher heat of combustion:

(a)  or

(b)  or

(c)  or

**2.20** Use a sawhorse representation to draw the three-dimensional arrangement in which optimal hyperconjugative stabilization is attained for each compound:

(a) propene   (b) 2-methyl-1-butene   (c) 1-pentene

**2.21** In each of the following structures, choose the bond that best matches the description:

(a) shortest carbon–carbon bond in

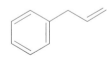

(b) longest carbon–carbon bond in

H—C≡C—CH₂—CH₃

(c) shortest carbon–carbon bond in

H₃C—C≡C—CH=CH₂

(d) shortest carbon–carbon bond in

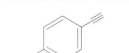

(e) longest carbon–carbon bond in

(f) longest carbon–hydrogen bond in

(g) shortest carbon–hydrogen bond in

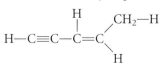

**2.22** Provide IUPAC names for the compounds shown in Problem 2.21.

**2.23** Draw valence bond representations of the other resonance contributors for each of the following chemical species:

(a)    (c)    (e)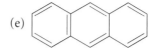

(b) 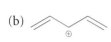   (d)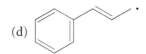

**2.24** On the basis of your knowledge of the relative stabilities of isomeric alkenes, predict which compound of the following isomeric pairs has the higher heat of hydrogenation.

(a)  or

(b)  or

(c)  or

(d) 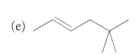 or

(e) 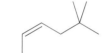 or

**2.25** Consider the structure of the cyclic polyene 10-annulene. Its synthesis by standard chemical methods has proven difficult.

**10-Annulene**

(a) From Hückel's rule, would you expect 10-annulene to be aromatic?

(b) If Hückel's rule were the only relevant factor, would 10-annulene be more or less stable than the acyclic conjugated polyene 1,3,5,7,9-decapentaene?

**1,3,5,7,9-Decapentaene**

**2.26** Draw a structure that corresponds to each of the following names:

(a) 1-isopropylnaphthalene

(b) 1-phenylethene
(also called styrene and vinylbenzene)

(c) 1,2-diphenylethane

(d) phenylethyne

(e) 2,6-dimethyl-4-octyne

(f) 3-phenylpropene

(g) *m*-bromotoluene

(h) 5-chloro-2-methyltoluene

(i) (*E*)-1-phenyl-1-butene

(j) 1,3-hexadiyne

**2.27** Assign the hybridization of the red atoms (labeled a–g) in dynemicin A.

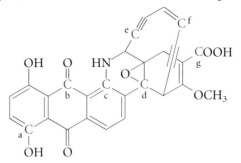

**Dynemicin A**

**2.28** For each of the following structures, draw and name a geometric isomer.

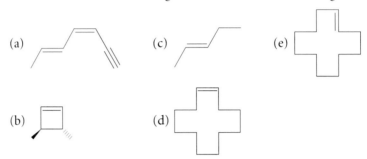

(a)

(b)

(c)

(d)

(e)

**2.29** Draw and name the structure that would be obtained from the catalytic hydrogenation of each of the following reactants.

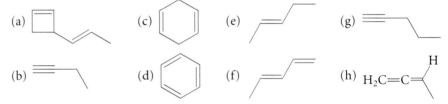

(a)

(b)

(c)

(d)

(e)

(f)

(g)

(h) $H_2C=C$

**2.30** How many electrons are shared between the carbon atoms in each of the following compounds?

(a) ethane     (b) ethene     (c) ethyne

**2.31** What is the structural feature that differentiates a saturated and an unsaturated hydrocarbon?

**2.32** Differentiate between $\sigma$ and $\pi$ bonds with respect to the following features:

(a) geometry                    (c) energy

(b) position of electron density     (d) ease of rotation about the bond

# Functional Groups Containing Heteroatoms

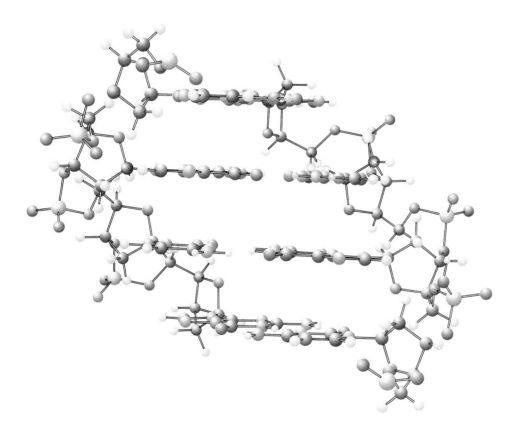

A short segment of DNA consisting of four base pairs. The intricate information-encoding system of DNA and RNA is built around hydrogen bonding, which would not be possible without the presence of the heteroatoms oxygen and nitrogen.

$\mathcal{M}$ost of the concepts of bonding developed for hydrocarbons in Chapters 1 and 2 also apply to organic molecules that contain other second- and third-row atoms. To differentiate these atoms from carbon, which is a constituent atom of all organic compounds, they are designated as *heteroatoms*. The chemical reactivity of an organic compound containing one or more heteroatoms usually differs significantly from that of an analogous compound that lacks these atoms. Thus, the part of the molecule containing the heteroatom constitutes a functional group.

In this chapter, we will consider how the presence of elements from the second and third rows of the periodic table provides unique properties to organic compounds. For example, we will compare the structure of $CH_4$ (methane) with those of $NH_3$ (ammonia) and $H_2O$ (water), and consider how the presence of a heteroatom affects the structures and reactivities of the organic derivatives of ammonia and water. We will also discuss some of the characteristics of families of organic compounds containing nitrogen, oxygen, or a halogen.

## 3.1

## Compounds Containing $sp^3$-Hybridized Nitrogen

**Ammonia** ($NH_3$) is the simplest member of a family of compounds built around nitrogen. Derivatives of ammonia in which one or more hydrogen atoms are replaced by alkyl or aryl groups are called **amines.**

### Ammonia: Hybridization and Geometry

Nitrogen has five electrons in its valence shell. Mixing the $2s$ and $2p$ atomic orbitals results in four $sp^3$-hybrid orbitals, as it did for carbon in Chapter 1.

$$N: \quad 1s \quad \underbrace{2s\,2p_x\,2p_y\,2p_z}$$

Four $sp^3$-hybrid
orbitals

There are seven protons in the nucleus of a nitrogen atom (one more than in a carbon nucleus), and a neutral nitrogen atom must have five valence electrons in addition to the two $1s$ electrons. Thus, two electrons must be accommodated in one of the four $sp^3$-hybrid orbitals. The two electrons accommodated in this filled orbital must have opposite spins. These electrons are referred to as a **lone pair** because they are associated with only one atom and do not take part in a covalent bond. The single electron in each of the three remaining $sp^3$-hybrid orbitals participates in a covalent bond with another atom. In this way, an octet electron configuration is achieved, satisfying nitrogen's valence requirement. As with an $sp^3$-hybridized carbon, each of these hybrid orbitals of nitrogen is directed toward the apex of a tetrahedron so as to minimize electron-pair repulsion. Thus, bond angles are approximately 109°, and the relative orientation of the three hydrogen substituents in ammonia ($NH_3$) is roughly the same as that of any three of the four hydrogen atoms in methane.

*CHEMICAL PERSPECTIVES*

INDUSTRIAL SYNTHESIS OF AMMONIA

More than 30 billion pounds of ammonia are produced industrially each year by means of the Haber–Bosch process. In this reaction, hydrogen and nitrogen are combined at very high pressures and temperatures (1000 atm, or 14,000 psi, and 700 °C). The hydrogen is obtained by passing very hot steam over heated coke. As a liquid, 30 billion pounds of ammonia would occupy a box measuring 270 meters (approximately three football-field lengths) on a side. One important use of ammonia is in fertilizers.

When each of the partly filled hybrid orbitals of nitrogen overlaps with the 1*s* orbital of a hydrogen atom, ammonia is formed. Because the electron density of the lone pair is closer to nitrogen's nucleus than is that of a nitrogen–hydrogen $\sigma$ bond, the lone pair exerts a somewhat greater repulsive force toward the electrons in the $\sigma$ bonds than do those electrons for each other. As a result, each H—N—H angle is slightly smaller (107°) than the expected tetrahedral bond angle, and the angle formed by a hydrogen, the nitrogen, and the lone pair is slightly larger (111°). Ammonia has three-fold symmetry because rotation by 120° (360° ÷ 3) about an axis passing through the nitrogen results in no change.

**Ammonia, NH₃**

**Deviation of Angles in the Ammonia
Molecule from Tetrahedral Angles**

Because only three substituents are attached to nitrogen in ammonia and similar compounds, this spatial arrangement is referred to as **pyramidal** (rather than tetrahedral) because the four atoms (nitrogen and three hydrogens) are located at the corners of a pyramid.

#05    Amines

## Amines

Methylamine, $CH_3NH_2$, has a methyl group (whose carbon is also *sp³*-hybridized) attached to nitrogen by a $\sigma$ bond; this bond replaces one of the nitrogen–hydrogen $\sigma$ bonds of ammonia. The geometry at nitrogen is approximately the same as in ammonia, and the geometry at carbon is similar to that in an alkane. The lone pair of electrons on the nitrogen atom is a site of high chemical reactivity and is therefore the functional group in methylamine.

Alkyl substituents other than a methyl group can be attached to nitrogen. The resulting compounds belong to the family referred to as **amines.** An amine can be named either as an alkyl derivative of ammonia (for ex-

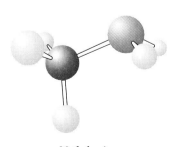

**Methylamine**

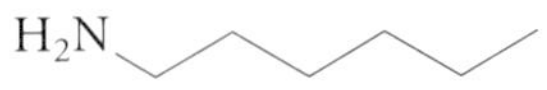

H₂N ∼∼∼∼

**1-Aminohexane**

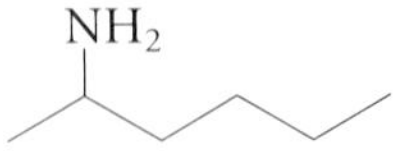

NH₂
∼∼∼∼

**2-Aminohexane**

NH₂
∼∼∼∼

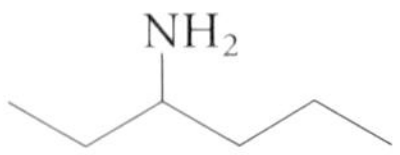

**3-Aminohexane**

**FIGURE 3.1**

Three primary amines. The designation *primary* means that only one carbon is attached to nitrogen.

ample, methylamine) or in accord with the IUPAC system as a nitrogen derivative of an alkane—that is, with the **amino group** ($NH_2$) as a substituent of an alkane (for example, aminomethane). The amino group can be placed at any position along a hydrocarbon chain. Figure 3.1 shows three amines in which the amino group is placed on the first, second, and third carbon of a six-carbon chain. These compounds are referred to as 1-aminohexane, 2-aminohexane, and 3-aminohexane.

*Primary, Secondary, and Tertiary Amines.* Methylamine and 1-, 2-, and 3-aminohexane are referred to as **primary amines** because nitrogen is connected to only one carbon substituent. Amines are classified by the number of carbons attached to nitrogen. (Keep in mind that all other functional groups are classified by the number of carbon substituents attached to the carbon atom bearing the heteroatom of the functional group.) A common convention (to emphasize the functional group, rather than the alkyl chain) is to designate a primary amine as $RNH_2$, where R represents any alkyl group.

**Secondary amines** such as dimethylamine have two carbon substituents on nitrogen, $R_2NH$. **Tertiary amines** such as trimethylamine have three carbon substituents on nitrogen, $R_3N$. (See Figure 3.2.) A fourth alkyl group can be attached to nitrogen, but doing so requires that both electrons of the lone pair be used to form a covalent bond. As a result, the nitrogen becomes positively charged. Such cations are referred to as **quaternary ammonium ions;** an example is the tetramethylammonium cation (also shown in Figure 3.2).

Note that the designation *primary, secondary,* or *tertiary* for an amine characterizes the degree of substitution at nitrogen, not carbon. Thus, both 1-aminohexane and 2-aminohexane are primary amines, despite the fact that the amino group is attached to a primary carbon in the first compound and to a secondary carbon in the second compound (refer to Figure 3.1).

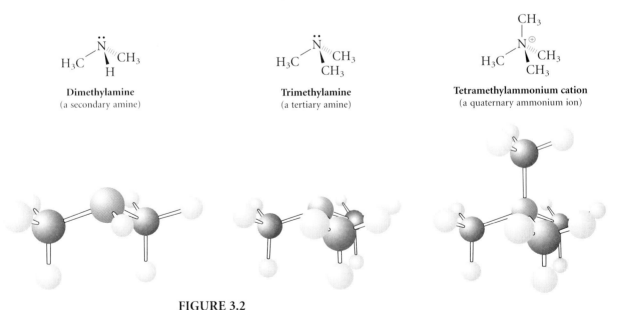

**Dimethylamine**
(a secondary amine)

**Trimethylamine**
(a tertiary amine)

**Tetramethylammonium cation**
(a quaternary ammonium ion)

**FIGURE 3.2**

A secondary amine, a tertiary amine, and a quaternary ammonium ion.

Classify each of the following as a primary, secondary, or tertiary amine or a quaternary ammonium salt:

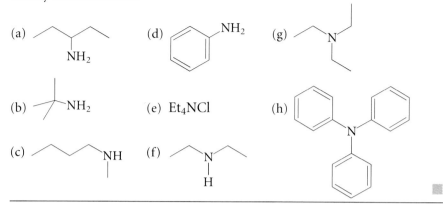

(a)

(b)

(c)

(d)

(e) Et₄NCl

(f)

(g)

(h)

**Formal Charges.** Trimethylamine and the tetramethylammonium ion can be represented by Lewis dot structures:

**Trimethylamine**

**Tetramethylammonium ion**

Formal charge on N:   $5 - (\%_2 + 2) = 0$          $5 - (\%_2 + 0) = +1$

As we did for carbon in Chapter 1, we calculate the formal charge on nitrogen in trimethylamine by comparing the number of valence electrons (5) with the sum of half the number of shared electrons ($\%_2 = 3$) plus the number of unshared electrons (2): thus, the nitrogen atom in trimethylamine bears a formal charge of zero $[5 - (3 + 2) = 0]$. In tetramethylammonium ion, the formal charge on nitrogen can be similarly calculated: here nitrogen bears a formal charge of +1.

**EXERCISE 3.2**

Determine the formal charge on carbon, nitrogen, and oxygen in each structure. Assume that each atom has a filled valence shell.

(a) $H_3C-\overset{\overset{\displaystyle CH_3}{|}}{\underset{\underset{\displaystyle CH_3}{|}}{N}}-O$

(b) $H_3C-C\equiv N-O$

SOME SIMPLE, NATURALLY OCCURRING AMINES

Compounds that contain nitrogen are pervasive in nature. Many of them have well-defined biological functions that are important—even essential—to life. As living materials decompose, many of these complex structures decompose to simple amines. For example, the odor of rotting fish is due to a mixture of amines (primarily trimethylamine). Putrescine, a diamine, is produced by bacteria during the decomposition of animal tissue. Its name provides an excellent description of its odor.

Very commonly, people associate bad odors with toxicity: things that smell bad are bad for you. In some cases, such as rotting flesh, this assessment is correct. However, putrescine occurs naturally in all cells, and compounds such as spermine are believed to be essential to cell division. Spermine, a tetramine with its own unique aroma, was first isolated as its acid–base salt with phosphoric acid from semen in 1678 by Anton von Leeuwenhoek, a Dutch chemist.

$$H_2N—CH_2—CH_2—CH_2—CH_2—NH_2$$

**Putrescine**

$$H_2N—CH_2—CH_2—CH_2—NH—CH_2—CH_2—CH_2—CH_2—NH—CH_2—CH_2—CH_2—NH_2$$

**Spermine**

---

### 3.2

## Polar Covalent Bonding in Amines

A covalent $\sigma$ bond between carbon and nitrogen is polar because of the greater electronegativity of nitrogen (3.1 versus 2.5 for carbon). The uneven charge distribution between atoms of unlike electronegativity that are connected by a polar covalent bond imparts unique properties to compounds containing such bonds. Thus, amines have different physical properties from alkanes having similar structures and molecular weights.

### ▣ Dipole Moments

Let's consider the effects of the presence of an amino group in an organic molecule. Hydrocarbons have only nonpolar covalent bonds, whereas the covalent bonding in amines results in regions of partial positive and partial negative charge within the molecule. Such separation of charge constitutes a dipole. A molecular dipole moment, $\mu$, exists when the resultant (the sum) of the individual dipoles of the bond polarities projected into three dimensions is not zero.

What structural features in amines produce a dipole moment? First, amines have a dipole due to the lone pair of nonbonding electrons on nitrogen. Furthermore, nitrogen is more electronegative than carbon and, as a result, more strongly attracts electrons toward itself and away from carbon (and hydrogen). Therefore, the carbon–nitrogen and hydrogen–nitrogen bonds in amines are polarized so that electron density in the bonds is shifted toward nitrogen, further enhancing the molecular dipole.

This shift of electron density can be indicated in several ways. In one method, an arrow pointed toward the center of partial negative charge indicates the direction of the shift in the $\sigma$ bond. Alternatively, a lowercase delta ($\delta$) indicates the development of a partial positive or negative charge on the atoms involved in polar covalent bonding. Both the carbon–nitrogen and hydrogen–nitrogen bond dipoles of methylamine combine with the dipole of the lone pair on nitrogen to produce the overall dipole of the molecule:

$$H_2N \overset{\longleftarrow|}{\phantom{x}} CH_3 \qquad H_2\overset{\delta-}{N} \overset{\delta+}{\phantom{x}} CH_3$$

**Bond dipoles of**       **Molecular**
**methylamine**             **dipole**

The presence of a dipole moment has important consequences for both a molecule's physical properties and its chemical reactivity. For example, because of the significant electron density on nitrogen, the negative end of a carbon–nitrogen dipole is attracted to nuclei of hydrogen atoms (the positive ends of carbon–hydrogen dipoles) on the surface of surrounding molecules. The influence of dipole–dipole interactions on intermolecular attractive forces is evident from the boiling points of isobutane, $(CH_3)_3CH$ ($-12$ °C), and trimethylamine, $(CH_3)_3N$ (3 °C), a difference of 15 °C (see Table 3.1 on page 94). The effect is small, however, because C—H bonds are only slightly polarized, and thus the attraction of these hydrogen atoms for the lone pair of electrons of nitrogen is small. The electrostatic attraction between the lone pair of electrons on nitrogen and a hydrogen atom nucleus is dramatically increased when the hydrogen is attached to a heteroatom.

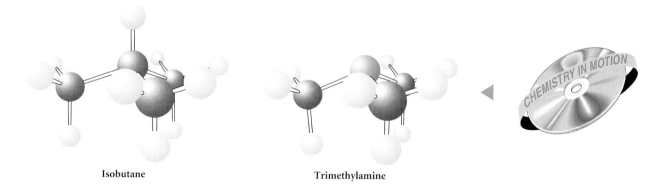

Isobutane                    Trimethylamine

## EXERCISE 3.3

For each of the following pairs, choose the molecule that is likely to have the larger dipole moment. Explain your reasoning.

(a)  $NH_3$ or $NF_3$

(b)  trimethylamine or 2-methylpropane

(c)  triphenylamine or triphenylmethane

**TABLE 3.1**

Boiling Points of Selected Hydrocarbons and Heteroatom-Containing Compounds

| Hydrocarbon | Boiling Point (°C) | Amine | Boiling Point (°C) | Alcohol or Ether | Boiling Point (°C) |
|---|---|---|---|---|---|
| $CH_4$ | −164 | $NH_3$ | −33 | $H_2O$ | 100 |
| $CH_3CH_3$ | −89 | $CH_3NH_2$ | −6 | $CH_3OH$ | 65 |
| $CH_3CH_2CH_3$ | −42 | $CH_3NHCH_3$ | 7 | $CH_3OCH_2CH_3$ | 11 |
| | | $CH_3CH_2NH_2$ | 16 | $CH_3CH_2OH$ | 78 |
| $CH_3CH_2CH_2CH_3$ | −0.5 | $CH_3CH_2CH_2NH_2$ | 48 | $CH_3CH_2CH_2OH$ | 97 |
| | | $CH_3NHCH_2CH_3$ | 37 | $CH_3CH(OH)CH_3$ | 82 |
| | | $CH_3CH(NH_2)CH_3$ | 33 | $(CH_3)_3COH$ | 82 |
| $(CH_3)_3CH$ | −12 | $(CH_3)_3N$ | 3 | $CH_3(CH_2)_3OH$ | 117 |
| $CH_3(CH_2)_3CH_3$ | 36 | $CH_3(CH_2)_3NH_2$ | 78 | $CH_3CH_2OCH_2CH_3$ | 35 |
| | | | | $CH_3(CH_2)_4OH$ | 138 |

## Hydrogen Bonding

Polar covalent $\sigma$ bonds are formed between hydrogen and highly electronegative atoms (those in the fifth, sixth, and seventh columns of the periodic table). The hydrogen of such a polar bond often participates in further association by hydrogen bonding. The partially positively charged hydrogen atom associates with a partially negatively charged center in another molecule. The weak attraction of a hydrogen atom bonded to an electronegative atom, X, for a lone pair of electrons on another electronegative atom, Y, is a **hydrogen bond:**

Hydrogen bonding causes a lengthening of the polar covalent bond between the heteroatom and hydrogen.

Let's consider the interaction between two ammonia molecules. Electrostatic attraction exists between the partial positive charge on a hydrogen of one ammonia molecule and the high electron density (partial negative charge) of the nitrogen lone pair of the other ammonia molecule (Figure 3.3). Thus, this hydrogen atom is linked with both the nitrogen to which it is covalently bonded and, more weakly, the lone pair on the other ammonia molecule through a hydrogen bond. A network is set up throughout the entire volume of a sample of liquid ammonia in which many such hydrogen bonds link individual ammonia molecules. When hydrogen bonds connect separate molecules, they are referred to as **intermolecular hydrogen**

$$\overset{\delta-}{H_2N}\!\!-\!\!\overset{\delta+}{H}\cdots\overset{\delta-}{:N}\!\!-\!\!\overset{\delta+}{H_3}$$

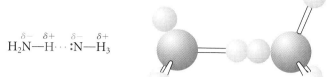

**FIGURE 3.3**

Intermolecular hydrogen bonding in ammonia. (The lone pairs of electrons on nitrogen are shown as small light blue spheres.)

**bonds;** when hydrogen bonds connect groups within the same molecule, they are called **intramolecular hydrogen bonds.**

Intermolecular hydrogen bonding is important wherever hydrogen is covalently bonded to such highly electronegative atoms as nitrogen, oxygen, sulfur, or a halogen. Hydrogen atoms attached to nitrogen, oxygen, or fluorine form the strongest hydrogen bonds. The valence requirement of the hydrogen atom participating in a hydrogen bond is fulfilled largely by the covalent bond. Thus, a hydrogen bond is substantially weaker than a typical covalent $\sigma$ bond. Hydrogen-bond strengths vary from 1 to 5 kcal/mole.

In primary and secondary amines, a hydrogen atom in the amino group forms a hydrogen bond to the lone pair on nitrogen in other amine molecules. The presence of hydrogen bonds between molecules of 1-aminopropane ($CH_3CH_2CH_2NH_2$) results in a boiling point 48 °C higher than that of the analogous hydrocarbon, butane ($CH_3CH_2CH_2CH_3$). Hydrogen bonding has even stronger effects on the physical properties of alcohols (discussed later in this chapter) because of the greater polarization of O—H bonds compared with N—H (and C—H) bonds. Boiling points increase uniformly from hydrocarbon to amine to alcohol, as, for example, in the series propane, ethylamine, and ethyl alcohol ($-42$ °C, 16 °C, and 78 °C, respectively).

**EXERCISE 3.4**

Draw each of the hydrogen bonds described as a dotted line:
(a) ammonia hydrogen bonded to another ammonia molecule
(b) ammonia hydrogen bonded to methylamine
(c) methylamine hydrogen bonded to another methylamine molecule
(d) ammonia hydrogen bonded to trimethylamine

**EXERCISE 3.5**

Determine how many atoms are included in the ring formed by intramolecular hydrogen bonding for each compound:
(a) 1,2-diaminoethane (also called ethylene diamine)
(b) putrescine, $H_2NCH_2CH_2CH_2CH_2NH_2$

## Solvation

Hydrogen bonding also occurs in mixtures between heteroatom-containing molecules. For example, consider the interaction of methylamine ($CH_3NH_2$) and water ($H_2O$), whose structure is considered in more detail later in this chapter. The hydrogen atoms of water are more electron-deficient than those of methylamine because oxygen is more electronegative than nitrogen. On the other hand, the lone pair of electrons of the amine is less tightly held by nitrogen than are the two lone pairs of water. Thus, the strongest hydrogen bond is formed when a hydrogen atom of water interacts with the lone pair of electrons of methylamine (Figure 3.4).

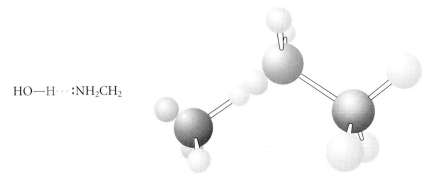

$$HO—H \cdots :NH_2CH_2$$

**FIGURE 3.4**

Intermolecular hydrogen bonding between water and methylamine. (Lone pairs of electrons are shown as pink or light blue spheres.)

With only one lone pair of electrons, the nitrogen atom of methylamine can engage in only one hydrogen bond with a water molecule. As a result, the electron density about nitrogen is decreased, with increased polarization of the two N—H bonds. Each of these hydrogens can then hydrogen bond with a lone pair of electrons on other water molecules (Figure 3.5).

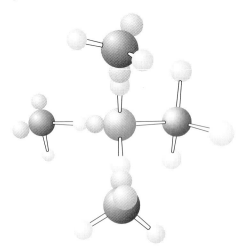

**FIGURE 3.5**

Orientation of three polar water molecules around methylamine. There is one hydrogen bond from a hydrogen of a water molecule to the lone pair of electrons of the amine and two hydrogen bonds from hydrogens of the amine to lone pairs of electrons on two other water molecules.

Hydrogen bonding is an important factor in accounting for the high solubility in water of methylamine (a gas at room temperature, with b.p. −6 °C). Water is an ideal compound for intermolecular hydrogen bonding. With two hydrogen atoms and two lone pairs of electrons, water molecules can form an almost infinite network in which each lone pair and each hydrogen atom is involved in hydrogen bonding. When other molecules (such as ammonia) dissolve in water, the hydrogen bonding network of water is disrupted. For dissolution to be energetically favorable, the hydrogen bonding network of water must be replaced by hydrogen bonding between water and the solute. The three water molecules shown surrounding methylamine in Figure 3.5 constitute all possible hydrogen-bonding motifs of methylamine. These associated solvent molecules together comprise the **inner solvation shell.** Additional water molecules can associate with those of the inner solvation shell because the latter have additional hydrogen atoms and lone pairs of electrons available for further hydrogen bonding. Each of the three water molecules can associate with as many as three more water molecules, forming a second solvation shell, as shown by the space-filling model in the margin. Note that the twelve water molecules (three in the inner shell and nine in the outer shell) completely encase the amino group of methylamine. The methyl group itself contributes nothing to the solubility of methylamine in water and, in fact, disrupts some hydrogen bonding that would be possible in its absence. As the size of the alkyl group increases, the solubility of a primary amine decreases.

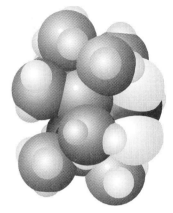

Methylamine with first and second solvation shells

## EXERCISE 3.6

For each of the following pairs, choose the compound that is likely to be more soluble in water. Explain your reasoning.

(a) ammonia or triethylamine

(b) methylamine or *n*-octylamine

(c) trimethylamine or *n*-propylamine

### Acidity and Basicity of Amines

A heteroatom bearing both a bond to hydrogen and a lone pair can act as either an acid or a base. The more commonly used definition of acids and bases is that originally suggested by Johannes Brønsted: an acid acts by transferring a proton to an acceptor. Therefore, a **Brønsted acid** is defined as a proton donor. For example, when HCl reacts with water to form $H_3O^{\oplus}$ and $Cl^{\ominus}$, a proton ($H^{\oplus}$) is donated by HCl to water:

$$H_2O + HCl \rightleftharpoons \begin{matrix} H \\ \diagdown \\ O^{\oplus}\!\!-\!\!H \\ \diagup \\ H \end{matrix} + Cl^{\ominus}$$

Similarly, a **Brønsted base** is a proton acceptor. In the ionization of water, one water molecule acts as a Brønsted acid, donating a proton, and the other acts as a Brønsted base, accepting a proton:

$$2\,H_2O \rightleftharpoons H_3O^{\oplus} + {}^{\ominus}OH$$

## QUININE: AN ALKALOID

Naturally occurring compounds with basic nitrogen atoms are referred to as *alkaloids.* Quinine, an alkaloid isolated from the bark of the cinchona tree, was the first compound found to be effective for treating malaria, a complicated disease state caused by a parasite. Extracts from cinchona bark were first used in Europe in the fifteenth century; the bark was brought to Europe by Spanish Jesuits, who learned of its medicinal properties from Peruvian Incas. The cinchona tree, which is native to the Amazon basin, grows only in tropical regions, and these trees are grown in plantations in the South Pacific.

**Quinine**

The control of many of the Pacific islands by the Japanese during World War II raised concerns about supplies of quinine to treat American troops operating in the Pacific. These concerns spurred interest in the synthesis of quinine. Two American chemists, Robert B. Woodward (Nobel Prize in Chemistry, 1966) and William von Eggers Doering, were the first to prepare quinine in the laboratory in 1944. This synthesis was the first preparation of such a complicated molecule and set the stage for a major revolution in how chemists viewed their ability to mimic nature.

These reactions can also be thought of in the Lewis acid–base sense: $Cl^{\ominus}$ accepts the electrons of the H—Cl bond that are freed by the interaction of hydrogen with the base (water). Simultaneously, water acts as an electron donor as one lone pair changes from being a nonbonding pair in water to participating in an O—H $\sigma$ bond in the hydronium ion.

The Brønsted and Lewis concepts of acidity are equally useful for mineral acids (such as HCl or $H_2SO_4$), but for organic acids, Brønsted acidity is often not relevant. For example, when the lone pair of electrons on the nitrogen of trimethylamine interacts with an electron acceptor such as boron trifluoride, the amine acts as an electron donor (a Lewis base), and $BF_3$ acts as an electron acceptor (a Lewis acid). Thus, a Lewis acid–base reaction takes place in this example even in the absence of an X—H bond in the reactants or the product.

$$(CH_3)_3N\colon + BF_3 \;\rightleftharpoons\; \overset{H_3C}{\underset{H_3C}{\diagup}}\!\!H_3C\!-\!\overset{\oplus}{N}\!-\!\overset{\ominus}{B}\!\overset{F}{\underset{F}{\diagdown}}$$

The presence of a lone pair of electrons (or a $\pi$ bond) is both necessary and sufficient for a compound to act as a Brønsted or Lewis base. When the lone pair on nitrogen in an amine interacts to form a covalent bond by donation of its electrons to an electrophile ($E^{\oplus}$), nitrogen is acting as a Lewis base.

$$R_3N\colon \qquad E^{\oplus} \;\rightleftharpoons\; R_3\overset{\oplus}{N}\!-\!E$$

**Lewis
base**

$$R_3N\colon \qquad H^{\oplus} \;\rightleftharpoons\; R_3\overset{\oplus}{N}\!-\!H$$

**Brønsted                                 Conjugate
base                                          acid**

When the electrophile is a carbon atom, a new carbon–nitrogen bond is formed, often irreversibly. When the electrophile is a proton ($H^{\oplus}$), nitrogen acts as a Brønsted base. Brønsted acid–base reactions (**proton transfer reactions**) are usually reversible. The product of the reaction of a Brønsted base with a proton is called the **conjugate acid** of the base.

Quantitative concepts of acidity and basicity will be developed in Chapter 6, but it will be useful to develop a general idea about the basicity of amines. A quick way to evaluate base strength is from the acidity of the conjugate acid: the stronger the conjugate acid, the greater is its propensity to give up a proton and, correspondingly, the weaker is the ability of the base to attract a proton. Typically the conjugate acids of amines (ammonium ions) have $pK_a$ values around 9. Thus, they are stronger acids than water ($pK_a$ 15.6) but substantially weaker than hydronium ion ($pK_a$ $-2$). Thus, the order of basicity is: $^{\ominus}OH > RNH_2 > H_2O$.

$$H_3CNH_2 + H^{\oplus} \;\rightleftharpoons\; H_3C\overset{\overset{\displaystyle H}{\displaystyle |}}{\underset{\oplus}{N}}H_2 \quad (pK_a\ 9)$$

$$HO^{\ominus} + H^{\oplus} \;\rightleftharpoons\; HO\!-\!H \quad (pK_a\ 15.6)$$

$$H_2O + H^{\oplus} \;\rightleftharpoons\; H_2\overset{\oplus}{O}\!-\!H \quad (pK_a\ -2)$$

Primary and secondary amines have polarized N—H bonds and can therefore also act as acids. However, the degree of polarization of an N—H bond is substantially less than that of an O—H bond. As a result, the N—H bond of an amine is only weakly acidic, with a typical $pK_a$ of 36. Amines are intermediate in acidity between compounds with O—H bonds, such as water and the alcohols ($pK_a$ 16–19), and hydrocarbons ($pK_a$ 40–60). In

Chapter 6 we will discuss some bases that are sufficiently strong to completely deprotonate primary and secondary amines.

$$(\text{p}K_a\ 36)\quad H_3C\overset{\overset{\displaystyle H}{|}}{N}H \rightleftharpoons H_3C\overset{\ominus}{N}H + H^\oplus$$

$$(\text{p}K_a\ 16)\quad H_3CO{-}H \rightleftharpoons H_3CO^\ominus + H^\oplus$$

$$(\text{p}K_a\ 48)\quad H_3C{-}H \rightleftharpoons H_3C^\ominus + H^\oplus$$

Amines are thus both bases and acids. The tendency of a compound or functional group to act as both an acid and a base is referred to as **ambiphilicity.** The requirements for ambiphilicity are the presence of a polarized X—H bond and a lone pair of electrons on the heteroatom, X. This dual acid–base reactivity is a property not only of amines, but also of alcohols (ROH), thiols (RSH), and phosphines ($RPH_2$), considered in later sections.

## EXERCISE 3.7

From your knowledge of the periodic table, predict whether each of the following compounds can act as a Lewis base. Why or why not?

(a) $Mg^{2+}$

(b) aluminum trichloride

(c) boron trifluoride

(d) triethylamine

(e) tin tetrachloride

## EXERCISE 3.8

Determine whether each of the following compounds can function as a Brønsted base. If so, write the product that would be formed by interaction with a Brønsted acid such as HCl.

(a) butane

(b) 1-butanol, $CH_3CH_2CH_2CH_2OH$

(c) methyl ether, $CH_3OCH_3$

(d) tertiary butanol, $(CH_3)_3COH$

(e) methyl sulfide, $CH_3SCH_3$

## 3.3

# Compounds Containing $sp^2$-Hybridized Nitrogen

Like carbon, nitrogen can participate in double and triple bonds. When a nitrogen atom is $sp^2$-hybridized, it can form a double bond (one $\sigma$ bond and one $\pi$ bond) by overlapping orbitals with those of another $sp^2$-hybridized atom.

### Double Bonding at Nitrogen

The mixing of atomic orbitals to form hybrid orbitals as described in Chapter 2 for carbon compounds also occurs with nitrogen to construct three $sp^2$-hybrid orbitals, leaving one $p$ orbital unhybridized:

$$\text{N:} \qquad 1s \qquad \underline{2s\ 2p_x\ 2p_y} \qquad 2p_z$$

$$\downarrow$$

Three      One $p$
$sp^2$-hybrid    orbital
orbitals

This $sp^2$-hybridized nitrogen can bond, for example, with an $sp^2$-hybridized carbon, forming a planar structure in which overlap of the $p$ orbitals at a right angle to the molecular plane forms a $\pi$ bond. The carbon–nitrogen and nitrogen–hydrogen $\sigma$ bonds and the nitrogen lone-pair hybrid orbital are coplanar, producing an **imine,** a compound containing a carbon–nitrogen double bond (Figure 3.6).

The geometries of imines are analogous to those of alkenes. For example, an imine contains two fewer hydrogen atoms than the corresponding amine, just as an alkene contains two fewer hydrogen atoms than the corresponding alkane. The two $\sigma$ bonds and the lone pair of electrons on nitrogen are coplanar and separated by approximately 120°, just like the three $\sigma$ bonds at carbon in an alkene. There is restricted rotation about a C=N bond in an imine, just as there is about a C=C bond in an alkene. Like a carbon–carbon $\pi$ bond, the $\pi$ bond of an imine undergoes catalytic hydrogenation, producing an amine.

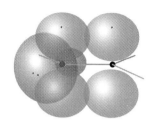

**FIGURE 3.6**

Overlap of $p$ orbitals (shown in blue) results in a $\pi$ bond between carbon and nitrogen in an imine, similar to that between two carbons in an alkene. The nitrogen atom bears a lone pair of electrons held in an $sp^2$-hybrid orbital (shown in gray) in the same plane as two other $sp^2$-hybrid orbitals that are involved in $\sigma$ bonding.

$$H_3C \quad H$$
$$\diagdown \quad \diagup$$
$$C=C \qquad \xrightarrow[\text{Pt}]{H_2} \qquad H_3C-CH_2-CH_2-CH_3$$
$$\diagup \quad \diagdown$$
$$H \quad CH_3$$

**An alkene**                **An alkane**

$$H_3C$$
$$\diagdown$$
$$C=N: \qquad \xrightarrow[\text{Pt}]{H_2} \qquad H_3C-CH_2-NH-CH_3$$
$$\diagup \quad \diagdown$$
$$H \quad CH_3$$

**An imine**                **An amine**

### Bond Strengths of Multiple Bonds

It is useful to compare the strengths of single and multiple bonds between various atoms. As we will see in later chapters, chemists are often able to understand why reactions occur by comparing the energies of the bonds that are broken in the starting material(s) with the energies of the bonds formed in the product(s). Table 3.2 (on page 102) gives average bond energies, obtained from heats of formation, for various types of bonds. Thus, for example, the entry for C—H is obtained as the heat required to convert methane into carbon and hydrogen—that is, for $CH_4 \rightarrow C + 4\ H$, $\Delta H°/4 =$ 99 kcal/mole. The entry for C—C is obtained by measuring the heat of formation of ethane and subtracting the bond energies of the six C—H bonds, assuming that each has the same value as a C—H bond in methane. The other bond energies listed are obtained by similar estimations. As we will see shortly, the C—H bond energies in methane and in ethane are *not* the same, and thus the values in Table 3.2 represent only approximations. However, these values are useful because the bond energies of multiple bonds (double and triple) are not otherwise available.

As you can see from Table 3.2, the energy required to break a double bond between two given atoms is greater than that needed to break a sin-

**TABLE 3.2**

Average Bond Energies (kcal/mole)

*Example:* $CH_4 \rightarrow C + 4\,H$; $\Delta H°/4 = 99$ kcal/mole

| C—H | C—C | C=C | C≡C | |
|---|---|---|---|---|
| 99 | 83 | 146 | 200 | |

| N—H | C—N | C=N | C≡N | |
|---|---|---|---|---|
| 93 | 73 | 147 | 213 | |

| O—H | C—O | C=O | C≡O | O=C=O |
|---|---|---|---|---|
| 111 | 86 | 179 | 257 | 225 (each) |

| H—H | N—N | N=N | N≡N | |
|---|---|---|---|---|
| 104 | 39 | 100 | 226 | |

| | O—O | $^3$(O=O) | |
|---|---|---|---|
| | 35 | 119 | |

| H—F | C—F | F—F |
|---|---|---|
| 136 | 108 | 38 |

| H—Cl | C—Cl | Cl—Cl |
|---|---|---|
| 103 | 81 | 58 |

| H—Br | C—Br | Br—Br |
|---|---|---|
| 87 | 68 | 46 |

| H—I | C—I | I—I |
|---|---|---|
| 71 | 51 | 36 |

gle bond, and the energy required to break a triple bond is greater than for a double bond. We can also see that an H—X bond becomes weaker in progressing down a column of the periodic table: H—F > H—Cl > H—Br > H—I. This trend is also observed when the halogens are bonded to carbon or to themselves.

With bond energies, we can calculate whether a given transformation is energetically feasible. For example, in the catalytic hydrogenation of an alkene to an alkane, the C=C bond is converted into a C—C bond, while an H—H bond in hydrogen gas is broken and two C—H bonds are formed.

By assigning positive values to the energies that must be supplied to break bonds and negative values to the energies released when bonds are formed, we can calculate $\Delta H°$ for this reaction:

$$\Delta H° = +146 + 104 - 83 - (2 \times 99) = -31 \text{ kcal/mole}$$

The negative value obtained here indicates that energy is released; that is, this is an energetically favorable—and therefore exothermic—reaction.

In the same way, we can calculate $\Delta H^\circ$ for the catalytic hydrogenation of an imine:

$$\Delta H^\circ = +147 + 104 - 99 - 93 - 73 = -14 \text{ kcal/mole}$$

Again, this reaction is predicted to be energetically favorable, which is consistent with the observation that amines are produced by catalytic hydrogenation. Table 3.2 is thus extremely valuable as a rough predictive tool for determining the feasibility of proposed reactions.

### EXERCISE 3.9

Using the bond energies in Table 3.2, predict whether the following proposed reactions are thermodynamically feasible:

(a)

(b) $N_2 \xrightarrow[\text{Catalyst}]{H_2} NH_3$

(c) $CH_3—C{\equiv}N \xrightarrow[\text{Catalyst}]{H_2} CH_3CH_2NH_2$

## Calculating Oxidation Levels

A method of "electron bookkeeping" is useful for describing conversions in which the number of multiple bonds or the number of bonds to heteroatoms is changed. For instance, comparing the oxidation levels of the atoms participating in such reactions helps us choose the type of reagent (and how much) to accomplish a particular chemical transformation. Reagents that can induce an oxidation or a reduction are called *redox reagents*. **Reduction** entails the addition of electrons and is always accompanied by the oxidation of a reaction partner. **Oxidation** is the loss of electrons.

By comparing the oxidation levels of starting materials and products, we can determine if an oxidation–reduction reaction has taken place. The **oxidation level** of an atom can be determined by a simple scheme. In a *mental* operation, you break all of the bonds, giving the electrons to the more electronegative partner. When the two atoms connected by a bond have the same electronegativity, the electrons are shared equally. A bond to a more electronegative atom results in a positive charge, a bond to a less electronegative atom results in a negative charge, and a bond to an atom with the same electronegativity contributes no charge. The sum of the charges on the atom is equal to the oxidation level of the atom. For example, the oxidation level of the carbon atom in methanol is $-2$. Correspondingly, the hydrogen atoms are at the $+1$ oxidation level.

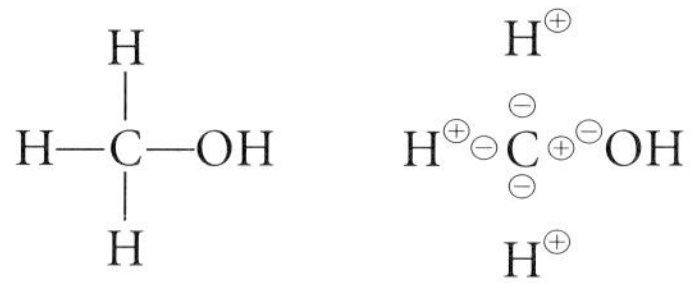

$$H-\overset{\overset{\displaystyle H}{|}}{\underset{\underset{\displaystyle H}{|}}{C}}-OH \qquad H^{\oplus}{\ominus}\overset{\overset{\displaystyle H^{\oplus}}{\ominus}}{\underset{\underset{\displaystyle H^{\oplus}}{\ominus}}{C}}{\oplus}{\ominus}OH$$

**Methanol**

Now let's examine the catalytic hydrogenation of ethene. The presence of the bonds to the two less electronegative hydrogen atoms contributes $-2$ to the oxidation level of the carbon atoms, and the two bonds ($\sigma$ and $\pi$) between the carbon atoms contribute 0. Thus, the oxidation level of the carbon atoms in this neutral molecule is $-2$. Conducting the same analysis for ethane (a product of catalytic hydrogenation), we find that each carbon atom is at the $-3$ oxidation level.

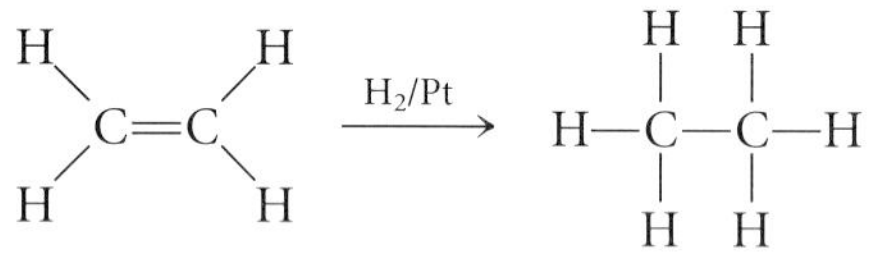

$$\overset{H}{\underset{H}{>}}C{=}C\overset{H}{\underset{H}{<}} \quad \xrightarrow{H_2/Pt} \quad H-\overset{\overset{\displaystyle H}{|}}{\underset{\underset{\displaystyle H}{|}}{C}}-\overset{\overset{\displaystyle H}{|}}{\underset{\underset{\displaystyle H}{|}}{C}}-H$$

**Ethene**            **Ethane**

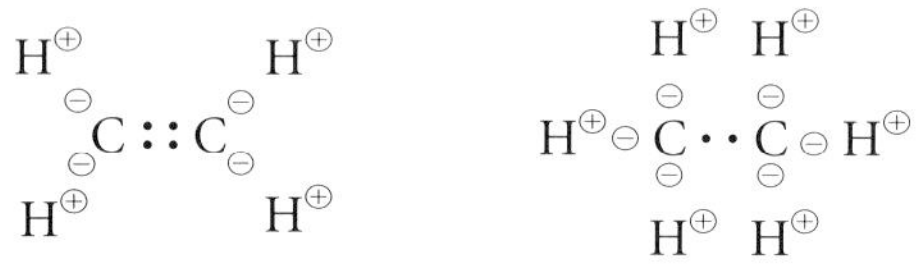

Thus, the formal oxidation level of carbon has been reduced (made more negative) by the addition of hydrogen to the molecule. In this example, each hydrogen atom in $H_2$ is oxidized (loses an electron) from the 0 to the $+1$ oxidation level; that is, the hydrocarbon is reduced as molecular hydrogen is oxidized.

Repeating the same analysis for ethyne ($C_2H_2$), we find that the carbon atoms are at a higher oxidation level than in ethene ($C_2H_4$). We conclude, as shown in Figure 3.7, that the greater the number of multiple bonds at a given carbon atom, the higher is its oxidation level. The introduction of

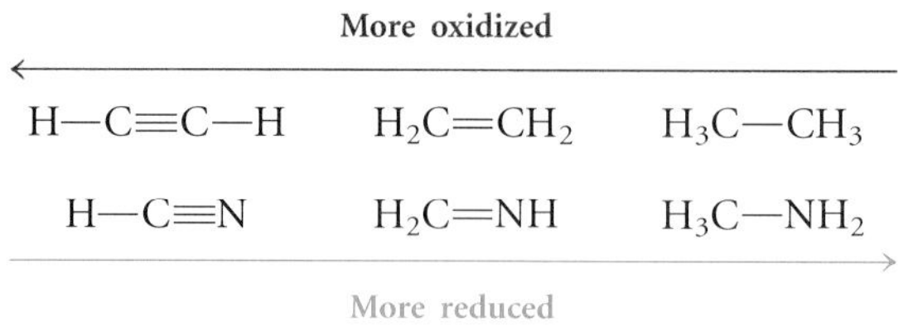

More oxidized

$$H-C{\equiv}C-H \qquad H_2C{=}CH_2 \qquad H_3C-CH_3$$

$$H-C{\equiv}N \qquad\quad H_2C{=}NH \qquad\quad H_3C-NH_2$$

More reduced

**FIGURE 3.7**

Relative oxidation levels of some hydrocarbons and nitrogen-containing compounds. The greater the degree of multiple bonding, the higher is the formal oxidation level of the atoms involved.

more multiple bonds requires an oxidizing reagent; to have fewer multiple bonds, we must use a reducing reagent. The same considerations also apply to compounds containing heteroatoms. The catalytic hydrogenation of an imine to the corresponding amine is similar to the reduction of an alkene to an alkane. As in an alkene, the carbon in an imine is at a higher oxidation level than the carbon in the amine formed by catalytic hydrogenation.

We can recognize a more highly oxidized functional group as being one with more multiple bonds or one with more bonds to heteroatoms. For example, the carbon atom in a C≡N triple bond (the structure is developed in more detail in the next section) is at a higher oxidation level than that in a C=N double bond, which, in turn, is more highly oxidized than the carbon atom in a C—N single bond. Thus, the compounds in Figure 3.7 go from a more highly oxidized state toward the left to a more highly reduced one on the right.

The same procedure used for calculating oxidation levels in hydrocarbons can be applied to heteroatom-containing organic compounds. For example, in the catalytic reduction of an imine to an amine, two hydrogen atoms change from an oxidation level of 0 in $H_2$ to a level of $+1$ in the product. We thus know that some atom or atoms must undergo reduction by a total change of $-2$.

An imine → (H₂/Pt) → An amine

By mentally breaking all the bonds, we see that the nitrogen atom has *not* changed ($-3$ in both imine and amine), whereas the carbon atom has changed from 0 in the imine to $-2$ in the amine.

This simple system for determining oxidation levels even works for charged atoms. For example, amines react with acids to form ammonium ions, as in the formation of methylammonium bromide from methylamine and HBr.

Methylamine → (HBr) → Methylammonium bromide

**TABLE 3.3**

Some Common Oxidizing and Reducing Agents

| Reducing Agents | Oxidizing Agents |
|---|---|
| $H_2$/Pt (catalytic hydrogenation) | Cr(VI) (chromate), often as $CrO_4^{2-}$ or $Cr_2O_7^{2-}$ |
| $NaBH_4$ (sodium borohydride) | Mn(VII) (permanganate), often as $MnO_4^-$ |
| $LiAlH_4$ (lithium aluminum hydride) | Cu(II) (cupric ion) |
| $NaB(CN)H_3$ (sodium cyanoborohydride) | Os(VIII) (osmate ion) |
| | Fe(III) (ferric ion) |

We know there is no change in oxidation level of the carbon atom, because its bonds are the same in starting material and product. Although the nitrogen atom has one additional bond (to hydrogen) in the product ammonium ion, this atom has not undergone an oxidation or reduction. Mentally disconnecting all of the bonds leaves a nitrogen atom with a *net* charge, and therefore an oxidation level, of $-3$, the same as in methylamine.

Note that the sum of the oxidation levels of all of the atoms in a molecule must equal its charge. We can use this observation to determine oxidation levels even when we are unsure of the bonding in a molecule, if we assume that certain elements are in their normal oxidation levels. For example, oxygen atoms are rarely found in oxidation states other than $-2$. Thus, for neutral $CrO_3$, the chromium must have an oxidation level of $+6$ to balance the three oxygens [$3 \times (-2) = -6$].

Once the transformation of one compound to another is identified as either an oxidation or a reduction, chemists can choose among possible reagents to effect the desired reaction (Table 3.3). In many common reagents used for reduction, hydrogen is present at an oxidation level lower than its usual $+1$ state—for example, 0 in hydrogen gas, $H_2$, or $-1$ in hydride, $H^-$, in complex metal hydrides such as $NaBH_4$ and $LiAlH_4$ (which will be discussed further in Chapter 12).

Oxidation reactions are accomplished with reagents that can change their formal oxidation levels by taking on additional electrons, thus undergoing reduction. Many oxidation reagents therefore contain a transition metal having two or more relatively stable oxidation states, such as chromium, manganese, copper, osmium, and iron. Several of these reagents are particularly convenient because their color changes when they are reduced, thereby allowing the reaction to be followed readily. For example, chromate reagents with chromium in the $+6$ oxidation state are red-orange but become green chromium ($+3$) salts when reduced by reaction with an organic substrate, which is oxidized. Manganese is purple as the permanganate ion, Mn(VII), in $KMnO_4$ and red-brown when reduced to Mn(IV) in $MnO_2$.

**EXERCISE 3.10**

Calculate the formal oxidation level of the carbon atoms in cyclohexane and cyclohexene.

Determine whether any of the carbon atoms are at different oxidation levels in each of the following pairs of compounds. If so, identify the structure—the left or the right—that contains the more highly oxidized atom.

(a) $H_3C-C\equiv N$ and $H_3C-C\overset{N-H}{\underset{H}{\diagup}}$

(b) $H_3C-\overset{O}{\overset{\|}{C}}-OH$ and $H_3C-\overset{O}{\overset{\|}{C}}-H$

(c) $CH_3CH_2NH_2$ and $H_3C-C\equiv N$

(d) $H_3C-\overset{O}{\overset{\|}{C}}-OH$ and $H_3C-\overset{O}{\overset{\|}{C}}-OCH_3$

(e) $H_3C-\overset{O}{\overset{\|}{C}}-OH$ and $H_3C-\overset{O}{\overset{\|}{C}}-Cl$

## 3.4

# Compounds Containing *sp*-Hybridized Nitrogen

The remaining mode for hybridization of nitrogen is the mixing of a $2s$ and a $2p$ atomic orbital to form two *sp*-hybrid orbitals, leaving two $2p$ orbitals unmixed.

N:     $1s$   $\underbrace{2s\ 2p_x}$   $\underbrace{2p_y\ 2p_z}$

Two
*sp*-hybrid
orbitals

Two *p*
orbitals

These remaining orthogonal *p* orbitals can participate in the same type of bonding as in alkynes. A compound containing an *sp*-hybridized nitrogen atom linked to a carbon by a triple bond is called a **nitrile.** This functional group is also called a **cyano group** when present in an organic molecule or a **cyanide ion** when it exists as a negatively charged ion ($^{\ominus}C\equiv N$) or in inorganic reagents. Acetonitrile, $H_3CC\equiv N$, is shown in the margin as both ball-and-stick and space-filling representations. As in the alkynes, the *sp*-hybrid orbitals participating in $\sigma$ bonding are oriented at 180° relative to each other. The *p* orbitals are orthogonal to the $\sigma$ bond, overlapping above and below and in front of and behind the linear array of atoms.

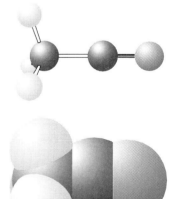

**Acetonitrile**

Isonitriles are similar in structure to nitriles, but are unusual in that both the carbon and the nitrogen joined by the triple bond are charged. Determine the formal charges on these atoms for the following isonitrile. Then draw an orbital picture of its structure, being sure to account for the lone pair of electrons on carbon.

$$H_3C\overset{\oplus}{-}\overset{\ominus}{N}\equiv C:$$

**An isonitrile**

## 3.5

## Compounds Containing $sp^3$-Hybridized Oxygen

Derivatives of water in which one of the hydrogen atoms is replaced by an alkyl or an aryl group are called *alcohols*. Compounds in which both hydrogen atoms of water are replaced are called *ethers*. The same structural features that characterize bonding in organic compounds containing nitrogen atoms are also encountered in oxygen-containing compounds.

### Water

Mixing of the $2s$ and $2p$ atomic orbitals of oxygen forms four $sp^3$-hybrid orbitals:

$$O: \quad 1s \quad \underline{2s\ 2p_x\ 2p_y\ 2p_z}$$

$$\downarrow$$

Four
$sp^3$-hybrid orbitals

Because these four orbitals must accommodate the six electrons of a neutral oxygen atom, they can accommodate only two additional electrons, and oxygen is limited to two covalent bonds before it reaches a filled-valence-shell electron configuration. When covalent bonds are formed with hydrogen atoms, the result is a water molecule, whose Lewis dot structure and geometry are shown in Figure 3.8. In the Lewis dot structure, the six valence electrons of oxygen are shown as red dots, and the two electrons contributed by the two hydrogen atoms as black dots. The valence requirement of each atom is satisfied, and the $sp^3$-hybrid orbitals are directed at ap-

H:Ö:H

**FIGURE 3.8**

Lewis dot structure and two three-dimensional representations (ball-and-stick and space-filling) of water. (The lone pairs of electrons on oxygen are shown as small pink spheres.)

proximately a tetrahedral angle. The repulsion between the lone pairs on oxygen in water is slightly larger than that between a lone pair and a $\sigma$ bond (as in ammonia), which, in turn, is larger than the bond–bond repulsion in methane. As a result, in water the angle between the lone pairs is expanded slightly, and the H—O—H angle somewhat compressed (to about 105°), from the ideal tetrahedral angle.

### Alcohols: R—OH

**Alcohols** are functional groups of the type R—O—H, where oxygen is bound on one side to carbon and on the other side to hydrogen. Methanol, in which a methyl group is bound to oxygen, is the simplest alcohol. Note the close similarity between the structures of methanol and methylamine (Section 3.1).

As in the amines shown in Figure 3.1, the OH group can be at various positions in an alcohol. Examples are the three isomeric six-carbon alcohols shown in Figure 3.9. Nomenclature for alcohols is in accord with the IUPAC rules for hydrocarbons (see Chapter 2), with the ending -**anol** used to indicate the presence of an OH group. A number preceding the name of the compound indicates the position along the chain at which the OH group is attached.

**Methanol**

**1-Hexanol**
(a primary alcohol)

**2-Hexanol**
(a secondary alcohol)

**2-Methyl-2-pentanol**
(a tertiary alcohol)

**FIGURE 3.9**

A primary, secondary, and tertiary alcohol.

*Primary, Secondary, and Tertiary Alcohols.* Alcohols are subclassified as primary, secondary, or tertiary according to the nature of the carbon to which the OH group is attached. This carbon is uniquely identified as the **carbinol carbon.** Unlike amines, in which one, two, or three carbons can be bonded to nitrogen, an alcohol can have only one carbon atom attached to oxygen and retains its functional-group identity. Thus, 1-hexanol is a **primary alcohol** because the OH group is attached to a primary carbon (that is, the carbinol carbon is itself bonded to only one other carbon atom), 2-hexanol is a **secondary alcohol** because the carbinol carbon is attached to two other carbon atoms, and 2-methyl-2-pentanol is a **tertiary alcohol** because the carbinol carbon is attached to three other carbons.

### EXERCISE 3.13

Classify each of the following alcohols as primary, secondary, or tertiary.

(a) OH

(b) OH

(c) OH

(d) OH

(e) OH

(f) OH

***Hydrogen Bonding.*** Like nitrogen, oxygen is more electronegative than both carbon and hydrogen. This leads to partial charge separation in the covalent bonds that oxygen forms with either carbon or hydrogen.

**Bond Polarization in Methanol**

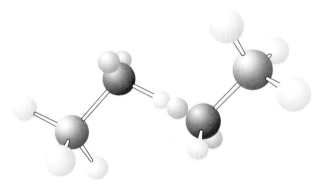

A methanol molecule forms hydrogen bonds with other methanol molecules (Figure 3.10) or with other polar molecules such as water (Figure 3.11). The boiling point of an alcohol is thus significantly higher than that of a hydrocarbon of similar molecular weight (see Table 3.1), and low-molecular-weight alcohols are significantly soluble in water and other po-

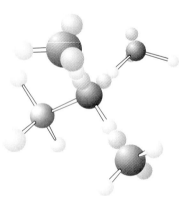

**FIGURE 3.10**

Hydrogen bonding between methanol molecules. (Lone pairs of electrons are shown as small pink spheres.)

**FIGURE 3.11**

Hydrogen bonding between methanol and water. Hydrogens of two water molecules (top) form hydrogen bonds with the two lone pairs of electrons on the oxygen of methanol, and the lone pair on the oxygen of a third water molecule (lower right) hydrogen bonds with the hydrogen of the OH group of methanol.

lar solvents. Indeed, methanol, ethanol, 1- and 2-propanol, and all of the butanols except *t*-butanol (2-methyl-2-propanol) are miscible (soluble in all proportions) with water. As the molecular weight of an alcohol increases, the polar OH group becomes a smaller fraction of the total molecular volume, and the solubility of the alcohol in water decreases. Most alcohols are miscible with all common organic solvents. The exception is methanol, which is not miscible with simple alkanes such as pentane and hexane.

## EXERCISE 3.14

Explain why the boiling point of ethanol (78 °C) is significantly higher than that of ethylamine (16 °C).

## Ethers: R—O—R

Ethers are another family of organic compounds that contains oxygen as the functional group. Unlike alcohols, **ethers** have two carbon atoms bound to an oxygen atom; the simplest example is methyl ether. A near-tetrahedral geometry is maintained at the *sp*³-hybridized oxygen atom, and the carbon–oxygen bonds are polarized such that the electron density on oxygen is higher than on carbon. Because of this molecular dipole, ethers are more polar than hydrocarbons and more soluble in polar solvents.

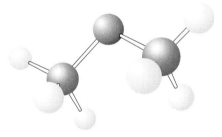

**Methyl ether**

Because ether molecules do not have an OH group, they cannot form hydrogen bonds with other ether molecules. (An ether molecule can participate in hydrogen bonding with a molecule that has a hydrogen bonded to a heteroatom.) The polarized carbon–oxygen bond enhances van der Waals attractions between ether molecules (compared with hydrocarbons), but without hydrogen bonding, ethers have weaker intermolecular interactions than do alcohols. Therefore, the boiling points of ethers are comparable to those of hydrocarbons of similar molecular weights but lower than those of the analogous alcohols.

The bond-dissociation energies of the carbon–oxygen bonds in ethers are similar to those of the carbon–oxygen bonds in alcohols, but heterolytic cleavage of that type of bond in an ether generally requires a very strong acid and rigorous heating. In general, ethers are very unreactive and therefore are useful as *solvents*. Good solvents interact with solutes well enough to dissolve them but remain chemically inert. Ethers are good solvents because the polar carbon–oxygen bonds participate in dipole–dipole interactions with other polar molecules. In addition, the lone pair of electrons on the oxygen of an ether participates in hydrogen bonding with molecules bearing a polar X—H bond. However, ethers are resistant to heterolytic cleavage reactions because they do not have an X—H bond and thus do not serve as proton donors, a reaction characteristic of alcohols. Although ethers are polar, they lack an acidic proton on a heteroatom. They are therefore aprotic solvents.

Ethers derive their names from the alkyl groups attached to oxygen—for example $CH_3OCH_2CH_3$ is methylethyl ether, and $CH_3CH_2OCH_2CH_3$ is ethyl ether. Note that names of ethers do not have a prefix (such as di-) to indicate two identical groups. When only one group is specified, as in ethyl ether, the compound is assumed to be a symmetrical ether.

### EXERCISE 3.15

Name each of the following ethers:

(a)

(d)

(b)

(e)

(c)

### 3.6

## Bond Cleavage

The transformation of one stable organic molecule into another is invariably accompanied by changes in bonding—existing bonds are broken and new bonds are formed. As bonds are broken, energy is consumed. **Homolytic cleavage** occurs when the electrons of a bond are distributed equally to the two atoms originally joined by that bond, and the heat consumed as this occurs is defined as the **bond-dissociation energy.** Homolytic cleavage yields two radicals, species with an unshared electron and thus one fewer electron than is required for a full valence shell. For example, homolytic cleavage of the C—O bond of methanol produces a methyl radical and an OH radical and consumes 92 kcal/mole:

$$H_3C-OH \longrightarrow H_3C\cdot + \cdot OH \qquad \Delta H° = 92 \text{ kcal/mole}$$

**Methanol**

The movement of one electron from a bond to an atom that occurs in homolytic cleavage is indicated with a **half-headed curved arrow** (see Chapter 2). All homolytic bond cleavages require the input of energy and are thus **endothermic.**

**TABLE 3.4**

Typical Bond Lengths and Bond Energies

| Bond | Bond Length (Å) | Bond Energy (kcal/mole) |
|---|---|---|
| —C—H | 1.10 | 93–105 |
| —C—C— | 1.54 | 84–90 |
| N—H | 1.01 | 91–103 |
| —C—N | 1.47 | 82–85 |
| —O—H | 0.97 | 102–109 |
| —C—O— | 1.43 | 80–94 |
| F—H | 0.92 | 136 |
| —C—F | 1.40 | 107–108 |

Bond-dissociation energies (as well as bond lengths) for several representative kinds of bonds are listed in Table 3.4. More complete lists of bond-dissociation energies are given in Tables 3.2 and 3.5, which are also reproduced inside the back cover of this book.

**EXERCISE 3.16**

Draw the structures of the radicals that would be produced by homolytic cleavage of each of the bonds shown in Table 3.4.

In **heterolytic cleavage,** the two electrons initially shared in a bond are distributed unequally—that is, both electrons of the bond go to one of the atoms. This process is more favorable for a polar $\sigma$ bond between atoms of unlike electronegativity than for a nonpolar bond. In any event, heterolytic cleavage does not occur easily because it requires considerable energy, not only to break the $\sigma$ bond but also to completely separate the resulting positive and negative charges. In a polar covalent bond, part of this additional energy cost has already been paid in achieving polarization of the bond; however, in the absence of other factors, heterolytic cleavage of a bond requires more energy than does homolytic cleavage. You will learn later how the energy required for heterolytic bond cleavage can be reduced substantially below that for homolytic cleavage in the presence of polar solvents that stabilize the ions produced.

Heterolytic cleavage can occur in polar molecules by two modes. As an example, consider the heterolytic cleavage of the C—O bond of methanol. In reaction 1, the two electrons in the C—O bond are shifted to oxygen, producing a methyl cation and a hydroxide ion.

$$H_3C\!\!-\!\!OH \longrightarrow H_3C^\oplus + \ ^\ominus OH \tag{1}$$

Methanol          Methyl          Hydroxide
cation          ion

$$H_3C\!\!-\!\!OH \longrightarrow H_3C^\ominus + \ ^\oplus OH \tag{2}$$

Methyl anion          Hydroxyl cation

Here, it is important to note that the movement of the two electrons is indicated by a **full-headed curved arrow.** The two electrons of the C—O bond are no longer in a bonding orbital; rather, they are localized on oxygen.

In the alternative mode of heterolytic cleavage shown in reaction 2, the two electrons in the C—O bond are shifted to carbon to produce a methyl anion and a hydroxyl cation. Again, the full-headed curved-arrow notation shows the flow of two electrons from the $\sigma$ bond to carbon. Both reaction 1 and reaction 2 require cleavage of the $\sigma$ bond between carbon and oxygen and separation of charge. However, the heterolytic cleavage in reaction 2 is more difficult than that in reaction 1 because it opposes the inherent electronegativity tendency in the polar $\sigma$ bond. In reaction 1, the electrons shift to the more electronegative oxygen atom, whereas reaction 2 reverses this flow of electrons and thus requires more energy.

### EXERCISE 3.17

Apply your knowledge of electronegativity to draw a full-headed curved arrow showing the preferred direction of electron flow when each of the following bonds is cleaved heterolytically:

(a) $CH_3$—$SCH_3$     (b) $CH_3S$—H     (c) $CH_3$—OH     (d) $CH_3O$—$CH_3$

### Homolytic Cleavage: Bond Energies and Radical Structure

The cleavage of the carbon–oxygen or oxygen–hydrogen bond of an alcohol can be, in principle, either homolytic or heterolytic. Let's consider the structural factors that might affect the cleavage mode.

The principal measure of the ease of homolytic cleavage is the bond-dissociation energy. Table 3.5 lists specific bond-dissociation energies for some common bonds encountered in organic chemistry. These values are not the same as the average bond energies in Table 3.2 because the values in the two tables are arrived at in different ways. The bond-dissociation energies in Table 3.5 indicate the energy required to break a specific bond in a particular molecule, whereas the average bond energies in Table 3.2 are calculated from a set of experimental data, assuming, for example, that all

**TABLE 3.5**

Bond-Dissociation Energies (kcal/mole)

| Bond | H | F | Cl | Br | I | OH | NH$_2$ | CH$_3$ |
|---|---|---|---|---|---|---|---|---|
| | | | | | X | | | |
| Ph—X | 111 | 126 | 96 | 81 | 65 | 111 | 102 | 101 |
| CH$_3$—X | 105 | 108 | 85 | 70 | 57 | 92 | 85 | 90 |
| CH$_3$CH$_2$—X | 100 | 108 | 80 | 68 | 53 | 94 | 84 | 88 |
| (CH$_3$)$_2$CH—X | 96 | 107 | 81 | 68 | 54 | 94 | 84 | 86 |
| (CH$_3$)$_3$C—X | 93 | — | 82 | 68 | 51 | 93 | 82 | 84 |
| PhCH$_2$—X | 88 | — | 72 | 58 | 48 | 81 | — | 75 |
| H$_2$C=CHCH$_2$—X | 86 | — | 68 | 54 | 41 | 78 | — | 74 |
| H—X | 104 | 136 | 103 | 87 | 71 | 119 | 107 | 105 |
| X—X | 104 | 38 | 59 | 46 | 36 | 51 | 66 | 90 |

C—H bonds have the same energy. Both types of values are useful: bond-dissociation energies provide an accurate assessment of the energy required to break a particular bond homolytically; average bond energies can be used to estimate changes in energy for transformations from one stable species to another, especially in cases where $\pi$ bonds are broken and made.

Bond-dissociation energies vary with the degree of substitution of the atoms involved in the bond being broken. Thus, the energy required to break a C—H bond decreases progressively in the series shown in Figure 3.12.

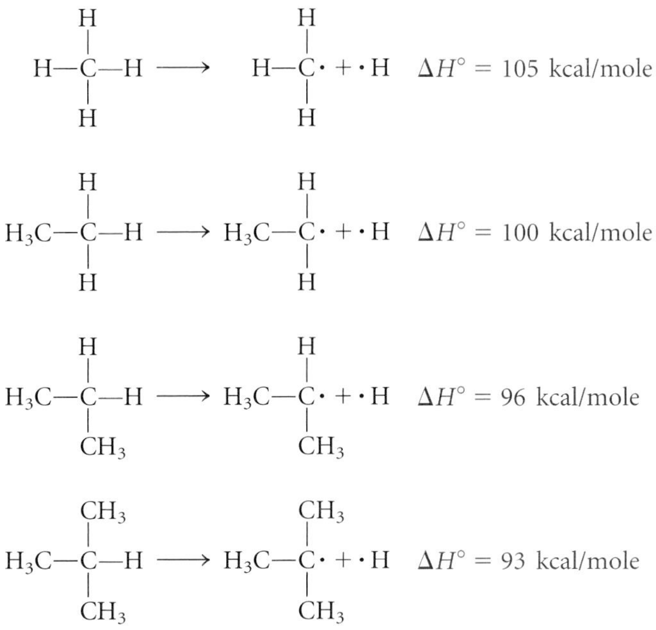

**FIGURE 3.12**

Bond-dissociation energies for a C—H bond in several simple hydrocarbons.

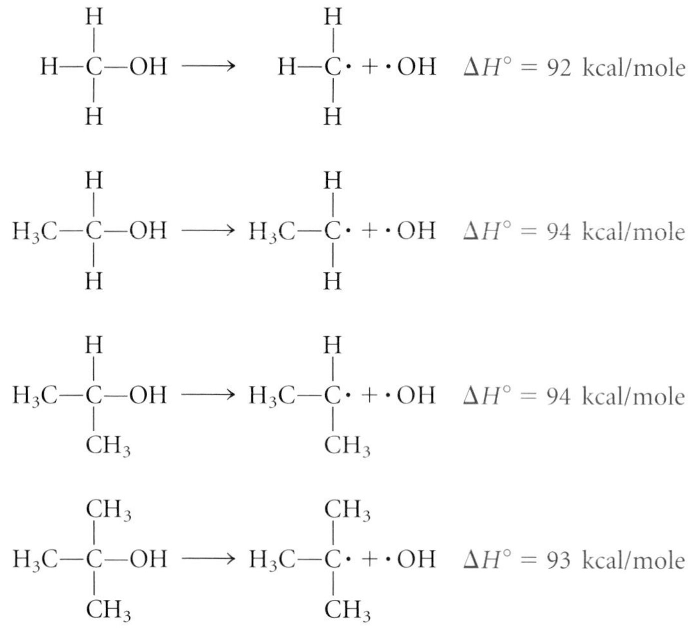

$$
\begin{array}{c}
\text{H} \\
| \\
\text{H} - \text{C} - \text{OH} \\
| \\
\text{H}
\end{array}
\longrightarrow
\begin{array}{c}
\text{H} \\
| \\
\text{H} - \text{C} \cdot + \cdot \text{OH} \\
| \\
\text{H}
\end{array}
\quad \Delta H^\circ = 92 \text{ kcal/mole}
$$

$$
\begin{array}{c}
\text{H} \\
| \\
\text{H}_3\text{C} - \text{C} - \text{OH} \\
| \\
\text{H}
\end{array}
\longrightarrow
\begin{array}{c}
\text{H} \\
| \\
\text{H}_3\text{C} - \text{C} \cdot + \cdot \text{OH} \\
| \\
\text{H}
\end{array}
\quad \Delta H^\circ = 94 \text{ kcal/mole}
$$

$$
\begin{array}{c}
\text{H} \\
| \\
\text{H}_3\text{C} - \text{C} - \text{OH} \\
| \\
\text{CH}_3
\end{array}
\longrightarrow
\begin{array}{c}
\text{H} \\
| \\
\text{H}_3\text{C} - \text{C} \cdot + \cdot \text{OH} \\
| \\
\text{CH}_3
\end{array}
\quad \Delta H^\circ = 94 \text{ kcal/mole}
$$

$$
\begin{array}{c}
\text{CH}_3 \\
| \\
\text{H}_3\text{C} - \text{C} - \text{OH} \\
| \\
\text{CH}_3
\end{array}
\longrightarrow
\begin{array}{c}
\text{CH}_3 \\
| \\
\text{H}_3\text{C} - \text{C} \cdot + \cdot \text{OH} \\
| \\
\text{CH}_3
\end{array}
\quad \Delta H^\circ = 93 \text{ kcal/mole}
$$

**FIGURE 3.13**

Bond-dissociation energies for a C—O bond in several simple alcohols.

Because these homolytic cleavages produce one fragment (the hydrogen radical) in common, the decreasing bond-dissociation energies in this series must result from a change in stability of the carbon radical formed or the ground-state strength of the C—H bond, or both.

For relatively nonpolar bonds (such as C—H and C—C bonds), the radical stabilization energy is important in determining the order of bond-dissociation energies. **Radical stability** follows the order tertiary > secondary > primary and influences the ease of homolytic cleavage. (Analogous to the subclassification of alcohols, the subclassification of radicals is based on the number of carbon atoms attached to the carbon bearing the unpaired electron.)

In contrast, bond-dissociation energies for bonds between carbon and more electronegative atoms such as oxygen and the halogens do not differ as much with the degree of substitution, as Figure 3.13. shows. The same primary, secondary, and tertiary carbon radicals are formed upon cleavage of the carbon–oxygen bond in this series as are produced in the breaking of the carbon–hydrogen bond in the series shown in Figure 3.12. However, the bond strength between carbon and oxygen increases as the degree of substitution increases, counteracting the effect of radical stability. The bond-dissociation energies of C—X bonds therefore do not easily conform to a degree-of-substitution trend, and individual entries in Table 3.5 must be used to determine $\Delta H^\circ$ for a reaction in which one of these bonds is broken or made.

## Radical Stabilization

Let's consider the structure of a carbon radical in order to understand why more substituted radicals are more stable. In the methyl radical ($\cdot \text{CH}_3$), there are three equivalent C—H bonds and one nonequivalent $p$ orbital bearing a single electron (Figure 3.14). In this configuration, the carbon atom has only seven valence electrons. Being one electron shy of a filled valence shell, carbon radicals are electron-deficient and highly reactive.

A radical has one unpaired electron, one fewer than required for a complete valence shell. Radicals are therefore electrophilic. Because the electron density of an $s$ orbital is closer to the nucleus than is that of a $p$ orbital, the lowest-energy arrangement of the methyl radical has as many electrons as possible in hybrid orbitals (with $s$ character). This is achieved when the single electron is held within a $p$ orbital, and the three hybrid orbitals are doubly occupied. (If the carbon were $sp^3$-hybridized, the three pairs of bonding electrons would be in orbitals with less $s$ character and thus farther from the nucleus.) The carbon of the methyl radical is therefore $sp^2$-hybridized, involved in three $\sigma$ bonds to hydrogens and having a singly occupied $p$ orbital. In this hybridization, the H—C—H angle is 120°, and the three C—H bonds are coplanar. Thus, when a methyl radical is formed by the homolytic cleavage of a $H_3C$—X bond, the carbon undergoes a geometric change from tetrahedral to planar and a rehybridization from $sp^3$ to $sp^2$.

Replacing one of the hydrogen atoms of the methyl radical by a methyl group yields the ethyl radical (Figure 3.15). An alkyl group is more polarizable than a hydrogen atom and can better satisfy the high electron demand of the electron-deficient $sp^2$-hybridized radical carbon. Furthermore, the carbon–hydrogen bonds of the methyl group can overlap with the singly occupied $p$ orbital, resulting in hyperconjugative stabilization similar to that seen in alkenes in Chapter 2. Any primary radical (such as the ethyl radical in Figure 3.15) has this hyperconjugative interaction, which is not possible in the simple methyl radical (Figure 3.14). Thus, the ethyl radical (or any other primary radical) is more stable than the methyl radical.

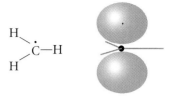

**FIGURE 3.14**

A three-dimensional representation of the methyl radical with a single electron in the $p$ orbital.

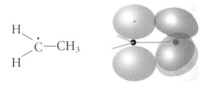

**FIGURE 3.15**

A three-dimensional representation of the ethyl radical. Note that it is possible to align the C—H bonds on the adjacent carbon atom with the singly occupied $p$ orbital, thus permitting hyperconjugative stabilization of the radical. (The $p$ orbital is shown in blue, the $sp^3$-hybridized orbitals of the methyl group in gray, and the hydrogen $s$ orbitals of the methyl group in off-white.)

The greater the number of alkyl groups directly connected to a radical carbon atom, the greater is the stabilization. In the isopropyl radical and the tertiary butyl radical, there are two and three alkyl groups, respectively, that provide hyperconjugative stabilization in the interaction with the electron-deficient carbon.

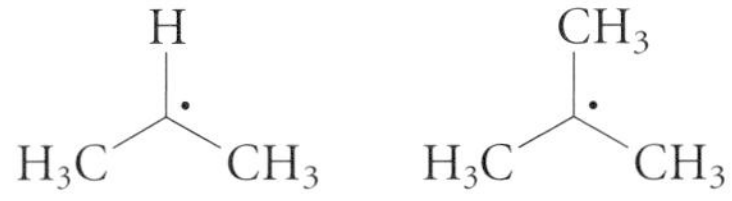

Isopropyl radical      *t*-Butyl radical

Thus, a tertiary radical is more stable than a secondary radical, which, in turn, is more stable than a primary radical, in part because of increasing stabilization by hyperconjugation along this series. This order of radical stability $(3° > 2° > 1°)$ is consistent with that observed from bond-dissociation energies for various carbon–hydrogen bonds.

### EXERCISE 3.18

Determine whether each of the following radicals is primary, secondary, or tertiary:

(a) 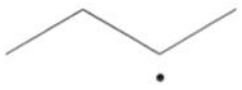 (b) 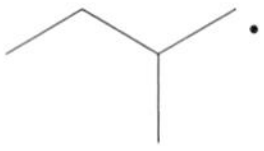 (c) 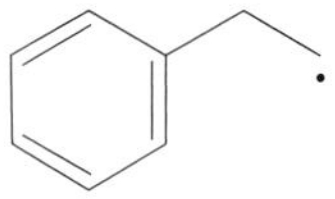 (d)

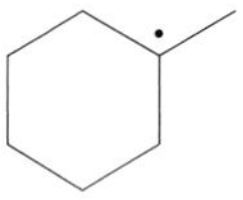

### EXERCISE 3.19

For each of the following pairs of compounds, choose the one that requires less energy for heterolytic cleavage of the C—X bond:

(a)  or

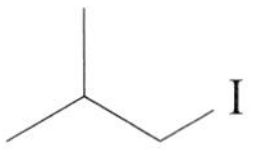

(b) 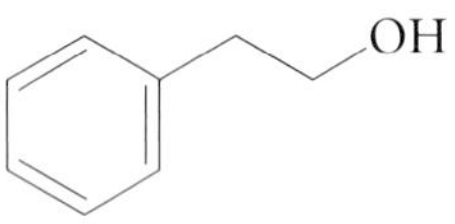 or

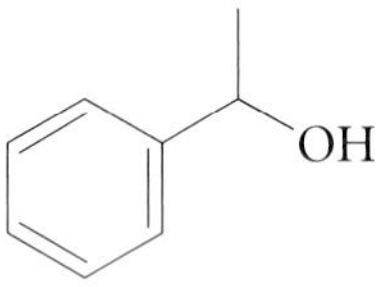

(c) 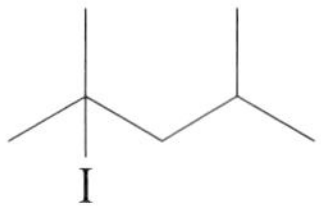 or

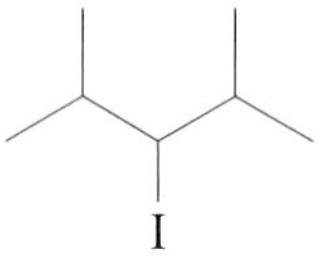

(d)  or

(e) 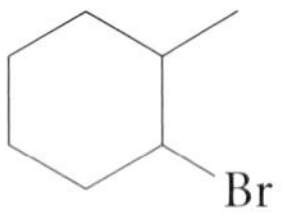 or

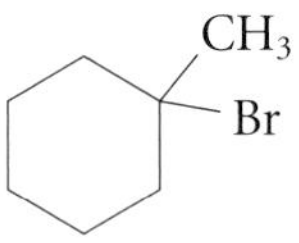

## Heterolytic Cleavage of C—OH Bonds: Carbocation Formation

In principle, heterolytic cleavage of a carbon–oxygen bond can occur in two ways that differ in the direction of the electron flow:

**Electron Flow to Oxygen**

$$H_3C\!-\!OH \longrightarrow H_3C^{\oplus} \quad {}^{\ominus}OH \tag{1}$$

**Electron Flow to Carbon**

$$H_3\overset{\frown}{C}\!-\!OH \longrightarrow H_3C^{\ominus} \quad {}^{\oplus}OH \tag{2}$$

Reaction 1 is more favorable than reaction 2 because the atom with the higher electronegativity (oxygen) takes on negative charge.

Spontaneous heterolytic cleavage of a carbon–oxygen bond is very difficult by either route because of the high energy cost of cleaving bonds and then separating charge to form ions. However, this reaction can be assisted significantly by the addition of a proton (from an acid) to one of the lone pairs of oxygen (we will reconsider this reaction in more detail in later chapters). For example, in the protonation of methanol, one of the lone pairs on oxygen must be shared between oxygen and the incoming proton, which is indicated by the curved-arrow notation.

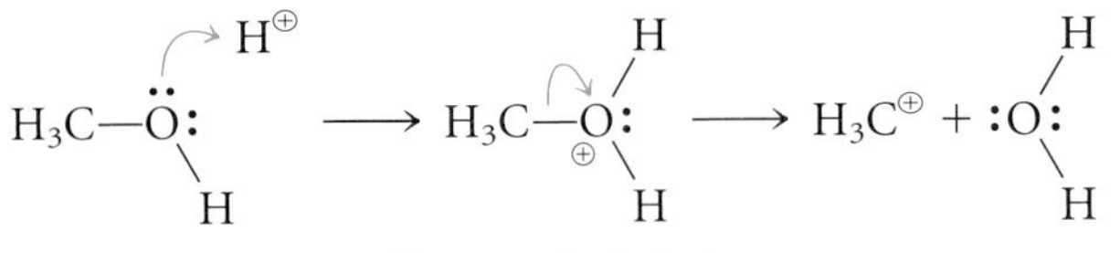

**Protonated alcohol**
(an oxonium ion)

The resulting **protonated alcohol** has a formal positive charge on oxygen, which bears three $\sigma$ bonds; this intermediate is called an **oxonium ion.** The ending *-onium* indicates positive charge, and the prefix *ox-* indicates that the charge resides substantially on oxygen. This positive charge induces even further polarization of the carbon–oxygen and hydrogen–oxygen bonds toward oxygen. Heterolytic cleavage of the carbon–oxygen bond in the second step of this reaction thus produces a methyl cation and a neutral water molecule. Because the oxonium ion is already positively charged, this cleavage does not require further charge separation and is therefore much more easily accomplished than heterolytic cleavage without acid.

In general, the ease of dehydration (loss of water) from an oxonium ion depends on the character of the alcohol. For various alcohols, the loss of water becomes progressively easier in the order methanol < primary alcohol < secondary alcohol < tertiary alcohol (Figure 3.16). Because water is

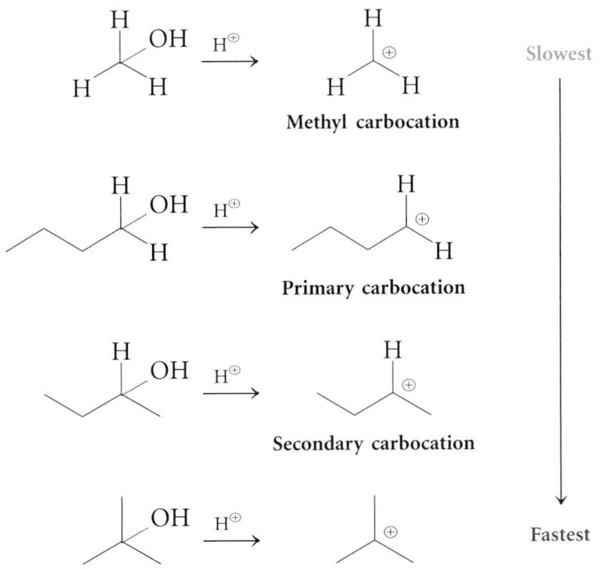

**Methyl carbocation**

Slowest

**Primary carbocation**

**Secondary carbocation**

**Tertiary carbocation**

Fastest

**FIGURE 3.16**

Dehydration of methanol and three isomeric butanols.

the common product, this order is largely governed by the stability of the resulting alkyl cation, called a **carbocation,** or **carbonium ion.** The order of stability of carbocations is the same as that of radicals. Because a carbocation has only six valence electrons, it is even more electron-deficient than a radical.

### Cation Stabilization

The structural factors that control radical stability and geometry also apply to carbocations. Thus, rehybridization of carbon from $sp^3$-hybridized in an alcohol to $sp^2$-hybridized in the analogous carbocation produces a planar cation with a vacant $p$ orbital orthogonal to the plane of the atoms. Because a carbon with a vacant $p$ orbital is even more electron-deficient than a radical (which bears a single electron in the $p$ orbital), the stability afforded to a carbocation by hyperconjugation (Figure 3.17) is even greater. Because of this hyperconjugative stabilization, the $t$-butyl cation is more stable than the isopropyl cation, which is more stable than the ethyl cation, which is more stable than the methyl cation. The order of stability of carbocations, like that of radicals, is therefore tertiary > secondary > primary > methyl.

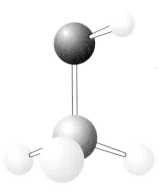

Methanol

Methyl carbocation

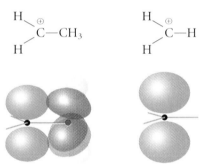

**FIGURE 3.17**

Hyperconjugation in the ethyl cation (left) affords stabilization not available to the methyl cation (right). (The $p$ orbital is shown in blue, the $sp^3$-hybridized orbitals of the methyl group in gray, and the hydrogen $s$ orbitals of the methyl group in off-white.)

### Ordering Alcohol Reactivity by Class

The observed reactivity of different classes of alcohols can be explained by the stability of the intermediate cations. Originally, cation stabilities were ordered based on experimental observations. One of the sources of experimental observations was the Lucas test, a classical qualitative test for distinguishing primary, secondary, and tertiary alcohols. When treated with Lucas reagent, a mixture of Brønsted (concentrated aqueous HCl) and Lewis ($ZnCl_2$) acids, alcohols undergo heterolytic cleavage, with the ultimate formation of alkyl halides (Figure 3.18). This conversion proceeds through an intermediary carbocation, and the rate of reaction parallels the ease of formation (stability) of that ion.

The alcohol initially dissolves in the Lucas reagent; however, with secondary and tertiary alcohols, a layer appears because the product alkyl halide does not have a polar O—H bond that can participate in hydrogen bond-

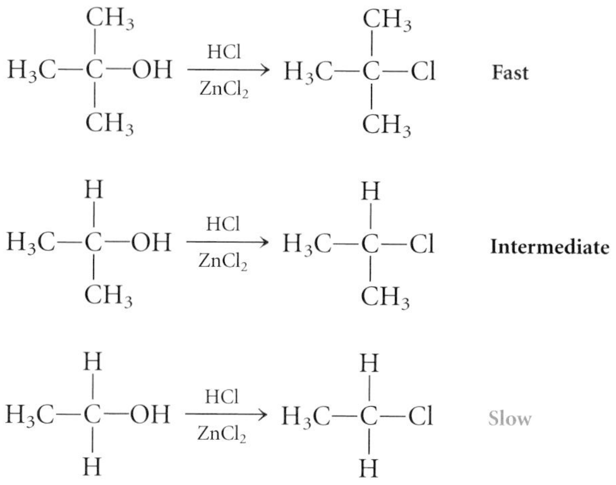

$$\underset{\underset{\displaystyle CH_3}{|}}{\overset{\overset{\displaystyle CH_3}{|}}{H_3C-C-OH}} \xrightarrow[\text{ZnCl}_2]{\text{HCl}} \underset{\underset{\displaystyle CH_3}{|}}{\overset{\overset{\displaystyle CH_3}{|}}{H_3C-C-Cl}} \quad \textbf{Fast}$$

$$\underset{\underset{\displaystyle CH_3}{|}}{\overset{\overset{\displaystyle H}{|}}{H_3C-C-OH}} \xrightarrow[\text{ZnCl}_2]{\text{HCl}} \underset{\underset{\displaystyle CH_3}{|}}{\overset{\overset{\displaystyle H}{|}}{H_3C-C-Cl}} \quad \textbf{Intermediate}$$

$$\underset{\underset{\displaystyle H}{|}}{\overset{\overset{\displaystyle H}{|}}{H_3C-C-OH}} \xrightarrow[\text{ZnCl}_2]{\text{HCl}} \underset{\underset{\displaystyle H}{|}}{\overset{\overset{\displaystyle H}{|}}{H_3C-C-Cl}} \quad \text{Slow}$$

**FIGURE 3.18**

Relative rates for appearance of an insoluble layer in a Lucas test.

ing. The product is therefore much less soluble in water than is the starting alcohol. How fast the layer appears depends on how quickly the alcohol reacts. As indicated in Figure 3.18, this layer appears almost immediately with tertiary alcohols, slowly with secondary alcohols, and virtually not at all (no reaction after 5 or 10 minutes) with primary alcohols. The tertiary alcohol, which can form a relatively stable tertiary carbonium ion, reacts much faster than the primary alcohol, whose corresponding carbonium ion is much less stable. Note that the —OH group is replaced by —Cl in this reaction, which is therefore called a **substitution reaction** (we will consider such reactions in much more detail in Chapters 7 and 8).

**EXERCISE 3.20**

For each of the following alcohols, predict whether the reaction with Lucas reagent will be immediate or slow or not take place at all:

(a) (structure: 2-pentanol with OH)

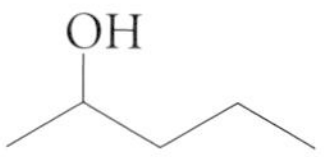

(c) (structure: 2-methyl-2-butanol with HO)

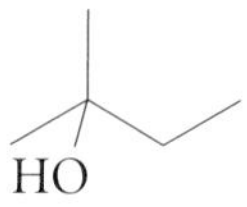

(b) (structure: 1-pentanol with OH)

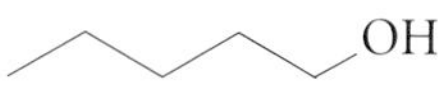

(d) (structure: 2-phenylethanol with OH and benzene ring)

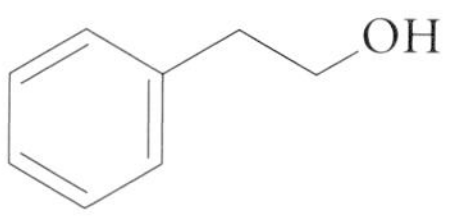

## Conjugation in Cations and Radicals

Let's now consider the interaction of the vacant $p$ orbital of a carbocation with an adjacent carbon–carbon double bond. Overlap of this $p$ orbital with the $\pi$ bond results in the conjugated system shown in Figure 3.19 (on page 122). The presence of easily polarized $p$ electrons adjacent to a vacant $p$ orbital allows for resonance interaction, which disperses positive charge to the atoms at opposite ends of this conjugated system of three carbon atoms. This shift of electron density is represented in the resonance

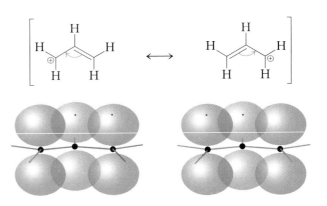

**FIGURE 3.19**

Three-dimensional representations of significant resonance contributors for an allyl cation.

structures in the figure by curved-arrow notation. (Recall from Chapter 2 that a resonance structure is a representation of electron distribution in a molecule in which atomic positions are fixed.) Whenever you can write two or more reasonable resonance structures, the molecule, ion, or radical being represented is unusually stable because of delocalization. In the example in Figure 3.19, partial positive charge is distributed over two carbons rather than localized on one. This primary allyl cation, containing two electrons in three adjacent $p$ orbitals, is almost as stable as a secondary alkyl cation.

A similar interaction takes place between a carbocationic center and a phenyl substituent. The resulting benzyl cation, which has a vacant $p$ orbital adjacent to an aryl ring (Figure 3.20), also has significant resonance stabilization and is about as stable as a tertiary alkyl cation. Thus, the order of stability presented earlier for carbocations should be expanded: benzylic ~ tertiary > allylic ~ secondary > primary > methyl.

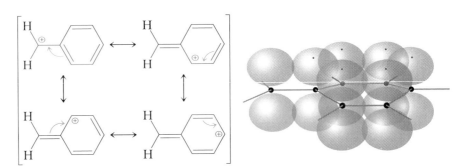

**FIGURE 3.20**

Resonance contributors (left) and an orbital picture (right) for a benzyl cation.

The same factors that are important in stabilizing allyl and benzyl cations affect the stability of allyl and benzyl radicals, as shown in Figure 3.21. Because the electron arrangement can be represented with several important resonance structures, these radicals are more stable than simple primary radicals. Thus, a more complete order of radical stability, inferred from the C—H bond-dissociation energies in Table 3.5, is: allylic > benzylic > tertiary > secondary > primary > methyl.

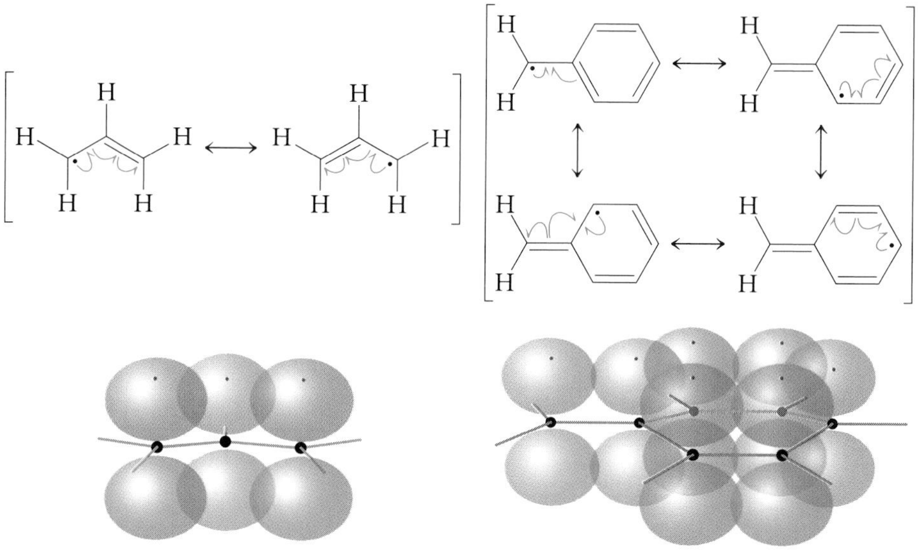

**FIGURE 3.21**

Resonance contributors for the allyl (left) and benzyl (right) radicals.

## EXERCISE 3.21

For each of the following alcohols, determine whether a primary, secondary, tertiary, allylic, or benzylic carbocation would be produced by protonation of oxygen followed by loss of water:

(a)

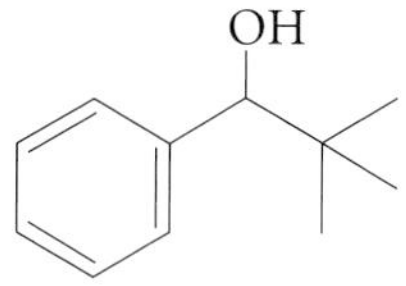

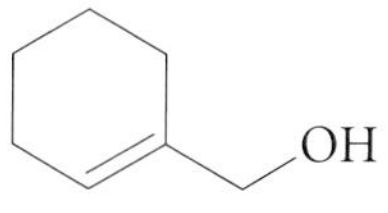

(b)

(c)

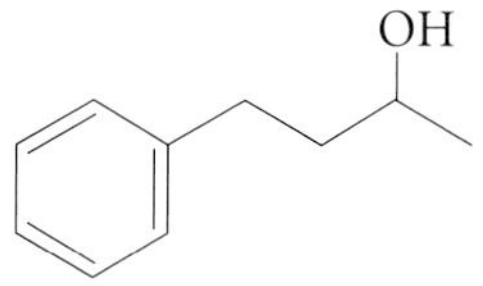

(d)

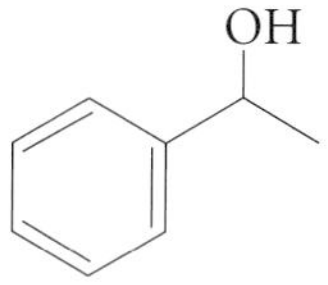

## EXERCISE 3.22

Rank each of the following sets of isomers in order of facility (from fastest to slowest) of acid-catalyzed dehydration. (*Hint:* Acid-catalyzed dehydration proceeds through a carbocation.)

(a)

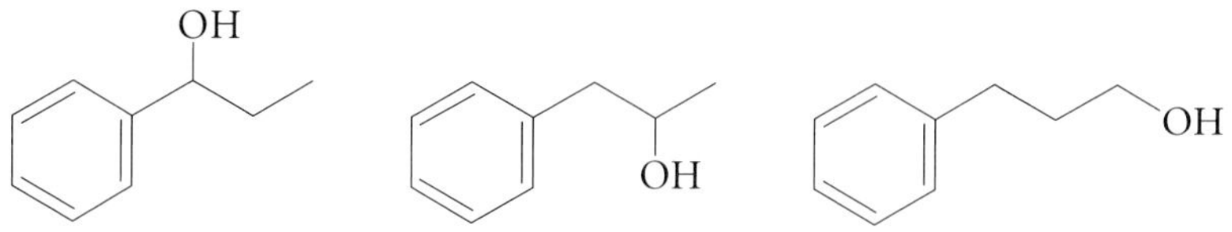

(b)

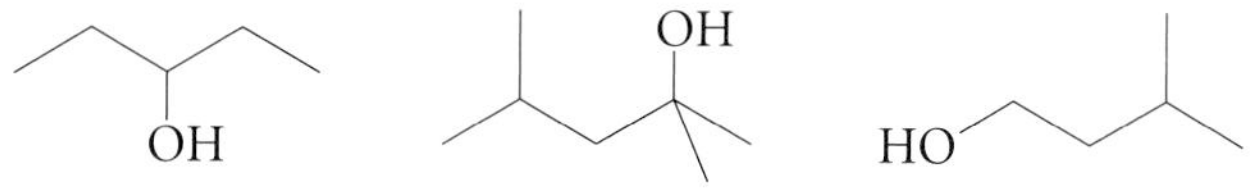

(c)

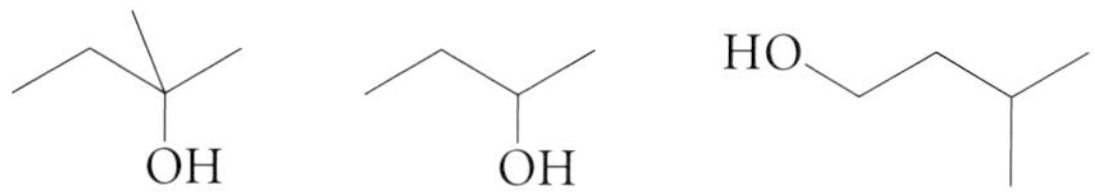

# Bond Formation: Nucleophiles and Electrophiles

Generally it is not possible to proceed from one stable organic molecule to another by bond cleavage alone—new bonds must also be formed so that all atoms in the product are at their normal valence level. For example, in the heterolytic cleavage of *t*-butanol in the presence of a Brønsted acid, the C—O bond is broken, forming water, in which both the oxygen and the hydrogen have filled valence shells, and *t*-butyl carbocation, in which the positively charged carbon atom has only six valence electrons.

*t*-Butanol          *t*-Butyl
                    carbocation

The tertiary carbocation can react with chloride ion to form the stable product, *t*-butyl chloride. The ions come together as an electrophile and a nucleophile:

Electrophile    Nucleophile

Electron-rich reagents such as chloride ion are referred to as **nucleophiles** because they seek centers of positive charge. The word *nucleophile* derives from the Greek *nucleo,* for "nucleus," and *philos,* for "loving." Electron-deficient reagents such as carbocations are referred to as **electrophiles.** The word *electrophile* is derived from the Greek *electros,* "electron," and *philos,* "loving."

Nucleophiles and electrophiles are not necessarily charged species. For a species to be a nucleophile, it is sufficient that it have an atom with a lone pair of electrons. For example, both $H_2O$ and $NH_3$ act as nucleophiles in organic reactions. Reaction of *t*-butyl cation with water produces *t*-butanol after loss of a proton from oxygen:

Electrophile    Nucleophile

Similarly, an electrophile need only have an atom that can accept electron density from a nucleophile.

**Nucleophilicity** can be defined as the tendency of an atom, an ion, or a group of atoms to donate electrons to an atom (usually carbon).

**Electrophilicity** can be defined as the tendency of an atom, ion, or group of atoms to accept electron density from some atom. A negatively charged ion (an anion) is thus more nucleophilic than the corresponding neutral species, and a positively charged ion (a cation) is more electrophilic than the corresponding neutral species. We will make extensive use of the concept of electrophiles and nucleophiles in later chapters. As we begin to explore organic reactions, you will become able to recognize reagents that are nucleophilic, those that are electrophilic, and those that can be either. Nucleophiles act as Lewis bases (electron-pair donors), and electrophiles act as Lewis acids (electron-pair acceptors).

## 3.8

## Carbonyl Compounds (Aldehydes and Ketones): $R_2C=O$

Oxygen atoms also participate in multiple bonding. The functional group consisting of oxygen doubly bonded to carbon is referred to as a **carbonyl group** ($C=O$), and the carbon involved in the double bond to oxygen is called a **carbonyl carbon.** Carbonyl compounds are often prepared by oxidation of alcohols (Figure 3.22), in a process that formally removes two hydrogen atoms, one from oxygen and one from carbon.

A carbonyl carbon that bears a hydrogen and an alkyl group is called an **aldehyde;** one that bears two alkyl groups is called a **ketone.** An aldehyde is produced by the oxidation of a primary alcohol, and a ketone is formed by the oxidation of a secondary alcohol. The fact that the conver-

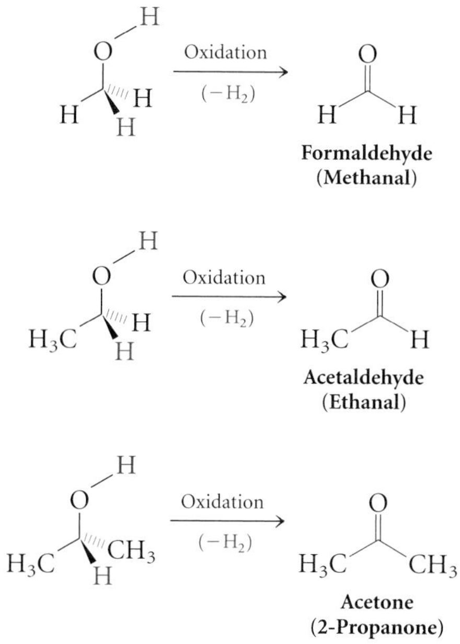

#03    Aldehydes/Ketones

**FIGURE 3.22**

Carbonyl compounds produced by oxidation of alcohols.

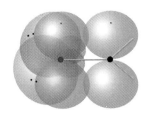

**FIGURE 3.23**

An orbital representation of a ketone. The $\pi$ bond is formed by overlap of $p$ orbitals on oxygen and carbon. The oxygen atom of a ketone has two lone pairs of electrons, held in $sp^2$-hybrid orbitals (gray) that lie in the plane of the $\pi$ bond.

sion of an alcohol to an aldehyde or a ketone is an oxidation can be verified by counting the number of bonds from carbon to the heteroatom. In an alcohol, there is one bond to oxygen; after oxidation, there are two. An aldehyde is named by adding the suffix **-anal** to the stem indicating the number of carbon atoms; a ketone is named by adding the suffix **-anone.**

The carbon atom of a carbonyl group is bonded to only three atoms and is thus $sp^2$-hybridized. The trigonal planar geometries of carbonyl compounds are therefore similar to those of imines, which contain C=N bonds. As shown in Figure 3.23, the planar carbonyl group has a C—C—O bond angle of about 120° and a $\pi$ bond formed by overlap of $p$ orbitals above and below the carbonyl carbon atom and the three attached substituents. Carbonyl compounds can be catalytically hydrogenated, yielding alcohols. However, the higher stability of the C=O $\pi$ bond compared with the C=N $\pi$ bond makes the reaction more difficult than hydrogenation of imines, and much more rigorous experimental conditions are required. Adjacent alkyl groups stabilize C=O bonds even more than C=C bonds.

### Resonance Structures

Oxygen, being more electronegative than carbon, attracts the electrons in the C=O bond more strongly. The electrons in a $p$ orbital are held less tightly than those in a $\sigma$ bond, and thus the degree of polarizability is greater for a $\pi$ bond than for a $\sigma$ bond. We can write a resonance structure in which the electrons initially shared between carbon and oxygen in the $\pi$ bond are shifted completely to oxygen.

**Resonance Structures of Acetone**

The resonance contributor at the right, when regarded as a Lewis dot structure, meets the valence requirement of oxygen, but not that of carbon. Furthermore, this structure has one fewer covalent bond and formal charge separation, with carbon bearing a positive charge and oxygen a negative charge. The structure at the right therefore contributes less to the real structure of the carbonyl group than does the one at the left. However, it does contribute to some degree, and the carbonyl group is polarized, making the carbon end of the C=O bond electron-deficient and the oxygen end electron-rich. Therefore, the carbon atom of a carbonyl group is readily attacked by nucleophiles, and the oxygen atom by electrophiles (these reactions will be covered in more detail in Chapters 12 and 13).

**Nucleophilic Attack at a Carbonyl Carbon**

**Electrophilic Attack at a Carbonyl Oxygen**

**EXERCISE 3.23**

Name the functional group present in each of the following oxygen-containing molecules:

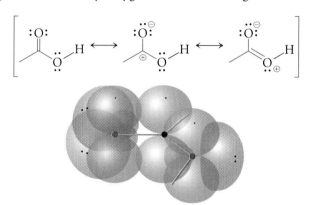

(a)  (b)  (c)  (d)

## Carboxylic Acids: $RCO_2H$

Because an oxygen atom needs only two covalent bonds to fulfill its valence requirement, it does not ordinarily participate in triple bonding. However, a carbon atom can bond to two oxygen atoms. Replacement of the hydrogen of an aldehyde with an OH group produces a class of compounds known as **carboxylic acids.** A carboxylic acid ($RCO_2H$) is easily named by the addition of the suffix **-anoic** to the root designating the appropriate hydrocarbon. For example, the three-carbon acid is called propanoic acid. The carbon atom of a carboxylic acid is $sp^2$-hybridized with a $\pi$ bond to oxygen forming a carbonyl group. Thus, the structure is trigonal planar. There are two additional resonance structures for a carboxylic acid, both with negative charge on the carbonyl oxygen, as shown in Figure 3.24.

HCHEM TV® II

#04   Carboxylic Acids/
Derivatives

**FIGURE 3.24**

Resonance contributors (top) and orbital picture (bottom) for a carboxylic acid.

One resonance structure has positive charge on carbon, whereas the other has this positive charge delocalized to the adjacent oxygen by the formation of a $\pi$ bond between the carbon and the oxygen. The electron density of the oxygen–hydrogen bond is thus shifted even farther toward oxygen than it is in an alcohol, and the partial positive charge facilitates the loss of a proton. Deprotonation of carboxylic acids is easier than deprotonation of alcohols; that is, carboxylic acids are more acidic than alcohols. We will see in Chapter 6 how resonance stabilization of the anion resulting from the loss of a proton from a carboxylic acid is important in enhancing the acidity of the OH group.

The ability to donate a proton (that is, to act as an acid) characterizes much of the chemistry of carboxylic acids. Carboxylic acids are substantially more acidic ($pK_a \sim 5$) than alcohols ($pK_a \sim 16$–19).

$$(pK_a\ 5) \quad H_3CCO_2H \rightleftharpoons H_3CCO_2^{\ominus} + H^{\oplus}$$

$$(pK_a\ 16) \quad H_3COH \rightleftharpoons H_3CO^{\ominus} + H^{\oplus}$$

The carbon atom of a carbonyl group in a carboxylic acid is at a higher oxidation level (three bonds to oxygen) than that in an aldehyde. Indeed, this carbon has an oxidation level of $+3$, the same as in a nitrile, in which carbon is triply bonded to nitrogen.

## Derivatives of Carboxylic Acids

There are other functional groups in which the carbon atom of the carbonyl group forms three bonds with heteroatoms. All of these compounds, considered to be derivatives of carboxylic acids, have the $+3$ oxidation level for the carbonyl carbon. Several of these functional groups are shown in Figure 3.25. You should become familiar with the names of all of these functional groups. **Esters** are named as alkyl derivatives with the suffix **-anoate, amides** as **-anoamides, acid chlorides** as **-anoyl chlorides,** and **acid anhydrides** as **-anoic anhydrides.** We will consider the chemistry of these functional groups in detail in Chapters 12 and 13, but for now it is sufficient to note that the carbonyl carbons are all at the same oxidation level and that each bears three bonds to heteroatoms.

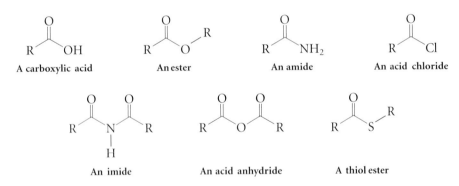

**FIGURE 3.25**

Derivatives of carboxylic acids.

*Resonance Effects: Hindered Rotation.* Resonance structures similar to those for the parent carboxylic acid can be written for carboxylic acid derivatives. Contributions from resonance structures analogous to the center and right-hand structures in Figure 3.24 significantly influence the physical properties of carboxylic acid derivatives. For example, rotation about the C—N bond in an amide is much more difficult than rotation about the C—N bond in an amine. The restricted rotation is caused by the partial double-bond character of the carbon–nitrogen bond in the amide; that is, it is due to a contribution from the zwitterionic resonance contributor.

**Two Resonance Structures of an Amide**

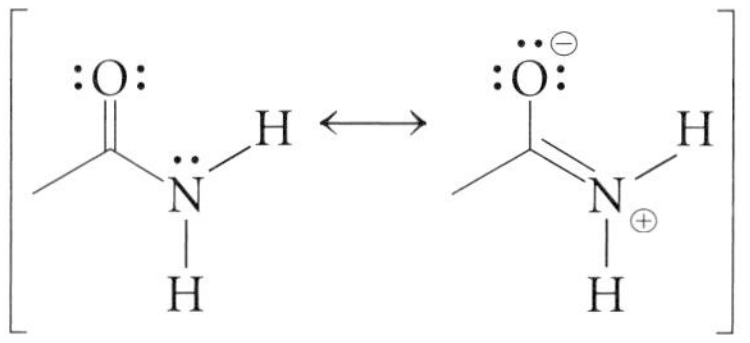

The orbital overlap between nitrogen and the carbonyl carbon in an amide is even greater than that between oxygen and the carbonyl carbon in a carboxylic acid: in amides, the barrier to rotation is ~18 kcal/mole. The overlap of the nitrogen lone pair of electrons with the $\pi$ system of the carbonyl group must contribute about 18 kcal/mole of bonding stabilization to amides. We will see in Chapter 16 that the partial double-bond character of the bond between nitrogen and the carbonyl carbon has important consequences for the physical properties of peptides and proteins, which contain many such amide groups.

## EXERCISE 3.24

Classify the functional group of each of the following compounds:

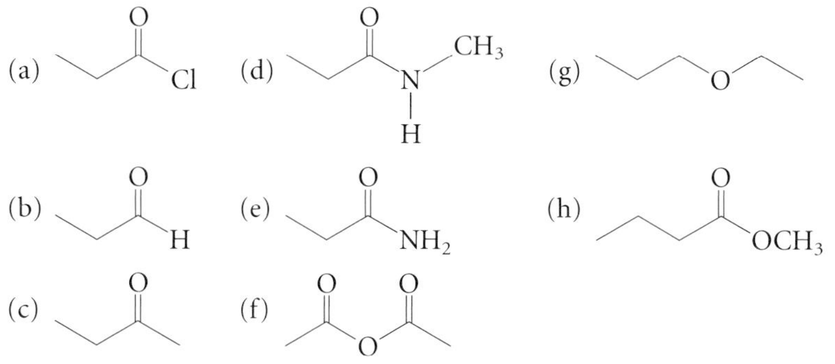

***Reactivity toward Nucleophiles.*** The ease with which a nucleophile attacks the carbonyl carbon of a carboxylic acid derivative is related to the amount of lone-pair delocalization. The contribution of resonance structures stabilizes carboxylic acid derivatives.

**Resonance Structures of an Amide**

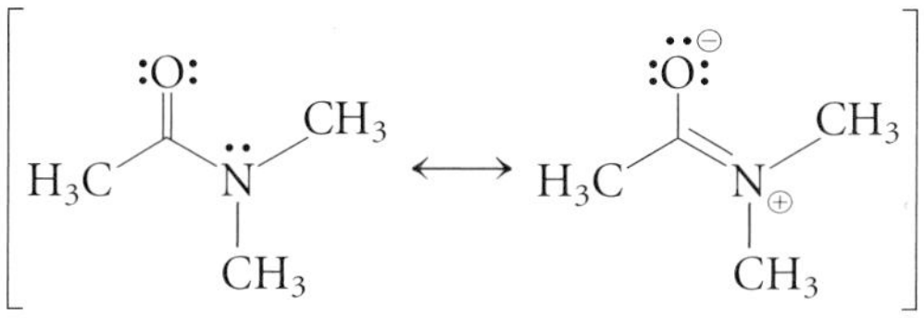

**Resonance Structures of an Ester**

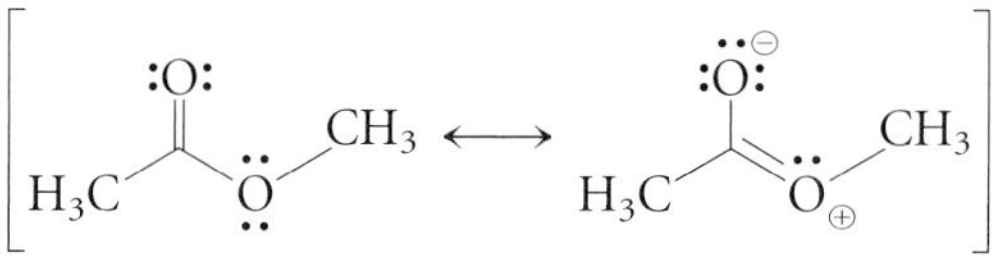

Addition of a nucleophile to the carbonyl carbon destroys the C=O $\pi$ bond and, along with it, the additional stabilization contributed by lone-pair de-

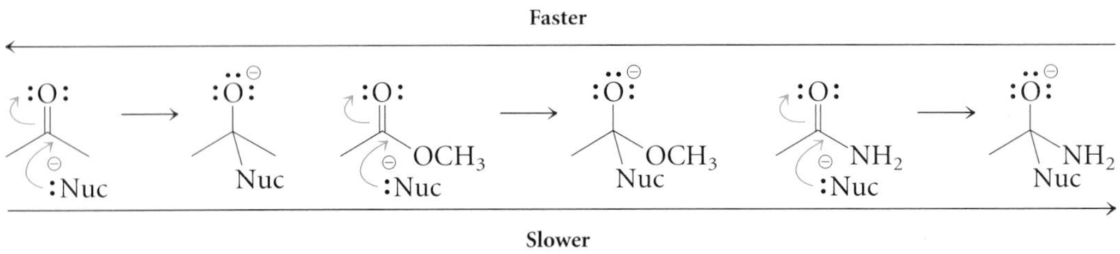

**FIGURE 3.26**

Relative rates of nucleophilic addition to ketones.

localization. Therefore, attack on carbon by a nucleophile is slower for amides than for ketones, as shown in Figure 3.26. Nitrogen, being less electronegative than oxygen, can more readily release electron density to the carbonyl oxygen. Resonance delocalization of a lone pair from nitrogen in amides is greater than that from oxygen in esters, and therefore amides are less reactive toward nucleophiles than are esters.

## EXERCISE 3.25

In each of the following pairs, choose the compound that would be more easily attacked by a nucleophile at the carbonyl carbon. Explain your reasoning.

(a) $CH_3CHO$ or $CH_3COCH_3$     (c) $CH_3COCH_3$ or $CH_3CON(CH_3)_2$

(b) $CH_3CHO$ or $CH_3CO_2CH_3$     (d) $CH_3CONH_2$ or $CH_3CO_2CH_3$

## Oxidation Levels

Oxidation levels of carbon atoms in oxygen-containing compounds can be determined using the same methods as for carbon atoms in nitrogen-containing molecules. Figure 3.27 summarizes the oxidation levels of carbons in compounds containing oxygen. Consider the oxidation levels of the two different carbons in ethanal ($CH_3CHO$) and the two identical carbons in ethyne ($HC{\equiv}CH$). In ethanal, the methyl carbon is at an oxidation level of $-3$, and the carbonyl carbon is at $+1$. In ethyne, both carbons are at an

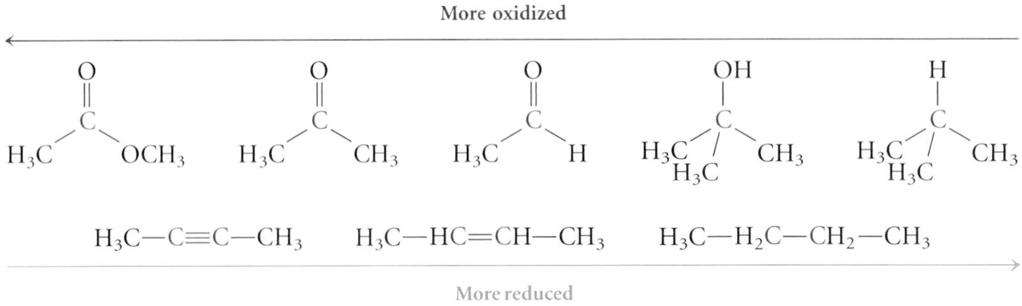

**FIGURE 3.27**

A comparison of the oxidation levels of carbons (in red) in oxygen-containing compounds with those in hydrocarbons.

oxidation level of $-1$. Adding the values for the carbon atoms of ethanal and those of ethyne separately gives the same total, $-2$. This correspondence is of chemical consequence, because it means that, overall, the interconversion of these two compounds involves neither oxidation nor reduction. Indeed, we will see in Chapter 10 that this reaction takes place by hydrolysis, the addition of water, not by treatment with a redox reagent.

The addition of water across a carbon–carbon double bond results in an alcohol:

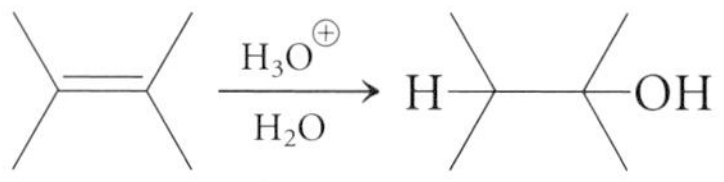

**An alkene**            **An alcohol**

This addition reaction does not change the oxidation levels of hydrogen or oxygen in water. However, a hydrogen atom is added to one carbon of the alkene (as a consequence, that carbon is reduced), and an OH group is added to the other carbon (an oxidation). These two processes—reduction and oxidation—exactly balance one another, and no overall change in oxidation level occurs when water is added to an alkene.

These two examples show, as we have seen before, that the presence of a multiple bond between carbon atoms means that the carbons are at a higher oxidation level than those in an alkane, and that a $\pi$ bond changes the oxidation level of a carbon atom to the same extent as does the introduction of a single bond to a heteroatom.

## EXERCISE 3.26

For each of the following reactions, indicate whether an oxidizing or a reducing agent is required. (Some reactions may require neither.)

(a) $H_3CCHO \longrightarrow CH_3CH_2OH$

(b) $H_3CCHO \longrightarrow H_2C{=}CH_2$

(c) $H_3CCHO \longrightarrow H_3CCH_3$

(d) $H_3CCH_2CHO \longrightarrow$

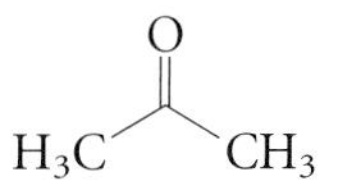

## 3.10

# Sulfur-Containing Compounds

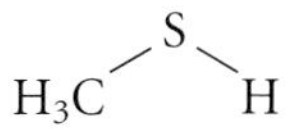

Oxygen and sulfur are in the same column of the periodic table and therefore have similar valence electron requirements. Just as oxygen forms hybrid orbitals from the combination of $2s$ and $2p$ atomic orbitals, sulfur forms hybrid orbitals from the combination of $3s$ and $3p$ atomic orbitals. These hybrid orbitals participate in covalent bonding. In a **thiol,** sulfur is bonded to one carbon atom and one hydrogen atom (analogous to the oxygen in an alcohol). When sulfur is bonded to two alkyl or aryl carbon atoms (analogous to the oxygen in an ether), the functional group is called a **thioether,** or an **alkyl sulfide.** A **thiol ester** is a compound in which an —SR group replaces the —OR group of an ester.

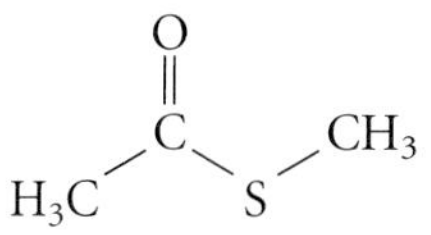

**A thiol**

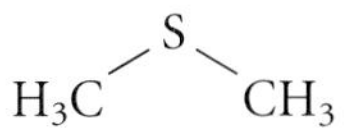

**A thioether**

**A thiol ester**

Many chemical properties of thiols are similar to those of alcohols, and many chemical properties of thioethers are similar to those of ethers. Most of the differences between these functional groups occur because sulfur has a third-level valence shell. Third-level orbitals are significantly larger than those of the second level, and the size mismatch between carbon second-level and sulfur third-level orbitals results in a carbon–sulfur covalent bond that is weaker than that between carbon and oxygen. (Compare the space-filling models of methyl ether and methyl thioether.)

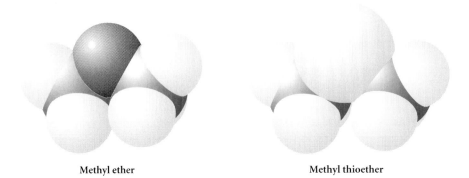

Methyl ether                              Methyl thioether

In addition, sulfur's electronegativity (2.5) is significantly lower than that of oxygen (3.5), and sulfur is more polarizable because its valence electrons are farther from the nucleus than are those of oxygen. Finally, because sulfur, in the third row of the periodic table, has access to $3d$ orbitals, its valence shell can expand beyond eight electrons: sulfur often participates in bonding with more than four atoms. As a result, the chemistry of sulfur compounds is somewhat more complex than that of oxygen compounds. For example, the formation of **sulfonic acids** is due to sulfur's capacity to form an expanded valence shell and has no analogy in oxygen chemistry, although sulfonic acids do form esters and amides that are analogous to carboxylic acid esters and amides.

$$
\begin{array}{ccc}
\overset{\displaystyle O}{\underset{\displaystyle O}{\overset{\|}{\underset{\|}{H_3C-S-OH}}}} & \overset{\displaystyle O}{\underset{\displaystyle O}{\overset{\|}{\underset{\|}{H_3C-S-OCH_3}}}} & \overset{\displaystyle O}{\underset{\displaystyle O}{\overset{\|}{\underset{\|}{H_3C-S-NH_2}}}} \\
\text{A sulfonic acid} & \text{A sulfonic acid ester} & \text{A sulfonamide}
\end{array}
$$

Some **sulfonamides** are potent antibacterial substances known as *sulfa drugs*.

---

**EXERCISE 3.27**

---

A thiol ester is analogous to an ester with the singly bonded oxygen replaced by sulfur. Would you expect a thiol ester to be more or less reactive than a simple ester toward nucleophilic attack at the carbonyl carbon? Explain your reasoning.

$$
\underset{H_3C}{\overset{\displaystyle O}{\overset{\|}{C}}}\overset{\displaystyle }{\underset{S}{}}\overset{CH_3}{}
$$

**A thiol ester**

---

DIMETHYL SULFOXIDE: A VERSATILE SOLVENT

Dimethyl sulfoxide (DMSO) is an odorless, dipolar, aprotic, organic solvent with unusual properties. By virtue of its highly polarized sulfur–oxygen bond, DMSO is miscible with water but also quite soluble in other, less polar organic solvents. It passes readily through the skin and will even carry other organic molecules with it. DMSO has been considered as a possible vehicle to deliver to the bloodstream drugs that are destroyed in the digestive system. This application of DMSO has not been commercialized, in part because of concern about possible toxic side effects of the solvent itself. Another complication is that DMSO is reduced in the body to methyl sulfide, a compound with a highly disagreeable odor.

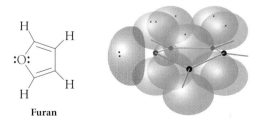

Dimethyl sulfoxide
(DMSO)                    Methyl sulfide

## 3.11

# Aromatic Compounds Containing Heteroatoms

You know from Chapter 2 that planar, cyclic, conjugated molecules containing $4n + 2$ electrons (where $n$ is an integer) are aromatic compounds of unusual stability. Aromatic molecules in which one or more carbon atoms are replaced by heteroatoms (usually oxygen, nitrogen, or sulfur) are **heteroaromatic molecules.** The stability of these compounds is similar to that of their all-carbon analogs. As a family, they are called **heterocyclic aromatics,** or **heteroaromatics,** because the heteroatom is one of the component atoms of the ring. These compounds have common, rather than systematic, names. Three examples with five ring atoms are **furan, pyrrole,** and **thiophene.** Each of these heterocyclic compounds can be represented by the cyclic array shown for furan (with blue $p$ orbitals and a gray $sp^2$-hybrid orbital containing a lone pair of electrons):

#06   Heterocycles

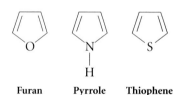

Furan        Pyrrole      Thiophene

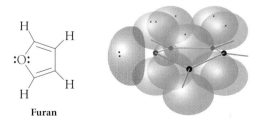

Furan

One lone pair of electrons on the heteroatom is held in a $p$ orbital perpendicular to the molecular plane and thus aligned for interaction with the $p$ orbitals of the carbon–carbon double bonds. These structures are analogous to the cyclopentadienyl anion (Figure 2.20) because each contains six electrons in a planar, cyclic, delocalized $\pi$ system, making them Hückel aromatics. The nitrogen–hydrogen bond of pyrrole is in the plane of the five-membered ring and orthogonal to the $\pi$ system. This position is occupied by a lone pair of electrons in both furan and thiophene.

**Pyridine** is the simplest example of a six-member heteroatomic aromatic compound (Figure 3.28). The aromatic $\pi$ system of pyridine is the same as that of benzene. Each of the six atoms of the ring provides a $p$ orbital and one electron, giving the total of six electrons needed for an aromatic compound. Thus, we can write Kekulé-like resonance contributors for pyridine in the same way we did for benzene. In contrast with pyrrole, in which the lone pair is part of the $\pi$ system, the lone pair on nitrogen in pyridine is contained in an $sp^2$ orbital in the plane of the six ring atoms and is orthogonal to the $\pi$ system formed from the $p$ orbitals.

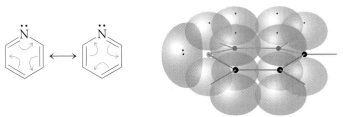

**FIGURE 3.28**

Pyridine has six overlapping $p$ orbitals (blue), which comprise the aromatic six-electron $\pi$ system, and an $sp^2$-hybrid orbital (gray) containing a lone pair.

### EXERCISE 3.28

Both pyridine and pyrrole have a lone pair of electrons on nitrogen, which can be protonated in an acid–base reaction. In view of Hückel's rule, which protonation will be easier? That is, will pyridine or pyrrole be the stronger base? Explain your reasoning.

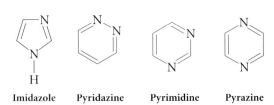

Pyridine          Pyrrole

## Biologically Important Heteroaromatics

A heterocyclic aromatic can contain more than one heteroatom, and each structure shown here represents a five- or six-member ring containing two nitrogen atoms.

Imidazole     Pyridazine     Pyrimidine     Pyrazine

## CHEMICAL PERSPECTIVES

### ANIMAL, VEGETABLE, MINERAL: THE COLOR OF THINGS

The world is full of wonderfully colored living creatures in almost endless variety. The marvelous colors of autumn leaves, seen especially in the northeastern United States, are the result of organic compounds in the leaves that are revealed once the normally dominant green color of chlorophyll disappears.

**Chlorophyll a**

Sometimes the colors of birds, beetles, butterflies, and other insects are the result of absorption of light by organic molecules. For example, xanthopterin is a yellow-orange pigment found in the wings of some insects.

**Xanthopterin**

However, many birds and insects use a trick of physics to produce color. They have layers of reflective material (usually inorganic salts) arranged at a precisely defined spacing that selectively reflects light of a specific wavelength. A physicist would call such a "device" a *quarter-wavelength interferometer*, which selectively reflects light of wavelength $\lambda$ from multiple layers with a thickness of $\lambda/4$. Many of you are taking physics at the same time as organic chemistry. Ask your instructor to explain how such a device operates. Having such knowledge, you may view the beauty of nature with even greater awe.

Three biologically important bases, **uracil, thymine,** and **cytosine,** are oxygen or nitrogen derivatives of pyrimidine.

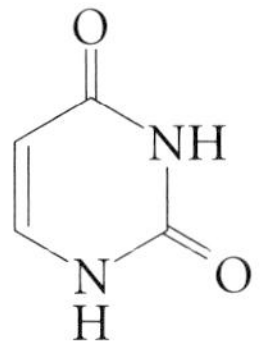

Uracil 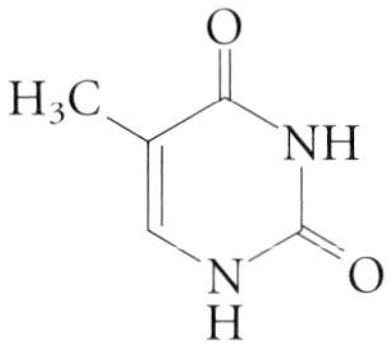      Thymine 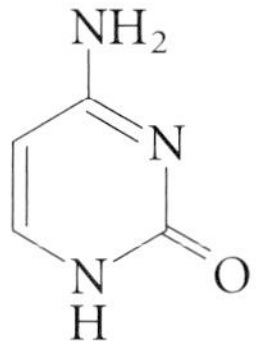      Cytosine

Heteroatoms are also found in fused-ring molecules (structurally similar to the polycyclic aromatic hydrocarbons). Following are three common fused-ring structures—**quinoline, pteridine,** and **purine**—and two purine derivatives, guanine and adenine. Purines are subunits of biologically important systems.

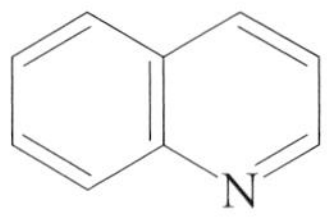

Quinoline 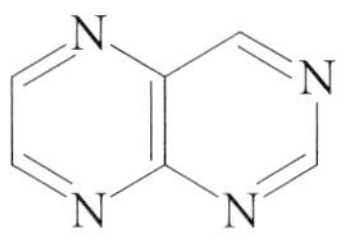      Pteridine 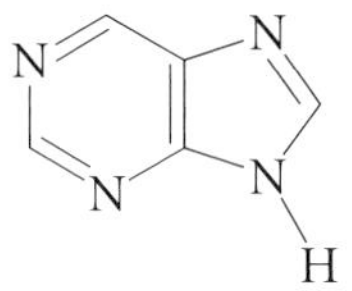      Purine

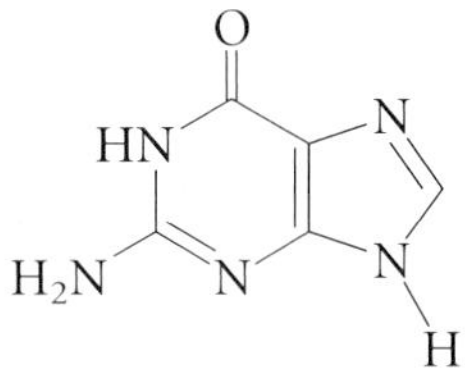

Guanine 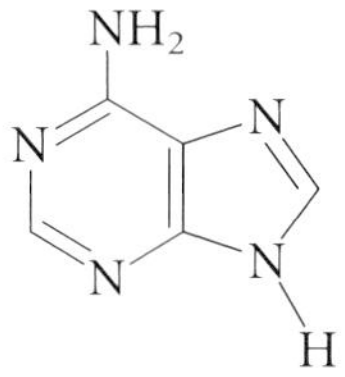      Adenine

**Guanine** and **adenine,** along with uracil, thymine, and cytosine, are aromatic bases. They are components of nucleotides, which constitute the chemical basis for genetic coding. Many derivatives of pteridine have been isolated from insects and are responsible for the bright and varied colors in butterfly wings.

### EXERCISE 3.29

For each nitrogen atom of purine, specify which type of orbital contains the lone pair of electrons and whether the lone pair is part of an aromatic, six-electron $\pi$ system:

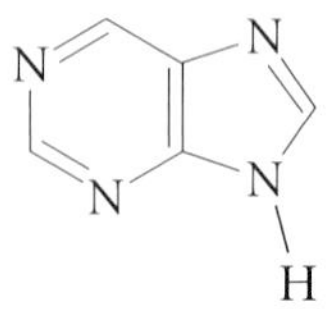

Purine

### Heteroatom-Substituted Arenes

In addition to the heteroaromatics, which have a heteroatom in the ring, a number of important compounds have a heteroatom attached to an all-

## CHEMICAL PERSPECTIVES

### CAFFEINE: A HETEROAROMATIC STIMULANT

Caffeine, a cyclic aromatic compound containing nitrogen, is present in both tea and coffee. Caffeine has a dramatic stimulating effect on people, and both tea and coffee have been consumed for centuries for this effect. In this century, caffeine has been marketed by itself and in combination with other ingredients for use as a stimulant by those who do not like coffee or cannot conveniently drink a beverage. Until recently, the caffeine sold in this way was prepared by adding a methyl group to theobromine, a related compound obtained from cocoa fruits. (Theobromine is also present in tea.)

**Caffeine**

**Theobromine**

However, the relatively large demand for decaffeinated coffee has resulted in large quantities of caffeine being available by extraction from coffee beans. At first, halogenated organic solvents were used to remove the caffeine from the bean, but concern about possible adverse health effects of these solvents has stimulated the development of alternative processes that use steam or supercritical carbon dioxide.

carbon aromatic ring. Because the ring contains only carbon atoms, such compounds are not called heteroaromatics. For example, as shown in Figure 3.29, in **aniline**, $sp^2$-hybridization of nitrogen produces the optimal geometry for the overlap of the lone pair of electrons in a $p$ orbital on the heteroatom with the aromatic array of $p$ orbitals on carbons in the ring. Although overlap of the heteroatom $p$ orbital with the ring $\pi$ system can also take place even without rehybridization, the arrangement of the bonds about the nitrogen atom of aniline and most of its derivatives is planar; therefore, the nitrogen is $sp^2$-hybridized in these compounds. We shall

**Aniline**

**FIGURE 3.29**

Three-dimensional representation of orbital interaction in aniline, where the $p$ orbital of the $sp^2$-hybridized nitrogen (at the left) overlaps with the $\pi$ system of the aromatic ring.

explore the effect of this extended conjugation on the chemical reactivity of some aromatic compounds in Chapter 11.

As we saw in Chapter 2, aromatic rings can have alkyl substituents, as in toluene. It is also possible for these substituents to bear heteroatoms. Several of these compounds that are encountered frequently and are usually referred to by common names are **benzoic acid, benzaldehyde, acetophenone,** and **anisole.**

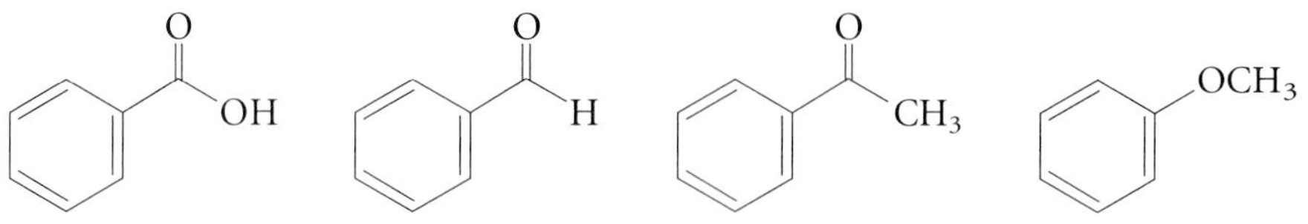

| Benzoic acid | Benzaldehyde | Acetophenone | Anisole |

## EXERCISE 3.30

Will contribution by the zwitterionic resonance structures shown below make aniline a stronger or weaker base than it would be if its structure could be represented by the resonance contributor at the left? Explain your reasoning.

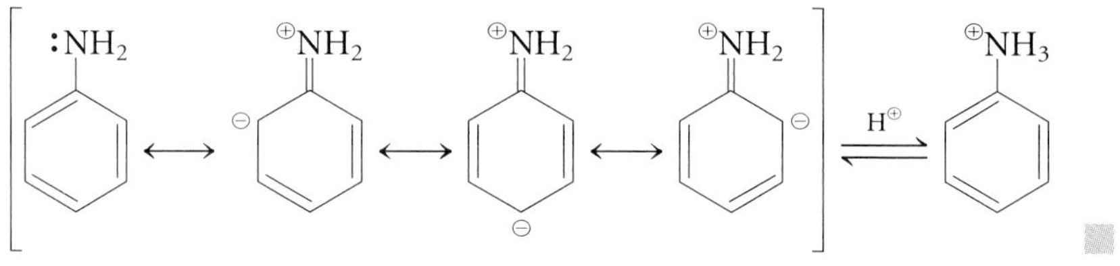

## 3.12

## Alkyl Halides

The next-to-last column on the right side of the periodic table contains the halogens. Fluorine requires only one $\sigma$ bond to satisfy its valence requirement. In hydrofluoric acid, fluorine contributes seven valence electrons and hydrogen contributes a single electron to satisfy the valence requirements of both atoms, making H—F a stable molecule.

**Three Representations of HF**

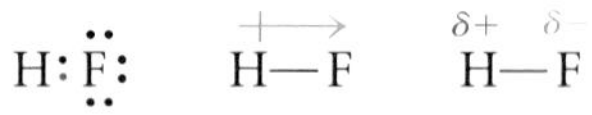

$$H : \ddot{F} : \qquad H—F \qquad \overset{\delta+}{H}—\overset{\delta-}{F}$$

The $\sigma$ bond is nonetheless highly polarized, because there is a substantial difference in electronegativity between hydrogen and fluorine. Carbon–fluorine bonds, such as that in methyl fluoride, are otherwise similar to carbon–nitrogen and carbon–oxygen bonds.

**Three Representations of H₃CF**

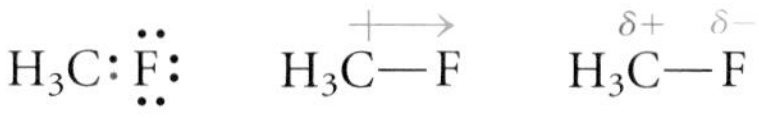

$$H_3C : \ddot{F} : \qquad H_3C—F \qquad \overset{\delta+}{H_3C}—\overset{\delta-}{F}$$

## CHEMICAL PERSPECTIVES

### CHEMICALLY INERT CARBON–FLUORINE BONDS

The bond between fluorine and carbon is much stronger than the bond between any other element and carbon. For example, a C—F bond is 25% stronger than a C—H bond. As a result, fluorocarbons are unusually stable. In fact, polymers such as Teflon, in which there are only C—F and C—C bonds, are almost completely inert to chemical reaction, except with strong reducing agents. Such polymers can therefore be used for applications where other organic materials are degraded, such as in coatings for heating utensils or as seals for containers of corrosive liquids.

Like alcohols, alkyl fluorides can be primary, secondary, or tertiary:

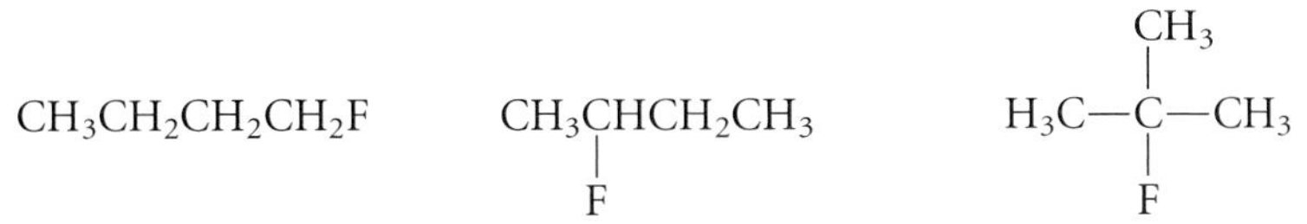

$CH_3CH_2CH_2CH_2F$      $CH_3CHCH_2CH_3$
                                    |
                                   F

$$CH_3CH_2CH_2CH_2F \qquad CH_3\overset{\displaystyle |}{\underset{\displaystyle F}{C}}HCH_2CH_3 \qquad H_3C-\overset{\displaystyle CH_3}{\underset{\displaystyle F}{C}}-CH_3$$

**1-Fluorobutane**       **2-Fluorobutane**       **2-Fluoro-2-methylpropane**
(a primary alkyl halide)    (a secondary alkyl halide)    (a tertiary alkyl halide)

Alkyl fluorides are members of the group known as alkyl halides, which are formed by bonding between carbon and a member of the halogen family (fluorine, chlorine, bromine, or iodine). The other alkyl halides have structures similar to those of alkyl fluorides. These compounds can be named either as halogenated alkanes (for example, bromoethane) or as alkyl halides (for example, ethyl bromide).

Comparing the bond-dissociation energies of alkyl fluorides, chlorides, bromides, and iodides (with the alkyl group constant) indicates that the σ bond becomes weaker as the difference in size between carbon and the halogen increases (Figure 3.30). In other words, homolytic cleavage is considerably easier for an alkyl iodide than for an alkyl fluoride.

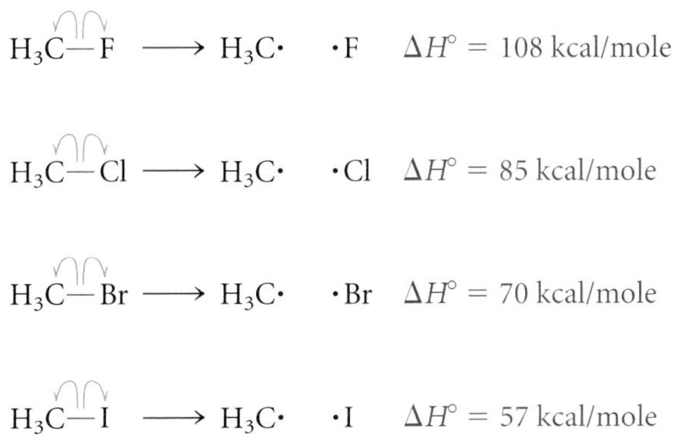

$$H_3C—F \longrightarrow H_3C\cdot \quad \cdot F \quad \Delta H° = 108 \text{ kcal/mole}$$

$$H_3C—Cl \longrightarrow H_3C\cdot \quad \cdot Cl \quad \Delta H° = 85 \text{ kcal/mole}$$

$$H_3C—Br \longrightarrow H_3C\cdot \quad \cdot Br \quad \Delta H° = 70 \text{ kcal/mole}$$

$$H_3C—I \longrightarrow H_3C\cdot \quad \cdot I \quad \Delta H° = 57 \text{ kcal/mole}$$

**FIGURE 3.30**

Bond-dissociation energies for methyl halides.

Heterolytic cleavage within a series of alkyl halides also becomes easier in the order fluoride < chloride < bromide < iodide. From the top to the bottom of the periodic table, the electronegativity of the halogen decreases, whereas the size—and hence the ability to respond to charge demand (polarizability)—increases. The stability of the anion (with the negative charge on $I^{\ominus}$ spread over a much larger area than that on $F^{\ominus}$) also is important. Therefore, the rates of C—X heterolytic cleavage increase as the R—X bond becomes weaker (Figure 3.31), a trend that parallels the acidity of the halogen acids (HX).

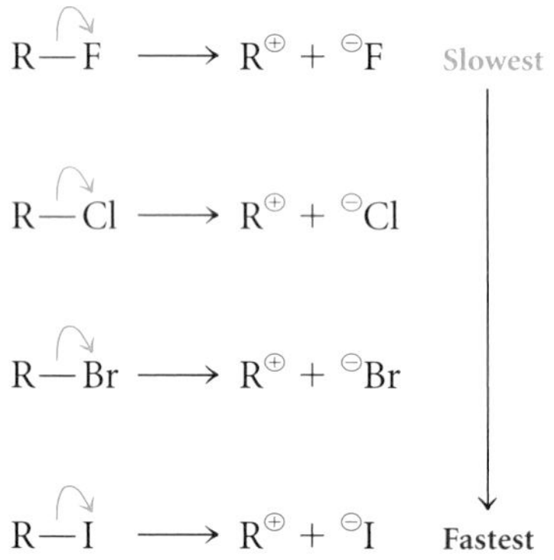

$$R\!-\!F \longrightarrow R^{\oplus} + {}^{\ominus}F \quad \text{Slowest}$$

$$R\!-\!Cl \longrightarrow R^{\oplus} + {}^{\ominus}Cl$$

$$R\!-\!Br \longrightarrow R^{\oplus} + {}^{\ominus}Br$$

$$R\!-\!I \longrightarrow R^{\oplus} + {}^{\ominus}I \quad \text{Fastest}$$

**FIGURE 3.31**

Relative rates of heterolytic cleavage of alkyl halides.

**EXERCISE 3.31**

Remembering that the dipole moment of a molecule is the vector sum of its bond dipoles, predict whether a molecular dipole exists in any of the following multi-halogen-substituted compounds. If so, draw (in three dimensions) the direction of the dipole.

(a) $CCl_4$     (b) $CHCl_3$     (c) $CH_2Cl_2$     (d) $CH_3Cl$     (e) $CBr_4$

## 3.13

## Solvents for Organic Chemistry

Organic chemists have many and varied uses for solvents in everyday laboratory operations. Solvents are used for separating and purifying organic compounds and for carrying out chemical reactions. **Solvents** are simple organic compounds that are liquids at room temperature; most have relatively low boiling points. Some common organic solvents are shown here.

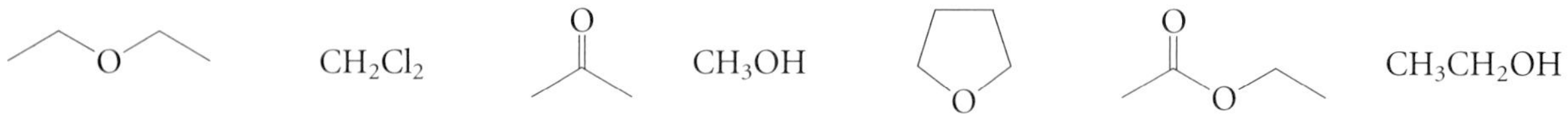

| | $CH_2Cl_2$ | | $CH_3OH$ | | | $CH_3CH_2OH$ |
|---|---|---|---|---|---|---|
| **Ethyl ether** | **Dichloromethane** | **Acetone** | **Methanol** | **Tetrahydrofuran** | **Ethyl acetate** | **Ethanol** |
| bp:   35 °C | 40 °C | 56 °C | 65 °C | 66 °C | 77 °C | 78 °C |

Solvents are of fundamental importance for separating organic compounds from inorganic materials. In general, organic compounds are considerably less soluble in water than they are in organic solvents. Thus, adding a mixture of inorganic and organic compounds to water and an organic solvent such as ether leads to a partitioning in which the inorganic compounds remain in the aqueous layer and the organic compounds are extracted into the ether layer. Usually the mixture is shaken to hasten equilibrium. This is followed by physically separating the layers and then evaporating the organic solvent to yield a residue consisting of the organic compounds.

Solvents are also useful for purifying organic compounds by recrystallization. A mixture of organic compounds is dissolved in a solvent with heating. When the solution is cooled to room temperature (or below), the less soluble organic compounds preferentially crystallize from the solution.

Solvents are also useful for conducting organic reactions. Most organic reactions are exothermic, so heat is released. Were two reactants simply mixed without solvent, the rate of heat release by the reaction might well exceed the loss of this heat through the walls of the containing vessel. The temperature of the reaction would then rise, leading to even faster reaction and an increased rate of heat release. Carrying out reactions in dilute solution allows the heat to be dissipated into the solution, whose temperature is controlled by external heating or cooling. Furthermore, if the solvent is heated to its boiling point, heat is effectively released by evaporation.

Solvents can also influence the rates of reactions. For example, heterolytic cleavage of a C—Cl bond is energetically unfavorable, because of the loss of the bond as well as the charge separation that accompanies formation of the carbocation and the chloride ion. Polar protic solvents greatly facilitate such heterolytic bond cleavage by stabilizing the ions through solvation. As we begin to discuss reactions in some detail starting in Chapter 8, we will revisit this role of solvents in organic reactions.

## 3.14

## Nomenclature for Functional Groups

Each heteroatom-containing functional group considered in this chapter is named in accord with the IUPAC rules presented in Chapter 1, except that the suffix is changed to identify the functional group. Table 3.6 (on pages 142–143) summarizes this nomenclature and presents the minimal representation needed to characterize the functional groups. Like hydrocarbons, these compounds are named by locating the longest continuous carbon chain that contains the functional group. The root designates the number of carbons, and the suffix designates the functional group. Substituents are assigned numbers to indicate their positions along the carbon skeleton.

For the low-molecular-weight members of some functional groups, common names are typically used rather than the IUPAC nomenclature. For example, formaldehyde ($H_2CO$), acetaldehyde ($CH_3CHO$), acetyl ($CH_3CO$—), acetic acid ($CH_3COOH$), and acetone ($CH_3COCH_3$) are used almost to the exclusion of the formal IUPAC names.

## TABLE 3.6

Nomenclature of Various Functional Groups

| Functional Group | Structure | Name | Example(s) |
|---|---|---|---|
| Acid chlorides | 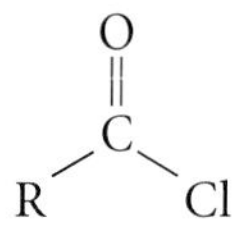 R—C(=O)—Cl | -anoyl chloride | 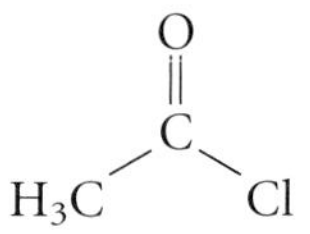 H$_3$C—C(=O)—Cl **Ethanoyl chloride (acetyl chloride)** |
| Alcohols | R—OH | -anol | CH$_3$CH$_2$OH **Ethanol** |
| Aldehydes | 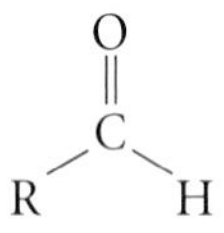 R—C(=O)—H | -anal | 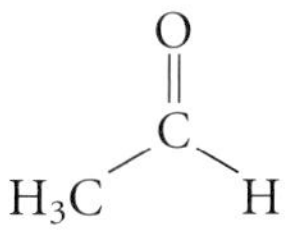 H$_3$C—C(=O)—H **Ethanal (acetaldehyde)** |
| Alkyl halides | R—X | haloalkane or alkyl halide | CH$_3$CH$_2$F **Fluoroethane (ethyl fluoride)** |
| Amides | 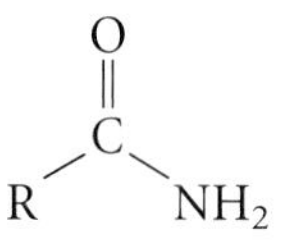 R—C(=O)—NH$_2$ | -anoamide | 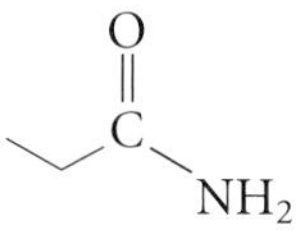 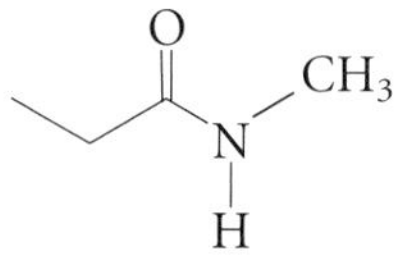 C(=O)—NH$_2$ **Propanoamide** ·· C(=O)—N(H)—CH$_3$ **N-Methyl propanoamide** |
| Amines | RNH$_2$, R$_2$NH, R$_3$N | alkylamine | H$_3$C—NH—CH$_2$CH$_3$ **Methylethylamine** |

---

## EXERCISE 3.32

Write an acceptable name for each of the following compounds:

(a) 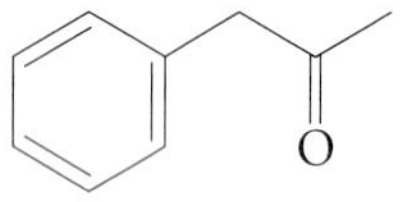   (b) 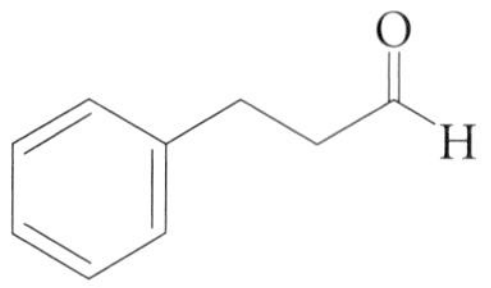   (c)

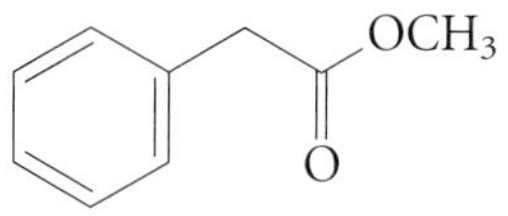

| Functional Group | Structure | Name | Example(s) |
|---|---|---|---|
| Anhydrides | 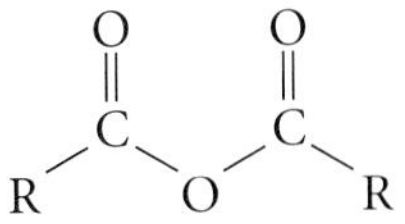 (structure: R—C(=O)—O—C(=O)—R) | -anoic anhydride | 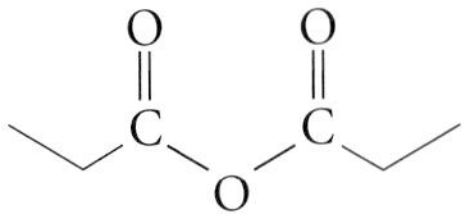 (structure) Propanoic anhydride |
| Carboxylic acids | 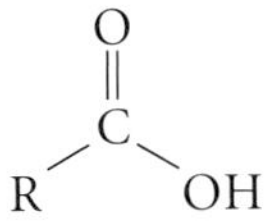 (structure: R—C(=O)—OH) | -anoic acid | 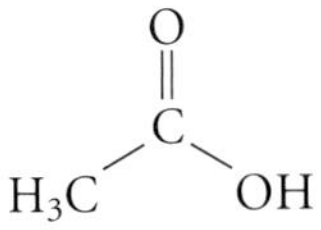 (structure: $H_3C$—C(=O)—OH) Ethanoic acid (acetic acid) |
| Esters | 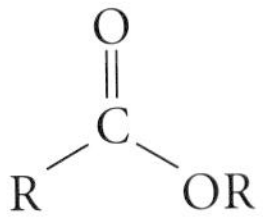 (structure: R—C(=O)—OR) | alkyl -anoate | 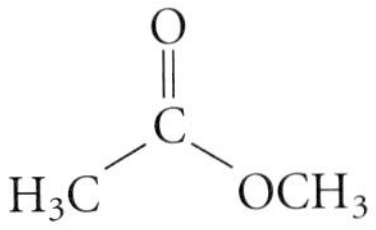 (structure: $H_3C$—C(=O)—$OCH_3$) Methyl ethanoate (methyl acetate) |
| Ethers | R—O—R | alkyl ether | $CH_3$—O—$CH_3$ Methyl ether |
| Imines | 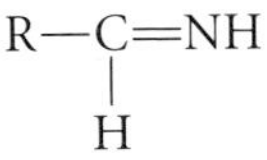 R—C(H)=NH | -anal imine | 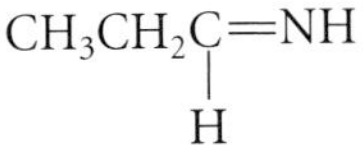 $CH_3CH_2C$(H)=NH Propanal imine |
| Ketones | 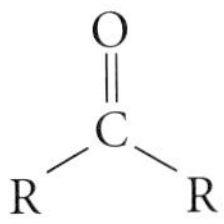 (structure: R—C(=O)—R) | -anone | 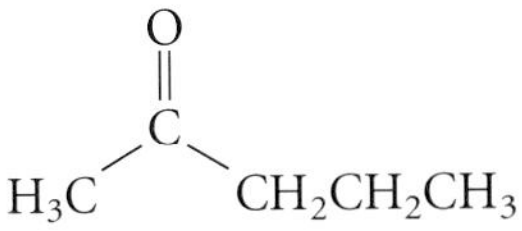 (structure: $H_3C$—C(=O)—$CH_2CH_2CH_3$) 2-Pentanone |
| Nitriles | R—CN | -anonitrile | $CH_3CH_2CN$ Propanonitrile |
| Thioethers | R—S—R | alkyl thioether | $CH_3SCH_2CH_3$ Methyl ethyl thioether (methyl ethyl sulfide) |

(d)

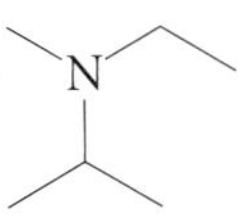

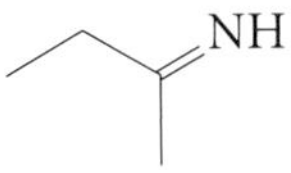

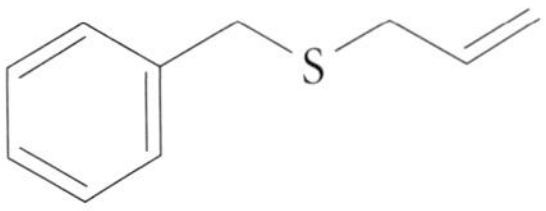

(e)

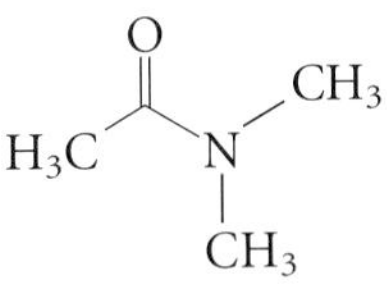

(f)

(g)

(h)

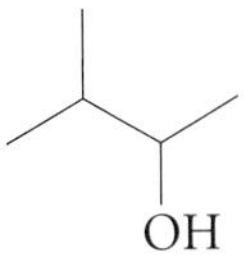

(i)

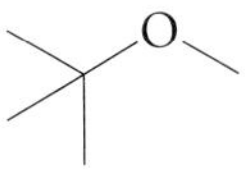

## EXERCISE 3.33

Draw a structure corresponding to each of the following IUPAC names:

(a)  butanone

(b)  2-hexanone

(c)  3-pentanone

(d)  4-methylpentanal

(e)  2-chloropropanoic acid

(f)  methyl propanoate

(g)  dimethylamine

(h)  propanoamide

(i)  butanoyl chloride

(j)  ethyl 2-bromopropanoate

## Summary

**1.** Considering the electronic configurations of nitrogen, oxygen, and fluorine (and atoms in the same columns of the periodic table) allows chemists to make important predictions about molecules containing them with respect to bond strength, geometry, and reactivity with nucleophiles and electrophiles.

**2.** The classification of heteroatom-containing molecules into subgroups is based on the degree of substitution of the carbon bearing the heteroatom, except in amines, for which the level of substitution on nitrogen is used. Thus, the terms *primary, secondary,* and *tertiary* applied to alcohols, ethers, and alkyl halides refer to the number of carbon substituents on the carbon bearing the oxygen or halogen substituent. When applied to amines, these designations refer to the number of alkyl groups attached to nitrogen.

**3.** Heteroatoms alter the structure of carbon compounds, because the presence of one or more lone pairs of electrons on an atom of higher electronegativity induces significant partial charge separation within the molecule. The heteroatom usually functions as a locus for chemical activity, that is, as the functional group in a molecule. Reactions of heteroatom-containing compounds usually take place at bonds to or near the heteroatom.

**4.** The difference in electronegativity between carbon and a heteroatom (X) to which it is bonded results in polarization of the C—X $\sigma$ bond. In many cases, the vectorial sum of such polar covalent bonds causes a net molecular dipole, which has consequences for the molecule's physical properties (greater reactivity toward charged reagents, higher melting and boiling points, higher solubility in polar solvents, and so forth). The presence of a dipole moment within a molecule often correlates with how easily it is attacked by nucleophiles and electrophiles. Nucleophiles attack molecules at centers of partial positive charge; electrophiles attack molecules at centers of partial negative charge.

**5.** A heteroatom bonded to both carbon and hydrogen can participate in hydrogen bonding. This interaction derives from polarization of the X—H bond, so that hydrogen is attracted to a lone pair of electrons on a heteroatom in another molecule or at another site within the same molecule. Hydrogen bonding has an important effect on the three-dimensional structure of a molecule (if intramolecular) and on solvation and association with other molecules (if intermolecular).

**6.** Multiple bonding between carbon and nitrogen occurs in imines (double bond) and nitriles (triple bond). Double bonding between carbon and oxygen is found in aldehydes, ketones, carboxylic acids, esters, amides, and other derivatives of carboxylic acids. Because oxygen's valence shell is filled if it participates in two bonds (and has two lone pairs of electrons), triple bonds to oxygen are not found in stable molecules.

**7.** Alcohols are functional groups in which oxygen has $\sigma$ bonds to both carbon and hydrogen. Ethers lack the OH group of alcohols and are therefore much less reactive than alcohols. The primary use of ethers in organic chemistry is as polar, aprotic, inert solvents.

**8.** Bond cleavage in organic molecules can be accomplished by homolytic or heterolytic pathways. In a homolytic cleavage, the two electrons initially shared between two atoms in a covalent bond are partitioned equally to the two radical fragments. In a heterolytic cleavage, both electrons of the covalent bond are transferred to one of the participating atoms, leaving the other atom with an electron deficiency. Radicals result from homolytic bond cleavage; ions, from heterolytic cleavage.

**9.** Carbocations (formed by heterolytic cleavage) and radicals (formed by homolytic cleavage) follow roughly the same relative order of stability: benzyl ~ tertiary > allyl ~ secondary > primary > methyl. This order of stability arises from resonance stabilization and greater hyperconjugation by alkyl groups.

**10.** A characteristic reaction of alcohols is the acid-catalyzed loss of water. The first step of this reaction is protonation to form an oxonium ion, from which water is lost to form a carbocation. The reactivity of an alcohol is dependent on the character of the carbon atom to which the OH group is attached. The Lucas test can be used to distinguish primary, secondary, and tertiary alcohols based on the rates at which they undergo conversion into alkyl halides.

**11.** Carbonyl groups are highly polarized. The carbonyl carbon bears appreciable partial positive charge, making it a potential site for nucleophilic attack.

**12.** Carboxylic acids form a number of derivatives. Resonance structures help explain the reactivity of these derivatives toward nucleophiles.

**13.** Sulfur-containing compounds are similar to those containing oxygen. The chemistry of thiols is similar to that of alcohols, and thioethers are similar to ethers. Thiol esters have some features of carboxylic acid esters, but because of mismatch in orbital size between the sulfur and the adjacent carbon, these carboxylic acid derivatives are not as stabilized by resonance as are their oxygen analogs. Enhanced reactivity toward nucleophilic attack at carbon results from this weaker interaction.

**14.** Heteroatoms can be incorporated in aromatic systems to which Hückel's rule applies. Several heteroaromatic compounds containing more than one heteroatom are important in nucleic acid chemistry.

**15.** Alkyl halides contain highly polar C—X bonds. Often such compounds have large dipole moments and are readily attacked by nucleophilic reagents at the carbon bearing the partial positive charge.

**16.** Nomenclature for compounds containing heteroatoms follows the IUPAC rules. A root designates the number of carbon atoms in the longest chain containing the functional group; suffixes designate the identity of the functional group; and prefixes and Arabic numerals designate the numbers and positions of substituents.

## Review of Reactions

Protonation of Amines

$$H_3CNH_2 + H^{\oplus} \rightleftharpoons H_3C\overset{\overset{\displaystyle H}{|}}{\underset{\oplus}{N}}H_2$$

Oxidation of Alcohols

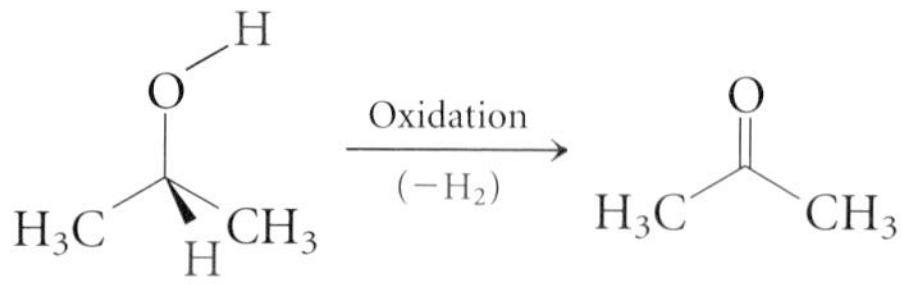

$$\xrightarrow[(-H_2)]{\text{Oxidation}}$$

Alcohol Substitution: Lucas Test

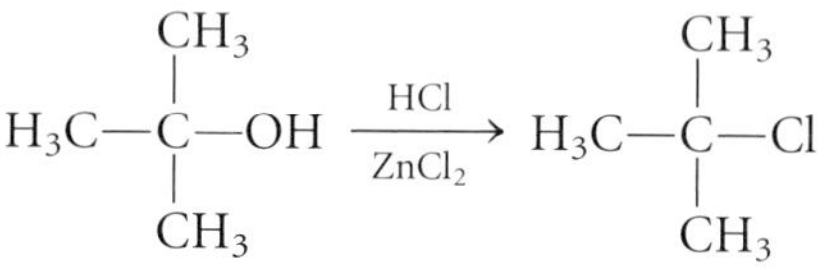

$$H_3C\overset{\overset{\displaystyle CH_3}{|}}{\underset{\underset{\displaystyle CH_3}{|}}{C}}{-}OH \xrightarrow[\text{ZnCl}_2]{\text{HCl}} H_3C\overset{\overset{\displaystyle CH_3}{|}}{\underset{\underset{\displaystyle CH_3}{|}}{C}}{-}Cl$$

Catalytic Hydrogenation of Imines

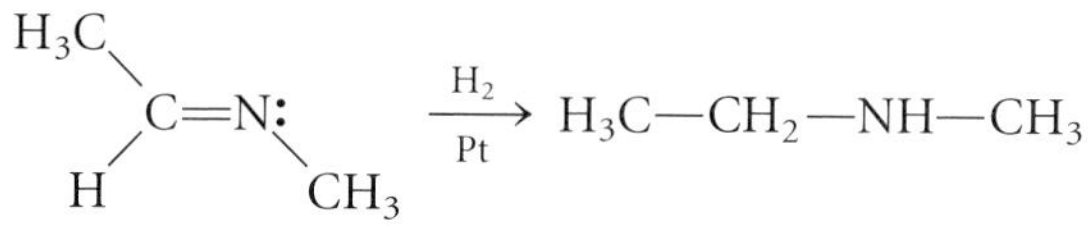

$$\xrightarrow[\text{Pt}]{\text{H}_2} H_3C{-}CH_2{-}NH{-}CH_3$$

## Review Problems

**3.1** Like alkenes, imines can exist as geometric isomers. Draw the *cis* and *trans* isomers of the imine of acetaldehyde. Would you expect interconversion of these isomers to be easier or harder than the *cis–trans* isomerization of 2-butene?

**3.2** Classify the following amines and alcohols as primary, secondary, or tertiary. Name each compound according to the IUPAC rules.

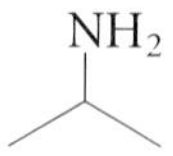

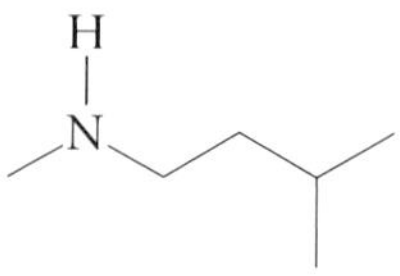

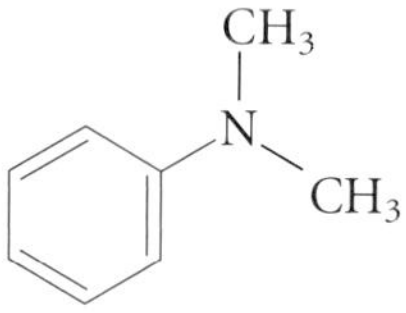

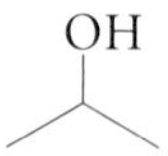

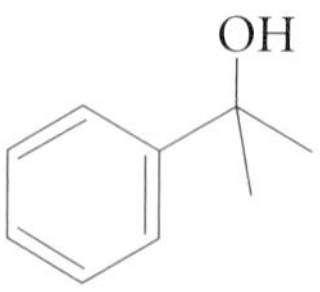

**3.3** Ethers, esters, aldehydes, and thioethers dissolve in concentrated sulfuric acid. Why?

**3.4** Environmentalists are greatly concerned about an atmospheric ozone hole centered on Antarctica and thought to be caused in part by the presence of chlorofluorocarbons in the atmosphere. Ozone, $O_3$, absorbs high-energy ultraviolet light (which is dangerous to plant and animal life), and the absence of ozone imperils many species.

(a) Draw a Lewis dot structure of ozone, $O_3$, being sure to indicate the formal charge on each atom.

(b) By drawing a resonance structure, explain how the two oxygen–oxygen bonds in ozone are of equivalent length.

(c) From the hybridization of the oxygen atoms in the structure you drew in part (b), predict whether ozone is linear or bent.

**3.5** The acidity of a sulfonic acid is due to the high stability of the conjugate base derived by deprotonation of the acid. Write significant resonance structures for the monoanion of benzenesulfonic acid ($C_6H_5SO_3H$), and use them to explain why the acidity of sulfonic acids is higher than that of carboxylic acids.

**3.6** Explain why iodomethane has a smaller dipole moment ($\mu = 1.62$ D) than fluoromethane ($\mu = 1.85$ D).

**3.7** Use resonance structures to explain why formaldehyde has a larger dipole moment than methanol.

**3.8** Calculate the formal oxidation level of carbon in

(a) ethyne        (b) acetonitrile ($CH_3CN$)        (c) ethyl amine

**3.9** Although ethyl ether has a substantially higher molecular weight than ethanol, ethanol has a higher boiling point. Explain.

**3.10** Derivatives of butane in which C—H bonds are replaced with C—Cl bonds can be obtained by exposing butane to chlorine gas in the presence of ultraviolet light.

(a) How many different monochlorobutanes are possible?

(b) How many dichlorobutanes are possible?

(c) How many trichlorobutanes are possible?

**3.11** Draw structures of all geometric isomers of each compound:

(a) 1,1,2-trichlorocyclopentane        (c) 1,2,4-trichlorocyclopentane

(b) 1,2,3-trichlorocyclopentane

**3.12** From what you know about intermolecular interactions, decide which compound in each of the following pairs has the higher boiling point.

(a) pentane ($C_5H_{12}$) or octane ($C_8H_{18}$)

(b) ethyl alcohol ($CH_3CH_2OH$) or methyl ether ($CH_3OCH_3$)

(c) ethylene glycol ($HOCH_2CH_2OH$) or ethyl alcohol ($CH_3CH_2OH$)

**3.13** Write structural formulas that correspond to the following descriptions:

(a) four esters with the formula $C_4H_8O_2$

(b) two aldehydes with the formula $C_4H_8O$

(c) a secondary alcohol with the formula $C_3H_8O$

(d) three ketones with the formula $C_5H_{10}O$

(e) a tertiary amine with the formula $C_4H_{11}N$

(f) a tertiary alkyl bromide with the formula $C_4H_9Br$

**3.14** Dimethyl sulfoxide ($H_3CSOCH_3$, often called DMSO), methylene chloride ($CH_2Cl_2$), dimethylformamide [$HCON(CH_3)_2$, called DMF], methanol ($CH_3OH$), ethyl ether ($CH_3CH_2OCH_2CH_3$, often called simply ether), and tetrahydrofuran [$—(CH_2)_4O—$, called THF] are common organic solvents. Classify each as dipolar aprotic, polar protic, or nonpolar. Identify the structural feature in each molecule from which its solvent classification derives.

**3.15** Identify each of the following as a nucleophile or an electrophile:

(a) triethylamine     (b) hydroxide ion     (c) Fe$^{3+}$     (d) methanethiol, $CH_3SH$

**3.16** Identify the functional group in each of the following compounds. Does the molecule act as a Lewis acid or base?

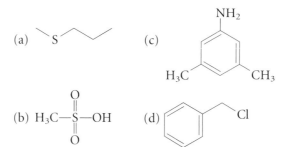

**3.17** What is the relation between the members of the following pairs of structures? Are they identical, positional (or structural) isomers, geometric isomers, or resonance contributors?

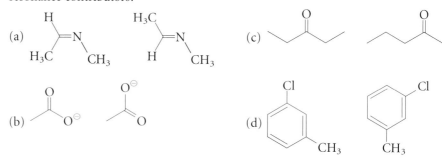

**3.18** The aromatic ring in phenol ($C_6H_5OH$) behaves as if it is particularly electron-rich.

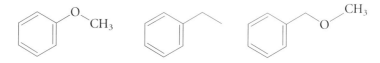

**Phenol**

(a) To explain this observation, draw resonance structures in which the electrons in one of the lone pairs on oxygen are shifted to another position.

(b) Rank the following compounds in order of decreasing ring electron density (most electron-rich first). Assume that resonance contributors parallel to those drawn for part (a) control the electron density of other aromatic rings.

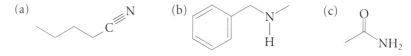

**Supplementary Problems**

**3.19** Provide an IUPAC name for each of the following structures:

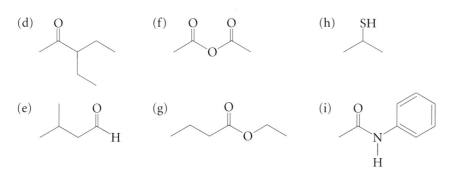

**3.20** Draw a structure that corresponds to each of the following IUPAC names:

(a) 2-aminobutane      (e) benzoyl chloride      (i) *N*-methylaniline

(b) methyl ethyl ether      (f) *N*-benzylethanoamide      (j) propanoic acid

(c) *t*-butyl allyl thioether      (g) *p*-bromoiodobenzene

(d) methyl pentanoate      (h) 2-methylthiophene

**3.21** Classify the following amines and alcohols as primary, secondary, tertiary, or quaternary.

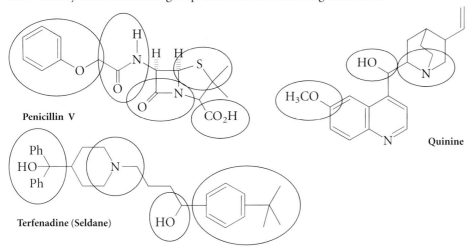

**3.22** Provide a valid name for each structure in Problem 3.21.

**3.23** Identify each functional group circled in the following structures:

Penicillin V

Quinine

Terfenadine (Seldane)

**3.24** Which compound in each of the following pairs is expected to be more soluble in hexane?

(a) 1-butanol or ethyl ether

(b) trihexylamine or 1-aminohexane

(c) octane or octanoic acid

(d) hexyl ethanoate or octanoic acid

**3.25** For each compound, which of the red covalent bonds is more easily broken by homolytic cleavage?

(a) H—CH$_2$O—H

(b) H—CH$_2$—CH$_3$

(c) H—CH$_2$NCH$_3$ (with H on N)

(d) H$_3$CCCH$_3$ (with H and F substituents)

**3.26** Determine whether each of the following compounds and ions is expected to act as a Lewis acid, a Lewis base, or neither.

(a) AlCl$_4^{\ominus}$     (b) FeCl$_3$     (c) TiCl$_4$     (d) HCl

**3.27** Use the average bond energies in Table 3.2 to calculate whether each of the following reactions is endothermic or exothermic.

(a) H$_3$C—C≡C—CH$_3$ + 2 H$_2$ ⟶ ⁀⁀⁀

(b) 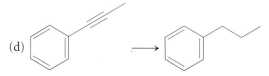 + H$_2$

(c) ⁀N⁀ + CH$_4$ ⟶ ⁀N⁀ (with H on N)

(d) ⁀C(=O)⁀ + HCl ⟶ ⁀C(OH)(Cl)⁀

**3.28** Determine whether each of the following conversions is an oxidation or a reduction, or neither.

(a) ⟶

(b) ⟶

(c) ⟶

(d) ⟶

**3.29** Classify the following radicals as primary, secondary, tertiary, benzylic, or allylic.

(a)    (b)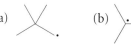

(c)    (d)

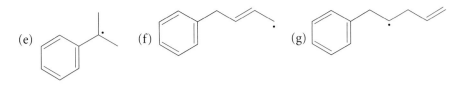

(e)    (f)    (g)

**3.30** Of the following pairs, choose the compound that is more easily dehydrated when treated with acid.

(a) [structure] or [structure]

(b) [structure] or [structure]

(c) [structure] or [structure]

(d) [structure] or [structure]

**3.31** For each pair in Problem 3.30, choose the compound from which a carbocation can be more easily obtained.

**3.32** Determine whether the asterisked carbon atom in each of the following compounds is more highly oxidized, more highly reduced, or at the same oxidation level as the carbonyl carbon of acetaldehyde.

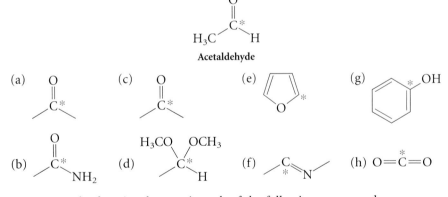

Acetaldehyde

(a) [structure]    (c) [structure]    (e) [structure]    (g) [structure]

(b) [structure]    (d) [structure]    (f) [structure]    (h) $O=\overset{*}{C}=O$

**3.33** Name the functional group in each of the following compounds:

(a) [structure]    (d) $H_3C—SH$    (g) [structure]

(b) $H_3C—\overset{O}{\underset{O}{S}}—NH_2$    (e) [structure]    (h) [structure]

(c) [structure]    (f) [structure]

**3.34** The following compounds exist in acid–base equilibria, differing only in the presence or absence of protons. For each equilibrium, identify any species that can act as a nucleophile. If more than one of the equilibrating species is a nucleophile, determine which is the more (or most) nucleophilic.

(a)

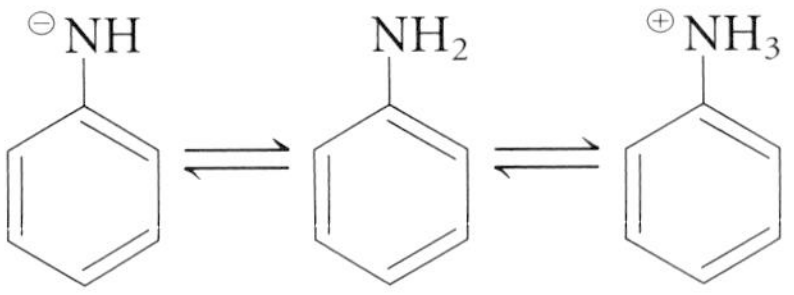

(b)

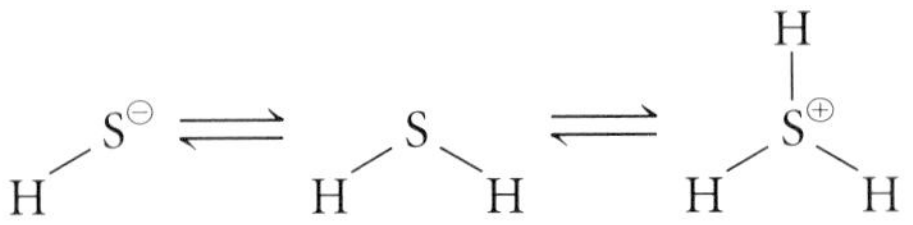

(c)

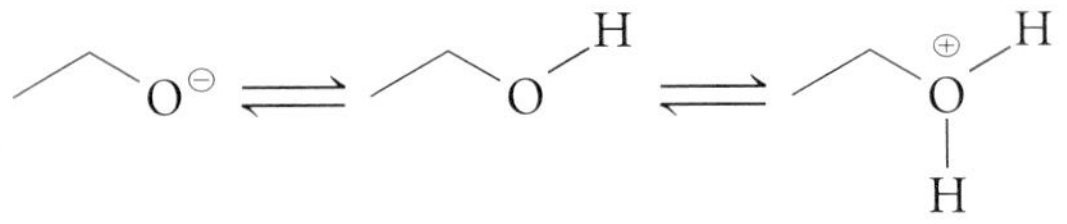

(d)

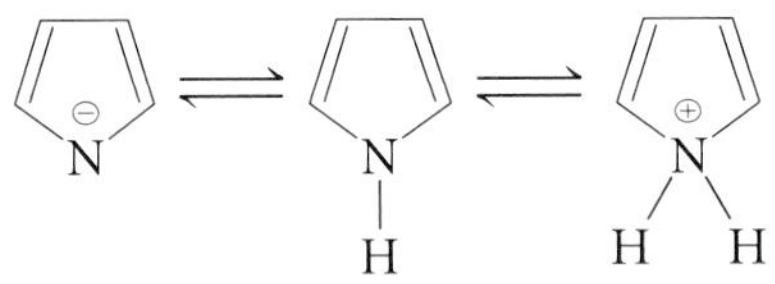

**3.35** For each of the following compounds, identify the principal functional group and define the hybridization at each carbon bonded to a heteroatom.

(a) R—C≡N

(b)

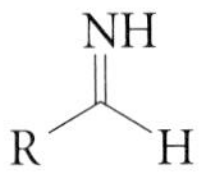

(c)

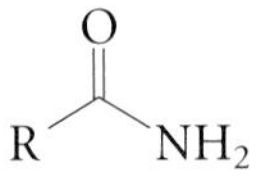

(d) R—CH$_2$—NH$_2$

(e)

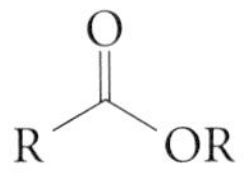

(f)

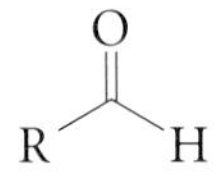

(g)

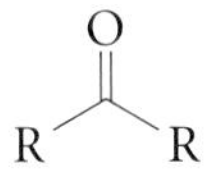

(h)

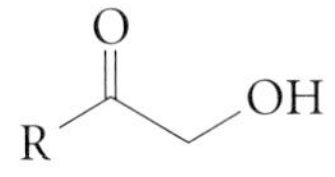

**3.36** Predict which of the two indicated C—H bonds in each of the following compounds would yield a more stable radical upon homolytic cleavage.

(a)

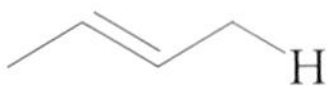

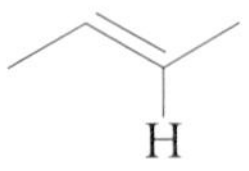

(b)

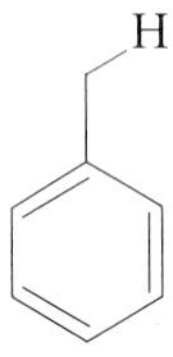

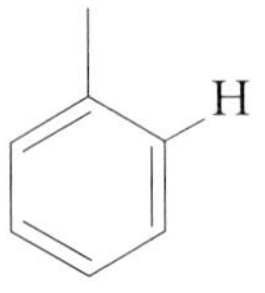

(c)

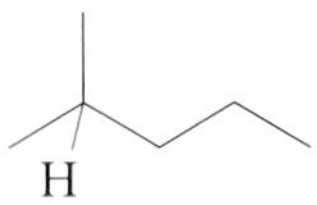

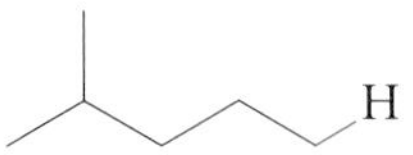

(d) 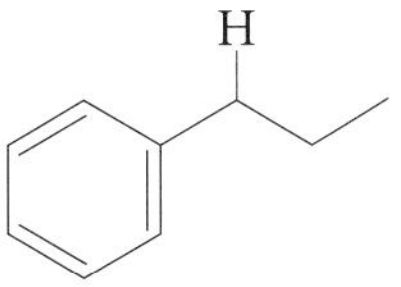 or

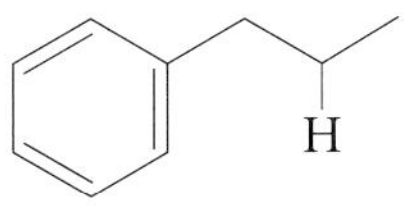

(e) 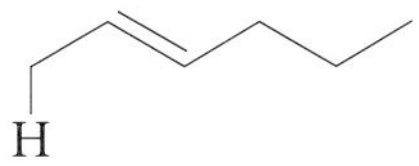 or

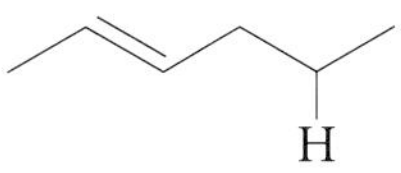

**3.37** Draw a three-dimensional representation of the relative orientation of the ring $\pi$ orbitals and the lone pair on the heteroatom in each molecule.

(a) 
Thiophene

(b) 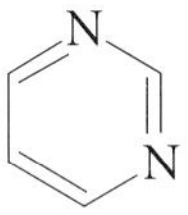
Pyrimidine

**3.38** Determine whether each heteroatom in uracil and cytosine will act as a proton donor or lone pair donor when participating in intermolecular hydrogen bonding.

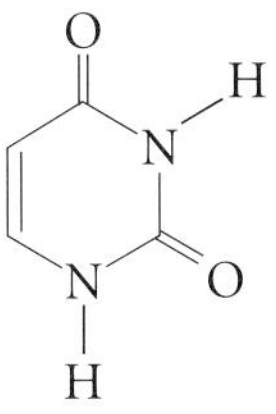

Uracil

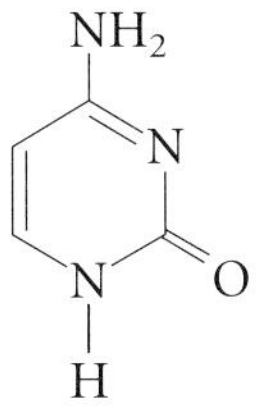

Cytosine

**3.39** Name each of the following compounds.

(a)

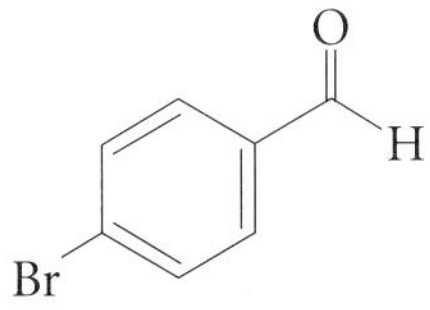

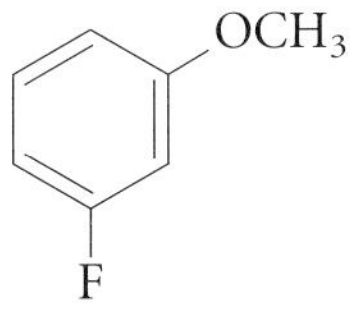

(b)

(c)

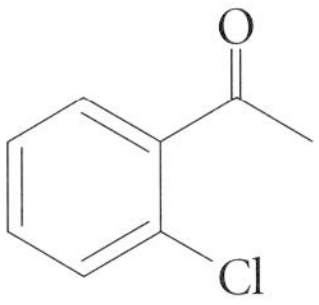

(d)

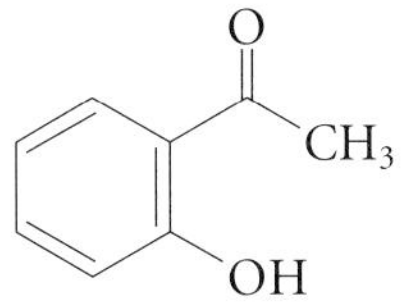

**3.40** Because of their low solubility in water, alkanes are often said to be *hydrophobic*. Explain why alkanes have such low affinity for water.

**3.41** From average bond energies (Table 3.2), calculate the heat of combustion of ethane (burning of ethane in oxygen to produce carbon dioxide and water).

**3.42** Compare the value calculated in Problem 3.41 with that measured experimentally (as listed in Table 1.4). If the values do not correspond, explain. (*Hint:* Recall how average bond energies are derived.)

**3.43** From average bond energies (Table 3.2), explain why C—C bond cleavage (causing the carbon skeleton to be broken) dominates over C—H bond cleavage when naturally complex mixtures of hydrocarbons are pyrolyzed at high temperatures in petroleum refining.

**3.44** Differentiate the structural features characteristic of an aromatic and a heteroaromatic compound.

**3.45** Draw a Lewis dot structure for each of the following heteroatom-containing compounds:

(a) hydrogen peroxide, HOOH

(b) phosphine, $PH_3$

(c) hydrogen sulfide, $H_2S$

(d) hydrazine, $H_2NNH_2$

**3.46** Write an electron dot structure for each of the following simple molecules, all of which contain at least one multiple bond.

(a) nitrogen gas, $N_2$

(b) "laughing gas," $N_2O$

(c) formaldehyde, $H_2CO$

(d) phosgene, $COCl_2$

**3.47** Identify and draw the structure of the lowest-molecular-weight compound that corresponds to each of the following descriptions:

(a) an enone

(b) a cyclic ether

(c) a cyclic amine

(d) an *N*-methylamide

**3.48** Explain why the H—X—H angle is about 107° in ammonia but about 105° in water, even though both nitrogen and oxygen are $sp^3$-hybridized.

**3.49** Explain why acetone, an organic ketone, is completely soluble in water.

# Chromatography and Spectroscopy

*Purification and Structure Determination*

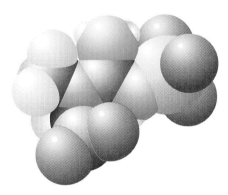

**Phosphocreatine**

$H_2O$

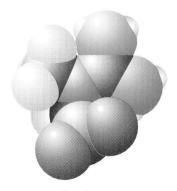

+

**Creatine**

**Inorganic phosphate**

*The conversion of phosphocreatine to creatine and inorganic phosphate ($P_i$) occurs in muscle tissue upon exertion and releases substantial energy. The decrease in concentration of phosphocreatine and the increase in concentration of $P_i$ in the forearm of a human subject can be followed using phosphorus NMR spectroscopy.*

*C*hapters 2 and 3 described how to identify the functional groups of organic compounds and how functionality influences physical properties. This chapter explores how physical techniques can be used to gather evidence about molecular structure. In later chapters, you will learn how to use these techniques with confidence to follow the course of reactions and to identify the products.

Among the interesting questions addressed in this chapter are these: How do chemists know whether a given sample is a pure compound or a mixture? How do they assign structure to a compound? What characteristics of the individual functional groups assist them in making correct structure assignments?

## Using Physical Properties to Establish Structure

When a student learns a new chemical reaction, the identities and structures of reactants, reagents, and products are given. However, the practicing chemist working in a laboratory usually knows only the structures of the reactant and the reagents and must demonstrate the structure of the product. Physical properties such as melting or boiling points can be used to help assign structure to a compound, provided that the compound has previously been prepared and these properties have been measured and recorded. The greater the number of physical properties that correspond to those of a known compound, the greater is the chemist's confidence that an assignment of structure is correct.

For example, if a compound thought to be 2-octanol exhibits a boiling point corresponding to that listed for this compound in reference books, this assignment would be reasonable. However, it is difficult to establish boiling points to closer than within 2 or 3 °C, and a perusal of even a simple reference such as the *Handbook of Chemistry and Physics* (or the *Merck Index*) will quickly convince you that many compounds have the same boiling-point range within 2 or 3 °C. If a second physical property (such as the melting-point range) also corresponds to that listed for 2-octanol, the structural assignment can be made with greater confidence because far fewer candidates will have both boiling and melting points that match those measured for the sample. You might be even more assured of the compound's identity if its reactivity also corresponds to that of 2-octanol. For example, suppose that you treat the compound with a strong oxidizing agent (say, chromic acid) and obtain a product mixture that can be distilled to give a clear liquid that boils at 173 °C:

(2-Octanol)
?

Product (bp 173 °C)

(2-Octanone)
?

This result corresponds to what would be obtained as the expected oxidation product, 2-octanone. Thus, more evidence has accumulated that the original assignment was correct.

Chemists working in the nineteenth century and the first half of the twentieth century spent a great deal of time investigating reactions. This involved not only carrying out the reaction of interest, but also transforming the products into known compounds simply to show in a convincing way that the suggested structural assignments for the products were correct.

Research productivity has dramatically increased in the past 50 years because of new techniques for isolating pure compounds from mixtures and new instrumentation for identifying the structures of organic compounds. Many of these instrumental methods provide direct evidence for the presence and spatial arrangement of a functional group. In this chapter, we consider the most common of these techniques: chromatography and spectroscopy.

## Purification of Compounds

Several different techniques are used to separate a mixture of organic compounds into its pure components; each has both advantages and limitations. Partitioning a mixture between two solvents—most commonly water and a water-immiscible solvent such as ethyl ether—is an effective method for separating compounds that differ in polarity. Recrystallization can be a powerful tool for small- to medium-scale separations if the compound to be purified is crystalline at a convenient temperature. However, because a solvent is involved, such separations can be prohibitive for the very large quantities required in the bulk chemical industry. Distillation is used extensively for large-scale separations, especially in the petroleum industry, where specialized distillation columns can separate compounds differing in boiling point by only a few degrees. However, only relatively volatile compounds can be purified by distillation.

**Chromatography** is a powerful method for separating the components of mixtures like those formed in chemical reactions. The technique is used both to obtain pure individual components of a mixture and to determine the ratio of these components. In chromatography, molecules are partitioned between two different phases, and separation is directly related to the difference in solubility that different molecules show in each phase. Furthermore, because compounds differ in mobility, chromatography can be used to demonstrate a correspondence between a compound and a reference sample of known structure.

## Determination of Structure

**Spectroscopy** constitutes a set of techniques that measure the response of a molecule to the input of energy. The resulting spectrum is a series of bands that show the magnitude of the response as a function of the wavelength of the incident energy. The energy source can be optical photons (as in ultraviolet spectroscopy, visible spectroscopy, and infrared spectroscopy) or radio-frequency energy (as in nuclear magnetic resonance spectroscopy).

In a somewhat different technique, mass spectroscopy, molecules are bombarded with high-energy electrons. In this case, the visual representation of the data is similar to the spectra obtained from interaction of molecules with energy of the electromagnetic spectrum. In this chapter, we deal with the physical basis of each of these methods and how characteristic spectra are interpreted to obtain structural information.

## 4.2

## Chromatography

At some time in your life, you have probably had an undesirable encounter with chromatography. For example, the ink on a neatly written homework paper may have become rain-soaked, causing the ink to "run," that is, to disperse into the component colors as they dissolve and flow at different rates across the paper surface. In fact, the word *chromatography* was first suggested by the Russian chemist Mikhail Tswett almost 100 years ago to describe the separation of pigments as "colored writing."

Chromatographic separations are usually accomplished by introducing organic compounds onto a **stationary phase** (paper) and then allowing a **mobile phase** (water) to flow past the mixture. Each component interacts with (adsorbs on) the stationary phase and dissolves in the mobile phase to a different extent. Components bound less tightly to the stationary phase and more soluble in the mobile phase travel farther than other components. The various methods of chromatography differ with respect to the mobile phase (a liquid or a gas), the stationary phase (paper, gel, or solid packing), and the driving force for the mobile phase (pressure, gravity, or an electric field).

Modern chromatographic techniques make use of the difference in solubility of different molecules in a mobile phase relative to a stationary phase. In **gas chromatography,** the mobile phase is a **carrier gas** (an inert gas such as argon), and the stationary phase is either a solid or a solid coated with a nonvolatile liquid. In **liquid chromatography,** the mobile phase (the *eluent*) is a liquid (any of a variety of aqueous and organic solutions), and the stationary phase is a solid composed of small particles around which the liquid can flow. Differences in the strength of interaction of the various components of a mixture with the stationary phase are a factor in all chromatographic techniques. Thus, it is important that the surface area of the stationary phase be as large as possible. The smaller the particle size, the larger is the surface area, so that very fine particles of a solid are often used for the most demanding separations. As a mixture of compounds passes over a solid support, the compounds exhibit differing *mobilities* (move at varying rates) due to the differences in their interactions with the stationary phase (*adsorption*) and the mobile phase.

Chromatographic techniques are limited in scale. Gas chromatography can be used to separate quantities up to hundreds of milligrams. Liquid chromatography is applicable for larger amounts and is often used in the synthesis of pharmaceutical compounds in quantities of multiple kilograms. Nonetheless, it is an expensive technique that involves silica gel as solid support (which is not reusable in many cases) and substantial quantities of solvent (which must be recovered and purified for reuse).

## Partitioning and Extraction

Chemists routinely use a simple form of selective partitioning when they extract organic molecules into an organic phase (such as ether) from water containing inorganic salts. In extraction, the separation occurs because organic compounds are generally more soluble in ether, whereas inorganic materials are more soluble in water. These solubility differences are usually very large, and it is often necessary to extract an aqueous layer only once to obtain most, if not all, of an organic material.

## Liquid Chromatography on Stationary Columns

A simple liquid chromatography column is constructed by packing a solid stationary phase (typically, alumina or silica gel) as a slurry into a burette (or other glass column with a restriction and a stopcock at one end). Both alumina and silica gel are polymers—arrays of large molecules composed of simple, repeating subunits (in this case, $Al_2O_3$ and $SiO_2$, respectively). As shown in Figure 4.1, a plug of cotton is usually inserted into the burette so that the stopcock does not become clogged with small particles of the solid phase. A layer of sand is added to form a more even surface, and then the solid phase is added. A second layer of sand is added at the very top of the column so that the solid phase is not disrupted as the mobile phase is added.

A concentrated solution of the mixture of compounds to be separated is applied at the top of the solid phase; then the mobile phase, or eluent, is added and allowed to flow toward the bottom under the influence of gravity. Components that adhere more tightly to the solid phase move down the column more slowly, requiring a greater volume of eluent before they reach the bottom. Thus, as the chromatography proceeds, a series of bands develops, and the mixture of compounds is separated into its components.

The motion of solute and solvent through the solid phase is called **elution.** The difference in the volume of solvent required for two different compounds to pass through a column is a measure of the degree to which each interacts with the solid phase. The ratio of solvent volumes required for elution represents the degree of separation of two components and is referred to as **alpha ($\alpha$).** For example, if the first compound elutes after 65 mL of solvent is collected and the second compound after 85 mL, $\alpha = 65 \div 85 = 0.76$. The time it takes for a compound to pass through the column is its **elution time.** As long as the rate of flow of the solvent remains constant, the separation factor $\alpha$ can also be determined from the ratio of elution times.

Both alumina and silica gel are polar metal oxides whose surfaces are covered with hydroxyl (OH) groups. As a result, molecules with higher polarity adsorb more strongly to these highly polar supports and are eluted more slowly than less polar molecules. Thus, the least polar component of the mixture usually elutes first from the column.

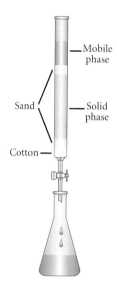

**FIGURE 4.1**

In column chromatography, a mixture of compounds placed at the top of a solid support slowly moves down the column as a liquid mobile phase flows over the stationary phase, partitioning the components of the mixture. The components elute separately as the eluent flows under the influence of gravity.

## EXERCISE 4.1

Which compound in each of the following pairs would be more likely to flow first from an alumina column if eluted with ethyl acetate?

(a) acetone or 2-propanol

(b) benzene or cyclohexane

(c) acetic acid or methyl acetate

(d) cyclohexyl chloride or cyclohexylamine

Liquid chromatography conducted in an open chromatographic column such as that shown in Figure 4.1 is often referred to as **column chromatography.** The degree of separation indicates the relative ease of elution of the components. The ease of separation, or **resolution,** in chromatography depends not only on the degree of separation but also on how much each component has spread while passing through the column. Band spreading decreases as the average size of the particles in the stationary phase is made smaller. However, with very small particle sizes, the flow of solvent induced by gravity all but stops. This problem is overcome with **high-performance liquid chromatography,** also referred to as **high-pressure liquid chromatography** (or **HPLC**), in which the mobile phase is driven through a sealed column by a mechanical pump. The same principles apply in HPLC as in simple column chromatography, but because smaller particles can be used, the degree of separation of compounds is better.

Let's consider how chromatography can be used to separate a mixture of compounds A, B, and C. Figure 4.2 shows how the three components resolve into bands and flow from the column. The least polar compound (A) elutes first. As additional solvent flows through the column, compounds B and C continue to move and are ultimately eluted, in turn, from the col-

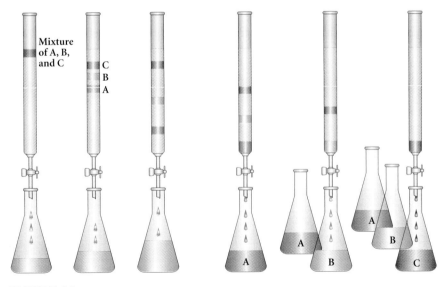

**FIGURE 4.2**

Separation of compounds A, B, and C by elution from a chromatographic column.

umn. Each component is collected in a separate flask. Thus, **chromatographic separation** is achieved. When each component is obtained separately, as here, chromatography is also a method of purification. Solvent can be removed from the eluent in each flask to yield the individual components of the original mixture in relatively pure form.

## Detectors

For efficient chromatographic separation, it is necessary to know when a component is eluting from the column. For this reason, a liquid chromatographic column is often coupled with a **detector** that responds to a change in some physical property when an additional compound is present in the eluting solvent. That is, the simple arrangement shown in Figures 4.1 and 4.2 is modified so that the eluent flows through a detector before being collected in a flask.

One very useful chromatographic detector is the **refractive index detector,** a so-called universal detector because it responds to virtually all compounds. The basis of its operation is that different materials have different indices of refraction. When a beam of light passes from one medium to another, the degree of bending it undergoes is related to the difference in the refractive indices of the two media. The refractive index of a solution differs from that of a pure solvent, and thus the eluent (containing a dissolved component) emerging from a chromatographic column has a different refractive index than does the pure solvent. In the detector, light passes from a compartment containing the solvent and sample to a compartment containing only the solvent. In the process, the beam of light is bent by an amount that depends on the difference in the refractive indices of the two liquids (Figure 4.3). The change in the path of the light is detected by comparing the intensity of the light received by two photocells positioned so that they "see" equal light intensities when the refractive index in both compartments is equal and thus there is no bending of the light. The refractive index of the liquid phase is monitored continually as it flows through the detector, and a change in refractive index indicates the presence of an additional component in the liquid.

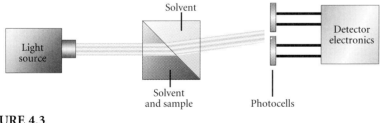

**FIGURE 4.3**

A refractive index detector. The magnitude of bending of the path of incident light indicates the presence of a compound having a refractive index different from that of the solvent.

The detector response can be plotted as a function of either the volume of eluent flowing through the column or the elution time. Such a plot, called a **chromatogram,** shows a peak for each component (A, B, and C, for example) as it flows from the column and through the detector. Most commercial HPLCs are equipped with detectors and automatic recording devices that trace a chromatogram as the chromatographic separation is being carried out. If the detector's response to each component is proportional to the amount of that compound in the eluting solvent, the ratio of the components in the original mixture can be determined from the areas under the peaks in the chromatogram. In the chromatogram in Figure 4.4, the ratio of the area under peaks A, B, and C is approximately 4:1:3. Because peaks B and C are not fully resolved, the ratio of these two components cannot be determined accurately. Complete separation of one component of a mixture from the others is referred to as **baseline separation.**

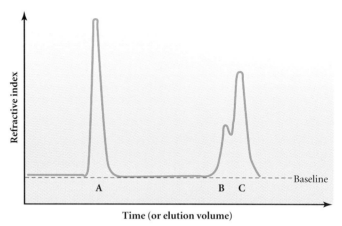

**FIGURE 4.4**

A chromatogram showing the elution of compounds A, B, and C from a chromatographic column. Because the refractive index returns to the level of pure solvent, this chromatogram provides evidence that A has been completely separated from B and C, which have not been completely separated (resolved).

Other changes in physical properties can also be used to detect the presence of a component in the eluting solvent. Commercially available chromatographic detectors employ a number of methods, such as ultraviolet absorption, fluorescence, and electrochemical conduction, to register the presence of a compound. Irrespective of the physical characteristic that is measured, however, such detectors are designed to indicate when the composition of the eluent has changed.

## Paper and Thin-Layer Chromatography

Other variants of liquid chromatography use sheets as the stationary phase rather than the particles in cylindrical columns employed in both column chromatography and HPLC. In **paper chromatography,** the mixture of compounds to be separated is applied as small drops of a solution near one edge of a sheet of chromatographic paper. This edge is immersed in a solvent that acts as the mobile phase, or eluent, which is pulled up the paper by capillary action. Alternatively, in **thin-layer chromatography** (Figure 4.5), a flat solid support such as a sheet of glass, plastic, or aluminum foil is coated with a thin layer of silica gel or alumina. As in paper chromatography, the solvent moves upward through the solid phase by capillary ac-

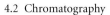

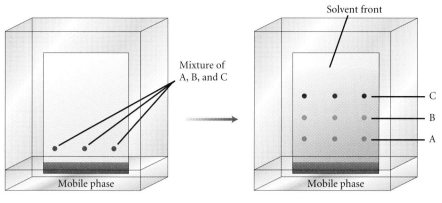

**FIGURE 4.5**

In chromatographic separation by thin-layer chromatography, the eluent moves up over a dry stationary phase by capillary action.

tion. This movement achieves the same separation as that accomplished by gravity in column chromatography.

The $R_f$ **value** is the ratio of the distance migrated by a substance compared with the farthest point reached by the solvent (the *solvent front*). The $R_f$ value is usually inversely proportional to the ratio of elution times observed in liquid column chromatography. Thus, the separation obtained by thin-layer chromatography parallels that obtained on a column, and it is common practice to employ thin-layer chromatography to find a solvent (or mixture of solvents) that will separate a mixture before carrying out column chromatography.

**EXERCISE 4.2**

Calculate $R_f$ values for the three separated compounds shown in the following thin-layer chromatogram:

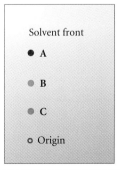

## Reverse-Phase Chromatography

Large biological molecules often have many polar functional groups, which bind too tightly to silica gel or alumina for column or thin-layer chromatography to be effective. For these compounds, a modified silica gel, in which a nonpolar organic molecule has been chemically bonded to the surface of the particles, is used as a stationary phase. This nonpolar phase binds the less polar compounds more tightly. The more polar compounds are carried more rapidly through the column by the solvent, which is often a mixture of a hydrocarbon (such as hexane) and a small amount of an alcohol (such as 2-propanol). Because the normal order of elution is reversed (with the more polar compounds eluting first), this technique is referred to as **reverse-phase chromatography.** The use of unmodified silica gel or alumina is sometimes called **normal-phase chromatography.**

## Gel Electrophoresis

We will see in later chapters that some biological molecules have many charged centers and are therefore called **polyelectrolytes.** Such a molecule can bear both positive and negative charges at various sites, giving either a net positive or negative charge, or, if the charges are exactly balanced, no overall charge. (Zwitterions are species that are neutral overall but have sites of both positive and negative charge.) Even in a neutral molecule, these ionic centers interact strongly with a stationary support. To separate molecules of this kind, a polar organic polymer such as polyacrylamide (properties of polymers will be discussed in more detail in Chapter 16) is used as the stationary phase. The polymer is saturated with water, causing it to swell, and ionic compounds move through this stationary phase under the influence of an electric field. Negatively charged ions migrate toward the positive pole when an electric field is applied, and positively charged ions migrate in the opposite direction. Molecules with a higher charge-to-mass ratio migrate faster, effecting separation of a mixture.

The migration of an ion under the influence of an electric field is known as **electrophoresis.** The use of an electric field to induce the movement of polyelectrolytes through a gel is referred to as **gel electrophoresis** (Figure 4.6). This technique is used extensively for separating and purifying biological macromolecules and for comparing an unknown with reference samples of known composition. The relative ease of migration of molecules across a gel depends on both the size and the charge: smaller molecules and those with higher charge move proportionally faster.

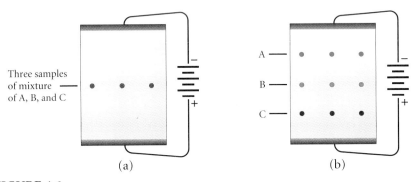

(a)      (b)

**FIGURE 4.6**

In gel electrophoresis, charged polyelectrolytes migrate across a gel under the influence of an electric field. Negatively charged ions migrate toward the positive electrode; positively charged ions migrate toward the negative electrode; neutral (including zwitterionic) molecules do not migrate. (a) Three identical samples of a mixture of A, B, and C will separate into components under the influence of an electric field. (b) Separation shown here would be attained if A were positively charged, B were neutral, and C were negatively charged.

In gas chromatography, a carrier gas (nitrogen or helium) sweeps a sample from a heated injector block onto and through a long chromatographic column heated in an oven. The gaseous effluent flows over a detector that registers the passage of each compound. Two types of detectors commonly are used. A **thermal conductivity detector** measures the difference in thermal conductivity between the pure carrier gas and the gaseous sample coming from the column. A **flame ionization detector** senses the presence of ions that are generated as the effluent from the column is burned in a hydrogen flame. A gas chromatograph is shown schematically in Figure 4.7.

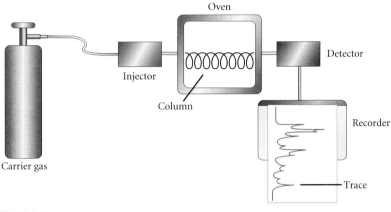

**FIGURE 4.7**

A schematic diagram of a gas chromatograph. An inert carrier gas moves under pressure over a column lined or filled with solid adsorbent. The mixture injected at the head of the column is fractionated and detected as the effluent gas passes through a detector.

Gas chromatography is used both in research laboratories and in routine analyses. It is the method of choice for analysis of trace amounts of compounds such as pesticides in foods or illicit drugs present in body fluids. Organic compounds are first extracted into an appropriate low-boiling organic solvent such as hexane or ethyl ether. The solution is then concentrated by fractional distillation so that only the solvent is removed.

The stationary phase in gas chromatography can be the walls of an empty column, solid packing within a column, or a polymeric liquid that coats either the wall or the porous solid packing. Typically, much longer columns (10–100 meters) are used in gas chromatography than in liquid column chromatography.

Just as for liquid chromatography, **retention times** for gas chromatography are influenced by the strength of the noncovalent interactions of the compounds being separated with the stationary phase. Roughly, these interactions can be considered to be governed by the effects of molecular polarity and van der Waals interactions, both of which also influence boiling points. It is common to find that the order of elution in gas chromatography approximates the order of the boiling points of the compounds separated. Because very long columns are used in gas chromatography, baseline separation is often achieved even for chemically similar compounds (Figure 4.8, on page 166).

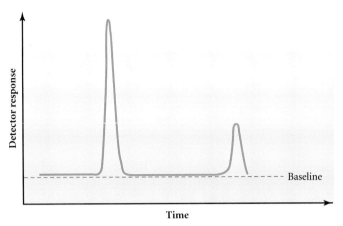

**FIGURE 4.8**

A typical gas chromatogram showing the baseline separation of two injected components.

## EXERCISE 4.3

Using your knowledge of how molecular structure influences physical properties, predict which compound in each of the following pairs will emerge first from a gas chromatography column.

(a)  $\diagup\!\!\!\diagdown\!\!\!\diagup\!\!\!\diagdown$  or  $\diagup\!\!\!\diagdown\!\!\!\diagup\!\!\!\diagdown\!\!\!\diagup\!\!\!\diagdown$

(b) or

(c) or

(d) $CH_4$ or $CCl_4$

## 4.3

# Spectroscopy

After a mixture has been separated into its components, spectroscopic techniques are often used to identify the individual compounds. Many spectroscopic techniques rely on the interaction of a compound with **electromagnetic radiation,** which can be considered as either a particle (called a photon) or a wave traveling at the speed of light. When regarded as a wave, light can be described by its wavelength ($\lambda$) or its frequency ($\nu$). *Wavelength* is the distance covered in one complete wave cycle. *Frequency* is the number of wave cycles that pass a fixed point in a defined time. (One **hertz, Hz,** equals one cycle per second.)

For some spectroscopic techniques, wavelength defines energy content. For others, frequency is used. These quantities are directly related, because the product of wavelength and frequency equals the speed of light, $c$ ($3 \times 10^{10}$ cm/sec). Thus,

$$\lambda = \frac{c}{\nu} \quad \text{and} \quad \nu = \frac{c}{\lambda}$$

The energy of a photon, $\epsilon$, can be easily calculated:

$$\epsilon = h\nu = \frac{hc}{\lambda}$$

where $h$ is Planck's constant ($6.6 \times 10^{-34}$ J/sec). (One joule, J, equals 4.186 calories. One calorie is the heat required to raise the temperature of 1 gram of water by 1 degree Celsius.) The energy of a photon increases with its frequency and is inversely proportional to its wavelength. Thus, high frequency or short wavelength means high energy. The regions of the electromagnetic spectrum are shown in Figure 4.9.

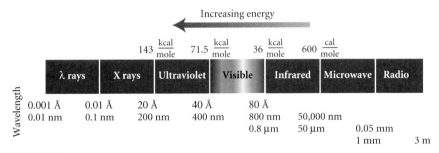

**FIGURE 4.9**

Energetic order of electromagnetic radiation. Human eyes are sensitive to radiation in the visible region of this spectrum.

In subsequent sections, we will consider the interaction of organic molecules with electromagnetic waves of increasing energy: radio frequencies in nuclear magnetic resonance spectroscopy, infrared photons in infrared spectroscopy, visible photons in visible absorption spectroscopy, and ultraviolet photons in ultraviolet absorption spectroscopy. Finally, we will consider the interaction of high-energy electrons with organic compounds in mass spectroscopy.

## Nuclear Magnetic Resonance (NMR) Spectroscopy

***Theoretical Background.*** Like electrons, both protons and neutrons have spin. A nucleus that contains an odd number of protons or neutrons (or both) has spin and is magnetically active. The smallest nucleus that meets this requirement is $^1$H, but so do $^{13}$C, $^{17}$O, $^{19}$F, and $^{31}$P. These nuclei behave as if they were spinning about an axis, and thus they have angular momentum. Because the nucleus is positively charged, this spinning motion causes it to behave as if it were a tiny magnet. In accord with the requirements of quantum mechanics, when a nucleus with a net spin is placed in a large magnetic field, quantized energy states for the nucleus are defined by its orientation with respect to the external magnetic field. In the case of nuclei with a spin of 1/2, such as $^1H$ and $^{13}C$, two orientations are possible: aligned with or against the external field. It is slightly more favorable for the spin of the nucleus to be aligned with the magnetic field than against it, and thus these two alignments are of different energy. As a result, the number of molecules whose nuclei are in parallel alignment will be slightly greater than the number with nuclei in antiparallel alignment (Figure 4.10). As will become evident, observing the transitions between these two spin states provides a wealth of information about the environment of nuclei in molecules.

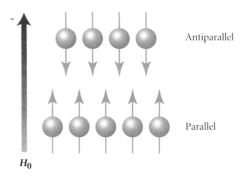

**FIGURE 4.10**

Parallel and antiparallel alignment of nuclear spins under an applied magnetic field, $H_0$.

Nuclei can be induced to jump from a lower- to a higher-energy spin state by electromagnetic energy of a frequency that matches the energy difference between the two states. Conversely, when a nucleus in the higher-energy state drops to the lower-energy state, electromagnetic energy of that frequency is emitted. The spin responsible for creating the two states is a property of the nucleus of the atom, and the technique is known as **nuclear magnetic resonance (NMR) spectroscopy.** (Note that *nuclear* has nothing to do with radioactivity in this context.) When nuclei of a sample are flipping rapidly between states, they are said to be **in resonance** with the applied electromagnetic radiation. Chemists can detect when this is happening in two ways: by measuring the energy absorbed from an applied electromagnetic signal by the nuclei as they jump to the higher-energy state, or by "listening" for the energy the nuclei radiate as they return to the lower-energy state. Today, most NMR spectrometers operate by detecting the radiant energy.

The frequency of the energy required to induce spin-state flipping of nuclei varies directly with the magnitude of the applied magnetic field. The greater the field strength, the larger is the difference between parallel and antiparallel spin states, and the higher is the energy of the signal required to induce the change. Commercial NMR spectrometers have very large magnets that employ superconducting wires to produce a magnetic field. With these field strengths, electromagnetic energy in the radio-frequency range is required to induce state flipping. NMR spectrometers are classified by the frequency used to change the spin state of magnetically active nuclei. The highest-field machines currently available from commercial instrument manufacturers operate at 750 MHz (1 megahertz, MHz, equals 1 million cycles per second). Instruments using signals from 100 to 300 MHz are much more common.

These frequencies actually correspond to very little energy: 100 MHz corresponds to only about $1 \times 10^{-5}$ kcal/mole, and this radio-frequency energy can be taken up only by nuclei that behave as magnets in an applied magnetic field. Hydrogen and carbon nuclei are of greatest interest to organic chemists. The most abundant isotope of hydrogen, $^1$H, has a net spin of 1/2, as does $^{13}$C. Although the latter isotope represents only 1.1% of the carbon present in normal samples and very sensitive instruments must be used to observe the spin-state changes of its nucleus, the wealth of information contained in $^{13}$C NMR spectra makes construction and use of these instruments worthwhile.

*Shielding.* The **effective field ($H_{eff}$)** felt by the nucleus differs from the **applied field ($H_0$)** because of a tiny **local magnetic field ($H_{loc}$)** set up by the circulating electron cloud surrounding the nucleus.

$$H_{eff} = H_0 - H_{loc}$$

The electron density about each atom in a molecule varies with the nature of the surrounding atoms and is slightly different for each nonequivalent atom. In essence, the nucleus of an atom experiences some degree of **shielding** from the external magnetic field. Thus, each unique nucleus experiences a different $H_{eff}$ and, as a result, emits energy at a different frequency. The result is a spectrum of different frequencies, on which each set of unique nuclei gives rise to a unique NMR signal. An **NMR spectrum** is a plot of signal intensity versus the frequency of the electromagnetic energy released by the various nuclei in a sample.

*Chemical Shifts.* Frequencies are reported as the difference (in parts per million, or ppm) between the signals recorded for a sample and that of a reference compound, tetramethylsilane, $(CH_3)_4Si$ (often called TMS), which is added to the sample for both proton ($^1$H) and carbon ($^{13}$C) spectroscopy. Thus, the signals are reported as **chemical shifts,** or changes, from this standard, on the delta ($\delta$) scale, where 1 $\delta$ equals 1 ppm and where the signal from tetramethylsilane is at 0:

$$\delta = \frac{\omega_{standard} - \omega_{sample}}{\omega_{standard}} \times 10^6$$

Because the $\delta$ scale is based on the ratio of the difference in frequency between a standard and a sample to the frequency of the standard, values are independent of the magnet's field strength. Signals at lower frequency (and lower field) than that of the standard have positive $\delta$ values. The majority of proton signals range between 0 and 12 ppm; the range for carbon is larger, from 0 to 250 ppm. Because the range in $^1$H spectroscopy is smaller than that for $^{13}$C, the accidental overlap of two nonequivalent signals is more likely to be found in a proton spectrum than in a carbon spectrum.

Almost all signals are at lower frequency than the standard and are said to be **downfield.** (This term carries over from early spectrometers where the magnetic field was varied while the frequency was held constant.) The

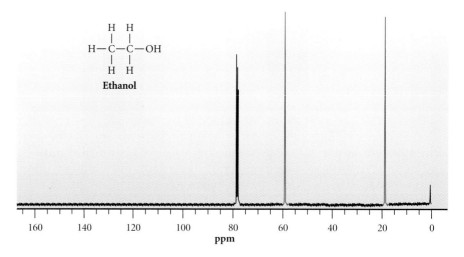

**FIGURE 4.11**

A $^{13}$C NMR spectrum of ethanol. The signal at $\delta$ 0 is that for the four identical methyl groups of the standard, tetramethylsilane. The signal at $\delta$ 17.9 is from the $CH_3$ carbon, and that at $\delta$ 57.3 is from the $CH_2$ carbon attached to oxygen. The three peaks centered at $\delta$ 77 are from solvent ($CDCl_3$).

silicon in TMS is responsible for the high field position of its protons and carbons relative to those of most organic compounds. Because silicon is less electronegative than carbon, the C—Si bond of TMS is polarized toward the carbon atoms of the methyl groups. This electron density shields the hydrogen and carbon nuclei of those groups from the applied magnetic field. Active nuclei that resonate at frequencies only slightly below TMS are said to appear in the *upfield region* of the NMR spectrum; those shifted to much lower frequencies, in the *downfield region*.

***Spectral Interpretation.*** What can chemists learn about molecular structure from an NMR spectrum? From the number of signals, they establish how many different types of nuclei are present; from the chemical shift of each signal, they learn details of the chemical environment of each type; and from the splitting of the signals, they can deduce how many protons are near each.

The $^{13}$C and $^{1}$H NMR spectra of ethanol are shown in Figures 4.11 and 4.12. The $^{13}$C NMR spectrum of ethanol shows two peaks because there are two types of carbon atoms present. The $^{1}$H NMR spectrum includes three groups of signals that result from the three kinds of hydrogen atoms. Note that the signals in the carbon spectrum are recorded as a series of single sharp lines, whereas those in the proton spectrum are split into symmetrical patterns. As we will see in the following sections, this splitting is the result of interactions with neighboring protons.

## $^{13}$C NMR Spectroscopy

Because $^{13}$C spectra are often simpler than $^{1}$H spectra, we will consider them first. A $^{13}$C NMR spectrum furnishes two basic pieces of information:

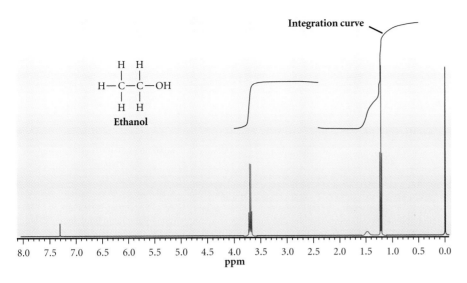

**FIGURE 4.12**

A 360-MHz $^1$H NMR spectrum of ethanol. The signal at $\delta$ 0 is from tetra-methylsilane. The broad singlet at $\delta$ 1.5 represents the OH proton. The signal at about $\delta$ 1.2 is split into a three-peak pattern (called a triplet) and is from the CH$_3$ group. The area under this peak is three times the area under the peak at $\delta$ 1.5. The signal at about $\delta$ 3.7 is split into a four-peak pattern (called a quartet) and is from the CH$_2$ protons. The area under this peak is twice that under the peak at $\delta$ 1.5. Splitting results directly from protons on adjacent atoms. (The proton of the OH group is moving from the oxygen of one molecule to the oxygen of another; as a result, it is not usually split by other protons and does not itself contribute to splitting.) The integration curve is explained on page 175. The small peak at about $\delta$ 7.3 is from CHCl$_3$, present as an impurity in the CDCl$_3$ solvent.

the number of distinct signals, corresponding to the number of different types of carbon atoms; and the chemical shift of each signal, which is determined by the molecular environment of each carbon. Under the usual instrumental conditions, each nonequivalent $^{13}$C nucleus is recorded as a distinct, sharp signal. The sharpness of the signal in $^{13}$C NMR spectroscopy is important for two reasons. First, a sharp signal is concentrated in a narrow frequency range and therefore can be distinguished more readily from random noise produced by the electronic circuitry. Second, the narrower the signal, the better is the *resolution,* the separation between signals that are close to one another.

Note, in Figure 4.11, that the two distinct carbons in ethanol are recorded as signals with different chemical shifts. In fact, as shown in Table 4.1 (on page 172), the chemical shifts of most distinct carbons are different and are characteristic of the type and number of carbon atoms and heteroatoms in the immediate molecular environment. In the table, an $\alpha$ substituent is an atom directly attached to the carbon being observed, a $\beta$ substituent is an atom one carbon removed down the chain, and a $\gamma$ substituent is an atom attached to a $\beta$ substituent. For example, the methyl carbon of ethanol has one $\alpha$ substituent (the carbon that bears the OH group) and one $\beta$ substituent (the oxygen of the OH group). Note that we do not consider hydrogen atoms in this analysis.

**TABLE 4.1**

General Effects on Carbon NMR Shifts

$$C—\alpha—\beta—\gamma$$

| | | 1° | 2° | 3° |
|---|---|---|---|---|
| For any $sp^3$ carbon, add to 0.0, for each: | $\alpha$ or $\beta$ substituent | | 8.0 | |
| In addition, add for each: | $\alpha$ oxygen substituent | | 38.0 | |
| | $\alpha$ nitrogen substituent | | 22.5 | |
| | $\alpha$ *trans* C=C | | 2.5 | |
| | $\alpha$ *cis* C=C | | −2.5 | |
| | $\alpha$ ester or acid | | −2.5 | |
| | $\alpha$ ketone | | 7.5 | |
| | $\alpha$ aldehyde | | 15.0 | |
| $\alpha$ chlorine | | 22.0 | 30.5 | 41.0 |
| $\alpha$ bromine | | 10.5 | 24.0 | variable |
| | $\alpha$ iodine | | variable | |
| | $\gamma$ carbon | | −2.0 | |
| | $\gamma$ oxygen | | −5.0 | |
| For 3° carbons, add for each: | $\beta$ substituent | | −1.5 | |
| For 4° carbons, add for each: | $\beta$ substituent | | −3.5 | |

$$
\begin{array}{c}
\beta—\alpha \diagdown \qquad \diagup \alpha'—\beta' \\
\qquad C=C \\
\beta—\alpha \diagup \qquad \diagdown \alpha'—\beta'
\end{array}
$$

| | | |
|---|---|---|
| For an $sp^2$ carbon of a C=C bond, add to 121.0 for each: | $\alpha$ or $\beta$ carbon substituent | 8.0 |
| | $\alpha'$ carbon substituent | −8.0 |
| | *cis* double bond | −1.0 |

The signal for the carbon that bears the OH group in ethanol is shifted significantly downfield from that for the methyl carbon group because of the presence of the electronegative oxygen. The $^{13}$C spectrum allows us not only to count the number of different carbons in a molecule of unknown structure, but also to have some idea of the immediate environment of each type of carbon atom. For example, using the values in Table 4.1, we can predict that the methyl carbon (CH$_3$—) of ethanol would resonate at about 16 $\delta$ because it has one $\alpha$ and one $\beta$ substituent (8 + 8), whereas the methylene carbon (—CH$_2$—) should appear near 54 $\delta$ because it has two $\alpha$ substituents, one of which has an added effect because it is oxygen (8 + 8 + 38). Both of these predictions are close to the observed values for the methyl and methylene carbons in ethanol: $\delta$ 17.9 and $\delta$ 57.3, respectively.

It is possible to count the number of different *types* of carbons present in a molecule from its $^{13}$C spectrum, but the intensity of each signal is only roughly related to *how many* carbon atoms produce that signal. The difference in peak size is caused by differences in the rate at which carbon atoms relax to the equilibrium distribution of their two energy states in the presence of a magnetic field. This rate is in turn influenced by the proximity of a given nucleus to other spin centers in the molecule. It is not necessary to define these factors precisely here, but you should realize that *$^{13}$C NMR signal intensity does not correlate accurately with the number of carbon atoms responsible for a given signal.* The signal resulting from the carbon of a methyl group is often somewhat weaker than that of methylene ($CH_2$) and methine (CH) carbon atoms, but stronger than that of quaternary carbons (those with no hydrogen atoms), which is quite weak.

Special techniques in $^{13}$C NMR spectroscopy permit the determination of the number of hydrogen atoms attached to each carbon atom. An understanding of how these methods work is well beyond the scope of this course. The information provided by the shift positions (a measure of the number of types of $\alpha$, $\beta$, and $\gamma$ substituents) combined with the assignment of each signal to a methyl, methylene, methine, or quaternary carbon atom can be invaluable in deducing structure from a $^{13}$C NMR spectrum.

## EXERCISE 4.4

For each of the following compounds, predict how many distinct carbon signals will be observed in its $^{13}$C NMR spectrum.

## EXERCISE 4.5

Using the values in Table 4.1, predict the chemical shift expected for each carbon in the following molecules. Then correlate each of your predictions with one of the observed signals so as to arrive at the smallest average error between prediction and experiment. What is the average error for your four predictions for part (a) and your five predictions for part (b)?

(a)

$\delta = 36.6, 29.5, 18.9, 11.6$

(b)

$\delta = 67.0, 41.6, 23.3, 19.1, 14.0$

## $^1$H NMR Spectroscopy

The signals from protons in $^1$H NMR spectra, like those from carbons in $^{13}$C NMR, are recorded as separate absorption peaks for nonequivalent nuclei. A $^1$H NMR spectrum provides four important pieces of information: the number of unique signals, the chemical shift, the splitting pattern, and the integrated signal intensity.

***Chemical Shifts.*** Interpreting a $^1$H NMR spectrum begins with the determination of the number of signals with distinct chemical shifts. For example, in Figure 4.12, there are three distinct signals corresponding to the three kinds of protons present in ethanol. In contrast with the sharp lines observed in the $^{13}$C NMR spectrum (Figure 4.11), the signal for each type of proton has a more complex pattern. The broad single peak that appears at 1.5 $\delta$ represents the proton on oxygen in the OH group. The signal for the CH$_2$ group at 3.7 $\delta$ is split into four lines, and the signal for the CH$_3$ group at highest field (1.2 $\delta$) is split into three lines. The center of each of ethanol's three signals (at 1.2, 1.5, and 3.7 $\delta$) defines the chemical shift for each type of hydrogen, which depends on the environment of that magnetically active nucleus. The data in Table 4.2 allow us to correlate the chemical shift of each $^1$H NMR signal with the molecular environment (the type and location of functional groups) of that type of nucleus in somewhat the same way as we did earlier for a $^{13}$C NMR spectrum, although less quantitatively. For example, the signal for protons closer to the more electronegative oxygen atom in ethanol is shifted farther downfield, as is the case for the carbon-bearing oxygen in the $^{13}$C NMR spectrum of ethanol.

## TABLE 4.2

Representative $^1$H Chemical Shifts

| Type of Proton | Chemical Shift ($\delta$) | Type of Proton | Chemical Shift ($\delta$) |
|---|---|---|---|
| —CH$_3$ | 0.7–1.3 | ![acetaldehyde CHO] | 9.5–10.0 |
| —CH$_2$— | 1.2–1.4 | ![carboxylic acid COOH] | 10.0–12.0 |
| —C—H | 1.4–1.7 | —C—OH | 1.0–6.0 (changes with solvent) |
| C=C(CH$_3$) | 1.5–2.5 | O—C—H | 3.3–4.0 |
| O=C—CH$_3$ | 2.1–2.6 | Cl—C—H | 3.0–4.0 |
| Ar—CH$_3$ | 2.2–2.7 | Br—C—H | 2.5–4.0 |
| C=C(H) | 4.5–6.5 | I—C—H | 2.0–4.0 |
| Ar—H | 6.0–9.0 | | |
| —C≡C—H | 2.5–3.1 | | |

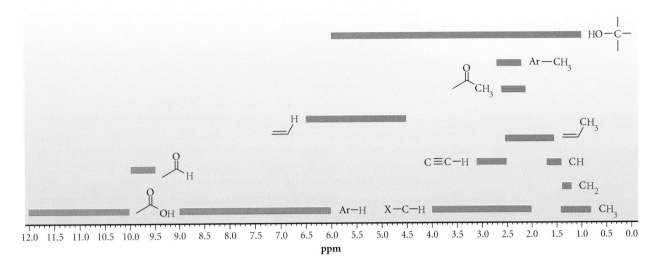

**FIGURE 4.13**

Charactertistic ranges on $^1$H NMR spectra where various types of protons absorb (X = halogen).

Because the chemical shifts of protons are influenced primarily by the atom to which they are attached and secondarily by other atoms in the immediate vicinity, the protons of common functional groups absorb in characteristic regions of $^1$H NMR spectra (Figure 4.13).

*Integration Curve.* The relative intensity of a signal in $^1$H NMR spectroscopy is proportional to the number of protons contributing to the signal. Proton NMR differs from carbon NMR in this respect. The curve superimposed on the signals in the NMR spectrum in Figure 4.12 is an **integration curve;** it is a measure of the area under each peak. The vertical rise in the stair step of an integration curve can be used to calculate the ratio of the number of hydrogens responsible for each signal. Thus, in Figure 4.12, the 1:2:3 ratio of the three peaks is proportional to the number of hydrogens responsible for each of the three different chemical shifts for ethanol. Because the absolute area under an integration curve depends on instrument sensitivity, not on the number of hydrogens, the integral gives a ratio of numbers of hydrogens, not the absolute number of each type. Integration information is reported along with the chemical shift of the signal—for example, 5.1 δ (3 H).

*Making Structural Assignments from Chemical Shift and Integration Data.* It is sometimes possible to make structural assignments for unknown compounds using only the chemical shift and integration information. For simple compounds, the number of possible isomers is small, and the spectral data can be used systematically to eliminate all structures but one. For example, an unknown compound, $C_3H_6O$, exhibits signals in its $^1$H NMR spectrum at 6.0 δ (1 H), 5.2 δ (2 H), 4.1 δ (2 H), and 2.9 δ (1 H). From the molecular formula, we can deduce that the compound contains one ring or one double bond and must be one of the following structures:

**175**

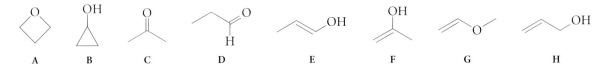

A B C D E F G H

Neither of the cyclic structures, A or B, would be expected to have absorptions lower field than about 4 δ (signals in that region of the spectrum arise from protons on $sp^2$- and $sp$-hybridized carbon atoms). Thus, the unknown structure must have a π bond. We can also rule out the ketone (structure C, acetone) because it, too, would have no signals in the downfield region. The aldehyde (structure D) would be expected to have a signal in the 9.5–10 δ region, so its structure is also inconsistent with the data. We can use the integration data to eliminate structures E and F, since each has only two protons on $sp^2$-hybridized carbon atoms, and the spectrum shows signals for three protons in the downfield region. We are left with structures G and H. Structure G has a methyl group and will therefore have a signal that would integrate as 3 H. No such signals are reported. Thus, by elimination, structure H must be the correct answer.

## EXERCISE 4.6

By assigning each set of protons to the appropriate signal, convince yourself that structure H (allyl alcohol) is consistent with the $^1$H NMR data given above.

## EXERCISE 4.7

The $^1$H NMR spectrum of a compound with the molecular formula $C_3H_6O$ exhibits signals at 9.7 δ (1 H), 2.5 δ (2 H), and 1.1 δ (3 H). Determine which of structures A through G above are possible for this compound by eliminating those that are inconsistent with this data.

***Spin–Spin Splitting.*** In $^1$H NMR spectra, each signal appears as a complex pattern rather than a single peak. The splitting of a signal often provides valuable information about the structure of the molecule because it is the direct result of interaction with neighboring protons. The number of lines in a **multiplet** reveals the number of hydrogens on adjacent carbons. The splitting patterns observed are caused by the interaction of the magnetic spin of each $^1$H nucleus with neighboring nuclei. This interaction is referred to as **coupling.**

Let's compare the spectrum of ethanol in Figure 4.12 with that of ethyl bromide in Figure 4.14. Both compounds have ethyl groups and exhibit the same general pattern of signals, although the chemical shifts observed are different. In both cases, the upfield $CH_3$ signal is split into a three-line multiplet (a **triplet**), and the methylene ($CH_2$) signal is split into a four-line multiplet (a **quartet**).

Let's examine the $CH_3$ groups in both compounds. The methyl hydrogens interact with the two hydrogens of the adjacent methylene group. As illustrated at the right in Figure 4.15, a sharp singlet is seen when the interaction with methylene hydrogens on the adjacent carbon atom is blocked. The observed methyl splitting pattern results from the way in which the interacting spins on adjacent hydrogens can align with or against the applied magnetic field. The two methylene hydrogens can be oriented in three

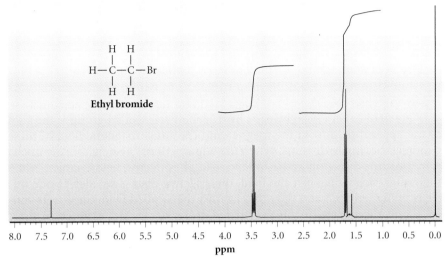

**FIGURE 4.14**

A 360-MHz $^1$H NMR spectrum of ethyl bromide. The sharp singlet at $\delta$ 0 is the TMS standard. The triplet from the CH$_3$ group appears at $\delta$ 1.7, and the quartet from the CH$_2$ group is at $\delta$ 3.4. (The small peak at $\delta$ 1.55 is due to contamination of the sample by H$_2$O and that at $\delta$ 7.3 is due to CHCl$_3$.)

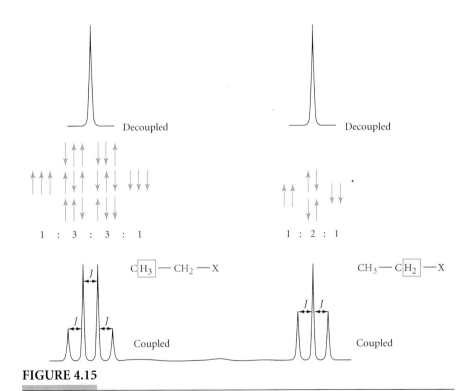

**FIGURE 4.15**

Coupling with protons on adjacent carbons is responsible for the observed splitting pattern ($J$ is the coupling constant). The protons under observation are in color; those splitting the observed signals are in boxes. The sets of arrows describe the possible alignments of adjacent nuclear spins with or against the applied magnetic field. The statistical abundance of each type determines the ratio of peak heights in the coupled spectrum.

possible ways: both aligned with, both against, and one with and one against
the applied field. There are two possible arrangements with identical energies for the last combination, and the arrows below the decoupled signal in
Figure 4.15 represent the alignments possible. The relative abundance of
each type of alignment gives the observed pattern: a quartet from three interacting neighboring hydrogens, a triplet from two interacting neighboring hydrogens.

The field experienced by each hydrogen of the methyl group is the sum
of the applied field from the magnet and the small magnetic field of these
two neighboring hydrogens. When the alignment of the neighboring nuclei
is in the same direction as the applied field, the effective field is larger. When
their alignment is against the field, the effective field is smaller. If one nucleus is aligned with and the other against the applied field, the effects cancel. Thus, the hydrogens on the methyl group are in fact exposed to three
different fields: one larger, one equal to, and one smaller than the applied
field. Each of these slightly different environments for the methyl hydrogen
exhibits its own unique resonance frequency. Thus, instead of seeing a sharp
singlet for the methyl group, we see a triplet resulting from the contribution of the spins of the methylene group hydrogens.

The **multiplicity** of a signal (that is, the number of peaks into which
the signal is split) observed in a $^1$H NMR spectrum provides valuable information about the number of hydrogens on adjacent positions. Figure
4.16 shows the shapes of simple multiplets that are derived by interaction

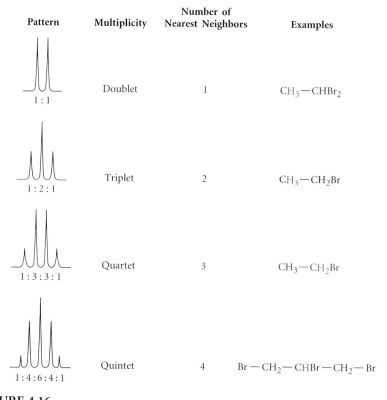

| Pattern | Multiplicity | Number of Nearest Neighbors | Examples |
|---|---|---|---|
| 1 : 1 | Doublet | 1 | $CH_3—CHBr_2$ |
| 1 : 2 : 1 | Triplet | 2 | $CH_3—CH_2Br$ |
| 1 : 3 : 3 : 1 | Quartet | 3 | $CH_3—CH_2Br$ |
| 1 : 4 : 6 : 4 : 1 | Quintet | 4 | $Br—CH_2—CHBr—CH_2—Br$ |

**FIGURE 4.16**

The splitting patterns in $^1$H NMR spectra of several compounds. The nuclei
for which the patterns are shown are highlighted in red.

with different numbers of nearest-neighbor nuclei. A doublet results when one magnetically active neighboring nucleus is present, a triplet when there are two, a quartet with three, and a quintet with four.

The degree of separation between the lines in a multiplet, referred to as **splitting,** varies directly with the effect of the neighboring hydrogens on the magnetic field. In general, the closer in space the hydrogens are, the larger the splitting. The magnitude of this splitting is referred to as the **coupling constant,** often designated simply as *J*. The magnitudes of the effects of two types of nuclei on each other will be equal. As a result, the coupling constants will be identical in the related multiplets: that is, the peak separations in the triplets in Figures 4.14 and 4.16 are exactly equal to the peak separations in the quartets. The *J* values for the methyl and methylene hydrogens of both ethanol and ethyl bromide are 7 Hz.

*Pascal's Triangle.* In the triplet pattern observed for the methyl group, the areas under the peaks are in the ratio 1:2:1. One-fourth of the signal intensity is upfield and one-fourth is downfield from the expected position; one-half of the signal intensity is at the center. The 1:2:1 ratio is the result of the probability distribution of the two spin states of the hydrogen atoms of the methylene group. There is only one way both spins can be aligned with the applied magnetic field, and only one way both can be aligned against this field. On the other hand, there are two ways in which one can be aligned with and one against the applied field. Because the hydrogen atoms are in identical environments, it does not matter which is aligned with and which against the magnetic field. We can quickly expand this analysis to three, four, five, and six (and more) neighboring hydrogen atoms using *Pascal's triangle*, a graphic construct devised by the mathematician Blaise Pascal to explain probability theory. The number at each point in the triangle is the sum of the numbers immediately above.

**Pascal's triangle**

Using Pascal's triangle, we can predict that the NMR signal for the methylene ($-CH_2-$) hydrogen atoms of ethanol or ethyl bromide will appear as a 1:3:3:1 quartet, because there are three hydrogen atoms on the adjacent methyl group. There are eight possible orientations of spin for the three neighboring hydrogen atoms in relation to the applied field, divided into four groups as they add to or subtract from the applied field. Thus, each hydrogen of the methylene group is exposed to four different effective field strengths, giving rise to four unique absorptions in a 1:3:3:1 ratio, a *quartet.*

***Spin–Spin Decoupling.*** The splitting by neighboring hydrogen nuclei can be selectively removed by a process called **spin–spin decoupling.** Application of a large radio-frequency signal that equals the resonance frequency of the methylene group induces these protons to flip rapidly from one spin state to the other. This process occurs so rapidly that the protons on the methyl group then experience an average of all three possible orientations of the methylene protons and resonate at a single frequency. Thus, the 1:2:1 triplet in the coupled spectrum at the right in Figure 4.16 is changed to a singlet in the decoupled spectrum.

Decoupling is normally done when obtaining $^{13}C$ NMR spectra so that the carbon signals are not split by attached and adjacent hydrogen atoms. This spectral simplification gives rise to single-line $^{13}C$ NMR spectra. In this routine technique, the entire proton region is irradiated while the $^{13}C$ NMR spectrum is recorded.

***Nonequivalent Nuclei.*** There are four possible ways to arrange the spins of the two hydrogen atoms of the methylene groups of ethanol and ethyl bromide. We see only three signals in the triplet for the adjacent methyl group, because two of these permutations are identical in energy.

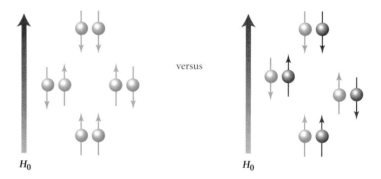

versus

$H_0$       $H_0$

**Styrene**

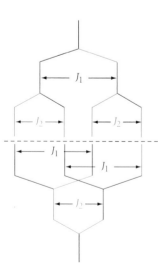

However, if the two hydrogen atoms causing such splitting are *not* identical, all four permutations are unique, and a four-line pattern is observed in the spectrum. (If the protons are different, it matters which one is aligned with and which against the applied magnetic field.) For example, the three vinyl hydrogen atoms of styrene (vinylbenzene) are all different. The values of the coupling constants between $H_a$ and each of the two other hydrogen atoms, $H_b$ and $H_c$, are quite different. The so-called *trans* coupling between $H_a$ and $H_c$ has a value of $J = 18$ Hz, whereas the *cis* coupling between $H_a$ and $H_b$ is only $J = 11$ Hz. Further, because $H_b$ and $H_c$ are not the same, they also couple with each other, with $J = 1$ Hz.

We can predict the coupling patterns to be expected when two nonequivalent adjacent hydrogen atoms are present by using a modification of Pascal's triangle. We start with a single, vertical line and branch down to the left and right, separating the two vertical lines by a distance proportional to one of the coupling constants. We then repeat the process, branching down from each of the two previous lines, separating the four new vertical lines by a distance proportional to the second coupling constant. It doesn't matter which coupling constant we use first and which second— exactly the same pattern results. The four signals will be of approximately the same intensity—that is, in a ratio of 1:1:1:1. This pattern is called a doublet of doublets (or simply dd) to distinguish it from the 1:3:3:1 quartet that arises from the presence of three adjacent and identical hydrogen atoms.

## EXERCISE 4.8

Use the graphical method described above for predicting coupling patterns to confirm that the appearance of each of the vinyl hydrogen atoms in styrene is as shown here.

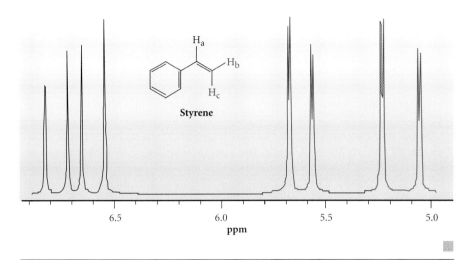

***Effect of Symmetry.*** The presence of symmetry in a molecule results in nuclei (hydrogens and carbons) that give rise to signals of the same frequency. Thus, the number of unique resonance signals for a symmetrical molecule is less than the number of individual atoms. For example, consider the following spectral data for two isomeric alcohols (d = double, t = triplet, q = quartet):

|  |  |
|---|---|
| **3-Pentanol** | **2-Pentanol** |

| $^{1}$H: H-1, H-5, t (6 H) | $^{13}$C: 73.8 | $^{1}$H: H-1 d (3 H) | $^{13}$C: 67.0 |
|---|---|---|---|
| H-2, H-4 d of q (4 H) | 29.7 | H-2 t of q (1 H) | 41.6 |
| H-3 quintet (1 H) | 9.8 | H-3 d of t (2 H) | 23.3 |
|  |  | H-4 t of q (2 H) | 19.1 |
|  |  | H-5 t (3 H) | 14.0 |

Only three signals are observed in the $^{13}$C NMR spectrum for the five carbons of 3-pentanol, whereas each of the five carbons in 2-pentanol gives rise to its own unique signal. Note that the symmetry in 3-pentanol makes the two methyl groups, at C-1 and C-5, as well as the two methylene groups, at C-2 and C-4, equivalent. As a result of this symmetry, the hydrogens and the carbons at C-1 and C-5 in 3-pentanol are identical and give rise to one signal in the $^{1}$H and $^{13}$C NMR spectra, respectively. The same can be said for the hydrogens and carbons at C-2 and C-4. However, unlike the methylene groups of ethanol and ethyl chloride, the C-2 (and C-4) hydrogens are split not only by the hydrogens on C-1 (and C-5), but also by the single hydrogen at C-3. The coupling constants of the C-2 hydrogens with these two different sets of adjacent hydrogens are not the same, resulting in a pattern that is a doublet (coupling with the C-3 hydrogen) of quartets (coupling with the three C-1 hydrogens). The signal for the C-4 hydrogens is at the same chemical shift as that for the C-2 hydrogens and is similarly split into a doublet (coupling with the C-3 hydrogen) of quartets (coupling with the three C-5 hydrogens). The signal for the hydrogen on C-3 appears as a quintet, being coupled with the four equivalent hydrogens on C-2 and C-4, and is downfield from the signals for the other hydrogen atoms because of the effect of the oxygen on C-3.

An NMR spectrum must be interpreted with care when there are fewer than the number of expected signals. In this situation, symmetry may be present, causing two or more nuclei to be in identical environments. On the other hand, it is always possible that the chemical shift difference between two carbon atoms (or two protons) is so small that they appear as a single signal. Such overlap happens more frequently in $^1H$ than in $^{13}C$ NMR spectra, where it rarely occurs.

See if you can explain the $^{13}C$ and $^1H$ signals and splitting patterns for 2-pentanol shown above. Clearly, the chemical shifts of hydrogens on C-1, C-3, C-4, and C-5, though not identical, do not differ dramatically. Instead of the simple pattern for 3-pentanol, a broad signal, a multiplet, is observed. Even though the proton spectrum of 2-pentanol is hard to interpret, its very complexity makes it distinguishable from its symmetrical isomer 3-pentanol. The relative numbers of signals in the $^{13}C$ NMR spectra make this assignment of isomers completely unambiguous.

### EXERCISE 4.9

Each of the $^1H$ spectra shown in parts (a) through (d) corresponds to one of the isomers accompanying it. Choose between the alternative compounds, and give reasons for your assignment.

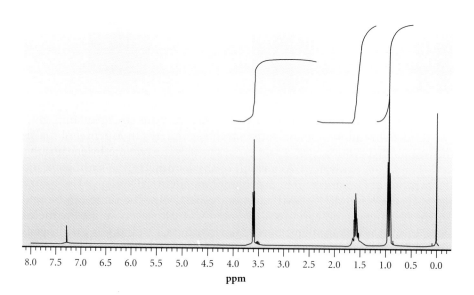

(b)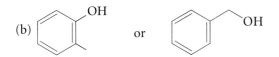

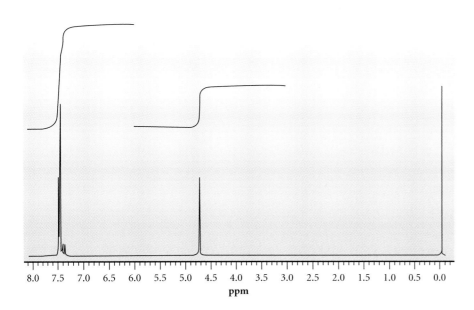

(c)

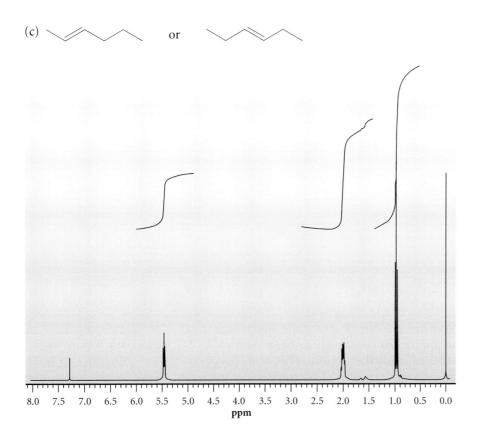

(d)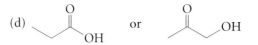

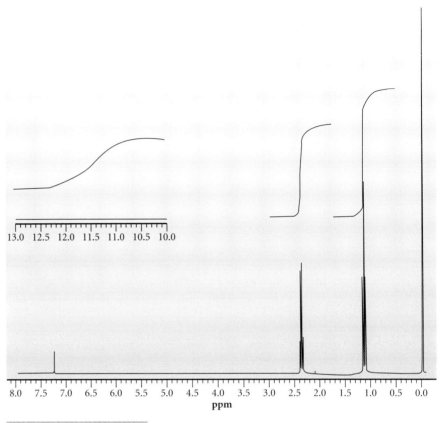

---

### EXERCISE 4.10

---

Predict the approximate $^1H$ NMR spectrum (number of signals, approximate chemical shift, multiplicity, and integration) for each of the following compounds.

(a) 1-butanol    (b) 1-butanal    (c) 2-butanol    (d) 2-butanone

---

In conclusion, $^1H$ NMR spectra can be interpreted on the basis of the number of signals, the chemical shifts, the splitting patterns, and integration. These spectral features provide valuable structural information about the nature of attached atoms (from chemical shifts) and the number of neighboring hydrogens (from splitting and integration). $^1H$ NMR spectroscopy is thus a very useful complement to $^{13}C$ NMR spectroscopy for assigning structure.

***The NMR Spectrometer.*** The basic components of an NMR spectrometer are shown schematically in Figure 4.17. The sample, dissolved in an appropriate solvent (most commonly $CDCl_3$), is placed in a thin-walled tube, which is then inserted between the poles of a superconducting magnet and spun rapidly. This spinning ensures that all nuclei experience the same applied magnetic field. Small variations in the magnetic field at different places between the poles of the magnet are averaged as each nucleus moves through all possible environments. A pulse of high-intensity, broad-

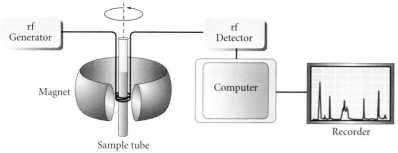

FIGURE 4.17

Schematic representation of the key features of a nuclear magnetic resonance spectrometer (rf is an abbreviation of radio-frequency).

spectrum radiation is applied through a coil that surrounds the tube. This pulse contains an even distribution of all frequencies to be observed, and nuclei are uniformly stimulated to jump from a lower- to a higher-energy state relative to the applied field. The pulse lasts for only a short time (~0.1 sec), and after it ends, a radio receiver is used to detect the emission of electromagnetic radiation as nuclei return to the lower-energy state. The process is repeated many times, and the result of each pulse is added to the sum of results from previous pulses, thus amplifying the very weak signals emitted by the nuclei.

When all of the nuclei being observed are identical—such as, for example, the six protons in benzene—the signal received has a single frequency that decreases exponentially as the normal distribution between the lower- and higher-energy states is re-established (Figure 4.18). This type of signal is referred to as a **free induction decay (FID) signal.**

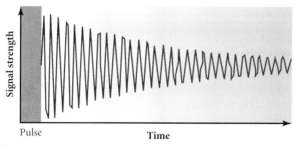

FIGURE 4.18

Free induction decay (FID) signal resulting from nuclei emitting a single frequency.

When nuclei have different chemical environments, they radiate different frequencies, and the result is a complex signal with many component contributions. Free induction decay signals resulting from two and three different frequencies are shown in Figure 4.19 (on page 186). These composite signals can be likened to the sound of a symphony orchestra (or a rock band): each instrument is producing a unique sound, yet a total, composite sound arrives at our ears, which our brains decode so that we recognize the presence of violins, pianos, drums (and guitars). Baron Jean

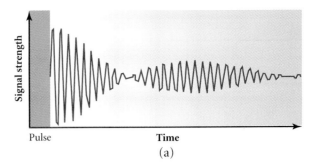

(a)

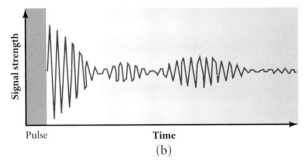

(b)

**FIGURE 4.19**

Free induction decay (FID) signals resulting from (a) two and (b) three component frequencies that differ by 5%.

Baptiste Joseph Fourier, a French mathematician and physicist who lived 200 years ago, devised a mathematical method, known as a *Fourier transform,* for the deconvolution of complex signals into their component parts. With the advent of high-speed computers, Fourier transformation of NMR signals became practical. The essential features of today's NMR spectrometers are a high-field, superconducting magnet and a high-speed computer. The Fourier transform of a free induction decay signal is plotted with signal intensity as the *y*-axis and time as the *x*-axis.

***Effect of Field Strength.*** The field strength of the magnet used in an NMR spectrometer significantly affects the quality and appearance of the spectra obtained. The possibility of accidental overlap of peaks in $^1$H spectroscopy is reduced by using an instrument with high field strength, because the frequency at which nuclei resonate increases with the field strength. Recall that the chemical shift ($\delta$) of a nucleus does not change with field strength. Thus, at 100 MHz, a difference in chemical shift between two nuclei separated by 1 ppm (1 $\delta$) is 100 Hz, whereas at 500 MHz, it is 500 Hz. As the field strength is increased, the frequency *difference* also increases. Thus, at higher field (and higher frequency), signals are more separated and therefore more readily distinguished. For example, two $^1$H spectra of linalool, at 90 and 360 MHz, respectively, are shown in Figures 4.20 and 4.21. In the spectrum obtained at 90 MHz, the signal at about 5 ppm is quite broad and difficult to interpret, because the individual components are quite close in frequency. In contrast, the spectrum obtained at 360 MHz exhibits sharp peaks, and although the patterns are complex, they can be readily interpreted.

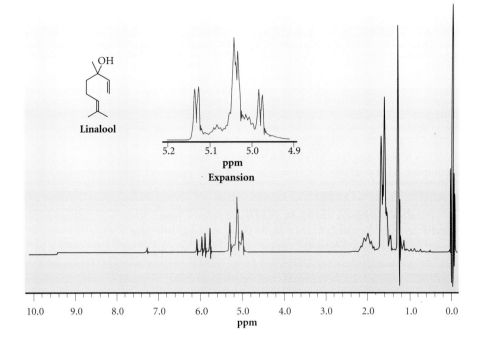

**FIGURE 4.20**

A 90-MHz $^1$H NMR spectrum of linalool. Expansion of the signal in the range from 4.9 to 5.3 ppm shows the significant peak overlap.

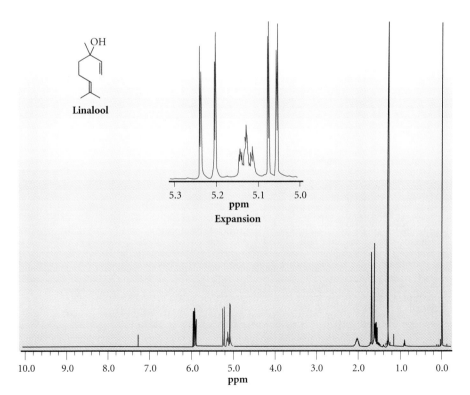

**FIGURE 4.21**

A 360-MHz $^1$H NMR spectrum of linalool. Expansion of the signal between $\delta$ 5.0 and 5.3 shows that this region can be resolved into two pairs of doublets and a triplet.

An added bonus of high-field spectrometers is enhanced sensitivity. With 300–500 MHz spectrometers, it is possible to obtain a reasonable $^{13}C$ spectrum with milligram quantities of sample. Because the natural abundance of $^1H$ is much larger than $^{13}C$, amounts under a milligram are sufficient for a proton spectrum.

### EXERCISE 4.11

Note that in the two free induction decay patterns in Figure 4.19, but not in that in Figure 4.18, there are nodes, or points along the time axis at which the signal strength is zero. Explain why nodes appear when two or more signals of different frequency are mixed.

*Medical Applications of NMR Spectroscopy.* NMR spectroscopy is used not only for identifying pure compounds, but also for detecting differences in relative abundances of magnetically active nuclei in solid samples and water-filled tissues. Very large NMR spectrometers are used for medical and biological research applications; plant or animal matter, or even a whole human body, can be inserted in the magnet and analyzed. Such a spectrometer is shown in Figure 4.22. This type of spectrometer produces a proton spectrum, from which a three-dimensional map of water concentration in an organ or other object of interest is made. Deviations from normal water concentrations and distributions indicate medical anomalies such as tumors. Representative three-dimensional images of human brains are shown in Figure 4.23.

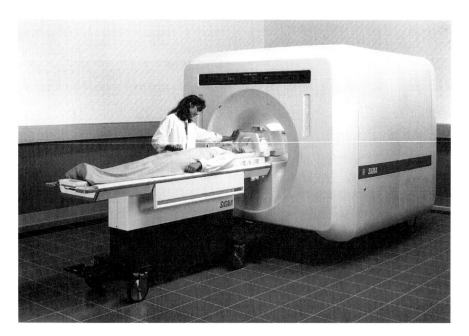

**FIGURE 4.22**

NMR spectrometer used for three-dimensional imaging of human bodies.

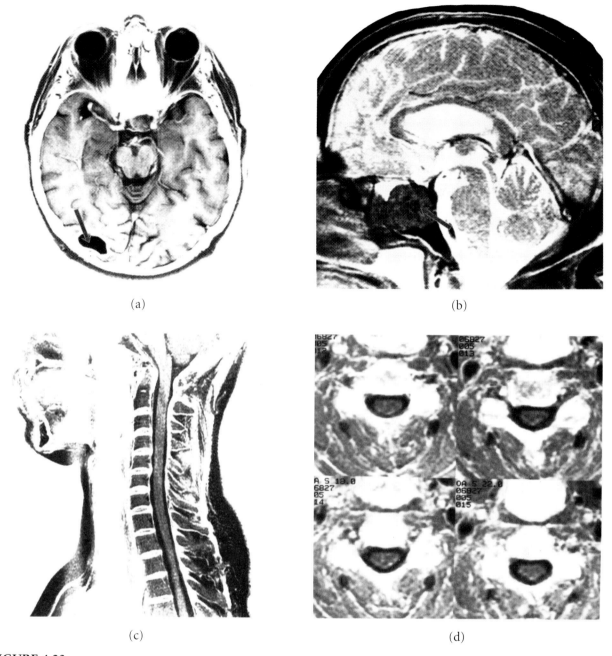

(a)

(b)

(c)

(d)

**FIGURE 4.23**

Three-dimensional NMR images (magnetic resonance imaging, MRI) of various parts of the human body: (a) a horizontal layer of a brain, clearly showing such features as the eyes, as well as a blood clot (arrow); (b) a vertical layer from the head of a patient with an enlarged pituitary gland (arrow); (c) a vertical layer of a spine; and (d) four horizontal slices. The images shown in parts (b) and (c) are similar to what would be shown in x-rays of the same areas of the body. On the other hand, the images in parts (a) and (d) are unique and allow a physician to examine internal body structures that cannot be visualized by x-ray techniques.

## CHEMICAL PERSPECTIVES

### NMR SPECTROSCOPY OF LIVING ORGANISMS

It is possible to "watch" organic compounds being digested by living creatures. Whole-body NMR spectrometers can track the concentrations of compounds that vary under different physiological conditions. For example, the conversion of phosphocreatine to creatine and phosphate ($P_i$) releases substantial energy.

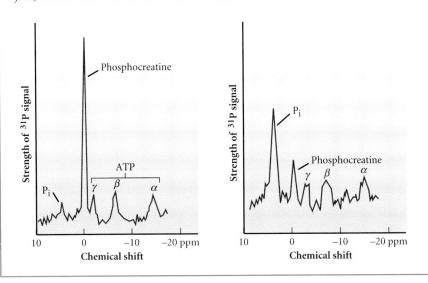

This conversion takes place in muscle tissue during exercise, depleting phosphocreatine and increasing the concentration of phosphate, as can be seen in the two phosphorus NMR spectra, taken on the forearm of a human subject, before and after 19 minutes of exercise.

An NMR spectrometer can also be used to search for weak points or irregularities in manmade objects. Figure 4.24 shows a picture of polystyrene tubing taken with an optical camera and a three-dimensional NMR image of a cross section of the tubing. The use of NMR is called noninvasive imaging by medical personnel and nondestructive testing by material scientists. The technique is considerably more sensitive (and at present more expensive) than x-ray imaging and does not damage tissue. Physicians usually use the expression *magnetic resonance imaging (MRI)* to differentiate this technique from x-ray imaging. (The word *nuclear* is omitted to avoid alarming those people who might otherwise connect the term—incorrectly—with nuclear fusion and fission.)

(a)                                    (b)

**FIGURE 4.24**

(a) Optical photograph and (b) three-dimensional NMR cross-sectional image of a piece of polystyrene tubing.

## Infrared (IR) Spectroscopy

The infrared (IR) region of the electromagnetic spectrum is just beyond the region that the human eye perceives as red light. The absorption of infrared light causes increases in the frequencies at which the bonds between atoms stretch and bend. The frequency at which a bond vibrates and bends is determined primarily by the mass of the atoms involved and the strength of the bond. The bonds that characterize functional groups have unique frequencies at which they absorb and characteristic **absorption bands** in the infrared region of the spectrum. Thus, absorption in the infrared region can be used to identify the types of functional groups present in a molecule.

*Theoretical Background.* The key principle of **infrared spectroscopy** is that infrared radiation is absorbed when there is a match between the radiant energy and the frequency of a specific molecular motion, usually bond bending or stretching. Atoms are not static within a molecule—they are constantly moving relative to each other, vibrating about the connecting bonds at constant frequencies. The bond lengths given in Table 2.2 represent the minimum-energy distance for the atoms involved. As they move closer to each other than that minimum-energy distance, repulsive forces increase, and as they move farther apart, attractive interactions decrease. This motion of alternately stretching and compressing resembles that of two spheres held together by a spring, as represented schematically for the O—H bond of $CH_3OH$. When atoms of unequal atomic mass are bonded, the lighter atom (in this case, hydrogen) moves farther than the heavier one. Absorption of infrared energy results in an increase in the frequency of vibration.

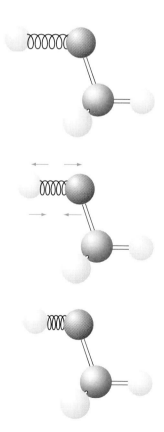

**Schematic representation of O—H stretching**

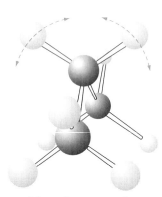

Schematic representation
of bending of C—H
bonds in propane

Atoms also display a motion that results in a constantly changing bond angle. This bond bending changes the relative positions of two atoms attached to a third. For example, in the methylene group of propane, the two hydrogen atoms are constantly moving closer together and then farther apart at a characteristic frequency. Again, absorption of infrared light of a characteristic frequency results in an increase in the frequency of this motion.

The representation of a bond between two atoms as a spring connecting two masses does not fully portray the actions of atoms in molecules. Classical Newtonian physics can be used to describe weights attached by a spring but fails when the scale is as tiny as atoms. Masses held together by a spring can vibrate at any speed and amplitude, whereas the vibrations of molecules are quantized. The atoms of a molecule can vibrate only at specific frequencies, known as **vibrational states.** Infrared light is absorbed by a molecule only when the energy of the photons is quite close to the energy gap between a vibrational state and the next higher one (Figure 4.25). The vast majority of molecules exist in the lowest-energy state, and the absorption of light that gives rise to an infrared spectrum is the result of raising molecules to the next higher state. Much less frequently, a photon of approximately twice the energy will be absorbed, promoting the molecule to a third energy state. The absorptions of these higher frequencies give rise to signals known as **overtones.** They are quite weak, but are often visible in the spectra of carbonyl compounds.

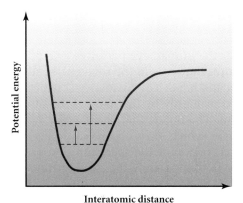

**FIGURE 4.25**

Quantized vibrational energy states of bonded atoms.

The absorption of infrared light by a molecule requires a dipole moment in the bond that will be stimulated to vibrate at a higher frequency. The intensity of the absorption of light is directly related to the magnitude of the dipole moment—the stronger the dipole moment, the stronger the absorption. Thus, an O—H bond absorbs more strongly than a C—H bond. Conversely, symmetrical bonds, such as the C=C bond in *trans*-2-butene, do not absorb at all.

Infrared spectrometers record spectra using electromagnetic radiation of wavelengths ranging from 2,000 to 15,000 nm. It is common practice to use the wavenumber scale ($\bar{\nu}$, 5000–700 cm$^{-1}$). The cm$^{-1}$ scale is a designation of frequency—that is, of how many waves will fit in 1 cm.

$$\bar{\nu} = \frac{1}{\lambda}$$

The characteristic absorption bands for almost all organic functional groups are found in the range 4000–800 cm$^{-1}$. Spectra are plotted with wavenumber as the x-axis (decreasing wavenumbers to the right) and either absorbance or percent transmittance as the y-axis.

*The Infrared Spectrometer.* A schematic representation of an infrared spectrometer is shown in Figure 4.26. Infrared radiation is emitted from a source (at the right), which consists of a heated ceramic rod. The infrared radiation from the source is split into two beams by mirrors. One beam passes through a cell containing a solution of the sample (typically in $CH_2Cl_2$), and the other beam goes through a cell containing only the solvent. Both beams are then directed to a chopper, a device that alternately passes one beam and then the other. The beam is then directed to a diffraction grating, where it is split into its component wavelengths. The grating is rotated, directing small samples of the spectrum through a narrow slit and onto a detector. The detector, a tiny coil of wire, is heated by the impinging radiation, increasing its resistance. Thus, the resistance of the detector varies with the intensity of the radiation that hits it. The action of the chopper results in alternation of the beam from the sample and from the reference cell falling on the detector. Electronic circuitry is used to compare these signals. The absorption by the solvent is the same in both cells, so the effect of the solvent can be subtracted, and the recorder receives only signals due to absorption by the sample.

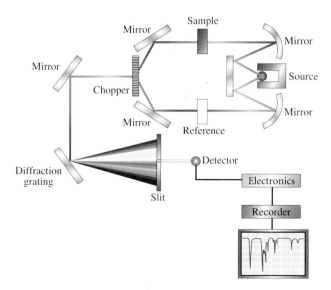

**FIGURE 4.26**

Schematic representation of an infrared spectrometer.

## TABLE 4.3

Typical Infrared (IR) Absorption Bands for Specific Functional Groups

| Functional Group | Band (cm$^{-1}$) | Intensity | Functional Group | Band (cm$^{-1}$) | Intensity |
|---|---|---|---|---|---|
| C—H | 2960–2850 | Medium | RO—H | 3650–3400 | Strong, broad |
| C=C—H | 3100–3020 | Medium | —C—O— | 1150–1050 | Strong |
| C=C | 1680–1620 | Medium | C=O | 1780–1640 | Strong |
| C≡C—H | 3350–3300 | Strong | R$_2$N—H | 3500–3300 | Medium, broad |
| R—C≡C—R′ | 2260–2100 | Medium (R ≠ R′) | —C—N— | 1230, 1030 | Medium |
| Ar—H | 3030–3000 | Medium | —C≡N | 2260–2210 | Medium |
| ⬡ | 1600, 1500 | Strong | RNO$_2$ | 1540 | Strong |

***Characteristic Absorptions of Functional Groups.*** The great utility of IR spectroscopy in assigning structures to organic molecules comes from the fact that each functional group exhibits a characteristic set of infrared absorptions. Modern IR spectrometers are sufficiently sensitive to be able to characterize a single layer of molecules on a surface, providing an invaluable tool for characterizing thin films of organic materials. Table 4.3 lists characteristic IR absorption bands for frequently encountered organic functional groups.

*Carbonyl groups* (aldehydes, ketones, esters, etc.) have intense absorptions in the region 1780–1640 cm$^{-1}$. Figures 4.27 through 4.31 show representative infrared spectra of an aldehyde, a ketone, a carboxylic acid, an ester, and an amide.

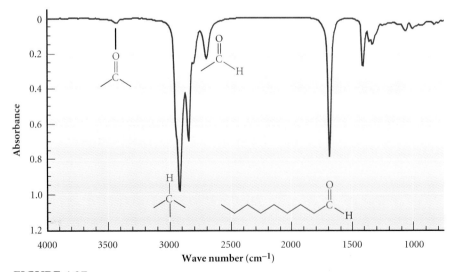

**FIGURE 4.27**

Infrared spectrum of nonaldehyde.

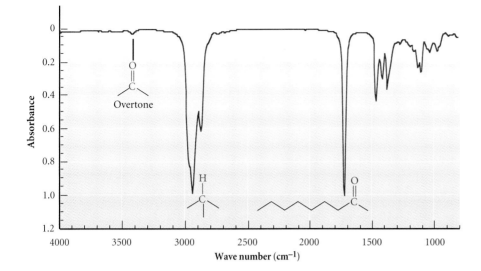

**FIGURE 4.28**

Infrared spectrum of
2-nonanone.

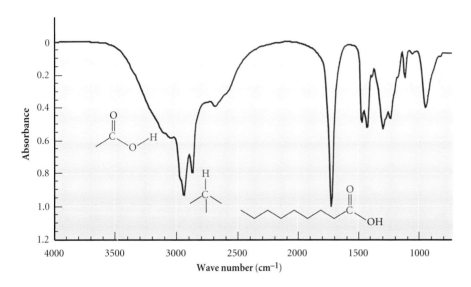

**FIGURE 4.29**

Infrared spectrum of
nonanoic acid.

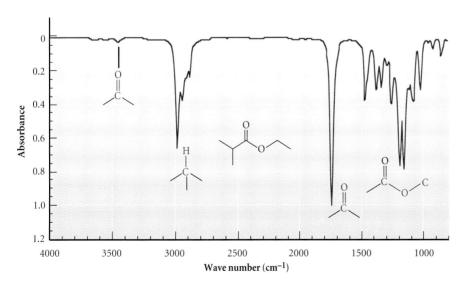

**FIGURE 4.30**

Infrared spectrum of ethyl
isobutyrate.

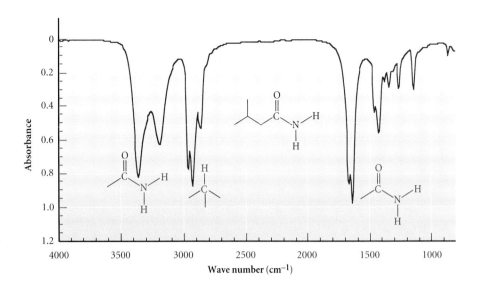

**FIGURE 4.31**

Infrared spectrum of
butyramide.

The infrared spectra of aldehydes, ketones, carboxylic acids, esters, and amides all have in common the strong absorption of the C=O bond. Note that in most cases the overtone at approximately twice the frequency of the primary absorption is visible, though small. In some cases, the frequency differences between the characteristic absorptions of functional groups are sufficiently large to be able to determine which group is present in a sample. In addition, each of these carbonyl functional groups (except a ketone) has additional absorptions that, when combined with the presence of the C=O band in the region 1780–1640 cm$^{-1}$, are quite characteristic of that group. These include the unique carbonyl C—H stretching absorption of aldehydes at 2800–2700 cm$^{-1}$, the O—H stretching of carboxylic acids (3500–3000 cm$^{-1}$), the C—O—C stretching of esters (1300–1100 cm$^{-1}$), and the N—H stretching of amides (3400–3100 cm$^{-1}$). Thus, a ketone can be characterized by the presence of the C=O band and the *absence* of the additional bands characteristic of the other carbonyl functional groups.

All of the spectra in Figures 4.27 through 4.31 have strong absorptions resulting from C—H stretching vibrational transitions. In comparing the intensity of these absorptions with those of the carbonyl groups, remember that there are many C—H bonds and only one carbonyl group in each of these examples.

### EXERCISE 4.12

An unknown compound with the molecular formula $C_3H_6O$ has strong infrared absorption at 1725 cm$^{-1}$. Draw the structures of all possible isomers with this formula. Which of these isomers are consistent with the data? How would you use $^1H$ NMR spectroscopy to decide among the various possibilities? How would you use $^{13}C$ NMR spectroscopy?

Absorptions of *X—H bonds,* bonds between hydrogen atoms and heteroatoms, dominate the high-wavenumber region of the infrared spectrum, in the order O—H, N—H, C—H, progressing from higher to lower wavenumber (left to right). Because virtually all organic compounds con-

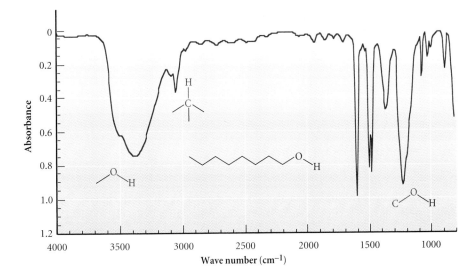

**FIGURE 4.32**

Infrared spectrum of
1-octanol.

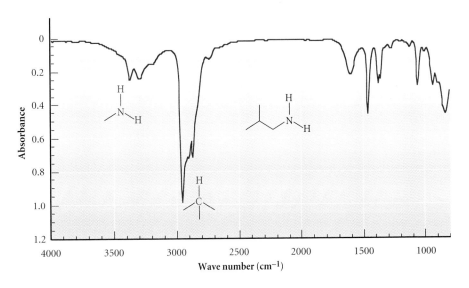

**FIGURE 4.33**

Infrared spectrum of 2-
methyl-1-aminopropane
(isobutylamine).

tain many C—H bonds, the infrared absorptions arising from their presence are not particularly useful in unraveling the structure of an unknown
compound. On the other hand, O—H and N—H absorptions are quite diagnostic for the presence of alcohols and amines and, as we have seen, for
carboxylic acids and amides (except tertiary amides, which lack N—H
bonds). The IR spectra of 1-octanol and 2-methyl-1-aminopropane are
shown in Figures 4.32 and 4.33, respectively.

The widths and intensities of the absorptions resulting from the presence of O—H and N—H bonds are quite sensitive to the structure of the
compound as well as to the conditions under which the spectrum is obtained. Although both of these functional groups can participate in intermolecular hydrogen bonding, these interactions are most significant in
primary alcohols and amines and less significant in secondary and especially tertiary alcohols and amines. Hydrogen bonding also decreases as the
concentration of the alcohol or amine in the solvent is decreased. The absorption bands for primary alcohols and amines are quite broad at high

concentrations, because many different species are present at equilibrium: dimers, trimers, tetramers, and so on, and each species absorbs at a different frequency.

**EXERCISE 4.13**

The O—H stretching absorption of an alcohol in solution in $CH_2Cl_2$ changes with concentration, appearing quite broad at high alcohol concentration and becoming sharper as the concentration decreases. Explain this observation.

Absorptions of *alkenes* in IR spectra resulting from stretching of C=C bonds are relatively weak (as shown for 1-hexene in Figure 4.34), because this functional group generally has no significant dipole moment. Indeed, no absorption is observed for symmetrical alkenes (such as *trans*-4-octene, Figure 4.35). Regardless of symmetry, this functional group is quite evident in both $^{13}C$ and $^1H$ NMR spectra (but in the latter only if vinyl protons, bonded to one of the doubly bonded carbons, are present). Thus, NMR

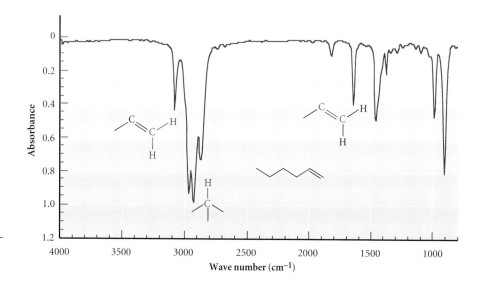

**FIGURE 4.34**

Infrared spectrum of
1-hexene.

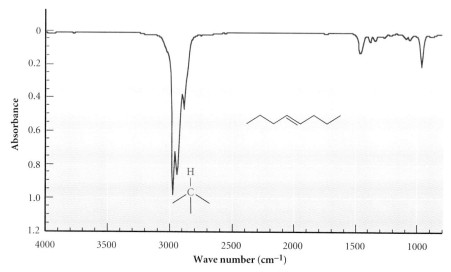

**FIGURE 4.35**

Infrared spectrum of
*trans*-4-octene.

spectroscopy represents a generally more reliable technique for establishing the presence of the C=C functional group.

## EXERCISE 4.14

The C=C stretch in the infrared spectrum of allyl alcohol is considerably stronger than is that of propene. Explain why this is reasonable.

**Allyl alcohol**          **Propene**

*Alkynes* and *nitriles* both have unique, characteristic absorptions in IR spectra. Because of the polarity of the nitrile group, its characteristic absorption around 2250 cm$^{-1}$ is quite strong, whereas the absorption for the carbon–carbon triple bond at around 2230 cm$^{-1}$ is much weaker, and is absent in symmetrical molecules. Representative spectra of an alkyne (1-pentyne) and a nitrile (isobutyronitrile) are shown in Figures 4.36 and 4.37.

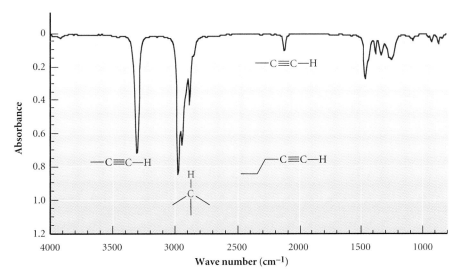

**FIGURE 4.36**

Infrared spectrum of 1-pentyne.

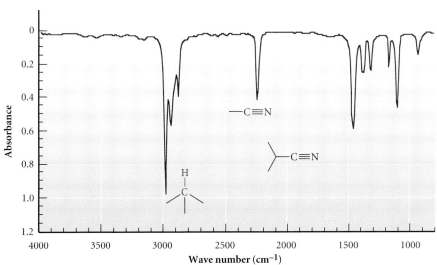

**FIGURE 4.37**

Infrared spectrum of isobutyronitrile.

The presence of a terminal alkyne (as in 1-pentyne) is characterized by a strong absorption for the C≡C—H stretch. This C—H bond is stronger than those to an $sp^3$-hybridized carbon atom and therefore has a higher frequency of infrared absorption.

*Aromatic compounds,* both benzene derivatives and other aromatic systems, have characteristic absorptions in the infrared region due to the presence of the cyclic $\pi$ system. Because a bond between a hydrogen atom and an $sp$-hybridized carbon atom is stronger than that with an $sp^3$-hybridized carbon atom, the former absorbs at somewhat higher frequency (recall the C≡C—H absorption in the spectrum of 1-pentyne, Figure 4.36). The stretching absorptions for aromatic C—H bonds just above 3000 cm$^{-1}$ are diagnostic of the presence of an aromatic ring, and can be seen in the IR spectrum of toluene (Figure 4.38). The four absorptions between 2000 and

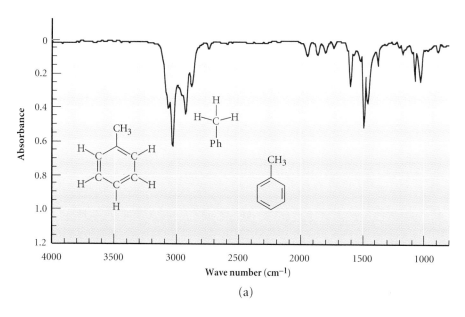

(a)

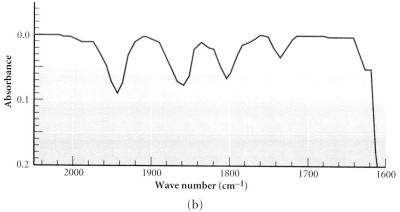

(b)

**FIGURE 4.38**

(a) Infrared spectrum of toluene. (b) Expansion of the region between 2000 and 1600 cm$^{-1}$ in the IR spectrum of toluene.

1600 cm$^{-1}$ are overtones arising from absorptions in the 1000–800 cm$^{-1}$ region of the spectrum. This pattern changes with the degree of substitution and with the relative orientation of the substituents on the aromatic ring. This region of the spectrum of toluene is enlarged in part (b) of the figure; it is characteristically seen in spectra of monosubstituted benzenes. The assignment of the substitution pattern for an unknown aromatic compound is usually done using NMR spectroscopy.

## EXERCISE 4.15

Indicate which features of the $^1$H and $^{13}$C NMR spectra would permit an unknown compound to be identified as one of the following.

Most bands characteristic of functional groups appear at frequencies higher than 1200 cm$^{-1}$. The frequencies of IR bands for functional groups are reasonably characteristic and are rarely found to vary from compound to compound; however, the intensity of the absorption and the width of the band do vary. Furthermore, interaction of functional groups can lead to changes in frequency and intensity of absorption. For example, simple alkenes have weak absorptions (if any at all) for the C=C bond because of the weak dipole, but the intensity of the absorption for this functional group is significantly increased in an $\alpha$, $\beta$-unsaturated carbonyl compound because the C=O group polarizes the C=C bond.

In the region from about 1200 to 700 cm$^{-1}$, complex bands characteristic of a specific molecule (rather than a functional group) are usually observed. This region is called the **fingerprint region.** A comparison of the IR spectrum of an unknown compound with a library of spectra of known compounds can often enable the unambiguous identification of both the functional group(s) present and the specific structure. In essence, there are so many possible patterns of absorption bands in the fingerprint region that it is highly improbable that the spectra of two different compounds would be the same in all details (including intensity).

You should learn to recognize the characteristic infrared absorptions for a small number of functional groups. For example, a C=O absorption of a carbonyl compound appears as a strong band between 1780 and 1640 cm$^{-1}$, and O—H and N—H stretches appear as bands in the range from 3600 to 3200 cm$^{-1}$. By using the information in Table 4.3, you can identify the presence of many of the common functional groups and eliminate from consideration those that are not present. For example, let's use infrared spectroscopy to distinguish 2-octanone from 2-octanol and, thus, determine whether the oxidation reaction discussed in Section 4.1 proceeded as expected. The starting material, 2-octanol, will show absorption characteristic of the OH functional group at about 3600 cm$^{-1}$. This absorption is absent in the spectrum of the product, 2-octanone, which will have a strong absorption in the region of 1700 cm$^{-1}$ resulting from stretching of the carbonyl group.

4.3 Spectroscopy

CHARACTERIZING MOLECULES IN INTERSTELLAR SPACE BY
INFRARED SPECTROSCOPY

The use of infrared spectroscopy is not limited to the chemistry laboratory; it has also been used in heat sensors and for remote sensing when the observer is at a distant location from the sample being analyzed. For example, infrared spectrometers aboard both *Voyager I* and *Voyager II* (unmanned spacecraft that have been exploring the solar system for more than 10 years) detected six simple hydrocarbons (ethyne, ethene, ethane, propane, propyne, and butadiyne) and three carbon-containing nitriles (HCN, $N{\equiv}C{-}C{\equiv}N$, and $HC{\equiv}C{-}C{\equiv}N$) in the atmosphere of Titan, a large moon of Saturn. Infrared spectrometry is sufficiently sensitive that $C_2N_2$ was detected at the parts-per-billion (ppb) level. It is interesting to speculate why these specific compounds are produced in an atmosphere whose major constituents are $N_2$ and $CH_4$.

The photo shows the design for a new IR telescope at the McDonald Observatory of the University of Texas in Austin. Using a large mirror constructed from many small ones, the telescope will be dedicated to spectroscopic analysis of deep space.

## EXERCISE 4.16

Which region of the IR spectrum might be used to distinguish between each pair of compounds?

(a) [structure with O] and [structure with OCH$_3$]

(b) [structure with OH] and [structure with O]

(c) $CH_3CH_2NH_2$ and $CH_3C{\equiv}N$

(d) [benzene ring] and [cyclohexane ring]

## Visible and Ultraviolet (UV) Spectroscopy

Radiation in the visible and ultraviolet regions of the spectrum has sufficient energy to promote electrons from lower- to higher-energy orbitals, especially in compounds with $\pi$ bonds. The energy separation of these orbitals is determined by the number of double bonds in conjugation in the $\pi$ system and by the nature of the substituents. Therefore, the frequency of radiation absorbed can be correlated with the structure of the $\pi$ system of an unsaturated compound.

***Theoretical Background.*** Proceeding from the infrared to the visible and ultraviolet regions of the spectrum increases the amount of energy of the photons. The energy in the ultraviolet region is large enough to perturb the electronic structures of many organic molecules. Even in the lower-energy visible region, some organic molecules can be excited.

In Chapters 1 and 2, you learned that the electronic structure of molecules can be represented by electrons located in molecular orbitals. The bonding ($\pi$) and antibonding ($\pi^*$) molecular orbitals formed, for example, by the interaction of two $p$ atomic orbitals are equally split about the zero point of energy (the energy when there is no interaction). Two electrons located in the $\pi$ orbital confer net bonding on the molecule. When a photon of sufficient energy ($h\nu$) interacts with a molecule, it is absorbed, promoting one of the electrons from a bonding to an antibonding orbital (Figure 4.39). After this process, called *photoexcitation,* the bonding and antibonding orbitals are each singly occupied (Figure 4.40).

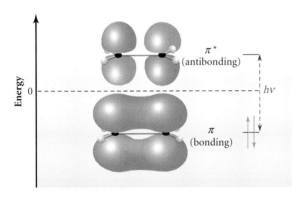

**FIGURE 4.39**

Photoexcitation in the visible or ultraviolet region consists of the absorption of a photon, which promotes an electron from a filled molecular orbital to a vacant one.

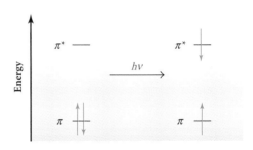

**FIGURE 4.40**

After photoexcitation, one electron is left in the bonding ($\pi$) molecular orbital; the second electron is located in the antibonding ($\pi^*$) orbital. The resulting state (shown at the right) is called a $\pi,\pi^*$ excited state.

Because an electron has moved from a $\pi$ to a $\pi^*$ molecular orbital, this change is called a $\pi,\pi^*$ (or $\pi \rightarrow \pi^*$) transition (read as "pi-to-pi-star"). The combined effect of one electron in a bonding molecular orbital and one electron in an antibonding molecular orbital is no net bonding interaction between the atomic $p$ orbitals. (In Section 5.1 we will see how this change in bonding forms the chemical basis for vision.) This electronic transition can occur only when the energy of the absorbed photon equals that required to raise a bonding electron to an antibonding orbital. A visible or ultraviolet spectrum is a plot of the intensity of absorption as a function of wavelength (corresponding to the excitation energy). Thus, in absorption spectroscopy, transitions between filled and vacant orbitals are measured as a function of wavelength. For ethylene, the $\pi, \pi^*$ transition requires high

MOLECULAR OXYGEN: A STABLE TRIPLET MOLECULE

The lowest-energy arrangement for molecular oxygen ($O_2$) is a triplet electronic state in which there is only one bond between the atoms and there are two unpaired electrons, one on each atom.

| Triplet | Singlet |
| oxygen | oxygen |

By interacting with other excited molecules (sensitizers), triplet oxygen ($^3O_2$) can be converted into singlet oxygen ($^1O_2$), a very unstable species that reacts rapidly with organic compounds. Virtually no organic molecules are stable for long in the presence of singlet oxygen. In fact, oxidation reactions involving this species have been implicated in aging and in other modes of cell damage. Long-chain polyenes having ten or more conjugated double bonds (many of which, like vitamin A, are found in nature) interact with singlet oxygen in a special way that results in the conversion of both species into the triplet state.

| Singlet alkene | Singlet oxygen | Triplet alkene | Triplet oxygen |

The triplet state of the polyene relaxes rapidly to the stable singlet state. Thus, polyenes serve as deactivators of $^1O_2$, and their presence protects other molecules from oxidation.

---

energy, and the absorption appears at about 171 nm (at the high-energy end of the ultraviolet spectrum). We can calculate the energy (in kcal/mole) of radiation of this wavelength using this formula:

$$E = \frac{2.86 \times 10^4 \text{ kcal} \cdot \text{nm} \cdot \text{mole}^{-1}}{\text{Wavelength (nm)}}$$

Thus, 200 nm corresponds to 143 kcal/mole, 300 nm to 95 kcal/mole, and 400 nm (the break between the ultraviolet and visible regions) to 72 kcal/mole.

***Molecular-Orbital Interpretation of UV Absorption.*** Excitation by light typically involves promotion of an electron from the **highest occupied molecular orbital (HOMO)** to the **lowest unoccupied molecular orbital (LUMO).** Let's consider the molecular-orbital picture for butadiene. Combination of four $p$ atomic orbitals yields four $\pi$ molecular orbitals: two $\pi$ bonding and two $\pi^*$ antibonding. The designations $\pi_1$, $\pi_2$, $\pi_3^*$, and $\pi_4^*$ indicate molecular orbitals of decreasing bonding (or increasing antibond-

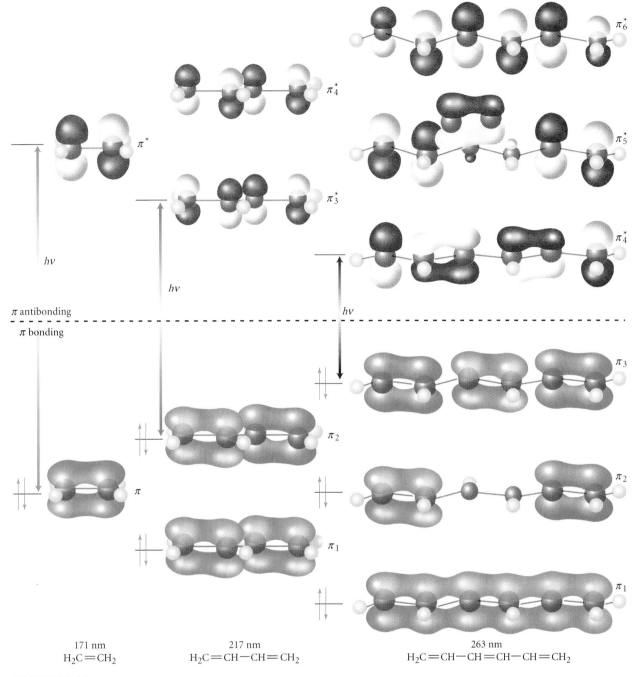

**FIGURE 4.41**

Molecular orbitals of ethene, butadiene, and hexatriene. The relative phasing of the lobes of the bonding orbitals is indicated by blue and green; that of the antibonding orbitals is shown by red and yellow.

ing) character. The two bonding orbitals ($\pi_1$ and $\pi_2$) are below and the two antibonding orbitals ($\pi_3^*$ and $\pi_4^*$) are above the zero of energy (Figure 4.41).

In the arrangement of lowest energy, bonding is continuous along a chain of carbon atoms ($\pi_1$). The next higher level, bonding orbital $\pi_2$, has

**205**

a node at the center of the chain, between C-2 and C-3. (Recall that, at a node, orbital phase inverts. Thus, orbital phasing, indicated by color in the figure, is reversed from one side of the node to the other.) In the antibonding orbital $\pi_3^*$, bonding is maintained only between C-2 and C-3, and there is no bonding between adjacent carbons in $\pi_4^*$.

In the ground state (the lowest-energy arrangement) of butadiene, the four $\pi$ electrons populate $\pi_1$ and $\pi_2$. Because the $\pi_2$ orbital of butadiene lies at a higher level of energy than the $\pi$ orbital of ethene, and the $\pi_3^*$ orbital of butadiene lies at a lower level than the $\pi^*$ orbital of ethene, the energy difference between the HOMO and the LUMO for butadiene is smaller than that for ethene. Thus, 1,3-butadiene absorbs light of longer wavelength and lower energy (217 nm) than does ethene (171 nm).

***Extended Conjugation.*** In general, the greater the extent of conjugation, the smaller is the energy difference between the HOMO and the LUMO, and the farther the absorption is shifted to longer wavelengths. Figure 4.41 also illustrates the bonding and antibonding molecular orbitals for 1,3,5-hexatriene. As for butadiene, the six molecular orbitals are arranged in order of increasing energy, which corresponds to an increasing number of nodes. The energy required for electronic excitation of hexatriene is even lower than for 1,3-butadiene, and the absorption for the triene is at 263 nm.

Conjugative effects are also important in aromatic rings. For example, benzene has an absorption maximum at 256 nm (not far from that for 1,3,5-hexatriene), whereas naphthalene absorbs at 286 nm. Some characteristic absorption maxima of other conjugated molecules are given in Table 4.4.

## TABLE 4.4

Representative Absorption Maxima for Typical Ultraviolet-absorbing Compounds

| Compound | $\lambda_{max}$ (nm) | Compound | $\lambda_{max}$ (nm) |
|---|---|---|---|
| $H_2C=CH_2$ | 171 | 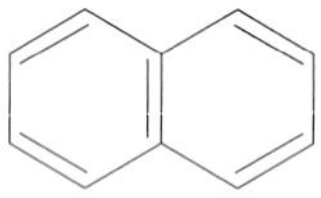 naphthalene | 286 |
|  | 182 |  |  |
|  | 217 | anthracene | 375 |
|  | 263 | acetone | (n, $\pi^*$) 279 ($\pi$, $\pi^*$) 188 |
| 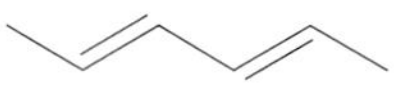  | 290 | 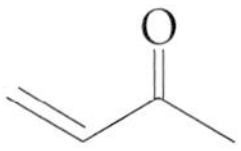 methyl vinyl ketone | (n, $\pi^*$) 315 ($\pi$, $\pi^*$) 210 |
|  | 256 | $-N=N-$ | ~ 350 |

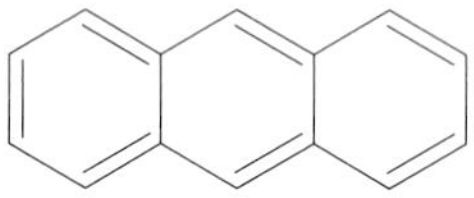

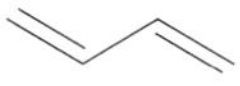

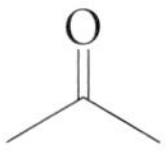

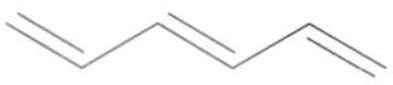

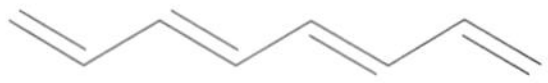

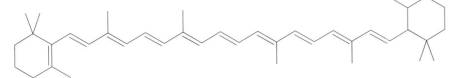

**β-Carotene**

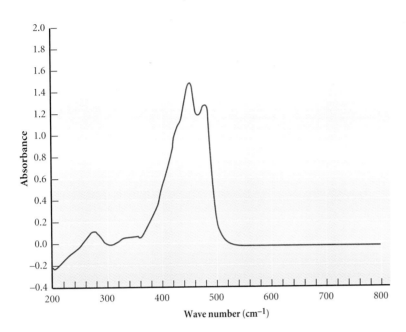

**FIGURE 4.42**

An absorption spectrum of β-carotene in the ultraviolet and visible regions.

Extended conjugation shifts the observed absorption maximum to longer and longer wavelengths. Ultimately, the absorption maximum shifts from the ultraviolet into the visible region. A compound that absorbs some wavelengths of visible light is perceived by the human eye as having color. For example, β-carotene has a long, conjugated hydrocarbon skeleton with many conjugated π bonds (Figure 4.42). Its absorption maximum is between 450 and 500 nm, which means that the energy difference between its HOMO and LUMO is small. This absorption maximum corresponds to blue light, and as a result of this absorption, β-carotene is perceived as having a bright yellow-orange color. This pigment stains your hands when you peel fresh carrots. Similar compounds, such as lycopene, are responsible for the color of other vegetables, such as tomatoes.

Lycopene (red)
(20 mg/kg of tomatoes)

## CHEMICAL PERSPECTIVES

### CONJUGATED OXYGEN-CONTAINING PLANT PIGMENTS

Anthocyanins are oxygen heterocycles that absorb strongly in the visible region of the spectrum. They occur naturally in plants and are responsible for purple, mauve, and blue colors. Although highly colored, these compounds have not been used commercially to any extent as dyes because they degrade (fade) upon prolonged exposure to ultraviolet radiation.

**Cyanidine**
(an anthocyanin)

### EXERCISE 4.17

Which compound in each of the following pairs exhibits an electronic transition to an antibonding orbital at the longer wavelength?

(a) or

(b) or

(c) or

(d) or

*Carbonyl Groups.* In addition to unsaturated hydrocarbons, other functional groups absorb ultraviolet light. For example, ketones and conjugated enones show weak absorption spectra in which the absorption maxima are shifted to longer wavelengths than would be expected for $C=C$ bonds. In addition to the promotion of an electron from the $\pi$ to the $\pi^*$ orbital (as we have seen for hydrocarbons), such compounds also show absorption resulting from the promotion of an electron from one of the non-bonded lone pairs of electrons on the oxygen of the carbonyl group to the $\pi^*$ orbital. For acetone, this $\mathbf{n,\pi^*}$ (or n → $\pi^*$) **transition** (read as "en-to-pi-star") results in a weak absorption band at about 279 nm.

## CHLOROPHYLL: A CONJUGATED NITROGEN-CONTAINING PIGMENT

Chlorophylls are intensely green pigments found in plants. These complex molecules, exemplified by chlorophyll a, are central to the conversion of light energy into chemical energy via the process known as *photosynthesis*. Upon absorption of a photon of light, an electron in the extended $\pi$ system of a molecule of chlorophyll is promoted to an antibonding molecular orbital, resulting in an excited state. In photosynthesis, an electron is transferred along a chain of molecules and ultimately effects the reduction of carbon dioxide. Reduction of carbon dioxide results in the incorporation of the carbon into carbohydrates, a process called *carbon fixation*.

Chlorophyll a

The position of the n,$\pi^*$ transition also has a molecular-orbital basis. The energy levels of the $\pi$ and $\pi^*$ orbitals of the carbonyl group are equidistant from zero energy (Figure 4.43). In contrast, the nonbonding electron pair on the carbonyl oxygen is located in an orbital that is only slightly below the zero of energy. The energy of the n,$\pi^*$ transition is therefore less than that of the $\pi,\pi^*$ transition. The absorption bands of n,$\pi^*$ transitions are usually at longer wavelengths (lower energies) than those of $\pi,\pi^*$ transitions and usually less intense. As we have seen earlier with hydrocarbons, further conjugation of the carbonyl group also results in a shift in the

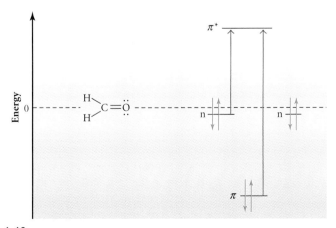

**FIGURE 4.43**

Energies of $\pi$ and $\pi^*$ molecular orbitals in formaldehyde.

absorption band to longer wavelengths. For 3-butene-2-one, the n,$\pi^*$
absorption band occurs at 315 nm and is accompanied by a stronger
$\pi,\pi^*$ band at 210 nm—that is, at a wavelength very similar to that seen for
butadiene.

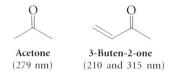

**Acetone**     **3-Buten-2-one**
(279 nm)    (210 and 315 nm)

## Mass Spectroscopy

Energies much higher than those required for electronic transitions can
cause the expulsion of an electron from a molecule. In a technique called
*mass spectroscopy (MS),* molecules are bombarded with high-energy electrons, resulting in the ejection of electrons from the molecule to form a
cation radical. These highly energetic species often break into smaller
cationic fragments.

$$RH \xrightarrow{e^{\ominus}} RH^{\oplus\cdot} + e^{\ominus}$$
$$\downarrow$$
$$Fragments^{\oplus\cdot}$$

The mass spectrometer measures the sizes and relative abundance of these
fragments and records the information as a mass spectrum. The mass spectrum can be used to identify a compound by comparison with a previously
identified sample, to determine the exact molecular weight (and thus the
molecular formula), and, by analyzing the fragmentation pattern, to establish the structure.

***The Mass Spectrometer.*** In the most common type of mass spectrometer (shown schematically in Figure 4.44), an ionizing electron beam
passes through a gaseous sample. The resulting ion mixture is swept by a
high vacuum into a strong magnetic field. The ions are deflected in a curved
path according to their mass-to-charge ($m/z$) ratio and then directed to a
detector that determines this ratio. The magnetic field is scanned, bringing
ions of successively higher $m/z$ ratio to the detector. The peak with the highest molecular weight represents the **molecular ion,** or **parent ion,** an unfragmented ion with the same mass as that of the starting material. Plotting
ion abundance as a function of $m/z$ gives a **fragmentation pattern** characteristic of that specific molecule. The fragmentation pattern results from
the breakdown of the molecular ion to give ions of lower molecular weight.

The accuracy with which the $m/z$ ratio can be measured is determined
by the quality of the magnet (especially its uniformity) and the sophistica-
tion of the detector. In a low-resolution mass spectrum, $m/z$ ratios are determined to about $\pm 0.2$ mass unit. For some purposes, this accuracy is sufficient for assigning molecular weights to the ions (assuming, as is generally
the case, that the ionic charge is $+1$).

***Molecular Weight Determination.*** High-resolution mass spectrometers are highly sophisticated instruments that have an accuracy of a small
fraction of a mass unit. With this level of accuracy, it is possible to distinguish between combinations of atoms that differ only slightly. For exam-

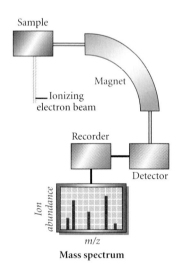

**FIGURE 4.44**

A schematic representation
of a mass spectrometer.

ple, carbon dioxide and propane have, to the nearest integer, the same molecular weight (44). However, this weight is based on naturally occurring distributions of $^{12}C$ and $^{13}C$, as well as $^{16}O$, $^{17}O$, and $^{18}O$ isotopes. In a mass spectrum, the molecular ion peak is that associated with the isotopes of highest abundance (in this case, $^{12}C$ and $^{16}O$). With these isotopes, $CO_2$ is calculated to have a mass of 43.9898 and propane ($C_3H_8$) a mass of 44.0626. These two masses can be readily distinguished by high-resolution mass spectrometry, which generally has an accuracy of $\pm 0.0001$ mass unit. There is usually only a single (reasonable) combination of atoms that corresponds to any given mass. It is possible, therefore, to calculate the formula for the ion from the exact mass of a parent ion found in a high-resolution mass spectrum.

Mass spectroscopy examines and determines the mass of individual ions and does not measure the average properties of a bulk sample. Thus, the atomic masses used in calculating an exact mass value are those of the individual isotopes (for example, 12.0000 for $^{12}C$, instead of 12.011), not those typically used—that is, atomic masses that represent the effect of all isotopes weighted for their relative abundances. Because individual ions are observed in a mass spectrometer, it is also easy to recognize the presence of elements such as chlorine, which exists as two abundant isotopes ($^{35}Cl$ and $^{37}Cl$) and is represented by two ions that are two mass units apart.

*Fragmentation Patterns.* An example of a low-resolution mass spectrum of hexane is shown in Figure 4.45. The peak with the highest molecular weight appears at $m/z = 86$. This peak represents the parent ion, ob-

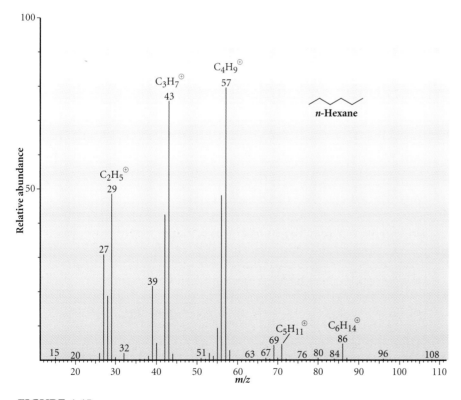

**FIGURE 4.45**

A low-resolution mass spectrum of *n*-hexane.

tained by the simple loss of an electron without fragmentation. The much smaller peak that appears at $m/z = 87$ represents the fraction of molecules containing a higher-weight isotope (either $^{13}C$ or $^{2}H$). The ratio of intensities of the parent and the parent +1 ions is defined by the natural abundance of $^{13}C$, $^{2}H$, and so forth, and thus by the probability that one of the higher-mass isotopes is incorporated in the ionized molecule. Lower-molecular-weight peaks also appear in this spectrum at $m/z = 71, 57, 43$, and 29. These peaks represent the sequential loss of a methyl group and then methylene groups along the straight chain of the parent ion. The most intense peak, at $m/z = 57$, is referred to as the **base peak.** Often, mass spectral data are reported as a series of peaks whose intensities are given as a fraction of this most intense base peak. Note that, even in this simple compound, some bonds are cleaved more readily than others. Thus, the base peak at $m/z = 57$ results from cleavage of the bond between C-2 and C-3, not from cleavage of the more abundant C—H bonds or one of the carbon–carbon bonds to the two methyl groups.

The bond cleavages that occur in a mass spectrometer are of radical cations in the gas phase and not of neutral molecules. Thus, it is not generally possible to rely on a knowledge of organic reactions in solution to predict which bonds will be cleaved most readily. For example, as shown in the mass spectrum of benzaldehyde (Figure 4.46), the parent peak (the peak with the highest $m/z$ value, 105) gives the molecular weight of the molecule from which the radical cation is produced. Other strong peaks in this

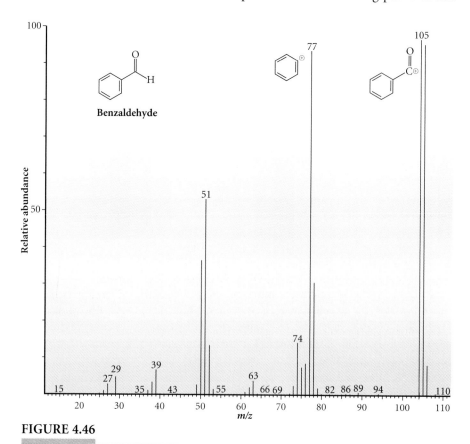

**FIGURE 4.46**

A low-resolution mass spectrum of benzaldehyde.

spectrum are for ions that result from cleavage on either side of the carbonyl group, with loss of H· or $C_6H_5$·. These fragmentations correspond to peaks showing greater abundance than those resulting from cleavage of C—H or C—C bonds in the aryl ring. None of these bond fragmentations represents a normal reaction of benzaldehyde in solution. (A detailed discussion of the reactions of radical cations is beyond the scope of this course.)

**E X E R C I S E   4 . 1 8**

For each of the following compounds, calculate the $m/z$ value of the parent peak and predict the mass of the base peak (major fragmentation) in the mass spectrum. (Exact mass values for the elements are included in the periodic table inside the front cover of the book.)

(a) $CH_3CH_2CH_2CH_2OH$

(b) $CH_3COCH_2CH_3$

(c) $CH_3CH_2CHO$

(d) $CH_3CHClCH_3$

## Summary

1. The identity of an organic compound can be determined by comparing its physical and chemical properties to those of known compounds. In addition to simple physical properties such as melting and boiling points and chemical reactivity, specific structural features in a molecule can be definitively characterized from spectroscopic evidence obtained through the use of instrumentation. Techniques for structure determination, including spectroscopic techniques, are most effective when used to analyze pure compounds.

2. Chromatography is a very versatile technique used to separate mixtures of compounds into individual components. Chromatographic separations take place because of the differential adsorption of a mixture of compounds on a solid or liquid stationary phase while a fluid mobile phase is flowing over this fixed support, eluting the compounds at different rates. Column chromatography employs a solid support (usually alumina or silica gel) through which a liquid solvent flows under the influence of gravity. In paper and thin-layer chromatography, the sample is applied to sheets of adsorbent (paper, or a thin layer of silica gel spread on a glass support) through which solvent moves by capillary action. In gel electrophoresis, a polar polyacrylamide gel is swollen with solvent and subjected to an electrical field. Different polyelectrolyte molecules migrate at different rates under the influence of this external electrical field, thus becoming separated from one another. In gas chromatography, the mixture to be analyzed passes through a very long column, which is either packed with a solid support or lined with a liquid adsorbent. An inert gas acts as a carrier to move the organic molecules through the column.

3. In chromatography, the time required for the elution of a desired compound is directly related to the degree of the compound's adsorption on the stationary phase. Strong polar interactions and van der Waals attractions increase adsorption, causing longer elution times and higher adsorptivity. The retention time is therefore a rough indicator of the polarity

and size of a given molecule. Chromatography can be used not only as a purification technique, but also as an analytical method for characterizing mixtures, because retention times are characteristic for individual compounds.

**4.** Spectroscopic techniques entail the interaction of some form of electromagnetic energy with molecules to produce a spectrum that can be interpreted to reveal the presence of characteristic groups and structural features.

**5.** Nuclear magnetic resonance (NMR) spectroscopy detects the nuclear spin flipping, induced by energy in the radio-frequency range, of a molecule placed in a high magnetic field. Because $^{13}C$ and $^{1}H$ are magnetically active nuclei, NMR spectroscopy can provide important structural information about organic molecules. The information derived from $^{13}C$ and $^{1}H$ NMR are complementary. In $^{13}C$ NMR spectroscopy, each unique carbon usually gives rise to a unique signal. The chemical shifts of these signals can be predicted with considerable accuracy. In $^{1}H$ NMR spectroscopy, chemical shifts, as well as the multiplicity and integration of signals, provide information on the type of protons, their number of nearest neighbors, and the number of protons responsible for an observed signal.

**6.** Infrared (IR) spectroscopy is a method for observing characteristic stretching and bending frequencies of bonds. Because the IR spectra of many common organic functional groups have characteristic absorptions, they can be used to identify functional groups in an unknown compound. In addition, the fingerprint region of an IR spectrum often provides a unique pattern for a molecule, and this feature can be used to identify a compound by comparison with a known sample.

**7.** Ultraviolet (UV), as well as visible, absorption spectroscopy probes electronic transitions from filled to vacant molecular orbitals. The absorption maxima of these bands provide information about the degree of conjugation and the types of electronic transitions possible within a compound.

**8.** In mass spectroscopy, high-energy electrons collide with a given molecule in the gas phase. The high-energy electrons effect ionization, producing a parent ion (a radical cation) and facilitating the determination of an accurate molecular weight. From isotopic abundances, high-resolution mass spectra also provide a molecular formula. Fragmentation patterns observed in the mass spectrum provide valuable information about the structure of a molecule.

## Review Problems

**4.1** Which compound in each of the following pairs will have the longer chromatographic retention time?

(a) 2-butanol or butanal

(c) cyclohexanol or benzene

(b) octadecane or octanoic acid

(d) pyridine or guanine

**4.2** Determine whether each pair of compounds in Problem 4.1 could be separated best by gas chromatography, high-pressure liquid chromatography, or gel electrophoresis. Explain.

**4.3** For each of the following compounds, the molecular formula, data from a $^1H$ NMR spectrum (s = singlet, d = doublet, t = triplet, q = quartet, m = multiplet) and the characteristic infrared bands are given. Propose one or more structures consistent with the data for each compound. If the data are consistent with more than one isomer, suggest another spectroscopic method that might distinguish them.

(a) $C_2H_3Cl_3$: δ 3.95 (d, 2 H), 5.77 (t, 1 H); 2950 and several below 850 cm$^{-1}$

(b) $C_2H_4O$: δ 2.20 (d, 3 H), 9.80 (m, 1 H); 1730 cm$^{-1}$

(c) $C_2H_4O_2$: δ 2.10 (s, 3 H), 11.37 (s, 1 H); broad band at 3200 and strong band at 1710 cm$^{-1}$

(d) $C_2H_4O_2$: δ 3.77 (broad s, 3 H), 8.08 (broad s, 1 H); 1745 and 1250 cm$^{-1}$

(e) $C_2H_6O$: δ 1.22 (t, 3 H), 2.58 (broad s, 1 H), 3.70 (q, 2 H); broad band at 3600 cm$^{-1}$

(f) $C_3H_5ClO_2$: δ 1.73 (d, 3 H), 4.47 (q, 1 H), 11.22 (s, 1 H); broad band at 3200, strong band at 1710, and several below 850 cm$^{-1}$

(g) $C_3H_5NO$: δ 3.47 (s, 3 H), 4.20 (s, 2 H); 2250 and 1100 cm$^{-1}$

(h) $C_3H_6O$: δ 2.72 (quintet, 2 H), 4.73 (t, 4 H); 1120 cm$^{-1}$

(i) $C_3H_6O$: δ 3.58 (s, 1 H), 4.13 (m, 2 H), 5.13 (m, 1 H), 5.25 (m, 1 H), ca. 6.0 (m, 1 H); broad bands at 3600, 3050, 2980, and 1420 cm$^{-1}$

(j) $C_6H_6ClN$: δ 3.60 (s, 2 H), 6.57 (d, 2 H), 7.05 (d, 2 H); broad bands at 3520 and 3400, 3050, 1490, 1590, and 910 cm$^{-1}$

**4.4** Give a structure for a compound with the formula $C_{10}H_{12}O_2$ that is consistent with the following spectra. Would the assignment of structure be unambiguous from the carbon spectrum alone?

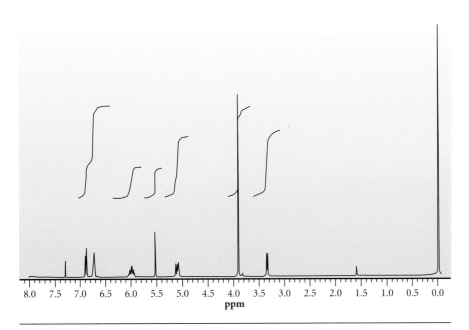

$^1H$ NMR spectrum of $C_{10}H_{12}O_2$

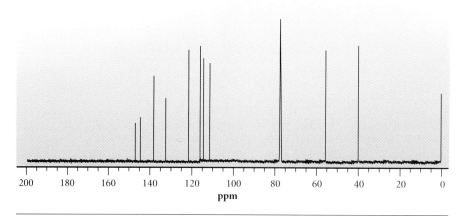

$^{13}$C NMR spectrum of $C_{10}H_{12}O_2$

**4.5** For each of the following compounds, how many peaks would you expect to find in the $^1$H NMR spectrum? What would you expect their splitting and integration to be? How many peaks would you expect to see in the $^{13}$C NMR spectrum? What characteristic peaks would you expect in the infrared spectrum?

(a) ethyl bromide

(b) propyne

(c) 2-propyne-1-ol

(d) allyl bromide

(e) 2-nitropropane

(f) *N,N*-dimethylformamide

(g) methyl ethyl sulfide

(h) 2-propanol

(i) vinyl acetate

(j) 2-bromobutanoic acid

(k) 2-butanone

(l) butanal

(m) 3-methoxybutanol

(n) toluene

**4.6** Suggest a spectroscopic method that will readily distinguish between members of the following pairs of compounds. Describe exactly what you would see for each compound using the chosen method.

(a) *n*-butylamine and *t*-butylamine

(b) *s*-butylamine and *t*-butylamine

(c) methylenecyclopentane and 1-methylcyclohexene

(d) *m*-methylphenol (*m*-cresol) and benzyl alcohol

(e) *p*-methylphenol (*p*-cresol) and anisole ($C_6H_5OCH_3$)

(f) styrene oxide ($C_6H_5CHCH_2O$) and acetophenone

(g) acetophenone and *p*-methoxybenzaldehyde (*p*-anisaldehyde)

(h) *m*-xylene and *p*-xylene

**4.7** Propose a structure for hydrocarbon X, whose mass spectrum shows a parent peak at $m/z = 86$, with a major peak 15 mass units lower than that, and whose $^{13}$C NMR spectrum shows peaks at $\delta$ 19.5 and $\delta$ 34.3. Explain how other isomeric structures can be definitely eliminated by these data.

**4.8** Isomers A and B have the molecular formula $C_6H_{12}$. Isomer A has a $^{13}$C NMR spectrum with peaks at $\delta$ 13.7, 17.8, 23.1, 35.0, 124.9, and 131.6. Isomer B has peaks at $\delta$ 12.7, 13.7, 23.1, 29.2, 123.9, and 130.7. How many double bonds are present in these compounds? Propose possible structures for each isomer, excluding structures specifically eliminated by the $^{13}C$ data. What additional information, if any, would be of use in making an unambiguous spectral assignment?

**4.9** From the ultraviolet absorption data in Table 4.4, roughly predict the wavelength(s) at which each of the following common solvents would absorb. Would any of these be an acceptable solvent for measuring the absorption spectrum of naphthalene?

(a) cyclohexane

(b) tetrahydrofuran

(c) toluene

(d) acetone

**4.10** Predict the mass of the parent ion and the major mass spectral fragments to be expected for each of the following compounds:

(a) pentanal

(b) acetophenone

(c) ethanol

(d) ethyl ether

## Supplementary Problems

**4.11** Each of the following isomers of $C_4H_{10}O_2$ has a unique $^{13}C$ NMR spectrum.

A    B    C    D

E    F    G    H

(a) Estimate the chemical shift values expected for each carbon of each isomer using the data in Table 4.1.

(b) Select the structure that best fits each of the following observed shifts:

(i) $\delta$ 72.3, 19.2

(ii) $\delta$ 101.4, 52.0, 18.8

(iii) $\delta$ 73.8, 66.3, 26.1, 10.0

(iv) $\delta$ 72.1, 66.6, 61.6, 15.0

**4.12** The compounds in Problem 4.11 can be divided into three groups: diols, monoethers–monoalcohols, and diethers. Explain how these structural features can be used to help make the assignments in part (b) of Problem 4.11.

**4.13** The observed shift values in part (b) of Problem 4.11 occur as sets of four, three, or two unique carbon atoms. How might the number of unique carbon resonances in the spectrum be used to determine which structure is consistent with this spectrum? (*Hint:* Which compounds in Problem 4.11 would be expected to have four resonances, which would have three, and which only two?)

**4.14** The NMR spectra shown were obtained for a compound with the molecular formula $C_4H_{11}NO$. First, determine the index of hydrogen deficiency to determine whether rings and/or double bonds are present. Then, suggest a structure consistent with all of the data. Note that the three resonances at approximately $\delta$ 77 in the $^{13}C$ NMR spectrum result from the solvent used, $CDCl_3$. (*Hint:* The resonance at $\delta$ 11 in the $^{13}C$ NMR spectrum could result only from a carbon with only one $\alpha$ and one $\beta$ carbon substituent, and $\gamma$ substituents that result in a significant upfield shift.)

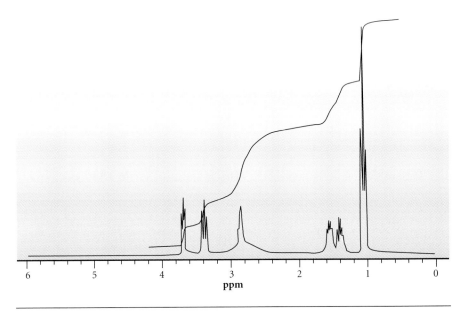

$^1H$ NMR spectrum of $C_4H_{11}NO$

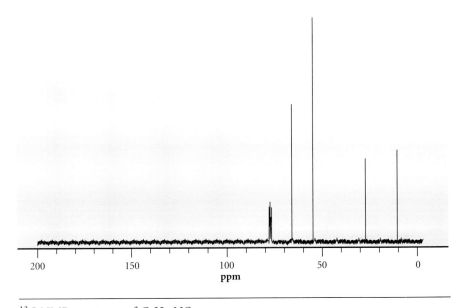

$^{13}C$ NMR spectrum of $C_4H_{11}NO$

**4.15** The NMR spectra shown were obtained for a compound with the molecular formula $C_8H_{10}O_2$. First, determine the index of hydrogen deficiency and if the answer is not zero, use the spectral data to determine whether rings and/or double bonds are present. Then, suggest a structure consistent with all of the data. Note that the three resonances at approximately $\delta$ 77 ppm in the $^{13}C$ NMR spectrum result from the solvent used, $CDCl_3$. (*Hint:* There are only four unique resonances in the $^{13}C$ NMR spectrum for a compound with eight carbon atoms. It is reasonable to presume that the compound is symmetrical.) Also, decide which atom is responsible for the resonances at $\delta$ 3.9 in the $^1H$ NMR spectrum and at $\delta$ 56 in the $^{13}C$ NMR spectrum.

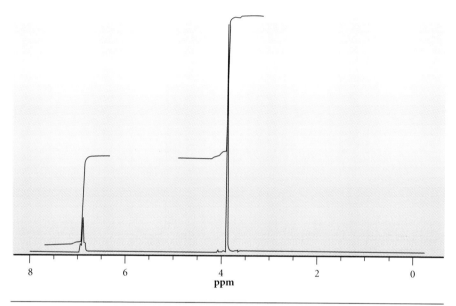

$^1H$ NMR spectrum of $C_8H_{10}O_2$

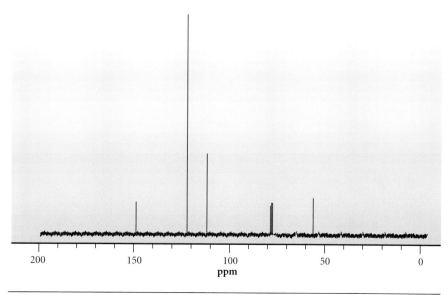

$^{13}C$ NMR spectrum of $C_8H_{10}O_2$

**4.16** The following spectra were obtained for a compound with the molecular formula $C_8H_{16}$.

(a) Determine the index of hydrogen deficiency. If the answer is not zero, use the spectral data to determine whether rings and/or double bonds are present.
(b) Suggest a structure that is consistent with all of the data presented for this compound. Note that the three resonances at approximately $\delta$ 77 in the $^{13}C$ NMR spectrum result from the solvent used, $CDCl_3$.

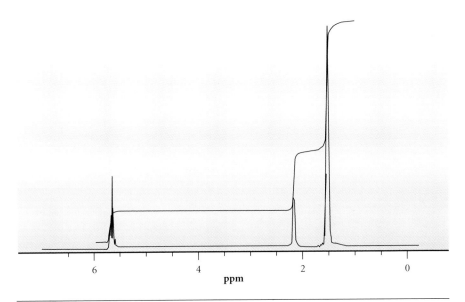

$^1H$ NMR spectrum of $C_8H_{16}$

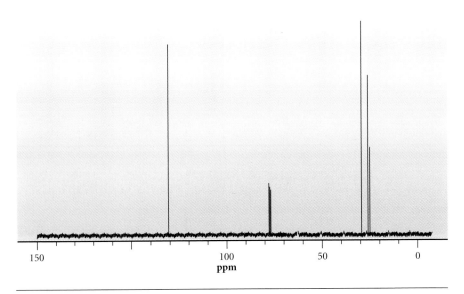

$^{13}C$ NMR spectrum of $C_8H_{16}$

**4.17** The NMR spectra shown were obtained for a compound with the molecular formula $C_4H_7N$. First, determine the index of hydrogen deficiency and if the answer is not zero, use the spectral data to determine whether rings and/or double bonds are present. Then, suggest a structure consistent with all of the data. Note that the three resonances at approximately $\delta$ 77 ppm in the $^{13}C$ NMR spectrum result from the solvent used, $CDCl_3$.

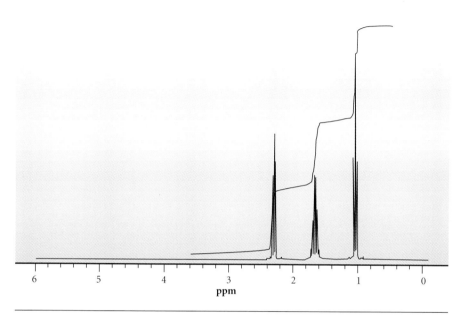

$^1H$ NMR spectrum of $C_4H_7N$

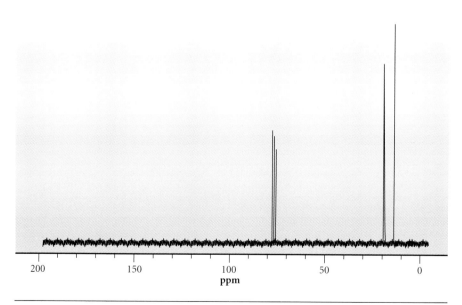

$^{13}C$ NMR spectrum of $C_4H_7N$

# Stereochemistry

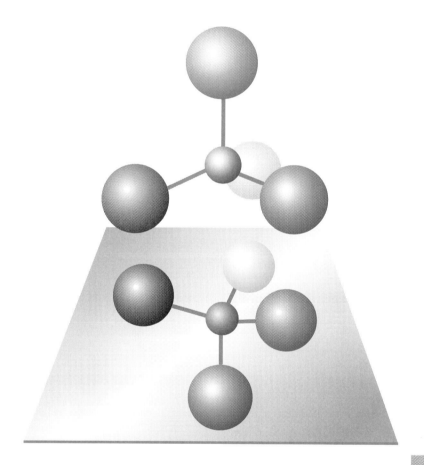

A three-dimensional, tetra-hedral object reflected in a mirror. The reflected image can-not, regardless of rotation, be superimposed on the original. The object (and its mirror image) are chiral.

$\mathcal{N}$ow that you understand the composition of the principal functional groups of organic chemistry, we will consider three-dimensional structure in more detail. Some of the compounds presented in Chapters 1 through 3 are **constitutional isomers,** which have the same molecular formula but their atoms are attached in different sequences; examples are methyl ether and ethanol:

Methyl ether    Ethanol

In this chapter, we will look at **stereochemical isomers,** or **stereoisomers,** which differ only in how their atoms are arranged in space, not in atomic connectivity. Stereoisomers can be considered as belonging to two classes: **conformational isomers** (those that can be interconverted by rotation about a $\sigma$ bond) and **configurational isomers** (those that can be interconverted only by the breaking and reforming of bonds). Two important subclasses of configurational isomers are **geometric isomers** (those in which restricted rotation in a ring or at a multiple bond determines the relative spatial arrangement of atoms) and **optical isomers** (those that differ in the three-dimensional relationship of substituents about one or more atoms). Figure 5.1 summarizes this classification of isomers.

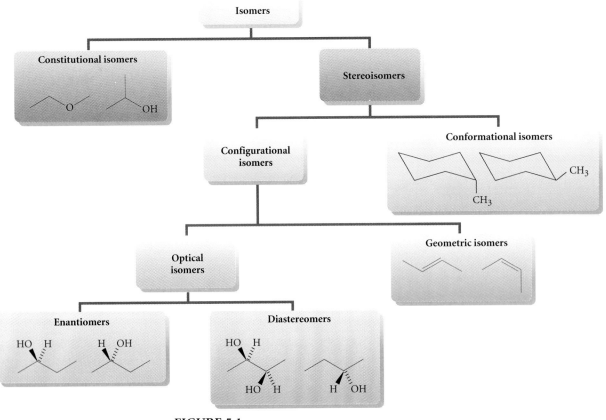

**FIGURE 5.1**

Isomers are divided into two main classes: constitutional and stereochemical. Stereoisomers are subdivided into configurational and conformational isomers. Configurational isomers comprise optical and geometric isomers.

In this chapter, we will discuss how to recognize different types of isomers, consider the energy relationships between various isomers, and correlate chemical and physical properties with the spatial arrangement of atoms. In doing so, we will address three topics that are crucial to an understanding of the structure and reactivity of molecules: geometric isomerization, conformational analysis, and chirality. Because many students have trouble visualizing molecules in three dimensions, we will look at these topics in order of increasing difficulty.

# Geometric Isomerization:
# Rotation about Pi Bonds

### ▨ Geometry of Alkenes

As you learned in Chapter 2, $\pi$ bonding requires the overlap of two (or more) $p$ orbitals on adjacent atoms. Because of the barrier to rotation about a double bond, geometric isomers such as *cis-* and *trans-*2-butene differ in how groups attached to the doubly bonded carbons are arranged relative to each other. The geometry of the $\pi$ bond between two $sp^2$-hybridized carbons forces them to be coplanar with the four attached substituent atoms. The degree of twisting about the $\pi$ bond is defined in terms of the **dihedral angle.** If C-1, C-2, and C-3 of 2-butene are viewed as forming a plane, the dihedral angle is the angle between this plane and that containing C-2, C-3, and C-4. In *cis-*2-butene, this dihedral angle is 0°, and in *trans-*2-butene, it is 180°:

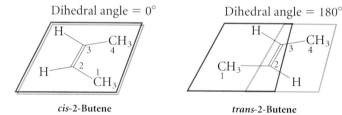

Dihedral angle = 0°    Dihedral angle = 180°

*cis-*2-Butene        *trans-*2-Butene

In both of these arrangements, the overlap of the $p$ orbitals of the $\pi$ bond is at a maximum. In contrast, when the dihedral angle is 90°, there is no $\pi$-bonding character.

### ▨ Energetics of Rotation about Pi Bonds

Rotating C-4 about the C=C bond changes the dihedral angle of 2-butene from 0° to 180°. We can plot the energy change as a function of the dihedral angle (see Figure 5.2 on page 226). This energy diagram shows three extremes: a maximum at 90° and two minima, at 0° and 180°. At the maximum, the $p$ orbitals on C-2 and C-3 are orthogonal, and there is no $\pi$ bonding. Thus, the difference in energy between the geometries at 0° and 180° (maximum overlap) and that at 90° (minimum overlap) is a rough measure of the $\pi$-bond strength.

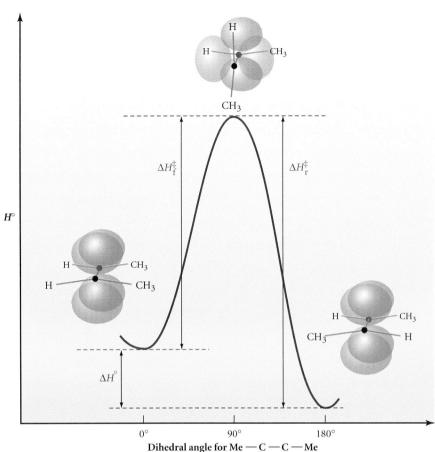

**FIGURE 5.2**

Energy changes induced by rotation about the C=C bond of 2-butene (Me is a methyl group). Activation energies for the forward (*cis → trans*: $\Delta H_f^\ddagger$) and back (*trans → cis*: $\Delta H_r^\ddagger$) reactions are indicated by the vertical arrows.

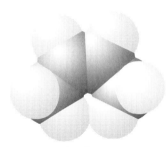

*cis*-2-Butene

*trans*-2-Butene

### ▣ Reaction Energetics

Plots such as that in Figure 5.2 can also be used to describe net reaction energetics. In fact, Figure 5.2 describes the net reaction energetics for the isomerization of *cis*- to *trans*-2-butene. The energy difference between a reactant and a product is called the **heat of reaction, $\Delta H°$.** In this case, the product, *trans*-2-butene, lies at a lower energy than the reactant, *cis-2*-butene, because the methyl groups in the *cis* isomer are too close to each other, causing steric strain. **Steric strain** results when atoms or groups of atoms separated by several bonds are brought into such close proximity that there is van der Waals repulsion. Therefore, the reaction converting the *trans* isomer of 2-butene to the *cis* isomer is energetically unfavorable.

The **free energy difference ($\Delta G°$)** for a reaction has contributions from both enthalpy ($\Delta H°$, heat content) and entropy ($\Delta S°$, disorder):

$$\Delta G° = \Delta H° - T\Delta S°$$

For isomerization reactions, it is usual to consider only the change in enthalpy, because the change in entropy is small.

The sign of $\Delta H°$ indicates whether an isomerization is endothermic ($+$) or exothermic ($-$). In Figure 5.2, $\Delta H°$ is negative, so the reaction is exothermic. However, knowing $\Delta H°$ provides no direct information about the **energy barrier,** $\Delta H^{\ddagger}$, which is the amount of energy required to reach the most unfavorable point along the path followed in the conversion of the reactant to the product. Describing the route by which a reaction takes place requires knowing not only the energy difference between the reactants and products, but also the energy of the least favorable arrangement through which the atoms must pass as the reaction proceeds. This arrangement having the highest energy is called the **transition state.**

In the geometric isomerization plotted in Figure 5.2, the transition state is the arrangement in which there is no stabilization from $\pi$ bonding. This transition state (with $p$ orbitals at a dihedral angle of 90°) can collapse to either the *cis* or the *trans* isomer without any additional energy barrier. The transition state is the least stable species in a reaction pathway, so no additional energy is required when it is converted into a more stable species.

The energy difference between the reactant and the high-energy transition state is referred to as the **energy of activation,** or **activation energy** ($\Delta H_f^{\ddagger}$ or $E_{act}$). In the specific transformation plotted in Figure 5.2, the activation energy for the forward reaction (*cis* $\rightarrow$ *trans*) is slightly less than that for the reverse reaction (*trans* $\rightarrow$ *cis*). Activation energies for the forward and reverse reactions always differ by precisely the value of $\Delta H°$ of the reaction.

---

## EXERCISE 5.1

For each of the following isomerizations, draw the transition state required for the conversion of the structure at the left into the structure at the right. Can you estimate from what you know about bond energies whether the activation energy for each of these reactions is larger or smaller than that for the isomerization of *cis*-2-butene to *trans*-2-butene?

(a) 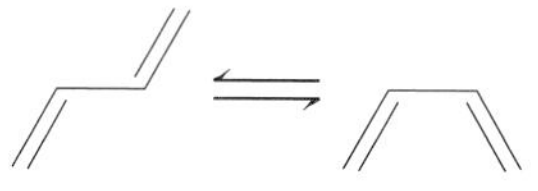     (c)

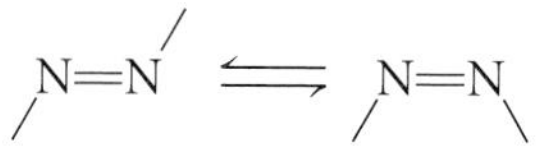

(b)

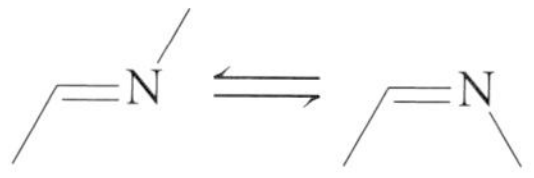

---

## Light-Induced Isomerization of Alkenes

The activation energy required for the interconversion of *cis* and *trans* isomers is larger than can usually be provided at typical reaction temperatures ($<300$ °C). However, you know from Chapter 4 that absorption of a photon of ultraviolet radiation by an alkene, diene, or triene is accompanied by promotion of a bonding electron to an antibonding orbital. In the simple case of an alkene with only one double bond, the resulting arrangement has one electron in the $\pi$ (bonding) and one in the $\pi^{\star}$ (antibonding) molecular orbital, and no net bonding between the $p$ atomic orbitals. In this excited state, rotation about the carbon–carbon bond occurs readily, even at very low temperatures. Ultimately, the electron in the energetically higher antibonding orbital returns to the bonding orbital, releasing most of the absorbed energy as heat or light.

The relatively free rotation about the carbon–carbon bond that occurs upon absorption of a photon has significant biological consequences: *cis–trans* isomerization forms the chemical basis for mammalian vision. Light-sensitive receptor cells in the eye contain 11-*cis*-retinal, which is chemically bound to the protein opsin (through an imine functional group), forming rhodopsin (Figure 5.3). Absorption of visible light by the 11-*cis* isomer (note its extended conjugation) results in an excited state that readily undergoes isomerization to the 11-*trans* isomer. The shape of the *trans* isomer is more extended than that of the *cis*, and the isomerization requires a change in the shape of the opsin protein. This shape change of opsin initiates the release of calcium ions, whose increased concentration triggers a nerve impulse that is interpreted by the brain as vision. (Thus, an electronic transition, whose energy is defined by the energy difference between the filled and vacant orbitals, provides a way for inducing new reactivity in the excited state of the bound polyene.)

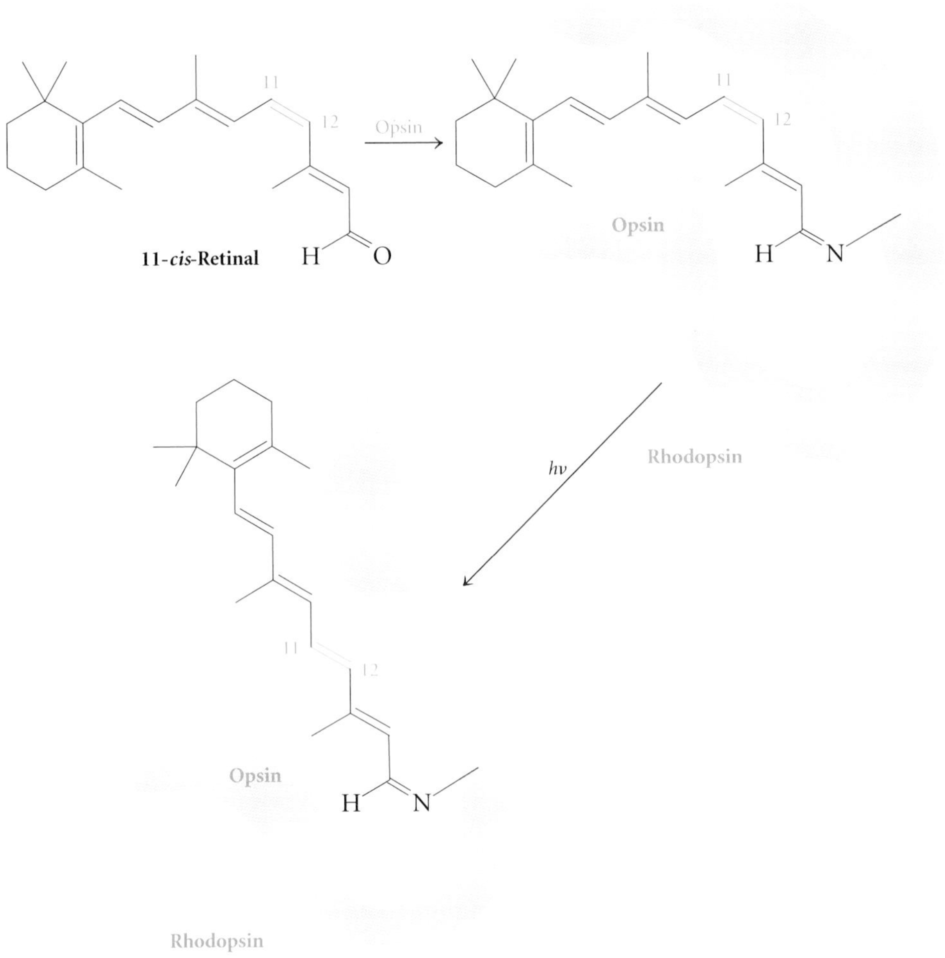

**FIGURE 5.3**

Geometric isomerism is induced in rhodopsin by the absorption of light.

Similar isomerizations of *cis* and *trans* alkenes can also be accomplished in the laboratory. The irradiation of an alkene with ultraviolet radiation of sufficient energy to promote an electron from a bonding to an antibonding $\pi$ orbital effectively breaks the $\pi$ bond, resulting in free rotation.

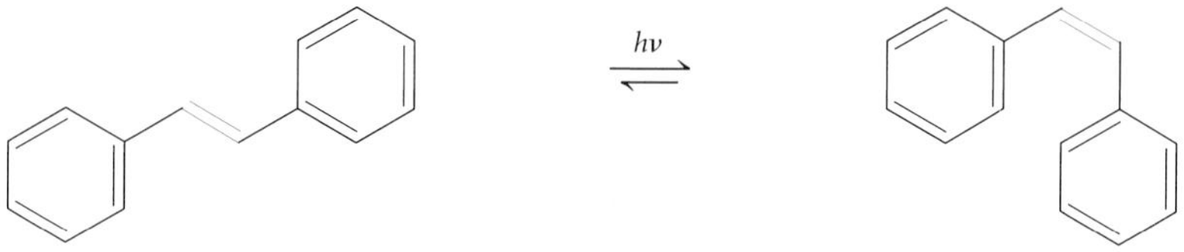

*trans*-Stilbene

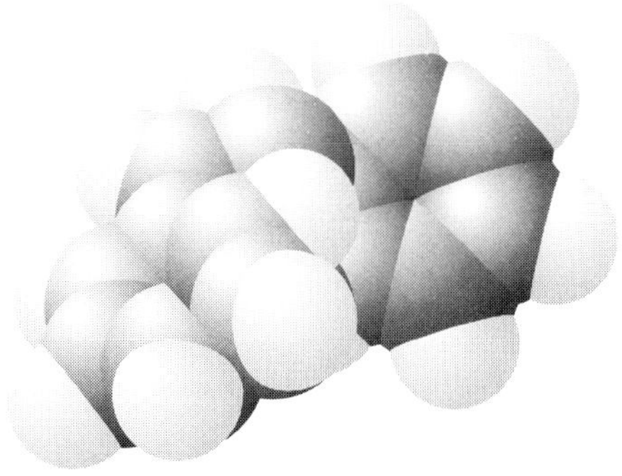

*cis*-Stilbene

## Geometric Isomerization to the Less Stable Isomer

When the ultraviolet absorption maxima of *cis* and *trans* isomers are sufficiently different, the *trans* isomer can be irradiated at a wavelength that the *cis* isomer absorbs to a lesser extent. In this case, the rate of conversion of the *trans* isomer to the *cis* is greater than that of *cis* to *trans,* and the *cis* isomer dominates the equilibrium. Such a reaction can thus be used to produce *cis* alkenes from the thermodynamically more stable *trans* isomers. For example, the ultraviolet absorption spectrum of *trans*-2-stilbene exhibits two maxima (296 and 305 nm), each approximately three times as strong as the single maximum in the spectrum of the *cis* isomer. These differences arise because of steric strain between the phenyl groups in the *cis* isomer, which prevents full planarity of the $\pi$ systems and causes a decrease in conjugation that increases the HOMO–LUMO gap and therefore the energy required to excite an electron from the HOMO to the LUMO.

## 5.2

# Conformational Analysis: Rotation about Sigma Bonds

Like geometric isomers, conformational isomers have the same skeletons but differ with respect to the relative positions of some atoms in three-dimensional space. In conformational isomers, these differences can be removed by rotation about one or more $\sigma$ bonds; that is, after rotation about at least one σ bond, the atomic arrangements become identical. The process known as **conformational analysis** describes the energetics of such conformational interconversion by relating the relative atomic positions during rotation about a σ bond to changes in potential energy.

You know from Chapters 1 and 2 that rotation about a single bond does not require bond cleavage and is therefore easier than rotation about a multiple bond, which requires the addition of sufficient energy to break the π bond. Thus, in the interconversion of conformational isomers, changes in energy are relatively small, and it is often said that there is "free" rotation about a σ bond.

#06 Ethane
Conformations

## Ethane

Let's consider rotation about the carbon–carbon bond of ethane. Although there are an infinite number of conformations, there are only two extremes (Figure 5.4). (You should build a model of this simple molecule so that you can better understand the analysis that follows.) In the structure at the left, called the **eclipsed conformation,** each C—H bond at C-1 is aligned with one at C-2 (the dihedral angle is 0°). In the structure at the right, called a **staggered conformation,** each C—H bond at C-1 is exactly between two C—H bonds at C-2 (the dihedral angle is 60°). In both structures, tetrahedral geometry at carbon is maintained. The structure at the left can be converted into the one at the right by simple rotation of 60° about the C—C σ bond.

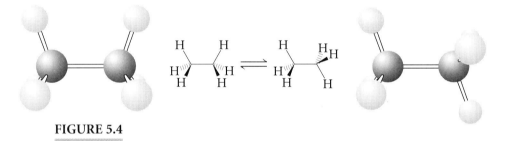

**FIGURE 5.4**

Two different orientations of substituents about the C—C bond of ethane.

*Newman Projections.* The relative positions of the hydrogen atoms on C-1 and C-2 in eclipsed and staggered ethane are clearly revealed by the representations in Figure 5.5. In the **Newman projections,** the C—C bond is directed away from the viewer, and both the carbon–carbon bond and C-2 are hidden behind C-1. The circle in these representations can be thought of as the electron density of the σ bond, with the front carbon implied at the junction of the three bonds and the back carbon hidden. Newman projections show the orientation of the hydrogen atoms on the front carbon atom relative to those on the back carbon and are quite useful for conformational analysis. (These representations are named in recognition of the chemist Melvin Newman, of Ohio State University, who first showed their utility in conformational analysis.)

Keep in mind that the bonds around the carbons do not form a plane and that, indeed, the Newman projections represent the same molecular geometry as is shown by the ball-and-stick models. In Newman projection A in Figure 5.5, all the hydrogen atoms of ethane are aligned with each other

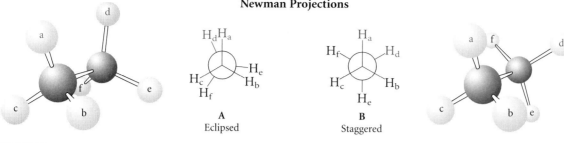

**Newman Projections**

A
Eclipsed

B
Staggered

**FIGURE 5.5**

Sawhorse representations of eclipsed and staggered ethane and their corresponding Newman projections. (In the Newman projection of the eclipsed conformation, the substituents are drawn slightly rotated from perfect alignment; otherwise, the hydrogen atoms bonded to the back carbon atom would be covered by those in the front.)

in the eclipsed conformation. Since each hydrogen atom has been identified with a subscript, you can see that the $C-H_a$ and the $C-H_d$ bonds are coplanar and on the same side of the carbon–carbon internuclear axis. The $\sigma$ bonds to $H_b$ and $H_e$ and to $H_c$ and $H_f$ are also coplanar. Rotation about the carbon–carbon $\sigma$ bond of this eclipsed conformation by 60° (moving the substituents on the back carbon) changes the relative positions of the hydrogen atoms in these pairs. In the resulting staggered conformation, shown by Newman projection B in Figure 5.5, the bonds to the front carbon are exactly between those of the back carbon.

## EXERCISE 5.2

Draw a Newman projection that illustrates each of the following descriptions:

(a) eclipsed conformation of 2,2,3,3-tetramethylbutane, viewed down the C-2—C-3 bond

(b) staggered conformation of 2,2,3,3-tetramethylbutane, viewed down the C-2—C-3 bond

(c) staggered conformation of propane, viewed down the C-1—C-2 bond

(d) eclipsed conformation of propane, viewed down the C-1—C-2 bond

***Torsional Strain.*** A change in energy is associated with the change in the relative positions of the atoms in eclipsed and staggered conformations. The electrons of the carbon–hydrogen bonds to the front and back carbons are closer to each other in the eclipsed conformation, resulting in greater electron–electron repulsion. In addition, because the hydrogen atoms are closer to one another, there are other electronic and nuclear interactions. The net result is that the eclipsed conformation is energetically less favorable than a staggered conformation. The total change in energy due to rotation from a staggered to an eclipsed conformation, which can be measured experimentally, is referred to as **torsional strain.**

For rotation about the $\sigma$ bond of ethane, we can draw a profile that relates the relative potential energy (degree of torsional strain) to the dihedral angle between a pair of hydrogen atoms, one on the front carbon and one on the back. In Newman projection A in Figure 5.5, $H_a$, C-1, and C-2

231

define one plane; C-1, C-2, and $H_d$ define a second plane; and the angle between these planes is the dihedral angle, $\Phi$. We begin with a high-energy conformation in which the carbon–carbon bond has an eclipsed arrangement, with $\Phi = 0$ (Figure 5.6). Keeping the front carbon stationary and rotating the back carbon clockwise by 60° yields a staggered conformation, which is of lower energy because torsional strain is relieved. Rotating the back carbon through another 60° results in another eclipsed conformation, in which $H_d$ aligns with $H_b$. Because all of the hydrogen atoms are identical, this eclipsed conformation is identical in energy with the first. Thus, the relative energy increases during rotation until it reaches the original value. A series of sequential 60° rotations until $H_a$ has returned to its original position provides a smooth energy profile for the interconversion of the eclipsed and staggered conformations.

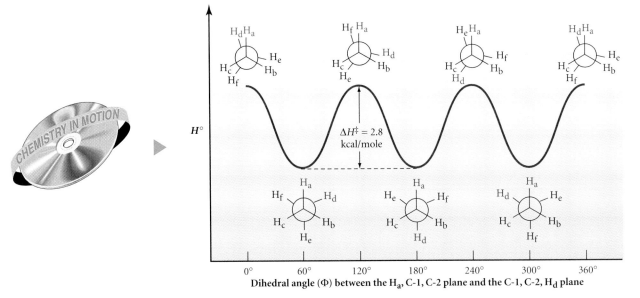

**FIGURE 5.6**

Energy changes resulting from rotation about the C—C bond of ethane.

The energy difference between the eclipsed and staggered conformations is about 2.8 kcal/mole (about 0.9 kcal/mole for each H—H torsional interaction). The staggered conformation is at the bottom of an energy well, whereas the eclipsed conformation is at the top of an energy hill. Indeed, no additional energy is required for further rotation of the eclipsed conformation.

The energy values used in conformational analysis are always relative values, generally comparing a less stable conformation with the conformation of lowest energy. For example, the torsional strain of 2.8 kcal/mole in ethane relates the less stable eclipsed conformation to the more stable staggered arrangement. Furthermore, it is often presumed that the energy values found experimentally for simple molecules such as ethane can be used to approximate the relative energies of the conformational isomers of more complicated molecules.

## Butane

Newman projections are also useful for the analysis of more complex hydrocarbons. For example, Newman projections for butane show that rotation about the central carbon–carbon bond results in several different eclipsed and staggered arrangements (Figure 5.7). These arrangements differ in energy. Conformation A in Figure 5.7 (in which two methyl groups are eclipsed) is energetically unfavorable—both because of interactions between the bonds (torsional strain) and because of the repulsive interaction resulting from two large methyl groups being in the same region of space.

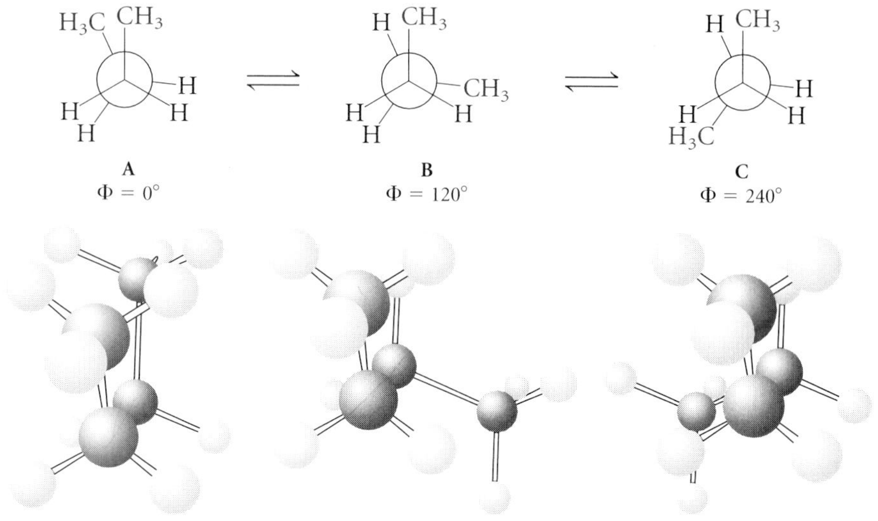

**FIGURE 5.7**

Newman projections and ball-and-stick representations of the eclipsed conformations of *n*-butane.

***Steric Strain.*** Destabilization resulting from van der Waals repulsion of groups that are close to each other is referred to as a **steric effect.** Thus, in conformation A in Figure 5.7, known as the *syn* eclipsed conformer of butane, there is both torsional strain (because the bonds are aligned) and steric strain (because the methyl groups are too close to each other).

In the other two eclipsed conformations in Figure 5.7, the methyl groups are eclipsed with hydrogen atoms. Because a hydrogen atom is smaller than a methyl group, there is less steric strain, and the total destabilization resulting from both torsional and steric strain in these two conformers of butane is only slightly larger than that for ethane. Thus, conformations B and C in Figure 5.7 are both more stable than the *syn* eclipsed conformer. The eclipsing interactions can be separated into those due to hydrogens eclipsing hydrogens and those due to eclipsing of hydrogens by methyl groups. By assuming that the pair of eclipsed hydrogens contributes the same degree of destabilization as in ethane ($2.8 \div 3 = 0.9$ kcal/mole), we can assign the remaining torsional strain ($3.4 - 0.9 = 2.5$ kcal/mole) to a contribution of 1.2 kcal/mole from each of the methyl–hydrogen eclipsing interactions.

The three staggered isomers of butane are shown in Figure 5.8 (page 234). Torsional strain is at a minimum in these staggered conformations.

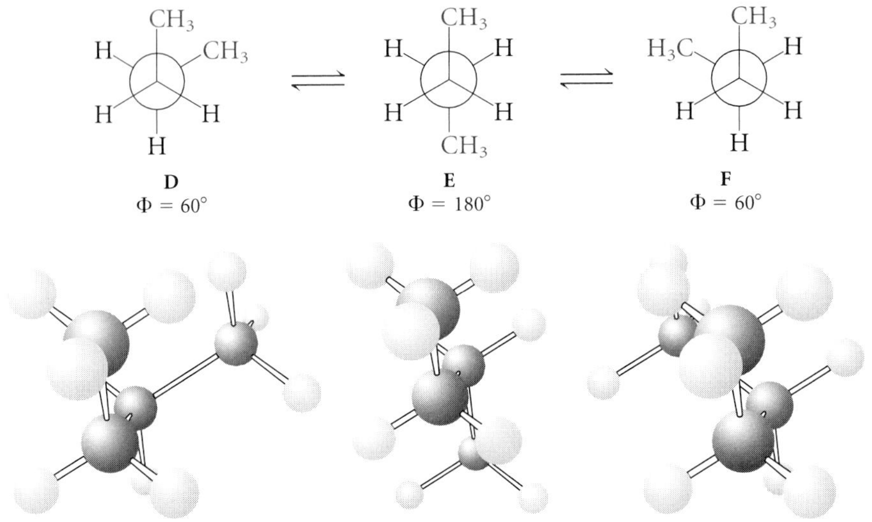

**FIGURE 5.8**

Newman projections and ball-and-stick representations of the staggered conformations of *n*-butane.

**Gauche *and* Anti *Conformers.*** Like the eclipsed conformations of butane, the staggered conformations of butane differ in energy. In conformation E in Figure 5.8, the dihedral angle between the methyl groups is 180°; in conformations D and F, this angle is 60°. There is some steric effect in structures in which the dihedral angle between the methyl groups is 60°. As a result, conformations D and F are higher in energy than is conformation E, in which these groups are as far away from each other as possible. Isomers bearing substituents near each other in a staggered conformation (that is, separated by a 60° dihedral angle) are referred to as *gauche* **conformers.** Isomers in which substituents are separated by a 180° dihedral angle are referred to as ***anti* conformers.** The two *gauche* conformers of butane are mirror images of each other and cannot be superimposed without rotation about the central carbon–carbon bond. (We will see this isomerism in other compounds later in the chapter.)

We can now construct an energy profile for rotation about the central carbon–carbon bond of butane, as shown in Figure 5.9. The energy difference between the *gauche* and *anti* conformers of butane is 0.9 kcal/mole, and the energy barrier for conversion of the *gauche* to the *anti* conformer (by way of the eclipsed conformer that represents the transition state) is about 3.4 kcal/mole. The energy of the *syn* eclipsed conformer, in which the methyl groups are aligned, is difficult to measure accurately but has been estimated to be about 5–7 kcal/mole higher than that of the *anti* isomer.

The energy cost for each conformational interaction for ethane and butane is given in Table 5.1. These values come from experimentation and can be assumed to be reasonable approximations for similar interactions in other molecules. The value for the methyl–methyl *gauche* steric strain is based on the energy difference between the *anti* and the *gauche* isomers of butane. The value for the methyl–hydrogen eclipsed interaction (steric and

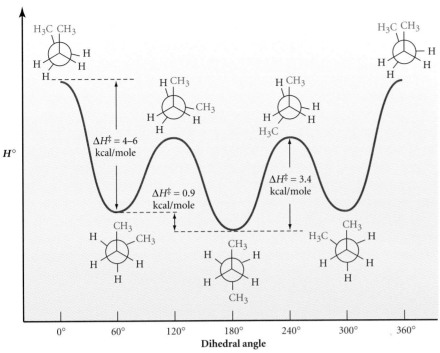

**FIGURE 5.9**

Energy changes induced by rotation about the C-2–C-3 bond of *n*-butane.

torsional strain) is derived from $\Delta H°$ between *anti*-butane (in which both steric and torsional strain are considered to be absent) and the eclipsed conformer (in which there is a 120° dihedral angle between the two methyl groups). The *gauche* strain thus obtained is corrected for the contribution of hydrogen–hydrogen eclipsing taken from the value for ethane (0.9 kcal/mole) and partitioned equally to the two *gauche* interactions. The estimate for the methyl–methyl *syn* eclipsed interaction is derived from $\Delta H°$ for the conformers at 0° and 180°, corrected for the two hydrogen–hydrogen eclipsing interactions.

**TABLE 5.1**

Approximate Energy Costs for Steric Interactions

| Type of Interaction | Energy, $\Delta H°$ (kcal/mole) |
|---|---|
| Hydrogen–hydrogen steric strain (*gauche*) | 0 |
| Methyl–methyl steric strain (*gauche*) | 0.9 |
| Hydrogen–hydrogen torsional strain (eclipsed) | 0.9 |
| Methyl–hydrogen steric and torsional strain (eclipsed) | 1.25 |
| Methyl–methyl steric and torsional strain (eclipsed) | ~3–5 |

Suppose that, instead of viewing butane down its C-2—C-3 bond, you view hexane down its C-3—C-4 bond. Draw the energy profile you obtain, and qualitatively compare the magnitudes of the hills and valleys in your profile with those in Figure 5.9.

---

*Equilibrium Ratios of* **Gauche** *and* **Anti** *Conformers.* The free-energy difference between two isomers determines their relative abundance at equilibrium, and an equilibrium constant can be calculated from the difference in free energy between the reactants and the products. That is, in the equilibration of two species, such as reactant A and product B, the relative amounts of A and B present at equilibrium depend directly on the size of the free-energy difference, $\Delta G°$, between them.

$$A \underset{k_r}{\overset{k_f}{\rightleftharpoons}} B$$

The concentrations of the species A and B (reactant and product) are related to the equilibrium constant $K$:

$$K = \frac{[\text{product}]}{[\text{reactant}]}$$

The equilibrium constant $K$ can also be calculated from the equation

$$\Delta G° = -RT \ln K \tag{1}$$

where $\Delta G° = \Delta H° - T\Delta S°$. Thus, equation 1 can be used to relate the change in free energy, $\Delta G°$, and the concentrations of reactants and products. Note that the dependence of the equilibrium constant on $\Delta G°$ (the difference in free energy between reactant and product) is exponential. Thus, small differences in $\Delta G°$ result in large changes in $K$.

In many cases, $\Delta G°$ is approximated by the difference in enthalpy, because entropy differences are often comparatively small. Thus, equation 1 becomes

$$\Delta H° \cong -RT \ln K$$

and

$$\ln K \cong -\frac{\Delta H°}{RT}$$

or

$$K = e^{-\Delta H°/RT} \tag{2}$$

Using equation 2 to calculate the ratio of the more stable to the less stable conformer as a function of the energy difference between them, we arrive at the values shown in Table 5.2. By interpolating from the values in

**TABLE 5.2**

Conformational Equilibrium Ratios
as a Function of Energy Differences

| $\Delta H°$ (kcal/mole) | Percentage of More Stable Isomer* (at 25 °C) |
|---|---|
| 0.0 | 50.0 |
| 0.65 | 75.0 |
| 1.3 | 90.0 |
| 1.7 | 95.0 |
| 2.7 | 99.0 |
| 4.1 | 99.9 |

*Calculated from $K = e^{-\Delta H°/RT}$.

the table, we find that the energy difference of 0.9 kcal/mole between the *gauche* and *anti* conformers of butane corresponds to an equilibrium in which the *anti* conformer is favored over either of the *gauche* conformers by a factor of about 4:1. Because there are two *gauche* conformers and only one *anti* one, the *anti*:*gauche* ratio is approximately 2:1. That is, the equilibrium mixture is composed of about 66% of the *anti* conformer and 34% of the two *gauche* conformers. The high-energy *syn* eclipsed conformation ($\Delta H° = 5$–7 kcal/mole) does not significantly contribute to the conformational equilibrium: the ratio of the staggered (*anti*) to the eclipsed conformer is greater than 1000:1.

Note in equation 2 that the equilibrium constant depends on the values of both $\Delta H°$ *and T*. Thus, the ratio of isomers in equilibrium varies with the temperature. We can arrive at a qualitative appreciation of the effect of temperature on the equilibrium constant by using values for the possible extremes of temperature. At very high temperatures, the exponent in equation 2 becomes very small, and $K$ thus approaches 1 ($e^0 = 1$). As the temperature approaches 0 K, the equilibrium constant goes to 0 ($e^{-\infty} = 0$). Thus, the ratio of conformations in equilibrium increases with decreasing temperature and decreases with increasing temperature. At higher temperatures, the contribution of less stable species to the conformational equilibrium increases.

**EXERCISE 5.4**

Using the data in Table 5.2, estimate the energy difference required to give the following equilibrium distributions of a more stable conformer A with a higher-energy conformer B at 25 °C:

(a) a 55:45 mixture of A and B

(b) a 70:30 mixture of A and B

(c) a 99.99:0.01 mixture of A and B

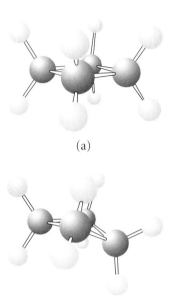

(a)

(b)

**FIGURE 5.10**

Ball-and-stick representations of (a) planar and (b) puckered cyclobutane. Minimization of torsional, steric, and angle strains results in the puckered form of cyclobutane.

## 5.3

## Cycloalkanes

The factors that control the conformational equilibrium of butane (namely, torsional and steric strain) also apply to cycloalkanes. The eclipsing interactions that induce torsional and steric strain favor staggered over eclipsed conformers and also play a role in the three-dimensional arrangements of small- and large-ring saturated hydrocarbons.

Two factors control conformational preference in small rings: angle strain (caused by distortion from the angle of maximum overlap dictated by the hybridization of the ring atoms) and torsional strain. (Steric strain does not play a role in small-ring hydrocarbons, because there are no non-bonding interactions not already accounted for by torsional strain.) In cyclopropane, the carbon atoms are coplanar because three points determine a plane, and the C—H bonds are eclipsed. The ring strain in cyclopropane is therefore caused both by angle strain (as described in Chapter 1) and by torsional strain. Similarly, planar cyclobutane is destabilized by angle strain, because the bond angles (90°) are smaller than the ideal tetrahedral angle. Furthermore, additional destabilization (torsional strain) results from the eclipsing of substituents of all four bonds at each carbon in planar cyclobutane (see Figure 5.10). Torsional strain resulting from eclipsing is reduced in the **puckered conformation** of cyclobutane in which one atom is moved out of the plane of the other three. This puckered form has less eclipsing and thus lower torsional strain, although angle strain is increased because the bond angles are smaller than 90°. The energy due to decreasing torsional strain and increasing angle strain is minimized in a conformation for cyclobutane that is substantially distorted from planarity.

A completely planar conformation of cyclopentane has all C—H bonds eclipsed (Figure 5.11), whereas a puckered form (the envelope conformation) in which one of the carbon atoms lies out of the plane defined by the remaining ring carbon atoms has less torsional strain. Of nearly equal energy to the envelope is another conformation of cyclopentane (called a half-chair) that has two atoms out of the plane of the other three, one above and one below. See if you can build it using your molecular models.

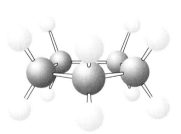

Planar

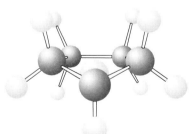

Envelope

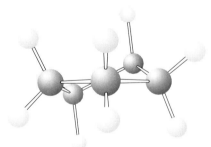

Half-chair

**FIGURE 5.11**

The envelope and half-chair conformations of cyclopentane are preferred because the sum of the strains is minimized.

For cyclic systems, the conformations in which all carbon atoms are in one plane have C—C—C bond angles dictated by the geometry of regular polyhedra, and any deviation from planarity reduces some or all of these angles. For four- and five-member rings, the planar conformations have angles (90° and 108°) smaller than tetrahedral. For a six-member planar ring, the angle dictated by geometry is 120°, clearly larger than the tetrahedral angle of 109.5° of $sp^3$-hybridized carbons. However, the nonplanar form of cyclohexane has angles close to 109.5°, smaller than those of the planar form. Similar nonplanar conformations also exist for even larger cycloalkane rings. Deviation from the plane thus decreases angle strain for large rings but increases it for small rings. For six-member and larger rings, deviation from planarity results in a conformation that is lower in energy because of a reduction in both angle strain and torsional strain.

For four- and five-member rings, deviation from planarity results in a conformation that balances a decrease in torsional strain with an increase in angle strain. As a result, four- and five-member rings must adopt conformations that are not ideal. Furthermore, the differences in energy between these low-energy conformations and the planar form are much smaller than for cyclohexane, and interconversions between the various conformations of cyclobutane or cyclopentane are much more rapid than for cyclohexane.

We can compare the effects of bond-angle strain and eclipsing interactions for different ring systems by examining the heat of combustion per methylene group, listed in Table 5.3. This value decreases with increasing ring size and reaches a minimum for cyclohexane (in which torsional and angle strains are relieved in a nonplanar conformation). The difference between the heat of combustion per methylene group for cyclobutane and that for cyclohexane is 7 kcal/mole. Because there are four $CH_2$ groups in cyclobutane, there is ring strain equal to 28 kcal/mole ($4 \times 7$ kcal/mole) in this four-member ring, the sum of the energetic costs of bond-angle distortions from the ideal 109.5° and torsional strain caused by partial eclipsing. The same calculation for cyclopropane yields a value of 30 kcal/mole of ring strain.

**TABLE 5.3**

Heat of Combustion per Methylene Group
in Cycloalkanes

| Compound | Heat Released per $CH_2$ (kcal/mole) |
|---|---|
| Cyclopropane | 167 |
| Cyclobutane | 164 |
| Cyclopentane | 159 |
| Cyclohexane | 157 |
| Cycloheptane | 158 |
| Cyclooctane | 159 |

Make models of cyclobutane and cyclopentane, and manipulate them so that they represent the conformations discussed in the text.

## 5.4

## Six-Member Carbon Rings

We have seen that there is a delicate conformational balance in four- and five-member rings between a preference for bond angles as near as possible to 109.5° (best in planar structures) and one for torsional angles of 60° (best in nonplanar structures). Six-member rings are different: both angle and torsional strains are minimized in nonplanar structures.

### Cyclohexane

Planar cyclohexane has all bonds eclipsed, as we have seen for planar three-, four-, and five-member rings. However, in contrast to the situation for smaller rings, the C—C—C bond angles of planar cyclohexane are *larger* (120°) than the ideal tetrahedral angle. Moving some of the carbon atoms out of the plane reduces these angles and thus reduces angle strain; the same movement also decreases torsional strain.

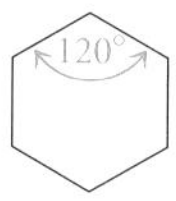

*Chair Conformation.* Moving one carbon atom up from the plane of the cyclohexane ring while moving the carbon atom at the other side down yields a conformation with ideal bond and torsional angles (Figure 5.12). Because line structures of these staggered conformations look roughly like the back, seat, and footrest of a chair, these conformations of cyclohexane are called **chair conformations.**

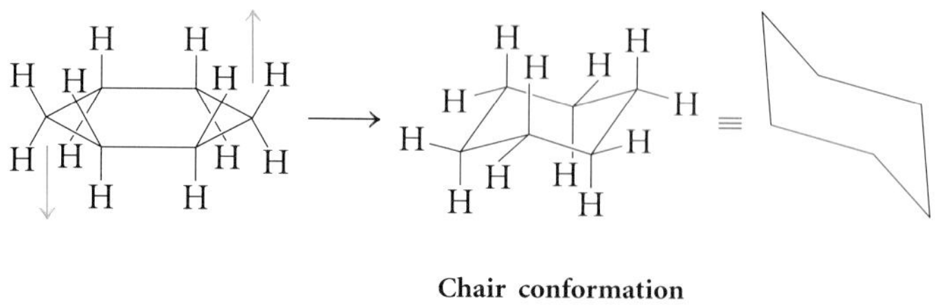

Chair conformation

**FIGURE 5.12**

Evolution of an unstable planar conformation of cyclohexane to the preferred chair conformation with bond angles near 109.5° and staggered C—C bonds.

We can draw a Newman projection that shows the view down two of the parallel bonds of chair cyclohexane (Figure 5.13).

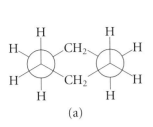

(a)          (b)

**FIGURE 5.13**

(a) Newman projection showing two of the bonds in the chair conformation of cyclohexane. (b) A ball-and-stick representation of this conformer.

    ***Boat and Twist-Boat Conformations.*** A different conformation of cyclohexane results from moving two of the carbon atoms simultaneously in the same direction (Figure 5.14). This arrangement is known as the **boat conformation** of cyclohexane because the line structure bears some resemblance to a boat.

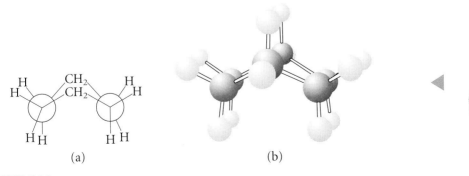

**Boat conformation**

**FIGURE 5.14**

A boat conformation is formed from planar cyclohexane by moving two carbon atoms in the same direction away from the plane.

    Although angle strain is relieved in the boat conformation, two bonds are still eclipsed (those forming the sides of the boat), as revealed by the Newman projection and ball-and-stick model in Figure 5.15.

(a)          (b)

**FIGURE 5.15**

(a) Newman projection showing two of the bonds in the boat conformation of cyclohexane. (b) A ball-and-stick representation of this conformer.

Some of the torsional strain of the boat conformation is relieved in the **twist-boat conformation** of cyclohexane. The twist boat is formed from the boat by grabbing and twisting the hydrogen atoms that point up on the frontmost and rearmost carbon atoms (called the **flagpole hydrogens,** in analogy to a boat), moving one to the left and the other to the right (Figure 5.16).

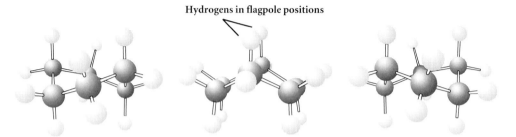

**FIGURE 5.16**

Boat (center) and twist-boat (left and right) conformations of cyclohexane.

### EXERCISE 5.6

Build a model of cyclohexane, including all twelve of the hydrogen atoms. Move the carbon atoms until you have a boat conformation. Twist the flagpole hydrogens as described in the text, and note how the torsional angles change. Do any bond angles change as the boat is converted to a twist boat?

*Interconversion of Chair, Boat, and Twist-Boat Conformations.* The three conformations of cyclohexane are all in equilibrium, although the energy barriers to interconversion are significantly higher than for acyclic and smaller-ring alkanes. The lowest-energy pathway for conversion of the chair to the boat and twist-boat conformations proceeds through the half-chair, a conformation in which five of the six carbon atoms are coplanar. If we start with a model of the chair conformation, we can arrive at the half-chair by taking any one carbon atom and moving it until it is in the plane of its four closest neighbors (Figure 5.17). (Alternatively, we can start with the planar conformation and move one of the atoms out of the plane of the other five.) By continuing to move this carbon atom in the same di-

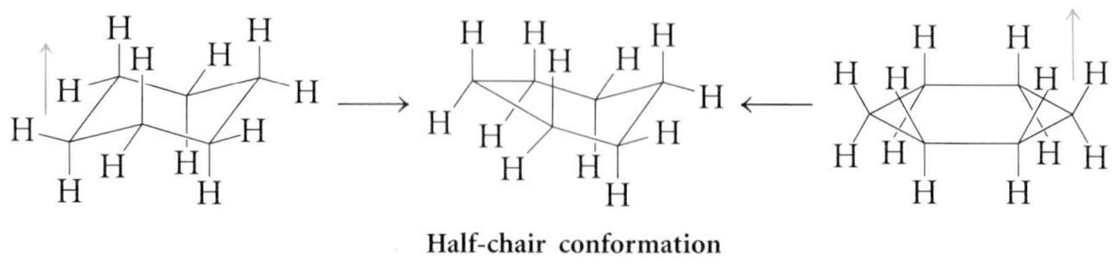

Half-chair conformation

**FIGURE 5.17**

Conversions of the chair (left) and planar (right) conformations of cyclohexane to the half-chair conformation.

rection and imparting a twist, we arrive at a twist-boat conformation. The twist boat can then interconvert with another twist boat by passing through the boat conformation as a transition state.

The activation energies and the energy differences between various conformations for the conformational interconversions of chair to twist boat and of twist boat to twist boat are summarized in Figure 5.18. There are several points worth noting in this energy profile. The energy of activation for the conversion of the chair conformation to the half-chair is 11 kcal/mole, substantially higher than any conformational barrier that we have encountered so far. Thus, this transformation of cyclohexane occurs about a million times more slowly than rotation about the carbon–carbon bond of ethane. On the other hand, the barrier to interconversion of the twist-boat conformations (through the boat) requires only 1.6 kcal/mole and is thus faster than rotation in ethane.

The energy difference between the chair and twist-boat conformations of cyclohexane is sufficiently large that the twist boat constitutes only a small fraction of the equilibrium concentration at room temperature. Chemists consider the typical conformation of cyclohexane to be the chair.

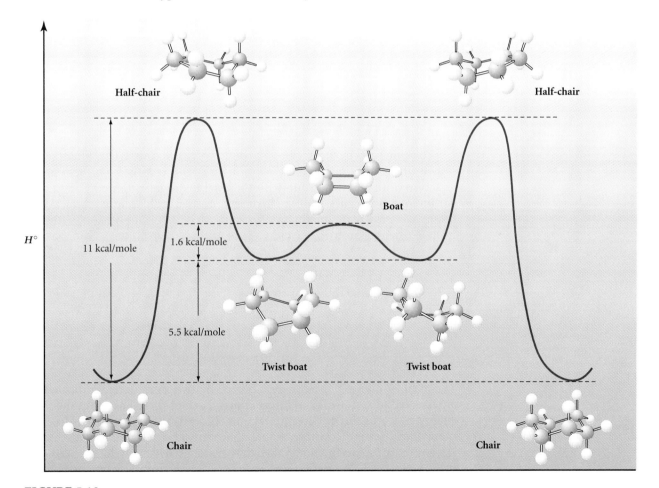

## FIGURE 5.18

An energy profile for the interconversions of the various conformations of cyclohexane.

With your molecular models, try to flip cyclohexane from one chair conformation to another and from one boat conformation to another. As you work with your model, it will become clear that it is possible to convert one chair to another by proceeding through intermediate twist-boat and boat conformations.

## Monosubstituted Cyclohexanes

Now let's consider the various conformations of a cyclohexane ring with a methyl group attached to C-1. There are two unique chair conformations of methylcyclohexane, which do not have the same energy. In one conformation, the methyl group points away from the "seat" of the chair, whereas in the other conformation, the methyl group is roughly coplanar with the seat. These two positions of the methyl group are called **axial** and **equatorial,** respectively. The two chair conformations can be interconverted simply by flipping the atoms of the ring back and forth. (Use a model to convince yourself that a ring flip converts each axial substituent into an equatorial position and changes each equatorial substituent into an axial one.)

Axial                    Equatorial

Looking down the C-1—C-6 bond of axial methylcyclohexane, as shown in the Newman projection in Figure 5.19, reveals a *gauche*-type interaction of the methyl group with the axial hydrogen atom on C-5 (a similar interaction occurs with the hydrogen on C-3). This steric interaction is not present when the methyl group is in the equatorial position, where its relative orientation to C-5 (and C-3) is *anti*.

Equatorial methylcyclohexane

The axial isomer is therefore destabilized by two steric interactions, which are referred to as **1,3-diaxial interactions.** Each of these has about the same energy as a *gauche* butane interaction. Indeed, the equatorial isomer of methylcyclohexane is more stable than the axial isomer by about 1.8 kcal/mole, an experimental value that matches what would be predicted from two *gauche* butane interactions ($2 \times 0.9$ kcal/mole). (Again, the use of molecular models will help you see that a 1,3-diaxial interaction is very similar to the interaction that destabilizes *gauche* butane.)

The chair conformations of methylcyclohexane, both axial and equatorial, lack the strong destabilizing eclipsing interactions of the boat and

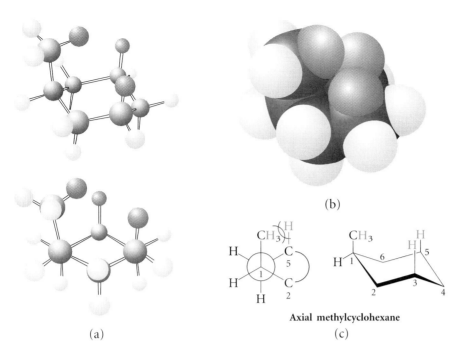

(b)

Axial methylcyclohexane

(a)                                    (c)

## FIGURE 5.19

(a) Ball-and-stick (from two perspectives) and (b) space-filling models of
axial methylcyclohexane. The steric repulsions between the hydrogen atoms
shown in green are responsible for the higher energy of the axial conformation compared with the equatorial isomer. (c) The Newman projection at
the lower center is obtained by visualizing down the C-1—C-6 bond, with
C-1 shown in the front and C-6 hidden from view by C-1. The overlapping
arcs between the methyl group and the hydrogen at C-5 represent repulsive
interaction between these groups. An identical interaction between the
methyl group and the axial hydrogen on C-3 is not shown.

twist-boat conformations. In the boat isomer, shown in Figure 5.20, there
is highly unfavorable steric interaction between the C-1 methyl group and
one of the hydrogen atoms on C-4 (both shown in green).

Because of 1,3-diaxial interactions, chair cyclohexanes with equatorial
substituents are more stable than those with axial substituents. In general,

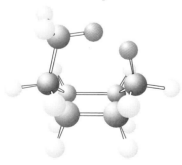

## FIGURE 5.20

Interaction of flagpole hydrogens (shown in green) in one boat conformation of methylcyclohexane.

## TABLE 5.4

Energy Cost of a Single 1,3-Diaxial Interaction

| 1,3-Diaxial Interaction | Energy (kcal/mole) |
|---|---|
| Hydrogen and methyl group | 0.9 |
| Hydrogen and ethyl group | 1.0 |
| Hydrogen and isopropyl group | 1.1 |
| Hydrogen and phenyl group | 1.5 |
| Hydrogen and t-butyl group | 2.7 |

either of these chair conformations is more stable than the other conformational possibilities—that is, the boat and twist-boat conformations. The energy difference between a conformation having a substituent in an axial position and one having the substituent in an equatorial position depends on the steric requirement, or size, of the substituent. The larger the substituent, the greater is the steric strain resulting from 1,3-diaxial interactions. Table 5.4 lists values representing the contribution of each 1,3-diaxial interaction to the relative destabilization of the axial conformation. (Recall from Table 5.2 how these energy differences affect the conformational equilibrium.) Some substituents are so large that they effectively act as conformational anchors, or **locks.** For example, the energy difference between an axial and an equatorial position for a *t*-butyl group is very large. Although ring flipping between the chair conformations is still rapid, the equilibrium is so strongly dominated by the conformer bearing the *t*-butyl group in the equatorial position that that conformation is considered "locked," with all other substituents fixed in axial or equatorial positions as determined by the preference of the anchoring *t*-butyl group for the equatorial orientation.

### EXERCISE 5.8

Note in Table 5.4 that the 1,3-diaxial interactions between a hydrogen atom and a methyl, ethyl, or isopropyl group are all approximately the same (0.9–1.1 kcal/mole), whereas the interaction of a hydrogen with a *t*-butyl group is much larger (2.7 kcal/mole). Use molecular models to explain why the *t*-butyl group is different from the other three alkyl substituents. (*Hint:* Don't forget that rotation is possible about the σ bond between the ring and the substituent.)

### Disubstituted Cyclohexanes

We can extend the ideas developed in considering the conformations of monosubstituted cyclohexanes to disubstituted cyclohexanes. In *cis*-1,4-dimethylcyclohexane, one methyl group is in an equatorial position and the other is axial. Chair–chair ring-flipping changes the position of each of these substituents, converting the axial position to an equatorial one, and moving the equatorial methyl group to an axial position. The pattern of axial and equatorial substituents is completely inverted by a ring-flip:

**Inversion of Axial (blue) and Equatorial (red) Substituents by Ring-Flipping**

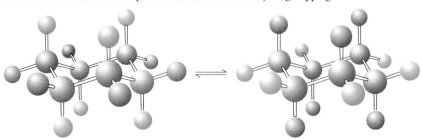

Thus, ring-flipping of *cis*-1,4-dimethylcyclohexane results in a conformation that is identical in all respects with the first (Figure 5.21). The sit-

cis-1,4-dimethylcyclohexane

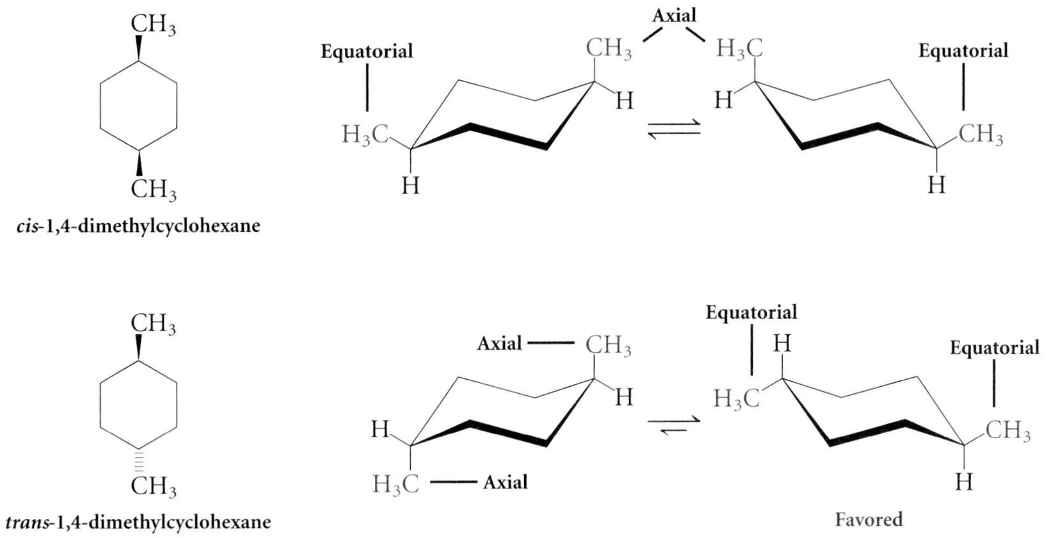

trans-1,4-dimethylcyclohexane

**FIGURE 5.21**

Chair conformations of *cis*- and *trans*-1,4-dimethylcyclohexane.

uation is quite different for *trans*-1,4-dimethylcyclohexane because both methyl groups must be either equatorial or axial. Thus, we would expect the energy difference between these two chair conformations to be large. The conformation with two equatorial methyl groups is clearly more stable.

In *cis*- and *trans*-1,3-dimethylcyclohexanes, the *trans* isomer has one axial and one equatorial substituent in both chair conformations (Figure 5.22). On the other hand, the *cis* isomer has a conformation in which both methyl groups are equatorial. This diequatorial conformer is more stable than the alternative, in which both methyl groups are axial (Figure 5.22).

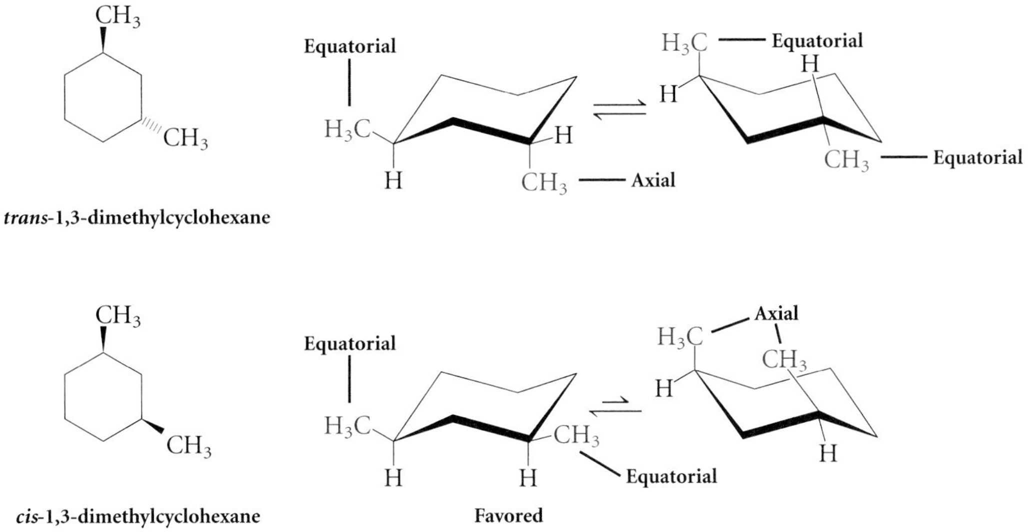

trans-1,3-dimethylcyclohexane

cis-1,3-dimethylcyclohexane

**FIGURE 5.22**

Chair conformations of *trans*- and *cis*-1,3-dimethylcyclohexane.

247

**EXERCISE 5.9**

Indicate which of the following pairs of isomers is conformationally more stable, and draw a line structure showing its preferred conformation.

(a) *cis*-1-*t*-butyl-2-methylcyclohexane or *trans*-1-*t*-butyl-2-methylcyclohexane

(b) *cis*-1,4-diisopropylcyclohexane or *trans*-1,4-diisopropylcyclohexane

(c) *cis*-1,3-dibromocyclohexane or *trans*-1,3-dibromocyclohexane

(d) *cis*-1-*t*-butyl-3-ethylcyclohexane or *trans*-1-*t*-butyl-3-ethylcyclohexane

## Fused Six-Member Rings: Decalins

Knowledge of the conformations of 1,2-dimethylcyclohexanes is useful in the conformational analysis of fused, saturated rings. For example, we can visualize *trans*-decalin, a hydrocarbon in which two cyclohexane rings have two carbon atoms in common, as being related to the most stable conformation of *trans*-1,2-dimethylcyclohexane by mentally extending the two methyl groups, with two additional carbons, into another ring (Figure 5.23). The *trans* ring fusion is clearly indicated by the relative positions of the hydrogen atoms at the **bridgehead positions,** that is, attached to the carbon

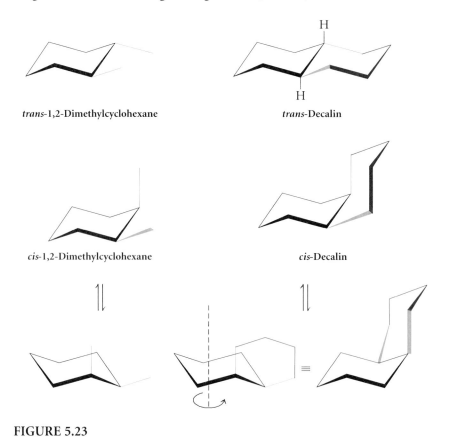

*trans*-1,2-Dimethylcyclohexane

*trans*-Decalin

*cis*-1,2-Dimethylcyclohexane

*cis*-Decalin

**FIGURE 5.23**

A comparison of the conformations of *trans*- and *cis*-decalin with the chair forms of *trans*- and *cis*-1,2-dimethylcyclohexane, respectively.

atoms common to both rings. Unlike dimethylcyclohexane, however, *trans*-decalin cannot flip to another, stable chair–chair form. Although the two additional carbons forming the second six-member ring of *trans*-decalin can bond to two adjacent equatorially oriented carbons on the first ring without strain, they cannot form sufficiently long links with two adjacent, axial carbons. (Use models to convince yourself that this is the case.) As a result, ring-flipping is blocked in *trans*-decalin.

In a similar fashion, we can visualize *cis*-decalin (Figure 5.23) as being formed from *cis*-1,2-dimethylcyclohexane by the addition of two carbons, which provide a chain long enough to reach between the axial and equatorial methyl groups. In *cis*-1,2-dimethylcyclohexane, one of the methyl groups is axial and the other equatorial. In this isomer, chair–chair ring-flipping takes place quite readily. The conformers of *cis*-decalin can undergo similar interconversions.

---

**EXERCISE 5.10**

Draw line structures for 2-methyl-*cis*-decalin and 3-methyl-*cis*-decalin (using the skeletal numbering scheme shown here) with the methyl groups *cis* to the hydrogen atom at C-1 and with the methyl groups *trans* to the hydrogen atom at C-1.

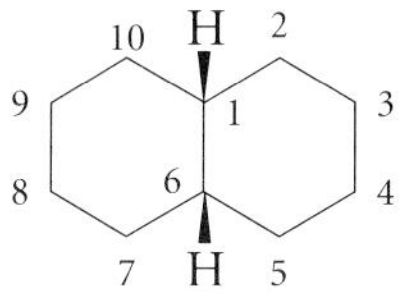

For all four structures, let both rings undergo a ring-flip, and draw the ring-flipped isomer. In each case, decide which ring-flipped structure is conformationally more stable.

---

## 5.5

# Chirality

A final type of stereoisomerism is found in molecules that are chemically and physically identical except for their interaction with polarized light. In these isomers, all the connectivities of the atoms are the same, but the isomers cannot be interconverted by bond rotation, the atoms in the two isomers are not superimposable on one another, and the shapes of the molecules are related as mirror images. Molecules (and other objects) having nonsuperimposable mirror images are said to be **chiral.** Chiral molecules are frequently encountered in nature.

One way of recognizing chirality in objects is to look for "handedness." Your left and right hands are clearly different even though each of the component parts (for example, the thumbs) appear to be the same (Figure 5.24, on page 250). A hand is chiral because one of its ends (the fingers) is different from the other end (the wrist), its thumb is different from its little finger, and its back is different from its palm. Thus, although your left hand is very similar to your right hand, the hands cannot be superimposed. Right-

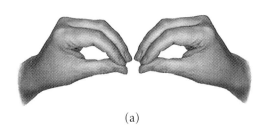

(a)              (b)

**FIGURE 5.24**

(a) Your left and right hands are not the same. (b) A clock is not identical to its mirror image. The hands on the clock run in a clockwise direction, whereas the hands on the mirror image run counterclockwise. Also note that the numbers are backwards in the mirror image. (Is there an achiral number?)

and left-handed gloves also are chiral, and thus different: a right-handed glove fits your right hand, not your left. The difference between your hands is maintained no matter how they are oriented—for example, with the thumbs up or down. Although your two hands are not the same, they are related: they are mirror images of each other. Gloves that are completely flat, like those worn by children, have a mirror plane and can be used on either hand. However, even then, if a glove is transferred from one hand to the other, the face of the glove that was on the palm of the left hand covers the back of the hand when worn on the right hand. Many everyday objects are chiral, such as the clock (and its mirror image) in Figure 5.24.

The presence of a **mirror plane** through an object assures that it will be superimposable on its mirror image. A mirror plane is a plane running through a three-dimensional object such that each part of the object on one side of the plane is mirrored by an identical part on the opposite side. If a human hand were completely flat and the back were identical with the palm, the hand could have a mirror plane through the palm and fingers. If hands were like this, the same glove would fit both the right and left hands identically. If there is no mirror plane through an object, it can be considered "handed"—that is, chiral.

Molecules (or objects) are either chiral or not. There is no in-between state. A molecule is considered **achiral** (not chiral) as long as at least one of its energetically accessible conformations has a mirror plane of symmetry, even if the others lack such a plane.

When an atom has substituents oriented in three dimensions such that there is no mirror plane through the atom, then (except in very special cases) it will not be possible to find a mirror plane for the molecule as a whole. Because of this **center of chirality,** the molecule is chiral. Such atoms are sometimes referred to as *chiral atoms,* or *chiral centers,* but chirality is a property only of a complete object, not of its parts. Therefore, the expression *center of chirality* is used in this book to emphasize that an atom contributes to the overall handedness of the molecule. (The term *stereogenic center* is also used instead of *chiral atoms,* but *center of chirality* is based on an established English word, *chirality,* whereas *stereogenic* is not.)

Any atom that is *sp*- or *sp*²-hybridized has a mirror plane (that containing the bonded atoms). Therefore, the carbons of an alkene or alkyne are not centers of chirality, irrespective of substituents. On the other hand, an *sp*³-hybridized carbon does have the three-dimensionality necessary to impart chirality to a molecule, provided that the four substituents are different.

Let's consider carbon atoms bonded to three or two identical atoms; examples of such compounds are shown in three dimensions in Figure 5.25 as pairs of mirror images. In both cases, the molecule at the right of the mirror can be superimposed on the molecule at the left by a 180° rotation about the vertical axis.

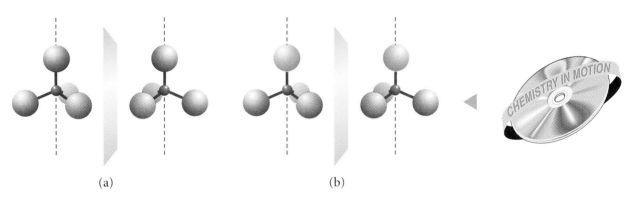

(a)                                         (b)

**FIGURE 5.25**

Mirror images of a carbon atom with (a) three and (b) two identical substituents (the gray spheres, perhaps hydrogen atoms).

Two molecules are **superimposable** if a conformation exists in which each of four substituents of one can be placed over the same substituent of the other. The atoms of the molecule are thus oriented in exactly the same way in space. (Use molecular models to confirm that this is so.) In the examples in Figure 5.25, mirror planes can pass *through* each molecule such that everything on one side of the plane has an exact counterpart on the other side. Any molecule (and any object) that has such a mirror plane is achiral.

**Achiral molecules have a mirror plane**

### Enantiomers

Any *sp*³-hybridized carbon atom that bears two identical substituents has a mirror plane through which one of the substituents is a reflection of the other. That plane contains the tetrahedral carbon and the two other substituents, X and Y. Thus, for an *sp*³-hybridized carbon atom to be a center of chirality, it is *necessary* and *sufficient* that four different groups be

#08   Intro to Stereochemistry

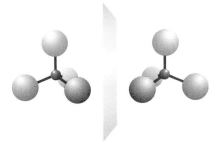

**FIGURE 5.26**

Mirror images of tetrahedral carbon bearing four different substituents. These molecules are not superimposable and are therefore chiral.

bound to carbon. Figure 5.26 shows a generalized example with four different substituents (represented by colored spheres) arranged at a tetrahedral center of chirality.

### EXERCISE 5.11

Locate one or more mirror planes in each of the following molecules:

(a) ethylene    (b) benzene    (c) *anti*-butane    (d) propyne

Each representation in Figure 5.26 shows a unique way that four groups can be oriented in three dimensions. These representations are mirror images of each other and cannot be superimposed. For example, if we place the two blue and green spheres on top of each other by rotation about the vertical axis, the red and white spheres will be in the wrong places. Indeed, no matter how we move and turn one of these images, we cannot overlap it with its mirror image. These two molecules are *stereoisomers,* molecules that differ only in the way in which the four substituent groups are oriented in space. Stereoisomers related to each other as nonsuperimposable mirror images are referred to as **enantiomers.**

Enantiomers can be interconverted only by switching the positions of two substituents, a process that requires the breaking and reforming of $\sigma$ bonds at the center of chirality. Specifically, enantiomers are not interconverted by rotations about $\sigma$ bonds and in this respect differ significantly from conformational isomers. In analyzing a molecule for the presence of a center of chirality, we are free to use any conformational representation (for example, eclipsed or staggered) to compare structures without worrying that we are changing the chiral center, because enantiomers are interconverted only by breaking $\sigma$ bonds.

### Representing Enantiomers in Two Dimensions

Describing three-dimensional molecules with two-dimensional drawings requires the use of certain conventions and styles. For example, we can use a wedge bond to emphasize that the bond is projecting out of the plane of the paper and toward the viewer, and a hatched line to indicate a receding bond, as for 2-butanol:

Human olfactory receptor sites are very sensitive to the shape of gaseous molecules. For example, a floral odor is caused by molecules that are spherical at one end and elongated at the other, somewhat like a miniature guitar, whereas a peppermint-like odor is produced by ellipsoidal molecules. For many components of perfume, shape seems to be more important than chemical composition: hexachloroethane, (+)-camphor, and cyclooctane have nearly the same odor, even though their molecular formulas are quite different. However, all of these molecules are roughly bowl-shaped, which allows a reasonable fit with the receptor site for what perfumers call "camphoraceous" molecules.

Hexachloroethane

Camphor

Cyclooctane

The ability to detect the presence of small quantities of molecules at various olfactory receptor sites varies. People gifted in this sense are well-respected (and highly paid) by the perfumers in the south of France and by wineries throughout the world.

Because stereochemistry significantly affects the shape of a molecule, a molecule's absolute configuration also strongly affects its odor. Enantiomers, for example, can elicit quite different responses: the characteristic aromas of oil of caraway and oil of spearmint are due to the separate enantiomers of carvone.

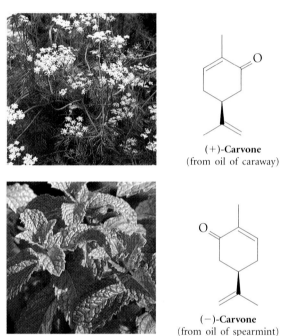

(+)-**Carvone**
(from oil of caraway)

(−)-**Carvone**
(from oil of spearmint)

2-Butanol

It is not necessary to use both a hatched and a wedge bond to indicate three-dimensional arrangements—one or the other is sufficient, as in the two representations at the right. These short-cuts are often used, but keep in mind that the clearest representation will be one that gives a "feeling" of three dimensions.

253

In each of the following molecules, indicate the location of a center of chirality with an asterisk.

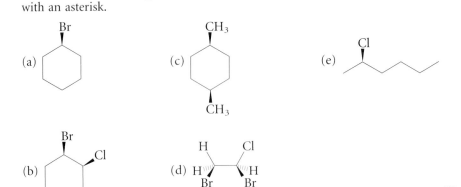

## 5.6

## Absolute Configuration

Because a molecule that has a center of chirality has two different enantiomers, chemists need to be able to refer uniquely to one or the other of the enantiomeric pair—just as you can specify, for example, a left or right shoe. An unambiguous method for specifying **absolute stereochemistry** was developed by three chemists and is known as the Cahn–Ingold–Prelog rules. (Vladimir Prelog was awarded the Nobel Prize in 1975 for his contributions to organic stereochemistry.) In contrast, **relative stereochemistry** refers only to the relation between two molecules; for example, saying that two molecules are enantiomers does not specify which is which.

The specification of absolute stereochemistry makes use of the same priority rules employed to describe $E$ and $Z$ isomers in Chapter 2. In applying these rules, we look first at the atoms directly attached to the center of chirality and assign priority on the basis of atomic number. In cases where two (or more) of these atoms have the same atomic number, we proceed along the chain until a difference is found. Thus, —$CH_2CH_3$ has higher priority than —$CH_3$, because the highest-priority substituent of the carbon atom of the ethyl group is C, whereas on the methyl group it is H:

$$-\!\!\!\!-CH_3 \qquad\qquad H_3C\!\!-\!\!H_2C\!\!-\!\!\!\!-$$

$$
\begin{array}{ccc}
\text{H} & & \text{C} \\
\text{H} & \text{versus} & \text{H} \\
\text{H} & & \text{H}
\end{array}
$$

When we encounter a double bond, we count the atom a second time, creating a "dummy" or "phantom" atom. Thus, an aldehyde carbon (—CHO) has higher priority than a primary alcohol (—$CH_2OH$):

$$\underset{}{\overset{O}{\underset{\|}{\phantom{}}}}$$

$\dashv CH_2OH \qquad HC\dashv$

| | |
|---|---|
| O | O |
| H | versus O |
| H | H |

In accord with these rules, priority is assigned to each of the four groups attached to the center of chirality, which are then uniquely defined as 1, 2, 3, and 4. Let's assume that the substituents A, B, C, and D have priorities that decrease in that order. We view the molecule by looking down the bond between the central carbon atom and the substituent of lowest priority (D), putting this substituent as far away as possible. This perspective for the isomer on the left in Figure 5.26 is shown in Figure 5.27. (Make a model to convince yourself that this is so.) When the assigned priorities (proceeding from highest to lowest) for the remaining three substituents are arranged in a counterclockwise direction (A → B → C), as in Figure 5.27, the isomer is designated **S** (from the Latin *sinister,* for "left"). The structure is referred to as the *S* isomer and as having the **S configuration** at its center of chirality.

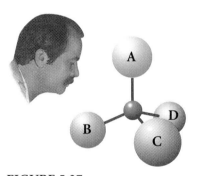

**FIGURE 5.27**

Orienting a center of chirality to assign absolute configuration with priorities as A > B > C > D. The center has the *S* configuration.

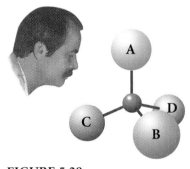

**FIGURE 5.28**

Orienting a center of chirality to assign absolute configuration with priorities as A > B > C > D. The center has the *R* configuration.

We assign a configuration to the structure on the right in Figure 5.26 in the same way, viewing down the bond from the center of chirality to substituent D, as shown in Figure 5.28. Here, the direction A → B → C is clockwise, and the isomer is assigned the stereochemical designation **R** (from the Latin *rectus,* "right") and has the **R configuration**.

Let's consider a specific example by assigning an absolute configuration to an isomer of 2-bromobutane.

$$\overset{Br\ \ H}{\underset{\underset{1\qquad\ 3}{\Large\diagup\!\!\!\diagup\!\!\diagdown\!\!\diagup}^{4}}{}}$$

**(R)-2-Bromobutane**

Our first task is to identify the center of chirality in the molecule. Carbon-1 bears three hydrogen atoms and cannot be a center of chirality. Carbon-2, which bears four different substituents, is a center of chirality, as indicated by the asterisk. Carbon-3 bears two hydrogen atoms and is therefore not a center of chirality, nor is carbon-4, which bears three hydrogen atoms. Using the atomic-number rule to assign priority, we find that the substituents at carbon-2 have the priorities $Br > CH_3CH_2 > CH_3 > H$. Because hydrogen is of lowest priority, we must visualize the molecule by looking down the C—H bond, that is, from C-2 toward hydrogen. From this orientation, the direction from bromine to ethyl to methyl is clockwise; thus, the isomer shown is the $R$ enantiomer.

**Assignment of priority
in ($R$)-2-bromobutane**

Now that we have a three-dimensional representation of ($R$)-2-bromobutane, it is easy to draw the $S$ isomer. We simply exchange any two substituents of the $R$ configuration, as shown in the following six three-dimensional representations of ($S$)-2-bromobutane.

($S$)-2-Bromobutane

Because the interconversion of enantiomers requires the breaking of $\sigma$ bonds, it is quite difficult to accomplish in most cases.

It is important for you to be able to assign absolute configuration to a given center of chirality, as this will aid your understanding of the relation between structures drawn in various ways.

### EXERCISE 5.13

Apply the Cahn–Ingold–Prelog rules to assign absolute stereochemistry to each center of chirality in the following molecules:

## EXERCISE 5.14

Draw a three-dimensional representation of each of the following stereoisomers:

(a) (*R*)-2-bromopentane

(c) (*R*)-2-fluoro-2-chlorobutane

(b) (*S*)-3-bromo-3-chlorohexane

(d) (*R*)-1-bromo-(*S*)-2-fluorocyclohexane

## 5.7

# Polarimetry

Because all the chemical bonds in a chiral molecule are also present in its enantiomer, two enantiomers might be expected to have identical physical properties. This is generally true except when chiral molecules interact with other chiral objects. The circularly polarized components of plane-polarized light are chiral, and the plane of polarization is rotated to the right by one enantiomer and to the left by the other. Figure 5.29 shows a schematic representation of the operation of a polarimeter capable of measuring this optical effect.

**FIGURE 5.29**

Schematic representation of a polarimeter. Normally, light is composed of rays that have electromagnetic fields oscillating in all directions. A polarizer inserted between the light source and the sample tube passes only light vibrating in one plane, as shown at the left. The direction of alignment of the polarizer defines the incident plane of the light as it passes into the cell containing the sample. As the plane-polarized light passes through an optically active sample, the light is rotated because of electronic interaction with the chiral molecules present, and the plane of polarized light emerging from the sample tube is rotated from its original plane of polarization. A second polarizer (at the right) placed at the sample-tube exit is rotated by the observer until the light intensity is greatest. The degree of rotation of the second polarizer relative to the first represents the measured rotation of the sample. (In practice, it is often easier to adjust the second polarizer until the light intensity is at a minimum. The relative rotation of the two polarizers must then be corrected by 90°.)

Ordinary light behaves like an electromagnetic wave that oscillates in all directions perpendicular to the path of propagation. When a light beam passes through a polarizer, the waves whose vibrations are not directionally aligned with the polarizer are absorbed (or reflected). The light beam that emerges from the polarizer has all electric and magnetic oscillations in the same plane. As this plane-polarized light passes through a chiral medium,

### SEPARATION OF ENANTIOMERS IN THE LABORATORY

The first resolution (separation of enantiomers constituting a racemic mixture) conducted in a laboratory was done by Louis Pasteur, whose contributions to microbiology (fermentation, pasteurization of milk, sterilization of surgical instruments, development of a rabies vaccine, and many others) are even better known than his contributions to chemistry. Pasteur decided to investigate the crystals that form in wine barrels during fermentation and aging (and on the corks of wine bottles). These crystals are called *racemic acid,* from the Latin word *racemus,* for "grapes." All the crystals were found to have the same molecular formula, which corresponded to a mixed sodium ammonium salt of tartaric acid,

$HO_2CCH(OH)CH(OH)CO_2H$. However, when viewed carefully under a microscope, the crystals appeared to consist of two different sets, differing in their three-dimensional shape. In fact, these shapes were mirror images of each other. With tweezers, Pasteur painstakingly separated the two different types of crystals and showed that they had exactly the same physical and chemical properties, except that in solution they rotated a plane of polarized light in opposite directions.

the asymmetrical nature of the chiral molecule causes the plane of vibration to rotate from its original position. A polarizer placed behind the sample is rotated so as to compensate for the induced rotation of the plane of polarization by the sample. The observed rotation will depend both on the rotating ability of the chiral molecules encountered by the light and on their number.

### ■ Optical Activity

Only objects that are chiral can rotate a plane of polarized light, and chiral molecules are often referred to as being **optically active.** The extent of rotation observed depends on the magnitude and asymmetry of a sample's electric field (which is characteristic of the particular molecule being measured), the wavelength of the light, and the number of optically active molecules in the sample. The **specific rotation** of an optically active compound is defined as the observed rotation divided by the concentration of the sample and the path length through which the light passes.

$$\text{Specific rotation} = [\alpha] = \frac{\alpha}{c \times l} \tag{3}$$

where $\alpha$ is the observed rotation (at the wavelength of the polarized light), $c$ is the concentration (in g/mL), and $l$ is the path length (in dm).

The yellow light emitted by a sodium lamp is often used as the polarized light, because the light source is inexpensive and the light can be easily filtered so that only a single wavelength of light (the D-line) can be used. The specific rotation induced by this light is called $[\alpha]_D$. Once the specific rotation has been measured for a pure, optically active compound, the anticipated rotation of a particular sample can be calculated:

$$\text{Observed rotation} = [\alpha] \times c \times l \tag{4}$$

The specific rotation can differ for dilute and concentrated solutions of the same compound, especially for molecules that self-associate through hydrogen bonding. Temperature can also influence the degree of association. For these reasons, it is common practice to report the concentration (in g/mL) and temperature at which a measurement was made.

$$[\alpha]_D^{25} = -13 \quad (c = 0.25)$$

Enantiomers differ only with respect to the sign of the specific rotation, that is, the direction of rotation of a plane of polarized light. A pure sample of one enantiomer of a pair rotates a plane of polarized light to a degree exactly equal to that of the other member but in the opposite direction. A 50:50 mixture of enantiomers does not show optical activity, because the two compounds have effects of equal magnitude but opposite direction. For an equimolar mixture of two enantiomers, the rotation induced by one enantiomer exactly cancels that of the other, and the mixture does not rotate the plane of polarized light. Such a 50:50 mixture of enantiomers is **optically inactive** and is referred to as a **racemic mixture,** a **racemic modification,** or simply a **racemate.** By definition, a racemic mixture is optically inactive, and its rotation is always 0°.

## Optical Purity

A mixture of enantiomers in a 75:25 ratio will have an observed rotation that is 50% of that for a single enantiomer (determined under identical conditions). For this reason, such a mixture is said to have an **optical purity** (o.p.) of 50%. In general, we can calculate the optical purity of a mixture of enantiomers as follows:

$$\text{o.p.} = \frac{\alpha}{[\alpha] \times c \times l} \times 100\% \tag{5}$$

where $\alpha$ is the measured rotation of a sample, and $[\alpha] \times c \times l$ is the observed rotation calculated from the specific rotation of a single enantiomer (equation 4, above).

Because optical purity is related to the *excess* of one enantiomer over the other, we can use it to determine **enantiomeric excess (e.e.).** For example, in a 75:25 mixture, there is 50% more of one enantiomer than of the other. In the past, optical purity was often equated with enantiomeric excess, but the latter has taken on a slightly different meaning with the advent of modern methods that allow the physical separation of enantiomers by chromatography and the observation of unique signals for each enantiomer in NMR spectra. Enantiomeric excess is defined as:

$$\text{e.e.} = \text{\% of major enantiomer} - \text{\% of minor enantiomer}$$

Note that for a racemic mixture, the enantiomeric excess is 0%. A racemic mixture has both an optical purity and an enantiomeric excess of 0%.

### EXERCISE 5.15

(a) Calculate the optical purity of an enantiomeric mixture in which the specific rotation of one enantiomer is $+100°$ and the observed rotation of the mixture is $+10°$.

(b) Calculate the optical purity of an enantiomeric mixture in which the specific rotation of one enantiomer is $+200°$ and the observed rotation of the mixture is $+50°$.

(c) Calculate the enantiomeric excess for a sample in which the ratio of enantiomers determined by chromatography is $3.5:1$.

## 5.8

## Designating Configuration

Enantiomers are stereoisomeric molecules that have nonsuperimposable mirror images. The members of the pair are separate compounds whose disposition in three-dimensional space must be individually defined. The use of the Cahn–Ingold–Prelog rules to do this has been covered in the preceding section. In this section, we will look at ways of specifying relative configuration and extend the method of assigning absolute configuration at a single center of chirality to specifying configuration when molecules contain more than one center of chirality.

### A Single Center of Chirality: Relative Configuration

As we have seen, the absolute configuration of a chiral molecule can be specified by applying the Cahn–Ingold–Prelog rules to designate the configuration as $R$ or $S$. This method specifies absolutely the direction of groups in space without relation to physical properties. An alternative way of referring uniquely to one member of the enantiomeric pair is to specify its **relative configuration,** which is based on the sign of its specific rotation. Thus, the enantiomer that rotates a plane of polarized light in a clockwise direction (when the observer looks at the light) is called the (+)-**isomer;** its mirror image, which rotates the plane of polarized light in a counterclockwise direction, is called the (−)-**isomer.** A (+) or (−) designation does not specify how the groups are arranged spatially but simply relates one structure to the sign of its specific rotation. *It is important to note that no simple relation exists between the sign of the optical rotation (±) and the absolute configuration (R,S) of an enantiomer.*

An equivalent representation is to use lowercase **d,** for **dextrorotatory** (from the Greek for "right rotating") to indicate the (+) enantiomer and

*CHEMICAL PERSPECTIVES*

THE ABSOLUTE CONFIGURATION
IN CHIRAL NATURAL PRODUCTS

Only in the 1930s did it become possible, by using x-ray crystallographic analysis, to determine the actual arrangement of atoms in three-dimensional space about a center of chirality. Long before, however, chemists were drawing three-dimensional representations of molecules based on whether they were related to (+)- or (−)-glyceraldehyde. A molecule was said to belong to one of these series if it could be converted into another compound already in that series by reactions that were not expected to change the stereochemistry. The original assignment of the arrangement for each series was made arbitrarily but ultimately was shown to be correct (a 50:50 chance).

lowercase *l*, for **levorotatory** (from the Greek for "left rotating") to indicate the (−) enantiomer. A third representation for relative configuration at centers of chirality is D or L, based on correspondence with naturally occurring glyceraldehyde, which is dextrorotatory.

## Multiple Centers of Chirality: Absolute Configuration

A molecule can have more than one center of chirality. Let's consider the possible configurations for 2-bromo-3-chlorobutane:

#09   Diastereomers

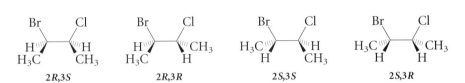

Note that this molecule contains two centers of chirality: one at C-2 and one at C-3. We can draw four different three-dimensional representations in which the carbon–halogen bonds are held in the plane of the paper. With every additional asymmetric center in a molecule, the number of possible stereoisomers doubles. Thus the *maximum possible number of stereoisomers for n centers of chirality is equal to 2 raised to the nth power, or $2^n$.*

Note that for 2-bromo-3-chlorobutane the $2R,3R$ and the $2S,3S$ isomers are an enantiomeric pair and the $2R,3S$ and the $2S,3R$ isomers are another enantiomeric pair. However, the $2R,3R$ isomer is not an enantiomer of the $2R,3S$ isomer, nor are the $2S,3S$ and the $2S,3R$ isomers enantiomers. Stereoisomers that are not mirror images are referred to as **diastereomers.** Thus, the relation between the $2R,3R$ and the $2R,3S$ isomers of 2-bromo-3-chlorobutane is diastereomeric, as is the relation between the $2S,3S$ and $2S,3R$ isomers. Although, except for the sign of their specific rotation, enantiomers have identical physical properties (melting points, boiling points, and so forth), diastereomers have different physical properties.

## ▓ Resolution of Enantiomers

It is not possible to separate enantiomers by the usual methods of purification. Because they have identical melting points, boiling points, and solubilities, enantiomers cannot be purified by recrystallization or the usual chromatographic techniques. On the other hand, because diastereomers do not have identical physical properties, they can be separated by various physical methods, including chromatography and recrystallization. By using a little ingenuity, chemists make use of these physical differences between diastereomers to separate the enantiomers. For example, a mixture of enantiomers can be converted into diastereomers by reaction with a single, optically active enantiomer as a reagent, as when the acid–base reaction of a racemic mixture of 2-chloropropanoic acid with one enantiomer of α-phenethylamine produces a mixture of diastereomeric salts.

In this acid–base reaction, the configuration does not change at the center of chirality in either the acid or the amine, because bonds are neither broken nor made at these carbon atoms. The salts formed from the racemic starting material retain the original configuration at each center of chirality.

The two salts ($R,R$ and $R,S$) formed are not mirror images and are therefore diastereomers. At this stage, they can be separated because diastereomers have different physical properties. Here, repeated recrystallization of the salts can yield a single diastereomer as a pure, crystalline solid. The separated diastereomers can then be reconverted into their components, the carboxylic acid and the amine. (How could you accomplish this process in the laboratory?)

This method of **resolution** (a technique for separating enantiomers) by forming and then separating diastereomers, followed by regeneration of the original reactants, makes use of a fundamental difference between stereoisomeric pairs that are diastereomers and those that are enantiomers. Diastereomers can be separated because they have different physical properties; enantiomeric pairs cannot be separated because their physical properties are identical except for the sign of optical rotation.

An alternative method for resolving a racemic mixture into individual enantiomers relies on diastereomeric interactions that take place when enantiomers are adsorbed on a chiral chromatography column. One enantiomer usually interacts more strongly with the chiral stationary phase than does the other, so that the less strongly adsorbed enantiomer elutes from the column first.

## *Meso* Compounds

The rule that there are $2^n$ stereoisomers for a compound with $n$ centers of chirality does not hold when two (or more) identically constituted centers are present. For example, let's consider 2,3-dibromobutane. We can draw three-dimensional representations of 2,3-dibromobutane that are analogous to those shown earlier for 2-bromo-3-chlorobutane.

The 2R,3R and 2S,3S isomers are nonsuperimposable mirror images—that is, they are enantiomers and, individually, they are optically active. When present in equal amounts, they constitute a racemate, which is referred to as a *d,l* pair. The representations of the 2R,3S and the 2S,3R isomers are analogous to (2R,3S)- and (2S,3R)-2-bromo-3-chlorobutane. They are not different isomers because they are superimposable and thus represent only a single stereoisomer. (Make a model to convince yourself that this is so.) This stereoisomer of 2,3-dibromobutane has a mirror plane of symmetry

## THALIDOMIDE: DISASTROUS BIOLOGICAL ACTIVITY OF THE "WRONG" ENANTIOMER

A dramatic and unfortunate consequence of absolute stereochemistry was revealed by the use of thalidomide, a drug produced as an antidepressant. Because of the keen insight of Frances Kelsey, a researcher at the U.S. Food and Drug Administration, thalidomide was never approved for use in the United States. However, this prescription drug was already in use in the 1950s in Canada and Europe, and, despite strong warnings against prescribing thalidomide for pregnant women or even women likely to become pregnant, it was being used to treat "morning sickness."

(*R*)-Thalidomide          (*S*)-Thalidomide

Unfortunately, thalidomide was marketed as a racemate. As the story unfolded, it became clear that one enantiomer acted as an antidepressant; the other was both a mutagen and an anti-abortive. The net result of the use of thalidomide was the birth of many very seriously deformed children, often having vestigial arms and legs. Curiously, the observation that Kelsey had used to hold back approval of thalidomide was that it caused abortions at high doses in rats. Clearly, human beings differ from rats in more ways than just size.

in the center of the molecule. Through this mirror plane (perpendicular to the C-2—C-3 bond), each center of chirality is reflected, *R* to *S* and *S* to *R*. As a consequence of this symmetry, there is no distinction between the designations 2*R*,3*S* and 2*S*,3*R*, because they describe the same molecule, differing only in the end of the carbon chain from which numbering begins.

Because this stereoisomer of 2,3-dibromobutane contains a plane of symmetry, it is optically inactive, despite the presence of centers of chirality. The term ***meso* compound** is used to designate such a stereoisomer. A *meso* compound has a mirror plane or center of symmetry interrelating centers of chirality in the molecule. A *meso* compound is, by definition, optically inactive. The enantiomers (the *d,l* pair 2*R*,3*R* and 2*S*,3*S*) and the *meso* compound (2*R*,3*S* ≡ 2*S*,3*R*) are not superimposable on one another, nor are they related as mirror images: they are thus diastereomers and can be separated by the usual methods of purification. However, members of the *d,l* pair are enantiomers and can be separated only by resolution. For compounds with *meso* stereoisomers, the number of possible stereoisomers is reduced from the maximum of $2^n$ by one for each *meso* compound.

Calculate the number of possible stereoisomers for each molecule in Exercise 5.12.

**EXERCISE 5.17**

Of the following compounds, identify those that are optically active and those that are *meso* compounds. (Methyl and ethyl groups are abbreviated Me and Et, respectively.)

(a) Br ... Br

(c) Cl ... Cl

(e) Cl Me / H Me Cl H

(g) Et OH / H HO Et H

(b) Br ... Br

(d) Cl ... Cl

(f) HO OH / H Me Me H

**EXERCISE 5.18**

For each stereoisomer in parts (e), (f), and (g) of Exercise 5.17, draw a Newman projection in which the hydrogen atoms at the centers of chirality are eclipsed.

## Fischer Projections

Stereoisomers with more than one center of chirality can often be recognized and compared through the use of a stick notation called a **Fischer projection,** which indicates absolute configuration. In a Fischer projection, the intersection of two orthogonal lines indicates the position of a chiral carbon. By convention, the horizontal lines indicate substituents directed toward the observer, and the vertical lines indicate substituents directed away from the observer. A prototype center of chirality bearing substituents A, B, C, and D is shown in the margin as a Fischer projection and as the equivalent hatch/wedge representation.

The carbon skeleton in a Fischer projection involving more than one carbon atom is usually arranged vertically with C-1 at the top, and the substituents are arranged horizontally. The three stereoisomers of 2,3-dibromobutane depicted earlier can be represented by the following Fischer projections. With the Fischer notation, it is easy to see the plane of symmetry between C-2 and C-3 in the *meso* compound.

$$
\begin{array}{ccc}
C & & C \\
A \!\!\!-\!\!\!\!\mid\!\!\!-\!\!\! B & \equiv & A \blacktriangleright\!\!\!\text{-}\!\!\!\text{B} \\
D & & D
\end{array}
$$

Fischer    Hatch/wedge
projection    representation

Me
Br——H
H——Br
Me
**2R,3R**

Me
H——Br
Br——H
Me
**2S,3S**

Me
H——Br
H——Br
Me
*Meso*

The structures at the left and center are enantiomers constituting a *d,l* pair. Both stereoisomers are chiral and are diastereomeric to the achiral *meso* compound at the right. Note that in a Fischer projection all of the bonds in sequential centers of chirality are eclipsed. The Fischer projection thus represents an unstable conformation, but is nonetheless useful for recognizing configurational isomers.

### EXERCISE 5.19

Determine whether the members of each of the following pairs of compounds are enantiomers, diastereomers, constitutional isomers, or identical.

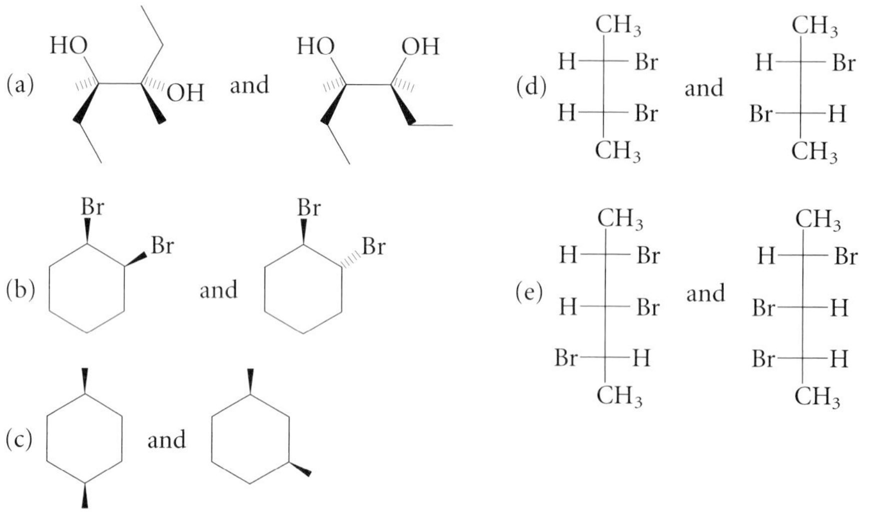

(a) and

(b) and

(c) and

(d)
| | CH$_3$ | | | | CH$_3$ | |
|---|---|---|---|---|---|---|
| H | — | Br | and | H | — | Br |
| H | — | Br | | Br | — | H |
| | CH$_3$ | | | | CH$_3$ | |

(e)
| | CH$_3$ | | | | CH$_3$ | |
|---|---|---|---|---|---|---|
| H | — | Br | and | H | — | Br |
| H | — | Br | | Br | — | H |
| Br | — | H | | Br | — | H |
| | CH$_3$ | | | | CH$_3$ | |

### EXERCISE 5.20

Draw Fischer projections that represent the compounds shown in parts (e), (f), and (g) of Exercise 5.17.

## 5.9

## Optical Activity in Allenes

In a *meso* compound, centers of chirality are present in a molecule that is itself achiral and optically inactive. Conversely, it is possible, although unusual, for molecules that lack chiral tetrahedral carbon atoms to be chiral and optically active. For example, the two isomers of 2,3-pentadiene (Figure 5.30) are not superimposable and are related as mirror images. Thus, they are enantiomers and are optically active. A necessary condition for chirality, and thus for optical activity, is the absence of a molecular mirror plane of symmetry, a feature that 2,3-pentadiene clearly lacks. In principle, therefore, one should look at the symmetry properties of the molecule, since they determine whether chirality is present. In practice, it is usually easier to recognize potential sites of chirality at tetrahedral carbon atoms.

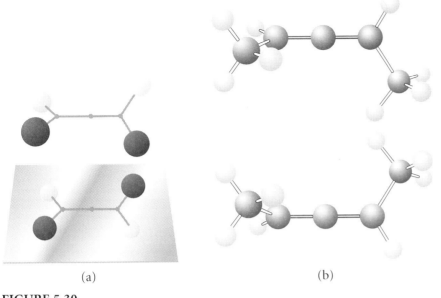

**FIGURE 5.30**

(a) Schematic representation of 2,3-pentadiene, $H_3CCH=C=CHCH_3$,
and its mirror image (the methyl groups are portrayed as black spheres).
(b) Ball-and-stick models of the enantiomers.

## 5.10

# Stereoisomerism at Heteroatom Centers

Elements other than carbon can also be centers of chirality. A nitrogen atom
with three different substituents exists in a pyramidal arrangement.
However, considering the lone pair of electrons to be a fourth group yields
a tetrahedral arrangement that is chiral.

A specific example of a chiral amine is ethylmethylamine. The lone pair
of electrons, a hydrogen atom, a methyl group, and an ethyl group represent four different groups. Simple amines cannot be resolved into separate
enantiomers because of the rapid inversion at nitrogen that proceeds
through a planar $sp^2$ arrangement. This process converts one enantiomer
into the other so rapidly that it is generally impossible to obtain neutral
amines in optically active form.

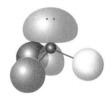

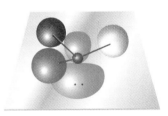

**A pyramidal amine
and its mirror image**

**Inversion of Configuration of Ethylmethylamine**

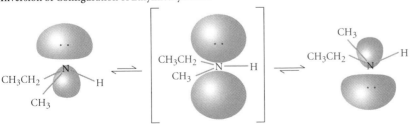

## CHEMICAL PERSPECTIVES

### WHY YOUR MOTHER TELLS YOU TO EAT YOUR BROCCOLI

The body is a marvelous chemical "factory." A vast array of chemical transformations required for life is constantly taking place. In addition, the body must deal effectively with unwanted and unneeded chemicals that are consumed, a task taken on in major part by the liver, where a complex series of oxidations and hydrolyses convert relatively nonpolar, lipophilic molecules into much more water-soluble products that can be easily excreted in the urine. Many different enzymes catalyze these reactions, but some enzymes are specifically responsible for degrading carcinogenic compounds. A compound known as sulforaphane, isolated from broccoli, has been shown to induce increased activity by these detoxification enzymes.

**Isothiocyanate**

**Sulforaphane**

Note that sulforaphane has a sulfoxide group and that four different groups are arranged about the sulfur (the oxygen, a four-carbon chain terminated by isothiocyanate, a methyl group, and the lone pair). Thus, the sulfur in this compound is a center of chirality. As with most naturally occurring chiral compounds, only one enantiomer ($R$) is found in the plant.

Quaternary ammonium ions with four different substituents are chiral and can be resolved into individual enantiomers. However, if one of the substituents is a hydrogen atom, a rapid sequence involving deprotonation, inversion of the tertiary center, and reprotonation will convert a single enantiomer of such a salt into a racemic mixture. Thus, ammonium ions can be obtained in optically active form only when none of the substituents is a hydrogen atom.

**A single enantiomer of a
quaternary ammonium ion**

Inversion is slower for third-row elements. Thus, phosphines ($R_3P$) and sulfoxides ($R_2S{=}O$) can be obtained in optically active form.

**Phosphine**

**Sulfoxide**

# Summary

**1.** Stereoisomers are isomers that differ not in atomic connectivity but rather in the three-dimensional disposition of atoms. Classes of stereoisomers include geometric isomers, conformational isomers, and configurational isomers.

**2.** The energy difference between geometric isomers is estimated as an enthalpy difference, $\Delta H°$. The energy required to accomplish an interconversion of geometric isomers is that required to overcome a barrier to reaching the highest-energy intermediate, the transition state, along the reaction coordinate. The energy of activation ($\Delta H^{\ddagger}$) defines the energy difference between the starting state and the transition state.

**3.** Conformational isomers are interconverted by rotations about $\sigma$ bonds.

**4.** Eclipsed and staggered conformations differ with respect to torsional strain caused by electron repulsion. The staggered conformation of ethane is more stable than the eclipsed conformation by about 2.8 kcal/mole.

**5.** The most stable staggered conformation of butane is favored by about 5–7 kcal/mole over the least stable, eclipsed conformation.

**6.** The energies of various staggered conformations may differ because of different steric interactions resulting from van der Waals repulsions. The actual energy of a conformation depends on the dihedral angle separating bulky substituents. A *gauche* isomer is one in which there is a 60° dihedral angle between carbon substituents; in an *anti* isomer there is a 180° dihedral angle. The energy difference between *gauche*- and *anti*-butane is about 0.9 kcal/mole, and the energy barrier for interconversion between these isomers is about 3.4 kcal/mole.

**7.** Conformational equilibria for cyclohexane and other cyclic saturated compounds are also governed by torsional and steric interactions. For cyclohexane, torsional strain is minimized in the chair conformation. The alternative boat and twist-boat conformations have fully or partially eclipsed bonds, and therefore appreciable torsional strain.

**8.** Ring-flipping from one chair conformation of cyclohexane to another has the effect of converting each axial substituent into an equatorial one, and vice versa. The axial substituents are destabilized by 1,3-diaxial interactions, whereas equatorial groups are not. Therefore, those conformers whose large substituents are in equatorial positions are most stable.

**9.** Chirality is a characteristic of molecules that lack a mirror plane. Such molecules usually have a center of chirality. An $sp^3$-hybridized atom bearing four different substituents constitutes a center of chirality.

**10.** Stereoisomers that are mirror images of each other are called enantiomers, and stereoisomers that are not mirror images are called diastereomers.

**11.** Enantiomers differ with respect to the direction of rotation of plane-polarized light and how they interact with other chiral molecules, but otherwise enantiomers have identical physical and chemical properties. Diastereomers have different physical and chemical properties.

**12.** Diastereomers can be separated by physical and chromatographic methods; enantiomers are usually resolved by a process that converts the enantiomers into diastereomers, which are then separated. Reversal of the reaction used to form the diastereomers then regenerates the starting materials in optically pure form.

**13.** A chiral molecule is optically active—that is, it rotates plane-polarized light by a value characteristic of that particular compound. This specific rotation can be used to gauge optical purity of a sample by comparison with the observed rotation.

**14.** There are several methods for uniquely identifying an enantiomer. The absolute configuration of a specific enantiomer is specified by the use of the Cahn–Ingold–Prelog rules. Alternatively, the designations (+) and (−) or $d$ and $l$ can be used to indicate the direction of rotation of the plane of polarized light for a given compound. However, there is no direct connection between the (±) designation or the $d,l$ designation and the absolute arrangement specified by the $R,S$ designation.

**15.** A stereoisomer bearing centers of chirality but having a mirror plane is optically inactive and is called a *meso* compound.

**16.** For a molecule with $n$ centers of chirality, there are in principle $2^n$ possible stereoisomers (but the number of *meso* compounds must be subtracted).

**17.** Two methods can be used for three-dimensional representations of molecules: Newman projections and Fischer projections. Newman projections show conformational relations (those resulting from rotation about $\sigma$ bonds) and can effectively illustrate the dihedral angles between substituent groups about $\sigma$ bonds. Fischer projections enable us to recognize the existence of mirror planes within molecules; they are a convenient means of representing the stereochemical relations of molecules containing multiple centers of chirality.

**18.** Although optical activity is encountered at atoms other than $sp^3$-hybridized carbon (for example, in allenes, quaternary ammonium salts, and sulfur and phosphorus compounds), its occurrence is relatively rare.

## Review of Reactions

Photochemical *trans–cis* Isomerization

Acid–Base Reaction

$$RCO_2H + RNH_2 \rightleftharpoons RCO_2^{\ominus} + RNH_3^{\oplus}$$

## Review Problems

**5.1** Crotonic acid, $CH_3CH{=}CHCO_2H$, a compound found in Texas clay and formed by the dry distillation of wood, exists as the $E$ isomer. Draw a correct geometric representation of this molecule.

**5.2** (a) Draw Newman projections, visualizing down the C-2—C-3 bond, of the most stable and least stable conformations of 2,3-dimethylbutane. (b) Using the values in Table 5.1, calculate the energy difference between these conformers, and estimate the equilibrium ratio of the most stable to the least stable.

**5.3** Draw an approximate potential-energy diagram to describe a 360° rotation about the C-2—C-3 bond of 2,2,3,3-tetramethylbutane.

**5.4** The preference of a methyl group for an equatorial rather than an axial position on cyclohexane is related to the number of *gauche* interactions in these two isomers. Draw a three-dimensional representation of 1-methylcyclohexane in its preferred conformation and in the conformation attained by flipping the ring. Then draw a Newman projection for each conformer, visualizing down the C-1—C-2 bond. From the structures, count the number of *gauche* and *anti* interactions that are like those considered in this chapter for butane.

**5.5** Explain in each case why the indicated isomer is the more stable one.

(a) *trans*-1,2-dimethylcyclohexane is more stable than the *cis* isomer

(b) *cis*-1,3-dimethylcyclohexane is more stable than the *trans* isomer

(c) *trans*-1,4-dimethylcyclohexane is more stable than the *cis* isomer

**5.6** For the following substituted cyclohexanes, label each substituent as axial or equatorial. If a ring-flip will produce a more stable conformation, draw a three-dimensional representation of that conformation.

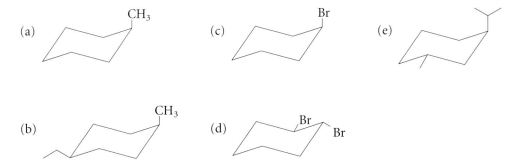

**5.7** Despite the usual preference for equatorial positions for substituents on chair cyclohexanes, 2-hydroxypyran exists predominantly in a chair conformation with the hydroxyl group axial. Draw this conformation, indicating the directionality of the lone pairs on oxygen.

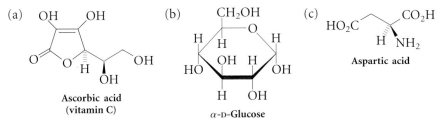

**2-Hydroxypyran**

**5.8** For each center of chirality in the following molecules, assign an *R* or *S* configuration according to the Cahn–Ingold–Prelog rules.

(a) Ascorbic acid (vitamin C)

(b) α-D-Glucose

(c) Aspartic acid

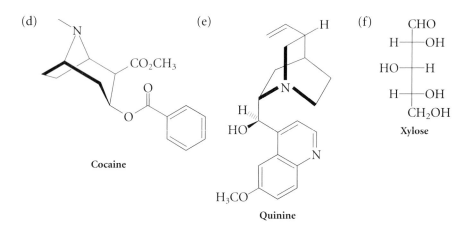

(d) Cocaine

(e) Quinine

(f) Xylose

**5.9** Draw a Fischer projection that represents each of the following compounds:

(a) (S)-2-pentanol

(b) (R)-serine, HOCH₂CH(NH₂)COOH

(c) (S)-glyceraldehyde (2-hydroxypropanal)

(d) (R)-3-methylheptane

(e) (2R,3R)-dihydroxybutane

(f) meso-2,3-dihydroxybutane

**5.10** For each of the following pairs of structures, identify the relation between them. Are they enantiomers, diastereomers, structural isomers, or two molecules of the same compound?

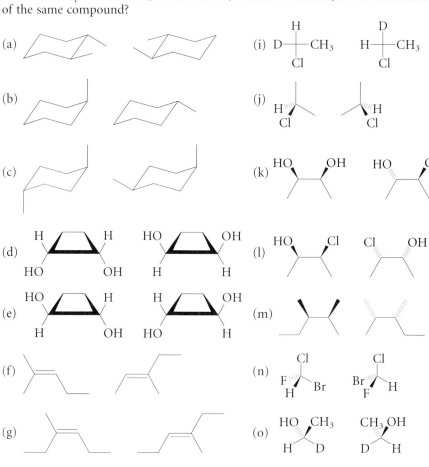

**5.11** The basic structure of cholesterol, the principal sterol found in all mammals, is shown here. Identify all centers of chirality, and calculate the number of possible stereoisomers.

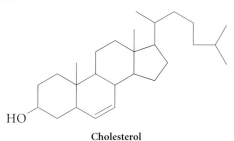

Cholesterol

## Supplementary Problems

**5.12** Use the following reaction profile for the conversion of compound A to compound C through B to determine:

(a) whether A, B, or C is the most stable chemical species

(b) whether the reaction requires or releases energy

(c) the activation energy for the reaction

(d) whether B is a reactive intermediate or a transition state

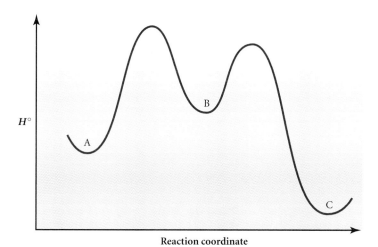

Reaction coordinate

**5.13** Name the compound represented by each of the following Newman projections:

(a)  (b)  (c)  (d)

**5.14** Estimate the energy difference between A and B as 0–0.5 kcal/mole, 0.5–1.5 kcal/mole, or > 2 kcal/mole for each equilibrium concentration ratio:

(a) $\dfrac{[A]}{[B]} = 1$    (b) $\dfrac{[A]}{[B]} = 3$    (c) $\dfrac{[A]}{[B]} = 10$    (d) $\dfrac{[A]}{[B]} = 100$

**5.15** Draw Newman projections to represent each of the following conformers:

(a) *gauche* conformer of hexane, visualized down the C-3—C-4 bond

(b) *anti* conformer of pentane, visualized down the C-2—C-3 bond

(c) eclipsed conformer of butane, visualized down the C-2—C-3 bond

(d) chair conformation of methylcyclohexane with the methyl group in an axial position

(e) boat conformation of methylcyclohexane with the methyl group in an equatorial position

**5.16** Draw a three-dimensional representation of each of the following compounds and its ring-flipped conformer:

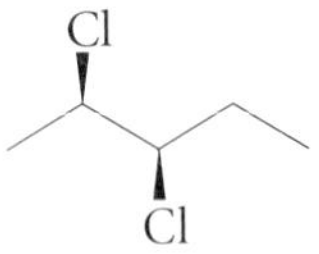

(a) *cis*-1,3-dimethylcyclohexane        (c) methylcyclobutane

(b) *trans*-1,2-dichlorocyclohexane

**5.17** Provide a complete name, including assignment of absolute configuration to any centers of chirality, for each of the following compounds:

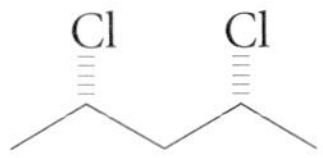

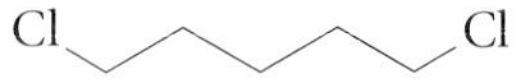

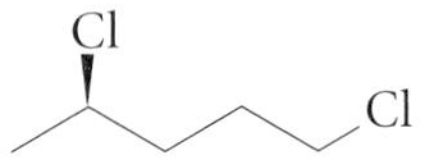

**5.18** Determine whether each compound in Problem 5.17 is optically active. Draw the enantiomer of each optically active compound. For each optically inactive compound, determine why it is inactive (no center of chirality or a *meso* compound).

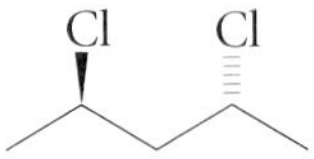

**5.19** For each of the following compounds, identify any centers of chirality, and calculate the number of possible optical isomers:

(a)

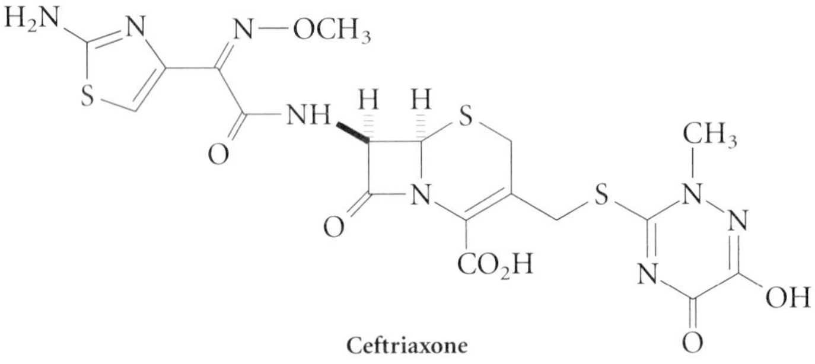

Ceftriaxone

(b)

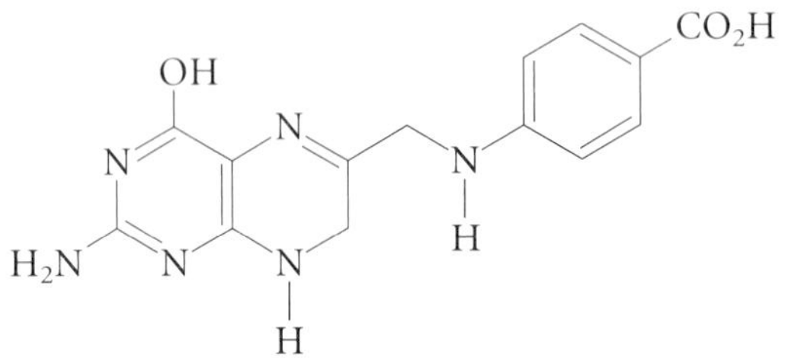

Dihydrofolic acid

**275**

Review Problems

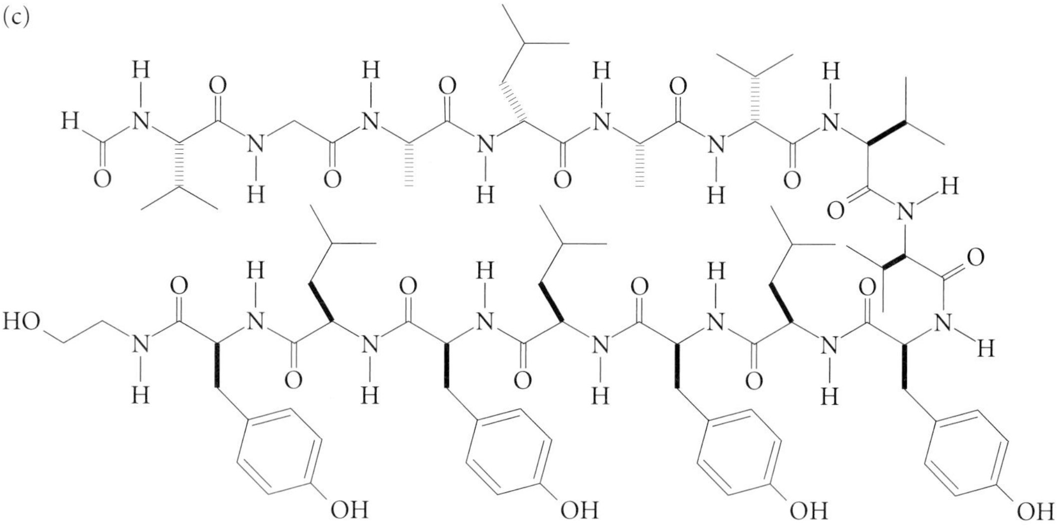

**Gramicidin A**

**5.20** For each of the following compounds, assign absolute configuration to each center of chirality:

(a) 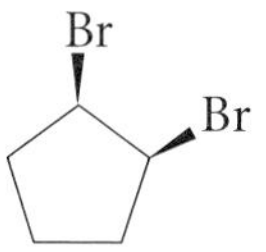  (b) 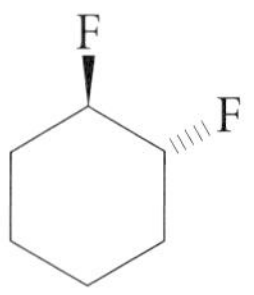  (c) 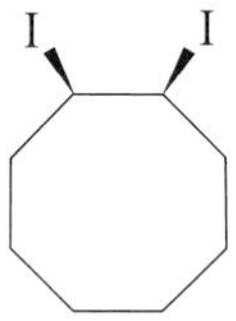  (d)

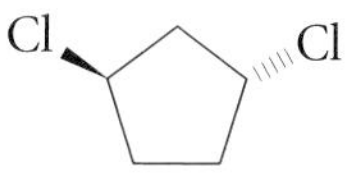

**5.21** Calculate the observed rotation expected in each of the following situations, assuming that a standard 1-dm cell is used.

(a) Ten grams of a 75:25 mixture of enantiomers is dissolved in 100 mL of solvent. The major enantiomer has a specific rotation of $+100°$.

(b) Ten grams of a 50:50 mixture of enantiomers is dissolved in 100 mL of solvent. The enantiomers have specific rotations of $+100°$ and $-100°$.

(c) Ten grams of a 75:25 mixture of enantiomers is dissolved in 100 mL of solvent. The major enantiomer has a specific rotation of $-50°$.

(d) Ten grams of a 9:1 mixture of enantiomers is dissolved in 100 mL of solvent. The major enantiomer has a specific rotation of $+200°$.

**5.22** Can the following pairs of compounds be separated in theory? By chromatography on an achiral support? Only by resolution? Not at all?

(a) 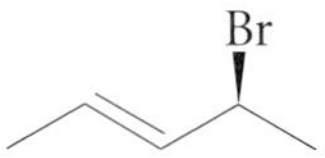   and

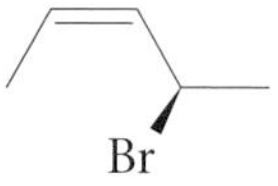

(b) 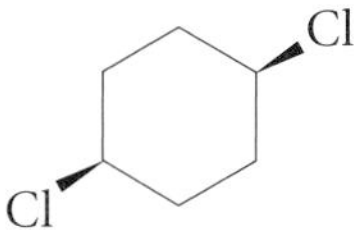   and

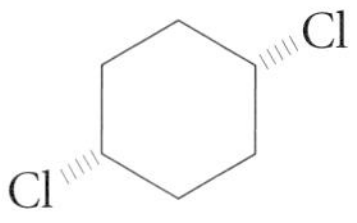

(c) 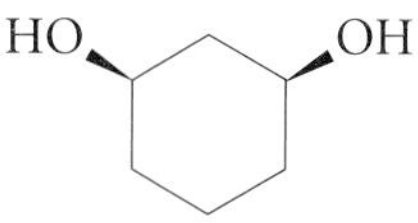   and

(d)

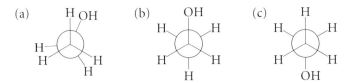

and

(e)

and

(f)

**5.23** Draw Newman projections viewed along the N—O bond to represent the possible conformations of hydroxylamine, $H_2N$—OH. From what you know about conformational analysis, speculate on the relative stabilities of these structures.

**5.24** Draw sawhorse representations of the following conformations of ethanol:

(a)

(b)

(c)

**5.25** At room temperature, the *anti* conformer of butane is more stable than the *gauche* conformer by a little less than 1 kcal/mole. This leads to an equilibrium constant of about 4.6 favoring the *anti* isomer. As the temperature is raised, will the value of this equilibrium constant increase or decrease? Explain.

**5.26** The NMR spectra on the opposite page were obtained for a compound with the molecular formula $C_8H_{18}O$. It is known to be one of the following compounds:

A

B

C

D

E

Select the structure that best corresponds to the spectral data. Then explain how the structure is consistent with the compound (that is, which structural subunits are responsible for which resonances). (*Hint:* It is often easier and quicker to decide which structures do *not* fit the data and exclude these.)

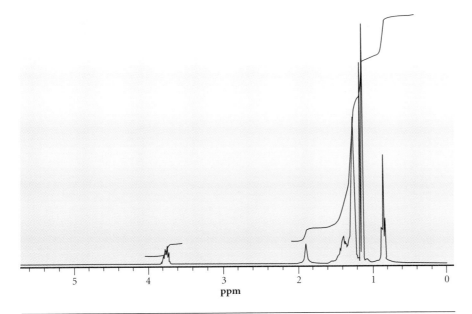

$^{1}$H NMR spectrum of $C_8H_{18}O$

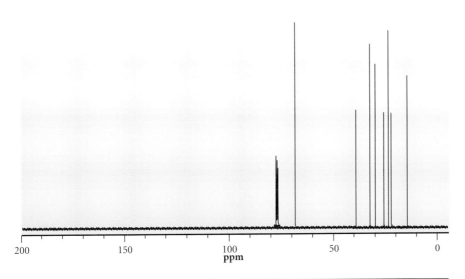

$^{13}$C NMR spectrum of $C_8H_{18}O$

# Understanding
# Organic Reactions

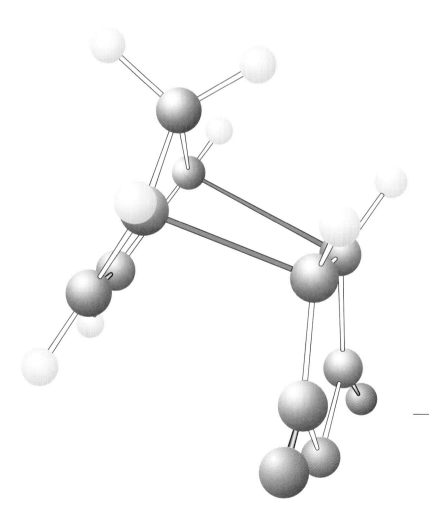

*In the Diels–Alder reaction, a diene and an alkene combine to form a single product, first passing through a highly organized transition state, like this, in which the new bonds between the starting materials are shown in green.*

*N*ow that you have some knowledge of the structure of organic molecules, we will turn to their reactions. In particular, we will consider some general principles that help explain why certain reactions proceed and others do not. We will also look at what controls the rates of reactions and how the structure and reactivity of a molecule are related. It is helpful to look at the driving forces that cause a given reaction to occur, such as the changes in energy content of products versus reactants (thermodynamics) and the pathway and rate by which the molecules become transformed from reactants to products (kinetics).

## 6.1

## Reaction Profiles (Energy Diagrams)

To visualize the progression from reactant to product, we can construct a reaction profile (also called an energy diagram) by plotting the change in potential energy as the reaction proceeds, just as we did for conformational interconversions in Chapter 5. The measure of how far a reaction has proceeded is called the *reaction coordinate*. For example, the reaction coordinate can follow how far a critical bond that is breaking in the reaction has stretched or how much a bond angle has expanded as an atom rehybridizes its orbitals from $sp^3$ to $sp^2$ as the reaction occurs.

### Free Energy

The energy content of a molecule is usually expressed as its free energy. **Free energy ($\Delta G°$)** is a measure of the potential energy of a molecule or group of molecules, and it is partitioned between enthalpy ($\Delta H°$) and entropy ($\Delta S°$):

$$\Delta G° = \Delta H° - T\Delta S° \tag{1}$$

Although free energy ($\Delta G°$) has both enthalpy ($\Delta H°$, bond energies) and entropy ($\Delta S°$, disorder) components, most organic reactions proceed with only very small entropy changes, when the number of moles of products equals the number of moles of reactants. Therefore, potential energy (free-energy) changes can often be approximated by enthalpy changes. Enthalpy relates to bonding (bond energies), whereas **entropy** relates to the degree to which a molecular system is ordered. As the amount of disorder increases, entropy increases. The entropy contribution to free energy depends on temperature, because $\Delta S°$ is multiplied by the temperature $T$ (in kelvins). As the temperature increases, the $T\Delta S°$ term becomes larger and can sometimes dominate over the $\Delta H°$ term.

### Thermodynamics: Initial and Final States

Thermodynamics and kinetics are important factors in describing how various energy components affect reactions. **Thermodynamics** focuses on the relative energies of the reactants and products, whereas **kinetics** describes the rate at which a reaction proceeds. For a reaction to be practical, thermodynamics must favor the desired product and the reaction rate must be fast enough, not only for the reaction to be complete within a reasonable time but also for it to win out over competing reactions.

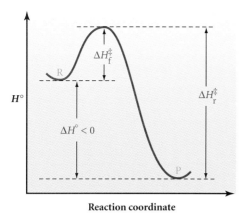

**FIGURE 6.1**

Potential energy diagram for an exothermic reaction.

We can use reaction profiles to compare the thermodynamic changes for various types of reactions—that is, compare the reactants and the products. When free energy is released in a reaction, the conversion is said to be **exergonic** and the reaction will proceed to product. When entropy changes are negligible (as is common in many organic reactions), the reaction is said to be **exothermic** if energy is released (if the total bond energy content of the products is lower than that of the reactants, Figure 6.1).

Several values occur repeatedly on energy diagrams. The change in energy from starting material to product is $\Delta H°$, and for exothermic reactions, $\Delta H° < 0$. Two other important values are: $\Delta H_f^{\ddagger}$, or the activation energy barrier for the forward reaction (that is, the energy difference between starting material and transition state), and **$\Delta H_r^{\ddagger}$**, or the activation energy barrier for the reverse reaction (the energy difference between product and transition state).

When free energy must be supplied to drive a reaction, the conversion is said to be **endergonic.** The term **endothermic** describes a reaction in which the bond energy content of the products is higher than that of the reactants (and the change in entropy is small compared with the change in enthalpy, Figure 6.2). A **thermoneutral reaction** is one in which the reactants and the products have the same energy content (Figure 6.3).

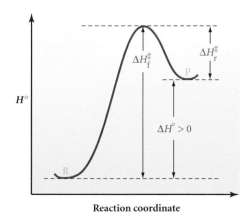

**FIGURE 6.2**

Potential energy diagram for an endothermic reaction.

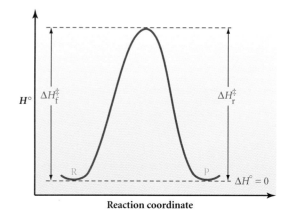

**FIGURE 6.3**

Potential energy diagram for a thermoneutral reaction.

## Kinetics: The Reaction Pathway

The rate of a reaction increases when heat is applied by increasing the temperature or energy is otherwise added to the system (a concept clear to anyone who has grilled a steak or baked a cake). For a very large number of reactions, the rate of reaction also increases as the concentration of the reactants increases. Kinetic theory is the result of the desire to explain these observations. The temperature (energy) dependence of reactions is explained in terms of *transition state theory,* described in Section 6.3; the dependence of reaction rates on the concentration of reactants is examined in Section 6.9.

An explanation of these experimental observations requires a consideration of the pathway—the molecular processes—by which reactants are converted to products. A reaction can proceed in one step from reactant to product—that is, as a **concerted reaction**—or through a sequence of steps that includes the formation of one or more intermediates. The nature of these species formed along the course of the reaction profile and the ease of their formation determine the rate at which a reaction proceeds.

## The Transition State

When a reactant is converted directly to product in one step, the reaction profile appears as a smooth curve connecting reactants and products. Thus, the energy diagrams in Figures 6.1 through 6.3 describe concerted reactions. The top of the smooth curve that connects reactants and products in a concerted reaction corresponds to the transition state of the reaction. A **transition state** is the species of highest energy along the reaction pathway. Because the transition state is at an energy maximum, it is very unstable and has only a transient existence.

The energy required to reach the transition state from the reactant energy minimum is defined as the **activation energy** (sometimes written as $E_{act}$). The activation energy is a free-energy term with contributions from both enthalpy and entropy components. However, the activation energy can often be approximated by the **enthalpy of activation ($\Delta H^{\ddagger}$)** encountered in reaching the transition state in reactions in which entropy changes are small. In the conversion of a reactant to a product, enough energy must be supplied to reach the transition state and thus to overcome the barrier separating these species. The higher the activation energy of a reaction, the more difficult it is to reach the transition state and the slower the reaction. The barrier is thus often called the **activation energy barrier,** an expression used interchangeably with activation energy. In a thermoneutral reaction, the activation energy barriers for the forward ($\Delta H_f^{\ddagger}$) and reverse ($\Delta H_r^{\ddagger}$) processes are identical, as shown in Figure 6.3. In an exothermic reaction (Figure 6.1), the activation energy barrier for the forward reaction is lower than for the reverse reaction; the reverse is true of an endothermic reaction (Figure 6.2).

For exothermic and thermoneutral reactions, there is not necessarily a connection between the enthalpy change in the reaction ($\Delta H°$) and the activation energy ($\Delta H^{\ddagger}$). On the other hand, we know that the activation energy of an *endo*thermic reaction *must* be at least as large as $\Delta H°$. This is an important relationship because it establishes a minimum energy of activation for an endothermic reaction, which cannot be done for thermoneutral and exothermic reactions.

## Reactive Intermediates

A reaction that takes place in several stages includes the formation of one or more **reactive intermediates**—species that exist at the bottom of a potential-energy well and thus have a real-time existence. Unlike a transition state, an intermediate must overcome some energy barrier, however small, before it can follow the reaction pathway and proceed onward to product (or backward to starting material). It is these barriers that give the intermediate a measurable lifetime and form the walls of the energy well in the reaction profile. The intermediate thus exists at an energy minimum.

In the reaction profile in Figure 6.4, a reactant, R, is converted to a product, P, through an intermediate, I, in a transformation that is exothermic overall. From this plot, we can follow the change in energy as the reaction progresses. In this case, there is an intermediate, I, of higher energy than either the reactant or the product. Because the entire reaction in Figure 6.4 (R $\rightarrow$ P) takes place in two steps (R $\rightarrow$ I and I $\rightarrow$ P), it can be broken down

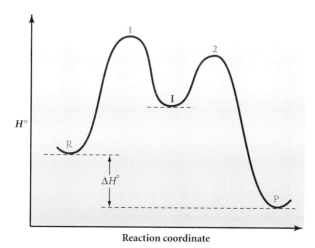

**FIGURE 6.4**

A step-by-step (nonconcerted) reaction, showing transition states 1 and 2.

into two simpler reactions: an endothermic step for R → I and an exothermic step for I → P. Neither step involves an intermediate, so each is a concerted reaction.

---

### EXERCISE 6.1

Draw a reaction profile representing each of the following situations:

(a) an exothermic reaction with a small activation barrier

(b) an exothermic reaction with a large activation barrier

(c) an endothermic reaction with a small activation barrier

(d) a thermoneutral reaction with a large activation barrier

---

## 6.2

## Thermodynamic Factors

Free energy has both enthalpy (bond energy) and entropy (disorder) components. Enthalpy changes are almost always important in chemical reactions, but entropy changes are usually significant in organic reactions only when the number of product molecules differs from the number of reactant molecules. Let's consider some examples of reactions that illustrate the significance of enthalpy and entropy changes.

### Enthalpy Effects: Keto–Enol Tautomerization

#10    Tautomerization

As you encounter new reactions, it will be important to understand why the reactions proceed in one direction—for example, from A to B—rather than vice versa.

$$A \underset{X}{\overset{\longrightarrow}{\longleftrightarrow}} B$$

The direction of a reaction is determined by thermodynamics—reactions proceed toward the more stable species. For a simple reaction, we can determine whether A or B is more stable by contrasting the strengths of the bonds that must be broken with those that are formed as a reactant is converted to product. When the relevant bond energies are known, the calculation of the heat of reaction is straightforward. If the new bonds formed in the product(s) are stronger than those that must be broken in the reactant(s), the reaction is exothermic and, consequently, thermodynamically favorable. The same kind of calculation can be used for multistep reactions, because only the relative energies of the starting material(s) and final product(s) determine the overall thermodynamics of the reaction. We cannot learn from these calculations anything about reaction rates, which are governed by the energies of transition states rather than the energies of reactants or products.

Let's consider a specific example. Upon treatment with base (often an alkoxide ion), a proton can be removed from the α position of a carbonyl compound:

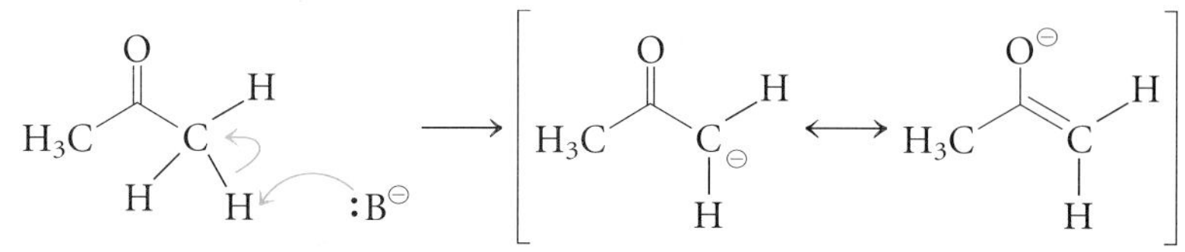

**Carbonyl compound**    **Base**          **Enolate anion resonance contributors**

In the **enolate anion** formed, two chemically different sites bear a partial negative charge, as indicated by the resonance contributors. It is therefore possible to reprotonate at either of these sites (that is, at carbon or at oxygen) to form two distinct products (Figure 6.5). When reprotonation occurs at carbon, the original carbonyl compound is formed again. When reprotonation occurs at oxygen, an **enol** is generated, so named because of the presence of a hydroxyl group attached to a doubly bonded carbon (alkene). An isomerization proceeding via deprotonation at the C—H bond adjacent to a carbonyl group and protonation at the carbonyl oxygen is called a **keto–enol tautomerization**. The keto and enol forms shown in Figure 6.5 are **tautomers** because they differ only with respect to the position of an acidic hydrogen.

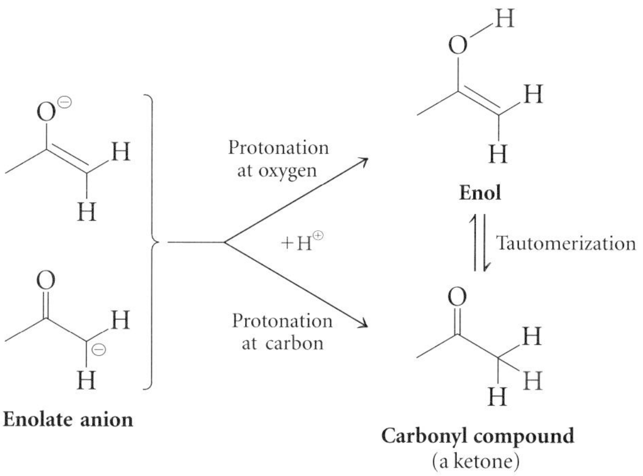

**FIGURE 6.5**

Protonation of an enolate anion forms either an enol (attachment of the proton to oxygen) or a carbonyl compound (attachment of the proton to carbon). The interconversion between a ketone and its enol form is tautomerization.

In this case, a hydrogen atom is shifted from one atom (carbon) to another atom (oxygen) that is two atoms away (a 1,3-shift), and the process is specifically referred to as **proton tautomerization.**

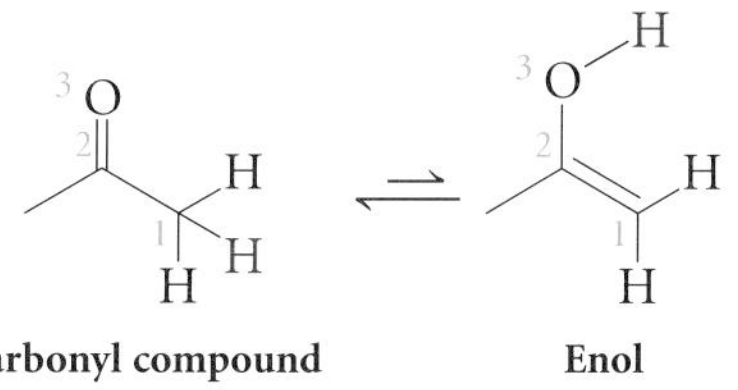

**Carbonyl compound**          **Enol**

Proton tautomerization can be induced by treatment with acid or base. With base, the α carbon is deprotonated in the first step, producing an enolate anion that is reprotonated at oxygen in a second step. With acid, the carbonyl oxygen is protonated in the first step, with removal of the α hydrogen occurring in the second step (Figure 6.6).

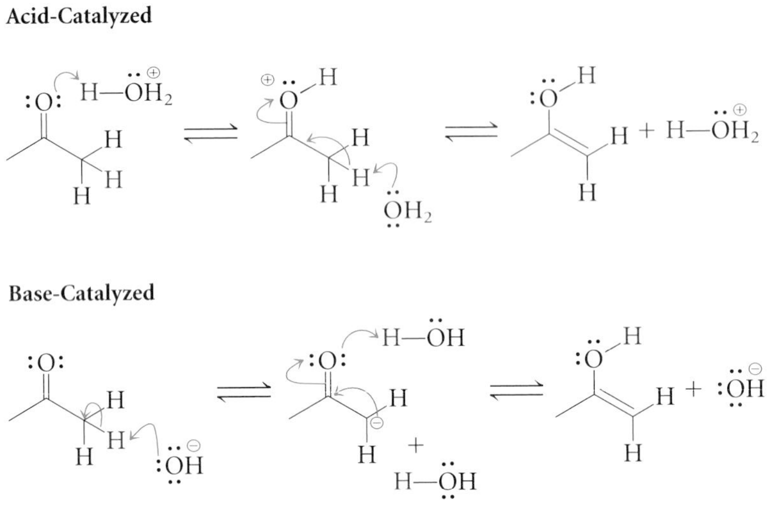

**Acid-Catalyzed**

**Base-Catalyzed**

**FIGURE 6.6**

Protonation is followed by deprotonation in the conversion of a ketone to an enol in the presence of acid. The sequence is reversed in the presence of base.

We can use bond strengths to determine whether the keto or enol form dominates this tautomeric equilibrium:

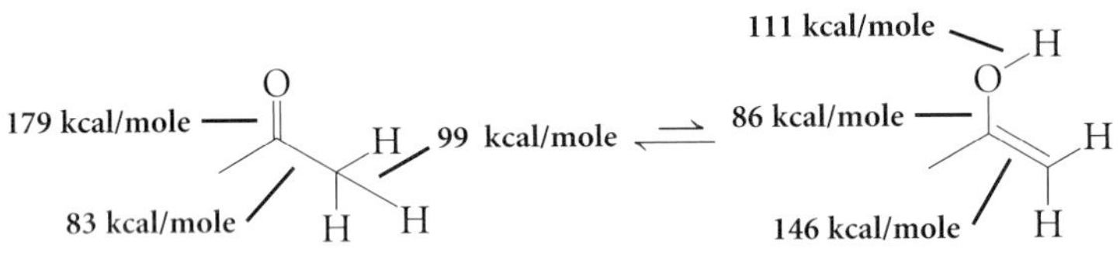

$$179 + 83 + 99 = 361$$
$$-343 = 86 + 146 + 111$$
$$+18 \text{ kcal/mole}$$

The ketone has a carbon–oxygen double bond, whereas the enol has a carbon–oxygen single bond (179 versus 86 kcal/mole, Table 3.2). The ketone has a carbon–carbon single bond, whereas the enol has a carbon–carbon double bond (83 versus 146 kcal/mole). The ketone has a carbon–hydrogen bond, whereas the enol has an oxygen–hydrogen bond (99 versus 111 kcal/mole). All other bonds are the same in the two species. In total, as shown above, the bonds of the ketone are stronger than those of the enol by 18 kcal/mole. Thus, the ketone is more stable and is the dominant species at equilibrium.

# EXERCISE 6.2

Referring to the bond strengths in Table 3.2, predict whether each of the following reactions is endothermic or exothermic:

(a) 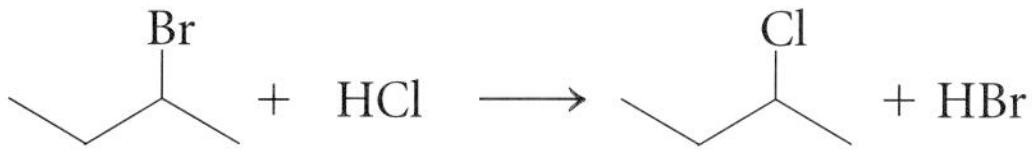 ![structure] + HCl ⟶ ![structure] + HBr

(b) 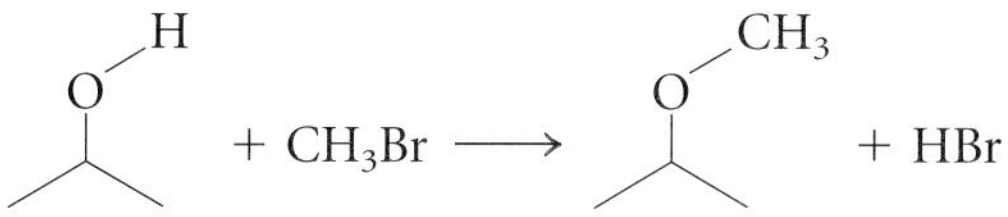 ![structure] + CH₃Br ⟶ ![structure] + HBr

(c) 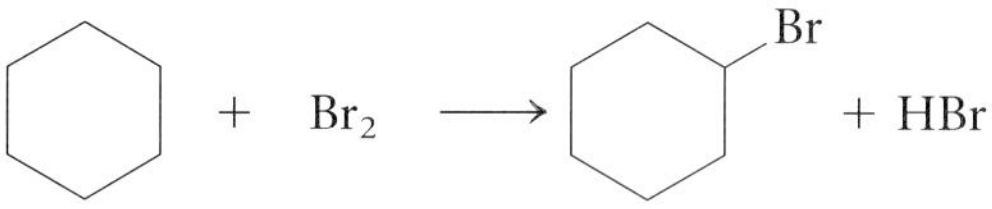 ![structure] + Br₂ ⟶ ![structure] + HBr

# EXERCISE 6.3

Write the structure of a tautomer for each of the following compounds:

(a) 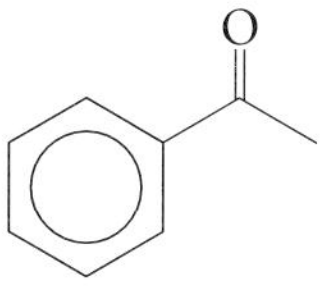 ![structure]

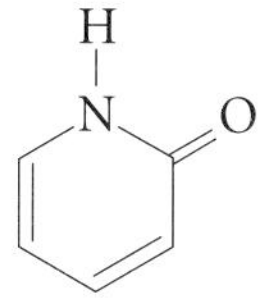

(b) 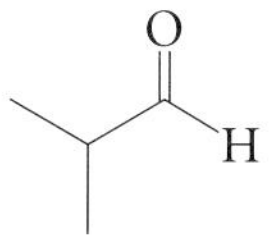 ![structure]

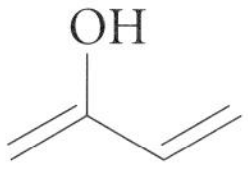

(c) 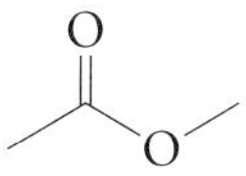 ![structure]

(d) ![structure]

(e) ![structure]

(f) 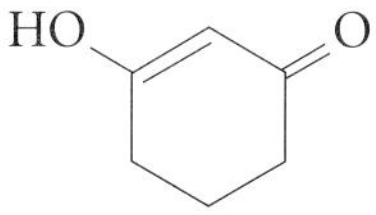 ![structure]

## Entropy Effects: The Diels–Alder Reaction

The entropy changes of most organic reactions are smaller than the enthalpy changes. As a result, the change in free energy, $\Delta G°$, can be approximated using only the enthalpy contribution, $\Delta H°$. One exception is the **Diels–Alder reaction,** in which two hydrocarbon reactants combine to form a cyclic product without proceeding through any intermediate that can be isolated, or even inferred from experimental data. In its simplest form, the Diels–Alder reaction combines butadiene with ethylene to form cyclohexene. However, you will learn in Chapter 10 that higher yields are obtained if one reactant is substituted with an electron-withdrawing group.

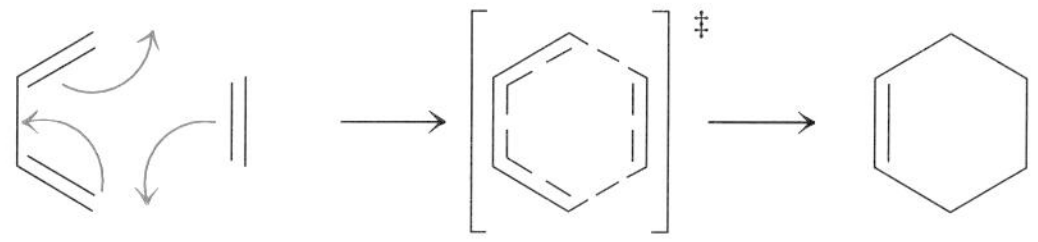

![reaction scheme]

**Butadiene   Ethylene**                              **Cyclohexene**

Partial bonds being formed and broken in the transition state are indicated by dashed lines. Because the Diels–Alder reaction is concerted, it does not include the formation of an intermediate. The structure in brackets, between the reactants and product, is a transition state that has no real-time existence. The progression of this reaction is indicated by the full-headed curved arrows showing the motion of each electron pair as the $\pi$ electrons of each reactant are delocalized in the cyclic, conjugated transition state. The rules for determining aromaticity explained in Chapter 2 also apply to transition states; because the transition state of the Diels–Alder reaction involves six electrons $(4n + 2)$, it is stabilized.

The reactants have a total of four carbon–carbon $\sigma$ bonds (three in butadiene and one in ethylene) and three carbon–carbon $\pi$ bonds (two in butadiene and one in ethylene). The product has six carbon–carbon $\sigma$ bonds in the carbon skeleton and one carbon–carbon $\pi$ bond. Therefore, two carbon–carbon $\pi$ bonds have been converted into two $\sigma$ bonds. Because the average bond strength of a carbon–carbon $\pi$ bond is lower than that of a carbon–carbon $\sigma$ bond by 20 kcal/mole (63 versus 83 kcal/mole), the change in bond energies, $\Delta H°$, is $-40$ kcal/mole. The Diels–Alder reaction is highly exothermic.

But what is the contribution of entropy to the free-energy change in the Diels–Alder reaction? Two molecules are converted into one, and thus the reaction is disfavored by entropy. The effect of entropy on the change in free energy is usually small compared to that of enthalpy ($\Delta G° = \Delta H° - T\Delta S°$); however, in this reaction, the relative rotational and translational freedoms of the two reactants do not exist in the product.

As a specific example, let's consider the reaction of cyclopentadiene with itself in a Diels–Alder reaction to form dicyclopentadiene, an important component of many polymers:

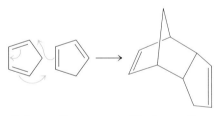

Cyclopentadienes          Dicyclopentadiene

The product is somewhat atypical of those formed in many Diels–Alder reactions, because there is significant strain in its tricyclic ring structure. Thus, the change in bond energies is only $-18.4$ kcal/mole, not the $-40$ kcal/mole that we just calculated based on the net conversion of two carbon–carbon $\pi$ bonds to two carbon–carbon $\sigma$ bonds. The entropy change that accompanies the reaction ($+34$ cal/K) contributes $+10.2$ kcal/mole at room temperature ($T\Delta S°$), and thus $\Delta G°$ is $-8.2$ kcal/mole ($-18.4$ kcal/mole $+ 10.2$ kcal/mole). Even with this relatively small value of $\Delta G°$, the product is favored in the equilibrium by $\sim 10^6 : 1$. The contribution of entropy to $\Delta G°$ is temperature-dependent and increases at higher temperatures. However, to increase the contribution of entropy to the point where it exceeds that of enthalpy—and thus invert the equilibrium in this Diels–Alder reaction—would require a temperature in excess of 270 °C.

## CHEMICAL PERSPECTIVES

### CRACKING DICYCLOPENTADIENE

A convenient method for preparing cyclopentadiene is from dicyclopentadiene, an inexpensive bulk chemical. Simply heating dicyclopentadiene in a
distillation apparatus to a temperature near its boiling point (170 °C) results
in a reverse, or retro, Diels–Alder reaction, and cyclopentadiene collects in
the receiver of the apparatus. The equilibrium between dicyclopentadiene
and cyclopentadiene favors the former at this temperature, but the reaction
is effectively reversed because of Le Chatelier's principle. Cyclopentadiene
rapidly distills from the dimer and is thus removed from the equilibrium,
because its boiling point (42 °C) is substantially below that of the distillation pot.

Dicyclopentadiene is prepared commercially from cyclopentadiene,
which itself is prepared by passing gaseous cyclopentane over transition metal
oxides on an alumina support. (The other product is $H_2$.)

Entropy's effect on the activation energy can be significantly larger than
its effect on the change in free energy of the reaction. For an exothermic
reaction such as the Diels–Alder reaction, the transition state resembles the
starting materials. (See the discussion of the Hammond postulate in the
next section.) Thus, there is little bond making and bond breaking at the
transition state, as can be seen by the long forming bonds (shown in green)
in the transition state for the Diels–Alder reaction of cyclopentadiene with
maleic anhydride. On the other hand, the organization of the two reactant
molecules required to form one product molecule is fully developed at the
transition state, and the negative contribution of entropy to the activation
energy is maximal and the same as that in the overall reaction ($\Delta S^\ddagger = \Delta S°$).

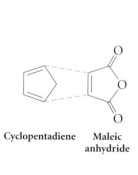

Cyclopentadiene    Maleic
anhydride

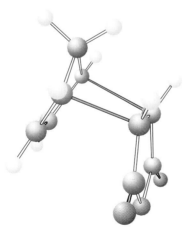

## EXERCISE 6.4

Calculate the entropy change (in cal/K) that would be necessary to reverse a reaction $A + B \rightarrow C$ for which $\Delta H°$ at room temperature is:

(a)  −1 kcal/mole      (b)  −3 kcal/mole      (c)  −10 kcal/mole

## 6.3

# Characterizing Transition States: The Hammond Postulate

A transition state has only a transitory existence because it represents the energy maximum along a reaction pathway. No additional energy is required for the changes that occur as the transition state proceeds to the product (or reverses to the starting material), and, indeed, energy will be released. Transition states have no real lifetime, and there are no physical techniques by which they can be directly characterized. As a result, chemists must infer the transition state's structure by relating it to the stable (or metastable) species to either side of it on the reaction pathway. The validity of this view cannot be rigorously proven, but it is logical. Bonds broken in the reaction must be only partially broken at the transition state, and the same is true for bonds formed. Thus, the transition state will have characteristics of the starting material imparted by the partial bonds still present and characteristics of the product imparted by the partially formed bonds.

The degree to which the starting material and product of a reaction contribute to the structure of the transition state was set forth by George Hammond, a noted American chemist. According to the **Hammond postulate,** *a transition state most closely resembles the stable species that lies closest to it in energy.* Thus, the Hammond postulate asserts that, in an endothermic reaction, the transition state is more similar to the product than to the reactant, and in an exothermic reaction, the transition state is more similar to the reactant. These transition states are referred to as *late* (product-like) and *early* (reactant-like), respectively, to indicate how far along the reaction coordinate each one lies.

Let's consider how the geometries of the two transition states in a two-step reaction like that depicted in Figure 6.4 might resemble that of either the reactant or the product. We start by dividing the reaction into the R → I and the I → P steps. Because the R → I conversion is endothermic, the transition state between R and I closely resembles the "product" of that step of the reaction—the intermediate, I. Because the second step is highly exothermic, the Hammond postulate tells us that its transition state more closely resembles the "reactant" of the second step—that is, the intermediate, I—than the final product, P. Thus, if we wish to understand the step-by-step conversion of R to P, it is essential that we understand intermediate I, because both transition states in this two-step reaction more closely resemble the intermediate, I, than either the reactant, R, or the product, P.

## EXERCISE 6.5

Consider the reaction of an alcohol, ROH, with a hydrogen halide, HX:

$$ROH + HX \longrightarrow RX + HOH$$

From the bond energies given inside the back cover of this book, calculate the energy changes for the reaction when X = (a) Cl, (b) Br, and (c) I. For each, state whether the reaction is endothermic, exothermic, or thermoneutral.

# Types of Reactive Intermediates

Knowledge of the molecular and electronic structures of the most common organic intermediates will help you understand how reactions occur. In some multistep reactions, the intermediate is sufficiently stable to be isolated. However, more frequently the intermediate is highly energetic and reactive. If it cannot be isolated, the intermediate must be characterized indirectly, either by spectroscopic methods or by piecing together inferences from a series of experiments. Common intermediates in organic reactions include carbocations, radicals, carbanions, carbenes, and radical ions.

Because a reactive intermediate is by definition relatively unstable, its concentration in the reaction medium is often quite low. Thus, most of the spectroscopic techniques covered in Chapter 4 are not appropriate for examining reactive intermediates. The exceptions are visible and ultraviolet spectroscopy, whose inherent sensitivity allows them to be used to detect very low concentrations. Also, the presence of ketyls (radical anions of ketones) can be detected readily by their intense absorption of visible light. For example, adding sodium to a solution of benzophenone in ether generates the ketyl, which is deep blue in color. Unfortunately, ultraviolet and visible spectroscopy do not provide the rich structural detail afforded by infrared and especially $^{13}$C and $^{1}$H NMR spectroscopy.

## Carbocations and Radicals

Recall from Chapter 3 that both a carbocation and a radical contain a carbon atom bearing three substituents in a trigonal planar arrangement (Figure 6.7). The $sp^2$-hybridized atom in a carbocation or radical is electron-deficient (compared with carbon's valence-shell requirement), and the stability of these intermediates is increased by substitution with alkyl groups. For both carbocations and radicals, the observed order of stability is tertiary > secondary > primary > methyl.

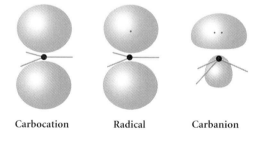

| Carbocation | Radical | Carbanion |

**FIGURE 6.7**

The three-dimensional structures of a carbocation, a radical, and a carbanion.

## Carbanions

Like a carbocation or a radical, a **carbanion** often has three $\sigma$ bonds, but it also bears an unshared electron pair (Figure 6.7) and is not electron-deficient because its valence shell is filled. [One way to prepare carbanions

### A FOUNTAIN OF YOUTH FOR FLEAS

Controlling fleas can be a constant nuisance to pet owners, especially those with cats and dogs that like to roam outside. And fleas aren't easy to kill manually, as their "skin" is really a tough exoskeleton made up primarily of chitin (a complex polysaccharide that also forms the outer shell of cockroaches). Pesticides, while effective in killing adult fleas, don't kill the eggs, and so repeated treatments must be made of both the animal and the home.

For some years, Sandoz (primarily a manufacturer of pharmaceuticals) has been marketing (*S*)-methoprene (Precor), a synthetic compound that mimics the action of a natural flea-growth regulator known as a *juvenile hormone*. Juvenile hormones are important in the development of many insects; such a hormone regulates the transition from one state to another—from larval to pupal stages, for example. When treated with Precor, the larvae never develop further and eventually starve to death. (Note that Precor is a doubly unsaturated ester.)

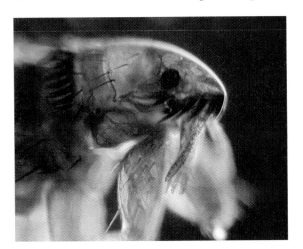

**(*S*)-Methoprene (Precor)**

Yet another solution to the flea problem has been developed by Ciba–Geigy, another major pharmaceutical company. Lufenuron is a synthetic compound that interferes with the production of chitin, and thus of the flea's exoskeleton, without which it cannot live.

**Lufenuron**

is to remove a proton, H$^\oplus$, from a C—H bond (**deprotonation**) by treating the starting compound with a strong base. The electrons originally in the C—H bond then become a nonbonding lone pair on carbon.] Although both the electron-deficient carbocation and the radical are $sp^2$-hybridized, a simple carbanion is pyramidal. With three $\sigma$ bonds and a lone pair, a carbanion is electronically similar to an amine; the carbanion and the amine are therefore said to be **isoelectronic.** Like an amine, a simple alkyl carbanion is $sp^3$-hybridized with a doubly occupied $sp^3$-hybrid nonbonding orbital directed at approximately a tetrahedral angle away from the bonding orbitals. Because the trigonal carbon of a carbanion bears a formal charge of $-1$, it is often strongly associated (ion-paired) with a positively charged metal counterion. When the carbanionic carbon is adjacent to a $\pi$ system, $sp^2$ hybridization is preferred so that the negative charge can be distributed by resonance throughout the $p$ orbital array. Resonance structures are used to describe the electron distribution in the resulting conjugated anion.

Because a simple carbanion is $sp^3$-hybridized, the carbanionic carbon can in principle be a center of chirality; in contrast, the carbon center of a carbocation or a radical, having symmetry about the plane of the carbon atom and its three substituents, cannot be a center of chirality. In most cases, however, inversion of configuration of carbanions is rapid and results in a racemic mixture.

## EXERCISE 6.6

Draw the structures of all significant resonance contributors for the anions formed by deprotonation of each of the following compounds. (The protons to be removed are shown in red.) Show the location of formal charge in each ion.

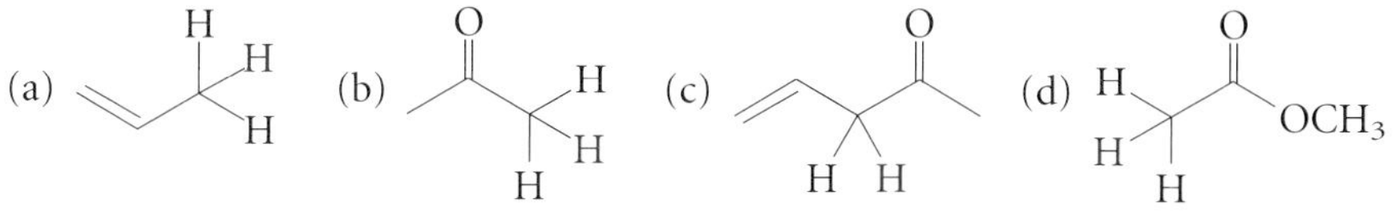

(a) (b) (c) (d)

## EXERCISE 6.7

Draw a three-dimensional structure for each of these reactive intermediates:

(a) allyl cation, $^{\oplus}CH_2CH{=}CH_2$     (c) cyclopropyl cation

(b) cyclopropyl anion     (d) allyl anion

## Carbenes

Carbocations and radicals are trigonal planar, having three $\sigma$ bonds at carbon, and carbanions are pyramidal, with three $\sigma$ bonds and a fourth hybrid orbital occupied by a lone pair of electrons. **Carbenes,** another class of neutral reactive intermediates, have only two $\sigma$ bonds. A carbon atom with two $\sigma$ bonds results in a geometry called *digonal*. For a digonal carbon to be neutral, two nonbonding electrons must be present in addition to those participating in the two covalent bonds. A neutral digonal intermediate, a carbene, is highly electron-deficient (despite its neutrality), because it lacks two electrons from the octet needed for an inert-gas configuration. Like cations and radicals, a singlet carbene (in which the nonbonded electrons are spin-paired) is $sp^2$-hybridized. In this hybridization, a $p$ orbital is oriented perpendicular to the plane containing the two $\sigma$ bonds and the additional $sp^2$-hybrid orbital (Figure 6.8). The electron pair is generally in the $sp^2$-hybrid orbital (rather than in the $p$ orbital). The greater $s$ character (33%) of an $sp^2$-hybrid orbital places the electrons closer to the positively charged nucleus than they would be in a $p$ orbital, making the former arrangement more stable.

A molecule in the singlet state must have all electrons paired, and generally two electrons of opposite spin are paired in each molecular orbital. (The spin state of a molecule is determined by adding one to the total number of electrons that are not spin-paired.) In a **singlet carbene,** the electrons of the $\sigma$ bonds are paired, as are the two electrons that occupy the nonbonding $sp^2$-hybrid orbital (Figure 6.8). This pair of electrons imparts nucleophilicity to a singlet carbene. On the other hand, there is a vacant $p$ orbital that is an electrophilic center. (The vacant $p$ orbital of the singlet carbene in Figure 6.8 can be compared to that of the carbocation in Figure

**Triplet carbene**

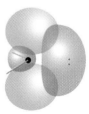

**Singlet carbene**

**FIGURE 6.8**

Orbital occupancy in a singlet and a triplet carbene.

## TRAPPING REACTIVE INTERMEDIATES

Although carbenes are highly reactive intermediates, they are sufficiently stable to be studied by spectroscopic means if they are isolated from each other and from other molecules with which they might react. When trapped in a frozen matrix of solid argon or other inert species, very reactive species can be characterized in the laboratory by absorption or infrared spectroscopy. Although cyclobutadiene can be prepared in the laboratory in the gas phase, it reacts rapidly with itself to form a dimer. However, cyclobutadiene can be frozen in argon immediately after it is formed.

The infrared spectrum of cyclobutadiene is consistent with a rectangular, planar structure. Thus, this cyclic collection of four $sp^2$-hybridized carbon atoms has distinct single and double carbon–carbon bonds, unlike benzene, in which all of the carbon–carbon bonds are identical.

The parent carbene is methylene, $:CH_2$. Methylene and several other very reactive molecules have been found in outer space, where very low tem-

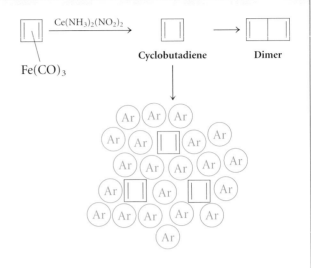

peratures and the low density of molecules inhibit intermolecular reactions.

---

6.7.) Carbenes are therefore reactive in an *ambiphilic* sense—that is, toward both electron-rich and electron-poor reagents.

In a **triplet carbene,** the two nonbonding electrons have the same spin. Therefore, these unshared electrons occupy different orbitals, with one electron in the $p$ and one in the $sp^2$-hybrid orbital. With two electrons in separate orbitals, triplet carbenes have radical-like reactivity.

## EXERCISE 6.8

Calculate the formal charge (see Chapter 1) of carbon in each carbene:

(a) $:CH_2$      (b) $:CCl_2$

## Radical Ions

Reactive intermediates can be formed by addition or removal of an electron. For example, an electron can be removed from the $\pi$ bond of an alkene, resulting in a species with only one $\pi$ electron. Because the resulting structure has one electron fewer than needed for neutrality, it is positively charged, and because it contains a single electron in one orbital, it is also a radical. We can write resonance contributors for this **radical cation** that localize the odd electron on either of the two atoms of the $\pi$ system, with the

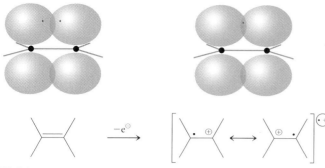

**FIGURE 6.9**

Removal of an electron from an alkene produces a radical cation.

positive charge borne formally at the other atom (Figure 6.9). (The symbol ⊙ is used to indicate the overall charge and odd electron of a radical cation.) The hybrid of these two resonance structures of the radical cation of an alkene has positive charge spread equally between the two carbon atoms and a single electron shared equally, in effect, as one-half of a $\pi$ bond. The $\pi$ system has unpaired-electron character (and therefore the species behaves as a radical); the species is also electron-deficient (like a carbocation) and can therefore act as an electrophile.

A **radical anion** is produced by adding an electron to an alkene, resulting in a structure in which three electrons must be accommodated in the $\pi$ system. Two of the three electrons fill the $\pi$ bonding orbital, and the third electron must be placed in the $\pi^\star$ antibonding orbital. Thus, we can no longer use a simple orbital picture (as we did for the radical cation, with $p$ orbital overlap representing a $\pi$ bonding orbital) to represent the three electrons of the radical anion, because we need two orbitals to hold the three electrons. Both the bonding and antibonding orbitals are required, as shown in Figure 6.10.

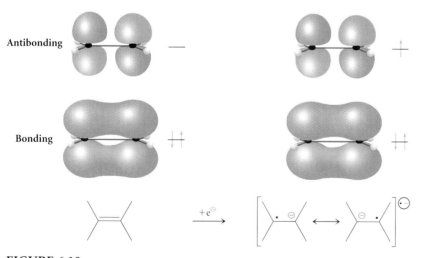

**FIGURE 6.10**

Addition of an electron to an alkene produces a radical anion with two electrons in the $\pi$ bonding orbital and one electron in the $\pi$ antibonding orbital.

The electron density in the anion radical is distributed equally between the two carbon atoms, with equal sharing of the bonding pair of electrons and the antibonding electron. The effect of the antibonding electron on the stability of the structure is opposite to that of one of the bonding electrons. Thus, the anion radical of an alkene has one-half of a $\pi$ bond between the carbon atoms, exactly like the cation radical.

When the $\pi$ system to which the electron is added is a carbonyl group, the resulting radical anion is known as a **ketyl**:

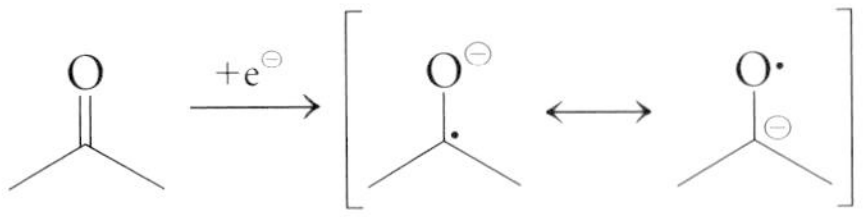

**A ketyl**

The two resonance structures are not the same, for in one the negative charge resides on the oxygen atom, and in the other it is on the carbon atom. Because oxygen is the more electronegative atom, these two contributors are not of equal energy, and the electron density in a ketyl is polarized toward oxygen. As a result of the unequal contribution of resonance structures for this radical anion, the oxygen atom has anionic character and the carbon atom has radical character.

### EXERCISE 6.9

Draw all significant resonance contributors for each radical:

(a) benzene cation radical

(b) naphthalene anion radical

(c) acetone anion radical (a ketyl)

## 6.5

# Kinetics: Relative Rates from Reaction Profiles

In a reaction that takes place without the formation of intermediates, there is only a single transition state, and the energy required to reach this point from the starting material(s) governs the rate of the reaction. In a reaction that takes place in more than one step, the step in which the transition state is of highest energy determines the overall reaction rate. This slowest step is called the **rate-determining**, or **rate-limiting, step** and is the "bottleneck" in a sequence of steps. For a multistep transformation to occur, the transition state of highest energy must be reached.

In the reaction profile shown in Figure 6.4, the formation of transition state 1 is the rate-determining step. According to the Hammond postulate, that transition state can be approximated best by the intermediate, I, because, as we have seen, the transition state lies close in energy to this intermediate. If a similar reaction proceeded via the same general pathway but through a more stable intermediate, the transition state would be lower in energy. Thus, the corresponding activation barrier would also be lower, and the reaction would proceed more rapidly. This can be seen graphically in

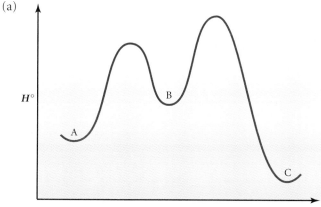

**FIGURE 6.11**

Activation energy barriers for the formation of intermediates of varying relative stabilities.

the three curves in the energy diagram in Figure 6.11. The more stable the intermediate, the more stable is the transition state from which it is formed. The reaction leading to the red intermediate is faster than that leading to the blue, which in turn is faster than that leading to the black. We will find exceptions to this trend that the more exothermic (or less endothermic) reaction will have the lower activation energy, but, in general, it is a good rule of thumb.

## EXERCISE 6.10

Draw energy diagrams for three similar reactions that proceed in a single *exothermic* step to three products, each of different energy. Using transition-state theory (the transition state is related to both starting material and product), determine the energy differences in the transition states and thus which reaction will be fastest and which slowest. Use the Hammond postulate to determine if the difference in rates will be the same, larger, or smaller than it would be if the reactions were all endothermic.

## EXERCISE 6.11

Identify the rate-determining step from each of the following reaction profiles:

(a)

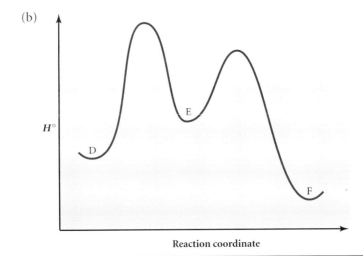

**Reaction coordinate**

## 6.6

## Kinetic and Thermodynamic Control

A study of the changes in bond energy that accompany the conversion of a starting material to a product provides information about $\Delta H^\circ$, but not about the energy of the relevant transition state or any possible reactive intermediates. Nonetheless, as we study various reactions, you will learn empirically (that is, from experimental observations) which reactions are rapid and which are slow. Such observations can be extrapolated (within limits) to predict rates for similar reactions.

Reactions are generally considered to take place under either kinetic or thermodynamic control. This distinction is based on the extent to which the reverse reaction (from product to reactant) takes place under the specified reaction conditions. If the reverse reaction is rapid, equilibrium is established quickly. If equilibrium is reached rapidly, the reaction is said to be under **thermodynamic control.** If the reverse reaction cannot occur (or does so very slowly) with the given reaction conditions, the reaction is said to be under **kinetic control.** The difference in the rates of the forward and reverse reactions is determined by the difference in activation energies for these two processes, which is always uniquely equal to the change in energy from starting material to product. Thus, for a reaction where $\Delta H^\circ = 0$, the forward and reverse rates are identical, and the reaction is under thermodynamic control. Conversely, for a highly exothermic reaction, where $\Delta H^\circ$ is very large, the reverse reaction will be very slow, and the reaction will be under kinetic control.

It is important to note that the distinction between kinetic and thermodynamic control is temporal: with sufficient time, all reactions can achieve thermodynamic control. For many reactions, it is possible to switch between kinetic and thermodynamic control by changing reaction conditions—either time or temperature, or both. For example, the Diels–Alder reaction of cyclopentadiene and furan proceeds at room temperature to produce the *endo* isomer, but at higher temperature and after a longer time, the product is the more stable *exo* isomer.

Cyclo-
pentadiene      Furan                               *endo* Isomer

Cyclo-
pentadiene      Furan                               *exo* Isomer

The distinction between kinetic and thermodynamic control is most relevant when a reaction can yield two products of different stabilities, as in the case of the Diels–Alder reaction just discussed. In general terms, a reaction proceeds under thermodynamic control when the difference in activation energies for two competing forward and reverse reactions is small; in this case, the energy difference between products governs which product predominates. The equilibrium thus established is governed by $\Delta G°$, the difference in free energy between the two products.

In contrast, a reaction proceeds under kinetic control when a large difference in activation energies allows one transition state to be reached more readily than the other. No equilibrium is established between products, and the preference for one product over another is determined by the relative heights of the activation energy barriers for the two processes. As stated in Section 6.1, in an exothermic reaction, the activation energy barrier for the forward reaction is less than that for the reverse reaction (by $\Delta H°$). For $\Delta H°$ values larger than a few kilocalories per mole, the rates of the forward and backward reactions become so different that equilibrium is generally difficult to establish within a reasonable period of time. Reactions with large $\Delta H°$ values are thus generally considered to be under kinetic control. Such reactions are finished, for practical purposes, once the energy barrier from the reactant to the transition state has been surmounted.

Let's construct a reaction profile to illustrate these two types of reaction control. Suppose that a reactant, R, can be converted into either of two products having different stabilities, $P_1$ or $P_2$. A specific example of this general type is the protonation of an enolate anion (R) to form either a ketone ($P_1$) or an enol ($P_2$), a reaction whose thermodynamics we considered earlier in this chapter.

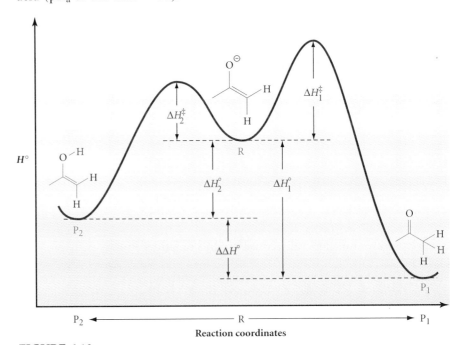

The ketone is more stable than the enol, but the conversion of the enolate anion into the ketone requires a higher activation energy than that needed to convert the enolate anion into the enol (Figure 6.12). If sufficient energy is supplied for the reactant to overcome both barriers, $\Delta H_1^{\ddagger}$ and $\Delta H_2^{\ddagger}$, then enough energy is available to interconvert the ketone, the enol, and the enolate, and so equilibrium is established. Under these reversible conditions, the more stable product (the ketone) is ultimately formed, with the distribution between the two products being governed by the enthalpy difference, $\Delta\Delta H° = \Delta H_2° - \Delta H_1°$ (assuming $\Delta\Delta S° \approx 0$). Because this difference is greater than a few kilocalories per mole, the product mixture is completely dominated by the more stable product. The reaction is then considered to proceed under thermodynamic control. Here, thermodynamics favors the weaker acid ($pK_a$ of the ketone $\approx 19$) over the stronger acid ($pK_a$ of the enol $\approx 10$).

**FIGURE 6.12**

Kinetic (R → P$_2$) and thermodynamic (R → P$_1$) control of the protonation of the enolate anion of acetone.

There are four simple, single-step reactions in Figure 6.12:

1. Deprotonation of the enol to form the enolate anion
2. Protonation of the enolate anion to form the ketone
3. Deprotonation of the ketone to form the enolate anion
4. Protonation of the enolate anion to form the enol

Of these, protonation of the enolate anion to form the enol has the lowest activation energy and, therefore, the fastest rate. If we could force this step of the reaction to be slow, we would be able to produce the enol from the enolate anion. Under these conditions (for example, at very low temperature), the reaction would be under kinetic control. If, on the other hand, we had chosen a higher temperature at which all four reactions proceed rapidly, an equilibrium would be established that favored the more stable ketone. For some reactions, it is possible to switch between kinetic and thermodynamic control. Conducting a reaction at low temperature generally favors kinetic control, but the specific temperature at which thermodynamics dominates depends on the specific reaction being considered. In the specific case of protonation of the enolate anion, it is not usually possible to effect kinetic control under reasonable conditions.

The distinction between kinetic and thermodynamic control is sometimes a qualitative one, because few reactions are so exothermic that equilibrium cannot be achieved under any conditions. In practice, the term *kinetic control* refers to reactions in which the conversion of reactant into one product that is less stable than other possible products can be driven essentially to completion under certain laboratory conditions before significant reverse reaction takes place.

### EXERCISE 6.12

Suggest a chemical reason why protonation of the enolate anion to form the enol may be faster than protonation to form the ketone. (*Hint:* Think about charge density in the resonance-stabilized enolate anion.)

### EXERCISE 6.13

From each of the following reaction profiles, indicate whether kinetic or thermodynamic control is more likely for the overall reaction forming $P_1$ and $P_2$ from R.

(a)

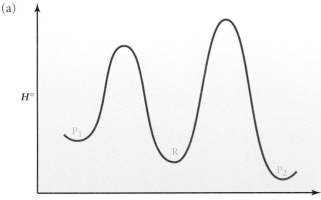

Reaction coordinate

(b)

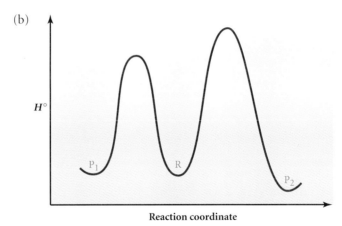

(c)

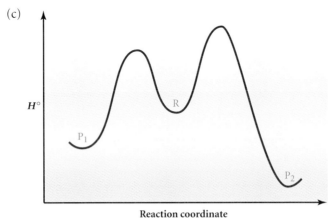

## 6.7

## Chemical Equilibria

### Relating Free Energy to an Equilibrium Constant

Equilibrium is defined as the state in which the forward and reverse reaction rates are equal. At equilibrium, the concentrations of starting materials and products do not change, and thus the ratio of these concentrations is constant.

Changes in free energy control chemical reactions—that is, the more exergonic a reaction (the more negative is $\Delta G°$), the larger is its equilibrium constant $K$ and the more the equilibrium favors the product:

$$\Delta G° = -RT \ln K = \Delta H° - T\Delta S° \qquad (2)$$

where $R$ is the ideal gas constant (1.987 cal/K), and $T$ is the temperature in kelvins. Because the contributions of enthalpy are usually more important than those of entropy to free-energy changes in organic reactions, changes in bond energies can often be related to equilibrium constants. Specifically,

$\Delta G°$ or $\Delta H°$ can also be related to the equilibrium constant $K$ by equation 2. For example, if we know that $K = 1000$ at 22 °C (295 K), we can solve for $\Delta G°$ (and, if $\Delta S°$ is negligible, the result of this calculation is also $\Delta H°$):

$$\Delta G° = -1.987 \text{ cal/K} \times 295 \text{ K} \times \ln(1000)$$
$$= -4050 \text{ cal/mole} = -4.05 \text{ kcal/mole}$$

Conversely, if we know $\Delta G° = +10$ kcal/mole, we can solve equation 2 for the equilibrium constant, $K$:

$$\ln K = \frac{-\Delta G°}{RT}$$

$$K = e^{-\Delta G°/RT}$$

$$= e^{-10,000 \text{ cal}/(1.987 \text{ cal/K} \times 295 \text{ K})}$$

$$= 3.9 \times 10^{-8}$$

For a reversible reaction between A and B to produce C and D (equation 3), the equilibrium constant, $K$, can be written either in terms of concentrations of reagents at equilibrium (equation 4) or as a ratio of the forward ($k_1$) and reverse ($k_{-1}$) reaction rate constants (equation 5).

$$A + B \underset{k_{-1}}{\overset{k_1}{\rightleftharpoons}} C + D \tag{3}$$

$$K = \frac{[C][D]}{[A][B]} \tag{4}$$

$$K = \frac{k_1}{k_{-1}} \tag{5}$$

## EXERCISE 6.14

For a general reaction, $A + B \rightleftharpoons C + D$, taking place at room temperature, calculate $K$ for each of the following conditions:

(a) $k_1 = 10^{10}$ (mole/L)$^{-1} \cdot$ sec$^{-1}$; $k_2 = 10^8$ (mole/L)$^{-1} \cdot$ sec$^{-1}$

(b) initial concentration of A = initial concentration of B; final concentrations of C and D = $0.5 \times$ initial concentration of A

(c) $\Delta G° = -1$ kcal/mole

(d) $\Delta G° = -10$ kcal/mole

(e) $\Delta G° = -30$ kcal/mole

## Acid–Base Equilibria

In **acid–base equilibria,** one of the principal types of chemical equilibria, a proton is transferred from an acid to a base. Cleavage of an H—X bond in an organic acid (to generate an anionic intermediate bearing a lone pair of electrons on $X^\ominus$) forms the **conjugate base** (the deprotonated form,

$X^{\ominus}$) of the acid. Such anionic intermediates are more reactive toward other functional groups than are their protonated precursors, and the initiation of reaction sequences by deprotonation of H—X to form $X^{\ominus}$ is a critical step in many organic transformations.

Bond cleavage of a general acid, HA, to generate an anion requires a *base*, a species active as a proton acceptor or as an electron-pair donor. The flow of electrons in the reaction of a base with an acid is as follows:

$$B\colon \quad H\!-\!A \rightleftharpoons {}^{\oplus}B\!-\!H + \colon\! A^{\ominus} \tag{6}$$

In equation 6, the anion, $A^{\ominus}$, is the conjugate base of the acid H—A. More specifically, an equilibrium for a general acid, HA, with water acting as a base, can be written as follows:

$$H_2O\colon \quad H\!-\!A \rightleftharpoons H_2\overset{\oplus}{O}\!-\!H + \colon\! A^{\ominus} \tag{7}$$

The equilibrium constant, $K$, is defined in the usual way (see equation 8, and compare with equation 4). In addition, because the concentration of water is constant in aqueous solution, we can define another equilibrium constant, $K_a$, as $K$ times the water concentration (equation 9).

$$K = \frac{[A^{\ominus}][H_3O^{\oplus}]}{[HA][H_2O]} \tag{8}$$

$$K_a = K[H_2O] = \frac{[A^{\ominus}][H_3O^{\oplus}]}{[HA]} \tag{9}$$

For example, applying this equation to acetic acid, $CH_3CO_2H$, gives $K_a = 10^{-5}$ (equation 10).

$$K_a(CH_3CO_2H) = \frac{[CH_3CO_2^{\ominus}][H_3O^{\oplus}]}{[CH_3CO_2H]} = 10^{-5} \tag{10}$$

A convention has been adopted to use a negative logarithm scale to describe acidity, in which **$pK_a$** is defined as the negative logarithm of $K_a$ (equation 11).

$$pK_a = -\log K_a \tag{11}$$

$$pK_a(CH_3CO_2H) = 5$$

Instead of saying that the $K_a$ of acetic acid is $10^{-5}$, we say that the $pK_a$ of acetic acid is 5. Thus, a small $K_a$ corresponds to a large $pK_a$. The acid dissociation constant, $K_a$, of acetic acid is quite small, but the $K_a$ values for most organic acids are even smaller. The smaller the $K_a$ of an acid, the larger is the positive value of its $pK_a$ and the less acidic it is.

## EXERCISE 6.15

Calculate the $pK_a$ values of acids having the following acid dissociation equilibria:

(a) $K = 4 \times 10^{-6}$      (c) $K = 1250$      (e) $K = 5$

(b) $K = 3 \times 10^{-40}$      (d) $K = 1$

# 6.8

## Acidity

### A Quantitative Measure of Thermodynamic Equilibria

A convenient way to describe the relative acidity of two organic compounds is to order them according to $pK_a$. Such a ranking of $pK_a$ values describes how easily heterolytic cleavage of an H—X bond can occur. On this scale, the less positive (or more negative) the $pK_a$, the stronger is the acid and the easier it is to cleave the H—X bond. The more stable the anion $X^-$, the more acidic is the hydrogen bonded to X. Thus, because acidity is directly related to the stability of the anion generated, this scale can be used to interrelate the relative reactivity of various anions as bases.

Mineral acids typically have negative $pK_a$ values, suggesting that their acid dissociation equilibria lie far to the right in the equilibrium described by equation 7, and that they are, therefore, very strong acids. Different functional groups have different characteristic $pK_a$ values. Those of carboxylic acids are typically around 5, those of phenols (aromatic alcohols) are at about 10, and those of aliphatic alcohols are around 16. More values are listed in Tables 6.1 and 6.2 (on page 307).

### TABLE 6.1

Approximate $pK_a$ Values of Organic and Inorganic Acids

| Compound | $pK_a$ | Compound | $pK_a$ | Compound | $pK_a$ |
|---|---|---|---|---|---|
| HOSO$_2$O—H<br>Sulfuric acid | −10 | CH$_3$COO—H<br>Acetic acid | 4.8 | ArO—H<br>A phenol | 10 |
| I—H<br>Hydroiodic acid | −10 | HOCOO—H<br>Carbonic acid | 5 | Acetone enol | 11 |
| Br—H<br>Hydrobromic acid | −9 | HS—H<br>Hydrogen sulfide | 7 | | |
| Cl—H<br>Hydrochloric acid | −7 | ArS—H<br>Thiophenol | 7 | Methyl acetoacetate | 11 |
| ArSO$_2$O—H<br>An arylsulfonic acid | −6.5 | H$_3$N$^{\oplus}$—H<br>Ammonium ion | 9 | | |
| H$_2$O$^{\oplus}$—H<br>Hydronium ion | −1.7 | N≡C—H<br>Hydrogen cyanide | 9 | Dimethyl malonate | 13 |
| O$_2$NO—H<br>Nitric acid | −1.5 | | 9 | | |
| F—H<br>Hydrofluoric acid | 3 | 2,4-Pentanedione | | CH$_3$O—H<br>Methanol | 15.5 |

(*continued*)

**TABLE 6.1**

Approximate $pK_a$ Values of Organic and Inorganic Acids (*continued*)

| Compound | $pK_a$ | Compound | $pK_a$ | Compound | $pK_a$ |
|---|---|---|---|---|---|
| HO—H<br>Water | 15.7 | N≡CCH$_2$—H<br>Acetonitrile | 25 | PhCH$_2$—H<br>Toluene | 41 |
| <br>Cyclopentadiene | 16 | <br>Methyl acetate | 25 | Ph—H<br>Benzene | 43 |
| CH$_3$CH$_2$O—H<br>Ethanol | 16 | CH$_3$C≡C—H<br>Propyne | 25 | <br>Propene | 43 |
| (CH$_3$)$_2$CHO—H<br>2-Propanol | 16.5 | <br>*N,N*-Dimethylacetamide | 30 | <br>Ethene | 44 |
| <br>Acetaldehyde | 17 | Ph$_3$C—H<br>Triphenylmethane | 32 | H$_3$C—H<br>Methane | 48 |
| <br>Acetamide | 17 | <br>1, 3-Pentadiene | 33 | CH$_3$CH$_2$—H<br>Ethane | 50 |
| (CH$_3$)$_3$CO—H<br>*t*-Butanol | 18 | H—H<br>Molecular hydrogen | 35 | <br>Cyclohexane | 51 |
| <br>Acetone | 19 | H$_2$N—H<br>Ammonia | 38 | | |
| | | ((CH$_3$)$_2$CH)$_2$N—H<br>Diisopropylamine | 40 | | |

In the following sections, we consider electronic and structural features that influence the $pK_a$ values of various molecules. The strength of an acid (its $pK_a$ value) depends directly on the extent to which a proton is transferred from the acid to a base, which in turn is determined by the stability of the resulting anion. First, we consider how acidity is influenced by changing the atom to which the acidic proton is attached. Then we examine more subtle effects by keeping constant the atom to which the acidic proton is attached, thereby minimizing the effect of electronegativity. This allows us to consider the effects on acidity of several other factors: bond energies, inductive and steric effects, hybridization, resonance stabilization, and aromaticity.

## TABLE 6.2

Inductive Effects on Acidity

| Compound | pK$_a$ | Compound | pK$_a$ |
|---|---|---|---|
| 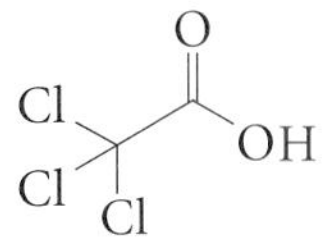 Trichloroacetic acid | 0.4 | 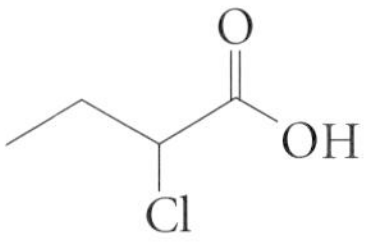 2-Chlorobutanoic acid | 2.9 |
| 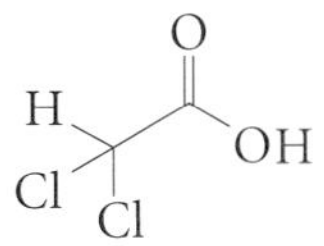 Dichloroacetic acid | 1.3 | 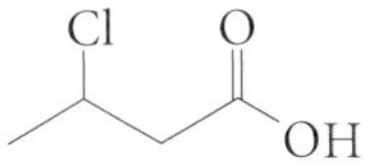 3-Chlorobutanoic acid | 4.1 |
| 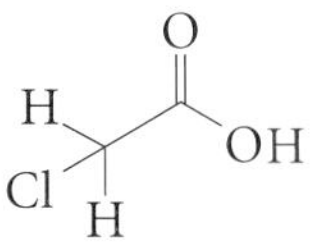 Chloroacetic acid | 2.9 | 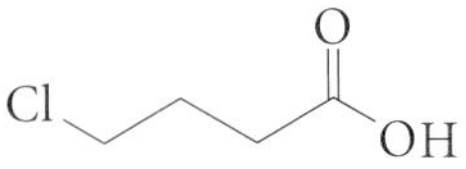 4-Chlorobutanoic acid | 4.5 |
| 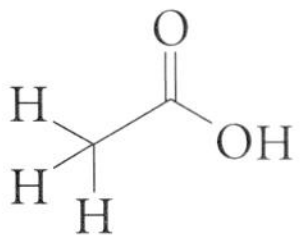 Acetic acid | 4.8 | 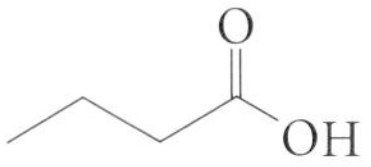 Butanoic acid | 4.9 |

## EXERCISE 6.16

Identify the most acidic hydrogen atom in each of the following molecules. Then use Table 6.1 to determine which member of each pair has the lower pK$_a$.

(a)

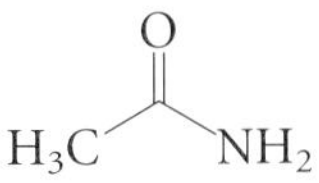

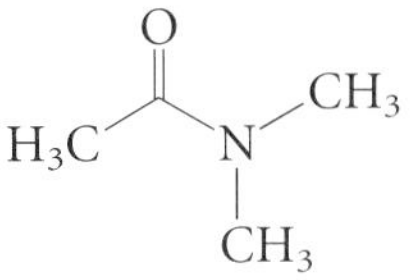

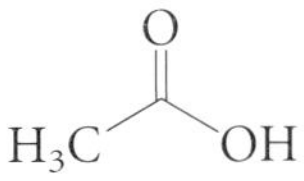

(b)

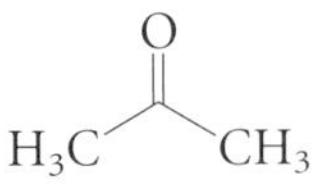

(c) H$_3$C—NH$_2$ or

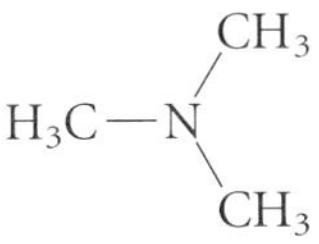

(d)

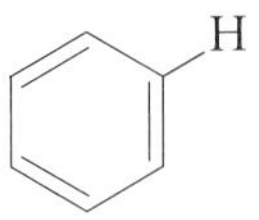

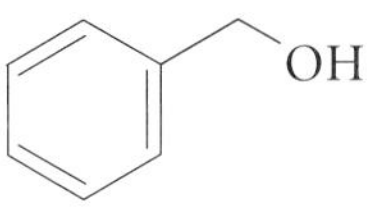

## Electronegativity

The bond connecting a proton to an electronegative atom is highly polar, usually making the molecule a strong acid. For example, the acidities of compounds containing second-row elements—$CH_4$ ($pK_a$ about 48), $NH_3$ (38), $H_2O$ (16), and HF (3)—increase steadily as the atom to which the proton is attached becomes more electronegative. (Remember that $pK_a$ values are logarithms; thus, HF is about $10^{45}$ times as acidic as $CH_4$.)

## Bond Energies

Within a single column of the periodic table, the trend in acidity is opposite to what is expected solely from electronegativity. As the atomic weight increases down a column, the ability of an atom to bear negative charge increases because of the atom's larger size, even though its electronegativity decreases. The H—X (and C—X) bond-dissociation energy also decreases in the progression down the periodic table from HF to HI because of the increasingly mismatched orbital sizes. Thus, the acidity of HI ($pK_a \approx -10$) is greater than that of HBr ($pK_a \approx -9$), which is greater than that of HCl ($pK_a \approx -7$), which, in turn, is greater than that of HF ($pK_a \approx +3$).

### EXERCISE 6.17

Predict which member of each of the following pairs of compounds is more acidic, and give reasons for your choice.

(a) $H_2S$ or $PH_3$        (d) $CH_4$ or $SiH_4$

(b) $H_2O$ or HCl          (e) $CH_3OH$ or $CH_3NH_2$

(c) $H_2O$ or $H_2S$        (f) HI or $H_2S$

## Inductive and Steric Effects

Acidity is also influenced by the presence of polar functional groups, which induce a shift of electron density within the molecule. We can see the effect of electron-donating and -releasing groups on acidity by comparing a series of similarly constructed acids. For example, we can see from the $pK_a$ values of butanoic acid, 4-chlorobutanoic acid, 3-chlorobutanoic acid, and 2-chlorobutanoic acid (Table 6.2) that the electronegative chlorine atom enhances acidity and that its effect is greater the closer it is to the acidic carboxyl group. The electronegative atom (chlorine) withdraws electron density through the series of $\sigma$ bonds. This **inductive effect**—a charge polarization through a series of $\sigma$ bonds—causes a shift of electron density from the acidic site and thus stabilizes the anion formed by deprotonation. An even stronger electron-withdrawing substituent, such as a nitro group ($-NO_2$), enhances the acidity of a carboxylic acid even more. The $pK_a$ of nitroacetic acid, $O_2NCH_2CO_2H$, is 1.68; that of chloroacetic acid, $ClCH_2CO_2H$, is 2.9.

Electron withdrawing by chlorine in 4-chlorobutanoic acid and its anion

Because an inductive effect is transmitted *through bonds,* the effect is greater when transmission is through fewer bonds. Thus, the effect is greater when the electronegative element is closer to the acidic site. Furthermore, the greater the number of electronegative atoms, the greater is the stabilization of the anion by electron withdrawal. Thus, acidity increases in the series: acetic acid < monochloroacetic acid < dichloroacetic acid < trichloroacetic acid (Table 6.2). Compared with changes due to the atom X involved in the H—X bond, these inductive effects are small, but nonetheless important. For example, trichloroacetic acid is ten times as acidic as dichloroacetic acid.

## EXERCISE 6.18

Predict which compound in each of the following pairs is more acidic. Give reasons for your choice.

(a) 

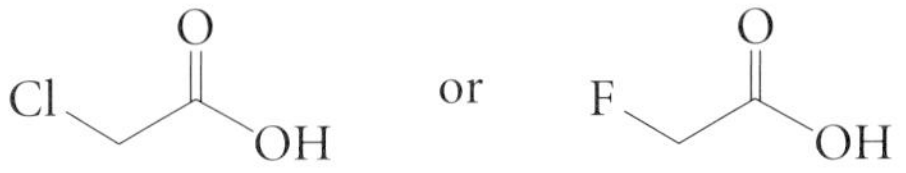

(b) 

(c) $F_3CCH_2$—OH    or    $H_3CCH_2$—OH

The acidity of alcohols decreases as the degree of substitution of the carbinol carbon increases. Thus, the solution-phase acidities of methanol, ethanol, 2-propanol, and *t*-butanol decrease by more than two $pK_a$ units as the carbon bearing the OH group becomes more fully alkylated:

$$pK_a$$

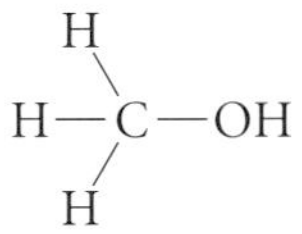

$$H-\underset{\underset{\displaystyle H}{|}}{\overset{\overset{\displaystyle H}{|}}{C}}-OH \qquad 15.5$$

**Methanol**

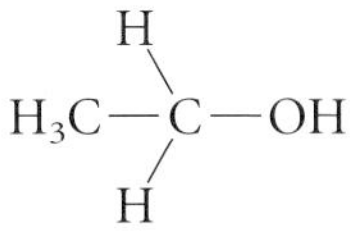

$$H_3C-\underset{\underset{\displaystyle H}{|}}{\overset{\overset{\displaystyle H}{|}}{C}}-OH \qquad 16$$

**Ethanol**

$$H_3C-\underset{\underset{\displaystyle H}{|}}{\overset{\overset{\displaystyle H_3C}{\diagdown}}{C}}-OH \qquad 16.5$$

**2-Propanol**

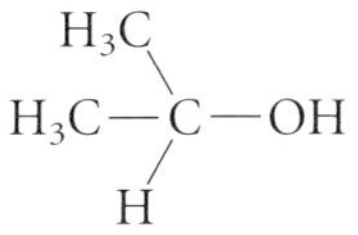

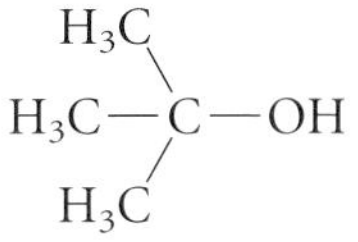

$$H_3C-\underset{\underset{\displaystyle H_3C}{\diagup}}{\overset{\overset{\displaystyle H_3C}{\diagdown}}{C}}-OH \qquad 18$$

**t-Butanol**

## ACIDITY AND WINE MAKING—ACIDS ON THE PALATE

The conversion of grape juice ("must" in wine-making nomenclature) into wine is a far more complex process than is the formation of alcohol from carbohydrates, which occurs by the action of yeasts in primary fermentation. For example, a second stage, referred to as *malolactic fermentation,* is desirable for nearly all red wines and often sought for white wines. In this fermentation, a bacterium belonging to the genus *Lactobacillus* converts malic acid to lactic acid. Malic acid is perceived as being softer and smoother in the mouth than lactic acid, even though there is very little difference in acidity between them. (The pH of wine varies from 2.8 to 3.8.)

Malic acid
(pK_a 3.40 and 5.11)

Lactic acid
(pK_a 3.08)

However, the order of acidity in solution (MeOH > EtOH > *i*-PrOH > *t*-BuOH), where solvation and intermolecular association are important, is reversed in the gas phase, where isolated molecules are observed. The $pK_a$ in solution must therefore be sensitive to intermolecular effects. The replacement of a hydrogen atom by a group of comparable electronegativity, but much larger size, stabilizes an ion in the gas phase but induces a pronounced destabilizing **steric effect** on the solvation of the same anion (conjugate base) in solution. The presence of alkyl substituents on the carbinol carbon pushes the solvent molecules away from the negatively charged oxygen atom of the alkoxide and thus inhibits the stabilizing interaction of the solvent with the anion (Figure 6.13).

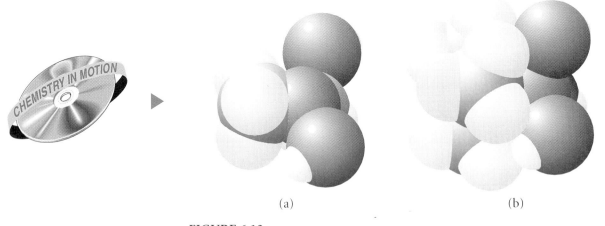

(a)                                        (b)

**FIGURE 6.13**

(a) Methoxide ion and (b) *t*-butoxide ion, each with the three water molecules that constitute the inner solvation shell. The additional alkyl groups in the *t*-butoxide ion interfere with solvation, increasing the energy of this anion. (Carbon is shown as black, hydrogen as off-white, and oxygen as red. Some of the hydrogen atoms on the water molecules are not visible.)

Potassium hydroxide (KOH) is very soluble in water (107 g/100 mL at 15 °C), quite soluble in methanol and ethanol (but less so than in water), and hardly soluble at all in *t*-butanol. Explain this trend in solubility. (*Hint:* First consider why an ionic compound such as KOH dissolves in water.)

## Hybridization Effects

The p$K_a$ values in Table 6.1 show that the hybridization of the carbon atom attached to a hydrogen greatly influences the acidity of that hydrogen—with more *p* character, the acidity of a C—H bond decreases appreciably. Thus, acidity increases appreciably from alkanes through alkenes to alkynes. The p$K_a$ of ethane (with an $sp^3$-hybridized C—H bond) is about 50, that of ethene (with an $sp^2$-hybridized C—H bond) is about 44, and that of ethyne (with an $sp$-hybridized C—H bond) is about 25.

The explanation for this order is that, in each case, the acid-dissociation equilibrium generates an anion whose lone pair of electrons is held in a different kind of hybridized orbital. We saw in Chapter 2 that $sp$-hybridized atoms are more electronegative than $sp^2$- or $sp^3$-hybridized atoms. Thus, a lone pair in an $sp^3$-hybrid orbital (25% *s* character) is held farther from the nucleus than one in an $sp^2$-hybrid orbital (33% *s* character), which is farther from the nucleus than one in an $sp$-hybrid orbital (50% *s* character). Because it is more favorable for the negative charge of an anion to be in an orbital closer to the positively charged nucleus, an $sp$-hybridized anion is more stable than an $sp^2$-hybridized anion, which is more stable than an $sp^3$-hybridized anion.

## Resonance Effects

We saw in Chapter 3 that allyl cations and radicals are stabilized by resonance.

Allyl cation          Allyl radical

The allyl anion, formed by the deprotonation of propene, is also stabilized. The electron pair released by deprotonation is accommodated in a *p* orbital formed as the original $sp^3$-hybridized center becomes $sp^2$-hybridized.

Propene          Allyl anion

Delocalization of negative charge along the three-atom system, together with equal contributions from the two resonance structures, appreciably enhances the stability of this anion. The stabilization accompanying this delocalization causes both the allyl and enolate ions to be planar because, in the flat geometry, orbital interaction is strongest. Thus, the $sp^3$-hybridized C—H bond in propene (p$K_a$ 43) is more acidic than that in propane (p$K_a$ 50). When the conjugation of the allyl anion is extended further, the additional resonance contributors stabilize the anion and make the protonated form more acidic. Thus, deprotonation of 1,3-pentadiene (p$K_a \approx 33$) is easier than that of propene, because of the additional resonance stabilization of the more highly conjugated anion derived from the diene.

Pentadienyl anion

There is also resonance stabilization of the anion that results from deprotonation of a C—H group adjacent to a carbonyl group. The resulting anion is stabilized not only because of delocalization like that encountered in the allyl anion, but also because one of the resonance contributors has negative charge on the more electronegative oxygen atom.

Ketone                              Enolate anion

This anion, referred to as an *enolate anion*, is one of the most important anions of organic chemistry. Despite the fact that deprotonation adjacent to the carbonyl group of an aldehyde or a ketone requires breaking a C—H rather than an O—H bond, the acidity of the $\alpha$ hydrogen is sufficiently enhanced that it is only about five p$K_a$ units less acidic than a hydrogen of an OH group.

## EXERCISE 6.20

Identify the most acidic hydrogen atom in each of the following molecules. Draw the significant resonance contributors for the enolate anion generated by deprotonation of each molecule.

Predict which hydrogen is most acidic in each of the following compounds:

(a)

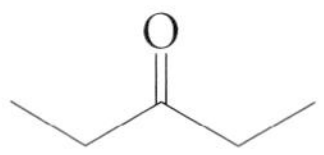

(c)

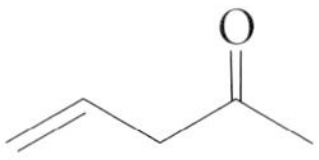

(b)

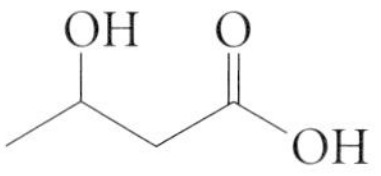

(d)

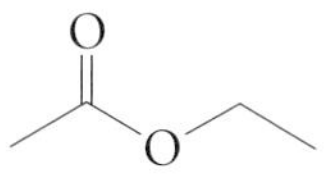

## Enolate Anion Stability

Let's consider in more detail how the structure of the carbonyl group influences the acidity of an $\alpha$ hydrogen. Typical $pK_a$ values for an aldehyde (17), a ketone (19), an ester (25), and a diketone (9) are listed in Table 6.1. The difference between values observed for aldehydes and ketones is the result of compensating factors—that is, an alkyl group attached to a carbonyl carbon of a ketone behaves as if it were electron-releasing, compared to a hydrogen atom, stabilizing the ketone relative to the aldehyde.

Indeed, the two bonds ($\sigma$ and $\pi$) between the carbon and oxygen of the carbonyl group in a ketone are together worth 179 kcal/mole, whereas those bonds in an aldehyde are worth 176 kcal/mole, and those in formaldehyde are worth only 173 kcal/mole.

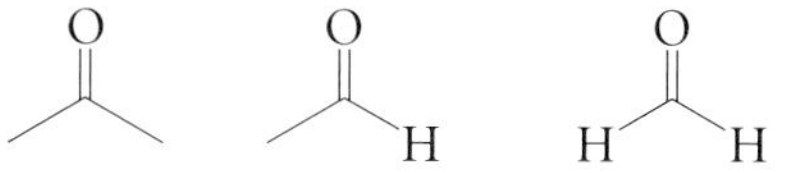

| Ketone | Aldehyde | Formaldehyde |

Although alkyl substituents also stabilize alkenes (Chapter 2), they do so by releasing electrons, which is much less important in the negatively charged enolate anion.

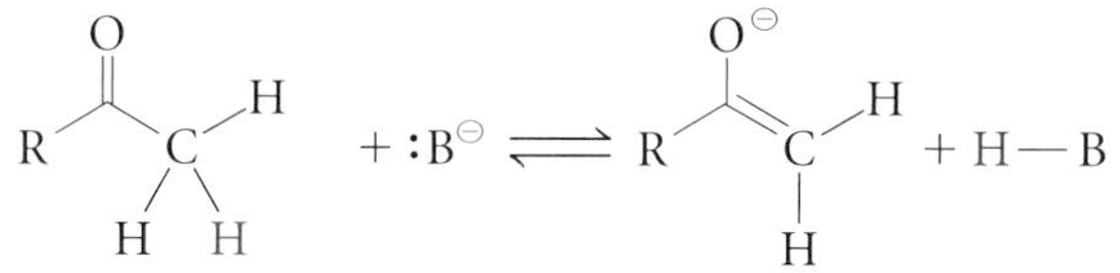

A ketone                        Enolate anion
(R = alkyl, aryl, or H)

To compare the acidity of an aldehyde or a ketone with that of a typical ester or amide, we must keep in mind that an acid derivative bears a heteroatom bonded to the carbonyl group. Oxygen or nitrogen can influence acidity by an inductive effect, but we must also consider the resonance interaction of the adjacent heteroatom with the carbonyl $\pi$ bond. Resonance contributors such as those shown at the far right for the ester and amide groups diminish the ability of the carbonyl oxygen to accommodate further charge, which is necessary for the stabilization of the enolate anion.

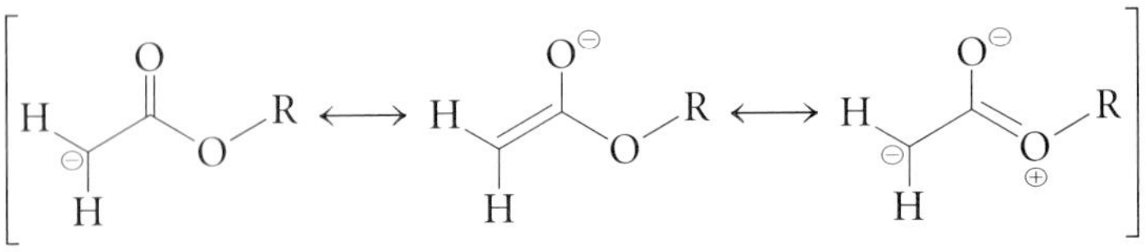

Resonance structures of an ester enolate anion

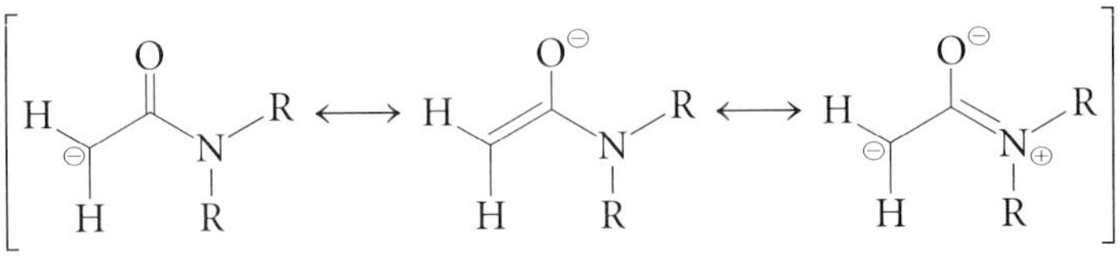

Resonance structures of an amide enolate anion

Thus, the acidities of $\alpha$ hydrogens of esters ($pK_a$ 25) and amides ($pK_a$ 30) are somewhat lower than those of $\alpha$ hydrogens of aldehydes ($pK_a$ 17) and ketones ($pK_a$ 19). Nitrogen (in an amide) is better able than oxygen (in an ester) to act as a $\pi$-electron donor to a carbonyl group; that is, electron donation from the less electronegative nitrogen atom induces greater partial double bond character in the C—N bond of an amide. The greater contribution of amide resonance compared to ester resonance (both of which act against enolate stabilization) is responsible for the lower acidity of the $\alpha$ hydrogen in an amide compared to that in an ester. (This argument applies only to tertiary amides because, in primary and secondary amides, the hydrogen on nitrogen is more acidic than the hydrogen $\alpha$ to the carbonyl group.)

When two carbonyl groups are adjacent to the same carbon atom in a 1,3 relationship, deprotonation results in an anion with charge delocalization over three atoms (two oxygen atoms and one carbon atom) in the same way as in the pentadienyl anion.

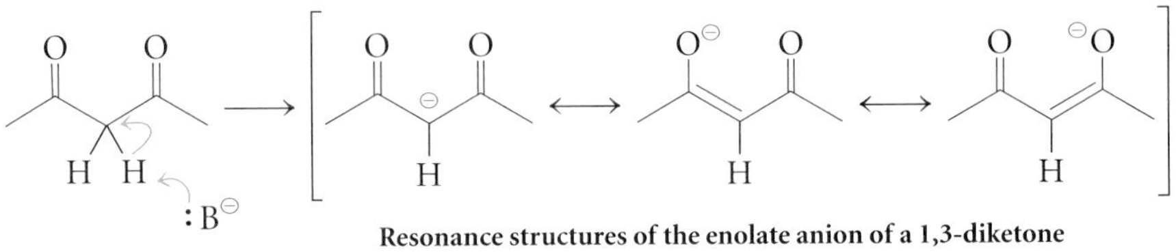

Resonance structures of the enolate anion of a 1,3-diketone

The anion resulting from deprotonation of a 1,3-diketone is more stable than a simple enolate anion, and as a result, 1,3-diketones (and other 1,3-dicarbonyl compounds) are more acidic than simple ketones. However, the effect of the second carbonyl group on acidity is not as great as that of the first.

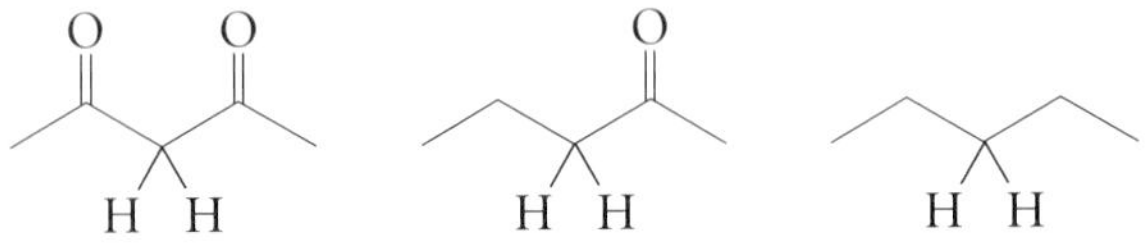

A diketone
($pK_a$ 9)

A ketone
($pK_a$ 19)

A hydrocarbon
($pK_a$ 51)

Aromaticity can contribute significantly to the stability of an anion. For example, deprotonation of cyclopentadiene generates the cyclopentadienyl anion, which contains six (that is, $4n + 2$) electrons in a Hückel aromatic system. The $pK_a$ of cyclopentadiene (16) is much lower than that of 1,3-pentadiene (33) and is, in fact, very close to that of water (15.7), despite the cleavage of a C—H rather than an O—H bond.

Cyclopentadiene

Cyclopentadienyl anion resonance contributors

Deprotonation of cycloheptatriene results in an anion for which we can draw seven identical resonance contributors:

Cycloheptatriene                    Cycloheptatrienyl anion

Nonetheless, cycloheptatriene has a $pK_a$ of about 40 and is thus significantly less acidic than cyclopentadiene. This decreased acidity is the direct result of the difference in the number of electrons in the two anions; six in cyclopentadienyl anion, corresponding to a Hückel aromatic system ($4n + 2$, $n = 1$), and eight in cycloheptatrienyl anion ($4n$, $n = 2$), a system that, though delocalized, is not a Hückel aromatic. The large difference in acidity ($10^{24}$) can thus be attributed directly to the effects of aromaticity on the stabilization of cyclopentadienide as a cyclic, conjugated, planar, delocalized aromatic anion.

## EXERCISE 6.22

Draw the other six resonance contributors for the cycloheptatrienyl anion:

## EXERCISE 6.23

Which member of each of the following pairs of compounds is more readily deprotonated?

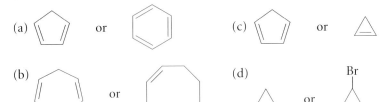

Thus, several factors govern the acidity of a given functional group and, as a result, the position of its chemical equilibrium with its conjugate base. Acidity is important not only as a concept used to illustrate thermodynamic equilibria, but also as a means for ranking the relative stabilities of anions and the thermodynamic feasibility of various reactions under acidic and basic conditions.

In summary, the following effects control relative acidity. Everything else being equal, HX is a stronger acid than HY if:

**1.** X is a more electronegative atom than Y.

$$HF > H_2O > NH_3 > CH_4$$

**2.** The H—X bond is weaker than the H—Y bond.

$$HI > HBr > HCl > HF$$

**3.** X bears more electronegative atoms closer to the site of negative charge in its conjugate base than does Y.

**4.** $X^{\ominus}$ is less sterically blocked from solvation than is $Y^{\ominus}$.

**5.** $X^{\ominus}$ has a greater fractional $s$ character than does $Y^{\ominus}$.

**6.** The negative charge in $X^{\ominus}$ can be delocalized over a larger number of atoms than it can in $Y^{\ominus}$.

**7.** The negative charge in $X^{\ominus}$ can be delocalized onto a more electronegative atom than it can in $Y^{\ominus}$.

**8.** The negative charge in $X^{\ominus}$, but not that in $Y^{\ominus}$, is stabilized by aromaticity.

Look carefully at Table 6.1 for other examples that illustrate these effects.

## 6.9

## Reaction Rates: Understanding Kinetics

The rate at which a reaction proceeds is governed by the energy of the highest-lying transition state. The best method for controlling reactivity depends on whether the rate-determining step involves a single species or whether it requires collision of two or more species. A reaction having only a single species in the rate-determining step is referred to as **unimolecular.** A reaction requiring a collision between two species in the rate-determining step is referred to as **bimolecular.** Reactions requiring collision between more than two species are rare.

### Unimolecular Reactions

Typically, a unimolecular reaction consists of either homolytic cleavage to radical fragments or heterolytic cleavage to ionic fragments, followed by fast conversion of these reactive intermediates into products. In either case, the rate is governed by the number of reactant molecules per unit time having sufficient energy to overcome the activation energy barrier that separates reactants and products. Estimation of the energy required for this transformation is derived from **transition-state theory,** which recognizes that a given reaction proceeds efficiently only if the energy necessary for a reactant to approach the transition state is available.

In general, the facility with which a unimolecular reaction takes place can be enhanced either by increasing the fraction of molecules that can pass over an activation energy barrier or by decreasing the barrier. The former is accomplished by changing the temperature; the latter is effected by altering the structure of the substrate undergoing the reaction or the way in which the reaction is conducted. (An example of a change in the way the reaction is conducted is an increase in solvent polarity in a reaction that includes the formation of ions or the introduction of a catalyst.)

The rate of a reaction is determined by the activation energy, because only molecules with kinetic energy equal to or greater than the activation energy can reach the transition state. The activation energy has contributions from both enthalpy and entropy:

$$\Delta G^{\ddagger} = \Delta H^{\ddagger} - T\Delta S^{\ddagger}$$

When entropy is negligible, the **Arrhenius equation** (equation 12) describes the relation between the observed rate constant, $k_{obs}$, and the activation energy, $\Delta H^{\ddagger}$:

$$k_{obs} = Ae^{-\Delta H^{\ddagger}/RT} \tag{12}$$

where $A$ is a fitting factor characteristic of the reaction, $R$ is the ideal gas constant, and $T$ is the temperature. Thus,

$$\ln\left(\frac{k_{obs}}{A}\right) - \frac{\Delta H^{\ddagger}}{RT}$$

This equation shows how activation energies can be determined in the laboratory: a plot of the logarithm of the observed rate constant (at various temperatures) against $1/T$ will have a slope equal to $-\Delta H^{\ddagger}/R$.

When the Arrhenius equation is rewritten as in equation 13, the form becomes parallel to equation 2, in Section 6.7:

$$\Delta H^{\ddagger} = -RT\ln\left(\frac{k_{obs}}{A}\right) \tag{13}$$

Thus, activation energies ($\Delta H^{\ddagger}$) relate to rate constants ($k_{obs}$) much like free-energy changes ($\Delta G^{\circ}$) relate to the equilibrium constant ($K$).

## EXERCISE 6.24

Calculate the relative rate ($k_1/k_2$) of two reactions, A → B and C → D, occurring at room temperature and having identical $A$ values, if the difference in activation energies is:

(a) 0    (b) 1 kcal/mole    (c) 2 kcal/mole    (d) 5 kcal/mole

## Boltzmann Energy Distributions

The distribution of energies of molecules is a function of temperature; the higher the temperature, the greater is the mean energy of a collection of molecules. (The **mean energy** is defined as that energy at which the number of molecules with energy greater than the mean equals the number with energy less than the mean.) Ludwig Boltzmann (1844–1906) was the first to set forth the mathematical relationship between temperature and molecular energy distribution. **Boltzmann distributions** for collections of molecules at three temperatures, 0 °C, 100 °C, and 200 °C, are shown in Figure 6.14, where the number of molecules (vertical axis) is plotted against enthalpy ($H°$, horizontal axis).

As the temperature increases, the Boltzmann distribution widens and the mean energy is somewhat higher. The mean value of the kinetic energy is given by:

$$\text{Mean energy (cal/mole)} = \frac{8RT}{\pi} \qquad (14)$$

where $R$ is the gas constant (1.987 cal/K · mole) and $T$ is the temperature (in K). Thus, the mean energy of a sample increases linearly with temperature. The mean energy at 0 °C is 1400 cal/mole (or 1.4 kcal/mole).

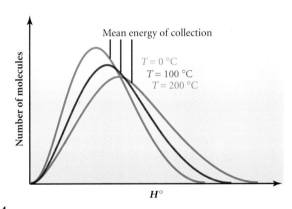

**FIGURE 6.14**

Distribution of enthalpies of a collection of molecules at 0 °C, 100 °C, and 200 °C (temperatures that span the range commonly used for organic reactions).

### EXERCISE 6.25

Use equation 14 to calculate the mean energy for $T = 100$ K and for $T = 200$ K.

The number of molecules that collide with energy of at least $H^{\ddagger}$ is represented by the area under the Boltzmann distribution curve to the right of that $H^{\ddagger}$ value. Most organic reactions have activation energies substantially higher than the mean energy at all reasonable temperatures. Thus, we must focus on the far right portion of the Boltzmann distribution curves to determine the fraction of molecules that collide with sufficient energy to overcome the activation energy barrier. Even for a quite low activation energy of 3.0 kcal/mole (approximately the rotation barrier in ethane), raising the temperature from 0 °C to 100 °C dramatically increases the fraction of molecules with $H° > H^{\ddagger}$, as can be seen in Figure 6.15.

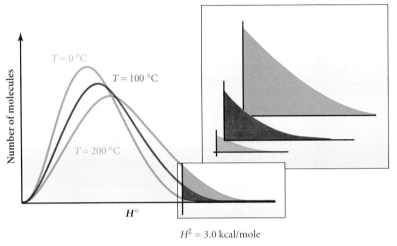

**FIGURE 6.15**

The fraction of molecules with sufficient free energy to react ($H° > H^{\ddagger}$) at 0 °C, 100 °C, or 200 °C is shown by the area under the blue, red, or green curve, respectively, to the right of $H^{\ddagger}$ on the horizontal axis. The insert shows an expansion of that region.

At higher activation energies, the effect of temperature on the reaction rate is even more dramatic. For a reaction with an enthalpy of activation ($H^{\ddagger}$) of 20 kcal/mole (a fast reaction), raising the temperature from 0 °C to 100 °C increases the rate by a factor of $4 \times 10^4$, even though the mean energy increases only from 1.4 to 1.9 kcal/mole. The same change in temperature for a reaction with an enthalpy of activation of 40 kcal/mole (quite slow) increases the rate by a factor of $2 \times 10^7$.

The contribution of entropy to the activation energy does not change with temperature, as can be seen by rewriting the Arrhenius equation:

$$\text{Rate} = e^{-[(\Delta H^{\ddagger} - T\Delta S^{\ddagger})/RT]} = e^{-\Delta H^{\ddagger}/RT} \times e^{-(-T\Delta S^{\ddagger})/RT} = e^{-\Delta H^{\ddagger}/RT} \times e^{\Delta S^{\ddagger}/R} \quad (15)$$

Thus, for reactions in which the activation energy is largely the result of entropy, temperature changes have relatively little effect on the rate. An example of a reaction in which the activation energy is dominated by entropy is proton transfer between heteroatoms (acid–base reactions):

| Acetic acid | Hydroxide ion | | Acetate ion | Water |

As a result, temperature has little effect on the rate of acid–base reactions, which can be conducted at very low temperatures ($-78$ °C).

Use the Arrhenius equation (equation 15) for the rate of reaction to evaluate:

(a) the effect on the rate of a reaction with an enthalpy of activation of 30 kcal/mole of changing the temperature from 0 °C to 100 °C

(b) the effect on the rate of the same reaction of changing the temperature from 0 °C to 10 °C

(*Hint:* Because you are comparing the same reaction at two different temperatures, all pre-exponential terms cancel and can be ignored.)

$$\text{Relative rate} = \frac{e^{-\Delta H^{\ddagger}/RT_1}}{e^{-\Delta H^{\ddagger}/RT_2}}$$

**EXERCISE 6.27**

How would the shape of a typical Boltzmann distribution curve change in each case?

(a) The temperature increases.

(b) The temperature decreases.

(c) Activation energy is increased at room temperature.

## Bimolecular Reactions

In a bimolecular (or higher-order) reaction, the rate is governed by three factors: the number of collisions between the reacting species per unit time, the energy of the colliding molecules, and the orientation of the reactants at the moment of collision. The number of collisions per unit of time is a function of the number of the reacting species per unit of volume. Increasing the number of molecules per volume—that is, increasing the concentration—proportionally increases the number of collisions and, therefore, the number of effective collisions. Thus, the rate of a bimolecular reaction is proportional to the concentration of each reactant. For a reaction between two molecules, A and B:

$$\text{Rate} \propto [A][B]$$

As in a unimolecular reaction, the complex formed by the collision of two reactants must have sufficient energy to overcome the activation energy barrier. Increasing the temperature increases the average kinetic energy of both reactants, and the probability increases that a collision between the two has sufficient energy to overcome the barrier and proceed toward the product. Thus, the effect of an increase in temperature is similar for unimolecular and bimolecular reactions (as long as entropy effects remain minor).

For a collision to induce a chemical change, two molecules must approach one another in the correct orientation that leads to the transition state. Because most organic molecules are not spherically symmetrical, the fraction of accessible geometries that can lead to product influences the fraction of productive collisions between molecules that have sufficient kinetic energy to surmount the activation barrier. Because the number of productive collisions increases in proportion to the total number of collisions, the reaction rates of bimolecular (or more complex termolecular) reactions can be controlled by increasing not only temperature, but also the relative concentration of each reactant required to form the transition state.

## Summary

**1.** A reaction profile (energy diagram) clearly conceptualizes the differences between endothermic, exothermic, and thermoneutral reactions.

**2.** A one-step reaction is referred to as concerted. A multistep reaction involves one or more reactive intermediates.

**3.** Reaction profiles illustrate enthalpy changes in the conversion of a reactant to a product and describe activation barriers in the reaction sequence. In most organic reactions, enthalpy changes have more effect on free energy than do entropy changes, although the importance of entropy depends on the reaction temperature and on the stoichiometry of the reaction.

**4.** The formation of reactive intermediates in multistep reactions is readily apparent in reaction profiles of the rate-determining step.

**5.** The important organic intermediates are carbocations, free radicals, carbanions, carbenes, radical cations, and radical anions.

**6.** The transition state represents an energy maximum in the conversion of a reactant to a product. According to the Hammond postulate, the transition state resembles the species—reactant or product—to which it is closest in energy. Thus, the transition state resembles the reactant in an exothermic reaction and the product in an endothermic reaction.

**7.** The position of a chemical equilibrium is determined by the relative stabilities of the reactant(s) and product(s). The thermodynamics of a reaction can be calculated from the measured equilibrium position. Under strictly reversible conditions, a reaction proceeds to the more stable product and is under thermodynamic control. A reaction is said to be under kinetic control when the rate of the reverse reaction is sufficiently slow that it is not observable, and the product distribution obtained is controlled by the difference in activation energy barriers (and therefore relative rates) rather than the difference in $\Delta G°$ (relative product stability).

**8.** Acid–base equilibria constitute an important class of chemical equilibria. The position of such an equilibrium is described by the $pK_a$, an indicator of acidity. The acidities of organic compounds are sensitively influenced by differences in electronegativity, hybridization, bond energies, resonance stabilization, aromaticity, and inductive effects in the anion formed by the deprotonation of a neutral organic molecule. A large positive $pK_a$ is indicative of a weak organic acid.

**9.** Rates of reaction are governed by activation energy barriers. At a given time, only a fraction of the individual molecules possess sufficient energy to overcome such a barrier. Such barriers are large for transition states in which bond breaking has been substantial and are smaller for transition states that have high degrees of bond making. According to collision theory, increasing the concentration of either reactant in a bimolecular reaction enhances the probability of a productive collision.

# Review of Reactions

Deprotonation $\alpha$ to a Carbonyl Group: Formation of an Enolate Anion

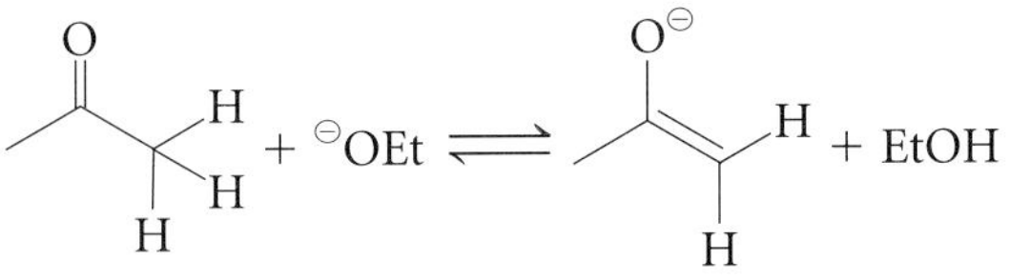

Keto–Enol Tautomerization

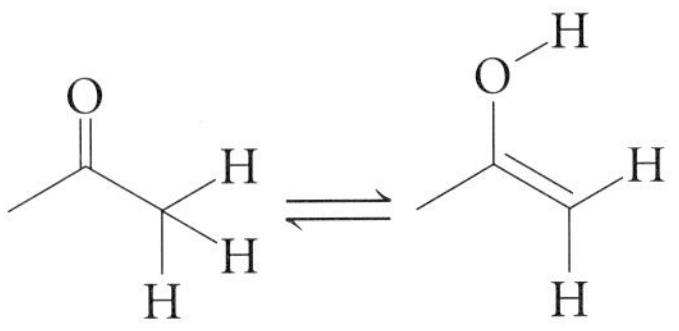

Diels–Alder Reaction

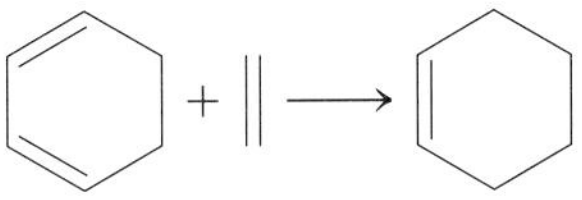

# Review Problems

**6.1** Draw a reaction profile that corresponds to each of the following descriptions:

(a) an exothermic concerted reaction

(b) an endothermic reaction occurring in two steps through a reactive intermediate

(c) an exothermic reaction occurring in three steps, where the second step is rate-determining

**6.2** Rank each set of intermediates according to stability (most stable first). Explain your choices.

(a) $CH_3CH_2CH_2\overset{\oplus}{C}H_2$, $CH_3\overset{\oplus}{C}HCH_2CH_3$, $(CH_3)_2\overset{\oplus}{C}CH_2CH_3$, $(CH_3)_3\overset{\oplus}{C}$

(b) $CH_3CH_2CH_2\overset{\cdot}{C}H_2$, $CH_3\overset{\cdot}{C}HCH_2CH_3$, $(CH_3)_2\overset{\cdot}{C}CH_2CH_3$, $(CH_3)_3C\cdot$

(c) $CH_3CH_2CH_2\overset{\ominus}{C}H_2$, $CH_3\overset{\ominus}{C}HCH_2CH_3$, $(CH_3)(C_6H_5)\overset{\ominus}{C}CH_2CH_3$, $(CH_3)_3\overset{\ominus}{C}$

(d) $:CH_2$, $CH_3\overset{\cdot\cdot}{C}H_2$, $C_6H_5C:$, $(C_6H_5)_2C:$

(e) $H_2CO^{\ominus}$, $(C_6H_5)_2CO^{\ominus}$, $H_2C{=}CH_2^{\ominus}$, $H_2C{=}C(C_6H_5)_2^{\ominus}$

(f) $(C_6H_5)^{\oplus}$, $p\text{-}NO_2(C_6H_5)^{\oplus}$, $p\text{-}CH_3(C_6H_5)^{\oplus}$, $p\text{-}Cl(C_6H_5)^{\oplus}$

**6.3** Draw resonance structures for each radical ion, indicating possible sites of formal charge or of the unpaired electron.

(a) naphthalene cation radical

(b) benzophenone anion radical

**6.4** Upon treatment with strong base at low temperature, *cis*-1,2-diphenylcyclo-propane forms an anion at the benzylic position. When quenched with $D_2O$, a mixture of 1-deutero-*cis*-1,2-diphenylcyclopropane and 1-deutero-*trans*-1,2-diphenylcyclopropane is formed. Draw three-dimensional structures of the intermediate anion and the product. Does the structure of the product allow you to say anything about whether the carbanionic carbon in the intermediate anion is a center of chirality?

**6.5** For each of the following reaction types, is the entropy change positive (favorable), negative (unfavorable), or near zero (negligible)?

(a) a large molecule fragments into three smaller ones

(b) a small molecule condenses with another small molecule to make a large one

(c) an ester is hydrolyzed by water to produce a carboxylic acid and an alcohol

**6.6** Which diene and dienophile would you choose to synthesize each of the following products by a Diels–Alder reaction?

(a) [structure with $CO_2CH_3$ and $CO_2CH_3$]

(c) [structure with $CO_2CH_3$]

(b) [structure with O, O, O]

(d) [structure with $CO_2CH_3$]

**6.7** Of the following proposed Diels–Alder reactions, reaction 1 proceeds but reaction 2 fails. Why?

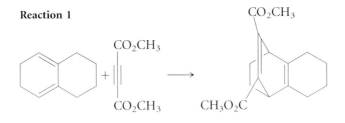

**Reaction 1**

**Reaction 2**

**6.8** Reorder the following sets of compounds according to increasing $pK_a$:

(a) cyclohexanol, phenol, cyclohexanecarboxylic acid

(b) 1-butyne, 1-butene, butane

(c) propanoic acid, 3-bromopropanoic acid, 2-nitropropanoic acid

(d) phenol, toluene, benzene

(e) dimethyl ether, ethanol, methyl acetate ($CH_3CO_2CH_3$)

(f) hexylamine, aniline, hexanoamide

(g) benzoic acid, *p*-chlorobenzoic acid, 2,4,6-trichlorobenzoic acid

(h) ethanoic acid (acetic acid), 1,2-ethanedioic acid (oxalic acid), 1,3-propane-dioic acid (malonic acid) (*Hint:* A carboxylic acid group acts as an effective electron-withdrawing group to the adjacent $\sigma$-bond system.)

(i) protonated forms of pyrrole, pyridine, *N*-methylpyrrole

**6.9** Crotonaldehyde, $CH_3CH{=}CHCHO$, has a p$K_a$ of 20, despite the fact that it lacks enolizable hydrogens $\alpha$ to the carbonyl group.

(a) Determine which hydrogen is removed by interaction with base.

(b) Write one or more resonance structures to account for the stability of the anion (that is, the conjugate base).

**6.10** Octylamine is insoluble in water but dissolves in dilute sulfuric acid. Octanoamide, $C_7H_{15}CONH_2$, does not dissolve in either water or dilute sulfuric acid. Rather, octanoamide dissolves in aqueous base. Propose an explanation for the contrasting solubilities of the amine and the amide.

**6.11** For each of the following molecules, determine which hydrogen is most acidic—that is, which one would be removed by treatment with one equivalent of base.

**6.12** Compound A is converted into a more stable product, B, upon heating without any additional reagent. The reaction profile shows the formation of one reactive intermediate.

(a) Draw an energy diagram that illustrates the key features of this reaction.

(b) Suppose the reaction is conducted at a higher temperature. Does this change the shape of the energy diagram?

(c) Does the rate of reaction depend on the concentration of A?

**6.13** In the following molecules, determine which hydrogen is most acidic—that is, which one would be removed by treatment with one equivalent of base?

**6.14** Consider a reaction in which reactants C and D combine in the rate-determining step. Determine which of the following statements applies to this reaction. Explain your reasoning.

(a) Doubling the concentration of C doubles the rate of the reaction.

(b) Doubling the concentration of D cuts the rate of reaction in half.

(c) Doubling the concentration of both C and D doubles the rate of the reaction.

(d) Increasing the temperature increases the rate of reaction.

**6.15** Draw an energy diagram for a two-step reaction passing through an intermediate that is less stable than both the starting material and the product, where the product is more stable than the starting material *and* the activation energy for proceeding from the intermediate to the product is higher than that for proceeding from the intermediate to the starting material.

**6.16** Answer the following questions for the reaction diagram you constructed in Problem 6.15.

(a) Which species does the first transition state resemble more closely, the starting material or the intermediate?

(b) Which species does the second transition state resemble more closely, the product or the intermediate?

(c) Is the first or the second transition state involved in the rate-determining step?

**6.17** Draw all significant resonance contributors for the following enone (methylvinyl ketone):

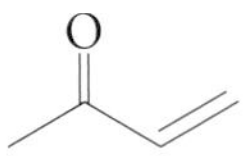

## Supplementary Problems

**6.18** Upon treatment with a Brønsted acid, methylenecyclohexane (A) undergoes isomerization to the more stable alkene methylcyclohexene (B). This reaction proceeds through a cation intermediate (I). Draw an energy diagram for this reaction, paying careful attention to the relative energies of the two transition states involved. (Recall the Hammond postulate.)

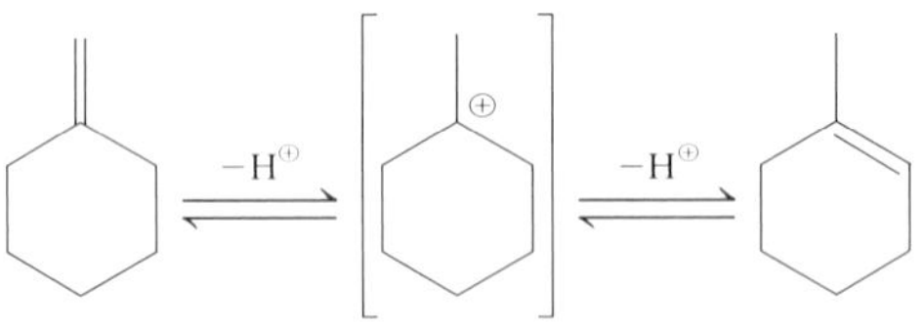

**6.19** Draw the most stable protonated form of each of the following anions, and provide the p$K$a of the resulting conjugate acid.

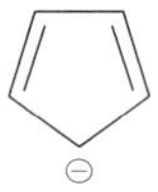

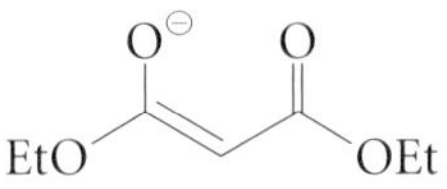

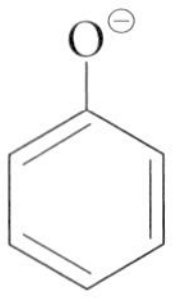

(d) 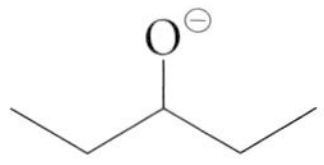  (e) 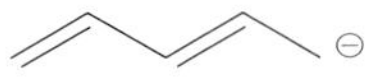  (f)

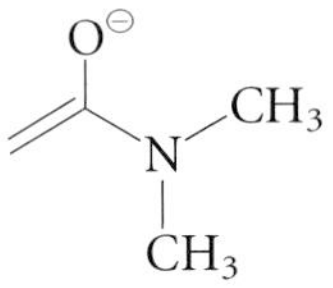

(g) 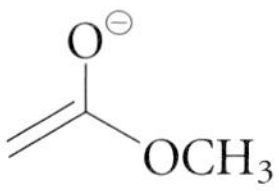  (h)   (i) Br⊖

**6.20** Draw the most stable protonated form of each of the following anions, and provide the p$K_a$ of the resulting conjugate acid.

(a) 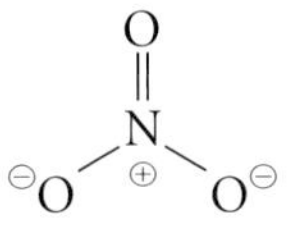  (d) $H_2O$  (g)

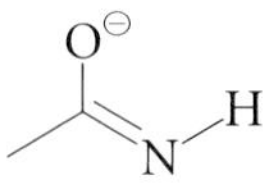

(b) 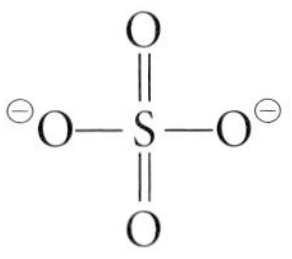  (e) ⊖C≡N  (h)

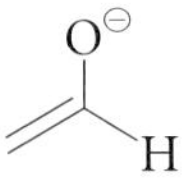

(c) 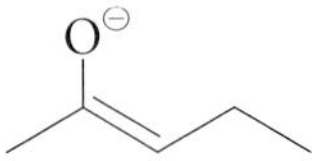  (f) 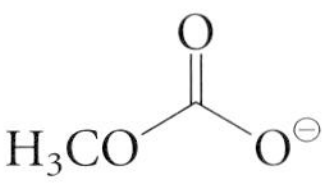  (i)

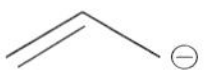

**6.21** Draw all proton tautomers of each of the following structures:

(a) 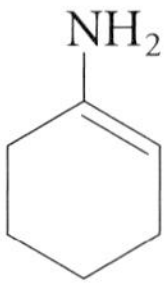  (c) 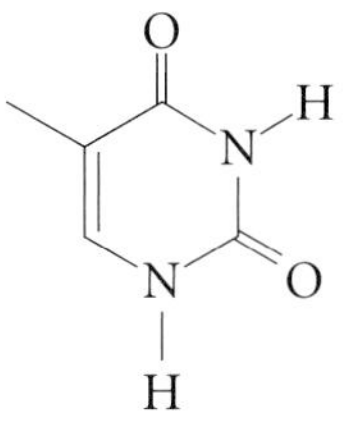  (e)

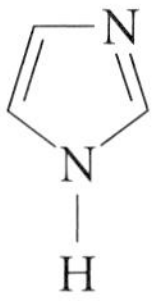

(b) 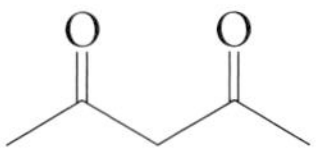  (d)

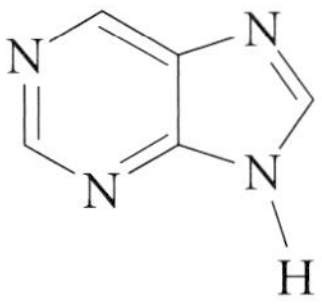

**6.22** Draw the two possible tautomers of theobromine. Determine whether each heteroatom in these molecules can act as a proton or lone pair donor or acceptor for hydrogen bonding.

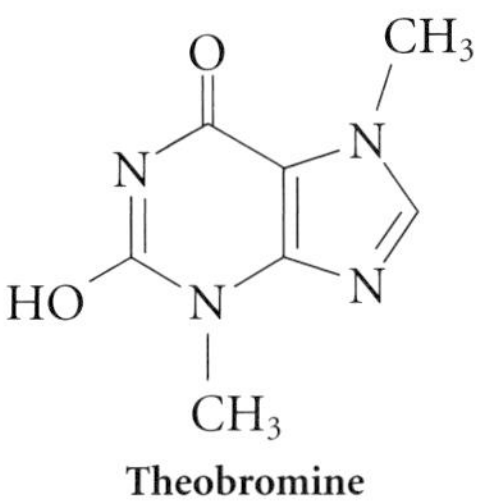

**Theobromine**

**6.23** Draw the structure of the Diels–Alder adduct expected from each of the following reactions:

(a)

(c)

(b)

(d)

**6.24** In each pair of compounds, choose the stronger base.

(a) $Cl^{\ominus}$   or   $F^{\ominus}$

(e) $Ph_3C^{\ominus}$   or

(b)    or

(f)    or

(c) $^{\ominus}OH$   or

(g)    or

(d) $NH_3$   or

(h) $H_2\overset{\ominus}{C}-C\equiv N$   or   $^{\ominus}C\equiv N$

**6.25** In each pair of compounds, choose the stronger acid.

(a)    or

(b) $H_2O$   or   $NH_3$

(c)    or

(d) $H_3C-OH$   or

(e)    or

**6.26** The following spectra were obtained for a compound with the molecular formula $C_4H_9Cl$.

(a) Determine the index of hydrogen deficiency. If the value obtained is not zero, use the spectral data to determine whether rings and/or double bonds are present.

(b) Suggest a structure consistent with all of the data presented for this compound. (The three resonances at approximately 77 in the $^{13}C$ NMR spectrum are due to the solvent, $CDCl_3$.)

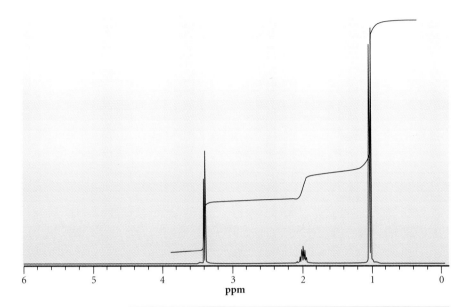

$^1H$ NMR spectrum of $C_4H_9Cl$

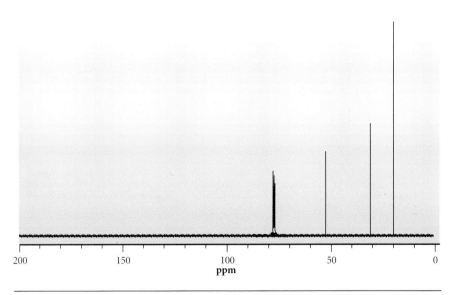

$^{13}C$ NMR spectrum of $C_4H_9Cl$

# Mechanisms of
# Organic Reactions

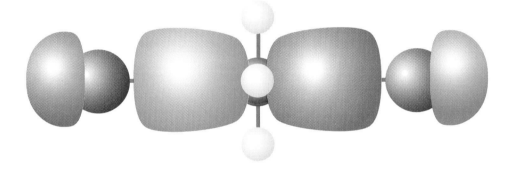

*The transition state for the displacement of bromide ion from methyl bromide by bromide ion. This molecular orbital is formed by overlap of a p orbital on each bromine (left and right, in red) with a p orbital on the central carbon atom (gray sphere in center, partially hidden by the off-white hydrogen atom in front), resulting in the four lobes shown in blue and green.*

*I*n Chapter 6, you learned that the first step in predicting chemical reactivity is to estimate whether a particular conversion is thermodynamically feasible. In this chapter, you will learn to group various chemical reactions according to reaction type and to describe how typical reactions take place. The specific sequence in which bonds are made and broken as a reactant is converted into a product is known as the *reaction mechanism.* Complete understanding of a chemical reaction requires following the flow of electrons in each step as they move to or from bonding orbitals. We will use curved arrows to indicate the electron movements in each step of several common organic reactions, and thus describe the reaction mechanisms.

We will analyze the mechanisms of several specific reactions representative of quite different reaction types: (1) a concerted nucleophilic substitution, (2) a multistep nucleophilic substitution that proceeds through an intermediate carbocation, (3) a multistep nucleophilic substitution in which two cationic intermediates are formed sequentially, and (4) a homolytic substitution proceeding through a free radical intermediate. But before we examine mechanistic details, we must understand what is taking place as a reaction proceeds. We will consider how to classify each reaction, determine its energetic feasibility, and represent the bonding changes that take place. This chapter focuses on how reaction mechanisms can be clearly defined by following electron flow. Later chapters will look at how specific transformations are used in synthesis.

## 7.1

## Classification of Reactions

In considering a new reaction, we first determine what is accomplished—whether the number of atoms in the product differs from the number in the reactant, whether any atoms in the product are different from those in the reactant, and whether the positions of any atoms in the product differ from their positions in the reactant. Depending on the answers to these questions, we then classify a given chemical conversion as one of seven major organic reaction types: addition, elimination, substitution, condensation, rearrangement, isomerization, or oxidation–reduction.

### Addition Reactions

In an **addition reaction,** two reactant molecules combine to form a product containing the atoms of both reactants. Two examples of addition reactions are **hydration** (addition of water) and **catalytic hydrogenation** (addition of two hydrogen atoms) of alkenes.

**Hydration of an Alkene**          **Catalytic Hydrogenation of an Alkene**

In the hydration reaction, water and cyclohexene combine to produce cyclohexanol. In the catalytic hydrogenation reaction, hydrogen is added to cyclohexene (in the presence of a metal catalyst) to form cyclohexane. These reactions will be treated more thoroughly in Chapters 10 and 12.

Some addition reactions require the presence of a catalyst, a substance that does not appear in the product. A **catalyst** is defined as a reagent that facilitates a reaction without itself ultimately forming chemical bonds in the product or appearing in the stoichiometric equation describing the reaction. For example, the addition of water to an alkene proceeds at a reasonable rate only in the presence of a strong acid, and the addition of hydrogen to an alkene occurs only when a metal surface is present. However, in neither case does the catalyst appear in the product; the catalyst remains unchanged. After the reaction, the catalyst is free to participate in another reaction cycle. Whether or not a catalyst is needed to accelerate the rate of an addition reaction does not influence the classification of the reaction.

## Elimination Reactions

An **elimination reaction** is the opposite of an addition. In an elimination reaction, a single complex molecule splits into two simpler products; the one reactant molecule contains all the atoms present in two product molecules. Two typical elimination reactions are **dehydrobromination** (loss of HBr) and **dehydration** (loss of water), both of which result in a carbon–carbon double bond. (These reactions will be treated more thoroughly in Chapter 9.)

**Dehydrobromination of an Alkyl Bromide**

**Dehydration of an Alcohol**

In the dehydrobromination reaction, cyclohexyl bromide can be induced to undergo elimination (loss of HBr) by treatment with a base. Under these conditions, HBr is not observed directly because, in the presence of base, it undergoes an acid–base reaction to form a salt. However, formally, HBr is lost from cyclohexyl bromide in forming cyclohexene, irrespective of its final form (here, in ethanol and bromide ion). It is because the HBr formed in the elimination is immediately converted into ethanol and bromide under the reaction conditions that HBr is shown in parentheses. In the dehydration reaction, the treatment of cyclohexanol with acid produces cyclohexene and water upon heating, accomplishing a reversal of the hydration addition reaction considered earlier.

### Substitution Reactions

In a **substitution reaction,** one atom or group of atoms in a molecule is replaced by another. For example, a hydrogen atom in cyclohexane is replaced by a bromine atom when the alkane is exposed to $Br_2$ in the presence of light or heat. (This reaction will be treated in more detail later in this chapter.)

In the products (cyclohexyl bromide and HBr), the bromine is substituted at a position previously occupied by a hydrogen in cyclohexane, and hydrogen takes the place of one of the two bromine atoms in molecular bromine. Another example of a substitution reaction is the treatment of cyclohexyl iodide with sodium bromide. Again, the positions of iodine and bromine in the reactants are interchanged in the products. (This substitution reaction is covered more thoroughly in Chapter 8.)

### Condensation Reactions

A **condensation reaction** consists of the interaction of two molecules of intermediate complexity to form a more complex product, usually with the loss of a small molecule. For example, the combination of a carboxylic acid with an alcohol in the presence of an acid catalyst produces an ester (a more complex molecule) and water (a small molecule).

In this reaction, two different organic reactants (an acid and an alcohol) combine to form an ester. The product ester has fewer atoms than the sum of those in the two reactants because water is formed as a by-product. (This reaction will be treated in Chapter 12.)

The aldol condensation reaction is an example of a condensation reaction in which a carbon–carbon bond is formed.

**An Aldol Condensation**

Two molecules of a single reactant (a ketone) combine to form a ketone of higher molecular weight, again with water formed as a by-product. (This reaction will be treated more thoroughly in Chapter 13.)

## Le Chatelier's Principle

Many condensation reactions are reversible, and the position of the equilibrium can often be controlled (that is, shifted toward product) if the small molecule is removed (for example, by distillation) as it is formed. This, in turn, shifts the equilibrium in accord with Le Chatelier's principle as the reacting system attempts to replenish the "missing" product. **Le Chatelier's principle** asserts that an equilibrium between A and B producing C and D can be shifted toward C and D by increasing the concentration of A or B, or both (pushing from the left), or by decreasing the concentration of C or D, or both (pulling from the right). The equilibrium can be shifted toward A and B by increasing the concentration of C or D, or both, or by decreasing the concentration of A or B, or both.

$$A + B \rightleftharpoons C + D$$

Le Chatelier's principle is applicable not only to condensation reactions, but also to many other equilibrium processes.

## Rearrangement Reactions

In a **rearrangement reaction,** the molecular skeleton is altered—that is, the sequence in which atoms are attached is changed. These reactions may also include other changes in the molecule; for example, one functional group may be converted into another.

Typically, rearrangement reactions have several steps, which makes these reactions both scientifically interesting and mechanistically complex. Examples are the Beckmann and pinacol rearrangements. In a rearrangement reaction, the atoms or groups present in the reactant are connected in a different fashion in the product. The reactant and product can have the same empirical formula, as in the Beckmann rearrangement, or different numbers and types of atoms, as in the pinacol rearrangement, in which water is formed as a by-product.

**Beckmann Rearrangement**

**Pinacol Rearrangement**

In the Beckmann rearrangement, an alkyl group originally attached to carbon becomes attached to nitrogen, an N—O bond in the reactant is broken, and a C=O bond appears in the product. In the pinacol rearrangement, a methyl group migrates from one carbon to the adjacent carbon, a C=O bond appears in the product, and a molecule of water is lost from the reactant. (These reactions are treated more thoroughly in Chapter 14.) The pinacol rearrangement could also be classified as an elimination reaction. In general, reactions that fall into more than one category are considered to be examples of the more complex process.

### Isomerization Reactions

An **isomerization** is a reaction in which species with the same molecular formula, but different structures, are interconverted. An isomerization differs from a rearrangement in that the carbon skeleton remains intact, but the disposition of substituents or functional groups in space is changed. In an isomerization, the molecular formulas of the reactant and product are always the same; in a rearrangement, they can be the same (as in the Beckmann rearrangement) or different (as in the pinacol rearrangement). There are two types of isomerization reactions: geometric and positional. In a **geometric isomerization,** all atoms in the product are attached to the same atoms as in the reactant, but the disposition in space of the bonds connecting them is changed. In a **positional isomerization,** the position (or positions) of one or more substituents or functional groups in the product differs from the original position(s) in the reactant. For example, the conversion of *cis*-2-butene to *trans*-2-butene is a geometric isomerization, and that of 2-butene to 1-butene or of *n*-butyl bromide to *s*-butyl bromide is a positional isomerization.

**Geometric versus Positional Isomerization of *trans*-2-Butene**

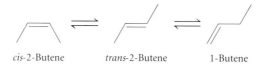

*cis*-2-Butene          *trans*-2-Butene          1-Butene

Geometric isomers differ only in the position of atoms or groups in space (*cis*- and *trans*-2-butene); positional isomers differ in the position of a functional group in the molecule (*trans*-2-butene and 1-butene).

### Oxidation–Reduction Reactions

In **oxidation–reduction reactions,** there is a net formal change in oxidation level of one or more carbon atoms in a molecule. Such reactions were discussed in Chapters 2 and 3, and further examples will not be given here. These reactions can often also be classified as substitutions (when the number of heteroatoms at a given carbon is changed), additions (when hydrogen is added across a multiple bond), or eliminations (when the elements of molecular hydrogen have formally been removed from adjacent atoms). Oxidation–reduction reactions are referred to as *redox reactions,* to emphasize the need to use oxidizing or reducing reagents to bring them about.

Classify each of the following conversions according to reaction type:

(a)

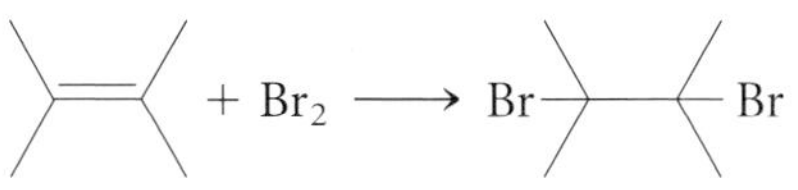

(b)

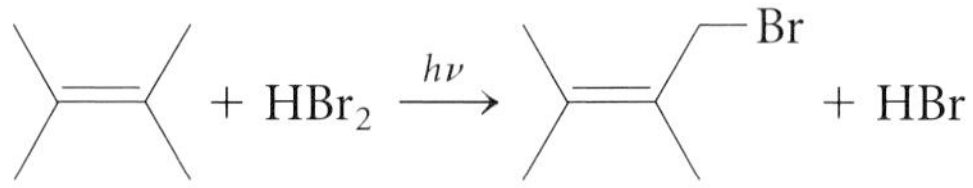

(c) $CH_3OH + HCl \longrightarrow CH_3Cl + H_2O$

(d)

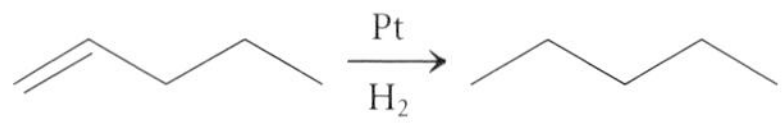

(e)

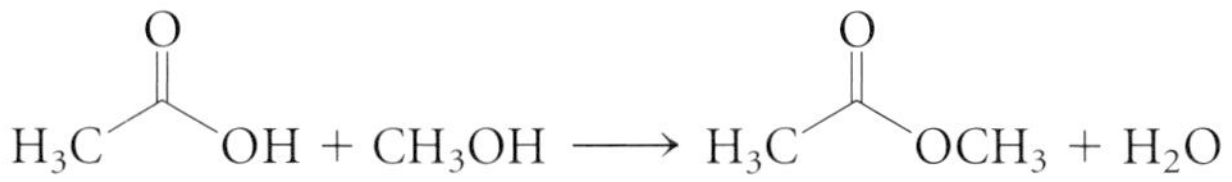

(f)

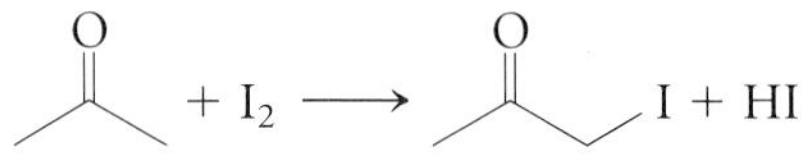

(g)

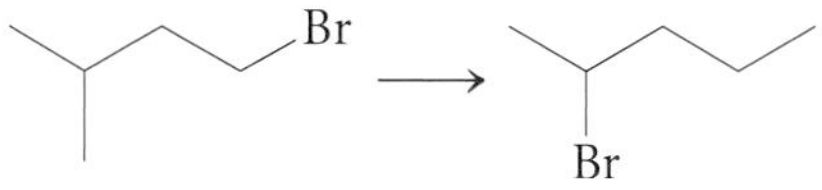

(h) 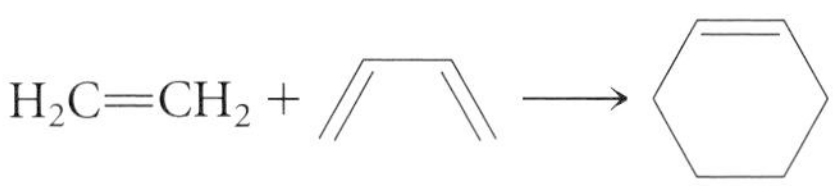 $H_2C{=}CH_2 +$

### Reaction Mechanisms

Correctly classifying a specific reaction as to type does *not* indicate *how* the reaction has taken place. For example, substitution could, in principle, occur either by direct replacement of groups or by a sequence of addition and elimination reactions. To illustrate this difference, let's consider a reaction sequence in which *trans*-stilbene reacts with molecular bromine to yield an addition product, 1,2-dibromo-1,2-diphenylethane. Then treatment with strong base induces elimination, producing a bromostilbene, (*Z*)-1-bromo-1,2-diphenylethene.

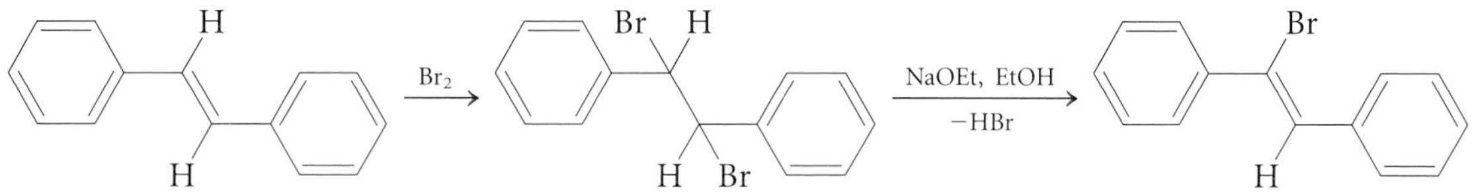

*trans*-Stilbene     1,2-Dibromo-1,2-diphenylethane     (*Z*)-1-Bromo-1,2-diphenylethene

Here a substitution product has been formally obtained from the original reactant (by replacement of H in *trans*-stilbene by Br) through a sequence of addition and elimination reactions. It is insufficient, therefore, to specify the kind of reaction without describing how the reaction proceeds, including a detailed description of electron flow and the identity of any intermediate formed—the **reaction mechanism.** Subsequent chapters will emphasize reaction mechanisms as an important means of intellectually organizing the organic reactions.

## 7.2

# Bond Making and Bond Breaking: Thermodynamic Feasibility

All chemical reactions entail bond making or bond breaking, or both. From Chapter 3, you know that a $\sigma$ bond between atoms A and B can sometimes be cleaved so that the two shared electrons are distributed equally (one to A and one to B), producing neutral species called *radicals*. As mentioned in Chapter 3, this process is referred to as **homolytic cleavage,** or **homolysis.**

$$\text{A--B} \longrightarrow \text{A} \cdot + \text{B} \cdot \qquad \Delta H^\circ = \text{Bond-Dissociation Energy (BDE) of A---B}$$

In a homolytic cleavage, the two electrons of the covalent bond are partitioned so that one electron is associated with each atom. A bond is broken, and two radicals are formed.

In the alternative mode of bond breaking, the two electrons of a $\sigma$ bond move as a pair to one of the initially bonded atoms. This process, called **heterolytic cleavage,** or **heterolysis,** produces a positive and a negative ion.

$$\text{A--B} \longrightarrow \text{A}^\oplus + \text{B}^\ominus$$

The atom that takes up the two electrons from the bond becomes an anion, and the atom that loses the two electrons becomes a cation. The enthalpy change for this bond cleavage is influenced by the bond strength and the solvation energy for the ions formed. The direction of the electron flow (to A or B) is governed by the relative electronegativity of these two atoms, with the more electronegative atom becoming negatively charged.

The convention employed to represent the two modes of electron movement is a half-headed arrow for a single electron (in a homolytic cleavage) and a full-headed arrow for an electron pair (in a heterolytic cleavage). To understand the reaction types in the remainder of this book, it is very important to recognize the precise meaning of this **arrow notation.** The use of half-headed or full-headed curved arrows to indicate motion of electrons is the best way to indicate clearly how a reaction occurs. Bear in mind that the curved arrows indicate movement of electrons, not atoms. When electrons move, atoms follow. The tail of the curved arrow marks the origin of the electron(s), and the head marks the site to which the electron(s) move.

### ▧ Energy Changes in Homolytic Reactions

Homolysis and heterolysis consist solely of bond breaking. However, in many chemical reactions, bond breaking is often accompanied by bond making, in which another reagent assists the cleavage. For example, an already available radical center can assist in homolytic cleavage.

$$\text{R} \cdot \qquad \text{A--B} \longrightarrow \text{R---A} + \text{B} \cdot \qquad \Delta H^\circ = \text{BDE of A---B} - \text{BDE of R---A}$$

$$\text{R} \cdot \qquad \text{B--A} \longrightarrow \text{R---B} + \text{A} \cdot \qquad \Delta H^\circ = \text{BDE of A---B} - \text{BDE of R---B}$$

In an assisted homolysis, some or all of the energy required to cleave the A—B bond is offset by the energy gained in simultaneously forming the R—A (upper reaction) or R—B (lower reaction) bond. Thus, when a radical, R·, with one unpaired electron interacts with an A—B σ bond, the electron that becomes accessible to A in the homolytic cleavage of the A—B bond enables A to form a new bond with R·; the second electron of the newly formed R—A bond is contributed by R·. Meanwhile, the other electron in the original A—B covalent bond becomes localized on a new radical, B·. If the radical, R·, attacked the other end of the molecule, the same kind of electron flow would have produced a bond between R and B and a localized electron on atom A, forming a radical. The relevant bond energies then would be those of A—B and R—B.

The enthalpy change of either of these reactions represents the balance between the bond-dissociation energy of A—B and the bond energy of the newly formed R—A or R—B. The bond energies of R—A and R—B need not be equal, and, as a result, the $\Delta H^\circ$ values for these two reactions can vary. Because the entropy change in such a reaction is small, the reaction proceeds along the more energetically favorable route; that is, R· will form a bond with A· or with B· according to which $\Delta H^\circ$ value is more negative (or less positive).

The coupling of bond making and bond breaking is very common. Many chemical bonds are very strong, and breaking them costs a lot of energy. When a part of this lost energy is regained in the formation of a new bond, the reaction proceeds much more easily.

Three half-headed arrows are used to indicate the motion of the three electrons in the assisted homolytic cleavage; only two are required to describe the motion of the two electrons in a simple homolysis. In the assisted reaction, the enthalpy difference includes not only the bond dissociation of A—B, as required for simple homolysis, but also the energy gained by the formation of a bond between R and A.

**Simple Homolysis**          **Assisted Homolysis**

A—B ⟶ A· + B·          R·          A—B ⟶ R—A + B·

## EXERCISE 7.2

From the bond energies listed in Table 3.5, calculate $\Delta H^\circ$ for each of the following reactions, and predict which C—H bond will be preferentially cleaved by interaction with a bromine radical:

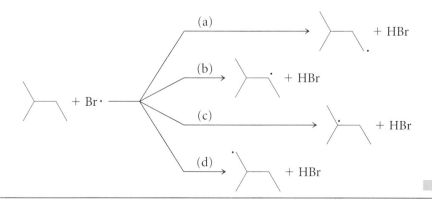

### Energy Changes in Heterolytic Reactions

Heterolytic cleavages are most likely to occur at polar $\sigma$ bonds. For example, in the reaction of an electrophile with a molecule containing a polar bond between A and B, the electrophile interacts more favorably with the electron-rich end of the A—B bond:

$$\overset{\delta+\ \ \delta-}{A—B} \quad E^\oplus \longrightarrow A^\oplus + B—E$$

If B is the more electronegative atom, the formation of a bond between B and the electrophile is facilitated by the polarization of the covalent bond between A and B, in which electrons are shifted toward the more electronegative atom B. Because this reaction entails the simultaneous movement of an electron pair (that is, of two electrons), a full-headed arrow represents this electron motion. Because ions are formed in heterolytic reactions, we cannot use bond-dissociation energies alone to calculate the enthalpy change but must also consider the energy needed to form and solvate the polar reactant and product ions.

In a similar way, and because of the same electrostatic factors, a nucleophile is attracted to the positive end of the polar A—B bond and can assist heterolytic cleavage by donating electrons to the developing positive charge at A, forming a Nuc—A bond as the A—B bond is cleaved. As this takes place, the two electrons of the A—B bond shift to B. Here, two electron pairs move, as indicated by two full-headed curved arrows:

$$Nuc:^\ominus \quad \overset{\delta+\ \ \delta-}{A—B} \longrightarrow Nuc—A + B^\ominus$$

Like homolytic reactions, heterolytic reactions can be assisted by other reagents, but the factors involved are different. How easily the ions are produced depends critically on the polarity of the solvent. In contrast, homolytic cleavages produce neutral radicals and are not greatly affected by the nature of the solvent. Solvation of the ions formed in a heterolytic cleavage provides a significant amount of energy, but the precise values are difficult to measure and depend on the specific reaction conditions. As a result, it is much more difficult to describe enthalpy changes for heterolytic cleavages than for homolytic cleavages.

Because it is difficult to describe enthalpy changes for heterolytic cleavages, an indirect means is usually used to predict the relative energies of heterolytic bond cleavage in a series of similar compounds in which parallel bond breaking occurs. For example, because a tertiary cation is more stable than a secondary one, reaction 1 costs less energetically than does reaction 2.

$$\text{(1)}$$

$$\text{(2)}$$

$$\Delta H_1^\circ < \Delta H_2^\circ$$

*CHEMICAL PERSPECTIVES*

CRACKING BREAKS BONDS IN CRUDE OIL

Petroleum, or crude oil, is a complex mixture of many different compounds, mostly aliphatic and aromatic hydrocarbons. As it comes from the well, petroleum contains many more compounds of high molecular weight than are needed. Various methods have been developed to degrade larger hydrocarbons into smaller ones; these processes are referred to collectively as *cracking*. In thermal cracking, hydrocarbons are heated to as high as 760 °C. At this temperature, sufficient energy is available to induce homolytic cleavage of carbon–carbon bonds when molecules collide, producing carbon free radicals. Alternatively, in the presence of an inorganic acid as catalyst, bonds are broken by heterolytic cleavage, with the generation of carbocations. In addition to producing hydrocarbons that are more suitable for use as fuels in internal combustion engines, cracking also yields propene and butene, which are used to make many products, including plastics.

Likewise, even without knowing the specific p$K_a$ values, we know that reaction 3 is less costly energetically than reaction 4 because an enolate anion is more stable than an alkyl anion.

$$\underset{\substack{H_3C}}{\overset{\displaystyle O}{\diagup}}\!\!\!\!\diagdown CH_2\!\!-\!\!M \longrightarrow \underset{\substack{H_3C}}{\overset{\displaystyle O}{\diagup}}\!\!\!\!\diagdown \overset{\ominus}{CH_2} + M^{\oplus} \tag{3}$$

$$CH_3CH_2CH_2\!\!-\!\!M \longrightarrow CH_3CH_2\overset{\ominus}{CH_2} + M^{\oplus} \tag{4}$$

$$\Delta H_3^\circ < \Delta H_4^\circ$$

Thus, the trends we have seen in the relative stabilities of intermediates (here, of carbocations and carbanions) serve us well in ordering reactivity in heterolytic reactions.

If we are concerned only with the net reaction (and not the intermediate ion-forming steps), we can use the table of bond-dissociation energies to calculate the enthalpy change ($\Delta H^\circ$) of a reaction, even one that proceeds through heterolytic steps. For example, even though reaction 5 proceeds through ions, we can nonetheless calculate the reaction enthalpy by subtracting the relevant bond energies of the products from those of the reactants. This approach assumes similar solvation energies for the reactants and products, a quite reasonable assumption for pairs of similar, neutral reagents.

$$(CH_3)_3C\!\!-\!\!Br + H\!\!-\!\!Cl \longrightarrow (CH_3)_3C\!\!-\!\!Cl + H\!\!-\!\!Br \tag{5}$$

In summary, reactions in which bonds are concurrently broken and formed (assisted homolytic and heterolytic reactions) proceed much more readily and with different energetic requirements than do reactions in which only bond cleavage takes place. Because of difficulties in measuring bond strengths, it is often difficult to predict quantitatively the thermodynamics of individual steps in heterolytic reactions, although trends in a series of similar compounds are predictable. In contrast, thermochemical calculations are often easily carried out for reactions involving homolytic cleavages.

### EXERCISE 7.3

Use the bond energies from Table 3.5 (also inside the back cover) to calculate $\Delta H^{\circ}$ for each of the following reactions:

(a) $(CH_3)_3C-Br + HCl \longrightarrow (CH_3)_3C-Cl + HBr$

(b) $H_3C-Br + HCl \longrightarrow H_3C-Cl + HBr$

(c) $H_3C-I + HCl \longrightarrow H_3C-Cl + HI$

## 7.3

## How to Study a New Organic Reaction

Once you have recognized its reaction type and have established that a proposed reaction is sensible thermodynamically, you can determine the conditions necessary for the proposed conversion. There are several ways to describe fully an organic reaction. A thorough knowledge of a particular reaction requires answering all of these four questions:

**1.** Given the reactant and reagents, together with a set of reaction conditions, what is the expected product(s)?

**2.** Given a reactant and a product, what reagents and conditions favor this transformation?

**3.** Given a product and a set of reagents, what is reasonable as starting material(s)?

**4.** What are the intermediates in the conversion (if any), and what electron flow accomplishes their formation and reaction?

### EXERCISE 7.4

From the reactions that you have learned so far, supply the missing information for the following reactions:

(a)

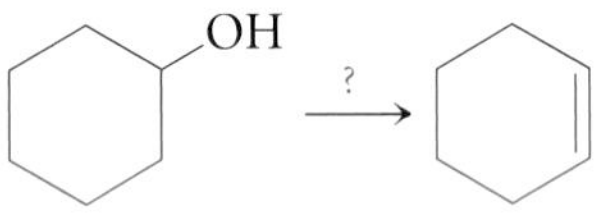

(b) 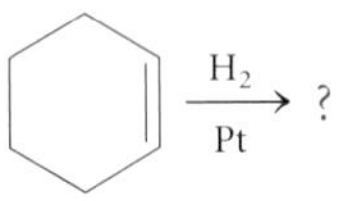 $\xrightarrow[\text{Pt}]{\text{H}_2}$ ?

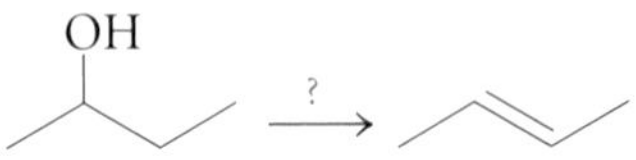

(c) 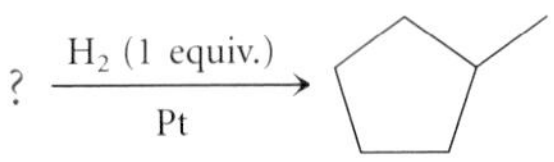 ? $\xrightarrow[\text{Pt}]{\text{H}_2 \text{ (1 equiv.)}}$

(d) $\xrightarrow{?}$

(e) $\xrightarrow{?}$

(f) 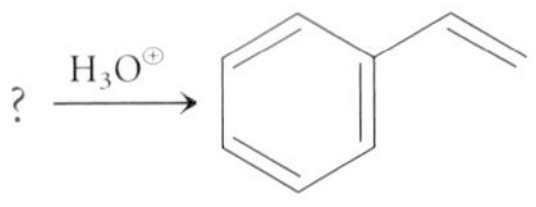 ? $\xrightarrow{\text{H}_3\text{O}^{\oplus}}$

Addressing the first two questions requires a familiarity with a range of reagents and reaction conditions. The third question is sometimes answered by proposing a series of reactions that would achieve in several steps what would be difficult to accomplish in one. To do this, chemists often use an approach known as **retrosynthetic analysis,** in which they work backward. Having chosen a target product, they choose a reasonable precursor, which in turn has a logical precursor, and so forth. Thus, when the analysis is finished, the result is a plan for building molecules of increasing complexity through a series of reactions, using readily available starting materials. This approach enables chemists to plan logically the construction of interesting new molecules or propose new synthetic routes to complex existing molecules (for example, natural products). This area, called **organic synthesis,** is a very important subfield of organic chemistry. It will be covered in more detail in Chapter 15.

The fourth question focuses on exactly how electrons (and thus atoms) move when a reactant is converted into a product. This detailed description, or *reaction mechanism,* is the underpinning of organic chemistry and constitutes much of the critical information that allows chemists to predict new reactions with confidence. A study of how organic reactions occur is called **mechanistic organic chemistry,** and the subfield of organic chemistry that relates structure to reactivity in explaining reaction mechanisms is called **physical organic chemistry.**

A major strength of organic chemistry is that the answers to these four questions for a small number of reactions can be generalized to other reactions of the same classification. It is not necessary to learn thousands of individual reactions. Instead, you will learn a few reactions *really well* and apply the knowledge to similar cases. (Read this paragraph to yourself three times—or however many it takes until it sinks in and you really believe it.)

You cannot extrapolate what you know about one organic reaction to another until you know its reaction mechanism. Typically, a chemical reaction consists of bond making and bond breaking as the reactant is converted into a product: a reaction mechanism is nothing more complicated than the sequence of elementary steps by which this occurs. These steps are represented by showing the flow of electrons (using curved arrows) as some bonds are broken and others formed. Reaction mechanisms, which describe how electrons move to make and break chemical bonds along a reaction coordinate, are therefore of *very great importance* to organic chemistry.

To write a reaction mechanism, we must establish the identities of all intermediates formed en route from reactant to product. If we know something about the energies of these intermediates (even if only roughly), we can approximate the structure and energy of the transition states leading to the formation of the intermediates and can predict relative reactivity for closely related reactions. In the following sections, we will consider the mechanisms of three representative kinds of reactions: *concerted reactions,* or those having no reactive intermediates; reactions involving *heterolytic cleavage,* and therefore ionic intermediates; and those involving *homolytic cleavage,* and thus radical intermediates. The goal is to show how *electron pushing* (using curved arrows to describe the movement of electrons as a reaction proceeds) helps define a reaction mechanism.

Although the focus of this chapter is on describing several types of reaction mechanisms, the examples used to illustrate these types are of additional interest because they accomplish chemical transformations that are useful in synthesis. You should begin now to apply the individual learning method most effective for you and assemble study aids that organize these reactions according to what they accomplish and how they proceed. Many students find it useful to prepare "flash cards" that summarize important features of each type of reaction.

### EXERCISE 7.5

For each of the following bond cleavages, use curved arrows to show the electron flow and classify each as homolysis or heterolysis:

(a) $CH_3O{-}OCH_3 \xrightarrow{\Delta} CH_3O\cdot + \cdot OCH_3$

(b) ⬡ + $E^{\oplus}$ ⟶ (cyclohexane with E and ⊕)

(c) (t-butyl bromide) ⟶ (carbocation)$^{\oplus}$ + $Br^{\ominus}$

(d) $\underset{H_3C \quad CH_3}{\overset{O}{\|}}$ + $^{\ominus}OH$ ⟶ $\underset{H_3C \quad CH_2^{\ominus}}{\overset{O}{\|}}$ + $H_2O$

## 7.4

## Mechanism of a Concerted Reaction: Bimolecular Nucleophilic Substitution (S$_N$2)

CHEM TV® I

#12    The S$_N$2
        Reaction

As defined in Section 6.1, a *concerted reaction* proceeds directly from reactant to product without forming any detectable intermediates, whether ionic or neutral. Thus, if chemists can find no evidence for the presence of a reactive intermediate, they conclude that the reaction is concerted. When charged intermediates (carbocations, carbanions, or radical ions) are formed in a reaction, the rate of reaction is significantly affected by changes in solvent polarity. Accordingly, if a reaction shows little change in rate when solvent polarity changes, chemists conclude that charged intermediates are not involved. Radicals react very rapidly with molecular oxygen—so much so that reactions involving these intermediates are greatly slowed by the presence of molecular oxygen, which siphons off the radicals and prevents the next cycle. Thus, if a reaction is insensitive to the presence or absence of oxygen, chemists conclude that radicals are not involved in the reaction pathway. If chemists rule out the involvement of any of the known charged and uncharged reactive intermediates, they conclude, in the absence of evidence to the contrary, that the reaction is concerted.

Concerted nucleophilic substitution reactions are referred to as **S$_N$2 reactions**. In this notation, S$_N$ describes the overall reaction (a nucleophilic substitution), and 2 is related to the molecularity of the rate-determining step (in which two species are involved).

Recall that one concerted addition reaction was presented in Section 6.2—the Diels–Alder reaction:

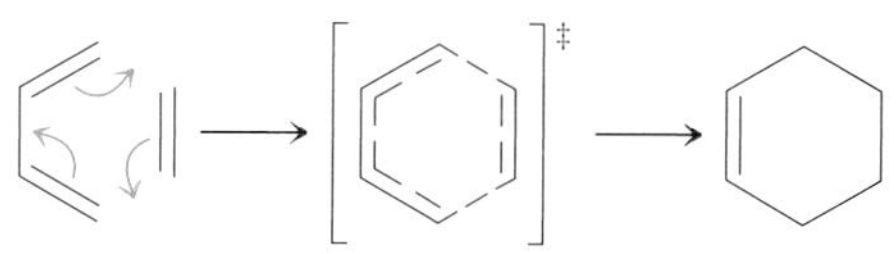

Another concerted reaction is a **self-exchange reaction,** an $S_N2$ reaction in which an incoming bromide ion interacts with methyl bromide, causing the carbon–bromine bond to break (displacing bromide) while forming a new carbon–bromine bond.

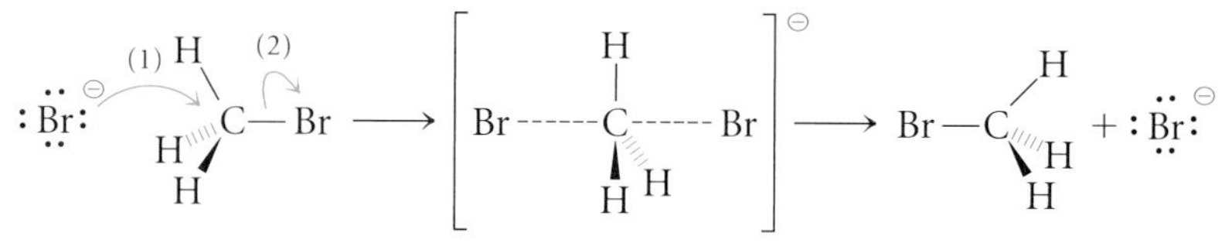

In this bimolecular nucleophilic displacement (an $S_N2$ reaction), the incoming bromide ion (the nucleophile) forms a bond to carbon as the carbon–bromine bond in the alkyl bromide reactant is broken. Because the bond making and bond breaking occur together in a single step without the formation of an intermediate, this reaction is *concerted.*

The reactants and products in this reaction are identical; that is, no chemical change is produced (unless the bromine atoms are isotopically different). Consequently, this reaction is thermoneutral. We use a full-headed curved arrow (1) to indicate that the two electrons of one of the lone pairs of the bromide ion move toward carbon to form what ultimately becomes a carbon–bromine $\sigma$ bond. However, the new bond cannot form without breaking the original carbon–bromine bond; otherwise, carbon would have to accommodate ten electrons in its valence shell. As a result, a second full-headed curved arrow (2) indicates that the two electrons originally in the carbon–bromine $\sigma$ bond of the starting material must move from a bonding orbital between these atoms to become a nonbonding lone pair on the product bromide ion as the bond is broken.

These two steps, (1) bond making and (2) bond breaking, are simultaneous; no intermediates are formed. In the transition state, carbon is partially bonded to both the incoming and the departing bromine atoms. This partial bonding is sometimes shown as dashed lines, with the structure enclosed in brackets to indicate that, as a transition state, it has no intrinsic stability and cannot be isolated. This dashed line notation is not nearly as precise as the curved-arrow notation and is not recommended as the primary way to think about mechanisms. Do not use the dashed-line notation unless it helps you see the electron flow. (And be very clear *not* to imply that you think the structures shown with dashed lines have more than a transient existence.)

Because the attacking reagent bears a nonbonding electron pair and is therefore nucleophilic and because a bond has been replaced by a new one, this reaction is called a **nucleophilic substitution.** (Recall that *nucleophilic* means "nucleus loving," and nucleophiles are attracted to a positive charge. Thus, anions are nucleophilic.) Two reagents (the starting material and the nucleophile) participate in the transition state of the rate-determining step, which makes this reaction type a **bimolecular nucleophilic substitution,** abbreviated $S_N2$. Because two reagents participate in bond making and bond breaking in the rate-determining step, the rate of this reaction is affected by the concentrations of both the substrate (LG stands for "leaving group") and the nucleophile.

$$\text{Rate} = k[\text{R—LG}][\text{Nuc}]$$

The same kind of $S_N2$ reaction can produce a net chemical change if the incoming and outgoing halide ions are different. For example, a bromide ion displaces iodine from methyl iodide, producing a carbon–bromine bond and an iodide ion:

$$: \overset{..}{\underset{..}{Br}} :^{\ominus} \quad H_3C—I \longrightarrow Br—CH_3 + : \overset{..}{\underset{..}{I}} :^{\ominus} \qquad \Delta H° < 0$$

$$: \overset{..}{\underset{..}{I}} :^{\ominus} \quad H_3C—Br \longrightarrow I—CH_3 + : \overset{..}{\underset{..}{Br}} :^{\ominus} \qquad \Delta H° > 0$$

The thermodynamics of a substitution reaction can be estimated from the bond energies of the reactant and the products. A nucleophilic substitution in which bromide ion displaces iodide ion from an alkyl halide is exothermic; one in which iodide ion displaces bromide ion is endothermic. Because a carbon–bromine bond is stronger than a carbon–iodine bond, this reaction is energetically favorable and thus exothermic; $\Delta H°$ is negative, assuming that other factors are equal. Here, iodide ion is the **leaving group,** pulling the electrons originally in the covalent C—I bond toward itself, and bromide ion is the nucleophile, donating an electron pair toward carbon. For the reverse reaction, in which iodide ion displaces bromine from methyl bromide to produce methyl iodide plus bromide ion, $\Delta H°$ is positive and the reaction is endothermic. In this unfavorable reaction, bromide ion is the leaving group and iodide ion is the nucleophile.

### The Transition State of an $S_N2$ Reaction

Let's consider the transition state for the reaction of bromide ion with methyl bromide:

$$: \overset{..}{\underset{..}{Br}} :^{\ominus} \quad H_3C—Br \longrightarrow Br—CH_3 + : \overset{..}{\underset{..}{Br}} :^{\ominus} \qquad \Delta H° = 0$$

In the self-exchange reaction, the carbon–bromine bond of methyl bromide must stretch in the transition state as a new carbon–bromine bond is being formed. (Otherwise, carbon would have to accommodate more than eight valence electrons.) Breaking a covalent bond is energetically costly, but forming a new bond is energetically favorable. Because the entering nucleophile and the leaving group are identical, there is no energetic preference for bonding to either the incoming or the outgoing bromide ion, so the transition state is symmetrical: the two partial bonds from carbon to bromine are identical, as is the angular relationship of these bonds to the carbon–hydrogen bonds. It is important to understand that species *in between* the starting material and the transition state have no measurable consequence for a reaction. Indeed, there are many different approaches that a nucleophile can take in attacking an electrophilic carbon, but all lead to the same transition state, defined as the maximum on the *lowest*-energy pathway. A smooth curve is used on reaction profiles to depict the transformation of starting material to transition state to product only for convenience; the molecules are under no such constraints.

In general, for an $S_N2$ reaction with $\Delta H° = 0$, the transition state resembles reactant and product equally. The partial bond with the entering nucleophile has the same strength as that with the departing leaving group (Figure 7.1). To attain this transition state, the incoming nucleophile ($Br^\ominus$) in the self-exchange reaction must attack from the carbon end of the C—Br bond—that is, from the side opposite that from which the leaving group ($Br^\ominus$) departs. This is called **back-side displacement.**

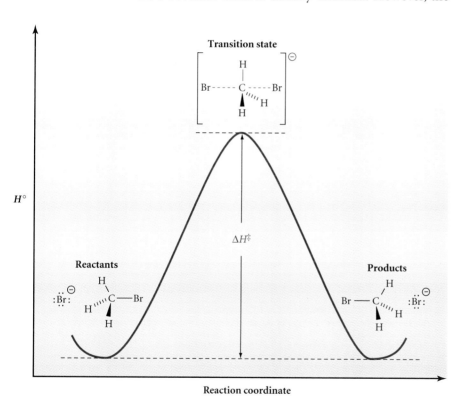

Alternatively, the bromide ion could approach the carbon atom from the same side as that where a bromine atom is already attached. However, the

**FIGURE 7.1**

A reaction profile for a self-exchange reaction is completely symmetrical. At the transition state, the strength of the bond being made to the incoming nucleophile is exactly equivalent to the strength of the bond being broken as the leaving group departs.

transition state for this front-side displacement would have the two electron-rich bromine atoms in close proximity:

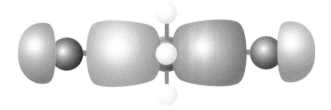

The first transition state, attained by back-side attack with inversion of configuration at carbon, is strongly favored over this second possible transition state.

According to the **principle of microscopic reversibility,** the reaction pathway in the forward and reverse directions must be the same. Both the forward and reverse reactions must proceed through the same transition state and the same intermediates, if any are involved in the reaction. The concept behind the principle of microscopic reversibility is that a reaction will follow the pathway that requires the least amount of energy to reach the transition state, and regardless of the direction in which the reaction is proceeding, there can be only one transition state of least energy.

For the single-step reaction of bromide ion with methyl bromide (and other single-step reactions for which the product is the same as the starting material), the transition state must be halfway between the structures of the starting material and product. At the transition state, it will not be possible to determine which bromine was originally bonded to carbon and which is the nucleophile. Thus, the bonds from carbon to, and the charges on, both bromine atoms must be identical. Furthermore, because the hydrogen atoms start on one side and finish on the other, these atoms must be coplanar with the carbon atom in the transition state. This can occur when the carbon atom is $sp^2$-hybridized in the transition state, and the partial bonds to the bromine atoms result from overlap of their orbitals with one of the lobes of the carbon's $p$ orbital:

### Inversion of Configuration

Back-side displacement causes an **inversion of configuration** at a center of asymmetry. Thus, when this displacement occurs at a center of chirality, the product will have the opposite three-dimensional arrangement of the substituents about the carbon atom. In most cases, the priority of the substituents will not change in an $S_N2$ reaction, and thus inversion of configuration changes the enantiomer $R$ to $S$, or vice versa. For example, reaction of $(R)$-2-bromobutane with bromide ion results in the formation of $(S)$-2-bromobutane:

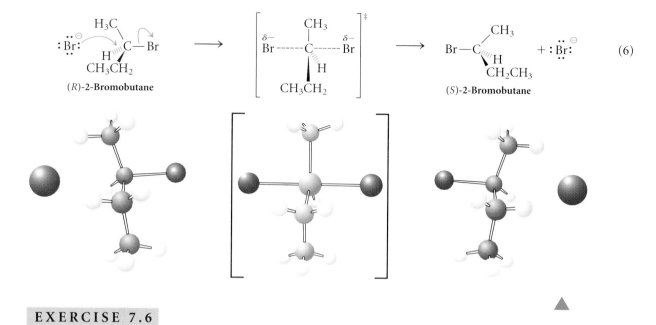

$$(6)$$

(R)-2-Bromobutane                                       (S)-2-Bromobutane

## EXERCISE 7.6

Could reaction 6 be used to make a pure, optically active sample of (S)-2-bromobutane starting from the pure R enantiomer? Explain your answer. (*Hint:* Consider whether the transition state is chiral.)

## EXERCISE 7.7

Using the Cahn–Ingold–Prelog rules you learned in Chapter 5, assign absolute configuration at each chiral center in the reactants and the products of each $S_N2$ reaction:

(a)    I, H        $+ Br^{\ominus} \longrightarrow$     H, Br

(b)    Br, H       $+ {}^{\ominus}SH \longrightarrow$     H, SH

(c)    Cl       $+ I^{\ominus} \longrightarrow$     I

## Nonsymmetrical $S_N2$ Transition States

Self-exchange reactions can help illustrate the nature of $S_N2$ reactions, but they are rarely of practical use. In general, the leaving group (LG) and the nucleophile (Nuc) are different, and the transition state need not be symmetrical. The Hammond postulate states that the transition state will resemble that species (starting material or product) to which it is closest in energy. Thus, for an exothermic $S_N2$ reaction, the transition state will be

**349**

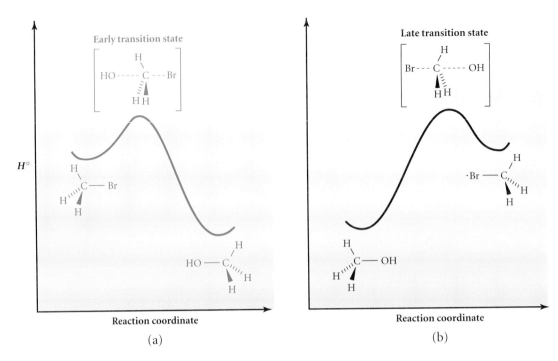

**FIGURE 7.2**

(a) For an exothermic $S_N2$ reaction, the transition state is early. (b) Conversely, the reverse (endothermic) reaction (which proceeds through the same transition state) is considered to have a late transition state.

*early,* with only a small degree of bond making between the nucleophile and the carbon atom and little breaking of the bond between the carbon and the leaving group. An example of an exothermic $S_N2$ reaction is the displacement of bromine by hydroxide ion, illustrated for methyl bromide in Figure 7.2(a).

In contrast, an endothermic $S_N2$ reaction has a transition state that is *late.* Thus, the transition state resembles the product, and the bond to the leaving group is nearly broken. This is illustrated by the displacement of hydroxide ion from methanol by bromide ion in Figure 7.2(b). Depending on the identities of the nucleophile and the leaving group, the range of $S_N2$ reactions spans a continuum from early to symmetrical to late transition states, allowing interconversion of functional groups.

### Factors Affecting the Rate of $S_N2$ Reactions

*Steric Hindrance in the Substrate.* Back-side displacement in an $S_N2$ reaction requires that the nucleophile approach carbon closely enough to permit partial bonding. Therefore, the approach of the incoming nucleophile is strongly affected by the bulkiness of the substituent groups present on the carbon bearing the leaving group. The ease with which displacement occurs is greatest for leaving groups bonded to primary carbon atoms; displacement occurs less readily at secondary carbons and even less so at tertiary ones.

Figure 7.3 illustrates a nucleophile's attack to break a carbon–bromine bond at a methyl, ethyl, isopropyl, and *t*-butyl center. The van der Waals

#14   Steric Factors
      in $S_N1$ Reactions

radii of the alkyl groups are drawn to show roughly the larger steric demand of an alkyl group over that of hydrogen. The ease of nucleophilic displacement within this series follows the order: methyl > ethyl (primary carbon) > isopropyl (secondary) >> *t*-butyl (tertiary). In fact, concerted displacements are so difficult at tertiary centers that other reactions occur instead.

The rate of S$_N$2 reactions is also reduced by the presence of bulky substituents on carbon atoms adjacent to the one undergoing substitution. For example, neopentyl bromide reacts only very slowly with most nucleophiles, even though the carbon atom bearing the leaving group is primary.

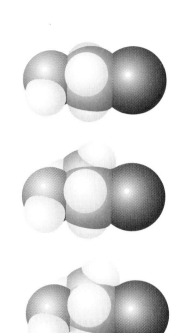

**Neopentyl bromide**

As in tertiary alkyl halides, the methyl groups in neopentyl bromide significantly interfere with the approach of the nucleophile, raising the energy of the transition state and increasing the activation energy.

***Electronic Effects in the Substrate.*** Some structural features increase the rate of S$_N$2 reactions. In particular, the presence of an adjacent $sp^2$- or $sp$-hybridized carbon atom results in a significant acceleration of bimolecular substitution reactions. Thus, allylic, benzylic, and propargylic halides, as well as $\alpha$-haloketones (and other $\alpha$-halocarbonyl compounds) are unusually reactive toward nucleophilic substitution.

**An allylic halide**   **A benzylic halide**   **A propargylic halide**   **An $\alpha$-haloketone**

In part, this increased reactivity results from *decreased* steric hindrance in the transition state, where there are fewer adjacent substituents and where those that are present are held farther from the nucleophile and leaving group—for example, in 1-propyl versus 3-propenyl bromide:

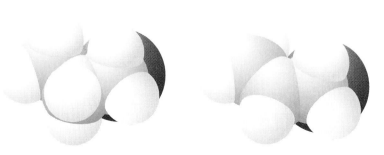

**1-Propyl bromide**                **3-Propenyl bromide**

## FIGURE 7.3

Back-side attack in an S$_N$2 reaction becomes more difficult when the carbon bears a larger number of bulky alkyl substituents. (Top to bottom) Transition states for the reaction of hydroxide ion with methyl, ethyl, isopropyl, and *t*-butyl bromide.

351

In addition, stabilization of the transition state is provided by overlap between a $p$ orbital of the carbon atom undergoing substitution and a $p$ orbital on the adjacent carbon atom.

In summary, a concerted nucleophilic displacement ($S_N2$) occurs when an organic substrate bearing a covalently bonded leaving group is attacked from the back side by an incoming nucleophile, causing inversion of configuration at carbon. A primary or secondary alkyl halide undergoes substitution by a nucleophile capable of forming a stronger bond with carbon than that between the carbon and the leaving group in the starting material, a general requirement for an exothermic reaction.

## EXERCISE 7.8

Which compound in each of the following pairs is more active toward a nucleophile under $S_N2$ conditions?

(a)

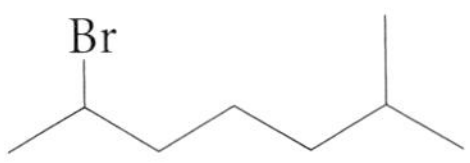

or

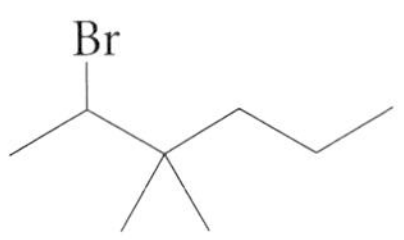

(b)

or

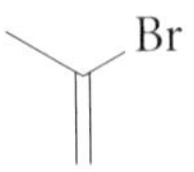

(c)

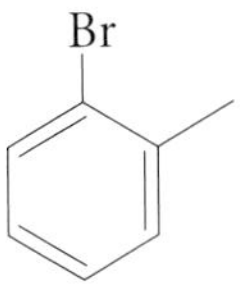

or

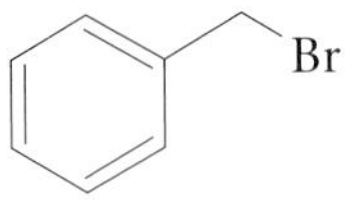

## EXERCISE 7.9

Arrange the following isomeric bromides in order of decreasing relative rate of their $S_N2$ displacement reactions.

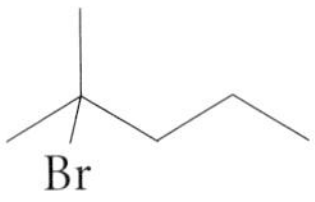

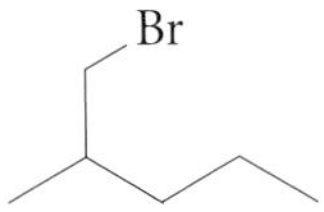

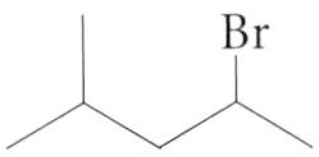

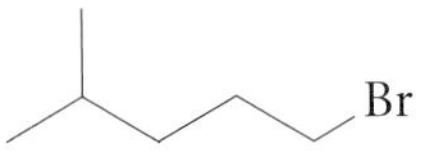

## EXERCISE 7.10

Which of the following substrates are good candidates for reaction with $NaN_3$ in acetone (typical $S_N2$ reaction conditions)?

(a)

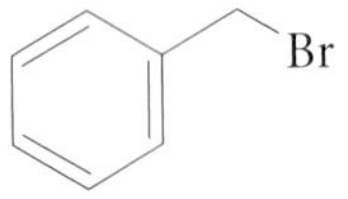

(b)

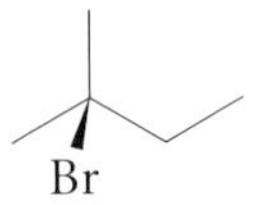

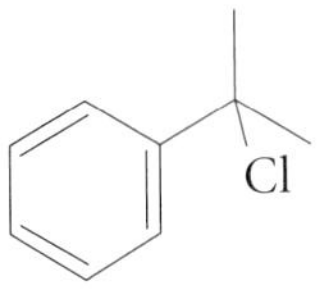

(c)

(d)

(e)

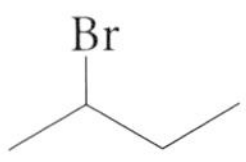

What reagent(s) are required for each of the following transformations?

(a)

(b)

(c)

(d)

(e)

*Nucleophilicity.* The fact that an effective nucleophile has high electron density means that there are two generally useful ways to compare the nucleophilicity of two species. First, an anion is more reactive as a nucleophile than the corresponding neutral species; for example, $^\ominus$OH is more nucleophilic than H$_2$O. Second, among species with the same charge, a less electronegative atom bearing a nonbonding electron pair is a better nucleophile than a more electronegative one, because it can more easily donate its electron pair in approaching the transition state. Within a single column or a single row of the periodic table, nucleophilicity increases with decreasing electronegativity. Thus, $^\ominus$OH is a better nucleophile than F$^\ominus$, because oxygen is less electronegative than fluorine. Similarly, I$^\ominus$ is more nucleophilic than F$^\ominus$ because iodine is less electronegative than fluorine, and NH$_3$ is a better nucleophile than H$_2$O. We cannot use this analysis to compare species in different rows *and* columns of the periodic table—for example, $^\ominus$OH and Cl$^\ominus$—but such information can be obtained empirically (that is, by experiment). The relative reactivity of some common nucleophiles is:

$$RS^\ominus > I^\ominus > {}^\ominus CN > {}^\ominus OH > {}^\ominus N_3 > Br^\ominus > Cl^\ominus > H_2O$$

*Leaving Group.* The rate of S$_N$2 reactions is also affected by the nature of the leaving group. The best leaving groups are those with weak bonds to carbon and those that can readily support negative charge. Chlorine, bromine, and iodine are all good leaving groups, with alkyl chlorides the least reactive toward S$_N$2 reactions, and alkyl iodides the most reactive. However, alkyl iodides are the most expensive and the most difficult to prepare.

***Solvent.*** The effect of the solvent on the rate of an $S_N2$ reaction varies with the nature of the nucleophile. If both the substrate and the nucleophile are uncharged, increasing solvent polarity increases the rate of the $S_N2$ reaction by stabilizing the transition state, where charge separation has developed.

$$(CH_3)_3N: \quad H_3C-Br \longrightarrow \left[ (CH_3)_3\overset{\delta+}{N}----\overset{H}{\underset{H\ \ H}{|}}----\overset{\delta-}{Br} \right]^{\ddagger} \longrightarrow (CH_3)_3\overset{\oplus}{N}-CH_3 + :\overset{\cdot\cdot}{\underset{\cdot\cdot}{Br}}:^{\ominus}$$

Conversely, reaction of a negatively charged nucleophile with a neutral substrate proceeds more slowly in polar solvents than in nonpolar solvents, because the localized charge in the starting nucleophile is stabilized by solvation to a greater extent than the more diffuse charge in the transition state.

$$N{\equiv}C:^{\ominus} \quad H_3C-Br \longrightarrow \left[ N{\equiv}\overset{\delta-}{C}----\overset{H}{\underset{H\ \ H}{|}}----\overset{\delta-}{Br} \right]^{\ddagger} \longrightarrow N{\equiv}C-CH_3 + :\overset{\cdot\cdot}{\underset{\cdot\cdot}{Br}}:^{\ominus}$$

### EXERCISE 7.12

Which reagent in each of the following pairs is the more reactive nucleophile?

(a) $^{\ominus}NH_2$ or $NH_3$      (c) $Cl^{\ominus}$ or $I^{\ominus}$

(b) $OH_2$ or $NH_3$      (d) $HS^{\ominus}$ or $HO^{\ominus}$

## Synthetic Utility of $S_N2$ Reactions

With the $S_N2$ reaction, it is possible to convert an alkyl halide into any of several different functional groups (Table 7.1). With different nucleophiles, alkyl halides will react to produce amines, azides, alcohols, ethers, thioethers, and other halides. These products, in turn, can be transformed

### TABLE 7.1

Preparation of Some Typical Functional Groups by $S_N2$ Displacement

$R-Br + {}^{\ominus}Nuc \longrightarrow R-Nuc$ (R = methyl, primary alkyl, or secondary alkyl)

| Source of Nucleophile | Product |
| --- | --- |
| $NH_3$ | Amine ($R-NH_2$) |
| $NaN_3$ | Alkyl azide ($R-N_3$) |
| $NaOH$ | Alcohol ($R-OH$) |
| $NaOCH_3$ | Ether ($R-OCH_3$) |
| $NaSCH_3$ | Thioether ($R-SCH_3$) |
| $NaCl$ | Alkyl chloride ($R-Cl$) |
| $KI$ | Alkyl iodide ($R-I$) |

into other materials. For example, we will see in the next chapter how alcohols can be converted into good leaving groups, so that alcohols also become starting materials for substitution reactions yielding a variety of functional groups. By learning the mechanism of the $S_N2$ reaction, you have also learned how each of these related transformations takes place. Because these reactions constitute useful methods for preparing each of these products, they are of both synthetic and mechanistic interest.

## 7.5

# Mechanism of Two Multistep Heterolytic Reactions: Electrophilic Addition and Nucleophilic Substitution ($S_N1$)

Reactions that include the formation of intermediates (Chapter 6) take place in distinct steps. Here, we consider the mechanisms of two reactions in which bond cleavage is heterolytic and the intermediates formed are cations. The first reaction, electrophilic addition of hydrogen chloride to an alkene, includes the formation of an intermediate carbocation. The second reaction, hydrolysis of alkyl bromide, involves the formation of two intermediates: a carbocation and then an oxonium ion. Because the rate-limiting step in the second reaction involves only the starting material, the rate of the reaction depends only on the concentration of starting material (rate = $k[R—LG]$), and the reaction is referred to as an **$S_N1$ reaction.** In this terminology, $S_N$ indicates the overall reaction (a substitution in which the substituent is a nucleophile), and 1 relates to the molecularity of the rate-determining step (unimolecular)—that is, the rate-determining step consists only of bond breaking in the substrate and does not involve the nucleophile.

As you will learn, the formation of cations in both of these reactions is inferred from three observations about the rate of reaction: (1) loss of stereochemistry as the reaction proceeds through a planar carbocation, (2) a correlation of relative reactivity with the stabilities of the intermediate carbocations, and (3) enhanced rates in polar solvents that stabilize the transition state for the formation of the intermediate carbocation itself.

#13    The $S_N1$ Reaction

### Electrophilic Addition of HCl to an Alkene

The addition of HCl to cyclohexene takes place in two steps:

Cyclohexene      Step 1      Step 2      Chlorocyclohexane

In the first step, the $\pi$ electrons of the double bond in cyclohexene are donated to the electrophile (HCl) to form a carbon–hydrogen $\sigma$ bond. The full-headed arrow indicates that the two electrons of the $\pi$ bond move to form a new C—H bond as the two electrons in the H—Cl bond shift to chlorine. The proton thus acts as an **electrophile.** Because the overall transformation is an addition, this reaction is called an **electrophilic addition.**

The protonation of one carbon converts the other carbon originally participating in the double bond into a cation. (As an exercise, calculate the formal charge of this carbon.) In this first step, more bonds are broken than are formed: both a carbon–carbon $\pi$ bond and a hydrogen–chlorine $\sigma$ bond are broken, whereas only a carbon–hydrogen $\sigma$ bond is formed, resulting in a carbocation and a chloride ion. In the gas phase, this reaction takes place only with great difficulty. However, in solution, the intermediate ions generated can be stabilized by interaction with polar solvent molecules, and, as a result, the reaction is accelerated by solvents such as water. Nonetheless, this solvation energy cannot compensate completely for the substantial bond breaking in this endothermic first step.

In the second step, the chloride ion formed in the first step reacts with the carbocation to form a carbon–chlorine $\sigma$ bond. This step occurs rapidly and easily because it consists only of bond making. Therefore, the endothermic first step is the slow step and is **rate-determining.** The transition state closely resembles the intermediate carbocation. A reaction profile is shown in Figure 7.4.

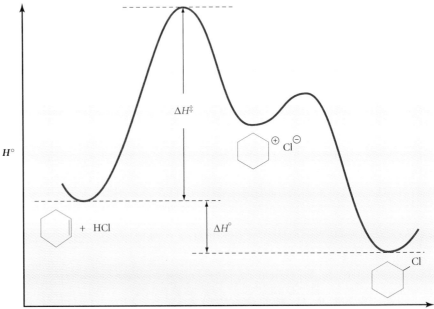

**FIGURE 7.4**

Reaction profile for a two-step electrophilic addition, the addition of HCl to cyclohexene. The reaction takes place by formation of a carbocation in the rate-determining step.

*Stabilization of Intermediate Cations.* The factors that stabilize the intermediate carbocation and chloride ion also stabilize the transition state. As the energy of the transition state is reduced, so is the required activation energy, resulting in a faster reaction. This reaction is accelerated by polar solvents because the critical intermediates are ions. The carbocation is stabilized by solvents containing heteroatoms involved in polar bonds, and the anion is stabilized by solvents having hydrogen atoms bonded to heteroatoms. Water has both of these features and is particularly good at stabilizing charged intermediates, both cations and anions.

You know that tertiary cations are more stable than secondary cations, which are, in turn, more stable than primary cations. As was explained in Chapter 3, the relative reactivity of alcohols is determined by the relative stability of the carbocations formed by cleavage of the carbon–oxygen bond. Thus, 2-methyl-2-propanol is cleaved by acid (ionized) more readily than is 2-butanol, which is more reactive than 1-butanol. The order of the rate of acid-induced ionization is

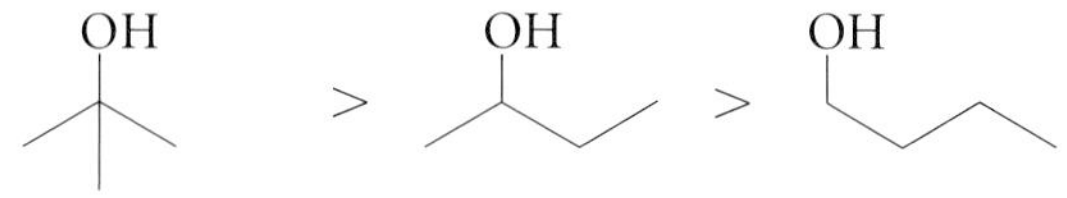

2-Methyl-2-propanol          2-Butanol          1-Butanol

This order of reactivities follows from, and is therefore the same as, the order of stability of the carbocations produced by loss of water: tertiary > secondary > primary.

***Regiospecificity: Markovnikov's Rule.*** A reaction that proceeds via an intermediate cation, such as that shown in Figure 7.5, is faster when the intermediate cation is tertiary than when it is secondary or primary. Therefore, in the addition of HCl to methylcyclohexene, where either a tertiary or a secondary cation can be formed as intermediate, the reaction proceeds through the more stable tertiary cation intermediate.

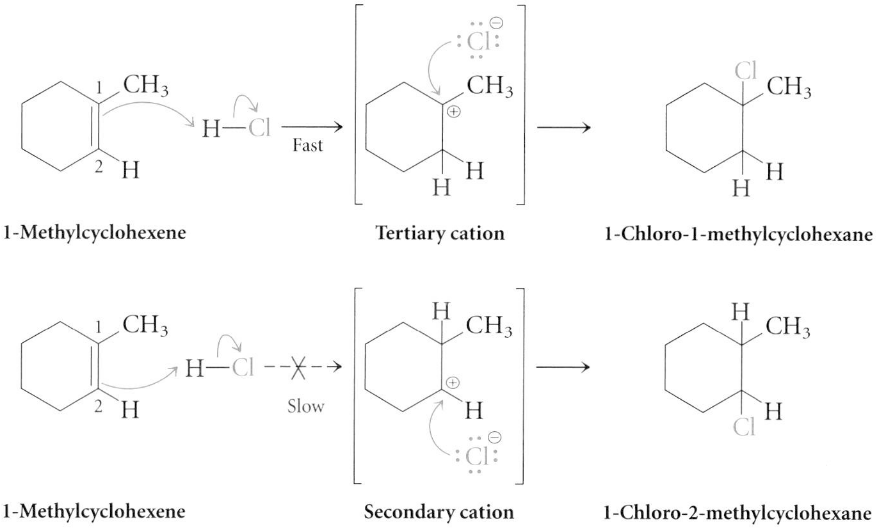

1-Methylcyclohexene          Tertiary cation          1-Chloro-1-methylcyclohexane

1-Methylcyclohexene          Secondary cation          1-Chloro-2-methylcyclohexane

**FIGURE 7.5**

Protonation of 1-methylcyclohexene at C-2 forms a tertiary cation, whereas protonation at C-1 produces a secondary cation. Because a tertiary carbocation is more stable than a secondary one, the upper reaction is thermodynamically favored over the lower one.

Note that the carbon atoms involved in the double bond in 1-methyl-cyclohexene are not equivalent. Protonation at C-2 gives the tertiary ion in the rate-determining step. Protonation at C-1 forms a secondary carbocation. Because secondary cations are less stable than tertiary ones, the formation of the secondary cation does not compete effectively with the formation of the tertiary cation, and 1-chloro-1-methylcyclohexane, the product derived from the tertiary carbocation, is formed preferentially.

These energy considerations are illustrated in Figure 7.6, which shows an energy profile very similar to that for the addition of HCl to cyclohexene. The reaction that forms the more stable, tertiary cation proceeds from the reactants in the center to the products at the left, whereas the slower process (forming the secondary cation) proceeds to the products at the right. Under either kinetic or thermodynamic control, 1-chloro-1-methylcyclohexane is formed: this product is more stable than the isomeric product at the far left, and it is formed through a pathway with a lower activation energy.

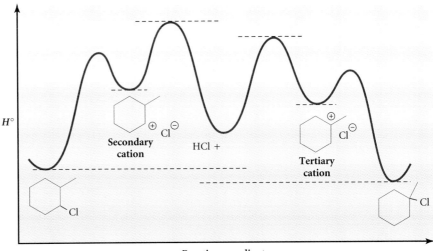

**Reaction coordinate**

**FIGURE 7.6**

Energy diagrams for the two possible outcomes for electrophilic addition of HCl to 1-methylcyclohexene.

Although the double bond in 1-methylcyclohexene is more stable than that in cyclohexene because of the methyl group on C-1, the effect of this substituent on the relative stabilities of the intermediate cations is even greater. Thus, the addition of a proton to 1-methylcyclohexene has a lower activation energy than the addition of a proton to cyclohexene, and it occurs more readily.

The addition of HX to a carbon–carbon double bond takes place in a stepwise fashion, with the positions of H and X in the product being governed by the stability of the intermediate cation. Electrophilic attack at the less highly substituted carbon gives the more stable carbocation. Therefore, protonation by acid occurs at the less highly substituted carbon atom of the double bond. This preferred orientation defines the **regiochemistry,** or positional isomerism, of the reaction. The regiochemistry of the addition of

HX to unsymmetrical alkenes will reflect the addition of a proton to the less substituted carbon. The Russian chemist Vladimir Markovnikov (1838–1904) was the first to make this observation, which is referred to as **Markovnikov's rule.** Markovnikov did not fully understand the chemical basis for his rule when he proposed it at the age of 31, for the mechanism was not uncovered until many years later. Nonetheless, he was an excellent scientist who generalized from his experimental observations to predict the course of new reactions.

### EXERCISE 7.13

Predict the preferred regiochemistry for the addition of HCl to each of the following compounds on the basis of carbocation stability:

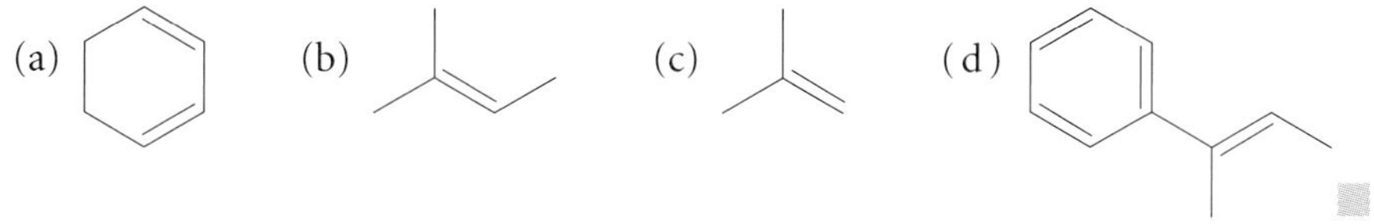

(a)   (b)   (c)   (d)

## Multistep Nucleophilic Substitution ($S_N1$): Hydrolysis of Alkyl Bromides

Another reaction that takes place through the formation of an intermediate cation is the conversion of an alkyl halide to an alcohol, with the replacement of the halogen by an OH group from water. Because this reaction proceeds through cationic intermediates, it is a multistep heterolytic substitution. In this hydrolysis reaction, the C—X bond is completely broken before a bond is formed with the nucleophile. Because the first step involves only the starting material, this nucleophilic substitution is unimolecular, an $S_N1$ reaction (rate = $k[R\text{—}LG]$).

Consider, for example, the hydrolyses of *t*-butyl bromide and isopropyl bromide:

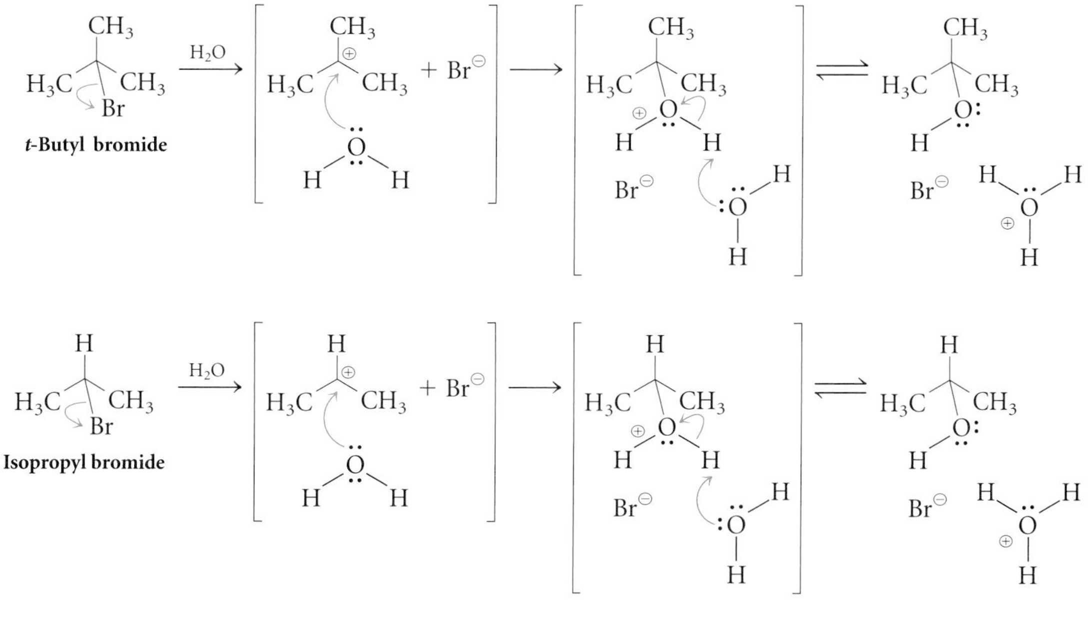

*t*-Butyl bromide

**Isopropyl bromide**

In the first step of an $S_N1$ reaction, bond breaking results in the formation of a carbocation; this step is rate-determining. Here, trapping of this cation by water in the second step is rapid, because only bond making is required. The oxonium ion formed is then deprotonated to produce the observed alcohol product.

### Rate-Determining Step: Formation of a Carbocation Intermediate.

These nucleophilic substitutions are mechanistically different from the $S_N2$ reactions considered in Section 7.4, because they take place in several steps. The first step is the heterolytic cleavage of the carbon–bromine bond to produce a carbocation and a bromide ion. (The $S_N2$ substitutions discussed earlier are concerted and do not proceed through a reaction intermediate.) After the carbocation is formed in the first, difficult, and rate-determining step, water attacks the cation in a rapid second step, forming a new carbon–oxygen bond. In this step, electron density flows from the lone pair on water's oxygen to the carbocationic carbon, forming a bond between carbon and oxygen. As a result, the oxygen now bears a formal positive charge in this intermediate, referred to as an *oxonium ion*. Loss of a proton restores neutrality to this oxygen in the last step of the overall process.

Because the first step of an $S_N1$ reaction consists only of bond breaking, it is endothermic. Thus, the transition state for cleavage of the carbon–bromine bond resembles the carbocation. The cation formed in the first step of the heterolytic cleavage of the carbon–bromine bond of *t*-butyl bromide is tertiary, whereas the cation derived from isopropyl bromide is secondary. Therefore, the first step of the reaction is less endothermic for *t*-butyl bromide than for *i*-propyl bromide. To the extent that a tertiary cation is more stable than a secondary one, the transition state leading to the former is favored, and a lower activation energy barrier is encountered. With a lower energy barrier, a greater fraction of the reactant molecules are able to reach the transition state (Chapter 6), and the reaction is faster. The order of reactivity of substrates in an $S_N1$ reaction is tertiary > secondary > primary >> methyl—the opposite of the order for an $S_N2$ reaction.

*t*-Butyl bromide

*i*-Propyl bromide

## EXERCISE 7.14

Which member of the following pairs of compounds is likely to react more rapidly under $S_N1$ conditions?

(a)

or

(b)

Br or Br

(c) Br or Br

(d) Cl or Cl

However, the carbocation is not the final product. Hydrolysis requires that this intermediate react with water to form an oxonium ion, deprotonation of which gives the final product, the alcohol (Figure 7.7). Of the two cations (the carbocation and the oxonium ion), the carbocation is less stable because its positively charged carbon lacks a full complement of valence electrons; in the oxonium ion, oxygen has access to eight valence electrons:

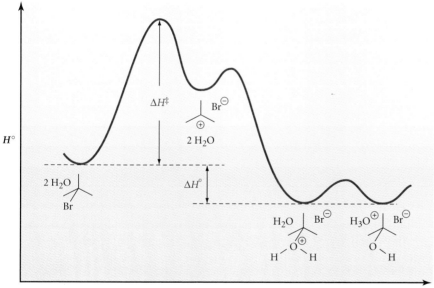

**FIGURE 7.7**

Complete reaction profile for the $S_N1$ hydrolysis of *t*-butyl bromide. The rate-determining step (carbocation formation) is followed by fast nucleophilic capture of the cation by water, forming an oxonium ion, which then transfers a proton to a neutral water molecule.

Formation of the carbocation, entailing only bond breaking, is thermodynamically unfavorable. Because this step has a large activation energy, it is undoubtedly rate-determining, and reasonable reaction rates are typically obtained only upon heating. Because the rate-determining step is unimolecular, the reaction rate depends only on the concentration of the reactant. The concentration of the nucleophile (water) does not appear in the reaction-rate expression because it does not enter into the reaction until after the rate-determining step.

***Reaction of the Carbocation Intermediate.*** Trapping of the carbocation by water in the second step of the hydrolysis reaction of an alkyl halide consists only of bond making and is exothermic. In the final step, a proton on the oxonium ion is transferred to a base—in this case, water. The deprotonation is very fast, as is generally true for reactions in which a proton is transferred from one heteroatom to another and in which the identity of the charged atom does not change (here, from an oxygen in the oxonium ion to an oxygen in the hydronium ion). Because the solvent acts as the nucleophile and traps the cationic intermediate, these $S_N1$ reactions are called **solvolysis reactions.** When water is the solvent, a solvolysis is called a **hydrolysis.**

We can also approach the carbocationic intermediate from the alcohol, via the reverse sequence of protonation, dehydration, and capture of the carbocation by the nucleophile:

Alcohol                                                                 Alkyl halide

In this way, an alcohol can be converted into an alkyl halide by an $S_N1$ reaction that is the reverse of the hydrolysis of an alkyl halide. In both reactions, formation of the carbocationic intermediate occurs in the rate-determining step. Protonation of the alcohol results in the formation of an oxonium ion, whose dehydration provides the cation. From the principle of microscopic reversibility, this carbocation is the same as that formed in the solvolysis of the corresponding alkyl halide. Therefore, the rate-determining step of both the forward and the reverse reactions is carbocation formation.

This example shows why it is very important to understand the structure and reactivity of intermediates in organic chemistry: when you really understand a reaction in one direction, you also understand its reverse. For example, the same carbocation can be trapped by halide ion to form alkyl halide or by water to form alcohol; the intermediate is the same whether the reaction is proceeding from halide to alcohol or from alcohol to halide. (As you continue the study of organic chemistry, you will find with increasing frequency that your understanding of reaction mechanisms begins to dovetail and that you will already know a great deal about the reactions of new functional groups.)

**EXERCISE 7.15**

363

7.5 Mechanism of Two
Multistep Heterolytic Reactions:
Electrophilic Addition and
Nucleophilic Substitution ($S_N1$)

Draw a complete energy diagram for the conversion of cyclohexanol to bromocyclohexane.

**EXERCISE 7.16**

Predict, on the basis of carbocation stability, which member of each of the following pairs is hydrolyzed at the faster rate:

(a) or

(b) or

(c) or

## Factors Affecting the Rate of $S_N1$ Reactions

Because the nucleophile is not involved in the rate-determining step, the rate of $S_N1$ reactions is affected by only three of the four factors that control the rate of $S_N2$ reactions: degree of substitution (as just discussed), identity of the leaving group, and polarity of the solvent. The effect of the leaving group on $S_N1$ reactions follows the same general trends as for $S_N2$ reactions. For example, alkyl iodides are more reactive than alkyl bromides or chlorides. However, in an $S_N1$ reaction of a neutral substrate, the polarity of the solvent affects the rate of reaction substantially more than in an $S_N2$ reaction and sometimes in the reverse direction. For example, calculations have shown that ionization of $t$-butyl chloride in the gas phase is endothermic by 150 kcal/mole, whereas the measured activation energy in water is only 20 kcal/mole. The calculated activation energy for the reaction of chloride ion with methyl chloride in water is about 25 kcal/mole; in the gas phase, the activation energy is less than 5 kcal/mole, and a complex between the starting materials is substantially more stable than methyl chloride and chloride ion separated at great distance.

## Rearrangements

As we saw in Chapter 3, alkyl groups donate electron density to carbocations, resulting in the following order of cation stability: tertiary $\sim$ benzylic > secondary $\sim$ allylic > primary > methyl. This difference in stability can provide a driving force for rearrangement reactions that result in an increase in the number of alkyl groups attached directly to the positively charged carbon of a carbocation.

*Hydrogen Shifts.* When 3-methyl-2-butanol is treated with hydrogen bromide, two isomeric bromides are formed: the major product, 2-bromo-2-methylbutane, and the minor product, 2-bromo-3-methylbutane:

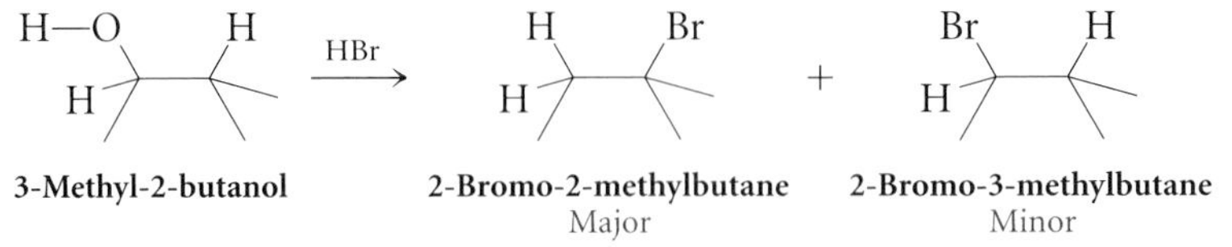

**3-Methyl-2-butanol**   **2-Bromo-2-methylbutane**   **2-Bromo-3-methylbutane**
                              Major                          Minor

In this reaction, protonation of oxygen is followed by loss of water to form a secondary carbocation (Figure 7.8). The secondary carbocation then re-arranges to a more stable tertiary carbocation by movement of a hydrogen atom *with* the pair of bonding electrons from C-3 to C-2. This fills the octet of C-2 and leaves C-3 with only six electrons. Thus, movement of the hydrogen atom and bonding electrons causes the center of positive charge to move from C-2 to C-3; that is, the initially formed secondary carbocation has rearranged to a more stable tertiary carbocation. Reaction of this re-arranged carbocation with bromide ion produces the tertiary alkyl bromide (2-bromo-2-methylbutane), whereas reaction of the initially formed secondary carbocation with bromide ion produces the secondary bromide (2-bromo-3-methylbutane), as a very minor product.

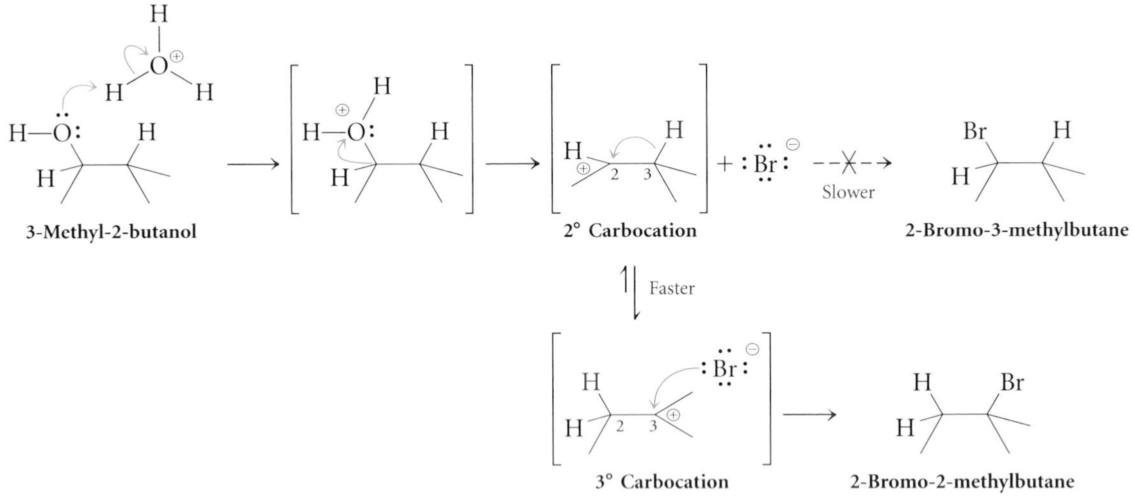

**FIGURE 7.8**

In the reaction of HBr with 3-methyl-2-butanol, the initially formed secondary carbocation rearranges by movement of a hydrogen atom (shown in red) and its associated bonding electrons. The resulting tertiary cation combines with bromide to form the major product.

Because the rearranged product is the major product observed, we can conclude that the rate of rearrangement is significantly faster than the rate of reaction of the secondary carbocation with bromide ion. Indeed, with primary substrates, rearrangement takes place simultaneously with loss of the leaving group, leading directly to the more substituted carbocation without formation of a primary one.

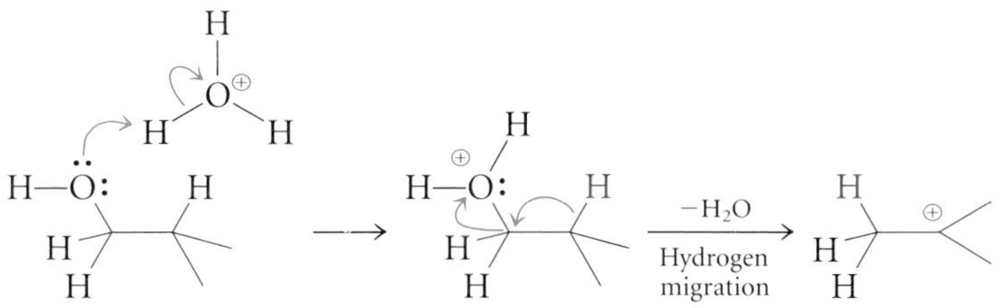

**Alkyl Shifts.** Alkyl groups are also observed to shift when a more stable cation is the result. When 3-methyl-2-butanol reacts with hydrogen bromide, either a hydrogen atom or a methyl group can be shifted. However, shifting the methyl group produces a secondary cation—a tertiary carbocation can be produced only by the shift of the hydrogen atom. When migration of more than one group is possible, the group that migrates is the one (here, a hydrogen atom) that will result in a more highly substituted carbocation. The preference for a shift of a hydrogen atom is not determined by the relative facility with which a hydrogen atom or an alkyl group can migrate:

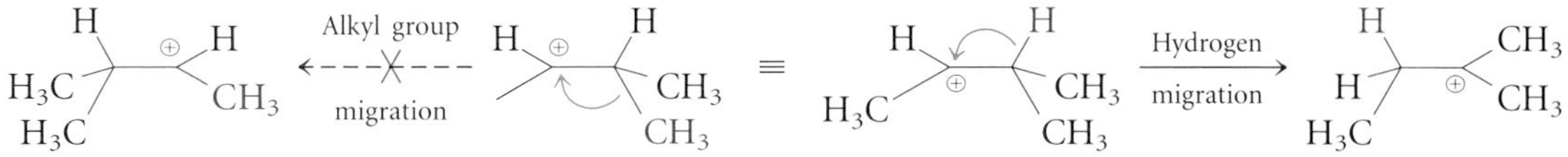

2° Carbocation                                                    3° Carbocation

When the shift of an alkyl group produces a more stable cation, this shift is also rapid. The change in structure that results from the migration of a carbon substituent is more profound than the one that accompanies the migration of a hydrogen atom, because the connectivity (and therefore the skeleton) of the structure changes. For example, treatment of 3,3-dimethyl-2-butanol with HCl produces 2-chloro-2,3-dimethylbutane:

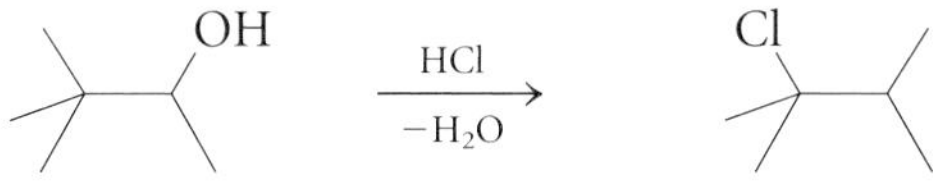

3,3-Dimethyl-2-butanol            2-Chloro-2,3-dimethylbutane

We will see additional examples of these rearrangements in Chapter 14.

## EXERCISE 7.17

Draw the expected alkyl halide that would be produced by treatment of each of the following alcohols with HBr. (In each case, a rearrangement reaction is involved.)

(a) 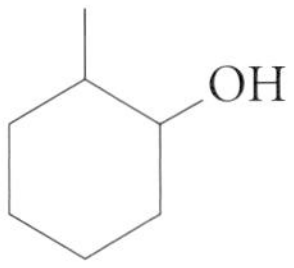    (c)

(b) 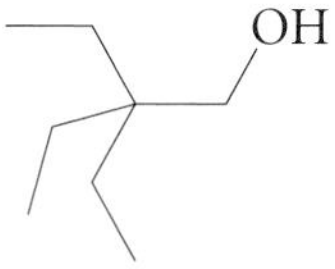    (d)

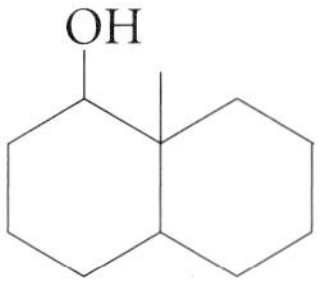

**7.6**

# Mechanism of a Multistep Homolytic Cleavage: Free-Radical Halogenation of Alkanes

One of the principal means of introducing functional groups into alkanes is homolytic substitution. In a homolytic reaction, radical intermediates are formed. The electrons of the $\sigma$ bonds undergoing cleavage do not remain paired, as they do in reactions in which ionic intermediates are formed. In representations of these homolytic substitution reactions, two half-headed arrows show the movement of single electrons.

## Energetics of Homolytic Substitution in the Chlorination of Ethane

In the **free-radical chlorination** of ethane, a C—H bond is replaced by a C—Cl bond.

$$\Delta H° = 100 \quad \Delta H° = 59 \qquad\qquad \Delta H° = -80 \quad \Delta H° = -103$$

$$CH_3\dot{C}H_2 \cdot H \quad Cl\cdot \cdot Cl \qquad \equiv \qquad CH_3\dot{C}H_2 \cdot Cl \quad H\cdot \cdot Cl$$

$$CH_3CH_2{-}H + Cl{-}Cl \xrightarrow{\;h\nu\;}$$

$$CH_3CH_2{-}Cl + H{-}Cl$$

$$\Delta H° = (100 + 59) - (80 + 103) = -24 \text{ kcal/mole}$$

This reaction proceeds through free-radical intermediates and is thus a **homolytic substitution.** We can estimate $\Delta H°$ for this reaction by comparing the energy of the bonds broken in the reaction (C—H and Cl—Cl) with the energy of the bonds formed (C—Cl and H—Cl). By using the energies from Table 3.5 (reproduced inside the back cover of the book), we can calculate $\Delta H°$ and find that this homolytic substitution is exothermic. To do this calculation, we "mentally" break all of the bonds present in the starting materials that are not present in the products and sum the bond energies. Putting the pieces together to form the products, we subtract the energies of the bonds that form in so doing. The result represents the change in bond energy for the reaction. Keep in mind that this sequence is merely a mental construct for the purpose of energy bookkeeping; the reaction does not actually happen by first breaking all of the bonds that disappear in the reaction and then making all of the new bonds that appear in the products.

### EXERCISE 7.18

Calculate $\Delta H°$ for each of the following reactions:

(a) ⬡ + $Br_2$ ⟶ ⬡Br + HBr

(b) ![structure] + $I_2$ $\longrightarrow$ ![structure]$^I$ + HI

(c) ![structure] + $Cl_2$ $\longrightarrow$ ![structure]Cl + HCl

## Steps in a Radical Chain Reaction

A **radical chain reaction** takes place in several steps, and we must evaluate each step to identify the one that is rate-determining.

***Initiation Step.*** A homolytic substitution reaction requires light or heat for initiation. In this **initiation step,** two free radicals are produced from a stable starting material:

$$Cl\!-\!Cl \xrightarrow{h\nu} Cl\cdot + Cl\cdot \qquad \Delta H° = +58\ \text{kcal/mole}$$

In an initiation step, a covalent bond is homolytically cleaved to produce two radicals. The energy needed to break the bond can be supplied as either light or heat.

The bond energy of the chlorine–chlorine $\sigma$ bond is relatively small, and so the energy required for its homolytic fission (the bond-dissociation energy) is not excessive. Thus, light or heat induces fission of this bond, producing two reactive radical fragments. This endothermic initiation step is critical in beginning the reaction, but it is not a part of the overall stoichiometry. Because initiation steps require that a bond be broken without

**367**

the simultaneous formation of another bond, they consume a substantial amount of energy. Fortunately, initiation steps need not be stoichiometric. Even a few radicals formed in an initiation step can begin a radical chain reaction leading to a large number of product molecules.

*Propagation Steps.* Although a homolytic substitution takes place through several steps, the propagation steps account for the bulk of product formation. In the propagation steps, the radicals produced in the initiation step react with neutral substrate to produce different radicals. In a **propagation step,** the number of product radicals is equal to the number of reactant radicals. In this case, a chlorine atom interacts with the alkane in a process in which a carbon–hydrogen bond is homolytically cleaved as a hydrogen–chlorine bond is formed.

$$CH_3\overset{\cdot}{C}H_2\!-\!H \quad \cdot Cl \longrightarrow CH_3\overset{\cdot}{C}H_2 + H\!-\!Cl \qquad \Delta H^\circ = 100 - 103 = -3 \text{ kcal/mole}$$

$$CH_3\overset{\cdot}{C}H_2 \quad Cl\!-\!Cl \longrightarrow CH_3CH_2\!-\!Cl + \cdot Cl \qquad \Delta H^\circ = 59 - 80 = -21 \text{ kcal/mole}$$

The propagation steps in a free-radical chlorination consume one radical while producing another. Both steps are exothermic and proceed efficiently. The product radical in one step is the reactant radical in the next, and so these reactions cycle repeatedly until the alkane or chlorine is almost completely consumed.

Again, referring to the table of bond-dissociation energies (Table 3.5), we find that the propagation step involving the chlorine atom is exothermic by 3 kcal/mole. Note that this step does not generate additional free radicals; it simply converts one reactant radical into a different product radical. One bond is broken as another is formed. The resulting alkyl radical then interacts with $Cl_2$. The chlorine–chlorine bond is broken at the same time that a carbon–chlorine bond is formed in a highly exothermic step ($\Delta H^\circ$ −21 kcal/mole). Like the first propagation step, this propagation step does not change the number of reactive intermediates—an alkyl radical is consumed as a chlorine atom is formed. Note that the reactive intermediate that initiates the first propagation step is formed as a product in the second propagation step. This chlorine atom can then serve as a reactant to repeat the first step. The propagation steps in this radical chain reaction alternate until the reactant alkane or chlorine, or both, are consumed.

*Termination Step.* When the initiation reaction is encouraged (by supplying continuous heat or light), the number of radicals increases until they begin to encounter each other, at least occasionally. Two of these radicals can combine exothermically to form a $\sigma$ bond in a process called a **termination step,** which is exactly the opposite of the initiation step. The following reactions convert two reactive radicals into one stable product:

$$Cl\cdot \quad \cdot Cl \longrightarrow Cl\!-\!Cl$$

$$CH_3\overset{\cdot}{C}H_2 \quad \cdot Cl \longrightarrow CH_3CH_2\!-\!Cl$$

$$CH_3\overset{\cdot}{C}H_2 \quad \overset{\cdot}{C}H_2CH_3 \longrightarrow CH_3CH_2\!-\!CH_2CH_3$$

In a termination step, a covalent bond is formed as each of two radicals donates its unpaired electron to form a σ bond. This bond formation releases energy and blocks further propagation steps by consuming a reactive free radical. In this way, termination reactions stop the radical chain reaction by consuming reactive intermediates without producing more.

The two different radicals produced by the two propagation steps of free-radical chlorination can combine with themselves or with each other. Therefore, the termination step of this chain reaction will produce not only small amounts of molecular chlorine (the combination of two chlorine atoms), but also small amounts of alkyl chloride (alkyl radical plus chlorine atom) and alkane (two alkyl radicals). Because termination steps begin to become important only when radical concentrations increase, which is usually when one of the reactant molecules is consumed, they do not contribute significantly to the observed product distribution.

***Net Reaction in a Radical Chain Reaction.*** The net stoichiometry and thermodynamics for a radical chain reaction derive from a consideration of the propagation steps alone; they are represented for the free-radical chlorination of ethane by the reaction profile in Figure 7.9.

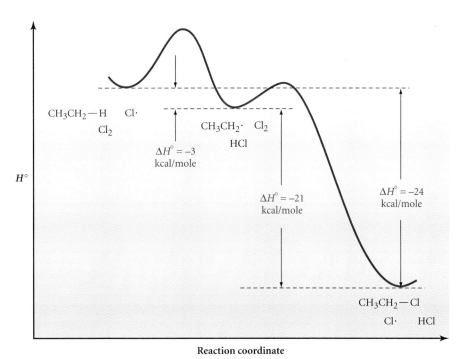

**FIGURE 7.9**

This reaction profile shows that two propagation steps are needed in a free-radical substitution reaction of ethane with chlorine.

**E X E R C I S E  7 . 1 9**

Calculate $\Delta H°$ for each of the following proposed propagation steps:

(a) ⌃⌄ + ·Br ⟶ ⌃⌄ + HBr

(b)

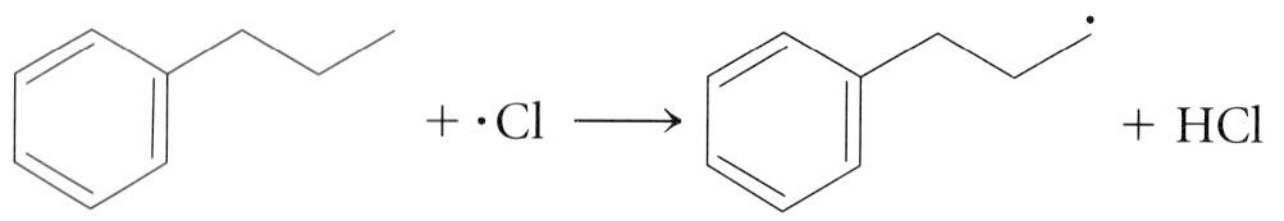

(c)

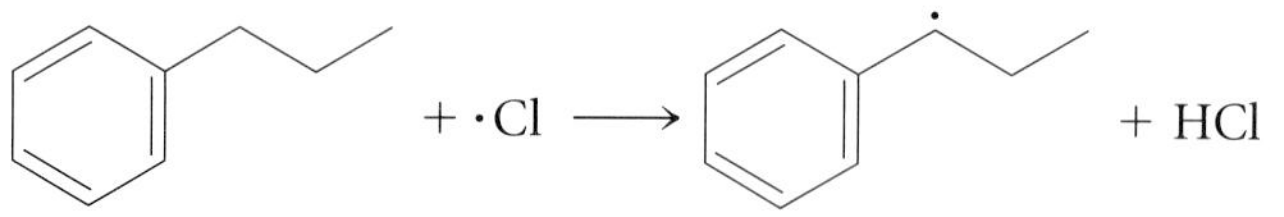

## Relative Reactivity of Halogens

Free-radical halogenation of alkanes is not limited to reactions with chlorine. For example, ethane reacts with bromine by a homolytic pathway essentially identical to that for chlorination.

*Transition States.* The energies of the various bonds broken and formed in free-radical chlorination and bromination are quite different, and these differences can significantly affect the usefulness of these reactions. Abstraction of hydrogen from ethane by a halogen atom is the rate-determining step in free-radical halogenation by either chlorine or bromine. This can be seen by comparing the energies of the two propagation steps for each halogen:

**Bromination**

$$CH_3CH_2\text{—}H \quad \cdot Br \longrightarrow CH_3\dot{C}H_2 + H\text{—}Br \qquad \Delta H° = 100 - 87$$
$$= +13 \text{ kcal/mole}$$

$$CH_3\dot{C}H_2 \quad Br\text{—}Br \longrightarrow CH_3CH_2\text{—}Br + \cdot Br \qquad \Delta H° = 46 - 68$$
$$= -22 \text{ kcal/mole}$$

**Chlorination**

$$CH_3CH_2\text{—}H \quad \cdot Cl \longrightarrow CH_3\dot{C}H_2 + H\text{—}Cl \qquad \Delta H° = 100 - 103$$
$$= -3 \text{ kcal/mole}$$

$$CH_3\dot{C}H_2 \quad Cl\text{—}Cl \longrightarrow CH_3CH_2\text{—}Cl + \cdot Cl \qquad \Delta H° = 59 - 80$$
$$= -21 \text{ kcal/mole}$$

Note that for chlorine, abstraction of hydrogen from ethane is slightly exothermic ($\Delta H° = -3$ kcal/mole), whereas the second propagation step is much faster ($\Delta H° = -21$ kcal/mole). For bromine, abstraction of hydrogen from ethane is endothermic ($\Delta H° = +13$ kcal/mole). The $\sigma$ bond is stronger in HCl than in HBr. Because of this difference, the transition state is early (reactant-like) in chlorination and is late (product-like) in bromination. In chlorination, the C—H bond is still mostly intact at the transition state; in bromination, this bond is substantially broken. The carbon atom undergoing substitution in bromination is therefore more radical-like (Figure 7.10), because the reaction involves a later transition state. As we will see in the next subsection, a late transition state that resembles the radical intermediate more than the starting material affords higher selectivity and better control of regiochemistry.

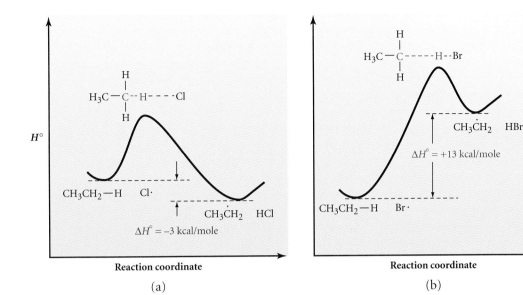

**FIGURE 7.10**

Reaction profiles for hydrogen abstraction from ethane by (a) chlorine and (b) bromine. Because the first step is exothermic for chlorination and endothermic for bromination, the transition state for chlorination is early and resembles the reactant alkyl halide rather than the intermediate radical, whereas that for bromination is late and has substantial radical character.

***Net Reaction Thermodynamics.*** Thermodynamics can reveal why radical chain reactions are not used for fluorination or iodination (Table 7.2). Free-radical fluorination is exothermic by approximately 109 kcal/mole, a value much too large to allow for safe and effective control of a self-propagating reaction without special precautions. In fact, free-radical

**TABLE 7.2**

Overall Enthalpy Changes ($\Delta H°$) for the Two Propagation Steps in Free-Radical Halogenation of Propane

| Halogen, X | $\Delta H°$ for Step 1 (kcal/mole) | $\Delta H°$ for Step 2 (kcal/mole) | $\Delta H°$ Overall (kcal/mole) |
|---|---|---|---|
| F | −39 | −70 | −109 |
| Cl | −7 | −23 | −30 |
| Br | +9 | −22 | −13 |
| I | +25 | −15 | +10 |

fluorination is so exothermic that extreme care must be exercised to prevent local heating and a violent reaction. Nonetheless, reasonable yields of fluorination products have been obtained under very carefully controlled conditions. Iodination, on the other hand, is unfavorable thermodynamically and, even if it were driven by the removal of HI by reaction with base, the first propagation step is so endothermic that it is impractically slow.

## Regiocontrol in Homolytic Substitution

Because most alkanes have a mixture of primary, secondary, tertiary, and quaternary carbon atoms, free-radical halogenation can give rise to a number of different products. As the size of the hydrocarbon increases, the number of possible isomers generally also increases. Even for the simple alkane pentane, there are three possible monochlorination products: 1-, 2-, and 3-chloropentane:

| Pentane | 1-Chloropentane | 2-Chloropentane | 3-Chloropentane |

With bromine, however, simpler mixtures are usually obtained.

Let's use the mechanism of free-radical halogenation to explain why bromination can often be used to obtain mostly one product, whereas chlorination is relatively nonselective and affords a mixture.

### EXERCISE 7.20

Draw structures for all of the possible monochlorination products for each of the following hydrocarbons.

(a)    (b)    (c)    (d)

*Selectivity in Free-Radical Chlorination and Bromination.* Let's

**373**

7.6 Mechanism of a Multistep
Homolytic Cleavage:
Free-Radical Halogenation
of Alkanes

consider free-radical chlorination and bromination of propane.

Two types of hydrogen atoms are present, primary and secondary. There are six primary and only two secondary hydrogen atoms. Thus, even if $\Delta H^o$ for abstraction of the two different hydrogen atoms were the same, there would be a statistical bias of 3:1 (6:2) favoring the formation of 1-chloropropane. However, because a primary C—H bond is stronger than a secondary C—H bond (100 versus 96 kcal/mole, Table 3.5 and inside the back cover), we can conclude that cleavage of the secondary C—H bond will be faster. How *much* faster is determined by the degree to which the bond is broken in the transition state. An early transition state, such as that in chlorination, will have undergone little bond breaking, and the strength of the C—H bond will not be a major influence on the activation energy or the rate of the reaction. Conversely, with a late transition state, in which substantial breaking of the C—H bond has occurred, the difference in the C—H bond energies becomes more important. This is the case for the abstraction of a hydrogen atom by a bromine atom, an endothermic reaction with a late transition state in which there is substantial breaking of the C—H bond. Thus, we expect the rate of bromination to differ significantly with the strength of the C—H bond and the transition state to look much like the radical intermediate.

We can see the effect of an early versus a late transition state from the experimental ratios of 1- to 2-halopropanes obtained in chlorination and bromination of propane:

44 : 56

4 : 96

**Propane**     **1-Halopropane**     **2-Halopropane**

The ratio of 1-bromopropane to 2-bromopropane (4:96) is far from that expected statistically (3:1), whereas the ratio of the chloropropanes is much closer to the statistical one. This "preference" (revealed by a different ratio from that predicted statistically) for one positional isomer over another is called **regiocontrol,** and these reactions are said to be **regioselective.** The fact that the amount of 2-bromopropane is much higher than expected from statistics is consistent with the greater stability of the secondary radical, which significantly affects the late transition state for abstraction of a hydrogen atom by a bromine atom.

It is not possible to perform a detailed analysis of the late versus early position of the transition states of most organic reactions, because the

## CHEMICAL PERSPECTIVES

### CHLOROFLUOROCARBONS IN THE ATMOSPHERE

Chlorofluorocarbons (CFCs) are chemicals with a variety of industrial applications, including use as refrigerants and degreasing solvents. They are named by a special system referred to as the *rule of 90*. For example, to arrive at the molecular formula of CFC-12, the numerical suffix, 12, is added to 90 (12 + 90 = 102). The result is read as one carbon atom (<u>1</u>02), no hydrogen atoms (1<u>0</u>2), and two fluorine atoms (10<u>2</u>). Any further atoms needed to complete the valence(s) of the carbon(s) are chlorine atoms (2). Thus, CFC-12 is $CCl_2F_2$.

Unfortunately, CFCs decompose on exposure to ultraviolet radiation in the upper atmosphere, generating (among other products) chlorine atoms. In turn, these chlorine atoms serve as catalysts for the decomposition of ozone into molecular oxygen:

$$Cl\cdot + O_3 \longrightarrow ClO\cdot + O_2$$

$$ClO\cdot + O_3 \longrightarrow Cl\cdot + 2\ O_2$$

Ozone in the upper atmosphere absorbs harmful ultraviolet radiation and thus plays a very important protective role (for example, against UV-induced skin cancer) for life on earth. A hole in the ozone layer near the South Pole has permitted higher-than-normal UV irradiation in the far southern regions of South America, and consequent blinding of entire flocks of sheep and of the Indians tending them. The blinding is thought to result from retinal damage induced by unfiltered UV light. Replacements for the CFCs, primarily hydrofluorocarbons that will not damage the ozone layer, are beginning to be produced on a large scale.

relevant bond energies are not available. In these cases, chemists use the degree of selectivity observed in reactions to decide whether the transition state of the step that determines the regiochemistry of the product is late or early.

By adjusting the observed 4 : 96 ratio for bromination of propane for the number of primary and secondary hydrogen atoms present, we can obtain an intrinsic reactivity ratio for *each* hydrogen atom that reflects the difference in activation energy ($\Delta H^{\ddagger}$) for abstraction of a primary or a secondary hydrogen atom. To do so, we divide 4 by the number of primary hydrogen atoms (6) and divide 96 by the number of secondary hydrogen atoms (2), and then express the result as a ratio:

$$\frac{\dfrac{96}{2}}{\dfrac{4}{6}} = \frac{48}{0.67} = 72 : 1$$

Thus, *each* secondary hydrogen atom is approximately 72 times more reactive than *each* primary hydrogen atom toward free-radical bromination.

We can use this relative reactivity to predict ratios of isomers for more complicated hydrocarbons. For example, butane has six primary and four secondary hydrogen atoms. The predicted ratio of 2- to 1-bromobutane is

$$\frac{4 \times 72}{6 \times 1} = \frac{288}{6} = 48:1$$

Converting this ratio to percentages gives

$$\frac{48}{48 + 1} \times 100\% = 98\%$$

and

$$\frac{1}{48 + 1} \times 100\% = 2\%$$

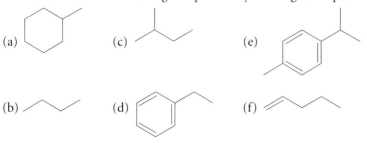

Butane   2-Bromobutane 1-Bromobutane

98 : 2 (predicted)
98 : 2 (observed)

## EXERCISE 7.21

Use the ratio of 1-chloro- to 2-chloropropane obtained from free-radical chlorination (44:56) to calculate the intrinsic reactivity difference between a primary and a secondary hydrogen atom of propane. Use this information to predict the expected ratio of 1-chloro- to 2-chlorobutane.

## EXERCISE 7.22

The intrinsic reactivity difference between primary and tertiary hydrogen atoms in free-radical chlorination is 5:1. Using this value, predict the ratio of 2-chloro-2-methylpropane to 1-chloro-2-methylpropane expected from the free-radical chlorination of isobutane.

## EXERCISE 7.23

Predict which hydrogen will be preferentially substituted in the free-radical bromination of each of the following compounds by drawing the expected product:

(a)    (c)    (e)

(b)    (d)    (f)

**EXERCISE 7.24**

Predict whether each of the following alkyl halides can be synthesized from the hydrocarbon shown by direct free-radical halogenation. If not, indicate the product that will be formed.

(a)  [cyclohexane] → [cyclohexyl chloride, Cl]

(b)  [neopentane] → [neopentyl bromide, Br]

(c)  [pentane] → [1-bromopentane, Br]

(d)  [pentane] → [2-iodopentane, I]

---

## 7.7

## Synthetic Applications

Alkyl halides are important synthetic intermediates because halogen can be easily replaced by other groups. We have seen that alkyl halides can be produced by three possible routes: addition of HX to an alkene, heterolytic substitution of an alcohol, or homolytic substitution of an alkane. Alkyl halides can be converted to other functional groups by the displacement reactions listed in Table 7.1.

As you learn new reactions, it is important to compare different routes that achieve a common synthetic objective. If you think about the unique characteristic of each reaction, you will learn how to use it in a discriminating fashion. (You will know that you have truly *arrived* in the world of organic synthesis when someone tells you that your synthesis is *elegant, innovative,* or *sophisticated!*) Table 7.3 is a way of compiling the reactions considered in this chapter. Refer to the section in the chapter that deals with each reaction to be sure you understand how it is best employed.

## Summary

**1.** Most organic reactions can be classified as one of seven major types: addition, elimination, substitution, condensation, rearrangement, isomerization, or oxidation/reduction.

**2.** Some reactions take place through reactive intermediates; others proceed directly from starting material to product in a single step. A reaction that includes the formation of one or more reactive intermediates is called a multistep reaction. Reactions in which there are no intermediates are referred to as concerted.

**3.** There are two distinctly different mechanisms for bond cleavage: homolytic and heterolytic. In homolytic cleavages, which are governed by bond-dissociation energies, single electrons move separately to form radical intermediates. In heterolytic cleavages, two electrons move as a pair, resulting in the formation of ions. Heterolytic cleavage reactions are much more affected by solvent polarity than are homolytic ones.

## TABLE 7.3

How to Make Various Functional Groups

| Functional Group | Reaction | Example |
|---|---|---|
| Alcohols | Acid-catalyzed hydration of an alkene | |
| | Hydrolysis of an alkyl halide | |
| Alkanes | Catalytic reduction of an alkene | |
| Alkenes | Acid-catalyzed dehydration of an alcohol | |
| | Dehydrohalogenation of an alkyl halide | |
| Alkyl azides | $S_N2$ displacement reaction of an alkyl halide with $NaN_3$ | |
| Alkyl halides | Free-radical halogenation (X = Br, Cl) | $R-H + X_2 \xrightarrow{h\nu} R-X + H-X$ |
| | Halogen exchange | $R-I \xrightarrow{KBr} R-Br$ |
| Amines | $S_N2$ displacement reaction of an alkyl halide with ammonia | |
| Ethers | $S_N2$ displacement reaction of an alkyl halide with an alkoxide | |
| Thioethers | $S_N2$ displacement reaction of an alkyl halide with thiolate anion | |

**4.** Reaction mechanisms are best represented (and understood) by the use of curved-arrow notation. A full-headed curved arrow indicates the motion of two electrons, and a half-headed curved arrow indicates the motion of one electron. The curved arrows indicate movement of electrons, not atoms. When electrons move, atoms follow.

**5.** Concerted nucleophilic substitution occurs through back-side attack by a nucleophile on the carbon attached to the leaving group. This reaction, called an $S_N2$ reaction, occurs with inversion of configuration, and its rate depends on the concentrations of both the substrate and the nucleophile.

**6.** The observed order of reactivity in an $S_N2$ reaction ($CH_3 > 1° > 2° >> 3°$) depends on steric access of the nucleophile to the reactive carbon atom.

**7.** The $S_N2$ mechanism effects several kinds of functional group interconversions.

**8.** Electrophilic addition takes place in two steps and proceeds through a carbocation intermediate. A hydrohalogenation reaction begins by protonation at one carbon of a $C{=}C$ bond, generating a carbocation at the other $sp^2$-hybridized carbon. The rate-determining step is the step leading to the cation, and the transition state is closely related to the cation. The product is formed in a second, fast step in which the cation is captured by the halide anion.

**9.** The regioselectivity observed for electrophilic addition to double bonds is governed by cation stability, which is consistent with Markovnikov's rule.

**10.** Like electrophilic addition to double bonds, multistep nucleophilic substitution through an $S_N1$ mechanism also involves formation of an intermediate carbocation. In the hydrolysis of an alkyl halide, this cation is captured by water to form a second intermediate, an oxonium ion, which loses a proton to form the final alcohol product. Thus, there can be more than one reactive intermediate along a reaction coordinate.

**11.** The rate-determining step of an $S_N1$ reaction is the step leading to the carbocation. The activation energy barrier for this reaction is affected by the energy needed for heterolysis of the C—X bond and by the stability of the carbocationic intermediate.

**12.** The reactivity order for $S_N1$ reactions ($3° > 2° >> 1°$) is governed by cation stability, and the reaction rate depends only on the concentration of the substrate (not on that of the nucleophile).

**13.** Free-radical halogenation also occurs by a multistep mechanism and is initiated by homolytic cleavage of a halogen molecule.

**14.** Radical chain reactions consist of three kinds of steps: initiation, propagation, and termination. The net stoichiometry of free-radical halogenation is controlled by the propagation steps.

**15.** Regiocontrol in homolytic substitution is governed by radical stability ($3° > 2° > 1° > CH_3$) and by whether the transition state is early or late. Bromine is more regioselective than chlorine, because bromine is less reactive and more selective in abstracting hydrogen through a later (more radical-like) transition state.

**7.1** Classify each of the following reactions as addition, elimination, substitution, condensation, rearrangement, isomerization, or oxidation–reduction.

(a)

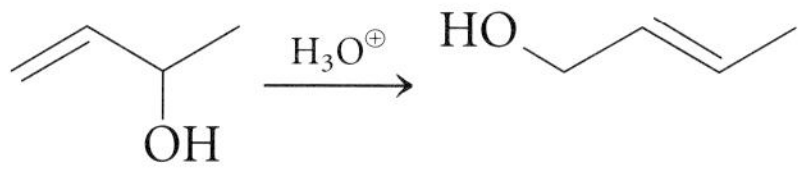

(b)

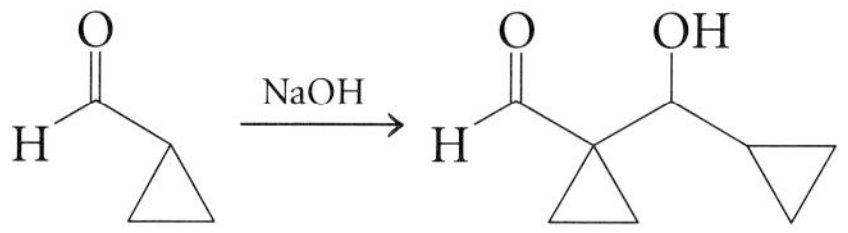

(c)

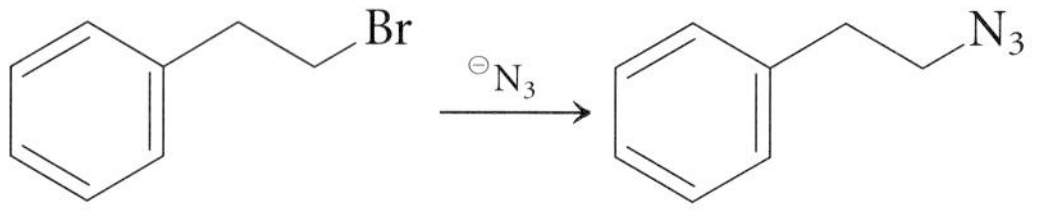

(d)

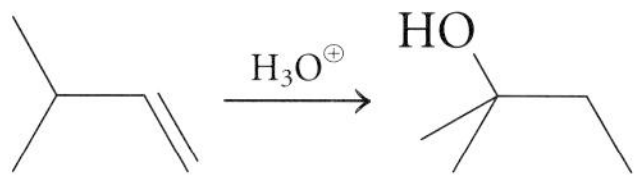

(e)

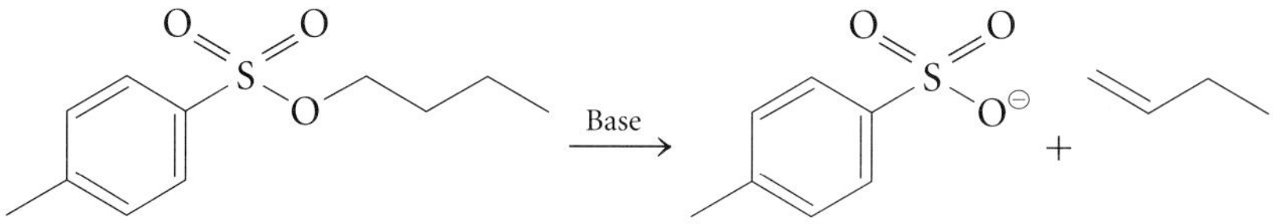

(f)

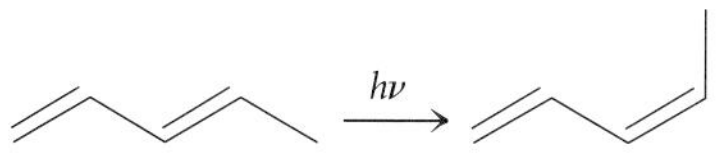

(g)

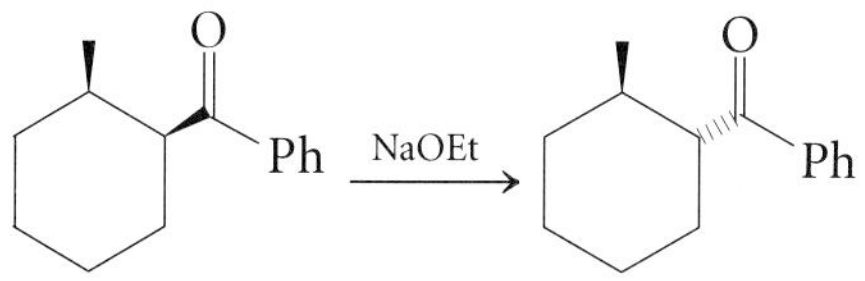

**7.2** The catalytic hydrogenation of an alkene is sometimes called an addition reaction. (Recall that it was also called a reduction in Chapter 2.)

(a) Explain why both classifications are reasonable.

(b) The conversion of 2-propanol into 2-propanone can be considered either an elimination or an oxidation. Explain why this reaction can be viewed as either type.

**7.3** Suppose you wished to make each of the following compounds by an $S_N2$ reaction. Identify the alkyl halide and the nucleophile you would need.

(a) $CH_3CH_2CH_2CH_2N_3$

(b) $CH_3CH_2CH_2CH_2CN$

(c) $CH_3OCH_3$

(d) tetrahydrofuran

(e) $CH_3CH_2CH_2CH_2SH$

(f) $CH_3CH_2CH_2SCH_2CH_3$

(g) $CH_3OSO_2Ph$

(h) $CH_3CH_2P^{\oplus}(C_6H_5)_3 \quad {}^{\ominus}Br$

**7.4** Epoxides can be formed through an intramolecular $S_N2$ reaction. Using what you know about p$K_a$ values, write a mechanism for the following reaction:

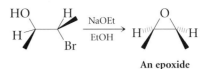

**An epoxide**

**7.5** The azide ion, $^{\ominus}N_3$, is known to react by an $S_N2$ mechanism thousands of times more rapidly with 2-bromopentane than with its isomer neopentyl bromide (1-bromo-2,2-dimethylpropane), despite the fact that the leaving group is at a secondary carbon in the former compound and at a primary carbon in the latter. Explain.

**7.6** To reach the conclusion that the reaction of azide ion with 2-bromopentane cited in Problem 7.5 did indeed occur as an $S_N2$ reaction, the chemists studying the reaction did several additional experiments:

(a) They used optically active (R)-2-bromopentane.

(b) They doubled the concentration of alkyl bromide.

(c) They doubled the concentration of azide ion.

Predict what they would have seen in each experiment if the reaction really took place via an $S_N2$ pathway.

**7.7** Choose the member of the following pairs of unsaturated hydrocarbons that is more reactive toward acid-catalyzed hydration, and predict the regiochemistry of the alcohols formed from that compound.

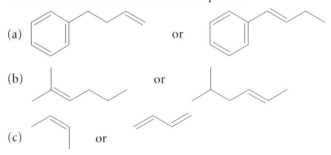

(a)                                    or

(b)                       or

(c)            or

**7.8** When allowed to stand in dilute aqueous acid, (R)-2-butanol slowly loses its optical activity. Write a mechanism that can account for this racemization.

**7.9** When the acid-catalyzed hydration of 3-methyl-1-butene is carried out in $D_2O$, the alcohol product does *not* have D and OD on adjacent carbons. Write a detailed mechanism, using curved arrows to show electron flow, that identifies the hydration product formed and explains this observation. (*Hint:* Carbocations can rearrange by shifting a hydrogen atom or alkyl group from an adjacent position if a more stable cation will be produced.)

**7.10** Rank the alcohols in each of the following sets according to their rates of reactivity toward treatment with HBr. Explain each ranking.

(a) *t*-butyl alcohol, *s*-butyl alcohol, *n*-butyl alcohol

(b) *p*-methoxybenzyl alcohol, *p*-nitrobenzyl alcohol, benzyl alcohol

(c) benzyl alcohol, *p*-methylphenol, $\alpha,\alpha$-dimethylbenzyl alcohol

**7.11** Under forcing conditions (such as hot concentrated sulfuric acid), ethyl ether reacts to form ethene and ethanol by a mechanism similar to the acid-catalyzed dehydration discussed in this chapter.

(a) Using curved arrows, write a mechanism by which ethyl ether is converted to ethylene and ethanol.

(b) Predict the product that would be obtained if tetrahydrofuran or dioxane (common organic solvents) were so treated.

**7.12** Using what you now know about the mechanism for acid-catalyzed dehydration, propose a detailed mechanism for the pinacol rearrangement mentioned (without detail) on pages 335–336.

**7.13** Heating many alkyl chlorides or bromides in water converts them to alcohols through an $S_N1$ reaction. Order each of the following sets of compounds with respect to this solvolytic reactivity:

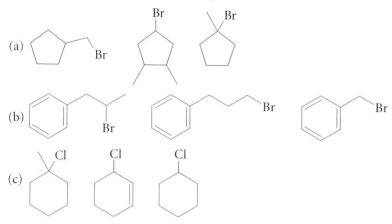

**7.14** When alkyl halides are treated with aqueous silver nitrate, silver halide precipitates and an alcohol is formed. From what you know about $S_N1$ reactions, propose a mechanism for the following conversion. (*Hint:* Consider a possible rearrangement to produce a more stable cationic intermediate.)

**7.15** Suppose the following reactions were proposed as routes for making the indicated products. Determine whether each reaction is likely to proceed as written. If not, write the expected product, and explain why the indicated reaction would not occur.

**7.16** Homolytic chlorination and bromination are effective means for producing alkyl halides from alkanes. Chlorine is somewhat less expensive than bromine and, if you were running a chemical plant, where it is important to keep the costs of reagents needed for large-scale (many tons) conversions as low as possible, it would be advantageous to use chlorine. For each of the following free-radical brominations, decide whether chlorine could be used instead of bromine to prepare the analogous alkyl chloride in good yield. Explain your reasoning.

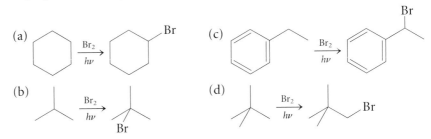

**7.17** Consider the following proposed reaction of cyclohexane with chlorine:

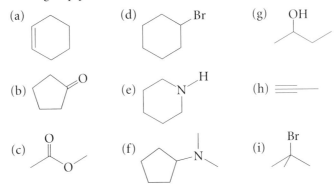

(a) Assuming that such a reaction would be initiated by the same route as in homolytic substitution (Cl—Cl ⟶ 2 Cl·), propose a mechanism by which the indicated reaction could proceed through a radical chain.

(b) For each propagation step you wrote for part (a) use the table of bond-dissociation energies (Table 3.5) to calculate the expected enthalpy change. (Assume that a C—C bond between secondary carbons is worth about 84 kcal/mole.) Does the calculation explain why this proposed reaction is not observed in the laboratory (that is, why cyclohexyl chloride is obtained instead)?

**7.18** In seeking a source for gasoline and other low-weight hydrocarbons (for example, butadiene, used in large quantities as a component of plastics and rubber, as described in Chapter 16), the petroleum industry runs large cracking towers in which complex mixtures of higher-molecular-weight alkanes are heated to very high temperatures. Under these conditions, the alkanes "crack" through the homolysis of C—C and C—H bonds. Consider propane to be a model for hydrocarbon cracking. Refer to the table of bond-dissociation energies (Table 3.6) to decide whether cracking would be more efficiently initiated by cleavage of a C—C, a primary C—H, or a secondary C—H bond.

## Supplementary Problems

**7.19** Classify each of the following compounds in terms of the principal functional group present (1° alcohol, 3° amine, etc.)

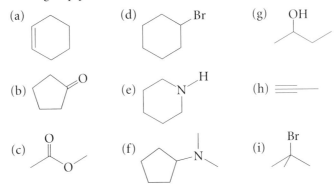

**7.20** Classify each of the following reactions as addition, elimination, substitution, condensation, rearrangement, isomerization, or oxidation–reduction.

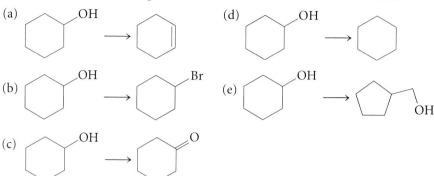

**7.21** Select the alkene in each pair that would be expected to undergo hydration in aqueous acid at the greater rate.

**7.22** Cyclohexene generally undergoes addition reactions at rates faster than cyclopentene does. Try to explain this observation. (*Hint:* Consider how the removal of the double bond might be influenced by the conformations of the starting materials and the products.)

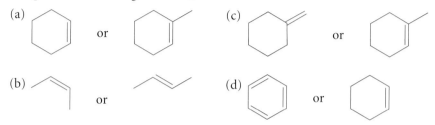

**7.23** Calculate $\Delta H°$ for each of the following reactions:

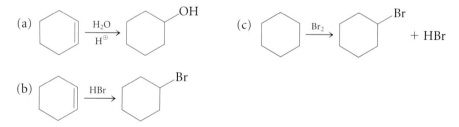

**7.24** Indicate which hydrogens in the following structure would be most reactive in a free-radical substitution reaction. (*Note:* Because each of these structures is symmetrical, each hydrogen atom is identical to one or more others.)

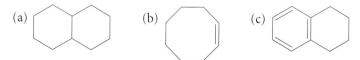

# Substitution by Nucleophiles at $sp^3$-Hybridized Carbon

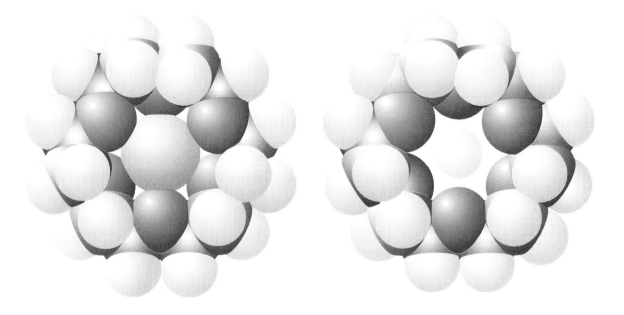

*H*omolytic substitution and nucleophilic substitution (two of the reactions considered in Chapter 7) are frequently used to change the functional groups in a molecule. Homolytic substitution can be used to introduce a chlorine or bromine substituent into a hydrocarbon, and then, in a subsequent step, a nucleophilic substitution reaction is employed to replace the halogen by another group. In this chapter, we look in more detail at the transformations that can be carried out by $S_N2$ and $S_N1$ reactions.

To use substitution reactions to form new carbon–carbon and carbon–heteroatom bonds at an $sp^3$-hybridized carbon, you must become familiar with sets of reactions in which reagents act as nucleophiles. These reactions fall into two classes: those that produce a new carbon–heteroatom bond in the product, and those that result in the formation of a new carbon–carbon bond. The first class represents a method for converting one functional group into another—for example, the conversion of methyl bromide to methanol.

$$HO^{\ominus} + H_3C—Br \longrightarrow HO—CH_3 + :Br^{\ominus}$$

(In this and subsequent equations in this chapter, nucleophiles are shown in blue and electrophiles in red.) The second class of reactions represents a valuable set of tools for increasing the number of carbon atoms—and therefore the complexity—of an organic molecule.

$$N≡C^{\ominus} + H_3C—Br \longrightarrow N≡C—CH_3 + :Br^{\ominus}$$

## Review of Mechanisms of Nucleophilic Substitution

Concerted ($S_N2$) and stepwise ($S_N1$) reactions at tetrahedral carbon are the two mechanistic extremes by which a new group can replace an existing substituent via nucleophilic substitution.

### $S_N2$ Mechanism

You learned in Chapter 7 that a concerted nucleophilic substitution takes place by **back-side attack:** an electron-rich nucleophile (Nuc:$^{\ominus}$) approaches from the side opposite to that from which the leaving group (LG) departs. A group can function as a leaving group only to the extent that it can accommodate the electrons that were originally in the C—LG bond. In most leaving groups, the atom directly connected to carbon is one of the more electronegative heteroatoms, often oxygen or a halogen. The $S_N2$ reaction results in inversion of configuration at the carbon atom undergoing substitution.

$S_N2$ **Reaction (for example, LG = Cl, Br, I)**

$$Nuc:^{\ominus} \quad \overset{R}{\underset{R}{\overset{|}{\underset{|}{H\cdots C—LG}}}} \longrightarrow Nuc—\overset{R}{\underset{R}{\overset{|}{\underset{|}{C\cdots H}}}} + :LG^{\ominus}$$

Because both the nucleophile and the substrate are partially bonded to carbon in the transition state, the reaction is bimolecular. (The 2 in $S_N2$ indicates that two molecules take part in the rate-determining step.) The rate of a bimolecular reaction depends on the concentration of both of these species, and second-order kinetics are observed.

$$\text{Rate} = k[\text{R—LG}][\text{Nuc}^{\ominus}]$$

## $S_N1$ Mechanism

At the other extreme of possible nucleophilic substitution mechanisms is the two-step $S_N1$ reaction (Figure 8.1). In the first step, the leaving group departs with the electrons from the C—LG bond, forming a trigonal, $sp^2$-hybridized carbocation. The resulting planar carbocation then reacts rapidly with the nucleophile. Because the attack of the nucleophile occurs on both faces of the planar carbocation at the same rate, both possible stereoisomers are formed. Therefore, even when the starting material is a single enantiomer (whose center of chirality is the carbon atom undergoing substitution), the $S_N1$ reaction produces a racemic product.

**$S_N1$ Reaction**

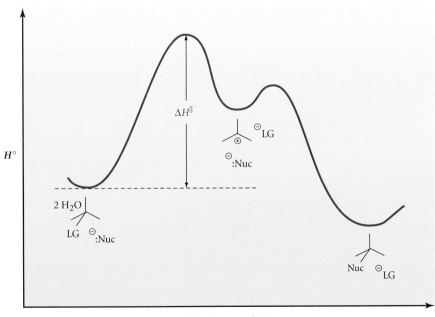

## FIGURE 8.1

Energy diagram for an $S_N1$ reaction. The two separate steps are formation of a carbocation and reaction of this intermediate with the nucleophile.

**Transition state for
formation of a carbocation**

The first step in the S$_N$1 mechanism consists only of bond breaking and undoubtedly is rate-determining. Loss of the leaving group is followed by a much faster step in which an external nucleophile ($:Nuc^{\ominus}$) provides a pair of electrons to form a new $\sigma$ bond to the carbocation. The first (rate-determining) step leading to the carbocation is endothermic, with a late transition state. Therefore, the C—LG bond is substantially broken at the transition state, which has appreciable carbocationic character. The facility of an SN1 substitution relates directly to the stability of the carbocation formed.

Because the rate-determining step of an S$_N$1 reaction involves only the substrate, this reaction is unimolecular. (The 1 in S$_N$1 indicates that only one molecule is involved in the rate-determining step.) The rate of an S$_N$1 reaction is not influenced by the concentration of the nucleophile, and first-order kinetics are observed. Thus, the reaction rate depends only on the concentration of the substrate.

$$\text{Rate} = k[\text{R—LG}]$$

## Solvents for Organic Reactions

Solvents can change the course of an organic reaction in several ways. In some cases, as in the conversion of $t$-butyl chloride to $t$-butanol in water, the solvent is also a reagent. These reactions are referred to as **solvolyses.** In most reactions, the solvent is not consumed but does perform several important functions. First, the solvent dissolves both reagent(s) and substrate so that they can react with each other. Second, the choice of solvent can influence the reaction pathway, or mechanism. Third, the solvent may control the temperature of the reaction.

The choice of solvent is governed to some extent by the solubility of the reactants. To dissolve ionic compounds such as NaOH, NaCN, and NaN$_3$, it is essential that the solvent interact strongly with (**solvate**) the ions in solution. Protic solvents (water and alcohols) are good solvents for salts (ionic compounds), because they engage in hydrogen bonding with the anions and have lone pairs of electrons on oxygen that interact with the cations.

Protic solvents facilitate S$_N$1 reactions by stabilizing the negatively charged leaving group. Conversely, aprotic solvents favor the S$_N$2 pathway. Since aprotic solvents do not stabilize negative ions, they do not facilitate S$_N$1 reactions; thus, by default, the S$_N$2 pathway is favored.

Many of the nucleophiles used in S$_N$2 reactions are not soluble in nonpolar aprotic solvents such as hydrocarbons or methylene chloride, so a polar aprotic solvent such as DMSO or DMF must be used. These solvents act as good cation stabilizers through interaction with oxygen, but have little influence on anions, which are free to take part in the reaction as nucleophiles.

**Dimethyl sulfoxide
(DMSO)**

**Dimethyl formamide
(DMF)**

## CHEMICAL PERSPECTIVES

### MOLECULAR "CROWNS"

A special class of ethers are the **crown ethers,** cyclic compounds with several ether functional groups. The name is derived from the resemblance of the complex to a crown. In 18-crown-6, a total of 18 nonhydrogen atoms are present, six of which are oxygens. Crown ethers form extremely tight complexes with metal cations of the right size (as shown at the beginning of this chapter).

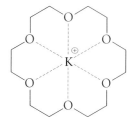

**18-Crown-6 ether–potassium complex**

The potassium ion is a near-perfect fit for the cavity of 18-crown-6, whereas the lithium ion is too small to span from one side to the other and consequently is bound much less tightly than potassium. Crown ethers have been used to dissolve otherwise insoluble salts in ether solvents. For example, potassium cyanide, KCN, can be dissolved in ethyl ether by the addition of 18-crown-6. The cyanide ion is much more nucleophilic (and basic) in ether than it is in water, where hydrogen bonding significantly stabilizes this anion.

Three chemists, Charles J. Pedersen (then at DuPont), Donald J. Cram (University of California at Los Angeles), and Jean-Marie Lehn (Paris and Strasbourg) shared the Nobel prize in 1987 for creating macrocyclic compounds that serve as strong ligands for metal cations.

#22   Crown Ethers

Ethers such as tetrahydrofuran (THF) also favor $S_N2$ reactions. These cyclic ethers are less polar than DMSO and DMF but still form tight complexes with alkali metal ions:

**Tetrahydrofuran–lithium complex**

(Because the alkyl portions of the cyclic ethers are tied back into a ring, the lone pairs of oxygen are able to bond with cations better than those of straight-chain ethers.) The association of the metal cation with the ether weakens the association between the metal cation and its negatively charged counterion. Ethers have a further desirable characteristic in that they are relatively inert to most reagents used for organic reactions.

Solvents can also control the temperature of a reaction. Most organic reactions carried out in the laboratory are exothermic, releasing heat. The mass of the solvent helps to moderate the temperature increase that ac-

**(a) Aprotic Solvents**

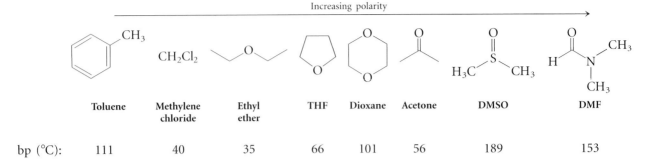

|  | Toluene | Methylene chloride | Ethyl ether | THF | Dioxane | Acetone | DMSO | DMF |
|---|---|---|---|---|---|---|---|---|
| bp (°C): | 111 | 40 | 35 | 66 | 101 | 56 | 189 | 153 |

**(b) Protic Solvents**

Increasing polarity

| | CH$_3$CH$_2$OH | CH$_3$OH | H$_2$O |
|---|---|---|---|
| | Ethanol | Methanol | Water |
| bp (°C): | 78 | 65 | 100 |

**FIGURE 8.2**

Boiling points of some common aprotic and protic solvents.

companies the release of heat. Moreover, the temperature of a reaction cannot exceed (significantly) the boiling point of the solvent. For a number of reasons, it is useful for a practicing chemist to commit to memory the boiling points of the more common organic solvents (see Figure 8.2).

## 8.2

# Competition between S$_N$2 and S$_N$1 Pathways

Whether a particular substrate follows the S$_N$2 or the S$_N$1 pathway for substitution is determined largely by the degree of substitution at the reactive center. Tertiary carbon atoms bearing a leaving group are too sterically hindered to enter into S$_N$2 reactions and by default follow the S$_N$1 pathway. Substrates with a leaving group on a primary carbon atom follow the S$_N$2 mechanism, because the formation of a primary carbocation is prohibitively high in energy and back-side attack is relatively unhindered. When the leaving group is on a secondary carbon atom, both S$_N$2 and S$_N$1 pathways are available. Which one is followed depends on more subtle features of the molecule undergoing reaction, as well as on the reaction conditions, such as the solvent and nucleophile used. For example, polar protic solvents (such as water) favor heterolytic cleavage of the C—LG bond to form an intermediate carbocation. Polar aprotic solvents (polar heteroatom-containing

**TABLE 8.1**

Effect of Substrate Structure and Reaction Conditions
on the Pathway of Substitution Reactions

|  | $S_N2$ | $S_N1$ |
| --- | --- | --- |
| **Substrate** | 1°, 2°, methyl | 2°, 3°, benzylic, allylic |
| **Nucleophile** | More nucleophilic | Less nucleophilic |
| **Solvent** | DMSO or acetone | $H_2O$ or ROH |

molecules that lack acidic hydrogen atoms that can participate in hydrogen
bonding—for example, dimethylsulfoxide, abbreviated DMSO) promote
$S_N2$ reactions.

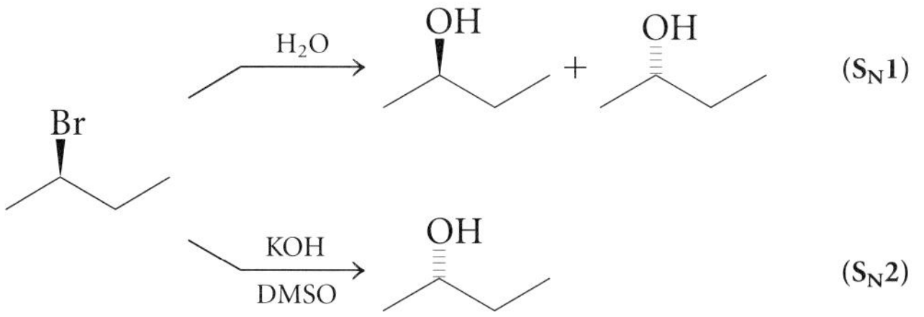

Except with simple substrates, yields are generally higher for an $S_N2$ reaction at a secondary carbon than for the comparable $S_N1$ reaction of the
same substrate. (You will see in subsequent chapters that the carbocations
involved as intermediates in $S_N1$ reactions also undergo elimination and rearrangement reactions, leading to the formation of other products mixed
with the substitution product.) Furthermore, with reactants that are single
enantiomers, $S_N2$ reactions result in inversion of configuration, whereas $S_N1$
reactions give rise to racemic products and, in general, to more complex
mixtures of products. Some aspects of $S_N2$ and $S_N1$ reactions are summarized in Table 8.1.

**EXERCISE 8.1**

Identify the carbon atom bearing a leaving group in each of the following compounds, and indicate if this carbon is primary, secondary, or tertiary.

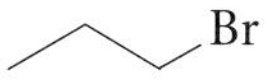

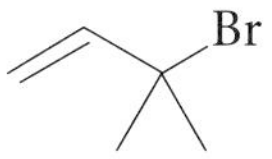

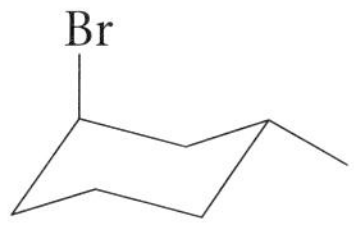

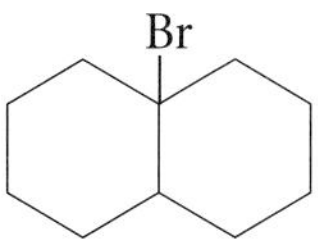

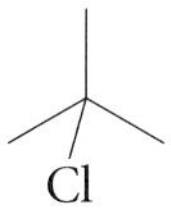

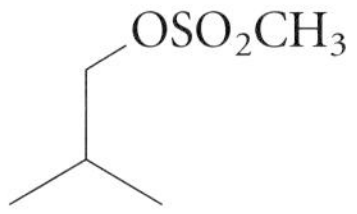

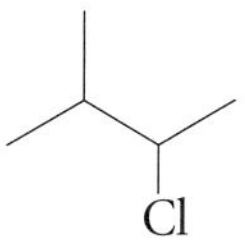

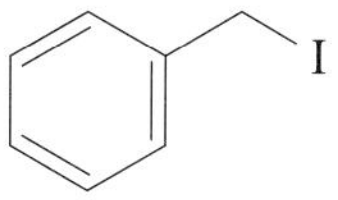

**8.3**

# Functional-Group Transformations through $S_N2$ and $S_N1$ Reactions

As noted at the beginning of this chapter, substitution reactions can be used to change one functional group into another. In this class of reactions, the nucleophilic center is a carbon atom bonded to a heteroatom: a halogen, oxygen, phosphorus, nitrogen, or sulfur. (Several of these nucleophiles were listed in Tables 7.1 and 7.3.) Both $S_N1$ and $S_N2$ pathways are followed for these reactions.

Recall that alkyl halides can be synthesized by free-radical halogenation of alkanes, a reaction that is one way to functionalize hydrocarbons:

$$R\!-\!H \xrightarrow[h\nu]{Br_2} R\!-\!Br$$

Once a carbon atom is functionalized as an alkyl halide, nucleophilic substitution reactions can be used to exchange the halogen atom for another substituent (Figure 8.3).

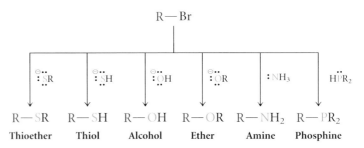

**FIGURE 8.3**

Substitution of the halogen of an alkyl halide by a variety of nucleophiles provides access to a number of functional groups.

## Substitution of Halogen to Form Alcohols by an $S_N2$ Mechanism

All classes of alkyl halides undergo substitution to form alcohols. Primary and secondary alkyl halides react with hydroxide ion in an $S_N2$ reaction to form primary and secondary alcohols.

$$H_3C\!-\!CH_2\!-\!Br \xrightarrow[H_2O]{Na^{\oplus}\ \ominus:\ddot{O}H} H_3C\!-\!CH_2\!-\!OH$$

**Primary**
**alkyl bromide**

**Substitution**
**product**

Generally, the reaction proceeds more rapidly with primary than with secondary alkyl halides because of increased steric hindrance in the latter.

Indeed, the lower rate of substitution of secondary alkyl halides makes competing elimination reactions more important, often decreasing the yield of the alcohol as some of the starting material is diverted to formation of an alkene. (Elimination reactions will be treated in depth in Chapter 9.)

A potential problem is associated with substitution reactions because all nucleophiles are also bases. Thus, there is a competition between substitution reactions, which depend on nucleophilic character, and elimination reactions, which require a base to remove a proton from an adjacent carbon atom. With secondary alkyl halides, the balance between substitution and elimination is often delicate: which reaction will dominate is determined by the nucleophile, various structural features of the reactant, and the details of the reaction conditions.

| Secondary alkyl bromide | Substitution product | Elimination product |

The balance between the nucleophilic and basic characters of a reagent is determined in large part by the size of the orbital containing the lone pair of electrons, and thus the polarizability of these electrons. Reagents with highly polarizable centers of electron density (for example, large halide ions) are better nucleophiles than bases, because they prefer to attack carbon rather than the considerably smaller proton. Thus, $I^{\ominus}$ is the most nucleophilic of the halide ions and also the least basic (and H—I is the strongest of the halogen acids). Conversely, the high concentration of electron density in $^{\ominus}NH_2$, $^{\ominus}OH$, $^{\ominus}C\equiv N$, and $^{\ominus}C\equiv C$—H makes these reagents good bases and only moderately active nucleophiles.

| | p$K_a$ |
|---|---|
| H—F | 3 |
| H—Cl | −7 |
| H—Br | −9 |
| H—I | −10 |

Although it is important that you recognize which reagents will provide the highest yield of substitution products along with the lowest level of elimination, the nature of the nucleophile is generally dictated by the desired outcome of the substitution reaction. If the desired conversion is of an alkyl halide to an alcohol, the nucleophile must be $^{\ominus}OH$ or $H_2O$. It is therefore sometimes difficult to achieve only substitution, without a competing elimination reaction.

When the reactive center of a secondary alkyl halide is a center of chirality, the back-side attack inherent in $S_N2$ reactions leads to an inversion of configuration at the functionalized carbon:

(R)-2-Bromobutane          (S)-2-Butanol

Substitution at a tertiary alkyl halide by an $S_N2$ mechanism is not possible, because the steric hindrance raises the energy of the transition state for substitution above that for the competing elimination reaction, and elimination occurs (Figure 8.4, on page 394).

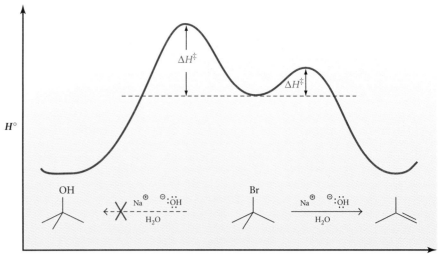

**FIGURE 8.4**

Reaction profile expected for the reaction of *t*-butyl bromide with sodium hydroxide. The product observed (2-methylpropene) is the result of elimination rather than substitution, because the activation energy required for the latter process is quite high as a result of steric hindrance.

### EXERCISE 8.2

Predict the product expected in each of the following substitution reactions if an $S_N2$ pathway is followed. Use $R,S$ nomenclature to specify absolute configuration at any centers of chirality in the reactants and products.

(a)

$$\xrightarrow[\text{Acetone}]{\text{NaI}}$$

(b)

$$\xrightarrow[\text{H}_2\text{O}]{\text{NaOH}}$$

(c)

$$\xrightarrow[\text{H}_2\text{O}]{\text{NaOH}}$$

### Substitution of Halogen to Form Alcohols by an $S_N1$ Mechanism

As noted in the previous section, tertiary alkyl halides undergo elimination reactions when treated with strongly basic nucleophiles such as hydroxide ion. However, when a tertiary alkyl halide is heated in water *in the absence of a strong base,* it loses halide ion to form a carbocation:

$$\xrightarrow[\text{H}_2\text{O}]{\text{Na}_2\text{CO}_3}$$

Tertiary alkyl halide                                        Tertiary alcohol

## CHEMICAL PERSPECTIVES

### A HIGHLY TOXIC ALKYL HALIDE

Mustard gas was used extensively in World War I as a chemical weapon and was stockpiled by many nations as a deterrent until several international treaties in the 1980s banned its use and called for its destruction. When this gas contacts the skin or lungs, the water present in the tissue rapidly displaces HCl, producing high local concentrations of acid, which cause extensive blistering, tissue destruction, and, in severe exposure, death.

**Mustard gas**

The nucleophilic substitution reaction with mustard gas is unusually fast because of the neighboring group effect of the sulfur atom.

The carbocation then reacts with water (a less basic nucleophile than hydroxide ion) to produce an alcohol. In this reaction, the alkyl halide is converted to the alcohol by a first-order substitution, an $S_N1$ reaction. This method for the preparation of tertiary alcohols by $S_N1$ substitution proceeds in high yield only for a limited number of alkyl halides. As you will learn in the next chapter, tertiary and especially secondary carbocations are prone to form a number of different products, of which alcohols are only one.

Although substitution at primary and secondary carbons generally proceeds with acceptable yield only under conditions that lead to an $S_N2$ reaction, there are notable exceptions to this rule when special structural features provide for an unusually stable carbocation. For example, the reactions of benzylic and allylic halides proceed through an $S_N1$ pathway, because the intermediate cations are stabilized by conjugation with the adjacent $\pi$ electron system (Section 3.6).

**Benzyl chloride**                    **Benzyl alcohol**

A weak base such as sodium carbonate or sodium bicarbonate is often added to absorb the acid produced in the reaction, especially if the product alcohol is not particularly stable in acid solution. (Recall from Chapter 3 that protonation of a hydroxyl group makes it easier to break the carbon–oxygen bond; that is, a protonated hydroxyl is a good leaving group.)

**Allyl bromide**                    **Allyl alcohol**

### EXERCISE 8.3

In principle, secondary alkyl halides can react by either an $S_N1$ or an $S_N2$ pathway.

(a) Predict the stereochemical consequence of each pathway in the hydrolysis of (2R,4R)-2-bromo-4-pentanol to a diol.

(b) The starting material in part (a) is optically active. Is the product also optically active?

(c) Would your answer to (b) be different if the starting material were (2S,4R)-2-bromo-4-pentanol?

### EXERCISE 8.4

Draw all significant resonance structures for the cations formed by heterolytic cleavage of the carbon–halogen bond in allyl bromide and benzyl bromide. In each case, indicate whether any of the resonance structures are identical in energy.

### EXERCISE 8.5

Indicate whether an $S_N2$ or an $S_N1$ mechanism is expected for each of the following reactions. Be sure to consider both the structure of the substrate and the reaction conditions.

(a)

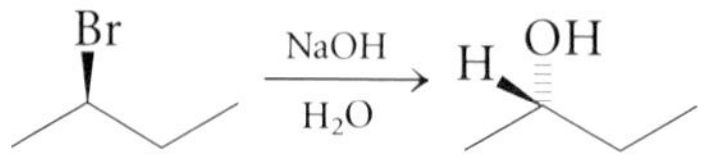

(b)

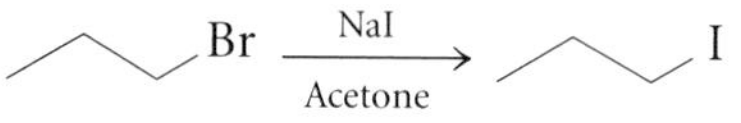

(c)

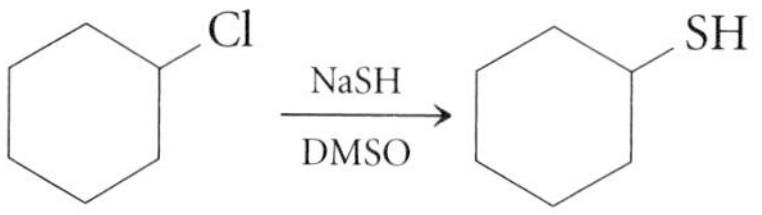

(d)

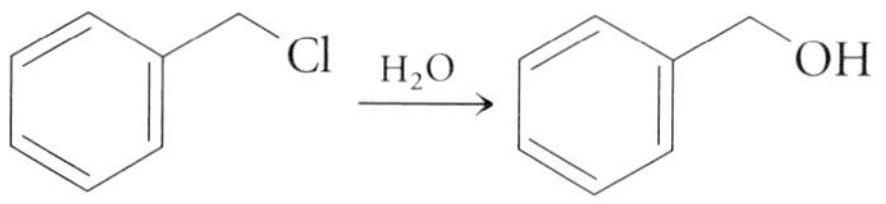

(e)

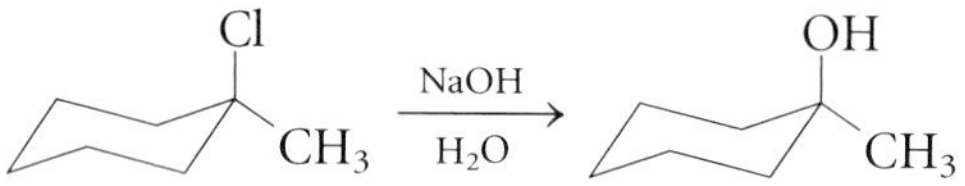

(f)

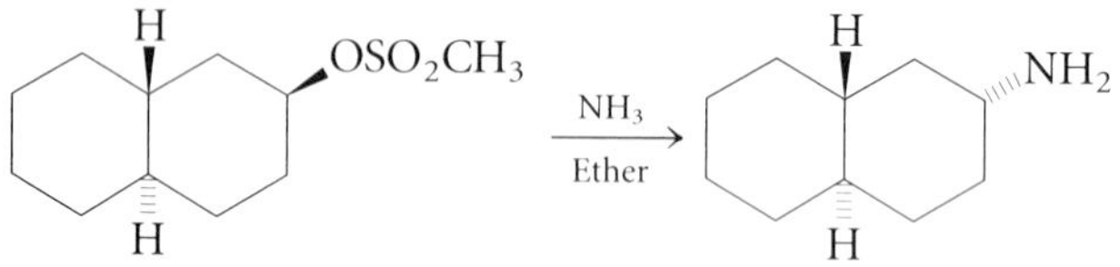

### Substitution of Halogen to Form Ethers: Williamson Ether Synthesis

Alkoxide ions are good nucleophiles and displace halide ions from alkyl halides, resulting in the formation of a new carbon–oxygen bond. Alkoxides are produced by treatment of alcohols with either a base or an alkali metal:

### ETHERS AS ANESTHETICS

Chemistry revolutionized medicine. Unlike the medical doctors of earlier centuries, today's physicians are much more likely to treat disease with chemicals than with a knife. The use of nitrous oxide and ethyl ether as anesthetics is one of the earliest examples of the profound effect of chemistry on the practice of medicine. But the practical utilization of early observations of the effects of these compounds on the perception of pain did not come quickly. Joseph Priestley was the first chemist to investigate the chemical properties of nitrous oxide (in 1772, two years before his discovery of oxygen). Some 26 years later, another English chemist, Humphry Davy, began exploring the medical uses of gases and tested nitrous oxide on himself by inhaling large quantities in a very short time. Apparently, he quite enjoyed the experience, but his suggestion that nitrous oxide be used in surgery was not followed.

It was not until well into the nineteenth century that dentists in Boston began demonstrating the effectiveness of nitrous oxide. However, in the first public demonstration in 1844, a tooth was pulled before the gas had taken effect, and the dentist, Horace Wells, was booed and hissed from the amphitheater. Two of his followers, William Morton and Charles Jackson, took up the use of nitrous oxide but also began experimenting with ethyl ether. These three—Wells, Morton, and Jackson—competed for the honor of being recognized for the discovery of anesthesia and even took their dispute to Congress for resolution (which, in typical fashion, failed to act). Meanwhile, a physician in the state of Georgia had been using ethyl ether in his practice since 1842. (The American Medical Association and the American Dental Association acknowledge Wells as the discoverer of anesthesia.)

**Ethylmethyl ether**

The ethers produced in this way have more carbon atoms than either of the starting materials and thus are more complex structures. This reaction, called the **Williamson ether synthesis,** is a straightforward application of an $S_N2$ reaction for construction of a complex organic molecule from simpler starting materials.

Because an $S_N2$ pathway is required for the Williamson ether synthesis, this reaction is useful only when the alkyl halide is primary or secondary. To synthesize *t*-butylethyl ether by this route requires the use of *t*-butoxide ion as the nucleophile and ethyl bromide as the electrophile.

*t*-Butoxide ion    Ethyl bromide            *t*-Butylethyl ether

Ethoxide ion    *t*-Butyl bromide        No ether

397

The alternative combination of ethoxide ion with *t*-butyl bromide would be unsuccessful because of crowding in the transition state. This steric hindrance of the $S_N2$ reaction at the tertiary carbon raises the energy of the transition state sufficiently that the competing elimination reaction is substantially faster. Thus, ethoxide ion reacts as a base with *t*-butyl bromide and effects elimination of HBr. With tertiary halides, the Williamson ether synthesis fails completely; with secondary halides, it often leads to low yields of ethers because of competing elimination reactions.

**EXERCISE 8.6**

Specify preferred reactants for preparing each of the following ethers. List all possible choices.

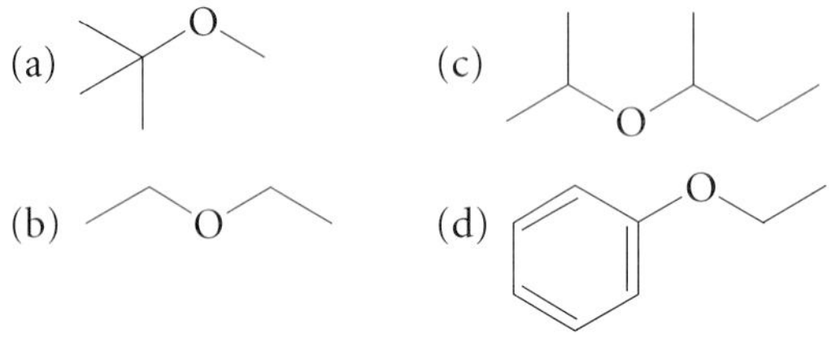

(a)

(c)

(b)

(d)

### Sulfonate Esters as Leaving Groups for Substitution Reactions

Sulfonate esters react very much like alkyl halides in both $S_N1$ and $S_N2$ reactions. The two most common sulfonate esters are methanesulfonate ester and *p*-toluenesulfonate ester, known respectively as **mesylate ester** and **tosylate ester** (these groups are commonly represented in structures as —OMs and —OTs, respectively). Either of these esters can be prepared readily from an alcohol by reaction with the appropriate sulfonyl chloride in the presence of a weak base such as pyridine or triethylamine:

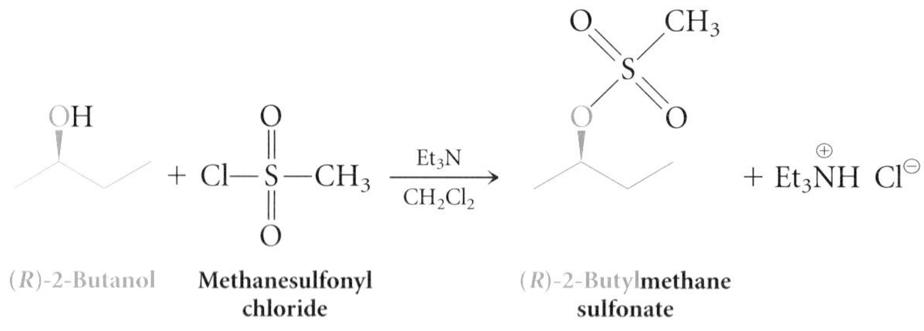

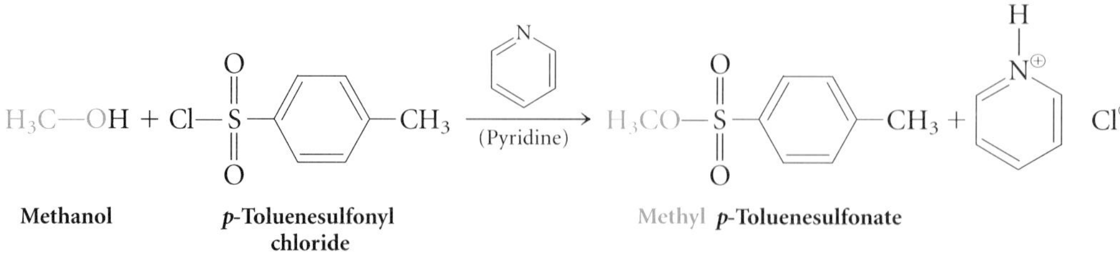

| (*R*)-2-Butanol | **Methanesulfonyl** **chloride** | | (*R*)-2-Butyl**methane** **sulfonate** |

| **Methanol** | **p-Toluenesulfonyl** **chloride** | | Methyl **p-Toluenesulfonate** |

Note that this reaction does *not* involve cleavage of the carbon–oxygen bond of the alcohol. Therefore, the absolute stereochemistry of (*R*)-2-butanol is retained in the product mesylate ester, (*R*)-2-butylmethane sulfonate.

Sulfonate esters react with hydroxide ion or alkoxide ions to form alcohols or ethers, respectively. At first, it might appear that there would be no point in converting an alcohol to a sulfonate ester and then reacting the ester with hydroxide ion to form an alcohol. Following this sequence with (R)-2-butanol gives the intermediate sulfonate ester having the same R configuration as the starting alcohol. However, reaction of the ester with hydroxide ion by an $S_N2$ reaction causes inversion of configuration, and the product is (S)-2-butanol. Thus, for a *chiral* alcohol, the overall sequence proceeding through a sulfonate ester produces *inversion* of configuration:

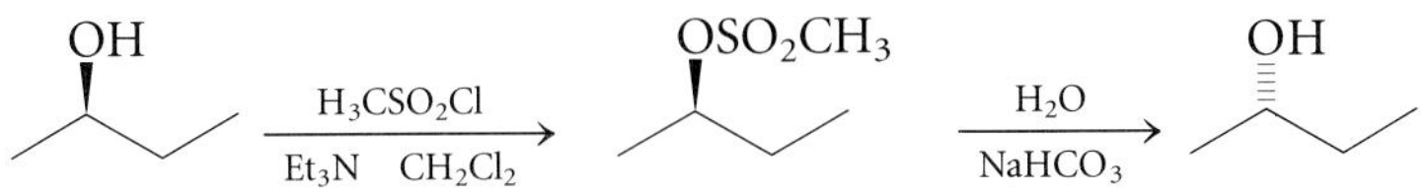

(R)-2-Butanol    (R)-2-Butylmethane
                      sulfonate          (S)-2-Butanol

Sulfonate esters can be used in the Williamson ether synthesis. For example, n-propyl ether can be prepared from n-propanol. Reaction of this alcohol with sodium produces the n-propoxide ion, and a separate reaction with tosyl chloride yields n-propyl tosylate. Combining these two reagents results in an $S_N2$ displacement, giving n-propyl ether.

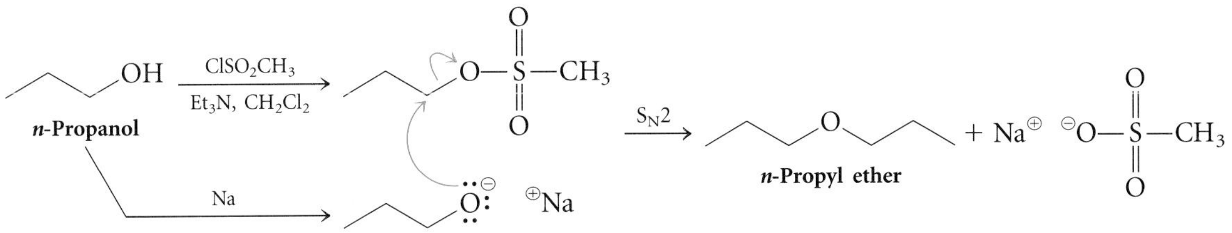

### EXERCISE 8.7

Reactions of sulfonate esters as electrophiles in nucleophilic substitution reactions produce the anions of sulfonic acids. One resonance structure of the anion is shown here. Draw all other possible resonance structures for this anion, and indicate which, if any, are equivalent in energy.

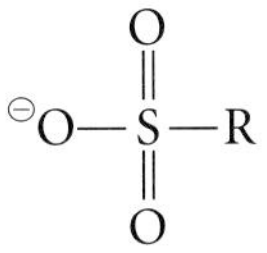

## Substitution of Alcohols to Form Alkyl Halides

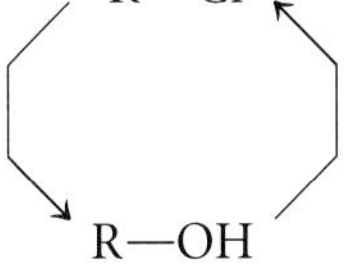

We will see in subsequent chapters that there are many methods for the preparation of alcohols, and most of them are generally superior to the substitution reaction of an alkyl halide. Indeed, alcohols are more often converted into alkyl halides than the other way around.

There are several methods for the conversion of alcohols to alkyl halides: via sulfonate esters, by reaction with thionyl chloride, or by treatment with concentrated halogen acids or phosphorus trihalides.

*Alkyl Halides from Alcohols via Sulfonate Esters.* As we saw in the preceding section, sulfonate esters are readily formed from primary and secondary alcohols by the action of the corresponding sulfonyl chloride in the presence of an amine. The sulfonate group can then be displaced by halide.

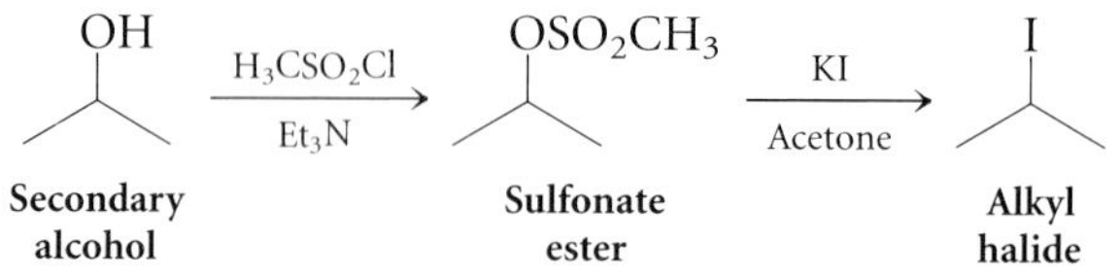

$$\underset{\substack{\text{Secondary} \\ \text{alcohol}}}{\text{OH}} \xrightarrow[\text{Et}_3\text{N}]{\text{H}_3\text{CSO}_2\text{Cl}} \underset{\substack{\text{Sulfonate} \\ \text{ester}}}{\text{OSO}_2\text{CH}_3} \xrightarrow[\text{Acetone}]{\text{KI}} \underset{\substack{\text{Alkyl} \\ \text{halide}}}{\text{I}}$$

The substitution of the sulfonate group by halide ion proceeds via an $S_N2$ mechanism; therefore the technique is applicable only to primary and secondary alcohols.

*Alkyl Chlorides from Alcohols by the Action of Thionyl Chloride.*
A simple method for the formation of alkyl chlorides that is effective for primary, secondary, and tertiary alcoholysis is the reaction of alcohols with thionyl chloride. A chlorosulfite ester formed as an unstable intermediate is converted into the alkyl chloride in a second step. (Note the similarity in structure between a chlorosulfite ester and a sulfonate ester—both are good leaving groups.)

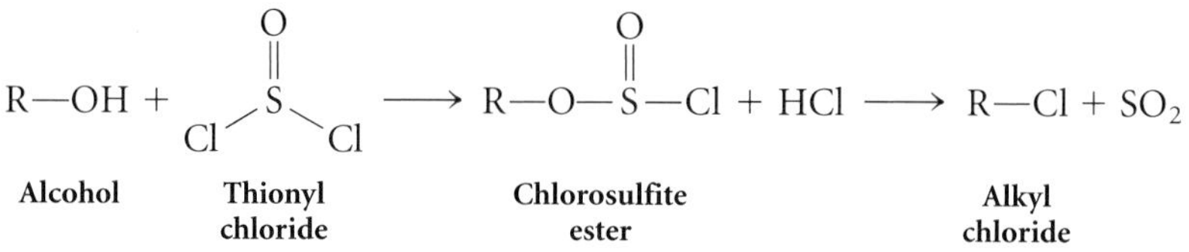

$$\underset{\text{Alcohol}}{\text{R—OH}} + \underset{\substack{\text{Thionyl} \\ \text{chloride}}}{\overset{\overset{\text{O}}{\|}}{\underset{\text{Cl}}{\text{S}}}} \longrightarrow \underset{\substack{\text{Chlorosulfite} \\ \text{ester}}}{\text{R—O—}\overset{\overset{\text{O}}{\|}}{\text{S}}\text{—Cl}} + \text{HCl} \longrightarrow \underset{\substack{\text{Alkyl} \\ \text{chloride}}}{\text{R—Cl}} + \text{SO}_2$$

The mechanism of the transformation of the chlorosulfite ester into the corresponding alkyl chloride depends on the degree of substitution. Primary alcohols undergo an $S_N2$ reaction, with chloride ion as the nucleophile and sulfur dioxide and chloride ion as leaving groups (Figure 8.5). Tertiary al-

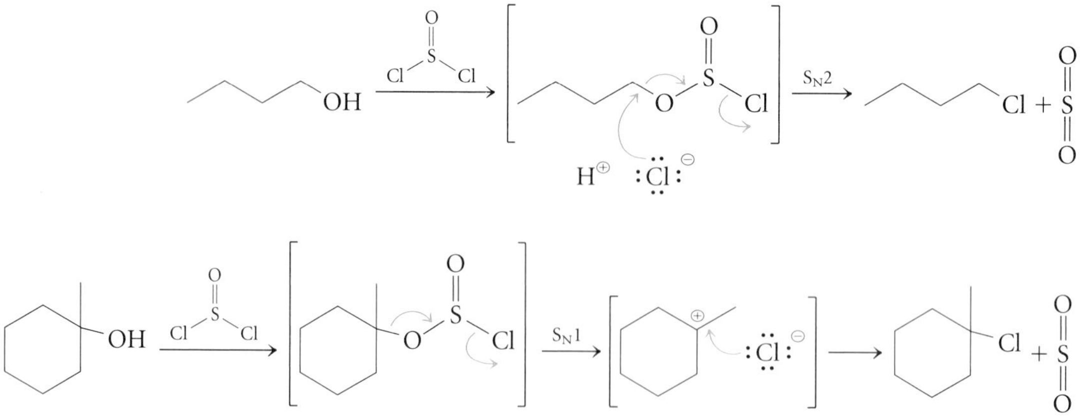

**FIGURE 8.5**

Reaction of alcohols with thionyl chloride produces chlorosulfite esters. These unstable intermediates are further transformed into alkyl chlorides. With primary alcohols, the substitution process occurs by an $S_N2$ mechanism. Tertiary alcohols follow an $S_N1$ pathway.

cohols follow an $S_N1$ pathway: first $SO_2$ and $Cl^{\ominus}$ are lost; then $Cl^{\ominus}$ reacts with the cation. (In some cases it has even been shown to be the same chloride ion that was lost.) Secondary alcohols follow both $S_N1$ and $S_N2$ pathways. An $S_N2$ pathway is favored when a base such as pyridine is added to the reaction mixture.

### EXERCISE 8.8

Write detailed, stepwise $S_N1$ and $S_N2$ mechanisms for the conversion of the chlorosulfite ester of 2-propanol to 2-chloropropane.

### EXERCISE 8.9

Explain why the rate of the $S_N2$ reaction of a chlorosulfite ester of a secondary alcohol increases when pyridine is added, whereas the rate of $S_N1$ substitution does not change.

***Alkyl Halides from Alcohols by Treatment with Concentrated Halogen Acids.*** Primary, secondary, and tertiary alcohols are also converted into alkyl chlorides by treatment with concentrated HCl. Hydrochloric acid serves two functions in these reactions: (1) it transfers a proton to the oxygen atom of the alcohol, generating an acceptable leaving group ($H_2O$); (2) it is a source of chloride ion, the nucleophile. As stated earlier, tertiary alcohols react by an $S_N1$ pathway; secondary alcohols, by both $S_N1$ and $S_N2$ pathways.

The reaction rate increases dramatically with the degree of substitution: tertiary alcohols are converted within seconds, and the reaction with primary alcohols is too slow to be of use. However, addition of the Lewis acid $ZnCl_2$ increases the rate and shortens the reaction time by changing the course of the reaction. Complexation of $^{\oplus}ZnCl$ with the oxygen of the hydroxyl group creates an even better leaving group than that formed by protonation of the oxygen. This complexation accelerates the reaction so that, with heating, primary alcohols can be converted slowly to the corresponding chlorides.

Alkyl bromides can be prepared from alcohols by treatment with HBr in a reaction that parallels the formation of alkyl chlorides using HCl:

***Alkyl Halides from Alcohols by Treatment with Phosphorus Trihalides.*** An alcohol can be converted to an alkyl bromide by treatment with phosphorus tribromide, $PBr_3$. (Phosphorus trichloride, $PCl_3$, can be used to make alkyl chlorides.) A simple $S_N2$ reaction of three equivalents of the alcohol with the reagent leads to the formation of a phosphite ester with three phosphorus–oxygen bonds. The high strength of the phosphorus–oxygen bond provides the driving force for this step. Bromide ion then effects $S_N2$ displacement, yielding the alkyl bromide. (Each of the three alkyl groups of the phosphite ester undergoes this reaction.)

**Phosphite ester**

### Substitution of Halogen to Form Thiols and Thioethers

***Thiols.*** Thiols (R—SH), also referred to as **mercaptans,** are the sulfur equivalents of alcohols and are important functional groups in biochemical systems. Thiols can be prepared from alkyl halides using reactions similar to those used to produce alcohols. Sulfur lies below oxygen in the periodic table, which means that sulfur is both less electronegative and more polarizable than oxygen. Therefore, functional groups containing sulfur are considerably more nucleophilic (and less basic) than the corresponding oxygen-based functional groups. Sodium hydrogen sulfide, NaSH, is an effective nucleophile and reacts readily with primary and secondary alkyl halides and sulfonate esters.

| **Alkyl halide** | **Benzyl mercaptan** | **Sulfonate ester** | **2-Propanethiol** |

***Thioethers.*** Thioethers can be synthesized by a pathway similar to that of the Williamson ether synthesis. A thiol is converted to the thiolate anion by reaction with sodium hydroxide. Thiols are significantly more acidic than alcohols or water, primarily because the S—H bond is substantially weaker than the O—H bond (82 versus 111 kcal/mole). The reaction of one equivalent of hydroxide ion with a thiol results in essentially complete conversion of the thiol into the thiolate anion (and of hydroxide ion to water).

$$H_3C-SH + {}^{\ominus}OH \rightleftarrows H_3C-S^{\ominus} + H_2O$$

**Thiol**
($pK_a$ 10)

**Thiolate anion**  ($pK_a$ 15.7)

## CHEMICAL PERSPECTIVES

### SULFUR–SULFUR BONDS

One of the reasons for the importance of thiols in biochemical systems is the facility with which they undergo oxidation to form disulfides. Even molecular oxygen is sufficient to convert thiols to disulfides. Conversely, even very mild reducing agents are able to reduce disulfides, forming two thiols:

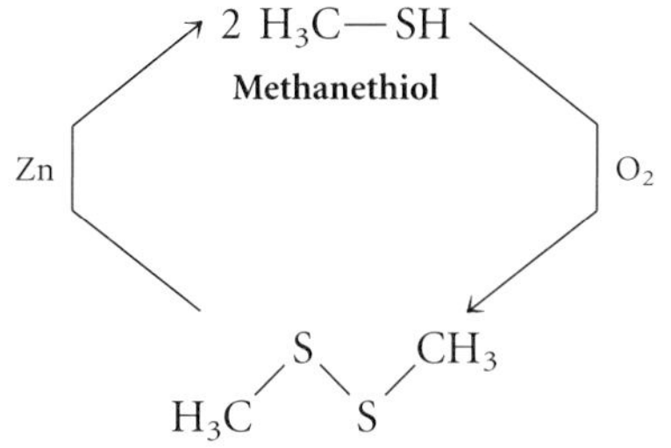

The formation of sulfur–sulfur bonds often imparts rigidity to the surrounding material. For example, a "permanent" wave is the result of treating hair with an oxidant that forms sulfur–sulfur bonds, linking one protein molecule to another. These cross-links help hold the hair in whatever shape it had during the oxidation process.

One theory of aging ascribes many of its physical changes to the formation of disulfide bonds, and many popular diets recommend taking antioxidants such as vitamins A, C, and E and coenzyme Q. Indeed, because of their antioxidant properties, vitamins C and E are used commercially to prolong the shelf life of foods. Whether the aging process can be slowed—or even reversed—by increasing antioxidants in the diet remains to be seen.

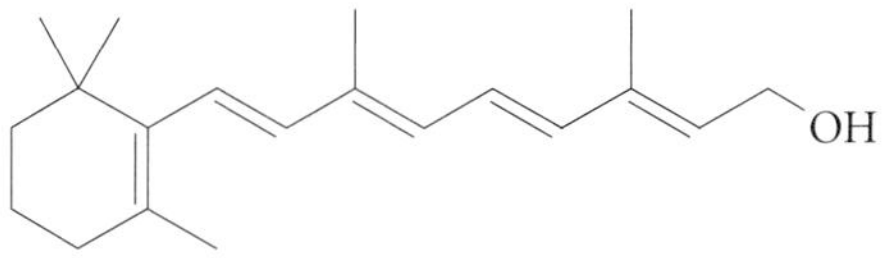

Vitamin A

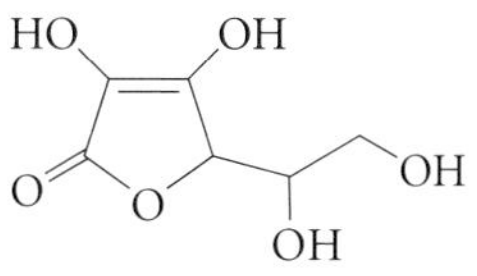

Vitamin C

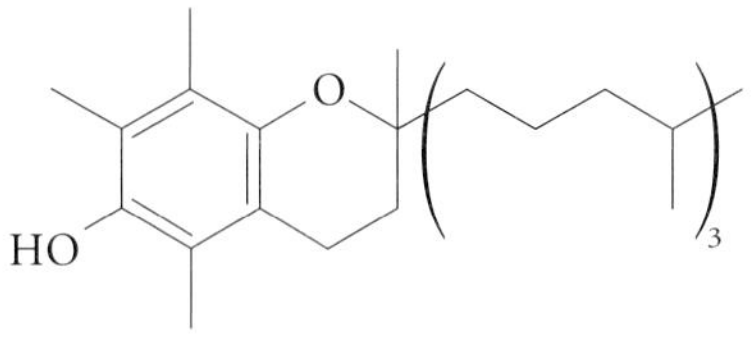

Vitamin E

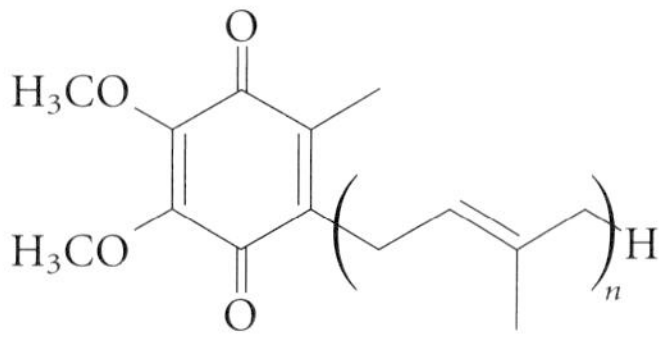

**Coenzyme Q** $(n = 6{-}10)$

Reaction of this nucleophile, the thiolate anion, with an alkyl halide or sulfonate ester yields a thioether (or sulfide).

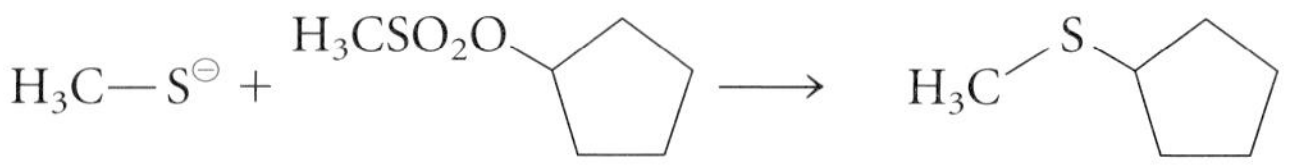

**Cyclopentyl methyl sulfide**
(a thioether)

***Controlling Substitution Reactions.*** Let's look again at the formation of thiols, using the reaction of NaSH with methyl bromide as a simple example.

$$H_3C{-}Br + Na^{\oplus}\ {}^{\ominus}SH \longrightarrow H_3C{-}SH + Na^{\oplus}\ Br^{\ominus}$$

As the reaction proceeds, the concentrations of the starting materials decrease and the concentrations of the products increase. Are there possible reactions of the product that might decrease the yield of methanethiol? Unfortunately, in this case and in many organic reactions, the answer is yes. For example, as you have just learned, thiols are readily converted into disulfides by mild oxidizing agents. This diversion of the desired product can be limited by carefully excluding oxygen (and other oxidizing agents). Another reaction of the product methanethiol is an acid–base reaction with NaSH to form methanethiolate anion. Like most acid–base reactions in which a proton is transferred, this reaction is very fast. The anion can then react with the starting material, methyl bromide, to form methyl thioether.

$$H_3C{-}SH + Na^{\oplus}\ {}^{\ominus}SH \rightleftharpoons H_3C{-}S^{\ominus}\ {}^{\oplus}Na + H_2S$$

$$H_3C{-}\overset{..}{\underset{..}{S}}{:}^{\ominus} + H_3C{-}Br \longrightarrow H_3C{-}S{-}CH_3 + Br^{\ominus}$$

This reaction illustrates a problem common to many organic reactions: a product is often capable of entering into reaction with one or more of the starting materials. A starting material, A, is converted to a product, B, which is further converted to another product, C.

$$A \xrightarrow{k_1} B \xrightarrow{k_2} C$$

The ability to obtain B in high yield depends on the relative rates $k_1$ and $k_2$. When $k_1 > k_2$, B can be obtained in good yield through control of the reaction time. However, if $k_1 < k_2$, then B is a transient intermediate in the conversion of A to C, and the yield of B is low.

*Effect of Reagent Concentration.* Chemists have some control over reactions of the type just described because the rate of an $S_N2$ reaction is concentration-dependent. For example, in the reaction of NaSH with methyl bromide, using an excess of NaSH results in an equilibrium in which the concentration of $HS^{\ominus}$ is much higher than that of $H_3CS^{\ominus}$. Therefore, the rate of reaction of $CH_3Br$ with $HS^{\ominus}$ will be higher than that with $H_3CS^{\ominus}$.

$$H_3C{-}SH + Na^{\oplus}\ {}^{\ominus}SH \rightleftharpoons H_3C{-}S^{\ominus}\ {}^{\oplus}Na + H_2S$$

$$\Big\downarrow Br{-}CH_3 \qquad\qquad \Big\downarrow Br{-}CH_3$$

$$H_3C{-}SH \qquad\qquad H_3C{-}S{-}CH_3$$

## EXERCISE 8.10

What effect does increasing the concentration of NaSH have on the equilibrium concentration of $H_3CSH$ and $H_3CS^{\ominus}$ ?

*Protecting Groups.* An alternative method for the formation of thiols is the reaction of the anion of thiolacetic acid with alkyl halides to form thiol esters.

Hydrolysis of the product thiol ester to the thiol is then effected with acid in a distinct and separate reaction. (We will deal with the details of this second reaction in Chapter 12.) Because the initial product (methyl thiolacetate) has no hydrogen atoms on sulfur, it is not possible to deprotonate it to form a sulfur anion. As a result, this product does not react with methyl bromide. Only after the second step is methanethiol formed, and at this point methyl bromide is no longer present. Using this sequence of two steps to form an alkanethiol allows the chemist to avoid entirely the complication of the formation of the thiol ether.

This reaction provides one example of the use of a protecting group in organic reactions. A **protecting group** is a functional group that masks the characteristic reactivity of another group, into which it can be converted. The formation of a thiol ester limits the alkylation reaction to the introduction of a single methyl group. Thus, the ester acts as a protecting group.

### EXERCISE 8.11

What reactant(s) and reagents are required to prepare each of the following sulfur compounds? All carbon–sulfur bonds must be formed in one or more of your steps. Otherwise, you may use any reagents and starting materials.

## Substitution of Halogen to Form Amines

Nitrogen lies to the left of oxygen in the periodic table, and therefore ammonia is more nucleophilic than water and alkylamines are more nucleophilic than alcohols. The difference between nitrogen and oxygen nucleophiles can be seen from the fact that ammonia and alkylamines react readily as neutral molecules with alkyl halides (and sulfonate esters) in an $S_N2$ displacement reaction, whereas the corresponding oxygen nucleophiles must first be converted into alkoxide anions, as in the Williamson ether synthesis. There is a further difference in the reaction of oxygen and nitrogen nucleophiles with alkyl halides. Reaction of alkyl halides with oxygen nucleophiles proceeds in two stages, from which the products are easily separated. For example, reaction of a primary or secondary alkyl halide with hydroxide ion yields the corresponding alcohol. The alcohol can then be converted to the alkoxide and treated with another equivalent of an alkyl halide to form an ether (Williamson ether synthesis). Reaction of alkyl halides with nitrogen nucleophiles is a more complex affair. Reaction of a primary or secondary alkyl halide with ammonia yields the corresponding primary amine, but this initial product is difficult to isolate because it also reacts with the alkyl halide.

***Stepwise Substitution on Nitrogen.*** Let's look in more detail at the alkylation of amines—that is, the nucleophilic displacement of alkyl halides by amines. The reaction of ammonia and methyl bromide illustrates this reaction. The initial product is an ammonium salt (methylammonium bromide), a species that does not have a lone pair of electrons on nitrogen and is thus not nucleophilic. However, proton exchange between this salt and ammonia rapidly yields an equilibrium mixture of ammonium bromide and the desired product, methylamine.

Methylamine has a lone pair of electrons and is similar in structure to ammonia. Indeed, methylamine is also a nucleophile and reacts with methyl bromide to form dimethylammonium bromide. As in the reaction forming methylammonium bromide, the dialkylammonium salt and ammonia establish an equilibrium with dimethylamine and ammonium bromide.

In proceeding from ammonia to methylamine to dimethylamine, these reactions replace first one and then a second hydrogen atom on nitrogen with a methyl group. These reactions can be repeated until all of the hydrogen atoms on nitrogen have been replaced by methyl groups, forming trimethylamine.

Dimethylamine reacts in the same way with methyl bromide, forming trimethylamine after equilibration with ammonia.

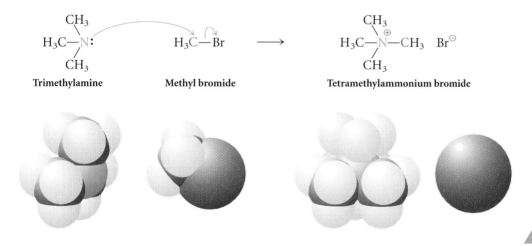

Trimethylamine also is susceptible to a substitution reaction, because it has a lone pair of electrons and is nucleophilic, like ammonia, methylamine, and dimethylamine. Reaction of trimethylamine with methyl bromide forms tetramethylammonium bromide.

$$
\underset{\text{Trimethylamine}}{\overset{\displaystyle CH_3}{\underset{\displaystyle CH_3}{H_3C-N{:}}}} \qquad \underset{\text{Methyl bromide}}{H_3C-Br} \qquad \longrightarrow \qquad \underset{\text{Tetramethylammonium bromide}}{\overset{\displaystyle CH_3}{\underset{\displaystyle CH_3}{H_3C-\overset{\oplus}{N}-CH_3 \;\; Br^{\ominus}}}}
$$

The ammonium salt does not have a lone pair of electrons and is therefore not nucleophilic. Note also that there are no protons on nitrogen in tetramethylammonium bromide, meaning that no nucleophilic amine can be produced by deprotonation. Thus, this sequence of substitution reactions that started with ammonia is finished.

The proton transfer steps in these reactions are quite fast and very much faster than the steps that result in the formation of carbon–nitrogen bonds. The overall sequence starting from ammonia and proceeding through a primary, then a secondary, and then a tertiary amine, ultimately forming a quaternary ammonium salt, can be summarized (and generalized).

## PRACTICAL USES FOR BAD-TASTING, NONTOXIC COMPOUNDS

Bitrex, a quaternary ammonium salt, has a very bitter taste but is nonmutagenic and nontoxic. Even when it is present in very low concentrations in ingested food, both people and animals find it repulsive. In the past, the U.S. Customs Agency required that foreign suppliers introduce Bitrex as a contaminant into imported sugar and alcohol to minimize smuggling and to assure that its removal was done only by legitimate processors who had paid a levied tax. A current practice among farmers is to apply a dilute solution of Bitrex to the backs of pigs to prevent them from biting each other and to add it to bait set out for deer who, having tasted the local wares, might be induced to move elsewhere for food. Bitrex is also sold in the United States as a solution to inhibit nail-biting and thumb-sucking by children.

**Bitrex**

NH3 $\xrightarrow[\text{NH}_3]{\text{R—X}}$ R—NH2 $\xrightarrow[\text{NH}_3]{\text{R—X}}$ R2NH $\xrightarrow[\text{NH}_3]{\text{R—X}}$ R3N—R $\xrightarrow[\text{NH}_3]{\text{R—X}}$ R4N$^{\oplus}$ X$^{\ominus}$

| Ammonia | Alkylamine | Dialkylamine | Trialkylamine | Tetraalkylammonium salt |

A $\xrightarrow{k_1}$ B $\xrightarrow{k_2}$ C $\xrightarrow{k_3}$ D $\xrightarrow{k_4}$ E

This general sequence of reactions represents the conversion of A through B, C, and D to the ultimate product, E. The overall conversion consists of four steps, each with its own rate of reaction: $k_1$, $k_2$, $k_3$, and $k_4$. (Again, we can ignore the proton transfer steps because they are very fast reactions.) This stepwise reaction is similar to one considered earlier for the conversion of A to B and then B to C. The rates of these alkylation reactions decrease progressively ($k_1 > k_2 > k_3 > k_4$) as the degree of substitution at nitrogen increases. However, the differences in these rates of reaction are not large, and it is difficult to control alkylation of nitrogen so as to obtain only (or mainly) the primary, secondary, or tertiary amine.

***Synthesis of Primary Amines.*** As we saw for the formation of thiols, the chemist must carefully control reaction conditions to obtain selective formation of one product to the exclusion of other possible products. To prepare primary amines ($RNH_2$), for example, a chemist could use an excess of ammonia in a fashion analogous to the use of excess NaSH for controlling the reaction of NaSH with methyl bromide. Because the concentration of ammonia will remain significantly higher than that of the product methylamine for the entire course of the reaction, the rate of alkylation of ammonia is much greater than the rate of alkylation of methylamine. This technique of increasing the concentration of one reagent is effective and efficient when that reagent is both inexpensive and readily separated from the desired product. In many organic reactions, the product can undergo a reaction similar to that by which it is formed. In most cases, these reactions are selective only if the rate of the first reaction is significantly faster than further transformation of the desired product.

*Gabriel Synthesis.* For the alkylation of nitrogen to be synthetically useful without the use of an excess of the nucleophile, the nitrogen atom of the product should not be nucleophilic. In the **Gabriel synthesis,** this is accomplished by the use of phthalimide instead of ammonia. The anion formed by deprotonation of phthalimide reacts as a nucleophile in an $S_N2$ displacement.

In phthalimide, the nitrogen atom is not nucleophilic because its lone pair is extensively delocalized into the extended $\pi$ bonding system that includes the two carbonyl groups and the benzene ring. However, the proton on the nitrogen of phthalimide is relatively acidic ($pK_a$ 9) when compared to simple amines such as ammonia ($pK_a$ 38), and phthalimide can be deprotonated with potassium hydroxide, generating the anion:

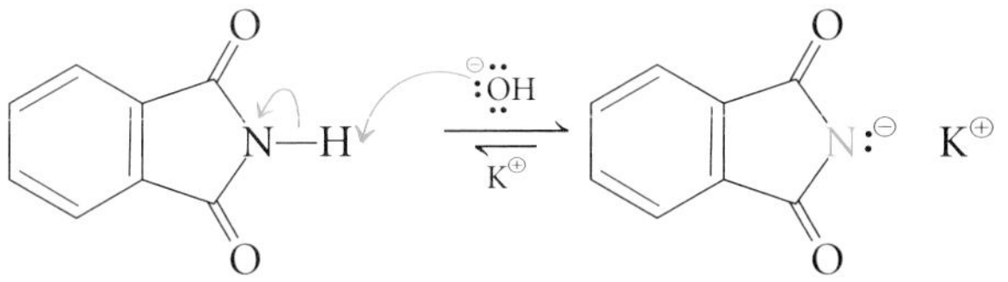

**Phthalimide**

In the anion, there are two lone pairs on nitrogen. One pair can be considered to be in a $p$ orbital and delocalized into the $\pi$ system, and the other to be localized on the nitrogen in an $sp^2$ orbital. The localized pair acts as a nucleophile in attacking an alkyl halide or tosylate (RX or ROTs), forming a new nitrogen–carbon bond. Note that the product of this alkylation of nitrogen has no protons on nitrogen. Therefore, an anion of the product cannot be formed, and the lone pair of electrons on the nitrogen in the alkylated phthalimide is delocalized into the $\pi$ systems of both carbonyl groups. As a result, the nitrogen atom in the alkylated product is essentially non-nucleophilic.

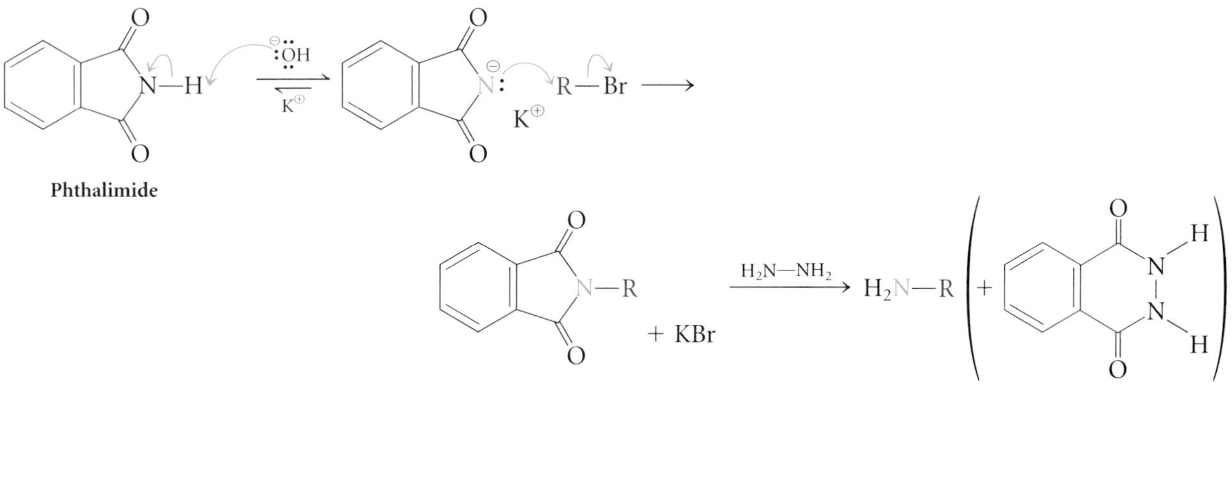

**Phthalimide**

## CHEMICAL PERSPECTIVES

### THE SMELL OF FISH

There are many naturally occurring amines, and chemists have so far been able to determine the function of only a few of them. Some amines serve a function that nature may have never intended. For example, when you buy fresh fish does the salesperson ever offer the fish to you to smell? It's a good test for freshness: the fresher the fish, the less "fishy" the smell. The bacteria that decompose the flesh of fish degrade the proteins to amino acids and then convert these amino acids into simple amines. Indeed, a "fishy" smell is nothing more than the odor of a mixture of amines, primarily methylamine.

Reaction of the *N*-alkylphthalimide with hydrazine, $H_2NNH_2$, a particularly good nitrogen nucleophile because of repulsion between the two lone pairs of electrons on its adjacent nitrogen atoms, releases the primary amine. The difficulty of controlling the alkylation of a simple amine is circumvented in the Gabriel synthesis by the construction of a system in which the product does not readily undergo the reaction by which it was formed. This synthesis of primary amines is analogous to the use of the anion of thiolacetic acid for the formation of alkanethiols, as discussed earlier, and the phthalic acid group can be considered to be a nitrogen-protecting group.

## EXERCISE 8.12

(a) Write resonance structures for the conjugate base of phthalimide. Do these structures explain phthalimide's high acidity (low p$K_a$)?

(b) Write resonance structures for phthalimide itself. Compare the structures in parts (a) and (b) and explain why it is necessary to deprotonate phthalimide to make it sufficiently active to enter into an $S_N2$ displacement. (In contrast, ammonia displaces bromide from methyl bromide without further activation.)

## EXERCISE 8.13

Show the sequence of reactions required to prepare each of the following amines in good yield without the use of a large excess of one component in any of the reactions.

(a) 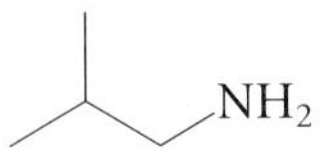


(c) 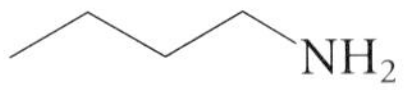


(b) 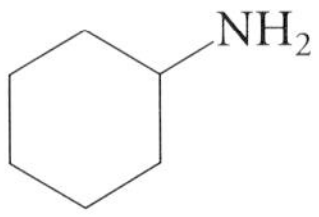


(d) 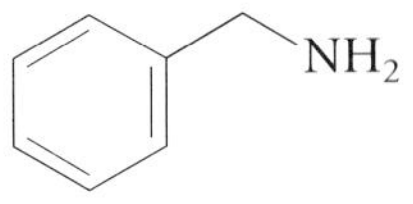

Phosphorus is in the same column and immediately below nitrogen in the periodic table and is thus both less electronegative and more polarizable than nitrogen. You should not be surprised that **phosphines,** $PR_3$, are also good nucleophiles. Indeed, phosphorus compounds generally have much higher reactivity as nucleophiles than their nitrogen counterparts. Most simple phosphines react vigorously with oxygen, and the parent phosphine, $PH_3$, as well as almost all simple mono-, di-, and trialkylphosphines, must be rigorously protected from air.

Aromatic substituents diminish the reactivity of phosphines, as exemplified by triphenylphosphine, a stable, crystalline material. It nonetheless retains sufficient nucleophilic character that it reacts readily with primary and secondary alkyl halides in an $S_N2$ displacement reaction. For example, reaction with methyl iodide produces methyltriphenylphosphonium iodide. (Each phenyl group in triphenylphosphine is represented by Ph. This convenient simplification is often used in structures when no chemical change takes place on the phenyl ring. When the phenyl ring bears one or more substituents, the abbreviation Ar is used, for "aryl.")

Triphenylphosphine

Methyltriphenylphosphonium iodide

Methylenetriphenyl phosphorane

The resulting **phosphonium salt,** $^{\oplus}PR_4$, can be deprotonated with a strong base such as *n*-butyllithium to form a zwitterion, methylenetriphenyl phosphorane.

*Phosphoranes.* The phosphonium salt resulting from alkylation of a phosphine is analogous to a tetraalkylammonium salt in that each lacks a lone pair of electrons on the heteroatom or a proton on the heteroatom that could be removed to yield such a lone pair. There are differences between these species, however. A phosphonium salt reacts with a strong base by loss of a proton at the $\alpha$ position of the alkyl group (the atom adjacent to phosphorus). The resulting zwitterion, a compound with both a positively charged and a negatively charged atom, is called a **phosphorane,** or a **phosphonium ylide.** Because the phosphorus atom in a phosphorane has unfilled $d$ orbitals, it is reasonable to write a resonance structure with a fifth bond to phosphorus, as shown on the right.

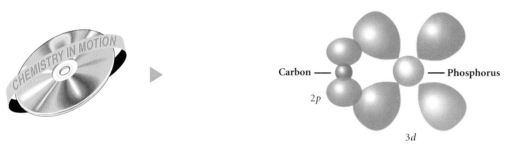

However, overlap of the relatively large $3d$ orbital on phosphorus with the smaller $2p$ orbital on carbon is not particularly effective because of the mismatch in orbital size. Thus, phosphoranes are better represented by a zwitterionic resonance structure. As expected, a phosphorane has significant negative charge on carbon and is a good nucleophile.

Phosphoranes are of interest because of the synthetic utility of their reactions. We will cover the nucleophilic reactions of phosphoranes in detail in Chapter 12.

## 8.4

## Preparation and Use of Carbon Nucleophiles

Section 8.2 focused on the $S_N2$ reaction of heteroatom nucleophiles with carbon electrophiles to form carbon–heteroatom bonds. The analogous reaction of carbon nucleophiles with alkyl halides, sulfonate esters, and epoxides (carbon electrophiles) affords a method for the formation of carbon–carbon bonds.

To make a new carbon–carbon bond by an $S_N2$ reaction requires a reactant with a carbon atom bearing a lone pair of electrons—that is, a carbon nucleophile. There are several important differences between the carbon nucleophiles we will be considering here and heteroatom nucleophiles. Carbon nucleophiles enter into relatively few $S_N2$ reactions with $sp^3$-hybridized carbons. This is, in part, because they are much stronger bases and bring about competing elimination reactions (to be discussed in Chapter 9). The more characteristic reactions of carbon nucleophiles are nucleophilic addition to $sp^2$-hybridized carbons (to be discussed in Chapter 13) and acid–base reactions (discussed later in this chapter). The anionic carbons considered in this chapter are sufficiently basic to be able to abstract not only relatively acidic protons, such as those attached to heteroatoms, but also the much less acidic protons attached to carbon. As a practical matter, most of the carbon nucleophiles considered here are highly reactive substances that are stable only in the absence of air and water. Their use, therefore, depends on the availability of more sophisticated laboratory techniques than are required for the manipulation of heteroatom nucleophiles.

We first consider the formation of $sp$-, $sp^2$-, and $sp^3$-hybridized carbon nucleophiles and then the use of these organometallic (or carbanionic) reagents as nucleophiles and bases.

## Carbon Nucleophiles

For a carbon atom to be nucleophilic, it must have significant negative charge. This is the case when carbon is bonded to an atom of significantly lower electronegativity, such as one of the alkali metals. The polarization of the bond between the metal and the carbon atom results in significant surplus of electron density on carbon, and with the most electropositive metals, the bond is ionic, such that the carbon bears a full negative charge and has a lone pair of electrons.

$$\overset{\delta-\quad\delta+}{H_3C\!-\!M} \qquad\qquad H_3C\!:^{\ominus}\quad M^{\oplus}$$

**Polar covalent bond**          **Ionic bond**

To the extent that this negative charge is available to enter into bond formation with an electrophile, the carbon is a nucleophile.

Progression through the alkali metals ($M^{\oplus} = Li^{\oplus}, Na^{\oplus}, K^{\oplus}, Cs^{\oplus}$) shows increasing differences in both size and electronegativity between carbon and the metal. The fraction of ionic character (that is, the negative charge on carbon) is thus larger in compounds containing sodium or potassium than in those containing lithium.

## General Methods for Preparation of Carbon Nucleophiles

Three ways are commonly used to produce nucleophilic carbanions. The first method consists of the removal of a proton from a carbon by a strong base in an acid–base reaction, as in the formation of an acetylide ion by the reaction of an alkyne with $NaNH_2$:

$$H_3C\!-\!C\!\equiv\!C\!-\!H \xrightarrow{\ NaNH_2\ } H_3C\!-\!C\!\equiv\!C\!:^{\ominus}\ ^{\oplus}Na$$

**Alkyne**                    **Acetylide ion**

A second method is the reaction of a zero-valent metal with a carbon–halogen bond to form a carbon–metal bond, as in the formation of $n$-butyllithium (and lithium chloride) by the reaction of lithium metal with $n$-butyl chloride. Such compounds have a very tight bond between the metal and the carbon atom and are known as **organometallic compounds.**

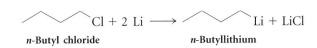

**n-Butyl chloride**      **n-Butyllithium**

Carbon nucleophiles can also be produced by *transmetallation,* in which one metal is exchanged for another, as in the formation of dialkylcuprates from alkyllithium reagents, as we will see shortly.

---

**EXERCISE 8.14**

---

Determine the oxidation level of the carbon atom bearing chlorine in *n*-butyl chloride and of the carbon atom bearing lithium in *n*-butyllithium. Does the conversion of an alkyl chloride to an alkyllithium represent an oxidation, a reduction, or no change in oxidation level?

---

## *sp*-Hybridized Carbon Nucleophiles: Cyanide and Acetylide Anions

***Preparation of Cyanide and Acetylide Anions.*** A hydrogen bonded to an *sp*-hybridized carbon atom is more acidic than a hydrogen bonded to an $sp^2$- or $sp^3$-hybridized carbon atom (Chapter 6). As a result, removal of a proton from an *sp*-hybridized carbon can be accomplished with relative ease. For example, $NaNH_2$ (prepared from sodium metal and ammonia) is sufficiently strong that treatment of a terminal alkyne ($R—C\equiv C—H$) with one equivalent of this base results in essentially complete conversion of the alkyne to the acetylide anion.

|  | $pK_a$ |
|---|---|
| $H_3C—CH_2—H$<br>*$sp^3$* | ~50 |
| $H_2C=CH—H$<br>*$sp^2$* | 44 |
| $HC\equiv C—H$<br>*$sp$* | 25 |
| $N\equiv C—H$<br>*$sp$* | 9 |

$$Na + NH_3 \longrightarrow$$

$$H_3C—C\equiv C—H \xrightarrow{Na^\oplus \,{}^\ominus NH_2} H_3C—C\equiv C:^\ominus \; Na^\oplus + H—NH_2$$
$$(pK_a \; 25) \qquad\qquad \textbf{Acetylide anion} \qquad (pK_a \; 38)$$

Hydrogen cyanide is similar in structure to an alkyne, having an *sp*-hybridized carbon atom bearing an acidic hydrogen. Because of the presence of the electronegative nitrogen atom, HCN is even more acidic ($pK_a$ 9) than a terminal alkyne and can be deprotonated with KOH. Hydrogen cyanide is extremely toxic. It is water-soluble and exists as a gas at room temperature. Therefore, chemists elect to use the stable salts, potassium or sodium cyanide, which, although toxic, are nonvolatile solids.

$$N\equiv C—H + KOH \longrightarrow N\equiv C:^\ominus \;{}^\oplus K + H—OH$$
$$(pK_a \; 9) \qquad\qquad\qquad\qquad (pK_a \; 15.7)$$

---

**EXERCISE 8.15**

---

Draw a Lewis dot structure for cyanide ion, $^\ominus CN$, where the negative charge is formally on carbon. There is an alternate resonance structure with the negative charge on nitrogen. Draw this resonance structure, first as a valence bond representation and then as a Lewis dot structure. Compare the two resonance structures, and explain why cyanide ion has greater negative charge on carbon than on nitrogen.

Demonstrate that both the carbon *and* the nitrogen atom in cyanide ion could be nucleophiles by showing the reaction with methyl bromide. (Use the first Lewis dot structure you drew for Exercise 8.15.) The product that results from nucleophilic attack by nitrogen is methyl isocyanide, $CH_3NC$, a compound that can be prepared by the reaction of methylamine with chloroform. What feature(s) of the structure of methyl isocyanide contribute to its being less stable than acetonitrile, $CH_3CN$, the product of $S_N2$ substitution by the carbon atom of cyanide ion?

*Reactions of Acetylide and Cyanide Anions.* Acetylide and cyanide anions are good nucleophiles, and each reacts with both alkyl halides and sulfonate esters, forming carbon–carbon bonds. For example, 2-pentyne can be prepared by the reaction of the mesylate of ethanol with the anion of propyne (derived by deprotonation with $NaNH_2$).

| Propyne anion | Mesylate of ethanol | 2-Pentyne |

Alternatively, the combination of cyanide ion with benzyl bromide results in the formation of phenylacetonitrile.

| Benzyl bromide | Cyanide ion | Phenylacetonitrile |

*Substitution versus Elimination with Cyanide and Acetylide Anions.* The reaction of secondary halides or sulfonates with cyanide ion and especially with acetylide ions results in significantly lower yields than the reaction with the corresponding primary electrophiles.

What goes wrong with these more hindered substrates? Because the additional substituent reduces the rate of substitution, another reaction becomes important: reaction of the anions as bases as well as nucleophiles, a complication we encountered earlier in this chapter in the reaction of hydroxide ion with alkyl halides. At this point, we do not need to concern ourselves with the details of this alternate reaction pathway—it is the subject of the next chapter. For now, it is important to realize that *all nucleophiles are also bases*. The balance between reactivity as a nucleophile and reactivity as a base is often very delicate, depending on many factors such as the nature of the nucleophile and electrophile, the solvent, the counterions, and even the temperature.

The reactions of acetylide ion and cyanide ion with electrophilic carbons are the first methods we have seen for the formation of C—C bonds, but we will encounter many more in our study of the organic reactions. These reactions are important to synthetic chemists, one of whose goals is to build more complex carbon skeletons from simpler precursors.

### EXERCISE 8.17

What starting material and reagents are appropriate for the formation of each of the following compounds? (You may use any starting materials and reagents as long as a carbon–carbon bond is formed in each case.)

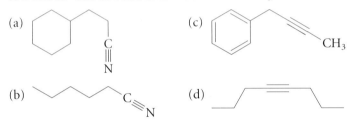

### EXERCISE 8.18

Which of the following alkyl halides would not be an appropriate substrate for an $S_N2$ reaction with cyanide ion? If the reaction will produce a nitrile, write the structure of that product. Explain what is wrong in any case in which the substrate will not react with cyanide ion in an $S_N2$ reaction.

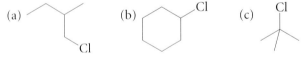

### $sp^2$- and $sp^3$-Hybridized Carbon Nucleophiles: Organometallic Compounds

***Formation of Organolithium and Organomagnesium Compounds.***
Organometallic compounds are formed by treatment of an alkyl halide or aryl halide with a zero-valent metal, usually lithium or magnesium. These reactions consist of reduction of the carbon atom by two electrons and simultaneous oxidation of the metal. For example, chlorobenzene reacts with

lithium metal to produce phenyllithium, and methyl bromide reacts with magnesium to produce methylmagnesium bromide.

In both of these reactions, there is a two-electron reduction of the carbon that bears the halogen in the starting material.

Lithium has only one valence electron, so that two atoms of lithium are required to effect a two-electron reduction of carbon. The bond between carbon and lithium in the resulting alkyllithium compound satisfies the valence requirement of both atoms. Magnesium has two valence electrons (is divalent). Only one magnesium atom is required to accomplish the two-electron reduction of the carbon atom, and the resulting organometallic species has both a carbon–magnesium bond and a magnesium–halogen bond. The metal atoms in both organomagnesium and organolithium compounds coordinate particularly well with the oxygen atoms of ethers. Therefore, ethers such as tetrahydrofuran and ethyl ether are commonly used as solvents for both the formation of organometallic reagents and their subsequent reactions.

Organomagnesium compounds are commonly called **Grignard reagents** in recognition of the French chemist François Auguste Victor Grignard, who was honored in 1912 with the Nobel prize for his discovery of this class of compounds. Note that the overall process results in insertion of magnesium between the carbon atom and the bromine atom to which the carbon was originally attached. Thus, the formation of an organometallic reagent in this way is often called an **insertion reaction.**

***Reactions of Organolithium and Organomagnesium Compounds with Epoxides.*** Organolithium and organomagnesium compounds act as nucleophiles in $S_N2$ reactions with epoxides. Organolithium and organomagnesium compounds are otherwise generally poor nucleophiles with respect to alkyl halides and alkyl sulfonate esters, typical $sp^3$-hybridized electrophilic carbon species. In part, this low nucleophilicity can be attributed to the tight bonding between the metal atom and the carbon atom. In addition, organolithium and organomagnesium compounds are very strong bases and promote elimination reactions.

When organolithium and organomagnesium compounds undergo reaction with epoxides, one of the carbon–oxygen bonds is cleaved, forming a new carbon–carbon bond and a metal alkoxide. For example, reaction of ethylene oxide with an organolithium or organomagnesium (Grignard) reagent forms a new carbon–carbon bond with opening of the three-membered ring. Protonation of the initially formed alkoxide by addition of water generates the product alcohol.

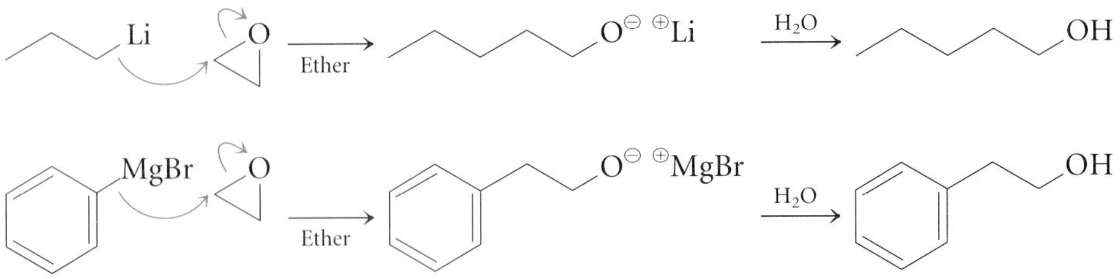


The C—O bonds of an epoxide are weaker than those of other ethers because of the strain inherent in a three-membered ring. These bonds are further weakened in the presence of organolithium and organomagnesium compounds by complexation between the metal acting as a Lewis acid and the oxygen atom of the epoxide acting as a Lewis base:

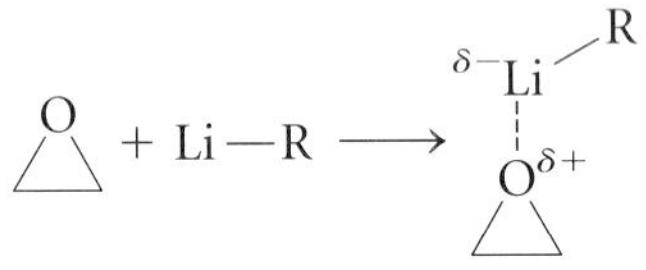


## EXERCISE 8.19

Draw the structure of the alkyl bromide that could be used to prepare each of the following alcohols by conversion into a Grignard reagent, followed by reaction with ethylene oxide.

(a) 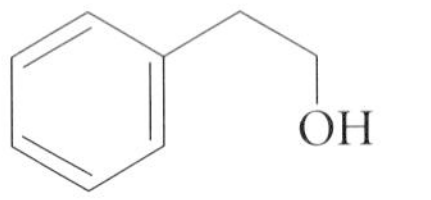


(c) 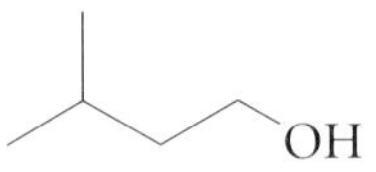


(b) 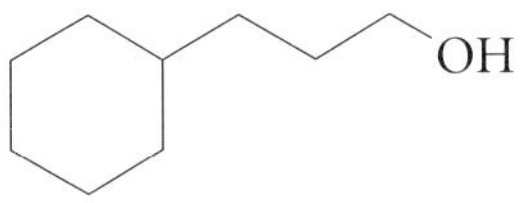


(d) 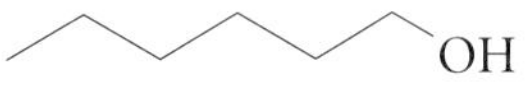


***Lithium Dialkylcuprates: Transmetallation.*** **Transmetallation** is the reaction of an organometallic compound with an inorganic salt, in which the carbon substituent is transferred from one metal to the other. An important example of transmetallation is the reaction of an alkyllithium with copper (I) iodide, CuI. (Note that, in CuI, copper is at the +1 oxidation level rather than the more stable +2 level.) When methyllithium is added to a suspension of CuI in an ether solvent, copper replaces lithium on the methyl group, producing methylcopper; at the same time, lithium iodide is formed.

$$\underset{\textbf{Copper (I) iodide}}{\mathrm{CuI}} \xrightarrow{\mathrm{H_3C{-}Li}} \underset{\substack{\textbf{Methylcopper}\\ \text{(bright yellow)}}}{(\mathrm{H_3C{-}Cu})_n} + \underset{\textbf{Lithium iodide}}{\mathrm{LiI}} \xrightarrow{\mathrm{H_3C{-}Li}} \underset{\substack{\textbf{Lithium dimethylcuprate}\\ \text{(nearly colorless)}}}{\mathrm{H_3C{-}\overset{\ominus}{Cu}{-}CH_3\ Li^{\oplus}}}$$

When a second equivalent of methyllithium is added to the suspension of methylcopper, lithium dimethylcuprate is formed, the simplest example of a lithium dialkylcuprate. Unlike the organolithium reagents from which they are formed, lithium dialkylcuprates are excellent nucleophiles in reactions with alkyl halides and alkyl sulfonate esters. Indeed, these reactions proceed with good yield even when both the alkyl group of the lithium dialkylcuprate and the alkyl halide are secondary.

Because the yields in these reactions range from good to excellent, this reaction represents a valuable tool with which to increase the number of carbon atoms and complexity of a compound.

The formation of a carbon–carbon bond through the use of a lithium dialkylcuprate is not restricted to reaction with alkyl halides. Aryl halides and vinyl halides also react with these reagents to form carbon–carbon bonds. Although these reactions are certainly not $S_N2$ reactions, researchers have not yet discovered exactly how they proceed, despite the fact that carbon–carbon bond formation via this type of reaction is a versatile and widely used synthetic method.

### EXERCISE 8.20

Several (and, in some cases many) different combinations of lithium dialkylcuprate reagents with alkyl halides can yield a particular hydrocarbon. Show all of the possible combinations of lithium dialkylcuprate reagents and alkyl bromides that can lead to 2-methylhexane.

**2-Methylhexane**

### Reaction of Organometallic Compounds as Bases

The most useful carbon–carbon bond-forming $S_N2$ reactions involving organometallic compounds are the reactions of organolithium or organomagnesium reagents with epoxides and the reactions of lithium dialkylcuprates with alkyl halides and alkyl sulfonate esters. The rest of the organometallic reagents considered so far have limited usefulness in $S_N2$

substitution reactions. All of the organometallic reagents we have considered are strong bases and react rapidly and irreversibly with water.

***Reduction of Alkyl Halides.*** Formation of an organometallic compound followed by protonation is a convenient method of reduction for alkyl halides. The overall process leading from an alkyl halide to a Grignard reagent and then, by protonation, to a hydrocarbon results in a net reduction of the carbon originally bearing the halogen substituent.

The base strength of most organometallic compounds requires that these reagents be protected from contact with water during their use as nucleophiles. Anhydrous ether solvents must generally be used, and neither the electrophile nor the organometallic reagent itself may contain acidic functional groups. The functional groups that readily act as acids toward organometallic reagents are carboxylic acids, alcohols, thiols, primary and secondary amines, and amides.

***Isotopic Labeling.*** The side reaction of an organometallic reagent with water often reduces the yield of the desired product. However, this reaction can be used to advantage by intentionally adding $D_2O$ to an organolithium or organomagnesium reagent, producing a compound with a single deuterium atom as a substituent on the carbon atom that originally held a halogen atom.

Alternatively, adding water that is enriched with tritium provides a hydrocarbon in which a specific carbon is partially substituted with tritium. (Tritium is very difficult to separate from deuterium and protium. Only a trace of tritium is generally used, but it can be easily detected because it is radioactive.)

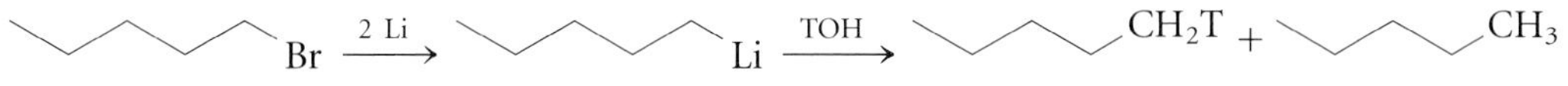

$$\text{(structure)}\;Br \xrightarrow{\text{2 Li}}\;\text{(structure)}\;Li \xrightarrow{\text{TOH}}\;\text{(structure)}\;CH_2T\;+\;\text{(structure)}\;CH_3$$

The pharmaceutical industry has done extensive research in labeling compounds with tritium. It is often desirable to determine the fate of a drug as it is degraded by the body. By feeding to test animals a compound labeled at a specific site with tritium and then analyzing the excretion products, pharmaceutical chemists can determine how a large molecule is converted biologically into smaller, soluble ones that can be excreted.

### EXERCISE 8.21

For each of the following compounds, state whether a stable Grignard reagent can be made by treating it with magnesium metal and anhydrous ethyl ether.

(a) 2-bromo-1-butanol

(b) 2-chloro-1-phenylpropane

(c) *p*-bromotoluene

(d) *p*-bromophenol

(e) 3-bromopropanoamide

### EXERCISE 8.22

If 10 mL of tetrahydrofuran is to be used as solvent, calculate the percentage of water (by volume) that it would have to contain to protonate 1 mmole of *n*-butyllithium. (Assume that *both* protons of water are effective and can be used to make butane.)

## 8.5

# Synthetic Methods:
# Functional-Group Conversion

As discussed in Chapter 7, it is useful to keep track of new reactions that alter functional groups. Table 8.2 (on pages 422 and 423) provides a compilation of the reactions presented in this chapter, which can be used to attach the carbon skeleton of an alkyl halide (or alkyl tosylate) to various other functional groups or to convert the halide into another functional group. In conceptually grouping reactions, you should take special note of those that form new carbon–carbon bonds.

As an exercise to complement the Review Problems at the end of the chapter, quiz yourself on each reaction listed in Table 8.2 and in the Review of Reactions (at the end of the chapter), according to the criteria set out in Section 7.3:

**1.** Can I predict the product (including stereochemistry and regiochemistry), given the reactants?

**2.** Do I know the reagent(s) and conditions necessary to convert the given functional group into the product?

**3.** For a desired product formed by a given reaction path, can I correctly choose an appropriate starting material?

**4.** Given the reactants, conditions, and products, can I write a detailed reaction mechanism showing the specific electron flow that describes how the reaction proceeds?

## TABLE 8.2

Using Substitution Reactions to Make Various Functional Groups

| Functional Group | Reaction | Example |
|---|---|---|
| Alcohols | Reaction of an alkyl halide with hydroxide ion or water | |
| | Opening of ethylene oxide by a Grignard reagent | |
| Alkanes | Coupling of an organocuprate with an alkyl halide | |
| | Protonation of Grignard (or organolithium) reagents | |
| Alkynes | Acetylide ion alkylation | |
| Alkyl bromides | Reaction of an alcohol with HBr | |
| | Reaction of an alcohol with PBr$_3$ | $H_3C—OH \xrightarrow{PBr_3} H_3C—Br$ |
| Alkyl chlorides | Reaction of an alcohol with HCl (or ZnCl$_2$, if primary) | |
| | Reaction of an alcohol with thionyl chloride | |
| Amines | Alkylation of ammonia (or an amine) | $H_3C—Br \xrightarrow[\text{(excess)}]{NH_3} H_3C—NH_2$ |

422

| Functional Group | Reaction | Example |
|---|---|---|
| Amines | Gabriel synthesis | |
| Ethers | Williamson ether synthesis | |
| Grignard reagents | Insertion of magnesium into carbon–halogen bond | $H_3C-Br \xrightarrow[Et_2O]{Mg} H_3C-Mg-Br$ |
| Nitriles | Reaction of an alkyl halide with cyanide ion | |
| Organocuprates | Transmetallation of an organolithium | $CuI \xrightarrow{H_3C-Li} H_3C-\overset{\ominus}{Cu}-CH_3\ Li^{\oplus}$ |
| Organolithiums | Insertion of lithium into carbon–halogen bond | |
| Phosphonium salts | Phosphine alkylation | |
| Phosphoranes | Deprotonation of a phosphonium salt | |
| Thioethers | Reaction of a sulfonate ester (or alkyl halide) with thiolate anion | |
| Thiols | Reaction of an alkyl halide (or sulfonate ester) with NaSH | |

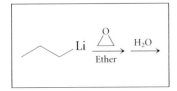

**Front**

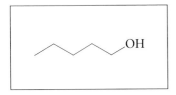

**Back**

**FIGURE 8.6**

Example of a reaction
flash card.

For synthetic utility, the third question is of utmost importance. Although the emphasis in this book has been on reaction mechanisms, practical organic chemistry requires the use of sequences of known reactions to make new molecules. This can be done only if you know reactions literally backward and forward. You can be sure of knowing these reactions well if you drill yourself on *both* the *mechanism* and the *functional-group interconversion* accomplished in each transformation. One effective way to study *what* reactions accomplish is by the use of reaction flash cards, with the reactant and reagents shown on the front of a card and the product on the back (Figure 8.6). You should make a card for each new reaction you encounter as you study organic chemistry. As you use the cards to test your knowledge of the reactions, do *not* separate those with which you have difficulty from those you know. Keep them together, and shuffle the deck and start over until you know them all. (A complete set of cards for the reactions in this textbook is available from the publisher, but constructing your own set can be a valuable learning experience.)

Only by understanding how various products can be obtained by several alternative pathways is it possible to plan syntheses intelligently. Having a variety of functional-group interconversions at hand makes it easier to integrate new reactions with those you already know.

## 8.6

### Spectroscopy

The spectral techniques covered in Chapter 4 are often very useful for determining whether an intended transformation has actually occurred as planned. Infrared spectroscopy can be used to check for the presence or absence of hydroxyl, amino, and carbonyl groups. The change in electron density at the carbon of a functional-group site results in different shifts for both the carbon (in the $^{13}C$ NMR spectrum) and any attached hydrogen atoms (in the $^1H$ NMR spectrum).

The reaction of alkyl halides with hydroxide ion (or with water in the case of tertiary, benzylic, and allylic substitution) to form alcohols can be followed by infrared spectroscopy. The product alcohols have distinctive and strong absorptions resulting from O—H stretching at 3650–3400 $cm^{-1}$ and C—O stretching at 1150–1050 $cm^{-1}$. Both of these absorptions are absent in the starting alkyl halides (C—Cl, C—Br, and C—I stretching vibrations are less useful as diagnostic tools for this class of compounds, because they often appear in the fingerprint region of the spectrum). Ethers lack the O—H stretching absorptions seen for alcohols, providing a simple way to determine that the reaction of the anion of an alcohol with an alkyl halide has resulted in the formation of an ether.

NMR spectroscopy also shows differences between alkyl halides and the corresponding alcohols. The greater electronegativity of oxygen compared with chlorine, bromine, or iodine results in a downfield shift of the carbon bearing oxygen relative to the carbon bearing halogen, as well as a downfield shift of hydrogen atoms, if any are present on this carbon. Representative examples of NMR shifts are provided in Table 8.3.

The replacement of a halogen substituent by nitrogen to form an amine is apparent in the infrared spectra for primary and secondary amines be-

## TABLE 8.3

$^{13}C$ and $^{1}H$ NMR Chemical Shifts for Alkyl Halides and Alcohols

| Alkyl Halide | NMR | | Alcohol | NMR | |
| | $^{13}C$ ($\delta$) | $^{1}H$ ($\delta$) | | $^{13}C$ ($\delta$) | $^{1}H$ ($\delta$) |
|---|---|---|---|---|---|
| $CH_3CH_2$—I | 20.3 | 1.85 | $CH_3CH_2$—OH | 58.0 | 3.66 |
| (Br, branched structure) | 67.3 | — | (OH, branched structure) | 69.1 | — |
| (H, Cl cyclohexyl) | 53.8 | 4.20 | (H, OH cyclohexyl) | 70.2 | 3.60 |

cause of the N—H stretching absorptions at 3500–3300 cm$^{-1}$. Nitriles have quite characteristic but weak stretching absorptions at 2260–2210 cm$^{-1}$. Alkynes also have unique absorptions at 2260–2100 cm$^{-1}$. However, the absorption resulting in a stretching interaction requires a dipole in the functional group, and all but terminal alkynes are sufficiently symmetrical that there is little, if any, dipole and therefore only very weak absorption in the infrared. The C—H bond of terminal alkynes shows a stretching absorption at 3350–3300 cm$^{-1}$.

NMR spectroscopy serves as an excellent tool for confirming that the reaction of a cuprate with an alkyl halide has indeed resulted in a larger carbon framework. The product of such a carbon–carbon bond-forming reaction almost always has a greater number of unique $^{13}C$ NMR absorption signals than does the starting alkyl halide.

$^{13}C$ NMR:     52.9, 37.6, 25.9, 25.2 $\delta$     35.8, 33.1, 26.7, 26.6, 22.9 $\delta$

There is also a change in the chemical shift of the carbon bearing a halogen when the halogen is replaced by an alkyl group. However, in some cases it is possible for the alkyl group to have nearly the same shift effect as the original halogen.

### EXERCISE 8.23

Which spectral techniques would be useful for determining how much of an alkene by-product was produced in the following substitution reaction?

# Summary

**1.** Nucleophilic substitutions are of two major types: (a) $S_N1$ reactions, in which only cleavage of the bond between carbon and a leaving group occurs in the rate-determining step; (b) $S_N2$ reactions, in which partial formation of a bond with the incoming nucleophile takes place simultaneously with cleavage of the bond to the leaving group. The $S_N1$ reaction takes place in a sequence of steps, with formation of a carbocation intermediate. The $S_N2$ reaction takes place in a single concerted step, with no intermediates.

**2.** There is inversion of stereochemistry in an $S_N2$ reaction because of the required back-side attack by the incoming nucleophile. In an $S_N1$ reaction, a planar, achiral, cationic intermediate is formed.

**3.** Because $S_N1$ reactions proceed through intermediate carbocations, they are complicated by competing elimination and rearrangement reactions. $S_N1$ reactions are usually observed at tertiary sites, and therefore are generally less useful for syntheses than $S_N2$ reactions, except with relatively simple substrates.

**4.** The $S_N2$ reaction proceeds through a pentacoordinate transition state and is therefore sensitive to steric effects, which dictate a reactivity order (primary > secondary > tertiary) opposite to that observed for an $S_N1$ reaction (tertiary > secondary >> primary).

**5.** The ease of cleavage of the bond between carbon and the leaving group in $S_N1$ and $S_N2$ reactions are influenced by the C—LG bond strength and the ability of the leaving group to accommodate negative charge.

**6.** A more reactive nucleophile more readily donates an electron pair to the electron-deficient carbon. Anionic nucleophiles (formed by deprotonation) are more reactive than their neutral precursors. A good nucleophile (which acts as an electron donor to an electron-deficient carbon) is therefore often a good base (an electron donor to an electron-deficient hydrogen).

**7.** Nucleophilic substitutions are employed as the critical step in several functional-group transformations, as well as in carbon–carbon bond formations. In such a reaction, an electron-rich reagent (often an anion) reacts with an organic molecule bearing a leaving group.

**8.** Carbon nucleophiles are generated in one of three ways: (a) by treatment of a compound bearing an acidic hydrogen atom with a base (specific examples covered in this chapter are the formation of cyanide and acetylide anions by deprotonation of the corresponding conjugate acids); (b) by treatment of an alkyl halide with a zero-valent metal such as lithium or magnesium to produce an organolithium compound or a Grignard reagent; (c) by transmetallation, where one metal is exchanged for another, as in the formation of a lithium dialkylcuprate from an alkyllithium reagent.

**9.** The carbon–metal bond in an organometallic compound is polarized, because the metal is electropositive and the carbon is electronegative. As a result, the carbon of an organometallic $\sigma$ bond has partial negative charge and reacts as a base with protons and as a nucleophile with carbon electrophiles.

**10.** The negatively charged carbon in organometallic compounds is associated with a counterion. The strength of the interaction between the positive and negative centers determines the type of bond between the metal and carbon, with free ions and pure covalent bonds presenting the extremes. Organolithium, organocopper, and Grignard reagents possess some covalent character in the carbon–metal bond.

## Review of Reactions

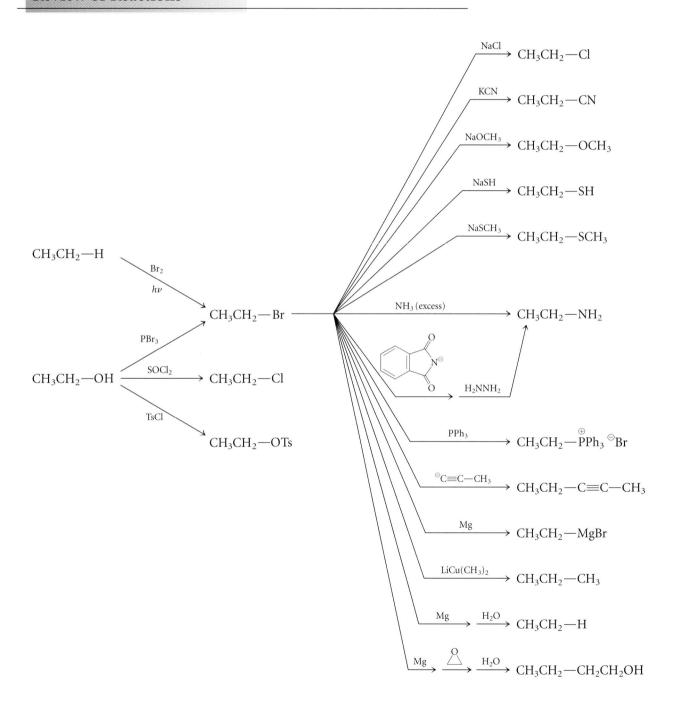

## Review Problems

**8.1** Assuming that the following reactions take place by an $S_N2$ displacement, choose the faster reaction of each pair and explain your reasoning:

(a) reaction of cyanide ion with *n*-iodoheptane or *n*-chloroheptane

(b) reaction of ethanol or sodium ethoxide with *n*-butyl bromide

(c) reaction of azide ion with *n*-butyl tosylate or *s*-butyl tosylate

(d) reaction of isopropoxide with ethyl bromide or reaction of ethoxide with 2-bromopropane

**8.2** For each of the following reactions, predict the expected product. Where a center of chirality is created or destroyed, indicate the expected absolute configuration.

(a) 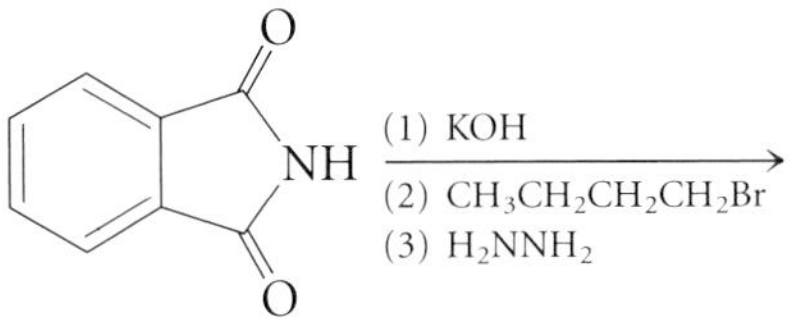
O
NH
O
(1) KOH
(2) CH3CH2CH2CH2Br
(3) H2NNH2

(b) 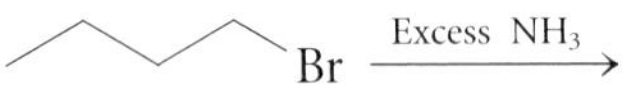
Br
Excess NH3

(c) 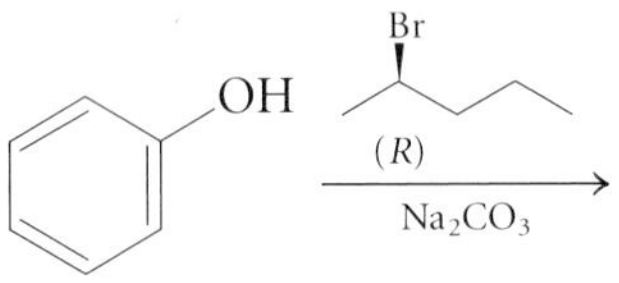
OH
Br
(R)
Na2CO3

(d) 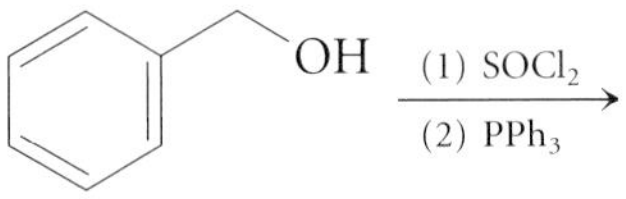
OH
(1) SOCl2
(2) PPh3

(e) 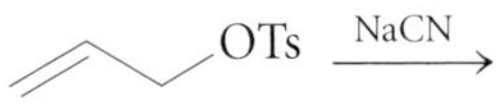
OTs
NaCN

(f) 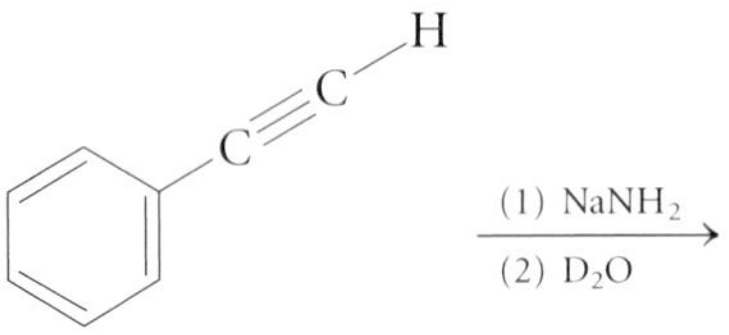
H
C
C
(1) NaNH2
(2) D2O

(g) 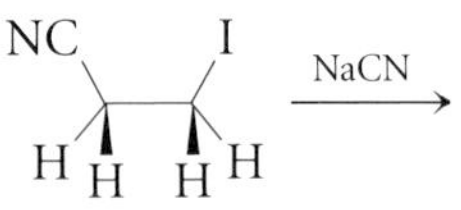
NC
I
H H H H
NaCN

(h) 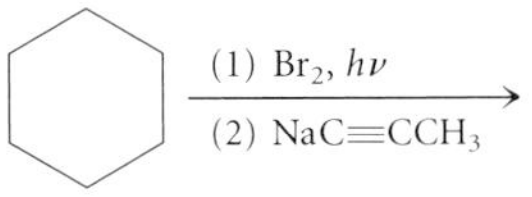
(1) Br2, hν
(2) NaC≡CCH3

(i) 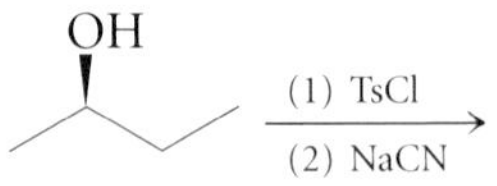
OH
(1) TsCl
(2) NaCN

(j) 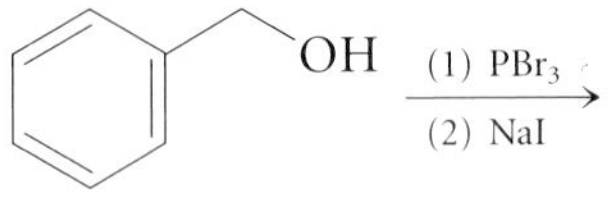
OH
(1) PBr3
(2) NaI

(k) 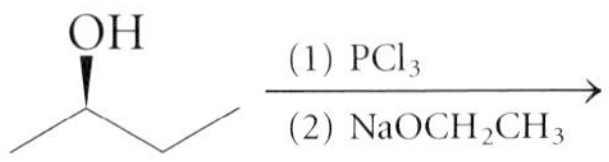
OH
(1) PCl3
(2) NaOCH2CH3

(l) 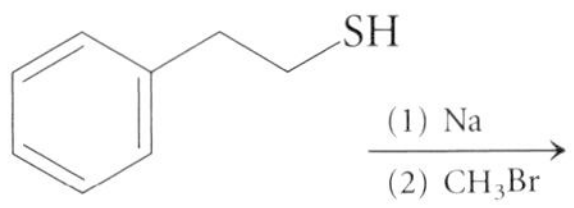
SH
(1) Na
(2) CH3Br

(m) 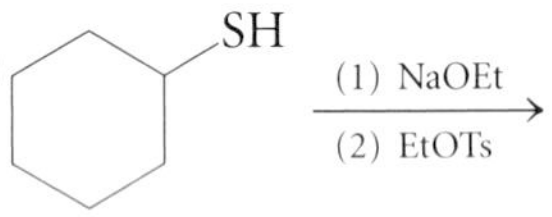
SH
(1) NaOEt
(2) EtOTs

(n) 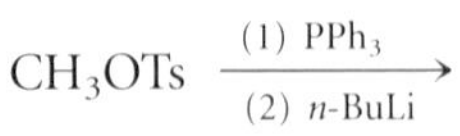
CH3OTs
(1) PPh3
(2) n-BuLi

**8.3** Using the reactions you learned in this and preceding chapters, suggest a route by which the carbon skeletons of the following compounds can be synthesized from a two- or three-carbon alkyne and any alkyl halide containing three carbons or fewer. (You may use any other reagents needed; more than one step may be required.)

(a) 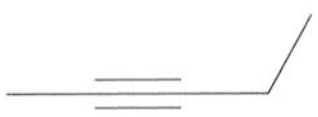

(b) 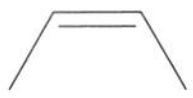

(c) 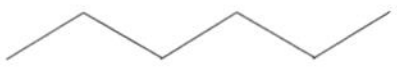

(d) 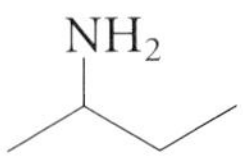
NH2

**8.4** Propose a sequence of reactions and the appropriate reagents that can be used to effect the following conversions. Specify any special conditions or solvents that are needed.

(a) benzyl chloride to benzyllithium

(b) toluene to benzylmagnesium bromide

(c) benzyl alcohol to lithium dibenzylcuprate

**8.5** Identify the organic halide that, after conversion to a Grignard reagent and treatment with ethylene oxide, would produce each of the following alcohols:

(a)

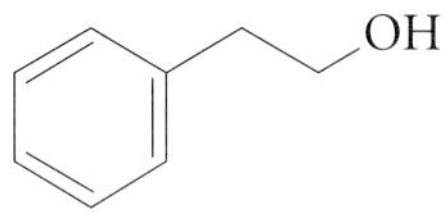

(b)

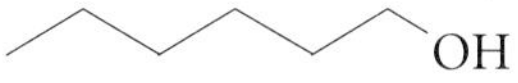

(c)

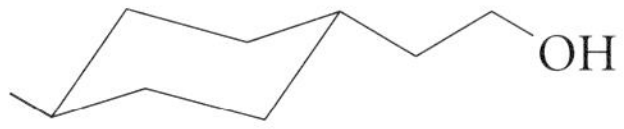

**8.6** $S_N1$ reactions proceed through planar carbocations, and only under special circumstances does such a reaction give a nonracemic product. For example, although the reaction rate depends only on the concentration of the reactant, the isolated product obtained when $(S)$-2-bromo-$n$-propylmethyl ether is hydrolyzed is $(S)$-2-hydroxy-$n$-propylmethyl ether. In addition, the solvolysis of this reactant is much faster than that of 2-bromopropane.

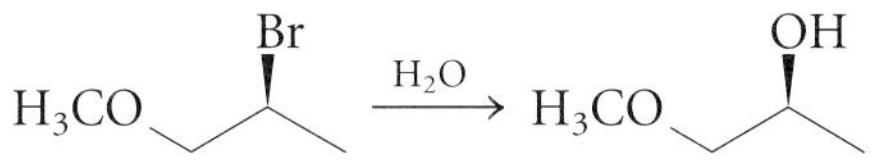

(a) Suggest an explanation for these observations.

(b) Predict the expected stereochemical course if sodium cyanide in acetone were used as the nucleophile instead of water.

**8.7** What starting materials and reagents are needed to prepare each of the following compounds by a nucleophilic substitution reaction? Do not concern yourself with making the starting materials.

(a)

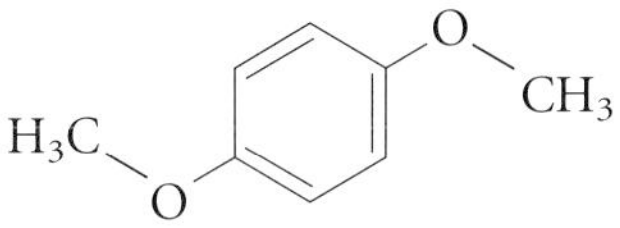

(b)

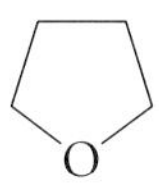

(c)

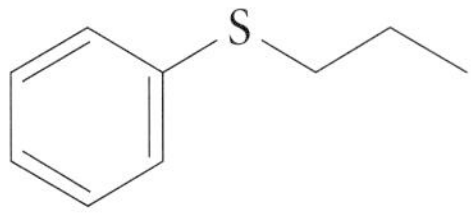

(d)

(e) $H_3CCH_2C\equiv N$

(f)

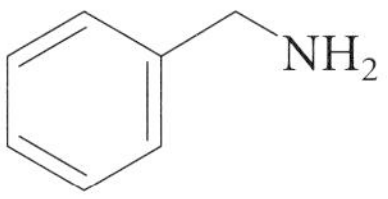

(g)

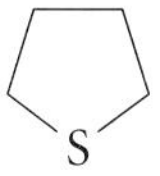

(h) $H_3CCH_2—NH_2$

**8.8** It is often possible to prepare symmetrical ethers by the reaction of a primary alkyl bromide with a limited amount of base in a small amount of water.

$$R—CH_2Br \xrightarrow{NaOH} R—CH_2—O—CH_2—R$$

(a) What starting material would be required to prepare tetrahydrofuran (THF) by this method?

THF

(b) Show the sequence of reaction steps with complete mechanisms for the conversion of the starting material chosen in part (a) into THF.

**8.9** Show how each of the following primary alcohols could be prepared by the reaction of an organometallic reagent with ethylene oxide.

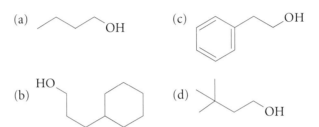

(a) OH

(c) OH

HO

(b)

(d) OH

**8.10** How can the following alcohol be prepared from bromocyclohexane and ethylene oxide as the only sources of carbon atoms? (*Hint:* Not counting proton-transfer reactions, five discrete chemical transformations are required.)

OH

**8.11** Show how each of the following primary amines can be prepared by the Gabriel synthesis.

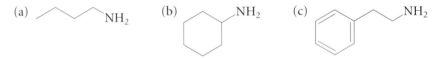

(a) NH$_2$    (b) NH$_2$    (c) NH$_2$

**8.12** Three of the following six amines can be prepared directly by the Gabriel synthesis. Identify the amines for which the Gabriel synthesis is not applicable, and briefly state why they cannot be prepared by this method.

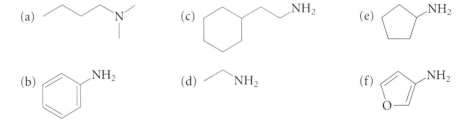

(a) N    (c) NH$_2$    (e) NH$_2$

(b) NH$_2$    (d) NH$_2$    (f) NH$_2$

**8.13** Provide an IUPAC name for each of the following compounds.

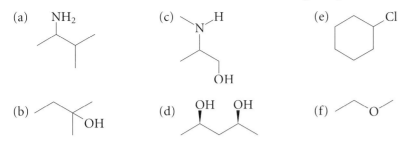

**8.14** Draw the structure that corresponds to each of the following names.

(a)  *N*-methylaniline

(b)  cyclohexylmethylamine

(c)  *N,N*-dimethylbenzylamine

(d)  *trans*-4-methylcyclohexanol

(e)  (*R*)-2-hexanol

(f)  cyclohexylmethyl ether

(g)  *trans*-2-aminocyclopentanol

**8.15** Tertiary nitrogen atoms at the bridgehead position of bridged bicyclic amines are often more reactive as nucleophiles (for example, in alkylations to form quaternary ammonium ions) than are noncyclic tertiary amines. Suggest a reason for this enhanced reactivity.

**A bridged bicyclic amine**

$\xrightarrow[\text{Faster}]{H_3C-I}$

$H_3C$      $I^{\ominus}$

$Et_3N$  $\xrightarrow[\text{Slower}]{H_3C-I}$  $Et_3\overset{\oplus}{N}-CH_3$   $I^{\ominus}$

**A noncyclic
tertiary amine**

**8.16** The nitrogen atom of aniline is significantly less reactive as a nucleophile than is the nitrogen atom of an aliphatic primary amine. What is the rationale for this observation?

$\xrightarrow[\text{Slower}]{H_3C-I}$

**Aniline**

$\xrightarrow[\text{Faster}]{H_3C-I}$

**8.17** Compound A is known by the trivial name "proton sponge," because it is significantly more basic than similar compounds with only one amino group—for example, compound B. Can you think of a reason for this difference in affinity for a proton? Would you expect the rate of alkylation of A to be faster or slower than that of similar monoamines? (Be careful here to consider the difference between thermodynamic and kinetic control on the outcome of reactions.)

**8.18** Tetrahydrofuran is generally a better solvent than ethyl ether for organometallic reagents (although there are some exceptions). What is the rationale for this observation?

Tetrahydrofuran          Ethyl ether

**8.19** What reagents are required to accomplish each of the following transformations, using the reactions described in this chapter?

**8.20** What reagents are required to accomplish each of the following transformations, using the reactions described in this chapter?

(b)

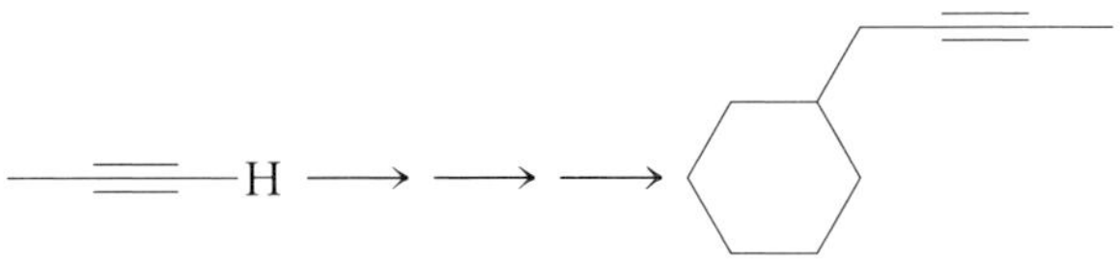

(c)

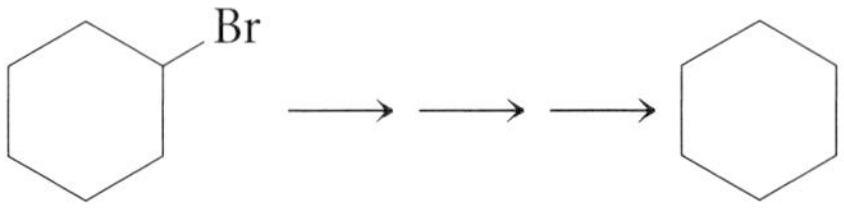

**8.21** In the reaction of an alkyl halide with hydroxide ion in water, the yield of alcohol increases as the concentration of the alkyl halide is decreased. Why should this happen? (*Hint:* Are there any reactions that the product alcohol might undergo under the reaction conditions?)

$$R\!-\!Br \xrightarrow[\text{H}_2\text{O}]{\ominus\text{OH}} R\!-\!OH$$

**8.22** Suggest a synthesis of the herbicide 2,4-dichlorophenoxyacetic acid (known as 2,4-D) starting from 2,4-dichlorophenol and any other starting material or reagents that you need.

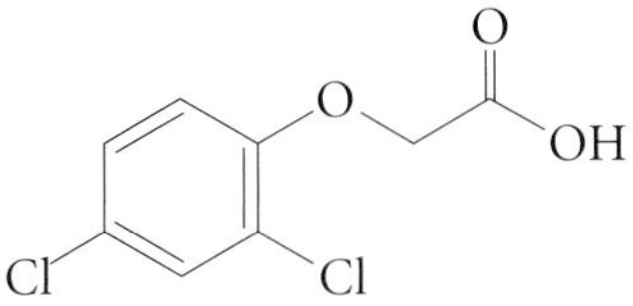

**2,4-D**

**8.23** Describe in a general way the differences that would be expected in the infrared, $^1$H NMR, and $^{13}$C NMR spectra of the starting materials and products of the following reactions.

(a)

(b)

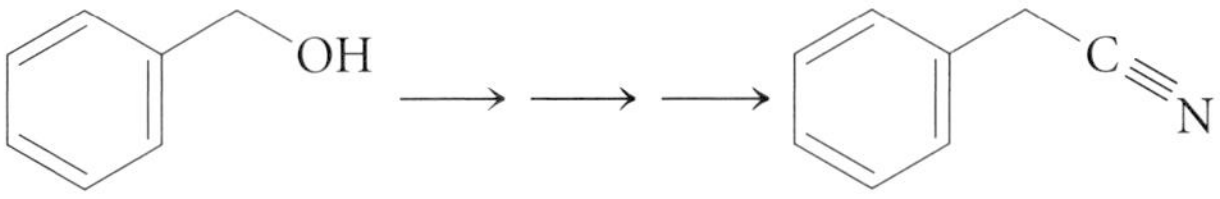

(c)

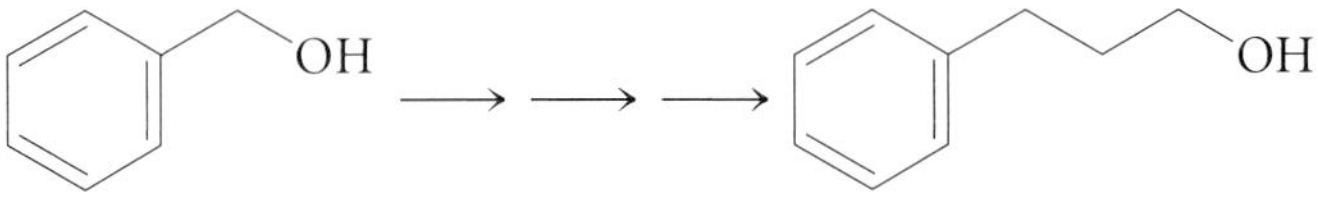

# 9

# Elimination
# Reactions

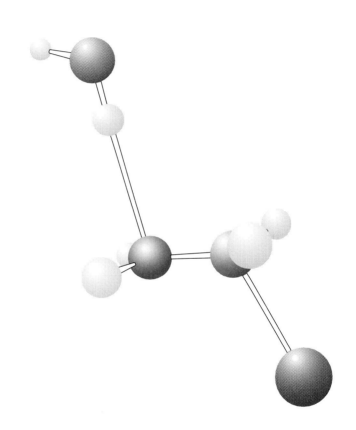

The ball-and-stick model represents the transition state for the reaction of
hydroxide ion with ethyl bromide to form ethylene:

*M* ost compounds that can undergo nucleophilic substitution (that is, those that bear a leaving group at an $sp^3$-hybridized atom) can also undergo elimination. In many elimination reactions, two groups on adjacent atoms are lost and a double bond is formed.

In a typical elimination, a substrate bearing two groups, A and B, on adjacent atoms undergoes cleavage of the two bonds connecting the carbon skeleton to A and to B:

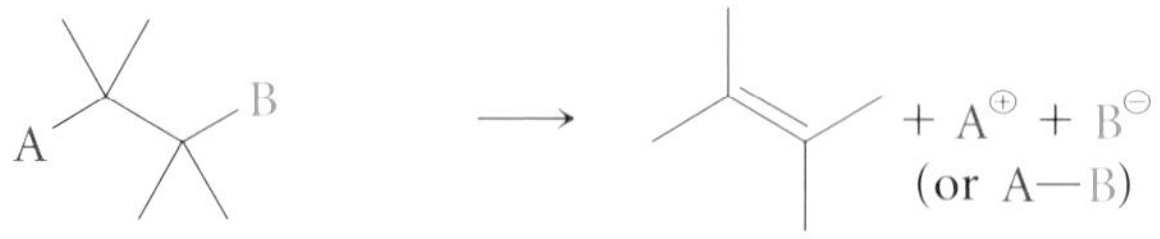

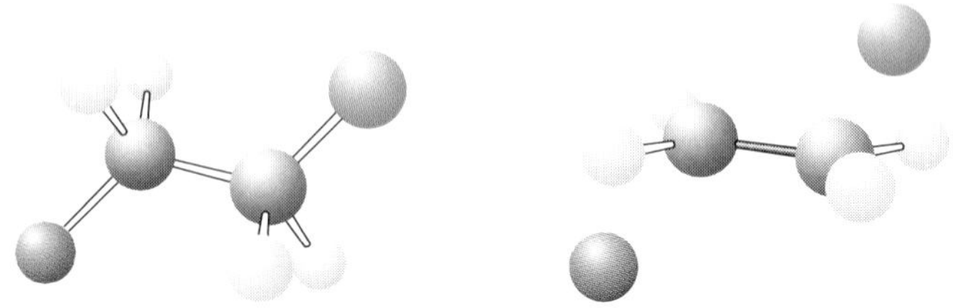

Two of the four electrons from these $\sigma$ bonds appear in the product as a $\pi$ bond. The remaining two electrons appear either as an electron pair localized on A or B, producing a cation and an anion, or as a covalent bond between the two fragments, forming a molecule A—B. At the same time, the two carbon atoms to which A and B were attached change from $sp^3$ to $sp^2$ hybridization.

In most organic elimination reactions, one of the eliminated groups is a hydrogen atom and the other is a leaving group like those involved in substitution reactions. Because hydrogen is more electropositive than most leaving groups (which usually have an electronegative atom at the point of attachment), the two eliminated products are usually $H^{\oplus}$ and a negatively charged leaving group, $LG^{\ominus}$.

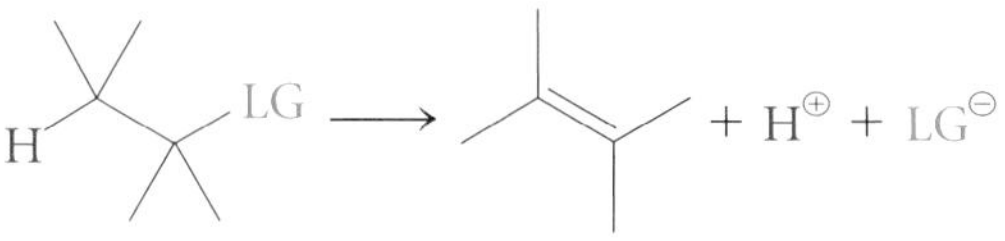

In this chapter, we will examine the mechanisms for several kinds of elimination reactions. Knowledge of these mechanisms enables chemists to choose starting materials incorporating structural features that lead to control of regiochemistry and stereochemistry in the product alkenes. Thus, elimination reactions can be valuable for the preparation of pure alkenes.

## 9.1

## Mechanistic Options for Elimination Reactions

There are three possible mechanisms for an elimination reaction, which differ in the timing of the cleavage of the two $\sigma$ bonds: (1) first C—LG, then C—H; (2) C—LG and C—H simultaneously; and (3) first C—H, then C—LG.

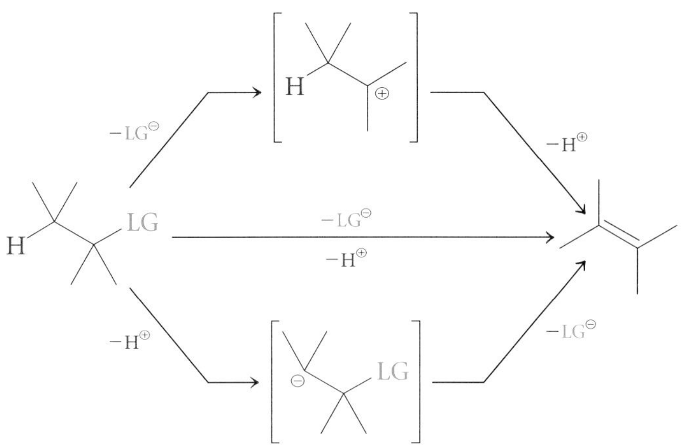

## ▨ E1 Mechanism: Carbocation Intermediates

In the first mechanistic option, the C—LG bond is broken heterolytically to form a carbocation intermediate and a leaving group $LG^{\ominus}$ (Figure 9.1). (This is the same step that initiates the $S_N1$ reaction discussed in Chapters 7 and 8.) This unimolecular, rate-determining step is followed by a second, rapid step, in which a proton is lost from an adjacent carbon atom to form a $\pi$ bond. The first step consists only of bond breaking, with no concomitant bond formation. In the second step, a carbon–hydrogen bond is cleaved at the same time as a carbon–carbon $\pi$ bond is formed; meanwhile, deprotonation is assisted by the transfer of the proton to a base, resulting in the formation of a covalent bond. As a result, the activation energy barrier for the second step is lower than that for the first, and so the loss of $LG^{\ominus}$ is the rate-determining step. If nucleophilic substitution is faster than loss of the proton, substitution (by an $S_N1$ mechanism) is observed instead of elimination.

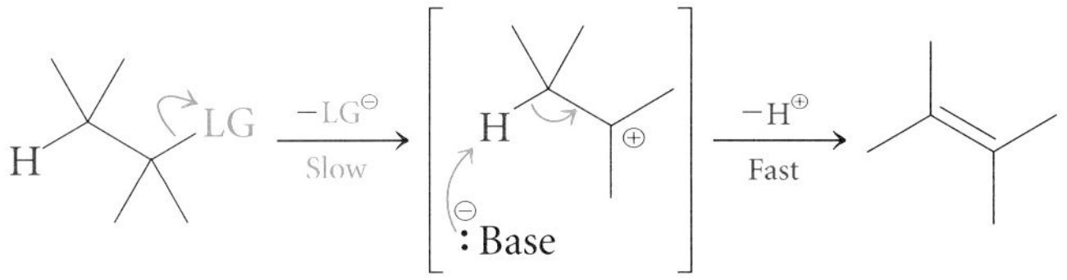

**FIGURE 9.1**

In an E1 elimination, the rate-determining step is the unimolecular loss of the leaving group, $LG^{\ominus}$, to form a carbocation. The elimination is completed by a second, fast step in which a proton is removed by a base.

Note that the rate-determining step of this pathway is unimolecular and endothermic, with a transition state closely resembling the intermediate carbocation. With terminology similar to that used for describing nucleophilic substitutions, this type of reaction is called an **E1 reaction.** The letter E indicates that the reaction is an elimination, and the number 1 indicates that the rate-determining (slow) step of the reaction—that is, heterolytic cleavage of the C—LG bond (with the formation of a carbocation)—is unimolecular. Because carbocations are formed as intermediates, E1 reactions are favored for compounds in which the leaving group is at a tertiary or secondary position.

***Elimination versus Substitution: E1 versus $S_N1$.*** Recall from Chapters 7 and 8 that $S_N1$ reactions also proceed through an intermediate carbocation and that rates of substitution are determined by the facility with which the carbocation is formed via loss of the leaving group.

Whether an elimination or a substitution ultimately occurs depends on the relative rates of deprotonation and of nucleophilic addition to the carbocation. Which pathway—elimination or substitution—is favored is determined mainly by reaction conditions. High concentrations of good nucleophiles favor substitution. Increasing the temperature of the reaction favors elimination.

Why should elimination be favored over substitution as the reaction temperature is increased? Recall from Chapter 6 that the contribution of enthalpy to the activation energy—and, therefore, to the rate of a reaction—varies with temperature. The contribution of entropy is invariant with temperature.

$$\text{Rate} \approx e^{-(\Delta H^{\ddagger} - T\Delta S^{\ddagger})/RT} = \underbrace{e^{-\Delta H^{\ddagger}/RT}}_{\substack{\textbf{Enthalpy} \\ \textbf{contribution}}} \times \underbrace{e^{\Delta S^{\ddagger}/R}}_{\substack{\textbf{Entropy} \\ \textbf{contribution}}}$$

For elimination reactions, both bond breaking and bond making occur, and progression from the cation to the transition state (Figure 9.2) requires a

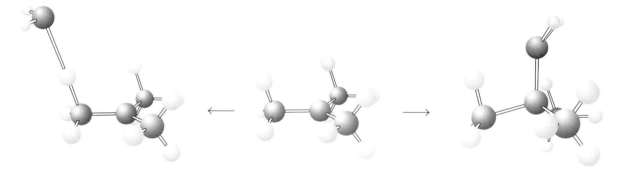

**FIGURE 9.2**

Reaction of water with *t*-butyl cation can lead either to overall elimination through the transition state at the left, or to overall substitution by bond formation between oxygen and carbon. Removal of a proton to form an alkene product involves both bond breaking and bond making, resulting in an activation energy with significant contribution from enthalpy, $\Delta H^{\ddagger}$. The substitution pathway proceeds through the transition state at the right with only bond making. The activation energy for this step is dominated by entropy ($\Delta S^{\ddagger}$).

significant change in both enthalpy and entropy. The reaction rate is there-fore temperature-dependent, and increasing the temperature increases the rate. This effect can be seen most easily in the simple example where $H^{\ddagger}$ is assumed to be zero. Then, regardless of the value of $T$, the enthalpy contribution becomes

$$\text{Enthalpy contribution} = e^{-0/RT} = e^0 = 1$$

and the rate is determined by the temperature-independent entropy term. Conversely, the larger the value of $\Delta H^{\ddagger}$, the greater is the effect of a change in temperature on the reaction rate for an elimination.

The reaction of the intermediate cation with a nucleophile leads to a substitution reaction, in which only bond making occurs. Thus, the energy required to reach the transition state is dominated by entropy. Because the entropy term does not vary with temperature, raising the temperature makes relatively little difference in the rate of a substitution reaction.

The second step of an E1 reaction involves both bond breaking and bond making as the proton is transferred to a base. Thus, $\Delta H^{\ddagger}$ will be relatively large. As the temperature of the reaction is increased, the rate of elimination increases and the rate of substitution remains relatively unchanged.

## E2 Mechanism: Synchronous Elimination

In the second mechanistic option for elimination, two $\sigma$ bonds are broken and a $\pi$ bond is formed simultaneously, with deprotonation by base occurring at the same time as the C—LG bond is broken (Figure 9.3).

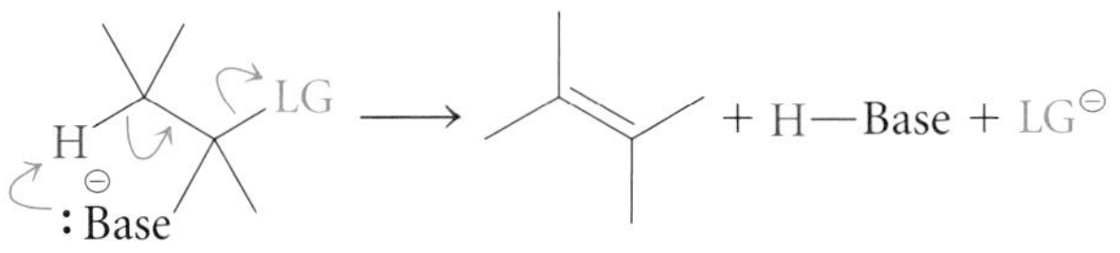

**FIGURE 9.3**

In an E2 elimination, $H^{\oplus}$ and $LG^{\ominus}$ are lost at the same time through a concerted, bimolecular, rate-determining step.

In this elimination, the conjugate acid of the base, a $\pi$ bond between the carbon atoms, and a negatively charged leaving group are all formed through a single transition state. The reaction is bimolecular, because both the base and the molecule bearing the leaving group participate in the transition state of the single, and therefore rate-determining, step. This concerted elimination is called an **E2 reaction.** In the transition state of lowest energy in an E2 reaction, the orbitals that form the $\pi$ bond in the product alkene are aligned for maximal overlap. In a cyclic compound, this alignment is best achieved when the leaving group is *trans* to the proton being eliminated. Unlike the E1 mechanism, the E2 pathway does not involve an intermediate carbocation. As a result, the rate of an E2 reaction is less sensitive than that of an E1 reaction to the degree of substitution—primary, secondary, or tertiary—at the carbon bearing the leaving group.

***Elimination versus Substitution: E2 versus S_N2.*** Under the basic conditions used for elimination of HX from a molecule, the possibility also exists for a bimolecular nucleophilic substitution. Which type of reaction occurs depends largely on the base chosen to carry out the elimination. If the base is also a good nucleophile, competing substitution can become a problem (discussed further later in this chapter).

## ▨ E1cB Mechanism: Carbanion Intermediates

In the third mechanistic option for an elimination reaction, the first step consists of the removal of a proton, $H^\oplus$, by a base—generating a carbanion. In the second step, the leaving group is lost from the intermediate anion (Figure 9.4), as the $\pi$ bond is formed.

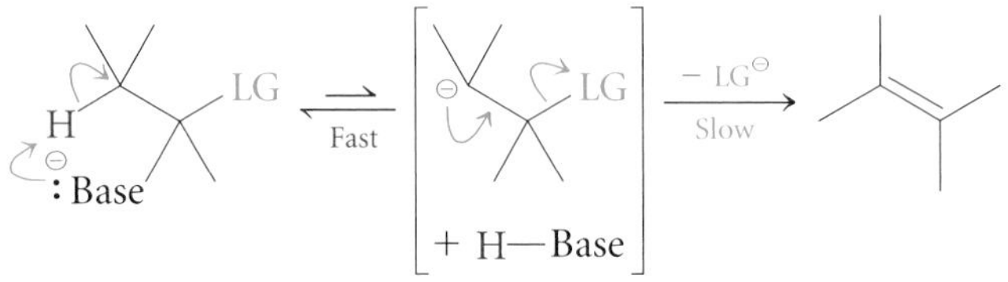

**FIGURE 9.4**

An E1cB elimination reaction involves the rapid and reversible formation of an intermediate carbanion.

In this elimination, an acid–base pre-equilibrium results in deprotonation of the neutral starting material to form its conjugate base. The loss of the leaving group, $LG^\ominus$, from the conjugate base in the second, unimolecular step is rate-determining. If the first step is significantly slower than the second, this two-step sequence cannot be distinguished kinetically from the concerted E2 reaction. When deprotonation is fast and reversible and the reaction rate is controlled by how fast the leaving group is lost from the intermediate carbanion, the reaction takes on unique characteristics. In this case, the loss of $LG^\ominus$ from the anion in the second, rate-determining step is unimolecular and does not involve the base. Therefore, the concentration of base does not directly affect the rate of reaction. This elimination pathway is called an **E1cB reaction.** (The letters cB in this notation refer to the intermediate, a *c*arbon *b*ase.) In the E1 mechanism, the leaving group is lost from the neutral substrate in the rate-determining step; in the E1cB mechanism, the leaving group is lost from the anionic conjugate base of the substrate in the rate-determining step. The E1cB mechanism for elimination is not common. It requires special features in the substrate, such as functional groups that can stabilize the intermediate carbanion. For example, in 3-bromocyclohexanone, the presence of the carbonyl group greatly enhances the acidity of the $\alpha$ protons.

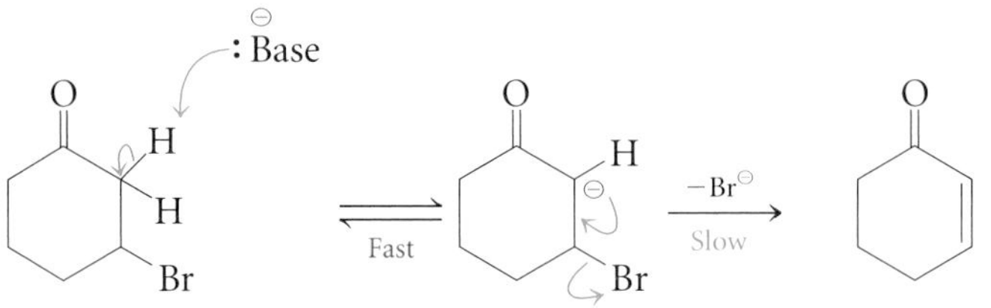

3-Bromocyclohexanone       2-Cyclohexenone

## CHEMICAL PERSPECTIVES

### RECORDING INFORMATION THROUGH A
### COLOR-PRODUCING ELIMINATION REACTION

When you sign a credit card receipt on carbonless paper, you are inducing an acid-catalyzed elimination reaction. The carbonless paper contains small microcapsules (3–8 micrometers in diameter) that hold a solution of a colorless compound. When you apply pressure (by writing), you break the capsules and allow the solution to come into contact with the acid-treated paper. As in the acid-catalyzed elimination reactions discussed here, a bond between carbon and oxygen is broken, producing a highly colored (blue, purple, or black) cation. One such dye is shown here.

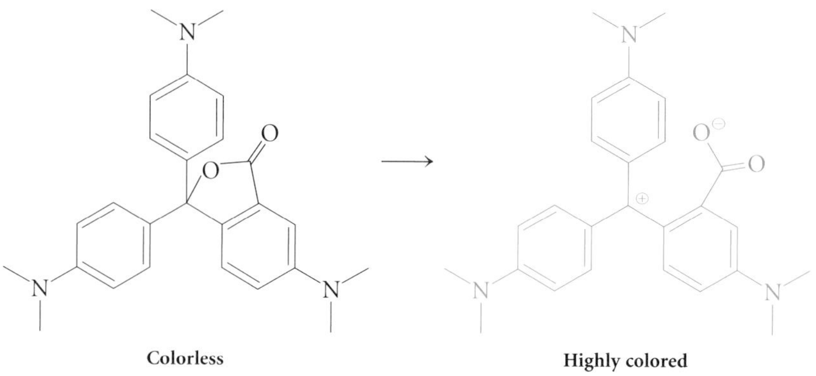

Colorless　　　　　　　　　　Highly colored

## EXERCISE 9.1

From what you know about the relative stability of cations and anions, which substrate in each of the following pairs undergoes elimination (of $H_2O$, HBr, or HCl) more easily through the mechanism indicated? Explain.

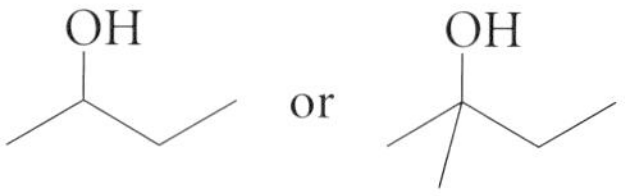

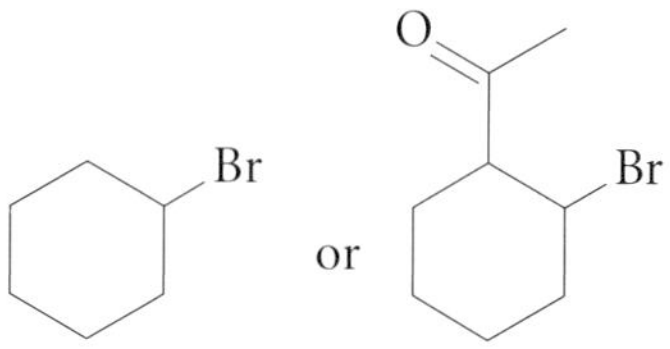

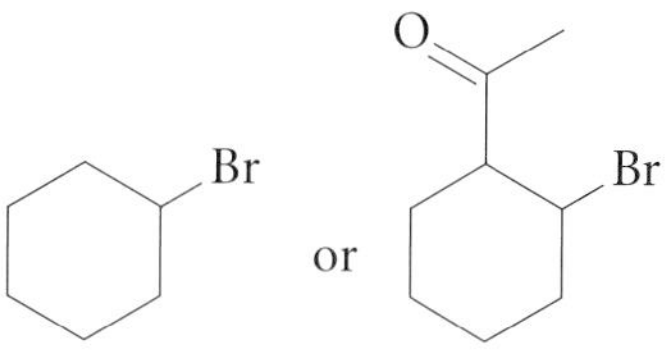

(d)    Br    or    Br    (E2)

(e)    Cl    or    Cl    (E2)

## EXERCISE 9.2

Draw an energy diagram for an E1cB reaction in which the first step (formation of the carbanion by deprotonation) is the slow step. Then draw an energy diagram for the opposite situation, in which loss of the leaving group from the carbanion is the slow step. Be sure to include structures for the starting materials, products, and any intermediates on both diagrams. By considering these energy diagrams, determine what factors will most influence the overall rate of the reaction.

## Transition States and Reaction Profiles for E1 and E2 Eliminations

In many cases, more than one alkene can be formed by an elimination. These isomers can differ in both regiochemistry and stereochemistry. An example is elimination of HBr from 2-bromobutane, for which there are three isomeric products:

Br

$\xrightarrow{-\text{HBr}}$

2-Bromobutane      *trans*-2-Butene    +    *cis*-2-Butene    +    1-Butene

The regiochemistry and stereochemistry of **dehydrohalogenation** (loss of HX) depend on which pathway (E1 or E2) is followed, but the outcome can be predicted from the nature of the substrate and the reaction conditions. Let's first consider the factors that lead to a preference for one stereoisomer over another and then look at those that control regiochemistry.

When geometric isomers are possible as products, the more stable alkene is generally favored in both E1 and E2 reactions—but for different reasons. In an E2 reaction, the geometry of the alkene produced is determined by the fact that both the hydrogen atom and the leaving group are lost in a single step (Figure 9.5). In contrast, in an E1 reaction, the geometry is fixed in the second, fast step, when a proton is removed from the intermediate carbocation.

E2 and E1 reactions differ significantly in the nature of the transition states that determine the regiochemistry of the product. The E2 pathway involves a transition state leading from starting material directly to product. The product-forming step of an E1 reaction is more exothermic than that of an E2 reaction. Thus, the E1 reaction has a relatively early transition state, closely resembling the carbocation formed in the rate-determining step.

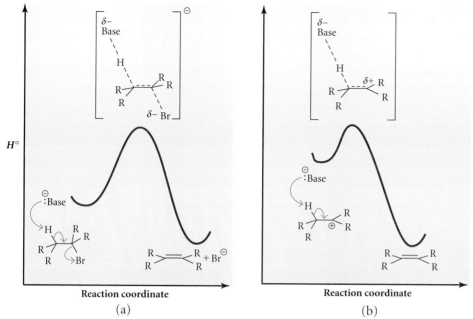

**FIGURE 9.5**

Energy diagrams showing the product-determining transition states for
(a) E2 and (b) E1 pathways for dehydrohalogenation.

## 9.2

# Dehydration of Alcohols

The dehydration of alcohols provides examples of all three elimination
mechanisms. The elimination of water (dehydration) from simple alcohols
requires acidic conditions so that the hydroxyl group, a relatively poor leaving group, can be transformed to $H_2O$, a moderately good one.

### Dehydration via an E1 Mechanism

Dehydration of secondary and tertiary alcohols under acidic conditions
follows the E1 pathway. As noted in Chapters 3 and 7, dehydration is facilitated by acid because protonation of the hydroxyl group effectively converts this leaving group from hydroxide ion to water. Because $H_3O^{\oplus}$ is a
stronger acid than $H_2O$, the conjugate base of the former ($H_2O$) is a better
leaving group than that of the latter ($^{\ominus}OH$). Whenever a relatively stable
carbocation is produced by dehydration of a protonated alcohol, an E1 elimination can occur.

### Dehydration via an E2 Mechanism

Because an unstable primary carbocation would be formed in the E1 dehydration of a primary alcohol, acid-catalyzed E1 elimination through such a carbocation is so slow that other pathways are followed. An E2 reaction occurs instead, in which a proton is lost from a carbon at the same time as water is lost from the adjacent carbon. This allows for the formation of an alkene without the intermediate formation of an unstable carbocation.

**A protonated primary alcohol**          **An alkene**

Dehydration is particularly easy when a conjugated double bond is the result. For example, an alcohol that bears a carbonyl group two carbons away (a $\beta$-hydroxyaldehyde or $\beta$-hydroxyketone) readily undergoes dehydration, yielding an $\alpha,\beta$-unsaturated carbonyl compound.

**$\beta$-Hydroxyketone**                    **$\alpha,\beta$-Unsaturated
carbonyl compound**

### Dehydration via an E1cB Mechanism

The position of the carbonyl group relative to the hydroxyl group in $\beta$-hydroxycarbonyl compounds opens the pathway for elimination under basic conditions by an E1cB mechanism. Here, the carbonyl group plays two critical roles: it stabilizes the intermediate carbanion, and it provides an additional driving force for elimination in giving enhanced stability to the conjugated product.

**$\alpha,\beta$-Unsaturated
carbonyl compound**

Indeed, the base-catalyzed loss of water from $\beta$-hydroxycarbonyl compounds is one of the few examples of an elimination involving an $sp^3$-hybridized carbon atom that follows the E1cB pathway.

In conclusion, acid-catalyzed dehydration of *secondary* and *tertiary* alcohols is usually accomplished by an E1 pathway and proceeds through a carbocation intermediate. This carbocation has two other reaction options (S$_N$1 substitution and rearrangement), which compete with simple elimination. Dehydration of *primary* alcohols under acidic conditions takes place by E2 elimination from the protonated alcohol. Alcohols do not undergo base-catalyzed dehydration because OH$^\ominus$ is a poor leaving group. However, an E1cB elimination of water occurs when $\beta$-hydroxycarbonyl compounds are treated with base.

---

### EXERCISE 9.3

Construct an energy diagram for the conversion of $\beta$-hydroxycyclohexanone to cyclohex-2-enone by an E1cB mechanism.

### EXERCISE 9.4

Write E1 and E2 mechanisms for the acid-catalyzed dehydration of cyclohexanol, using curved arrows to indicate electron movement. Show the sequence of any intermediates. What is the rate-limiting step in each mechanism? Why doesn't cyclohexanol undergo elimination following an E2 pathway under basic conditions?

---

## 9.3

## E2 Elimination Reactions: Dehydrohalogenation of Alkyl Halides

In order to look at the E2 mechanism in more detail, we will consider the elimination of HX from an alkyl halide (**dehydrohalogenation**).

#15   The E2 Reaction

### Transition State for E2 Elimination: *Anti*-periplanar Relationship

Let's look at the elimination of HBr from 2-bromobutane under basic conditions:

In an E2 elimination, the C—H and C—LG bonds are broken simultaneously as the reaction proceeds to the transition state. At the same time, the carbon atoms bearing the hydrogen atom and the leaving group undergo rehybridization from $sp^3$ to $sp^2$. The $\pi$ bond is formed at the transition state only to the extent that the developing $p$ orbitals on the two carbons overlap. This key factor influences the stereochemical and regiochemical outcome of an E2 reaction. Overlapping of the *p* orbitals requires that the C—H and C—LG bonds be coplanar in the transition state. The necessary spatial relationship between these bonds can be achieved in only two ways, referred to as ***anti*-periplanar** and ***syn*-periplanar** (Figure 9.6). Most E2 elimination reactions take place through an *anti*-periplanar transition state in which the C—H and C—LG bonds are staggered. This is partly due to the fact that the same factors that destabilize an eclipsed conformation also destabilize the transition state for elimination from the *syn*-periplanar conformation of the starting material.

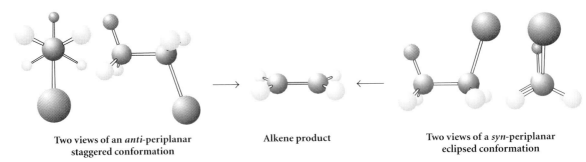

<table>
<tr><td align="center">Two views of an *anti*-periplanar<br>staggered conformation</td><td align="center">Alkene product</td><td align="center">Two views of a *syn*-periplanar<br>eclipsed conformation</td></tr>
</table>

**FIGURE 9.6**

In the *anti*-periplanar arrangement (left), the C—H bond (hydrogen shown as small green sphere) and the C—LG bond (leaving group shown as red sphere) are in a staggered conformation. In the *syn*-periplanar arrangement (right), these two bonds are eclipsed.

### Stereochemistry of E2 Elimination Reactions

Only two of the three possible staggered conformations of 2-bromobutane meet the requirement for an *anti*-periplanar arrangement in the transition state of the concerted elimination. In the third staggered arrangement (Figure 9.7, right), a methyl group is *anti*-periplanar to the leaving group, and no elimination is possible. The transition state resulting from the *gauche* conformation (Figure 9.7, left) produces *cis*-2-butene. Steric interactions between the methyl groups destabilize both the starting *gauche* conformation of 2-bromobutane and the product *cis*-2-butene. Indeed, the magnitude of these two interactions is nearly the same: *gauche* 2-bromobutane is 0.8 kcal/mole less stable than the *anti* conformer, and *cis*-2-butene is 1.0 kcal/mole less stable than *trans*-2-butene. In contrast, interaction between the methyl groups is absent in the transition state leading from the conformer with the methyl groups *anti* to *trans*-2-butene (Figure 9.7, center). Thus, the energy of the interaction of the methyl groups in the transition state leading from the conformer with methyl groups *gauche* to *cis*-2-butene should be somewhere between 0.8 and 1.0 kcal/mole. Indeed, the typical ratio of *trans* to *cis* alkenes formed in E2 elimination reactions is 80:20, quite close to that predicted based on an energy difference of 0.8 to 1.0 kcal/mole between the competing transition states. (Recall that ratio = $e^{\Delta H/RT}$.)

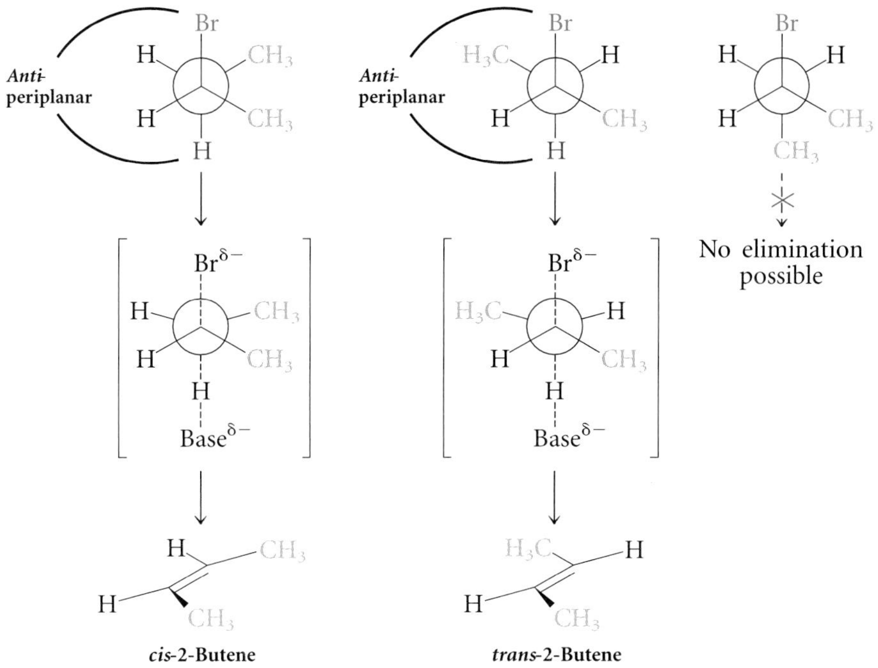

**FIGURE 9.7**

In the *gauche* conformer at the left, elimination of HBr through an *anti*-periplanar transition state produces *cis*-2-butene. In the *anti* conformer in the center, E2 elimination leads to *trans*-2-butene. No elimination is possible from the *gauche* conformer at the right, because there is no hydrogen atom in an *anti*-periplanar arrangement with the bromine atom.

## EXERCISE 9.5

For each structure, indicate which hydrogen atoms are (or can be) *anti*-periplanar to bromine (the leaving group).

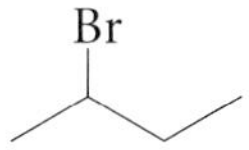

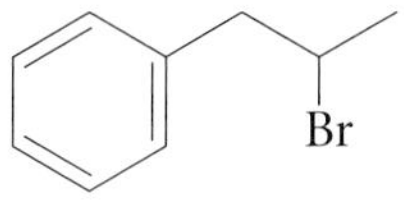

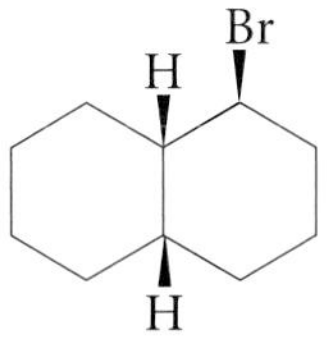

## Regiochemistry of E2 Elimination Reactions

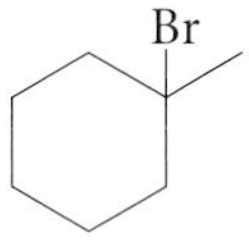

Now let's look more closely at the regiochemical outcome of E2 reactions. Again using 2-bromobutane as an example, we see that elimination of HBr can produce both 1-butene and 2-butene (as a mixture of stereoisomers). These sets of less and more substituted alkenes constitute regioisomers.

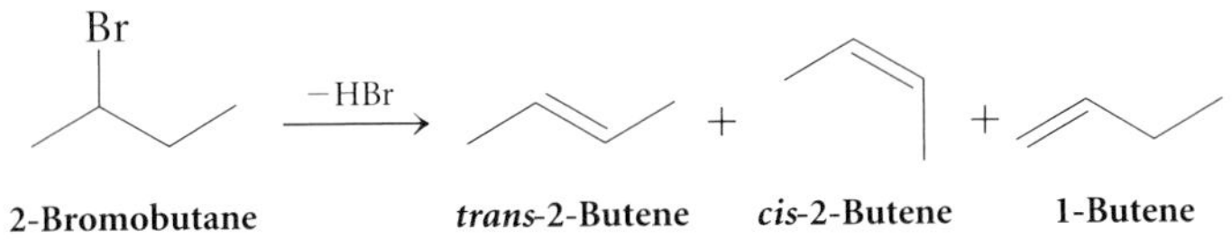

2-Bromobutane      trans-2-Butene    cis-2-Butene    1-Butene

The greater stability of more substituted alkenes means that the 2-butene isomers should be formed in preference to 1-butene in an elimination reaction. Recall from Chapter 2 that the relative stabilities of alkene isomers can be evaluated from heats of hydrogenation (Table 2.2).

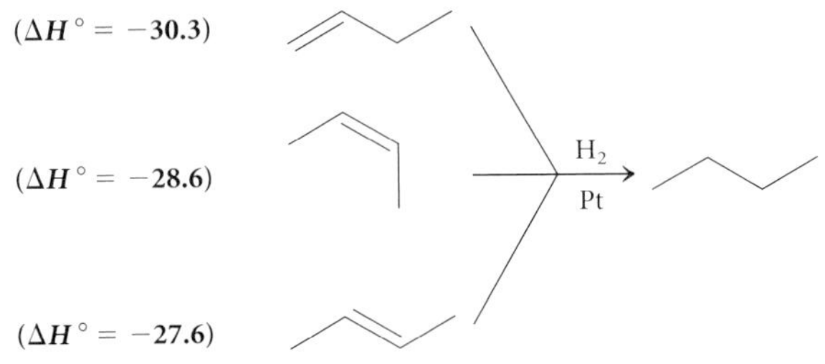

$(\Delta H^\circ = -30.3)$

$(\Delta H^\circ = -28.6)$

$(\Delta H^\circ = -27.6)$

## Effect of Reaction Conditions on Regiochemistry in E2 Reactions

The stability of the product alkene is an important factor in determining the rate of an E2 reaction. However, it is not always the main determinant of the product regiochemistry, and a less stable product is sometimes formed more rapidly.

***The Effect of Base Structure on Elimination Reactions.*** Because there is significant interaction between the base and the hydrogen in the transition state, the nature of the base has a significant effect on E2 eliminations. For example, in the E2 reaction of 2-bromobutane, the reaction can follow two distinct pathways that result in different regioisomers. The base can approach a proton at either the C-1 or the C-3 position, with simultaneous C—H bond cleavage, $\pi$-bond formation, and elimination of bromide ion.

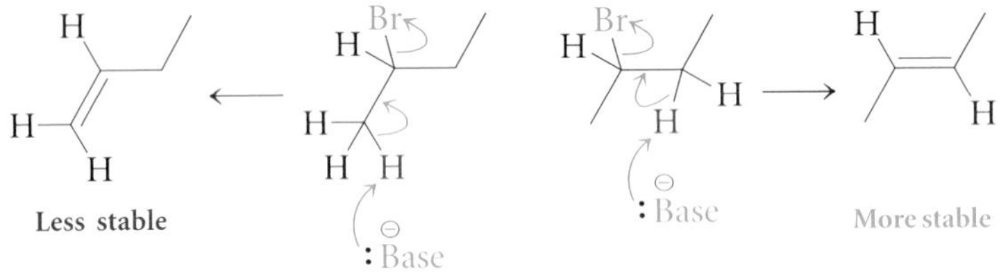

Less stable                 More stable

Recall from Chapter 6 that the affinity of an anion (or a nucleophile) for a carbocation (or other electrophile) is referred to as *nucleophilicity*. Because *basicity* is a measure of the affinity of an anion or a nucleophile for a proton, it is not surprising that trends in these two properties—nucleophilicity and basicity—are often parallel. In general, for species containing the same element, the stronger the base, the better is the nucleophile (and the poorer the leaving group). For example, methylamine is more basic than aniline. In aniline, the electron density of the nitrogen atom is decreased because of delocalization into the aromatic ring. Methylamine is also a better nucleophile than aniline. There is one important distinction between nucleophilicity and basicity: in general, nucleophilicity is related to rate of reactions, because most reactions involving nucleophiles are irreversible under the conditions used; basicity (or acidity) is related to thermodynamic stability because measuring acidity requires consideration of equilibrium processes.

Because all bases can also be nucleophiles, it is important to choose the base for an E2 reaction carefully, so as to avoid competing substitution. Elimination is favored by the use of a reagent that is a strong base but a poor nucleophile. It is also necessary to keep in mind the effect of the base on the regiochemistry of elimination.

*Effect of Steric Hindrance on Nucleophilicity.* Because a proton is very small, basicity is relatively unaffected by steric interactions. On the other hand, as discussed in Chapters 7 and 8, steric hindrance greatly affects the rate at which a transition state is reached in a bimolecular nucleophilic substitution ($S_N2$) reaction. For example, although *t*-butoxide ion is a much stronger base (by a factor of over 100) than methoxide ion, the latter is a better nucleophile because it is less sterically hindered.

*Effects of Charge Density on Basicity.* Another factor that influences the balance between basicity and nucleophilicity is the degree of charge localization. Anions with highly concentrated charge are generally better bases than they are nucleophiles, in part because the small size of the anion improves overlap with the small $1s$ orbital of a proton. For example, hydroxide ion is a good nucleophile *and* a good base, and its use often leads to mixtures of both substitution products. On the other hand, hydrogen sulfide anion, $HS^{\ominus}$, is a much better nucleophile than a base because of the large size of the third-level valence orbitals of sulfur (third row of the periodic table).

For many E2 elimination reactions, potassium *t*-butoxide is the base of choice, because it is a moderately strong base and a relatively poor nucleophile. It is readily prepared by adding potassium metal to *t*-butyl alcohol and is soluble in solvents such as dimethylsulfoxide. However, it is too expensive for use in large-scale reactions, for which sodium ethoxide is often used instead.

## EXERCISE 9.6

Suggest a method for the preparation of sodium ethoxide, and write a balanced equation for the reaction.

## EXERCISE 9.7

In each of the following pairs of compounds, which one is likely to be the stronger base? The better nucleophile?

(a) $^{\ominus}NH_2$    or    $NH_3$

(b) $^{\ominus}OH$    or    $H_2O$

(c) $^{\ominus}OH$    or    $^{\ominus}SH$

(d) 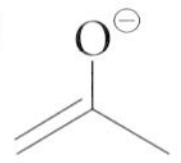 [structure: acetone enolate O$^{\ominus}$]    or    [structure: phenolate O$^{\ominus}$]

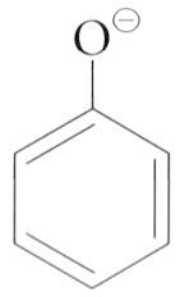

(e) 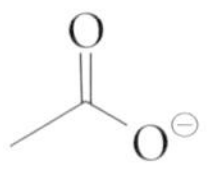 [structure: acetate] $^{\ominus}$    or    [structure: ethoxide] $^{\ominus}$

**Hofmann Orientation.** There are three hydrogen atoms on C-1 in 2-bromobutane, but only two on C-3:

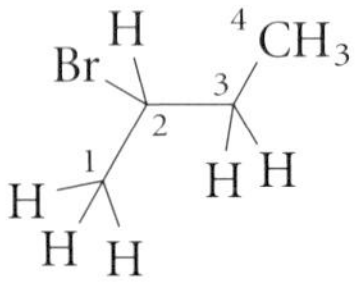

2-Bromobutane

Statistically, therefore, it is more favorable to remove a proton from C-1 (to form 1-butene) than from C-3 (to form 2-butene). Furthermore, the steric congestion around the primary hydrogen atoms on C-1 is lower than that around the secondary hydrogen atoms on C-3. Thus, a base encounters less steric repulsion in the transition state leading to 1-butene. This factor becomes important—and can even dominate—when very large bases, such as potassium *t*-butoxide, are used. In such cases, the less substituted (and less stable) alkene dominates the product mixture. A reaction that gives the less substituted alkene as the major product is said to follow **Hofmann orientation.** Hofmann orientation occurs most frequently in E2 reactions.

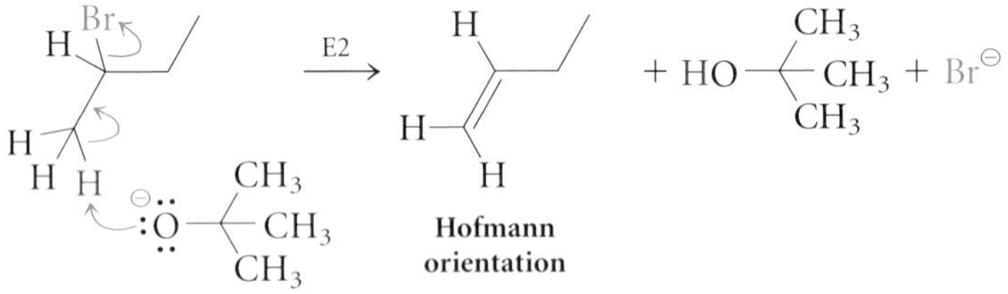

Hofmann orientation is named after the German chemist August Wilhelm Hofmann, who discovered that treatment of quaternary ammonium halides with $Ag_2O$ leads to an elimination reaction in which the less substituted alkene is highly favored. This reaction is referred to as a **Hofmann elimination** and is said to be **regioselective** (favoring one possible regioisomer).

**Hofmann elimination**

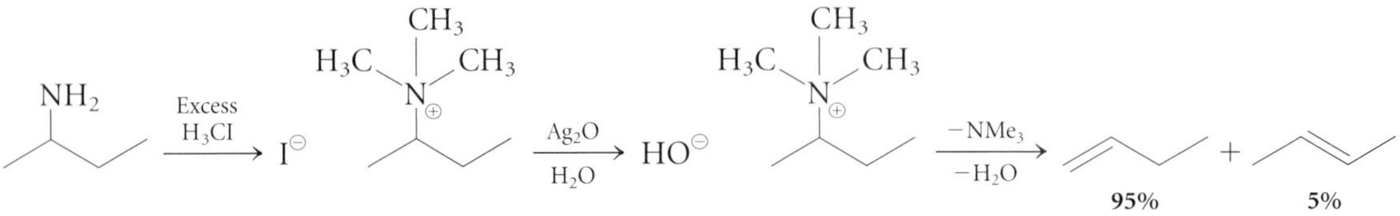

***Zaitsev Orientation.*** At the other extreme, a small base such as sodium hydroxide is less sensitive to steric interactions than a larger one such as potassium *t*-butoxide. With a small base, the thermodynamic stability of the product becomes a more important factor in determining the stability of the transition state, and the two isomeric 2-butenes are formed preferentially. When the more stable and more substituted alkene predominates, the reaction is said to follow **Zaitsev orientation.** (E1 reactions generally exhibit Zaitsev orientation.)

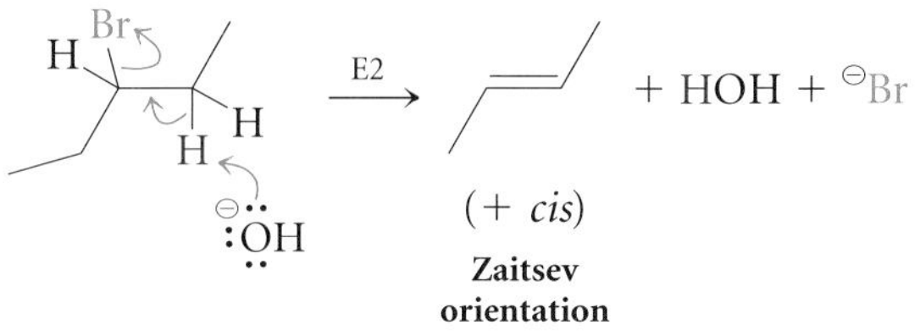

A. N. Zaitsev was a Russian chemist born in 1841 who studied elimination reactions and formulated the general rule (Zaitsev's rule) that the more substituted alkene product is favored.

To summarize, use of a larger base favors the Hofmann orientation (for example, 1-butene), whereas use of a smaller base favors the Zaitsev orientation (for example, *cis*- and *trans*-2-butene).

## EXERCISE 9.8

For each of the following elimination reactions, indicate whether the regiochemistry is of the Hofmann or Zaitsev type:

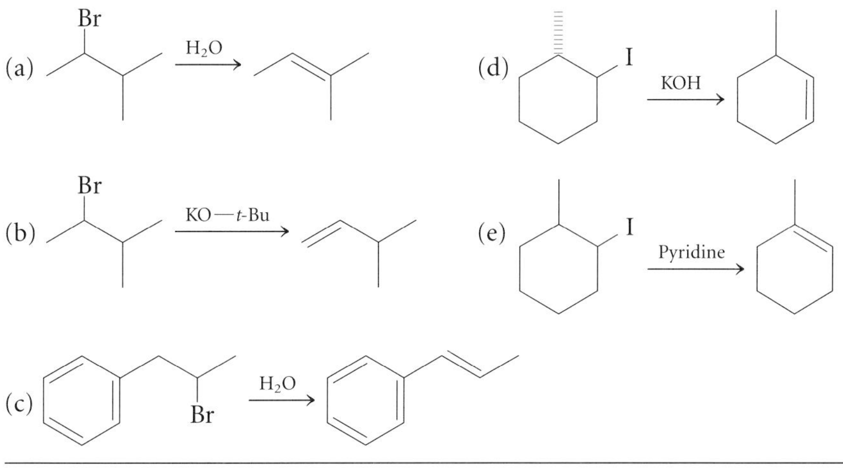

## Effect of Substrate Structure on the Regiochemistry of E2 Elimination in Cyclohexane Rings

For small cycloalkyl halides, the stereochemistry of the product is often constrained by the ring system. In addition, the regiochemistry is often determined by the stereochemical requirement for an *anti*-periplanar arrangement of the hydrogen atom and the leaving group. Consider the dehydrobromination of *trans*-2-methylbromocyclohexane, an unsymmetrically substituted secondary alkyl halide:

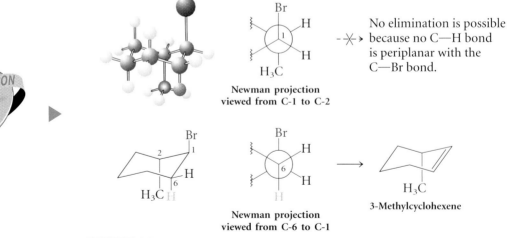

*trans*-2-Methylbromocyclohexane · 3-Methylcyclohexene 100% · 1-Methylcyclohexene 0%

There are two possible regioisomeric elimination products, 1-methylcyclohexene and 3-methylcyclohexene. (Note that only the *cis* isomer can be formed in either case because of the ring size.) Only 3-methylcyclohexene is obtained when the starting alkyl bromide is treated with a strong base (conditions that favor an E2 mechanism).

In all small and moderate size rings (three to seven atoms), an *anti*-periplanar arrangement can be achieved only when the hydrogen atom that is removed and the leaving group are *trans*. Let's first consider the chair conformation of *trans*-2-methylbromocyclohexane with the bromine in an axial position, as shown in the three-dimensional representations in Figure 9.8. Examination of a Newman projection viewed along the bond from C-1 to C-2 reveals that no elimination is possible between the atoms on C-1 and C-2 because the C—Br bond is coplanar with the C—CH$_3$ bond, not with the C—H bond, as required for the E2 transition state. Therefore, elimination cannot take place so as to produce a $\pi$ bond between C-1 and C-2. As the Newman projection viewed along the bond from C-6 to C-1 shows, the C—Br bond is coplanar with the axial hydrogen on C-6. Elimination is possible with this *anti*-periplanar arrangement, giving rise to the formation of 3-methylcyclohexene.

**FIGURE 9.8**

In the chair conformer at the left, bromine is *anti*-periplanar to a hydrogen (green) at C-6 but *gauche* to a hydrogen at C-2. Concerted elimination can produce a double bond only between C-1 and C-6, yielding 3-methylcyclohexene.

Now consider the situation with bromine in an equatorial position. (Recall from Chapter 5 that axial and equatorial conformers of substituted cyclohexanes can be interconverted by a ring flip.) There are no C—H bonds in proper alignment for concerted elimination (coplanar with the C—Br bond) on either C-2 or C-6 (Figure 9.9). E2 elimination cannot occur from this conformation because it specifically requires a periplanar transition state.

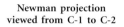

No elimination is possible because no C—H bond is periplanar with the C—Br bond.

Newman projection viewed from C-1 to C-2

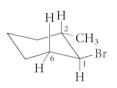

No elimination is possible because no C—H bond is periplanar with the C—Br bond.

Newman projection viewed from C-6 to C-1

**FIGURE 9.9**

A ring flip of the chair conformation shown in Figure 9.8 produces the conformation shown at the left with the bromine atom in an equatorial position where it is *anti*-periplanar to the C—C bonds of the ring rather than to any C—H bond. Elimination is therefore impossible from this ring-flipped conformer, even though it is the more stable conformer and is present in higher abundance at equilibrium.

The requirement for periplanar alignment in a transition state is an example of **stereoelectronic control** of regiochemistry. Even though the axial conformer of *trans*-2-methylbromocyclohexane in Figure 9.8 constitutes only a small fraction of the equilibrium mixture, it is the only chair conformer from which elimination can occur, because all of the others lack the required *anti*-periplanar alignment of the C—H and C—Br bonds.

*Elucidation of Mechanism by Isotopic Labeling.* The requirement for a periplanar alignment of bonds in the transition state for an E2 reaction has been verified experimentally by using an isotopically labeled substrate. In 2,2-dimethyl-*trans*-6-deuterocyclohexyl bromide, the elimination must take place through the formation of a π bond between C-1 and C-6, because of the presence of two methyl groups at C-2. Thus, either HBr or DBr will be eliminated.

The isolated product obtained upon treatment with strong base is that re-sulting from the loss of DBr, not HBr. The deuterium located *trans* to bromine in the starting material can assume an *anti*-periplanar alignment suitable for the E2 elimination, whereas this is not possible for the *cis* hydrogen. That is, the requirement for a periplanar arrangement in the transition state means that the groups lost must be *trans*, not *cis*, to each other.

## Summary of E2 Elimination

In summary, elimination reactions that take place via an E2 mechanism proceed through a transition state in which the C—H and C—LG bonds are in a periplanar arrangement. For cyclohexane derivatives, only the *anti*-periplanar arrangement can be achieved readily. For the bonds to hydrogen and the leaving group to have this arrangement, both groups must be axial.

### EXERCISE 9.9

Predict the product expected from an E2 elimination through an *anti*-periplanar transition state for each of the following substrates:

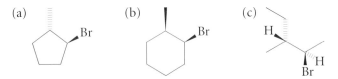

(a)      (b)      (c)

---

## 9.4

# E1 Elimination Reactions

## Intermediate Cations in E1 Elimination Reactions

The intermediate cation formed in an E1 elimination is the same as that formed in the $S_N1$ reaction (see Chapter 7). The factors that affect the rate of the $S_N1$ reaction, therefore, also affect the rate of E1 elimination. Whereas the rate of the E1 elimination depends on the nature of the transition state in the rate-determining first step, the nature of the products depends on the transition state for the second, faster step. It is in this second, fast step of the reaction that both the stereochemistry and the regiochemistry of elimination are fixed, and thus this step is **product-determining.**

Reactions that typically proceed by an E1 mechanism are the acid-catalyzed dehydration of alcohols and the dehydrohalogenation of secondary and tertiary alkyl halides.

## Stereochemistry of E1 Elimination Reactions

Let's consider the options available to the intermediate carbocation in an E1 reaction. Figure 9.10 shows Newman projections of the carbocation formed by loss of bromide ion from 2-bromobutane. To proceed to prod-

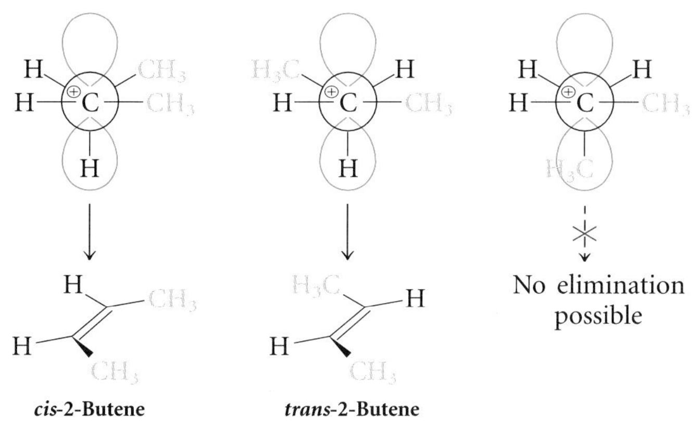

*cis*-2-Butene      *trans*-2-Butene

**FIGURE 9.10**

Newman projections of the 2-butyl cation in 2-butene viewed down the bond between C-2 and C-3. With C-2 held steady, C-3 (the back carbon atom) is rotated counterclockwise by 120° to step from the left conformer to the center and then to the right.

uct by loss of a proton, a C—H bond and the *p* orbital of the carbocation must overlap in the transition state. Just as for the E2 elimination reaction, only two of the three unique conformations of the intermediate carbocation (Figure 9.10, left and center) have C—H bonds so aligned.

The Newman projections in Figure 9.10 are obtained by viewing down the bond from C-2 to C-3, with C-2 (the positively charged carbon) in the foreground, with its $\sigma$ bonds orthogonal to the vacant *p* orbital. With C-2 fixed in this position, the substituents on the $sp^3$-hybridized C-3 are rotated to obtain the various projections. To form a $\pi$ bond, the C—H bond that is broken must be aligned with the vacant *p* orbital of the carbocation. Only in this way can electron density flow into the developing $\pi$ bond as C-3 re-hybridizes from $sp^3$ to $sp^2$. E1 reactions favor the formation of the more stable (*trans*) alkene for the same reasons as E2 reactions do: the interactions between methyl groups in the products are also present in the transition states leading from the carbocation to the alkene, and these methyl–methyl interactions in the transition state cause formation of the *cis* isomer to be slower than that of the *trans* isomer.

## Factors Affecting Regioselectivity in E1 Reactions

Like the stereoselectivity of an E1 elimination, the regioselectivity of such a reaction is determined by the second, faster step, in which the intermediate carbocation is transformed into the product alkene. There is no longer a bond to the leaving group in the intermediate carbocation. Thus, the only requirement for formation of an alkene is that a proton be oriented such that the C—H bond aligns with the vacant *p* orbital of the carbocation.

***E1 Elimination in Cyclohexane Rings.*** The available options for elimination of a proton from the carbocation can be illustrated by examining the loss of HBr from 1-bromo-1-methylcyclohexane:

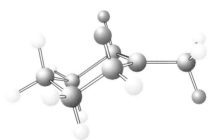

The loss of bromide occurs in the rate-determining step to yield a tertiary carbocation. Fast deprotonation of the cation from either of the adjacent positions completes the reaction. The cation is symmetrical; thus, loss of a proton from either of the carbons adjacent to the positively charged one leads to the same product, 1-methylcyclohexene, in which the double bond is in the ring (**endocyclic**). On the other hand, loss of a proton from the methyl group yields methylenecyclohexane, in which the double bond is outside the ring (**exocyclic**). As stated in Chapter 2, the more highly substituted alkene is generally more stable.

The ball-and-stick representation of the tertiary carbocation shows that it is the axial hydrogen atoms (shown in green) adjacent to the positively charged carbon that have the correct orientation to be lost, thus forming an alkene.

Tertiary intermediate carbocation

Similarly, only one of the three methyl-group hydrogen atoms is correctly oriented for elimination. Thermodynamics favors the formation of the more stable alkene—in this case, the trisubstituted alkene formed when the double bond is endocyclic rather than the disubstituted alkene with the double bond exocyclic to the ring. Thus, the regiochemical preference for the formation of 1-methylcyclohexene is related to product stability. Note that the isomer ratio of the products formed is determined in the second step, which is *not* the rate-determining step of the overall reaction. *Thus, the factors that influence the rate of the reaction may play little or no role in determining the ratio of the isomeric alkenes produced.*

Another example of an E1 reaction is the elimination of HBr from 1-bromo-1,2-dimethylcyclohexane; three regioisomeric alkenes are possible as products:

CH₃ ... CH₃ CH₃ —Br⁻ ... H CH₃ CH₃ ... —H⁺ ... CH₃ CH₃ + CH₃ CH₃ + CH₃ CH₂

Br H H

**Major**     **Minor**     **Minor**

Loss of bromide ion from the starting alkyl bromide results in a carbocation in which there are three adjacent carbon atoms bearing hydrogen atoms. Because these three carbon atoms are unique, there are three possible alkene products: one in which the double bond is exocyclic, and two in which the double bond is endocyclic. Again, the product distribution obtained in an E1 elimination is related to the relative stability of the possible alkenes. The major product is the most stable isomer, having the more highly substituted double bond—in this case, 1,2-dimethylcyclohexene.

**E1 Elimination in Acyclic Systems.** Similar considerations apply to E1 elimination in acyclic compounds, except that, in addition to regioisomers, geometric isomers are also possible. For example, consider 2-bromobutane, in which the leaving group (Br⁻) is attached to a secondary carbon. Elimination can give two possible products. An E1 elimination pathway results in ionization to form an *s*-butyl cation, a species with nonequivalent hydrogen atoms on adjacent carbons.

Br —Br⁻ Hₐ (a) + —H⁺ **Major** Hᵦ (b)

Loss of proton Hₐ from C-3 in the carbocation (path a) yields 2-butene as a mixture of *cis* and *trans* isomers. Loss of proton Hᵦ from C-1 in the carbocation (path b) yields 1-butene. The three alkenes formed are not equally stable. The most stable product, *trans*-2-butene, is formed in the highest yield because path a is favored over path b—that is, the rates of product formation reflect the order of stability of the products.

**EXERCISE 9.10**

Write the major elimination products to be expected from each of the following alkyl halides under E1 reaction conditions:

(a) [structure with Br]

(b) [bicyclic structure with Br]

(c) [structure with Cl]

(d) [cyclohexane structure with Cl]

## 9.5

# Competing Rearrangements in E1 Reactions

E1 reactions proceed through cationic intermediates. We saw in Chapter 3 that tertiary carbocations are more stable than secondary ones, which are in turn more stable than primary ones. Secondary and especially primary carbocations are prone to rearrangement when a more stable cation can be produced by shifting a hydrogen or a carbon atom from an adjacent atom. This propensity of carbocations to undergo rearrangement was noted in Chapter 7 with reference to $S_N1$ reactions. Such rearrangements can also influence the outcome of E1 elimination reactions.

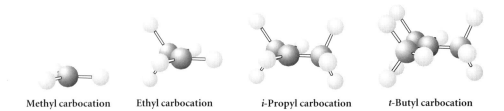

Methyl carbocation     Ethyl carbocation     *i*-Propyl carbocation     *t*-Butyl carbocation

Elimination of water from secondary and primary alcohols is often complicated by competing rearrangement reactions. For example, treatment of 3,3-dimethyl-2-butanol with acid produces mainly 2,3-dimethyl-2-butene and only a small amount of 3,3-dimethyl-1-butene. The major product results from a rearrangement of the initially formed secondary carbocation to a more stable tertiary carbocation. We will encounter other examples of rearrangements of this type in Chapter 14.

**3,3-Dimethyl-2-butanol** →[H₂SO₄]

**3,3-Dimethyl-1-butene**
Minor

**2,3-Dimethyl-2-butene**
Major

### EXERCISE 9.11

Treatment of neopentyl alcohol with sulfuric acid yields 2-methyl-2-butene. Write a mechanism for this reaction, being sure to note that a rearrangement of the carbon skeleton is involved.

**Neopentyl alcohol** →[H₂SO₄] **2-Methyl-2-butene**

## E1 versus E2 Elimination

The major determinants of the mechanism followed in an elimination reaction are the structure of the starting material and the nature of the leaving group. Primary and secondary alkyl halides react primarily by an E2 mechanism under basic conditions; tertiary alkyl halides undergo E1 elimination because of the greater stability of the tertiary cation intermediate. Dehydration of secondary and tertiary alcohols proceeds via an E1 mechanism because the nature of the leaving group requires that the reaction be carried out under acidic conditions, which are unfavorable for the E2 pathway.

## Substrate Structure

In the discussion of $S_N1$ reactions in Chapter 7, it was pointed out that, for molecules capable of forming a tertiary carbocation, substitution occurs via the $S_N1$ mechanism (involving a carbocation intermediate) rather than via the $S_N2$ mechanism. Similarly, for elimination reactions, if the leaving group is at a tertiary center, the elimination proceeds preferentially by the E1 mechanism. Eliminations at primary centers usually proceed by the E2 mechanism. Eliminations at secondary centers usually proceed by the E2 mechanism, but may proceed by the E1 route.

## Leaving Groups

How does the identity of the leaving group affect the efficiency of heterolytic bond cleavage in an E1 or E2 elimination? In both E1 and E2 mechanisms, the C—LG bond is broken in the rate-determining step. Clearly, the weaker the C—LG bond, the easier is its cleavage, because less energy is needed to reach the transition state in which this bond is substantially broken. A comparison of the facility of ionization within a series of $t$-butyl halides (Figure 9.11) reveals that the rate decreases with increasing strength of the carbon–halogen bond. Because the same carbocation is generated in each case, the order of reactivity must result from differences in the halide. Of significance here, in addition to bond strength, is the stability of the halide anion generated in the reaction.

R—I (51 kcal/mole) < R—Br (68 kcal/mole) < R—Cl (82 kcal/mole) < R—F (106 kcal/mole).

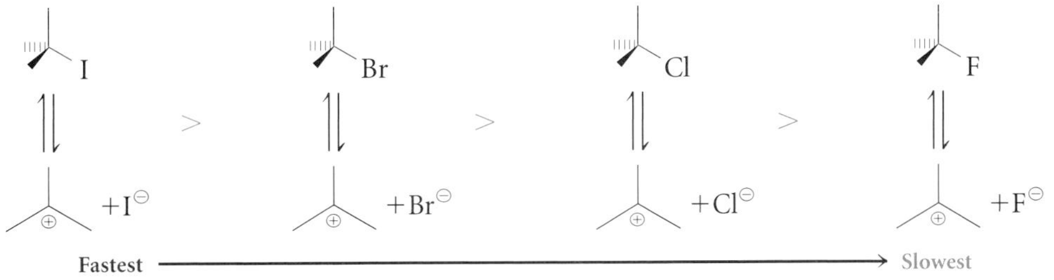

Fastest ——————————————————————————→ Slowest

**FIGURE 9.11**

The rate of cleavage of a C—X bond decreases in progressing from the bottom to the top of the halogen column in the periodic table. This trend is parallel to the progression of bond-dissociation energies for these bonds.

Recall from Chapter 6 that HF is the weakest halogen acid and HI the strongest, and that the H—F bond is the strongest and the H—I bond the weakest. The effect of the stabilities of the anions on the rates of cleavage of alkyl halides is analogous to their effect on the acidity of the acid: HI is the most acidic of the halogen acids, and alkyl iodides are the most reactive of the alkyl halides. The ability of a given halogen to act as an effective leaving group (readily departing with the two electrons that initially constituted the $\sigma$ bond) is thus roughly inversely related to the basicity of the ion formed: the weaker the base (and the stronger the conjugate acid, HX), the better the leaving group.

In an E2 elimination, both the C—H and the C—LG bonds are broken in the rate-determining step. Because the C—LG bond is broken in the transition state of both the E1 and E2 reactions, its bond strength affects the rates of both cleavages in roughly the same way. Thus, although the identity of the leaving group affects the rate of elimination of HX, it has little effect on whether the elimination reaction proceeds via the E1 or the E2 mechanism.

### EXERCISE 9.12

Compare hydroxide ion with the halide ions as a leaving group. Consider C—O versus C—X bond strength, as well as the basicity of the ions as measured by the acidity of the corresponding acids ($H_2O$ and HX).

## 9.6

## Elimination of $X_2$

Elimination reactions are not restricted to loss of HX or $H_2O$. Elimination of two electronegative atoms on adjacent atoms—for example, in a vicinal dihalide—can also be accomplished by treatment with an active metal. The formation of the alkene is often accompanied by formation of a halide salt of the metal:

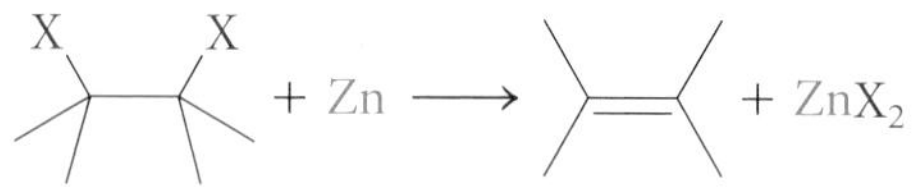

Here, the net conversion includes the oxidation of zinc and the reduction of the two carbons of the vicinal dihalide. The metals most useful for this elimination reaction are those that can easily support a +2 change in oxidation level—usually zinc or tin. The formation of an alkene from a vicinal dibromide frequently occurs through an organozinc intermediate (Chapter 8), in which the metal has been inserted into the carbon–bromine bond.

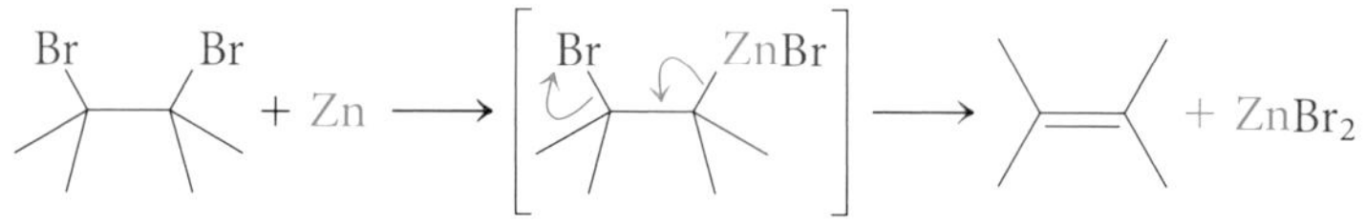

## EXERCISE 9.13

When the dibromide shown here is treated with zinc, only *cis*-stilbene is produced. First, write a mechanism for this transformation, and then explain why only the *cis* alkene is formed. If there are chiral centers in the starting dibromide, determine their *R,S* configuration. Is the starting dibromide chiral?

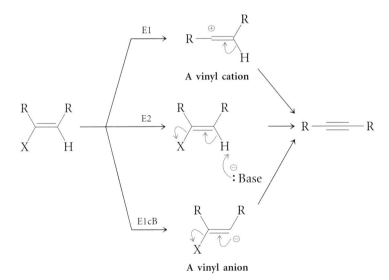

---

### 9.7

# Elimination of HX from Vinyl Halides

The elimination of HX from a vinyl halide produces an alkyne. Very strong bases, such as sodium amide, $NaNH_2$, are generally employed for this purpose. As for elimination reactions involving $sp^3$-hybridized atoms, there are three possible pathways for elimination reactions at $sp^2$-hybridized atoms (Figure 9.12). Because the product alkyne has no stereochemical features, it is difficult to find experimental details to verify which mechanism is actually followed. However, the E1 route is unlikely because the vinyl cation is so unstable, and energetic considerations favor the concerted E2 and the E1cB elimination mechanisms.

**FIGURE 9.12**

Elimination reactions at the $sp^2$-hybridized carbon atom of a vinyl halide can proceed through three possible mechanisms: (upper) an E1 mechanism through an intermediate vinyl cation, (center) a concerted loss of HX in an E2 mechanism, or (lower) an E1cB mechanism through a vinyl anion.

## E1 Elimination of HX from Vinyl Halides

Elimination of HX via the E1 route involves loss of a halide ion and formation of a vinyl cation. If such a cation is *sp*-hybridized, it is particularly unstable. The positively charged carbon has access to only six valence electrons; four of these are accommodated in the C—R and C—C *sp*-hybridized bonds, while the remaining two valence electrons are held in a $\pi$ orbital and are farther from the nucleus than if the carbon were $sp^2$-hybridized (Figure 9.13). This orbital picture explains why an E1 elimination is more difficult in a vinyl halide than in an alkyl halide and is only rarely observed.

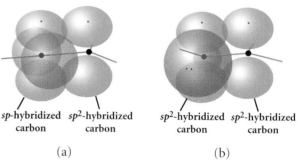

| *sp*-hybridized carbon | $sp^2$-hybridized carbon | $sp^2$-hybridized carbon | $sp^2$-hybridized carbon |

(a)             (b)

**FIGURE 9.13**

(a) In the vinyl cation, the cationic carbon is *sp*-hybridized, bearing two $\sigma$ bonds, a $\pi$ bond from overlap of two *p* orbitals (shown in blue), and a vacant *p* orbital (light gray). The electrons are accordingly held at a position farther from the nucleus than in an $sp^2$-hybridized trigonal cation. The vinyl cation is therefore less stable than an $sp^2$-hybridized carbocation. (b) In the vinyl anion, the anionic carbon is $sp^2$-hybridized with a lone pair of electrons in an $sp^2$-hybrid orbital (shown in dark gray). Because the lone pair is held in an $sp^2$-hybrid orbital, a vinyl anion is more stable than a simple carbanion in which the electrons are in an $sp^3$-hybrid orbital. As in the vinyl cation, there is a $\pi$ bond formed by overlap of two atomic *p* orbitals (in blue).

## E1cB and E2 Elimination of HX from Vinyl Halides

In the E1cB elimination route, deprotonation takes place first, generating a vinyl anion:

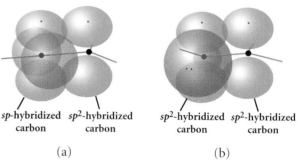

The carbon atom bearing negative charge in a vinyl anion is $sp^2$-hybridized, bearing two bonds formed from hybrid orbitals, a lone pair in a hybrid orbital, and a *p* orbital participating in $\pi$ bonding. The unshared electron pair in the vinyl anion is held in an $sp^2$-hybrid orbital and is therefore closer to the nucleus than the electron pair in $sp^3$-hybridized anions, considered earlier in this chapter. As a result, vinyl anions are unusually stable (Chapter 6), as indicated by the acidity of the vinyl C—H bond (p$K_a$ 44). Therefore, initial deprotonation is more favorable in this elimination than at an $sp^3$-hybridized atom.

The lone pair in the hybrid orbital of a vinyl anion is orthogonal to the $\pi$ system and coplanar with the C—X bond. This coplanarity facilitates the loss of $X^\ominus$ in the next step, completing the elimination. At the extreme, the loss of $X^\ominus$ is concerted with deprotonation, giving rise to an E2 elimination:

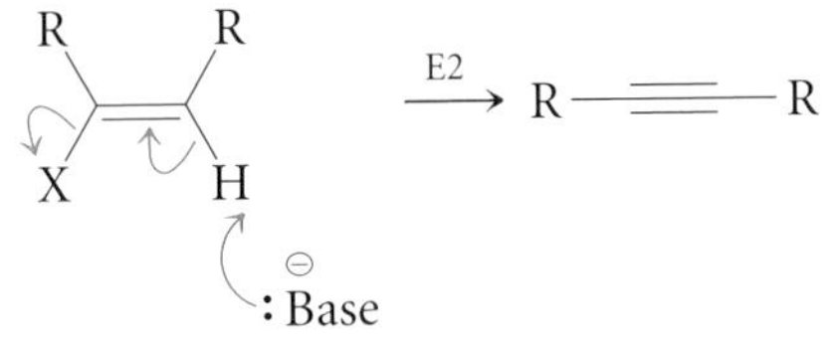

## Preparation and Use of Vinyl Halides

Vinyl halides can be prepared by elimination of HX from geminal or vicinal dihalides, which bear two halogens either on the same carbon atom or on adjacent ones.

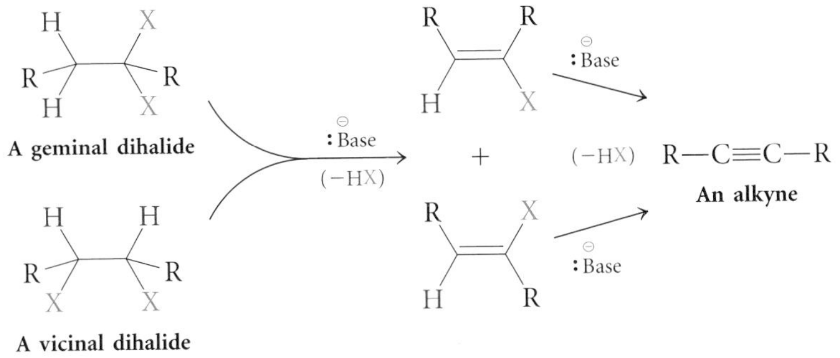

When alkoxide bases are used, it is possible to stop with the loss of one equivalent of HX. With a stronger base such as sodium amide, NaNH$_2$, a second equivalent of HX is lost, producing an alkyne.

Vicinal dibromides are prepared by treating alkenes with Br$_2$ in CCl$_4$, a reaction that will be discussed in more detail in Chapter 10. The loss of two equivalents of HBr from the resulting vicinal dibromide constitutes a method by which an alkene can be converted to an alkyne.

## EXERCISE 9.14

In the elimination reaction induced by treating ($Z$)-2-bromo-2-butene with NaNH$_2$ in NH$_3$, 1-butyne is formed. Suggest a mechanism for this transformation, and give reasons for the observed regiochemistry.

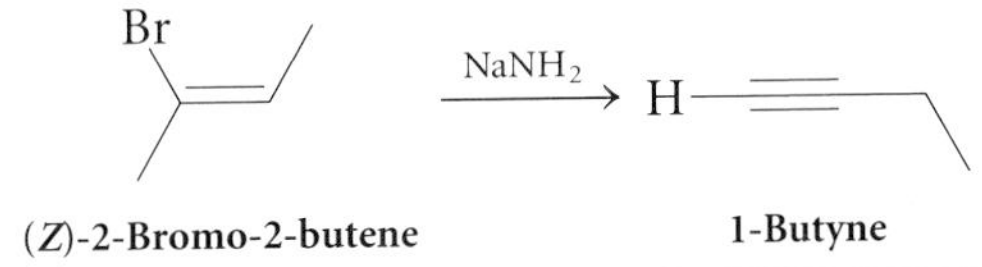

(*Z*)-2-Bromo-2-butene                    1-Butyne

# Elimination of HX from Aryl Halides: Formation and Reactions of Benzyne

## Mechanisms of Elimination from Aryl Halides

There are three possible pathways for the elimination of HBr from bromobenzene (Figure 9.14). An E1 mechanism would proceed with the loss of bromide ion to form a phenyl cation, a species that is destabilized by the same hybridization effects that destabilize a vinyl cation (see Figure 9.13). Because the positively charged carbon of a phenyl cation is constrained within a six-member ring, this intermediate is even further destabilized by angle strain. The phenyl cation is therefore even less stable than the vinyl cation and is an even less likely intermediate. Indeed, no direct evidence has been obtained for the formation of a phenyl cation in the course of this elimination.

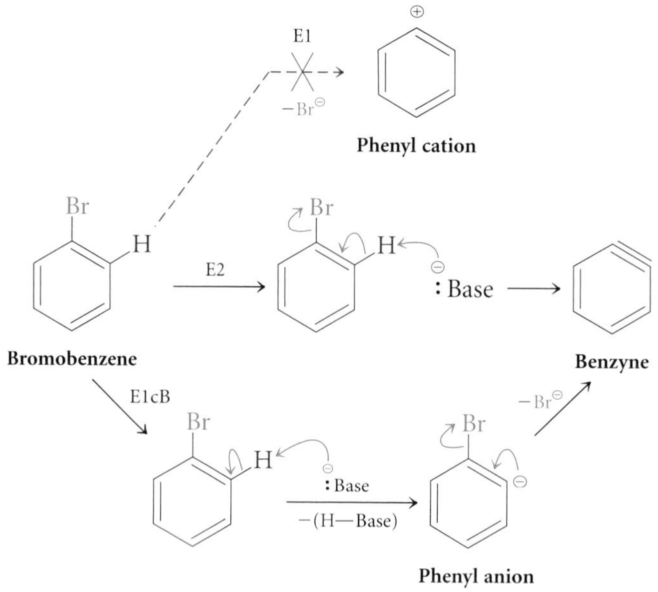

**FIGURE 9.14**

The first possible mechanism, the E1 route proceeding through a phenyl cation, is unlikely because of the great instability of this intermediate. Both the E2 and E1cB are reasonable pathways, and which is followed can be hard to determine experimentally.

As with vinyl halides, elimination from aryl halides is much more likely through a concerted E2 pathway or through a phenyl anion (via an E1cB pathway). Whether elimination takes place in one concerted step (as an E2 mechanism) or in two steps (as an E1cB mechanism) is unclear, but the same product, benzyne, is formed by either route. The phenyl anion formed by deprotonation of bromobenzene has the same hybridization as a vinyl anion. The $sp^2$-hybrid orbital containing the lone pair of the phenyl anion is coplanar with the carbon–bromine bond, and elimination occurs easily in the next step.

## Structure of Benzyne

The structure of benzyne is interesting because it contains a formal triple bond within a six-member ring. The orbitals that form the second $\pi$ bond must be perpendicular to the $\pi$ system of the aromatic ring, as in all alkynes (Chapter 2). In benzyne, however, the preferred angle for $sp$-hybridization (180°) is not possible, because it would require four of the six atoms of the ring to be arranged in a colinear fashion (Figure 9.15).

**FIGURE 9.15**

Two $sp^2$-hybrid orbitals overlap in benzyne to form a $\pi$ bond. The orbitals are shown at the left, and the resulting $\pi$ orbital at the right. Note that the large lobes of the $sp^2$-hybrid orbitals are pointed away from each other, fixed in this relative position by the underlying carbon framework. As a result, overlap of these orbitals is significantly less than with two $p$ orbitals.

## Reactions of Benzyne

The weak $\pi$ bond in benzyne formed by overlap of two $sp^2$-hybrid orbitals is much less stable than the $\pi$ bond formed by overlap of two $p$ orbitals. As a result, benzyne undergoes addition reactions very readily. For example, the addition of ammonia or water to form aniline or phenol, respectively, occurs very rapidly.

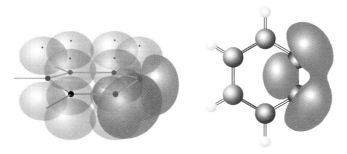

| Aniline | Benzyne | Phenol |

The mechanisms of addition reactions will be treated in greater detail in Chapter 10. For now, you should simply recognize that the elimination reaction forming benzyne followed by an addition reaction constitutes a route by which an aryl halide can be converted to an aryl alcohol or aryl amine, via a net substitution.

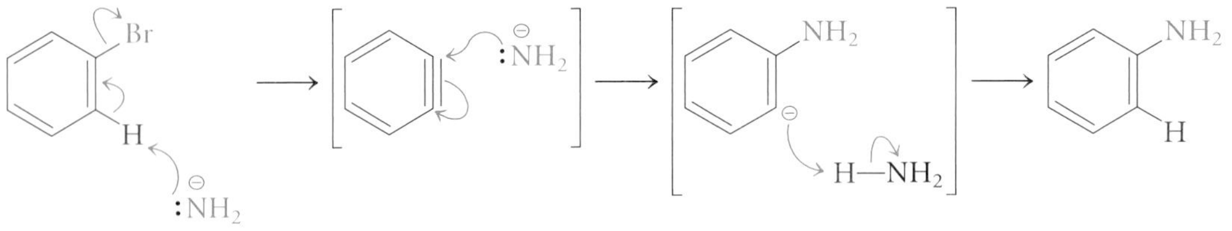

Bromobenzene          Benzyne                                    Aniline

Because this substitution takes place through an elimination–addition sequence, the simple classification as a substitution is mechanistically inadequate.

### EXERCISE 9.15

Write a mechanism for the formation of phenol from chlorobenzene upon treatment with concentrated KOH at high temperature and pressure. (This conversion is the basis for one industrial preparation of phenol.)

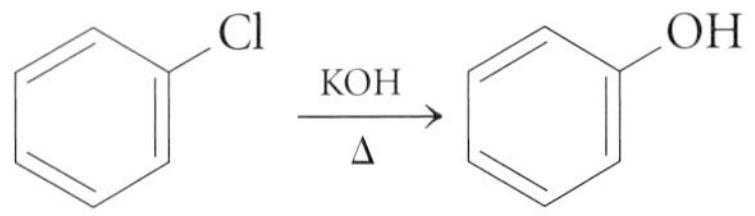

Chlorobenzene                    Phenol

## 9.9

## Oxidation

The oxidation reactions of organic compounds may be formally considered to be dihydrogen eliminations, illustrated here by oxidations of alcohols to aldehydes or ketones.

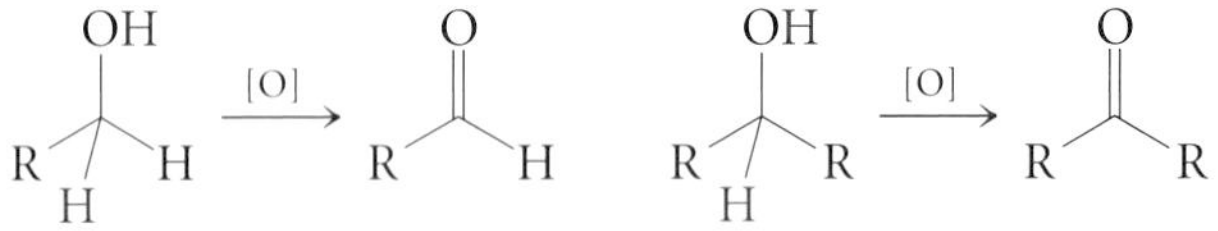

In reality, oxidation reactions of organic compounds rarely occur by direct loss of $H_2$. Rather, the *elements* of $H_2$ are eliminated as:

$$H^{\oplus} + H^{\oplus} + 2\,e^{\ominus} \qquad \text{two protons and two electrons, or}$$

$$H\cdot + H\cdot \qquad \text{two hydrogen atoms, or}$$

$$H^{\ominus} + H^{\oplus} \qquad \text{a hydride ion equivalent and a proton}$$

The nature of the functional group being reduced and of the oxidant taking part in the reaction usually determines which pathway is followed.

466

The oxidation of primary alcohols to aldehydes and of secondary alcohols to ketones using chromate ion ($Cr^{6+}$) and other transition metals proceeds as loss of two protons and two electrons. In these reactions, two electrons find their way to chromium, and two protons are lost to a weak base such as water. Other reagents that can be used to accomplish such conversions are listed in Table 3.3. Like $Cr^{6+}$ ion, these reagents have in common a stable, lower oxidation level of the metal ion. Chromium reagents are commonly used in small- to moderate-scale reactions to prepare carbonyl compounds from the corresponding alcohols. Unfortunately, it is not feasible to use these reagents for large-scale production (multiton quantities) of ketones and aldehydes, because of chromium's toxicity and the high cost of reagents in which it has a high oxidation level.

***Chromium Species Used in Oxidation Reactions.*** Aqueous solutions of metals with relatively high oxidation states often contain a number of species in equilibrium. For example, the addition of chromium trioxide, $CrO_3$, to water results in immediate hydration to form chromic acid, $H_2CrO_4$, a species resembling sulfuric acid, $H_2SO_4$. When the pH of a solution of chromic acid is increased by the addition of NaOH or KOH, the chromate ion, $CrO_4^{2-}$, is generated. This species is relatively unstable and exists in equilibrium with the dichromate ion, which can be precipitated as $Na_2Cr_2O_7$ (or $K_2Cr_2O_7$). An aqueous solution of $Cr^{6+}$ can be prepared from $CrO_3$, $Na_2Cr_2O_7$, or $K_2Cr_2O_7$. Regardless of the specific reagent used, these reactions are called chromate or chromic acid oxidations.

| Chromium trioxide | Chromic acid | Chromate ion | Dichromate ion |

***Mechanism of Chromate Oxidation.*** Treatment of an alcohol with chromic acid produces an alkyl chromate ester:

| An alcohol | Chromic acid | A chromate ester ($Cr^{6+}$) | A ketone | ($Cr^{4+}$) |

Because the chromium atom in a chromate ester has a +6 oxidation level, it is highly electron-deficient and can readily change its oxidation level by accepting additional electron density. Cleavage of the oxygen–chromium bond with simultaneous loss of a proton from the carbinol carbon and formation of a carbon–oxygen $\pi$ bond produces the product ketone, in what is effectively an E2 elimination. In this reaction, water acts as a base to remove a proton from the carbon to which the OH group was originally attached. The electrons originally in the carbon–hydrogen bond form a carbon–oxygen double bond as the electrons in the oxygen–chromium bond are shifted to chromium, reducing its oxidation level from +6 to +4. A number of secondary reactions then ensue, because chromium ultimately appears as chromium(III). However, the significant step is the conversion, in an E2-like process, of the chromate ester intermediate to a chromium(IV) species, a reduction that accompanies the two-electron oxidation of the alcohol to the ketone.

In oxidation of alcohols with $Cr^{6+}$, the initial oxidation level results in a two-electron reduction of chromium. The resulting $Cr^{4+}$ species is quite unstable and has not been observed. Disproportionation presumably occurs, with one $Cr^{4+}$ undergoing a one-electron reduction to $Cr^{3+}$, the ultimate end point for chromium in these oxidations, while the other is oxidized to $Cr^{5+}$. This latter species is also unstable and provides two electrons to oxidize another molecule of alcohol, while being reduced to $Cr^{3+}$.

### Functional Group Conversions Using Chromate Oxidations

Chromate oxidations convert secondary alcohols to ketones and primary alcohols to aldehydes and then to carboxylic acids (Figure 9.16). It is

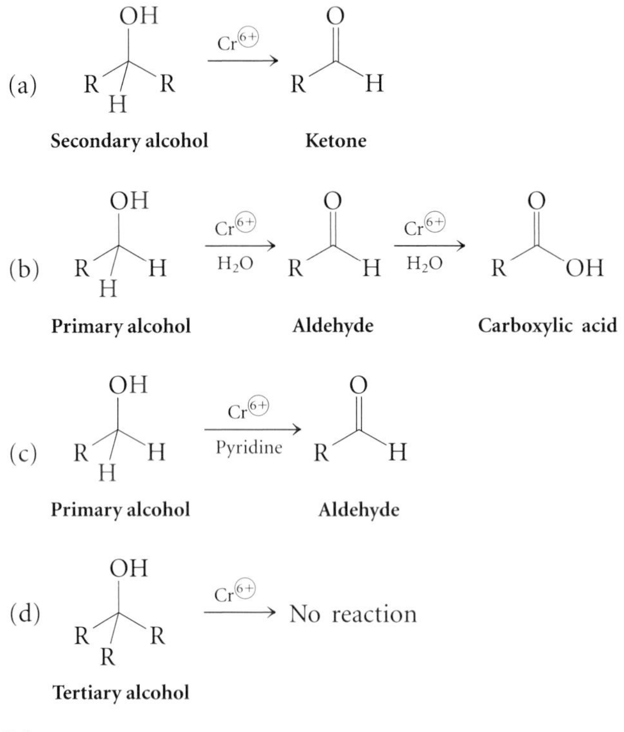

**FIGURE 9.16**

(a) Chromic acid oxidations produce a ketone from a secondary alcohol. (b) Oxidation of a primary alcohol with aqueous chromic acid produces a carboxylic acid. (c) Oxidation with pyridine produces an aldehyde. (d) No oxidation is observed with tertiary alcohols when a chromic acid oxidation is attempted at room temperature.

## CHEMICAL PERSPECTIVES

### HIGHWAY SAFETY THROUGH CHEMISTRY

The change in color from red-orange to green upon reduction of chromium(VI) to chromium(III) is the basis for the Breathalyzer test. Studies have shown that the alcohol concentration in the blood correlates well with the concentration in air exhaled from the lungs. Passing a defined volume of air through a tube containing chromate ion causes oxidation of ethanol to acetic acid and reduction of the chromium to the +3 oxidation level. The greater the concentration of alcohol in the breath, the farther the green color progresses down the tube. (The green color of jade also is due to the presence of $Cr^{3+}$ salts.)

$$CH_3CH_2OH + Cr^{6+} \longrightarrow CH_3CO_2H + Cr^{3+}$$
$$\text{Orange} \qquad\qquad\qquad \text{Green}$$

difficult to stop the oxidation of a primary alcohol at the aldehyde stage in aqueous solution. However, if the reaction is conducted in pyridine, the oxidation does stop at the aldehyde. A tertiary alcohol is resistant to chromate oxidation because it lacks a hydrogen on the carbinol carbon.

Chromate oxidation constitutes a useful chemical test for the presence of an oxidizable substrate (alcohol or aldehyde), because the orange-red chromium(VI) reagent is converted to deep green chromium(III) as the oxidation proceeds.

$$Cr(VI) \xrightarrow{+2e} [Cr(IV)]$$

$$2\ Cr(IV) \longrightarrow [Cr(V)] + Cr(III)$$

$$Cr(V) \xrightarrow{+2e} Cr(III)$$

## EXERCISE 9.16

For each of the following substrates, write the product expected (if any) from a chromate oxidation in $H_2O$.

(a) [cyclohexane with OH]

(b) [CH₃CH₂CH₂CH₂OH structure]  OH

(c) [cyclohexane with C(OH) ethyl group]

(d) [sec-butyl alcohol] OH

(e) $CH_3CH_2CH_2CO_2H$

(f) [cyclohexanone]

## EXERCISE 9.17

The oxidation of an alcohol to an aldehyde and then to a carboxylic acid in the presence of water probably includes oxidation of the hydrate of the aldehyde (a geminal diol). This hydrate is produced by the addition of water to the aldehyde in a step that does not involve chromium.

| Alcohol | Aldehyde | Geminal diol | Carboxylic acid |

Write a mechanism by which propanol can be converted to propanoic acid using chromic acid in water. (You need not concern yourself with the details of the formation of chromium–oxygen bonds or of the hydrate.)

## Biological Oxidations

In the chromate oxidation of alcohols, the carbinol hydrogen is removed as a proton from the chromate ester by a weak base (water). Such metal-centered oxidations differ significantly from biological oxidations, in which the bond between the carbinol carbon and hydrogen is broken with the opposite polarity—that is, by the loss of hydrogen as hydride ion, $H^{\ominus}$.

Protonated nicotinamide

In biological oxidations, the electron flow in the alcohol is the reverse of that in a chromic acid oxidation. The electrons from the O—H bond form the C=O bond, releasing the electrons from the C—H bond. Hydrogen is therefore transferred as hydride to an electron-deficient site, such as that found in protonated nicotinamide. Hydride ion is a nucleophilic rather than an electrophilic species, and the inverse of these oxidation reactions (namely, the addition of hydride as a nucleophile to a carbonyl group) is also important.

## 9.10

# Oxidation of Hydrocarbons: Dehydrogenation

## Direct Dehydrogenation

The conversions of alkanes to alkenes (and of alkenes to alkynes) can also be considered oxidations in which two hydrogens are eliminated. Although these reactions have high activation energies, they are routinely accomplished under conditions of high temperature—for example, in a cracking tower for the processing of petroleum. In a refinery, a gaseous stream of hydrocarbons is heated to some temperature significantly above 300 °C as it passes over alumina, causing cleavage of C—H and C—C bonds and yielding a product mixture rich in alkenes. However, these direct dehydrogenations are not usually carried out in the laboratory because they tend to be nonselective, producing a varied mixture of products. An exception is the reaction of cyclohexane, because this molecule is symmetrical:

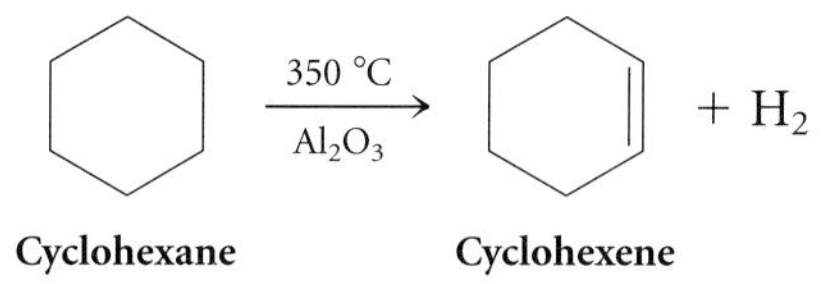

Cyclohexane     Cyclohexene

## Laboratory-Scale Dehydrogenation

***Indirect Dehydrogenation via a Series of Steps.*** Oxidation of hydrocarbons in the laboratory can be accomplished with greater control by using a series of steps. For example, the conversion of cyclohexane to cyclohexene can be achieved in good yield in two steps: free-radical halogenation followed by dehydrohalogenation.

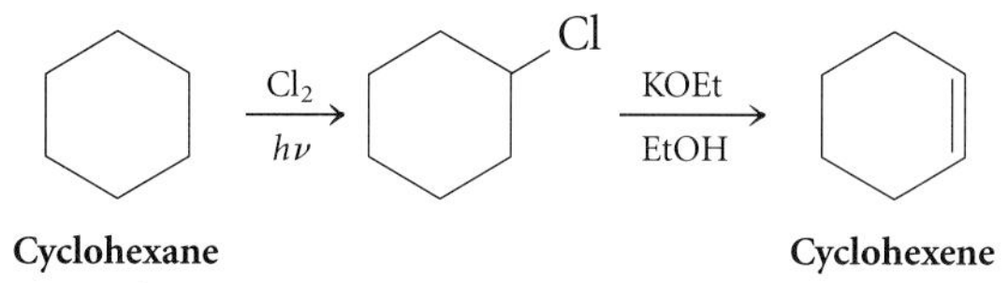

Cyclohexane     Cyclohexene

***Direct Dehydrogenation to Aromatic Compounds.*** Certain laboratory reagents can be used to effect the direct oxidation of hydrocarbons when the products are aromatic. Quinones are powerful oxidizing agents because, through reduction, they are converted to the more stable aromatic hydroquinones:

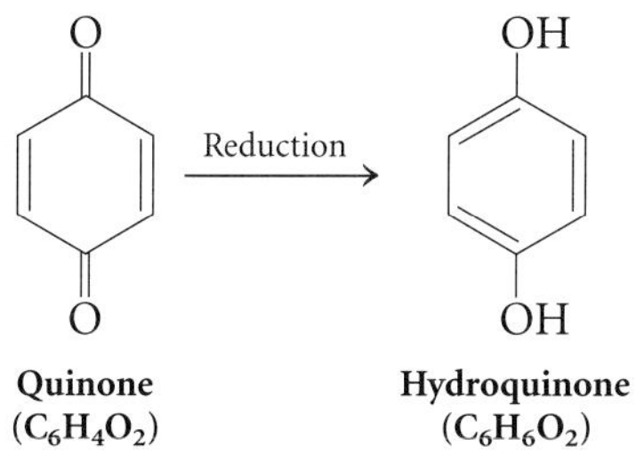

**Quinone**     **Hydroquinone**
$(C_6H_4O_2)$     $(C_6H_6O_2)$

Quinones are especially powerful oxidizing agents when electron-withdrawing substituents are present, as in tetracyanoquinone (TCNQ) and dichlorodicyanoquinone (DDQ). Oxidations using quinones involve radical intermediates, but stepwise mechanistic details are not available.

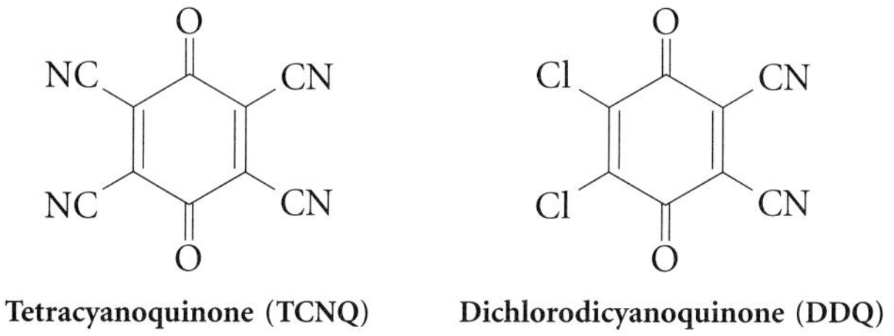

**Tetracyanoquinone (TCNQ)**     **Dichlorodicyanoquinone (DDQ)**

## HYDROQUINONES IN PHOTOGRAPHY

The chemistry of photography is fascinating, in part because many aspects are still not very well understood. When a photon of light is absorbed by the surface of a crystal of a silver halide (AgBr, for example), the energy of the photon is consumed in the reduction of $Ag^{\oplus}$ to $Ag^0$. This conversion creates a defect site that makes that face of the crystal particularly susceptible to further reduction during development of what is known as the *latent image*. One common reducing agent used to reduce the remainder of the $Ag^{\oplus}$ is hydroquinone, which in the process is oxidized to quinone. This oxidation–reduction reaction is quite exothermic, and the heat released propagates the reduction throughout the crystal, leaving behind a trail of silver metal and creating part of the image seen on a photographic print.

In effect, the single photon that created the original defect by converting one Ag(I) to Ag(0) is amplified as all the remaining silver atoms in the crystal are also reduced. The larger the original crystal, the more silver metal is produced and the darker the image. Thus, high-speed films intended for use in low-light situations have relatively large crystals of silver halide. The larger the crystal, the larger is the trail of silver metal created as the reduction takes place. High-speed films therefore yield "grainy" images, with relatively large clusters of silver metal being produced for each photon absorbed on the plate. High-definition films have very small crystals of silver halide and show fine grain. However, each crystal produces substantially less silver metal, and so such films require substantially more light to activate a large number of the smaller crystals.

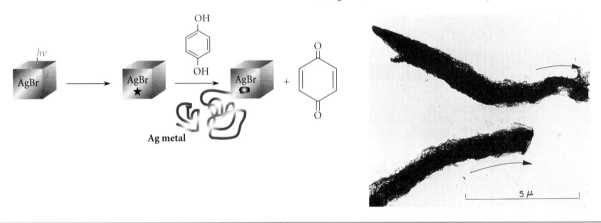

Each of these quinones can be used to oxidize hydrocarbons and other functionalized compounds, often in a controlled fashion. However, the relatively high cost of these reagents precludes their use in large-scale oxidations.

## QUINONES AND PHENOLS IN FOODS AND PLASTICS

Phenols, like hydroquinones, are readily oxidized, and compounds such as butylated hydroxytoluene (BHT) are used as preservatives in foods, as well as in other products such as rubbers and plastics. These readily reduced aromatic compounds inhibit the oxidation of organic compounds by molecular oxygen. They do this by acting as "radical scavengers," combining with hydroperoxide radicals, which are the chain-propagating species in the process by which organic compounds are oxidized by molecular oxygen.

Quinones are also essential components of the biological oxidation–reduction (redox) processes that are essential, for example, for muscle function. Ubiquinones occur in most aerobic organisms, including bacteria, plants, and animals—the prefix *ubi*- stands for "ubiquitous." Coenzyme $Q_{10}$ has been used clinically to treat cardiac insufficiency, a disease state in which there is insufficient blood flow through the heart, presumably because the body's own production of $Q_{10}$ has become insufficient. Coenzyme $Q_{10}$ is a popular item in the vitamin section of health food stores.

**Butylated hydroxytoluene (BHT)**

**Coenzyme Q**
($n = 6$–$10$)

---

## EXERCISE 9.18

Suggest a sequence of steps to accomplish each of the following transformations:

(a)

(b)

(c)

---

## 9.11

# Synthetic Methods

Elimination reactions can be grouped according to the functional-group conversion accomplished. Table 9.1 provides a summary. This table is intended to draw your attention to the synthetic applicability of each reaction. In planning syntheses, it is useful to have several possible ways to make a given functional group. Furthermore, you thoroughly understand a reaction only if you can recognize the reactant needed to make a specific product. As before, you may find it useful to write these reactions on flash cards and use them for study drill.

TABLE 9.1

Using Elimination Reactions to Make Various Functional Groups

| Functional Group | Reaction | Example |
|---|---|---|
| Aldehyde | Chromate oxidation of a primary alcohol in pyridine | |
| Alkene | Dehydrohalogenation of an alkyl halide | |
| | Dehydration of an alcohol | |
| | Debromination of a vicinal dibromide | |
| | Radical halogenation of an alkane, followed by dehydrohalogenation | |
| | High-temperature dehydrogenation of an alkane | |
| Alkyne | Dehydrohalogenation of a vinyl halide | |
| | Dehydrohalogenation of a geminal dihalide | |

## Summary

1. In an elimination reaction, two bonds are cleaved from adjacent positions, forming a $\pi$ bond.

2. When the groups eliminated are hydrogen and a halide ion, two mechanisms are commonly encountered: (a) the rate-determining loss of the leaving group to generate a carbocation (E1 elimination), and (b) a concerted reaction in which a base abstracts a proton while a double bond is being formed as the leaving group leaves with its electron pair (E2 elimination). The E1cB mechanism, in which an E2-type elimination occurs from the conjugate anion of the substrate, is normally encountered only in spe-

| Functional Group | Reaction | Example |
|---|---|---|
| Alkyne | Dehydrohalogenation of a vicinal dihalide | |
| Aniline | Ammoniation of benzyne | |
| Benzyne | Dehydrohalogenation of an aryl halide | |
| Carboxylic acid | Chromate oxidation of a primary alcohol in water | |
| | Chromate oxidation of an aldehyde in water | |
| Ketone | Chromate oxidation of a secondary alcohol | |
| Phenol | Hydration of benzyne | |

cial cases—for example, in loss of water from $\beta$-hydroxycarbonyl compounds and in vinyl halides.

**3.** Dehydration of primary and secondary alcohols via an E1 mechanism occurs readily under acid-catalyzed conditions. Under the same conditions, primary alcohols undergo dehydration via an E2 mechanism. Under basic conditions, the dehydration of $\beta$-hydroxycarbonyl compounds by the E1cB pathway takes place with ease because of the formation of a conjugated double bond.

**4.** The E2 pathway is observed for elimination of HX from primary and secondary alkyl halides under basic conditions and from primary alcohols under acidic conditions.

475

**5.** The transition state for the E2 elimination involves an *anti*-periplanar relationship between the two atoms that are lost, hydrogen and the leaving group.

**6.** The *anti*-periplanar geometry required by the E2 transition state dictates the stereochemistry of the products, usually resulting in formation of the most stable stereoisomer.

**7.** The *anti*-periplanar geometry also affects the regiochemistry of the E2 elimination product. However, in the E2 elimination, the size of the base is an additional factor. With small bases, the most stable product is formed (Zaitsev's rule), but with larger bases the less substituted, less stable isomer can predominate (Hofmann elimination).

**8.** E1 eliminations commonly take place at tertiary halide centers, less commonly at secondary centers, and almost never at primary centers.

**9.** In the two-step E1 elimination, the first step is rate-determining, and the second step is product-determining. The rate of the first step, in which a carbocation is formed, is fastest for formation of a tertiary carbocation and slowest for formation of a primary carbocation.

**10.** The steroechemistry of an E1 elimination favors formation of the most stable geometric isomer. The regiochemistry of the reaction also favors formation of that isomer (Zaitsev orientation).

**11.** Because the E1 elimination proceeds through formation of an intermediate carbocation, it is complicated by side reactions characteristic of that intermediate—that is, substitutions and rearrangements.

**12.** Dehydrohalogenation of vinyl bromides is unlikely to proceed through an E1 mechanism because of the instability of a vinyl cation. Instead, alkynes are formed either in a concerted pathway or through a vinyl anion. Alkyne formation takes place upon treatment with strong base ($NaNH_2$).

**13.** The elimination of HX from an aromatic halide produces benzyne, an interesting compound in which ring aromaticity is maintained while an orthogonal $\pi$ bond is formed between two of the carbons in the ring. The geometric distortion from the ideal 180° angle for *sp*-hybridized atoms makes benzyne exceedingly reactive, and an elimination–addition sequence provides a route for aromatic substitutions that convert aryl bromides into phenols and anilines.

**14.** The elimination of $X_2$ from vicinal dihalides can be accomplished by treatment with an easily oxidizable zero-valent metal such as zinc or tin.

**15.** Although oxidation reactions constitute formal elimination of $H_2$, the mechanisms by which they occur are often complex.

**16.** Metal-centered oxidations of alcohols (chromate oxidations) take place through a simultaneous change in the oxidation level of the metal ion or complex.

**17.** The direct elimination of two hydrogen atoms from alkanes is not a practical laboratory method for the oxidation of hydrocarbons. In the laboratory, hydrocarbons are usually oxidized by sequences of functional-group manipulations.

## Review of Reactions

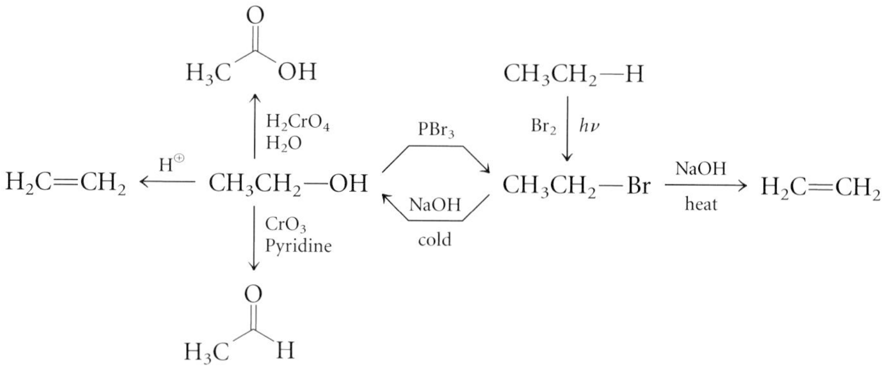

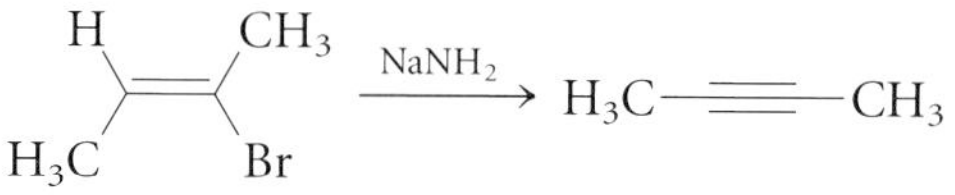

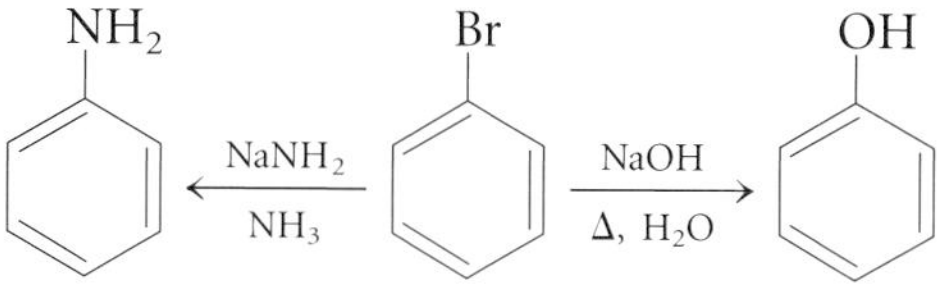

## Review Problems

**9.1** Using curved arrows to show the electron flow, write the preferred reaction mechanism for each of the following eliminations. Briefly discuss the reasoning that led you to choose an E1, E1cB, or E2 mechanism.

(a) 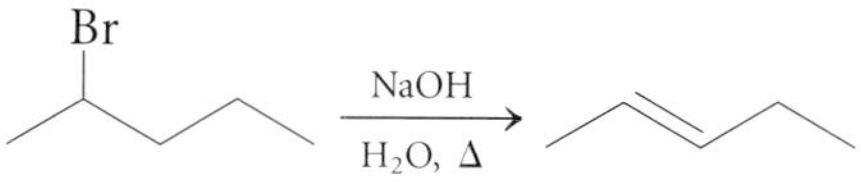

(b) 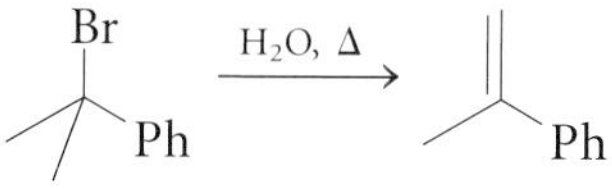

(c) 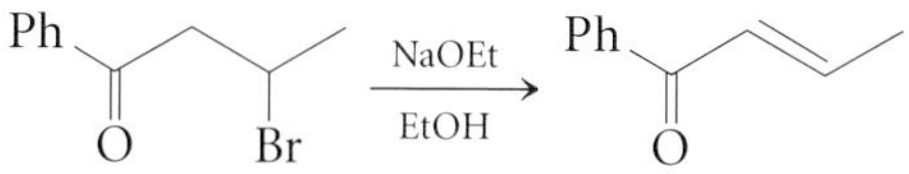

**9.2** For each of the following pairs of compounds, choose the one that better fits the description.

(a) Follows Zaitsev orientation in an E2 reaction:

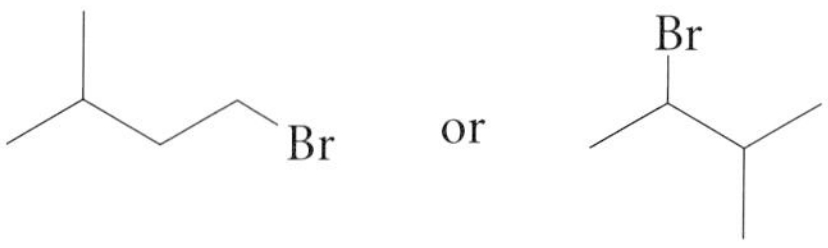

(b) Reacts more rapidly with cold aqueous HBr:

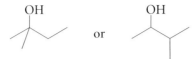

or

(c) Undergoes elimination with less competing nucleophilic substitution upon treatment with HBr:

OH

Ph   or   HO ⟍⟍⟍ Ph

(d) Gives a mixture of two alkenes in E1 elimination induced by sulfuric acid:

OH

or

OH

(e) Gives exactly three alkenes in E2 elimination:

Cl

or

Cl

**9.3** Predict the major product expected for each of the following reactions:

(a) Ph, Br, H₂O, Δ

(d) OTs, NaOEt, EtOH

(b) Ph, OH, Cold HBr

(e) I, NaOH, H₂O

(c) OH, Δ, H₂SO₄

(f) O, Cl, NaOEt, EtOH

**9.4** Predict whether the amount of Hofmann elimination observed in each of the following reactions is larger, smaller, or unchanged from that observed in the treatment of 2-bromobutane with sodium ethoxide in ethanol:

(a) 2-chlorobutane with sodium ethoxide in ethanol

(b) 2-iodobutane with sodium ethoxide in ethanol

(c) 2-methyl-2-bromobutane with hot water

(d) 2-bromobutane with sodium t-butoxide

(e) 2-bromobutane with sodium hydroxide

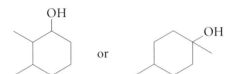

**9.5** Assume that treatment of the following cyclohexyl halides with base effects an elimination through an E2 mechanism. Draw Newman projections to represent the transition states for all possible products, and predict the preferred geometry of the product.

(a)

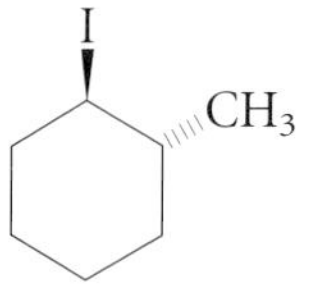

(b)

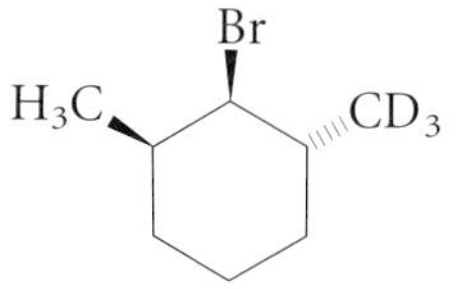

**9.6** If the starting material is labeled with deuterium as indicated, predict how many deuterium atoms will be present in the major elimination product and where they will appear.

(a)

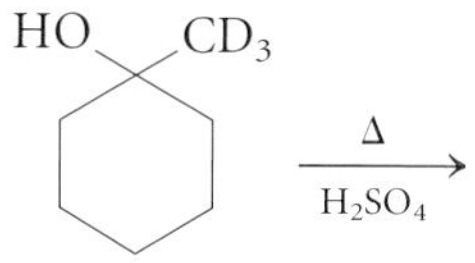

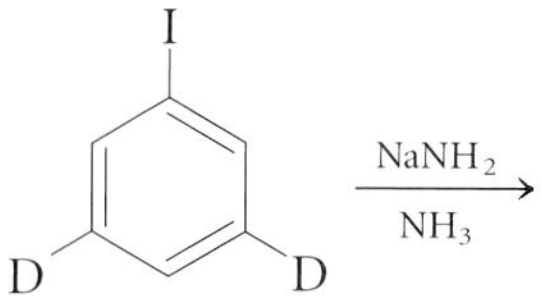

(b)

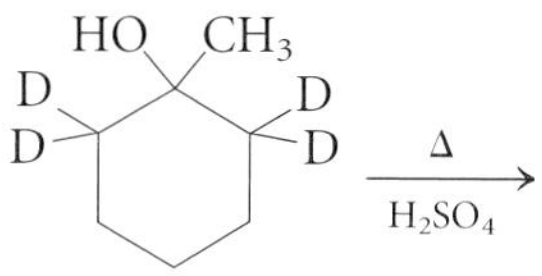

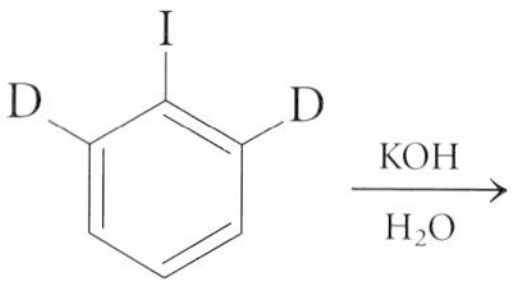

(c)

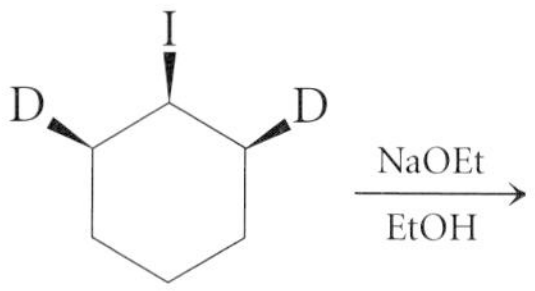

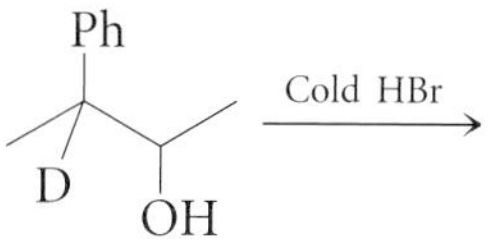

(d)

(e)

(f)

(g)

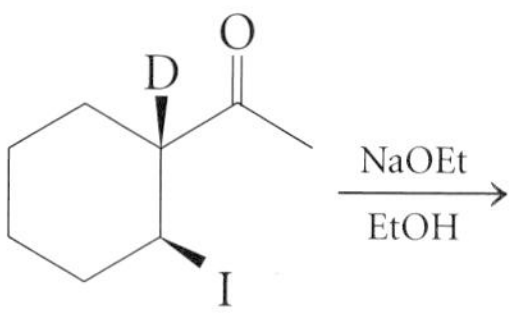

**9.7** In the following reactions, a rearranged skeleton is observed in the principal product. Write a mechanism for each reaction that leads to the observed product. Use curved arrows to indicate electron flow.

(a)

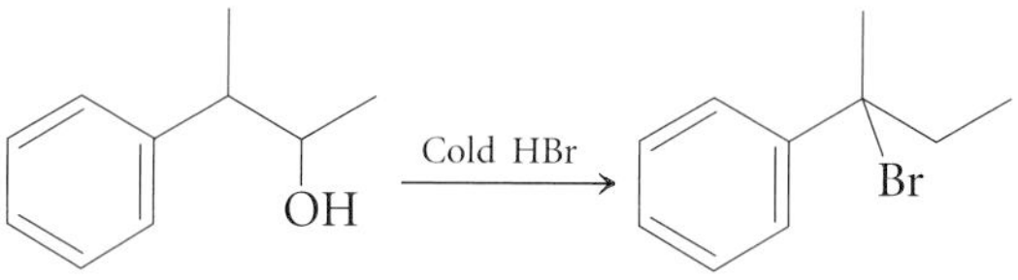

(b)

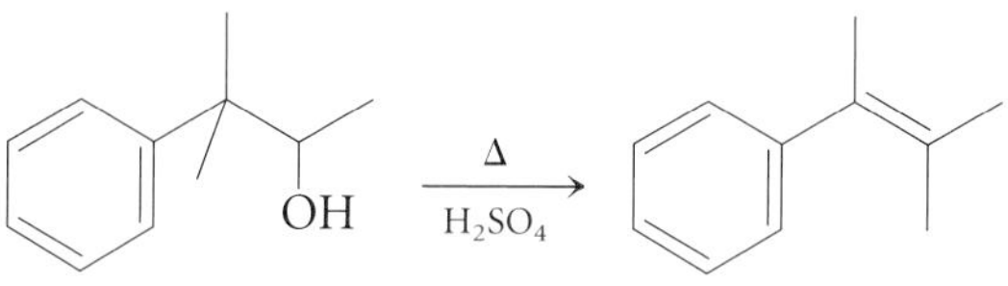

**9.8** Draw the structure of the major product expected when 2-butanol is treated with each of the following sequences of reagents:

(a) (1) $PBr_3$; (2) NaOEt, EtOH

(b) (1) hot $H_2SO_4$; (2) $Br_2$, $CCl_4$; (3) $NaNH_2$, $NH_3$

(c) (1) cold HBr; (2) KO-$t$-Bu, $t$-BuOH

(d) (1) $SOCl_2$ in pyridine; (2) NaOH in EtOH; (3) $Br_2$ in $CCl_4$; (4) Zn in $Et_2O$

**9.9** For each of the following compounds, describe what, if anything, you would see upon treatment with aqueous chromic acid. What would you see with Lucas reagent (Chapter 3)?

(a) 2-octanol

(b) 3-octanol

(c) cycloheptanol

(d) 1-methylcycloheptanol

(e) 3-methyl-1-octanol

(f) 3-methyl-3-octanol

(g) 3-octanone

(h) octanal

**9.10** Propose three routes, employing different starting materials, by which you could synthesize 2-butyne.

**9.11** In each of the following reactions, two or more regioisomers are found as products. Draw the structures of the isomers.

(a)

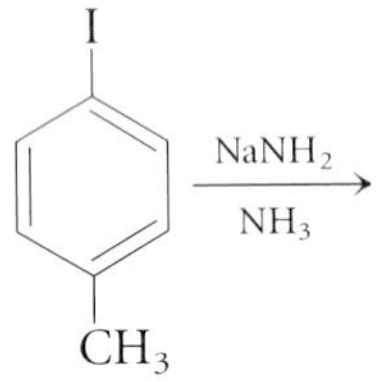

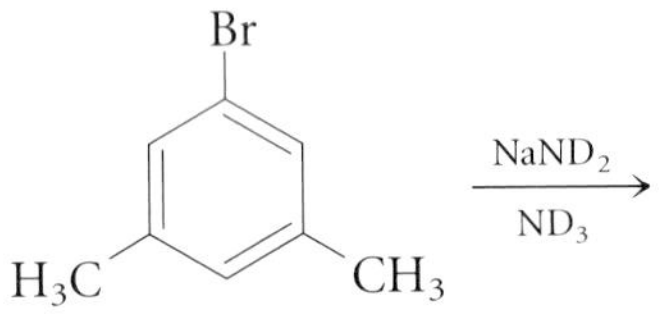

(b)

(c)

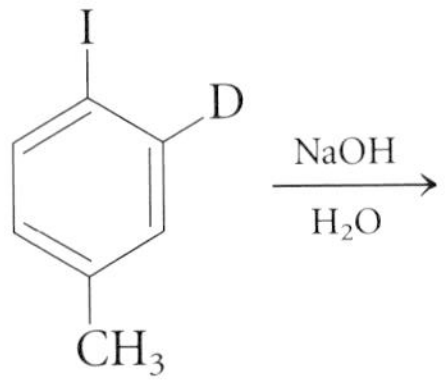

(d)

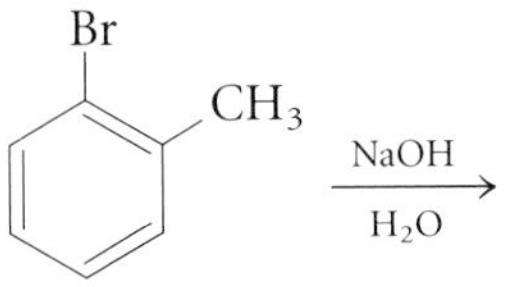

**9.12** One of the hardest predictions to make before a reaction is run is how important elimination will be as competition for the desired substitution. Shown here are $S_N1$ and $S_N2$ reactions. Draw the structures of the elimination products that might be competitively formed. Suggest how you would use spectroscopy to determine whether the major product isolated is a substitution or an elimination product. Write a reaction mechanism that shows how each elimination product might be formed.

(a)

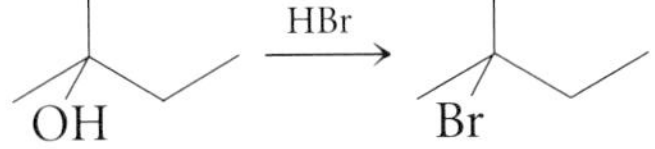

(b)

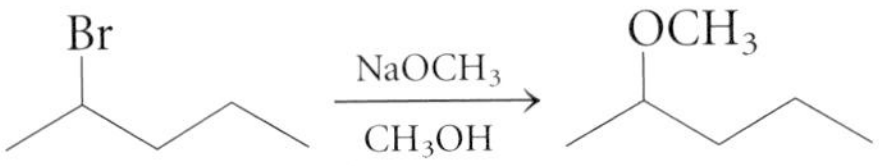

## Supplementary Problems

**9.13** Provide an IUPAC name for each of the following unsaturated compounds. Suggest an alkyl halide from which each could be prepared by an elimination reaction.

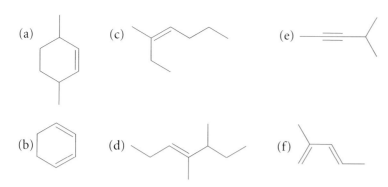

**9.14** Provide a structure that corresponds to each of the following names. Suggest an alkyl halide from which each compound could be prepared by an elimination reaction.

(a) 3-chloro-1-butene

(b) *trans*-4,4-dimethyl-2-heptene

(c) 1,8-dimethyl-*cis*-cyclooctene

(d) 2,5-dimethyl-3-hexyne

**9.15** E2 elimination reactions generally involve *anti* elimination. However, there are exceptions—particularly in the case of compounds with the bicyclo[2.2.1]heptane skeleton. Can you provide an explanation for why *syn* elimination is often preferred in this system?

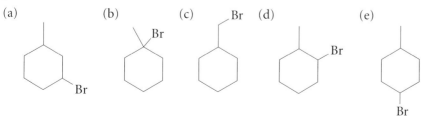

**9.16** The preparation of 2,4,5-trichlorophenoxyacetic acid involves the treatment of 2,4,5-trichlorophenol with sodium hydroxide and chloroacetic acid. During this process, some of the phenol is converted to a small amount of the carcinogen 2,3,7,8-tetrachlorodibenzo-*p*-dioxin. Account for the formation of both products by writing detailed reaction mechanisms. (*Hint:* The reaction involves a substituted benzyne intermediate.)

**9.17** Draw structures for all possible alkenes that could be formed by loss of HBr from each of the following alkyl halides. In each case, predict which will predominate based on Zaitsev's rule.

**9.18** Three of the five alkyl bromides in Problem 9.17 can exist in two stereoisomeric forms. Identify these compounds. Indicate for each how the stereoisomers might behave differently when treated under E2 reaction conditions.

**9.19** Two of the alkyl bromides in Problem 9.17 are chiral, but one produces an achiral alkene as one of the elimination products. Which alkyl bromide fits this description? Which of the alkyl bromides is achiral but produces a chiral alkene (as a racemic mixture) upon elimination of HBr?

**9.20** Elimination reactions almost invariably produce by-products in which the base has acted as a nucleophile. For example, an alcohol is produced along with an alkene upon treatment of an alkyl halide with hydroxide ion. Rank the alkyl halides in Problem 9.17 in decreasing order of ease of $S_N2$ substitution. In the three alkyl bromides that can exist as two stereoisomers, consider which stereoisomer would undergo a greater degree of substitution.

**9.21** Treatment of both $\alpha$- and $\beta$-bromoketones with base results in loss of HBr to form $\alpha,\beta$-unsaturated ketones. However, the former react much more slowly and require much stronger bases. Account for this difference in reactivity by providing a complete analysis of both reaction mechanisms.

**9.22** Treatment of vicinal dibromides with a strong base such as $NaNH_2$ generally leads to the formation of alkynes, although lesser amounts of dienes and allenes are also produced.

The formation of the alkyne requires regiospecific removal of protons in both steps: first, from the carbon bearing one of the bromine atoms, and second, from the vinylic carbon of the intermediate vinylic bromide. Explain why in each case deprotonation with the regiochemistry required for formation of the alkyne is preferred.

**9.23** The E stereochemistry of the vinylic bromide shown as an intermediate in Problem 9.22 is determined by the stereochemistry of the starting vicinal dibromide (not shown in the problem). Determine which diastereomer of 2,3-dibromobutane is required to produce (E)-2-bromo-2-butene. Draw a Newman projection of this diastereomer in an appropriate conformation for entering into the reaction to form the alkene. Again using a Newman projection, show how the other diastereomer would yield (Z)-2-bromo-2-butene.

**9.24** Treatment of 1,8-bis-(bromomethyl)naphthalene with butyllithium results in the formation of the tricylic product shown. Provide a detailed reaction mechanism for this conversion.

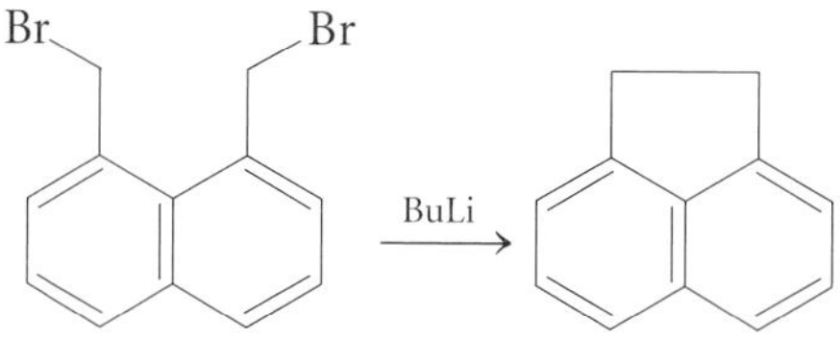

**9.25** How might you convert the product of Problem 9.24 to the corresponding alkene?

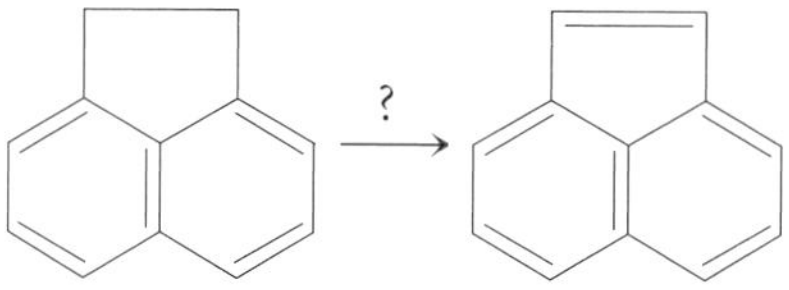

**9.26** Under sufficiently mild conditions, diols such as the two shown here undergo selective monodehydration to form alkenols. Suggest a reason for the selective formation of the indicated products.

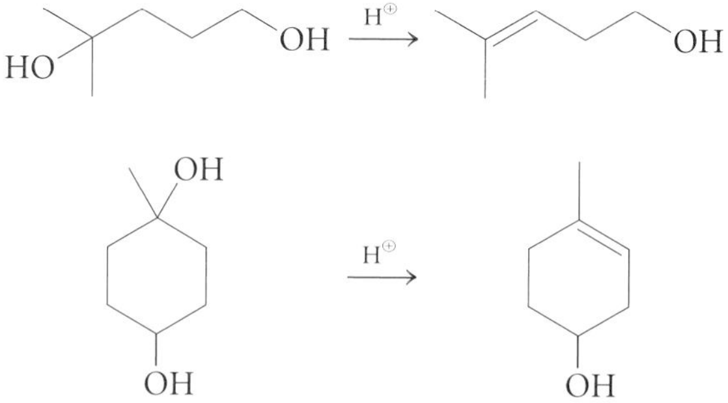

**9.27** Treatment of a vicinal dibromide with strong base generally results in the formation of an alkyne.

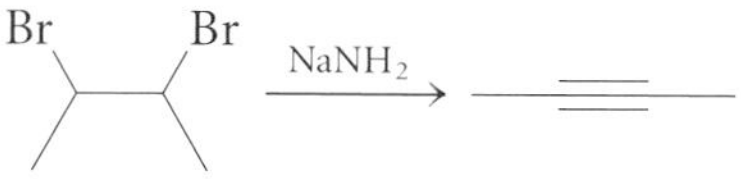

Nevertheless, yields of the alkyne are only moderate because of the formation of other products.

(a) What other compounds might be formed by the double dehydrohalogenation of a vicinal dibromide?

(b) Can you think of a vicinal dibromide for which formation of an alkyne might be highly unfavorable?

**9.28** Upon treatment with acid, 2,2-dimethylcyclohexanol undergoes dehydration to form a mixture of 3,3-dimethylcyclohexene and another alkene that exhibits only four unique signals in the $^{13}C$ NMR spectrum yet has a mass of 110, corresponding to the formula $C_8H_{14}$. Assign a structure to this product, and account for its formation with a detailed reaction mechanism.

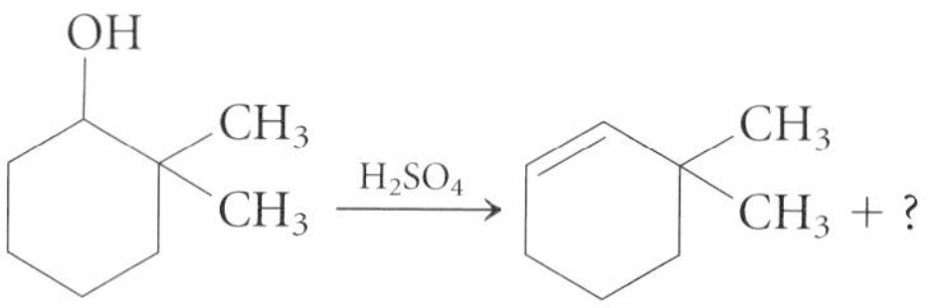

**9.29** Estimate the chemical shifts of the carbons of the alkenes produced from the alkyl bromide in part (c) of Problem 9.17. Note which signals could be used to make an assignment of the isomer that predominates in the reaction product mixture.

**9.30** How would the $^1$H NMR spectra for 1- and 3-methylcyclohexene differ? How would their $^{13}$C NMR spectra differ?

**9.31** How would you use proton NMR spectroscopy to distinguish among the three hydrocarbon products in Problem 9.22 (an alkyne, an allene, and a diene)?

# 10

# Addition to Carbon–Carbon Multiple Bonds

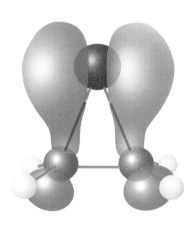
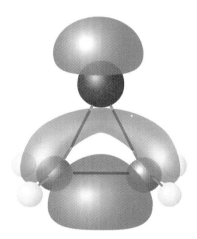

*The two molecular orbitals for the pair of C—Br bonds in the bromonium ion formed upon the addition of bromine to ethylene:*

**Bromonium ion**

*I*n Chapter 9, we discussed the elimination of two groups from adjacent carbon atoms to form a multiple bond. In this chapter, we will consider the inverse reaction: the addition of an electrophile to one carbon atom of a multiple bond and a nucleophile to the other carbon atom. Some of these addition reactions follow pathways that are the reverse of those for elimination reactions.

We will also deal in more detail with the Diels–Alder reaction (introduced in Chapter 6) and related reactions—all of which are characterized by a concerted mechanism, proceeding from starting material to product without the formation of intermediates.

## 10.1

## Electrophilic Addition of HCl, HBr, and H₂O

An **electrophilic addition** is called an *addition* because the product incorporates all atoms of both reactants; it is called *electrophilic* because it is initiated by interaction of the $\pi$ bond with an electrophile.

The characteristic carbon–carbon multiple bonds of alkenes and alkynes constitute sites of high electron density, which are readily attacked by electron-deficient reagents called **electrophiles.** In contrast, alkenes and alkynes are relatively unreactive with the nucleophiles and bases that attack $sp^3$-hybridized carbon atoms, discussed in Chapters 7 through 9. In this chapter, we focus on addition reactions initiated by electrophilic attack on alkenes, alkynes, and dienes. In electrophilic addition to an alkene, the carbon–carbon $\pi$ bond is replaced by two $\sigma$ bonds to new substituents. Typically, one substituent is an electrophile and the other a nucleophile.

486

The electrons in the $\pi$ bond of an alkene are less tightly held than those in $\sigma$ bonds, where the atomic orbitals involved ($sp$, $sp^2$, or $sp^3$) have significant $s$ orbital character. These $\pi$ electrons are available to begin the process of addition by bond formation with an electrophile.

## Mechanism of Electrophilic Addition

***Electrophilic Addition: First Step.*** Let's consider a typical example of electrophilic addition:

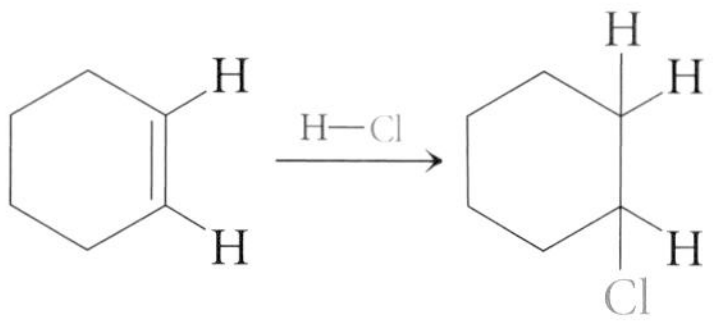

This reaction is initiated by the addition of an electrophile (often a proton) to a C=C bond, as in the electrophilic addition of HCl to cyclohexene, discussed briefly in Chapter 7. In this step, an electron-deficient reagent, $E^{\oplus}$ (often a proton, $H^{\oplus}$), approaches the $\pi$ cloud of a carbon–carbon double bond. The electrons of the $\pi$ bond flow toward the electrophile, forming a C—E $\sigma$ bond and a carbocation. In this step, a carbon–carbon $\pi$ bond and the $\sigma$ bond between electrophile and nucleophile are broken.

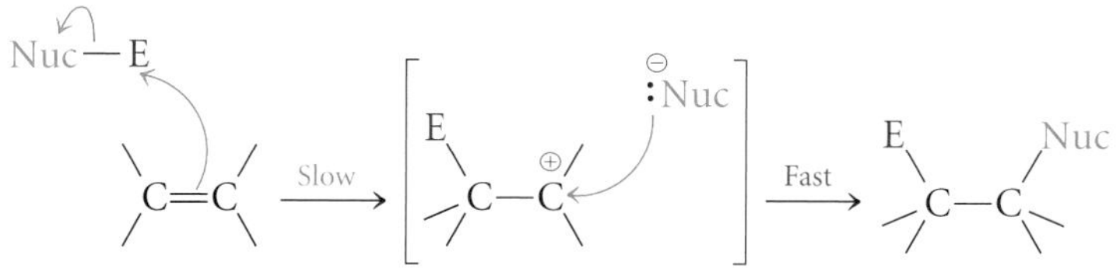

***Electrophilic Addition: Second Step.*** The carbocation formed in the first step is easily attacked in a second step by a reagent that acts as a nucleophile. When the nucleophile attacks this carbocation, it donates its lone pair of electrons to form a new $\sigma$ bond between itself and carbon, completing the addition. Overall, in an electrophilic addition, two new $\sigma$ bonds are formed to the two carbon atoms that originally participated in the carbon–carbon $\pi$ bond: one $\sigma$ bond with an electrophile and one with a nucleophile. Because the second step of electrophilic addition consists only of bond making, it is faster than the first. Therefore, the first step—the step in which the carbocation is formed—is rate-determining.

***Thermodynamics of Electrophilic Addition.*** Is an electrophilic addition to an alkene thermodynamically favorable? Overall, the $\pi$ bond of the alkene and a $\sigma$ bond between the electrophile and the nucleophile are broken. In their place, two new $\sigma$ bonds are formed: one between the electrophile and one of the $\pi$-bonded carbons and the other between the nucleophile and the other carbon. Calculation of the enthalpy change accompanying these transformations requires knowledge of the bond energies for the specific bonds being broken and made. For example, in the addition of HCl to cyclohexene, both the $\pi$ bond of the alkene and the $\sigma$ bond of HCl are broken and replaced by a C—H and a C—Cl bond in the addition product.

$$\Delta H = (63 + 103) - (99 + 81) = -14 \text{ kcal/mole}$$

However, even without the bond-energy information, we can draw some qualitative conclusions about the thermodynamics of such reactions. Most $\sigma$ bonds to carbon are stronger than either a typical carbon–carbon $\pi$ bond (Table 3.2) or the common electrophile–nucleophile bond. (The $\pi$ bond contribution in an alkene is 63 kcal/mole, which is the difference between the $\pi$ and $\sigma$ bonds: 146 kcal/mole − 83 kcal/mole.) As a result, the combined strength of the two bonds formed in electrophilic addition (C—E and C—Nuc) usually exceeds that of the bonds consumed (C—C $\pi$ and E—Nuc), and most electrophilic additions are exothermic.

### EXERCISE 10.1

Calculate $\Delta H$ for each of the following addition reactions using the bond energies provided in Table 3.2.

(a)

(b)

(c)

*Kinetics of Electrophilic Addition.*  An understanding of the controlling features of the two-step pathway involved in electrophilic addition to alkenes depends on knowing which step is rate-determining. The rate-determining step is that in which the transition state of highest energy is formed, and the factors that stabilize (or destabilize) this transition state directly affect the rate of reaction and the relative reactivities of different alkenes.

In the first step of electrophilic addition, a C—C $\pi$ bond and an E—Nuc $\sigma$ bond are broken as a C—E $\sigma$ bond is formed. Because two bonds are being broken while only one is being formed, this step is generally endothermic. In contrast, the second step consists only of bond making and is highly exothermic. Therefore, the first step, producing a carbocation, is likely to be rate-determining. Because the stability of the transition state for an endothermic reaction is substantially influenced by the product, the transi-

## CHEMICAL PERSPECTIVES

### A HYPERVALENT CARBOCATION

The strengths of acids, as measured by $pK_a$ values, range from very weak ones such as hydrocarbons to acids that are much stronger than sulfuric acid. Acids with acidities equal to or greater than sulfuric acid are called *super acids*. Some of them are so acidic that they can donate a proton to an alkane, even to methane to form the $CH_5^{\oplus}$ ion, first prepared by the American chemist George Olah. Such cations are said to be *hypervalent* because they formally possess "extra" bonds beyond those normally formed by second-row atoms. Thus, they differ in character from the trivalent carbocations encountered as intermediates in electrophilic addition. There can be no "classical" formalism for a hypervalent ion, because it does not conform to the rules of valency ($C = 4$, $H = 1$). In one possible "nonclassical" structure, the added proton is associated side-on with the electron density of one of the C—H bonds of methane. These hypervalent ions are unstable and undergo loss of $H_2$.

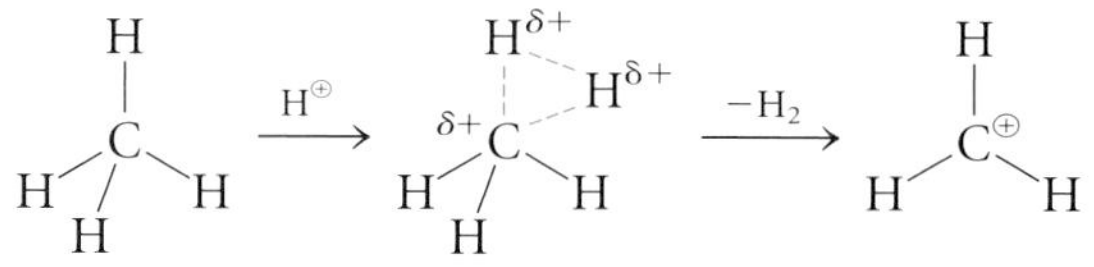

tion state of the first step in this sequence resembles the cationic intermediate. Thus, the rates of electrophilic addition reactions increase with increasing stability of the intermediate cation.

### EXERCISE 10.2

Use your knowledge of normal bond polarization to identify the part of each of the following reagents that can act as an electrophile:

(a) H—Br

(e) Br₂

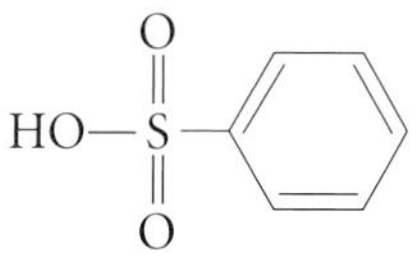

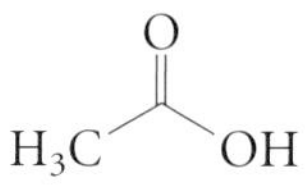

(b)

(f) HOCl

(c)

(g)

(d) Cl₂

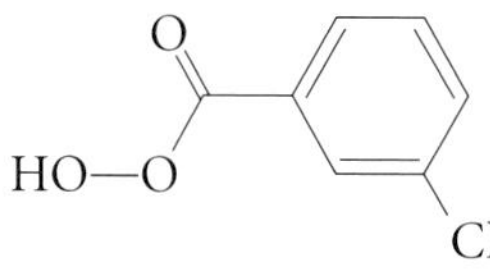

(h) H₂O

### Addition of HCl

Let's consider a specific example, the addition of hydrogen chloride to *cis*-2-butene to produce 2-chlorobutane, first in the gas phase and then in solution.

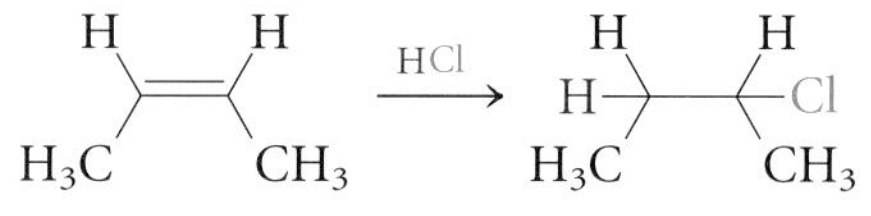

*cis*-2-Butene        2-Chlorobutane

**Gas-Phase Reactivity.** Because chlorine is more electronegative than hydrogen, the $\sigma$ bond connecting these atoms in HCl is polarized toward chlorine, leaving the hydrogen electron-deficient and therefore electrophilic. Thus, the proton assumes the role of electrophile in initiating the addition (Figure 10.1).

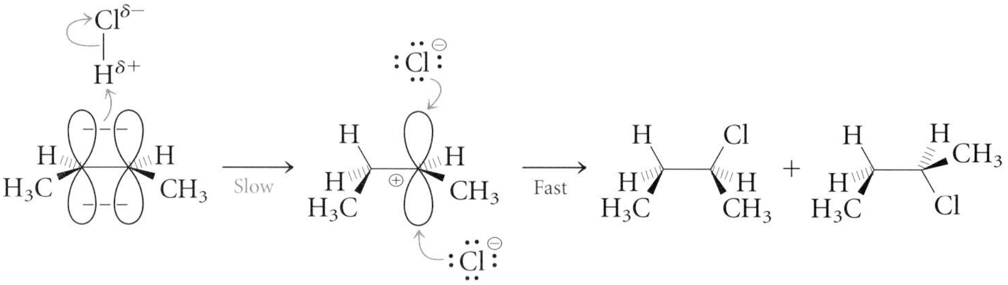

**FIGURE 10.1**

The electrostatic attraction between the partial positive charge on hydrogen in the polar H—Cl $\sigma$ bond and the electron cloud of the alkene $\pi$ bond brings these molecules into a geometry in which electrons can flow from the $\pi$ orbital to form a $\sigma$ bond between carbon and hydrogen. Chloride ion, formed by taking up the electrons of the H—Cl $\sigma$ bond, then approaches the vacant $p$ orbital of the carbocation. Because the cation is $sp^2$-hybridized and planar, this approach is equally easy on the top or bottom face. Equal amounts of each enantiomer are therefore formed at the new center of chirality.

The partially positively charged end of H—Cl approaches the $\pi$ bond (either above or below the carbon plane), because this is the region where the electron density is greatest. The proximity of the electrophile causes electrons to flow from the $\pi$ bond to form a $\sigma$ bond. The flow of the two electrons, originally in the polar covalent H—Cl bond, toward the chlorine atom results in formation of a chloride ion. As a $\sigma$ bond is formed between one carbon and the proton, the other carbon takes on carbocationic character. This ion pair is less stable in the gas phase than in solution, because of the absence of solvent interactions with both the cation and the anion. In a rapid second step, chloride ion reacts with the carbocation to form a second $\sigma$ bond. (Note that approach of chlorine from above and below the carbons results in two enantiomers as product.)

**Solution-Phase Reactivity: Effect of Solvation on Generation of the Carbocation Intermediate.** The character of the reaction of HCl with an alkene is changed dramatically when water is used as a solvent, and solvation plays a major role in the generation of electrophiles from acids in aqueous solution. Water stabilizes cations through interaction with the lone pairs of electrons on oxygen. Indeed, HCl dissolves in water with essentially complete dissociation to form a hydronium ion and a chloride ion:

$$H_2O + HCl \rightleftharpoons H_3O^{\oplus} + Cl^{\ominus}$$

In water, it is the hydronium ion that donates a proton to the alkene; in the gas-phase reaction, it is HCl. This distinction leads to a dramatic difference in reaction rates.

In the gas phase, the first step of electrophilic addition yields a carbocation and a chloride ion, and the resulting separation of charge makes achieving the transition state highly unfavorable energetically. When the reaction of HCl with an alkene takes place in water, addition of HCl to water also results in charge separation, with the formation of $H_3O^\oplus$ and $Cl^\ominus$. Nonetheless, this reaction is energetically favorable because of the solvation of both ions by water (Figure 10.2).

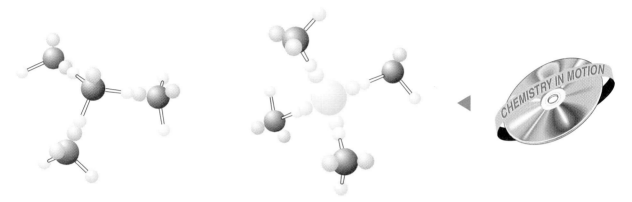

**FIGURE 10.2**

The hydronium ion (left) and chloride ion (right) formed by the reaction of HCl with $H_2O$ are both solvated in water. This solvation more than compensates for the accompanying charge separation. (The chloride ion and its lone pairs are light green, oxygen is red, hydrogen is off-white, and the lone pairs of electrons on oxygen are small pink spheres.)

The subsequent transfer of the proton from the hydronium ion to the alkene involves no additional charge separation in the transition state, because one cation is changed into another.

$$\begin{array}{c}\text{H}\qquad\text{H}\\\diagup\!\!=\!\!\diagdown\end{array} + \text{H}_3\text{O}^\oplus + \text{Cl}^\ominus \xrightarrow[\text{H}_2\text{O}]{} \text{H} \begin{array}{c}\text{H}\qquad\text{H}\\\diagup\;\;\oplus\;\diagdown\end{array} + \text{H}_2\text{O} + \text{Cl}^\ominus$$

Thus, solvation facilitates the formation of the carbocation-like transition state, and electrophilic additions of H—X are much faster when conducted in water.

### EXERCISE 10.3

Assuming that cation stability determines the barrier for protonation in addition of HX, predict which compound in each of the following pairs will be more rapidly hydrochlorinated in a polar solvent.

(a) $H_2C{=}CH_2$    or

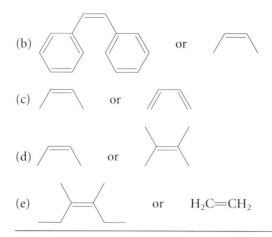

(b)

(c)   or

(d)   or

(e)   or   $H_2C{=}CH_2$

***Solution-Phase Reactivity: Competition between Nucleophiles.*** In addition to its role in the solvation of ions, water can also act as a nucleophile. Although the rate of addition of a proton to an alkene can be accelerated by using water as a solvent, the reaction medium in that case contains two nucleophiles (chloride ion and water), and two products (an alkyl chloride and an alcohol) are possible. If chloride ion acts as the nucleophile, the same product is formed as in the gas-phase reaction. When water (rather than chloride ion) acts as the nucleophile, an oxonium ion is formed. The positive charge in this ion can be relieved by deprotonation, a process in which a proton is transferred to the solvent—in this case, water. This cleavage of an oxygen–hydrogen $\sigma$ bond completes the formation of an alcohol, and the overall addition is called **hydration.**

Thus, in contrast to the gas-phase reaction, electrophilic addition of HCl in water has two possible products. The relative amounts of alkyl chloride and alcohol produced are determined by the relative nucleophilicities of chloride ion and water, as well as their relative concentrations. You know from Chapter 6 that chloride ion, being negatively charged and polarizable, is a much better nucleophile than a neutral water molecule. Therefore, alkyl chloride formation dominates, although significant amounts of alcohol are also produced.

***Hydration.*** In solution, HCl dissociates, and the hydronium ion ($H_3O^\oplus$) acts as an acid in transferring a proton to an alkene:

Electron flow from the C=C $\pi$ bond to $H_3O^\oplus$ forms a C—H $\sigma$ bond, freeing neutral water. (As with HCl, the initial attack by $H_3O^\oplus$ takes place with equal ease from both the top and the bottom faces of the $\pi$ bond.) The planar, $sp^2$-hybridized carbocation formed in this rate-determining step is then captured by nucleophilic attack by either chloride or water. Although chloride, an anion, is a more reactive nucleophile, it is present in much lower concentration than water, the solvent. Both the alkyl chloride and the alcohol are racemic, because attack is equally easy on the top and bottom faces of the planar carbocation.

The alcohol can be made the dominant product by using a strong acid such as sulfuric acid, H$_2$SO$_4$, for which the conjugate base is a relatively weak nucleophile. When added to water, H$_2$SO$_4$ dissociates, forming H$_3$O$^{\oplus}$ and HSO$_4^{\ominus}$ (bisulfate ion), which is a poor nucleophile. Addition of an alkene to this solution results in the formation of the alcohol by hydration.

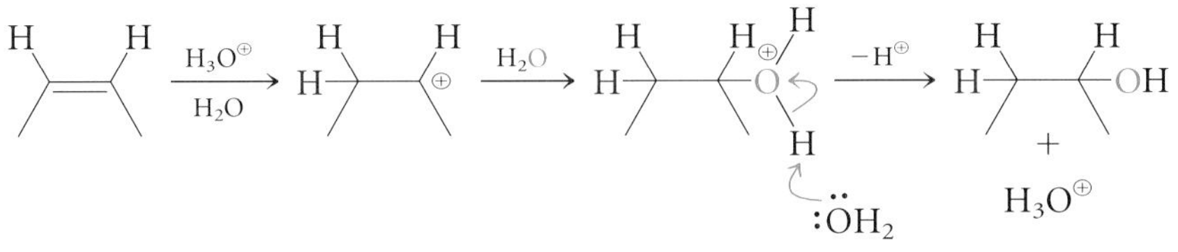

In the last step of hydration, a proton is transferred from the oxonium ion to water, replacing the hydronium ion consumed in the first step. Thus, the proton is not consumed, and the formation of alcohol by hydration is *acid-catalyzed*. An acid is required to initiate the reaction but is regenerated as the product is formed. The hydration of an alkene thus requires acid, but only water and the alkene are consumed in forming the product.

***Addition of HX.*** In general, acids of the type HX, where X$^{\ominus}$ is a moderate to good nucleophile, are effective reagents for electrophilic addition to alkenes, as generalized in Figure 10.3. When H and X add across a C=C $\pi$ bond, the result is hydrohalogenation if X is a halide (X = Cl, Br, or I) and hydration if X is OH. Both reactions are initiated by protonation (by a hydronium ion) in the initial electrophilic attack, which results in the breaking of the carbon–carbon double bond. This step is the slower one, and thus the reaction rate is determined by the stability of the resulting carbocation.

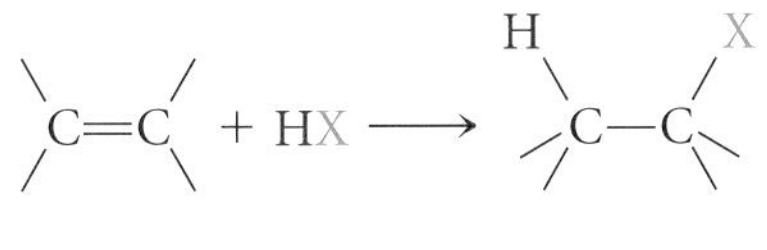

**FIGURE 10.3**

Hydrohalogenation (where X = Cl, Br, or I) and hydration (where X = OH) both proceed as electrophilic addition through an intermediate carbocation.

## Regiochemistry of Electrophilic Addition

Let's now consider what happens when 1-butene rather than 2-butene undergoes hydration with H$_2$SO$_4$ in water (Figure 10.4, on page 494). Unlike the double bond in 2-butene, that in 1-butene is not symmetrical. Protonation of 1-butene at C-1 produces a secondary carbocation, whereas protonation at C-2 generates a primary carbocation. As you know, a sec-

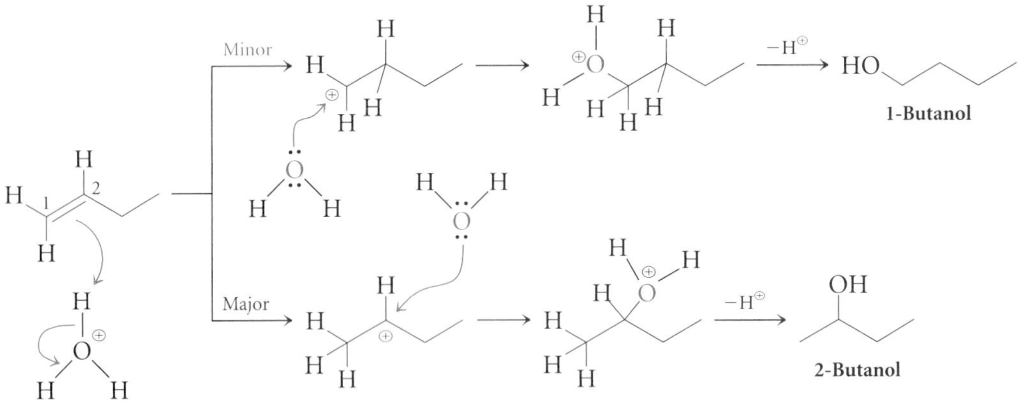


**FIGURE 10.4**

Electrophilic protonation of 1-butene at C-1 produces a secondary carbocation at C-2 (bottom reaction), whereas protonation at C-2 leads to a primary cation at C-1 (top reaction). Because a secondary cation is more stable than a primary one, the lower route is the major pathway.

ondary cation is more stable than a primary one. Because carbocation formation is the rate-determining step of an electrophilic addition, the greater stability of the secondary carbocation dictates that protonation at C-1 is dominant. Capture of this cation by water generates 2-butanol, whereas capture of the less stable primary cation leads to 1-butanol. Preferential formation of the more stable secondary cation thus leads to the secondary alcohol as the major *adduct* (addition product).

Cation stability also controls the regiochemistry of addition to other substrates. For example, the addition of HCl to 2-methylpropene leads predominantly to 2-chloro-2-methylpropane through the more stable tertiary cation:

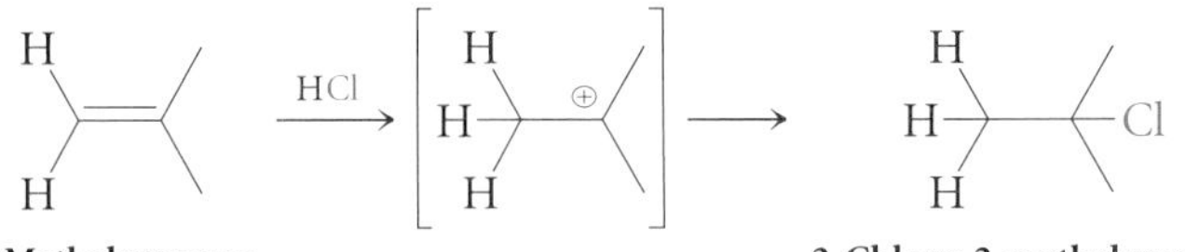


**2-Methylpropene** **2-Chloro-2-methylpropane**

Similarly, the addition of HCl to styrene produces 1-chloro-1-phenylethane through a highly stabilized benzylic cation:

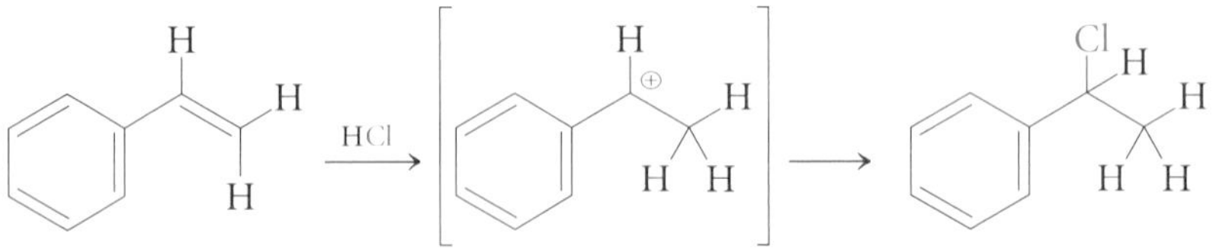


**Styrene** **1-Chloro-1-phenylethane**

Recall from Chapter 7 that, according to **Markovnikov's rule,** when an alkene undergoes electrophilic addition, the less highly substituted position is attacked by the electrophile. This is due to the fact that the rate-determining step is formation of the carbocation, and the order of carbocation stability is tertiary ≈ benzylic > allylic ≈ secondary > primary ≈ vinyl > phenyl.

*CHEMICAL PERSPECTIVES*

PEOPLE VERSUS COCKROACHES

People tend to draw a correlation between "strong" reagents such as sulfuric acid and danger. Certainly sulfuric acid should be treated with care and respect. Nonetheless, how vigorously such reagents react depends on the nature of the substance with which they are reacting. Human tissue is made up of many different hydrophilic molecules, containing virtually all the common functional groups. These functional groups react rapidly with concentrated sulfuric acid, which is why human skin is destroyed relatively rapidly by concentrated sulfuric acid. On the other hand, some insects, such as cockroaches, are coated with a mostly hydrocarbon layer that is relatively inert to sulfuric acid. Indeed, a cockroach will swim about, apparently merrily, on the surface of concentrated sulfuric acid. (Please wear safety glasses if you try this—roaches splash terribly when doing the backstroke.)

## EXERCISE 10.4

What carbocation is formed preferentially as an intermediate in the acid-catalyzed hydration of each of the following compounds? What is the structure of the product alcohol?

(a) 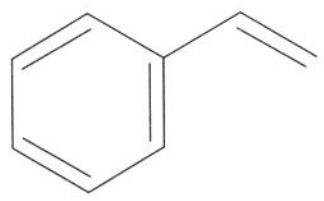  (b) 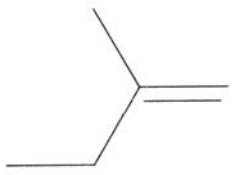  (c) 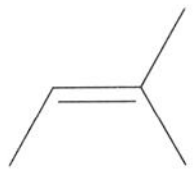  (d)

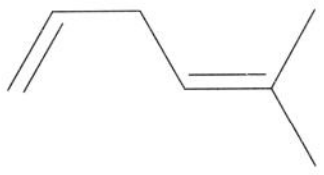

## Addition to Conjugated Dienes

The addition of HX to 1,3-butadiene, a conjugated diene, proceeds by the addition of a proton to one of the terminal carbon atoms, generating an allylic secondary carbocation as intermediate. As shown in Figure 10.5,

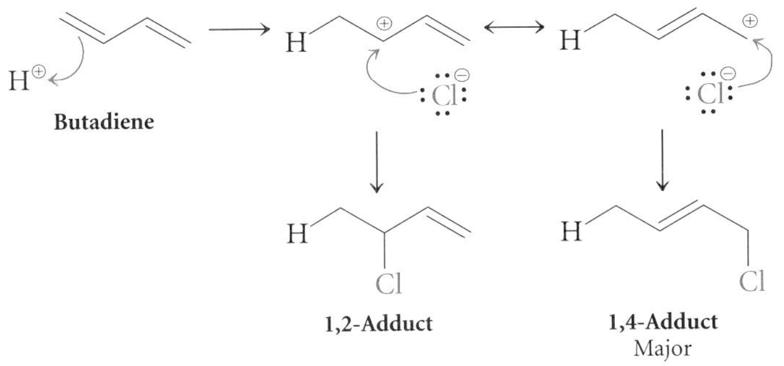

**FIGURE 10.5**

Protonation of butadiene at C-1, the least substituted position, produces a resonance-stabilized allylic cation. Nucleophilic capture of the cation at the two sites that bear formal positive charge leads to two products—the 1,2- and 1,4-adducts.

Markovnikov's rule predicts that the electrophile will attack at the end of the conjugated system to form an allylic cation. Resonance delocalization of this allylic cation places positive charge at C-2 and C-4. Trapping of the allylic cation by chloride at each of these positions leads to 3-chloro-1-butene and 1-chloro-2-butene, the 1,2- and 1,4-adducts, respectively. (This terminology identifies the positions of the added H and X with respect to the carbon skeleton.)

The stabilities of the two products are not equal: the double bond in 1-chloro-2-butene is disubstituted, and that in 3-chloro-1-butene is monosubstituted. Because both addition products are allylic chlorides, they are unusually reactive and can undergo loss of $Cl^{\ominus}$ followed by loss of $H^{\oplus}$, generating the original diene:

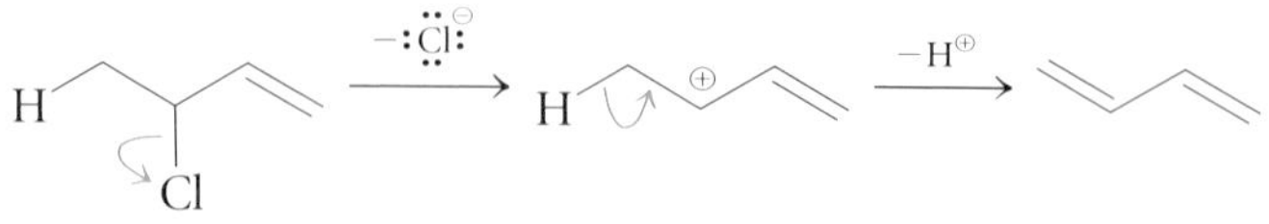

When the reaction takes place at room temperature, where the reverse reaction is reasonably rapid, the more stable alkene, the 1,4-adduct, predominates, and the reaction is under thermodynamic control. Conversely, at low temperature, the products can be isolated before significant back-reaction occurs. Under these conditions, the reaction is under kinetic control and the 1,2-adduct predominates. The addition of HX across a four-carbon system (as in forming the 1,4-adduct) is an example of **conjugate addition.**

### EXERCISE 10.5

Predict the regiochemistry of both the 1,2- and the 1,4-adducts formed by treatment of each of the following dienes with HBr:

(a)      (c)

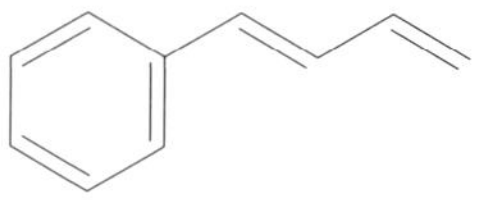

(b)

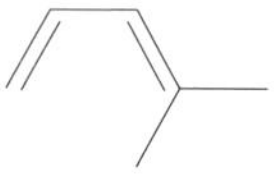

### Stereochemistry of Electrophilic Addition

Electrophilic addition of HX to many alkenes results in the formation of a new center of chirality. Even if chiral centers are absent in the starting alkene, the products will be racemic because the addition of $X^{\ominus}$ to the carbocation occurs with equal facility from either of the two faces. For example, hydration of styrene results in a mixture of equal amounts of (S)- and (R)-α-phenylethyl alcohol (Figure 10.6), because water attacks the top and bottom lobes of the vacant $p$ orbital in the intermediate carbocation with equal facility. The result is a racemic mixture. This addition makes C-2 a center of chirality, despite the fact that this carbon was not chiral in the reactant. Thus, the carbocation is said to be **prochiral,** meaning that the carbon atom is not a center of chirality but becomes one as the reaction proceeds from the intermediate to the product.

## FIGURE 10.6

Protonation of styrene produces a benzylic cation. Attack by water is equally easy on either face of the intermediate cation, leading to racemic α-phenylethyl alcohol.

The presence of a center of chirality in the starting alkene can result in an unequal mixture of product diastereomers. For example, acid-catalyzed addition of water to 1,4-dimethylcyclohexene proceeds through a tertiary carbocation in which the two faces are *not* identical:

**1,4-Dimethylcyclohexene**

**Major product**

Addition of water to the carbocation from the axial direction is hindered by the two axial hydrogen atoms on that face of the ion and is therefore slower than addition to the equatorial face. Therefore, addition of water to the equatorial face dominates, and the major product is the diastereomer where the hydroxyl group is *trans* to the methyl group at C-4.

## EXERCISE 10.6

Draw the structures of the intermediate carbocation and the product alcohol that will result from treatment of each of the following alkenes with $H_2SO_4$ in water. If more than one diastereomeric product is possible, indicate which is likely to predominate.

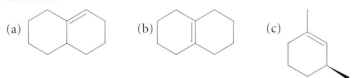

(a)    (b)    (c)

### Rearrangements

The Markovnikov orientation observed in electrophilic addition is a direct consequence of differences in carbocation stability. Another consequence of such differences is the possibility for rearrangement. A group at an adjacent carbon rapidly migrates to the carbocationic center whenever a thermodynamic driving force exists for such a migration—as it does, for example, in the conversion of a secondary carbocation into a tertiary one.

By using these principles, we can predict the course of the hydrobromination of 3-phenyl-1-propene:

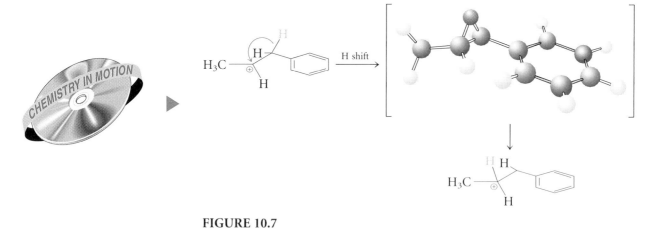

Protonation of 3-phenyl-1-propene takes place at C-1 to produce the more stable secondary carbocation (rather than the primary one). When a more stable carbocation is produced by a shift of a group from an adjacent atom, a rearrangement occurs rapidly. Here, a hydrogen originally bonded to C-3 migrates with its electrons to this secondary center to produce an even more stable benzylic carbocation (Figure 10.7). Capture of this planar carbocation by bromide takes place with equal ease on either face, leading to equal amounts of the two enantiomeric bromides. This bond shift produces a more stable, rearranged carbocation that is benzylic as well as secondary. Capture of this cation by bromide gives rise to the observed racemic product.

#### FIGURE 10.7

The transition state for the hydrogen shift has the hydrogen atom that migrates (shown here in green) partially bonded to both C-3 and C-2.

Because of this rearrangement, bromine is found in the product at a carbon atom that did not originally participate in $\pi$ bonding. Thus, there are two pieces of evidence for a carbocationic rearrangement: (1) the isomerization of the carbon skeleton, and (2) the presence of the nucleophile in the product at a position that was not part of the original reactant's double bond. For example, you can recognize that a rearrangement has taken place with the addition of HCl to 3,3-dimethyl-1-butene because (1) C-6 has moved from C-3 to C-2 in the product, and (2) Cl is attached to a carbon atom in the product (C-3) that was not part of the double bond in the starting material:

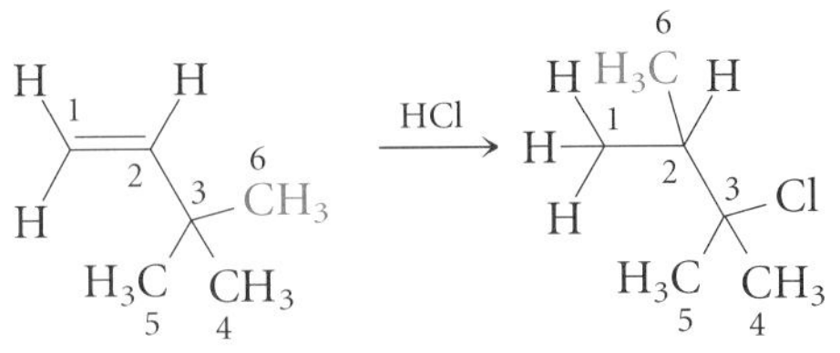

**3,3-Dimethyl-1-butene**

## EXERCISE 10.7

Write a complete mechanism for each of the following transformations, and explain the driving force for the formation of the rearranged skeleton.

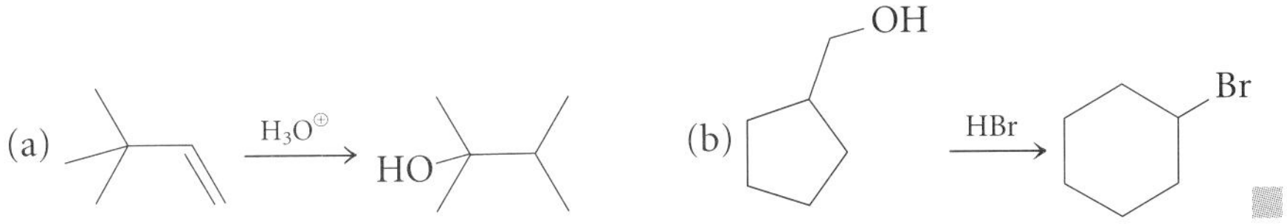

### Addition of HX to Alkynes: Formation of Geminal Dihalides

Because the $\pi$ system of an alkyne closely resembles that of an alkene, addition of HX to triple bonds seems likely to occur by a similar mechanism. Using knowledge of the mechanism of electrophilic addition, let's work through the intermediates formed in the electrophilic addition of HBr to an alkyne to understand the relative reactivity of alkynes and alkenes.

*Step 1: Formation of Vinyl Halides.*   The electrophilic addition of HBr to an alkyne (Figure 10.8, on page 500) proceeds through protonation, using the $\pi$ electrons of one of the bonds in the triple bond. The resulting **vinyl cation** is digonal and *sp*-hybridized, bearing $\sigma$ bonds to two, rather than three, substituents (as is the case for cations discussed earlier in this chapter). As stated in Chapter 9, the vinyl cation is appreciably less stable than a comparably substituted trigonal $sp^2$-hybridized carbocation.

The vinyl cation has a stability somewhat like that of a primary $sp^2$-hybridized cation. Therefore, its formation by protonation of an alkyne is more difficult than the corresponding protonation of an alkene to form a secondary, or more stable, cation, and the rate of protonation of an alkyne is somewhat slower. In fact, this reaction probably would not take place at all if the electron density of a triple bond was not higher than that of a double bond. However, as in electrophilic addition to alkenes, the second step (capture of the vinyl cation by bromide ion) is particularly facile, generating

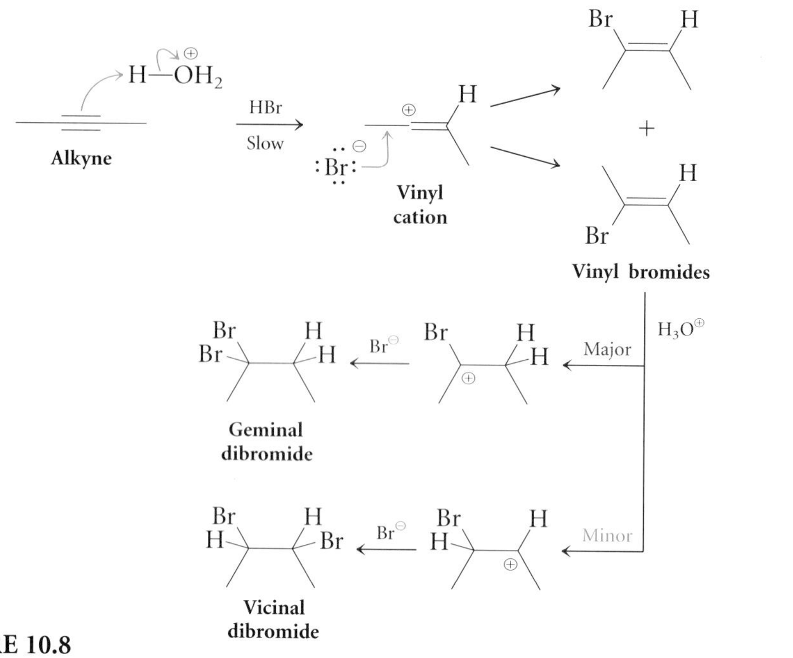

**FIGURE 10.8**

Protonation of an alkyne produces a vinyl cation, a particularly unstable
cationic intermediate. This vinyl cation is rapidly captured by a nucleophilic
bromide anion to form a vinyl bromide. Further protonation is preferred at
the carbon atom of the alkene not bonded to bromine. When the resulting
cation is trapped by a second bromide ion, a geminal rather than a vicinal
dibromide is formed.

a vinyl bromide. Because the vinyl cation is linear, it is easily attacked on
either side of the remaining alkenyl double bond, generating a mixture of
both *cis-* and *trans-*vinyl halides.

**Step 2: Formation of Geminal Dihalides.**  The vinyl halide formed in
the first addition step is an alkene derivative, and is also subject to electrophilic attack by HBr. Electron withdrawal by the electronegative bromine
atom leaves less electron density in the $\pi$ bond of a vinyl bromide than is
found in that of a simple alkene, and further electrophilic attack is slower.
There are two regiochemical possibilities for protonation: at the carbon
bearing hydrogen or the carbon bearing bromine. Both cations are secondary. However, for the one bearing the positive charge adjacent to the
bromine atom, an additional resonance contributor can be written by employing one of bromine's lone pairs to interact with the vacant $p$ orbital,
stabilizing this cation relative to the alternative cation, in which the positively charged carbon is bonded only to carbon and hydrogen.

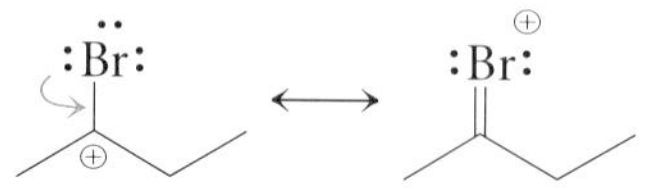

As in all electrophilic additions, the more stable cation is formed preferentially, dictating the observed regiochemistry. Capture of this cation by bromide gives rise to a geminal dibromide (as shown in Figure 10.8).

The observed regiochemistry for an addition yielding a geminal halide follows from the preferential formation of the more stable cation. Once again, the product is that predicted by Markovnikov's rule, in which hydrogen is added to the carbon with more bonds to hydrogen atoms. With a terminal alkyne, protonation takes place so as to produce the more stable secondary vinyl cation, leading to Markovnikov addition. Again, the second addition takes place so as to produce a geminal dihalide, in which both bromine atoms are bonded to the same carbon atom.

$$H_3C\!\!=\!\!=\!\!H \xrightarrow{HBr} \qquad \xrightarrow{HBr}$$

---

### EXERCISE 10.8

For each of the following additions of HX, predict the regiochemistry of the adduct(s). Also, state whether the product mixture is optically active. If it is not, determine whether the inactivity results from the absence of chiral centers, the formation of equal amounts of enantiomers, or the formation of a *meso* compound (recall Chapter 5).

(a) $\xrightarrow{DCl}$  (b) $\xrightarrow{HBr}$  (c) $\xrightarrow{D_3O^{\oplus}}$

---

### 10.2

## Addition of Other Electrophiles

Up to this point, this discussion of electrophilic additions has focused on the role of protons (from HX) as the electrophilic species. Other positively charged or partially positively charged reagents can also fill this role.

### Oxymercuration–Demercuration

The hydration of an alkene can be accomplished by a two-step sequence known as **oxymercuration–demercuration:**

Oxymercuration                    Demercuration

$$\text{C}=\text{C} \xrightarrow[\text{H}_2\text{O}]{\text{Hg(OAc)}_2} \xrightarrow{\text{NaBH}_4}$$

$$\left( \text{OAc} = \underset{O}{\overset{O}{\diagdown}} \right)$$

In the first step, $^{\oplus}$Hg(OAc), formed by dissociation of Hg(OAc)$_2$ (mercuric acetate), acts as a Lewis acid, adding to the alkene to form a cyclic intermediate called a **mercuronium ion.** In this intermediate, mercury bridges

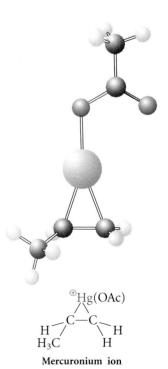

$^{\oplus}$Hg(OAc)

**Mercuronium ion**

between two carbons, forming a three-member ring. The ring is broken by the attack of water at one of the two carbons bonded to mercury; the transition state for this process resembles that for a simple $S_N2$ reaction.

**Oxymercuration**

When the starting alkene is unsymmetrical (as is propene), water preferentially attacks the *more* substituted carbon of the mercuronium ion. Thus, hydration via oxymercuration follows Markovnikov's rule, adding an OH group to the same carbon as in simple acid-catalyzed hydration, for which cation stability controls regiochemistry. In oxymercuration, the orientation of addition is dictated by the relative strengths of the two C—Hg bonds: the bond to the more substituted carbon is weaker, and so the transition state involving breaking of this bond is lower in energy.

In the second step of oxymercuration–demercuration, the C—Hg(OAc) group is replaced by C—H upon treatment with $NaBH_4$. The details of this reaction are not well understood, but it is likely that radicals are involved.

**Demercuration**

Overall, oxymercuration–demercuration effects hydration of an alkene with Markovnikov orientation—the same outcome as with acid-catalyzed hydration. However, because carbocations are *not* involved as intermediates in oxymercuration, rearrangements do not occur. For example, oxymercuration–demercuration of 3,3-dimethyl-1-butene yields 3,3-dimethyl-2-butanol; addition of water in the presence of acid to the same alkene is accompanied by rearrangement, forming 2-hydroxy-2,3-dimethylbutane:

2-Hydroxy-2,3-dimethylbutane      3,3-Dimethyl-1-butene      3,3-Dimethyl-2-butanol

### EXERCISE 10.9

For each of the following reactants, predict the product that would be formed by oxymercuration–demercuration and by acid-catalyzed hydration.

(a)      (b)      (c)      (d)

# Hydration of Alkynes

The hydration of alkynes can also be achieved by electrophilic addition, typically via treatment with a mercuric salt in an aqueous acidic solution:

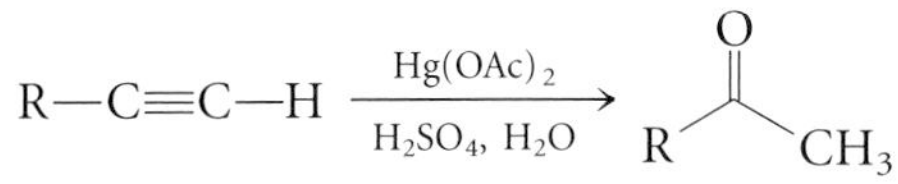

$$R-C\equiv C-H \xrightarrow[\text{H}_2\text{SO}_4, \text{ H}_2\text{O}]{\text{Hg(OAc)}_2} \underset{R}{\overset{O}{\|}} CH_3$$

 **CHEM TV® II**

#11   Alkyne Hydration

The role of the mercuric ion is the same as in oxymercuration of an alkene— that is, mercuric ion serves as a Lewis acid in the first stage of the reaction to produce a bridged mercuronium ion (Figure 10.9). Although hydration of an alkyne can be effected by aqueous acid without Hg(OAc)$_2$, the conditions required are much more vigorous because an unstable vinyl cation is involved.

The reaction continues via attack by water on the mercuronium ion at the more highly substituted carbon. The resulting oxonium ion then loses a proton to produce an enol, which also contains a carbon–mercury bond. This C—Hg bond is then replaced by a C—H bond. Tautomerization of the resulting enol to the more stable keto form is rapid in aqueous acid, and a ketone is produced. With a terminal alkyne, the hydrogen atom ends up on the less substituted carbon, in accordance with Markovnikov's rule, and after enolization, a methyl ketone is formed. With a nonterminal and unsymmetrical alkyne, mixtures of the two possible ketones are usually obtained, because there is only a small difference in energy between the two possible vinyl cations formed in the initial electrophilic attack.

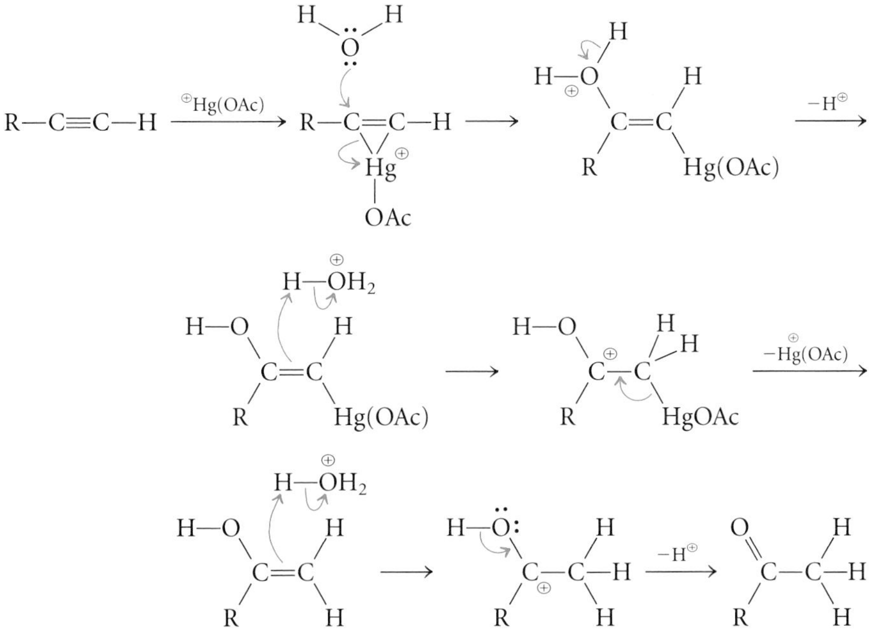

**FIGURE 10.9**

In the hydration of a terminal alkyne, mercuric ion binds to C-1 and C-2 to form a bridged mercuronium ion. Nucleophilic attack by water, followed by deprotonation of the resulting oxonium ion, produces a mercurated enol. Protonation followed by loss of $^{\oplus}$Hg(OAc) produces an enol, whose acid-catalyzed tautomerization yields the observed methyl ketone.

### EXERCISE 10.10

Draw the structure(s) of the ketone product(s) expected from hydration of each of the following alkynes:

(a) 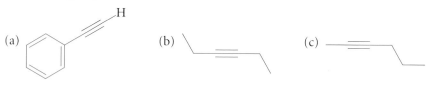   (b)   (c)

### EXERCISE 10.11

Write a detailed mechanism for the conversion of propyne to acetone, catalyzed by $H_2SO_4$ in $H_2O$:

$$H_3C\!-\!\!\!\!\equiv\!\!\!\!-\!H \xrightarrow[H_2O]{H_3O^\oplus} \text{(acetone)}$$

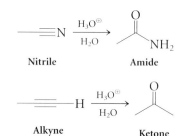

Nitrile → Amide

Alkyne → Ketone

### Hydration of Nitriles

The addition of water to a nitrile to produce an amide is similar to hydration of a terminal alkyne to produce a ketone. An organic nitrile contains a carbon atom with the same oxidation level as that of a carboxyl carbon (count the number of bonds to heteroatoms). Organic nitriles are formed either by using cyanide as a nucleophile in an $S_N2$ displacement from an alkyl halide (Chapter 8) or by treating a primary amide with a strong dehydrating reagent such as $POCl_3$:

$$R\!-\!X \xrightarrow{^\ominus C\equiv N} R\!-\!C\equiv N \xleftarrow{POCl_3} R\!-\!\!\!\overset{\overset{\displaystyle O}{\parallel}}{C}\!\!-\!NH_2$$

Acid-catalyzed hydrolysis of nitriles converts these species first into amides (Figure 10.10) and then into the corresponding carboxylic acids. These, in

**FIGURE 10.10**

Acid-catalyzed hydration of a nitrile produces an amide. This reaction begins by protonation of the nitrile nitrogen, thus activating the nitrile carbon to nucleophilic attack by water. Proton exchange and tautomerization lead to the amide.

turn, can be converted into other acid derivatives. Thus, nitriles are versatile substrates that can be converted into a range of other functional groups.

Hydrolysis of nitriles takes place via protonation on nitrogen (Figure 10.10), followed by attack of water on the carbon bonded to nitrogen in the resulting cation. A primary amide is then obtained after a series of protonation–deprotonation steps. The overall reaction is

$$H_3C-C\equiv N \xrightarrow[\text{H}_2\text{O}]{\text{H}_3\text{O}^{\oplus}} H_3C-C\overset{\displaystyle O}{\underset{\displaystyle NH_2}{\big\|}}$$

## Addition of Halogens

Molecular bromine and chlorine ($Br_2$ and $Cl_2$) also function as electrophiles in addition reactions, rapidly reacting with alkenes to produce vicinal dihalides in high yield. The reaction of bromine with alkynes produces tetrahalides.

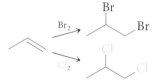

***Bromine Test for Unsaturation.*** A solution of $Br_2$ in $CH_2Cl_2$ is deep red-brown, and the organic dibromides formed by the addition reaction are colorless. This rapid color change that accompanies addition of bromine to double and triple carbon–carbon bonds can be used to test for the presence of unsaturation.

#19 Bromination of Alkenes

***Mechanism of Br₂ Addition.*** As $Br_2$ (or $Cl_2$) approaches an alkene, the halogen–halogen bond becomes polarized by interaction with the electron-rich $\pi$ cloud of the carbon–carbon double bond. Reaction of $Br_2$ with an alkene proceeds through an intermediate cyclic **bromonium ion.** Attack by bromide ion on the bromonium ion yields a **vicinal dibromide,** bearing two bromine atoms on adjacent carbon atoms:

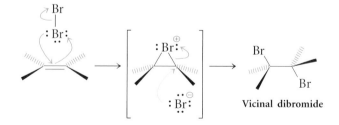

Vicinal dibromide

The bromine molecule becomes polarized as it approaches a $\pi$ cloud, and so the nearer bromine atom can act as an electrophile. Sigma bonds are formed between the two carbon atoms of the alkene and this bromine, as the electrons of the $\pi$ bond are donated toward the partially positively charged bromine atom. The result is the cyclic bromonium ion, in which bromine bears formal positive charge. (The molecular orbitals representing the two C—Br $\sigma$ bonds are shown at the beginning of this chapter.)

The bromonium ion is a reactive intermediate and is attacked by bromide ion. The transition state for this second step in the overall sequence resembles that for the back-side attack in $S_N2$ reactions, discussed in Chapters 7 and 8. As bromide ion attacks, one of the carbon–bromine bonds of the cyclic bromonium ion is broken, and the electrons are returned to bromine as a lone pair. The resulting adduct is a vicinal dibromide.

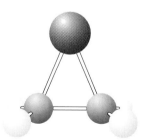

A simple bromonium ion

In the past, carbon tetrachloride was often used as a solvent for the addition of bromine to an alkene. Unfortunately, this solvent has been shown to have detrimental effects on the health of those exposed to it over a long period of time. Dichloromethane, $CH_2Cl_2$, can be used in place of $CCl_4$, although it, too, is suspected of causing cancer (based on the effect of long-term, high-level exposure on laboratory animals bred to be susceptible to carcinogenic compounds).

***Stereochemistry of $Br_2$ Addition to Cyclic Compounds.*** The stereochemical outcome of the *anti* addition of $Br_2$ can be seen in the reaction with cyclohexene. In the product, *trans*-1,2-dibromocyclohexane, the two bromine atoms are oriented on opposite sides of the ring.

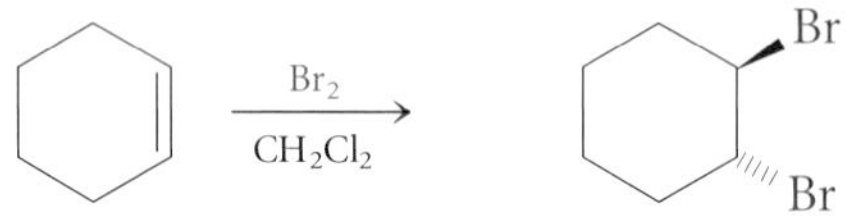

Cyclohexene    *trans*-1,2-Dibromocyclohexane

Even though the *anti* addition of bromine produces two centers of chirality, the product is an optically inactive mixture. This result is the same as that for the other electrophilic additions we have considered so far, although the process by which it is achieved is somewhat different. Formation of a cyclic bromonium ion in the first step of electrophilic bromination produces two centers of chirality. Both carbons bonded to bromine in the bromonium ion derived from cyclohexene are stereocenters. A mirror plane is possible for this species (through the bromine atom), and consequently the two stereocenters are of opposite configuration (one *R* and one *S*):

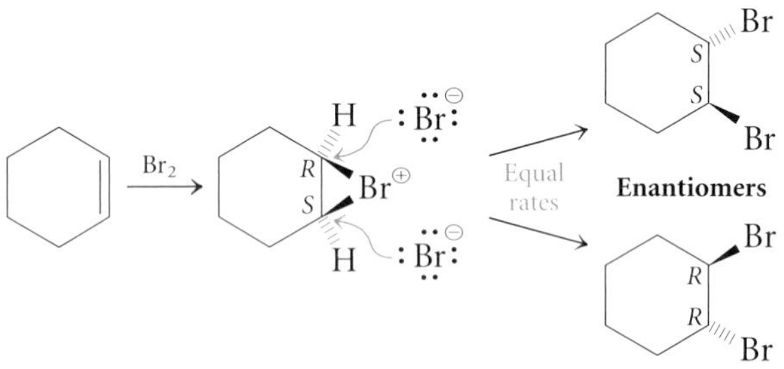

Back-side nucleophilic attack by bromide ion, which results in opening of the bromonium ion, is equally likely at either of the two stereocenters of the bromonium ion. Because the process is an $S_N2$ reaction, inversion of configuration occurs. Reaction at the carbon with *S* configuration in the bromonium ion leads to the *R,R* dibromide, and attack at the carbon with *R* configuration leads to the *S,S* dibromide. These two stereoisomers constitute the mirror-image components of the racemic mixture produced. (Note that, by this pathway, generation of the *R,S* diastereomeric dibromide, which is the *cis* isomer and a *meso* compound, is not possible—and, indeed, is not observed.)

***Stereochemistry of $Br_2$ Addition to Acyclic Compounds.*** This type of *anti* addition, in which two substituents assume a *trans* relation in cyclic systems, also has significance for acyclic compounds. *Anti* addition of

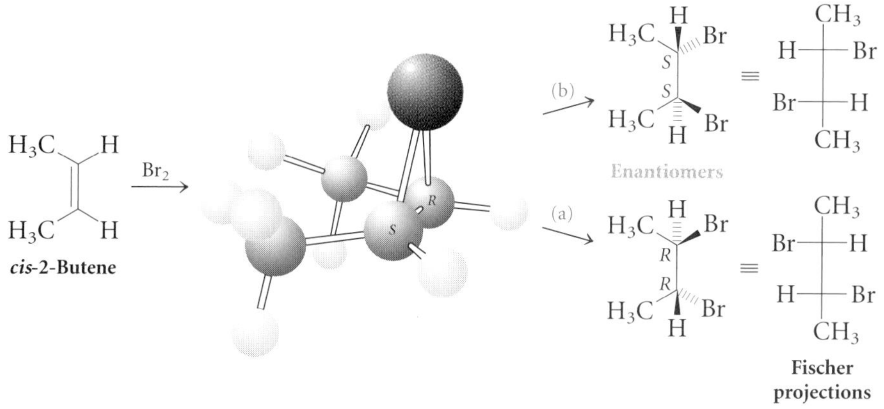

**FIGURE 10.11**

Delivery of an electrophilic bromine from the top face of the $\pi$ bond produces the cyclic bromonium ion shown. Nucleophilic attack at the $R$ center (path b) leads to the $S,S$ enantiomer; attack at the $S$ center (path a) leads to the $R,R$ enantiomer. These two compounds are formed in equal amounts, so the resulting mixture is optically inactive.

bromine to *cis*-2-butene produces the $R,R$ and $S,S$ isomer pair rather than the $R,S$ (*meso*) compound (Figure 10.11). This is seen more easily in the Fischer projections at the right in the figure.

## EXERCISE 10.12

Write a detailed mechanism for the bromination of *trans*-2-butene. Identify whether the product mixture is a *meso* compound or a racemic pair. (Be careful in assigning the stereochemistry of the bromonium ion: the situation here is different from the bromination of *cis*-2-butene.)

*Addition of Cl$_2$.* Similar cyclic halonium ions are also formed with chlorine and iodine. Apart from the higher electronegativity and smaller size of chlorine compared with bromine, the chloronium ion is quite analogous to the bromonium ion. Reaction of alkenes with $Cl_2$ yields vicinal dichlorides. The stereochemistry of chlorination is the same as bromination: the chlorine atoms are added in an *anti* fashion. For example, the reaction of cyclopentene with $Cl_2$ produces *trans*-1,2-dichlorocyclopentane:

Cyclopentene       *trans*-1,2-Dichlorocyclopentane

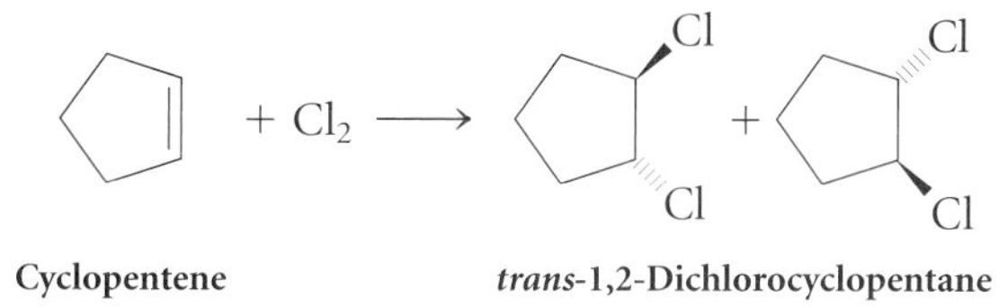

*Reactivity of I$_2$ and F$_2$.* Iodine does not react with simple alkenes to form diiodoalkane products. This lack of reactivity is not due to kinetics or the inability to form a bridged ion; rather, it arises because the reaction is endothermic. Although the strengths of the two C—I $\sigma$ bonds of the product are slightly more than the sum of the strengths of the C=C $\pi$ bond and the I—I bond, the reaction is entropically unfavorable, because two molecules are joined as one product and $\Delta G°$ ($= \Delta H° - T\Delta S$) is positive.

Fluorine is much more reactive than the other halogens and reacts even with C—H bonds, resulting in complex mixtures of products. Therefore, the dihalogen addition through cyclic halonium ions is important only for $Br_2$ and $Cl_2$.

### EXERCISE 10.13

Calculate $\Delta H°$ for the reactions of $F_2$, $Cl_2$, and $Br_2$ with cyclohexene using the average bond energies from Table 3.2 (also inside the back cover).

### EXERCISE 10.14

Predict the structure of the intermediate cation and the stereochemistry of the final product in the reaction of each of the following reagents with cyclohexene.

(a) ICl     (b) PhSBr     (c) BrCl     (d) HOCl

***Synthetic Utility of Vicinal Dihalides: Protection of Double Bonds and Preparation of Alkynes.*** The halogenation of an alkene generates a vicinal dihalide, which, as discussed in Chapter 9, can lose the two vicinal halogen atoms by treatment with zinc. The high yields obtained in such elimination reactions make bromination–debromination a means of reversibly protecting a double bond.

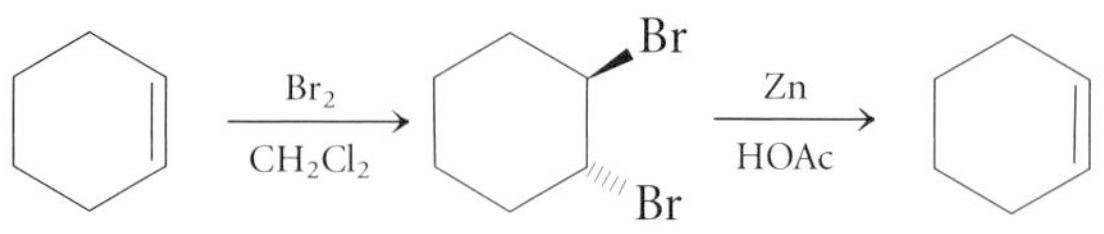

**Cyclohexene**                                    **Cyclohexene**

For example, cyclohexene can be brominated by treatment with $Br_2$ in $CH_2Cl_2$. Because the dibromide lacks the characteristic reactivity of the original double bond, it is possible to use a number of reagents that would ordinarily attack a double bond to alter functionality in other parts of this molecule. The double bond can then be regenerated by removing the two bromides via treatment with Zn. The ease of this reverse reaction allows the vicinal dibromide to be regarded as a **protected alkene,** and the bromines are referred to as a **protecting group.** Here, a reactive functional group (the double bond) is "protected" from reagents used to achieve transformations in other parts of the molecule. The concept of the protection of functional groups, as well as its applications, will be described in more detail in Chapter 15.

Because vicinal dihalides can be twice dehydrohalogenated by treatment with base (Chapter 9), halogenation also provides an intermediate through which an alkene can be converted to an alkyne, as shown here for the transformation of *cis*-2-butene to 2-butyne:

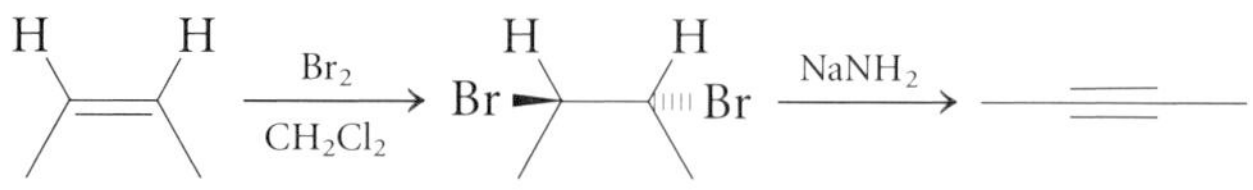

*cis*-**2-Butene**                                    **2-Butyne**

The carbocation formed by the protonation of an alkene is itself an electrophilic species and can attack the starting alkene if the concentration of the original alkene is high. For example, the protonation of styrene generates a carbocation with a regiochemistry consistent with Markovnikov's rule (Figure 10.12). This species can itself be attacked by another molecule of styrene, forming a second carbon–carbon $\sigma$ bond.

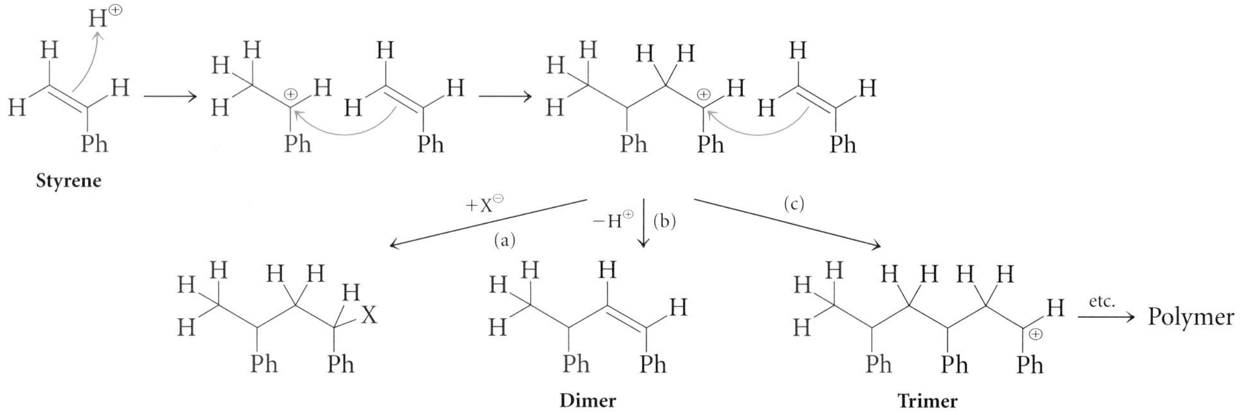

**FIGURE 10.12**

The carbocation generated by protonation of styrene can act as an electrophile to attack another molecule of styrene. The cation formed by the combination of the two alkenes can be captured by a nucleophile, $X^{\ominus}$ (path a), lose a proton to form a neutral dimer (path b), or attack a third molecule of alkene (path c).

Several fates are possible for this more complex structure: as in the simple addition of HX to an alkene (Section 10.1), the cation can be trapped by an anion to form a simple adduct (path a). This product has the formal elements of HX added across a **dimer** of the starting alkene. Alternatively, the more complex cation can lose a proton (path b), as in the second step of an E1 elimination, discussed in Chapter 9. The dimeric cation can also act as an electrophile, attacking yet another molecule of starting material (path c) to make an even longer chain. In the **trimer** shown in Figure 10.12, three molecules of styrene are bound together. This reaction can recur again and again, ultimately forming a **polymer**—in this case, **polystyrene**. This important process is discussed in detail in Chapter 16.

## 10.3

## Radical Additions

Recall from Chapters 2 and 6 that carbon radicals, as well as carbocations, are electron-deficient and are stabilized by electron donation from alkyl groups. For this reason, radicals and cations have the same order of stability: tertiary > secondary > primary. Thus, the reactivity of radicals and cations toward double bonds is similar.

### Radical Addition of HBr: Reversing Markovnikov Regiochemistry

Although Markovnikov's rule applies to the electrophilic addition of HBr to carbon–carbon double bonds in most organic solvents, the reverse regiochemistry is sometimes observed when the reaction is conducted in ether:

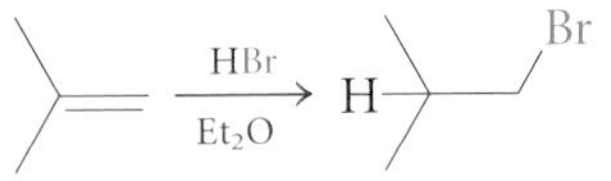

Because the mechanism determines the regiochemistry of a reaction, this switch in regiochemistry means that the anti-Markovnikov hydrobromination product is formed by a different mechanism from that discussed in Section 10.2:

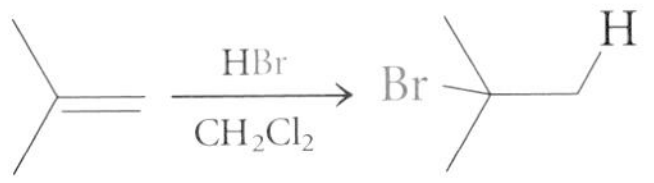

*Initiation of Addition.* Ether solvents are notorious for containing small amounts of **peroxides**, ROOR, or hydroperoxides, ROOH, which are formed by partial decomposition of the ether upon standing. Peroxides and hydroperoxides are characterized by a weak oxygen–oxygen bond that can be cleaved homolytically, either by gentle warming or by light, to initiate a radical chain reaction like those described in Chapter 7.

$$RO\!-\!OH \longrightarrow RO\cdot + \cdot OH$$

**A hydroperoxide**

Homolytic cleavage of the peroxide bond generates alkoxy and hydroxy radicals that can serve as initiators for a radical chain reaction, whose regiochemistry is opposite that observed for a reaction that proceeds through cationic intermediates.

Interaction of an alkoxy (or hydroxy) radical with HBr results in the rapid abstraction of a hydrogen atom, generating a reactive bromine atom that then attacks an alkene to begin a radical chain propagation.

$$RO\cdot \quad H\!-\!Br \longrightarrow ROH + \cdot Br$$

*Radical Propagation Steps.* In the first radical propagation step, the bromine atom produced by reaction with the alkoxy radical adds to the alkene's double bond, forming a C—Br $\sigma$ bond and the more stable carbon free radical. In forming the C—Br $\sigma$ bond, one electron is contributed by the bromine radical and the other by the $\pi$ system of the reactant alkene. The second electron of the $\pi$ bond remains in a $p$ orbital on the adjacent carbon atom.

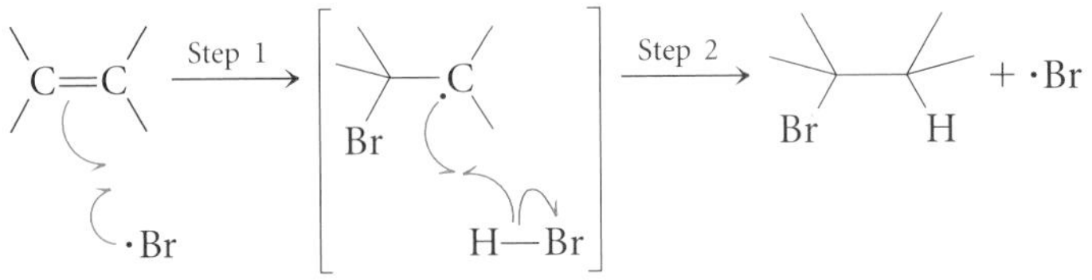

In this step of the reaction, a radical is consumed and one is simultaneously generated. This sequence constitutes chain propagation because a reactant free radical is converted to a product that is also a free radical: the number of reactive radicals on the left and right sides of the equation for Step 1 is equal.

In the second radical propagation step, the alkyl free radical, produced by the earlier reaction of the bromine radical with the double bond, in turn attacks another molecule of HBr, abstracting hydrogen. In this transformation, the formal hydrohalogenation of the C=C bond is completed, and a bromine atom is regenerated. This bromine atom can then attack a second alkene, and the cycle can be repeated again and again. Once again, chain propagation occurs, because as one radical is consumed, another radical is generated.

*Regiochemistry of Addition.* The two propagation steps consume one equivalent of alkene and one equivalent of HBr to generate one equivalent of the adduct. The first step, which generates a carbon free radical, is thermodynamically less favorable than the second. The second step, which consumes the alkyl free radical and HBr, generating a C—H $\sigma$ bond and a bromine radical, is faster.

The intermediate formation of the alkyl radical determines the regiochemistry of this reaction. Because radicals and cations follow the same order of reactivity (3° > 2° > 1°) and because the identity of the attacking reagent is reversed ($H^{\oplus}$ in electrophilic hydrobromination and Br· in radical hydrobromination), the regiochemistry of this free radical addition is also the opposite of that normally observed in an electrophilic addition (and predicted by Markovnikov's rule). This free-radical hydrobromination is therefore said to have occurred with an **anti-Markovnikov orientation.**

*Stereochemistry of Addition.* To gain some insight into the stereochemistry of free radical addition of HBr, let's consider the addition of HBr to 1-methylcyclohexene. The alkoxy radical formed by the decomposition of trace amounts of peroxide abstracts hydrogen from HBr to produce a bromine radical, which adds at C-2 to generate the more stable tertiary radical. The alternative secondary radical is not formed to an appreciable extent because it is considerably less stable. (Note that the regiochemistry of the addition is fixed at this point.)

Although the order of radical stability is the same as that observed for carbocations, the identity of the attacking species has changed from a proton, $H^{\oplus}$, to a bromine free radical, Br·. Thus, the anti-Markovnikov product is formed.

The reaction is completed in the second propagation step as the tertiary radical, whose reactive center is planar, abstracts hydrogen from a second molecule of HBr. Hydrogen abstraction takes place to nearly equal extent on the two faces of the planar radical intermediate, and so al-most equal amounts of *syn* and *anti* addition of HBr result.

Conducting HBr additions in the presence of peroxides circumvents the usual regiochemical preference of such additions, but, as with the Markovnikov addition of HBr, it does not lead to controlled stereochem-istry. Also, radical addition is restricted to hydrobromination and fails with other hydrogen halides.

## EXERCISE 10.15

To obtain each of the following products, would you use a solvent (such as ethanol and water) that favors polar intermediates or one (such as ether) that favors radi-cal chain reactions?

(a)
(b) 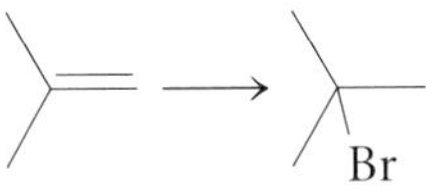
(c)
(d)

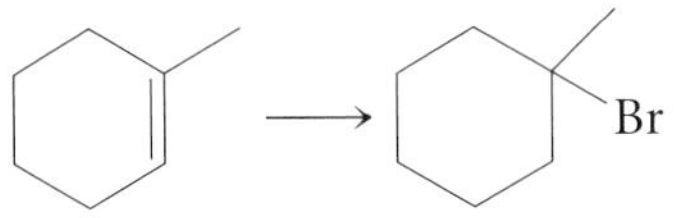

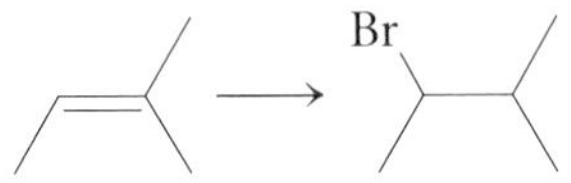

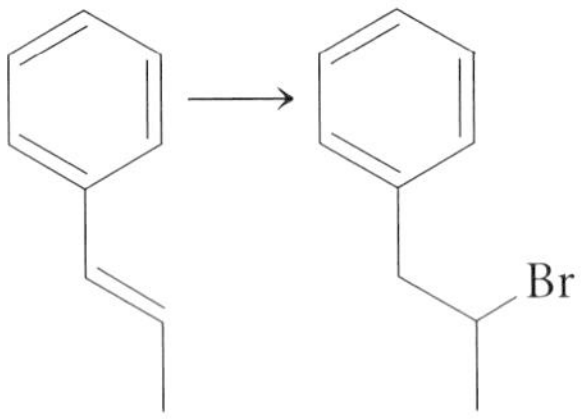

## Radical Polymerization

The electrophilic attack by a carbocation is very similar to the attack by a carbon free radical on a double bond. The only significant difference is that radical intermediates rather than cations are formed. Bearing these facts in mind, it should not be too surprising to find that the free radical formed by interaction of an alkene with an initiator can attack another mol-ecule of alkene, leading to the sequence of reactions shown in Figure 10.13. The dimer radical formed by $\sigma$ bonding of a carbon free radical to a mol-

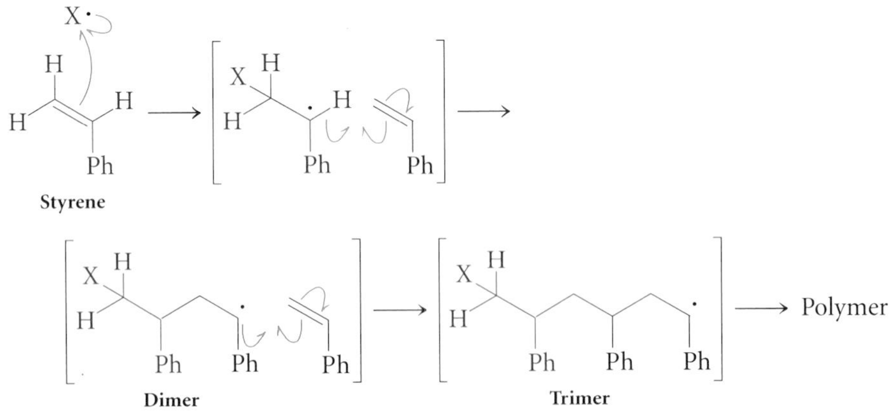

**FIGURE 10.13**

Free-radical polymerization of styrene, yielding polystyrene.

ecule of the starting alkene has several avenues open to it. It can combine with another radical, terminating the reaction sequence; it can lose a hydrogen atom to form a neutral dimer; or it can attack a third molecule of alkene. Such a series of reactions parallels the polymerization sequence initiated by carbocations. Thus, polystyrene can be prepared by either radical or cationic polymerization pathways. However, radical polymerization can be terminated by radical coupling, a pathway not open to cationic polymerization.

## EXERCISE 10.16

Radical polymerization of styrene gives a high yield of a regular polymer with a phenyl group on every other carbon atom:

Provide a rationale for this observation.

## 10.4

# Cycloaddition Reactions

A number of reactions of alkenes involve simultaneous formation of new bonds to both $sp^2$-hybridized carbon atoms of the double bond and, at the same time, breaking of the $\pi$ bond. These reactions are called **cycloadditions.** We have already discussed one such reaction in Chapter 7—the Diels–Alder reaction. In this section, we will consider other cycloaddition reactions, which proceed through transition states with three-, four-, and five-membered rings, and we will also cover the Diels-Alder reaction in more detail.

**Diels–Alder Reaction**

### Synthesis of Cyclopropanes

Cyclopropane rings can be generated by the reaction of carbenes or carbenoids, which are electron-deficient, with alkenes. The mechanisms of these reactions are considered below. A simple carbene, dichlorocarbene, can be prepared by $\alpha$-elimination of HCl from chloroform upon treatment with base.

***Singlet Carbenes.*** As described in Chapter 6, a singlet carbene has a vacant $p$ orbital that is perpendicular to the plane containing the two $\sigma$

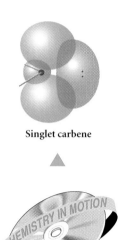

**Singlet carbene**

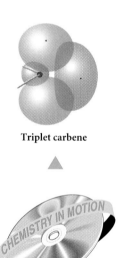

bonds and the lone pair. This vacant *p* orbital makes the carbene electrophilic, allowing it to attack double bonds in the same way as other electrophiles considered in this chapter do. At the same time as the electrons flow from the carbon–carbon $\pi$ bond toward the vacant carbene *p* orbital, forming a new carbon–carbon $\sigma$ bond, the pair of electrons of the carbene interacts with the other carbon atom of the alkene, forming a second carbon–carbon bond. As in the formation of halonium ions, both new $\sigma$ bonds are formed simultaneously—that is, the formation of a cyclopropane derivative is concerted, and there are no intermediates.

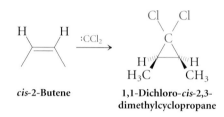

*cis*-2-Butene      1,1-Dichloro-*cis*-2,3-dimethylcyclopropane

Because there is no free carbocation, the relative stereochemistry of the substituents on the starting alkene is maintained. Thus, when dichlorocarbene adds to *cis*-2-butene, the *cis* relation between the methyl groups is found in the product. This retention of stereochemistry is excellent evidence for the concerted nature of carbene addition to the double bond.

**Triplet carbene**

**Triplet Carbenes.** Note that a concerted addition (parallel to the formation of a cyclic bromonium ion) is possible only with a singlet carbene. Because a triplet carbene has two singly occupied orbitals, it is more likely to act as a biradical than as an electrophile. A triplet carbene leads to formation of a biradical intermediate (Figure 10.14), because there is a single electron on each of the two developing centers of reactivity. (Recall from Chapter 2 that a biradical is a reactive intermediate in which two noninteracting radical sites are present within a single molecule.) This biradical can exist long enough for rotation to occur about the $\sigma$ bond joining the two carbons that were originally joined by a double bond, which means that

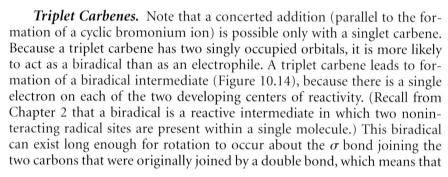

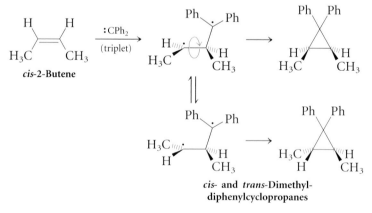

*cis*- and *trans*-Dimethyl-diphenylcyclopropanes

**FIGURE 10.14**

The addition of a triplet carbene to an alkene proceeds in two steps through a biradical intermediate. Free rotation about the C—C bond of what was the alkene leads to formation of both *cis*- and *trans*-1,1-diphenyl-2,3-dimethyl-cyclopropanes.

stereochemistry is lost. In a slower, second step, the biradical collapses, yielding a mixture of *cis* and *trans* cyclopropanes. This result is unlike that obtained with singlet carbenes, for which retention of the original stereochemistry about the alkene bond is observed.

The stereochemistry of carbene addition depends on whether the carbene involved is a singlet or a triplet. The multiplicity of a carbene is determined by the way in which it is formed. Both singlet and triplet carbenes can be formed through photochemical routes. Further, some singlet carbenes convert to triplet carbenes over time, and the stereochemical outcome of both singlet and triplet carbene addition to alkenes can be observed in the same reactions. The subtleties of carbene reactions are the subject of more advanced courses in chemistry.

*Carbenoids.* In the **Simmons–Smith reaction,** an alternative method for the formation of cyclopropane rings, $CH_2I_2$ is reacted with Zn metal, forming a **carbenoid,** a carbene complexed with a metal:

$$CH_2I_2 \xrightarrow{Zn} I-CH_2-Zn-I$$

**A carbenoid**

Despite the presence of the metal, carbenoids react like singlet carbenes, with retention of the geometric relationship present in the alkene. Thus, *trans*-dimethylcyclopropane is obtained by reaction of the carbenoid with *trans*-2-butene:

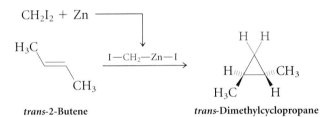

*trans*-2-Butene          *trans*-Dimethylcyclopropane

## EXERCISE 10.17

Predict the product(s) obtained from addition of singlet dichlorocarbene, :CCl₂, to each of the following compounds, and indicate whether they are *meso* compounds or pairs of stereoisomers.

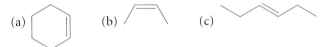

(a)          (b)          (c)

## Epoxidation

The reaction of peracids with alkenes is the primary laboratory method for the synthesis of **epoxides** (or **oxiranes**), three-member rings containing oxygen.

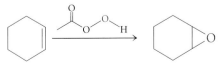

Cyclohexene oxide
(an epoxide)

#20   Alkenes to Glycols
(*anti* addition)

**Peracids** are oxygenated relatives of carboxylic acids and are represented by the formula $RCO_3H$. Because of their stability and storability as laboratory reagents, *meta*-chloroperbenzoic acid (MCPBA) and peracetic acid are the most widely used peracids. Both of these reagents convert alkenes into epoxides.

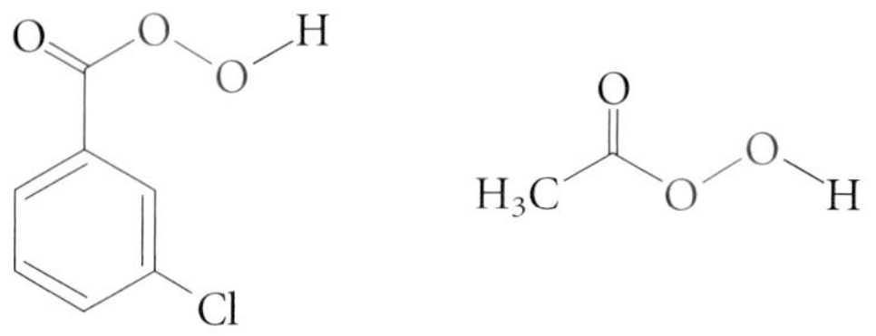

*m*-Chloroperbenzoic acid
(MCPBA)

Peracetic acid

The mechanism of peracid epoxidation is believed to include a cyclic, concerted transition state, as shown in Figure 10.15, in which oxygen is transferred to the alkene at the same time as the carbon–carbon $\pi$ bond is broken and the proton is transferred to the carbonyl oxygen.

Peracid

+

Epoxide

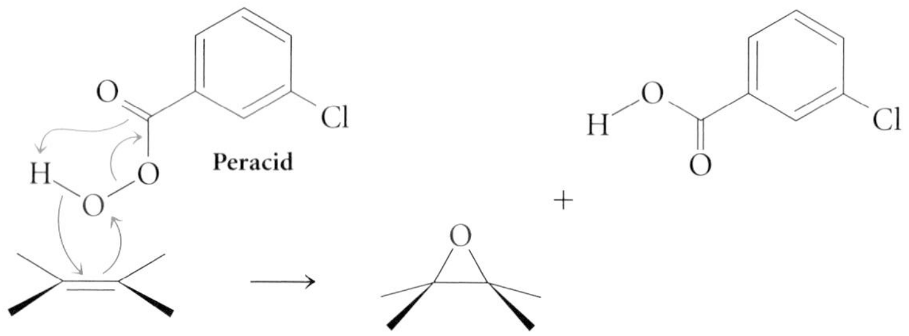

**FIGURE 10.15**

The hydroxyl oxygen in a peracid initiates electrophilic attack on an alkene. As $\pi$ electrons flow from the carbon–carbon double bond to form a $\sigma$ bond between one of the doubly bonded carbons and oxygen, another bond is formed to the second doubly bonded carbon as electron density from the O—H bond flows toward oxygen and the proton is transferred to the carbonyl oxygen atom.

### EXERCISE 10.18

Write a mechanism for the formation of an epoxide by treatment of *cis*-2-butene with peracetic acid. Predict whether the product will be optically active.

Epoxides, like cyclic halonium ions, can undergo ring opening through back-side attack by a nucleophile:

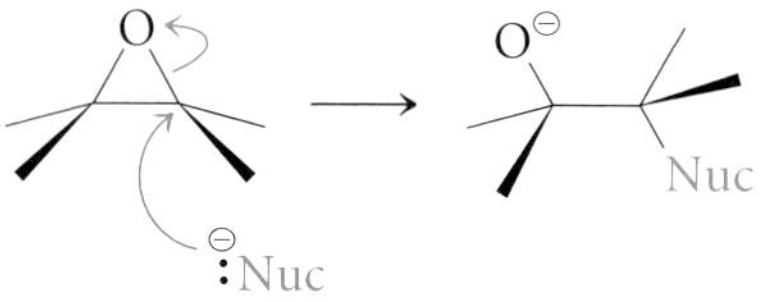

*CHEMICAL PERSPECTIVES*

ETHYLENE OXIDE: THE SIMPLEST EPOXIDE

Ethylene oxide, the simplest epoxide, is produced in very large quantities via oxidation of ethylene using $O_2$ and proprietary catalysts containing copper and silver. Reagents such as peracetic acid are far too expensive for use in the bulk production of chemicals. In 1995, 7.6 billion (!) pounds of ethylene oxide were produced worldwide.

Ethylene oxide is incorporated into a number of products, including the antifreeze ethylene glycol and a variety of plastics. In addition, ethylene oxide is an effective sterilizing agent for medical equipment and a fumigant for food and clothing. Because of its low boiling point (11 °C), it is readily removed after application.

**Ethylene oxide**          **Ethylene glycol**

Ring opening of epoxides is quite slow unless a Lewis acid (or a proton) is complexed with oxygen. For example, the opening of epoxides with NaOH in $H_2O$ requires elevated temperatures, as in the commercial synthesis of ethylene glycol from ethylene oxide:

Ethylene          Ethylene
oxide          Ethylene
glycol

In the reaction of Grignard reagents with epoxides, the epoxide ring is activated by complexation between magnesium and the epoxide oxygen; therefore, the reaction occurs at room temperature and below. The product alcohol contains a new carbon–carbon bond (Chapter 8).

(1)

**Grignard reagent**

Epoxides are relatively reactive under biological conditions, and some molecules containing more than one of these functional groups have been demonstrated to induce cancer in laboratory test animals. Although the mode of biological action is not absolutely clear, carcinogenicity is thought to result from sequential reactions with both strands of double-stranded DNA.

## EXERCISE 10.19

Suggest a sequence of reactions that could be used to achieve each of the following overall conversions:

(a)

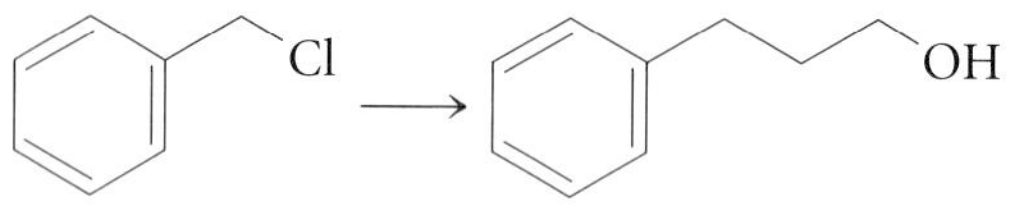

(b)

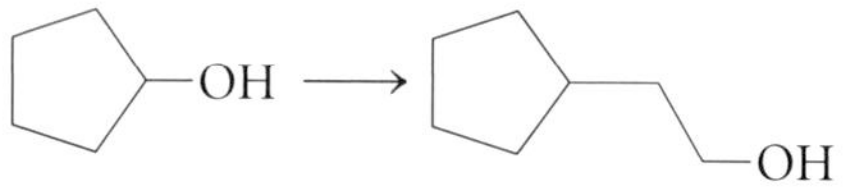

(c)

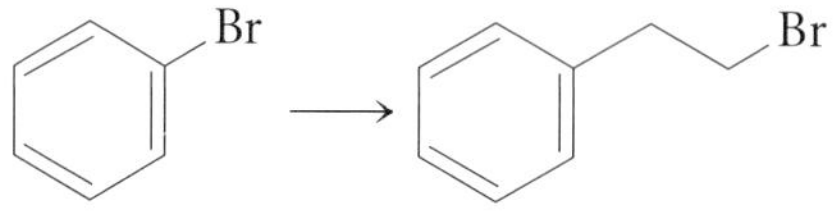

## EXERCISE 10.20

Suggest a mechanism for the opening of the ethylene oxide ring by the phenyl Grignard reagent, as shown in reaction 1.

### Four-Member Cyclic Transition State: Hydroboration–Oxidation

The reaction of diborane, $B_2H_6$, with an alkene, followed by oxidation with alkaline hydrogen peroxide, results in net addition of water across the double bond. Reaction of an unsymmetrical alkene in this two-step sequence of **hydroboration–oxidation** yields an alcohol having anti-Markovnikov orientation (with the hydroxyl group on the less substituted carbon). Thus, products with opposite regiochemistry are formed by the hydration of 1-methylcyclohexene in dilute acid and by the hydroboration–oxidation sequence. Further, the product of hydroboration–oxidation of 1-methylcyclohexene is *trans*-2-methylcyclohexanol. Both regio- and stereochemical control are observed in this sequence.

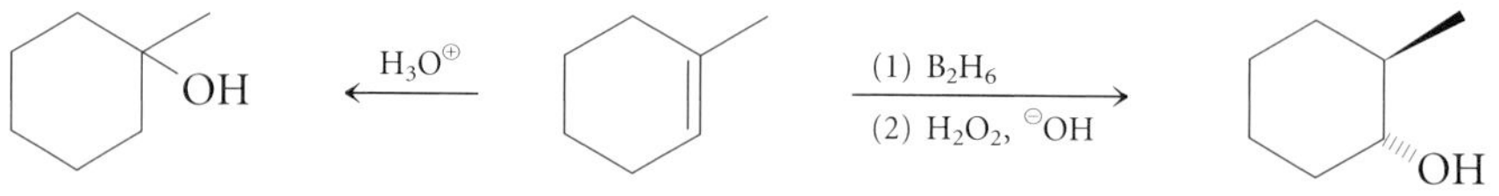

*cis*-1-Methylcyclohexanol        1-Methylcyclohexene        *trans*-2-Methylcyclohexanol

***Mechanism of Hydroboration:* syn *Addition.*** Let's consider hydroboration more closely. Diborane, $B_2H_6$, exists in equilibrium with small amounts of borane, $BH_3$.

$$B_2H_6 \rightleftharpoons 2\ BH_3$$

Because boron has only three valence electrons, it can form only three covalent bonds and still remain neutral. Therefore, borane is highly electron-deficient and acts effectively as a Lewis acid to coordinate with virtually any source of electron density. (This is why it exists predominantly as a dimer rather than a monomer.) When brought into contact with a double bond, a boron–hydrogen bond can interact directly with the alkene *p* orbital to form the four-membered cyclic transition state shown in Figure 10.16. Because the carbon–boron and carbon–hydrogen bond are formed on the same face of the double bond, this first step, referred to as **hydroboration,** is a *syn* addition. Hydroboration has proved to be a very important reaction in both organic and inorganic chemistry. For his extensive and pioneering studies of organoboron chemistry, Herbert C. Brown of Purdue University was awarded a Nobel prize in chemistry in 1979.

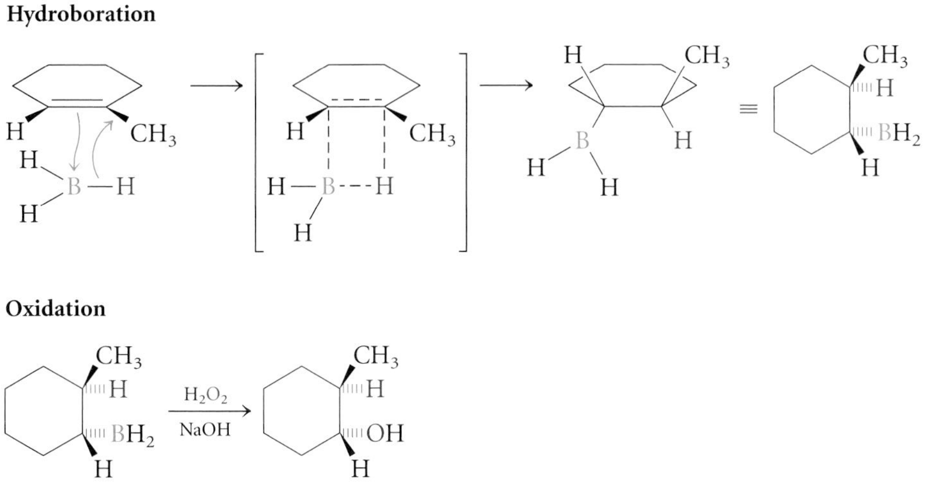

**Hydroboration**

**Oxidation**

**FIGURE 10.16**

Because hydroboration is a concerted reaction in which hydrogen and $BH_2$ are delivered simultaneously to one face of the alkene, *syn* addition is observed. When the $BH_2$ group is replaced by an OH group in the oxidation step, the configuration at the center of chirality is retained, and the same stereochemical relation (*syn*) is maintained between the H and OH groups.

*Oxidation of Organoboranes.* Although we will not consider the mechanistic details of the next step, a peroxide-induced oxidation, carbon–boron bonds can commonly be replaced by carbon–oxygen bonds upon treatment with alkaline hydrogen peroxide. When this occurs, the product of the hydroboration is converted to an alcohol. The orientation of the C—O bond formed in the oxidation step is identical with that of the original C—B bond. The *syn* addition achieved through hydroboration thus leads to *syn* addition in the formation of the final alcohol product.

*Regiochemistry of Hydroboration–Oxidation.* The regiochemistry of the hydroboration step controls the regiochemistry of the overall sequence, because in the oxidation step the boron is replaced by a hydroxyl group, on the same carbon and with the same orientation, to give the product alcohol. The less substituted alcohol is produced, because in the reaction of the alkene with borane, addition of boron to the less substituted carbon atom is favored. In most cases, this regiochemistry is dictated by steric factors, and the transition state with hydrogen added to the more substituted carbon atom is lower in energy than that with the reverse regiochemistry.

## EXERCISE 10.21

For each of the following reactions, predict whether the desired stereochemistry and regiochemistry can be attained with acid-catalyzed hydration, hydroboration–oxidation, or not at all.

(a)

(d)

(b)

(e)

(c)

(f)

## Five-Member Cyclic Intermediates

**Oxidation by Osmium Tetroxide or Potassium Permanganate.**
**Osmium tetroxide** and **potassium permanganate** are both oxidants that convert alkenes into 1,2-glycols. In both cases, the addition of the two hydroxyl groups is *syn*, as can be seen in the reactions of cyclohexene and *trans*-2-butene with these reagents. These addition reactions are called ***cis*-hydroxylations.**

Osmium tetroxide

Potassium permanganate

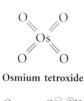

CHEM TV® I

#20   Alkenes to Glycols
(*syn* addition)

520

Cyclohexene    1,2-Glycol

(Quinuclidine)

*trans*-2-Butene    1,2-Glycol

There is some evidence that both of these reagents add to alkenes in a concerted process, simultaneously forming the two C—O bonds present in the product 1,2-glycols. The cyclic **osmate ester** and **manganate ester** are converted to the glycols under the reaction conditions, although the details of how the Os—O and Mn—O bonds are cleaved are not known.

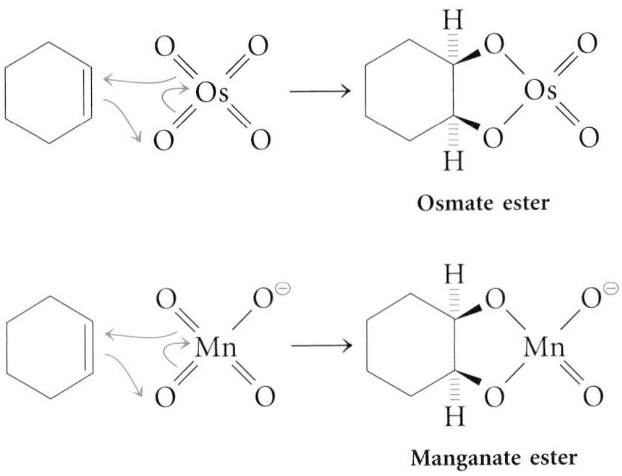

Osmate ester

Manganate ester

There are disadvantages to the use of these reagents. Osmium tetroxide is both expensive and toxic, although it can be used in a catalytic amount when another oxidizing agent—for example, $K_3Fe(CN)_5$—is present. (The tertiary amine base quinuclidine greatly increases the reaction rate.) Potassium permanganate is significantly less toxic and is even used as a topical antibacterial agent. However, the permanganate ion is a more powerful oxidizing agent than osmium tetroxide, and yields of the glycol are generally low (50%) because of overoxidation. Indeed, potassium permanganate is used in acid solution to cleave both the $\pi$ and $\sigma$ bonds of an alkene. Depending on the substitution pattern of the carbon atoms of the alkene, permanganate oxidizes the $sp^2$-hybridized carbons of the double bond to ketone or carboxylic acid groups (Figure 10.17).

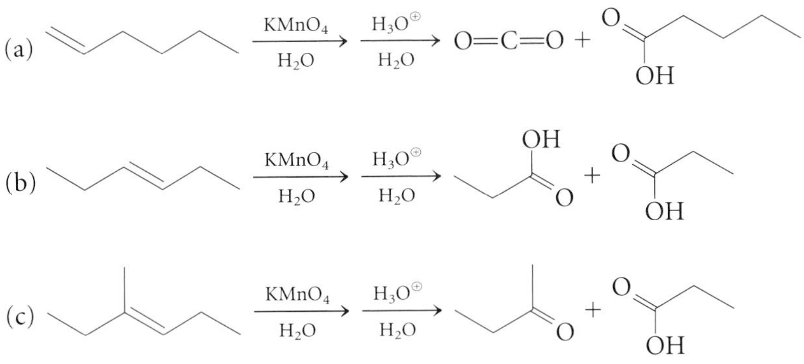

**FIGURE 10.17**

The oxidation level of the products of oxidative degradation with hot aqueous $KMnO_4$ is determined by the degree of substitution of the $sp^2$-hybridized carbon atoms of the starting alkene. (a) When the carbon bears two hydrogen atoms (as in a terminal alkene), $CO_2$ is produced. (b) When the carbon bears one hydrogen atom, a carboxylic acid is produced. (c) When the carbon has two substituents (and no hydrogen atoms), a ketone results.

When the alkene carbon is bonded to a hydrogen and an alkyl group, the initial product of permanganate oxidation is an aldehyde, but this is rapidly oxidized to a carboxylic acid by permanganate. In fact, when two such hydrogen atoms are present, $CO_2$ is formed (see, for example, the oxidation of 1-hexene in Figure 10.17). Oxidation of a trisubstituted alkene produces a ketone and a carboxylic acid. Permanganate oxidation of cyclic alkenes results in difunctional products:

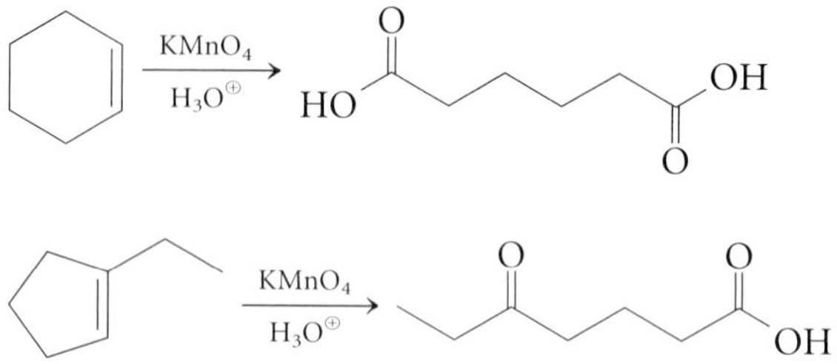

***Permanganate Test for Oxidizable Functional Groups.*** Permanganate is purple, and its reduction product, $MnO_2$, is brown. The fading of the purple color of the permanganate ion is indicative of oxidative degradation and is another color test that can be used as a quick indicator of the presence of oxidizable functional groups. Alkenes, alkynes, alcohols, and aldehydes give positive permanganate oxidation tests. The simple rule for permanganate oxidations is that all carbon–carbon multiple bonds will be completely cleaved (two bonds for alkenes and three for alkynes), and the original $sp^2$-hybridized carbons will appear in the products at the highest possible oxidation level that can be achieved without cleavage of another carbon–carbon $\sigma$ bond.

### EXERCISE 10.22

Draw the structure of an alkene that would yield each of the following compounds upon treatment with potassium permanganate in acid:

(a)

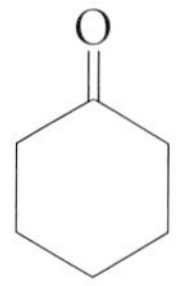

(c)

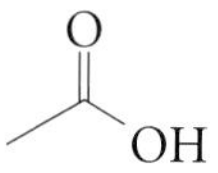

(b)

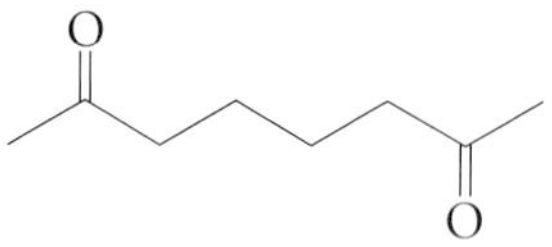

(d)

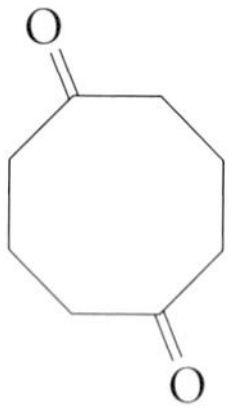

***Ozonolysis.*** Ozone, $O_3$, also acts as an electrophilic agent. It exists in a zwitterionic form in which the central oxygen formally bears positive charge. (Recall from Chapter 2 that a *zwitterion* is a neutral molecule that bears two oppositely charged centers in at least one significant resonance structure.) Two resonance contributors make the terminal oxygens of ozone partially negatively charged, but the formal oxygen–oxygen multiple bond confers part of the electrophilicity of the central positively charged oxygen on the terminal oxygens.

This zwitterion can thus initiate electrophilic attack. Electrons from the carbon–carbon $\pi$ bond of an alkene are polarized toward the terminal oxygen, allowing a shift of electrons:

Molozonide       Ozonide

Electrophilic attack by ozone proceeds through a six-electron transition state to form a **molozonide,** a compound containing two weak oxygen–oxygen bonds. This compound is relatively unstable and rearranges to an **ozonide** by routes that need not be considered here. This rearrangement accomplishes the breaking of both the $\sigma$ and the $\pi$ bonds between the carbons originally joined by a double bond, replacing each with a single bond to oxygen. That is, in the ozonide, each of the original $sp^2$-hybridized carbon atoms is joined to oxygen by two $\sigma$ bonds. Upon treatment with zinc in acetic acid, the ozonide is reduced to the corresponding carbonyl compounds.

The ozonolysis of alkynes, followed by treatment with zinc and acetic acid, leads to carboxylic acids at each of the originally $sp$-hybridized atoms:

Alkyne       Carboxylic acids

Ozonolysis effects the net conversion of an alkene into two carbonyl compounds (aldehydes or ketones). This occurs by cleavage of the carbon skeleton in a process known as **oxidative degradation.** Ozonolysis is a technique commonly used to simplify complex structures by replacing a C=C bond with two carbonyl groups. This method has proved very effective for determining structures of complex, naturally occurring compounds. Ozonolysis degrades a complex molecule to smaller fragments that can be more easily analyzed.

Here are simple examples of oxidative degradation:

*trans*-2-Hexene       Acetaldehyde       Butanal

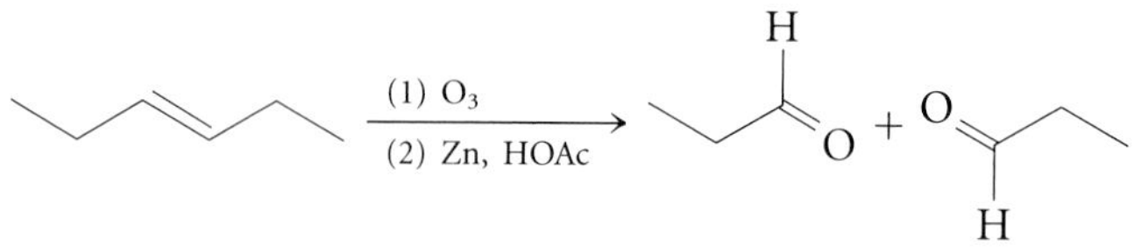

***trans*-3-Hexene** **Propanal**

One can distinguish *trans*-2-hexene from *trans*-3-hexene, despite their close chemical structures and reactivities, on the basis of the products formed upon ozonolysis. *Trans*-2-hexene gives rise to two products (acetaldehyde plus butanal), whereas *trans*-3-hexene gives rise to only one (two equivalents of propanal). Ozonolysis is particularly important in the degradation of functionally complicated, naturally occurring molecules. The structures of such complex molecules can be more readily deduced by first degrading them to the smaller and less complex corresponding carbonyl compounds, whose structures can be analyzed more easily.

Ozonolysis of cyclic alkenes can also be used to prepare molecules containing two carbonyl groups. For example, ozonolysis of 1,2-dimethylcyclohexene affords a diketone, an ozonolysis of 1-methylcyclooctene produces a ketoaldehyde. We will see in Chapter 13 how these dicarbonyl compounds can be further converted to cyclic compounds with ring sizes different from the starting alkene.

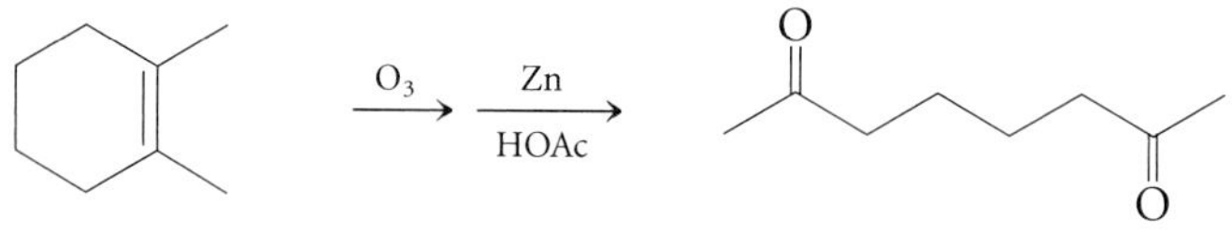

**1,2-Dimethylcyclohexene** **Diketone**

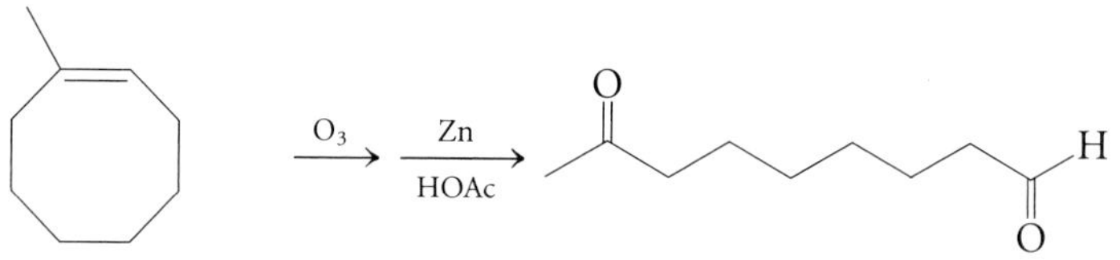

**1-Methylcyclooctene** **Ketoaldehyde**

**EXERCISE 10.23**

Predict the product(s) expected from ozonolysis of each of the following compounds:

(a) 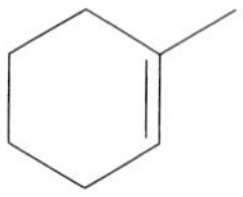 (c) 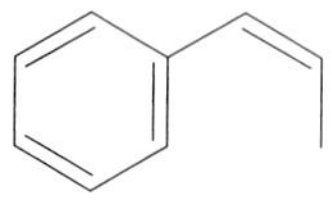 (e) 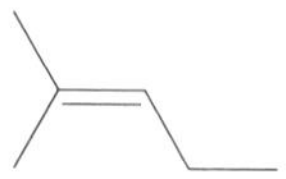

(b) 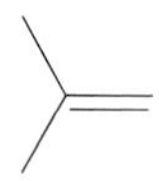 (d) 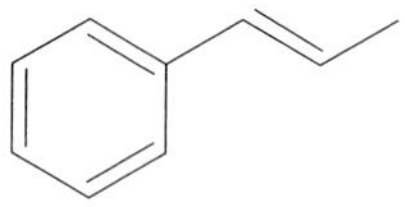

## Formation of Six-Member Rings: The Diels–Alder Reaction

CHEM TV® II

#7 Diels–Alder Reaction

The Diels–Alder reaction is a versatile synthetic tool for the construction of six-member rings. However, there are certain restrictions on the nature of the two components, the **diene** and the **dienophile.**

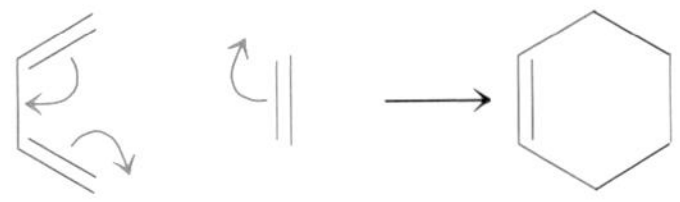

Diene Dienophile

The most simple example of the Diels–Alder reaction—the reaction of butadiene with ethylene—requires fairly rigorous conditions and gives unsatisfactory yields of cyclohexene because of two competing reactions. First, butadiene reacts with itself, at a rate comparable to its reaction with ethylene; one molecule of butadiene serves as the diene and another as the dienophile, producing a substituted cyclohexene:

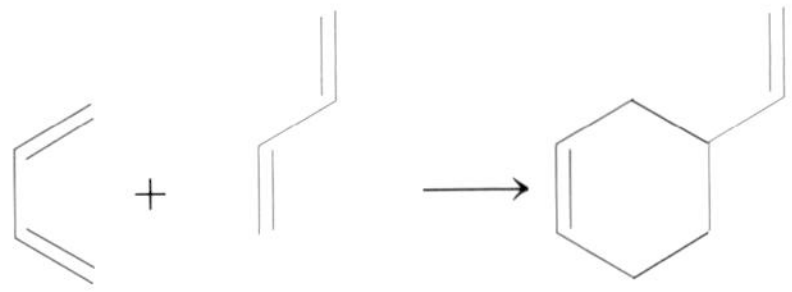

Diene Dienophile

Second, the cyclohexene formed from the reaction of butadiene and ethylene has a reactivity comparable to that of ethylene:

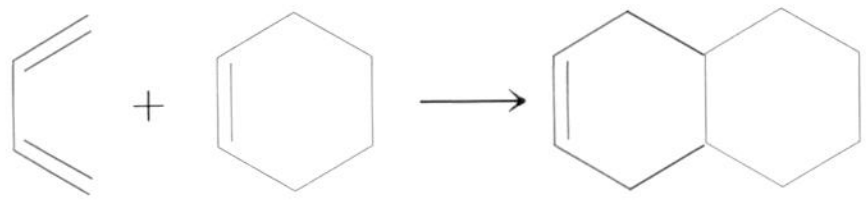

Diene Dienophile

Furthermore, the product of each of these two competing reactions is itself a dienophile and can participate in further Diels–Alder reactions with butadiene.

Fortunately, the presence of electron-withdrawing substituents in the dienophile offers the means of controlling the Diels–Alder reaction. First, electron-withdrawing groups substantially increase the rate of the reaction. For example, methyl acrylate and maleic anhydride both participate in Diels–Alder reactions at rates that are orders of magnitude faster than those of simple alkenes.

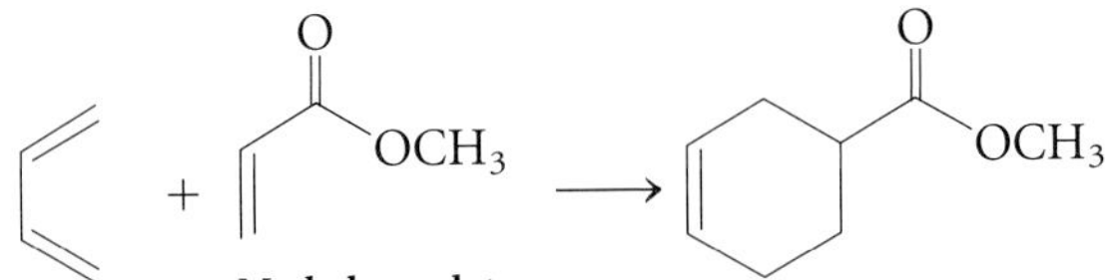


Methyl acrylate

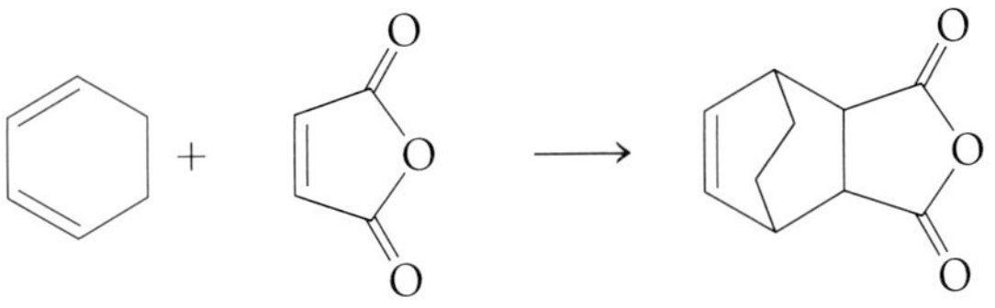

Maleic anhydride

Second, although the products retain the electron-withdrawing groups of the starting dienophiles, these groups are *not* substituents of the $sp^2$-hybridized carbon atoms of the products. Thus, the double bonds of the products react very slowly with the starting dienes, and further reaction is avoided.

**EXERCISE 10.24**

Provide the structures of the starting diene and dienophile that would produce each of the following compounds by a Diels–Alder reaction:

(a)

(b)

(c)

CO₂CH₃

## 10.5

# Reduction of Multiple Bonds

The addition of hydrogen across multiple bonds (also called *reduction*) can be accomplished by two fundamentally different methods, as illustrated for the reduction of an alkene:

**Catalytic Hydrogenation**

$$\text{+ (2 H·)} \xrightarrow[\text{Pt}]{\text{H}_2}$$

**Dissolving-Metal Reduction**

$$+ 2\ e^{\ominus} + 2\ H^{\oplus} \longrightarrow$$

These two methods—catalytic hydrogenation and electron transfer reduction–protonation (dissolving-metal reduction)—are general methods for achieving the reduction of unsaturated organic compounds. Neither method can be described mechanistically using full-headed curved arrows, because each takes place through steps in which only one electron is transferred.

### Catalytic Hydrogenation

#18 Catalytic Hydrogenation
of Alkenes

*Catalysts.* Catalytic hydrogenation requires the activation of molecular hydrogen, H₂, through interaction with the surface of a noble metal. A **noble metal** is one that is very stable at the zero oxidation level. Among the noble metals most commonly used in catalytic hydrogenation are platinum and palladium, but finely divided nickel and other metals can also be used. Typically, noble metals used as catalysts are in the form of highly dispersed powders on a support with a large surface area (carbon or alumina). Sometimes, the zero-valent metal is generated *in situ* by reduction of the corresponding oxide. For example, treating platinum oxide, $PtO_2$, with hydrogen generates water and finely divided platinum, which is highly active in catalytic hydrogenation. The interaction of molecular hydrogen with the surface of platinum results in the rupture of the hydrogen–hydrogen bond and the formation of two metal–hydrogen bonds.

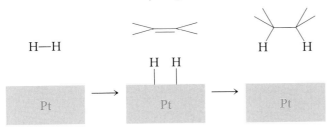

A multiple bond (for example, the C=C bond of an alkene) interacts with this activated form of hydrogen, which is then transferred, resulting in the net addition of $H_2$ to the C=C $\pi$ bond, with formation of two C—H $\sigma$ bonds. This conversion effects reduction of the double bond, while the metal catalyst is regenerated in its initial form, ready to interact with another molecule of hydrogen.

The metal surface is absolutely critical for these reactions, but it is not consumed in the net chemical transformation. It is said, therefore, to act as a catalyst—that is, as a species that accelerates the rate of a reaction without itself being consumed. For this reason, these reactions are called *catalytic hydrogenations.*

***Hydrogenation of Alkenes.*** In the catalytic hydrogenation of an alkene, both hydrogens are delivered to the same face of the molecule. This reaction is thus a stereospecific *syn* addition—that is, the hydrogens add to give *cis* products. This result can be observed in the addition of one equivalent of $H_2$ to an alkyne:

$$-C\equiv C- \xrightarrow[\text{Pt}]{H_2} \underset{\textbf{A } \textit{cis-}\textbf{alkene}}{\overset{H}{\underset{}{C}}=\overset{H}{\underset{}{C}}}$$

**An alkyne**          **A *cis*-alkene**

Catalytic hydrogenation of alkenes normally yields the product expected from *syn* addition, but this is not always the case. For example, the following reduction of a bicyclic alkene affords a 9:1 mixture of the *trans* and *cis* products:

Here, the *trans* product arises because the rate of reduction of the double bond decreases with increased alkyl substitution, and positional isomerization of the starting alkene occurs at a comparable rate. Reduction of the isomerized starting material yields the *trans* product. In practice, the observation that catalytic reductions proceed by the *cis* addition of hydrogen is usually of consequence only for the reduction of disubstituted alkynes.

*Hydrogenation of Alkynes.* The ease with which a multiple bond is reduced by catalytic hydrogenation is related to the strength of the $\pi$ bond. For example, because of electron–electron repulsion, the two $\pi$ bonds of an alkyne are individually weaker than the $\pi$ bond that remains in the product alkene after the addition of two hydrogen atoms.

Thus, the rate of catalytic hydrogenation of an alkyne is faster than that of the resulting alkene, and, with care, the process can be limited to the addition of one equivalent of hydrogen. If the alkene is the desired product, it is often useful to deactivate the catalyst somewhat—for example, by adding small amounts of an amine (such as pyridine or quinoline) that can bind to the surface of the metal. (These species are referred to as *catalyst poisons*.) Even in these circumstances, further reduction can be accomplished, and whether an alkene or the corresponding alkane is the final product is often determined by the reaction time. The progress of catalytic hydrogenation can be monitored by the uptake of hydrogen from the gas phase as the reduction proceeds.

*Hydrogenation of Aromatic Compounds.* Catalytic reduction of aromatic compounds is also possible, but such reactions are quite slow. The addition of one equivalent of hydrogen to an aromatic compound such as benzene requires disruption of the aromatic system and is considerably slower than the addition of hydrogen to the remaining $\pi$ bonds. In contrast to the reduction of alkynes, in which it is possible to produce an alkene, it is not possible to isolate any partially reduced products—a cyclohexadiene, for example—at an intermediate oxidation level.

*Hydrogenation of Heteroatom Functional Groups.* Carbon–oxygen and carbon–nitrogen double bonds are also reduced with hydrogen and a metal catalyst. The $\pi$ bond of a carbonyl group is considerably stronger than that of an alkene (93 versus 63 kcal/mole), and catalytic hydrogenation of aldehydes, ketones, and (especially) esters requires the use of high temperature and pressure to increase both the hydrogen concentration in the solution and the reaction rate. The rates of reduction of these stronger and more polarized bonds are much slower than those for carbon–carbon double bonds.

Imines undergo reduction to form amines. The carbon–nitrogen $\pi$ bond in an imine is weaker than the carbon–oxygen $\pi$ bond in a carbonyl group but stronger than the carbon–carbon $\pi$ bond in a simple alkene. Therefore, imines are reduced at rates somewhere between those for the other two functional groups. However, amines complex with noble metals and greatly reduce their effectiveness as catalysts. (Quinoline, for example, is a catalyst poison.) Thus, the reduction of an imine is best carried out in the presence of acid (for example, by using acetic acid as a solvent) so that the product amine is protonated, blocking surface complexation.

Molecules containing both an alkene and a carbonyl group undergo catalytic hydrogenation at room temperature and atmospheric pressure only at the carbon–carbon $\pi$ bond:

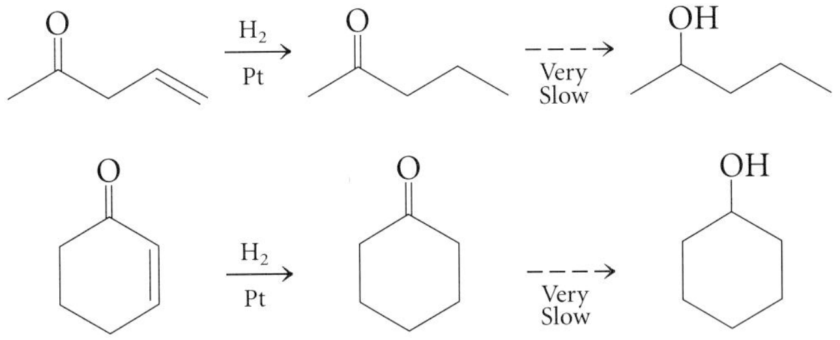

For hydrogenation of a carbonyl group or imine, *syn* addition cannot be demonstrated because the position of the hydrogen atom on the hetero-atom is not fixed with regard to that of the hydrogen atom on the carbonyl or imine carbon. Since no isomers can be formed, it is not possible to use the structure of the product to determine the stereochemistry of addition.

***Selective Hydrogenation.*** The relative ease with which catalytic hy-drogenation reduces the various functional groups discussed so far is shown in Figure 10.18. Alkynes are hydrogenated most easily, and esters are the most difficult to reduce. Even though the catalytic hydrogenation of esters and aromatic hydrocarbons requires high temperatures and pressures and special catalysts, it is often the preferred method for accomplishing indus-trial-scale reductions. Factors that seem undesirable for small-scale reac-tions in the laboratory are often advantageous when applied to large quan-tities. For example, in the case of catalytic hydrogenation, excess reagent (hydrogen) and the insoluble metal catalyst are easily removed, and no waste by-products are formed from the reagent, as is the case in reduction with complex metal hydrides.

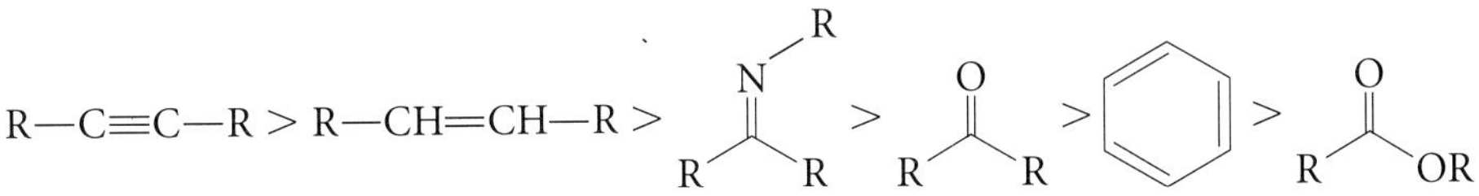

**FIGURE 10.18**

The ease of catalytic hydrogenation (with $H_2$ and a metal catalyst) depends on the functional group being reduced.

## ▪ Dissolving-Metal Reductions

Another method for effecting reduction provides an electron source (often a zero-valent alkali metal) capable of generating a radical anion or dianion intermediate, which is then protonated *in situ.*

***Reduction of Carbonyl Groups.*** The addition of an electron to a carbonyl group generates a radical anion (a ketyl), which is ion-paired with the alkali metal cation formed when the electron is transferred from the neutral metal.

In the presence of a proton source (typically, *tert*-butanol), the radical anion is protonated, producing a carbon radical that is, in turn, further reduced by the addition of an electron from a second atom of sodium. Protonation of the resulting carbanion leads to the fully reduced product. In some cases, the ketyl formed on the surface of sodium is further reduced by the addition of another electron before it is protonated. The resulting dianion then adds two protons, resulting in the same product.

Although the sequence may be different, both routes involve adding two electrons and two protons, giving rise to a net reduction.

Because sodium can react directly with an alcohol (generating hydrogen gas), these reactions are usually conducted in an ether solvent or in liquid ammonia containing only enough alcohol to serve as a proton source.

$$ROH + Na \longrightarrow RO^{\ominus} + Na^{\oplus} + {}^{1}/_{2}\, H_2 \uparrow$$

In the course of these reactions, the sodium metal disappears as it reacts and is converted into soluble sodium salts. Such reactions are therefore called **dissolving-metal reductions.** In principle, these reactions require not a metal, but rather a source of electrons. Indeed, the reactions can also be carried out by the addition of electrons in an electrochemical cell.

***Reduction of Alkynes.*** The facility with which dissolving-metal reductions proceed depends on the ability of the organic compound to accommodate an extra electron to form a radical anion. These reductions work well for carbonyl groups and alkynes but not for simple alkenes, because alkali metals do not transfer electrons efficiently to alkenes. Therefore, it is possible to reduce an alkyne to an alkene without further reduction to the alkane. When a dianion intermediate is formed, the negative charges are *trans* to each other in order to minimize electron repulsion. Protonation of the dianion leads to *trans* addition. This stereochemical course differs from that of catalytic hydrogenation, where *cis* addition of hydrogen to an alkyne occurs.

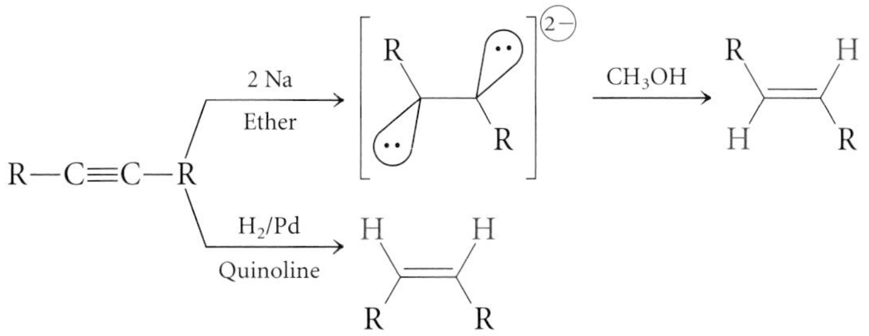

***Biological Reductions.*** Dissolving-metal reductions have particular biological relevance, although in living systems, less rigorous reducing agents than alkali metals are used to form the radical anion intermediates.

### EXERCISE 10.25

Write the structures of the products, if any, expected from each of the following reductions:

(a) 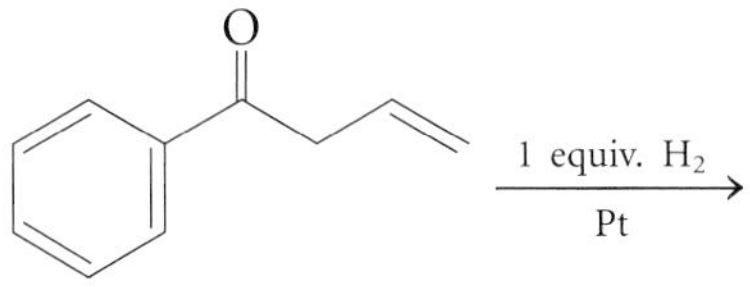 $\xrightarrow[\text{Pt}]{\text{1 equiv. } H_2}$

(b) 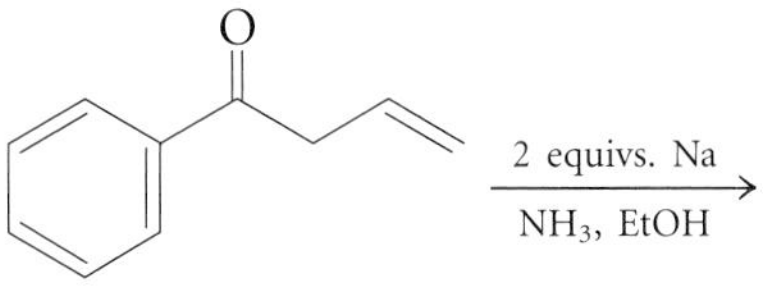 $\xrightarrow[\text{NH}_3,\ \text{EtOH}]{\text{2 equivs. Na}}$

(c)  $\xrightarrow[\text{NH}_3,\ \text{EtOH}]{\text{2 equivs. Na}}$

(d)  $\xrightarrow[\text{Quinoline}]{\text{1 equiv. } H_2,\ \text{Pt}}$

(e) 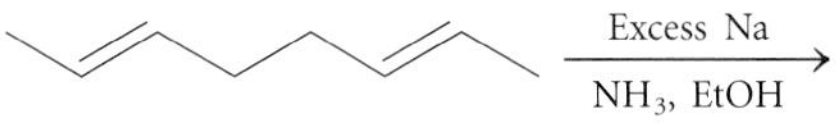 $\xrightarrow[\text{NH}_3,\ \text{EtOH}]{\text{Excess Na}}$

(f) 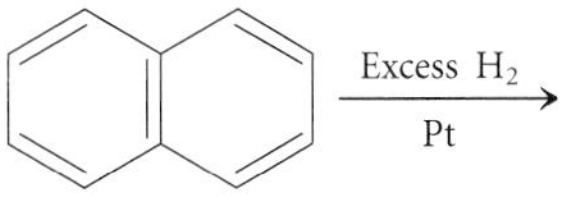 $\xrightarrow[\text{Pt}]{\text{Excess } H_2}$

## 10.6

## Synthetic Methods

Like other types of reactions, addition reactions can be grouped according to the functional-group conversion accomplished. Table 10.1 provides such a summary. When combined with the methods presented in Chapter 9 for preparing alkenes, these reactions are powerful techniques for interconverting a number of functional groups.

**TABLE 10.1**

Using Addition Reactions to Make Various Functional Groups

| Functional Group | Addition Reaction | Example |
|---|---|---|
| Alcohol | Acid-catalyzed hydration of an alkene | |
|  | Oxymercuration–demercuration | |
|  | Hydroboration–oxidation | |
| Aldehyde | Ozonation–reduction (ozonolysis) of an alkene | |
| Alkane | Catalytic reduction of an alkene | |
|  | Catalytic reduction of an alkyne | |
| Alkene | Partial hydrogenation of an alkyne (*cis*) | |
|  | Dissolving-metal reduction of an alkyne (*trans*) | |
| Alkyl halide | Electrophilic hydrohalogenation | |
|  | Peroxide-initiated radical hydrobromination | |
| Carboxylic acid | Hot KMnO$_4$ oxidation of alkenes | |
| Cyclohexene | Diels–Alder reaction | |

| Functional Group | Addition Reaction | Example |
|---|---|---|
| Cyclopropane | Carbene addition to alkenes | :CCl$_2$ |
| Dihalide, vicinal | Electrophilic bromination or chlorination | Br$_2$ / CH$_2$Cl$_2$ |
| Dihalide, geminal | Hydrohalogenation of an alkyne | HBr ... HBr |
| Epoxide | Peracid oxidation of an alkene | |
| 1,2-Glycol | Osmium tetroxide oxidation of an alkene | OsO$_4$ (cat.) / K$_3$Fe(CN)$_5$ (Quinuclidine) |
| | Basic permanganate oxidation of an alkene | KMnO$_4$ / NaOH |
| Ketone | Acid-catalyzed hydration of an alkyne | R—C≡C—H, Hg(OAc)$_2$ / H$_2$SO$_4$, H$_2$O |
| | Ozonation–reduction (ozonolysis) of an alkene | (1) O$_3$ (2) Zn, HOAc |
| | Hot KMnO$_4$ oxidation of an alkene | KMnO$_4$ / H$_2$O, H$_3$O$^\oplus$ / H$_2$O |
| Polymer | Cationic polymerization | H$^\oplus$ |
| | Free-radical polymerization | X· |

# Summary

1. Electrophiles attack regions of electron density, particularly $\pi$ orbitals in alkenes, dienes, and alkynes, so as to form the more stable of two possible carbocations.

2. Electrophilic addition is a two-step process. In the first, rate-determining step, the electrons of the $\pi$ bond move toward the electrophile, forming a new bond between one of the carbons and the electrophile. In the second step, the other carbon of the original double bond becomes bonded to a nucleophile.

3. Because carbocations are the intermediates in electrophilic addition, the regiochemistry of these reactions is controlled by the formation of the more stable cation. Consistent with Markovnikov's rule, the order of carbocation stability (tertiary > secondary > primary) dictates that the more highly substituted carbon becomes positively charged as the carbocation intermediate forms in electrophilic addition.

4. Carbocations are planar, and so the addition of a nucleophile in the second step of electrophilic addition occurs equally easily on either face. Thus, both stereoisomers are present in a product mixture when stereocenters are formed.

5. The carbocation intermediates formed in electrophilic addition reactions undergo skeletal rearrangements when a driving force exists for doing so. For example, when secondary carbocations are formed in electrophilic addition reactions, they often rearrange to more stable tertiary cations.

6. Reversal of the normal (Markovnikov) orientation is accomplished by changing reagents, and thus the mechanism of addition. (a) Hydrobromination of alkenes occurs in polar solvents according to Markovnikov's rule; in ether in the presence of a free-radical initiator, the opposite regiochemistry is obtained. (b) Anti-Markovnikov orientation is achieved by hydroboration–oxidation proceeding through a *syn* addition of borane in the first stage.

7. Addition to 1,3-dienes (conjugated dienes) occurs with a regiochemistry consistent with the formation of the most stable cationic intermediate (an allylic cation). This intermediate cation adds nucleophiles so as to form both 1,2- and 1,4- (conjugated) adducts.

8. Stereochemical control is afforded in some electrophilic additions to alkenes because of the intermediacy of bridged cationic intermediates: (a) The cyclic bromonium ion formed as an intermediate in the halogenation of an alkene is responsible for the overall *trans* addition. (b) Stereochemistry is also controlled in singlet carbene additions forming cyclopropanes and in peracid oxidations forming epoxides. Both are concerted reactions that proceed without the formation of intermediates.

9. The first step of ozonolysis is concerted, forming a molozonide. However, rapid rearrangement followed by reduction ultimately produces two carbonyl compounds, with the cleavage of both the $\sigma$ and $\pi$ bonds of the original alkene.

10. The addition of bromine to an alkene and oxidative cleavage of an alkene with permanganate are used as characteristic color tests for the presence of carbon–carbon double or triple bonds.

11. Reaction of alkenes with peracids provides a practical method for preparing epoxides. Epoxides react with powerful nucleophiles by ring opening via an $S_N2$ mechanism.

12. The Diels–Alder reaction forms six-member rings with the simultaneous formation of two C—C $\sigma$ bonds.

13. Catalytic hydrogenation adds two hydrogen atoms to sites of unsaturation with a *syn* orientation. Alkenes, alkynes, aromatic rings, carbonyl groups, and imines can be reduced by this method.

14. Dissolving-metal reductions with a proton source in inert solvent add hydrogen to carbonyl compounds and alkynes. Addition of $H_2$ to alkynes gives *trans* stereochemistry.

## Review of Reactions

### Alkenes

Alkenes *(continued)*

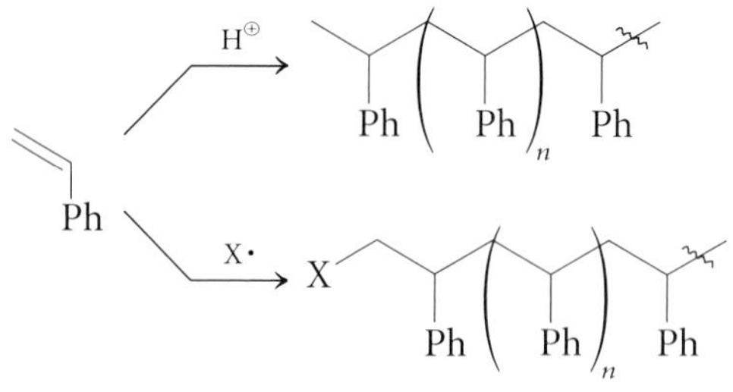

Alkynes

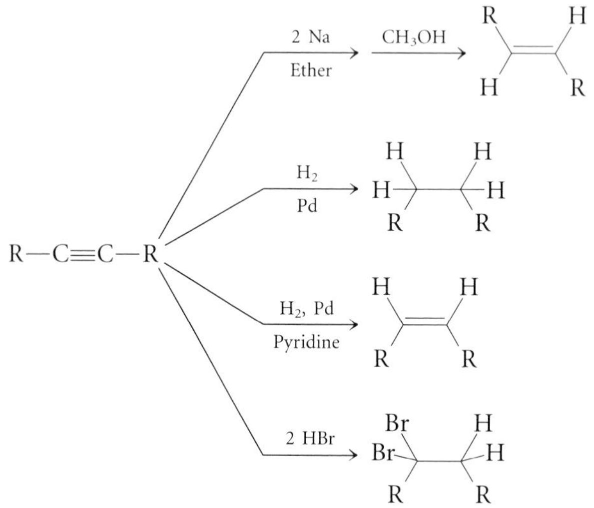

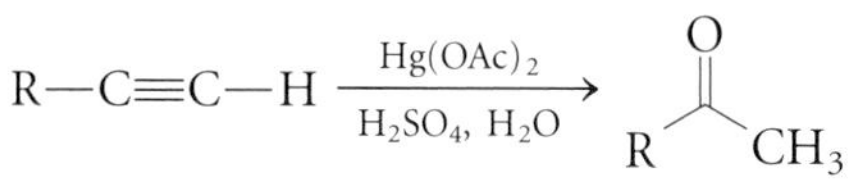

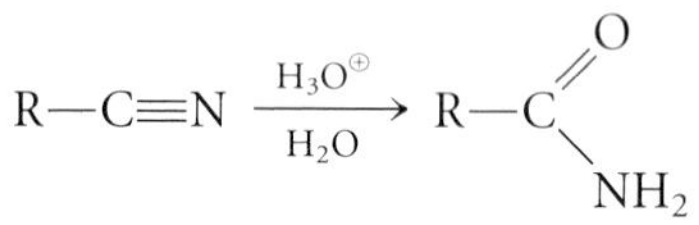

## Review Problems

**10.1** What product(s) would you expect when 1-methylcyclohexene is treated with each of the following reagents?

(a) HBr in MeOH

(b) HI in $H_2O$

(c) HBr in ether

(d) (1) $B_2H_6$; (2) $H_2O_2$, NaOH

(e) (1) $Hg(OAc)_2$; (2) $NaBH_4$

(f) $Br_2$ in $CH_2Cl_2$

(g) :$CH_2$ (singlet)

(h) *m*-chloroperbenzoic acid

(i) (1) $O_3$; (2) Zn, HOAc

(j) $Cl_2$ in $CH_2Cl_2$

(k) hot $KMnO_4$

(l) dilute aqueous $H_2SO_4$

**10.2** What product(s), if any, do you expect from the reaction of 2-butyne with each of the following reagents?

(a) Pt, excess $H_2$

(b) one equivalent of HBr

(c) two equivalents of HBr

(d) one equivalent of $Br_2$ in $CH_2Cl_2$

(e) two equivalents of $Br_2$ in $CH_2Cl_2$

(f) aqueous $HgSO_4$

(g) (1) $O_3$; (2) Zn, HOAc

(h) hot $KMnO_4$

**10.3** What major product(s), if any, do you expect from the reaction of 1,3-butadiene with each of the following reagents?

(a) excess $H_2$, Pt

(b) one equivalent of $Br_2$

(c) dilute aqueous $H_2SO_4$

(d) HCl

**10.4** What reagent (or series of reagents) can transform 1-butene into each of the following compounds?

(a) 1-bromobutane

(b) 2-bromobutane

(c) 1-butanol

(d) 2-butanol

(e) butane

(f) 1-butyne

(g) 2-butyne

(h) *s*-butyllithium

(i) propanoic acid

(j) 2-aminobutane

(k) 2-heptanone

(l) hexanoic acid

(m) 1-hexanol

**10.5** Suggest a sequence of reagents that can convert 1-pentanol into each of the following products:

(a) 1-chloropentane

(b) 1-pentene

(c) 2-pentanol

(d) 2-pentene

(e) 2-pentanone

(f) pentanoic acid

(g) pentanal

(h) 2-bromopentane

**10.6** Determine the correct starting material required to prepare each of the following products by using the indicated reagents:

(a) $Br_2$ in $CH_2Cl_2$ to prepare *meso*-2,3-dibromobutane

(b) $D_2$/Pt to prepare (*d,l*)-2,3-dideuterobutane

(c) $Cl_2$ in $CH_2Cl_2$ to prepare (*d,l*)-2,3-dichlorobutane

(d) HBr in $CH_2Cl_2$ to prepare racemic 2-bromobutane

(e) methylene (:$CH_2$) to prepare (*d,l*)-*trans*-dimethylcyclopropane

(f) *m*-chloroperbenzoic acid to prepare this *meso*-epoxide:

**10.7** Arrange each of the following series of alkenes in order of reactivity toward acid-catalyzed hydration, and explain your reasoning.

(a) 1-hexene, 2-methyl-1-pentene, 2-hexene

(b) 2-methylpropene, *cis*-2-butene, *trans*-2-butene

(c) 1-phenyl-1-butene, 1-phenyl-2-butene, 2-phenyl-1-butene

**10.8** The conditions commonly used for acid-catalyzed hydration of alkenes can sometimes lead to competing positional isomerization. Propose a mechanism by which 1-pentene can be converted to *trans*-2-pentene by treatment with acid, justifying at each step any relevant stereochemistry or regiochemistry.

**10.9** When HBr is added to a simple alkene in the presence of an ether solvent containing peroxides, the regiochemistry obtained is the reverse of that observed in a polar solvent. From your knowledge of the radical addition mechanism, what product would you expect from treatment of 1,3-butadiene with HBr in peroxide-containing ether?

**10.10** Chlorine adds to *trans*-2-butene in methylene chloride to give *meso*-2,3-dichlorobutane. However, if the reaction is conducted in water, a chlorohydrin (a compound bearing OH and Cl on adjacent carbons) is obtained. Using curved arrows to indicate electron flow, suggest a mechanism for this reaction that also predicts its stereochemical course.

**10.11** Explain why 2-butyne is less reactive than *trans*-2-butene toward most electrophiles such as bromine, and why it is nonetheless possible to stop after a single equivalent of bromine has been added to the alkyne.

**10.12** Draw three-dimensional structures for all products of addition of singlet diphenylcarbene to *trans*-2-butene. Are there centers of asymmetry in the product(s)? Compare the product(s) with that obtained from treatment of *trans*-2-butene with *m*-chloroperbenzoic acid.

**10.13** Write a detailed mechanism for each of the following reactions. Explain any relevant regio- or stereochemistry.

(a) 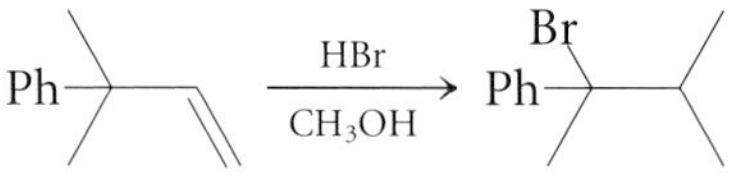 Ph $\xrightarrow[\text{CH}_3\text{OH}]{\text{HBr}}$ Ph—⟨Br⟩

(b) 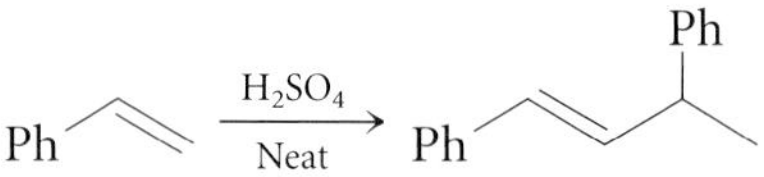 Ph $\xrightarrow[\text{Neat}]{\text{H}_2\text{SO}_4}$ Ph—⟨Ph⟩

(c) 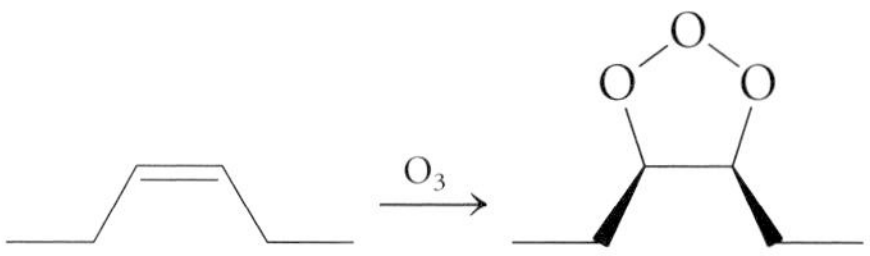 $\xrightarrow{\text{O}_3}$

(d) 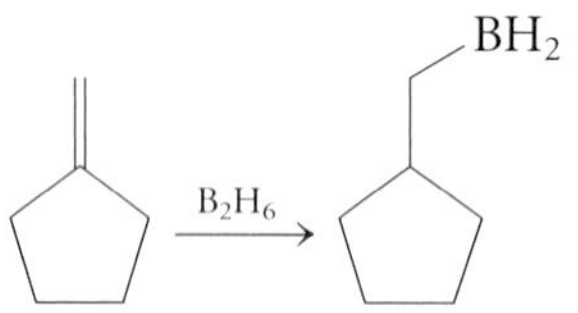 $\xrightarrow{\text{B}_2\text{H}_6}$ BH$_2$

(e) 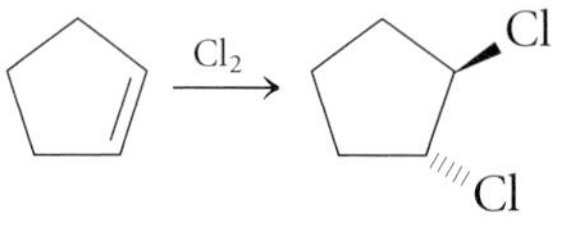 $\xrightarrow{\text{Cl}_2}$ Cl / Cl (and enantiomer)

(f) 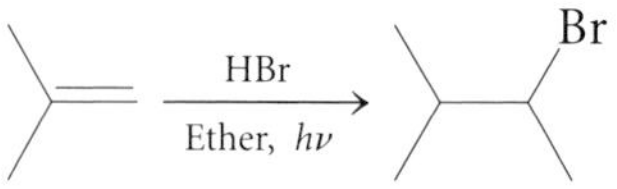 $\xrightarrow[\text{Ether, } h\nu]{\text{HBr}}$ Br

(g) 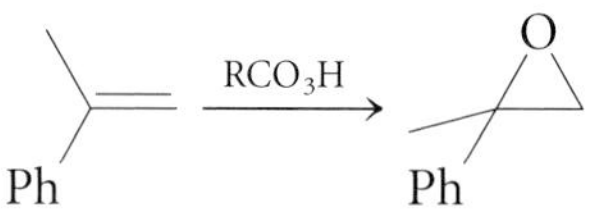 Ph $\xrightarrow{\text{RCO}_3\text{H}}$ Ph—O

(h) H$_3$C━━━H $\xrightarrow[\text{H}_2\text{O}]{\text{H}_3\text{O}^\oplus}$ H$_3$C—C(=O)—CH$_3$

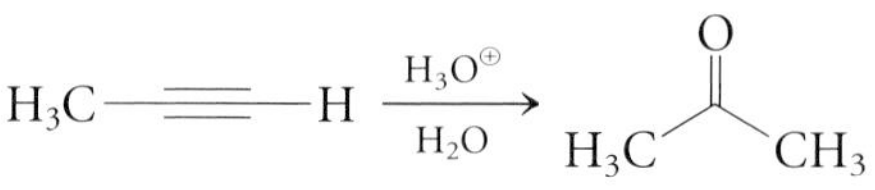

**10.14** Choose a chemical and a spectroscopic method that can be used to distinguish between each pair of compounds. Describe what you would see for each compound with each technique.

(a) cyclohexane and cyclohexene

(b) 1-hexene and 1-hexyne

(c) 2-hexene and 1-hexanol

(d) 2-hexene and 2-bromohexane

(e) 2-hexene-1-ol and 1-hexanol

(f) 2-hexanol and 2-bromohexane

(g) hexanal and hexane

**10.15** Osmium tetroxide can add to an alkene, giving an intermediate osmate ester that is hydrolyzed stereospecifically to a 1,2-diol. For example, when cyclohexene is treated with $OsO_4$, *meso*-1,2-cyclohexanediol is ultimately formed. Assuming that the hydrolysis occurs with retention of configuration, determine whether the stereochemistry for the formation of the osmate ester is *syn* or *anti*. Write a mechanism for this formation that is consistent with your answer.

**10.16** Limonene, $C_{10}H_{16}$, is a naturally occurring hydrocarbon that gives lemons their odor (a "citrusy" smell). When treated with excess hydrogen in the presence of a platinum catalyst, limonene takes up two equivalents of hydrogen. When treated with ozone, followed by zinc in acetic acid, it forms one equivalent of formaldehyde and one equivalent of a tricarbonyl compound:

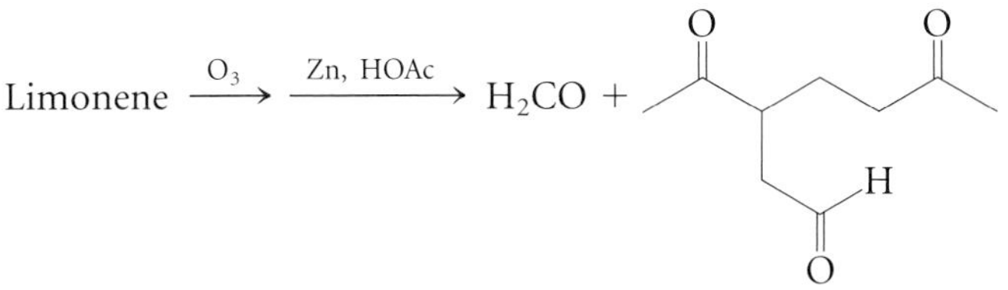

Propose one or more structures for limonene that are consistent with these data. If you cannot distinguish between two or more structures on the basis of the data given, propose a chemical or a spectroscopic method, or both, by which the structures could be distinguished. How many unique peaks would appear in the $^{13}C$ NMR spectrum of each structure you proposed?

**10.17** Hydroboration–oxidation of an alkyne proceeds through the same *syn* addition mechanism as for an alkene. The C—B bond of the resulting vinyl borane is also replaced, with retention of configuration in the oxidation step. From what you know about the mechanism of hydroboration–oxidation of alkenes, what products will be formed by hydroboration–oxidation of each of the following alkynes? (*Hint:* Recall that enols tautomerize easily to carbonyl compounds.)

(a) 1-hexyne    (b) 3-hexyne    (c) phenylacetylene

## Supplementary Problems

**10.18** Draw an energy diagram for the following transformation. Be sure to include any intermediates and to place multiple transition states at the appropriate relative energies.

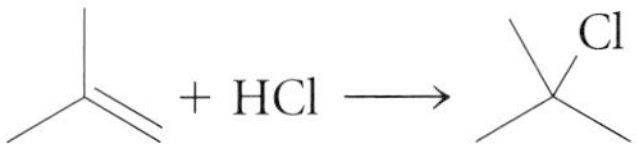

**10.19** Provide reagents that could be used to effect the transformation of methylenecyclohexane to each of the following compounds.

**Methylenecyclohexane**

(*Note:* Some transformations may require two separate reaction steps.)

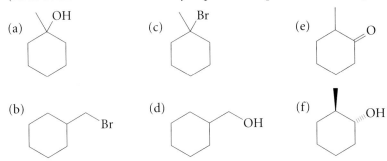

(a)  OH

(c)  Br

(e)  O

(b)  Br

(d)  OH

(f)  OH

**10.20** Bromine in solution is in rapid equilibrium with $Br_3^{\oplus}$ and $Br^{\ominus}$:

$$2\ Br_2 \rightleftharpoons Br_3^{\oplus} + Br^{\ominus}$$

$Br_3^{\oplus}$ is considerably more reactive in providing $Br^{\oplus}$ to an alkene than is $Br_2$. How might this affect the rate of bromination of an alkene as the concentration is lowered?

$Br^{\oplus}$  ←$\dfrac{Br_2}{\text{Slower}}$ — $\dfrac{Br_3^{\oplus}}{\text{Faster}}$→  $Br^{\oplus}$

**10.21** Treatment of a vicinal dibromide with strong base generally results in the formation of an alkyne.

Br   Br   $\xrightarrow{\text{NaNH}_2}$

Nevertheless, yields of the alkyne are only moderate because of the formation of other products.

(a) What other compounds might be formed by the double dehydrohalogenation of a vicinal dibromide?

(b) Can you think of a vicinal dibromide for which formation of an alkyne might be highly unfavorable?

**10.22** From what alkene might the following carbonyl compounds be prepared by reaction with ozone followed by reduction with zinc in acetic acid?

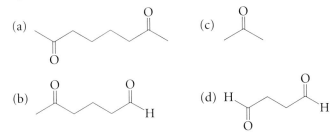

(a)

(c)

(b)

(d) H

# 11

# Electrophilic Aromatic Substitution

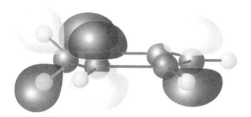

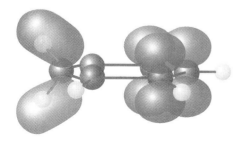

*Protonation of benzene produces an arenium ion, typical of all of the intermediates in electrophilic aromatic substitution:*

Benzene → An arenium ion (pentadienyl cation)

*Shown here are three of the five π molecular orbitals of the arenium ion of benzene. Each of the bonding orbitals (green and blue) is occupied by two electrons. The antibonding orbital (red and yellow) is empty but can accept electron density from substituents on the ring at positions where this orbital has density.*

The $\pi$ cloud of an aromatic ring, like that of a carbon–carbon double bond, makes it a potential chemical target for electrophiles. However, with aromatic molecules, electrophilic attack leads to substitution products, whereas electrophilic attack on alkenes generally leads to addition products. This difference in reactivity follows from the energetic importance of aromaticity in planar, cyclic, conjugated compounds that contain $4n + 2$ electrons. In this chapter, we examine electrophilic substitution on aromatic rings. The groups introduced as ring substituents in these substitution reactions can be chemically modified to other substituents that cannot be introduced directly by electrophilic substitution. Some of these transformations were presented in earlier chapters and some are new. This chapter will cover the production of the active electrophiles needed for aromatic substitution, the identity of the products formed when these electrophiles attack aromatic rings, and the mechanisms of these substitution reactions. The examination of the mechanisms of substitution reactions will explain why substituents alter the reaction rate and why they direct additional substitution to specific sites on the aromatic ring.

A study of the chemistry of aromatic compounds leads to deeper understanding of the effects of electronic and steric factors on reactivity, which will be of benefit in the further study of chemistry and biochemistry.

## 11.1

# Mechanism of Electrophilic Aromatic Substitution

#08    Electrophilic Aromatic Substitution

In an **electrophilic aromatic substitution,** an electrophile reacts with an aromatic ring to give overall replacement of a hydrogen atom with another substituent. In nitration, for example, the nitronium ion ($^{\oplus}NO_2$) replaces a proton, producing nitrobenzene:

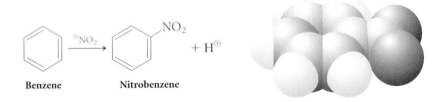

Benzene                    Nitrobenzene

Electrophilic substitution occurs in two steps: first, the addition of the electrophile; second, the loss of a proton. The first step is generally rate-determining. Therefore, significant insight into the overall reaction process can be gained by considering the structure of the intermediate cation formed upon addition of an electrophile to benzene. Because this step is quite endothermic, the transition state will closely resemble this intermediate.

### Step 1: Addition of the Electrophile and Formation of a Pentadienyl Cation as Intermediate

Although the $\pi$ electrons in an aromatic compound such as benzene are delocalized, they are still available for reaction with an electrophile, just like the $\pi$ electrons of an alkene (Chapter 10). Addition of an electrophile to benzene results in a cation in which electrons are also delocalized, called a **pentadienyl cation,** or an **arenium ion.** We can draw three significant resonance structures for this cation:

A pentadienyl cation

This cation bears significant positive charge on three of the ring carbon atoms, as can be seen in the three significant resonance structures that contribute to the hybrid. Two of the three charged carbon atoms are adjacent to the site of electrophilic attack, and one is on the other side of the ring.

Although this cation is significantly stabilized by delocalization, the loss of aromaticity from the addition of an electrophile to an aromatic ring represents a substantial energy cost, making the first step in an electrophilic aromatic substitution highly endothermic.

## Step 2: Loss of a Proton

In the second step of electrophilic aromatic substitution, a proton is lost, forming the ultimate substitution product. This step is quite exothermic as the aromatic $\pi$ system is reformed.

In theory, the arenium ion intermediate could react with a nucleophile, just as in electrophilic addition to alkenes. However, if this did happen, the product would lack the aromatic stabilization derived from cyclic conjugation. The alternative route, whereby a proton is lost, restores the $4n + 2$ aromatic system and is thus highly favored. In electrophilic addition to carbon–carbon double bonds, reaction of the cation intermediate with a nucleophile is usually observed. In those cases, addition of the nucleophile results in a C—Nuc bond that is generally stronger than the $\pi$ bond that would be formed by loss of a proton. The stability of the aromatic system accounts for the difference observed when electrophiles react with aromatic systems versus nonaromatic systems (e.g., alkenes). The former systems undergo substitution; the latter, addition.

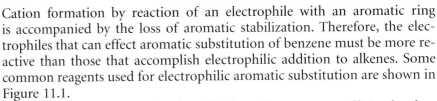

Loss of a proton from the arenium ion does not affect the overall rate of electrophilic aromatic substitution because the first step is rate-determining. Furthermore, only one proton can be lost to form a stable product, and therefore only one product can be formed from the arenium ion. In considering various factors that affect the rate and outcome of electrophilic aromatic substitution, we can safely focus on the first step—formation of the arenium ion intermediate.

## 11.2

### Introduction of Groups by Electrophilic Aromatic Substitution: Activated Electrophiles

Cation formation by reaction of an electrophile with an aromatic ring is accompanied by the loss of aromatic stabilization. Therefore, the electrophiles that can effect aromatic substitution of benzene must be more reactive than those that accomplish electrophilic addition to alkenes. Some common reagents used for electrophilic aromatic substitution are shown in Figure 11.1.

Reagents such as molecular chlorine, $Cl_2$, are not sufficiently electrophilic to bring about a practical rate of electrophilic aromatic substitution with benzene. However, prior interaction of $Cl_2$ with a strong Lewis acid such as $AlCl_3$ or $FeCl_3$ produces a complex that has greatly enhanced reactivity:

**Complex of Cl$_2$ and AlCl$_3$**

$$Cl-Cl + AlCl_3 \longrightarrow Cl-\overset{\oplus}{Cl}-\overset{\ominus}{AlCl_3}$$

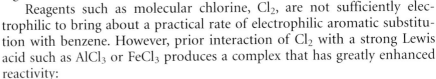

**FIGURE 11.1**

Some of the common reagents used for electrophilic aromatic substitution. In some cases, the reactivity of an electrophile is increased by interaction with either a Lewis acid such as $AlCl_3$ or a strong mineral acid such as $H_2SO_4$.

It is often difficult to achieve a sufficiently reactive electrophile for electrophilic aromatic substitution; thus, most electrophiles are produced in the reaction medium from a neutral species and a Lewis or Brønsted acid.

## Halogenation

The mechanism for electrophilic aromatic halogenation is shown in Figure 11.2 for the production of chlorobenzene from benzene. The reaction of $Cl_2$ with the Lewis acid $AlCl_3$ (acting as an electron acceptor) results in a complex that is significantly more reactive than molecular chlorine alone. The terminal chlorine in this complex is a very reactive electrophile because the Cl—Cl bond is strongly polarized toward the bridging, positively charged chlorine. In part, the formation of the Cl—Al bond in the complex weakens the Cl—Cl bond that must be broken. To some extent, the charge separation in the ion pair consisting of the arenium ion and the counterion is already partially developed in the complex. Loss of the proton from the carbon that has bonded to the electrophile returns a pair of electrons to the $\pi$ system, restoring aromaticity.

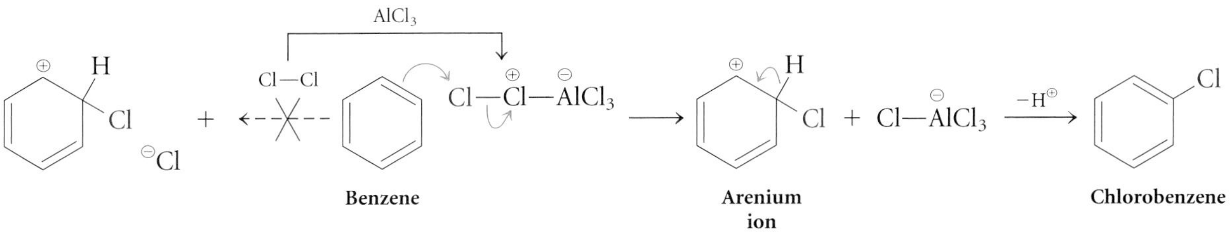

**FIGURE 11.2**

Complexation of $Cl_2$ with $AlCl_3$ (a strong Lewis acid) results in polarization of the Cl—Cl bond. Electrons are then donated from the aromatic $\pi$ system to form a C—Cl $\sigma$ bond with the terminal chlorine as the Cl—Cl bond is broken. Upon deprotonation of the intermediate cation, two electrons from the C—H $\sigma$ bond are released to restore ring aromaticity in the substitution product, chlorobenzene.

Bromination of benzene follows a pathway similar to that for chlorination, using a combination of $Br_2$ and $FeBr_3$. The corresponding reactions of $F_2$ and $I_2$ fail, however. We will see later in this chapter how fluorine and iodine can be introduced as substituents on an aromatic ring.

### EXERCISE 11.1

Write a detailed mechanism for the formation of bromobenzene by treatment of benzene with $Br_2$ and $FeBr_3$:

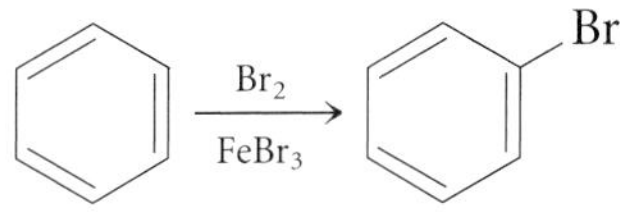

## TOXICITY OF CHLORINATED AROMATICS

Dioxins are a group of chlorinated aromatic hydro-carbons that are formed in trace amounts during production of many chlorinated compounds. For example, 2,4,5-trichlorophenoxyacetic acid (2,4,5-T, also known as Agent Orange) is an effective her-bicide that kills many different kinds of plants. It was widely used as a defoliant during the Viet-nam War. In the manufacture of 2,4,5-T from 2,4,5-trichlorophenol and chloroacetic acid, small amounts of the dioxin 2,3,7,8-TCDD also are gen-erated. This compound is extremely toxic in very small doses and has been shown to cause mutations in laboratory animals. Some of the health problems that later developed in soldiers exposed to Agent Orange in Vietnam are related to contact with this dioxin.

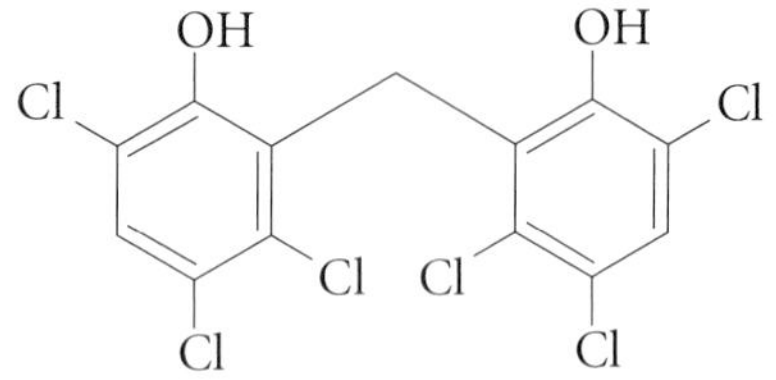

**Hexachlorophene**

Hexachlorophene, an effective topical antibacte-rial agent, has a structure similar to that of 2,3,7,8-TCDD. It has been shown to cause neurotoxic symp-toms in laboratory animals when given in very large doses. Because of the possible risks associated with long-term exposure, hexachlorophene has been banned for most uses.

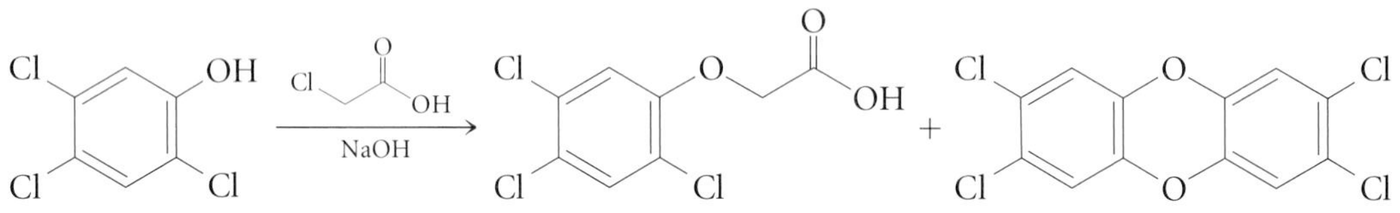

**2,4,5-Trichlorophenol**      **2,4,5-Trichlorophenoxyacetic acid (2,4,5-T, Agent Orange)**      **2,3,7,8-TCDD** (a dioxin)

## Nitration

Nitration of benzene using nitric acid involves the nitronium ion, $^{\oplus}NO_2$, as the active electrophile. This ion is generated from nitric acid by treatment with sulfuric acid; protonation is followed by loss of water:

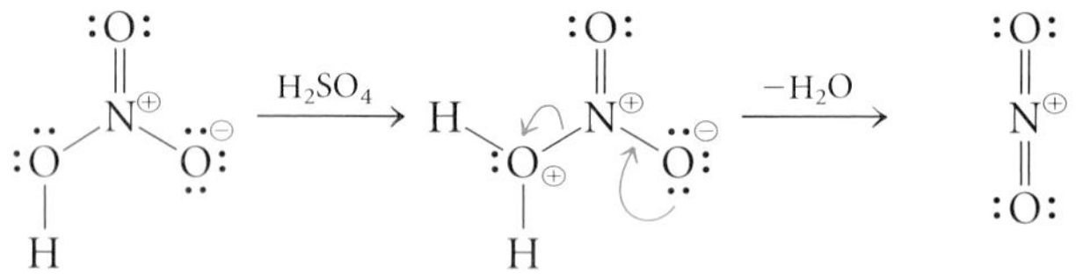

**Nitric acid**      **Nitronium ion**

The highly reactive nitronium ion can attack an aromatic hydrocarbon to form a $\sigma$ bond to carbon in the rate-determining step of the electrophilic aromatic substitution. Deprotonation of the resonance-stabilized cation re-stores aromaticity to the ring, yielding nitrobenzene:

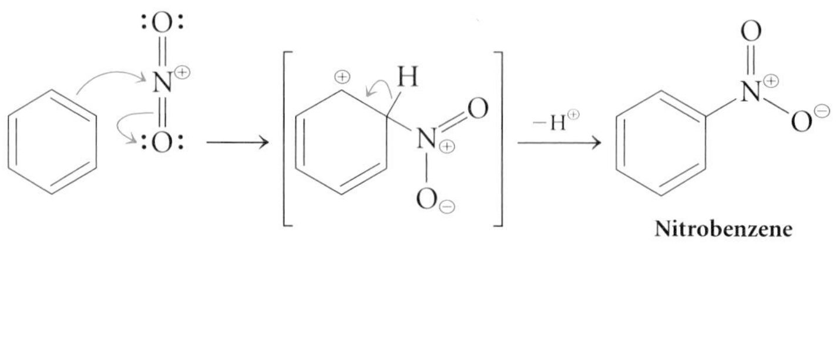

**Nitrobenzene**

**EXERCISE 11.2**

**547**

11.2 Introduction of Groups
by Electrophilic Aromatic
Substitution: Activated
Electrophiles

Draw a Lewis dot structure for $^{\oplus}NO_2$. Compare it with the Lewis dot structure for $CO_2$. From what you know about hybridization, do you expect the nitronium ion to be linear or bent?

## Sulfonation

Electrophilic aromatic sulfonation can be induced by the use of either concentrated sulfuric acid or sulfuric acid in which $SO_3$ (as much as 30%) has been dissolved, a mixture called *fuming sulfuric acid* (the word *fuming* is used because of the tendency of $SO_3$ to escape into the air where it mixes with water vapor, forming microdroplets of sulfuric acid). Actually, many different species present in sulfuric acid act as electrophiles. These include $SO_3$ and protonated $SO_3$. The reaction of $SO_3$ with benzene results in an intermediate zwitterion:

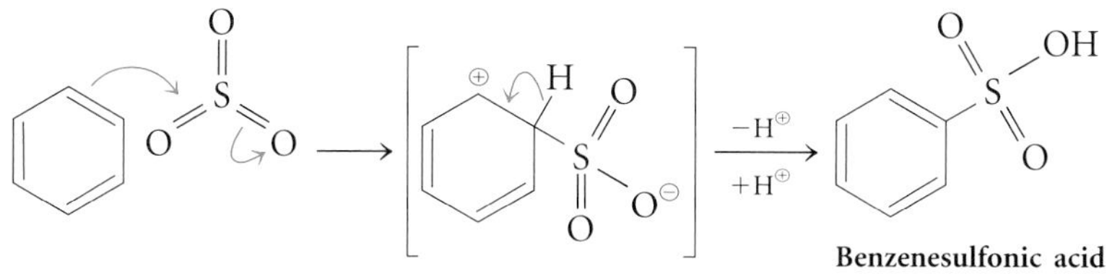

**Benzenesulfonic acid**

Loss of a proton from carbon regenerates the aromatic system, and protonation of the sulfonate group forms the product, benzenesulfonic acid.

## Friedel–Crafts Alkylation

Reaction of an alkyl halide with an aromatic compound in the presence of a Lewis acid results in replacement of a hydrogen by an alkyl substituent.

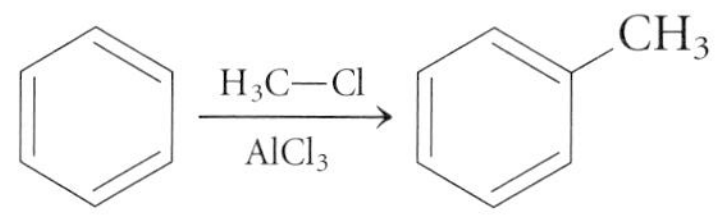

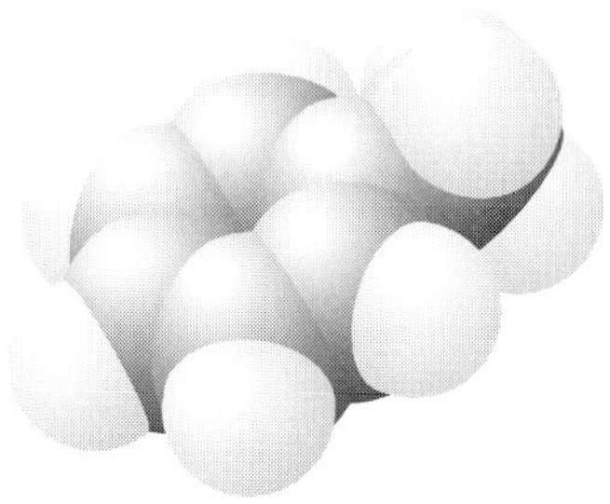

**Toluene**

This net carbon–carbon bond-forming reaction is referred to as a **Friedel–Crafts alkylation,** in acknowledgment of the contributions of two chemists, Charles Friedel, a Frenchman, and James Crafts, an American, who discovered this reaction in 1877. The mechanism is quite similar to that shown in Figure 11.2 for the chlorination of benzene.

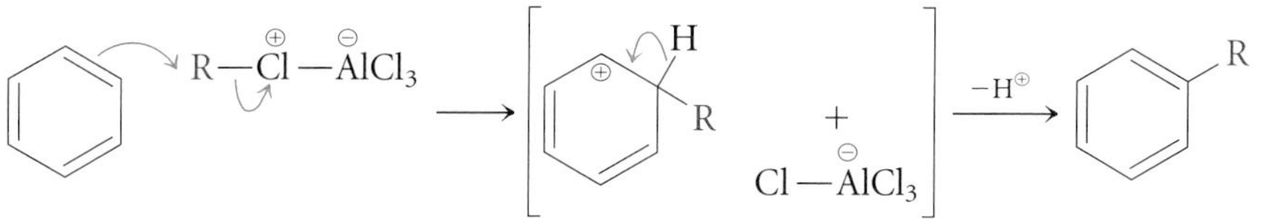

Complexation of $AlCl_3$ with the chlorine atom of an alkyl chloride weakens the carbon–chlorine bond and induces charge separation, just as in molecular chlorine. Reaction of the complex of R—Cl and $AlCl_3$ with benzene produces an arenium ion that, upon loss of a proton, forms an alkylbenzene, R—Ph.

### *Complications in Friedel–Crafts Alkylation: Carbocation Formation and Rearrangement.*

Primary alkyl halides react in the form of Lewis acid complexes, but tertiary and some secondary alkyl halides form a carbocation (and $^{\ominus}AlCl_4$) by cleavage of the C—Cl bond of the complex between the alkyl halide and $AlCl_3$. Then the free carbocation reacts with benzene.

*tert*-Butylbenzene

We can understand this difference between primary and tertiary alkyl halides on the basis of cation stability. (Recall that 3° cations are much more stable than 2° ones, and that 1° cations are rarely, if ever, formed.) The rate of formation of the carbocation for a tertiary halide is faster than reaction of the complex of the alkyl halide and $AlCl_3$ with benzene. Note also that the formation of the cation from the complex is a unimolecular reaction, whereas transfer of the alkyl group from the complex to benzene is bimolecular. Because unimolecular reactions are not affected by concentration as bimolecular reactions are, whether the attacking species is an alkyl halide–Lewis acid complex or a carbocation will depend to some extent on reaction conditions.

An important limitation of Friedel–Crafts alkylations is the occurrence of cationic rearrangements with primary and secondary alkyl halides. Indeed, virtually all primary alkyl halides undergo rearrangement upon treatment with $AlCl_3$, presumably as a result of the formation, in the complex, of appreciable positive charge on the primary carbon. The driving force behind these rearrangements is the greater stability of secondary and tertiary carbocations compared with primary ones.

Because of the facility with which such rearrangements occur, it is not possible to introduce a straight-chain alkyl group via Friedel–Crafts alkylation. For example, reaction of benzene with 1-chlorohexane and $AlCl_3$ produces 2-phenylhexane:

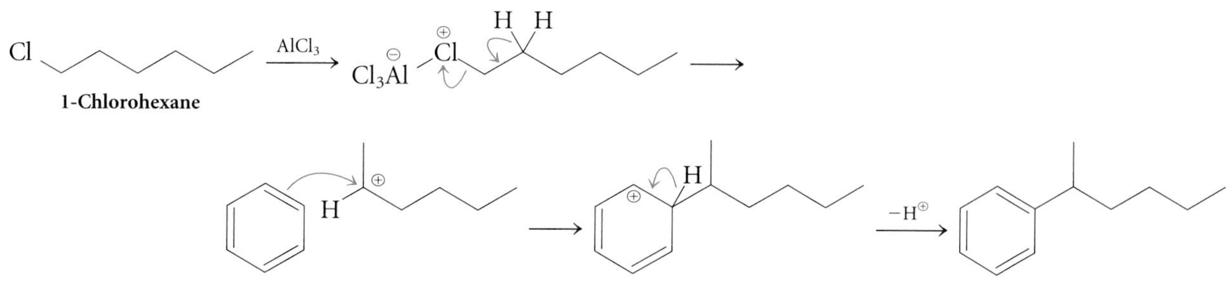

1-Chlorohexane

2-Phenylhexane

**Starting Materials for the Alkyl Side Chain.** It is possible to use bromo- and iodoalkanes in a Friedel–Crafts alkylation. However, there is no particular advantage to doing so, and these halogens and their derived alkyl halides are more expensive than molecular chlorine and alkyl chlorides. On the other hand, alkylation of benzene can be accomplished with a number of species that can give rise to $sp^2$-hybridized carbocations—such as alkenes or alcohols in the presence of Lewis and Brønsted acids. For example, cyclohexene reacts with benzene in the presence of $AlCl_3$ to produce cyclohexylbenzene:

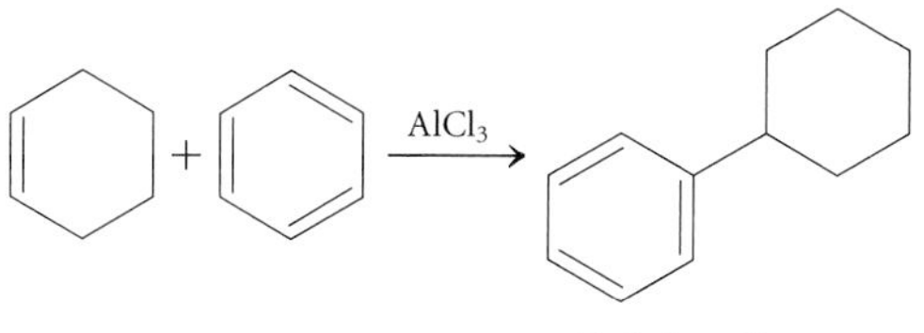

Cyclohexylbenzene

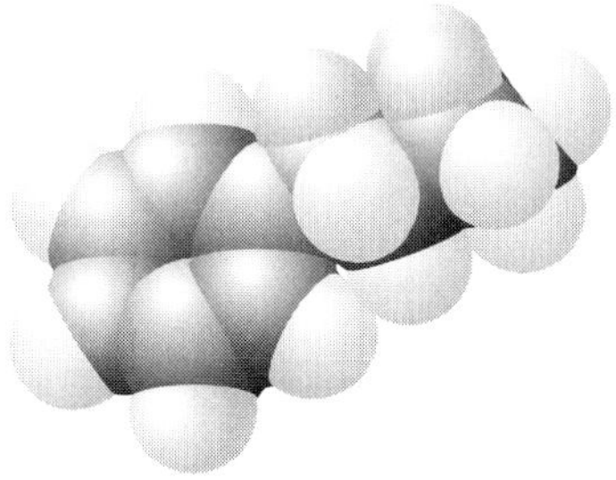

Reaction of terminal alkenes with benzene in the presence of acid gives the same products as are obtained from primary alkyl halides. Alkenes are substantially cheaper than alkyl chlorides, and their use in Friedel–Crafts alkylations is of significant industrial importance.

**EXERCISE 11.3**

Write a complete and detailed mechanism for the following reaction:

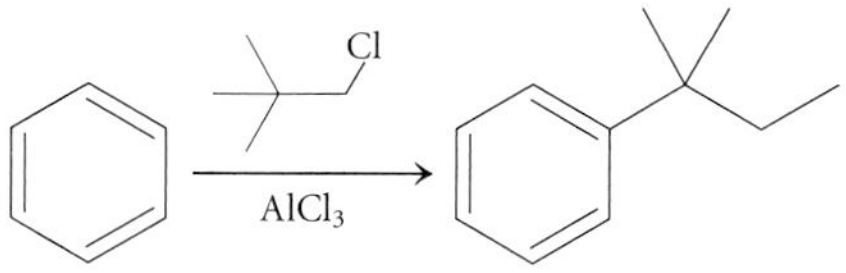

From the point of view of the alkyl halide, Friedel–Crafts alkylation is a nucleophilic substitution reaction—$S_N1$ for tertiary and secondary alkyl halides, and $S_N2$ for secondary and primary alkyl halides. Recall that nucleophilic substitution does not occur with aryl halides and vinyl halides, in which the carbon bearing the halogen leaving group is $sp^2$-hybridized. Thus, the group that becomes attached to the ring in a Friedel–Crafts alkylation must be derived from an $sp^3$-hybridized alkyl halide.

**549**

## RESEARCHERS DISCOVER THAT SQUIRRELS ARE NOT FOND OF MEXICAN FOOD

Hot peppers are the mainstay of Mexican and Southwestern American cooking. Capsaicin is the primary constituent of peppers such as the jalapeño (top photo) and the habañero (bottom photo). (Peppers are rated for "hotness" on the Scoville scale. The most intense pepper known is the habañero, rated at 100,000–300,000 Scoville heat units. The jalapeño's heat rating is only 2,500–5,000.)

Capsaicin has several uses, including use in personal defense sprays and as a component (0.025%) in a topical analgesic cream to relieve arthritis pain. When first applied to skin, capsaicin causes local heating and irritation. However, after repeated use, the area becomes desensitized to pain without loss of the sense of touch.

Researchers have uncovered another possible application for this naturally occurring compound. It appears that squirrels avoid bird seed treated with capsaicin (although so far they have not told investigators why). There are difficulties to be overcome before truly squirrel-proof bird seed is available. It takes 11 grams of capsaicin—the extract from 30 habañero peppers—per pound of bird seed to completely turn off squirrels.

**Capsaicin**

## EXERCISE 11.4

What product will be formed in each of the following Friedel–Crafts alkylations?

(a) [benzene] + [CH₃CH₂CH₂CH₂Cl] $\xrightarrow{AlCl_3}$

(b) [benzene] + [CH₃CH₂CH(Cl)CH₃] $\xrightarrow{AlCl_3}$

(c) [benzene] + [benzyl chloride] $\xrightarrow{AlCl_3}$

(d) [benzene] + [2-bromo-3-methylbutane] $\xrightarrow{AlCl_3}$

(e) [benzene] + [cyclohexanol] $\xrightarrow{H_2SO_4}$

(f) [benzene] + [1-bromo-2-methylbutane] $\xrightarrow{AlBr_3}$

**Friedel–Crafts acylation** (named for the same chemists who invented Friedel–Crafts alkylation) produces a phenyl ketone from reaction of benzene with a carboxylic acid chloride in the presence of $AlCl_3$:

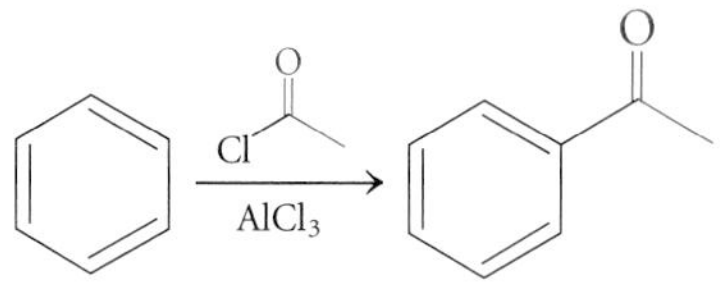

**Acetophenone**

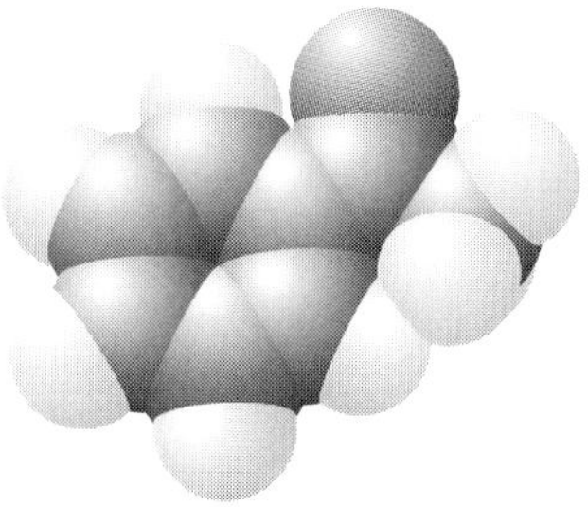

The reaction proceeds by a pathway very similar to that for alkylation, with complexation between a Lewis acid and the chlorine atom of the acid chloride:

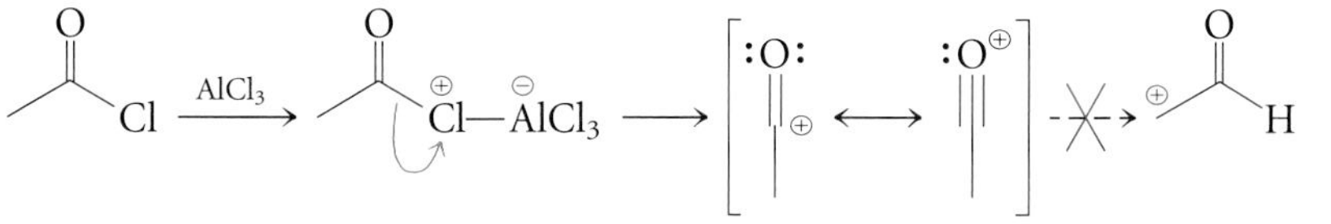

$$ \text{(1)} $$

**Acylium ion**

An **acylium ion** is then formed by cleavage of the C—Cl bond (in analogy with the formation of a carbocation intermediate in Friedel–Crafts alkylation). This ion bears formal positive charge on carbon and is stabilized by a resonance contributor that uses one of the lone pairs of oxygen to "backbond," forming a triple bond between carbon and oxygen. The acylium ion is a highly reactive electrophile.

Unlike simple carbocations, acylium ions do *not* undergo rearrangement. The resonance-stabilized acylium ion is considerably more stable than the cation (shown at the right in equation 1) that would result from the 1,2-shift of a hydrogen atom. As a result, an aryl ketone is always produced from Friedel–Crafts acylation. Overall, the mechanism is

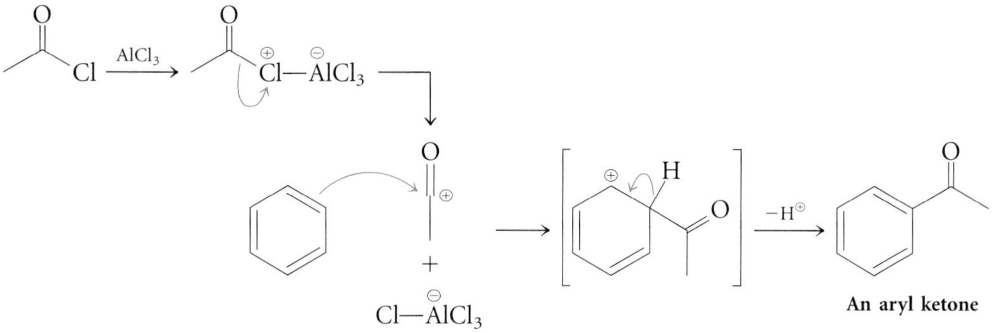

**An aryl ketone**

***Comparison of Friedel–Crafts Alkylation and Acylation.*** There is one practical difference between Friedel–Crafts alkylation and acylation. The aryl ketone that is the product of acylation is sufficiently basic that it forms a complex with the Lewis acid used to form the acylium ion, effectively removing the Lewis acid from further reaction:

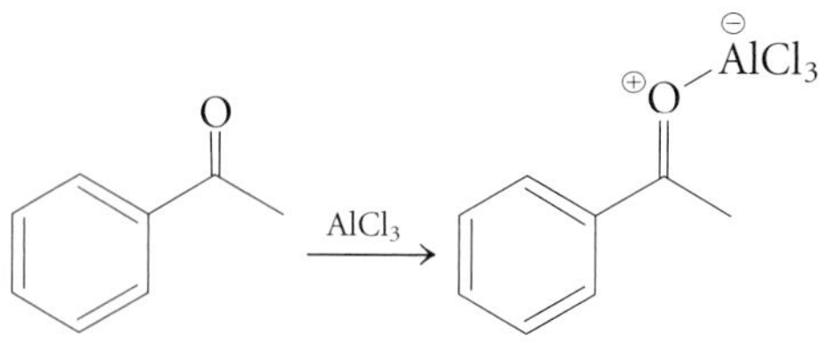

Thus, at least one equivalent of Lewis acid must be used in order to effect complete conversion. Furthermore, this complex bears positive charge adjacent to the benzene ring, and the ring is deactivated toward further reaction; a second acylation reaction does not occur. This is not the case with alkylation, where the presence of an alkyl group in the product actually increases the rate of further alkylation reactions. (We will discuss the effect of substituents on the rate of electrophilic aromatic substitution in a later section.)

*Synthetic Utility of Friedel–Crafts Acylation.* Friedel–Crafts acylation is important because it results in the attachment of a straight-chain (unbranched) carbon fragment to an aromatic ring. (This is not always possible with Friedel–Crafts alkylation because of fast skeletal rearrangements.) The Friedel–Crafts acylation attaches an unbranched acyl group to the ring, producing an aryl ketone. It is relatively easy to reduce aryl ketones under either acidic or basic conditions to the corresponding hydrocarbon (that is, to convert a C=O bond into a $CH_2$ group). Therefore, it is possible to convert the straight-chain ketone substituent attached via a Friedel–Crafts acylation to the corresponding unbranched hydrocarbon chain. This reduction can be accomplished in several ways, as shown in Figure 11.3. Under acidic conditions, reduction can be carried out by treatment with zinc in HCl, a reaction called the **Clemmensen reduction.** Reduction can be carried out under neutral conditions with $H_2$ and Pd as a catalyst at high temperature, although this reaction is accelerated by acid. Under basic conditions, reduction can be carried out by treatment with hydrazine, $NH_2NH_2$, and KOH at elevated temperature, a reaction known as the **Wolff–Kishner reduction.**

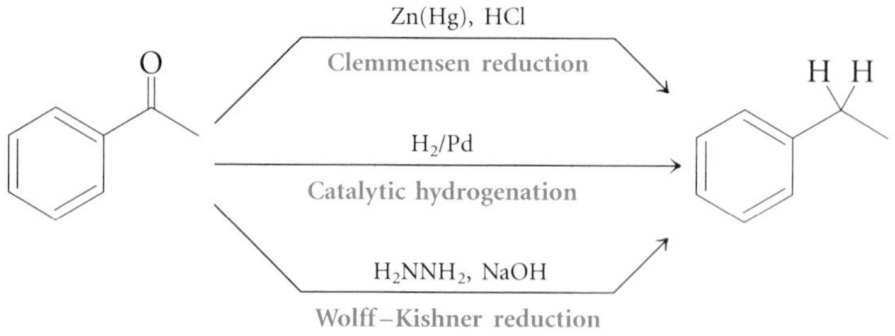

**FIGURE 11.3**

The carbonyl group of a ketone can be converted into a methylene group by reduction under acidic (Clemmensen reduction), neutral (catalytic hydrogenation), or basic (Wolff–Kishner reduction) conditions.

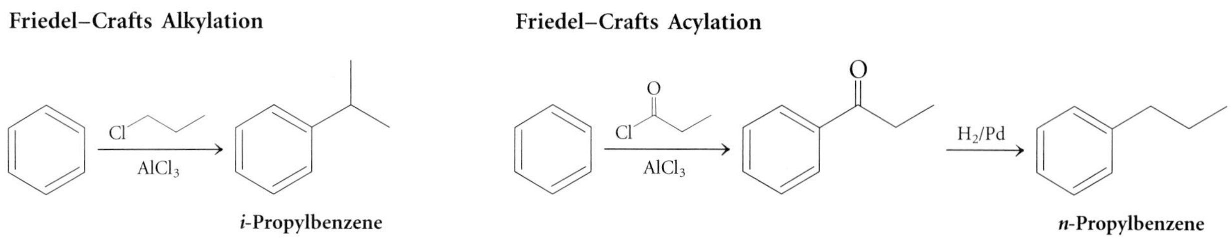

**Friedel–Crafts Alkylation**                    **Friedel–Crafts Acylation**

*i*-Propylbenzene                                                                    *n*-Propylbenzene

## FIGURE 11.4

Straight-chain alkyl groups cannot be attached directly to a benzene ring by Friedel–Crafts alkylation because rearrangements occur. However, straight chains can be attached indirectly to a benzene ring by Friedel–Crafts acylation followed by reduction.

The sequence of Friedel–Crafts acylation followed by reduction makes it possible to attach long-chain hydrocarbons to a ring without the rearrangements that accompany Friedel–Crafts alkylation. For example, Friedel–Crafts alkylation with 1-propyl chloride produces isopropylbenzene, whereas Friedel–Crafts acylation by propanoyl chloride, followed by reduction, produces *n*-propylbenzene (Figure 11.4).

## 11.3

# Reactions of Substituents and Side Chains of Aromatic Rings

Just as the acyl group introduced by Friedel–Crafts acylation can be converted to an alkyl group, other substituents introduced by electrophilic aromatic substitution can be converted to different functional groups.

### Reduction of Nitro Groups to Primary Amines

A nitro ($-NO_2$) group can be reduced to the corresponding amino ($-NH_2$) group by treatment with a reducing metal such as tin or zinc in the presence of acid. Sodium bicarbonate is used to neutralize the acid from the first step. This two-step sequence is an important route to aniline.

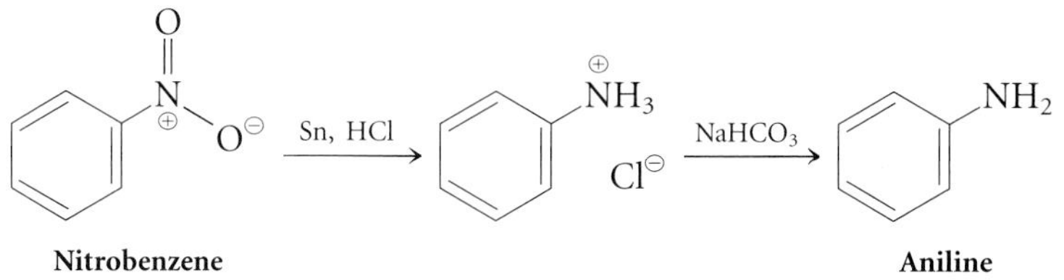

Nitrobenzene                                                                    Aniline

For the introduction of a primary amino group, this sequence (nitration followed by reduction) is superior to the alternative route in which bromobenzene is treated with sodium amide to produce a benzyne intermediate. The benzyne intermediate subsequently reacts with ammonia, as described in Chapter 9.

## Diazotization

The treatment of a primary aniline with nitrous acid, $HNO_2$, produces a **diazonium salt,** in a reaction called **diazotization.** Diazonium salts are important synthetic intermediates that can undergo coupling reactions to form azo dyes and substitution reactions to effect functional group conversions on aromatic rings.

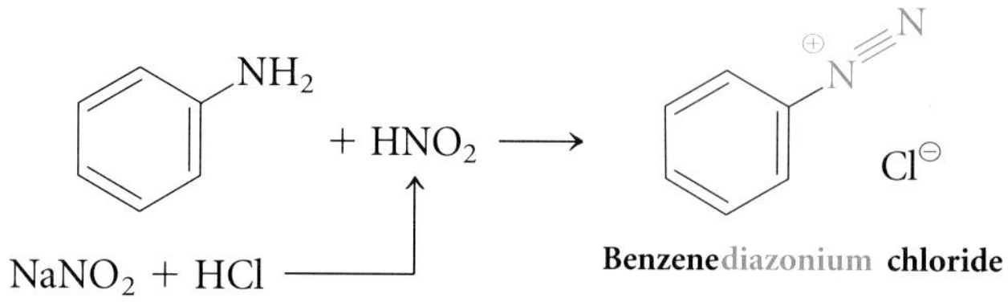

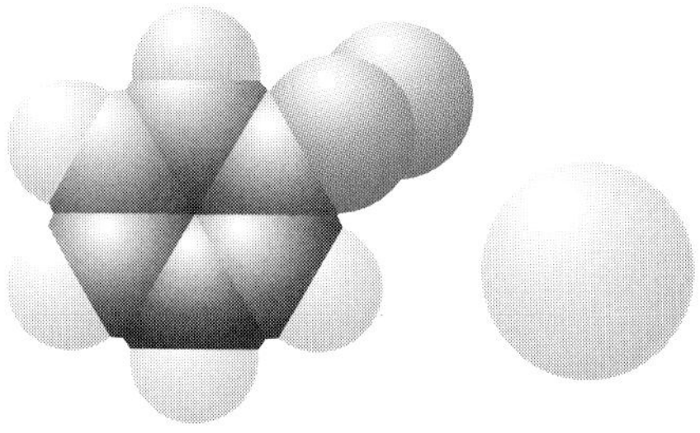

**Benzene**diazonium **chloride**

*Diazo Coupling.* Anilines are important because they constitute the starting material for the production of azo dyes. The diazonium salts obtained from anilines by diazotization are cations, which, as active electrophiles, can attack other, electron-rich aromatic rings. For example, benzenediazonium chloride reacts with *N,N*-dimethylaniline in the *para* position to produce *p*-(dimethylamino)azobenzene, a bright yellow solid that was, at one time, used as a colorant in margarine (Figure 11.5).

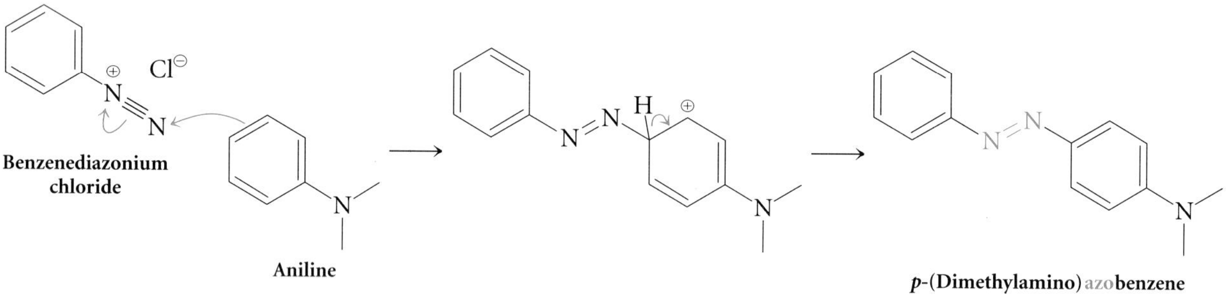

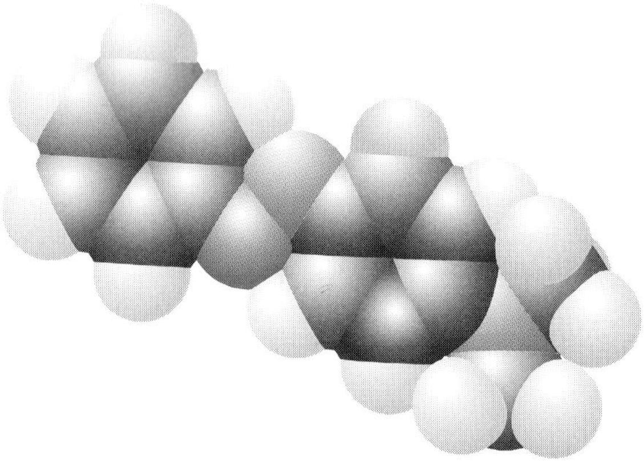

**Benzenediazonium
chloride**

**Aniline**

*p*-**(Dimethylamino)**azo**benzene**

**FIGURE 11.5**

Electrophilic attack by the terminal nitrogen atom of a diazonium salt forms a C—N $\sigma$ bond to an electron-rich aromatic ring. Deprotonation of the resonance-stabilized cation produces the highly colored azo compound.

*CHEMICAL PERSPECTIVES*

A WARNING FOR THOSE WHO ENJOY COCKTAILS
WITH FUNNY UMBRELLA HATS

Red No. 2, an azo compound, is a commercial dye now used only for dyeing wool and silk. Before its use was restricted in 1976 by the U.S. Food and Drug Administration, this dye was used for coloring food, especially maraschino cherries. However, Red No. 2 was shown to be a mutagen in one especially sensitive test. The battle raged for years between those who argued that there was no rationale for using, even at very low levels, any food color that might be dangerous and those who contended that no one consumed enough cherries to be at significant risk. Finally, a compromise banned Red No. 2 but permitted existing stocks of colored cherries to be sold.

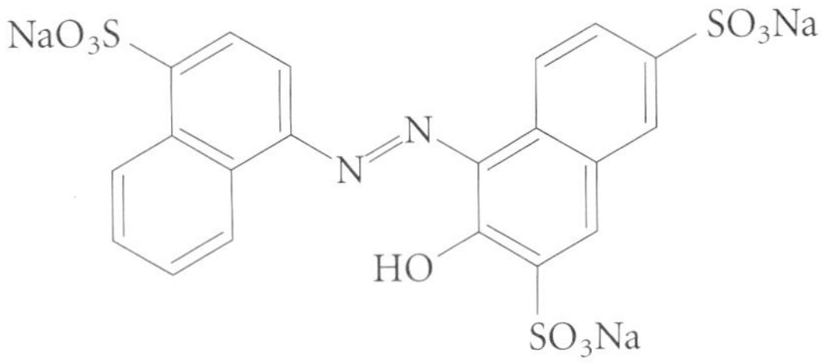

Red No. 2

Azo compounds are highly colored because the —N=N— linkage between aromatic rings extends the conjugation in the $\pi$ systems, resulting in strong absorption in the visible region. With various substituents on the two aromatic rings, azo compounds of nearly every color have been prepared. Azo dyes were among the first synthetic colorfast agents to be used for dyeing wool and cotton.

***Diazo Substitution: Functional Group Conversion (Sandmeyer Reaction).*** The aryl diazonium functional group can be converted into a number of other functional groups, some of which are difficult to introduce onto an aromatic ring in other ways.

*Nitriles.* Reaction of benzenediazonium chloride with $Cu_2(C{\equiv}N)_2$ and $KC{\equiv}N$ at 50 °C yields benzonitrile:

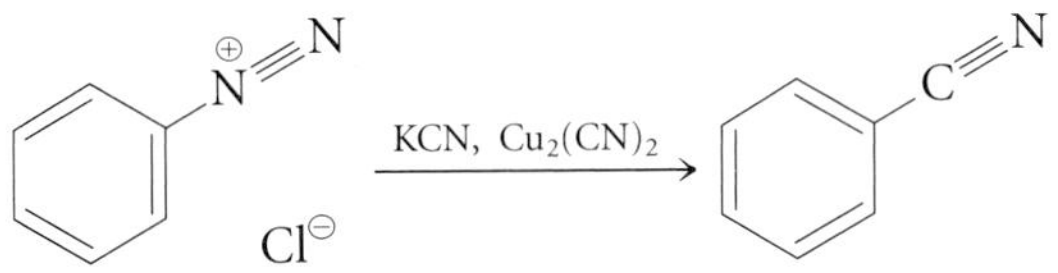

**Benzenediazonium chloride**

**Benzonitrile**

This reaction, discovered by Traugott Sandmeyer in 1884, is known as the **Sandmeyer reaction.** Its mechanism (and those of other reactions that replace the diazonium group) is not known, although there is sufficient evidence to rule out loss of nitrogen to form a phenyl cation as an intermediate. Radical intermediates are probably involved in most of these reactions.

*Halides.* In reactions similar to those of diazonium salts with cyanide ion (Sandmeyer reactions), aryl halides can be produced by the treatment of aryl diazonium salts with $Cu_2Cl_2$, $Cu_2Br_2$, or KI to yield aryl chlorides, bromides, or iodides:

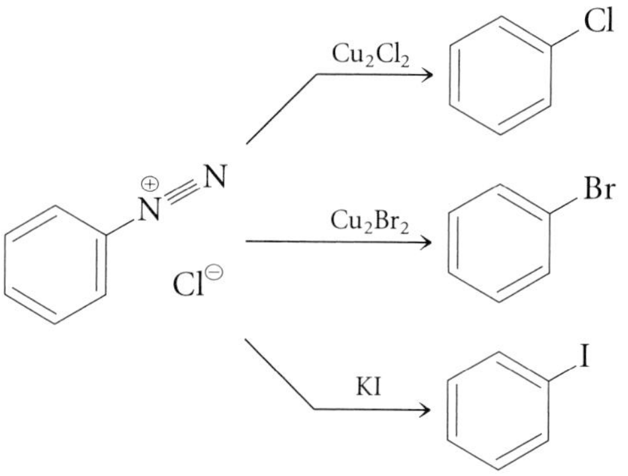

The formation of aryl iodides via this reaction is especially important because in the reaction of $I_2$ and $FeI_3$ with benzene, the equilibrium favors the starting materials and the aryl iodide is not obtained. Aryl fluorides can be prepared by heating aryl diazonium tetrafluoroborates, which are prepared by reaction of an aryl amine with $NaNO_2$ and fluoroboric acid, $HBF_4$:

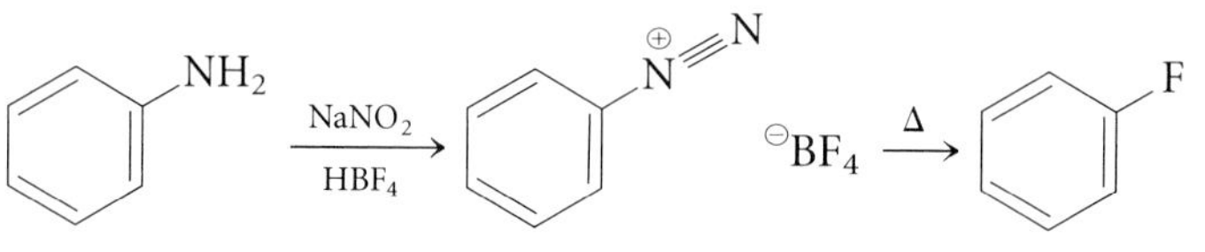

*Phenols.* Reaction of diazonium salts with water at 100 °C represents a valuable method for preparing phenols. The reaction is carried out in the presence of acid to suppress the reaction of the product phenol with the starting diazonium salt. This route to phenol is superior to the alternative discussed in Chapter 9, the treatment of bromobenzene with NaOH.

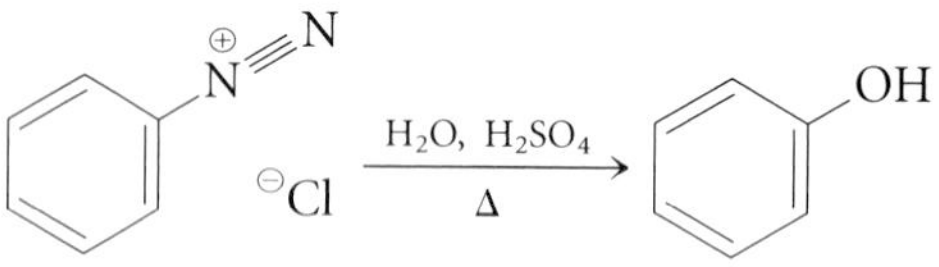

## EXERCISE 11.5

Show all of the steps, including intermediates and reagents, required to convert benzene into the following diazo compound:

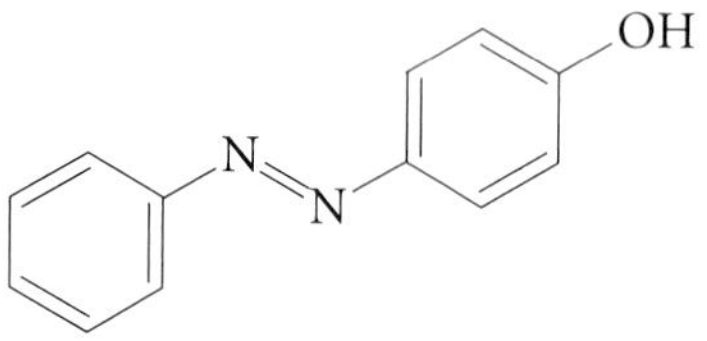

*Replacement by Hydrogen.* The diazonium group can be replaced by a hydrogen atom upon treatment with hypophosphorous acid, $H_3PO_2$:

1,3,5-Tribromobenzene

This might at first seem of little value—why introduce a nitro group, reduce it to an amine, convert it to a diazonium salt, and then remove the nitrogen to arrive where you started? However, in Exercise 11.15, we will see how this replacement of a nitrogen substituent by hydrogen can be of value in synthesis.

## Oxidation of Carbon Side Chains

The oxidation of side chains on aromatic rings is another means of imparting functionality. The vigorous oxidation of alkyl-substituted aromatic rings with hot aqueous $KMnO_4$ results in oxidative cleavage of the side chain, forming a carboxylic acid group, irrespective of the length or branching of the side chain. For example, both alkyl groups in *p*-diethylbenzene are cleaved by hot $KMnO_4$ to form the diacid terephthalic acid:

*p*-Diethylbenzene                    Terephthalic acid

Various benzoic acids can be prepared by Friedel–Crafts alkylation or acylation followed by oxidative degradation of the attached chains.

## Reactions of Aryl Side Chains

The presence of an aryl group can modify the reactivity of side-chain functional groups. For example, alkylbenzenes produced by electrophilic aromatic substitution exhibit special reactivity at the benzylic position. An alkyl chain introduced by electrophilic substitution can undergo free-radical bromination at the benzylic position, as discussed in Chapter 7:

### ASPIRIN: A SIMPLE AROMATIC PHARMACEUTICAL COMPOUND

Aspirin, or acetylsalicylic acid, is the world's most widely used remedy for reducing pain and lowering fever. Although aspirin's first description in the medical literature in 1899 was for the treatment of rheumatic fever, its phenolic precursor, salicylic acid, had been described earlier (1876) as being effective for controlling fever and treating gout and arthritis. Still earlier (1763), homeopathic medical practitioners reported that the chewing of willow (*Salix*) bark was effective in treating malaria. Only later was it determined that extraction of willow bark yields salicin, a compound that can be hydrolyzed and oxidized to salicylic acid:

Salicin [O] H₃O⊕ → Salicylic acid → Acetylsalicylic acid (aspirin)

This reaction is regiospecific for C—H bonds adjacent to the aromatic ring because of the enhanced radical stability of benzylic radicals. The bromide can be dehydrobrominated by treatment with a strong base, as described in Chapter 9.

### EXERCISE 11.6

Devise a synthesis for each of the following compounds starting from benzene and any other reagents needed.

(a)

(b)

(c)

(d)

(e)

(f)

(g)

(h)

To this point we have considered only benzene as a substrate for electrophilic aromatic substitution reactions. However, the resulting monosubstituted benzene derivatives (list them in your mind) can undergo further substitution. In this way, disubstituted aromatics can be prepared and can, in turn, serve as starting materials for more highly substituted aromatic compounds.

A ring substituent has two important effects on further electrophilic aromatic substitution reactions: (1) it affects reactivity, that is, the rate of the reaction compared with that of benzene, and (2) it affects orientation, that is, the regiochemistry of the substitution. Both factors are illustrated by the following examples:

**Faster than nitration of benzene**

**Slower than nitration of benzene**

**Slower than nitration of benzene**

There are two possible effects on the rate (substitution is either faster or slower relative to the same reaction with benzene) and two possible effects on regiochemistry (the orientation is either *ortho* and *para*, or *meta*). In this section, we will discuss why some substituents are *ortho, para* directors and others are *meta* directors. Furthermore, we will see why some substituents accelerate the rate of substitution (**activating substituents**), and others decrease the rate (**deactivating substituents**). Finally, we will explore why there are both activating and deactivating *ortho, para* directors, whereas all *meta* directors are deactivating toward electrophilic aromatic substitution.

It is important to keep in mind that relative reactivity is evaluated by comparing the rate of reaction of a substituted benzene with that of benzene itself. In contrast, directing effects are determined by comparing the intermediates resulting from *ortho, meta,* and *para* attack on a monosubstituted benzene derivative. Therefore, there is no necessary relationship between activation/deactivation and orientation.

## Weakly Activating Substituents: Alkyl Groups

Toluene is 600 times more reactive toward electrophilic aromatic bromination than is benzene, because the methyl group of toluene stabilizes the transition state (and the arenium cation) more than does the hydrogen atom of benzene.

***Orientation.*** The electrophilic bromination of toluene affords mostly *para*-bromotoluene, not a statistical mixture of the three possible bromine substitution products.

|  | *ortho* | *para* | *meta* |
|---|---|---|---|
| Statistical: | 40 | 20 | 40 |
| Observed: | 40 | 60 | <1 |

In the absence of other factors, a purely statistical distribution of products would give a 2 : 2 : 1 mixture of *ortho, meta,* and *para* substitution, reflecting the number of hydrogen atoms available for substitution. (Because there are two *ortho* and *meta* positions but only one *para,* statistics favors *ortho* and *meta* by a factor of 2.) However, in the electrophilic bromination of toluene, the ratio of *ortho* and *para* products to *meta* products is substantially greater than statistically predicted.

Electrophilic bromination takes place in the same way as electrophilic chlorination, following the mechanism shown in Figure 11.2. Complexation of $Br_2$ and a Lewis acid (in this case, $FeBr_3$) gives an activated electrophile that can deliver the equivalent of $Br^\oplus$ to the aromatic ring. Attack by this electrophilic reagent on toluene can occur at the *ortho, meta,* and/or *para* positions. In each case, a resonance-stabilized cation results, and three resonance contributors can be written for each of these cations (Figure 11.6).

For the cations produced by attack at the *ortho* and *para* positions, one of the contributing resonance structures bears positive charge at the site substituted by the methyl group. Because an alkyl substituent stabilizes a cation (recall that tertiary cations are more stable than secondary or primary ones), these structures are particularly stable. The existence of a resonance contributor having unusual stability stabilizes the delocalized cations formed from *ortho* and *para* attack and facilitates their formation in the rate-determining step. Deprotonation, as described earlier for benzene, then gives rise to the substitution products. Note that none of the carbocations resulting from attack at the *meta* position is particularly stable.

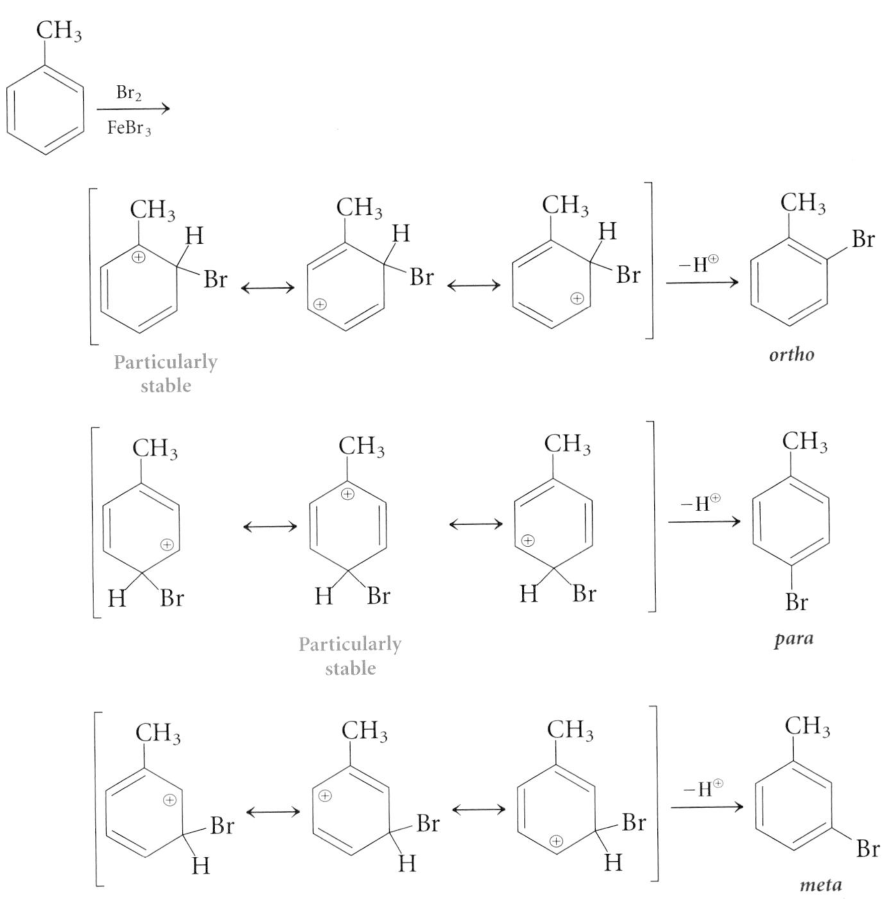

**FIGURE 11.6**

The resonance-stabilized cations formed by attack at the *ortho* or *para* positions bear positive charge at a site to which the methyl group of toluene is attached. When attack is at the *meta* position, the positive charge in the resonance-delocalized cation appears on an unsubstituted carbon atom. Because more highly substituted cations are more stable, the cations formed by attack at the *ortho* and *para* positions are favored over the cations produced by *meta* attack.

The observed preference for the *ortho* and *para* products is the result of electron release from the substituent methyl group, which stabilizes the intermediate carbocations for the *ortho* and *para* (but not the *meta*) products. Thus, the transition states leading to these isomers are lower in energy. If this electronic effect were the only important one, two-thirds of the product would be *ortho*-substituted and one-third would be *para*-substituted. The fact that the fraction of *para*-substituted product is larger than expected means that some additional factor is working against formation of the *ortho*-substituted product.

The proximity of an alkyl side chain makes the transition state for *ortho* attack more crowded than that leading to *para* substitution. For example, the electrophilic sulfonation of *t*-butylbenzene produces almost exclusively *para* product because the bulky alkyl group effectively blocks *ortho* attack. Thus, electronic factors favor *ortho* and *para* attack, and steric hindrance further favors *para* substitution.

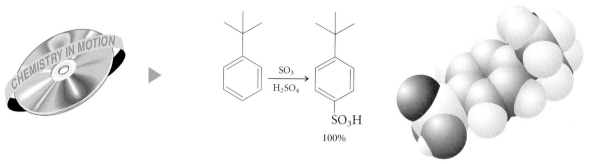

100%

*Activation.* Not only does the methyl group of toluene control regiochemistry, by directing the attack of the electrophile to the *ortho* and *para* positions, but it also accelerates the reaction. The bromination of toluene is faster than the bromination of benzene, because the methyl group stabilizes the transition state leading to the intermediate cation more than does the hydrogen atom in benzene.

More stable          Less stable

More stable          Less stable

Recall that electrophilic substitution of toluene leads to *ortho* and *para* products, because the transition state leading to the intermediate cation is more stable when the methyl group is *ortho* or *para* to the site of electrophilic attack than when it is *meta*. As noted earlier, it is important to keep in mind that relative reactivity is evaluated by comparing the reaction of a substituted benzene with benzene itself, whereas directing effects are determined by comparing the intermediates resulting from *ortho, meta,* and *para* attack. Thus, it is possible that a deactivating group can direct *ortho–para* substitution because it destabilizes these intermediates *less* than it does the *meta.* The halogens fit into this category of substituents that are *ortho, para* directors but are also deactivating. On the other hand, *all* activating groups are also *ortho, para* directors.

The activating nature of alkyl groups creates a problem in Friedel–Crafts alkylation: the product is more reactive than the starting material. Thus, simply mixing one equivalent each of benzene and an alkyl halide with $AlCl_3$ will lead to a mixture of mono-, di-, and trisubstitution products.

This polyalkylation is an unavoidable consequence of the activating nature of the alkyl groups introduced by Friedel–Crafts alkylation. However, this problem is unique to alkylation. All of the other electrophilic aromatic substitution reactions we have discussed introduce substituents that deactivate the aromatic ring toward further reaction.

## EXERCISE 11.7

The three xylenes (dimethylbenzenes) do not undergo electrophilic aromatic substitution at the same rate. Indeed, *meta*-xylene undergoes chlorination 200 times faster than *para*-xylene and 100 times faster than *ortho*-xylene. Explain these rate differences using resonance structures for each of the intermediate arenium ions.

## EXERCISE 11.8

Aryl substituents, like alkyl groups, are activating *ortho, para* directors. Thus, biphenyl undergoes bromination to give predominantly *p*-bromobiphenyl.

Rationalize the directing effect of a phenyl group as a substituent, using resonance structures of the intermediate arenium ion involved in this reaction.

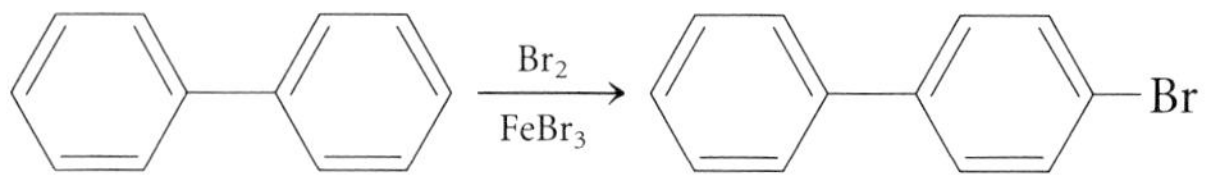

## Strongly Activating Heteroatom Substituents

Nitrogen and oxygen attached directly to an aromatic ring are effective stabilizers of the intermediate arenium cations leading to *ortho* and *para* substitution products. Indeed, they are even more effective than alkyl groups in activating the aromatic ring toward substitution and directing the orientation of substitution.

*Hydroxyl Group.* The effectiveness of the hydroxyl group in activating the aromatic ring toward electrophilic substitution and directing the orientation of bromination can be seen from the bromination of phenol (Figure 11.7). Carried out using bromine in the absence of a Lewis acid catalyst, at temperatures near 0 °C, this reaction leads almost exclusively to *para*-bromophenol.

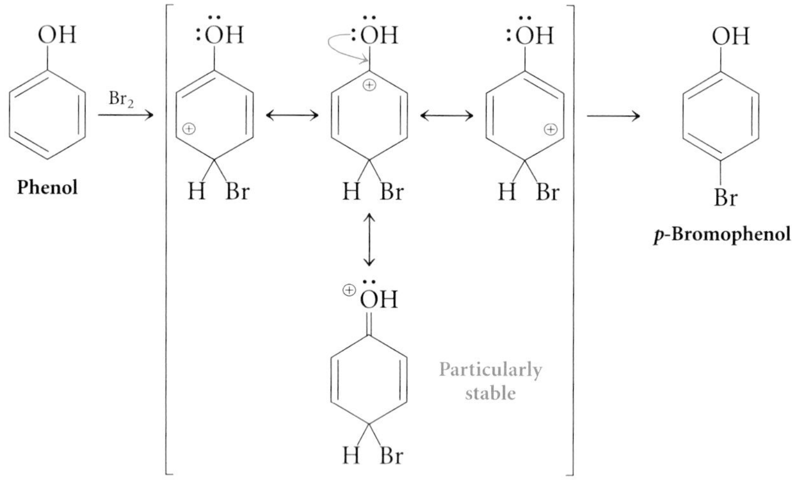

**FIGURE 11.7**

Bromination of phenol proceeds through the cation shown here. A fourth resonance structure involving donation of lone-pair electron density to the arenium ion contributes significantly to the hybrid.

The effectiveness of oxygen and nitrogen substituents in directing the orientation *ortho, para* is due to the contribution of a fourth resonance structure to the stability of the cation. At first it might appear that this fourth resonance structure would *not* be particularly stable because it bears positive charge on oxygen. Nonetheless, this structure is the dominant contributor because it has one more bond (a carbon–oxygen $\pi$ bond) than the other three structures. Moreover, all the atoms in this structure have a filled valence shell, whereas in each of the others, the carbon bearing formal positive charge has only six electrons. Indeed, stabilization of the cation by oxygen is so significant that phenol reacts with bromine in the absence of a Lewis acid catalyst.

The intermediate cation that would lead to *m*-bromophenol cannot be stabilized by lone-pair electron donation, because the oxygen substituent is not located on one of the three carbon atoms that bears positive charge:

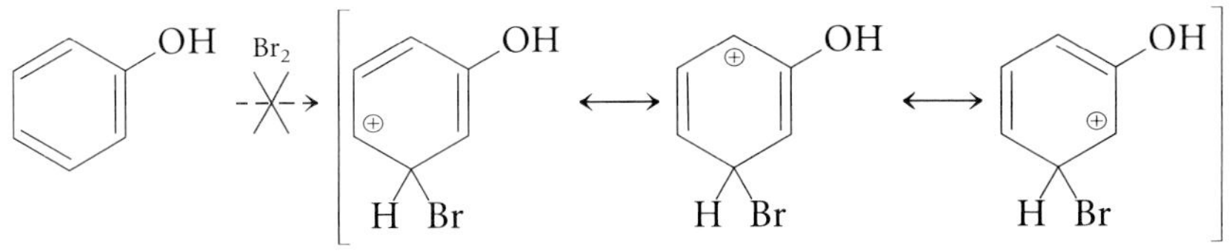

Without lone-pair donation, the oxygen substituent, because of its electron-withdrawing inductive effect, *destabilizes* this cation relative to that formed in the bromination of benzene. Thus, a free hydroxyl group is a strongly activating *ortho, para* director.

Introduction of a bromine substituent on phenol produces a product (*p*-bromophenol) that is *less* reactive than phenol toward electrophilic aromatic substitution. As you will learn shortly, halogens are deactivating substituents. Thus, bromination of phenol can be controlled to monobromination by adjusting the amount of $Br_2$ added, because each successive reaction is slower.

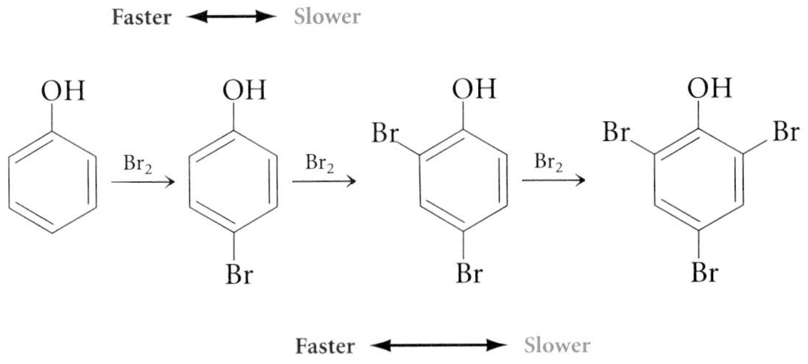

## EXERCISE 11.9

Unlike electrophilic bromination, chlorination and nitration of phenol both lead to nearly equal mixtures of *ortho* and *para* products. Draw all reasonable resonance structures for the cationic intermediate from which *ortho*-chlorophenol is produced.

*Amino Group.* The nitrogen atom of aniline is also able to donate lone-pair electron density to the intermediate cations formed in electrophilic aromatic substitution. Treatment of aniline with excess $Br_2$ leads to 2,4,6-tribromoaniline, in which all possible positions *ortho* and *para* to the amino substituent have undergone substitution. The reaction rate decreases significantly as the reaction proceeds, because the HBr produced protonates nitrogen to produce an ammonium ion. The nitrogen atom of the ammonium ion does not have a lone pair of electrons to donate to the arenium ion. Furthermore, it bears formal positive charge and thus is a deactivating substituent. Bromination takes place on the small concentration of free amine present in equilibrium with the salt.

2,4,6-Tribromoaniline

## Moderately Activating Heteroatom Substituents

Amides and esters act as electron donors, directing *ortho, para,* but they are only moderate activators of electrophilic aromatic substitution. Because the lone pair of electrons on the heteroatom attached to the ring also takes part in a resonance interaction with a carbonyl group, the substituent is less able to release electrons to the ring. In phenyl acetate and *N*-phenylacetamide, for example, the existence of resonance structures reduces the availability of the lone pair of electrons on the nitrogen atom for stabilization of the arenium ion intermediate:

Phenyl acetate          *N*-Phenylacetamide (acetanilide)

In many electrophilic aromatic substitution reactions, it is preferable to use these acyl derivatives rather than the free phenol or free amine. Hydroxyl and amino groups are so strongly activating that their substitution reactions can be difficult to control.

Alkyl groups, lacking a delocalizable lone pair on the atom attached to the benzene ring, stabilize the carbocationic intermediates to a somewhat lesser degree than do amide or ester groups. Alkyl groups cannot stabilize the transition state through $\pi$-electron release in the way these strong or moderate electron donors can.

## Moderately Deactivating Substituents: The Halogens

The halogens are moderately deactivating, as can be seen from the fact that the nitration of chlorobenzene is approximately 50 times slower than nitration of benzene:

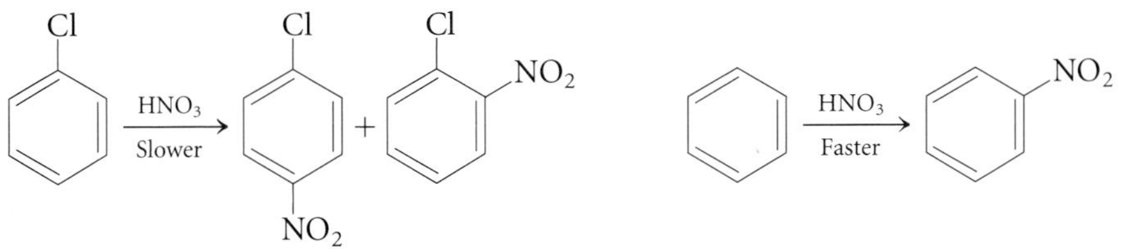

Although the halogens are moderate deactivators, they do direct *ortho, para,* just like other substituents with lone pairs of electrons. The intermediate cations leading to *ortho* and *para* substitution are stabilized *relative* to that leading to *meta* substitution by the same overlap of lone pair and $\pi$ electrons as occurs with oxygen and nitrogen substituents. These interactions are less effective than those with second-row elements such as nitrogen or oxygen, however.

The halogen atom of an aryl halide has three lone pairs of electrons, but donation of this electron density to stabilize the arenium ion intermediate requires overlap between the second-level *p* orbital of carbon with a substantially larger third-level (Cl), fourth-level (Br), or even fifth-level (I) orbital of the halogen atom (Figure 11.8). Because the orbital overlap is not sufficient to compensate for the high electronegativity of the halogens, they are moderate deactivators, providing less electron density to stabilize the intermediate arenium ion than does a hydrogen atom. Although fluorine has approximately the same orbital sizes as carbon, its high electronegativity prevents significant electron donation. (Recall that all four halogens are more electronegative than hydrogen.)

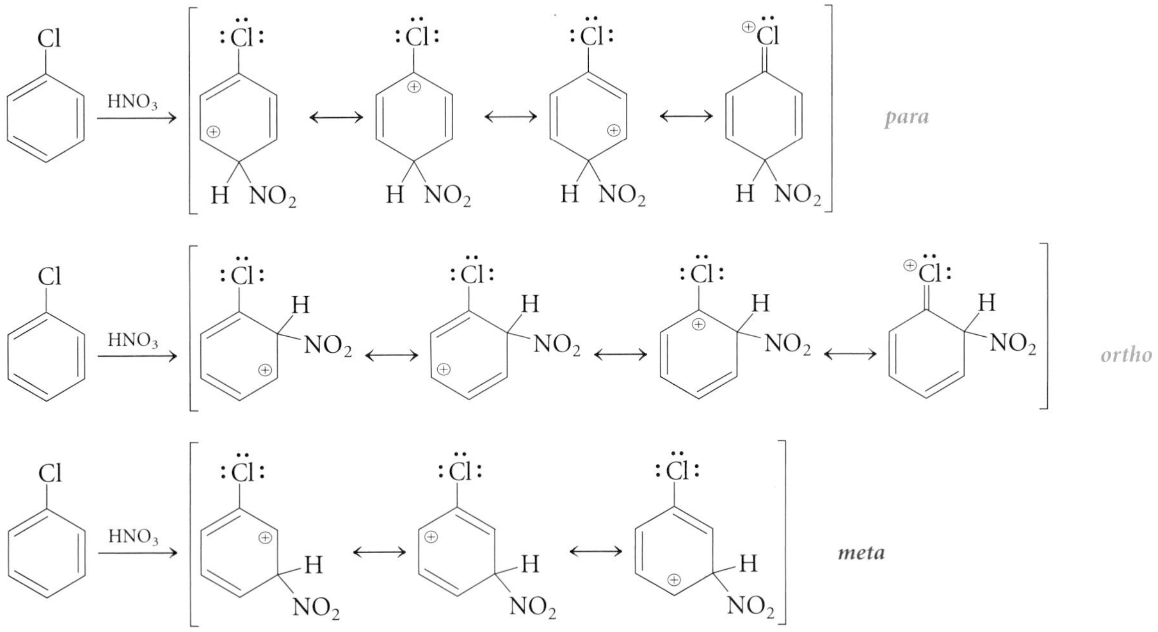

**FIGURE 11.8**

Halogen substituents are deactivating because they provide less electron density to the arenium ion than does a hydrogen atom. They are *ortho, para* directors because the *meta* intermediate is destabilized by the presence of the halogen substituent to a greater extent than are the *ortho, para* intermediates.

In the cation leading to *meta* product, the major effect of the halogen is to withdraw electron density (relative to hydrogen) through the $\sigma$ bond. In the cations leading to *ortho* and *para* substitution, this electron withdrawal is partially offset by lone-pair $\pi$ donation. Thus, the *ortho* and *para* intermediates are more stable than the *meta* intermediate but *less* stable than the intermediate arenium ion for benzene itself.

Because the halogens are deactivating substituents, the product of substitution is less reactive than the starting material, and the extent of substitution can be readily controlled by adjusting the ratio of reagent to substrate. For example, to obtain monochlorination of isopropylbenzene, one equivalent of $Cl_2$ is used. To obtain dichlorination, two equivalents of $Cl_2$ are used.

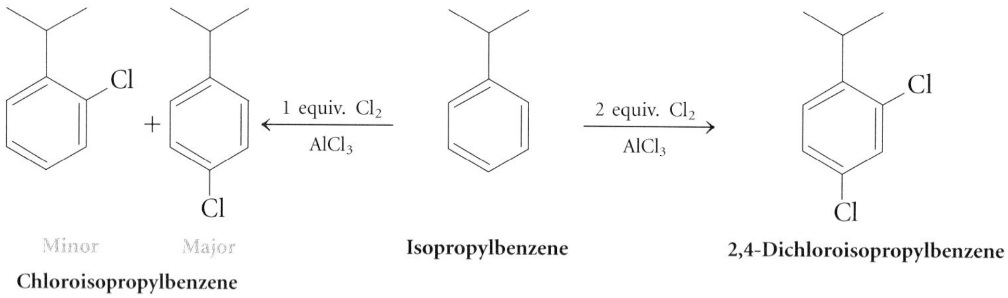

Minor    Major
Chloroisopropylbenzene

Isopropylbenzene

2,4-Dichloroisopropylbenzene

## Moderately and Strongly Deactivating Substituents

The substituents shown in Figure 11.9 all decrease the stability of the intermediate arenium ion and are thus deactivating substituents. Most bear significant positive charge because of polarization of the $\pi$ system toward a heteroatom.

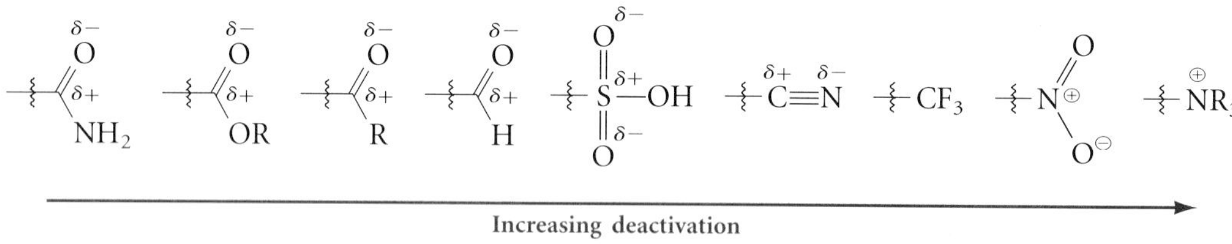

Increasing deactivation

**FIGURE 11.9**

Electron-withdrawing substituents deactivate an aromatic ring toward electrophilic aromatic substitution.

Two of the substituents, the ammonium ion and the nitro group, are attached to the ring through a nitrogen atom with a formal positive charge and are strongly deactivating. The trifluoromethyl group is unique because it deactivates as a result of the polarization of the C—F $\sigma$ bonds. None of these substituents bears a lone pair of electrons at the atom directly attached to the ring (as the halogens do). As a result, all of these deactivating substituents are also *meta* directors (Figure 11.10), because they lack the ability to serve as $\pi$-electron donors to the arenium ion intermediate.

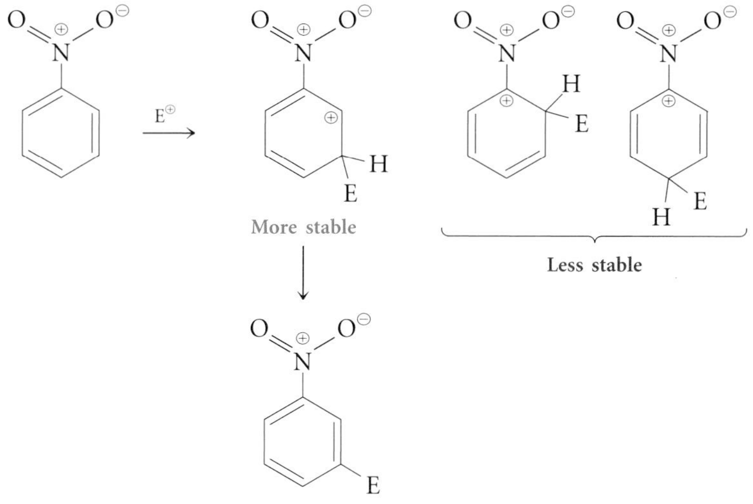

**FIGURE 11.10**

A nitro group strongly deactivates an aromatic ring toward electrophilic aromatic substitution. Destabilization is greatest for the intermediate arenium
ions that lead to the *ortho* and *para* products because one of the three resonance structures for the arenium ion has positive charge on the carbon atom
bearing the positively charged nitrogen atom of the nitro group. The intermediate leading to *meta* substitution is also destabilized, only less so because
none of its three resonance structures has this proximity of positive charges.

For example, nitration of nitrobenzene gives a mixture of *ortho-, para-,*
and *meta*-dinitrobenzene, with *meta* predominating.

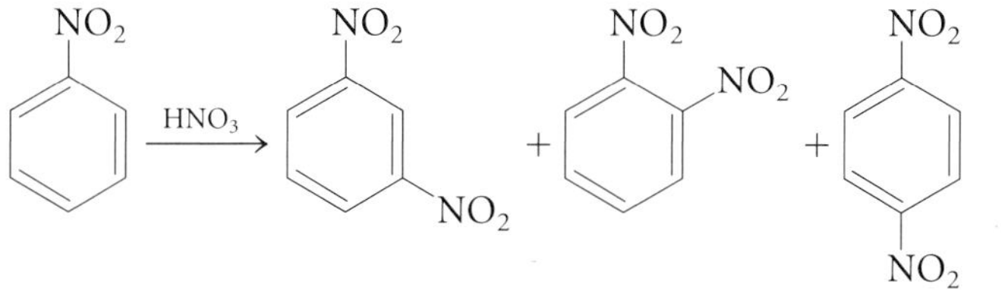

Product ratio = 93 : 6 : 1

The effects of an electron-withdrawing group are sometimes so significant
that the reaction fails—for example, it is impossible to conduct a
Friedel–Crafts alkylation or acylation on a ring deactivated by the presence
of a moderately or strongly electron-withdrawing group such as —NO$_2$,
—C≡N, or —CO$_2$R.

## Summary of Substituent Effects

Table 11.1 (on page 570) summarizes the substituents discussed here,
organized into groups based on activation/deactivation effects. The groups
that bear a lone pair of electrons on a second-row element directly attached
to the ring are most effective in stabilizing the intermediate cation and are
therefore the strongest activators. Donation of lone-pair electrons can occur only in the intermediate cations leading to *ortho* and *para* substitution;

## TABLE 11.1

Reactivity and Orientation Effects of Substituents in Aromatic Compounds

Strongly activating; *ortho, para* directors

$+\overset{\cdot\cdot}{\underset{\cdot\cdot}{O}}{:}^{\ominus}$  $+\overset{\cdot\cdot}{N}H_2$  $+\overset{\cdot\cdot}{N}R_2$  $+\overset{\cdot\cdot}{\underset{\cdot\cdot}{O}}H$  $+\overset{\cdot\cdot}{\underset{\cdot\cdot}{O}}R$

Moderately activating; *ortho, para* directors

Weakly activating; *ortho, para* directors

$+$Alkyl     $+$Aryl

Weakly deactivating; *ortho, para* directors

$+$F     $+$Cl     $+$Br     $+$I

Moderately and strongly deactivating; *meta* directors

thus, these substituents are *ortho, para* directors. With the larger halogen substituents—Cl, Br, and I—the stabilizing effect of lone-pair donation is diminished by the mismatch in orbital size. The halogens are therefore moderate deactivators but *ortho, para* directors, because they destabilize the *ortho* and *para* intermediates less than the *meta* intermediate. The groups that bear positive charge (or partial positive charge) at the atom attached directly to the ring deactivate the ring toward electrophilic attack and direct substitution toward the *meta* position.

### EXERCISE 11.10

Only one resonance structure is shown in Figure 11.10 for each of the three intermediate cations. Draw the remaining significant resonance structures for each intermediate.

### EXERCISE 11.11

Write the resonance structures for an electrophilic attack at the *para* position of fluorobenzene. Justify why fluorobenzene undergoes electrophilic substitution at the *ortho* and *para* positions.

### EXERCISE 11.12

Predict the regiochemistry of the monosubstitution product expected in each case:

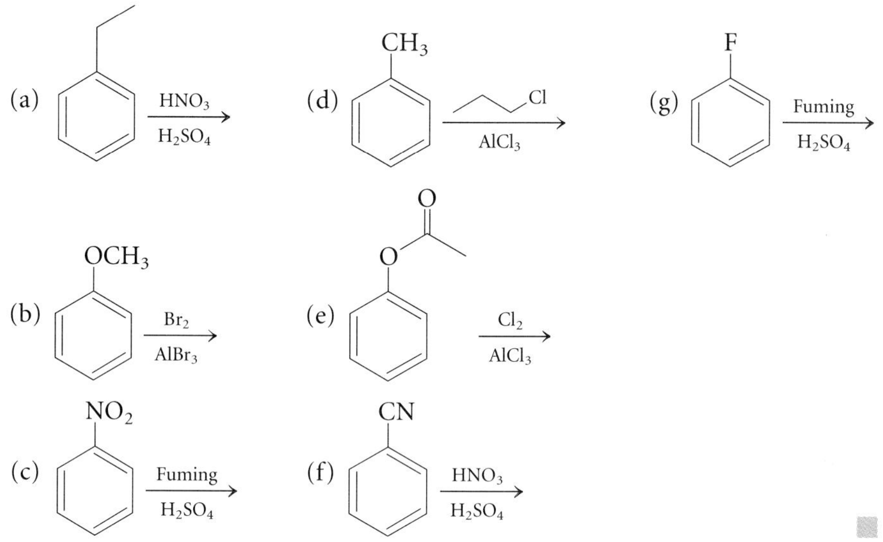

(a) $\xrightarrow[H_2SO_4]{HNO_3}$

(d) $\xrightarrow{AlCl_3}$

(g) $\xrightarrow[H_2SO_4]{Fuming}$

(b) $\xrightarrow[AlBr_3]{Br_2}$

(e) $\xrightarrow[AlCl_3]{Cl_2}$

(c) $\xrightarrow[H_2SO_4]{Fuming}$

(f) $\xrightarrow[H_2SO_4]{HNO_3}$

## Multiple Substituents

When more than one group is attached to a benzene ring, the effect of
the stronger activator (or weaker deactivator) prevails. In *o*-methylanisole,
for example, substitution is directed by the strongly activating —OCH$_3$
group to the *ortho* and *para* positions, rather than being directed by the
more weakly activating methyl group.

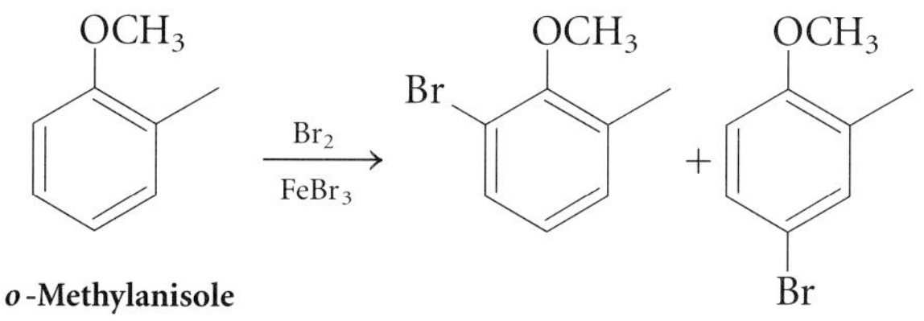

*o*-Methylanisole

## EXERCISE 11.13

Predict the regiochemistry of the monosubstitution product expected in each case:

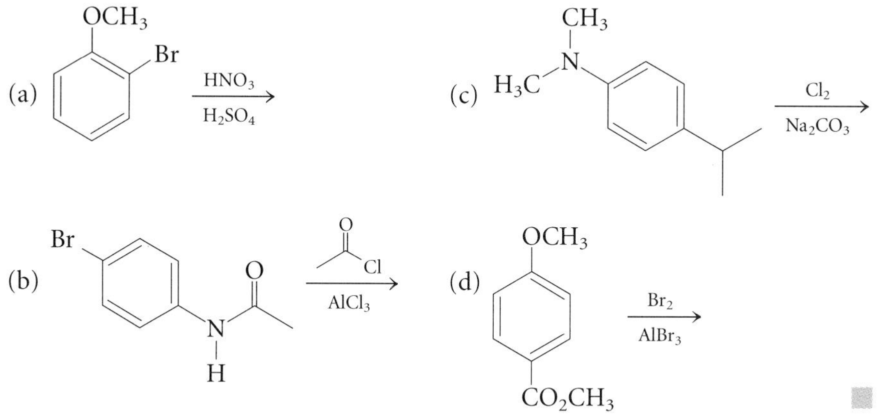

(a) $\xrightarrow[H_2SO_4]{HNO_3}$

(c) $\xrightarrow[Na_2CO_3]{Cl_2}$

(b) $\xrightarrow[AlCl_3]{}$

(d) $\xrightarrow[AlBr_3]{Br_2}$

## Using Substituent Effects in Synthesis

As explained in Section 11.3, many substituents possess characteristic chemical reactivity that enables them to be converted from electron-withdrawing groups to electron-releasing ones, or vice versa. For example, an electron-withdrawing —NO$_2$ group can be reduced to an electron-releasing —NH$_2$ group, and an electron-releasing alkyl group can be oxidized to an electron-withdrawing —CO$_2$H group. The order in which substituents are introduced onto an aromatic ring can control the regiochemistry of the isomeric products, and the ability to alter the substituents further enhances the chemist's control over regiochemistry.

As an example, let's consider the preparation of *m*-chlorobenzoic acid and *p*-bromobenzoic acid from benzene. First, we recognize that, for both products, we must replace (1) a ring hydrogen atom by a halogen and (2) a second ring hydrogen by a —CO$_2$H group. The carboxyl group can be introduced by the oxidation (with hot KMnO$_4$) of an alkyl chain introduced by a Friedel–Crafts alkylation. If we chlorinate benzene first, the chlorine substituent will direct further substitution to the *ortho* and *para* positions:

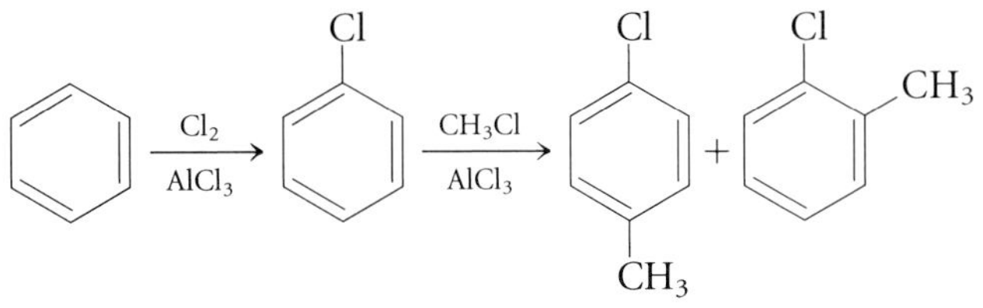

A nonstatistical mixture of *o*- and *p*-chlorotoluene is formed, because $\sigma$ inductive withdrawal and steric hindrance are greater at the positions *ortho* to the bulky chlorine substituent. Thus, Friedel–Crafts alkylation of chlorobenzene cannot place the alkyl group at the *meta* position. If we reverse the order, first alkylating to yield toluene and then chlorinating, we obtain the same mixture, because —CH$_3$ also is *ortho, para* directing.

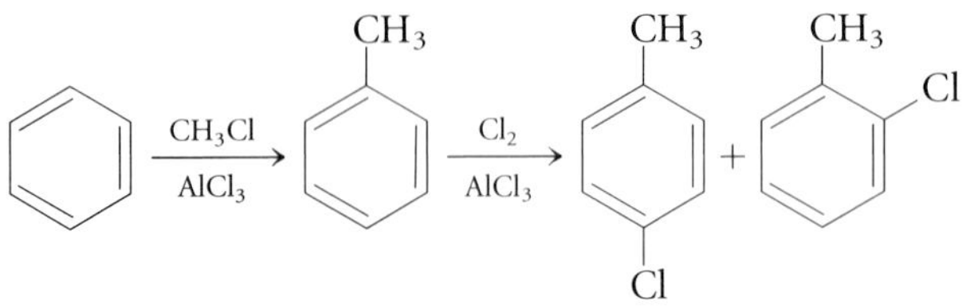

However, if we first introduce the methyl group by a Friedel–Crafts alkylation and then oxidize the methyl group, we obtain a *meta*-directing —CO$_2$H group. Chlorination of benzoic acid then gives the desired product.

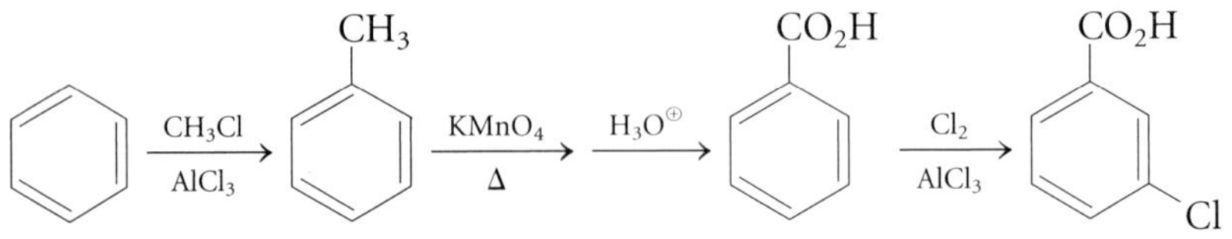

Two routes to the second synthetic target, *p*-bromobenzoic acid, are shown in Figure 11.11. The choice between these two routes would be made on secondary grounds—for example, the relative ease with which the mixtures produced could be separated to give a pure product. For example, separation of the liquid *o*-bromotoluene (mp −26 °C) from the solid *p*-isomer (mp 28 °C) could be effected relatively easily by crystallization. The *meta* isomer is also low-melting, at −40 °C. (The more symmetrical isomer almost always has the higher melting point.) On the other hand, the boiling points of the three bromotoluenes do not differ sufficiently for easy separation by distillation (*ortho*, 181 °C; *meta*, 183.7 °C; *para*, 184.5 °C). Alternatively, either toluene or bromobenzene may be available to be used as a starting material, allowing a step in one of the sequences to be omitted. For our purposes, it is important to note that several routes are possible for this synthesis. An important element of designing chemical synthesis is knowing not only what reactions are needed, but also the order in which they can be used. This is especially important in the synthesis of compounds with multiple substituents on aromatic rings.

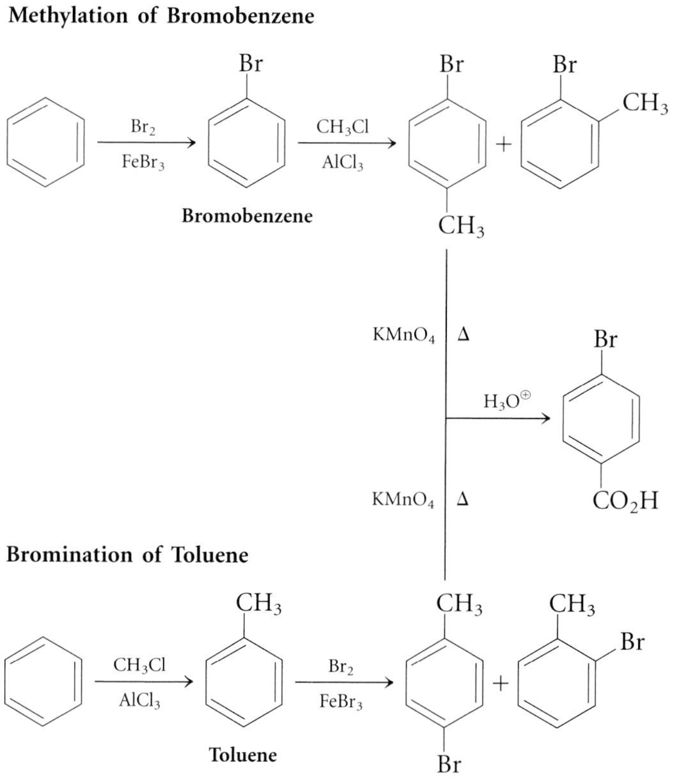

**Methylation of Bromobenzene**

**Bromination of Toluene**

**FIGURE 11.11**

Methylation of bromobenzene (formed by electrophilic bromination of benzene) by Friedel–Crafts alkylation affords a mixture of *o*- and *p*-bromotoluene (upper sequence). The same mixture is formed by bromination of toluene (lower sequence). The *ortho* and *para* isomers of bromotoluene can be separated, and *p*-bromotoluene can then be oxidized to *p*-bromobenzoic acid.

## EXERCISE 11.14

Identify the reagents required and the order in which they should be used to generate each of the following products from benzene:

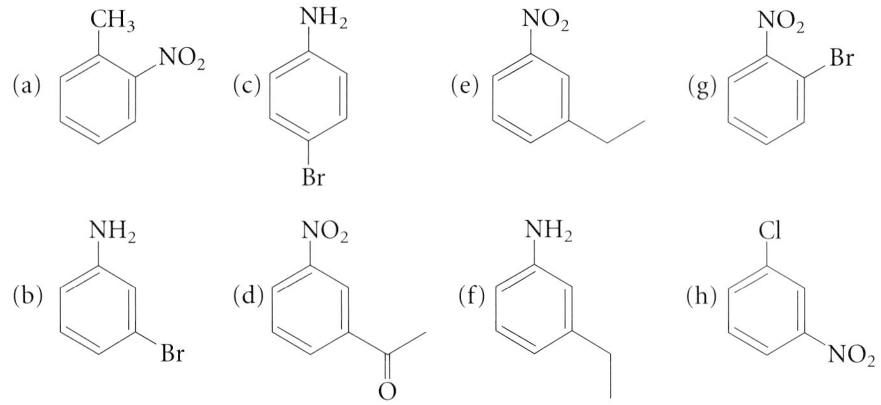

## EXERCISE 11.15

Recall that it is possible to remove an amino group from a benzene ring by reaction with $HNO_2$ and treatment of the resulting benzenediazonium ion with $H_3PO_2$. Thus, it is possible to use the directing influence of a nitro or an amino group to introduce substituents onto benzene and then completely remove the directing group. For example, 1,3,5-tribromobenzene can be prepared from aniline by bromination followed by removal of the amino group:

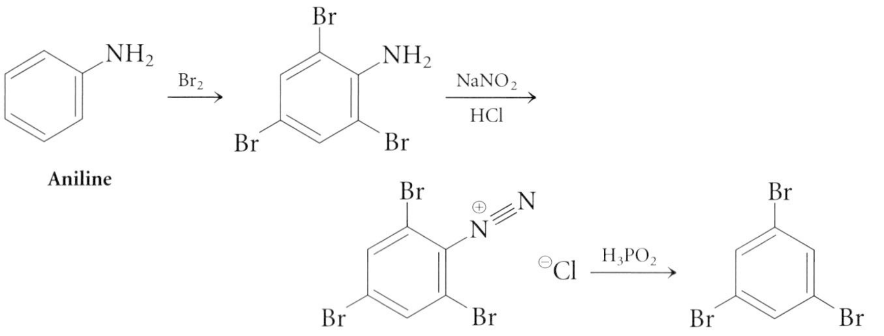

**Aniline** **1,3,5-Tribromobenzene**

The diazonium group can also be replaced by a halogen or a nitrile group via the Sandmeyer reaction. Using these reactions, devise a good synthesis of each of the following compounds starting with benzene. (*Hint:* You may use more than one amino or nitro group, or a combination of both.)

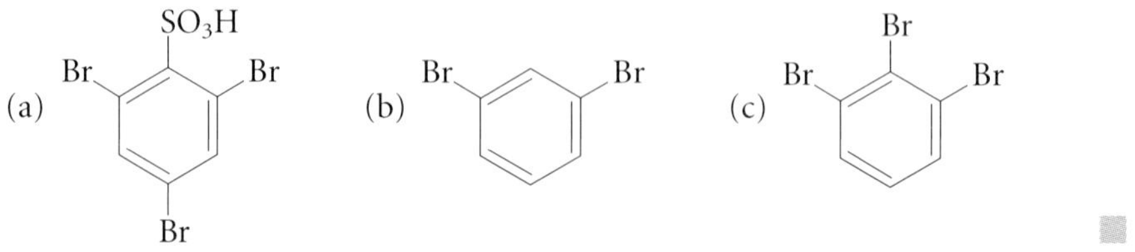

## 11.5

# Electrophilic Attack on Polycyclic Aromatic Compounds

The same rationale used to explain substituent directive effects (resonance stabilization of transition states leading to the most stable cationic intermediate) can also explain the regiochemistry of electrophilic attack on polycyclic aromatics. This electrophilic reactivity is important because it is believed to be related to the carcinogenicity associated with these compounds. (Recall from Chapter 2 that benzo[a]pyrene is a known carcinogen.)

The reaction of naphthalene with electrophiles illustrates these $\pi$ resonance effects. Figure 11.12 shows the cationic intermediates formed by electrophilic attack at the $\alpha$ and $\beta$ positions of naphthalene. In both cases,

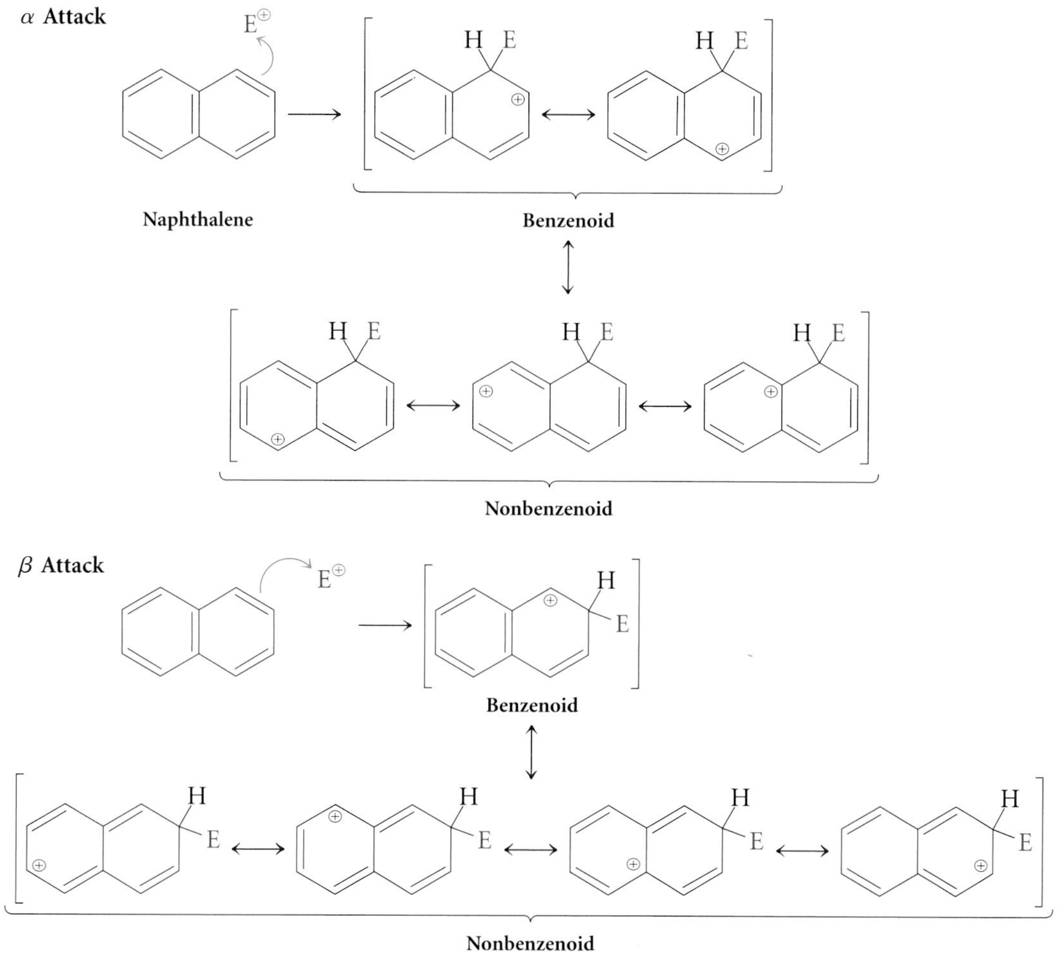

**FIGURE 11.12**

Electrophilic attack at either the $\alpha$ or $\beta$ position of naphthalene produces a cation with five significant resonance contributors. In the cation formed by $\alpha$ attack, two of these structures are benzenoid, whereas in the cation formed by $\beta$ attack, only one is benzenoid.

*CHEMICAL PERSPECTIVES*

### PERKIN'S MAUVE: THE DYE THAT NATURE FORGOT

William Perkin was just 18 in 1856, when he decided to spend his vacation from London's Royal College of Chemistry working in his home laboratory, somewhat naively trying to make quinine. In one attempt, he tried to attach an allyl group to a mixture of aniline and toluidine (methylaniline), producing a mixture that gave a black sludge after being oxidized with chromic acid. The residue, which contained the first synthetic dye, dissolved in alcohol and dyed cloth purple. It was fortunate for Perkin that this dye was purple: since ancient times, purple was the color of royalty. Only kings could afford purple or lavender cloth that did not fade, because the only source of color-fast purple dye was a small shellfish that lived in the Tyrian Sea.

Perkin's mauve

With a method in hand to make a mauve dye, Perkin interrupted his education, obtained a patent on the process, and persuaded his father to lend him money to build a factory to manufacture the several dyes he quickly learned to synthesize. Eighteen years later, at age 36, Perkin sold his factory, leaving him sufficiently wealthy that he was free to do research without financial concern. He later discovered a number of important organic reactions. The highest award of the American Chemical Society is named in his honor.

five resonance contributors can be drawn in which the positive charge is delocalized throughout the fused rings. However, with $\alpha$ attack, two of these structures retain three formal double bonds within a single six-member ring. This structural element is called a *benzenoid ring*; in accord with Hückel's rule (Chapter 2), it exhibits particular stability and contributes significantly to the resonance stabilization of the cation. The transition state leading to the cation with a larger number of benzenoid contributors is more stable. Thus, the cation formed by $\alpha$ attack is more stable than the one formed by $\beta$ attack, and electrophilic attack on naphthalene is easier at the $\alpha$ position. For example, nitration of naphthalene occurs exclusively at the $\alpha$ position:

Naphthalene     1-Nitronaphthalene

This extra stabilization of the transition state leading to the cationic intermediate results in acceleration of the electrophilic aromatic substitution of naphthalene relative to that of benzene.

The reactions involving derivatives of benzene can also be used with comparably substituted derivatives of naphthalene. For example, $\alpha$-naphthyl amine can be prepared by reduction of $\alpha$-nitronaphthalene with Fe and HCl, and $\alpha$-naphthoic acid can be prepared by reacting $\alpha$-bromonaphthalene with Mg, followed by reaction of the Grignard reagent with $CO_2$.

## EXERCISE 11.16

Predict the monosubstitution product formed in each of the following reactions:

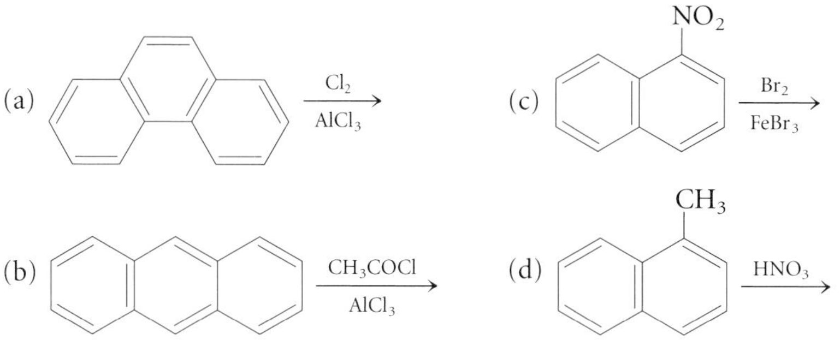

(a) $\xrightarrow[\text{AlCl}_3]{\text{Cl}_2}$

(b) $\xrightarrow[\text{AlCl}_3]{\text{CH}_3\text{COCl}}$

(c) $\xrightarrow[\text{FeBr}_3]{\text{Br}_2}$

(d) $\xrightarrow{\text{HNO}_3}$

## 11.6

# Synthetic Applications

Table 11.2 (on pages 578–579) is a summary of the reactions discussed in this chapter, grouped according to their synthetic utility. These reactions can be used in various sequences to achieve conversions not specifically listed. An important intellectual challenge is the integration of these new reactions with those presented in earlier chapters.

## 11.7

# Spectroscopy

Both $^1$H and $^{13}$C NMR spectroscopy are useful tools for assigning structures to the products of the reactions covered in this chapter. For example, p-chlorotoluene can be readily distinguished from the *ortho* and *meta* isomers by $^{13}$C NMR, because there are only four unique $sp^2$-hybridized carbon atoms in the symmetrical *para* isomer. There are six signals in the aromatic region for *o*- and *m*-chlorotoluene but only four for *p*-chlorotoluene.

In principle, the presence or absence of symmetry should also impact the $^1$H NMR spectra of aromatic compounds. Unfortunately, it is not always the case that the chemical-shift differences in the proton spectra of aromatic compounds are sufficiently definitive so as to distinguish each unique proton or group of identical protons. For example, there are two different types of aromatic hydrogen atoms in *p*-cymene (*p*-isopropyl-

## TABLE 11.2

Using Electrophilic Aromatic Substitution to Introduce Various Functional Groups

| Functional Group | Reaction | Example |
|---|---|---|
| Alkyl aromatics | Friedel–Crafts alkylation | |
| | Wolff–Kishner reduction of an aryl ketone | |
| | Clemmensen reduction of an aryl ketone | |
| Anilines | Reduction of an aryl nitro compound (Zn, HCl) | |
| Aryl carboxylic acids | Oxidation of an alkylbenzene | |
| Aryl halides | Halogenation of an aromatic hydrocarbon (not for I or F) | |

| Functional Group | Reaction | Example |
|---|---|---|
| Aryl halides (*continued*) | Substitution reaction of a diazonium salt | |
| Aryl ketones | Friedel–Crafts acylation | |
| Aryl nitriles | Sandmeyer reaction | |
| Aryl nitro compounds | Nitration of an aromatic hydrocarbon | |
| Aryl sulfonic acids | Sulfonation of an aromatic hydrocarbon | |
| Azo compounds | Treatment of an electron-rich aromatic with a diazonium salt | |
| Diazonium salts | Diazotization of aniline | |

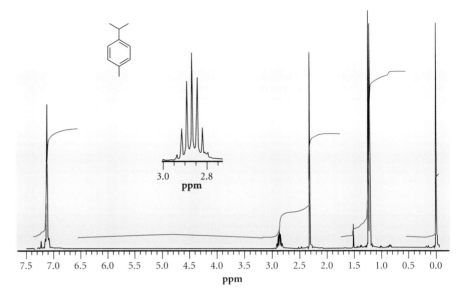

**FIGURE 11.13**

¹H NMR spectrum of *p*-cymene.

methylbenzene), but they occur as what appears to be a broad singlet (Figure 11.13).

Spectroscopy has also been of value in confirming the existence of the arenium ion that exists as an intermediate in all the electrophilic aromatic substitution reactions covered in this chapter. For example, treatment of benzene with a mixture of HF, $SbF_5$, $SO_2ClF$, and $SO_2F_2$ (an exotic and strongly acidic mixture) at $-134$ °C led to protonation of benzene, producing the simplest arenium ion that could be studied by $^{13}C$ NMR spectroscopy. Signals for the carbon atoms *ortho* and *para* to the site of protonation are shifted dramatically downfield because of delocalization of positive charge to these positions. The Nobel prize in chemistry was awarded in 1994 to George Olah of the University of Southern California in part for his contributions to the characterization of cations by spectroscopic techniques.

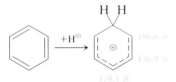

## Summary

**1.** Despite the accompanying loss of aromaticity, a highly active electrophile can attack an aromatic ring to form a new $\sigma$ bond and a resonance-stabilized cation.

**2.** The rate and regiochemistry of electrophilic attack are determined by electron density in the $\pi$ system of the aromatic ring and resonance stabilization of the intermediate arenium ion.

**3.** The resonance stabilization available in a re-aromatized six-member ring dictates the ensuing deprotonation of the cationic intermediate, restoring aromaticity and producing net substitution rather than addition.

**4.** Electron-donating substituents stabilize the arenium ion intermediates and accelerate the rate of electrophilic substitution. Such substituents also direct further substitution to the *ortho* and *para* positions in substituted benzenes.

**5.** Electron-withdrawing substituents inhibit electrophilic attack. This inhibition is most intense at the *ortho* and *para* positions, allowing *meta* substitution to become dominant in compounds bearing electron-withdrawing groups.

**6.** Halogenated aromatic compounds combine $\sigma$-electron withdrawal with $\pi$-electron donation, leading to slower electrophilic substitution and *ortho*, *para* direction.

**7.** Resonance and inductive effects similar to those seen in substitution reactions of benzene influence the regiochemistry of electrophilic substitution in fused aromatic rings.

**8.** Substituents introduced onto an aromatic ring retain their characteristic reactivity. For example, alkyl side chains can be oxidized, reduced, halogenated, and so forth.

**9.** Oxidation–reduction reactions often reverse the electron-donating or electron-withdrawing character of a substituent, altering its directive effect on further substitution.

## Review of Reactions

## Review Problems

**11.1** Predict the preferred regiochemistry for the product(s) expected from treatment of each of the following compounds with chlorine and $AlCl_3$:

(a) toluene                    (e) *p*-chlorophenol

(b) nitrobenzene               (f) acetophenone

(c) acetanilide                (g) *o*-chlorotoluene

(d) anisole

**11.2** Draw the structure of the major product obtained from each of the following sequences:

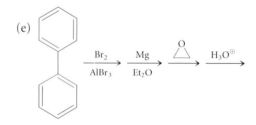

**11.3** Predict the major product expected from the reaction of $Br_2$ and $AlBr_3$ with phenyl benzoate. Be sure to explain why one ring is more active than the other.

**11.4** Suggest a sequence of reagents for converting toluene into each of the following compounds:

(a) *p*-nitrobenzoic acid        (c) *p*-nitrobenzyl alcohol

(b) *m*-nitrobenzoic acid        (d) *p*-toluenesulfonic acid (HOTs)

**11.5** Write a mechanism that accounts for the formation of each of the following products. Be sure to explain any important regiochemical and stereochemical control elements.

**(a)**

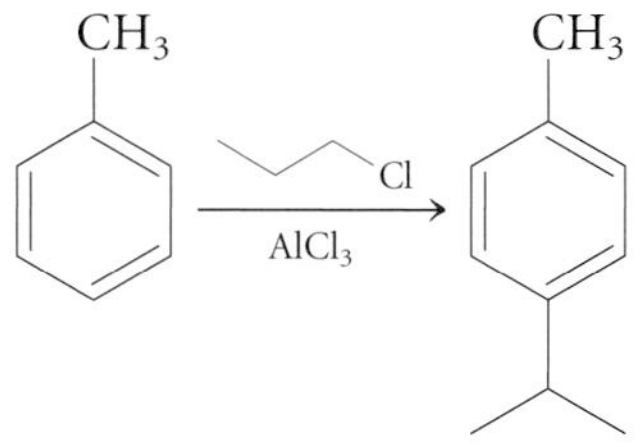

**(b)**

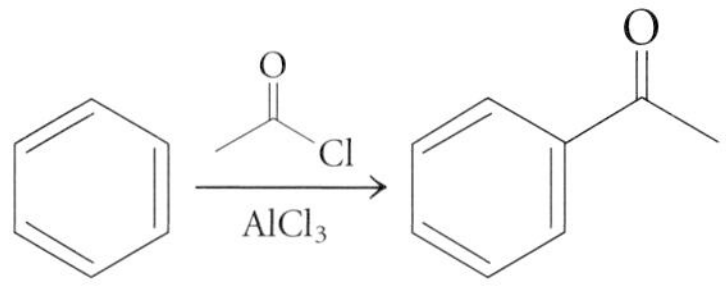

**(c)**

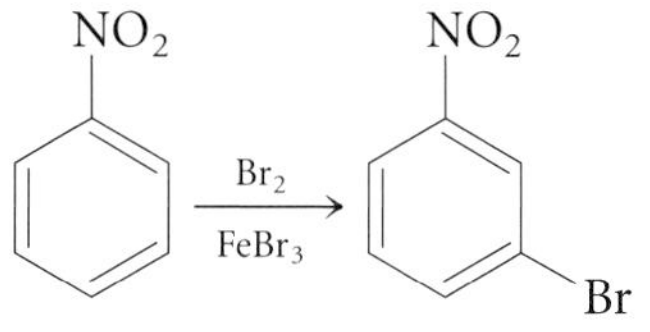

**(d)**

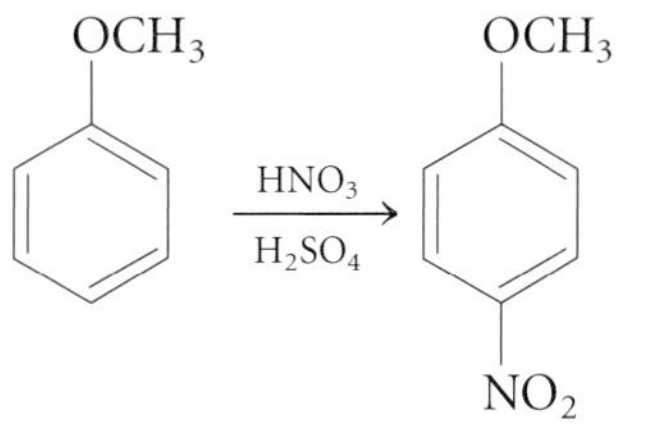

**(e)**

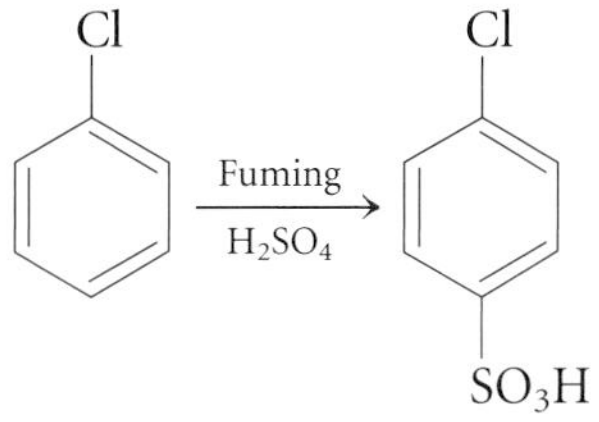

**11.6** Predict the major product (or products) obtained from nitration of each of the following compounds:

(a) *N*-phenylacetamide     (c) fluorobenzene          (e) 1-methylnaphthalene

(b) methyl benzoate          (d) *n*-propylphenylether

**11.7** Propose a synthesis for each of the following compounds from either benzene or toluene. You may use any inorganic reagent(s) and any organic compound containing up to four carbons.

(a)

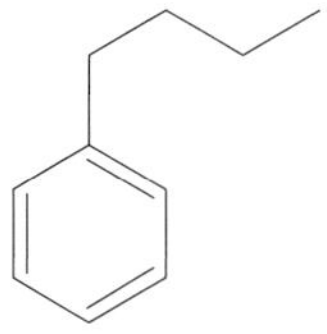

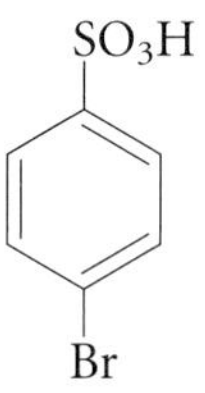

(e)

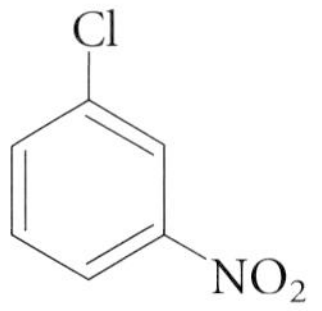

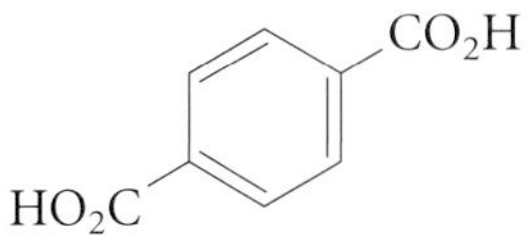

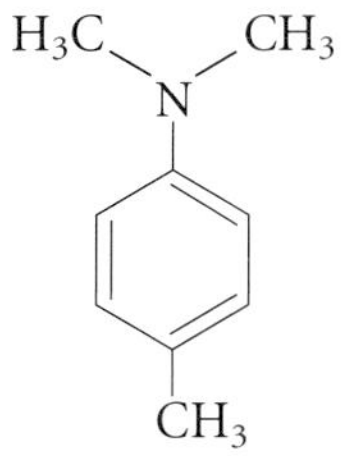

(i)

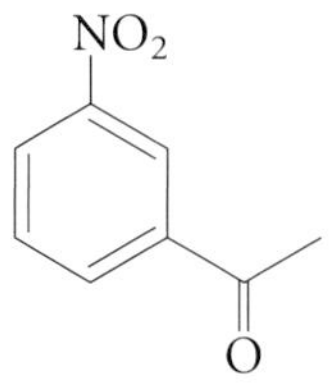

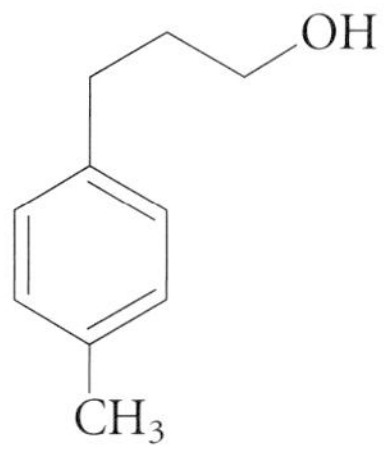

(b)

(f)

(j)

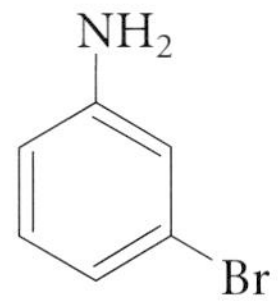

(c)

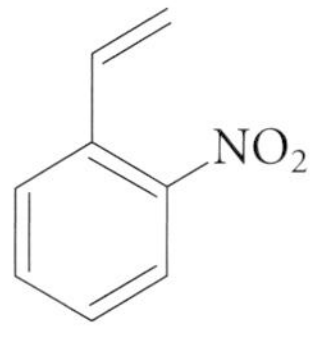

(g)

(d)

(h)

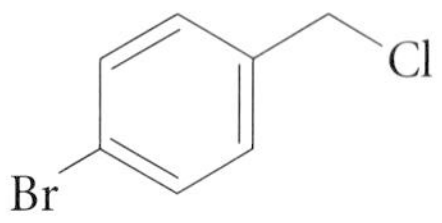

**11.8** Identify the starting material needed to prepare each of the following compounds from the given reagents.

(a) 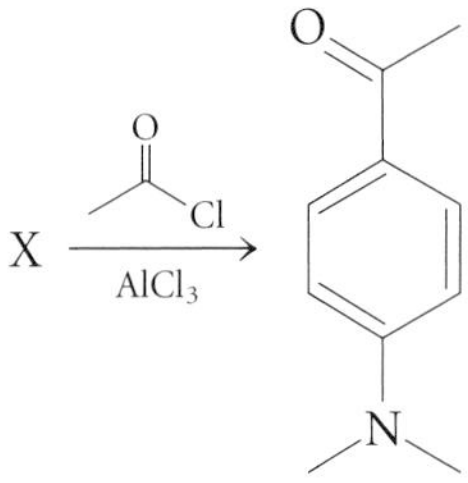 X $\xrightarrow[\text{AlCl}_3]{}$ (acetyl chloride)

(b) 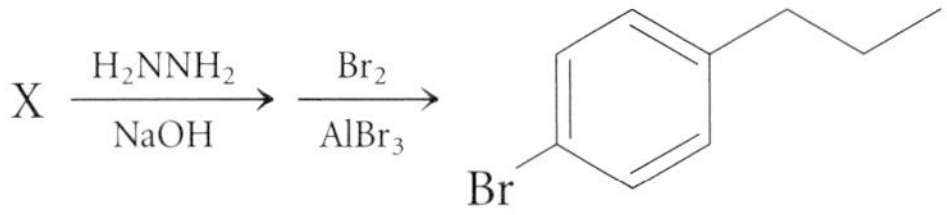 X $\xrightarrow[\text{NaOH}]{\text{H}_2\text{NNH}_2}$ $\xrightarrow[\text{AlBr}_3]{\text{Br}_2}$

(c) 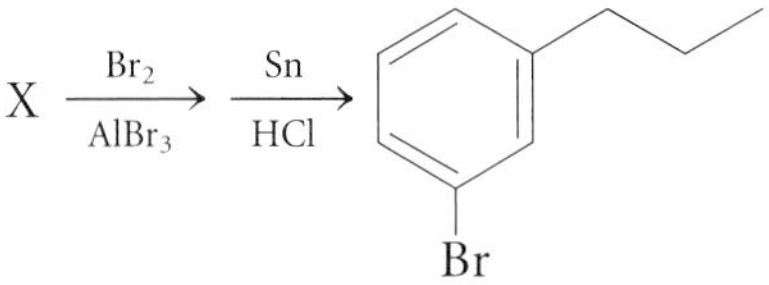 X $\xrightarrow[\text{AlBr}_3]{\text{Br}_2}$ $\xrightarrow[\text{HCl}]{\text{Sn}}$

(d) 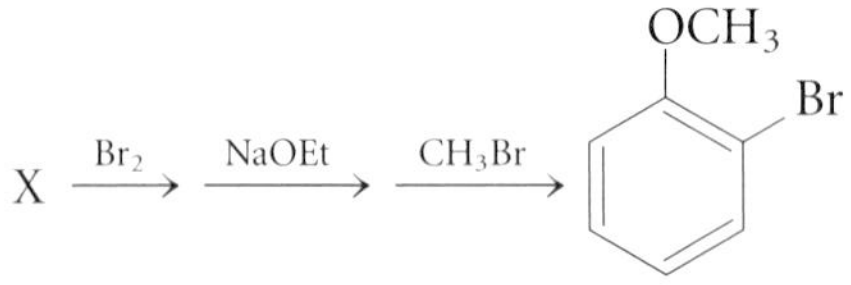 X $\xrightarrow{\text{Br}_2}$ $\xrightarrow{\text{NaOEt}}$ $\xrightarrow{\text{CH}_3\text{Br}}$

(e) 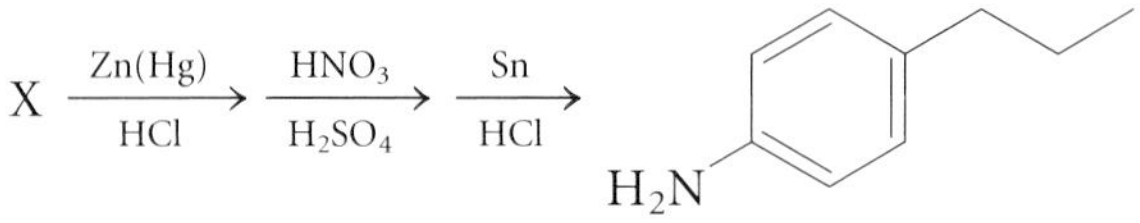 X $\xrightarrow[\text{HCl}]{\text{Zn(Hg)}}$ $\xrightarrow[\text{H}_2\text{SO}_4]{\text{HNO}_3}$ $\xrightarrow[\text{HCl}]{\text{Sn}}$

**11.9** Write a mechanism, using curved arrows to indicate electron flow, for the Friedel–Crafts acylation of naphthalene at the $\alpha$ and $\beta$ positions. Explain why the major product is that formed by $\alpha$ substitution.

**11.10** BHT, the major antioxidant used as a food preservative in the United States, is a mixture of positional and structural isomers of butylated hydroxytoluene. The major component in BHT is 2,6-di-$t$-butyl-4-methylphenol, which is made industrially from $p$-methylphenol and isobutene (2-methylpropene). Explain the preferred regiochemistry of the major product. Propose a mechanism for the formation of 2,6-di-$t$-butyl-4-methylphenol from these reagents in acidic methanol.

### Supplementary Problems

**11.11** Provide an IUPAC name for each of the following aromatic compounds:

(a)

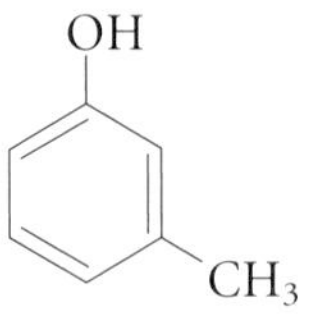

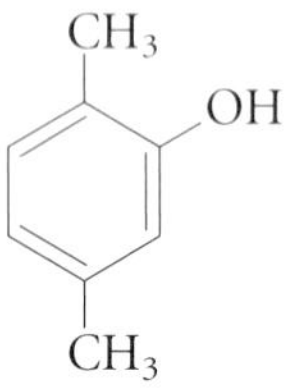

(b)

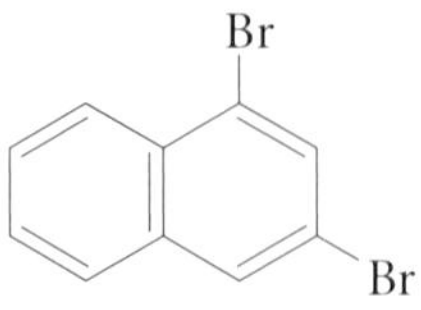

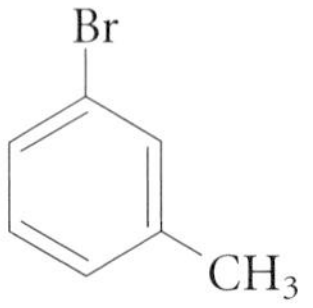

(c)

(d)

(e)

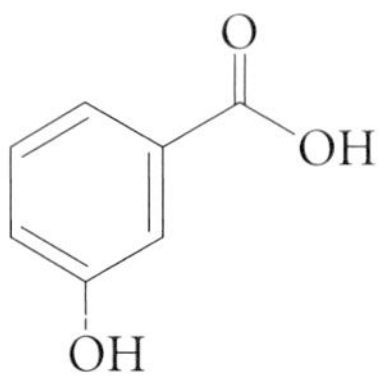

(f)

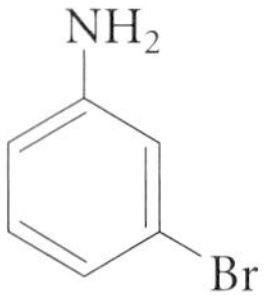

**11.12** Draw a structure corresponding to each of the following names:

(a) 2-chloro-3-bromophenol    (c) 3,4,5-trimethoxybenzoic acid

(b) *p*-divinylbenzene        (d) 3,4-dichlorobenzaldehyde

**11.13** Phenolphthalein sodium is used as a pH indicator because this red compound becomes colorless upon protonation. Phenolphthalein sodium is prepared from phenolphthalein, which, in turn, is produced by the reaction of phthalic anhydride and phenol in the presence of $ZnCl_2$. Account for this reaction by writing a detailed, stepwise mechanism.

Phenolphthalein                        Phenolphthalein sodium

**11.14** Amino groups are *ortho, para* directing and activating toward electrophilic aromatic substitution, whereas nitro groups are *meta* directing and deactivating. The nitroso group, whose nitrogen is intermediate in oxidation level between that of an amino and that of a nitro group, is *ortho, para* directing but *de*activating. Account for this difference in reactivity and directive effect.

*o,p*-Directing,     *o,p*-Directing,     *m*-Directing,
activating         deactivating     deactivating

**11.15** An aryl substituent is weakly activating and *ortho, para* directing. Thus, chlorination of biphenyl results in the formation of two monochlorobiphenyls:

Biphenyl                   Monochlorobiphenyls

Account for the directing influence of the phenyl ring on electrophilic aromatic substitution by examining the three possible cationic intermediates that result from addition of the electrophile at positions *ortho, meta,* and *para* to the phenyl substituent.

**11.16** Chlorination of alkylbenzenes produces a mixture of *ortho* and *para* isomers from which further reaction yields mainly the 2,4-dichloro isomer. This compound is converted to the 3,5-dichloro isomer upon heating in the presence of $AlCl_3$ for an extended period. Account for this isomerization with a detailed mechanism.

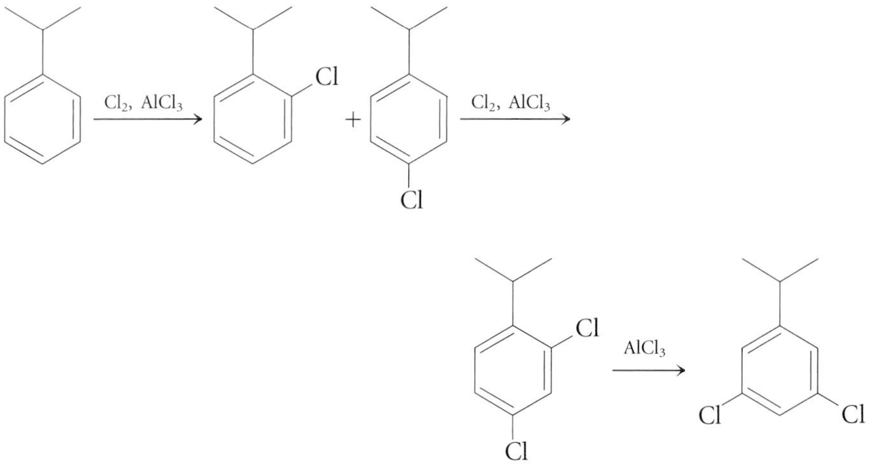

**11.17** Reaction of dimethylaniline with benzaldehyde in the presence of HCl results in the formation of compound X ($C_{25}H_{30}N_3$), which is oxidized by $PbO_2$ to form the intensely colored dye malachite green. (This dye is named for the mineral malachite, basic copper carbonate, which it resembles in color.) Propose a structure for the intermediate X, and account for its formation with a detailed reaction mechanism.

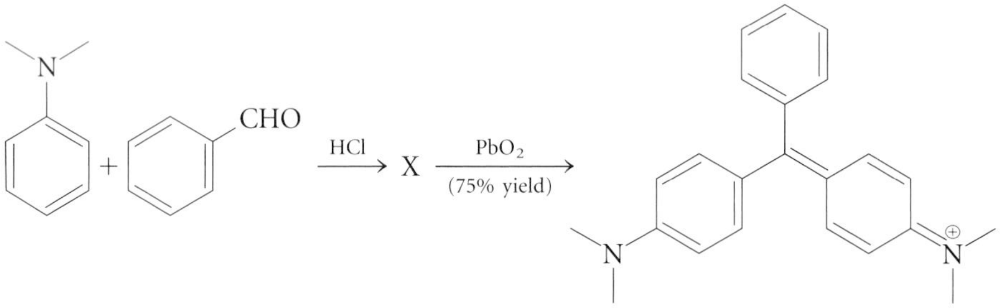

**Malachite green**

**11.18** Predict the major product expected when each of the following compounds is treated with the indicated reagent(s).

(a)

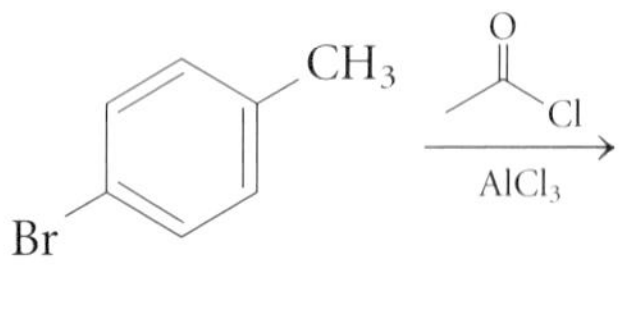

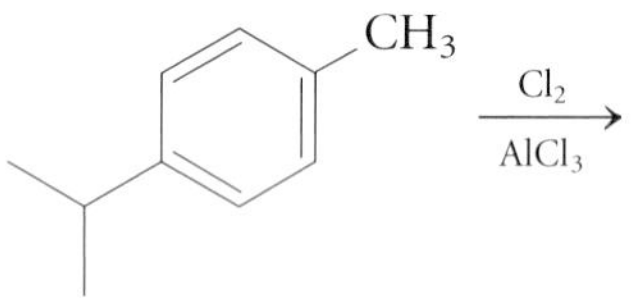

(b)

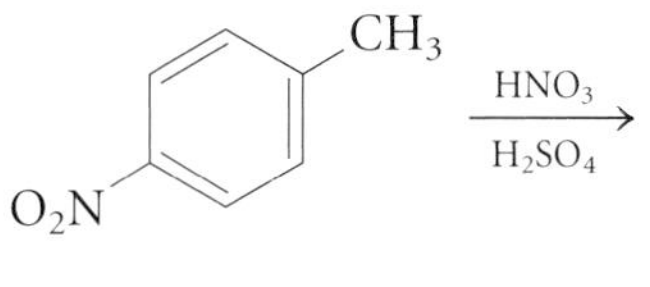

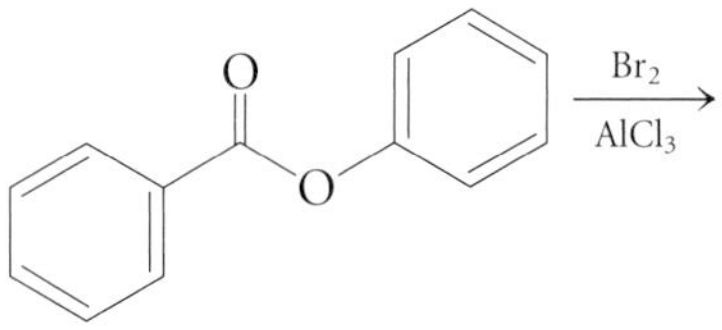

(c)

(d)

(e)

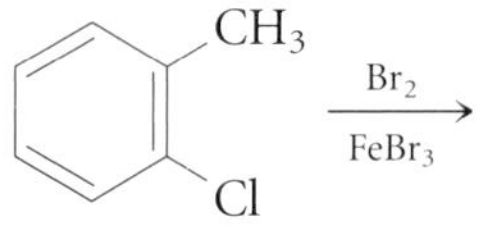

**11.19** Explain why the reaction of acetanilide with bromine is considerably slower than the reaction of aniline with bromine.

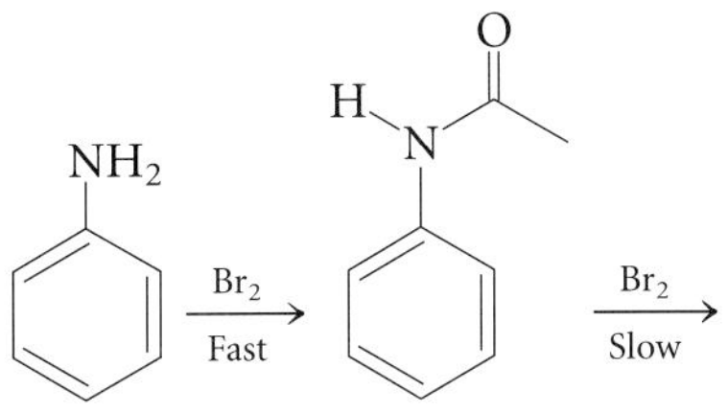

**11.20** Bisphenol A, a component of several plastics, is synthesized industrially by the following reaction. Propose a detailed mechanism for the formation of bisphenol A.

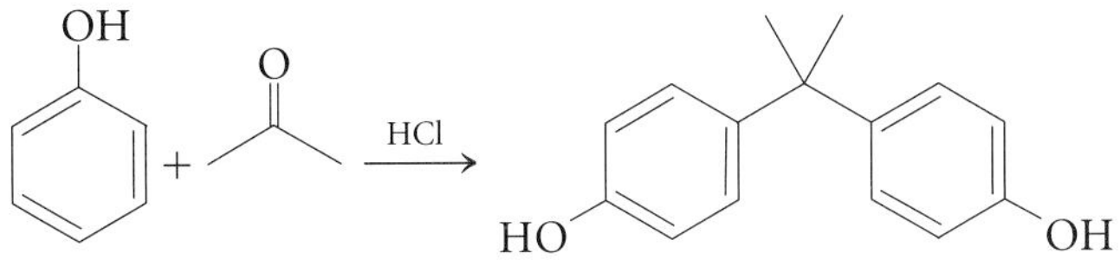

**Bisphenol A**

**11.21** There are a number of different possible ways that this tertiary alcohol could be prepared from benzene:

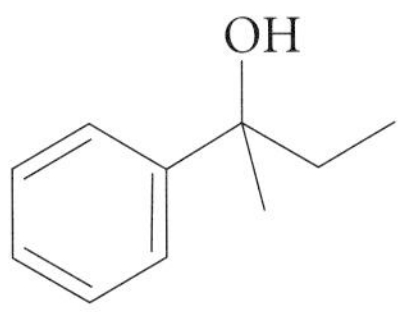

Devise three syntheses of this compound using:

(a) Friedel–Crafts acylation

(b) Friedel–Crafts alkylation

(c) electrophilic aromatic bromination

**11.22** Reaction of $H_3CCl$ with *p*-xylene in the presence of $AlCl_3$ at 25 °C gives 1,2,4-trimethylbenzene. However, when this product is treated further with $AlCl_3$ at 100 °C, a 63% yield of 1,3,5-trimethylbenzene is obtained.

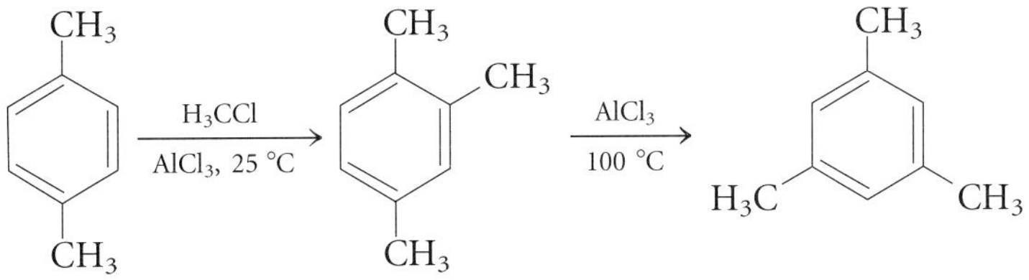

(a) Which isomer, 1,2,4- or 1,3,5-trimethylbenzene, is the kinetic isomer in this reaction, and which is the thermodynamic isomer?

(b) What factors lead to the greater stability of the thermodynamic isomer of trimethylbenzene?

(c) Devise a mechanism for the conversion of 1,2,4-trimethylbenzene to the 1,3,5-isomer.

**11.23** When either *o*- or *p*-xylene is heated at 50 °C with 5% AlCl$_3$ in the presence of HCl, a mixture of all three xylenes in the ratio shown is obtained.

**Ratio found in lab = 17 : 62 : 21**
**(Ratio found in coal tar = 15 : 61 : 24)**

(a) Can you devise a mechanism for the transformation of either *o*- or *p*-dimethylbenzene to the *m*-isomer?

(b) Why is the *meta* isomer favored in the equilibrium?

(c) Why might the ratio of the three isomers formed in the experiment described be nearly the same as that found in oil obtained from coal (coal tar)?

**11.24** Treatment of *m*-xylene with 80% H$_2$SO$_4$ at 25 °C results in sulfonation to form (mostly) 2,4-dimethylbenzenesulfonic acid. On the other hand, both the *ortho* and *para* isomers of xylene are unreactive under these conditions but do react when 84% H$_2$SO$_4$ is used.

(a) What other isomer(s) might be formed in the sulfonation of *m*-xylene, and why is 2,4-dimethylbenzenesulfonic acid the major product?

(b) Why is the *meta* isomer more reactive than the *ortho* and *para* isomers?

(c) Why does the relatively small change in the concentration of sulfuric acid produce a significant rate increase in sulfonation? (*Hint:* The percent composition is by volume, and the remainder is H$_2$O.)

**11.25** Draw structures for all of the possible isomers of tetrachlorobenzene. One is found in higher concentration in the mixture of tetrachlorobenzenes formed when benzene is treated with excess chlorine in the presence of AlCl$_3$. Which isomer is it? Reaction of this tetrachlorobenzene with NaOH at elevated temperatures produces 2,4,5-trichlorophenol, an intermediate in the synthesis of hexachlorophene (see Problem 11.26). Propose a mechanism for this conversion. Why does this reaction not proceed to replace all of the chlorine substituents with hydroxyl groups?

**11.26** Propose a synthesis for hexachlorophene starting with 2,4,5-trichlorophenol.

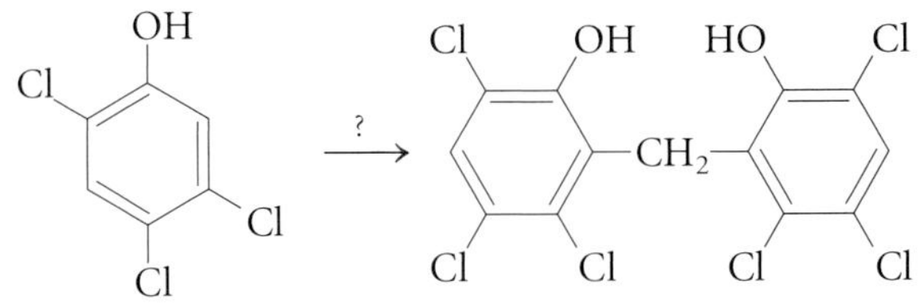

2,4,5-Trichlorophenol                    Hexachlorophene

**11.27** The presence of a nitro group increases the acidity of both phenol and protonated aniline. First, rationalize this observation. Then explain why the effect in both cases is greater when the nitro group is *ortho* or *para* to the acidic group rather than *meta*.

| | pKₐ | | pKₐ | | |
|---|---|---|---|---|---|
| | | | *ortho* | *meta* | *para* |
| 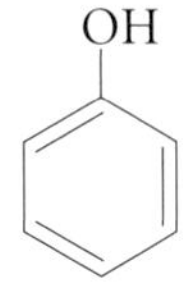 Phenol | 10.0 | 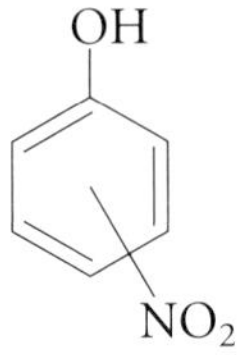 (NO₂ phenol) | 7.2 | 8.0 | 7.2 |
| 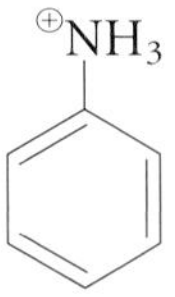 Protonated aniline | 4.6 | 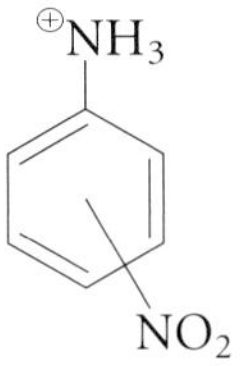 (NO₂ aniline) | −0.7 | 2.6 | 1.0 |

**11.28** During the industrial production of trichlorophenol (see Problem 11.25), a by-product known as a dioxin is produced. (This class of compounds has been implicated as the causative agent of a number of serious ailments.) Propose a mechanism for the formation of the following dioxin:

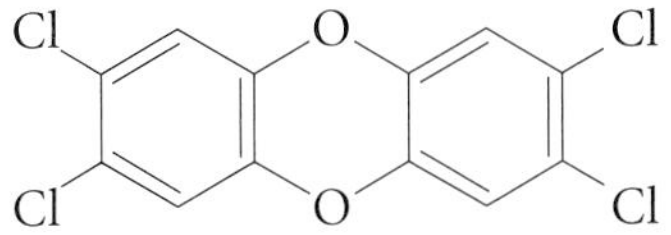

**A dioxin**

**11.29** Using the bond energies provided in Table 3.5, calculate $\Delta H°$ for electrophilic chlorination of benzene with molecular chlorine in the presence of catalytic amounts of AlCl₃.

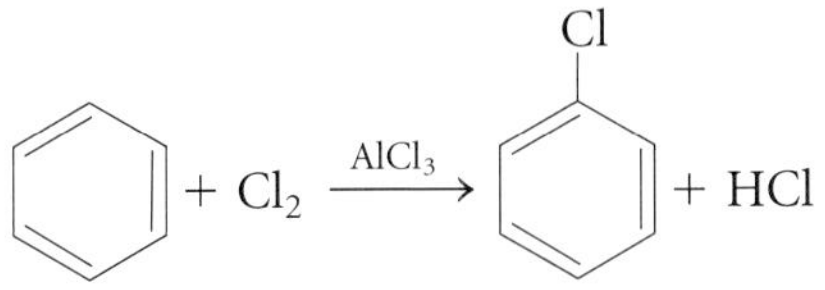

**11.30** Devise a synthesis of 1,3,5-benzenetricarboxylic acid starting from benzene and any reagents and other sources of carbon needed. (*Hint:* See Problem 11.22.)

# 12

# Nucleophilic Addition and Substitution at Carbonyl Groups

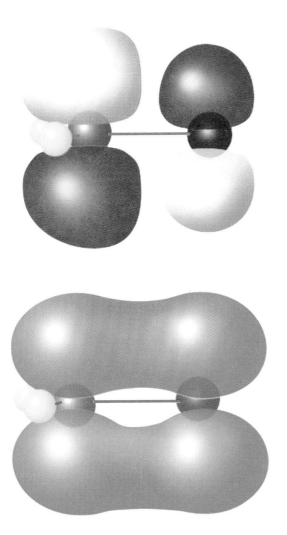

*The π bonding (lower) and π antibonding (upper) molecular orbitals of formaldehyde, $H_2CO$, are shown here. When a nucleophile approaches the carbonyl carbon, electron density is added to the π antibonding orbital as the electrons in the π bonding orbital shift toward oxygen. These electrons ultimately become the third lone pair of the negatively charged oxygen of the alkoxide ion.*

Many of the reactions described in Chapters 10 and 11 proceed through the formation of a carbocationic intermediate derived from an organic compound containing one or more multiple bonds. In Chapter 10, we saw how an electrophile interacts with an electron-rich carbon–carbon multiple bond, initiating electrophilic addition. By contrast, as described in Chapter 11, electrophilic attack on the electron-rich aromatic $\pi$ system results in net substitution. In this chapter, we consider reactions in which addition is initiated by the attack of a nucleophile on the C=O bond of a carbonyl group. In some cases, the initial product undergoes further reaction, with loss of the oxygen atom of the original carbonyl group. Whether the final product is the result of addition or of addition followed by substitution depends on the nature of the nucleophilic reagent and the carbonyl group.

## 12.1

## Nucleophilic Addition to a Carbonyl Group

#03    Aldehydes/Ketones

### The Carbonyl Group

Carbonyl groups have unique chemical properties. The $\pi$ bond between carbon and oxygen is polarized, because of oxygen's greater electronegativity (3.5 for oxygen versus 2.5 for carbon). This polarization in the carbonyl group can be viewed as resulting from significant contributions from two resonance structures, one in which the C=O $\pi$ bond is intact, and a second with only a $\sigma$ bond between carbon and oxygen and formal positive charge on carbon and negative charge on oxygen:

**Hybrid**

The hybrid has significant contributions from both of these resonance structures, and as a result, there is significant positive charge on the carbon end and significant negative charge on the oxygen end of the carbonyl group.

The polarization of the carbonyl group enhances the electrophilic character of its carbon atom. Addition of a nucleophile to a carbonyl group leads to an intermediate with negative charge on the oxygen atom:

($pK_a$ 16–18)

592

In contrast, if the same nucleophilic addition occurred to a simple alkene, a species with negative charge on carbon would be formed:

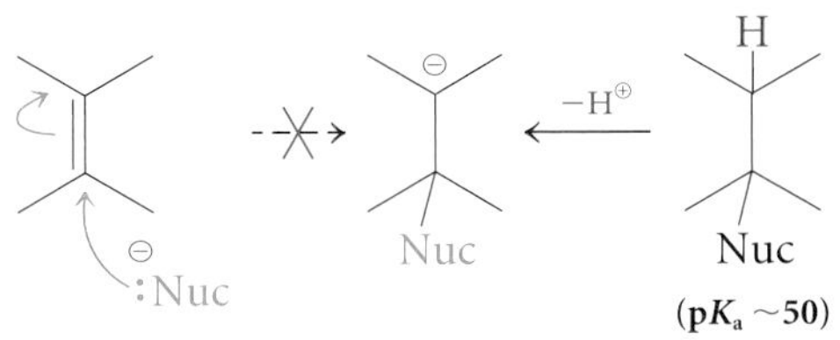

The significance of this difference is revealed by the differing acidities of a simple hydrocarbon and a typical alcohol ($pK_a \sim 50$ versus 16–18).

The interaction of a nucleophile with a carbonyl carbon results in the formation of a C—Nuc $\sigma$ bond that provides energetic compensation for the accompanying rupture of the carbonyl $\pi$ bond. The two electrons originally in that $\pi$ bond shift to the more electronegative oxygen atom, placing surplus electron density on the atom best able to accommodate negative charge. Polarization of the carbonyl group is increased by prior coordination of the carbonyl oxygen with a proton or metal ion.

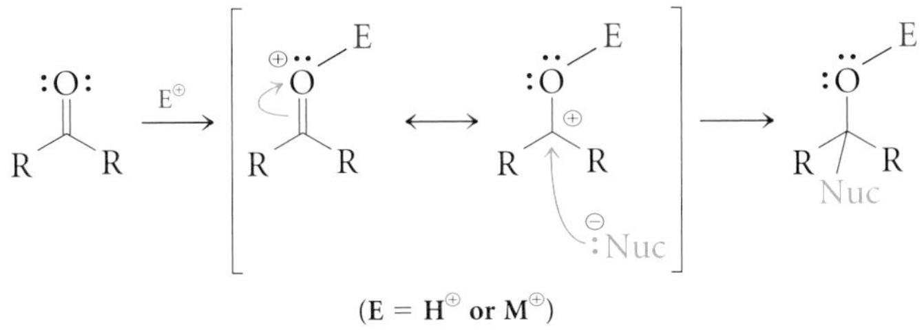

$$(E = H^{\oplus} \text{ or } M^{\oplus})$$

Nucleophilic addition is facilitated by this complexation, because the nucleophile interacts with an intermediate bearing a full positive charge. As a result, the product formed is neutral (when the attacking nucleophile is anionic).

## Possible Reactions of a Nucleophile with a Carbonyl Group

The reactions of nucleophiles with carbonyl compounds can be divided into two important classes: those that result in a bond between carbon and a heteroatom, and those in which a new carbon–carbon bond is formed. Each class can result in net addition or net substitution. When addition occurs, the nucleophile becomes bonded to carbon and an electrophile becomes bonded to oxygen. With net substitution, the nucleophile replaces either the carbonyl oxygen or another heteroatom. The four possibilities— addition or substitution by a heteroatom nucleophile, and addition or substitution by a carbon nucleophile—are illustrated in Figure 12.1 (on page 594). We will examine all of these reactions in this chapter.

Regardless of whether the final product is the result of net addition or substitution, the first step is the addition of the nucleophile to the carbonyl group's carbon atom. Additions and substitutions of heteroatom nucleophiles result in the conversion of one functional group to another.

**Addition of a heteroatom nucleophile—for example, hydrate formation**

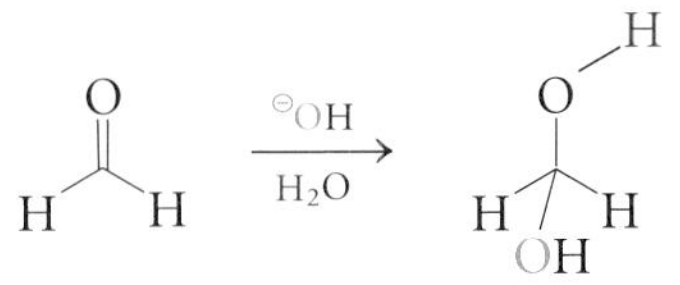

**Substitution by a heteroatom nucleophile—for example, imine formation**

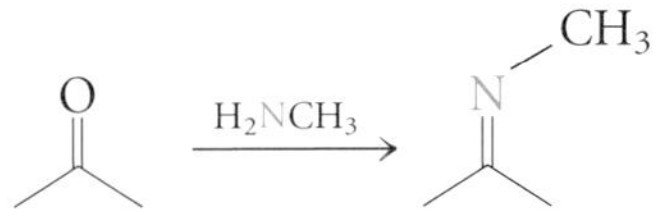

**Addition of a carbon nucleophile—for example, cyanohydrin formation**

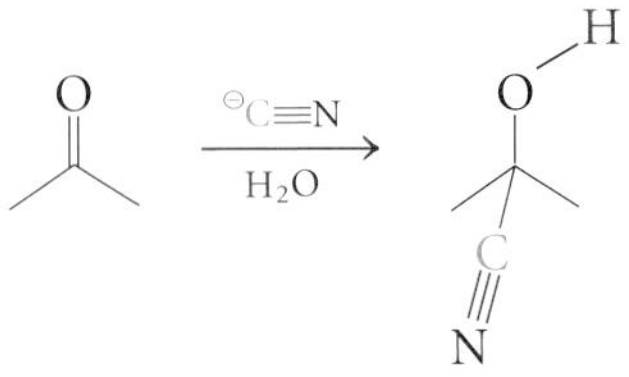

**Substitution by a carbon nucleophile—for example, Wittig reaction**

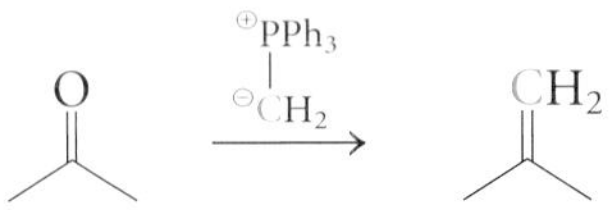

**FIGURE 12.1**

The four classes of reactions of nucleophiles with carbonyl groups. Reactions from each class can occur with aldehydes, ketones, and carboxylic acid derivatives.

Additions and substitutions of carbon nucleophiles result in the formation of products with larger, and often more complex, carbon skeletons.

An important goal of organic chemists is to make new molecules that have interesting properties or that duplicate the structural features of naturally occurring molecules having significant biological activity. To understand how such molecules can be constructed both in the laboratory and in nature, you must know how to manipulate functional groups and how to build molecules of greater structural complexity from simple precursors. This chapter will expand your repertoire of synthetic reactions to include the interactions of heteroatom and carbon nucleophiles with the $sp^2$-hybridized carbon atoms in carbonyl groups.

## Anions as Nucleophiles

Negatively charged reagents act as nucleophiles and can add to the carbon end of the C=O group. Let's consider how the reactions of carbonyl compounds depend on the activity of the nucleophile.

As you know from the trends discussed in Chapter 6, the acidity of an acid, HX, increases as the position of X in the periodic table progresses from

## SIMPLE CARBONYL COMPOUNDS AND CHEMICAL COMMUNICATION BY INSECTS

Pheromones are chemicals used for communication between individual members of a species. These compounds function as sex attractants, trail markers, or alarms. When you see ants marching in a line across your kitchen, know that they are following a chemical trail to food or water mapped by a successful explorer that excreted a specific compound to help her coworkers. One species of ant (*Iridomyrmex priunosus*) uses 2-heptanone to alert other members of the hill to danger. The placement in traps of molecules that have been identified as sex attractants has made it possible to control both the voracious gypsy moth (which threatened New England's forests) and the screwworm (which created serious problems in Texas cattle) without resorting to widespread spraying of insecticides.

To be effective, a pheromone must be both highly potent and narrowly specific, so that only one species will respond. Most known sex attractants contain 10–17 carbon atoms—a range that permits sufficient complexity to create a molecule that is unique to a given species, but also allows for easy biosynthesis. Many pheromones are simple carbonyl compounds or carboxylic acid derivatives. For example, 9-ketodecenoic acid is the sex attractant used by a honeybee queen in her nuptial flight; it also develops ovaries in worker bees that ingest it.

**9-Ketodecenoic acid**
(honeybee queen substance)

Do human pheromones exist? Studies indicate that mammals do not give an automatic, standardized response to chemicals the way insects do. However, two steroids have been used commercially to induce the mating stance in sows, facilitating the artificial insemination of pigs. Folk medicine in Africa and Asia abounds with examples of substances purported to be human aphrodisiacs; however, none of those substances has been proved objectively to be effective. One of them, xylomollin, was first synthesized in the laboratory by one of the authors of this book.

**Xylomollin**

---

left to right. Therefore, the trend in acidity of binary hydrogen compounds with first-row elements is $CH_4 < NH_3 < H_2O < HF$. This also means that basicity, the affinity of an atom or anion for a proton, decreases in the same left-to-right progression ($NH_3$ is a stronger base in water than is HF). A strong base has a high affinity for a proton, and the conjugate acid of a strong base is thus a weak acid. For example, $NH_3$ is a strong base, and $^{\oplus}NH_4$ is a weak acid:

$$H_3N\!: \quad H\!-\!A \rightleftharpoons H_3\overset{\oplus}{N}\!-\!H + A^{\ominus}$$

**Strong**        **Weak**
**base**          **acid**

**Nucleophilicity** refers to the affinity of an atom or an anion for an electrophilic carbon atom. Basicity, the affinity of an atom or anion for a proton, sometimes correlates well with nucleophilicity. Within a row in the periodic table, the elements farther to the left are less electronegative and, as a result, more nucleophilic. Ammonia is more nucleophilic than water, for example. This relation also holds in the progression from top to bottom of a column of the periodic table. The larger, less electronegative halide ions (for example, iodide ion) are more nucleophilic than the smaller, more electronegative fluoride ion; sulfur anions are more nucleophilic than oxygen anions. This is because differences in polarizability (the larger atoms are more polarizable), as well as in electronegativity, contribute to differences in nucleophilicity within a column of the periodic table. Because two factors, electronegativity and polarizability, affect nucleophilicity, such comparisons must be restricted to the same row or the same column of the periodic table.

In Chapter 8, we saw that halide ions are effective nucleophiles for $S_N1$ and $S_N2$ reactions at $sp^3$-hybridized carbon atoms. However, halide ions do *not* effect nucleophilic addition or substitution at carbonyl groups. Let's look at the mechanism to see why this is so. In the reaction of chloride ion with an aldehyde, the aldehyde is favored over the addition product, for two reasons.

First, the carbon–chlorine $\sigma$ bond formed upon addition is weaker than the carbonyl $\pi$ bond of the aldehyde that is lost (81 versus 91 kcal/mole). Second, chloride ion is a much weaker base than the alkoxide ion generated by the addition (compare the acidity of HCl, $pK_a$ $-7$, to that of a typical alcohol, $pK_a$ 16–18), and simple acid–base chemistry favors the weaker acid and weaker base in an equilibrium reaction. Thus, although the addition of halide ion to an aldehyde (and other carbonyl functional groups) can and does take place, there is no net consequence because the position of equilibrium favors the carbonyl compound.

## EXERCISE 12.1

Use the bond energies in Table 3.5 (page 115) and the $pK_a$ values in Table 6.1 (page 305) to evaluate the addition of bromide ion and iodide ion to an aldehyde. Rank the three halides—chloride, bromide, and iodide—according to how much addition product will be present in equilibrium with the aldehyde.

## EXERCISE 12.2

Using curved arrows to represent the flow of electrons, write a mechanism that shows how an acid chloride can be converted into a carboxylate anion upon treatment with aqueous base. Explain why the reverse reaction, the conversion of a carboxylic acid into an acid chloride by treatment with NaCl, does not occur.

# Nucleophilic Addition of Hydrogen to Carbonyl Groups

Among the simplest possible nucleophilic reagents is the hydride ion, $H^\ominus$. However, because of the low solubility of alkali metal hydrides (LiH, NaH, and KH) in organic solvents and their strong basicity, these reagents cannot be used as sources of nucleophilic $H^\ominus$. On the other hand, complex reagents in which hydride is bonded to boron or aluminum are soluble in organic solvents and can provide the equivalent of the hydride ion as a nucleophile.

## ▓ Complex Metal Hydride Reductions

The two most common sources of hydride are sodium borohydride, $NaBH_4$, and lithium aluminum hydride, $LiAlH_4$ (abbreviated as LAH). Both have appreciable solubility in organic solvents. Indeed, 35–40 g of $LiAlH_4$ will dissolve in 100 mL of ethyl ether. In the space-filling models of these two reagents, note the greater size of sodium compared with lithium and of aluminum compared with boron. The addition of a hydride ion to a carbonyl group effects net reduction.

**Sodium borohydride**

**Lithium aluminum hydride**

***Reduction of Aldehydes and Ketones.*** When a complex metal hydride delivers the equivalent of $H^\ominus$ to the C=O bond, a new C—H bond is formed at the carbonyl carbon. In this way, an aldehyde is converted to a primary alcohol, and a ketone to a secondary alcohol.

| | | |
|---|---|---|
| **Butanal** | ***n*-Butanol** | **Cyclohexanone**     **Cyclohexanol** |

Let's look more closely at the reaction of acetone with sodium borohydride in ethanol. Sodium borohydride, $NaBH_4$, is sufficiently stable that protic solvents such as alcohols can be used for this reaction. In alcohol, ionization occurs to form $Na^\oplus$ and $^\ominus BH_4$. As this negatively charged hydride complex approaches the polar C=O bond, a bond between carbon and hydrogen forms at the same time as the $\pi$-bond electron pair shifts onto oxygen and the bond between the hydrogen and boron breaks. When the reaction is carried out in ethanol, the reduction is facilitated by hydrogen-bonding interactions between the carbonyl oxygen and the acidic hydroxyl group of ethanol.

A borate

The ethoxide produced reacts in a Lewis acid–base fashion with $BH_3$, forming a borate with an O—B $\sigma$ bond. This borate has three remaining B—H bonds, each of which, in turn, acts as a hydride equivalent for the reduction of another molecule of the starting ketone.

Upon treatment with aqueous acid, the borates decompose, forming boric acid, $B(OH)_3$. Overall, one mole of $NaBH_4$ reduces four moles of ketone. (In practice, there is also some side reaction between the reagent and the solvent, producing hydrogen gas. Thus, a small excess of $NaBH_4$ is typically used.) The product alcohol—in this case, 2-propanol—is formed by a reduction of the C=O bond. The reducing agent $NaBH_4$ thus effects a nucleophilic addition of hydride to the C=O bond.

Lithium aluminum hydride also effects nucleophilic addition of hydride to carbonyl groups. For example, reduction of acetone with $LiAlH_4$ results in the delivery of a hydride equivalent through the same pathway as for reduction with $NaBH_4$:

An alkyl aluminate | A tetra-alkyl aluminate | 2-Propanol

Lithium aluminum hydride is considerably more reactive than sodium borohydride and must be used in aprotic solvents such as tetrahydrofuran (THF) or another ether. In the absence of a proton source in the solvent, the neutral $AlH_3$ complexes with the negatively charged oxygen, producing an aluminate containing an oxygen–aluminum bond. This species still contains three aluminum–hydrogen bonds and can provide three additional hydride equivalents. By further reaction, all of the hydrogen atoms of $^{\ominus}AlH_4$ can be delivered to the C=O bonds of additional molecules of the ketone being reduced, ultimately producing the tetra-alkyl aluminate shown. The aluminum–oxygen bond has appreciable polar character, much like the ionic bond in a sodium alkoxide. Therefore, like the boron–oxygen bond of the borates produced with $NaBH_4$, it is easily hydrolyzed by aqueous acid, forming the alcohol product by protonation of the oxygen. [Any unreacted or partially reacted aluminum hydride is rapidly protonated to form $Al(OH)_3$ and hydrogen gas.] Overall, the stoichiometry is the same for $NaBH_4$ and $LiAlH_4$. In the absence of side reactions, one mole of reagent provides four hydride equivalents, which can reduce four moles of aldehyde or ketone. Primary alcohols are produced from the reduction of aldehydes, and secondary alcohols from that of ketones.

The reduction with lithium aluminum hydride and the second step, in which the aluminate salt is decomposed by acid, must be conducted as two separate steps. Otherwise, an acid–base reaction intervenes, decomposing the hydride reagent and generating hydrogen faster than hydride can be added to the carbonyl carbon.

$$LiAlH_4 \xrightarrow{H_2O} Al(OH)_3 + LiOH + 4\ H_2 \uparrow$$

Reduction of unsymmetrical ketones such as 2-butanone with $NaBH_4$ and $LiAlH_4$ produces the alcohol as a racemic mixture. The two faces of the carbonyl group do not appear different to these achiral reducing agents, and the reaction rates are the same for reduction leading to ($R$)- and ($S$)-2-butanol.

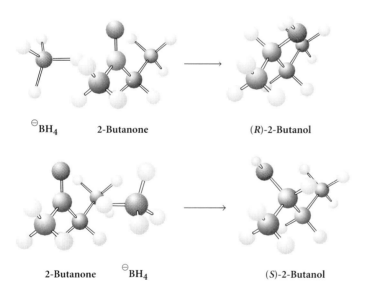

$^{\ominus}BH_4$     **2-Butanone**           **($R$)-2-Butanol**

**2-Butanone**    $^{\ominus}BH_4$        **($S$)-2-Butanol**

## EXERCISE 12.3

Using curved arrows to show all electron flow, write a complete mechanism for the reduction of 2-butanone with $LiAlH_4$.

***Reduction of Derivatives of Carboxylic Acids.*** All ketones, aldehydes, and carboxylic acid derivatives have carbonyl groups, but the reactions of carboxylic acid derivatives with hydride reagents differ from those of aldehydes and ketones in a significant way. Once hydride ion has been added to the carbonyl group of a carboxylic acid derivative, the intermediate loses a heteroatom substituent, a reaction that is not possible in the reduction of an aldehyde or ketone. This loss of the heteroatom substituent paves the way for delivery of a second equivalent of hydride ion.

*Esters.* In the reaction of lithium aluminum hydride with an ester such as ethyl acetate, an aluminate is initially formed, just as in the reduction of an aldehyde or ketone.

In the ester, however, there is a leaving group ($^{\ominus}$OR) bonded to the carbonyl carbon. Electron density flows from the oxygen–aluminum bond to regenerate a $\pi$ bond as the leaving group takes up the electrons originally present in the $\sigma$ bond between it and the carbonyl carbon. The aldehyde produced in this step is more reactive toward hydride reducing agents than is the ester from which it was derived, and it is reduced more rapidly by aluminum hydride. An alcohol is produced when the aluminate is treated with acid. Thus, reduction of an ester with lithium aluminum hydride results in the formation of a primary alcohol, with the OR group of the original ester lost as the corresponding alcohol.

*Amides.* The reduction of a tertiary amide by lithium aluminum hydride begins in the same way as the reduction of an ester:

The initial adduct has one oxygen and one nitrogen substituent rather than the two oxygen groups present in the tetrahedral intermediate formed in reduction of an ester. Oxygen is more electronegative than nitrogen, and the carbon–oxygen $\sigma$ bond is more easily broken than the carbon–nitrogen $\sigma$ bond. The lone pair of electrons on nitrogen forms a carbon–nitrogen double bond as the carbon–oxygen bond is broken, resulting in an **iminium ion.** The iminium ion is highly electrophilic and rapidly undergoes hydride reduction by a second equivalent of $LiAlH_4$, producing an amine with all C—C and C—N bonds originally present still intact. Thus, the reduction of an amide with $LiAlH_4$ has the net effect of replacing the carbonyl oxygen and the C—O $\sigma$ and $\pi$ bonds with the two carbon–hydrogen bonds in a $—CH_2—$ unit. The degree of substitution on nitrogen does not change, and the tertiary amide is reduced to form a tertiary amine.

Primary and secondary amides are also reduced by $LiAlH_4$, to form, respectively, primary and secondary amines:

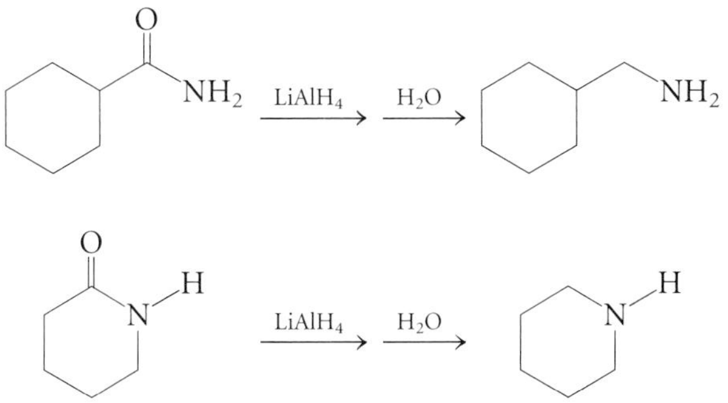


The pathway for the reduction of primary and secondary amides is similar to that for tertiary amides, as illustrated by the reduction of the secondary amide *N*-methylacetamide:

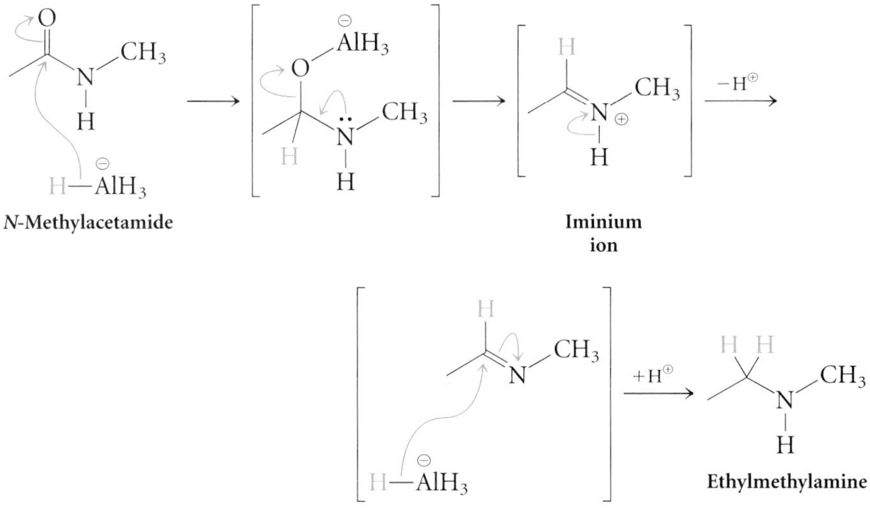


***N*-Methylacetamide** **Iminium ion** **Ethylmethylamine**

However, the intermediate iminium ion in this case has a proton on nitrogen that is rapidly lost, forming a neutral imine. The C═N bond in an imine is polarized in the same direction as the C═O bond in a carbonyl compound, with carbon bearing partial positive charge and the heteroatom bearing partial negative charge. As a result, hydride attack at the carbon end of the C=N bond is rapid, effecting reduction and the formation of a new C—H $\sigma$ bond, with a shift of the $\pi$ electron density onto nitrogen. The resulting negatively charged nitrogen atom coordinates with $AlH_3$, producing a complex that, after treatment with mild acid, leads to the free amine. Thus, metal hydride reductions of all three classes of amides generate the corresponding amine, as the carbonyl group is converted to a methylene unit.

### EXERCISE 12.4

Write a detailed mechanism for the $LiAlH_4$ reduction of acetamide:

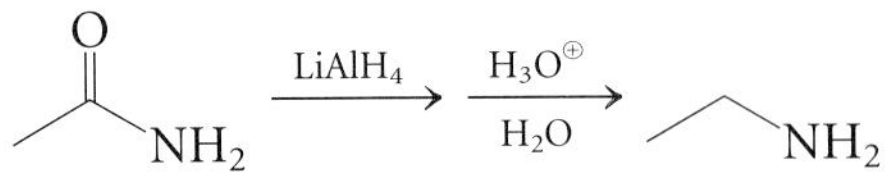

### EXERCISE 12.5

Draw the structures of all possible amides from which each of the following amines could be produced by reduction with $LiAlH_4$:

(a)

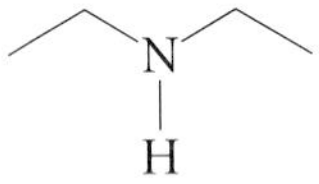

(c)

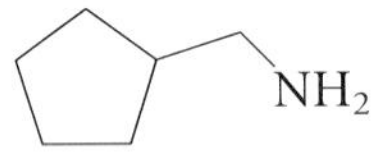

(b)

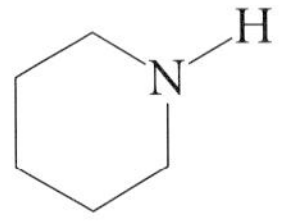

(d)

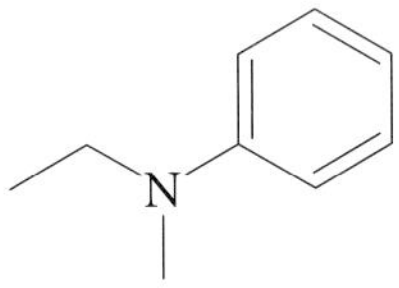

## Relative Reactivity of Carbonyl Compounds toward Hydride Reducing Agents

The relative reactivity of various carbonyl compounds toward nucleophilic attack roughly correlates with the magnitude of the positive charge density on the carbonyl carbon. In a nucleophilic hydride transfer reaction, the $\pi$ bond of the carbonyl group is destroyed. The rate of hydride reduction is affected by factors that increase or decrease the stability of the $\pi$ bond of the carbonyl group in the starting material and by factors that affect the stability of the intermediate tetrahedral anion. For example, carboxylic acid esters react more slowly than ketones and aldehydes, because donation of electrons from the oxygen of the —OR group stabilizes the ester and shifts electron density toward the carbonyl group (see Chapter 3).

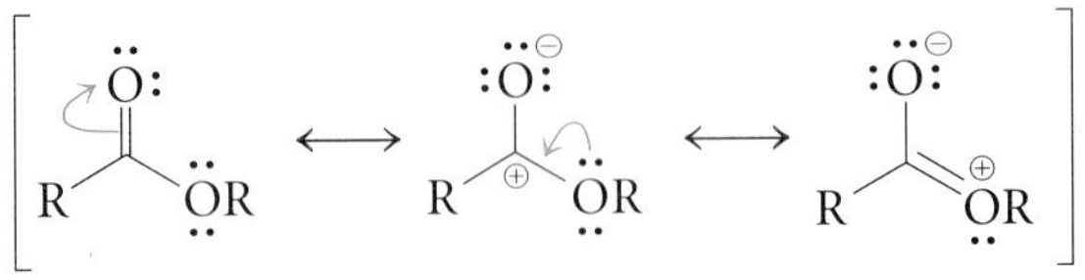

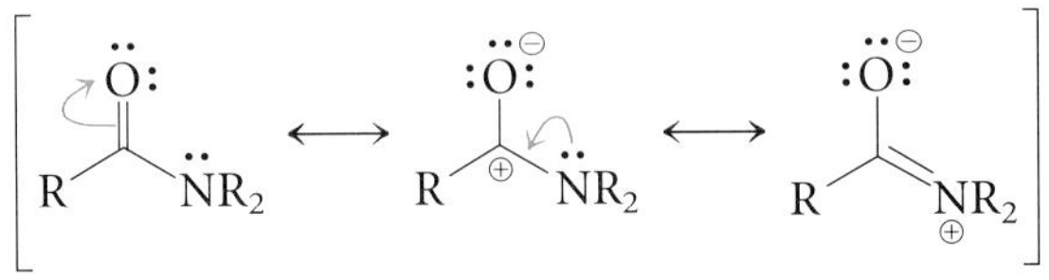

Resonance stabilization is even greater in amides because nitrogen is less electronegative than oxygen. On the other hand, resonance stabilization is weaker in thiol esters than in simple esters. This stabilization can be viewed as a contribution from the zwitterionic ester and amide resonance structures. Because this resonance stabilization is lost as the $\pi$ bond is broken and the nucleophile is added to the carbonyl carbon, resonance-stabilized reactants have higher activation energies and, consequently, are slower to react with nucleophiles (including complex metal hydrides) than are aldehydes or ketones.

The order of reactivity toward nucleophiles of several functional groups containing C=O bonds is shown in Figure 12.2. Aldehydes are more reactive than ketones, and both are more reactive than carboxylic acid derivatives. Because the hydrogen atom attached to the carbonyl carbon in an aldehyde is smaller than the second alkyl group of a ketone, and because hyperconjugative electron release from the additional alkyl group of a ketone stabilizes its carbonyl double bond relative to that of an aldehyde, activation energy barriers for nucleophilic attack are somewhat larger for ketones than for aldehydes. Therefore, aldehydes are more reactive than ketones. The relative reactivity of carboxylic acid derivatives is governed by how effectively the heteroatom bonded to carbon can release electrons to the carbonyl $\pi$ system. The third-level orbitals of sulfur are significantly larger than the second-level orbitals used by carbon for bonding. As a consequence, there is little overlap between the $2p$ and $3p$ orbitals and little resonance stabilization in a thiol ester, which is more reactive than a simple ester.

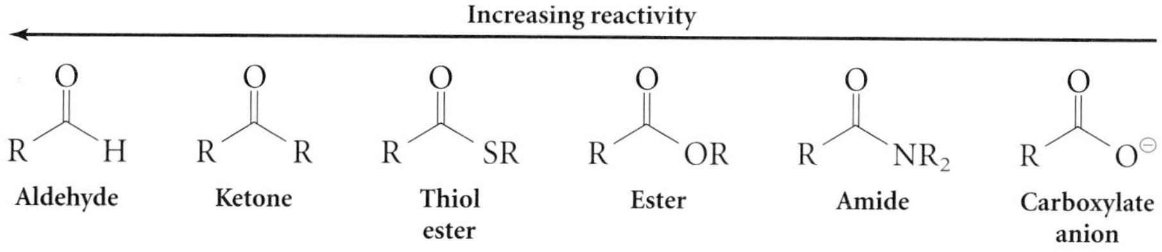

**FIGURE 12.2**

Order of reactivity of carbonyl compounds toward nucleophiles.

The relative reactivity of the various carbonyl groups has important consequences. Both sodium borohydride and lithium aluminum hydride reduce aldehydes and ketones, but the more reactive reagent, lithium aluminum hydride, is required for the reduction of thiol esters, esters, and amides.

**EXERCISE 12.6**

Recall that complex metal hydrides react with acids as weak as water to generate hydrogen in an acid–base reaction. Why are carboxylic acids much more difficult to reduce with $LiAlH_4$ than esters are?

**EXERCISE 12.7**

For each of the following pairs of compounds, consider whether it would be possible to reduce the first compound in a mixture of the two. If so, what reducing agent could be used to effect the selective reduction?

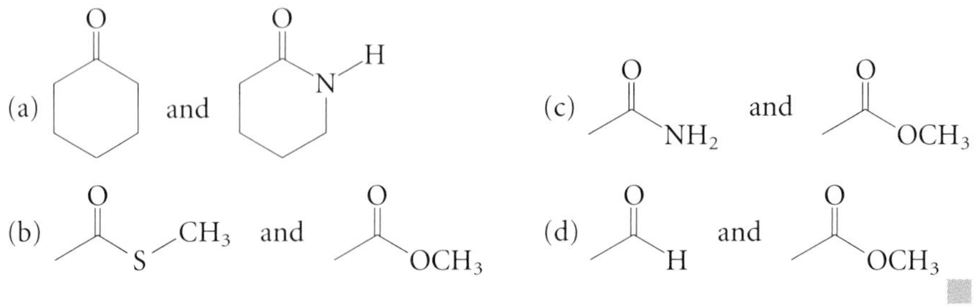

(a) [structure: cyclohexanone] and [structure: piperidin-2-one with N—H]

(c) [structure: acetamide, CH₃C(=O)NH₂] and [structure: methyl acetate, CH₃C(=O)OCH₃]

(b) [structure: CH₃C(=O)S—CH₃] and [structure: CH₃C(=O)OCH₃]

(d) [structure: acetaldehyde, CH₃C(=O)H] and [structure: methyl acetate, CH₃C(=O)OCH₃]

## 12.3

## Oxygen Nucleophiles

Unlike halide ions, oxygen nucleophiles productively attack carbonyl compounds. Common oxygen nucleophiles that effect such nucleophilic additions include water, hydroxide ion, and alcohols. In this section, we consider how these nucleophiles react with aldehydes and ketones to produce, respectively, hydrates, disproportionation products (through the Cannizzaro reaction), and acetals and ketals.

### Addition of Water: Hydrate Formation

For the reasons given in Section 12.5, halide ions do not effect nucleophilic addition to carbonyl groups. However, elements to the left of the halogens in the periodic table form stronger bonds to carbon than do the halogens and are less stable as free anions than are halide ions. These factors combine to increase the productivity of attack by oxygen nucleophiles.

*Base-Catalyzed Hydration.* Attack by hydroxide ion (an oxygen nucleophile) on an aldehyde leads to a **hydrate**:

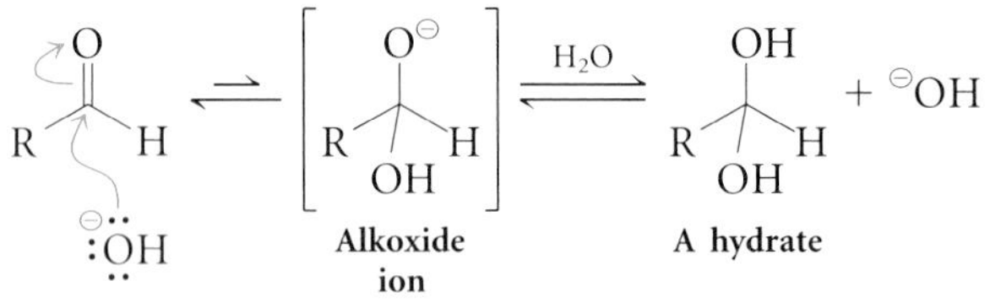

When hydroxide ion attacks the carbonyl group, the carbonyl oxygen becomes negatively charged, and an alkoxide ion is formed. There is little difference in the stability of the two negatively charged species, the tetrahedral alkoxide ion and the nucleophilic hydroxide ion. Although the carbon–oxygen $\pi$ bond of the aldehyde is still somewhat stronger (91 kcal/mole) than the carbon–oxygen $\sigma$ bond that replaces it (86 kcal/mole), the energy difference is smaller than it would be if a halide ion were the attacking nucleophile. Protonation of this intermediate alkoxide ion by water produces a geminal diol (a hydrate) and hydroxide ion. The overall reaction can thus be viewed as the addition of water across the $\pi$ bond, catalyzed by hydroxide ion. In fact, this hydration of an aldehyde (as well as that of a ketone) is very rapid in alkaline solution. Nonetheless, the product hydrate is generally less stable than the starting carbonyl compound,

## CHEMICAL PERSPECTIVES

### MICKEY FINN: WHO WAS HE?

The presence of electron-withdrawing groups on the carbon adjacent to a carbonyl group destabilizes the $\pi$ system. With sufficient destabilization, the energy of the aldehyde or ketone can be raised until the hydrate is more stable. For example, chloral (trichloroacetaldehyde) reacts exothermically with water to form the crystalline hydrate:

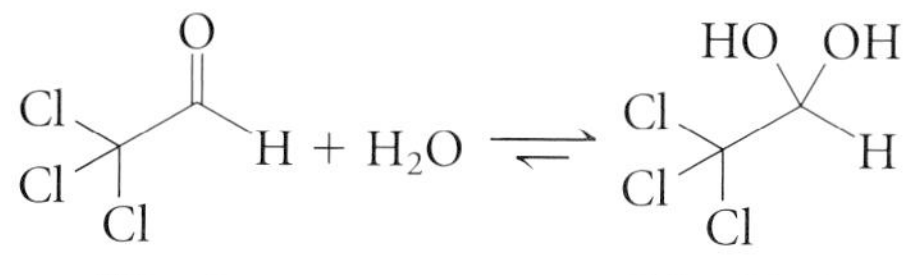


**Chloral** **Chloral hydrate**

Chloral hydrate is both a sedative and a hypnotic and has been used medicinally for these purposes. It has also been used for more sinister purposes: when secretly added to another person's drink, it yields a so-called Mickey Finn, a surreptitious knock-out. For this reason, chloral is a controlled substance.

and the equilibrium for this reversible reaction lies on the side of the carbonyl compound. For most aldehydes, there is no net chemical consequence of this reversible addition. However, the addition can be detected, and its rate can be measured by the use of water enriched in $^{17}O$ or $^{18}O$, since the heavy isotope of oxygen finds its way into the carbonyl group.

## EXERCISE 12.8

Propose a mechanism by which normal acetone (in which the oxygen is $^{16}O$) is labeled with $^{18}O$ upon treatment with $H_2O^{18}$ in the presence of acid.

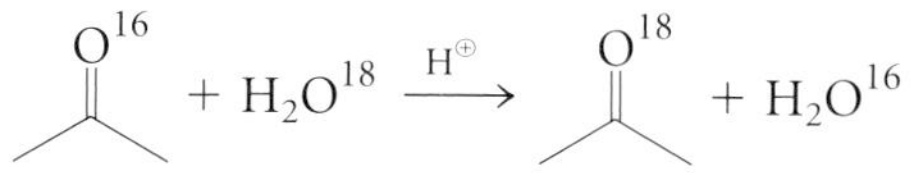


***Acid-Catalyzed Hydration.*** In acidic water (at low pH), the concentration of hydronium ion increases, and the concentration of hydroxide ion is correspondingly reduced. (Remember that $[H^{\oplus}][^{\ominus}OH] = 10^{-14}$.)

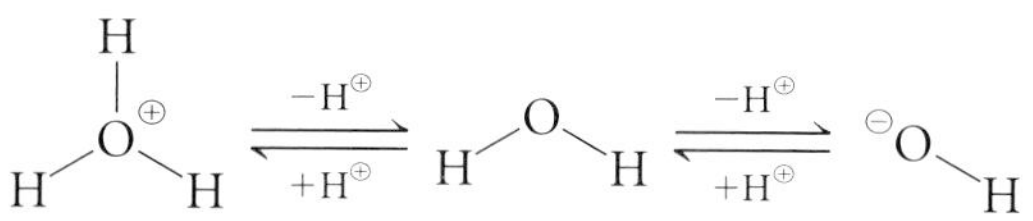


**Water**

Low pH High pH

The nucleophilicity of a neutral water molecule is much lower than that of a hydroxide ion, which should result in a correspondingly slower attack on a carbonyl group by water. (The hydronium ion is not nucleophilic because the oxygen bears a formal positive charge.) However, in acidic solution, water does add to the C=O bond.

**Aldehyde** — **Hydrate**

Here, the carbonyl group is rapidly and reversibly converted by protonation into a significantly better electrophile, one sufficiently reactive to be attacked by the less nucleophilic neutral water molecule. Thus, nucleophilic attack occurs in acidic solution, even with a weak nucleophile such as water, to generate an oxonium ion. Deprotonation of this species gives the hydrate. Here again, the net reaction is addition of water across the $\pi$ bond, but in a process catalyzed by acid.

An important point to be inferred from this comparison of two means of addition of water to an aldehyde is that nucleophilic attack can be accelerated under both basic and acidic conditions. In the presence of base, the nucleophile is deprotonated, resulting in an anion with enhanced nucleophilicity. In the presence of acid, the nucleophile is neutral (and thus less reactive than the anion), but the carbonyl compound can be activated toward reaction with the nucleophile by protonation on oxygen.

High pH

Low pH

**Hydrate**

***Equilibria Involving Carbonyl Compounds and Their Corresponding Hydrates.*** Each step in the addition of water is reversible and rapid, and thus an equilibrium is established between the carbonyl compound and its hydrate. This process can be catalyzed by either acid or base. By definition, a catalyst is not consumed in a reaction and can have no effect on the overall energetics of the reaction. The equilibrium position of hydration is therefore unaffected by whether an acid or a base is the catalyst. The position of this equilibrium is governed by the stability of the hydrate adduct relative to that of the starting carbonyl compound. Structural features in the starting material that stabilize the carbonyl group include the presence of electron-donating groups or, conversely, the absence of electron-withdrawing substituents. The relative amounts of four carbonyl compounds and their hydrates present at equilibrium are as follows (the significance of these equilibrium processes will become apparent when we turn our attention to sugars in Chapter 16):

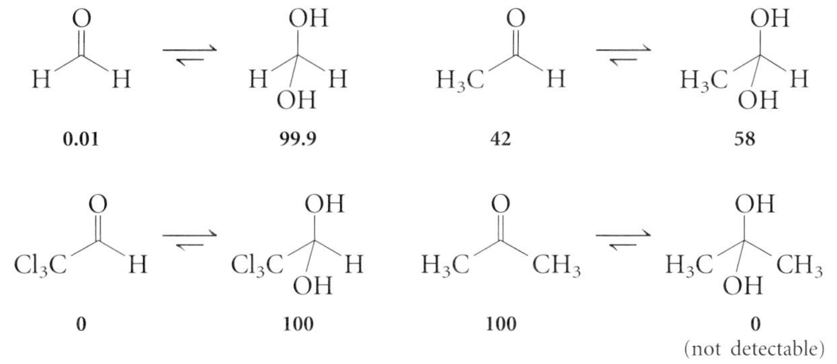

0.01       99.9       42       58

0       100       100       0
(not detectable)

## EXERCISE 12.9

For each of the following carbonyl compounds, predict whether the amount of hydrate present at equilibrium in aqueous acid is larger or smaller than that present when acetone is dissolved in the same acidic medium. Explain your reasoning clearly.

(a)       (b)       (c)

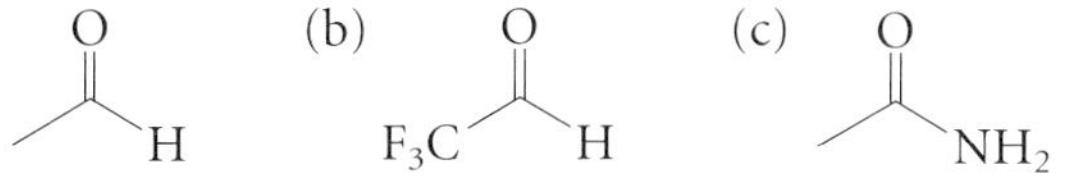

## Addition of Hydroxide Ion: The Cannizzaro Reaction and Hydride Transfer

Certain aldehydes, upon treatment with sodium or potassium hydroxide, are converted to equal amounts of the corresponding carboxylate anion and alcohol. This is known as the **Cannizzaro reaction.**

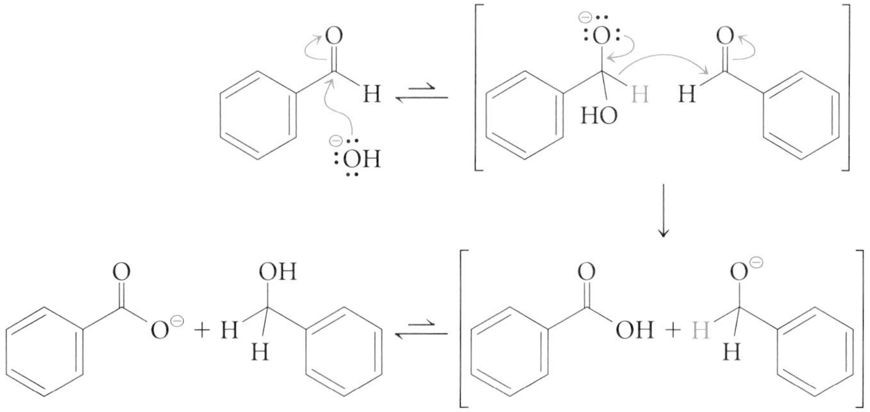

Carboxylate       Alcohol

This reaction works only with aldehydes (such as benzaldehyde) that lack $\alpha$ hydrogen atoms. For such aldehydes, $\alpha$-deprotonation by hydroxide ion cannot lead to an enolate anion, for which other reactions are possible, as we will see in Chapter 13. Instead, hydroxide ion adds to the carbonyl carbon. Reversal of this addition is quite rapid, but the reverse reaction simply reforms the starting material. In an alternative pathway, the carbonyl group can be regenerated from the tetrahedral intermediate if hydride ion (instead of hydroxide ion) is lost. Simple loss of hydride ion is not possible because this ion is very unstable as a result of its concentrated charge. However, hydride can be transferred simultaneously to an electrophile—in this case, benzaldehyde—by a route very similar to that for the complex metal hydride reductions presented in Section 12.2. Thus, as the carbonyl group is reformed from the tetrahedral intermediate derived from one molecule, hydride is transferred to a second molecule of starting material, so that a carboxylic acid and an alkoxide ion are formed. Proton transfer between these products generates a carboxylate anion and a neutral alcohol. This hydride transfer mechanism results in the oxidation of aldehyde that was initially subject to hydroxide attack and reduction of the aldehyde group to which the hydride was transferred.

The hydroxide ion–induced conversion of two molecules of benzaldehyde into one molecule of carboxylic acid and one of alcohol is a **disproportionation,** a reaction in which a species of intermediate oxidation level (an aldehyde) is both oxidized (to an acid) and reduced (to an alcohol). In the Cannizzaro reaction, this conversion is catalyzed by hydroxide ion.

***Hydride Transfer under Biological Conditions.*** In the Cannizzaro reaction, an aldehyde is reduced by the transfer of a hydride ion equivalent from a C—H bond. The biological cofactor nicotinamide adenine dinucleotide (NADH) is also a carbon-based reducing agent.

NADH
(reduced form)

NAD$^+$
(oxidized form)

Like the intermediate adduct in the Cannizzaro reaction, NADH delivers a hydride equivalent to the carbon of a C=O bond in the presence of a catalyst (the enzyme alcohol dehydrogenase). The resulting pyridinium ion (NAD$^+$) is aromatic, providing a driving force for the transfer of H$^\ominus$. The reverse transfer of a hydride from an alcohol to NAD$^+$ accomplishes alcohol oxidation. Note that this process is quite different from that discussed in Chapter 8 for metal-centered redox reactions.

### EXERCISE 12.10

Predict which of the following compounds can undergo a Cannizzaro reaction. If the reaction is possible, write structures for the expected products.

(a)

(b)

(c)

(d)

## Addition of Alcohols

The addition of an alcohol across the carbonyl $\pi$ bond of an aldehyde or ketone takes place by a pathway essentially identical to that for the addition of water. These additions can be catalyzed by either base or acid.

#12   Hemiacetal/Acetal
        Formation

*Formation of Hemiacetals and Acetals.* Hemiacetals and acetals are formed by reaction of alcohols with aldehydes. **Hemiacetals** are unstable substances that cannot be isolated; they exist only in solution. Their formation can be catalyzed by either base or acid. **Acetals** are stable, isolable compounds. Their formation from hemiacetals is acid-catalyzed.

*Base Catalysis of Hemiacetal Formation.* Let's first consider the reaction of an alcohol and an aldehyde in the presence of base. In this reaction, the attacking nucleophile is an alkoxide ion:

$$\text{ROH} \xrightarrow{\text{Li (or Na or K)}} M^{\oplus} \; {}^{\ominus}\!\!:\!\ddot{\text{O}}\text{R} + \text{H}_2 \qquad (M^{\oplus} = \text{Li}^{\oplus}, \text{Na}^{\oplus}, \text{or } \text{K}^{\oplus})$$

As we have seen, alcohols are acidic, and treatment with an alkali metal results in the formation of an alkoxide anion. (The reaction with potassium metal is so exothermic that dangerous conditions can result when potassium salts are made from primary or secondary alcohols in this way. The heat generated can cause the hydrogen that is formed to combine explosively with oxygen, forming water in a highly exothermic reaction.) Alternatively, smaller concentrations of alkoxide ions can be generated *in situ* by treatment of the neat alcohol with strong base, such as solid sodium hydroxide or potassium hydroxide. Alkoxide ions are effective nucleophiles and rapidly add to a carbonyl carbon. However, as in the reaction of an aldehyde with hydroxide, this first step is readily reversible, and the addition product is less stable than the starting carbonyl compound both because the bonds of the hemiacetal are weaker than those of the aldehyde and alcohol and because entropy favors the two molecules of starting material over the single product molecule.

*Acid Catalysis of Hemiacetal and Acetal Formation.* In contrast to catalysis by base, catalysis by acid does produce an observable product. Treatment of an aldehyde with an alcohol in the presence of acid leads to the formation of an acetal and water in a multistep process:

$$\underset{\text{Aldehyde}}{\text{O}} + 2\; \text{H}_3\text{COH} \xrightarrow{\text{H}_3\text{O}^{\oplus}} \underset{\text{Acetal}}{\overset{\text{H}_3\text{CO} \quad \text{OCH}_3}{\diagup\hspace{-0.3em}\backslash}} + \text{H}_2\text{O}$$

Aldehyde      Alcohol                              Acetal

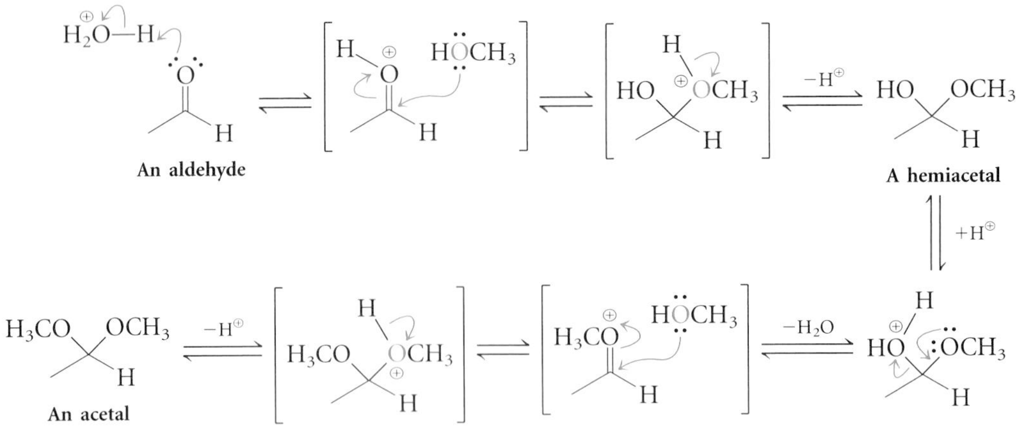

FIGURE 12.3

Mechanism for the acid catalyzed reaction of an aldehyde and an alcohol. The first stage produces a hemiacetal, and the second stage produces an acetal.

Because water is produced as a product, it is possible to "pull" the reaction toward the acetal by removal of the water as it is formed. The steps in the formation of an acetal (Figure 12.3) are quite similar to those for formation of a hydrate under acidic conditions.

In the presence of acid, protonation of the carbonyl oxygen activates the carbonyl group toward nucleophilic attack. An alcohol can therefore attack, producing a hemiacetal after deprotonation. Then a hydroxyl group of the hemiacetal is converted to a good leaving group by protonation by a relatively strong acid. The loss of water from the protonated hemiacetal is assisted by donation of a lone pair from the alkoxy oxygen, and an intermediate oxonium ion analogous to a protonated carbonyl group is formed.

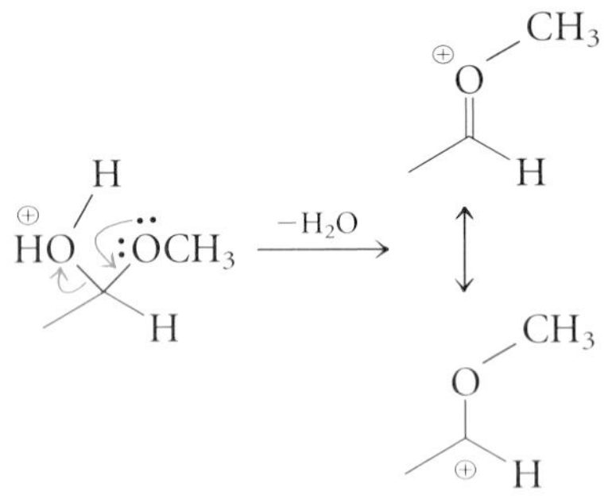

Addition of a second molecule of alcohol to the carbonyl carbon results in the formation of an oxonium ion intermediate. The sequence is finished by deprotonation, producing an acetal—a tetrahedral carbon with two geminal alkoxy groups. Hemiacetals, acetals, and analogous functional groups are of great importance in the chemistry of sugars and nucleic acids.

***Structure and Nomenclature of Carbonyl–Alcohol Adducts.*** A hemiacetal, like a hydrate, is thermodynamically less stable than the starting material. However, under acidic conditions, further transformations ultimately convert the aldehyde and two equivalents of alcohol into an acetal and water. A hemiacetal is a functional group with both an alkoxy and a hydroxy group attached to the same carbon atom; an acetal has two alkoxy groups on the same carbon atom. The corresponding products derived from a ketone are called a **hemiketal** and a **ketal.** Shown here are ball-and-stick representations of the methyl hemiacetal and methyl acetal of acetaldehyde and the methyl hemiketal and methyl ketal of acetone.

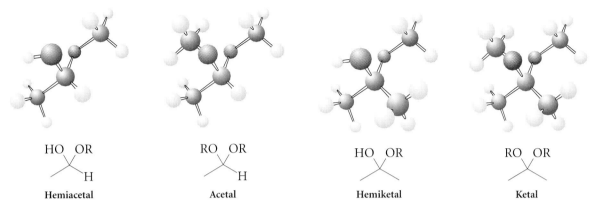

| HO   OR | RO   OR | HO   OR | RO   OR |
| :---: | :---: | :---: | :---: |
| H | H | | |
| **Hemiacetal** | **Acetal** | **Hemiketal** | **Ketal** |

Acetals and ketals are often named as derivatives of the underlying carbon skeleton rather than by terms that refer to acetal or ketal. For example, the dimethyl acetal of acetaldehyde is called 1,1-dimethoxyethane, and the dimethyl ketal of acetone is called 2,2-dimethoxypropane. The key feature that helps in recognizing acetals and ketals is the presence on the same carbon of two alkoxy substituents.

*Manipulating the Point of Equilibrium in Alcohol Addition.* All of the reactions involved in hemiacetal and acetal formation occur under equilibrium conditions—that is, the energy difference between reactants and products is sufficiently small that both forward and reverse reactions occur with ease. Using tabulated bond energies, we can calculate the enthalpy change for the formation of an acetal. Comparing the bonds present in the reactants (the carbonyl compound and two molecules of alcohol) with the bonds present in the products (acetal and water), we find the same change in bonding as occurs in the hydration of an aldehyde: the net change is that one carbon–oxygen $\pi$ bond is replaced by a new carbon–oxygen $\sigma$ bond. Because the $\pi$ bond is stronger than the $\sigma$ bond in most aldehydes and ketones, the formation of either the hydrate or an acetal from the carbonyl compound is an endothermic process.

$$\text{(acetaldehyde)} + H_2O \rightleftharpoons \underset{\textbf{Hydrate}}{\text{HO\ OH}}$$

$$\text{(acetaldehyde)} + 2\ ROH \rightleftharpoons \underset{\textbf{Acetal}}{\text{RO\ OR}} + H_2O$$

| **Bonds Lost** | | **Bonds Gained** | |
| :--- | :--- | :--- | :--- |
| C—O $\pi$ | 90 (176 − 86) | C—O $\sigma$ | 86 |
| 2 O—H | 222 | 2 O—H | 222 |
| | 312 | | 308 |

$\Delta H° = +4$ kcal/mol

The formation of the acetal differs from the formation of the hydrate in two important ways. First, the formation and hydrolysis of an acetal require acid catalysis, and the reaction can be stopped by the addition of base. Second, water is formed in addition to the acetal, and removal of the water physically (by azeotropic distillation, for example) will "pull" the endothermic formation of the acetal toward the right (recall Le Chatelier's principle). Alternatively, a reagent can be added that reacts exothermically with water to compensate for the energy required in acetal formation. Coupling of such a reaction with acetal formation makes the overall process exothermic. Examples of such reagents are $Na_2SO_4$ and molecular sieves, both of which react exothermically with water to form hydrates. *Molecular sieves,* also called *zeolites,* are mixed salts of silicon and aluminum oxides with $Li^{\oplus}$, $Na^{\oplus}$, and/or $K^{\oplus}$ as counterions. They crystallize from water as hydrates from which the water can be driven upon heating. The crystalline structure remains intact, leaving "holes" that are the right size for water, as in this simple zeolite with a framework of silicon (purple) and oxygen (red):

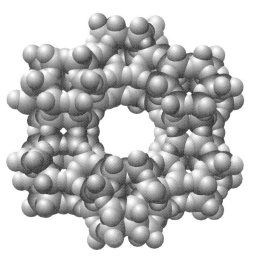

**A zeolite**

Under such conditions, aldehydes can be converted into acetals in good yield. The formation of the acetal is also favored when the alcohol is used as solvent. Under these conditions, the high concentration of starting material "pushes" the reaction toward the products.

The idea that reactions can be "pushed" or "pulled" toward the desired product is very important in biochemical transformations. Active living systems cannot be at thermodynamic equilibrium: an influx of starting materials (food) and expulsion of waste products are required for activity. As new starting materials are ingested, they are pushed toward product by the temporary increase in their concentrations. In turn, these products are starting materials for reactions. Many of these reactions are not highly exothermic (and some are even endothermic), but the constant flux of concentrations of reactants and products drives the various reactions involved in metabolic processes, which produce the energy required by living systems.

***Formation of Hemiketals and Ketals.*** The reactions just described for aldehydes also take place with ketones, forming the structurally analogous hemiketal and ketal functional groups:

The formation of acetals and ketals is reversible in the presence of aqueous acid. The mechanisms for the conversion of a ketal into a hemiketal and of a hemiketal into a ketone are identical with those for the conversion of an acetal into a hemiacetal and of a hemiacetal into an aldehyde.

*Acetals and Ketals as Protecting Groups.* Formation of an acetal from an aldehyde temporarily masks the characteristic reactivity of the carbonyl group and protects it from nucleophilic addition under basic conditions. After reaction of a nucleophile with other functional groups, the aldehyde can be regenerated by treating the acetal with aqueous acid. (Recall the bromination–debromination sequence for the protection of alkenes discussed in Chapter 10.)

**EXERCISE 12.11**

Write a complete mechanism for the formation of a ketal from acetophenone and ethylene glycol, $HOCH_2CH_2OH$. Then write a mechanism for the hydrolysis of the ketal to reform the starting materials.

## 12.4

# Nitrogen Nucleophiles

Nitrogen-containing compounds that bear a lone pair of electrons on nitrogen are active nucleophiles. Common nitrogen nucleophiles that attack carbonyl groups to effect nucleophilic addition include ammonia, primary and secondary amines, hydrazine derivatives, and hydroxylamine. In this section, we consider how these nucleophiles react with aldehydes and ketones to produce various nitrogen derivatives.

## Amines

Nitrogen functional groups are more basic and more nucleophilic than their comparable oxygen counterparts (for example, $NH_3$ versus $H_2O$), because nitrogen is less electronegative than oxygen. Therefore, the reactions of nitrogen nucleophiles with carbonyl compounds take place under less stringent conditions than are required for reactions of oxygen nucleophiles.

*Imine Formation.* The reaction of ammonia or a primary amine with an aldehyde or a ketone forms an **imine,** also known as a **Schiff base.** The reaction of ammonia with an aldehyde or a ketone begins with nucleophilic attack on the carbonyl carbon by a pathway similar to that for the addition of water to a carbonyl $\pi$ bond. However, the higher nucleophilicity of ammonia makes it unnecessary to employ either acid or base catalysis to initiate the reaction.

The lone pair of nitrogen attacks the carbonyl carbon to generate a tetrahedral zwitterionic intermediate. Rapid deprotonation at nitrogen and reprotonation at oxygen lead to a neutral species, but because this intermediate is in a protic medium, it can be protonated again, on either oxygen or nitrogen. Protonation on nitrogen is not productive, but protonation on oxygen sets the stage for loss of water and formation of a C=N bond using nitrogen's lone pair. The resulting iminium ion loses a proton to form a neutral **imine.** Imines of aldehydes and those derived from ammonia are particularly unstable. Not easily isolated, they are readily converted into the starting ketone and amine in the presence of water.

### EXERCISE 12.12

Write a full mechanism, using curved arrows, to illustrate how an imine is formed when acetone is treated with *n*-butylamine and mild acid:

*Reductive Amination.* Imines are reduced to amines by hydride reagents or by catalytic hydrogenation. An electron-deficient derivative of $NaBH_4$—namely, sodium cyanoborohydride, $NaBH_3CN$—is frequently used as the complex metal hydride in this reaction. The conversion of a carbonyl group to an amine through an intermediate imine is called **reductive amination.** Although a reducing agent is required only for the second step, the amine and metal hydride are added together so that the imine can be reduced as it is formed.

The presence of the electron-withdrawing cyano group on boron in sodium cyanoborohydride makes this reagent less reactive than sodium borohydride. Sodium cyanoborohydride does not reduce aldehydes and ketones. Indeed, it is the protonated imine that undergoes reduction in these reductive amination reactions and not the imine, which is also unreactive with this reducing agent.

**Sodium cyanoborohydride**

## EXERCISE 12.13

Suggest a mechanism for the reduction of acetone imine to 2-aminopropane by sodium cyanoborohydride.

***Enamines.*** The reaction of a secondary amine with an aldehyde or ketone forms an enamine:

$$\text{O} + \text{HNR}_2 \longrightarrow + \text{H}_2\text{O}$$

The reaction mechanisms for the formation of imines and enamines are quite similar. Addition of a primary or secondary amine to an aldehyde or ketone, followed by loss of water, results in formation of an iminium ion. The iminium ion derived from a primary amine can lose a proton either from nitrogen to form an imine or from the α-carbon to form an enamine:

**An imine**     **Iminium ion**     **An enamine**

With a secondary amine, the resulting iminium ion has no protons on nitrogen, so a proton is lost from carbon to form an enamine.

***Imine–Enamine Tautomerization.*** Imines that bear a hydrogen at the carbon α to the C=N bond (such as the imine derived from acetone) can tautomerize to generate an enamine, by protonation–deprotonation. In an enamine, a proton α to the original carbonyl carbon has been lost.

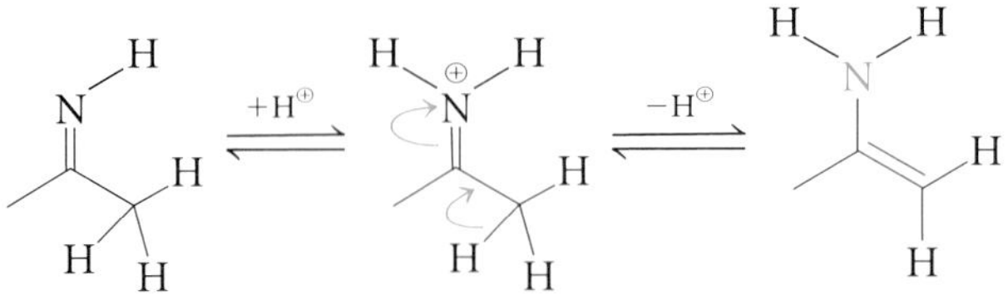

Imine            Enamine

This imine–enamine tautomerization is similar to the keto–enol tautomerization discussed in Chapter 3 and is catalyzed by either acid or base. The iminium ion is an intermediate in the interconversion of an imine to an enamine, and vice versa, a process that is quite rapid in the presence of acid. Except when there are very large alkyl groups on nitrogen, the imine tautomer is favored for the same reasons that the keto form is favored in a keto–enol tautomerization (see Chapter 3).

Ketones react with ammonia and primary and secondary amines to form imines (also known as Schiff bases) and enamines (Figure 12.4). An enamine is an important intermediate in that the α carbon bears significant negative charge (or electron density), much like an enol or enolate anion, and enamines undergo reactions as carbon nucleophiles.

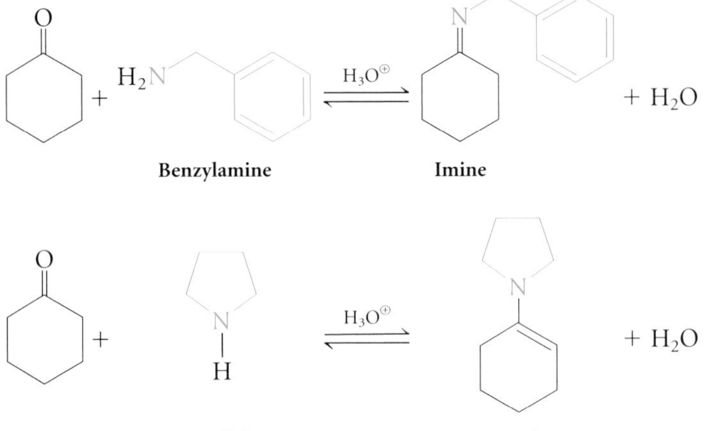

Benzylamine            Imine

Pyrrolidine            Enamine

**FIGURE 12.4**

Reaction of a ketone with a primary amine such as benzylamine yields an imine. Reaction of a ketone with a secondary amine such as pyrrolidine yields an enamine.

### ▨ Other Nitrogen Nucleophiles

***Formation of Derivatives.*** A number of other nitrogen-containing nucleophiles form derivatives of ketones. Before spectroscopic methods were available, solid, sharp-melting derivatives were valuable aids to the identification of organic compounds. Hydrazine, $H_2N\!-\!NH_2$, reacts with a ketone such as cyclohexanone to form a hydrazone by a mechanism identical to that for the formation of imines:

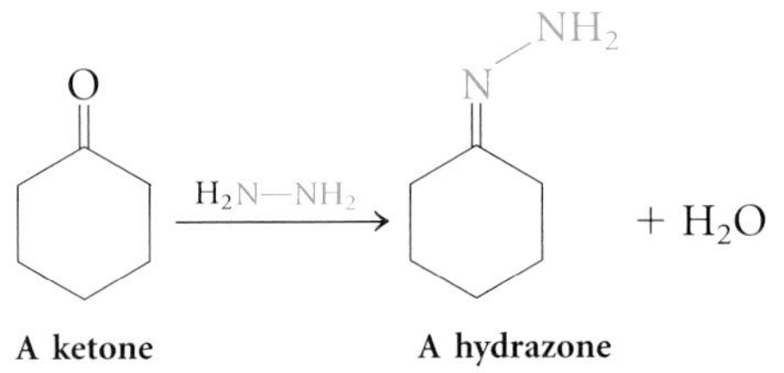

A ketone → A hydrazone

Hydrazones are often solids, and so the reaction with hydrazine is useful for converting a liquid carbonyl compound into a solid derivative that can be more easily characterized—for example, by its melting point.

Phenylhydrazines are used to make phenylhydrazones of ketones and aldehydes (Figure 12.5). The product of the reaction of a ketone with 2,4-dinitrophenylhydrazine is a 2,4-dinitrophenylhydrazone (2,4-DNP), a brightly colored (often red-orange) solid, whose formation is frequently used as a qualitative test for the presence of an aldehyde or ketone. Oximes and semicarbazones are derivatives (often solids) formed by the treatment of an aldehyde or ketone with hydroxylamine or semicarbazide.

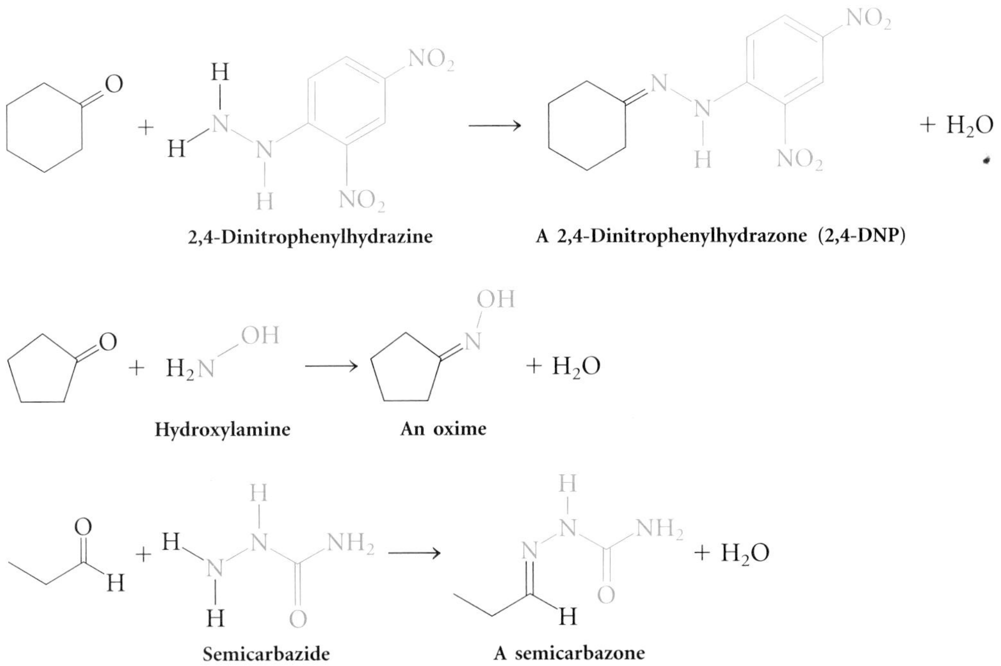

2,4-Dinitrophenylhydrazine → A 2,4-Dinitrophenylhydrazone (2,4-DNP) + $H_2O$

Hydroxylamine → An oxime + $H_2O$

Semicarbazide → A semicarbazone + $H_2O$

**FIGURE 12.5**

Ketones and aldehydes react with phenylhydrazines, hydroxylamine, and semicarbazide to form substitution products.

***Variations in Nucleophilicity of Nitrogen Nucleophiles.*** A number of the reagents used for derivatization (for example, hydrazine or semicarbazide) have more than one possible nucleophilic site available for reaction with a carbonyl group. In 2,4-dinitrophenylhydrazine, the terminal nitrogen (in blue) is more nucleophilic than the other amino nitrogen:

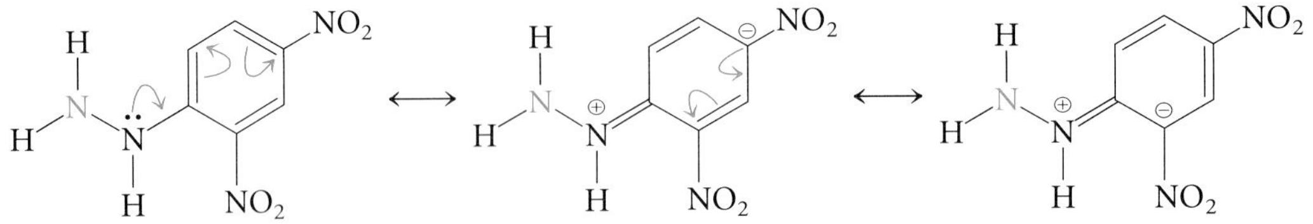

**2, 4-Dinitrophenylhydrazine**

The amino nitrogen attached to the aromatic ring has greatly diminished nucleophilicity because of delocalization of the lone pair of electrons into the aromatic $\pi$ system. This is evident in the two resonance contributors shown at the right. In these forms, the nitrogen atom adjacent to the ring bears positive charge, and negative charge is localized on the carbons bearing the nitro groups. (The nitrogen atom in an —$NO_2$ group is formally positively charged, as described in Chapter 11, and is not nucleophilic.) Because the electron density at the terminal amino group is not as greatly affected by the polarization in the aromatic ring, this group can more effectively serve as an active nucleophile.

Similarly, in semicarbazide, the two nitrogen atoms directly attached to the carbonyl carbons are resonance donors to the carbonyl oxygen. Therefore, the remaining nitrogen (shown in blue) is the more nucleophilic atom and ends up bonded to carbon in the semicarbazone.

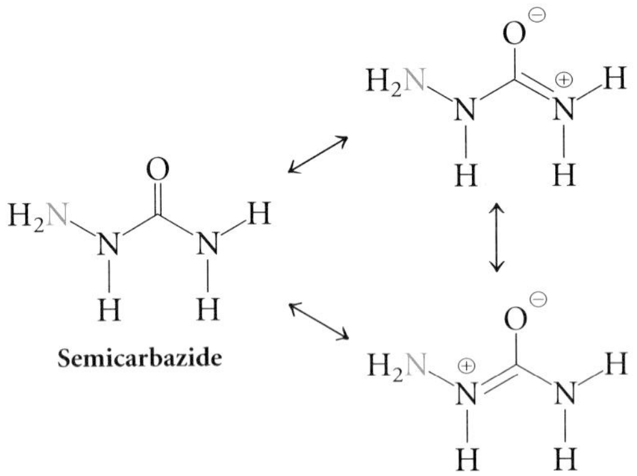

**Semicarbazide**

In hydroxylamine, both the oxygen and the nitrogen bear lone-pair electron density, and either could react as a nucleophile.

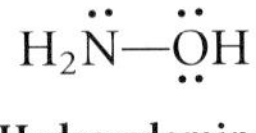

$$H_2\ddot{N}—\ddot{O}H$$

**Hydroxylamine**

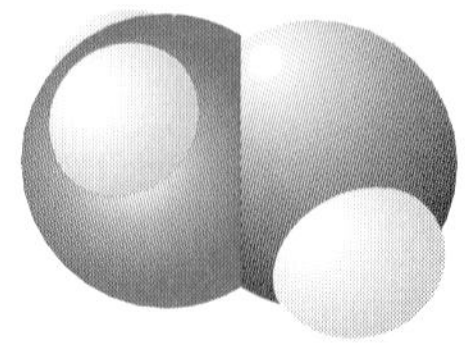

However, the less electronegative atom within a row of the periodic table is generally more basic and more nucleophilic. In fact, it is the nitrogen atom of hydroxylamine that reacts as a nucleophile, combining with aldehydes and ketones to form oximes (Figure 12.5).

618

Write a full mechanism, in parallel with that for imine formation, to illustrate how a hydrazone is formed when acetone is treated with hydrazine:

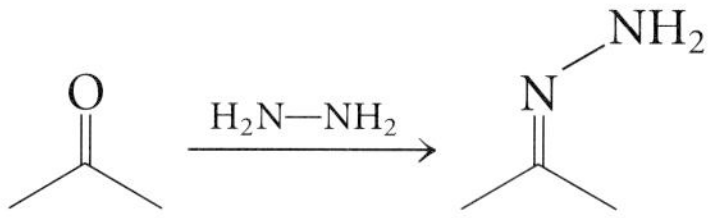

## 12.5

# Nucleophilic Acyl Substitution of Carboxylic Acids and Derivatives

In several families of carboxylic acid derivatives, a heteroatom is bonded to a C=O group. Examples include carboxylic acid chlorides, anhydrides, esters, thiol esters, and amides. Let's consider a general scheme in which a heteroatomic nucleophile, $X^{\ominus}$, attacks the carbonyl carbon of a carboxylic acid derivative whose heteroatom is represented by Y:

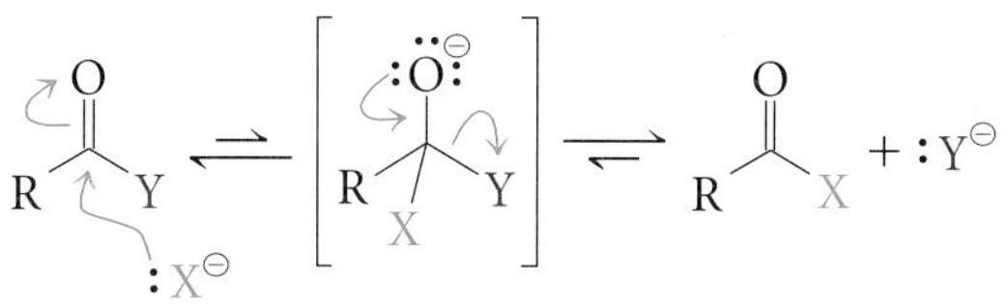

#04   Carboxylic
       Acids/Derivatives

In this reaction, one carboxylic acid derivative, RCOY, is converted to another, RCOX, by a process called **nucleophilic acyl substitution.** (Recall that —COR is an acyl group.) These conversions can be accomplished either by the use of a good nucleophile, as is implied by the negatively charged species $X^{\ominus}$, or by enhancing the electrophilicity of the starting material by protonation (Figure 12.6). Each step in a nucleophilic acyl substitution

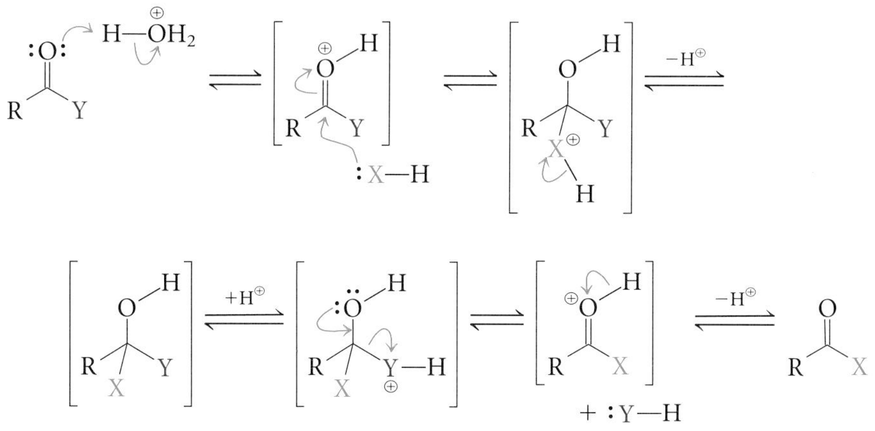

**FIGURE 12.6**

Protonation of the carbonyl oxygen of a carboxylic acid derivative converts it to a cation that is more easily attacked by nucleophiles.

reaction is potentially reversible, and the position of the overall equilibrium is determined mainly by the relative stabilities of the reactant and product carboxylic acid derivatives. Other important factors are the relative stabilities of $X^\ominus$ and $Y^\ominus$ and the relative strengths of the H—X and H—Y bonds formed when the reaction takes place in an acidic medium.

## Relative Stability of Carboxylic Acid Derivatives

It is useful to review the order of stability of various carboxylic acid derivatives (Figure 12.7). With regard to nucleophilic attack, the most reactive derivative is the acid chloride; the least reactive is the carboxylate ion. In all cases, reactivity is determined by the degree of electron delocalization from the heteroatom (for example, chlorine or nitrogen) into the carbonyl $\pi$ system. For the acid chloride, this delocalization is of minor importance, both because chlorine is relatively electronegative and because it is larger than carbon (in a different row in the periodic table). Consequently, the relevant orbitals are mismatched in size. Sulfur is about the same size as chlorine, although less electronegative; thiol esters, therefore, are much more stable than carboxylic acid chlorides. The central oxygen of a carboxylic acid anhydride must interact equally with each of the adjacent carbonyl groups; therefore, the donation of its lone pairs is less extensive than in an ester. Carboxylic acid esters have a slightly greater resonance stabilization than do neutral carboxylic acids, because the alkyl group of an ester releases electron density to the ester oxygen by hyperconjugation. Amides are quite stable, because nitrogen is similar in size to carbon and is less electronegative than oxygen. The carboxylate ion is the most stable of all carboxylic acid derivatives, because its two resonance forms have identical energy.

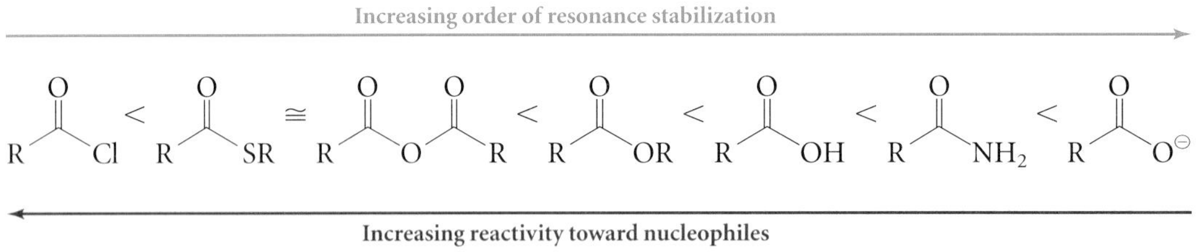

FIGURE 12.7

Order of stability of the various carboxylic acid derivatives.

### EXERCISE 12.15

Draw all significant resonance structures for each species:

(a) a carboxylic acid chloride

(b) a primary amide

(c) a carboxylate anion

It is possible to convert a more reactive derivative into a less reactive one by simple nucleophilic acyl substitution under equilibrium conditions, but the reverse conversion is not possible under the same conditions. For example, an amide can be prepared easily from an acid chloride, but the conversion of an amide into the corresponding acid chloride cannot be accomplished directly. Similarly, an ester can be converted into an amide, but the reverse conversion cannot take place directly. However, each of these acid derivatives, can be hydrolyzed under alkaline conditions, producing a carboxylate ion. We will see shortly how a carboxylate ion can be converted to an acid and then to a carboxylic acid chloride, allowing a cycle of interconversion through the entire range of carboxylic acid derivatives.

## Interconversion of Carboxylic Acid Derivatives

*Hydrolysis of Carboxylic Acid Derivatives.* All carboxylic acid derivatives can be hydrolyzed under acidic conditions to carboxylic acids and under basic conditions to carboxylate ions. In most cases, the equilibrium is driven toward the carboxylic acid (or carboxylate ion) by the greater resonance stability of the product formed. As an example of this class of reactions, let's consider the hydrolysis of a thiol ester under basic and under acidic conditions.

*Base Hydrolysis of Thiol Esters.* In base, hydroxide ion attacks the carbonyl carbon of the thiol ester, forming a negatively charged tetrahedral intermediate. In the space-filling model of methyl thiol acetate, note the large size of the sulfur (yellow) as compared with the neighboring carbon:

Methyl thiol acetate

The tetrahedral intermediate can either revert to starting material by rupture of the C—O bond just formed or it can proceed to product by cleavage of the C—S linkage. However, the reaction is not complete at this point, because the leaving group, the thiolate anion, is sufficiently basic to convert the carboxylic acid into a carboxylate anion almost quantitatively.

| Thiol ester | Tetrahedral intermediate | Carboxylic acid | Carboxylate anion |

An energy diagram for the steps involved in the process of hydrolysis is shown in Figure 12.8 (on page 622). The carboxylic acid lies lower in energy than the thiol ester, and deprotonation of the acid forms the carboxylate ion, which is further stabilized by resonance delocalization. Thus, under basic conditions, thiol esters are converted essentially quantitatively into carboxylate ions. This is not a base-catalyzed reaction because the original nucleophilic species, the hydroxide ion, is consumed: for each equivalent of thiol ester produced, one equivalent of hydroxide must be used. Such reactions are referred to as **base-induced reactions.**

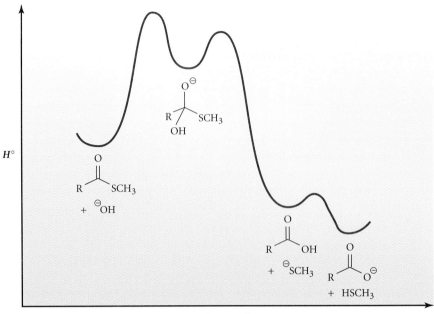

**FIGURE 12.8**

Energy diagram for base-induced hydrolysis of a thiol ester. The process is
highly exothermic because of the high stability of the carboxylate anion.

*Acid Hydrolysis of Thiol Esters.* Under acidic conditions, the major features of the hydrolysis reaction of a thiol ester are the same as under basic
conditions, although the details differ (Figure 12.9). The reaction is initiated by the transfer of a proton from the acidic medium to the carbonyl
oxygen of the thiol ester. Reaction with neutral water as a nucleophile then

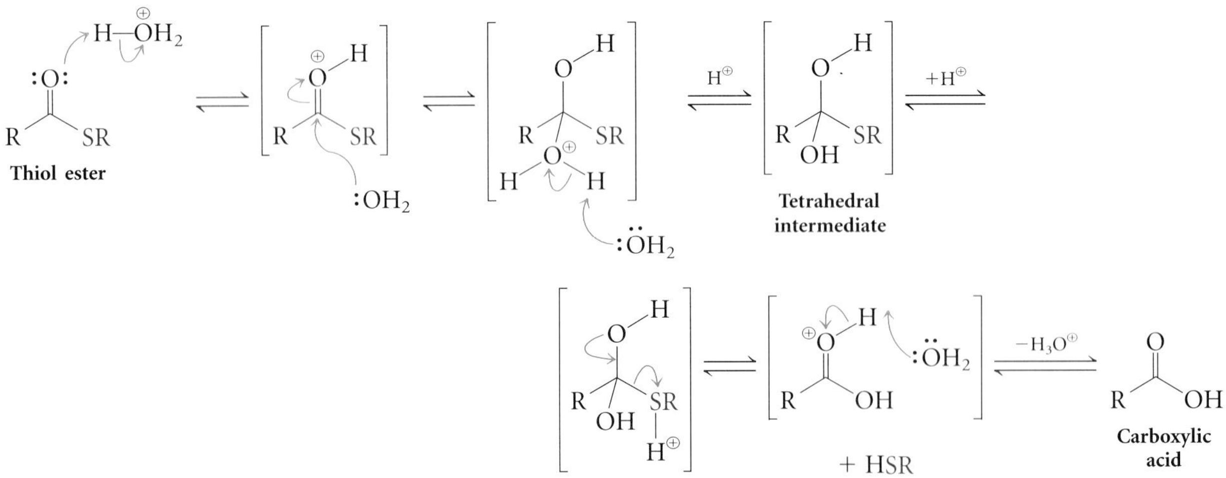

**FIGURE 12.9**

Acid-catalyzed hydrolysis of a thiol ester to a carboxylic acid. The steps include protonation of the carbonyl group, formation of a tetrahedral intermediate, and reprotonation and deprotonation.

leads to a positively charged tetrahedral intermediate, which rapidly loses a proton to form a neutral species. Reprotonation on sulfur provides an opportunity to form the carbonyl group once again by expulsion of a thiol. Loss of a proton from the initially formed species produces the carboxylic acid. Although this sequence includes a number of protonation and deprotonation steps, they balance overall, so that there is no net consumption of acid. Therefore, the process is acid-catalyzed, in contrast with the similar base-induced reaction.

***Interconversion of Carboxylic Acids and Esters.*** The hydrolysis of a thiol ester to a carboxylic acid is an example of a reaction in which the products are more stable than the reactants, in part because of greater resonance stabilization in the carboxylic acid compared with the thiol ester. Analogous interconversions between other acid derivatives in which the reactant and product are more similar in energy (for example, between a carboxylic acid ester and the corresponding carboxylic acid) are reactions in which the position of the equilibrium can be controlled by the use of an excess of one reagent (the acid is favored in water) or the removal of one product (the ester is favored when water is at low concentration). Under different conditions, it is possible to hydrolyze an ester or esterify a carboxylic acid.

*Shifting the Equilibrium in Acid-Catalyzed Ester Hydrolysis.* Let's consider an acid-catalyzed hydrolysis of an ester to a carboxylic acid—the reaction of methyl acetate and water to form acetic acid and methanol:

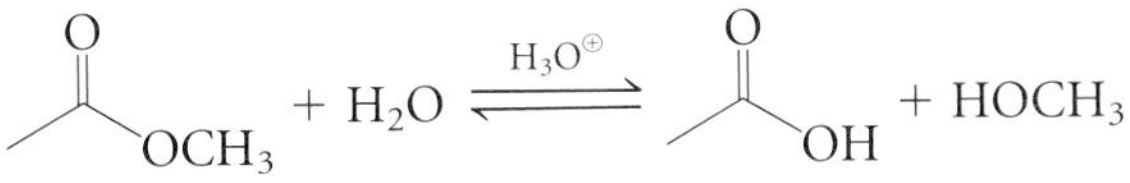

Methyl acetate          Acetic acid     Methanol

Because the difference in resonance stabilization between the reactant and product is small, the equilibrium can be shifted by the application of Le Chatelier's principle. As discussed in Chapter 6, an equilibrium describes a state in which the rate of conversion of starting material to product exactly equals the rate from product back to starting material. These rates are influenced not only by the relative activation energies, but also by the concentrations of the species required for the forward and reverse reactions. Thus, the equilibrium can be shifted toward the carboxylic acid by using water as solvent, and the equilibrium of the reverse reaction can be shifted toward the ester by using the alcohol as solvent. The same acid-catalyzed pathway can therefore be used for both ester hydrolysis and acid esterification.

*Base Hydrolysis: An Irreversible Reaction.* Nucleophilic attack by hydroxide ion on an ester produces a carboxylic acid under alkaline conditions. However, the acid thus produced is rapidly converted to a carboxylate anion by reaction with the alkoxide ion simultaneously produced or with a second hydroxide ion:

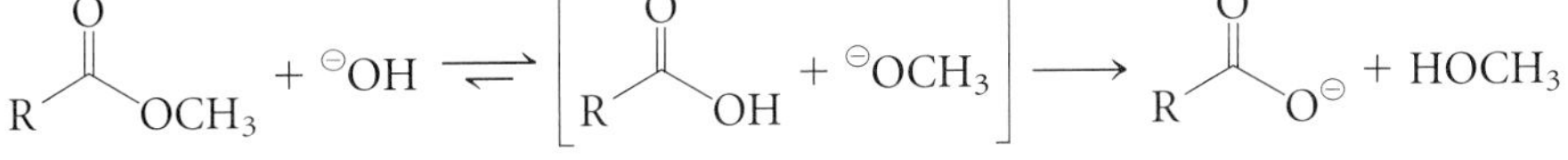

Ester                    Carboxylic   Alkoxide        Carboxylate   Methanol
                         acid         ion             anion

The carboxylate anion has the greatest resonance stabilization of all of the carboxylic acid derivatives, because there are two equivalent resonance contributors to the hybrid structure. Therefore, the equilibrium greatly favors the carboxylate anion and methanol. Indeed, the energy difference between the ester and the carboxylate anion is sufficiently large that this reaction is often described as irreversible. The difference in stability between a carboxylate anion and the other carboxylic acid derivatives is sufficiently large that it is not possible to proceed from the carboxylate anion to an ester or any other carboxylic acid derivative under most standard conditions. Basic conditions can thus be used to hydrolyze an ester but not to esterify an acid.

### EXERCISE 12.16

One method for driving an acid-catalyzed esterification reaction is to remove water by the use of a Dean–Stark trap. In this apparatus, the reaction mixture is heated under reflux, but the condensing vapors do not return directly to the flask. Instead, they are diverted to a side-arm in which the condensed solvent is collected before it is returned to the distillation pot. When a water-immiscible solvent that forms a low-boiling azeotrope with water is used (an example is benzene), water is removed from the flask and transferred to the bottom of the side-arm. (Recall that benzene is less dense than water.) Explain how Le Chatelier's principle applies to the use of this apparatus.

*Transesterification.* The interconversion of one carboxylic acid ester into another is called **transesterification.** For example, a methyl ester can be converted to an ethyl ester under acidic conditions:

$$
\underset{\textbf{Methyl ester}}{R\!\!-\!\!\overset{O}{\overset{\|}{C}}\!\!-\!\!OCH_3} \; + \; \underset{\textbf{Ethanol}}{HOCH_2CH_3} \; \underset{}{\overset{H^{\oplus}}{\rightleftharpoons}} \; \underset{\textbf{Ethyl ester}}{R\!\!-\!\!\overset{O}{\overset{\|}{C}}\!\!-\!\!OCH_2CH_3} \; + \; \underset{\textbf{Methanol}}{HOCH_3}
$$

Because the two esters are of comparable energy, the principal factor affecting the equilibrium position is the relative concentrations of the two alcohols. When one alcohol is more volatile than the other, the course of the reaction can be directed by removal of the low-boiling alcohol, once again applying Le Chatelier's principle. Because methanol boils at a lower temperature than does ethanol, methyl esters can be efficiently converted to ethyl esters by carrying out the reaction in ethanol at a temperature at which methanol boils off as it is produced.

### EXERCISE 12.17

Write a detailed, stepwise mechanism for the transesterification of ethyl acetate with methanol in acid to form methyl acetate and ethanol:

$$
\underset{}{CH_3\!\!-\!\!\overset{O}{\overset{\|}{C}}\!\!-\!\!OCH_2CH_3} \; \underset{CH_3OH}{\overset{H_3O^{\oplus}}{\longrightarrow}} \; \underset{}{CH_3\!\!-\!\!\overset{O}{\overset{\|}{C}}\!\!-\!\!OCH_3} \; + \; CH_3CH_2OH
$$

Transesterification can also be accomplished, in principle, by base catalysis. However, acid conditions are preferred for practical reasons: under al-

## WHAT AIRPORT BEAGLES KNOW ABOUT ESTERIFICATION

The structural differences between morphine, codeine, and heroin are relatively minor. Morphine has a phenolic —OH group and a secondary allylic —OH group. In codeine, the phenolic —OH group has been converted to —OCH₃. In heroin, both —OH groups are acetylated. Morphine occurs naturally in the opium poppy and accounts for as much as 40% of the dried weight of sap collected from the seed pods. (The common poppy flower has none of this alkaloid.) Illicit drug laboratories prepare heroin by treating morphine with acetic anhydride. Because acetic acid is produced as a by-product, drug enforcement agents can seek out these covert laboratories using dogs specially trained to recognize the characteristic pungent odor of acetic acid, even in very low concentrations.

Morphine

Codeine

Heroin

kaline conditions, esters undergo not only nucleophilic addition reactions, but also α-deprotonation and complicating side reactions. (We will explore the latter reactions in Chapter 13.)

*Amide Hydrolysis.* Amides can be hydrolyzed under both acidic and basic conditions. However, because of the high stability of the amide functional group, hydrolysis is quite slow. For practical purposes, it is often best to employ acidic conditions; the reaction is driven by protonation of the amine generated, forming an ammonium ion. Because acid is consumed in this process, hydrolysis of amines under acidic conditions is acid-induced, not acid-catalyzed.

Methylamine          *N*-Methylacetamide          Acetic acid     Methylammonium ion

The resistance of carboxylic acid amides to hydrolysis is an important characteristic of the amide functional group. Proteins and peptides are large molecules made up of smaller ones joined by amide linkages, and the stability of the C(O)—NH bond is of great biochemical significance.

## EXERCISE 12.18

Write a mechanism for the conversion of *N*-methylacetamide to acetic acid and methylammonium chloride by treatment with HCl in water:

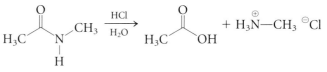

N-Methylacetamide

---

### EXERCISE 12.19

As in all nucleophilic acyl substitution reactions, addition of hydroxide ion to an amide produces a tetrahedral intermediate. Draw this species for acetamide. Then write the reaction for the further transformation of this intermediate to the carboxylic acid and the reaction for the reversion of this intermediate to the starting ester by loss of hydroxide ion. Which reaction is faster? Explain your reasoning. Construct an energy diagram for the conversion of acetamide to acetate ion in aqueous sodium hydroxide.

---

#14    Acetyl Chloride and
       Methanol

## Carboxylic Acid Chlorides

***Substitution Reactions of Acid Chlorides.*** Because carboxylic acid chlorides are the most reactive (and least stable) of the carboxylic acid derivatives, they can be converted readily into any of the other derivatives. For example, carboxylic acid chlorides react with water at neutral pH to form carboxylic acids and HCl. Unlike other carboxylate derivatives, the acid chloride is sufficiently reactive that neither protonation of the carbonyl oxygen atom nor use of a highly nucleophilic species such as hydroxide ion is required to bring about hydrolysis by nucleophilic acyl substitution. Similarly, acid chlorides are easily converted into esters upon treatment with alcohol and into amides by treatment with ammonia or a primary or secondary amine.

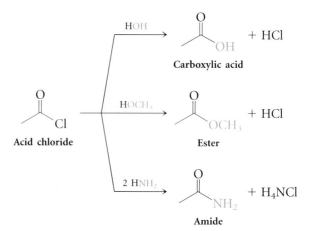

Because of the ease with which they can be converted into other carboxylic acid derivatives, acid chlorides play an important role in carboxylic acid chemistry. If it is desirable to avoid strongly acidic conditions, a weak, poorly nucleophilic base such as pyridine can be incorporated into the reaction medium for the formation of a carboxylic acid amide from an acid chloride and an amine:

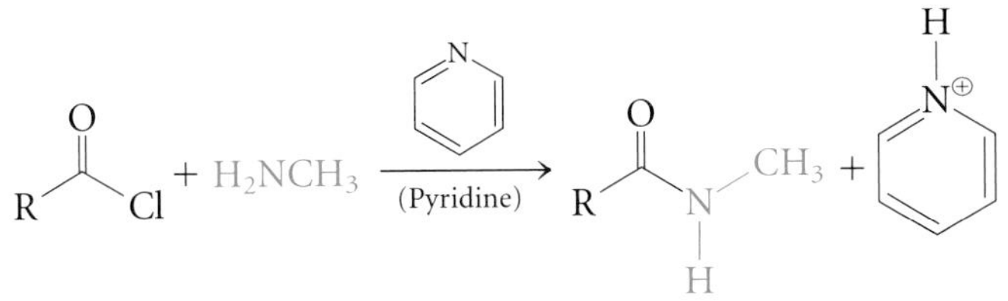

Without pyridine, the formation of an amide from an amine and an acid chloride would require two equivalents of the amine, because one would be consumed in neutralizing the HCl produced, forming the ammonium salt.

***Preparation of Acid Chlorides.*** Carboxylic acid chlorides are the least stable of the carboxylic acid derivatives; therefore, they cannot be formed directly from other acid derivatives by simple nucleophilic acyl substitution. Instead, a highly activated sulfurous acid derivative is often employed. One of the most common techniques for preparing carboxylic acid chlorides is treatment of a carboxylic acid with thionyl chloride, $SOCl_2$. (Recall that this reagent is also useful for converting alcohols into alkyl chlorides.)

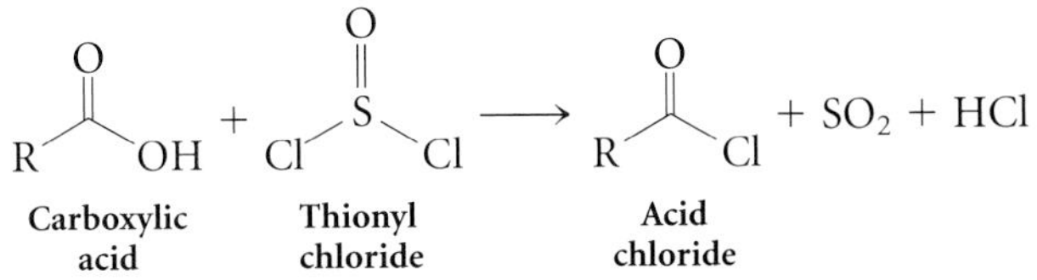

Let's consider the mechanism of the reaction of a carboxylic acid with thionyl chloride. In the first step, the carbonyl oxygen acts as a nucleophile, attacking the sulfur of thionyl chloride and forming a tetravalent sulfur intermediate (Figure 12.10). The intermediate is quite similar to the tetrahe-

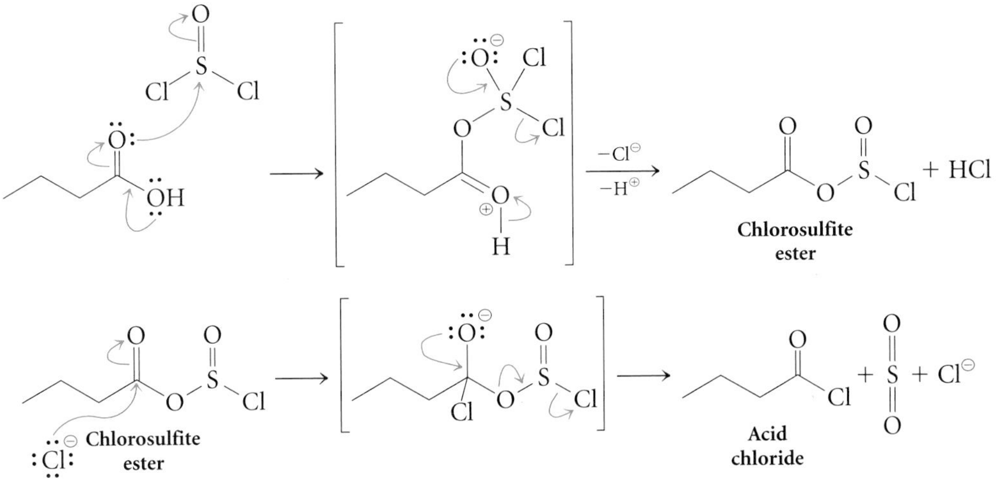

**FIGURE 12.10**

The reaction of a carboxylic acid with thionyl chloride yields an acid chloride. The reaction proceeds through a sulfur intermediate, a chlorosulfite ester, and an acyl chlorosulfite anhydride intermediate.

dral intermediate formed in the reactions of carboxylic acid derivatives. In an analogous fashion, chloride ion is lost from this intermediate, reforming the sulfur–oxygen $\pi$ bond and producing a second intermediate, an acyl chlorosulfite anhydride. This attack achieves the overall replacement of a chlorine atom in thionyl chloride by the acyloxy unit of the carboxylic acid. In the next step, the freed chloride ion acts as a nucleophile, attacking the activated carbonyl group. Unlike the attack by a halide ion on an aldehyde or ketone, which simply reverses, this attack forms a tetrahedral intermediate in which an even better leaving group ($-SO_2Cl$) is present. Collapse of this tetravalent intermediate forms the acid chloride, sulfur dioxide, and another chloride ion. Overall, this transformation takes a relatively stable carboxylic acid species (the acid itself) to its most reactive and least stable derivative, the acid chloride.

When used in conjunction with the other reactions of carboxylic acid derivatives, this reaction can be used to convert any carboxylic acid derivative into another, even if the product is less stable. For example, by a sequence of reactions, a more stable carboxylic acid amide can be converted into a less stable thiol ester. First, the amide can be hydrolyzed to a carboxylic acid, which can be converted into the carboxylic acid chloride by treatment with thionyl chloride. Finally, the reaction of the acid chloride with an alkyl thiol produces a thiol ester.

Overall, the conversion of an amide into the corresponding thiol ester is an uphill process. The driving force comes from the quite exothermic conversion of thionyl chloride into sulfur dioxide and HCl as the acid chloride is produced.

***Energetics of Acid Chloride Preparation.*** Thionyl chloride is structurally similar to a carboxylic acid chloride and is highly reactive. The conversion of a carboxylic acid to an acid chloride by the action of thionyl chloride is accompanied by the formation of two very stable, small molecules, $SO_2$ and HCl, both of which are gases at room temperature. Both entropy and enthalpy provide a strong energetic driving force for the otherwise thermodynamically unfavorable direct conversion of the carboxylic acid into its chloride. Furthermore, because $SO_2$ rapidly escapes from the reaction as a gas, the reverse reaction is impossible. Thus, an unfavorable reaction (conversion of the carboxylic acid into the acid chloride) is driven by the simultaneous conversion of thionyl chloride into $SO_2$ and HCl. These two reactions are not independent, and one cannot occur without the other. Chloride ion is required for the first, and water for the second. However, we can write two partial, or "half," reactions:

**Net reaction:**

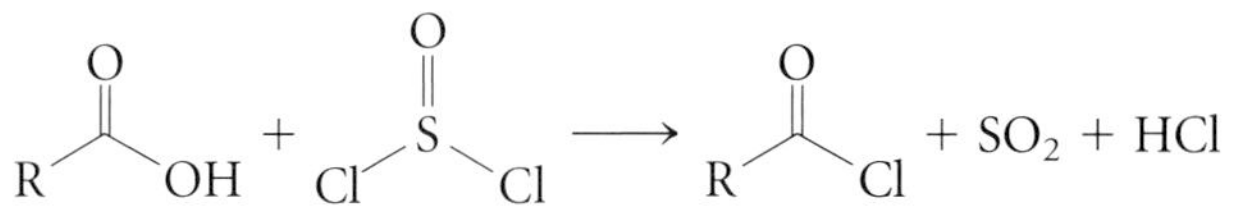

$$R-COOH + Cl-S(=O)-Cl \longrightarrow R-CO-Cl + SO_2 + HCl$$

The first reaction is quite endothermic, and the second reaction is exothermic. We can view these two reactions as coupled reactions for the formation of the acid chloride: the exothermic reaction provides sufficient energy to enable the endothermic reaction to occur. Coupled reactions are often depicted with intersecting, curved reaction arrows (*not to be confused with electron-flow arrows!*):

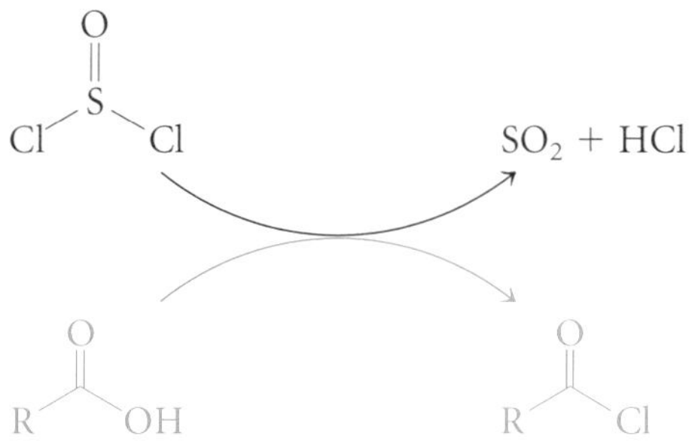

Coupled reactions are important in considering the energetics of biochemical transformations.

### EXERCISE 12.20

For each of the following pairs, choose the substrate that is more readily hydrolyzed. Give reasons for your choices.

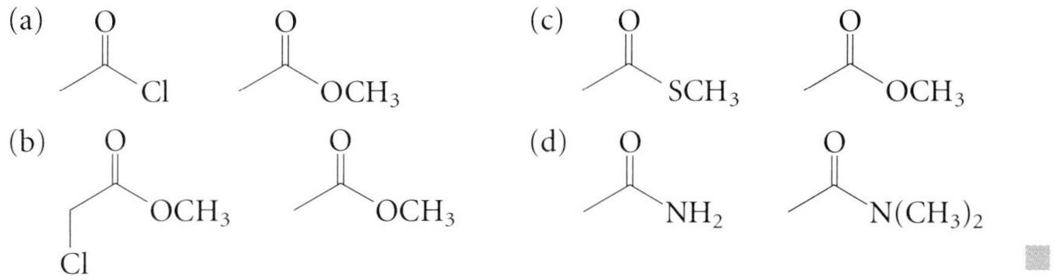

(a) [acetyl chloride] [methyl acetate]

(b) [methyl chloroacetate] [methyl acetate]

(c) [S-methyl thioacetate] [methyl acetate]

(d) [acetamide] [N,N-dimethylacetamide]

### Reactions of Acid Anhydrides

An acid anhydride is similar in reactivity to an acid chloride and can be formally derived from the dehydration of two moles of a carboxylic acid.

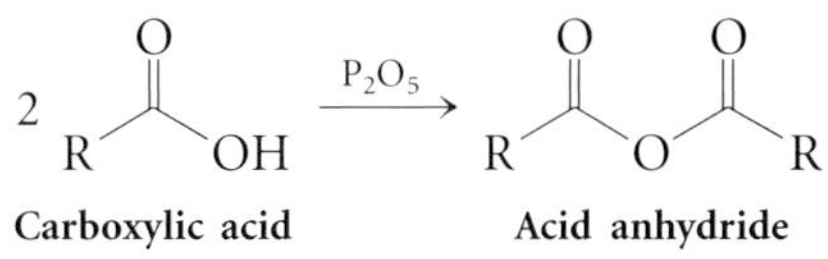

$$2\ R-COOH \xrightarrow{P_2O_5} R-CO-O-CO-R$$

Carboxylic acid          Acid anhydride

Although $P_2O_5$ is generally an effective desiccant and is sometimes used for this purpose, acyclic anhydrides are more difficult to prepare than the correspondingly activated acid chlorides. (Acetic anhydride is one of only a handful of acyclic anhydrides that are commercially available.) The difficulty in preparing anhydrides is caused by the high exothermicity of the hydrolysis of an anhydride linkage.

$$R-\overset{O}{\underset{}{C}}-O-\overset{O}{\underset{}{C}}-R \xrightarrow{H_2O} 2 \ R-\overset{O}{\underset{}{C}}-OH$$

Similar hydrolyses of the structurally analogous phosphoric acid anhydrides are important biologically and constitute a major method by which energy is stored in living organisms.

### Hydrolysis of Nitriles to Carboxylic Acids

Nitriles are not formally carboxylic acid derivatives because they lack a carbonyl group. However, you should not forget that the carbon atom of a nitrile is at the same +3 oxidation level as that of a carboxylic acid. Indeed, nitriles can be readily converted to carboxylic acids and ammonia by acid-catalyzed addition of water (Chapter 10). Hydrolysis first yields an amide; then further treatment with aqueous acid, under more vigorous conditions, results in hydrolysis of the amide to form the acid:

$$H_3C-C\equiv N \xrightarrow[H_2O]{H_3O^{\oplus}} H_3C-\overset{O}{\underset{}{C}}-NH_2 \xrightarrow[H_2O]{H_3O^{\oplus}} H_3C-\overset{O}{\underset{}{C}}-OH$$

Recall that nitriles can be prepared from alkyl halides by an $S_N2$ reaction with cyanide ion (Chapter 8). Combining this reaction with the hydrolysis of the nitrile to form the carboxylic acid provides a method for adding one carbon, as a carboxylic acid group, to an alkyl halide.

### EXERCISE 12.21

Each method for making carboxylic acids from alkyl halides has unique advantages, as well as inherent limitations. For each of the following carboxylic acids, only one of the two methods is useful. In each case, indicate which method is applicable, and explain why the other does not work.

(a)

(c)

(b)

(d)

### EXERCISE 12.22

Suggest a mechanism for the acid-catalyzed hydrolysis of acetonitrile ($CH_3C\equiv N$) to acetic acid:

$$H_3C-C\equiv N \xrightarrow[H_2O]{H_3O^{\oplus}} H_3C-\overset{O}{\underset{}{C}}-OH$$

## 12.6

# Derivatives of Sulfonic and Phosphoric Acids

### Sulfonic Acid Derivatives

Derivatives of sulfonic acids are similar to those of carboxylic acids except that they contain hexavalent sulfur rather than tetravalent carbon. There are several derivatives of sulfonic acids; among the most important are sulfonyl chlorides, sulfonic esters, and sulfonamides.

*Sulfonyl Chlorides.* Like carboxylic acids, sulfonic acids can be converted into sulfonyl chlorides by treatment with thionyl chloride. Sulfonic acid derivatives undergo nucleophilic substitution by reaction pathways that parallel those for carboxylic acid derivatives. For example, the sulfonyl group of benzenesulfonyl chloride is polarized in the same way as the carbonyl group of a carboxylic acid chloride and is easily attacked by an oxygen or nitrogen nucleophile:

Benzenesulfonyl chloride

A sulfonic ester

*Sulfonate Esters.* Reaction of a sulfonyl chloride with an alcohol produces a sulfonate ester through a sequence parallel to that for the formation of a carboxylic acid ester from an acid chloride. Upon nucleophilic attack by the oxygen atom of an alcohol, a zwitterionic intermediate is produced; it loses chloride ion to form a sulfonate ester. For example, when *p*-toluenesulfonyl chloride (tosyl chloride) is treated with an alcohol, a **tosylate ester** is formed:

Methyl *p*-toluenesulfonate

In this sulfonate ester, there is appreciable polarization in the carbon–oxygen (R—O) bond as a result of strong electron withdrawal by the two S=O groups. Furthermore, this same electron withdrawal is effective in stabilizing the negative charge in the $ArSO_3^{\ominus}$ ion. As a result, this anion is a very effective leaving group, and the R—O bond exhibits a reactivity similar to that of a C—Br bond. The $p\text{-MeC}_6\text{H}_4\text{SO}_3^{\ominus}$ group, called tosylate and often abbreviated as $TsO^{\ominus}$, is useful in effecting the substitution of alcohols by an $S_N2$ pathway (bimolecular back-side displacement). The hydroxyl group is a poor leaving group, and in order to carry out an $S_N2$ displacement at this center, the hydroxyl group must be converted to a better leaving group. One way to do this is to convert it to a halide. Unfortunately, conversions of primary and secondary alcohols to halides are often accompanied by acid-catalyzed dehydration and/or rearrangement reactions. This difficulty can be avoided by transforming the hydroxyl group into a tosylate ester, a good leaving group. For primary and secondary alcohols, the reaction with tosyl chloride proceeds under mild conditions and with minimal side reactions:

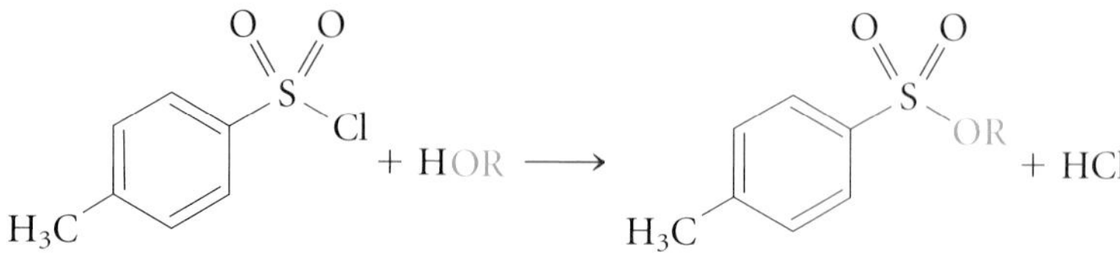

*p*-Toluenesulfonyl chloride
(tosyl chloride = TsCl)

Tosylate ester
(= TsOR)

### EXERCISE 12.23

Suggest a reagent (or a series of reagents) that can be used to accomplish each of the following conversions:

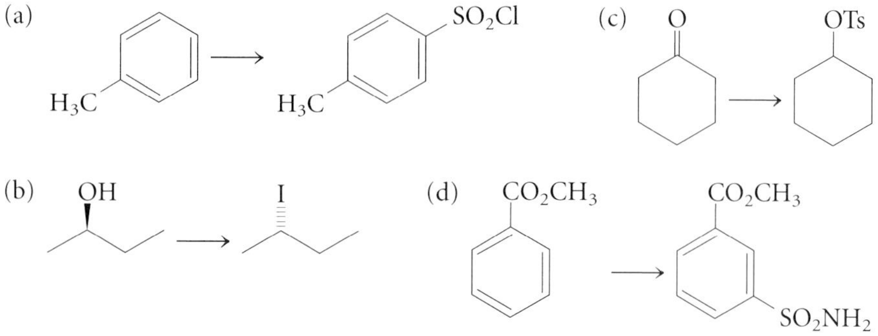

***Sulfonamides.*** Amine nucleophiles react with sulfonyl chlorides, resulting in the substitution of chlorine by nitrogen and the formation of sulfonamides. For example, treatment of benzenesulfonyl chloride with dimethylamine produces a benzenesulfonamide:

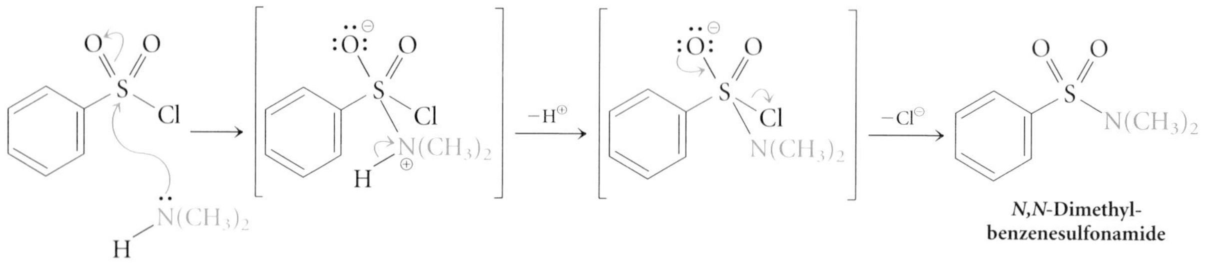

Benzenesulfonyl
chloride

*N,N*-Dimethyl-
benzenesulfonamide

## CHEMICAL PERSPECTIVES

### SULFONAMIDES: NOT ONLY ANTIBIOTICS

The biological activity of sulfonamides is not limited to human use: Glean (chlorsulfuron) is used as a herbicide to control broad-leaved weeds that plague cereal crops. After killing the competing weed by inhibiting an enzymatic transformation, chlorsulfuron is hydrolyzed to a soluble chlorosulfonic acid, urea (a fertilizer), and a cyanuric acid derivative (a triazole that is also a fungicide).

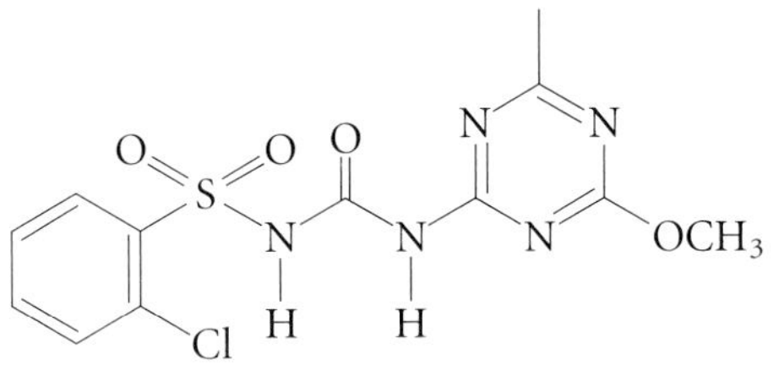
O
S
O
O
N
H
N
H
N
N
N
OCH3
Cl

**Glean**

The mechanism of this reaction is essentially identical to the one that produces an amide from an amine and a carboxylic acid chloride or a sulfonic acid ester from a sulfonic acid chloride. Sulfonamides are potent antibiotic agents that kill bacteria chemically without damaging the cells of the mammalian host.

## Phosphoric Acid Derivatives

Reaction of an alcohol with a halogenated derivative of phosphoric acid is an alternative method for converting the alcohol into the corresponding alkyl bromide or chloride. The reaction takes place via nucleophilic attack by the oxygen of the alcohol at the phosphorus atom of a phosphoryl halide ($POCl_3$ or $POBr_3$):

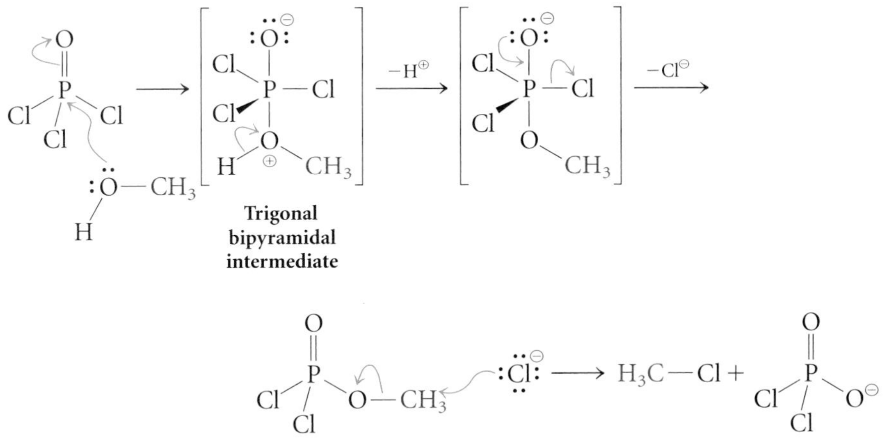
−H⊕
−Cl⊖
Trigonal bipyramidal intermediate
H3C—Cl +

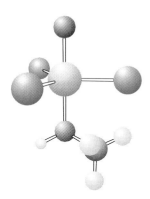

**Trigonal bipyramidal phosphorus intermediate**

Like the S=O bond, the P=O bond is highly polarized, with phosphorus bearing appreciable partial positive charge; this facilitates nucleophilic addition and consequent rapid loss of chloride ion. The intermediate formed upon addition of the alcohol is similar to the tetrahedral intermediate in nucleophilic acyl substitution reactions. Because the substitution is occurring at phosphorus, there are five bonds to the central phosphorus atom, arranged in a trigonal bipyramid.

The oxygen–carbon bond in the resulting intermediate is even more highly polarized than that in a tosylate ester, and nucleophilic substitution by chloride ion follows rapidly. For primary and secondary alcohols, the reaction follows an $S_N2$ pathway; for a tertiary alcohol, a multistep $S_N1$ reaction is required. Although the carbon–chlorine bond in the product is weaker than the carbon–oxygen bond in the starting material, this difference is more than offset by the high strength of the phosphorus–oxygen bond formed in the inorganic anionic by-product. Once again, an energetically unfavorable conversion of one organic material into another is driven by coupling it with an exothermic transformation.

### EXERCISE 12.24

An alkyl chloride can be prepared from the corresponding alcohol and thionyl chloride by a reaction that resembles that of an alcohol with $POCl_3$. Write a detailed mechanism for the conversion of ethanol into ethyl chloride using $SOCl_2$.

## 12.7

## Carbon Nucleophiles

In Chapter 8, we discussed the nucleophilic substitution reactions of a number of carbon nucleophiles. These carbon nucleophiles also add to the carbonyl group of aldehydes or ketones. In this section, we consider a number of nucleophilic addition reactions: of cyanide ion to make cyanohydrins, of Grignard reagents or organolithium reagents to make a variety of oxygen-containing functional groups, and of phosphonium ylides to prepare alkenes in the Wittig reaction.

$$K^{\oplus} {}^{\ominus}C \equiv N \qquad \overset{\delta^-}{H_3C}\overset{\delta^+}{-}MgBr \qquad \overset{\delta^-}{H_3C}\overset{\delta^+}{-}Li \qquad Ph_3\overset{\oplus}{P}\overset{\ominus}{-}CH_3$$

| Potassium cyanide | Methylmagnesium bromide | Methyl lithium | A phosphonium ylide |

### Cyanide Ion

One of the simplest carbon nucleophiles is cyanide ion, $^{\ominus}CN$. (Recall the reaction of cyanide ion with alkyl halides and sulfonate esters.) The addition of $^{\ominus}CN$ to a carbonyl group results in the formation of a new carbon–carbon bond in the product α-cyanoalcohol, called a **cyanohydrin.** Cyanide is both a good nucleophile and a reasonable leaving group. Therefore, cyanohydrin formation is usually reversible, as shown here for the reaction with acetone:

## CHEMICAL PERSPECTIVES

### TOXICITY OF A NATURALLY OCCURRING CYANOHYDRIN DERIVATIVE

Amygdalin (also called laetrile) is a naturally occurring cyanohydrin that can be isolated from bitter almond seeds and peach and apricot pits. For some years, it was touted as an anticancer drug and, although not approved for use in the United States, was administered to many cancer patients who went abroad for chemotherapy. Unfortunately, it has been shown to be not only highly toxic but also ineffective as a cancer treatment.

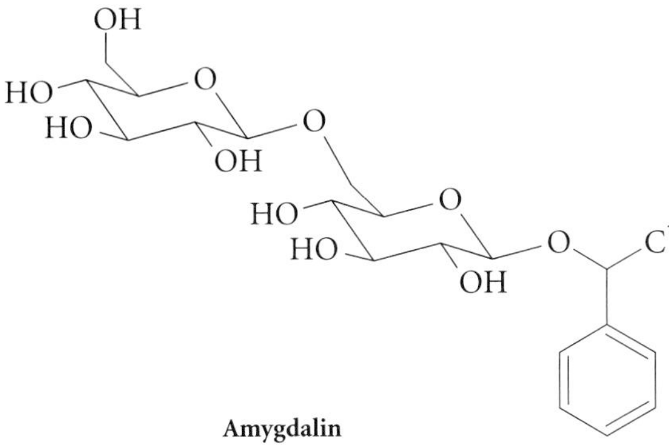

**Amygdalin**

Amygdalin's high toxicity derives from its nonselective release of hydrogen cyanide, HCN, under physiological conditions. In the laboratory, in a reversal of cyanohydrin formation, treatment of amygdalin with acid produces HCN, benzaldehyde, and two equivalents of glucose.

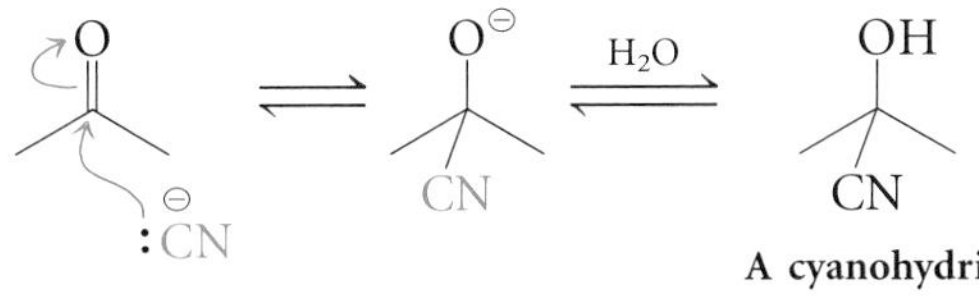

**A cyanohydrin**

In a variation of cyanohydrin formation known as the *Strecker synthesis,* an aldehyde is treated with ammonium chloride in the presence of potassium cyanide to form an $\alpha$-aminonitrile:

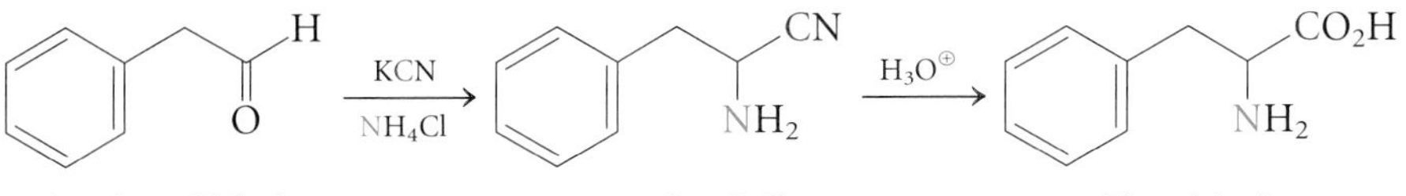

| Phenylacetaldehyde | An $\alpha$-aminonitrile | Phenylalanine (an $\alpha$-amino acid) |

Ammonia (present in equilibrium with $NH_4Cl$) converts the aldehyde into an imine. This reaction begins in the same way as the aminations presented earlier in this chapter—that is, by the formation of an imine via the reaction of ammonia (present in equilibrium with its conjugate acid, the ammonium ion) and an aldehyde. Nucleophilic attack by cyanide ion on the imine produces an $\alpha$-aminonitrile. Hydrolysis of this nitrile leads first to a primary amide by addition of water across the C≡N bond (Chapter 10) and then to an $\alpha$-amino acid by nucleophilic acyl substitution. In this example, phenylacetaldehyde is converted into the natural amino acid phenylalanine, one of the important $\alpha$-amino acids that are the "repeat" units in proteins and peptides. The rate of addition of cyanide ion to the two faces of the carbonyl group is the same; thus, the $\alpha$-aminonitrile intermediate and the product $\alpha$-amino acid are racemic.

### EXERCISE 12.25

Provide a rational mechanism, using curved-arrow notation, for the reactions shown here (the Strecker synthesis of amino acids).

### Grignard Reagents

A space-filling model of a Grignard reagent, $CH_3CH_2$—MgBr, is shown here.

The electron density in the carbon–magnesium bond of a Grignard reagent is highly polarized toward carbon because of the large difference in electronegativity between these elements.

***Addition to Carbonyl Groups: Synthesis of Alcohols.*** As a result of the polarization of the carbon–magnesium bond in a Grignard reagent, the carbon is quite nucleophilic and adds readily to aldehydes or ketones.

The Grignard reagent forms a $\sigma$ bond between its carbon and the carbonyl carbon while electrons from the C=O $\pi$ bond are shifted onto oxygen. After this nucleophilic addition is complete, protonation of the newly formed alkoxide salt by water (or dilute acid) produces an alcohol. The addition of Grignard reagents to carbonyl groups is quite general: these reagents add to formaldehyde, other aldehydes, and ketones, providing synthetically useful routes for primary, secondary, and tertiary alcohols, respectively.

Nucleophilic attack by a Grignard reagent on an ester forms a tetrahedral intermediate that is readily transformed into a ketone by loss of alkoxide ion—methoxide ion in the case shown here:

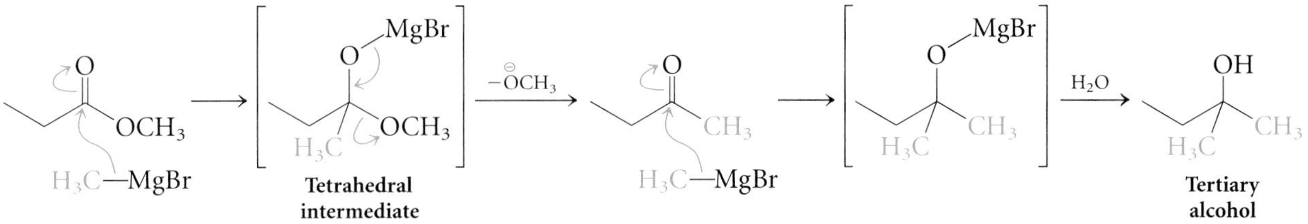

Because ketones are more reactive toward nucleophilic attack than are esters, it is not possible to stop the reaction at the ketone, which reacts rapidly with a second equivalent of Grignard reagent. Protonation of the alkoxide salt with dilute acid affords a tertiary alcohol. Note that the tertiary alcohol formed by the reaction of a Grignard reagent with an ester must have two identical groups attached to the carbinol carbon, but when the tertiary alcohol is formed by reaction with a ketone, all three alkyl groups can be different.

***Addition to Carbon Dioxide: Synthesis of Carboxylic Acids.*** Grignard reagents react similarly with the carbon atom in carbon dioxide, producing a resonance-stabilized carboxylate anion that, after acidification, gives a carboxylic acid:

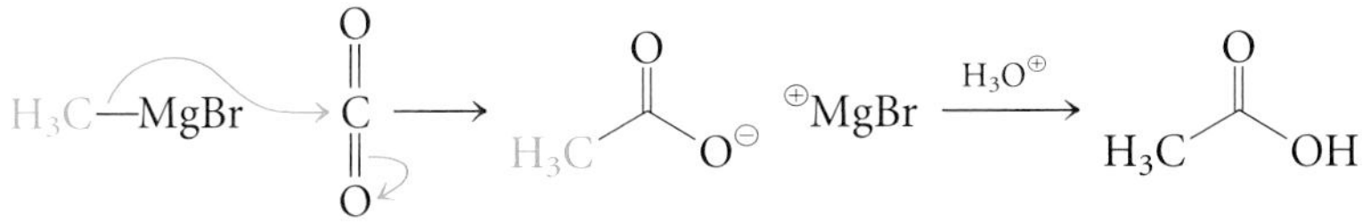

This reaction is quite useful for the synthesis of carboxylic acids. Starting from an alkyl halide, reaction with magnesium metal forms a Grignard reagent. Reaction of this organometallic reagent with carbon dioxide and acidification of the initial product produces a carboxylic acid with one more carbon than the starting alkyl halide.

***Synthetic Utility of Grignard Reagents.*** Recall from Chapter 8 that Grignard reagents react with ethylene oxide to form a new carbon–carbon $\sigma$ bond in the product, a primary alcohol. In this reaction, two carbons are added to the carbon skeleton in the Grignard reagent.

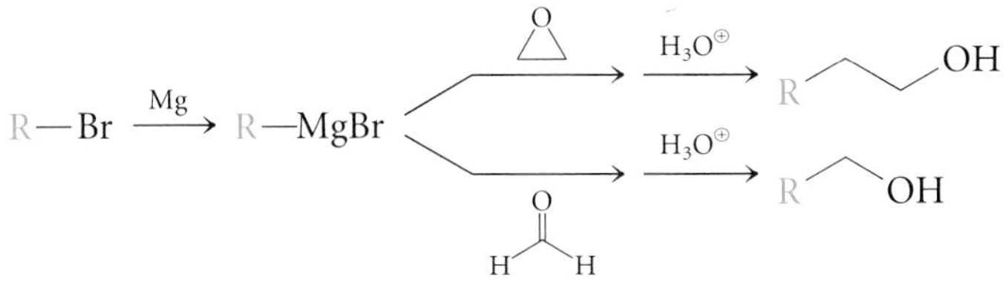

In contrast, the reaction of a Grignard reagent with formaldehyde produces a primary alcohol that has only one more carbon than did the alkyl halide from which the Grignard reagent was prepared.

**TABLE 12.1**

Synthetic Utility of Grignard Reactions

| Reactants | | Product |
|---|---|---|
| R—MgX + | 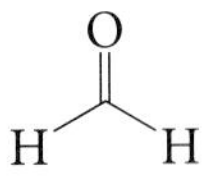 O<br>‖<br>H⌍H<br>**Formaldehyde** | 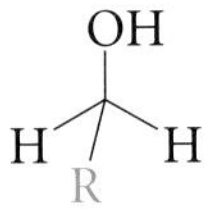 OH<br>\|<br>H—C—H<br>\|<br>R<br>**Primary alcohol** |
| R—MgX + | O<br>‖<br>R⌍H<br>**Aldehyde** | 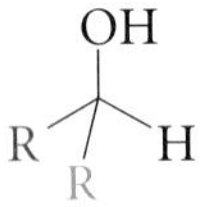 OH<br>\|<br>R—C—H<br>\|<br>R<br>**Secondary alcohol** | (R groups can be the same or different) |
| R—MgX + | O<br>‖<br>R⌍R<br>**Ketone** | 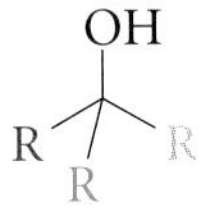 OH<br>\|<br>R—C—R<br>\|<br>R<br>**Tertiary alcohol** | (R groups can be the same or different) |
| R—MgX + | O<br>‖<br>R⌍OR<br>**Ester** | 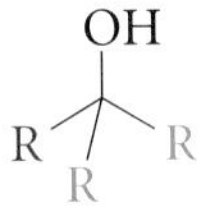 OH<br>\|<br>R—C—R<br>\|<br>R<br>**Tertiary alcohol** | (two of the R groups must be the same) |
| R—MgX + | $CO_2$ | 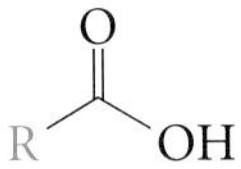 O<br>‖<br>R⌍OH<br>**Carboxylic acid** |
| R—MgX + |  O<br>△<br> |  R⌍⌍OH<br>**Primary alcohol** |
| R—MgX + | $H_2O$ | R—H<br>**Hydrocarbon** |

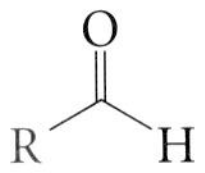

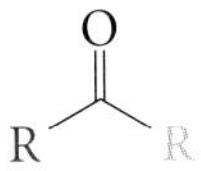

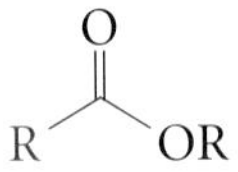

The synthetic utility of Grignard reagents is summarized in Table 12.1, which shows that the reaction of such a reagent with the appropriate electrophile leads to a primary, secondary, or tertiary alcohol, as well as to a carboxylic acid or alkane. A reaction that extends the carbon skeleton by C—C bond formation and that retains an alcohol functional group for subsequent manipulation is a valuable synthetic tool for the construction of complex molecules.

**EXERCISE 12.26**

For each of the following targets, several different synthetic routes that employ a Grignard reagent are possible. Provide at least two different sets of Grignard reagent and substrate that will lead to each of the observed products. Describe the appropriate reaction conditions.

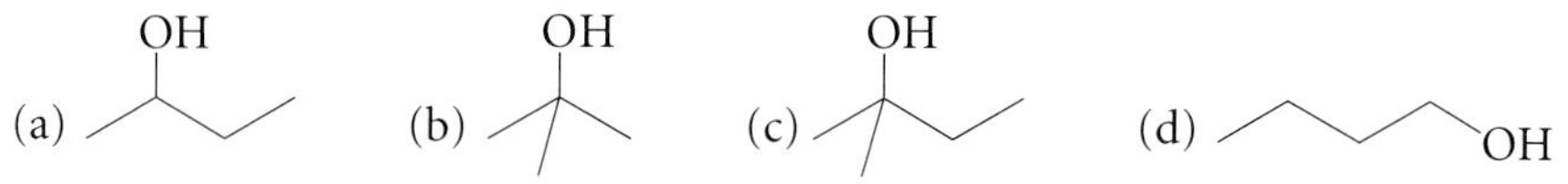

(a)  (b)  (c)  (d)

## EXERCISE 12.27

The yield in Grignard reactions is often decreased by a competing reduction reaction, and with sterically hindered ketones, this competing reaction often dominates. Suggest a mechanism for the reduction of adamantanone by Et—MgBr:

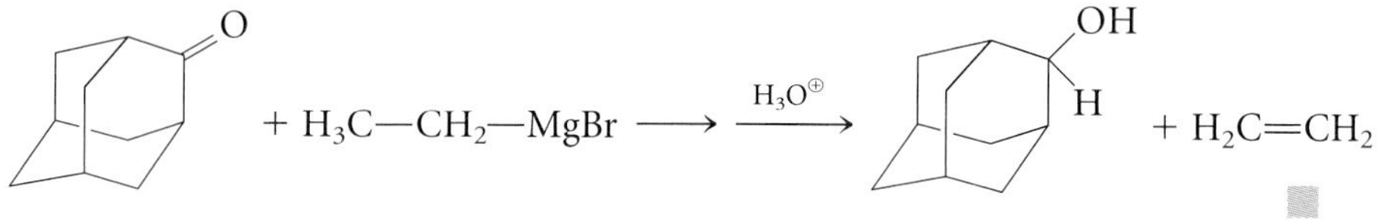

### Organolithium Reagents

Organolithium reagents react with carbonyl-containing compounds in virtually the same way as Grignard reagents do. Typical examples are the reaction of benzaldehyde with butyllithium or a lithium acetylide:

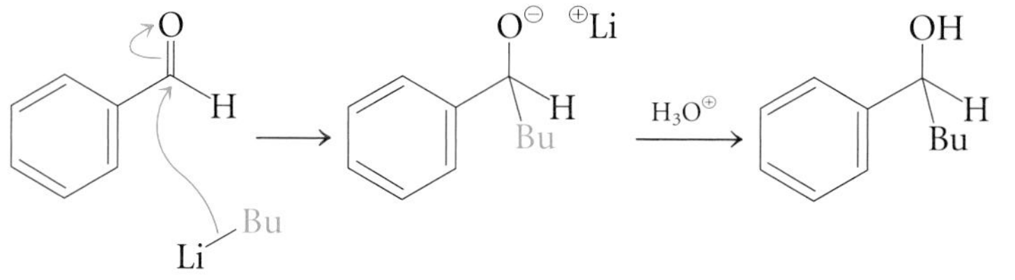

Benzaldehyde and
butyllithium

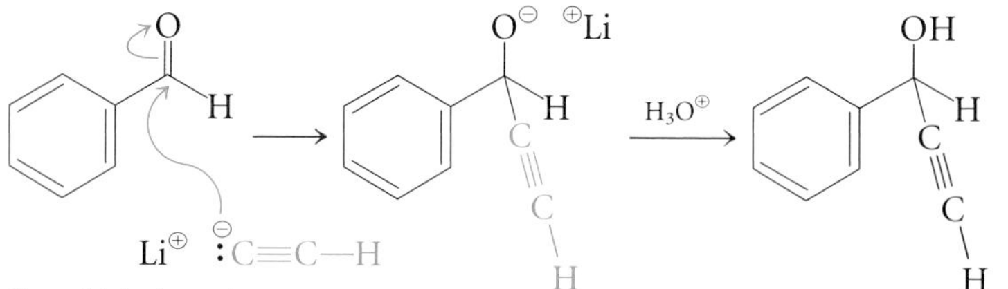

Benzaldehyde and
lithium acetylide

You should not forget that an unshared electron pair—the feature that imparts nucleophilic character to these reagents—also makes them basic. However, in reactions with carbonyl compounds (aldehydes, ketones, and esters), both Grignard reagents and alkyllithium compounds react preferentially as nucleophiles.

### The Wittig Reaction

Carbon nucleophiles can be formed by deprotonation of a carbon adjacent to an electron-withdrawing group. In Chapter 8, we looked at the reaction of trialkylphosphines with alkyl halides to form phosphonium salts.

## ALKYNES AS CONTRACEPTIVE AGENTS

Chemists at pharmaceutical companies have developed synthetic mimics of natural steroid hormones for use as contraceptive agents. Progesterone is produced naturally during the menstrual cycle to prepare the uterine lining for implantation of a fertilized egg and to suppress ovulation. Norethindrone and norethynodrel are synthetic compounds that have the same anti-ovulatory effect on the ovaries as progesterone.

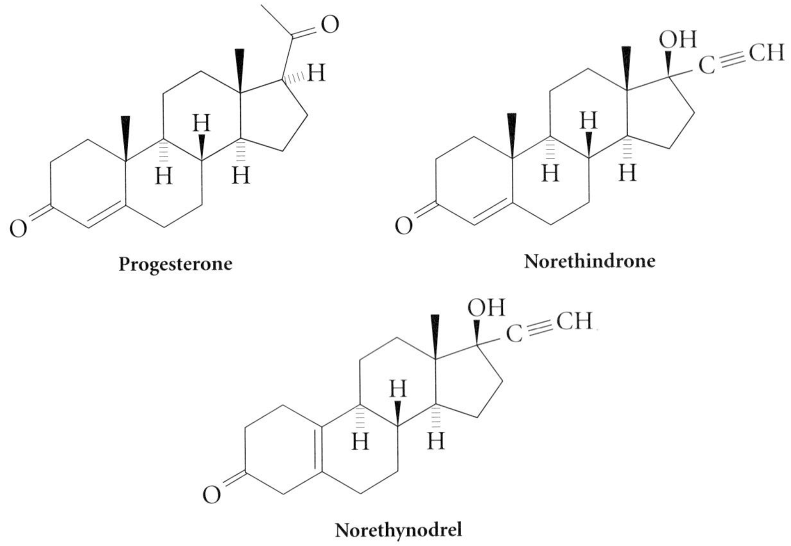

Progesterone

Norethindrone

Norethynodrel

The presence of the positively charged phosphorus atom significantly enhances the acidity of protons on attached carbon atoms, and treatment of phosphonium salts with base produces the corresponding anions, called *phosphoranes.* These anions are good nucleophiles. Consider the ylide formed by deprotonation of the phosphonium salt formed by nucleophilic ($S_N2$) substitution of methyl iodide by triphenylphosphine:

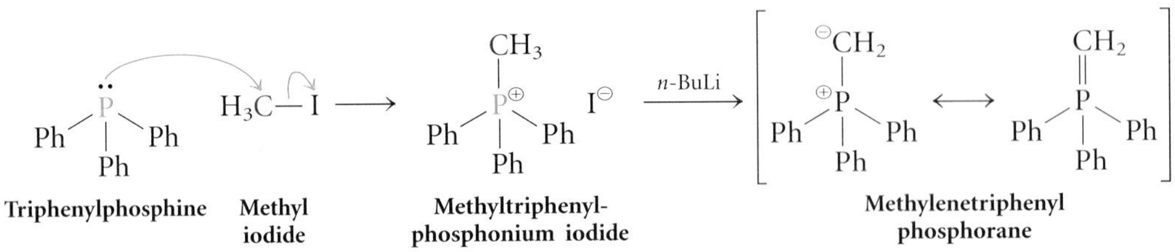

Triphenylphosphine    Methyl iodide    Methyltriphenyl-phosphonium iodide    Methylenetriphenyl phosphorane

Deprotonation requires a strong base such as a lithium dialkylamide or *n*-butyllithium. Because the carbon center in the resulting ylide bears significant negative charge, it is nucleophilic and adds to the carbonyl carbon of a ketone such as acetone, as shown in Figure 12.11. The resulting inter-

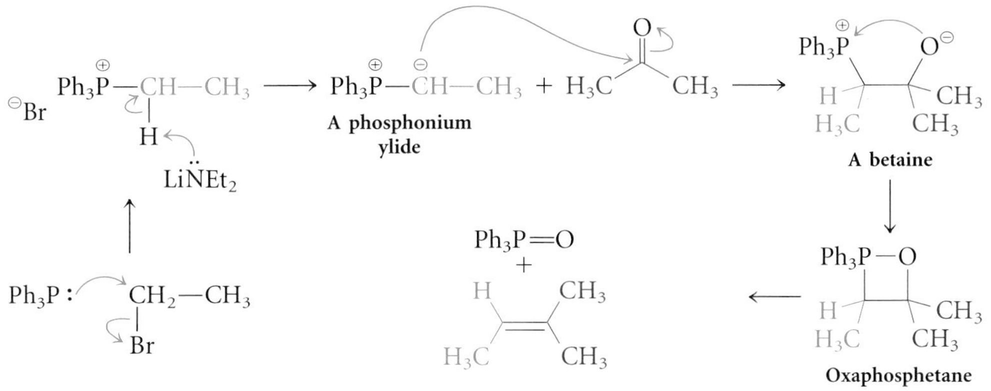

**FIGURE 12.11**

Removal of a hydrogen from the carbon adjacent to the positively charged phosphorus atom of a phosphonium ion produces a nucleophilic ylide. Reaction with an aldehyde or ketone in the Wittig reaction forms an alkene.

mediate, called a **betaine,** is zwitterionic, with a negatively charged oxygen atom separated by two atoms from a positively charged phosphorus atom. Betaines are named for the naturally occurring compound betaine (be ′tā-en′) found in sugar beets and other plants.

In the next step, these charged oxygen and phosphorus atoms bond, forming a four-member ring in a species called an **oxaphosphetane;** a simple example of an oxaphosphetane is shown in Figure 12.12. In the last step of the reaction sequence, the four-member ring opens, generating two fragments, one with a carbon–carbon double bond and the other with a phosphorus–oxygen double bond.

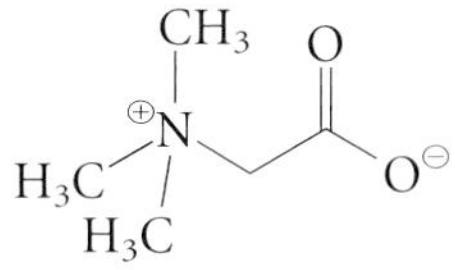

**Betaine**

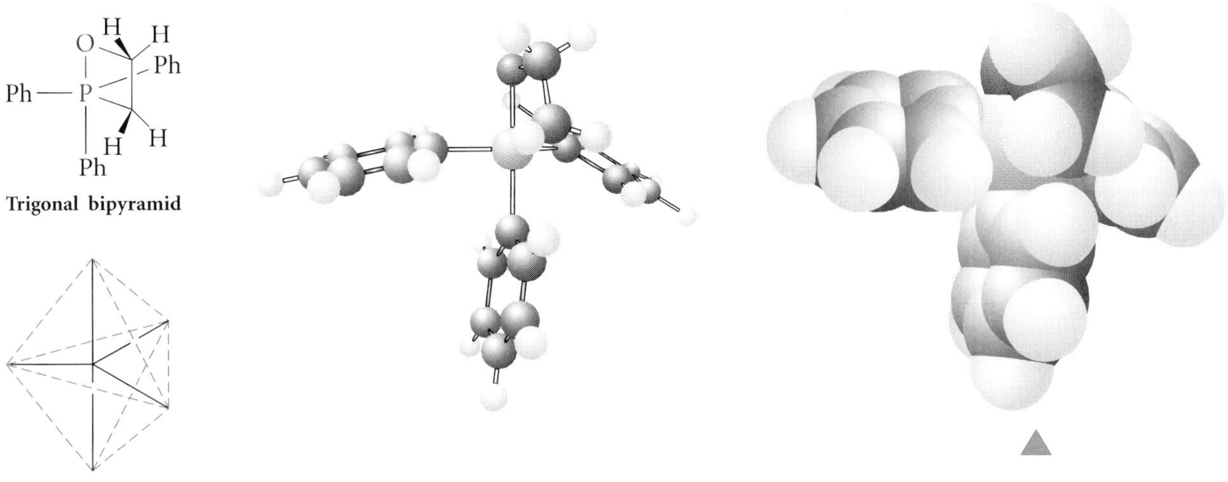

**FIGURE 12.12**

Ball-and-stick and space-filling representations of a simple oxaphosphetane. Note that the phosphorus atom (purple) has five ligands arranged in a trigonal bipyramid and that the carbon–phosphorus and oxygen–phosphorus bonds are longer than the others.

This transformation is called the **Wittig reaction** in recognition of its discovery by Georg Wittig, for which he received the Nobel prize in chemistry in 1979. Overall, this reaction converts an aldehyde or ketone into an alkene. The double bond of the product joins two fragments that were starting materials, one from the carbonyl compound, the other from the phosphonium ylide. We can backtrack one step and think of the sequence as originating in an alkyl bromide and a ketone. In the overall conversion of the alkyl bromide and the ketone to the alkene, two other transformations are involved—the conversion of triphenylphosphine to triphenylphosphine oxide and of butyllithium to butane:

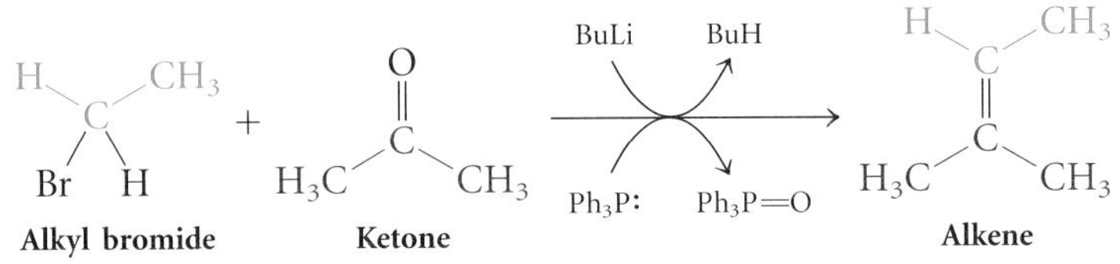

The use of coupled arrows in representing this reaction summarizes the principal organic conversion and implies that these processes are not independent. Furthermore, these reactions consist of several steps and cannot take place by simultaneously mixing all required reagents. This method of showing transformations as **coupled reactions** is convenient for summarizing a net conversion, especially when the molecules are complicated. We will encounter coupled reactions more frequently in later chapters where biochemical transformations are described.

### EXERCISE 12.28

Identify two possible combinations of organic reagents from which each of the following alkenes could be prepared via a Wittig reaction:

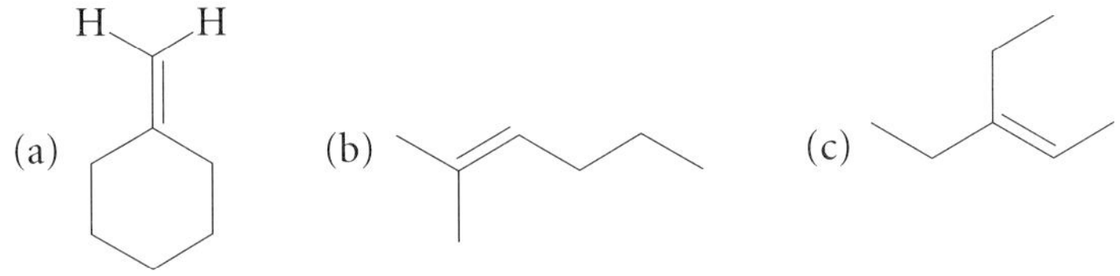

### EXERCISE 12.29

Each of the alkenes in Exercise 12.28 could also be prepared by two unique combinations of a Grignard reagent with an aldehyde or ketone, followed by dehydration. Draw the structures of the starting Grignard reagent and carbonyl compound for each of these possibilities. What alkene would be produced in highest yield upon acid-catalyzed dehydration of each of the six alcohols? Compare these results with the outcomes of the Wittig reactions in Exercise 12.28.

### 12.8

## Synthetic Applications

The new reactions considered in this chapter are summarized in Table 12.2 according to their usefulness for interconverting various functional groups.

Using Nucleophilic Additions and Substitutions (and Related Reactions)
to Make Various Functional Groups

| Functional Group | Reaction | Example |
|---|---|---|
| Acetals | Reaction of an aldehyde with an alcohol in the presence of acid | |
| Acid chlorides | Treatment of an acid with $SOCl_2$ | |
| Acids | Nitrile hydrolysis | |
| | Cannizzaro disproportionation of an aldehyde | $2\ Ph\!-\!CHO \xrightarrow{\ NaOH\ } Ph\!-\!CH_2OH + Ph\!-\!CO_2H$ |
| | Amide hydrolysis | |
| | Acid chloride hydrolysis | |
| | Ester hydrolysis | |
| | Thiol ester hydrolysis | |
| | Acid anhydride hydrolysis | |
| Alcohols | Complex metal hydride reduction of an aldehyde or ketone | |
| | Lithium aluminum hydride reduction of an ester | |
| | Cannizzaro disproportionation of an aldehyde | $2\ Ph\!-\!CHO \xrightarrow{\ NaOH\ } Ph\!-\!CH_2OH + Ph\!-\!CO_2H$ |

*(continued)*

TABLE 12.2

| Functional Group | Reaction | Example |
|---|---|---|
| Alcohols | Reaction of organometallic reagent with formaldehyde, an aldehyde, a ketone, an ester, or ethylene oxide | (see Table 12.1) |
| Alkenes | Wittig reaction | |
| Alkyl bromides | Treatment of an alcohol with POBr$_3$, PBr$_3$, or PBr$_5$ | $R{-}OH \xrightarrow[\text{(or PBr}_3 \text{ or PBr}_5)]{\text{POBr}_3} R{-}Br$ |
| Alkyl chlorides | Treatment of an alcohol with POCl$_3$ or SOCl$_2$ | $R{-}OH \xrightarrow[\text{(or SOCl}_2)]{\text{POCl}_3} R{-}Cl$ |
| Amides | Reaction of an acid chloride with an amine | |
| | Reaction of an ester with an amine | |
| | Hydrolysis of a nitrile | |
| Amines | Lithium aluminum hydride reduction of an amide | |
| | Reductive amination of an aldehyde or ketone and reduction of the imine | |

| Functional Group | Reaction | Example |
|---|---|---|
| Esters | Reaction of an acid chloride with an alcohol | |
| Hydrates | Hydration of an aldehyde or ketone | |
| Hydrazones | Reaction of an aldehyde or ketone with hydrazine | |
| Imines | Reaction of a primary amine with an aldehyde or ketone | |
| Ketals | Treatment of a ketone with alcohol in the presence of acid | |
| Oximes | Reaction of an aldehyde or ketone with hydroxylamine | |
| Semicarbazones | Reaction of an aldehyde or ketone with semicarbazide | |
| Sulfonate esters | Reaction of a sulfonyl chloride with an alcohol | $R{-}SO_2Cl + ROH \longrightarrow R{-}SO_2OR$ |
| Sulfonamides | Reaction of a sulfonyl chloride with an amine | $R{-}SO_2Cl + H_2N{-}R \longrightarrow R{-}SO_2NHR$ |

## 12.9

# Spectroscopy

Many of the transformations discussed in this chapter result in significant differences between the infrared spectra of starting materials and those of the products, because of the highly characteristic absorptions of the various carbonyl functional groups. Indeed, many reactions of carbonyl compounds lead to products that do not have carbonyl groups—for example, the reduction of a ketone to an alcohol and the formation of a ketal from a ketone. In the conversion of 1,4-cyclohexanedione to the monoketal, the intensity of the absorption due to the carbonyl group in the product

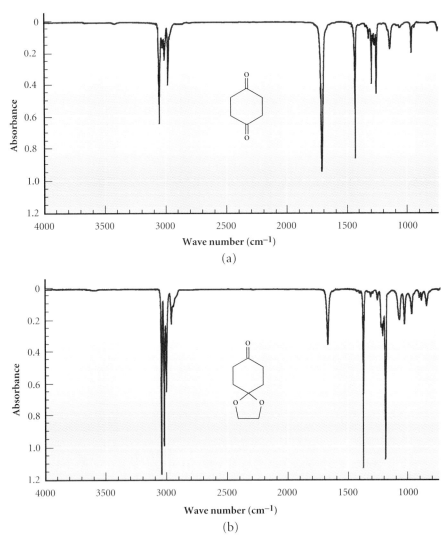

**FIGURE 12.13**

(a) Infrared spectrum of 1,4-cyclohexanedione. (b) Infrared spectrum of the monoketal of 1,4-cyclohexanedione.

TABLE 12.3

Characteristic Infrared Absorption Maxima for Various Carbonyl Functional Groups

| Functional Group | Frequency (cm$^{-1}$) | Functional Group | Frequency (cm$^{-1}$) |
|---|---|---|---|
| (acetyl chloride) | 1800 | (thioester) | 1690 |
| (anhydride) | 1820 and 1760 | (amide, NH$_2$) | 1690 and 1600 |
| (carboxylic acid, OH) | 1760 (monomer) 1710 (dimer) | (amide, NH) | 1680 and 1530 |
| (ester, O) | 1735 | (amide, N) | 1650 |
| (aldehyde, H) | 1725 | | |
| (ketone) | 1715 | | |

monoketone clearly declines (relative to the C—H stretching absorptions) whereas there are new absorptions in the region 1300–1000 cm$^{-1}$ resulting from C—O stretching (Figure 12.13).

Infrared spectroscopy is also a useful tool for analyzing the change of one carbonyl functional group into another because the frequency of the C=O stretch is often different. Table 12.3 lists the infrared absorption maxima for the most common carbonyl functional groups.

## EXERCISE 12.30

For each of the reactions shown in Review of Reactions on pages 649–650, indicate what difference(s) you would expect to see between the infrared spectrum of the starting material and that of the product.

## Summary

**1.** Nucleophilic addition to an $sp^2$-hybridized carbon usually occurs only if a heteroatom is part of the $\pi$ system—as in a C=O or C=N bond. Upon nucleophilic addition to such compounds, the negative charge shifts onto the more electronegative oxygen or nitrogen atom.

**2.** The reactivity of various carbonyl compounds is determined by their relative stabilities, which is affected by such factors as resonance stabilization and electron release from substituent groups. The order of reactivity toward nucleophiles is aldehydes > ketones > esters > amides > carboxylate anions.

**3.** Nucleophilic attack on a carbonyl group can result in either net addition or net substitution. Addition takes place if a poor leaving group is bonded to the carbonyl carbon, as in aldehydes and ketones. Substitution takes place if a good leaving group is present, as in acid chlorides, anhydrides, esters, amides, and nitriles.

**4.** Whether the overall reaction is characterized as addition or substitution, the first step in nucleophilic attack at the carbonyl carbon is the formation of a tetrahedral intermediate. In nucleophilic addition, this tetrahedral intermediate is trapped, usually by protonation at oxygen. In nucleophilic acyl substitution, one of the $\sigma$ bonds to the carbonyl carbon is fragmented as the electrons on the carbonyl oxygen in the tetrahedral intermediate reform the C=O bond.

**5.** Nucleophilic addition by complex metal hydrides is equivalent to transfer of a hydride ion and results in reduction. Primary alcohols are produced from aldehydes, and secondary alcohols from ketones. With an ester or amide, nucleophilic attack by a hydride produces an intermediate carbonyl compound or imine, respectively, which is then attacked by a second hydride equivalent. The net reduction transforms an ester to a primary alcohol and an amide to an amine. Nucleophilic addition of a hydride equivalent is characteristic of biological reductions. For example, biological reductions are accomplished with the cofactor NADH, which delivers the equivalent of a hydride ion to a reducible substrate.

**6.** Nucleophilicity, the affinity of an atom or group of atoms for a partially positively charged carbon, often parallels basicity, the affinity for a proton. For example, oxygen nucleophiles are less reactive than nitrogen nucleophiles, which in turn are less reactive than carbon nucleophiles—following the order of basicity. Because nucleophilicity is inversely related to leaving-group ability, the opposite order is followed when the leaving group is being displaced. However, care must be taken in using this analogy, because basicity is generally measured as an equilibrium value, whereas nucleophilicity always refers to relative reactivity (or the rate of reaction).

**7.** The nucleophilic addition of water to a ketone or aldehyde occurs rapidly and reversibly with mild acid or base catalysis to form a hydrate, whereas the addition of alcohols yields a ketal or acetal (or a hemiketal or hemiacetal). The positions of these equilibria are controlled by the concentration of water or alcohol and the inherent stability of the carbonyl compound.

**8.** Hydroxide ion attack on an aldehyde lacking an $\alpha$-hydrogen results in disproportionation through the Cannizzaro reaction. In this sequence, the tetrahedral intermediate acts to deliver a hydride equivalent to a second molecule, thereby simultaneously forming an oxidized (carboxylic acid) and a reduced (alcohol) derivative of the starting aldehyde.

**9.** Nitrogen nucleophiles also add to carbonyl compounds. Depending upon the substitution at nitrogen, either imines (from primary amines) or enamines (from secondary amines) are formed. When imines are formed in the presence of a reducing agent, amines are produced through reductive amination.

**10.** Nucleophilic attack on a carbonyl group by hydrazine (or a substituted hydrazine) gives rise to a hydrazone. Because such derivatives are often intensely colored solids, their formation can be used as a qualitative indicator of the presence of an aldehyde or a ketone.

**11.** Carboxylic acid derivatives are interconverted by nucleophilic acyl substitution. When an oxygen nucleophile attacks an ester, transesterification or hydrolysis to a carboxylic acid results. Thiol esters are more reactive than their oxygen counterparts because of the mismatch in orbital size between sulfur and the carbonyl $\pi$ system. Acid anhydrides and acid chlorides are readily attacked by oxygen nucleophiles (water or alcohols), producing the corresponding acids or esters. The high reactivity of these substrates derives from $\sigma$-electron withdrawal by Cl in an acid chloride and by the carboxyl substituent in an anhydride.

**12.** Nucleophilic attack by water on nitriles effects hydrolysis to carboxylic acids.

**13.** Nucleophilic substitution reactions also occur with sulfonic acid esters of alcohols. Tosylates are alcohol derivatives that are reactive toward nucleophilic substitution. Sulfonamides are amides of sulfonic acid; they are analogous to carboxylic acid amides, and some are biologically active as antibiotics.

**14.** Certain phosphorus and sulfur reagents ($POCl_3$, $PI_3$, $PBr_3$, $PCl_5$, and $SOCl_2$) are useful for converting alcohols to alkyl halides.

## Review of Reactions

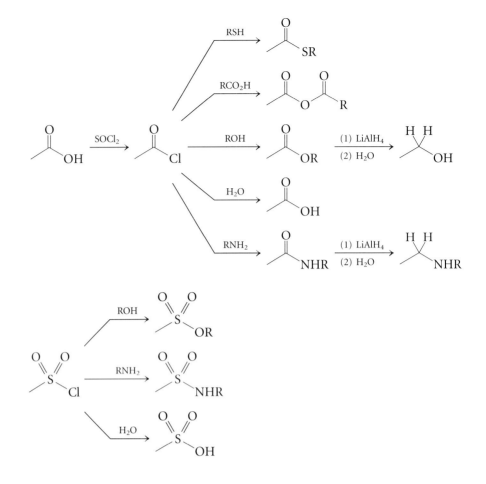

## Review Problems

**12.1** Determine the structure of the product(s) formed, if any, when pentanal is treated with each of the following reagents:

(a) NaBH$_4$ in EtOH

(b) LiAlH$_4$ followed by H$_2$O

(c) *n*-propylamine

(d) phenylhydrazine

(e) CH$_3$CONH$_2$

(f) HOCH$_2$CH$_2$OH, H$_3$O$^{\oplus}$

(g) aqueous NaCl

(h) SOCl$_2$

(i) NH$_2$OH

(j) cold KMnO$_4$

**12.2** Determine the structure of the product(s) formed, if any, when methyl pentanoate is treated with each of the following reagents:

(a) NaBH$_4$ in EtOH

(b) LiAlH$_4$ followed by H$_2$O

(c) *n*-propylamine

(d) aqueous NaCl

**12.3** Determine the structure of the product(s) formed, if any, when pentanoamide is treated with each of the following reagents:

(a) NaBH$_4$ in EtOH

(b) LiAlH$_4$ followed by H$_2$O

(c) aqueous ethanol, H$_3$O$^{\oplus}$

(d) *n*-propylamine

(e) aqueous NaCl

**12.4** Determine the structure of the product(s) formed, if any, when acetyl chloride is treated with each of the following reagents:

(a) $H_2O$

(b) *n*-propanol, acid

(c) $(CH_3)_2NH$

(d) $NH_3$

(e) $C_6H_6$ and $AlCl_3$

(f) $CH_3CH_2SH$, pyridine

(g) $CH_3COO^{\ominus} Na^{\oplus}$

(h) $C_6H_5OH$, pyridine

(i) $H_2$, Pt

**12.5** Determine the reagent needed to effect each of the following conversions:

(a) $CH_3CH_2COCl$ to $CH_3CH_2COOH$

(b) $CH_3CH_2COOH$ to $CH_3CH_2COCl$

(c) $CH_3CH_2COOCH_3$ to $CH_3CH_2CONH_2$

(d) $CH_3CH_2COOCH_3$ to $CH_3CH_2COOCH_2CH_3$

(e) $CH_3CH_2COOH$ to $CH_3CH_2CONH_2$

(f) $CH_3CH_2CN$ to $CH_3CH_2COOH$

**12.6** Using curved arrows to indicate electron flow, propose a detailed mechanism for each of the following conversions.

(a)

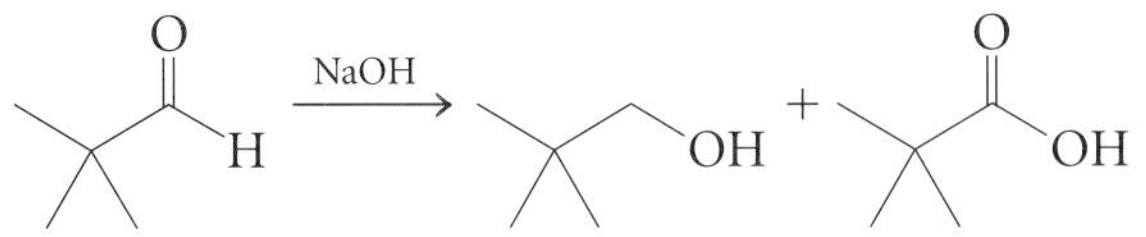

(b)

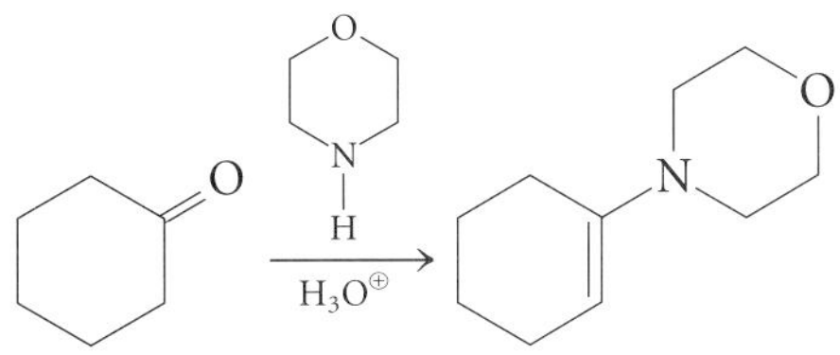

(c)

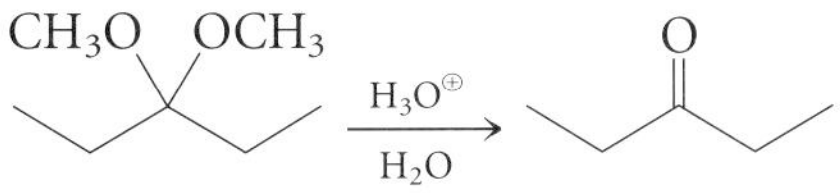

(d)

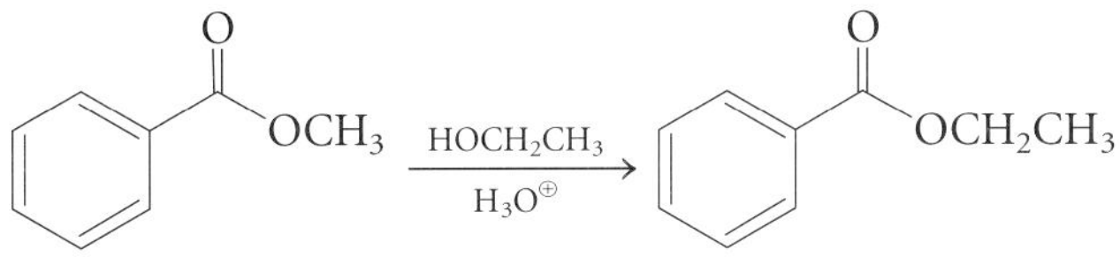

(e)

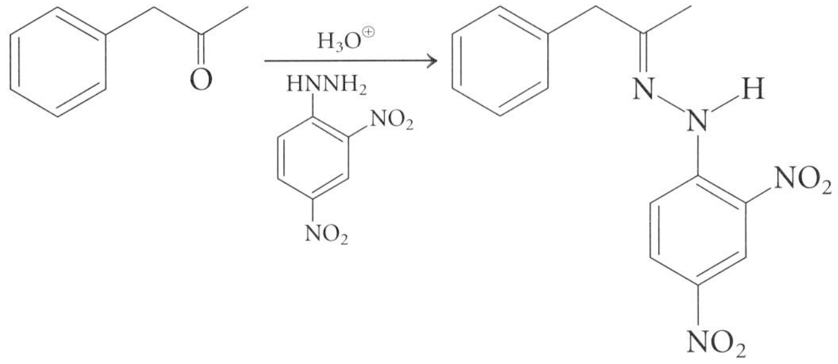

(f)

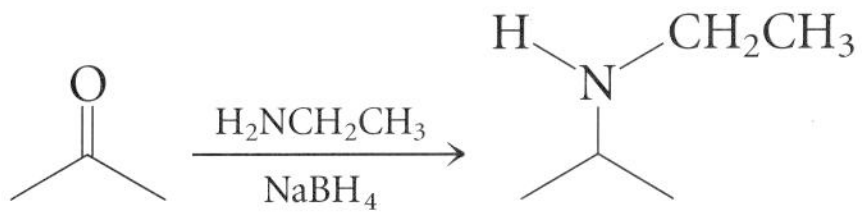

(g)

**12.7** Predict the product, if any, expected from each of the following reactions.

(a) $\xrightarrow{\text{(1) LiAlH}_4}$ $\xrightarrow{\text{(2) H}_3\text{O}^{\oplus}}$

(k) $\xrightarrow[\text{(2) POCl}_3]{\text{(1) conc. H}_2\text{SO}_4}$ $\xrightarrow{\text{(3) CH}_3\text{CH}_2\text{NH}_2}$

(b) $\xrightarrow[\text{(2) H}_3\text{O}^{\oplus}]{\text{(1) NaBH}_4}$

(l) $\xrightarrow[\text{(2)POCl}_3]{\text{(1) NaBH}_4}$

(c) $\xrightarrow[\text{(2) H}_3\text{O}^{\oplus}]{\text{(1) LiAlH}_4}$

(m) $\xrightarrow{\text{PhNHNH}_2}$

(d) $\xrightarrow[\text{H}_3\text{O}^{\oplus}]{\text{CH}_3\text{CH}_2\text{OH}}$

(n) $\xrightarrow{\text{PhNH}_2}$

(e) $\xrightarrow[\text{(2) H}_3\text{O}^{\oplus}]{\text{(1) LiAlH}_4}$

(o) $\xrightarrow[\text{(2) CH}_3\text{CH}_2\text{NH}_2]{\text{(1) SOCl}_2}$

(f) $\xrightarrow[\text{H}_3\text{O}^{\oplus}]{\text{CH}_3\text{CH}_2\text{OH}}$

(p) $\xrightarrow[\text{CH}_3\text{CH}_2\text{OH}]{\text{H}_3\text{O}^{\oplus}}$

(g) $\xrightarrow{\text{NaOH}}$

(q) $\xrightarrow{\text{Et}_3\text{N}}$

(h) $\xrightarrow[\text{H}_3\text{O}^{\oplus}]{}$

(r) $\xrightarrow[\text{H}_3\text{O}^{\oplus}]{\text{CH}_3\text{CH}_2\text{OH}}$

(i) $\xrightarrow{\text{H}_3\text{O}^{\oplus}}$

(s) $\xrightarrow[\text{AlBr}_3]{\text{CH}_3\text{Br}}$

(j) $\xrightarrow{\text{H}_3\text{O}^{\oplus}}$

(t) $\xrightarrow{\text{MgBr}}$ $\xrightarrow{\text{H}_3\text{O}^{\oplus}}$

**12.8** The hydrolysis of a ketal in the presence of acid and water to form a ketone (and an alcohol) follows a reaction path that is exactly the reverse of that for ketal formation. Write this mechanism without consulting the text. (*Hint:* The first step is protonation of one of the oxygen atoms.)

$$\text{H}_3\text{CO} \quad \text{OCH}_3 \xrightarrow[\text{H}_2\text{O}]{\text{H}^{\oplus}} \quad \text{O} \quad + \text{ 2 HOCH}_3$$

**12.9** The rate-limiting step for the conversion of a ketal to a ketone and an alcohol is almost always the cleavage of the first carbon–oxygen bond.

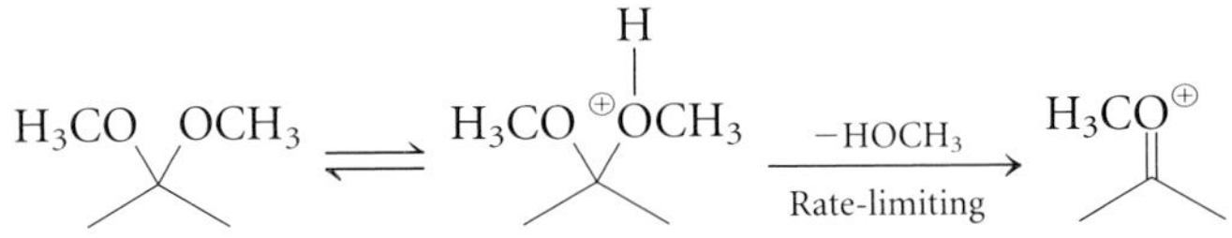

$$H_3CO \underset{}{\overset{OCH_3}{\bigvee}} \rightleftharpoons H_3CO \overset{\overset{\displaystyle H}{|}}{\underset{}{\overset{\oplus}{O}CH_3}} \xrightarrow[\text{Rate-limiting}]{-HOCH_3} H_3CO^{\oplus}$$

Based on this knowledge, predict which ketal in each of the following pairs will undergo hydrolysis more rapidly.

(a)  $H_3CO \overset{OCH_3}{\bigvee}$  versus  $H_3CO \underset{H}{\overset{OCH_3}{\bigvee}}$

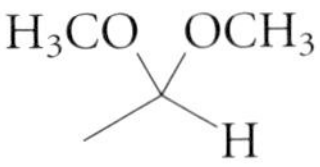

(b) 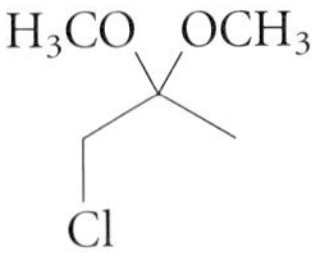 $H_3CO \overset{OCH_3}{\bigvee}_{CH_2Cl}$  versus  $H_3CO \overset{OCH_3}{\bigvee}$

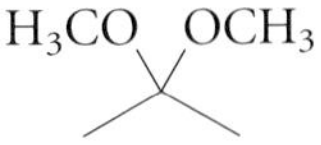

(c) 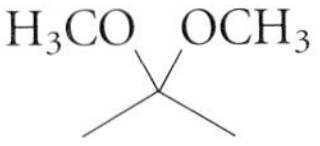 $H_3CO \overset{OCH_3}{\bigvee}$  versus  $H_3CO \overset{OCH_3}{\bigvee}Ph$

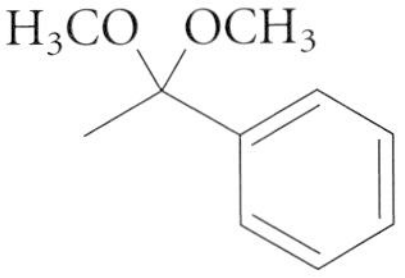

**12.10** The hydrolysis of thioacetals and thioketals is considerably slower than the same reaction of the corresponding acetals and ketals. Based on your answer to Problem 12.9, suggest a reason for this difference in reactivity.

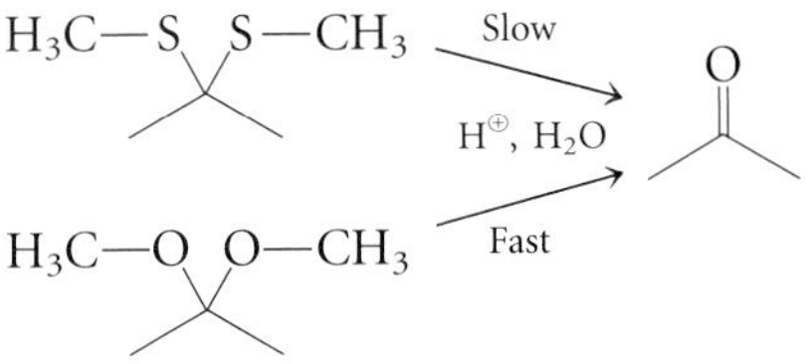

$$H_3C-S \overset{S-CH_3}{\bigvee} \xrightarrow{\text{Slow}}$$

$$\xrightarrow{H^{\oplus}, H_2O}$$

$$H_3C-O \overset{O-CH_3}{\bigvee} \xrightarrow{\text{Fast}} \overset{O}{\underset{}{\bigvee}}$$

## Supplementary Problems

**12.11** Provide an IUPAC name for each of the following structures. If chiral centers are present, assign the absolute stereochemistry as *R* or *S*.

(a) 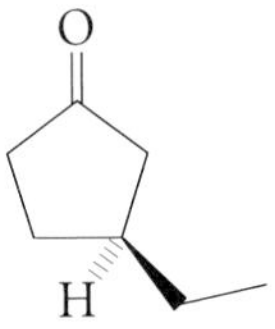  (d) 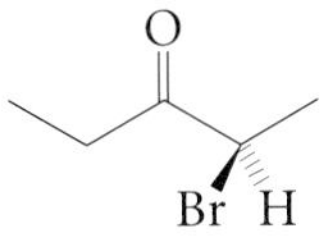  (g)

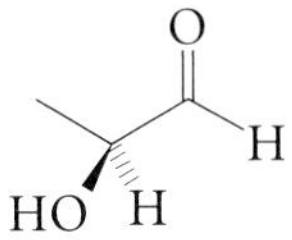

(b) 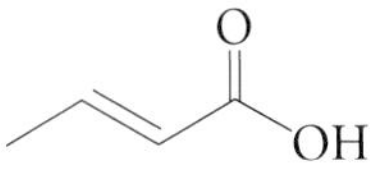  (e) 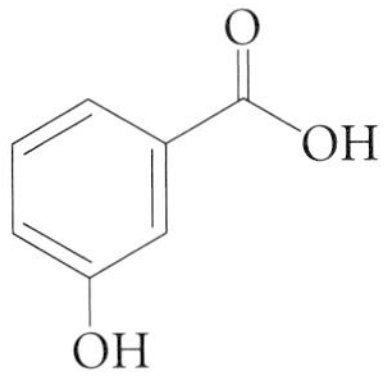  (h)

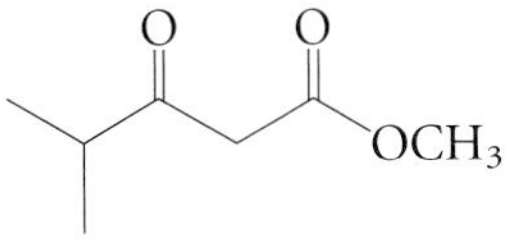

(c) 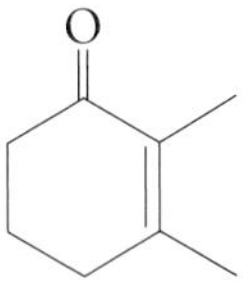  (f) 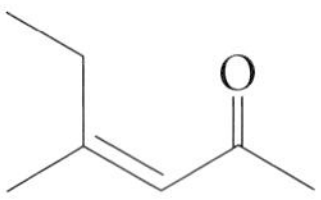  (i)

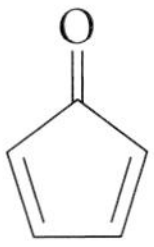

**12.12** Write a structure corresponding to each of the following IUPAC names:

(a) (*S*)-2-methylcyclohexanone

(b) *cis*-pent-2-enol

(c) (*R*)-2-hydroxypentan-3-one

(d) *trans*-2-pentenoic acid

(e) 4-nitrobenzaldehyde

(f) 2,4-pentanedione

(g) (*S*)-4-hydroxy-2-cyclopentenone

(h) (*R*)-2-methylpentanal

**12.13** The following sequence of reactions has been used successfully as a route for the conversion of aldehydes to ketones.

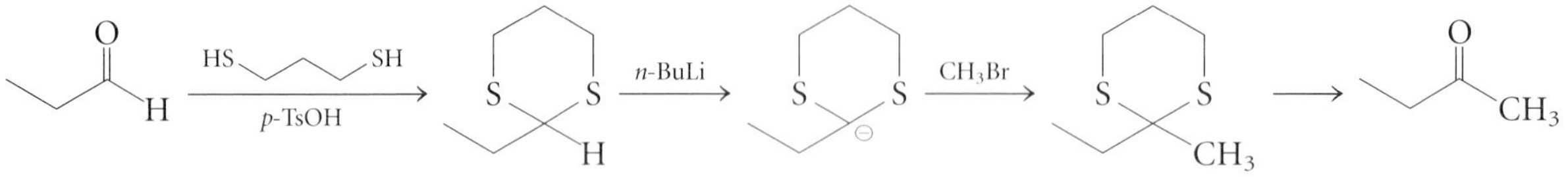

(a) Write a mechanism for the formation of the thioacetal in the first step. Is this reaction faster or slower than the formation of the corresponding acetal with ethylene glycol, HOCH$_2$CH$_2$OH?

(b) A key step in this sequence is the deprotonation in the second step. Explain why this deprotonation occurs with the thioacetal shown here but fails to occur with a normal acetal.

(c) Based on your knowledge of acetal chemistry, propose a reagent that could be used to induce the last step of this sequence.

**12.14** In many biological redox reactions, nicotinamide is an important oxidant because its alkylated form is very easily reduced—that is, it easily accepts electrons to generate an anion that is then protonated. Consider the following reaction a laboratory model for the biological reduction of an alkylated nicotinamide:

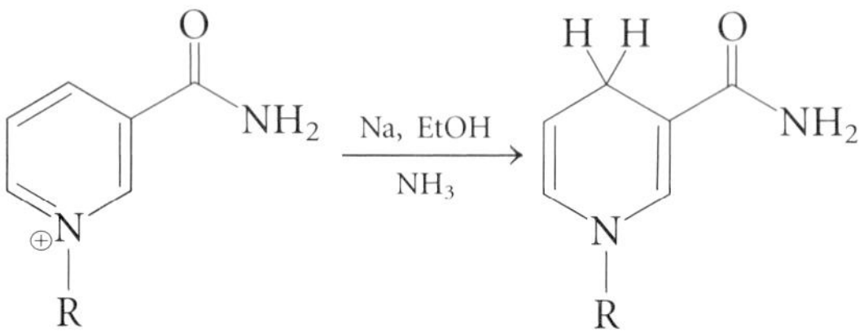

Is the reduction of this nicotinamide easier or harder than the reduction of the related alkylated pyridinium salt?

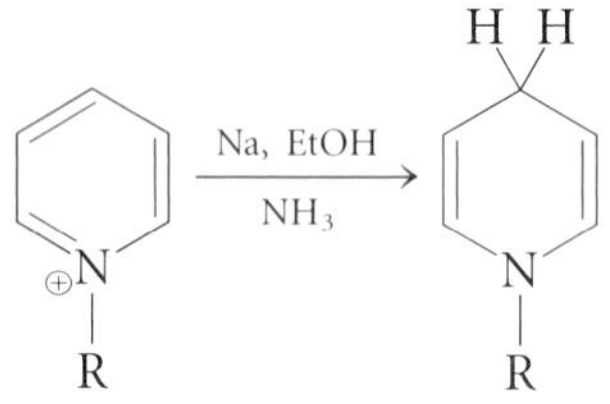

(*Hint:* Write resonance structures for the products obtained from each cation.)

**12.15** Nucleophiles in which another heteroatom is directly attached to the nucleophilic heteroatom are often quite reactive. Examples include $^{\ominus}$O—OH, which is more nucleophilic than $^{\ominus}$OH, and H$_2$N—NH$_2$, which is more nucleophilic than NH$_3$. Suggest a reason for the enhanced nucleophilicity resulting from the presence of the adjacent heteroatom.

**12.16** The formation of an imine requires the addition of a primary amine (or ammonia) to an aldehyde or a ketone. In a second and equally important step, the elements of water must be lost:

Both stages could conceivably be accelerated by acid. In the first step, protonation of the starting carbonyl group oxygen increases the electrophilicity of the carbonyl carbon. In the second step, protonation of oxygen again provides a better leaving group (water versus $^{\ominus}$OH).

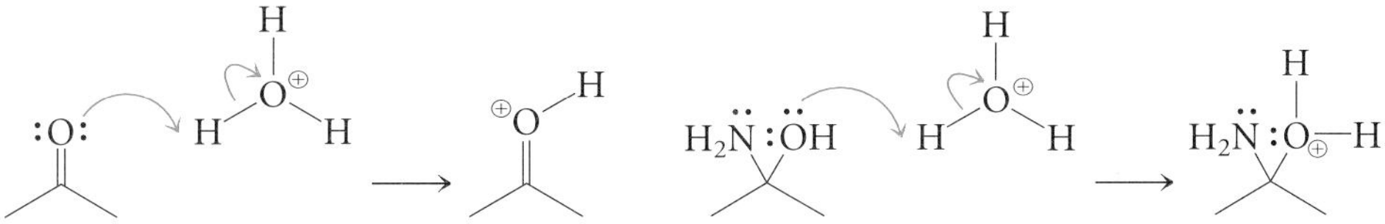

However, excess acid results in a slower rate of imine formation, with the maximum rate being observed at approximately pH 5. Can you explain why increasing the acidity below pH 5 should result in a slower rate of reaction?

**12.17** The reaction of ammonia or a primary amine with a ketone usually results in the formation of an imine rather than an enamine.

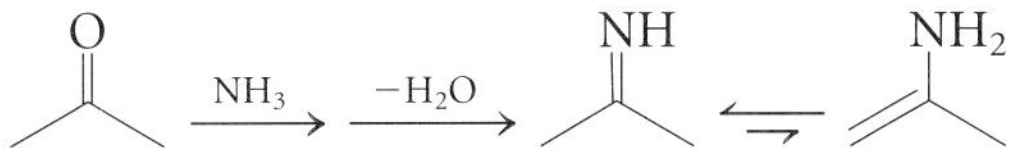

However, in the case of a β-ketoester, the amount of enamine tautomer is increased and often dominates the equilibrium with the imine. Why should the enamine of a β-ketoester be unusually stable?

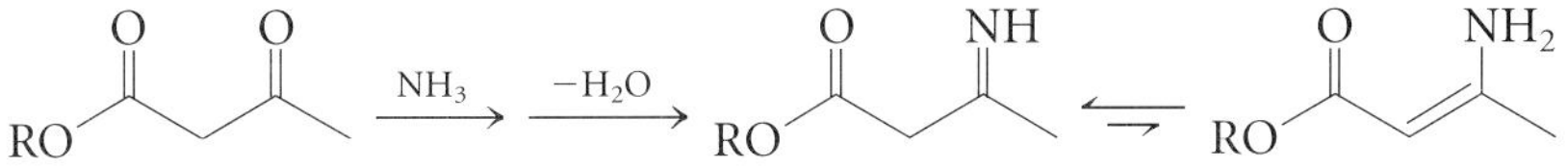

**12.18** Carboxylic acid chlorides can be prepared readily from carboxylic acids using thionyl chloride, $SOCl_2$. An alternative reagent, oxalyl chloride, can also be used and has advantages in certain cases.

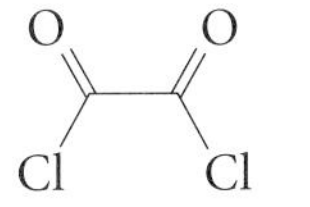

**Oxalyl chloride**

Write a mechanism for the reaction of oxalyl chloride with a carboxylic acid to form a carboxylic acid chloride. (*Hint:* The details of the mechanisms are very similar; only the structural details of the reagents differ.)

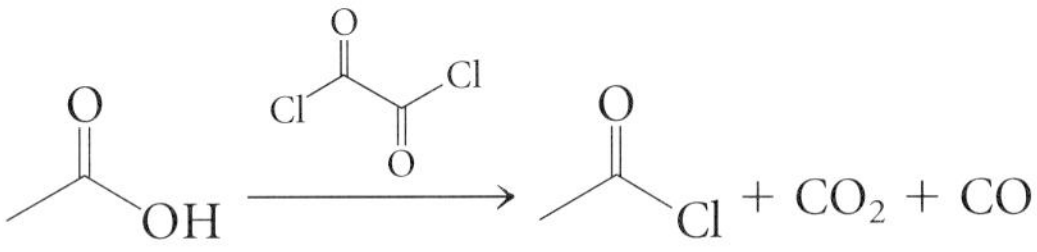

$$+ CO_2 + CO$$

**12.19** Use the bond energies provided in Table 3.5 to calculate $\Delta H°$ for each of the following reactions. Be sure to indicate whether the reaction is exothermic (negative $\Delta H°$) and therefore favorable in the direction written, or endothermic.

(a)

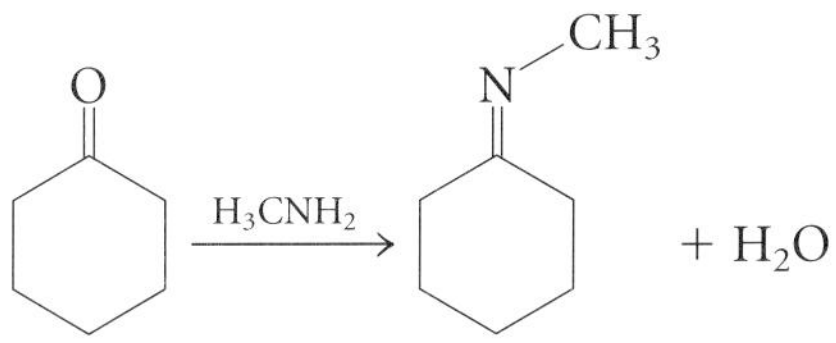

$$+ H_2O$$

(b)

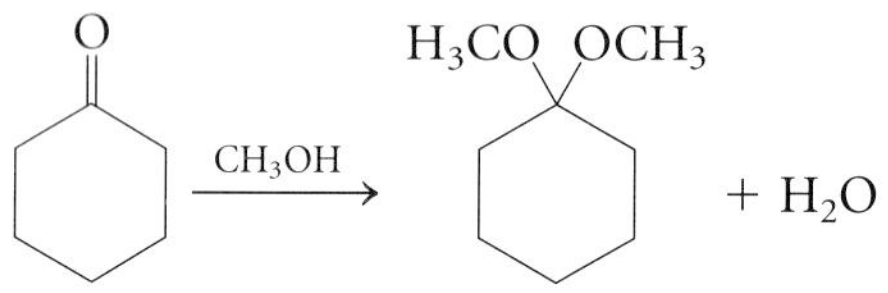

$$+ H_2O$$

(c)

$$+ H_2O$$

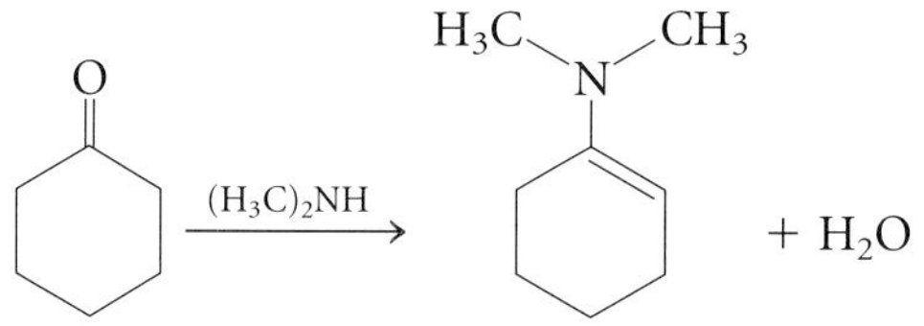

**12.20** Ketals and acetals are often used as protecting groups (derived functionalities that temporarily hide the characteristic reactivity of the carbonyl group). Formation of cyclic ketals from diols also hides the characteristic reactivity of the hydroxyl group:

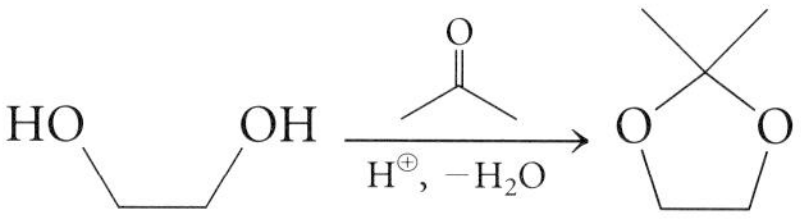

For each of the following glycols, indicate whether reaction with acetone is expected to lead to the formation of a ketal. Draw a clear representation of each ketal that is formed.

(a)

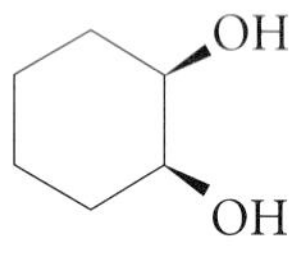

(c)

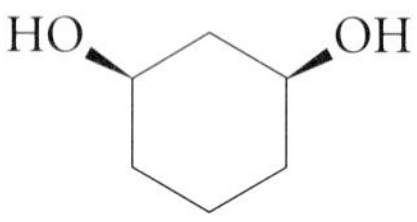

(e)

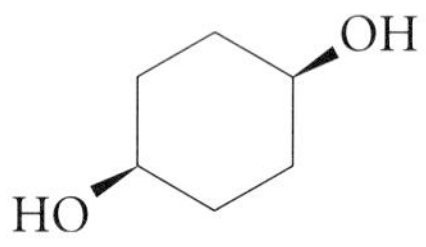

(b)

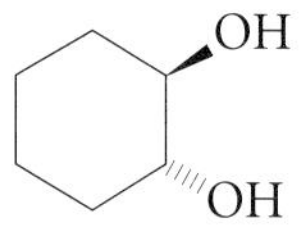

(d)

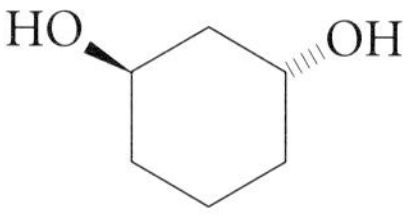

(f)

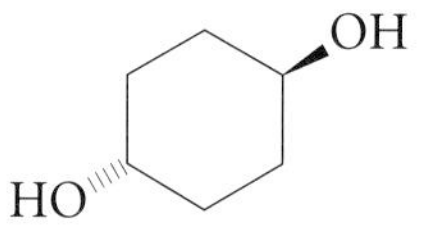

**12.21** What starting materials are required to prepare each of the following amines by reductive amination?

(a)

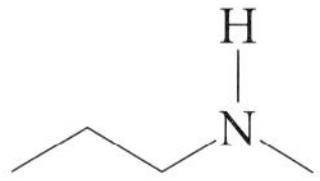

(c)

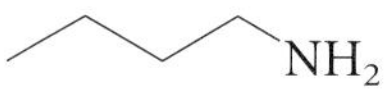

(e)

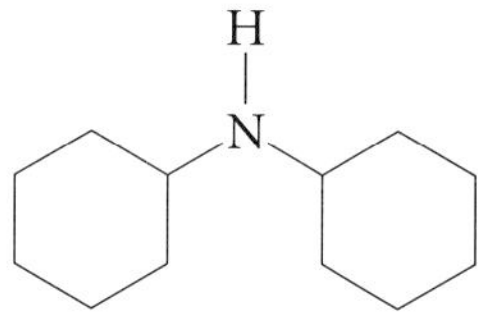

(b)

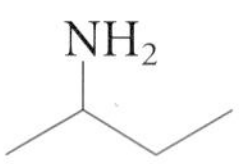

(d)

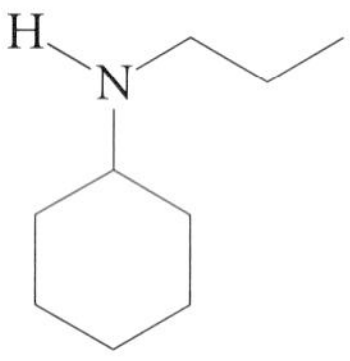

(f)

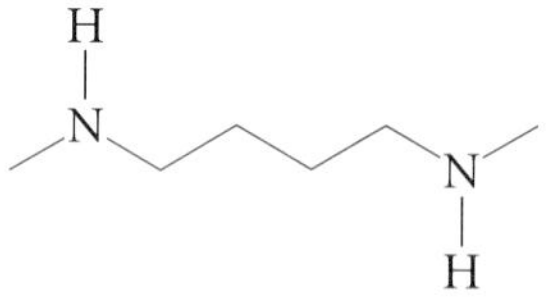

**12.22** Reaction of certain dialdehydes with primary amines in the presence of reducing agents such as sodium cyanoborohydride leads to the formation of cyclic amines. Propose a detailed, stepwise mechanism for this transformation:

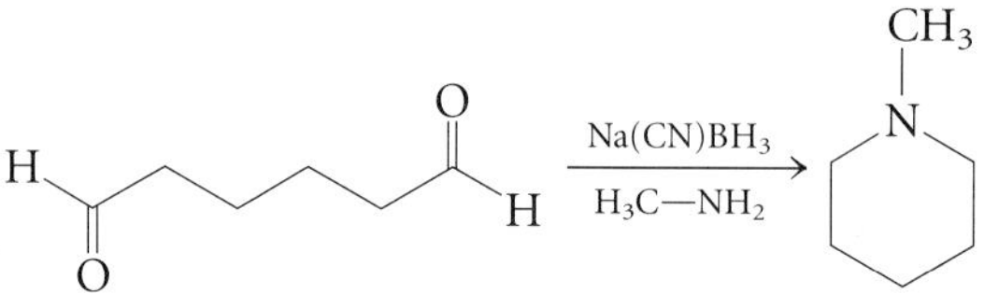

**12.23** Specify the starting materials required to prepare each of the following amines by reduction of a carboxylic acid amide:

(a)

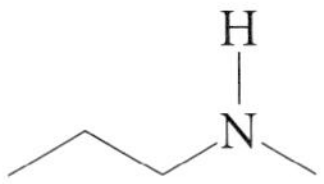

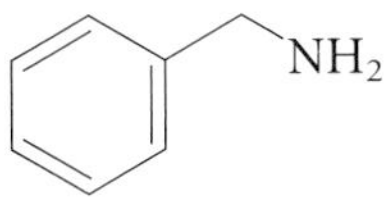

(b)

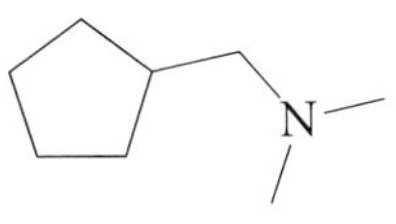

(c)

(d)

(e)

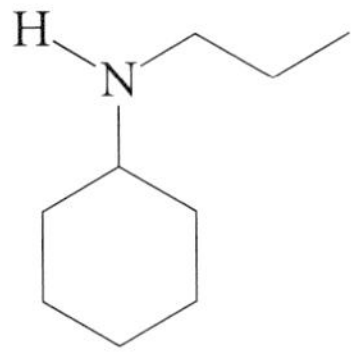

(f)

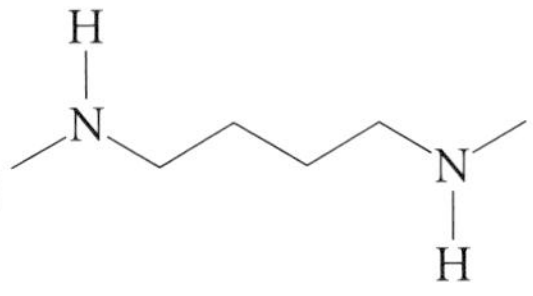

**12.24** Would you expect an N-aryl amide (an amide derived from aniline) to be more or less reactive toward nucleophilic addition to the carbonyl carbon than an amide bearing an alkyl group? Explain your answer.

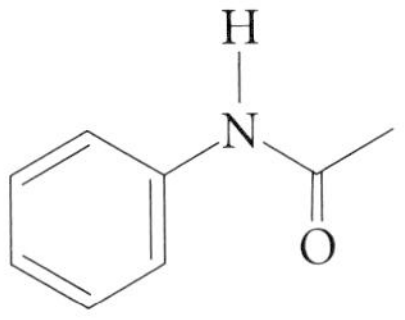

versus

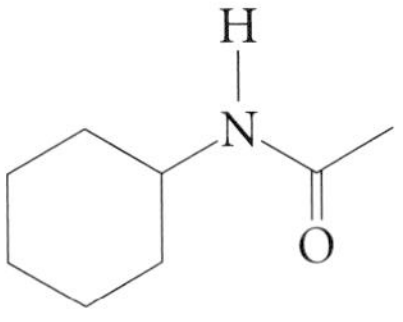

**12.25** Specify the reagents necessary to prepare the following carboxylic acid derivatives from acetic acid. (More than one step may be required.)

(a)

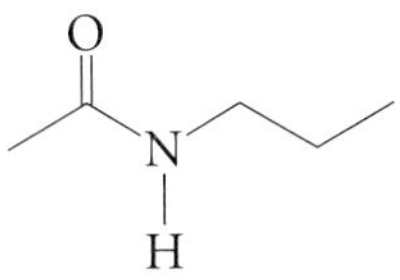

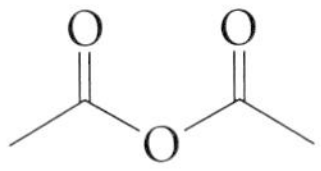

(b)

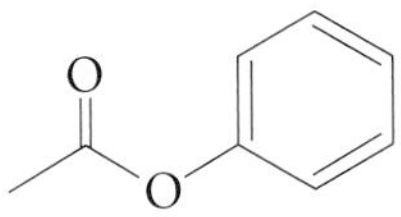

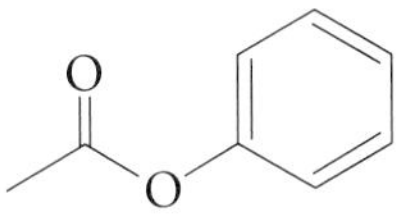

(c)

(d)

(e)

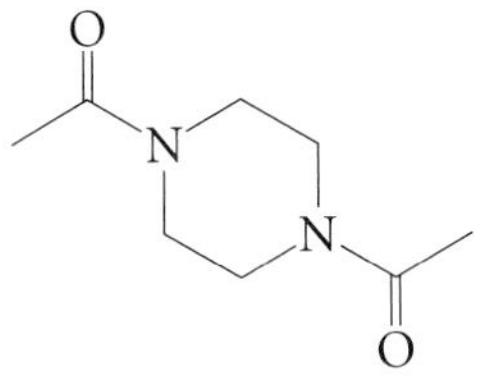

(f)

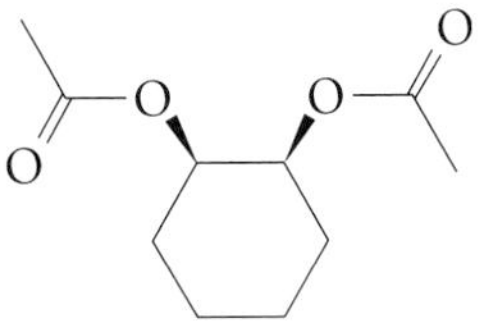

**12.26** Esters derived from phenol are considerably more reactive than those derived from cyclohexanol. Can you think of an explanation for this difference in reactivity?

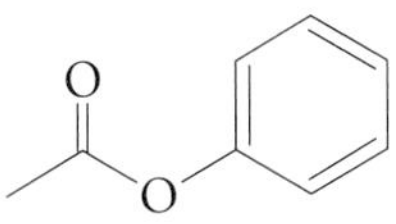

versus

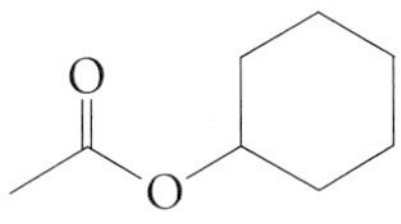

**12.27** Propose a chemical test and a spectroscopic method to distinguish the compounds in each of the following pairs:

(a) 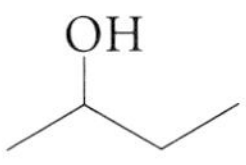 and

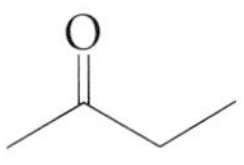

(b) 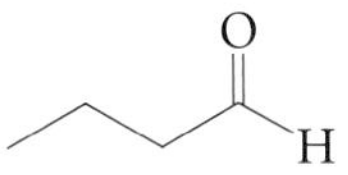 and

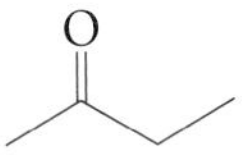

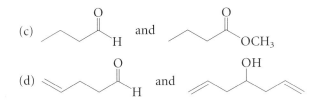

(c) and

(d) and

**12.28** Each of the following compounds can be prepared in at least two unique ways. In each case, two separate reactions are required, one being the addition of a Grignard reagent to a carbonyl compound. Show one viable two-step sequence that could be used to prepare each of the following products. (You may use any source of carbon atoms and any needed reagents as long as the sequence includes a Grignard reaction.)

(a)

(c)

(b)

(d)

**12.29** The reaction of acetone with cyanide ion forms a cyanohydrin. This reaction represents the first step in one industrial synthesis of methyl methacrylate, an important olefin used in production of the polymer plastics Plexiglas and Lucite (Chapter 16). The conversion of the cyanohydrin to methyl methacrylate is carried out by treatment with acid and methanol. Write a detailed mechanism for this transformation. (*Hint:* The first step is an acid-catalyzed loss of water.)

Acetone     Cyanohydrin     Methyl methacrylate

# Substitution Alpha to Carbonyl Groups

*Enolate Anions and Enols as Nucleophiles*

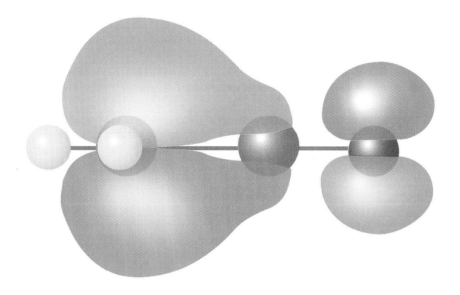

*The highest occupied π molecular orbital of an enolate anion. The electron density of lower-lying orbitals is greater near the more electronegative oxygen atom (right). The electron density in this π molecular orbital is distributed mainly between the two carbon atoms. When the overlapping orbitals of the π system and the nucleophile are of similar size, orbital overlap (and therefore bond formation) can be maximized. Thus, larger electrophiles react most rapidly at the carbon end of this orbital (the carbon alpha to the carbonyl group), whereas small electrophiles—especially protons—react most rapidly at the oxygen end.*

In Chapters 8 through 12, we examined in detail some of the many important reactions of organic chemistry. They are classified by their mechanisms as substitutions, eliminations, additions, and so on. Because this is a course in organic chemistry, we usually focus on the organic portion of a reaction, and in most cases the assignment of reaction class has been unambiguous. For example, the reaction of hydroxide ion with methyl bromide to form methanol is classified as an $S_N2$ reaction, because we are focusing on what is happening at the carbon of methyl bromide, not on what is happening to the oxygen atom of hydroxide ion.

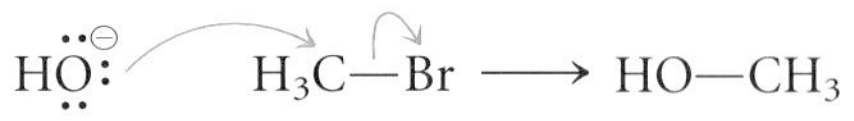

$$HO:^{\ominus} \qquad H_3C\!-\!Br \longrightarrow HO\!-\!CH_3$$

However, in many reactions, two organic compounds are joined. One example is the reaction of an alkyl halide with the anion derived by deprotonation of a terminal alkyne:

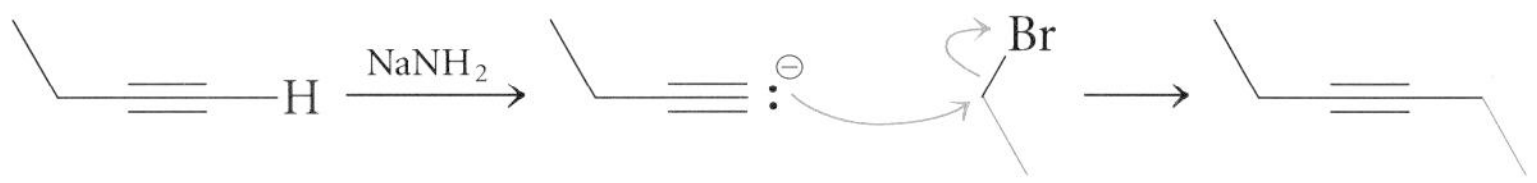

We classified this bond formation as an $S_N2$ reaction in Chapter 8, focusing on the alkyl halide—but we could just as easily take the view that the reaction is a substitution of a different kind–namely, of the hydrogen of the alkyne by an alkyl group. And, of course, it is both.

This chapter brings together a large and diverse group of reactions under the theme of the replacement of a hydrogen at the position alpha ($\alpha$) to a carbonyl group. Recall that relative positions of functional groups within a molecule are designated with Greek lowercase letters. For example, 2-bromocyclohexanone is an $\alpha$-bromoketone; 3-ketocyclohexanone is a $\beta$-diketone:

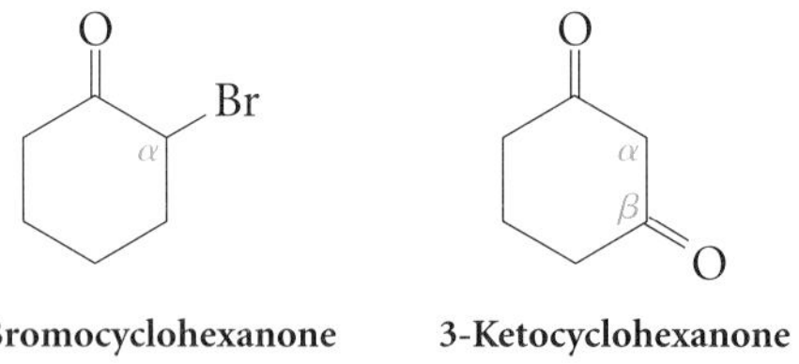

2-Bromocyclohexanone        3-Ketocyclohexanone

Specifically, we will look at replacement of an $\alpha$ hydrogen by a halogen atom or a carbon substituent. A typical example is the two-step process in which deprotonation of cyclohexanone by a base is followed by reaction of the resulting anion with methyl iodide to form 2-methylcyclohexanone:

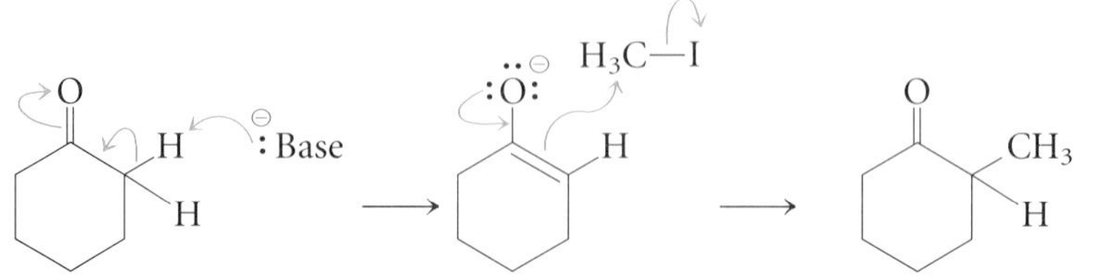

Cyclohexanone                                           2-Methylcyclohexanone

A variety of electrophilic partners can take part in such reactions, including aldehydes, ketones, acid chlorides, and esters, as well as alkyl halides (and sulfonate esters). In most cases, an enolate anion plays a crucial role as a reactive intermediate.

# Formation and Reactions of Enolate Anions and Enols

Ketones, aldehydes, and carboxylic acid derivatives (esters, amides, etc.) that have protons on the carbon $\alpha$ to the carbonyl group can be deprotonated to form anions known as **enolate anions.** These anions are substantially more stable than simple carbanions, and this increased stability is reflected in the acidity of the carbonyl compounds themselves (see Table 6.1). Enols are formed from enolate anions by protonation on oxygen:

An enolate anion          An enol

## Molecular Orbitals of Enolate Anions

The enhanced acidity of a hydrogen $\alpha$ to a carbonyl group compared with that of a hydrogen in a simple alkane is a direct result of the enhanced stability of the enolate anion.

| | | | |
|---|---|---|---|
| p$K_a$ | 17 | 19 | 25 | 50 |

These anions are stabilized by resonance delocalization of electron density such that both oxygen and carbon bear significant negative charge in the hybrid (Figure 13.1, on page 662). Each of the two $\pi$ molecular orbitals is filled with two electrons. The lower-lying orbital has electron density on three atoms, but the density is concentrated on the central carbon atom. The HOMO (highest occupied molecular orbital) has electron density distributed so as to avoid the electron density of the underlying $\pi$ molecular orbital. The combined electron density of these two orbitals distributes greater charge to the oxygen end than to the carbon end of the enolate anion. However, it is the HOMO that provides electron density for bonding between an enolate anion and an electrophile.

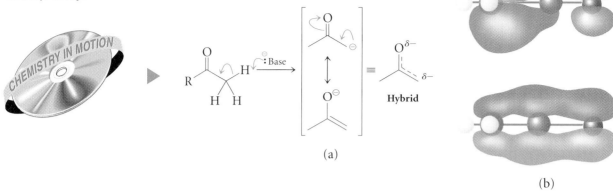

(a)

(b)

**FIGURE 13.1**

(a) The enolate anion is stabilized by resonance delocalization of electron
density—in the hybrid, both oxygen and carbon bear partial negative charge.
(b) The orbital representations depict the two filled $\pi$ molecular orbitals.
The upper one (also shown at the beginning of the chapter) is the highest
occupied molecular orbital (HOMO) of the enolate anion. (Not shown is
the third orbital formed by the combination of the three atomic $p$ orbitals.
In an enolate anion, this third orbital has no electrons.)

## Structure of Enolate Anions

The structure of an enolate anion is influenced by the associated counterion. For enolate anions derived from ketones and esters, the use of lithium
as the counterion generally leads to higher yields in subsequent reactions.
Thus, we will see many examples of lithium enolate anions. The structures
of these lithium enolates are complex, often involving aggregates of several
molecules joined together by common bonds to lithium. Figure 13.2 shows
the crystal structure of the lithium enolate formed by deprotonation of
$t$-butyl methyl ketone in tetrahydrofuran.

## Protonation of Enolate Anions

Protonation of an enolate anion on oxygen gives the corresponding
enol; protonation on carbon yields the corresponding tautomeric ketone.
In most cases, the considerably greater strength of a C=O $\pi$ bond compared with a C=C $\pi$ bond results in an equilibrium that favors the carbonyl tautomer.

**Enolate**        **Enol**        **Ketone**

Protonation of an enolate anion proceeds more rapidly on oxygen, producing an enol. Why does protonation of the enolate anion to form the less
stable enol occur more rapidly than protonation to form the ketone? The
rate of proton-transfer reactions is usually controlled by the degree to which
electron density can effectively overlap with the small $1s$ orbital of a pro-

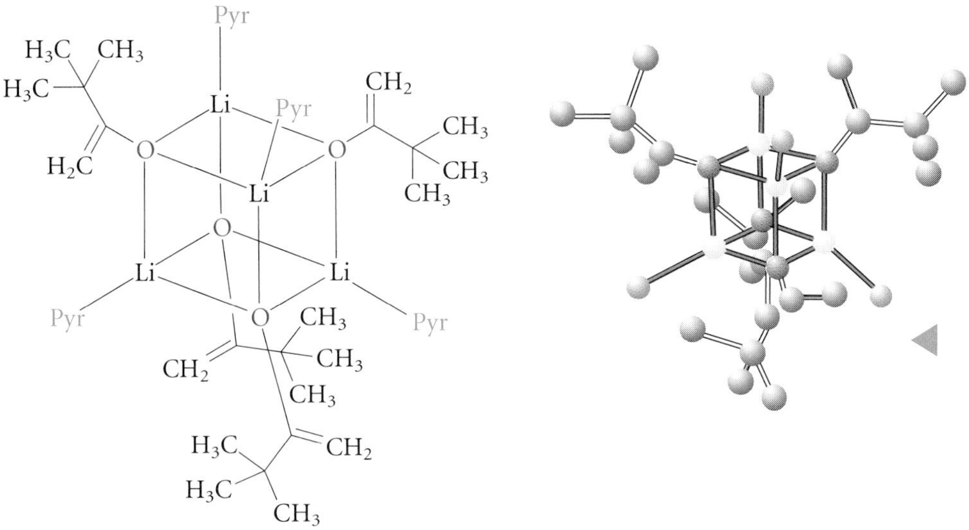

**FIGURE 13.2**

Crystal structure of the lithium enolate formed by deprotonation of *t*-butyl methyl ketone in tetrahydrofuran in the presence of pyridine (Pyr). There are four lithium enolates arranged in cubic fashion. Each of the four lithium atoms (white) is at one of the eight corners of the cube—the lithium atoms and oxygen atoms (red) are at alternating corners. Each lithium atom is surrounded by four ligands: three oxygen atoms from enolate anions and the nitrogen atom (blue) of a pyridine. (For clarity, only the nitrogen atoms of the pyridines are shown in the ball-and-stick representation; the other atoms have been omitted, along with the hydrogens of the methyl and methylene groups.)

ton. The electron density in the HOMO of an enolate anion (Figure 13.1) is quite compact at oxygen but large and diffuse at carbon. Thus, orbital overlap with a proton is most effective at oxygen. However, regardless of where the proton is first attached, proton tautomerization rapidly establishes the equilibrium concentrations of enol and ketone. In the case of acetone, the ketonic form is favored by a factor of more than $10^6$.

## EXERCISE 13.1

Ketones and enols are rapidly interconverted in protic solvents in the presence of low concentrations of acid or base. (a) Write a detailed reaction mechanism for the conversion of a ketone to an enol catalyzed by base ($^{\ominus}$OH). (b) Write a mechanism for the reaction in the presence of catalytic acid ($H_3O^{\oplus}$).

## Halogenation Alpha to Carbonyl Groups

*Halogenation of Ketones under Basic Conditions: A Sequence Out of Control.* Enolate anions generated in aqueous or alcoholic solution can displace halide ion in an $S_N2$ reaction with molecular iodine, bromine, or chlorine, effecting the halogenation of the enolate anion at the $\alpha$ position. As we shall see, this reaction initiates a cascade of halogenation reactions, each proceeding more rapidly than the preceding one.

663

***Halogenation of Methyl Ketones under Basic Conditions: The Iodoform Reaction.*** With methyl ketones and iodine, this $S_N2$ reaction is the first step of a sequence called the **iodoform reaction,** in which a ketone with three $\alpha$ hydrogens is converted to a carboxylic acid and iodoform ($CHI_3$). Let's consider the reaction of acetophenone in aqueous base in the presence of three equivalents of iodine.

*Sequential Halogenation.* Carbon–halogen bonds are formed $\alpha$ to a carbonyl group by sequential $S_N2$ displacement reactions of iodide ion from $I_2$. Because acetophenone is a considerably weaker acid than water, only a very small concentration of the enolate anion is formed. But because the enolate anion is so reactive, it is removed from the equilibrium by reaction and quickly reformed in an acid–base pre-equilibrium. As iodine approaches the negatively charged enolate, the I—I bond becomes polarized, allowing nucleophilic displacement (Figure 13.3).

The $\alpha$ hydrogen atoms in the resulting $\alpha$-iodoacetophenone are even more acidic than those in acetophenone itself, as a result of inductive electron withdrawal by iodine. Thus, in base, deprotonation occurs rapidly to form an iodoenolate. Nucleophilic displacement of $I^{\ominus}$ from $I_2$ with the io-

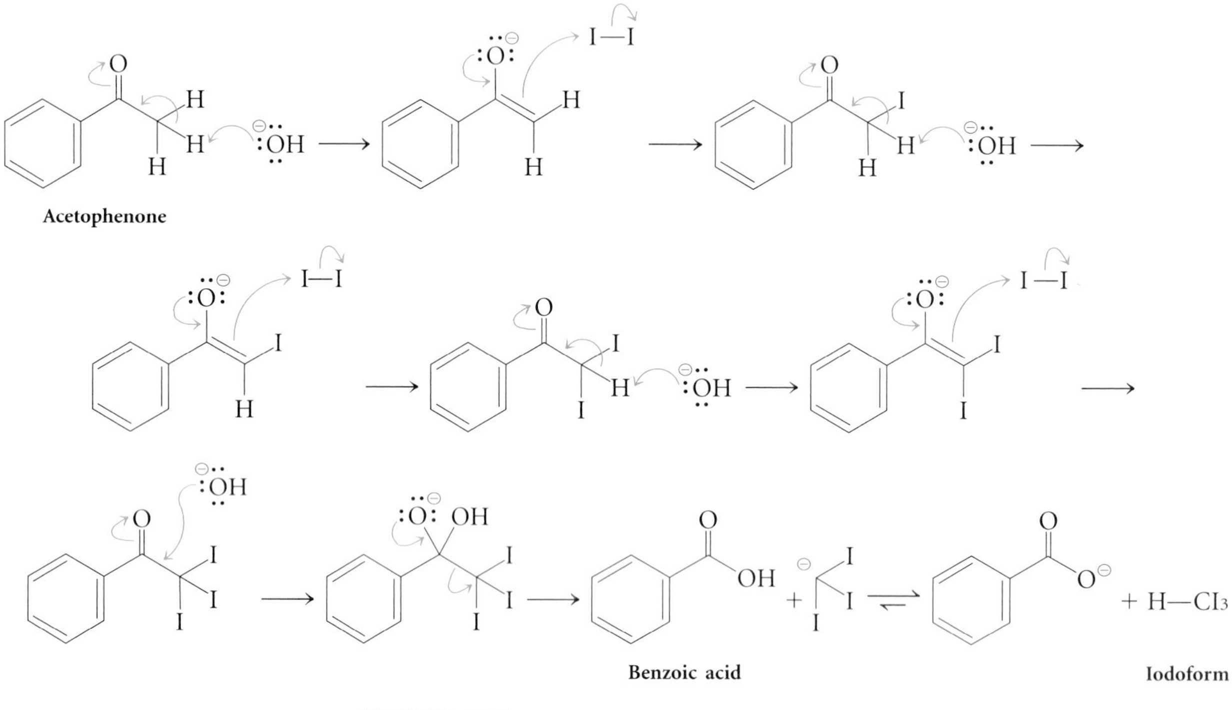

**FIGURE 13.3**

The iodoform reaction begins with the formation of an enolate anion from a ketone of the general type $RCOCH_3$ (illustrated here by acetophenone). The enolate anion acts as a nucleophile to displace $I^{\ominus}$ from $I_2$ by an $S_N2$ route. The resulting $\alpha$-iodoketone produced has two acidic hydrogen atoms, and enolization takes place again. The resulting iodoenolate anion reacts again by an $S_N2$ pathway, producing a diiodoketone. Upon repetition of this sequence, a triiodoketone is formed. Largely because of the stability of the triiodomethyl anion, the triiodoketone is converted by base into a carboxylic acid and iodoform.

doenolate as nucleophile results in a diiodoacetophenone, in which the remaining carbon–hydrogen bond is even more acidic. Repetition of the deprotonation and displacement reactions produces triiodoacetophenone.

*Cleavage of the Triiodoketone.* At this point, there are no $\alpha$ hydrogen atoms available for deprotonation, and hydroxide ion now assumes another role. Acting as a nucleophile, hydroxide ion adds to the carbon end of the C—O bond dipole, as in the first step in nucleophilic acyl substitution (covered in Chapter 12), to form a tetrahedral intermediate. The C=O bond is re-formed by expulsion of $^\circ CI_3$, a particularly stable carbanion because of the three attached polarizable halogens and therefore a good leaving group. Protonation produces iodoform ($CHI_3$, mp 120 °C), which forms a bright yellow precipitate because it is not very soluble in water. (It is also very heavy, with a density of 4.1 g/mL.) The formation of this precipitate is considered a positive **iodoform test.** The iodoform test is diagnostic for the presence of a methyl ketone, because $CHI_3$ cannot be formed unless there are three replaceable hydrogen atoms on one $\alpha$ carbon atom.

*The Haloform Reaction.* This type of reaction also works with $Br_2$ or $Cl_2$ and is referred to in general as the **haloform reaction.** The products of the haloform reaction with acetone are the trihaloacetones; the space-filling models of these molecules show the difference in size between chlorine, bromine, and iodine:

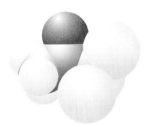

Trichloroacetone

Tribromoacetone

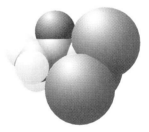

Triiodoacetone

However, because bromoform and chloroform are colorless liquids, the reactions with $Br_2$ and $Cl_2$ are not useful as diagnostic tests. The key feature of the haloform reaction is the increased acidity of the remaining $\alpha$ hydrogens with each successive halogenation.

For methyl ketones with no other $\alpha$ hydrogens (for example, acetophenone), the haloform reaction can be used to prepare carboxylic acids in good yield. With a methyl ketone that has other $\alpha$ hydrogens—and with ketones and aldehydes in general—indiscriminate halogenation occurs at the carbons $\alpha$ to the carbonyl group, and the reaction is not synthetically useful.

### EXERCISE 13.2

Which of the following molecules will give a positive iodoform test?

(a) 3-pentanone      (c) pentanal      (e) acetic acid

(b) 2-pentanone      (d) acetophenone

*Halogenation of Ketones under Acidic Conditions.* Halogenation of ketones can also be accomplished under acidic conditions—for example, using bromine in acetic acid solution. Under such conditions, without a base present, halogenation of a ketone produces a mole of the corresponding halogen acid (HX) for every mole of halogen consumed. The rate of halogenation increases with the increase in acid concentration as the reaction proceeds.

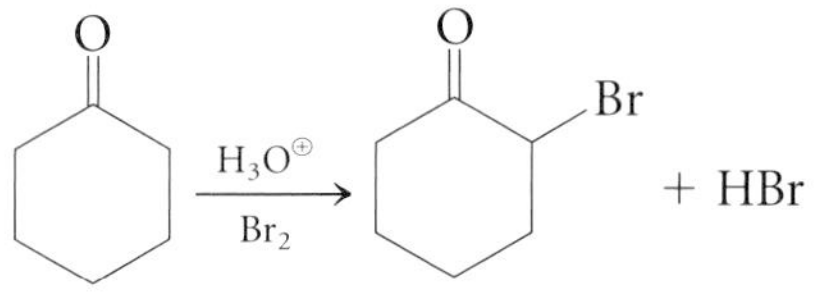

Acid-catalyzed halogenation proceeds through the enol of the ketone, which is generated by protonation of the carbonyl group oxygen atom, followed by loss of a proton from the adjacent carbon. Formation of the enol is the rate-limiting step in the sequence. The enol reacts rapidly with bromine, and subsequent loss of a proton forms the product $\alpha$-bromoketone.

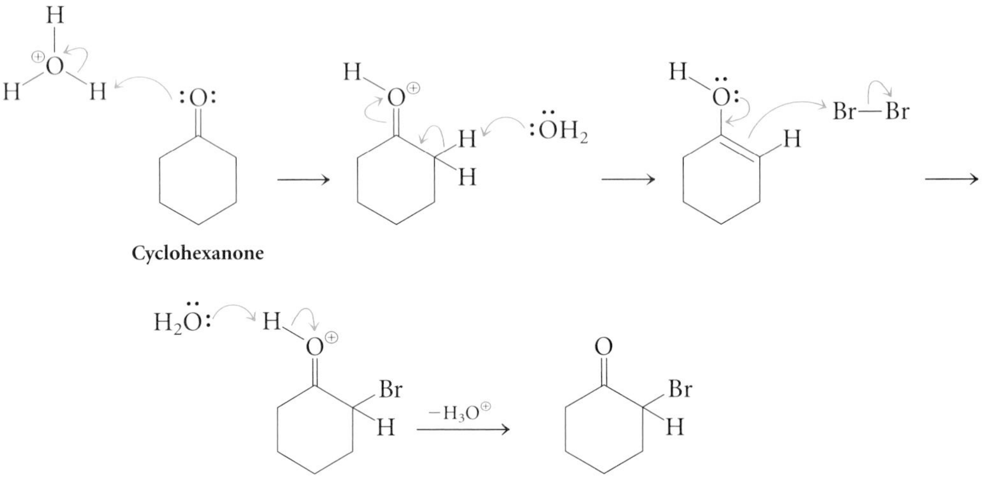

Cyclohexanone

$\alpha$-Bromocyclohexanone

The rate of formation of the enol of a ketone in acidic solution *decreases* as $\alpha$ hydrogen atoms are replaced by halogen atoms, possibly because the presence of an $\alpha$ halogen atom decreases the basicity of the carbonyl group. Thus, using only one equivalent of chlorine or bromine allows the monohaloketone to be obtained in good yield.

***The Hell–Volhard–Zelinski Reaction.*** Monohalogenation $\alpha$ to a carboxylic acid group can be achieved by treatment of a carboxylic acid bearing $\alpha$ hydrogen atoms with bromine in the presence of phosphorus tribromide, a reaction referred to as the **Hell–Volhard–Zelinski reaction.** In phenylacetic acid, for example, $\alpha$-bromination takes place with high yields. The mechanism is an $S_N2$-like pathway similar to that for the iodoform reaction (see Figure 13.3).

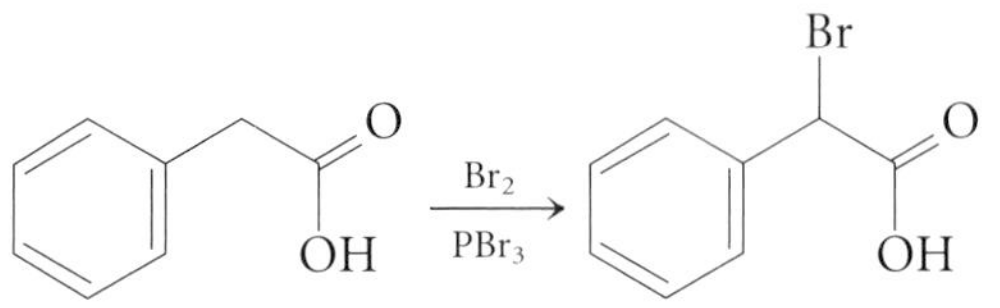

The mechanism of this reaction is less important than its synthetic utility—for example, $\alpha$-bromoacids are important intermediates in the synthesis of $\alpha$-amino acids.

In the first step of the Hell–Volhard–Zelinski reaction, the carboxylic acid is converted to an acid bromide (the bromine analog of an acid chloride). The acid bromide then undergoes enolization, with the enol displacing bromide from $Br_2$.

(a) Draw the structure of the enol obtained from the acid bromide of acetic acid, and show, with curved arrows, the flow of electrons that accomplishes $\alpha$-bromination.

(b) Explain why a second $\alpha$-bromination does not take place as readily as the first.

## Kinetic versus Thermodynamic Deprotonation of Carbonyl Groups

The iodoform reaction proceeds through a multistep sequence that ultimately replaces all three $\alpha$ hydrogen atoms of a methyl ketone with iodine. The initial deprotonation of the methyl ketone by hydroxide ion gives only a low concentration of the enolate anion, because water is more acidic than the ketone:

$$^{\ominus}OH + \underset{\substack{R \quad CH_3 \\ pK_a \sim 19}}{\overset{O}{\|}} \rightleftharpoons H_2O + \underset{\substack{R \quad CH_2 \\ pK_a\ 15.7}}{\overset{O^{\ominus}}{\|}} \xrightarrow{I_2} \underset{R \quad CH_2-I}{\overset{O}{\|}}$$

Nonetheless, because the small amount of enolate anion is consumed as it reacts rapidly with $I_2$, additional methyl ketone is deprotonated by hydroxide ion, maintaining the equilibrium concentration of enolate anion.

Thus, in the iodoform reaction, the initial product, $\alpha$-iodoketone, is formed in a solution rich in hydroxide ion. The $\alpha$-iodoketone is deprotonated by hydroxide ion and halogenated repeatedly until the triiodoketone ultimately results. This cascade of reactions resulting in multiple substitution is desirable for the iodoform reaction; however, in most cases, a chemist sets out to replace only one of the available hydrogen atoms.

*Quantitative Deprotonation.* Ketones and esters can be converted essentially completely to the corresponding enolate anions by using only one equivalent of a base that is significantly stronger than hydroxide ion.

$$\underset{\substack{H_3CO \quad CH_3 \\ pK_a\ 25}}{\overset{O}{\|}} + \underset{\substack{\text{Lithium diisopropyl-}\\ \text{amide}}}{Li-N} \rightleftharpoons \underset{\substack{H_3CO \quad CH_2 \\ pK_a\ 40}}{\overset{O^{\ominus}\ Li^{\oplus}}{\|}} + H-N$$

This process is often referred to as **quantitative deprotonation,** because virtually all of the ketone or ester is deprotonated to form the enolate anion.

**Lithium diisopropylamide**

A number of bases are used for quantitative deprotonation of ketones and esters. Of these, lithium diisopropylamide, abbreviated LDA, is representative, and we will concentrate on its use. LDA and other lithium dialkylamide bases are readily formed by combining an alkyllithium (commonly *n*-butyllithium) and a secondary amine in an ether solvent:

$$n\text{-Bu}-\text{Li} + \text{H}-\text{N(CHMe}_2)_2 \xrightarrow{\text{THF}} n\text{-Bu}-\text{H} + \text{Li}-\text{N(CHMe}_2)_2$$
$$\text{p}K_a\ 40 \qquad\qquad\qquad\qquad\qquad \text{p}K_a\ 50$$

***Quantitative Deprotonation of Unsymmetrical Ketones.*** Quantitative deprotonation of unsymmetrical ketones often leads to formation of a preponderance of one of the two possible enolate anions, as in the examples shown here.

**Methylcyclohexanone**       **99 : 1**

**2-Butanone**       **80 : 22**

Methylcyclohexanone gives essentially only one isomer; with 2-butanone, the predominance of one isomer over the other is less striking.

This regioselective deprotonation removes a proton from a methyl group rather than a methylene group or from a methylene group rather than a methine group, in each case preferentially forming the less substituted enolate anion. Although the origin of this regioselectivity is not fully understood, it is clear that the formation of the less substituted enolate anion is the result of the *rate* of deprotonation rather than the *stability* of the enolate anion. In almost all cases, the more substituted enolate anion predominates at equilibrium.

**Kinetic versus Thermodynamic Deprotonation of Unsymmetrical Ketones**

**Kinetic enolate**       **Thermodynamic enolate**

**1 : 10**

The greater stability of more substituted enolate anions is analogous to the greater stability of more substituted alkenes, and the origin of the preference may be the same in both cases. Because quantitative deprotonation of an unsymmetrical ketone by a strong base often forms the less stable enolate anion, the process is under kinetic control. Thus, quantitative deprotonation is also referred to as *kinetic deprotonation*, and less substituted enolate anions are called *kinetic enolate anions*.

The formation of an equilibrium mixture of both enolate anions from the less sub-stituted, kinetic enolate anion can be accomplished by the addition of a small amount of a proton source such as water. Write a detailed, stepwise reaction mech-anism for the conversion of the less substituted enolate anion of 2-butanone to the more substituted anion.

## 13.2

## Alkylation of Ketones and Esters: $S_N2$ Reaction with Alkyl Halides

Enolate anions are good nucleophiles that react with a variety of elec-trophiles—for example, molecular iodine in the iodoform reaction. Such anions react with $sp^3$-hybridized electrophiles in $S_N2$ reactions, as illustrated here for the enolate anions of cyclohexanone and ethyl acetate:

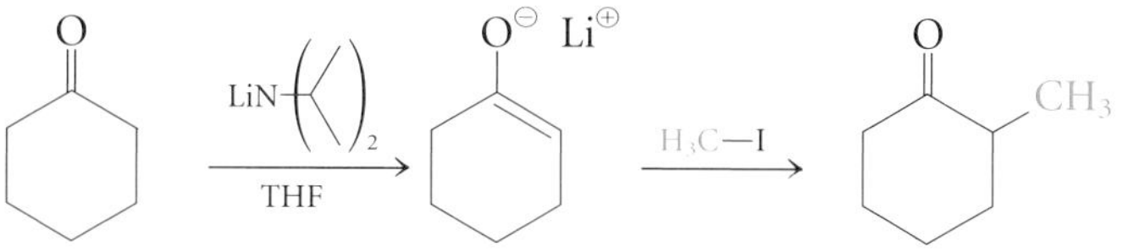

Cyclohexanone                    Methylcyclohexanone

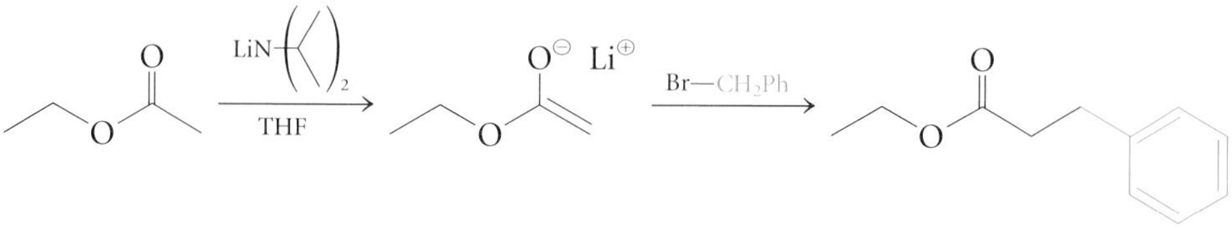

Ethyl acetate                    Ethyl 3-phenylpropanoate

The nucleophilic anion produced by deprotonation of the ketone or ester reacts with an electrophilic carbon species, such as an alkyl halide, to form a carbon–carbon bond. The process replaces an $\alpha$ hydrogen atom of a ke-tone or an ester with an alkyl group: it is therefore referred to as *alkylation*. Substitution can be limited to the introduction of a single alkyl group by the use of one equivalent of base to form the enolate anion quantitatively in the first step. Under these conditions, no dialkylamide base is present during the alkylation step, and the product ketone or ester is formed un-der conditions where further deprotonation followed by alkylation does not occur to a significant extent. Therefore, monoalkylation products predom-inate.

Although we are viewing these reactions as substitution reactions at the $\alpha$ position of carbonyl groups, they are also $S_N2$ reactions of the electrophile. Consequently, they are subject to the restrictions discussed in Chapter 8 with regard to nucleophilic substitution of alkyl halides under basic condi-tions. Enolate anion alkylation reactions afford the highest yields with pri-mary (and methyl) alkylation agents. Alkylation with secondary groups is satisfactory, but side products resulting from elimination are also obtained. The alkylation reaction fails completely with tertiary alkyl halides, and the only products derived from the alkyl halide are those due to elimination.

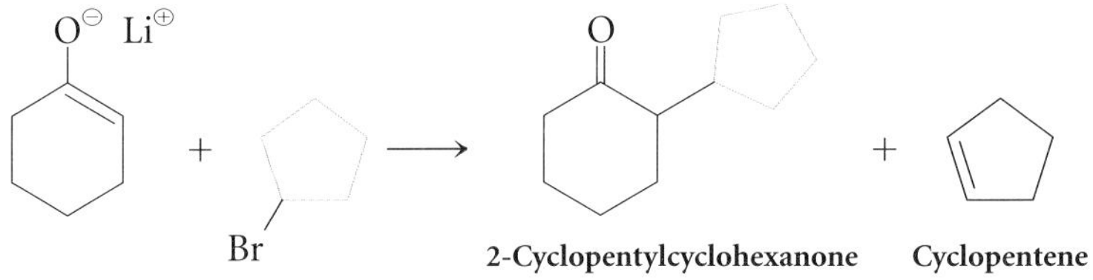

2-Cyclopentylcyclohexanone      Cyclopentene

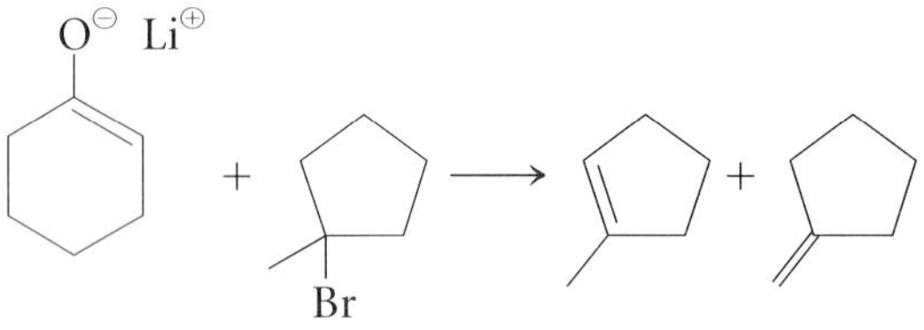

**EXERCISE 13.5**

Identify the starting materials required to prepare each of the following carbonyl compounds by alkylation of a kinetic enolate anion:

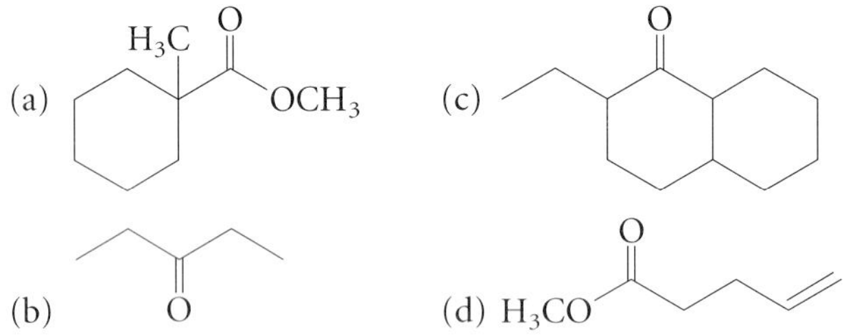

(a)    (c)

(b)    (d) $H_3CO$

## 13.3

# Aldol Reaction, Aldol Condensation, and Related Reactions: Nucleophilic Addition of Enolate Anions to Carbonyl Groups

### The Aldol Reaction

We have seen how enolate anions function as nucleophiles in halogenation and alkylation reactions. In the aldol reaction, enolate anions function as effective nucleophiles in reacting with the electrophilic carbonyl group in ketones and aldehydes. The reaction of acetaldehyde in base is one example. The enolate anion of acetaldehyde effects a nucleophilic addition to the carbonyl group of another molecule of acetaldehyde, forming an intermediate alkoxide ion. The combination of two molecules of acetaldehyde, one as a nucleophilic enolate anion and the other as an electrophilic carbonyl group, results in the formation of a new carbon–carbon bond (highlighted here in green):

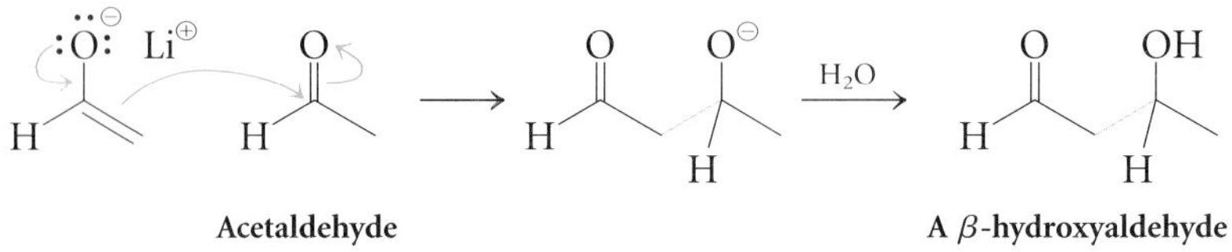

Acetaldehyde                    A $\beta$-hydroxyaldehyde
                               (an aldol)

Protonation of the intermediate alkoxide ion yields the product, a β-hy-droxyaldehyde known as an **aldol** because of the presence of both aldehyde and alcohol functional groups.

In this reaction, a single carbonyl compound serves as the source of both the electrophile and the nucleophile in the reaction that forms a new carbon–carbon bond. This reaction can be carried out by transforming one equivalent of the ketone or the aldehyde to the enolate anion–for example, by quantitative deprotonation—and then adding a second equivalent of the carbonyl compound. However, it is easier simply to add to the carbonyl compound a small amount of a relatively weak base such as hydroxide ion. In this way, a pre-equilibrium is established between the carbonyl com-pound and a small amount of the enolate anion in a solution rich in the carbonyl compound, a good electrophile.

**671**

13.3 Aldol Reaction, Aldol Condensation, and Related Reactions: Nucleophilic Addition of Enolate Anions to Carbonyl Groups

## The Aldol Condensation

If the aldol product is subjected to more vigorous reaction conditions—for example, by increasing the base concentration or the temperature, or both—it undergoes further reaction with hydroxide ion through a reversible deprotonation. Removal of a proton from the carbon atom between the car-bonyl and hydroxyl groups sets the stage for loss of hydroxide ion, pro-ducing an α,β-unsaturated aldehyde. This two-stage reaction is referred to as an **aldol condensation.**

CHEM TV® II

#13   The Aldol Reaction

Acetaldehyde

Crotonaldehyde
(an α,β-unsaturated aldehyde)

Hydroxide ion is a catalyst for both steps of the aldol condensation. At low temperatures (5–10 °C) and low concentrations of base, aldehydes form al-dol products via the aldol reaction. At higher temperatures (80–100 °C) and greater concentrations of base, the reaction proceeds to the α,β-unsaturated aldehyde via the aldol condensation.

## EXERCISE 13.6

In theory, two isomeric aldol condensation products could be formed when ac-etaldehyde is treated with base. However, only the α,β-unsaturated ketone is ob-served, not the β,γ-isomer. This preference can be rationalized on both kinetic and thermodynamic grounds. Provide reasons, based on both kinetics and thermody-namics, for the observed regioselectivity.

Not formed

### Aldol Reaction and Aldol Condensation of Ketones

The aldol reaction of simple ketones such as acetone is energetically unfavorable by a few kilocalories per mole. Therefore, in the presence of base, the equilibrium is unfavorable for the formation of a $\beta$-hydroxyketone (a **ketol**). However, when the temperature is raised, the reaction proceeds through the intermediate aldol to form the aldol condensation product, an $\alpha,\beta$-unsaturated ketone.

Acetone

Aldol
condensation product
(an $\alpha,\beta$-unsaturated ketone)

$+ \, H_2O$

### Intramolecular Aldol Reaction and Aldol Condensation

Both the aldol reaction and the aldol condensation proceed with excellent yields when both carbonyl groups, the electrophile and the nucleophile, are part of the same molecule. For example, treatment of 2,5-hexanedione or 2,6-heptanedione with base leads to the formation of a cyclic $\alpha,\beta$-unsaturated ketone. In the first case, the aldol reaction forms a five-member ring; in the latter, a six-member ring. The dehydration that introduces a double bond into a five-member ring is often considerably faster than the analogous reaction resulting in a cyclohexene product. Thus, treatment of a dicarbonyl compound that can form a cyclopentene product generally produces the $\alpha,\beta$-unsaturated condensation product, even at low temperature. In contrast, when the reaction forms a six-member ring, aldol products obtained at low temperature are converted to $\alpha,\beta$-unsaturated ketones at higher temperature in the presence of base.

2,5-Hexanedione

3-Methyl-2-cyclopentenone

2,6-Heptanedione

3-Methyl-2-cyclohexenone

The aldol condensation of 2,5-hexanedione could have produced a product with a three-member ring and that of 2,6-heptanedione could have produced a product with a four-member ring. However, because of ring strain, aldol products with three- and four-member rings are not observed.

## NATURALLY OCCURRING ALDOLS AND KETOLS

Many compounds produced by plants and animals have β-hydroxy- and α,β-unsaturated ketones or aldehydes. One example is cycloheximide, a β-hydroxyketone isolated from *Streptomyces griseus*. Cycloheximide is used as a fungicide and plant growth regulator. Upon treatment with mild base, cycloheximide undergoes a retro-aldol reaction, producing 2,4-dimethylcyclohexanone, a quite fragrant ketone.

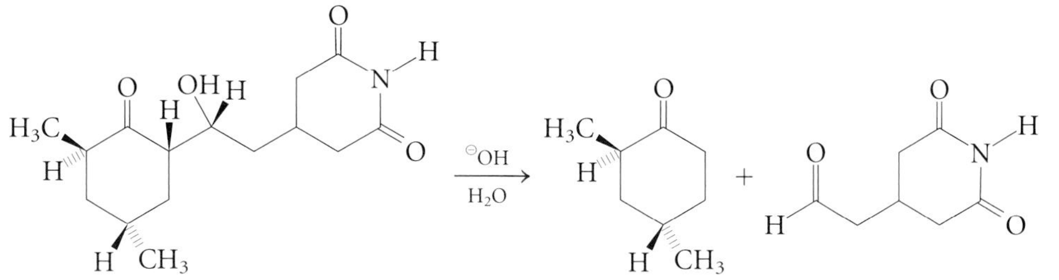

Cycloheximide

2,4-Dimethyl-
cyclohexanone

## EXERCISE 13.7

Draw the structure of the aldol condensation product for each of the following diketones:

(a)

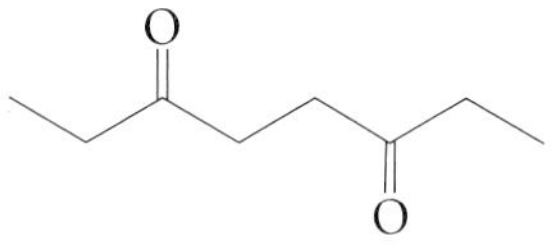

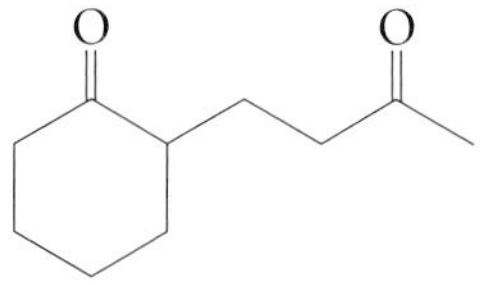

(b)

(c)

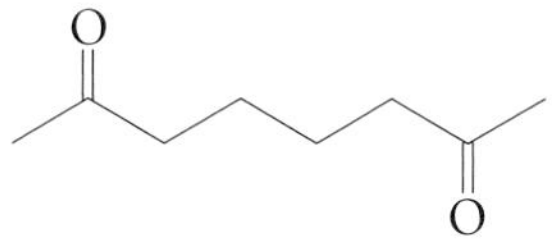

(d)

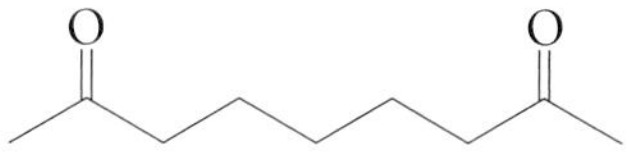

## EXERCISE 13.8

The aldol condensation can also be carried out in the presence of acid; in this case, the enol of the aldehyde or ketone is the nucleophilic species. Write a detailed, stepwise mechanism for the following aldol condensation in acid:

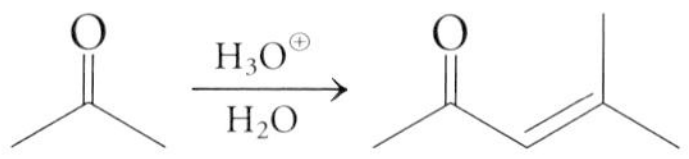

## Crossed Aldol Reaction

The synthetic utility of the aldol reaction would be greatly enhanced if the enolate anion of one substrate could be used to attack the carbonyl group of another substrate, forming a crossed condensation product.

However, when two different carbonyl compounds are mixed with base, a complex product mixture is typically obtained. Even when each carbonyl compound is symmetrical and can generate only a single enolate anion, four possible cross products can be formed because each component can serve as either a nucleophile or an electrophile. The situation becomes even more complicated when one or both of the starting materials can form two different enolate anions. For example, when acetone and 2-butanone are treated with base, all of the products shown in Figure 13.4 are formed. The complexity of the mixture, together with the difficulty of separating such chemically similar products, makes this reaction of little synthetic value. Furthermore, only one of these starting ketones (2-butanone) can form two different enolate anions. An even more complex product mixture would result from the reaction of, say, 2-butanone and 2-pentanone.

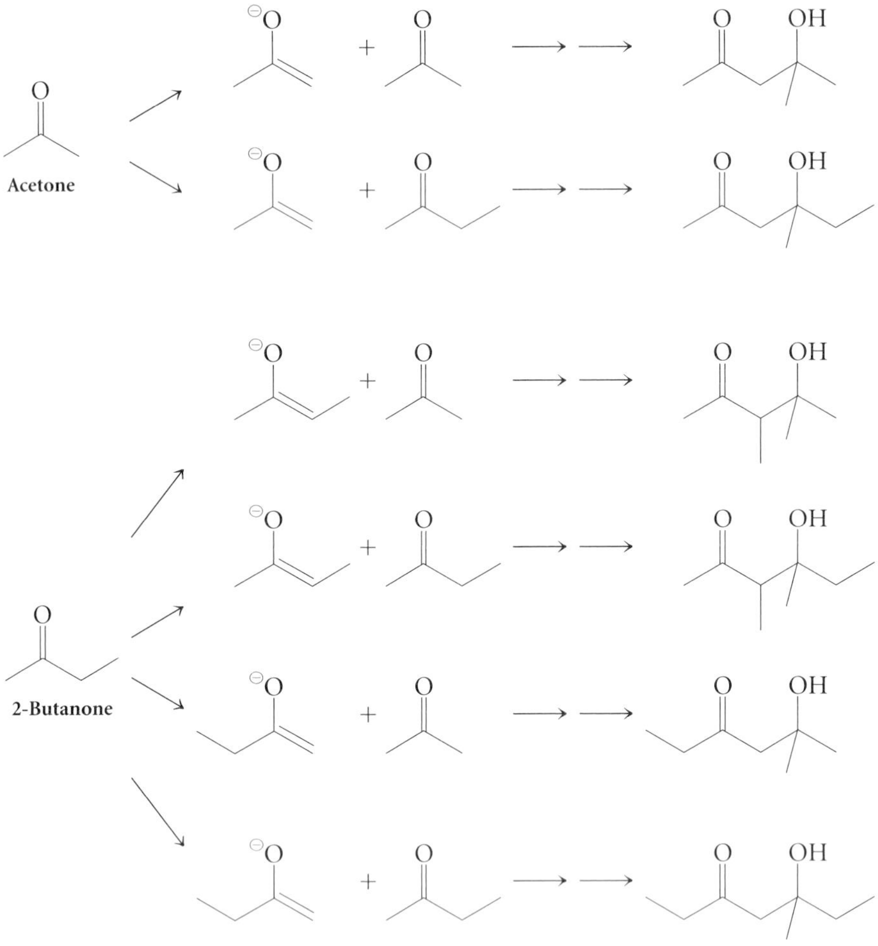

**FIGURE 13.4**

A very complex array of products is formed when a crossed aldol reaction is attempted with a mixture of two different ketones. With acetone and 2-butanone, three enolate anions are formed in the first deprotonation step. Each enolate anion can attack either starting ketone to produce six possible $\beta$-hydroxyalcohols.

## EXERCISE 13.9

**675**

13.3 Aldol Reaction, Aldol
Condensation, and Related
Reactions: Nucleophilic
Addition of Enolate Anions to
Carbonyl Groups

Draw the structures of all of the crossed aldol products that would result when a mixture of 2-methylcyclohexanone and 2-butanone is treated with aqueous sodium hydroxide.

A crossed aldol reaction is practical only when one of the carbonyl components is an aldehyde that lacks α hydrogen atoms and the other is a ketone. In this case, it is possible to form the enolate anion of the ketone, thereby limiting the number of possible products. The greater electrophilic reactivity of aldehydes compared with ketones (recall Chapter 12) also favors the crossed aldol product. To illustrate these points, let's look at the reaction of acetone with benzaldehyde:

Acetone    Benzaldehyde

*trans*-4-Phenylbut-3-ene-2-one

Here, the enolate anion of acetone attacks benzaldehyde, forming a carbon–carbon bond and ultimately producing an aldol condensation product. Because benzaldehyde has no α hydrogen atoms, it can act only as an electrophile, and the enolate of acetone is the only anion formed. The reactivity of an aldehyde toward nucleophilic attack is higher than that of a ketone (Chapter 12); thus, even though an attack by the enolate anion of acetone on acetone is also possible, it is slower than the attack on the more reactive aldehyde. Thus, the reaction of the enolate anion of acetone with acetone is slower than the crossed aldol reaction, and the major product observed is that obtained by attack of the enolate anion of acetone on benzaldehyde. The loss of water from the aldol product is especially easy, because of extended conjugation in the resulting condensation product.

In later chapters, we will see how the problem of multiple products in a crossed aldol reaction is solved in biological systems by enzymes that catalyze the formation of only one product. Enzymes exert a level of chemical selectivity that is not usually possible in simple chemical reactions in the laboratory.

**EXERCISE 13.10**

Draw the structures of the starting carbonyl compounds that could be used to synthesize each of the following $\alpha,\beta$-unsaturated ketones by a crossed aldol reaction:

(a)

(c)

(b)

(d)

## Nucleophilic Addition to $\alpha,\beta$-Unsaturated Carbonyl Groups: Conjugate Addition

***Conjugation in $\alpha,\beta$-Unsaturated Carbonyl Compounds.*** The aldol condensation gives rise to $\alpha,\beta$-unsaturated carbonyl compounds. In these molecules, the $\pi$ system extends over the carbonyl group and the alkene portion. The $p$ orbital overlap—and thus the stability—of these systems is greatest when the $\pi$ systems of the carbonyl group and the alkene are coplanar. This conjugation has important consequences for both the structure and reactions of enones. In 4-phenylbut-3-en-2-one, for example, the $\alpha,\beta$-unsaturated carbonyl system is further conjugated with an aromatic ring. For such compounds, the most stable conformation is also that in which the $\pi$ systems are coplanar.

**4-Phenylbut-3-en-2-one**

Because of conjugation, $\alpha,\beta$-unsaturated carbonyl compounds have two electrophilic sites. Both the carbonyl carbon and the $\beta$ carbon of an enone are reactive as electrophiles, and both can form a bond with a nucleophile. Which site is attacked depends on the nature of the nucleophile.

***1,2- versus 1,4-Addition of Nucleophiles to $\alpha,\beta$-Unsaturated Carbonyl Compounds.*** In general, nucleophiles with concentrated negative charge (such as alkyllithium and Grignard reagents) add to the carbonyl carbon of an $\alpha,\beta$-unsaturated carbonyl compound, and those with diffuse, delocalized charge (such as enolate anions) add to the $\beta$ carbon. Nucleophilic addition at the $\beta$ carbon of such systems is known as **conjugate addition.**

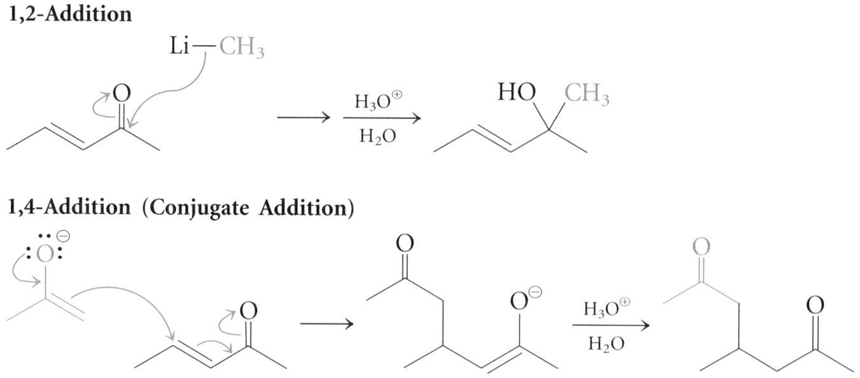

677

13.3 Aldol Reaction, Aldol
Condensation, and Related
Reactions: Nucleophilic
Addition of Enolate Anions to
Carbonyl Groups

**1,2-Addition**

**1,4-Addition (Conjugate Addition)**

***Nucleophilic Addition at the $\beta$ Carbon.*** Resonance structures help explain the electrophilic character of the $\beta$ carbon atom of an enone:

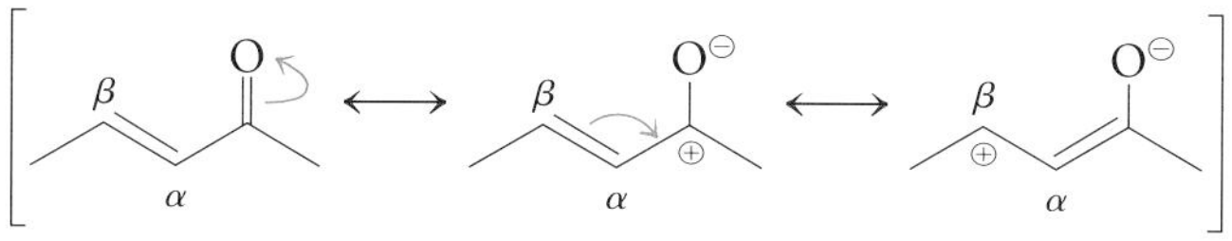

To the extent that the resonance structure at the right contributes to the hybrid, the $\beta$ carbon has electrophilic character—the partial positive charge on this carbon enhances its reactivity toward a nucleophile. Thus, an $\alpha,\beta$-unsaturated carbonyl compound can be attacked by a nucleophile either at the carbonyl carbon (as in simple aldehydes, ketones, and esters) or at the $\beta$ carbon.

When a nucleophile attacks the $\beta$ carbon of an $\alpha,\beta$-unsaturated ketone and a $\sigma$ bond is formed to that carbon, the electron pair of the carbon–carbon $\pi$ bond simultaneously shifts toward oxygen, ultimately forming an enolate anion.

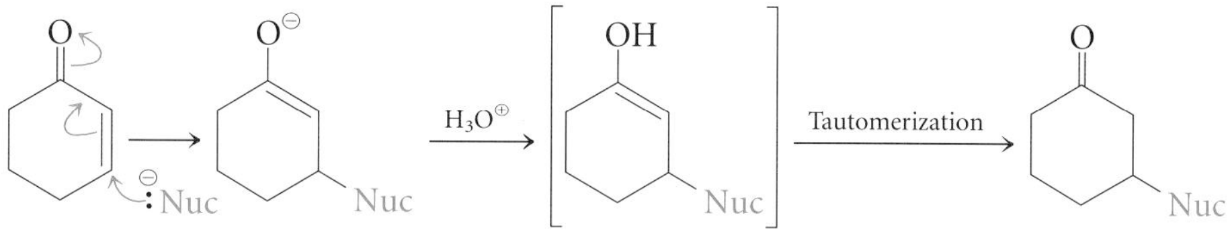

**1,4-Adduct**

In conjugate addition, the initially formed enolate anion is protonated on oxygen, as we would expect from the discussion earlier in this chapter, to produce an enol. It is for this reason that the reaction is called a 1,4-addition. Tautomerization of the enol to the more stable carbonyl group follows, completing the sequence. The final product corresponds to that expected from addition of the hydrogen atom and the nucleophile across the carbon–carbon double bond. You must not forget, however, that this conjugate addition requires a carbonyl group: nucleophiles do not add to an isolated C=C bond because of the instability of the simple carbanion that would result from this reaction:

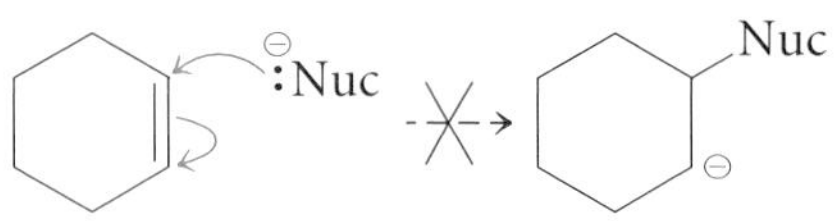


The critical role of the carbonyl group in the enone is to provide stabilization (through resonance delocalization) of the negatively charged intermediate formed upon attack by the nucleophile at the $\beta$ position.

***Alkylation of α,β-Unsaturated Compounds.*** Dialkylcuprate reagents (such as lithium dimethylcuprate) add in a conjugate (1,4) sense, providing a convenient method for preparing a wide range of complex carbon skeletons. With lithium dialkylcuprates, the nucleophile approaches the $\beta$ carbon, causing a shift of negative charge to oxygen, to form an enolate anion. Protonation on oxygen, followed by keto–enol tautomerization, gives the observed adduct. In contrast, Grignard reagents add in a 1,2 fashion to enones.

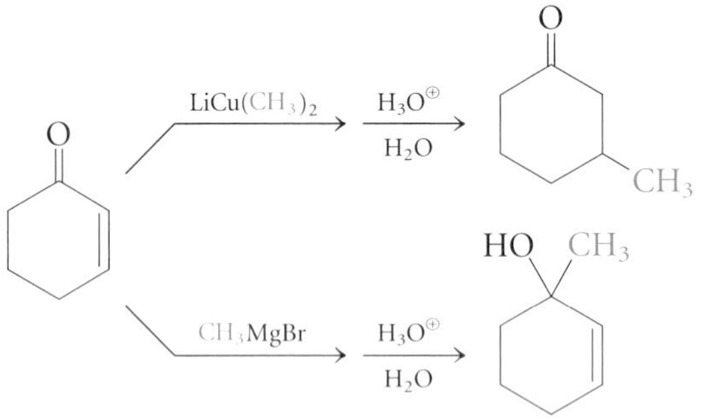


**Conjugate (1,4) Addition**

**1,2-Addition**

Why there is a preference for conjugate addition of alkyl groups from cuprate reagents is not entirely clear. Certainly, the negative charge of a cuprate, which resides on the copper atom, is more diffuse than that of a Grignard or alkyllithium reagent. Grignard reagents will add a carbon nucleophile in a conjugate sense in the presence of a catalytic amount of a copper (I) salt such as CuI. In this case, it is possible that a cuprate is formed from the Grignard reagent and the copper salt. (In the absence of a copper salt, Grignard reagents add to enones in the 1,2 sense.)

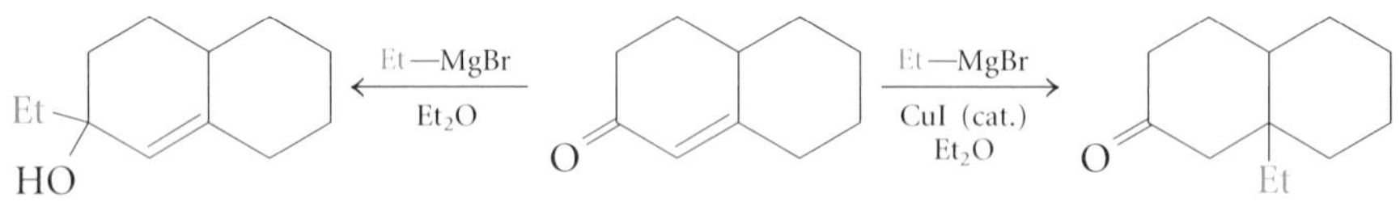


**1,2-Adduct** **1,4-Adduct**

***Michael Addition.*** When anions that are resonance-stabilized by two carbonyl groups (for example, anions of diketones) are used as nucleophiles in a conjugate addition, the reaction is called a **Michael addition.** (Indeed, the term *Michael addition* is often used loosely to refer to all conjugate addition reactions.) In such nucleophiles, charge dispersal is extensive (spread over three atoms), as in the anion derived by deprotonation of dimethyl malonate. The Michael addition takes place by nucleophilic attack of the resonance-stabilized malonate anion at the $\beta$ position of methyl vinyl ketone.

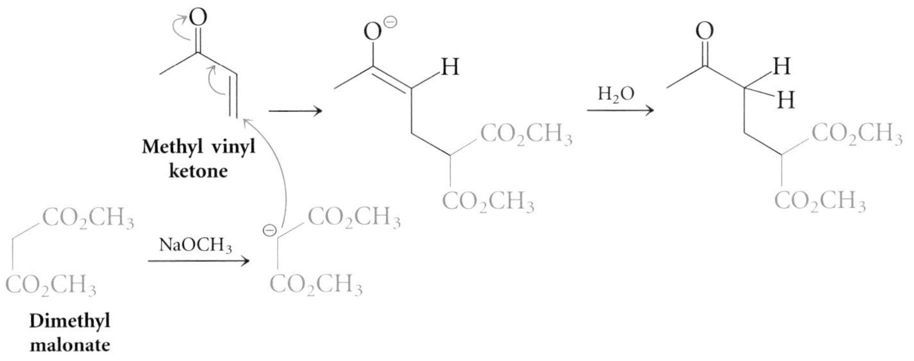

**679**

13.3 Aldol Reaction, Aldol
Condensation, and Related
Reactions: Nucleophilic
Addition of Enolate Anions to
Carbonyl Groups

The resulting enolate anion is protonated on carbon to form the ketone product.

## EXERCISE 13.11

Predict whether 1,2- or 1,4-addition is more likely with each of the following reagents. Draw the product expected from the reaction of each reagent with cyclohex-2-enone.

(a) $CH_3CH_2MgBr$     (b) $(CH_3CH_2)_2CuLi$     (c) $CH_3Li$

*Robinson Ring Annulation.* An intramolecular aldol reaction is an essential part of the **Robinson ring annulation,** a process that fuses a cyclohexanone ring onto an existing ketone (almost always a cyclic ketone). The process begins by addition of the enolate anion of a ketone to methyl vinyl ketone. The enolate anion adds in a conjugate sense to the $\beta$ carbon of the enone, forming a new carbon–carbon bond.

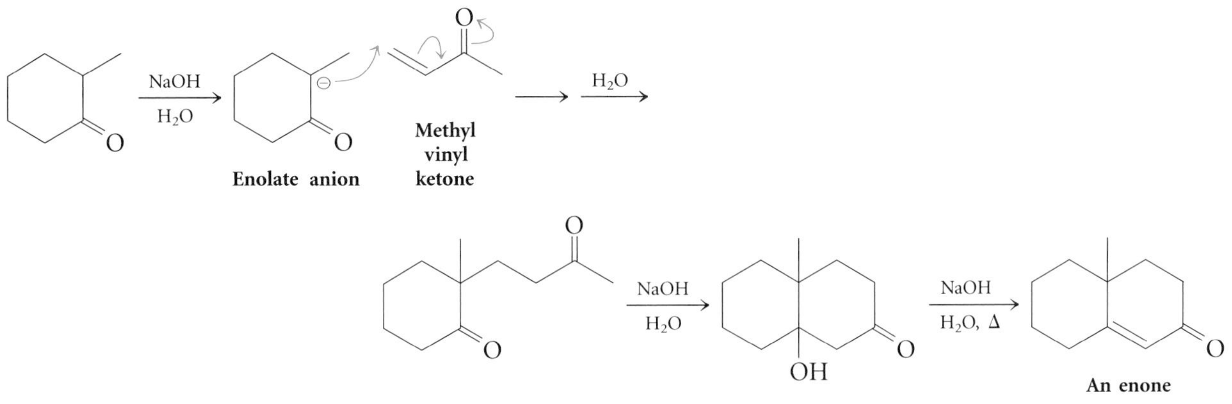

A second carbon–carbon bond results from an intramolecular aldol reaction. Under the reaction conditions (presence of base), the diketone product of the first step is further converted by an intramolecular aldol reaction to a cyclic $\beta$-hydroxyketone. Then, further treatment with either base or acid at a higher temperature converts the initially formed aldol to the enone, the aldol condensation product.

The following Robinson ring annulation involves a crossed aldol reaction:

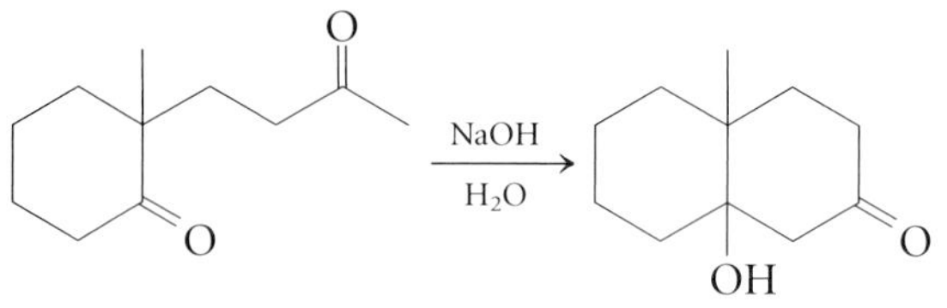

In theory, three aldol products might be formed, because both carbonyl groups are electrophilic and three different enolate anions can be formed by deprotonation of one or the other of the carbonyl groups. Draw the structures of the other possible aldol products. Explain why only the one shown is observed.

## 13.4

# The Claisen Condensation and Related Reactions: Acylation of Esters

### The Claisen Condensation

In the aldol reaction, a new carbon–carbon bond is formed between an enolate anion and a carbonyl carbon. In a **Claisen condensation,** both the enolate anion and the carbonyl carbon belong to an ester. The enolate anion generated by a deprotonation of an ester is sufficiently nucleophilic to react with the carbonyl group of another equivalent of the ester in a nucleophilic acyl substitution. This reaction, known as the Claisen condensation, results in the formation of a **β-ketoester.**

The Claisen condensation begins with the generation of the enolate anion of an ester.

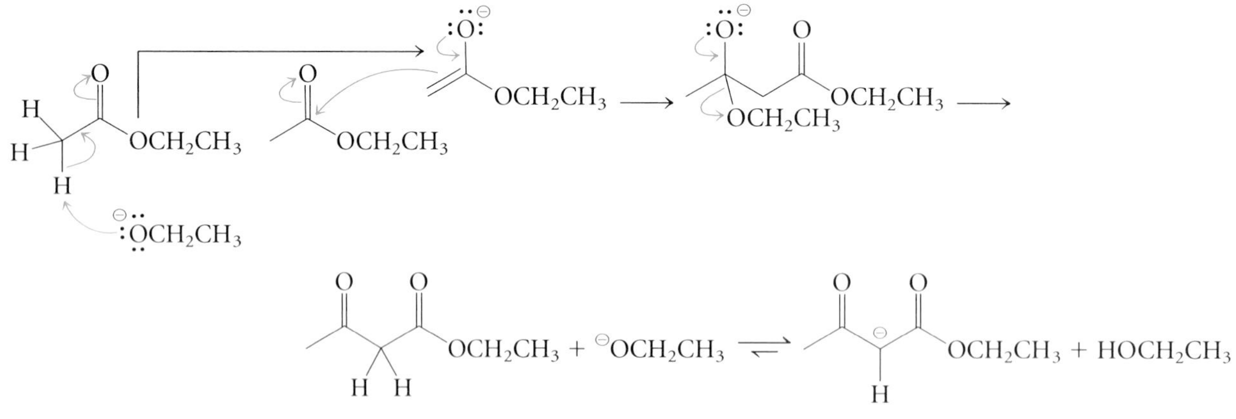

β-Ketoester

An ester enolate anion is formed when an ester with α hydrogen atoms is treated with alkoxide. Like an enolate anion derived from an aldehyde or a ketone in the aldol reaction, the ester enolate anion attacks a second neutral carbonyl carbon–in this case, of an ester—to form a negatively charged, tetrahedral intermediate. This intermediate loses an alkoxide ion with simultaneous re-formation of the C=O bond to form the neutral β-ketoester, just as in the nucleophilic acyl substitutions discussed in Chapter 12. As noted in Chapter 6 (Table 6.1), the hydrogen bonded to the carbon between the two carbonyl groups of a β-ketoester is especially acidic ($pK_a$ 11). Thus, deprotonation by an alkoxide ion of the β-ketoester formed in the Claisen condensation is thermodynamically favorable. Indeed, this acid–base reaction constitutes a principal driving force for the Claisen condensation, which, as a result, is best carried out with a full equivalent of base.

## EXERCISE 13.13

Draw the structure of the Claisen condensation product expected when each of the following esters is treated with KOEt in EtOH:

(a) $H_3CCO_2CH_2CH_3$

(c) $(H_3C)_2CHCH_2CO_2CH_3$

(b) $H_3CCH_2CO_2CH_2CH_3$

(d) $PhCH_2CO_2Ph$

## The Dieckmann Condensation

An intramolecular variant of the Claisen condensation is known as the **Dieckmann condensation.** Similar to an intramolecular aldol reaction, an intramolecular Claisen condensation can take place when a molecule bears two ester groups. With the following symmetrical diester, for example, deprotonation at the position α to either ester group produces the same ester enolate anion:

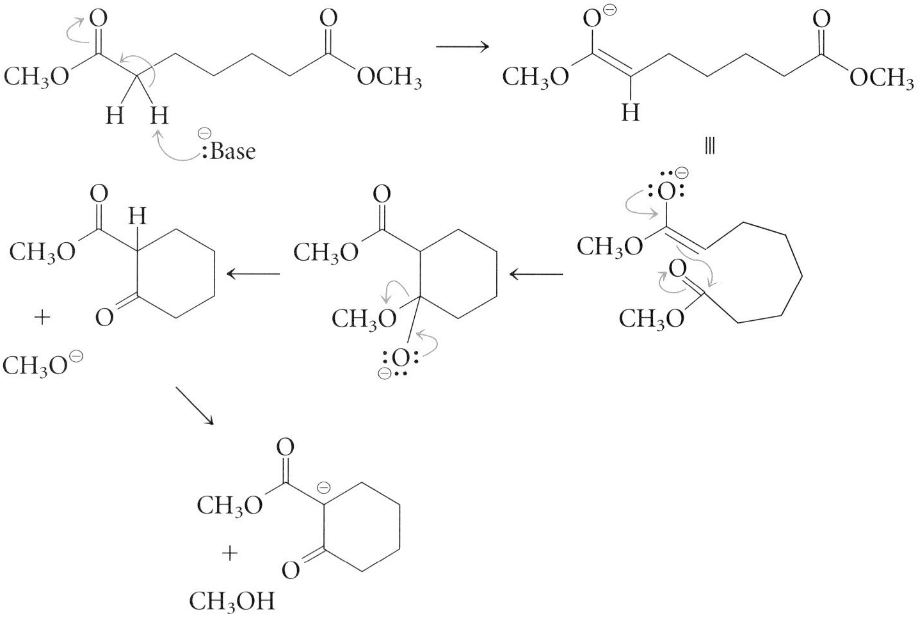

The ester enolate anion then attacks the second ester group intramolecularly, through a six-member transition state. The resulting tetrahedral intermediate loses methoxide ion to produce the β-ketoester expected in a Claisen condensation.

A primary driving force of the Dieckmann condensation, as with the Claisen condensation, is the formation of the anion of the product β-ke-

toester. As a result, the Dieckmann condensation of an unsymmetrical
diester is thermodynamically favorable only in the direction that results in
a β-ketoester with an acidic hydrogen between the two carbonyl groups:

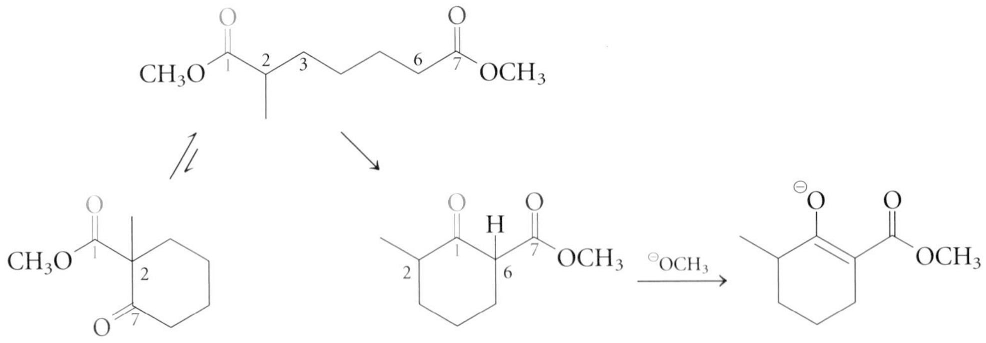

Although an ester enolate anion can be formed at either C-2 or C-6, only
the C-6 enolate anion leads to product, because the β-ketoester formed in
the cyclization of the C-6 enolate anion (at the right) bears an acidic proton at the α position, and the β-ketoester formed from the C-2 enolate anion (at the left) does not. Therefore, the β-ketoester at the right is deprotonated to a resonance-stabilized anion under the basic conditions of the
reaction.

## EXERCISE 13.14

Draw the structure of the ester enolate anion needed to form each of the two possible Dieckmann condensation products of the following reaction:

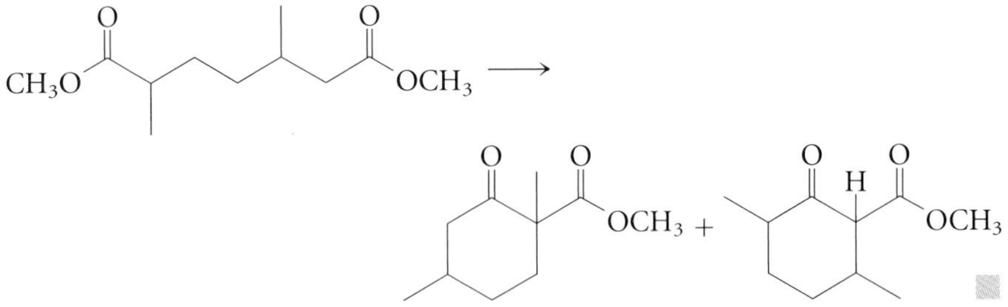

## Crossed Claisen Condensation

The reaction of an ester enolate anion with a different ester presents
the same problem that arises in the crossed aldol condensation. A crossed
Claisen condensation proceeds cleanly only if one of the esters lacks α hydrogen atoms (and therefore cannot form an enolate anion). Such an ester
should also be a more reactive electrophile than the other. An example of
a crossed Claisen condensation is the reaction of the enolate anion of methyl
acetate with methyl formate:

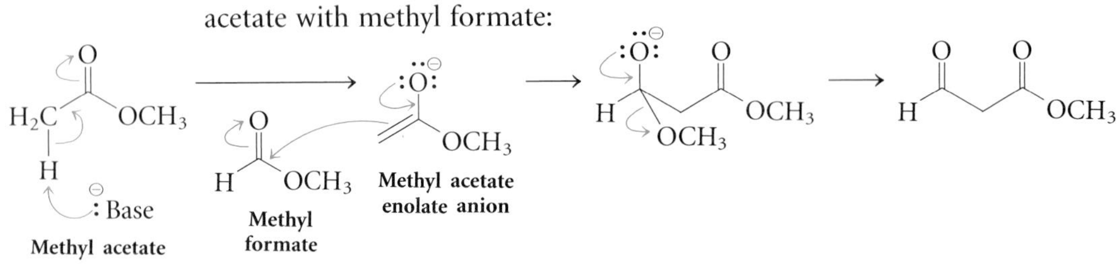

No enolate anion is possible from a formate ester. In addition, formate esters are especially reactive toward nucleophiles for both electronic and steric reasons (recall the differences between aldehydes and ketones).

Crossed Claisen condensations can also be accomplished with carbonate, oxalate, and benzoate esters, all of which lack C—H bonds $\alpha$ to the carbonyl group.

| Formate | Carbonate | Oxalate | Benzoate |
| ester | ester | ester | ester |

Aliphatic esters completely substituted at the $\alpha$ position (such as methyl pivalate) also have no $\alpha$ hydrogen atoms, but are relatively poor electrophiles because formation of the required tetrahedral intermediate is so sterically hindered. As a result, they are rarely useful in Claisen condensations.

**Methyl pivalate**

## EXERCISE 13.15

Predict the product of Claisen condensation of the enolate anion of methyl acetate, $H_3CCO_2CH_3$, with (a) carbonate, (b) oxalate, (c) benzoate, and (d) pivalate esters.

## EXERCISE 13.16

Propose appropriate starting materials for the synthesis of each of the following compounds by a route employing a Claisen condensation:

(a)

(c)

(b)

### The Reformatsky Reaction

The reaction of two ketones (or aldehydes) in an aldol condensation or of two esters in a Claisen condensation yields a complex mixture of products, unless one reactant lacks $\alpha$ hydrogen. For instance, simply treating a mixture of a ketone and an ester with base does not lead to a synthetically

useful product, because the ketone is both more acidic and more electrophilic than the ester. Thus, this combination leads only to the aldol condensation of the ketone. The **Reformatsky reaction** addresses this problem
by having an ester enolate anion act as a nucleophile to attack a ketone or
an aldehyde. The ester enolate anion is formed first, in the absence of the
ketone, by reduction of an $\alpha$-bromoester with zinc. (The Hell–Volhard–
Zelinski reaction discussed earlier in this chapter can be used to prepare an
$\alpha$-bromoacid; this acid can then be esterified to form the starting material
for the Reformatsky reaction.)

A zinc
enolate

In the reduction with zinc, a zinc enolate is formed. This enolate attacks an
aldehyde or a ketone more rapidly than it can attack its ester precursor. The
crossed condensation product, a $\beta$-hydroxyester, is thus produced in good
yield.

## EXERCISE 13.17

How might the Reformatsky reaction be used to synthesize the following $\alpha,\beta$-un-
saturated ester? *(Hint:* Recall the aldol and aldol condensation reactions.)

## 13.5

# Alkylation of $\beta$-Dicarbonyl Compounds

## $\beta$-Dicarbonyl Compounds

Compounds with carbonyl groups that are 1,3 to each other have unique
chemical properties and undergo reactions that are difficult or impossible
in the absence of this special relationship of functional groups. Malonate
esters, $\beta$-ketoesters, and $\beta$-diketones that have hydrogens on the carbon between the carbonyl groups are quite acidic, and can be converted essentially
completely to enolate anions with bases as mild as hydroxide ion.

**Dimethyl malonate**
(a malonic acid diester)
p$K_a$ 13

**Ethyl acetoacetate**
(a $\beta$-ketoester)
p$K_a$ 11

**Acetylacetone**
(a $\beta$-diketone)
p$K_a$ 9

Generally, the enolate anions are formed using a solution of an alkoxide ion in the alcohol that corresponds to the alkoxy group of the ester (for example, sodium methoxide in methanol for dimethyl malonate or sodium ethoxide in ethanol for ethyl acetoacetate).

Both β-ketoesters and β-diketones have an unusually high content of their enol tautomers in equilibrium. For example, at equilibrium, the enol of acetone constitutes only about 1 ppm (part per million), but both ethyl acetoacetate and acetylacetone have significant percentages of the enol. Indeed, in the latter case, the enol dominates the equilibrium.

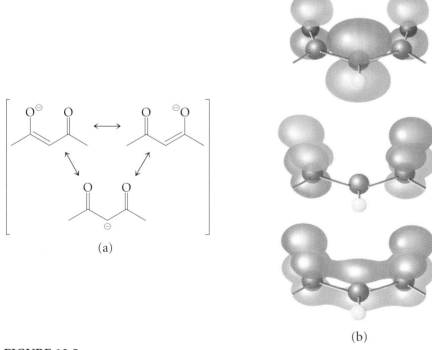

The enolate anions of β-dicarbonyl compounds are more stable than those derived from simple ketones and esters, because the negative charge is delocalized over three atoms—two oxygen and one carbon (Figure 13.5).

## FIGURE 13.5

(a) The enolate anion of acetylacetone has negative charge distributed over three atoms—two oxygen and one carbon. (b) The molecular orbitals of the π system, with the HOMO at the top. It is this orbital, which bears significant negative charge on the central carbon atom, that provides electron density for reaction with an electrophile. (For clarity, the methyl groups have been omitted from the orbital picture.)

### Alkylation of β-Ketoesters

Because of charge delocalization, enolate anions derived from β-dicarbonyl compounds are less reactive as nucleophiles than those derived from simple carbonyl compounds. However, these stabilized anions do react with primary and secondary alkyl halides in S$_N$2 reactions. For example, ethyl acetoacetate can be alkylated with methyl iodide in the presence of sodium ethoxide. Treatment of ethyl acetoacetate with sodium ethoxide in ethanol generates the enolate anion. Addition of methyl iodide as an electrophile results in alkylation of the carbon atom situated between the two carbonyl groups of the β-ketoester.

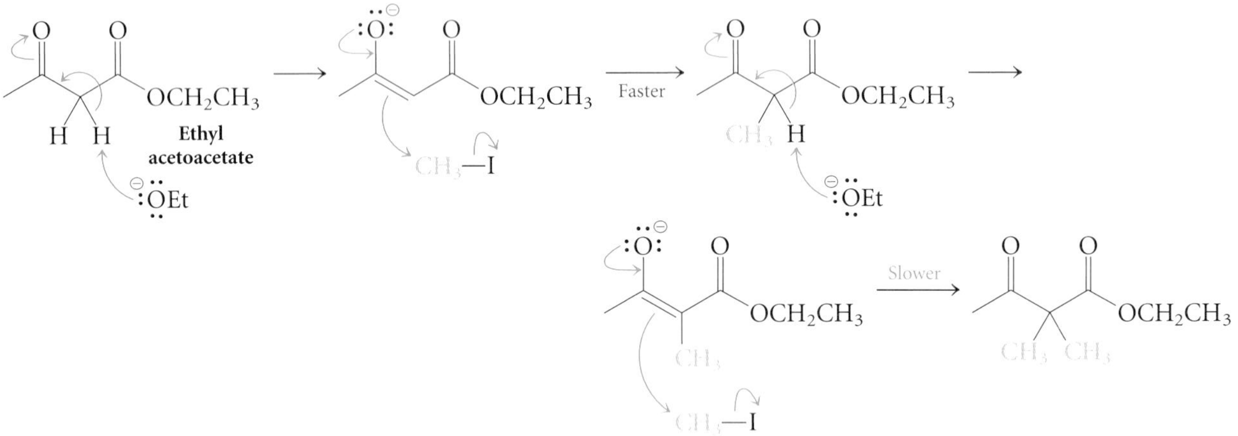

The alkylation product still has an acidic α hydrogen, which can also be removed by ethoxide ion. A second S$_N$2 alkylation can occur to form a second carbon–carbon bond at the carbon between the carbonyl groups. This second alkyl group can be the same as the first (as shown here) or different.

Because the enolate anion formed from the alkylated product is substantially less reactive than the unsubstituted enolate anion, monoalkylation of β-ketoesters can be achieved by the use of one equivalent of alkylating agent. Using two equivalents of both the base and the alkylating agent brings about replacement of both hydrogen atoms originally present on the carbon between the two carbonyl groups. The second alkylation can be carried out with a different alkyl halide, leading to a β-ketoester substituted with two different alkyl groups. When two different groups are to be introduced, a higher yield is achieved when the less substituted alkyl halide is used in the second alkylation. As a typical example, alkylation of ethyl acetoacetate, first with n-propyl iodide and then with methyl iodide, leads to the introduction of two different alkyl groups at the central carbon atom that is α to both carbonyl groups:

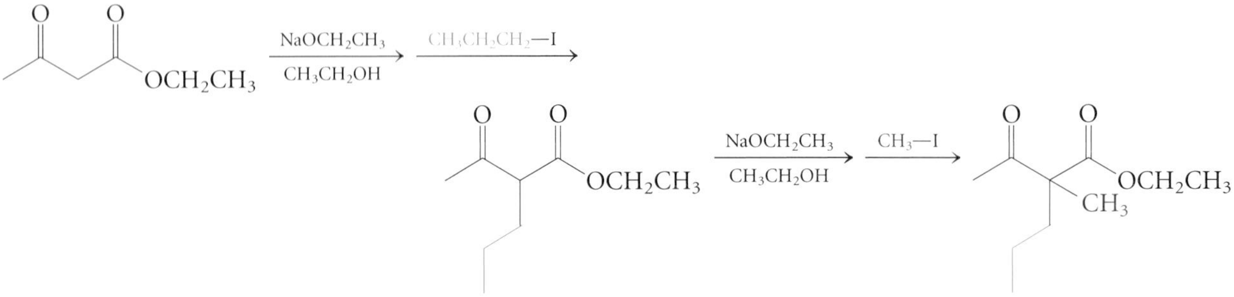

## Alkylation of Malonic Acid Diesters

Malonic acid diesters are also β-dicarbonyl compounds. They can be converted to enolate anions that undergo alkylation in the same way as those of acetoacetic acid esters. The second alkylation is substantially slower than the first. Thus, alkylation of malonic acid diesters can be limited to a single alkyl group, or two different alkyl groups may be added sequentially. For example, a two-step alkylation of dimethyl malonate can be carried out under basic conditions:

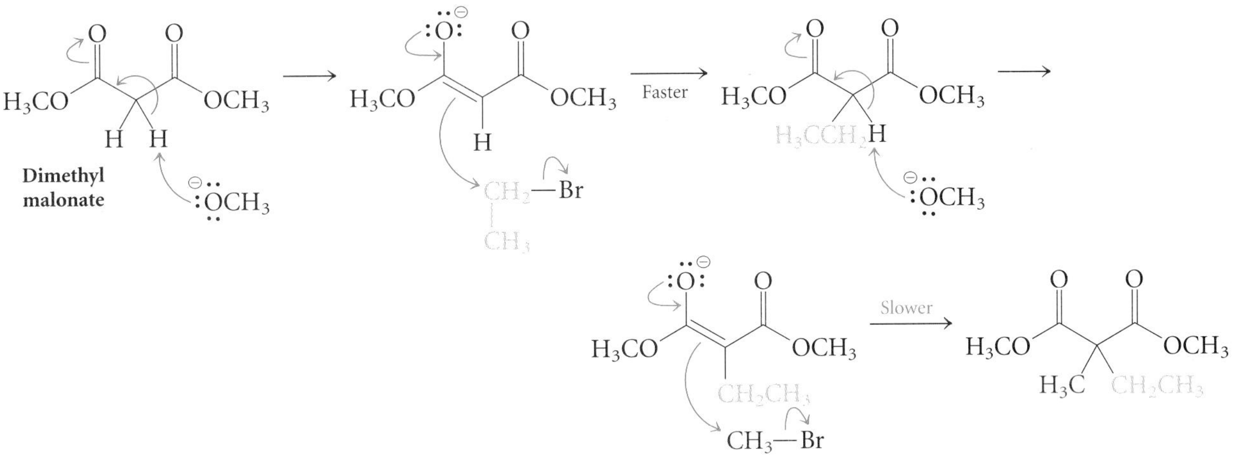

Dimethyl malonate is converted to its enolate anion upon treatment with methoxide ion. This enolate anion reacts with ethyl bromide in an $S_N2$ reaction to form a new carbon–carbon bond. The sequence can be repeated (reaction first with methoxide ion and then with methyl bromide) to introduce a second alkyl group. The alkyl groups introduced to dimethyl malonate by this sequence can be either the same or different.

## Hydrolysis and Decarboxylation of β-Ketoesters and Malonic Acid Diesters

The two carbonyl groups in β-ketoesters and in malonic acid diesters undergo the typical reactions of ketones and esters. For example, treatment of acetoacetic ester or malonic acid diester with sodium hydroxide in water effects hydrolysis to a carboxylate ion:

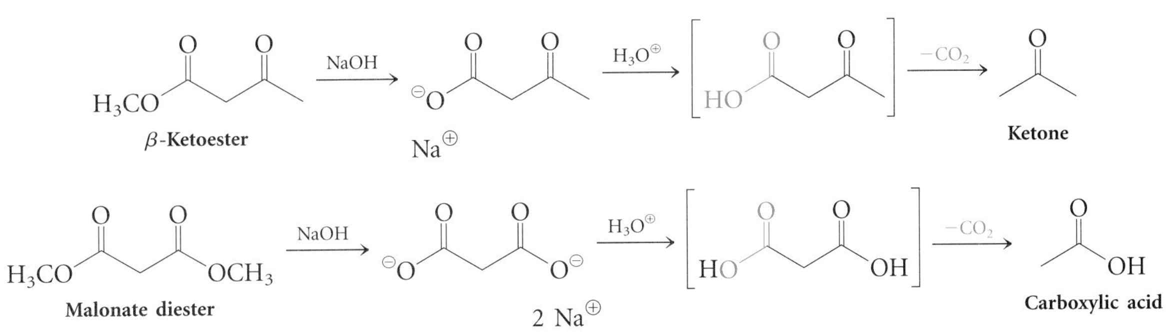

This hydrolysis of β-ketoesters and malonic acid diesters proceeds in the normal fashion by nucleophilic acyl substitution to produce carboxylate ions. However, the corresponding $\beta$-ketoacids and 1,3-diacids formed by acidification are not stable and undergo spontaneous loss of carbon dioxide at, or slightly above, room temperature. In this way, $\beta$-ketoesters are converted to ketones, and malonic acid diesters are converted to simple carboxylic acids.

Because of the facility with which this decarboxylation takes place, it is believed to involve a six-member cyclic transition state in which the proton of the acid is being transferred to the oxygen of the other carbonyl group. The enol produced by decarboxylation undergoes rapid proton tautomerization to yield the product ketone or carboxylic acid.

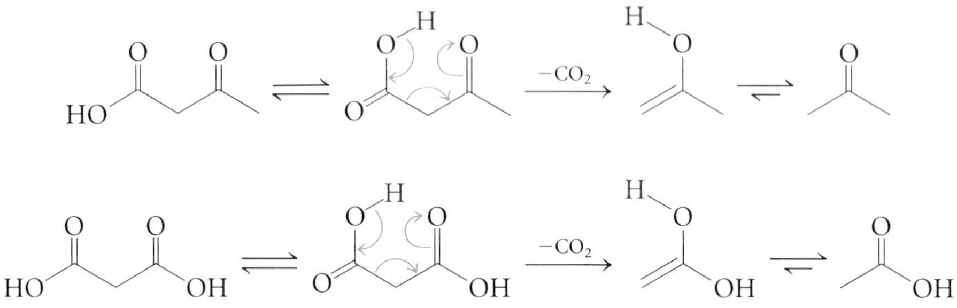

Applying this sequence of hydrolysis and decarboxylation to the $\beta$-ketoesters produced by both Claisen and Dieckmann condensations is a useful synthetic pathway to simple ketones, especially those with five- and six-member rings.

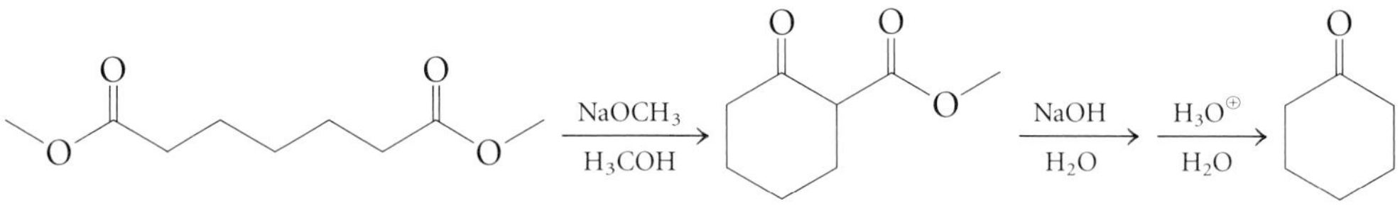

### EXERCISE 13.18

Review your knowledge of nucleophilic acyl substitution by writing a detailed mechanism for the conversion of methyl acetoacetate to acetoacetic acid.

### EXERCISE 13.19

Write a detailed, stepwise mechanism for the conversion of each enol to acetone or acetic acid:

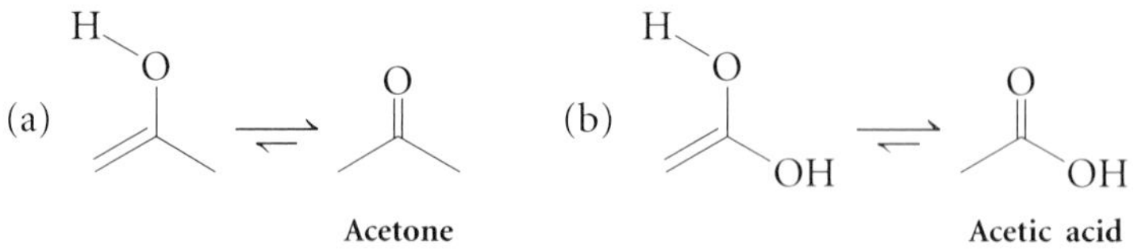

(a)    Acetone

(b)    Acetic acid

### EXERCISE 13.20

Unlike $\beta$-ketoacids, $\alpha$-ketoacids do not undergo decarboxylation on heating. Explain why.

# EXERCISE 13.21

Show the starting materials that could be used in either a Claisen or a Dieckmann condensation to produce each of the following ketones after hydrolysis and decarboxylation:

(a)

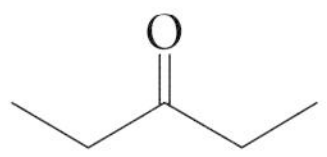

(b)

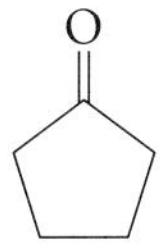

## Acetoacetic Ester and Malonic Ester Syntheses

Hydrolysis and decarboxylation of substituted acetoacetic esters and malonic acid diesters provide a convenient method for the preparation of α-monosubstituted and α,α-disubstituted ketones and carboxylic acids. The conversion of acetoacetic esters to ketones is known as the **acetoacetic ester synthesis,** and the conversion of malonic acid diesters to substituted acetic acids is referred to as the **malonic ester synthesis.**

The preparation of 2-heptanone, for example, can be carried out using the acetoacetic ester synthesis. Alkylation of the enolate anion of ethyl acetoacetate with 1-bromobutane, followed by hydrolysis and decarboxylation, gives the ketone:

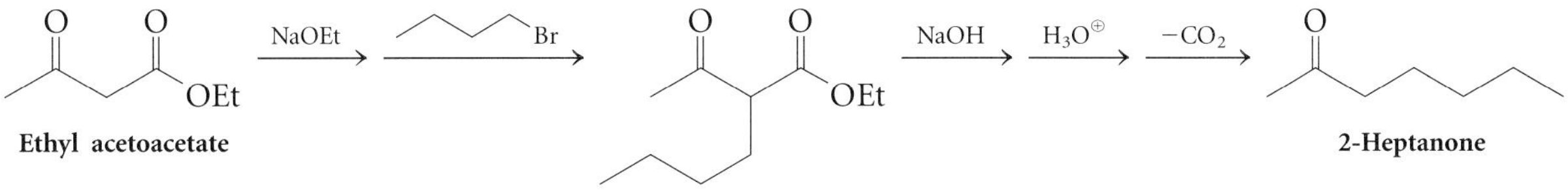

**Ethyl acetoacetate**    →    NaOEt    →    **2-Heptanone**

Similarly, the preparation of 2-methylpentanoic acid can be accomplished by the malonic ester synthesis:

**Dimethyl malonate**

**Dimethyl methylpropyl malonate**    →    **2-Methylpentanoic acid**

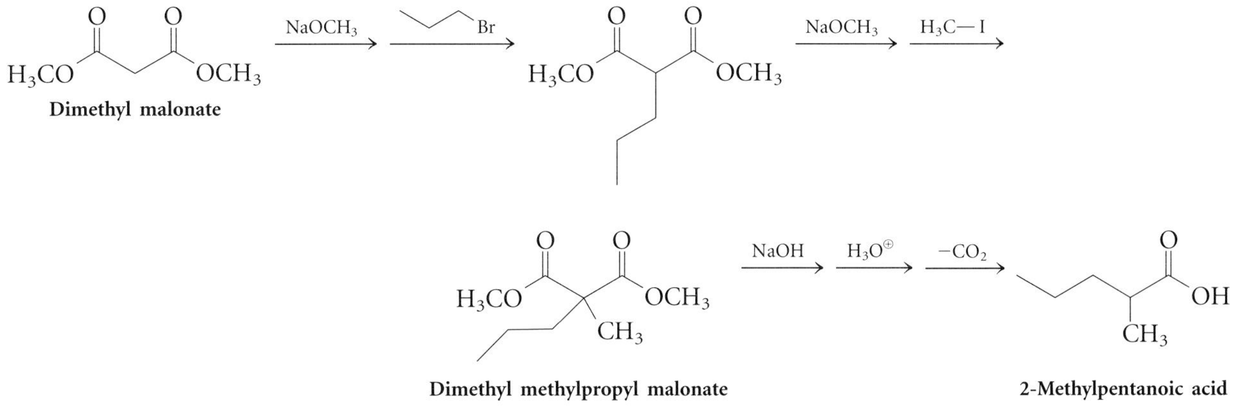

Stepwise alkylation of the enolate anion of dimethyl malonate with 1-bromopropane is followed by alkylation of the monoalkylation product with methyl iodide, yielding dimethyl methylpropyl malonate. Hydrolysis of the diester, followed by decarboxylation, produces 2-methylpentanoic acid.

Choosing any reagents you need, show how 2-methylbutanoic acid and 3-methyl-2-pentanone can be prepared using the acetoacetic ester synthesis or the malonic ester synthesis.

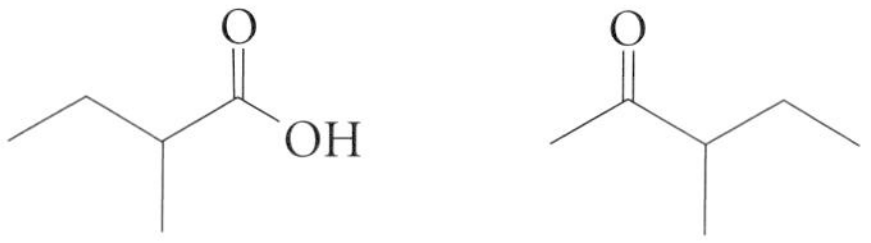

**2-Methylbutanoic acid**          **3-Methyl -2-pentanone**

## Formation of Carbocyclic Rings Using Acetoacetic Ester and Malonic Ester Syntheses

Both the acetoacetic ester and malonic ester syntheses can be used to make carbocyclic rings by employing dihaloalkanes in the alkylation. For example, formation of the enolate anion of dimethyl malonate using sodium methoxide, followed by alkylation with 1,3-dibromopropane, results in monoalkylation:

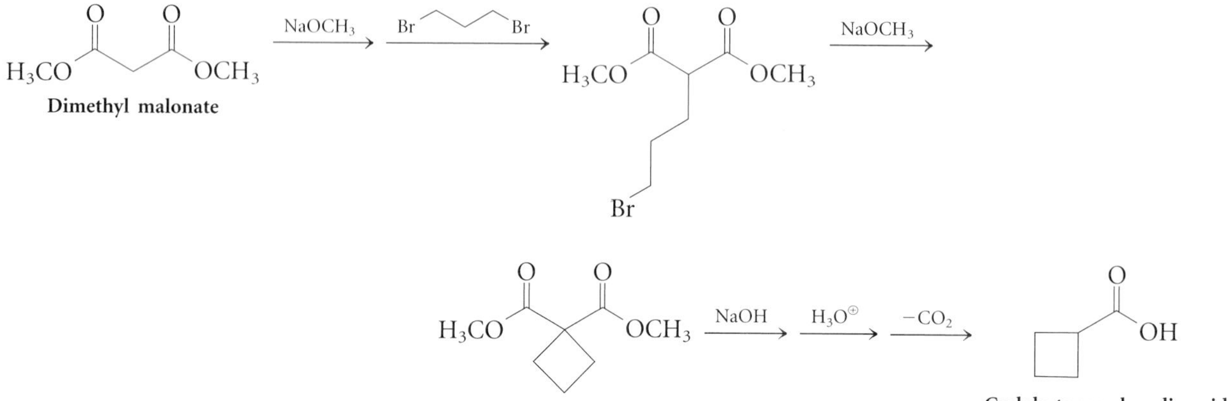

**Dimethyl malonate**

**Cyclobutanecarboxylic acid**

Treatment of this product with an additional equivalent of sodium methoxide generates an alkylated enolate anion, which undergoes an intramolecular nucleophilic substitution, leading to the formation of a four-member ring. Hydrolysis and decarboxylation yield cyclobutanecarboxylic acid. Unlike the aldol and Dieckmann reactions, the malonic ester synthesis can be used to form four-member and even three-member rings because the alkylation reactions are not reversible.

**EXERCISE 13.23**

Show the steps involved in synthesizing acetylcyclohexane from ethyl acetoacetate and any other reagents needed.

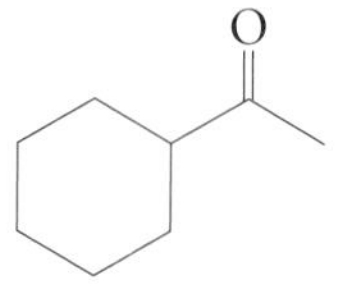

**Acetylcyclohexane**

BARBITURATES

Reaction of dimethyl malonate with urea in the presence of sodium ethoxide forms a cyclic derivative known as barbituric acid:

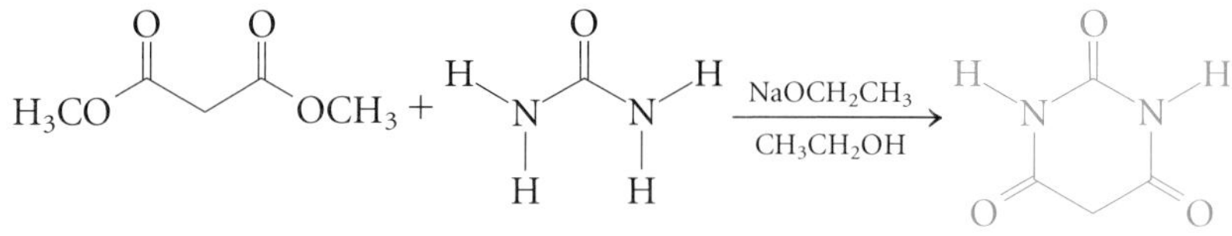

Dimethyl malonate    Urea    Barbituric acid

Derivatives of this parent compound are known as barbiturates and have been used since the beginning of the twentieth century as sleep inducers (soporifics). There are many different barbiturates; three are shown here.

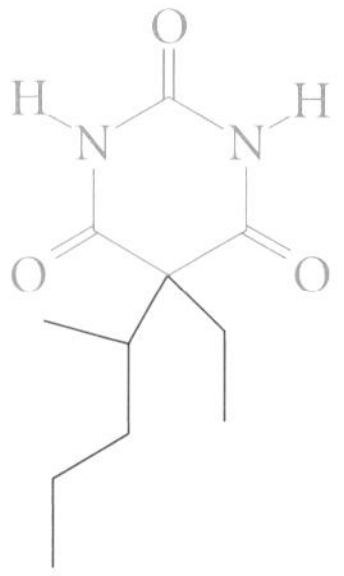

Pentobarbital 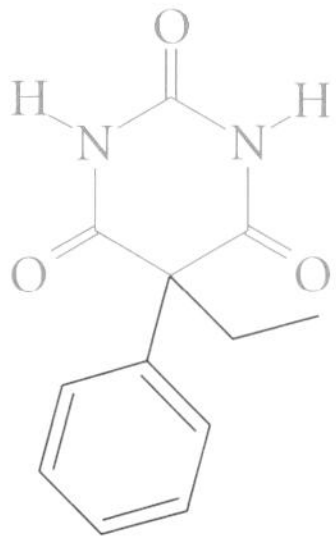    Phenobarbital 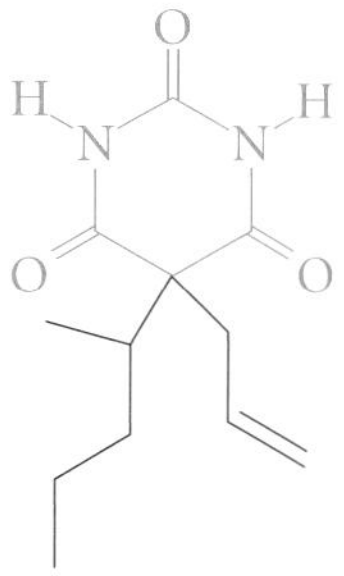    Seconal

## 13.6

# Synthetic Methods

The reactions discussed in this chapter are particularly important as synthetic methods because they produce new carbon–carbon bonds. Applications of these reactions in making structurally more complex carbon skeletons with various functional groups are listed in Table 13.1 (on pages 692–693).

## 13.7

# Spectroscopy

Both the starting materials and the products for the reactions in this chapter have carbonyl groups. Although there are differences in the infrared absorption frequencies for the carbonyl groups of ketones and esters, the difference between a $\beta$-ketoester and a ketone or between an unsubstituted and an alkylated ketone is often not sufficient to make an unambiguous

**TABLE 13.1**

Using Enolate Anions and Enols to Introduce Various Functional Groups

| Functional Group | Reaction | Example |
|---|---|---|
| Acid | Iodoform reaction | |
| | Malonic ester synthesis | |
| $\alpha$-Bromoacid | Hell–Volhard–Zelinski reaction | |
| Alcohol | Aldol reaction ($\beta$-hydroxycarbonyl compound) | |
| Alkene | Aldol condensation ($\alpha,\beta$-unsaturated aldehyde) | |
| Ester | Alkylation of esters | |
| | Reformatsky reaction ($\beta$-hydroxyester) | |
| | Claisen condensation ($\beta$-ketoester) | (see $\beta$-ketoester, next page) |

assignment. On the other hand, distinctive new absorptions appear in the products of the aldol reaction and the aldol condensation. For the aldol reaction, the product's hydroxyl group will result in O—H stretching absorption in the region 3650–3400 cm$^{-1}$. The aldol condensation reaction results in an $\alpha,\beta$-unsaturated enone, in which the positions of both the C=O and C=C absorptions are shifted as a result of the interaction of

| Functional Group | Reaction | Example |
|---|---|---|
| Ester (*continued*) | Dieckmann cyclization (cyclic β-ketoester) | (see β-ketoester, below) |
| Ketone | Alkylation of ketone | |
| | Claisen condensation (β-ketoester) | (see β-ketoester, below) |
| | Dieckmann cyclization (cyclic β-ketoester) | (see β-ketoester, below) |
| | Aldol reaction (β-hydroxyketone) | |
| | Aldol condensation (α,β-unsaturated ketone) | |
| | Acetoacetic ester synthesis | |
| β-Ketoester | Claisen condensation (β-ketoester) | |
| | Dieckmann cyclization (cyclic β-ketoester) | |

these two functional groups. For an α,β-unsaturated carbonyl compound, the C=O stretching absorption is at 1675 cm$^{-1}$, and there is a strong absorption for the C=C bond in the region 1650–1600 cm$^{-1}$.

Both proton and carbon NMR spectroscopy are useful tools for assessing the changes from starting materials to products in the reactions in this chapter. Alkylation of the carbon α to a carbonyl group adds absorptions

**693**

for the protons and carbons in the introduced alkyl group. In addition, at least one additional carbon substituent is added to the carbon $\alpha$ to a carbonyl group, resulting in a downfield shift in the $^{13}C$ NMR spectrum for the $\alpha$ carbon.

The presence of the double bond in the product of the aldol condensation will also be clear in the $^{13}C$ NMR spectrum, where unsaturated and saturated carbons are readily differentiated. In addition, polarization of the carbon–carbon double bond of an enone toward the carbonyl group results in a significant shift difference for the $\alpha$ and $\beta$ carbons.

Any protons on the unsaturated carbons of the aldol product will be readily visible in the $^1H$ NMR spectrum.

## Summary

**1.** The collection of reactions that form a new bond at the $\alpha$ position to a carbonyl group represents one of the most important for the construction of complex molecules. In all of these reactions, the carbonyl group plays the critical role of enhancing the acidity of hydrogens on the adjacent ($\alpha$) carbons. Treatment with base results in removal of a proton to yield an enolate anion, a species with a carbon with significant nucleophilic character. When treated with acid, a ketone is in rapid equilibrium with a low concentration of the corresponding enol. In the Reformatsky reaction, the action of zinc on an $\alpha$-bromoester provides the reactive equivalent of an enolate anion.

**2.** Enolate anions can be formed under either equilibrium or kinetic conditions. Under equilibrium conditions (use of hydroxide or alkoxide), the most stable (usually the most substituted) enolate anion is formed. Under kinetic conditions (for example, quantitative deprotonation by lithium diisopropylamide), the less stable (less substituted, least sterically hindered) enolate anion may predominate.

**3.** Enolate anions and enols are nucleophilic and react with a range of electrophiles, including halogens, alkyl halides, ketones and aldehydes, and esters. Examples of such reactions include the halogenation of ketones (specifically the iodoform reaction), the alkylation of ketones and esters, the aldol reaction and aldol condensation, the Claisen condensation, the Dieckmann cyclization, and the acetoacetic and malonic ester syntheses.

**4.** Nucleophilic addition to an $\alpha,\beta$-unsaturated carbonyl group can occur in a 1,2 or 1,4 sense. Grignard and alkyllithium reagents tend to add in a 1,2 sense; lithium dialkylcuprates and enolate anions add in a 1,4 sense.

**5.** The aldol and Claisen condensations are valuable tools for the construction of carbon–carbon bonds between two carbonyl groups. When the two carbonyl groups are in separate molecules, a significant increase in molecular weight results from these reactions, providing an organic product of increased size and complexity. When the carbonyl groups are in the same molecule, a cyclization occurs, forming a five- or six-member ring.

**6.** The Dieckmann condensation and the Robinson annulation are versatile methods for the formation of five- and six-member rings. Such rings can also be formed by the acetoacetic and malonic ester syntheses, using suitable alkyl dihalides.

**7.** Hydrolysis of acetoacetic or malonic acid diesters yields the corresponding $\beta$-ketoacids or malonic acids, which undergo decarboxylation via a six-member cyclic transition state to yield the corresponding ketones or acids. These steps form the final part of the acetoacetic and malonic ester syntheses, which yield substituted ketones and carboxylic acids, respectively.

## Review of Reactions

*(continued)*

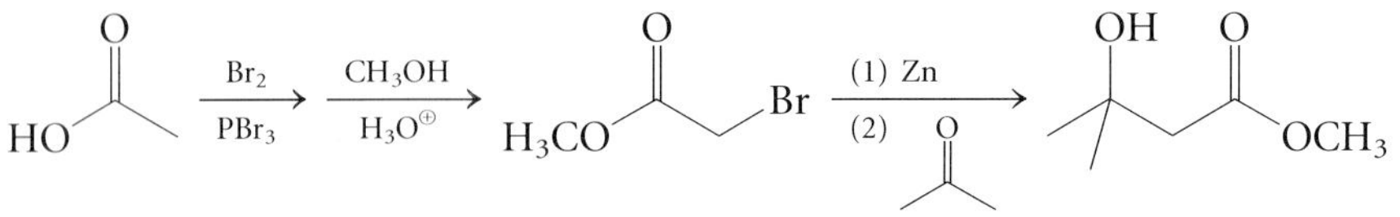

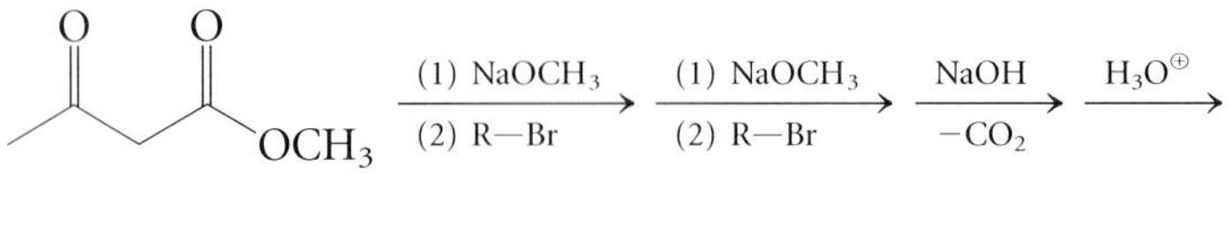

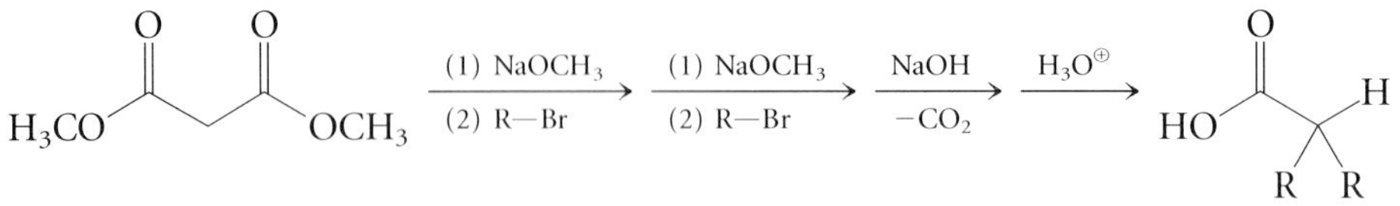

## Review Problems

**13.1** Predict the major product expected when 2-pentanone is treated with each of the following reagents:

(a) NaOH, I$_2$

(c) (1) LiN($i$-Pr)$_2$; (2) EtI

(e) NaOH, 80 °C

(b) NaOH, room temp.

(d) Br$_2$, HBr

**13.2** Predict the major product expected when methyl pentanoate is treated with each of the following reagents:

(a) MeMgBr     (b) NaOCH$_3$     (c) H$_3$O$^{\oplus}$, H$_2$O

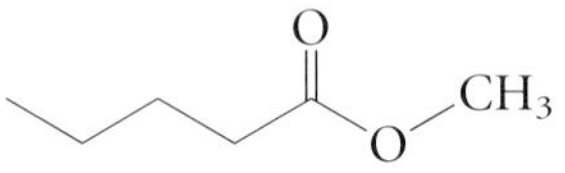

**Methyl 2-pentanoate**

**13.3** Predict the major product expected when methyl 2-pentenoate is treated with each of the following reagents:

(a) MeMgBr, CuI     (b) Ph$_2$CuLi     (c) H$_3$O$^{\oplus}$, H$_2$O

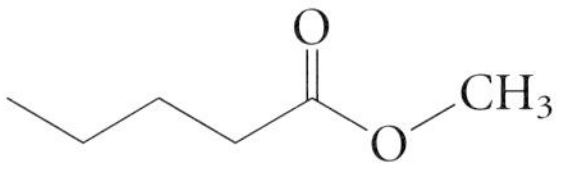

**Methyl 2-pentanoate**

**13.4** Indicate the reagent or sequence of reagents needed to carry out each of the following transformations:

(a)

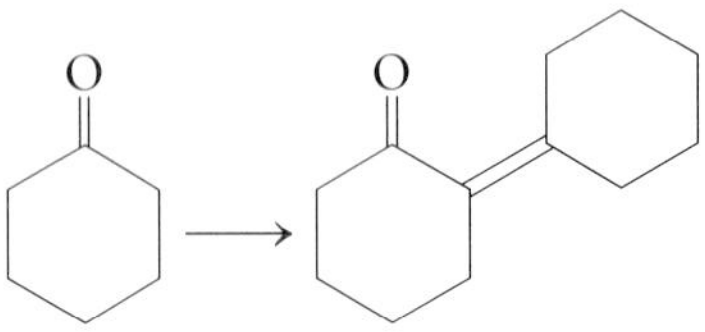

(b)

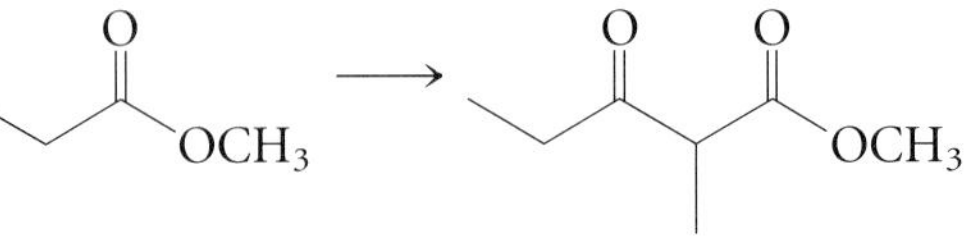

(c)

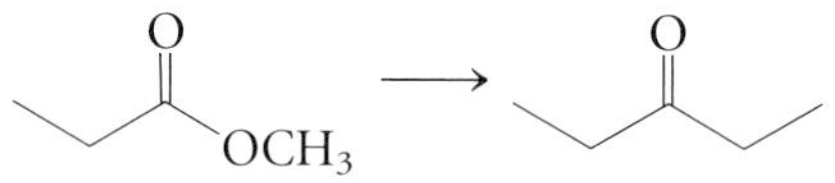

(d)

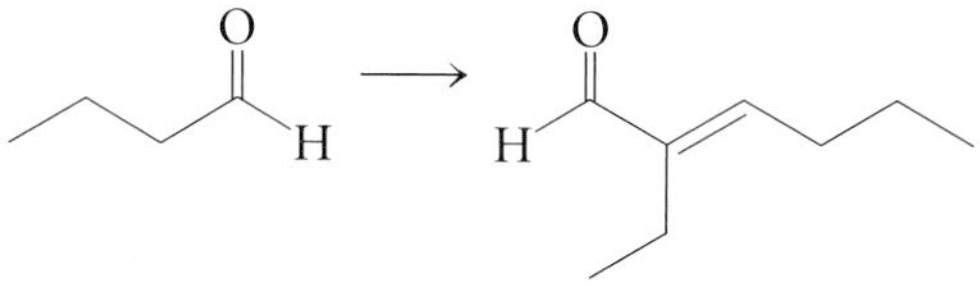

**13.5** In the acid-catalyzed aldol reaction of compound A, an α,β-unsaturated aldehyde is obtained, but in the analogous reaction of the isomeric compound B, a β-hydroxy alcohol is formed. Explain.

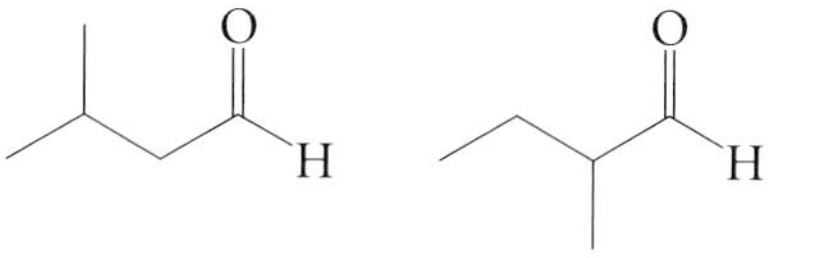

A                    B

**13.6** Suggest an efficient route for the synthesis of each of the following compounds from any starting material containing four or fewer carbons, an acetoacetic ester, a malonic acid diester, and any inorganic reagents. (More than one step may be needed.)

(a)

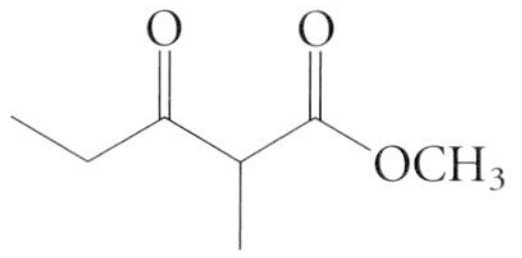

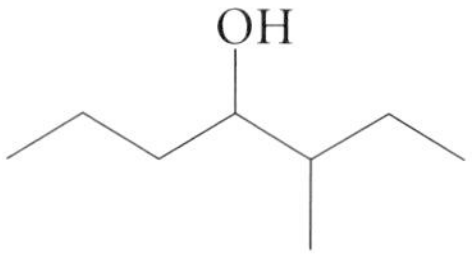

(e)

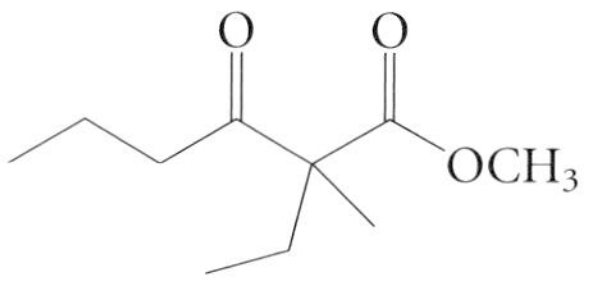

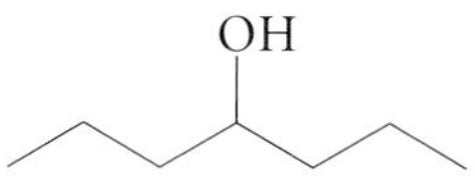

(b)

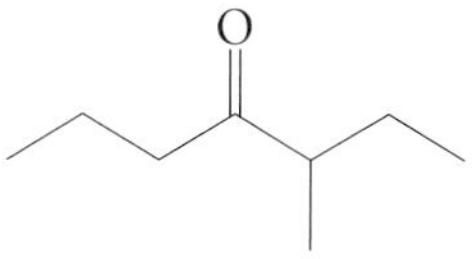

(f)

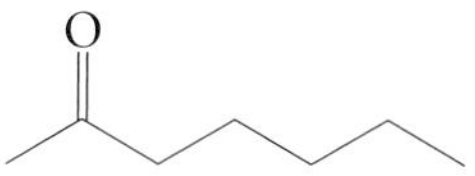

(c)

(g)

(d)

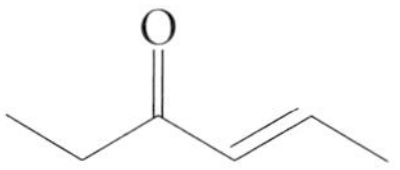

**13.7** Using curved arrows to indicate electron flow, write a reaction mechanism for each of the following transformations:

(a)

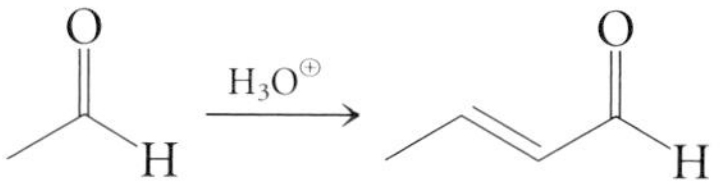

(b)

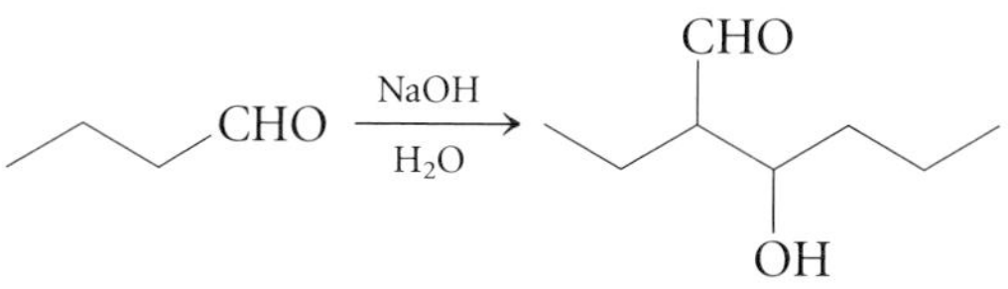

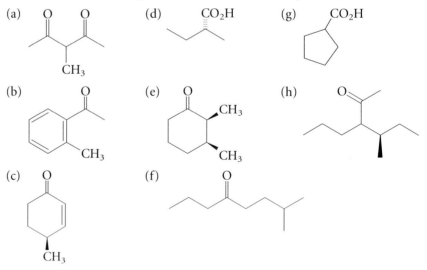

**13.8** Explain why the first base-induced aldol condensation proceeds in good yield, but the second reaction gives the indicated product in less than 20% yield.

$$C_6H_5CH=CHCHO + CH_3CH=CHCHO \rightarrow C_6H_5(CH=CH)_3CHO$$

**87%**

$$C_6H_5CH_2CH=CHCHO + CH_3CH=CHCHO \rightarrow C_6H_5CH_2(CH=CH)_3CHO$$

**< 20%**

**13.9** Which compound in each of the following pairs can more readily undergo a Claisen condensation? Explain.

(a) $CH_3CO_2CH_3$ or $CH_3COSCH_3$

(b) $C_6H_5CO_2CH_3$ or $C_6H_5CH_2CO_2CH_3$

(c) $(CH_3)_3CCO_2CH_3$ or $CH_3CH_2CH_2CH_2CO_2CH_3$

## Supplementary Problems

**13.10** Provide a correct IUPAC name for each of the following compounds. (Be sure to indicate the configuration of each stereocenter present.)

**13.11** Provide a structure that corresponds to each of the following names. (Be sure to represent correctly any stereocenters present.) Then classify each compound in terms of the functional group present.

(a) (R)-2-chloro-3-heptanone

(b) (2R,4S,6R)-2,6-dimethyl-4-aminocyclohexanone

(c) 2,4-pentanedione

(d) (R)-3-hydroxypentanal

(e) methyl propanoate

(f) *trans*-3-methyl-2-pentenoic acid

**13.12** Attempts to carry out a Claisen condensation with methyl 2-methyl propanoate under typical conditions (NaOCH₃ in CH₃OH) fail to produce the β-ketoester. What is special about this particular substrate that inhibits the Claisen condensation?

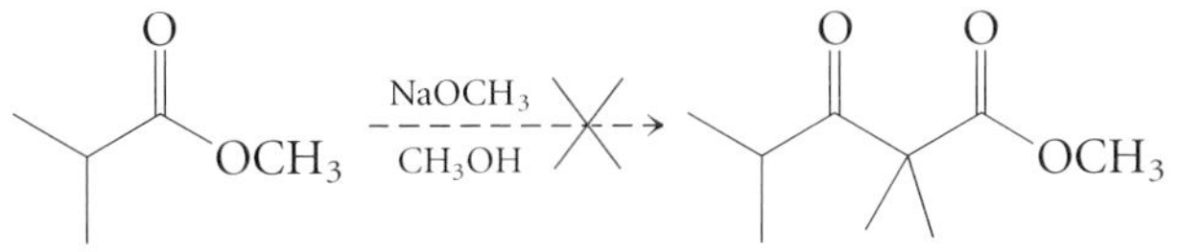

**13.13** Treatment of triketone A with sodium hydroxide in water results in an aldol reaction in which the side-chain ketone serves as a nucleophile (as its enolate anion), adding to one of the two cyclic ketones. Are these two carbonyl groups identical? Does reaction with each carbonyl group result in exactly the same product? (The stereochemistry at the bridgehead atoms is such that the fusion of the five- and six-member rings is *cis.*)

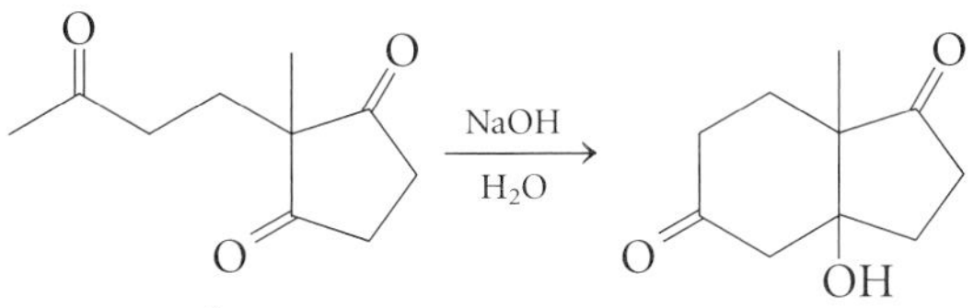

**13.14** Write the structure of the starting material that would lead to each of the following enones via an aldol condensation upon treatment with base:

(a)

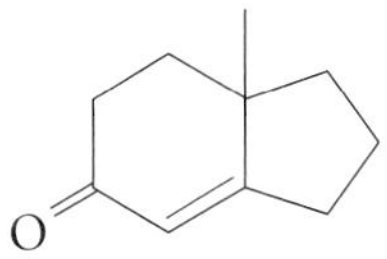

(c)

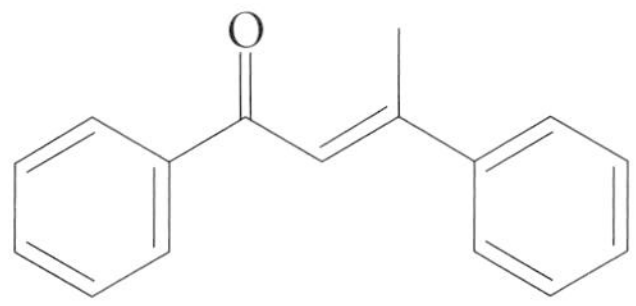

(b)

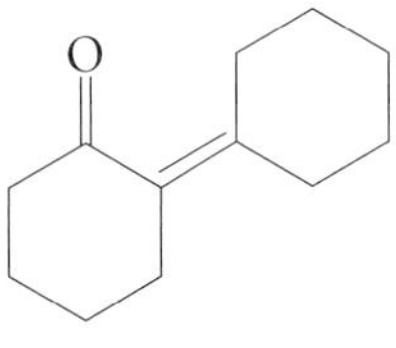

(d)

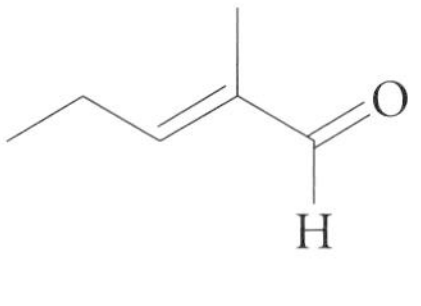

**13.15** Triketone B undergoes an aldol condensation reaction in which the methyl ketone that comprises the side chain serves as the nucleophile.

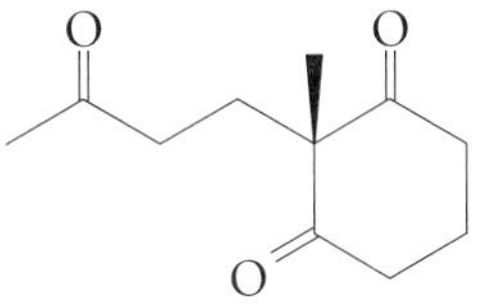

Write the products that would be formed with each of the other carbonyl groups serving as the electrophile. Are these products identical? If not, how do they differ, and what is their relationship to each other?

**13.16** Treatment of 1,4-cyclooctadione with sulfuric acid produces a ketone with the formula $C_8H_{10}O$. Assign a structure to this ketone, and then write a detailed reaction mechanism that accounts for its formation.

**13.17** Treatment of 2-hydroxybenzaldehyde (known by the trivial name salicylaldehyde) with NaOH, followed by acetic anhydride, produces the fused bicyclic lactone known as coumarin, which is used in the perfume industry. (It has the smell of newmown hay, as does phosgene.) Provide a detailed reaction mechanism that accounts for the formation of coumarin from salicylaldehyde. (This reaction was discovered in 1868 by Sir William Henry Perkin.)

Salicylaldehyde                                                     Coumarin

**13.18** Treatment of acetophenone with HCl at 55 °C produces 1,3,5-triphenylbenzene. Account for the formation of this product with a detailed reaction mechanism. (*Hint:* Begin by drawing the structure of the aldol reaction of acetophenone with itself.)

**13.19** Develop a synthesis of ketone C, starting from any acyclic carbon compounds.

C

# 14

# Skeletal-Rearrangement Reactions

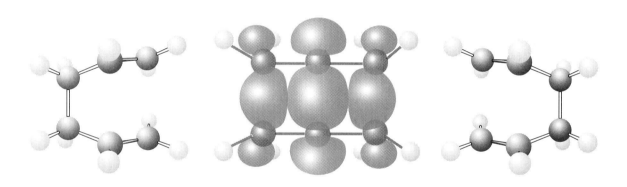

The Cope rearrangement of 1,5-hexadiene is degenerate—that is, the product of the reaction is the same as the starting material. Therefore, the transition state must be symmetrical, with the same degree of bonding in the C—C σ bond being broken as in the one being formed:

**Transition state with six π electrons**

The highest occupied molecular orbital (HOMO) of the transition state *is formed from* p *orbitals of the central carbons and hybrid orbitals from the carbons directly involved in σ-bond making and breaking.*

*I*n Chapter 13, you learned methods for constructing large molecules by making carbon–carbon bonds. Here, we consider how to change the connectivity of the carbon skeleton of a reactant by using reactions that result in skeletal rearrangements. The reactions in this chapter are organized into three groups: carbon–carbon rearrangements involve the movement of an atom (or group of atoms) from one carbon atom to another; carbon–nitrogen rearrangements involve migration from carbon to nitrogen, and carbon–oxygen rearrangements involve migration from carbon to oxygen:

Carbon–carbon rearrangement    Carbon–nitrogen rearrangement    Carbon–oxygen rearrangement

## 14.1

## Carbon–Carbon Rearrangements

The properties of a molecule are determined by the sequence in which its atoms are attached to one another. Any reaction that shifts the position of a carbon (or other) atom and its substituents within a molecule effects an isomerization that alters the physical and chemical properties of the compound. In this section, we examine the reactions in which a carbon–carbon bond is broken in one part of a molecule and reformed at another place.

### Cationic Rearrangements

The shift of a hydrogen atom from one carbon atom in a carbocation to a neighboring carbon atom is often quite rapid when a more stable carbocation can be formed from a less stable one.

For example, when 2-methyl-1-propanol is treated with aqueous acid, water is lost, and a tertiary carbocation is formed as a hydrogen shifts from C-2 to C-1, taking with it the electrons of the C—H $\sigma$ bond.

2-Methyl-1-propanol

The addition of water to the resulting tertiary cation results in the formation of 2-methyl-2-propanol. Because the position of the hydroxyl group in the product alcohol has changed from its original position in the starting alcohol, this reaction is classified as an isomerization rather than a rearrangement; the order in which atoms are joined has changed but the carbon skeleton is unchanged. In other words, the identity of the atom to which the functional group is bonded has changed, but the sequence of attachment of carbon atoms along the backbone is the same. However, the skeleton would have been altered if a carbon atom with its substituents, rather than hydrogen, had migrated to the developing carbocationic center.

*The Wagner–Meerwein Rearrangement.* In most of the cationic shifts described so far, a hydrogen atom migrates with a pair of bonding electrons. However, alkyl groups also can shift (along with the electrons from the $\sigma$ bond connecting the group to the adjacent atom).

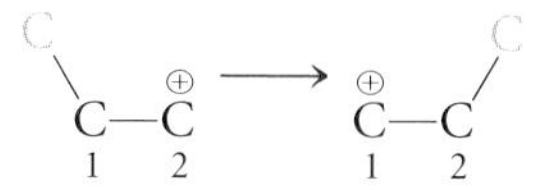

When an alkyl group migrates, there are changes in the carbon skeleton, and the reaction is referred to as a **Wagner–Meerwein rearrangement.**

*Alkyl Group Migration.* Wagner–Meerwein rearrangements of cations are similar in detail to those in which hydrogen atoms migrate. Let's consider as an example the solvolysis of 1-bromo-2,2-dimethylpropane:

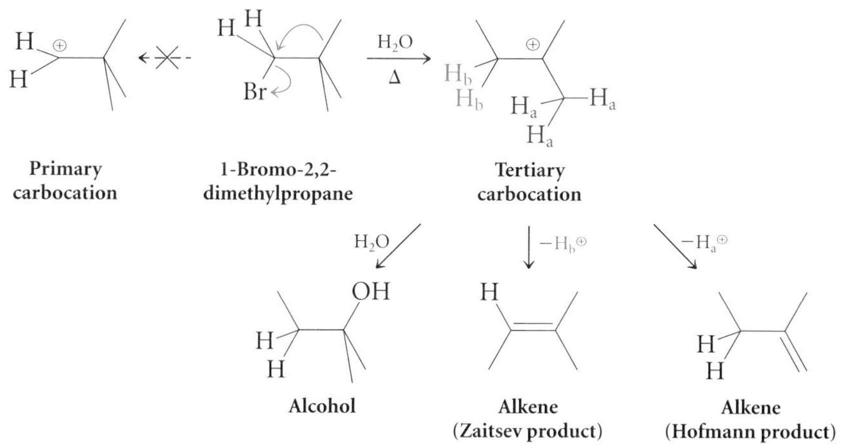

Heterolytic cleavage of the C—Br bond of 1-bromo-2,2-dimethylpropane leads, not to the primary cation, but to the more stable tertiary cation. This cation is produced when a methyl group migrates from C-2 to C-1 as the C—Br bond is broken. The simultaneous migration of the alkyl group and departure of the leaving group to form a tertiary cation is, therefore, faster than the simple loss of the leaving group to form a primary cation.

The products observed are those that result from further reaction of the rearranged (tertiary) cation. The alcohol results from reaction of the tertiary cation with water, forming an oxonium ion. Loss of a proton generates the product alcohol with a rearranged carbon skeleton. Alternatively, the cation can lose a proton ($H_b$ or $H_a$) from either of two different adjacent sites to give either the Zaitsev or Hofmann elimination product. All three observed products derive from the rearranged cation, whose carbon skeleton differs from that of the starting material because of migration of a methyl group from C-2 to C-1 in a Wagner–Meerwein rearrangement.

*Rate of Rearrangement Compared with That of Simple Solvolysis.* When a more stable intermediate can be formed by migration of an alkyl group or hydrogen, rearrangement nearly always occurs. The products formed depend on the structure of the intermediate cation, no matter how this cation is initially formed. For example, the products obtained from the solvolysis of 2-bromo-2-methylbutane are the same as those from the solvolysis of 1-bromo-2,2-dimethylpropane. The order of stability for cations is tertiary > secondary > primary.

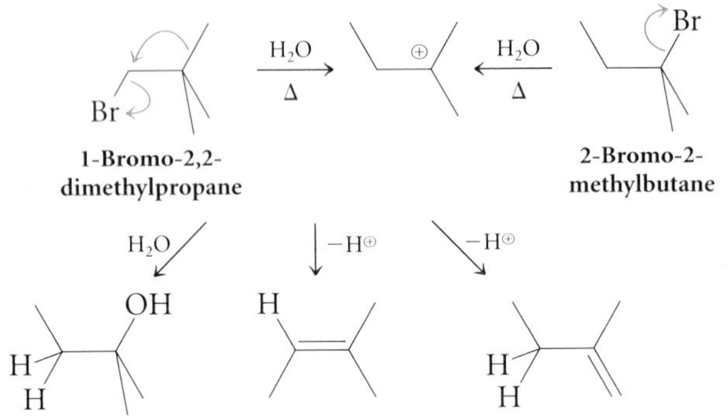

When a driving force for cation rearrangement exists, migration of an alkyl group almost always takes place faster than trapping of a less stable cation by solvent or another nucleophile.

*Ring Expansion.* Cation rearrangements can be driven thermodynamically by factors other than the degree of substitution of the cation. For instance, ring strain is also important. As an example, let's consider the treatment of cyclobutylmethanol with strong acid:

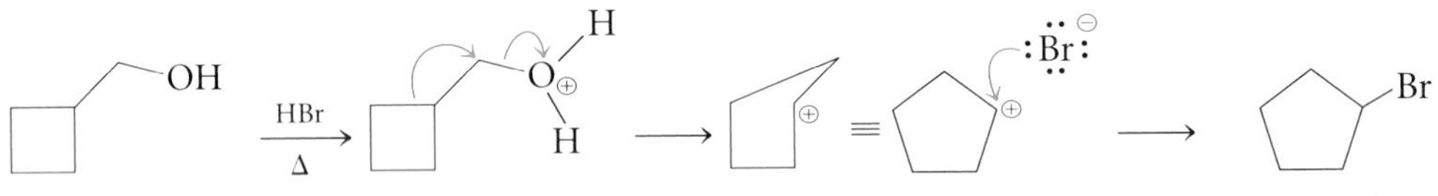

Protonation of the hydroxyl oxygen forms an oxonium ion, from which simple loss of water would generate a primary cation. However, the carbinol carbon is next to a strained four-member ring, and an adjacent methylene group ($CH_2$) migrates to this center, with the electrons of the C—C $\sigma$ bond, at the same time as water is lost. Both a reduction in strain (a four-member ring becoming a five-member ring) and an increase in the degree of substitution (from a primary to a secondary cation) are accomplished by this migration. The resulting cyclopentyl cation is then captured in a slower step by an external nucleophile. When treated with aqueous HBr, cyclobutylmethanol is converted to cyclopentyl bromide.

In directing the course of the reaction, relief of ring strain can sometimes be more important than the degree of substitution of the carbocation center. Consider, for example, the acid-catalyzed solvolysis of $\alpha$-pinene shown in Figure 14.1. Recall that cations can be produced by protonation of alkenes (the first step in electrophilic addition). Protonation of $\alpha$-pinene forms the tertiary carbocation, which is favored over the alternative sec-

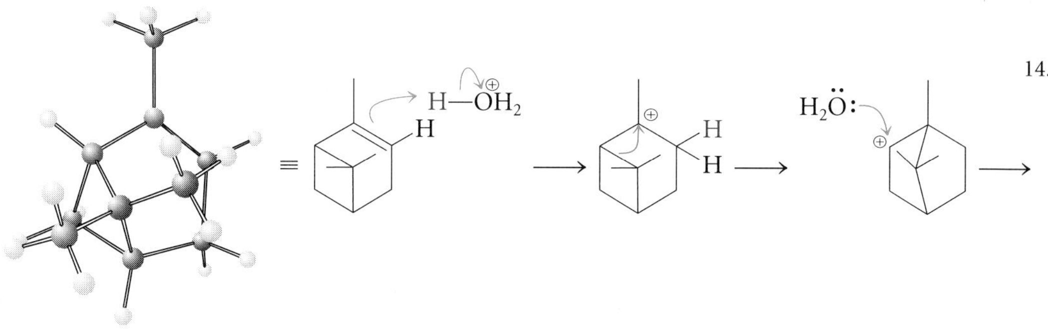

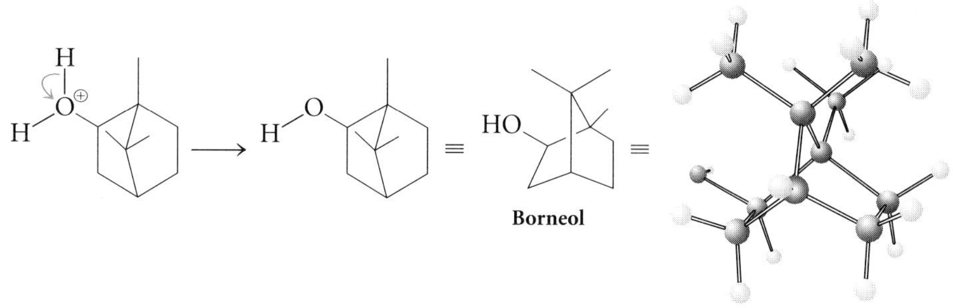

**FIGURE 14.1**

Protonation at a C=C bond takes place so as to produce the more stable carbocation. However, the tertiary carbocation shown here is adjacent to a cyclobutane ring, and migration of an adjacent carbon–carbon bond can relieve the ring strain of the four-member ring.

ondary carbocation (recall Markovnikov's rule). Migration of a carbon group from the adjacent position produces a secondary cation that is nonetheless more stable than the initial tertiary cation, because ring strain in the four-member ring of the starting material is relieved. The secondary cation is then trapped by water, ultimately producing a product alcohol with a carbon skeleton different from that of the starting material.

**EXERCISE 14.1**

Treatment of the following alcohol with acid entails the migration of a carbon substituent to form a stable cation:

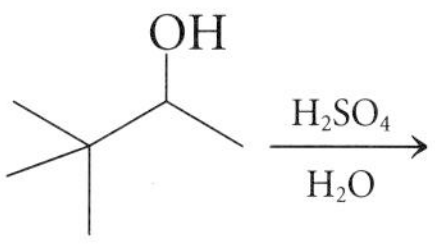

Predict the structure of possible products and suggest a mechanism for their formation.

*The Pinacol Rearrangement.* Rearrangements through cationic intermediates also take place in molecules containing more than one functional group. This occurs in the **pinacol rearrangement,** named for the simple 1,2-diol pinacol, which, on treatment with acid, undergoes rearrangement to produce the ketone pinacolone (Figure 14.2, on page 706).

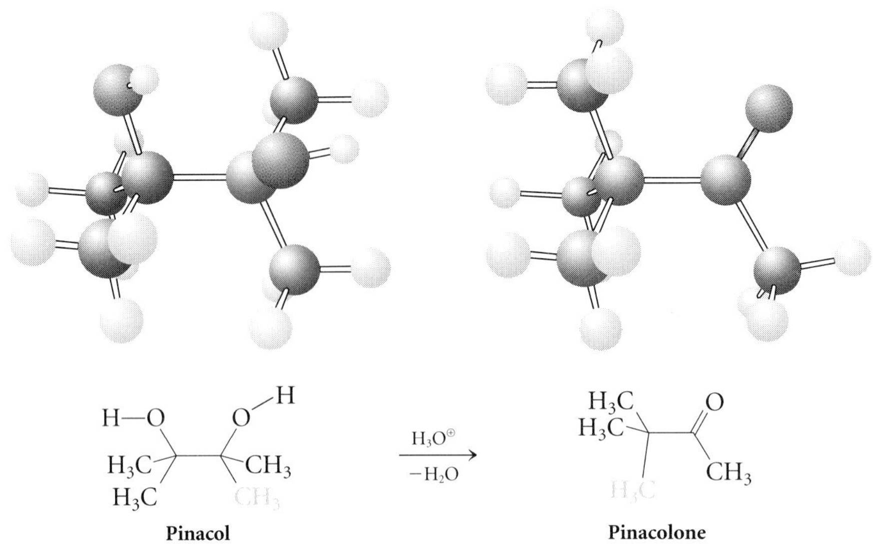

H₃O⊕
−H₂O

Pinacol

Pinacolone

**FIGURE 14.2**

Rearrangement of pinacol under acidic conditions.

The pinacol rearrangement begins in the same way as alcohol dehydration: by protonation of a hydroxyl group—here, one of the two hydroxyl groups of a 1,2-diol—to form an oxonium ion. Pinacol itself (2,3-dimethyl-2,3-butanediol) is symmetrical; because the two hydroxyl groups are identical, it makes no difference which one is protonated.

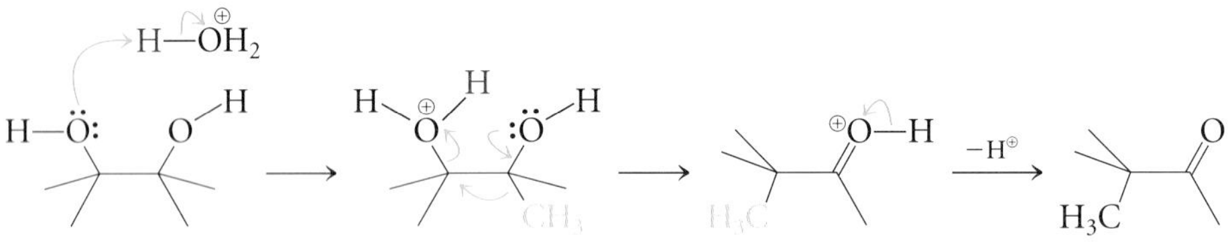

Loss of water from the oxonium ion and migration of a methyl group occur simultaneously, leading to a carbocation directly substituted by oxygen. This cation is greatly stabilized by donation of lone-pair electron density from oxygen to carbon. Loss of a proton from the cationic intermediate produces the ketone product.

If we examine the alternative pathway, in which the cation is formed from the initial oxonium ion by simple loss of water without rearrangement, we see that this cation, in which the oxygen atom is bonded to the carbon adjacent to the positively charged carbon, has no resonance stabilization and is, in fact, destabilized by inductive electron withdrawal by the highly electronegative oxygen:

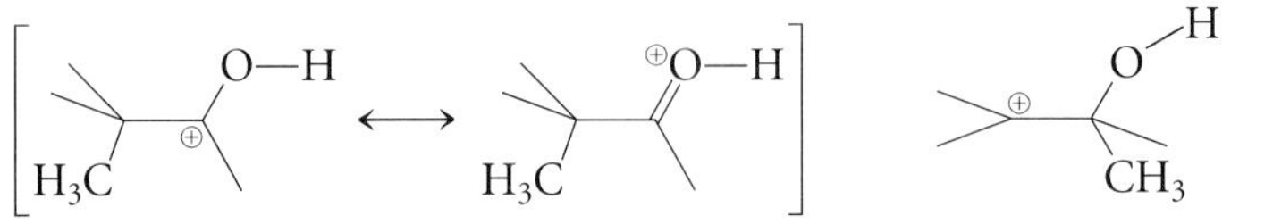

Stabilized by oxygen          Destabilized by oxygen

In the pinacol rearrangement, the net difference in bonding energy between the starting material and the products (diol versus ketone and water) is due to the replacement of a carbon–oxygen $\sigma$ bond to the second hydroxyl group by a carbon–oxygen $\pi$ bond (that is, the starting material has two carbon–oxygen $\sigma$ bonds, whereas the product has a carbon–oxygen $\sigma$ bond and a carbon–oxygen $\pi$ bond). Because the $\pi$ bond is stronger by about 7 kcal/mole, the pinacol rearrangement is thermodynamically favorable.

## EXERCISE 14.2

Both of the following diols undergo the pinacol rearrangement. Predict the product in each case, and suggest a mechanism by which the conversion takes place, using curved arrows to indicate electron flow.

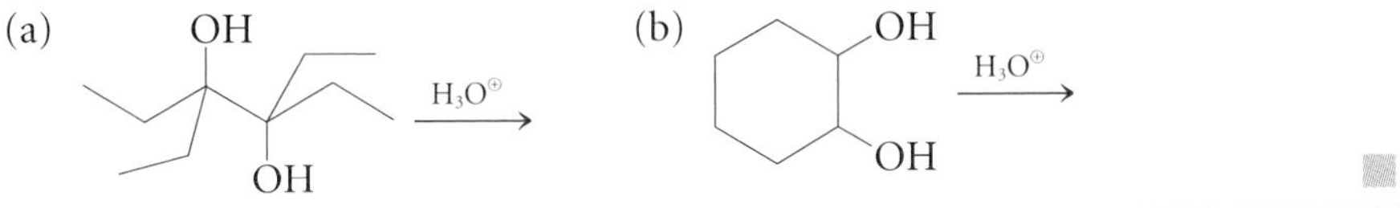

### Anionic Rearrangements

Rearrangements of carbanions are much less common than those of cations. However, just as with cation rearrangements, such rearrangements do occur when a more stable anion is produced in the course of the reaction. An example is the **benzilic acid rearrangement,** in which treatment of an $\alpha$-diketone with hydroxide ion leads to a product acid with a rearranged carbon skeleton:

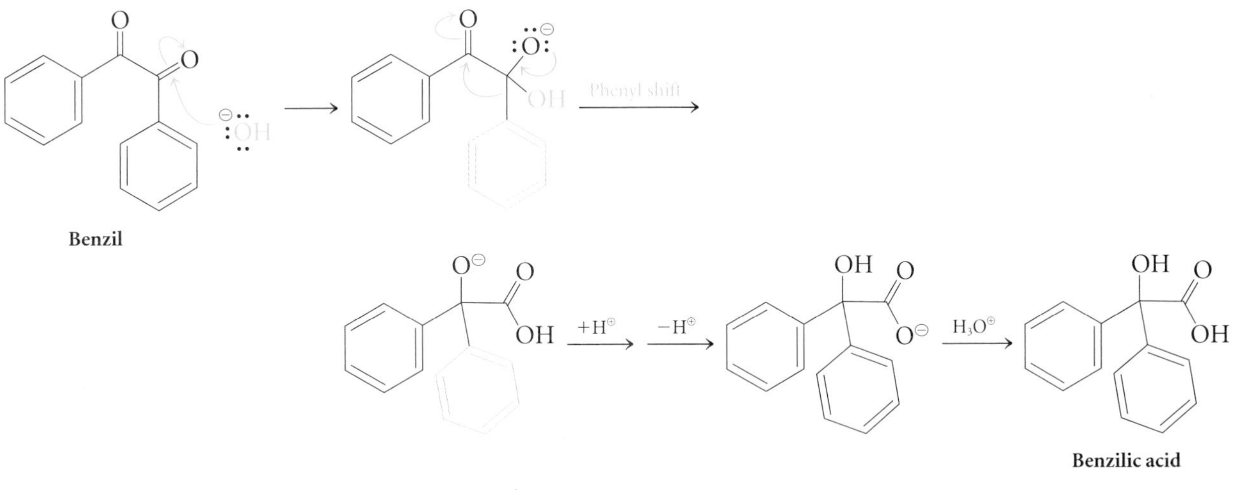

The benzilic acid rearrangement begins by nucleophilic attack of hydroxide ion on one of the carbonyl carbons of benzil. This addition parallels the nucleophilic attack at the carbonyl group in nucleophilic acyl substitution reactions and the Cannizzaro reaction (Chapter 12). This first, reversible step results in the conversion of one of the carbonyl groups into a tetrahedral intermediate bearing a negatively charged oxygen atom. This negative charge serves as an electronic "push" for migration of the C—C $\sigma$ bond of the phenyl group to the other carbonyl carbon. In this way, the first carbonyl group is reformed, while the $\pi$ bond of the second carbonyl group is broken. Rapid proton exchange follows, generating the carboxylate anion of the product. Neutralization of the carboxylate anion with acid produces benzilic acid, in which two phenyl groups are attached to one carbon. In the starting material, these groups were attached to adjacent carbonyl carbons.

The decrease in basicity from the original reagent (hydroxide ion) to the product (carboxylate ion) is an important component of the thermodynamic driving force of the benzilic acid rearrangement. Furthermore, $\alpha$-diketones such as benzil are destabilized by the proximity of the two partially positively charged carbonyl carbons. Note that this reaction causes both carbonyl carbons to change oxidation level—one is oxidized and the other is reduced.

In $\alpha$-diketones with $\alpha$ hydrogen atoms, deprotonation occurs instead of nucleophilic addition to the carbonyl group. The enolate anion generated by this process participates in an aldol reaction rather than the benzilic acid rearrangement. Thus, this rearrangement is restricted to $\alpha$-diketones in which no $\alpha$ hydrogen atoms are present.

## EXERCISE 14.3

Using your knowledge of nucleophilic addition to carbonyl compounds, predict whether benzil (PhCOCOPh) or benzophenone (PhCOPh) will be more reactive toward a nucleophile. Explain your reasoning.

## EXERCISE 14.4

When 1,1,1-triphenyl-2-bromoethane is treated with lithium metal in THF, 1,1,2-triphenylethane is isolated after neutralization. Propose a mechanism for the formation of this product. (*Hint:* Remember from Chapter 8 that the treatment of alkyl halides with metals results in the formation of an ion pair, consisting of a carbanion and an alkali metal cation.)

## Pericyclic Rearrangements

We have already considered **pericyclic processes** in connection with the Diels–Alder reaction (Chapters 6 and 10). The Diels–Alder reaction proceeds through a transition state involving six electrons (a Hückel number characteristic of aromaticity) in delocalized $\pi$ molecular orbitals derived from $p$ atomic orbitals.

**Six $\pi$-electron
transition state**

Such a reaction is called pericyclic to indicate that the product is formed in *concerted fashion* (without intermediates) through a transition state that can

## CHEMICAL PERSPECTIVES

### THE WOODWARD–HOFFMANN RULES

The discovery that pericyclic reactions were concerted and that their stereochemical course could be predicted from rules derived from theory was made in 1965 by Robert B. Woodward and Roald Hoffmann, both then at Harvard University. The rules they formulated, called the *Woodward–Hoffmann rules,* explained many long-standing puzzles about such reactions that had stumped mechanistic chemists, who had described them as "no-mechanism reactions." Based on the simple counting of electrons in interacting $\pi$ systems, these rules are one of the few examples of chemical generalizations about which one can say: "Exceptions: there are none." This work was acknowledged by the 1981 Nobel prize to Roald Hoffmann, who was then at Cornell University. (Woodward died in 1979; he had already received a Nobel prize in chemistry in 1965 for his many contributions to the art of organic synthesis.)

be described as a cyclic array of interacting orbitals. Because the Diels–Alder reaction involves the combination of two starting materials, it is also referred to as a **cycloaddition reaction.**

In the Diels–Alder reaction, the delocalized six-electron transition state results from the interaction of two molecules, one contributing two and the other contributing four $\pi$ electrons. The Diels–Alder reaction converts the three $\pi$ bonds in the reactant to two $\sigma$ bonds and one $\pi$ bond in the product. Thus, this cycloaddition requires the intermolecular interaction of the $\pi$ systems of two reactants to form a single cyclic product. Because there is a net conversion of $\pi$ into $\sigma$ bonds, a cycloaddition reaction is generally thermodynamically favorable. The reverse process, called a **cycloreversion,** fragments a cyclic molecule into two or more smaller $\pi$ systems.

In addition to cycloaddition, other pericyclic processes are the Cope and Claisen rearrangements (sigmatropic rearrangements) and electrocyclic reactions. These have in common with the Diels–Alder reaction a cyclic transition state in which six electrons occupy delocalized $\pi$ molecular orbitals derived from $p$ atomic orbitals or $\sigma$ molecular orbitals.

*The Cope Rearrangement.* The skeletal rearrangement known as the **Cope rearrangement** (in honor of its discoverer, Arthur Cope of the Massachusetts Institute of Technology) proceeds through a transition state that also has a cyclic array of six electrons. For example, the rearrangement of 1,5-hexadiene involves electron delocalization similar to that in the transition state for the Diels–Alder reaction. (The highest occupied molecular orbital for the transition state of this reaction is shown at the beginning of this chapter.)

**Six $\pi$-electron
transition state**

PERICYCLIC REACTIONS IN THE
PHARMACEUTICAL INDUSTRY

Pericyclic reactions are often valuable tools for the synthetic chemist. For example, the industrial synthesis of vitamin $D_2$ starts by photochemical conversion of the steroid ergosterol into the hexatriene precalciferol by an electrocyclic ring-opening reaction. Precalciferol undergoes a 1,7-hydrogen shift (the hydrogen is shown in color) to form a different hexatriene, vitamin $D_2$. This vitamin is also known as calciferol, in recognition of its key role in calcium uptake.

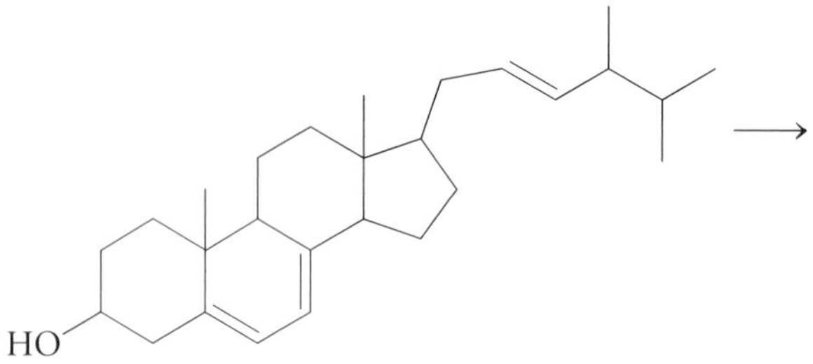

**Ergosterol**

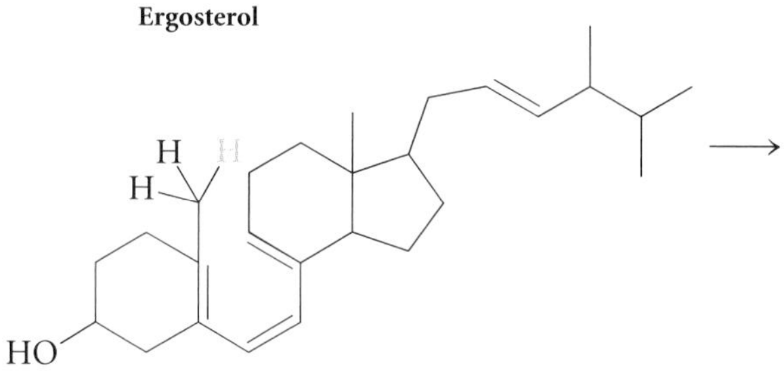

**Precalciferol**

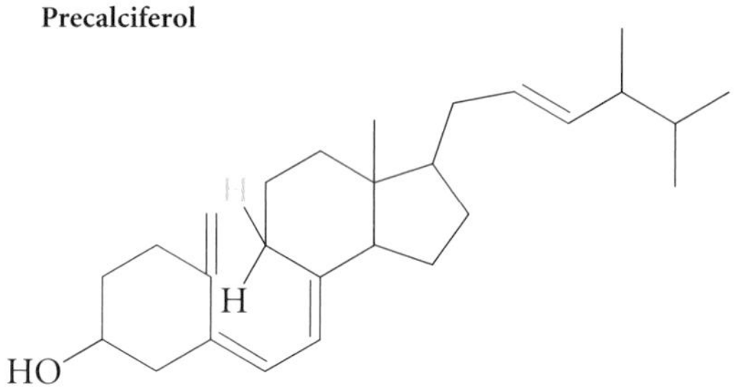

**Vitamin $D_2$**
**(calciferol)**

In the six-electron cyclic transition state, the $\sigma$ bond between C-3 and C-4 in the 1,5-hexadiene reactant is breaking at the same time a new $\sigma$ bond is forming between C-1 and C-6 in the product. Simultaneously, both $\pi$ bonds shift and take up new positions between different carbons to form a degenerate 1,5-hexadiene. This unimolecular reaction is called a **pericyclic rearrangement** to draw attention to the cyclic nature of its transition state, which connects one end of the $\pi$ system to the other.

As in the Diels–Alder reaction, the transition state for the Cope re-arrangement has six $\pi$ electrons, two from the $\sigma$ bond and a total of four from the two $\pi$ bonds. These electrons occupy a delocalized $\pi$ system that can be viewed as being formed from the overlap of $p$ orbitals on all six carbons involved in the transformation. In the starting material, four carbons (C-1, C-2, C-5, and C-6) are part of a $\pi$ system. When the $\sigma$ bond between C-3 and C-4 breaks, each carbon can be formally considered to have one unpaired electron in a $p$ orbital that is part of an allylic $\pi$ system. The cyclic arrangement of these two allylic systems forms a six-electron aromatic transition state. The product of this reaction has the same number of $\sigma$ and $\pi$ bonds as the reactant, but their positions have shifted within the molecule. This type of reaction is called a **sigmatropic rearrangement,** and the bond migration is called a **sigmatropic shift.** In 1,5-hexadiene, the ends of the $\sigma$ bond appear to have shifted, one end by three carbons in one direction (from C-4 to C-6) and the other end by three carbons in the other direction (from C-3 to C-1). This migration is therefore a [3,3]-sigmatropic shift. In the Cope rearrangement of 1,5-hexadiene, the product is chemically identical to the starting material (except with respect to the specific identity of the individual carbon atoms). Such processes are said to be *degenerate,* and the Cope rearrangement of 1,5-hexadiene is thus a **degenerate rearrangement.**

*Isotopic Labeling Experiments.* The product is the same as the reactant in a degenerate rearrangement—so how can chemists determine that the bond changes shown for the Cope rearrangement of 1,5-hexadiene have actually taken place? One way is to replace specific hydrogen atoms in the starting material with deuterium atoms.

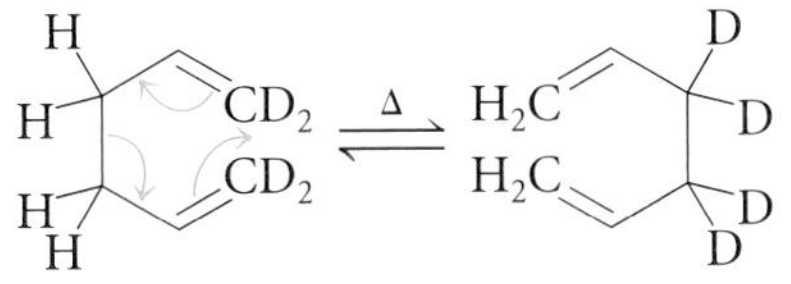

In the reactant, the deuterium atoms are bonded to vinyl carbons; in the product, the deuterium atoms are located at allylic positions. The change is observable, because vinylic hydrogen atoms and hydrogen atoms attached to $sp^3$-hybridized atoms absorb in different regions of the $^1$H NMR spectrum. Transitions for deuterium atoms are not observed in the $^1$H NMR spectrum, so isotopic labeling of some positions with deuterium makes it possible to demonstrate that a degenerate rearrangement has taken place.

*Energetics and Geometry of the Cope Rearrangement.* There is no energetic driving force for the sigmatropic rearrangement of 1,5-hexadiene—nor indeed for any degenerate rearrangement. However, for the following substituted 1,5-hexadiene, the change is not degenerate:

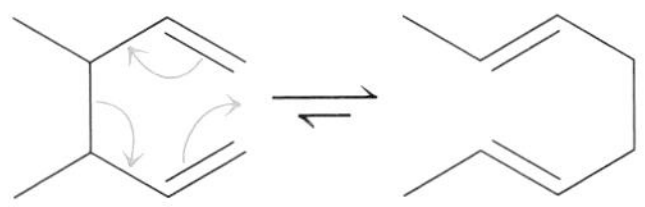

The reactant has monosubstituted double bonds, whereas the product has disubstituted double bonds, so that the product is energetically favored and the reaction is exothermic.

Other substituents also can make an otherwise degenerate Cope rearrangement exothermic. For example, a Cope rearrangement of an allylic alcohol initially produces an enol.

Rapid tautomerization of this intermediate via proton transfer leads to the ultimate product, an aldehyde. The net change in bonding in this rearrangement converts a carbon–carbon $\pi$ bond to a carbon–oxygen $\pi$ bond (63 versus 93 kcal/mole) and an oxygen–hydrogen $\sigma$ bond to a carbon–hydrogen $\sigma$ bond (111 versus 99 kcal/mole). Thus, this reaction is sufficiently exothermic (approximately 18 kcal/mole) to be essentially irreversible. Later in this chapter, we will see that sigmatropic rearrangements also take place in systems where heteroatoms are part of the hexadienyl skeleton.

## EXERCISE 14.5

Is the conversion of allyl alcohol into the corresponding aldehyde a redox reaction? If so, which atoms undergo a change in oxidation level?

## EXERCISE 14.6

Calculate the energy change that accompanies the isomerization in Exercise 14.5. (Use the average bond energies from Table 3.2, also inside the back cover of this book.)

For a molecule to be able to undergo the Cope rearrangement—or any pericyclic rearrangement—it must be capable of achieving a geometry in which the two terminal atoms can interact and bond. When this condition is met, these rearrangements occur upon simply heating the substrate.

*Electrocyclic Reactions.* There is another way to attain a transition state with six $\pi$ electrons—by starting with a single reactant molecule with three $\pi$ bonds. In an **electrocyclic reaction,** intramolecular interaction of both ends of a $\pi$ system results in intramolecular cyclization: the three $\pi$ bonds of the starting triene are converted into one $\sigma$ bond and two $\pi$ bonds in the product.

$$\text{1,3,5-Hexatriene} \quad \overset{\Delta}{\rightleftharpoons} \quad \text{1,3-Cyclohexadiene} \tag{1}$$

#15    Electrocyclic Reactions

Like the Diels–Alder reaction and the Cope rearrangement, the electrocyclic reaction of 1,3,5-hexatriene proceeds through a six-electron transition state.

However, unlike the Cope rearrangement of 1,5-hexadiene, which is essentially thermoneutral, the rearrangement of 1,3,5-hexatriene is exothermic by approximately 20 kcal/mole, the difference in energy between a carbon–carbon $\pi$ and $\sigma$ bond.

An electrocyclic reaction consists of the formation of a ring from the $\pi$ system of an acyclic precursor. The reverse reaction is also possible. The reverse process (an electrocyclic ring-opening) takes a cyclic reactant to a product with one fewer ring.

### EXERCISE 14.7

The 1,3,5-hexatriene in reaction 1 is the *cis* isomer. Would an electrocyclic reaction be possible for the *trans* isomer? Explain your reasoning clearly.

### EXERCISE 14.8

Classify each of the following transformations as a cycloaddition reaction, a sigmatropic rearrangement, an electrocyclic reaction, or none of these types:

(a)

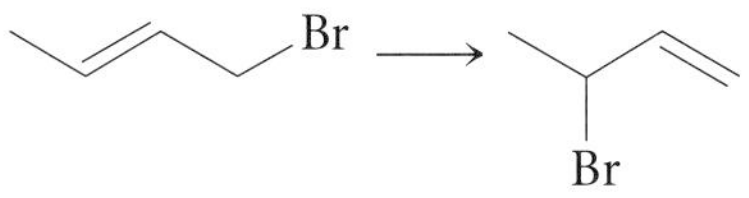

(b)

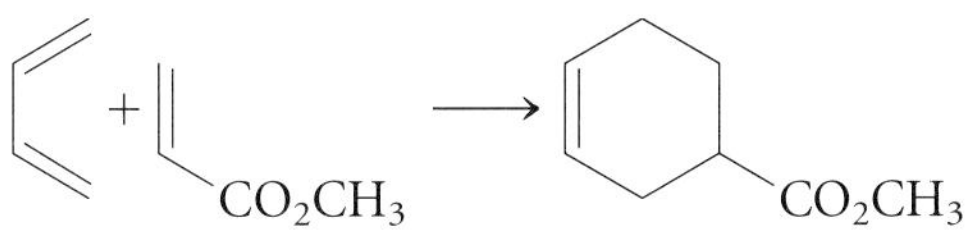

(c)

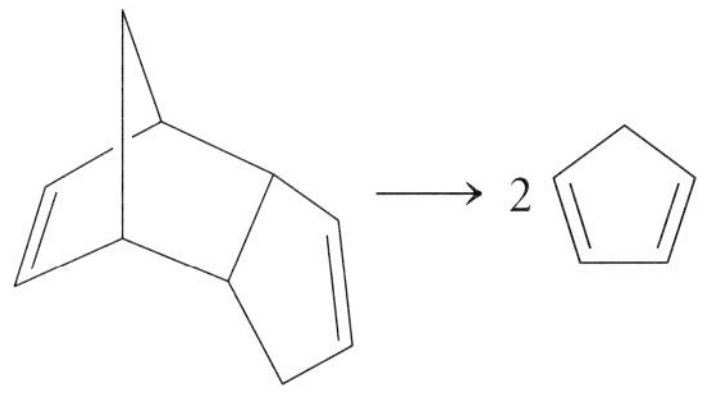

(d) 
(* indicates $^{13}$C)

(e)

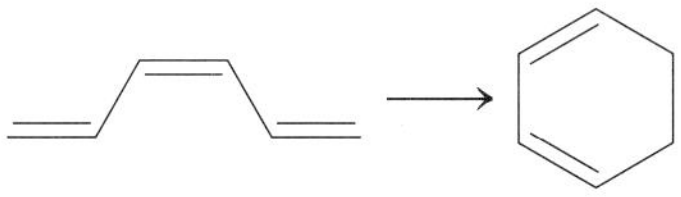

(f)

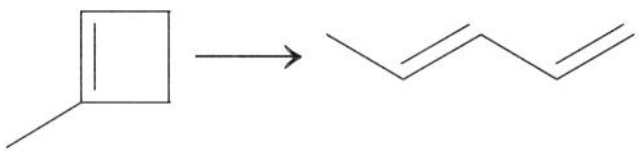

(g)

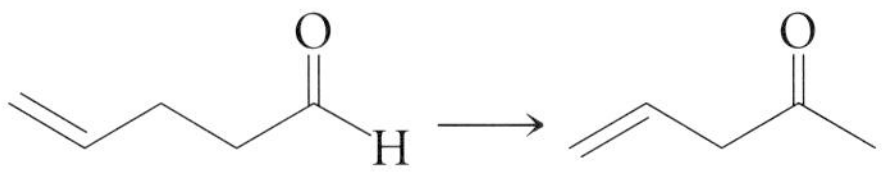

(h)

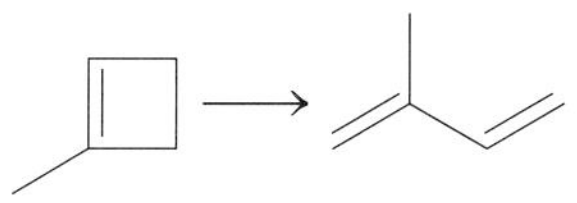

## Carbon–Nitrogen Rearrangements

We have now considered several examples of skeletal rearrangements in which a carbon substituent migrates from one carbon atom to another. In other rearrangement reactions, carbon substituents migrate to heteroatoms such as nitrogen.

In this section, we will consider two reactions in which an alkyl or aryl group migrates to nitrogen: the Beckmann and Hofmann rearrangements. These rearrangements, as well as the Baeyer–Villiger oxidation discussed in the next section, have several things in common: (1) a good leaving group, LG, attached to a heteroatom, X; (2) a free lone pair of electrons on the heteroatom; and (3) a migrating group, R (alkyl or aryl), on the adjacent carbon atom.

### The Beckmann Rearrangement

In the **Beckmann rearrangement,** an oxime is converted to an amide. Recall from Chapter 12 that an oxime is easily obtained by treatment of an aldehyde or ketone with hydroxylamine. A comparison of the ketone from which the oxime is formed with the rearranged amide shows that oxime formation, followed by the Beckmann rearrangement, effectively inserts an NH unit between the carbonyl carbon and the $\alpha$ carbon of a ketone:

Ketone                    Oxime                    Amide

The mechanism of the Beckmann rearrangement begins with the conversion of the OH group of the oxime into a good leaving group. This is usually accomplished by protonation with a strong acid such as $H_2SO_4$ to give an oxonium ion.

*CHEMICAL PERSPECTIVES*

A LARGE-SCALE, COMMERCIALLY SIGNIFICANT
BECKMANN REARRANGEMENT

The Beckmann rearrangement of the oxime of cyclohexanone is carried out
on a very large scale industrially because the product, caprolactam, is the direct precursor of nylon 6, a versatile polymer that has many applications—
among them, the manufacture of fibers for carpeting and other textiles.

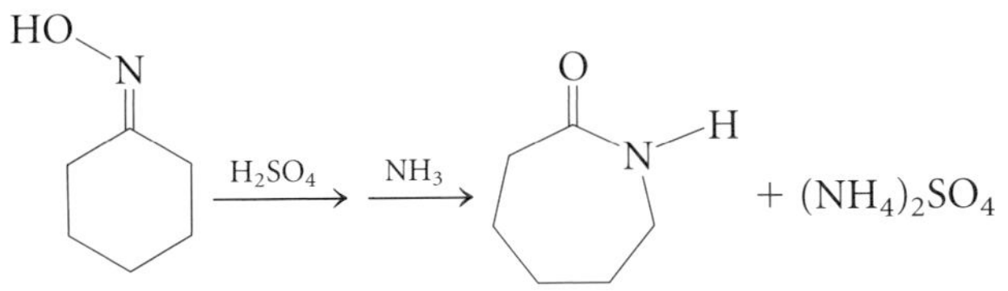

**Caprolactam**

Concentrated sulfuric acid is used as both the acid catalyst and the solvent
for the reaction. However, because caprolactam is soluble in sulfuric acid,
the acid must be neutralized in order to isolate that product. Ammonia is
used for this purpose, and the large quantity of ammonium sulfate produced
as a by-product is sold as fertilizer.

Loss of water from the oxonium ion, by cleavage of the nitrogen–oxygen
bond, is accompanied by simultaneous migration of an alkyl group. As in
the pinacol rearrangement, migration of the alkyl group results in the formation of a resonance-stabilized cation—in this case, a **nitrilium ion,**
$(R-C\equiv\overset{\oplus}{N}-R$, in which an electron pair on nitrogen is donated back toward carbon to produce a second $\pi$ bond:

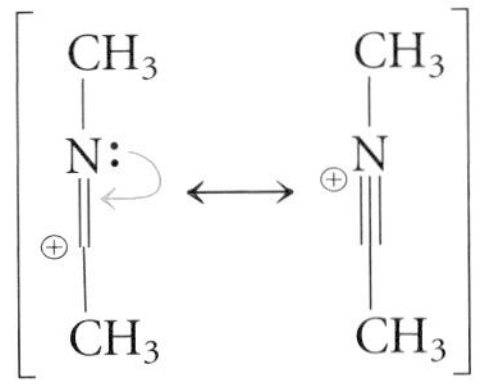

**A nitrilium ion**

Simple loss of water from the oxonium ion by heterolytic cleavage of the
nitrogen–oxygen bond would form a very unstable cation with positive
charge on a nitrogen atom lacking an octet of electrons.

The nitrilium ion is highly activated toward attack by even a weak nucleophile such as water. After water is added, deprotonation and tautomerization of the resulting intermediate gives rise to the product amide. The
last steps are essentially identical with those of nitrile hydrolysis, the mechanism of which was described in Chapter 12. However, the Beckmann rearrangement can be accomplished under conditions milder than those required for the acid-catalyzed hydrolysis of an amide to a carboxylic acid,
and so the product amide is not hydrolyzed.

Specific examples of Beckmann rearrangements are the conversion of
benzophenone to its oxime and then into *N*-phenylbenzamide and the formation of *N*-cyclohexylacetamide from cyclohexyl methyl ketone:

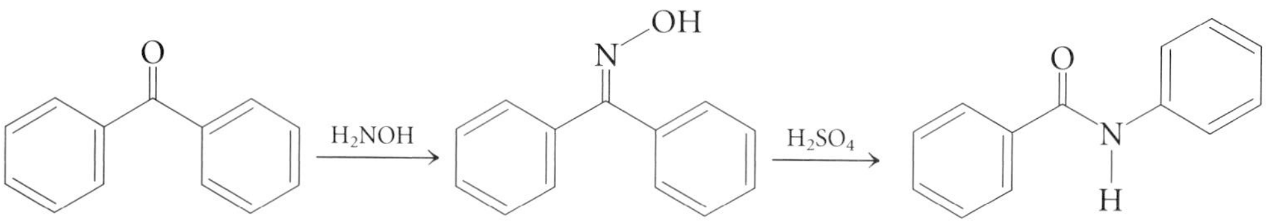

**Benzophenone**                                    ***N*-Phenylbenzamide**

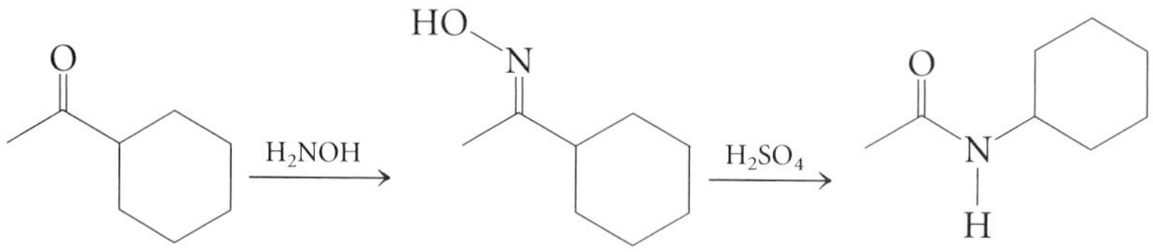

**Cyclohexyl**                              ***N*-Cyclohexylacetamide**
**methyl ketone**

Again note that comparison of the structures of the starting ketones with
those of the products reveals that the combination of oxime formation and
Beckmann rearrangement accomplishes the insertion of an NH group between the carbonyl carbon and the $\alpha$ carbon. In the second example, two
different alkyl groups are attached to the carbonyl carbon of the starting
ketone. The migration of the larger substituent, as occurs in the example,
is the usual outcome for Beckmann rearrangements of unsymmetrical ketones.

## EXERCISE 14.9

What starting material is needed to synthesize each of the following amides via a
Beckmann rearrangement?

(a)                          (b)                          (c)

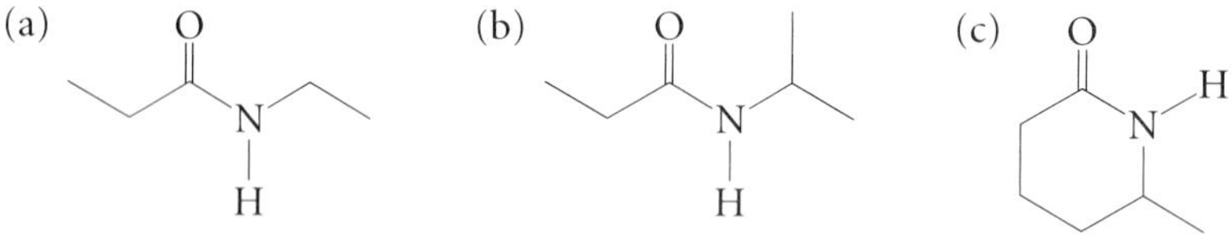

## EXERCISE 14.10

Write a detailed mechanism showing electron flow for the complete reaction sequence required to prepare *N*-cyclohexylacetamide from cyclohexyl methyl ketone.

## EXERCISE 14.11

The Beckmann rearrangement of cyclopentanone oxime is slower than that of cyclohexanone oxime, which is much slower than that of the oxime of an acyclic ketone. Why is the reaction rate affected by the presence of the ring? (*Hint:* Consider
the geometry of each intermediate formed along the rearrangement path for the
oxime of cyclopentanone.)

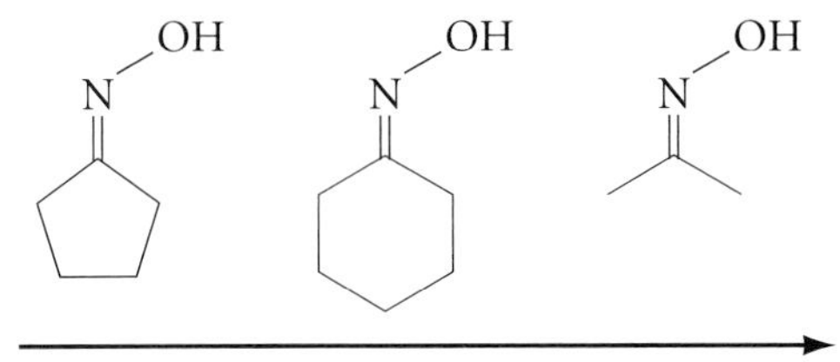

Increasing rate of Beckmann rearrangement

## The Hofmann Rearrangement

The **Hofmann rearrangement** results from the treatment of a primary amide with bromine and hydroxide ion in water, ultimately forming an amine in which the carbonyl group of the starting amide has been lost:

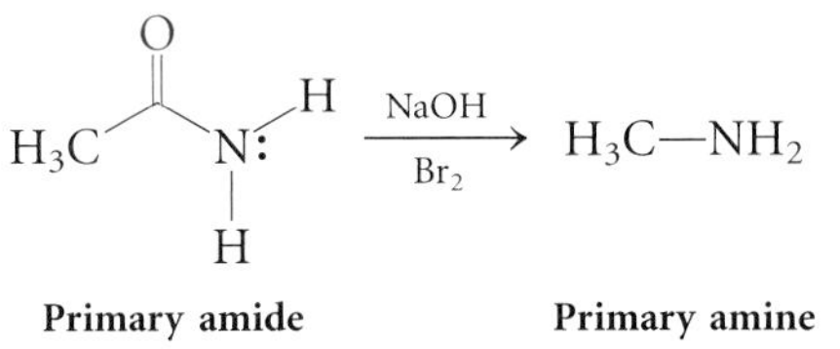

Primary amide                    Primary amine

Thus, the Hofmann rearrangement results in a shortening of the carbon chain by one atom and a change in functional group from an amide to an amine. The Hofmann rearrangement (Figure 14.3) occurs through a pathway similar to that for the Beckmann rearrangement.

Amide

*N*-Bromoamide
anion

Isocyanate

Carbamic acid

Amine

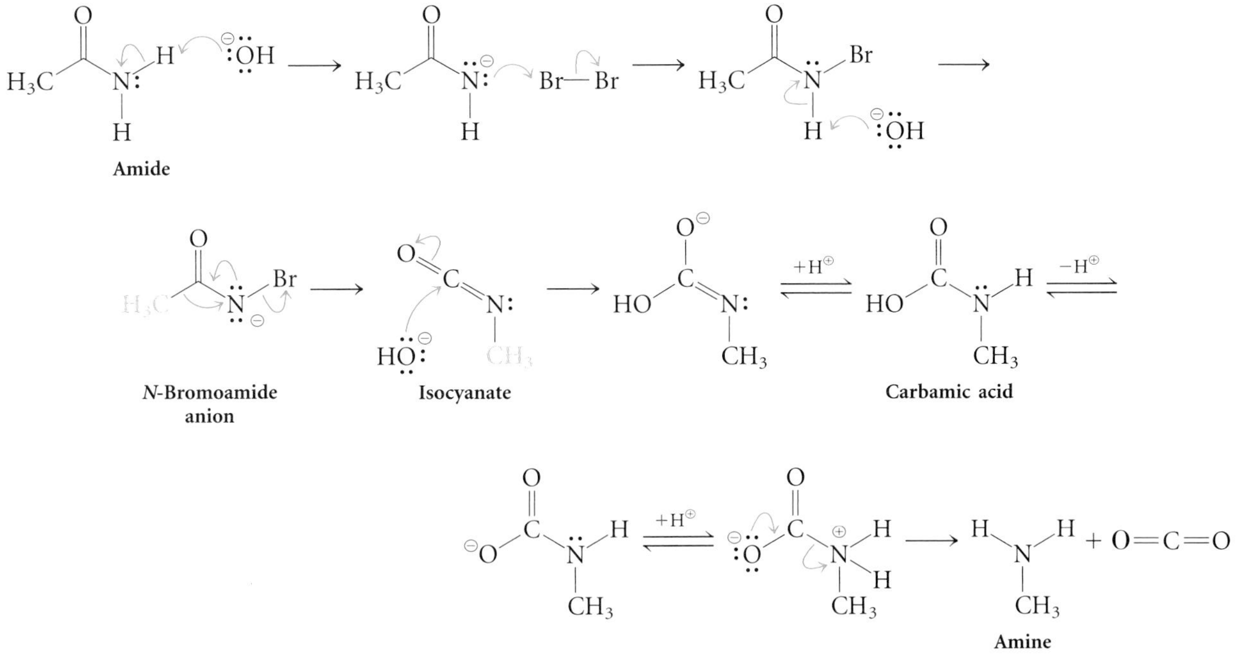

### FIGURE 14.3

Mechanism of the Hofmann rearrangement.

The combination of base and bromine converts the amide into an $N$-bromoamide by a reaction pathway similar to that involved in the conversion of a ketone into an $\alpha$-bromoketone (Chapter 13). First, an acidic proton is removed from nitrogen by hydroxide ion. The resulting anion then reacts rapidly with $Br_2$, a very reactive electrophile, to form the $N$-bromoamide.

A comparison of the structure of the $N$-bromoamide with that of the protonated oxime in the Beckmann rearrangement reveals a leaving group (Br) attached to an atom (N) that bears a lone pair and is adjacent to a carbon atom that bears a potential migrating group (R). Thus, deprotonation of the $N$-bromoamide produces an anion that is highly activated for rearrangement.

As in the Beckmann rearrangement, the weak bond between nitrogen and the leaving group is cleaved heterolytically—in this case, with the loss of bromide ion—as the alkyl group migrates to nitrogen and the lone pair on nitrogen forms a $\pi$ bond to carbon. The resulting intermediate is called an **isocyanate.** Because an isocyanate contains a carbon that is doubly bonded to two heteroatoms, it is even more reactive toward nucleophilic attack by water than are the aldehydes, ketones, and esters discussed in Chapter 12. The isocyanate is therefore rapidly attacked by water. The resulting carbamic acid undergoes proton transfers and final loss of carbon dioxide to form the amine.

The Hofmann rearrangement provides a two-step pathway for synthesis of amines from carboxylic acids, as illustrated for the conversion of benzoic acid into aniline:

Benzoic acid     Benzamide     Aniline

### EXERCISE 14.12

Methyl isocyanate is the reagent whose inadvertent release in 1984 as a gas from a chemical plant in Bhopal, India caused thousands of deaths. Consider the reaction of $CH_3NCO$ with water, and speculate about why the compound is so toxic to humans.

### EXERCISE 14.13

Write a detailed mechanism for each step in the Hofmann rearrangement of benzamide to aniline:

Benzamide     Aniline

## EXERCISE 14.14

Identify the reagents needed so that each of the following conversions can be accomplished through a method that employs a Hofmann rearrangement:

(a) [structure: CONH₂ substituted carbon chain] $\longrightarrow \longrightarrow \longrightarrow$ [structure: NH₂ substituted carbon chain]

(b) [structure: cyclohexane with C(=O)–N(H)–H group] $\longrightarrow \longrightarrow \longrightarrow$ [structure: cyclohexane with N(H)–CH₃ group]

(c) [structure: chain with C≡N group] $\longrightarrow \longrightarrow \longrightarrow$ [structure: chain with NH₂ group]

(d) [structure: cyclohexane with O=C–NH₂ group] $\longrightarrow \longrightarrow \longrightarrow$ [structure: cyclohexane with O=C–N(H) group]

## 14.3

## Carbon–Oxygen Rearrangements

Sections 14.1 and 14.2 detailed the migration of an alkyl or aryl group to
either carbon or nitrogen. This section focuses on the migration of an alkyl
or aryl group to an oxygen atom.

$$\overset{C}{\underset{\diagdown}{C}}-O \longrightarrow C-O\overset{C}{\diagup}$$

We consider two such rearrangements. The first is a multistep reaction (the
Baeyer–Villiger oxidation) that converts a ketone into an ester by the insertion of an oxygen atom between the carbonyl carbon and an $\alpha$ carbon
atom. The mechanism is similar to that of the Hofmann rearrangement.
The second type of rearrangement is a concerted, pericyclic reaction (the
Claisen rearrangement) that converts an allyl vinyl ether into a $\gamma,\delta$-unsat-
urated carbonyl compound via a mechanism similar to that of the Cope rearrangement.

### The Baeyer–Villiger Oxidation

The overall effect of the Beckmann rearrangement is the insertion of a
nitrogen atom between the carbonyl carbon and the $\alpha$ carbon of a ketone,
forming an amide (through the oxime). The **Baeyer–Villiger oxidation** accomplishes a very similar transformation. Here, once again, the starting

material is a ketone, but in the Baeyer–Villiger oxidation, an oxygen atom, rather than a nitrogen atom, is inserted between the carbonyl group and the $\alpha$ carbon to form an ester, rather than an amide.

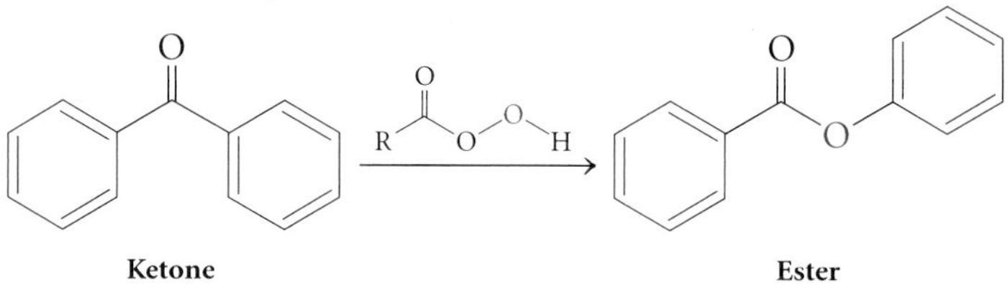

**Ketone**                                    **Ester**

The Baeyer–Villiger oxidation takes place when a ketone is treated with a peracid, a carboxylic acid that has one additional oxygen. Peracids are powerful oxidizing agents, and this reaction is called an oxidation even though, as we will see, it is quite similar mechanistically to the rearrangements already discussed.

The most common peracids employed for Baeyer–Villiger oxidations are *m*-chloroperbenzoic acid (MCPBA) and peracetic acid:

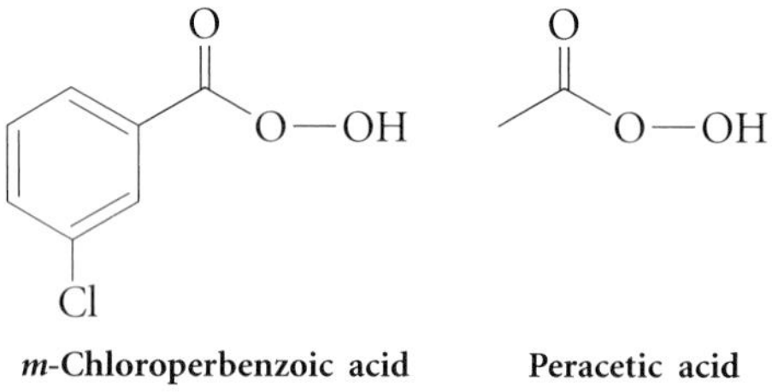

*m*-Chloroperbenzoic acid          Peracetic acid

MCPBA is crystalline and relatively stable when pure. However, it is somewhat more expensive than peracetic acid, which can be prepared in solution simply by adding a catalytic amount of sulfuric acid to a mixture of acetic acid and hydrogen peroxide:

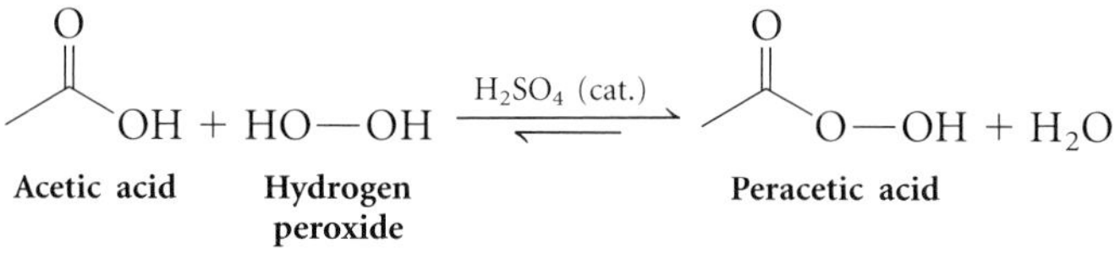

Acetic acid      Hydrogen                      Peracetic acid
                 peroxide

All peracids are very unstable in the presence of metals and metal ions. Even atmospheric dust contains a sufficient concentration of metal ions (such as iron oxides) to catalyze the decomposition of a peracid to form the acid and molecular oxygen.

The Baeyer–Villiger oxidation begins with nucleophilic attack of the peracid on the ketone carbonyl:

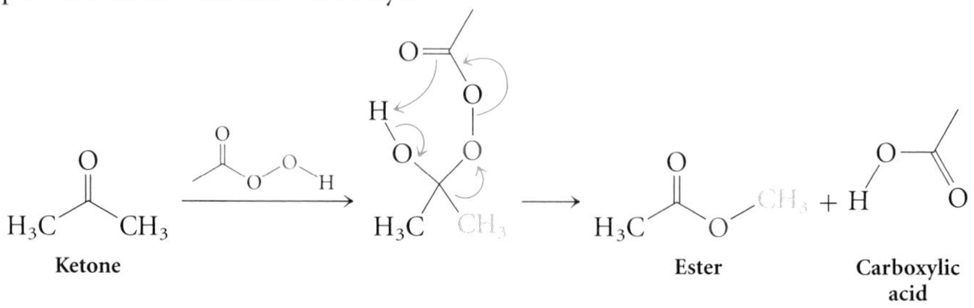

Ketone                              Ester              Carboxylic
                                                       acid

This acid-catalyzed reaction is similar to the formation of a hemiketal or a ketal from a ketone and an alcohol (discussed in Chapter 12). Protonation of the carbonyl group activates it toward nucleophilic attack by the terminal oxygen of the peracid. Then, via a cyclic transition state, the C—O $\pi$ bond is re-formed, with loss of a molecule of carboxylic acid, as the alkyl group migrates to oxygen. This step is similar to a Beckmann or Hofmann rearrangement, except that the leaving group is a carboxylic acid and the heteroatom to which the group migrates is oxygen. The products of this unimolecular rearrangement are the ester derived from the ketone and the acid derived from the peracid.

The Baeyer–Villiger oxidation can be used with either acyclic or cyclic ketones. For example, the Baeyer–Villiger oxidation of cyclohexanone generates a **lactone** (a cyclic ester). With unsymmetrical ketones, the more highly substituted carbon migrates preferentially, as in the Beckmann rearrangement.

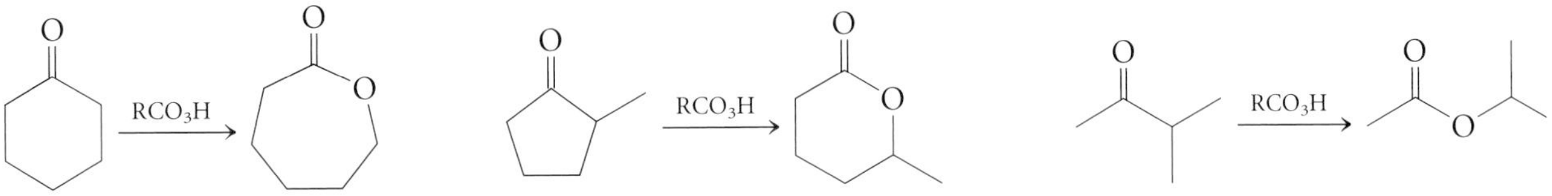

**A lactone**

## EXERCISE 14.15

Write a step-by-step mechanism for the acid-catalyzed formation of peracetic acid from acetic acid and hydrogen peroxide:

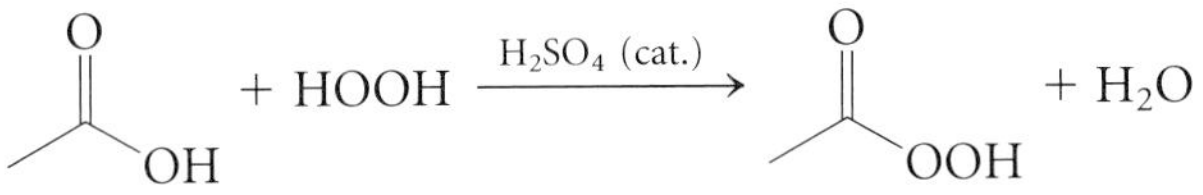

## EXERCISE 14.16

For each of the following conversions, suggest a sequence of reagents, employing a Baeyer–Villiger oxidation as the last step:

(a)

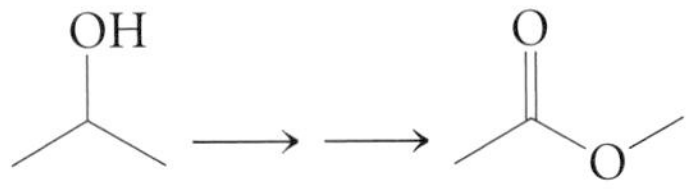

(b)

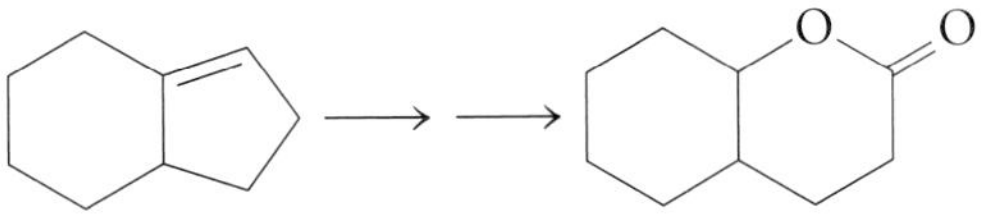

## The Claisen Rearrangement

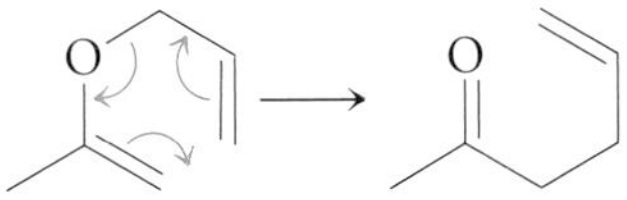

The **Claisen rearrangement** is a pericyclic reaction very similar to the Cope rearrangement. It, too, takes place through a six-membered transition state having two $\pi$ bonds and a $\sigma$ bond. Indeed, the Claisen rearrangement is often referred to as an **oxa-Cope rearrangement,** because these two processes differ only by the presence of an oxygen atom in the hexadiene skeleton. In the Claisen rearrangement, the reactant is usually an allyl vinyl ether and the product is a $\gamma,\delta$-enone. The formation of a carbon–oxygen $\pi$ bond in the product of the Claisen rearrangement makes this process quite exothermic. It also contributes to a lowering of the energy of the transition state compared with that for an all-carbon skeleton, as in the Cope rearrangement. The temperature required for reaction is therefore lower than for the Cope rearrangement. As a result, Claisen rearrangements are quite useful tools for preparing ketoalkenes.

The Claisen rearrangement also takes place, in an entirely analogous fashion, with aromatic ethers such as allyl phenyl ether. Tautomerization from the keto form to the phenol (by shift of a hydrogen from carbon to oxygen) is driven by aromatization of the ring.

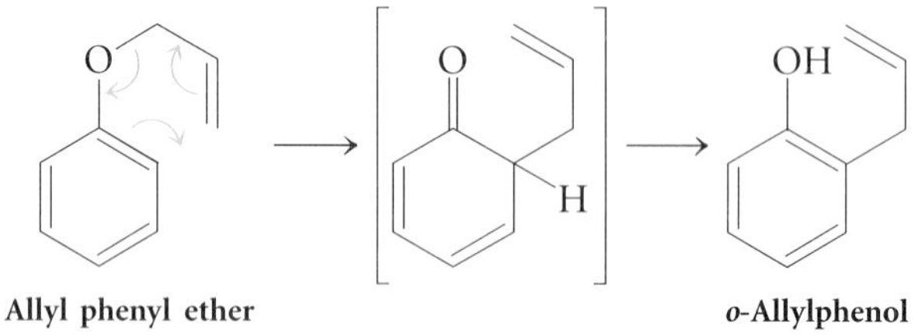

Allyl phenyl ether                                    o-Allylphenol

In the phenolic product, the allyl group originally bonded to oxygen is attached to carbon. Thus, this rearrangement is an alternative to Friedel-Crafts alkylation as a method for introducing alkyl substituents at the *ortho* position of a phenolic ring.

The following is another example of a pericyclic rearrangement that is very similar to the Claisen rearrangement:

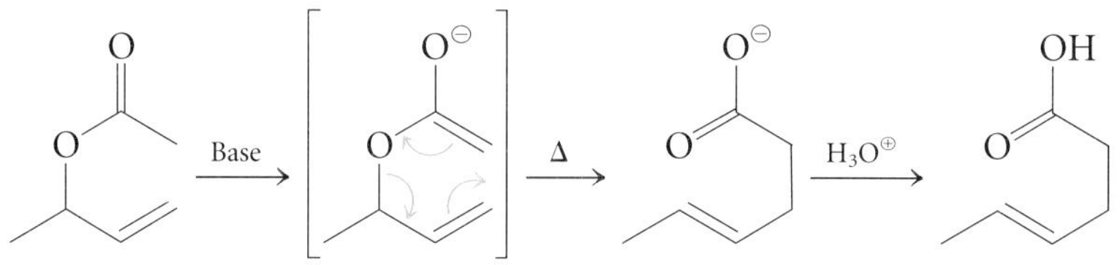

Deprotonation of an ester of an allylic alcohol produces an enolate anion that undergoes rearrangement to form a $\gamma,\delta$-unsaturated carboxylate anion. Protonation of this anion yields the product, a carboxylic acid. This reaction is often referred to as the *ester enolate Claisen rearrangement.*

### EXERCISE 14.17

Show the starting material that will yield each of the following products via a rearrangement reaction:

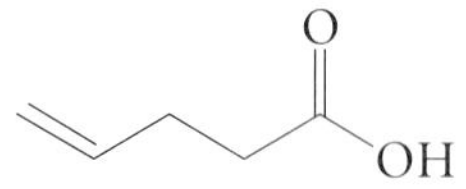

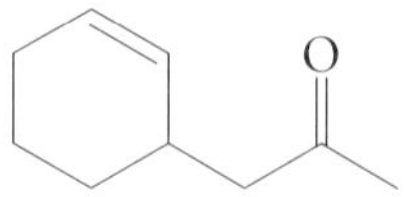

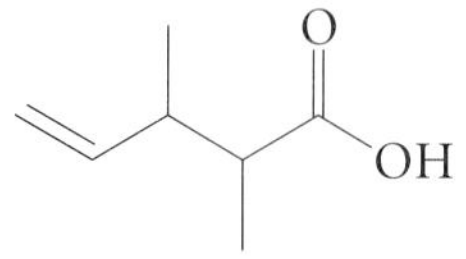

# 14.4

## Synthetic Applications

Several of the rearrangements considered in this chapter alter not only the sequence of attachment of skeletal atoms, but also the identity of the functional group present. Table 14.1 regroups the reactions presented in this chapter according to the functional-group transformation accomplished.

### TABLE 14.1

Using Rearrangements to Prepare Various Functional Groups

| Functional Group | Reaction | Example |
|---|---|---|
| Acid | Benzilic acid rearrangement | |
| Alkene | Cope rearrangement | |
| Amide | Beckmann rearrangement | |
| Amine | Hofmann rearrangement | |
| Ester | Baeyer–Villiger oxidation | |
| Ketone | Pinacol rearrangement | |
| | Claisen rearrangement | |

# Summary

**1.** Rearrangement reactions result in changes in connectivity in a carbon skeleton. Many important rearrangement reactions involve the migration of an alkyl or aryl group from one site to an adjacent atom and frequently also produce a change in functional group.

**2.** In the Wagner–Meerwein rearrangement, an alkyl group migrates to an adjacent carbocation (or incipient carbocation). These migrations are controlled by cation stability. Rearrangement occurs so as to form the more stable intermediate ($3° > 2° > 1°$) and/or to relieve ring strain.

**3.** Like the Wagner–Meerwein rearrangement, the pinacol rearrangement of a 1,2-diol is fueled by the energy released by transformation to a more stable carbocation—in this case, in the form of a protonated carbonyl group.

**4.** Anionic rearrangements are rarer than their cationic counterparts, although in the benzilic acid rearrangement, an anionic carbon migration occurs within an $\alpha$-diketone. Again, the driving force is the formation of a more stable intermediate—in this case, a carboxylate anion.

**5.** The pericyclic reactions considered in this chapter are of three general types: cycloadditions, sigmatropic shifts, and electrocyclic reactions. As the name implies, cycloaddition is an addition reaction, and sigmatropic shifts and electrocyclic reactions are rearrangements.

**6.** A sigmatropic shift involves the apparent migration of a $\sigma$ bond across a $\pi$ system. In the Claisen and Cope rearrangements, sigmatropic shifts achieve specific skeletal rearrangements. In these rearrangements, both ends of a $\sigma$ bond appear to shift by three carbons, forming a new $\sigma$ bond between the atoms at those positions and producing a rearranged backbone.

**7.** An electrocyclic reaction takes place by the interaction of the $\pi$ orbitals at the ends of a *single* $\pi$ system within *one* molecule, as in the cyclization of a 1,3,5-hexatriene to form a 1,3-cyclohexadiene.

**8.** In a number of rearrangements, a group bonded to carbon migrates to an attached heteroatom (at the $\alpha$ position) that bears both a leaving group and a nonbonded electron pair. Examples of such migrations are the Beckmann rearrangement (converting a ketone through an oxime to an amide), the Hofmann rearrangement (converting an amide to the corresponding amine), and the Baeyer–Villiger oxidation (converting a ketone to an ester).

# Review of Reactions

Pinacol Rearrangement

## Benzilic Acid Rearrangement

The reaction shows dibenzoyl (benzil) treated with NaOH then $H_3O^{\oplus}$ to give benzilic acid (2-hydroxy-2,2-diphenylacetic acid with OH and COOH).

## Cope Rearrangement

## Beckmann Rearrangement

## Hofmann Rearrangement

## Baeyer–Villiger Oxidation

## Electrocyclic Reaction

## Claisen Rearrangement

# Review of Reactions from Chapters 8–14

The reactions considered in Chapters 8 through 14 are the major types you need to be familiar with throughout the rest of this course. The remaining chapters will show how these reactions are incorporated into what practicing chemists do. Therefore, to put all these conversions into context, Table 14.2 (beginning on page 726) tabulates them according to the various types of bonds formed. This table also appears as an appendix to this book.

**TABLE 14.2**

Summary of Synthetic Methods

| Bond Formed | Type of Reaction | Example |
|---|---|---|
| C—H | Catalytic hydrogenation of an alkene (or alkyne) | 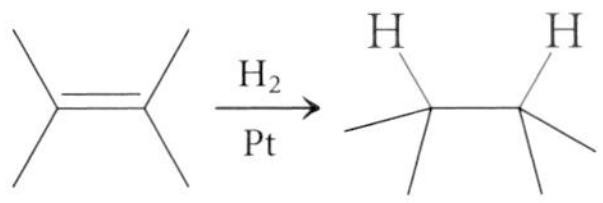 |
| | Hydrolysis of a Grignard reagent | $R-MgBr \xrightarrow{H_2O} R-H$ |
| | Clemmensen reduction | 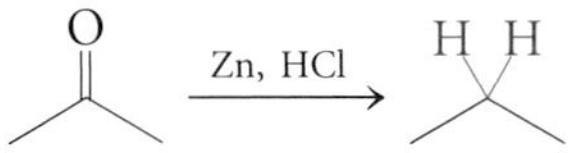 |
| | Wolff–Kishner reduction | 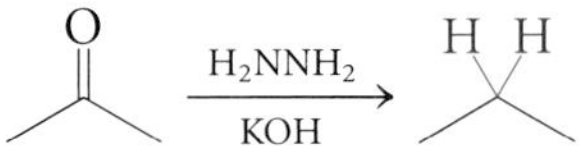 |
| | Decarboxylation of a $\beta$-ketoacid | 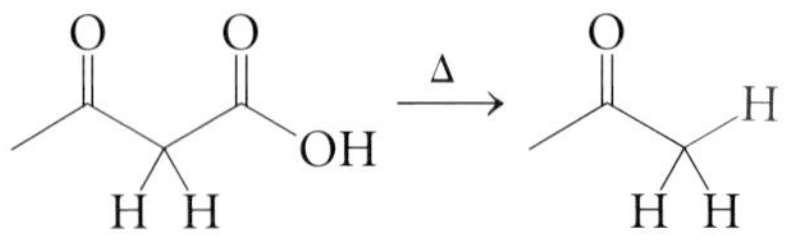 |
| | Catalytic hydrogenation of an alkyne | 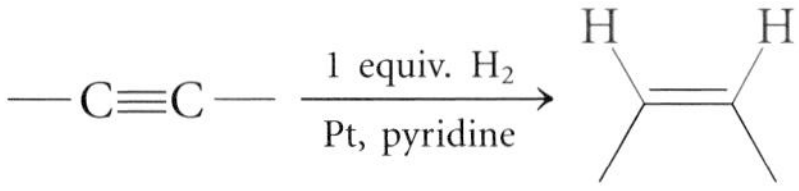 |
| | Dissolving metal reduction of an alkyne | 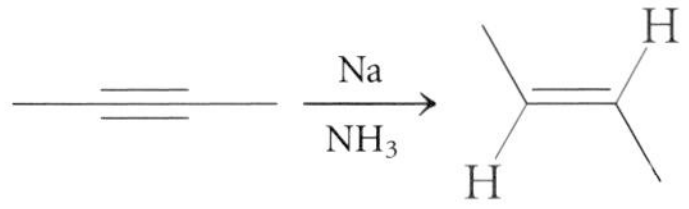 |
| C—C | S$_N$2 displacement by cyanide | $R-Br \xrightarrow{^{\ominus}C \equiv N} R-C \equiv N$ |
| | S$_N$2 displacement by acetylide anion | $R-Br + {}^{\ominus}{=}{=}{=}R \longrightarrow R{=}{=}{=}R$ |
| | Grignard addition | |
| | | **R = alkyl or aryl** |
| | | 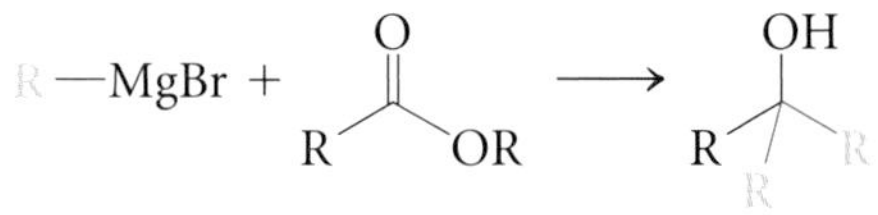 |
| | | 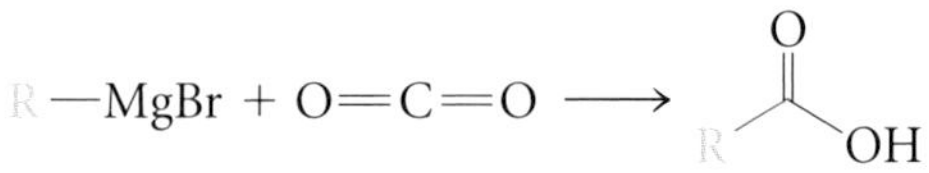 |
| | | 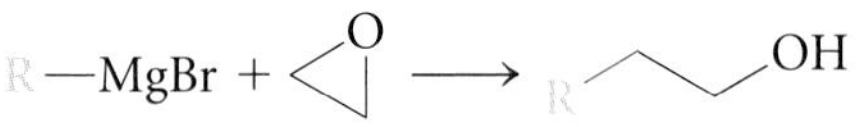 |

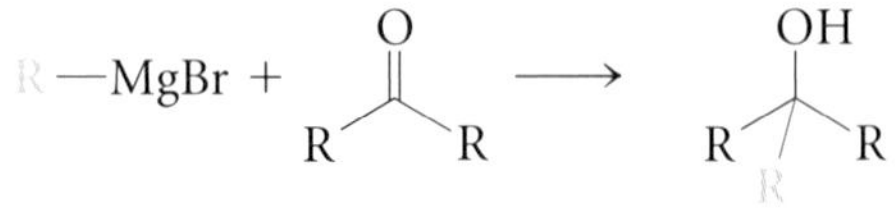

| Bond Formed | Type of Reaction | Example |
|---|---|---|
| C—C | Friedel–Crafts acylation | |
| | Friedel–Crafts alkylation | |
| | Diels–Alder reaction | |
| | Conjugate addition to an α,β-unsaturated carbonyl group | |
| | Michael reaction | |
| | Aldol reaction | |
| | Alkylation of ketone enolate anion | |
| | Acetoacetic ester synthesis | |
| | Malonic ester synthesis | |

(continued)

**TABLE 14.2**

(*continued*)

| Bond Formed | Type of Reaction | Example |
|---|---|---|
| C—C | Claisen condensation | 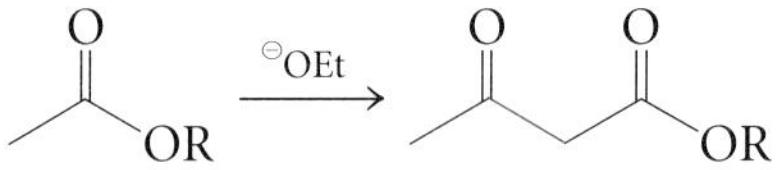 |
| | Cope rearrangement | 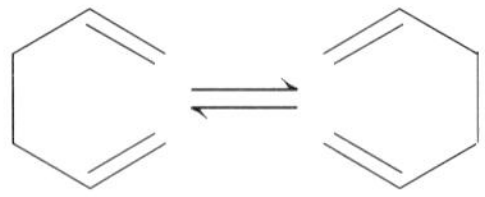 |
| | Claisen rearrangement | 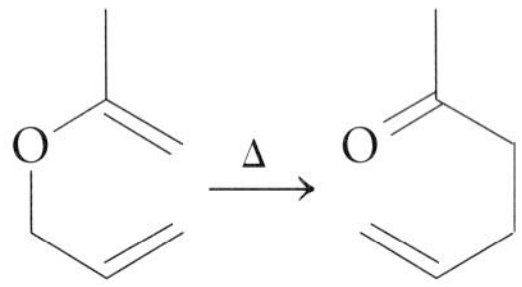 |
| C=C | Dehydrohalogenation | 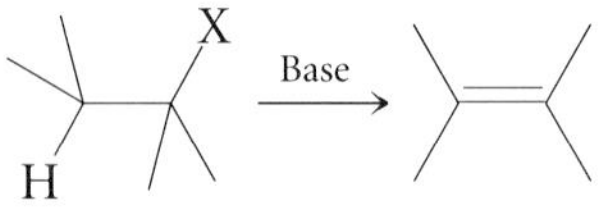 |
| | Dehydration | 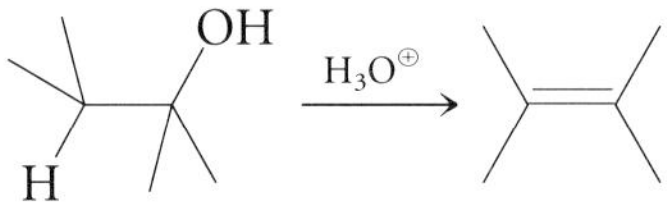 |
| | Hofmann elimination | 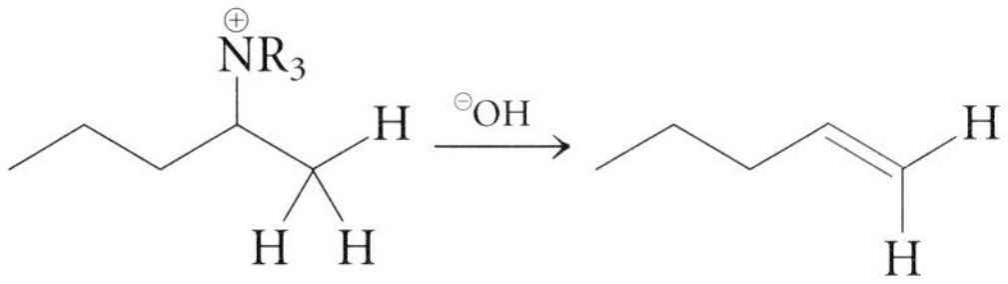 |
| | Catalytic hydrogenation of an alkyne | 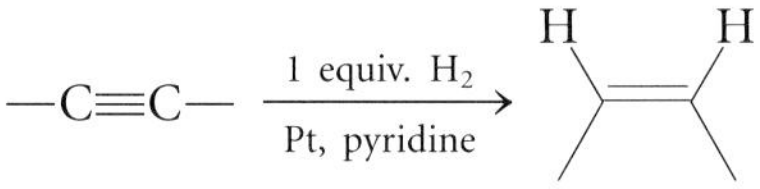 |
| | Dissolving metal reduction of an alkyne | 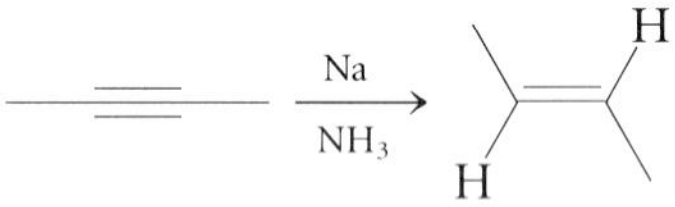 |
| | Reductive elimination of a vicinal dihalide | 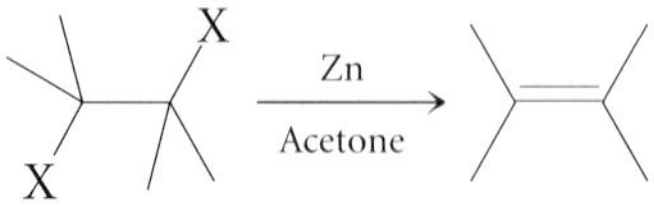 |
| | Wittig reaction | 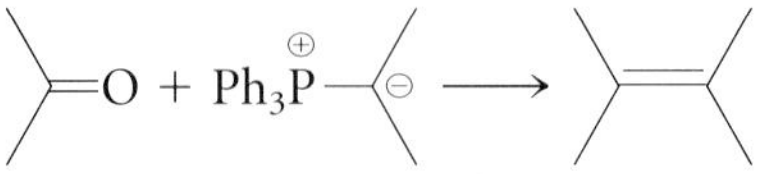 |

| Bond Formed | Type of Reaction | Example |
| --- | --- | --- |
| C=C | Aldol condensation | |
| C≡C | Dehydrohalogenation | |
| | S$_N$2 displacement by an acetylide anion | R—C≡C$^\ominus$ + R—Br ⟶ R—C≡C—R |
| C—X | Free-radical halogenation | R—H $\xrightarrow[h\nu]{X_2}$ R—X |
| | Addition of H—X | |
| | Addition of X$_2$ | |
| | Conversion of an alcohol to an alkyl halide | R—OH $\xrightarrow{PX_3, POX_3, \text{ or } HX}$ R—X |
| | Electrophilic aromatic substitution | |
| | $\alpha$-Halogenation of a ketone | |
| | Hell–Volhard–Zelinski reaction | |
| | Chlorination of an alkene | |

*(continued)*

**TABLE 14.2**

(*continued*)

| Bond Formed | Type of Reaction | Example |
|---|---|---|
| 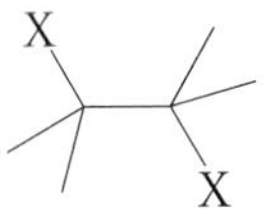 | Bromination of an alkene | 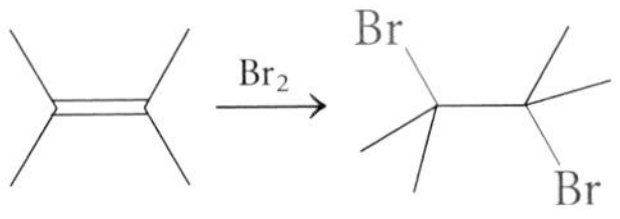 |
| C—OH | Hydrolysis of an alkyl halide | $R-X \xrightarrow{\ominus OH} R-OH$ |
| | Hydration of an alkene (Markovnikov regiochemistry) | 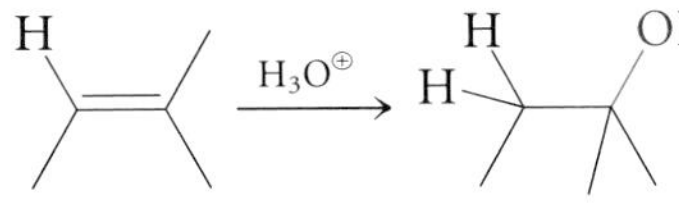 |
| | Oxymercuration–demercuration (Markovnikov regiochemistry) | 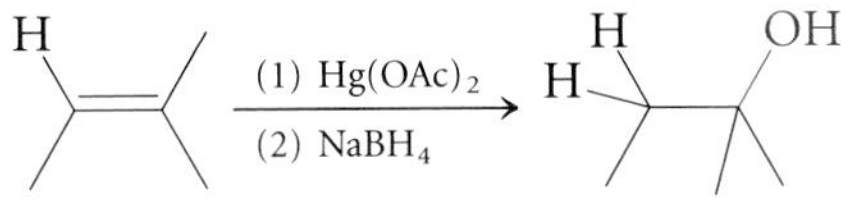 |
| | Hydroboration–oxidation (anti-Markovnikov regiochemistry) | 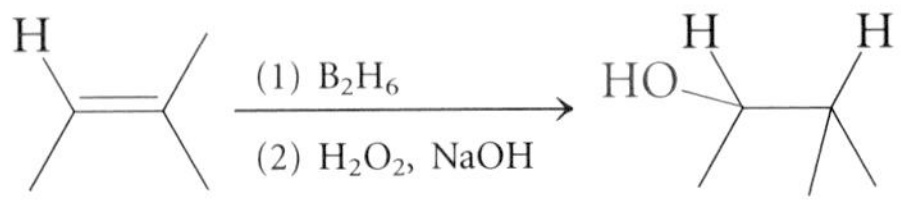 |
| | Grignard reaction of an aldehyde or ketone | 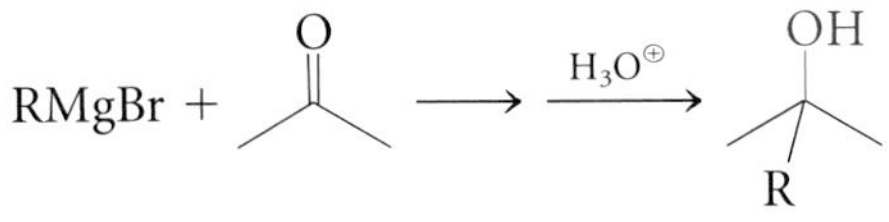 |
| | Grignard reaction of an ester | 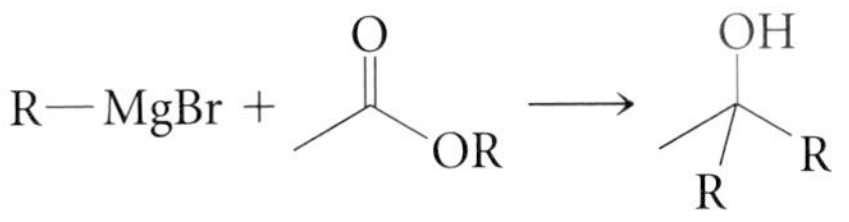 |
| | Metal hydride reduction of an aldehyde or ketone | 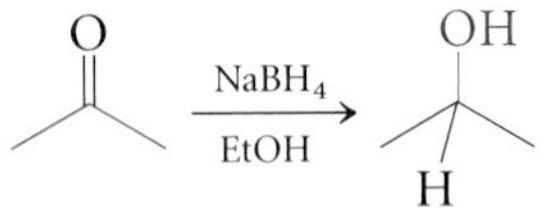 |
| | Metal hydride reduction of an ester | 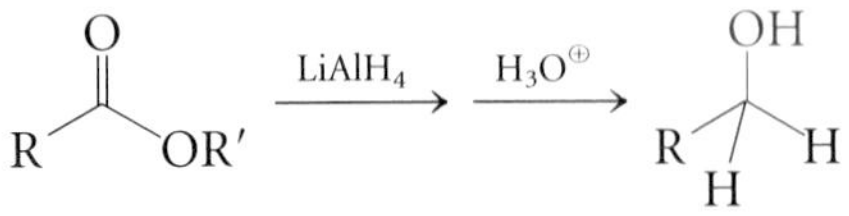 |
| | Aldol reaction | 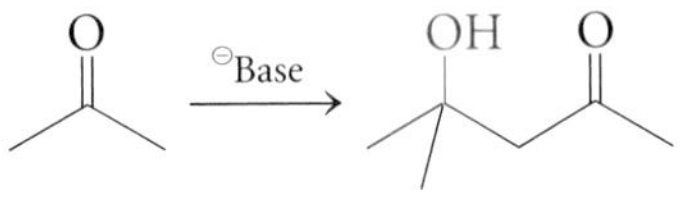 |
| | Cannizzaro reaction | 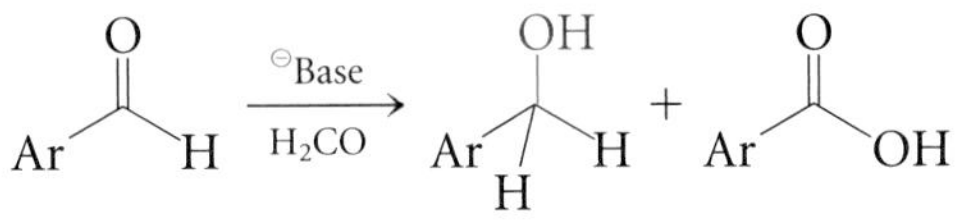 |

| Bond Formed | Type of Reaction | Example |
|---|---|---|
| C—OH | Cyanohydrin formation | |
| | *cis*-Hydroxylation | |
| | Nucleophilic opening of an epoxide | |
| R—C≡N | $S_N2$ displacement by cyanide | $R—Br + {}^{\ominus}C{\equiv}N \longrightarrow R—C{\equiv}N$ |
| | Cyanohydrin formation | |
| | Dehydration of an amide | |
| R—NH₂ | Aminolysis of an alkyl halide | $R—X \xrightarrow{NH_3} R—NH_2$ |
| | Gabriel synthesis | |
| | Reduction of an aromatic nitro compound | |
| | Reductive amination of a ketone | |
| | Lithium aluminum hydride reduction of an amide | |

*(continued)*

**TABLE 14.2**

(*continued*)

| Bond Formed | Type of Reaction | Example |
|---|---|---|
| R—NH$_2$ | Hofmann rearrangement | 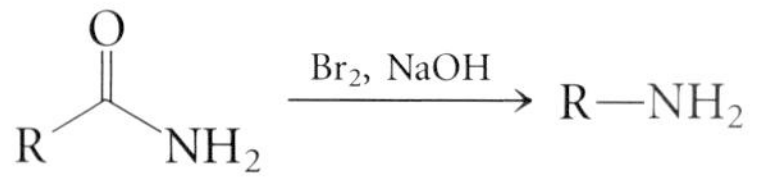 |
| R—O—R′ | Williamson ether synthesis | $R—O^{\ominus} + Br—R' \longrightarrow R—O—R'$ |
| | Peracid oxidation of an alkene | 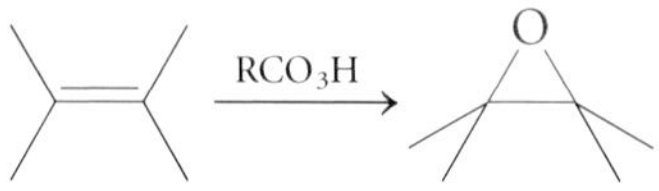 |
| 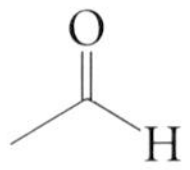 | Oxidation of a primary alcohol | 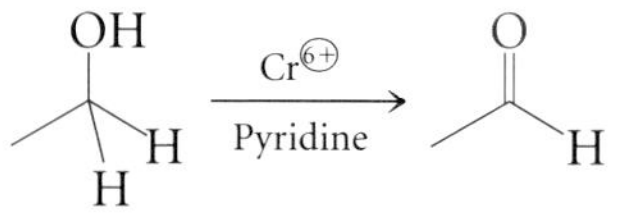 |
| | Ozonolysis of an alkene | 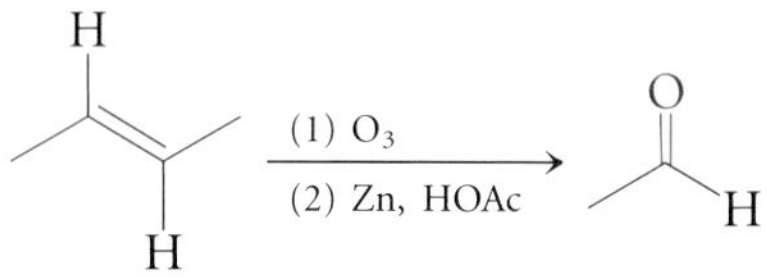 |
| | Hydrolysis of an acetal | 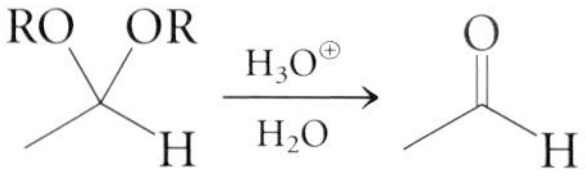 |
| 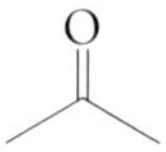 | Hydrolysis of a ketal | 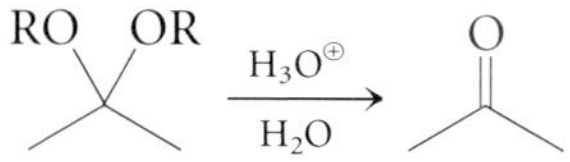 |
| | Hydrolysis of a terminal alkyne | 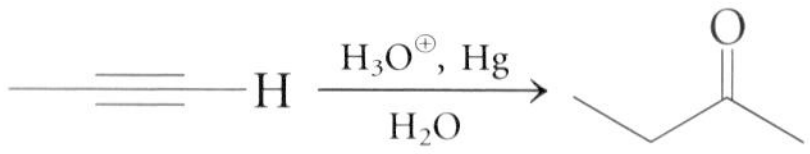 |
| | Chromate oxidation of a secondary alcohol | 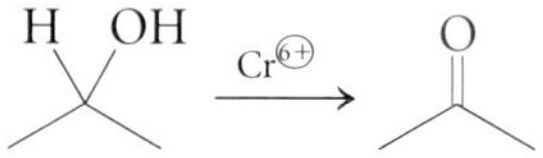 |
| | Friedel–Crafts acylation | 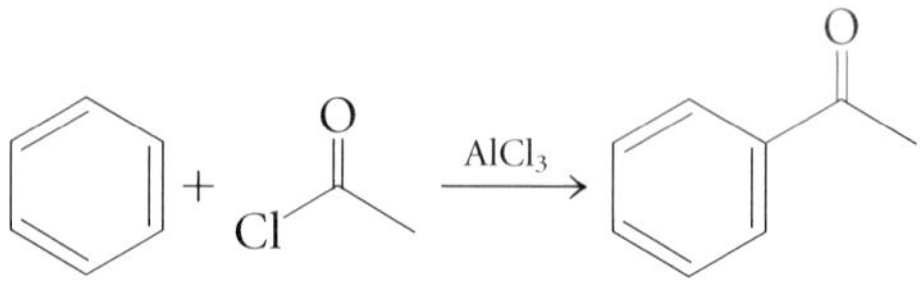 |
| | Claisen condensation | 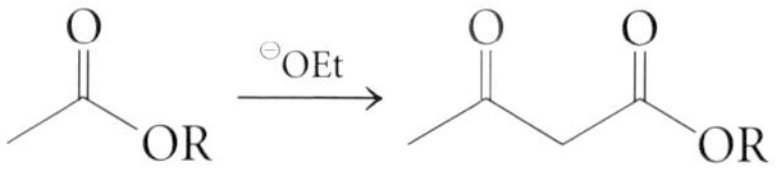 |

| Bond Formed | Type of Reaction | Example |
|---|---|---|

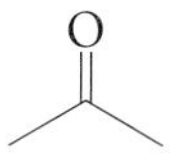

**Bond Formed:** (ketone C=O structure)

Pinacol rearrangement

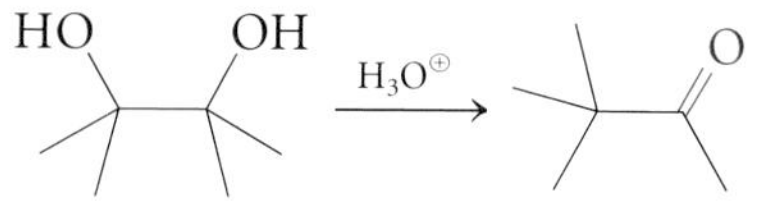

Claisen rearrangement

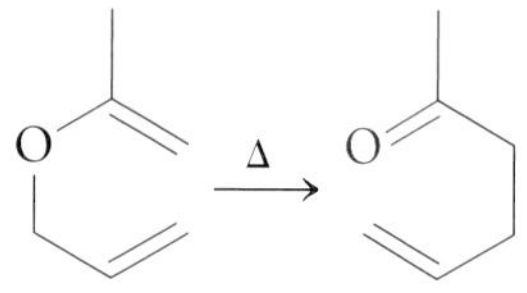

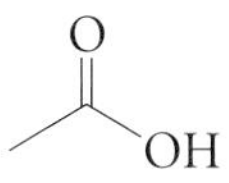

**Bond Formed:** (carboxylic acid structure)

Hydrolysis of a carboxylic acid derivative

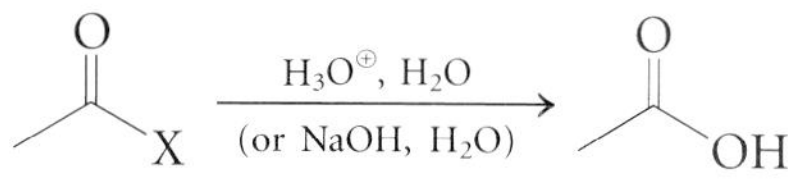

X = **Cl, OR, OAc, NR$_2$, SR**

Hydrolysis of a nitrile

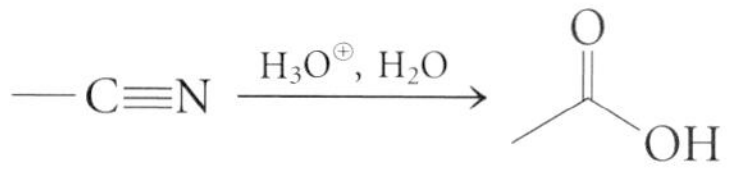

Oxidation of a primary alcohol

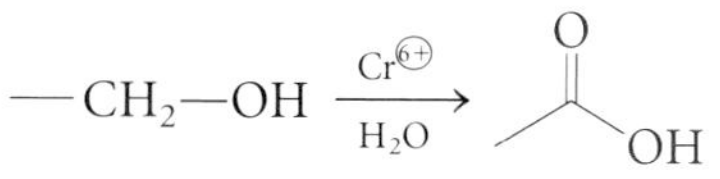

Oxidation of an aldehyde

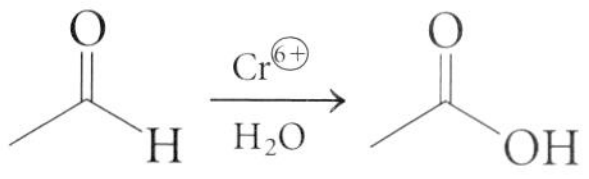

Permanganate oxidation of an alkyl side chain of an arene

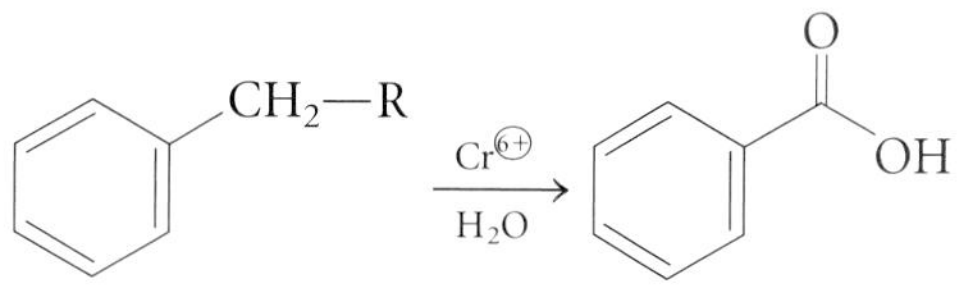

Iodoform reaction

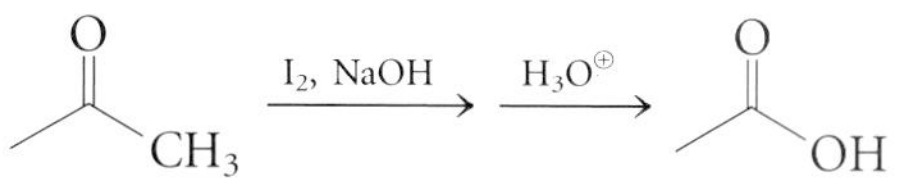

Carboxylation of a Grignard reagent

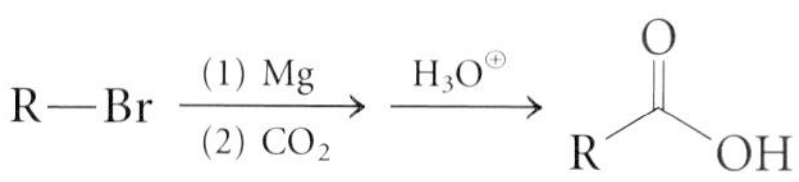

Benzilic acid rearrangement

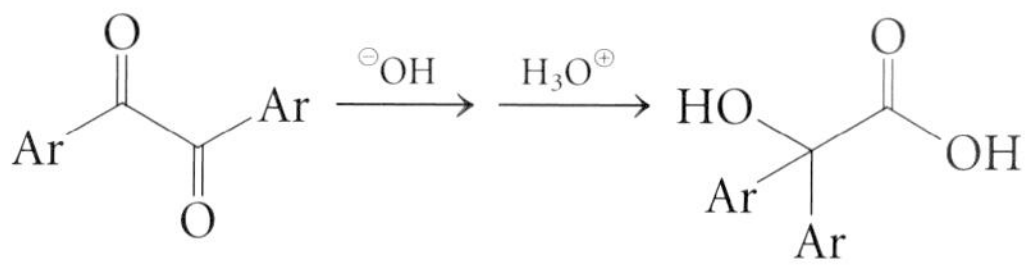

*(continued)*

## TABLE 14.2

*(continued)*

| Bond Formed | Type of Reaction | Example |
|---|---|---|
| (acid chloride) | Treatment of an acid with thionyl chloride | $\xrightarrow[\text{(or PCl}_3 \text{ or PCl}_5\text{)}]{\text{SOCl}_2}$ |
| (anhydride) | Acid dehydration | $\xrightarrow{\Delta}$ |
| (amide) | Amidation of a carboxylic acid derivative | $\xrightarrow{\text{HNR}_2}$    X = Cl, OR, OAc |
| | Beckmann rearrangement | $\xrightarrow[\text{(2) H}_2\text{SO}_4]{\text{(1) H}_2\text{NOH}}$ |
| (ester) | Esterification of a carboxylic acid | $+ \text{HOR} \xrightarrow{\text{H}_3\text{O}^{\oplus}}$ |
| | Transesterification | $+ \text{HOR}' \xrightarrow{\text{H}_3\text{O}^{\oplus}} \quad + \text{HOR}$ |
| | Baeyer–Villiger oxidation | $\xrightarrow{\text{RCO}_3\text{H}}$ |
| (ketal) | Ketal (acetal) formation | $+ \text{ROH} \xrightarrow{\text{H}_3\text{O}^{\oplus}}$ |
| (imine) | Imine formation | $+ \text{H}_2\text{N}-\text{R} \longrightarrow$ |
| (enamine) | Enamine formation | $+ \text{H}-\text{N} \longrightarrow$ |

**14.1** For each of the following reactions, predict the major product expected when the reactant is treated with the given reagent:

(a) 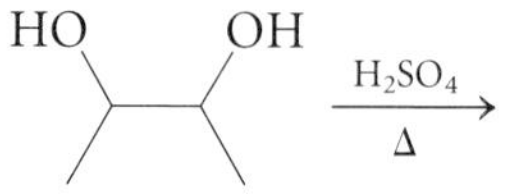
HO    OH
[structure] $\xrightarrow[\Delta]{H_2SO_4}$

(b) 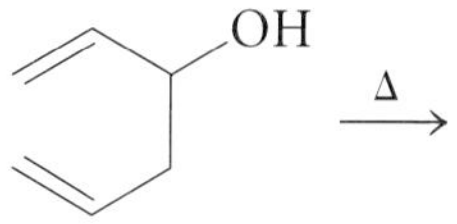
[structure with OH] $\xrightarrow{\Delta}$

(c) 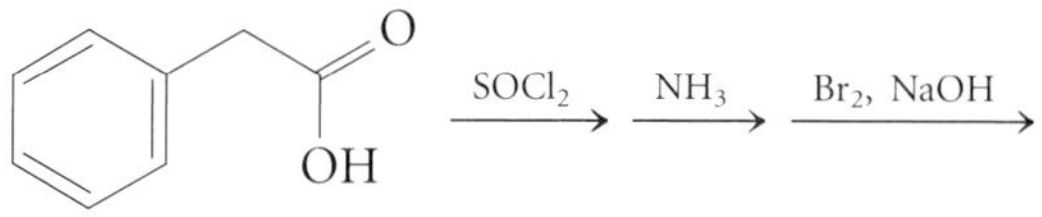
[structure with O and OH] $\xrightarrow{SOCl_2} \xrightarrow{NH_3} \xrightarrow{Br_2, NaOH}$

(d) 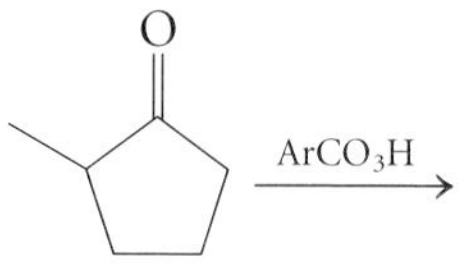
[cyclopentanone structure] $\xrightarrow{ArCO_3H}$

(e) 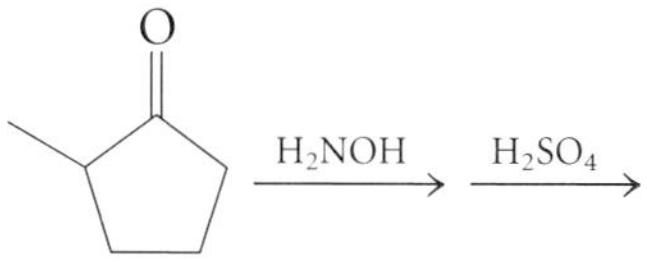
[cyclopentanone structure] $\xrightarrow{H_2NOH} \xrightarrow{H_2SO_4}$

(f) 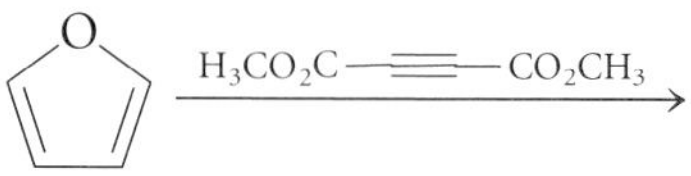
[furan structure] $\xrightarrow{H_3CO_2C \equiv CO_2CH_3}$

(g) 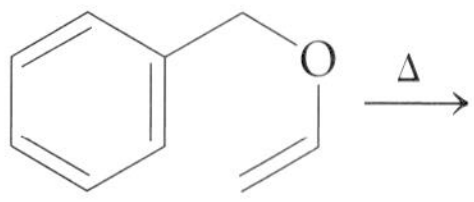
[benzyl allyl ether structure] $\xrightarrow{\Delta}$

**14.2** Specify the reagent (or sequence of reagents) and conditions required to convert cyclohexanone to each of the following products:

(a) 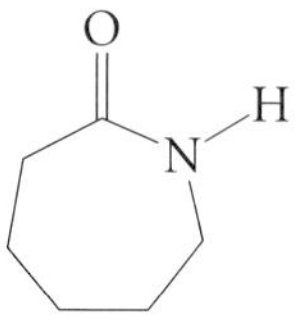
[lactam structure with O, N-H]

(d) 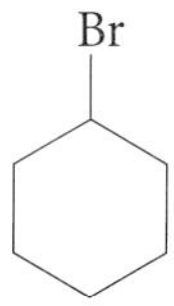
[cyclohexane with Br]

(g) 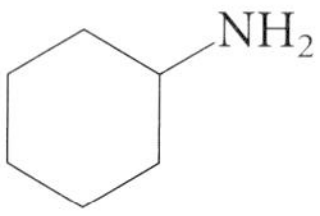
[cyclohexane with NH_2]

(b) 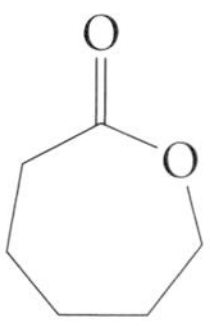
[lactone structure with O, O]

(e) 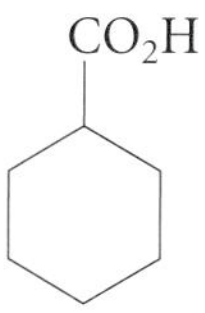
[cyclohexane with CO_2H]

(c) 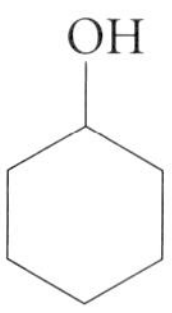
[cyclohexanol with OH]

(f) 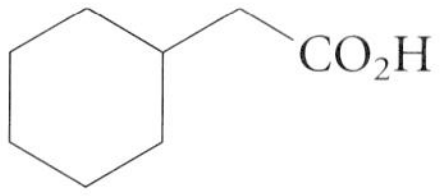
[cyclohexane with CH_2CO_2H]

**14.3** Identify the starting materials and reagents required to prepare 1-butylamine using each of the following reactions:

(a) a Gabriel synthesis

(b) a Hofmann rearrangement

(c) a Beckmann rearrangement

**14.4** Write a detailed mechanism, using curved arrows to indicate electron flow, for each of the following reactions:

(a)

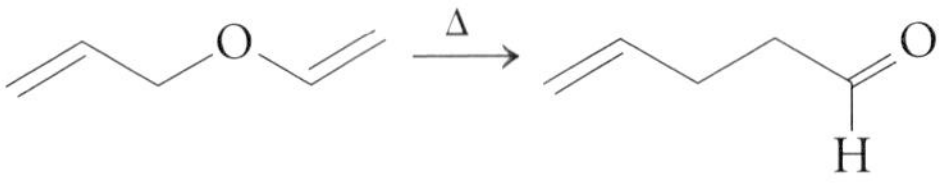

(b)

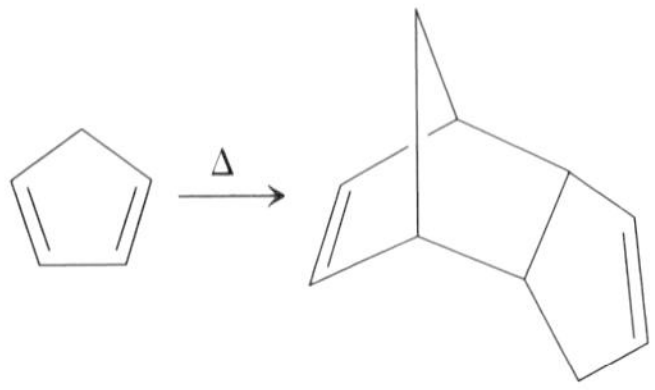

(c)

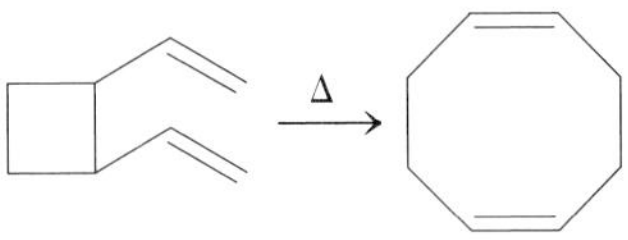

(d)

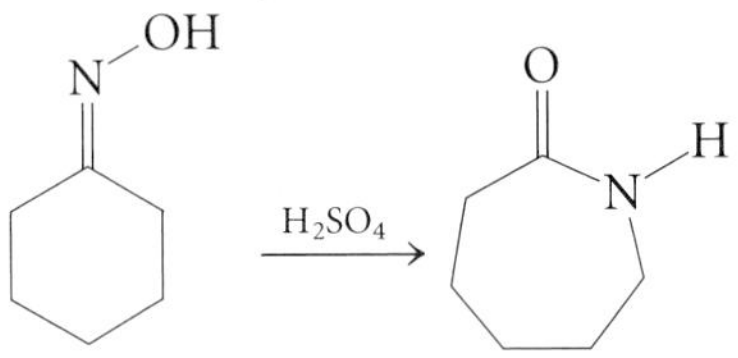

(e)

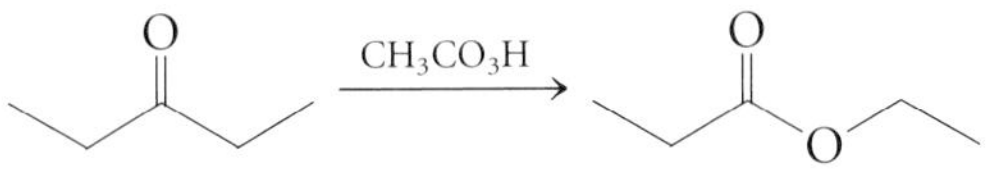

(f)

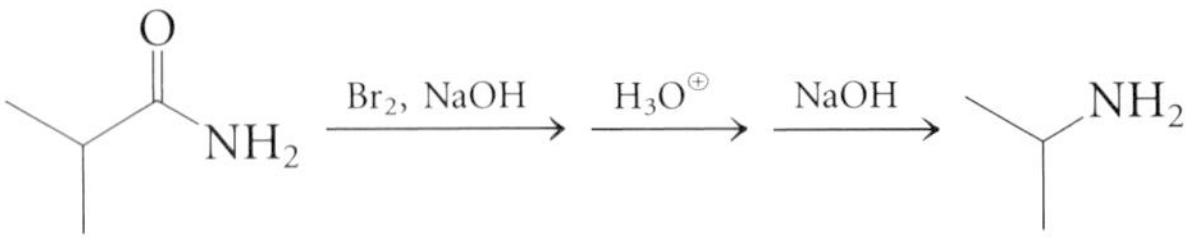

(g)

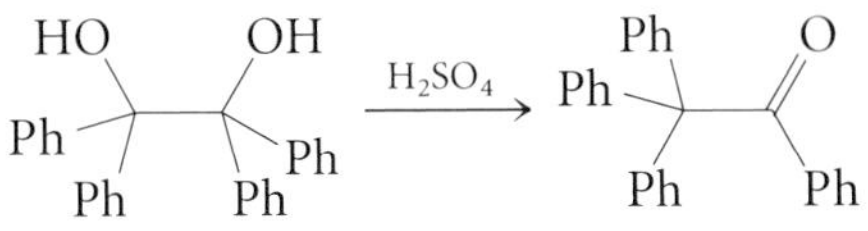

**14.5** The usual method for the attachment of alkyl chains to aromatic rings is electrophilic aromatic substitution. (Recall the Friedel–Crafts acylation and alkylation in Chapter 11.) However, an allyl group can be attached to the aromatic ring of a phenol via a sequence in which a Williamson ether synthesis is followed by a Claisen rearrangement. With this sequence in mind, write a mechanism for the following reaction:

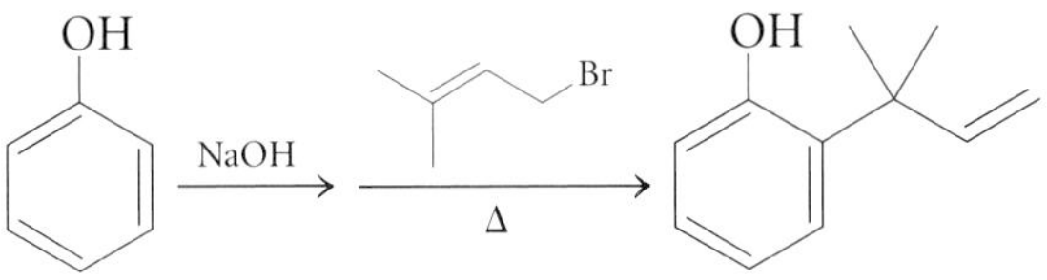

**14.6** In each of the following rearrangements, at least one functional group present in the molecule is altered. As each reaction proceeds, describe the changes, if any, that will occur in (1) the infrared spectrum, (2) either the $^{13}C$ or $^1H$ NMR spectrum, and (3) the mass spectrum.

(a) 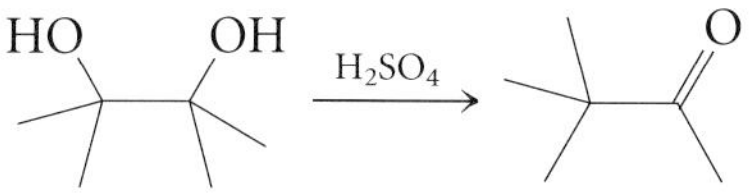 $\xrightarrow{H_2SO_4}$

(b) 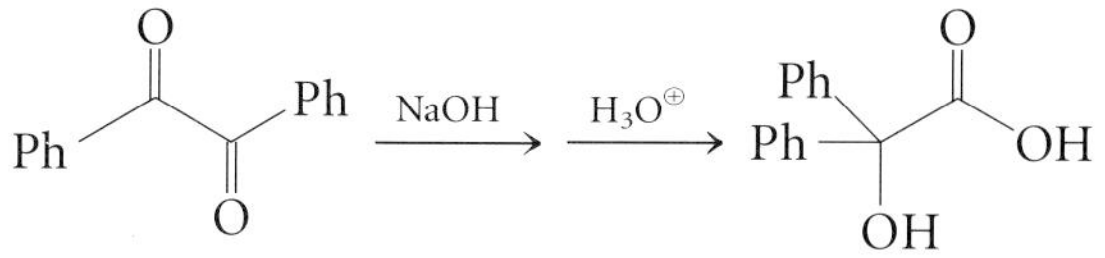 $\xrightarrow{NaOH}$ $\xrightarrow{H_3O^{\oplus}}$

(c) 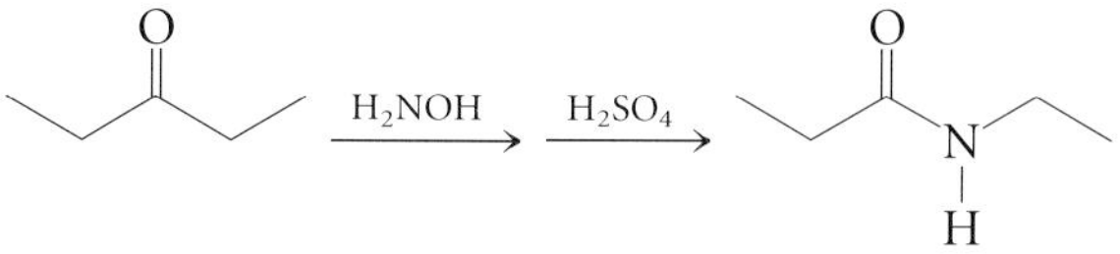 $\xrightarrow{H_2NOH}$ $\xrightarrow{H_2SO_4}$

(d) 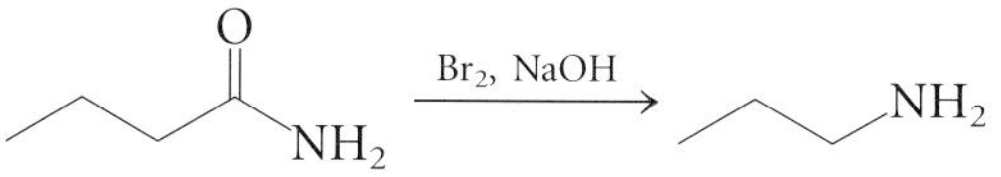 $\xrightarrow{Br_2, NaOH}$

(e) 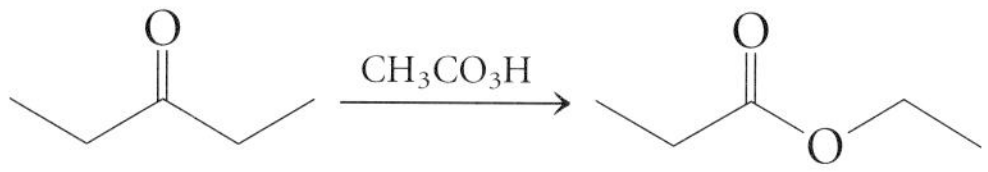 $\xrightarrow{CH_3CO_3H}$

(f) 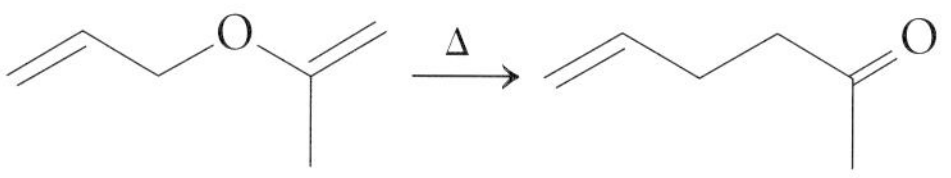 $\xrightarrow{\Delta}$

**14.7** Specify the reagents and conditions required to convert 2-butanol into each of the following compounds:

(a)

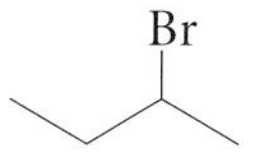

(b)

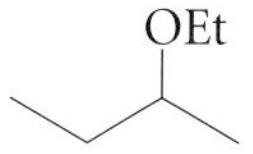

(c)

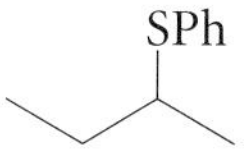

(d)

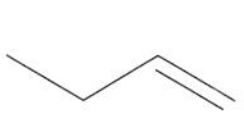

(e)

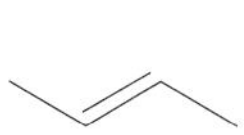

(f)

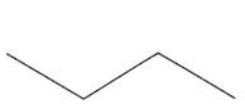

(g)

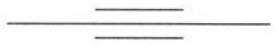

(h)

(i)

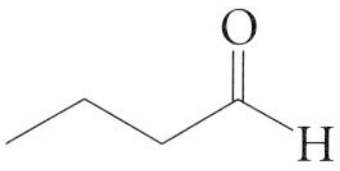

(j)

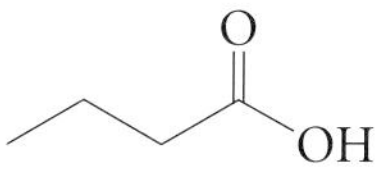

(k)

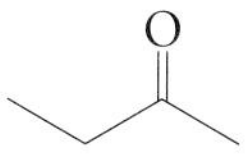

(l)

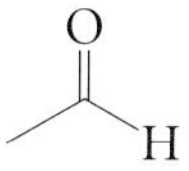

(m)

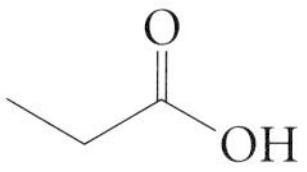

(n)

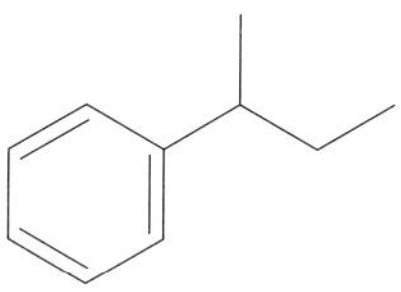

(o)

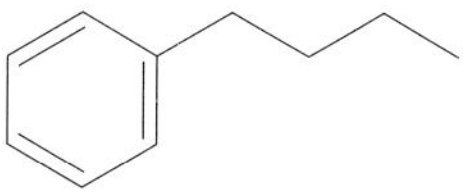

(p)

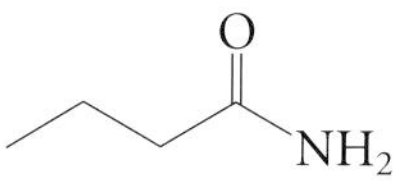

(q)

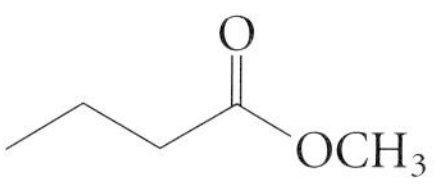

(r)

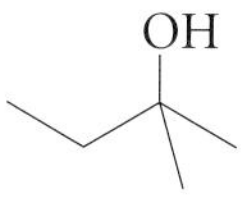

(s)

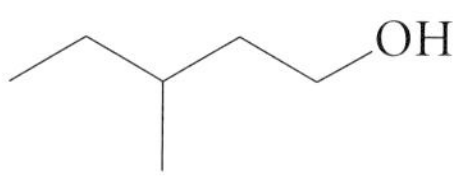

(t)

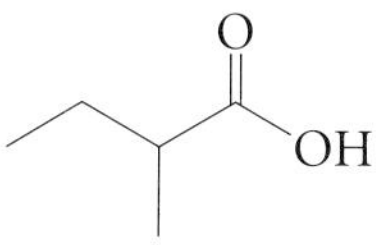

## Supplementary Problems

**14.8** Epoxides undergo a rearrangement reaction in the presence of acid to form ketones, as shown here for the epoxide of 1-methylcyclopentene. The reaction involves a hydrogen shift analogous to those described in Section 14.1. Write a detailed reaction mechanism for this transformation.

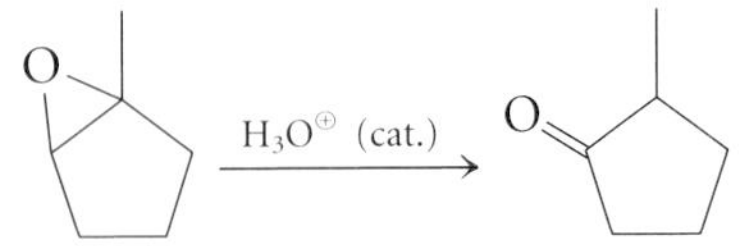

**14.9** What reagent(s) could be used to carry out each of the following transformations? (*Hint:* More than one step may be required.)

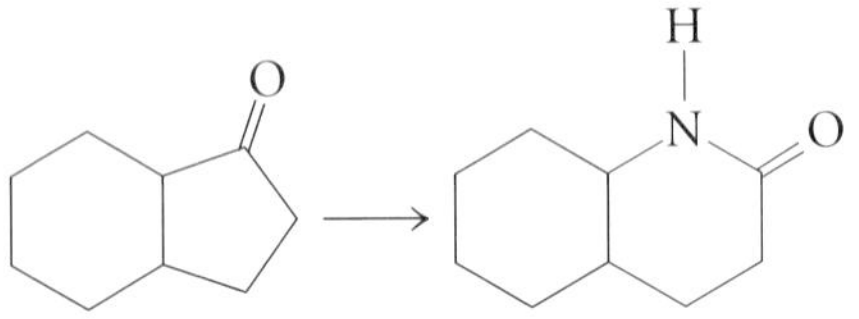

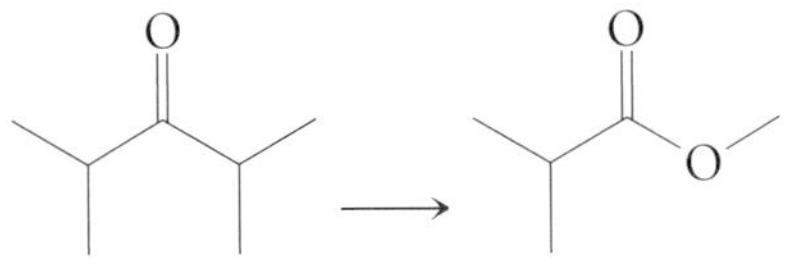

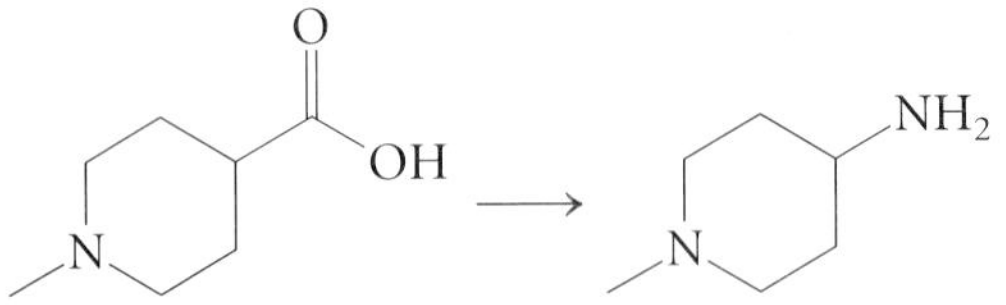

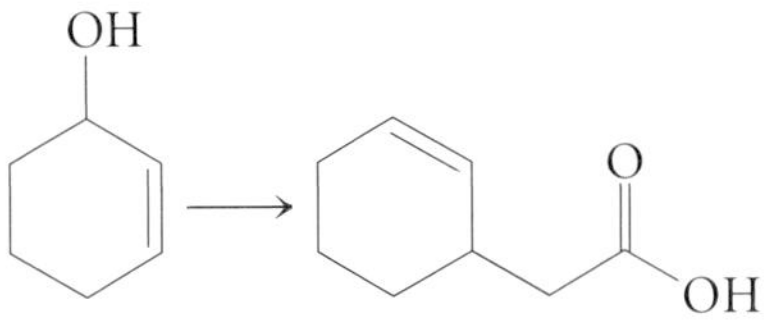

**14.10** Each of the following amines can be prepared by a Beckmann rearrangement of the oxime of a ketone followed by one additional step. Draw the structures of the starting ketone, the Beckmann rearrangement product, and the reagents necessary to form the product amine.

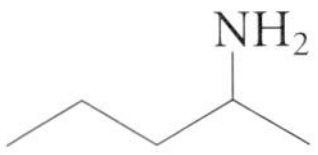

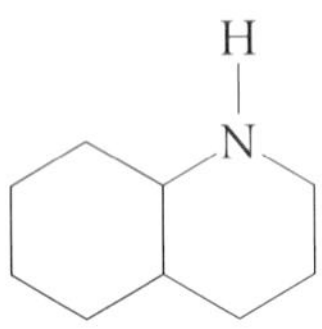

**14.11** Each of the following compounds can be prepared by a Baeyer–Villiger oxidation of a ketone followed by one additional step. Draw the structures of the starting ketone, the Baeyer–Villiger oxidation product, and the reagents necessary to form each product.

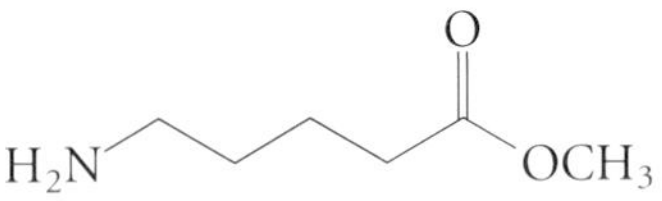

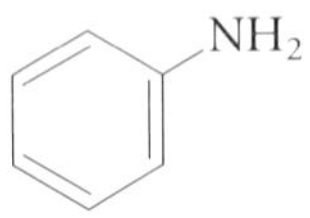

(a)

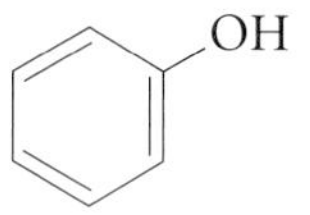

(c)

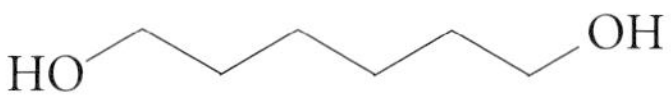

(b)

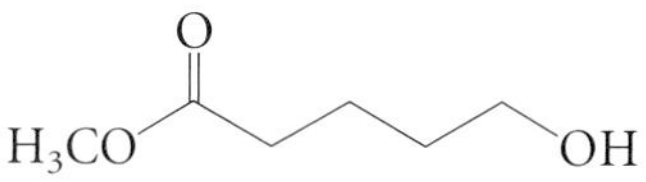

(d)

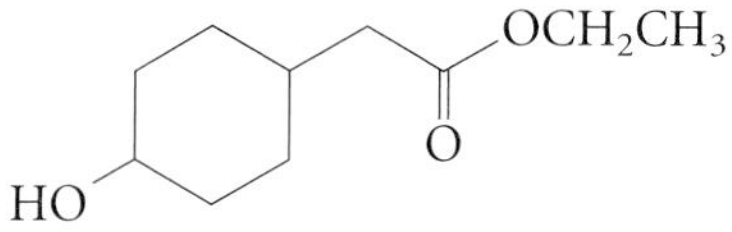

**14.12** The reaction of a Grignard reagent with $CO_2$ followed by acidification results in a carboxylic acid that can then be converted to a carboxylic acid amide. In turn, the amide can be converted to an amine using the Hofmann rearrangement.

$$R—Br \longrightarrow R—MgBr \longrightarrow R—CO_2H \longrightarrow R—CONH_2 \longrightarrow R—NH_2$$

Use this sequence to prepare each of the following amines, showing all reagents needed for each step:

(a) 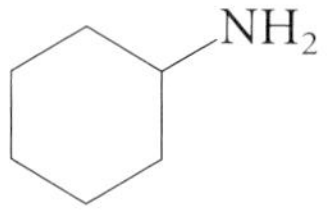   (b)

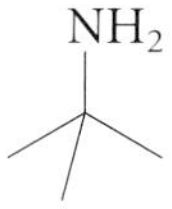

**14.13** The Beckmann rearrangements of oximes derived from unsymmetrical ketones generally result in migration of the more substituted carbon atom to nitrogen. What features(s) of the $^1H$ and $^{13}C$ NMR spectra would be especially useful in establishing that isomer A, rather than B, is formed in the Beckmann rearrangement of the oxime of 2-methylcyclohexanone?

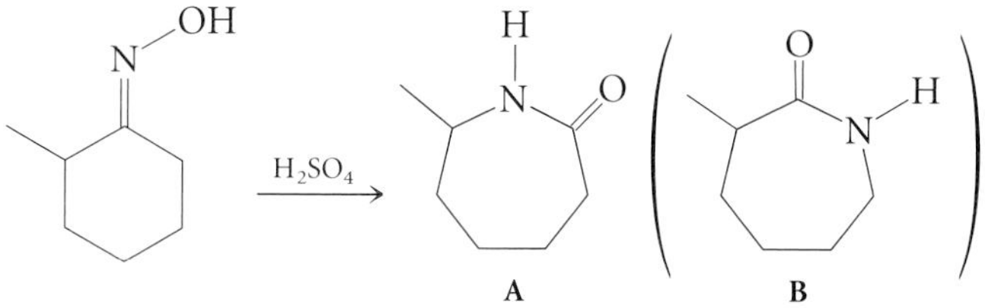

**14.14** It is often difficult to visualize the connection between starting material and product in a rearrangement reaction. For each of the following reactions, indicate which bond(s) are broken in the starting materials and which bonds are formed in the products:

(a)

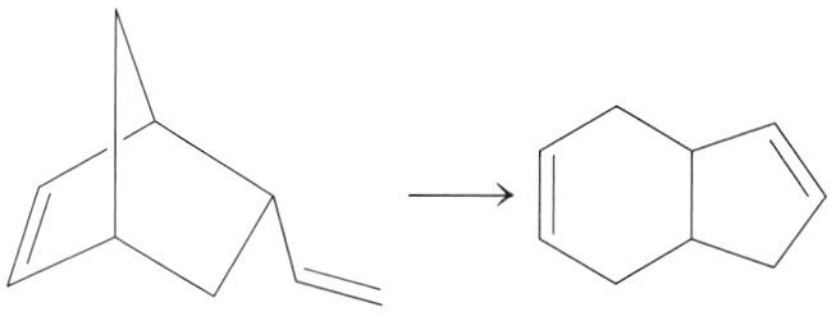

(b)

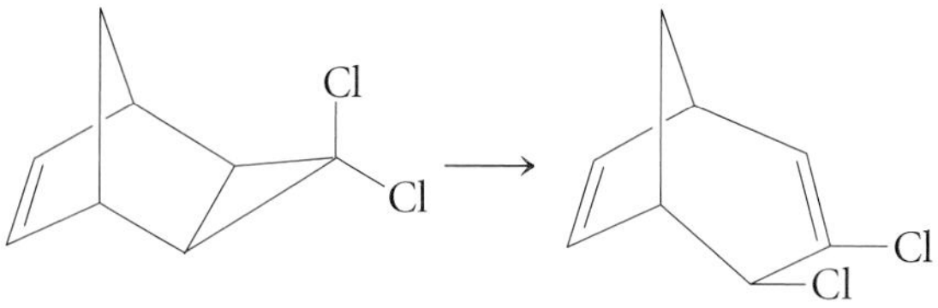

**14.15** The 1,2-shift of a carbon or hydrogen atom represents one of the fundamental rearrangement reactions, driven in most cases by the conversion of a less to a more stable carbocation. Occasionally, a sequence of 1,2-shifts will move the

site of positive charge far from its original position. Such a sequence of shifts occurs in the biosynthesis of cholesterol. Although the details are not known, it is
clear that cation C is converted to cation D. Write a mechanism for each of the
1,2-shifts that must be involved in this transformation.

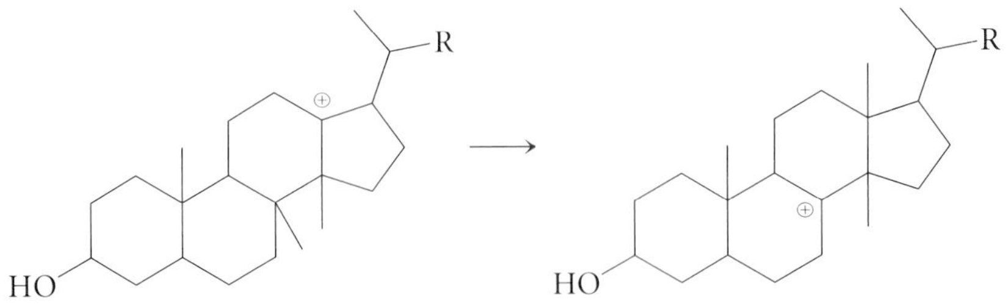

**14.16** Treatment of $\alpha$-pinene with hydrochloric acid yields bornyl chloride.
Write a detailed mechanism for this reaction. (*Hint:* Note that a skeletal rearrangement is involved.)

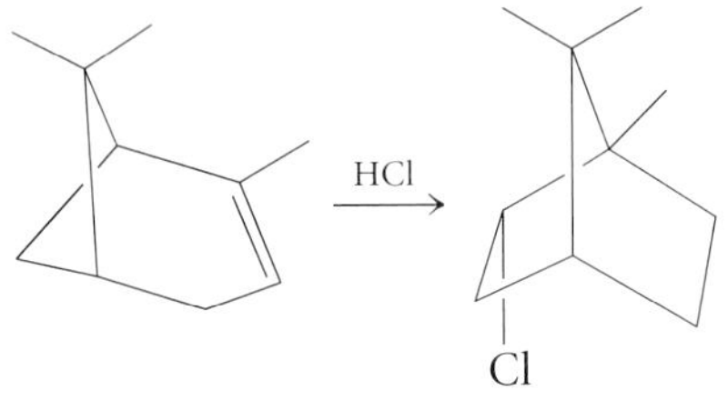

$\alpha$-Pinene      **Bornyl chloride**

**14.17** Reduction with NaBH$_4$ of the $\alpha$-bromoketone shown here produces two
bromoalcohols. On treatment with NaOH, one bromoalcohol is converted to an
epoxide and the other to a ketone. Assign structures to the two bromoalcohols.
Provide a detailed reaction mechanism for the formation of the epoxide, and
specify which isomer gives the epoxide.

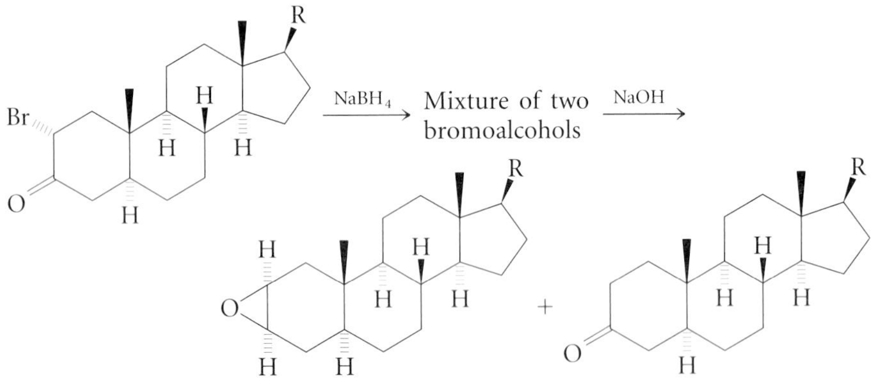

**14.18** Provide a detailed, stepwise mechanism for the following reaction. What is
the driving force for the conversion of the starting material into the product?
What type of rearrangement is involved in this reaction?

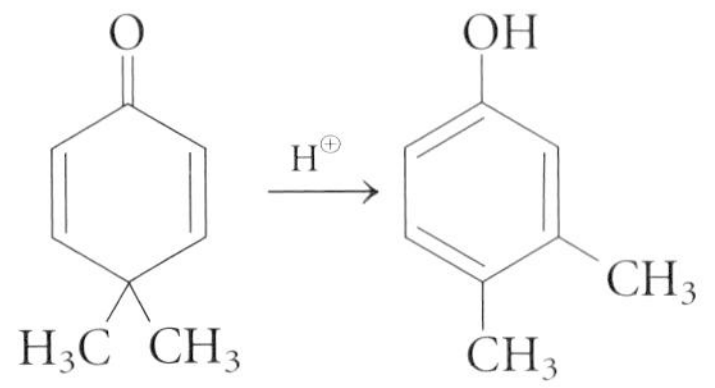

**14.19** Normally, treatment of a tertiary alcohol with a chromium(VI) reagent under mild conditions does not result in oxidation. However, allylic alcohols such as the following example do undergo oxidation with rearrangement:

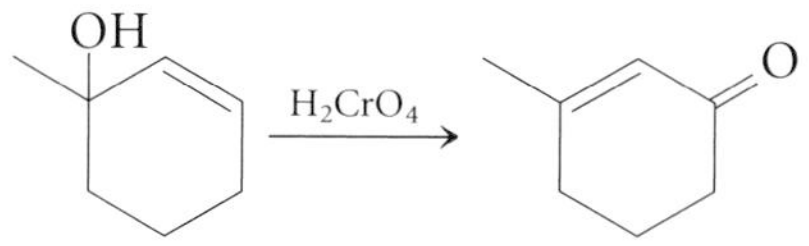

Devise a detailed reaction mechanism for this process. (*Hint:* Start by drawing the structure of the chromate ester formed from the alcohol by direct reaction with aqueous $H_2CrO_4$.) What type of rearrangement is involved?

**14.20** One important method for the industrial production of phenol and acetone is as follows:

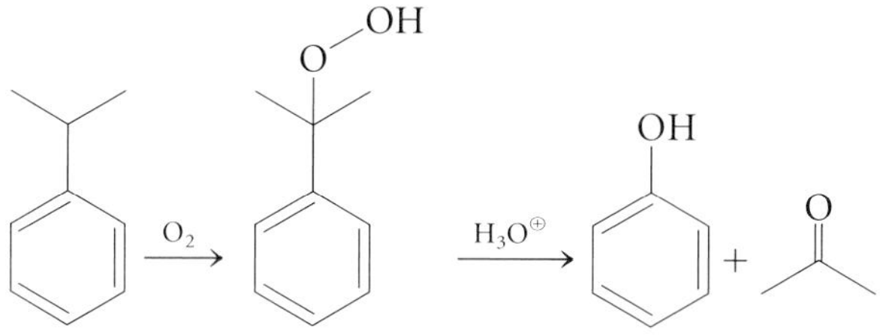

| Cumene | Cumene hydroperoxide | Phenol | Acetone |

The second step, the conversion of cumene hydroperoxide to the desired products, involves a migration reaction. Write a detailed mechanism for this second step.

**14.21** The following rearrangement is similar to the Claisen rearrangement but differs in some details. Propose a mechanism for this transformation. Then describe the driving force for the reaction by calculating $\Delta H^\circ$ using the bond energies in Table 3.5.

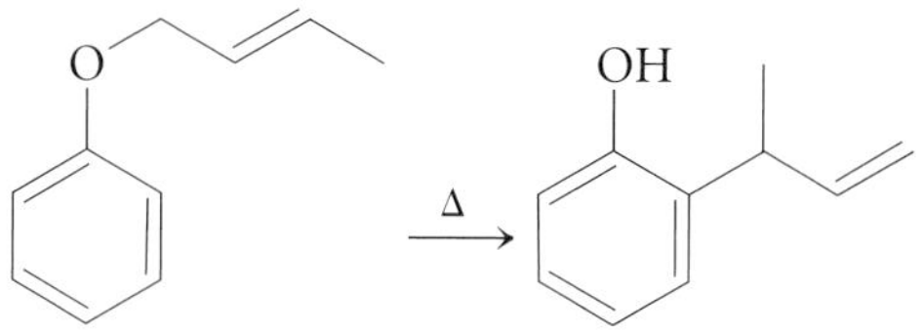

**14.22** The hydrocarbon prismane is an unusual species, having two three-member and three four-member rings:

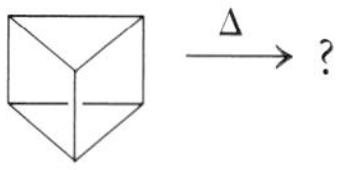

**Prismane**

Because of the strain resulting from these small rings, prismane is not stable. Upon heating, it is converted to another hydrocarbon, also with the formula $C_6H_6$. The product hydrocarbon has only a single absorption signal in the $^{13}C$ NMR spectrum (at $\delta$ 128.3) and only one peak in the $^1H$ NMR spectrum (at $\delta$ 7.33). Propose a structure for this hydrocarbon. Write a detailed mechanism for the transformation of prismane.

# 15

# Multistep Syntheses

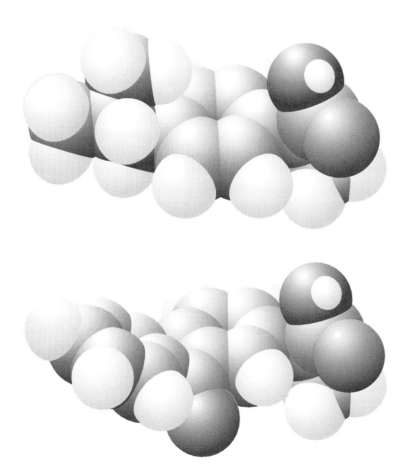

*Ibuprofen (top) and ketoprofen (bottom) are two over-the-counter drugs widely used for the relief of pain. Their skeletal structures are similar, and their three-dimensional shapes and sizes are also quite similar.*

**Ibuprofen**

**Ketoprofen**

You have now studied a wide range of reactions and their mechanisms in detail. With this information, you are in a position to view these reactions as processes that transform one species into another and, thus, as tools for chemical synthesis. Although you are familiar with the functional-group transformations that these reactions can accomplish, each transformation by itself may not be an impressive change. On the other hand, when a number of these transformations are carried out in sequence, the structural resemblance of the ultimate product to the initial starting material may be extremely slight.

As an example, consider the transformation of 2-propanol into 2-butanone:

2-Propanol                    2-Butanone

Even though we have not covered a single reaction that can induce this specific conversion, you will see shortly that a series of known reactions can be used to achieve it. In this chapter, you will learn to recognize clues provided by the starting material and the product that can direct you toward the appropriate reaction choices.

Why might we be interested in combining reactions? The field of organic chemistry owes its diversity to the almost unlimited number of possible structures based on carbon. Because each chemical reaction generally makes only a relatively minor change in structure, chemists must use several of them in sequence to prepare complex molecules such as those found in nature from simple and readily available molecules such as acetone, ethanol, and ethyl acetate.

## 15.1

## Grouping Chemical Reactions

We have considered many different kinds of chemical transformations in the preceding chapters, grouping them according to their mechanisms. For example, the reactions of methyl iodide with hydroxide ion and with the enolate anion of acetone both take place by $S_N2$ reaction pathways:

$$HO^{\ominus} + CH_3-I \longrightarrow CH_3-OH$$

Yet, when these reactions are viewed from the perspective of what they accomplish in terms of structural transformation rather than how they occur, they can be put into entirely different categories. For the purpose of combining reactions into sequences to construct complex molecules from simple ones, it is convenient to classify reactions in three categories:

1. Carbon–carbon bond-forming processes
2. Oxidation–reduction reactions
3. Functional-group transformations

This classification scheme is a natural one in terms of the need to organize and remember chemical transformations for synthetic purposes. Because all sequences that result in the transformation of a small organic molecule into a larger one require carbon–carbon bond formation, reactions of the first category are particularly important to synthesis. However, many carbon–carbon bond-forming reactions require carbonyl functional groups, so it is also important to know both how to make and how to remove these functional groups—because they are often not present in the desired product.

Let's return to the conversion of 2-propanol into 2-butanone to consider one possible solution:

| 2-Propanol | Acetone | Enolate anion | 2-Butanone |

Here we have one possible sequence (of three steps) that effects the transformation: oxidation of the alcohol to acetone, formation of the enolate anion, and alkylation with methyl iodide to form 2-butanone. With regard to the overall transformation, the most important process is the carbon–carbon bond-forming reaction because that reaction, in effect, extends the smaller molecule to yield a larger one. This is not to say that the other reactions are unimportant: the oxidation yields the carbonyl functional group necessary for the carbon–carbon bond-forming reaction. Yet, the oxidation step serves only to provide the requisite functional group for the key reaction.

Before we consider methods for creating sequences of transformations, it is useful to review some of the reactions from the preceding chapters, placing them in the three categories listed above. Table 15.1 (page 746) gives the major carbon–carbon bond-forming reactions; Table 15.2 (page 747), the oxidation–reduction transformations; and Table 15.3 (page 748), several important functional-group transformations. Included in a separate section of Table 15.2 are reactions such as the addition of water to an alkene, in which there is no net change in oxidation but one carbon undergoes a reduction that is balanced by the oxidation of another carbon. Although there is no net redox change, the oxidation and reduction levels of some atoms within a molecule change in the course of these reactions.

It is often the case that a specific reaction fits more than one category. For example, the oxidation of an alcohol to a ketone is both an oxidation of carbon and a functional-group transformation.

**TABLE 15.1**

Carbon–Carbon Bond-Forming Reactions

As a second example, the addition of a Grignard reagent to a ketone (followed by protonation) fits all three categories: it is a carbon–carbon bond-forming reaction, a reduction of a carbonyl carbon, and a functional-group transformation of a ketone into an alcohol.

When a reaction can be placed in more than one category, it is viewed as belonging to the category that is higher on the list of three. For example, a reaction that is both a functional-group transformation and an oxidation–reduction reaction is classified as the latter, and the addition of a Grignard reagent to a ketone is considered a carbon–carbon bond-forming process.

# TABLE 15.2

Oxidation–Reduction Reactions

| Oxidations | Reductions |
|---|---|

## Reactions with Simultaneous Oxidation of One Carbon and Reduction of Another

X = Cl, Br, I, OH  X = Cl, Br, I, OH

## TABLE 15.3

### Functional-Group Transformations

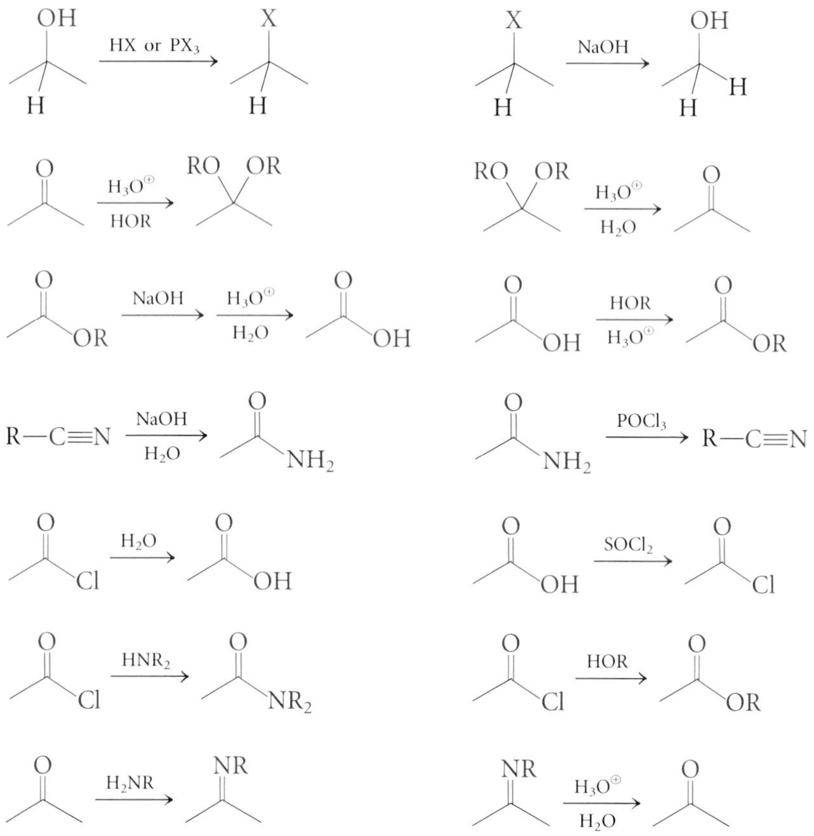

## EXERCISE 15.1

Classify each of the following reactions as a carbon–carbon bond-forming process, an oxidation–reduction reaction, or a functional-group transformation.

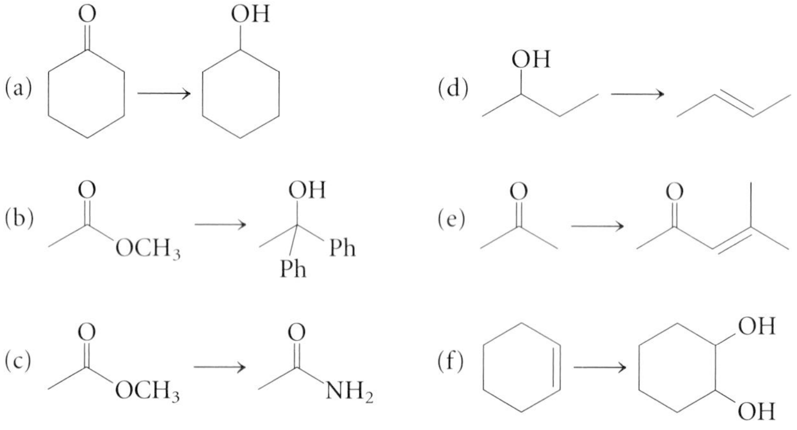

# Retrosynthetic Analysis

## Designing a Synthesis by Working Backward

What general method can we use to solve problems such as the conversion of 2-propanol into 2-butanone? In other words, what process will lead us to a workable series of reactions?

We might start by "trying" (as a thought experiment) various reactions on the starting material, 2-propanol. Although this appears to be a major task, we can narrow the search by noting that at some point we must form a carbon–carbon bond, because there are four carbon atoms in the product and only three in the starting material. Thus, we systematically examine the carbon–carbon bond-forming reactions in Table 15.1 to see whether any apply to the starting material. In this case, none is directly applicable.

Rather than immediately giving up, we look at the product, 2-butanone, to see what chemical transformation will produce it. Again, if possible, we look for a reaction that forms a carbon–carbon bond, because we know that this must occur at some point in the sequence. From Table 15.1, we see that alkylation of a simple ketone accomplishes the needed transformation. Furthermore, we see that the necessary starting materials are acetone (as its enolate) and an alkylating agent such as methyl iodide. (Recall from Chapter 7 that one way of analyzing a reaction is to try to identify the correct starting materials for forming the desired product under defined conditions.)

At this point, we realize that the number of carbons in acetone (the immediate precursor for 2-butanone by this analysis) and in the given starting material, 2-propanol, is the same. Thus, we have no further need of carbon–carbon bond-forming reactions and can limit the rest of the analysis to categories 2 and 3 (oxidation–reduction and functional-group transformation). An oxidation of 2-propanol to acetone enables us to connect the starting material, 2-propanol, with the ultimate product, 2-butanone, through the intermediate formation of acetone. Rather than following each possible reaction of 2-propanol through many steps, we have greatly simplified the analysis by starting with the final product and thinking about how to make the most logical precursor for that compound. Because this approach begins with the product and goes back, step-by-step, to the starting material(s), it is called **retrosynthetic analysis.**

## Rationale for Retrosynthetic Analysis

Why work backward? Although it may seem unnatural because what we want to accomplish, in fact, is the transformation in the forward direction, there is a very simple answer. In any sequence that progresses from smaller to larger molecules, the number of options rises dramatically in the forward direction but diminishes in the backward direction. Diagrammatically, the retrosynthetic approach looks much like a Christmas tree: any one of a number of reactions can apply to the starting material, and many additional reactions can apply to each initial product.

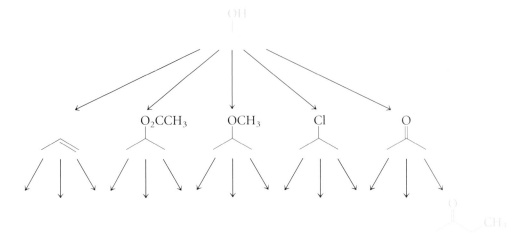

Although it is certainly possible that more than one of the branches will ultimately lead to the desired product, some may never reach this goal. On the other hand, by thinking in the backward direction and utilizing the knowledge that the product has more carbon–carbon bonds than the starting material does, we follow a path that, in a sense, "funnels" along a useful sequence.

This way of thinking of transformations may be confusing at first because reactions are taught and learned in the forward rather than the backward direction. It is critical that you "relearn" the reactions you are familiar with so that you think of how the *product* can originate, by one of the three classes of transformations, from an intermediate that can be linked through a series of reactions to a given starting material.

### EXERCISE 15.2

The following transformations take place via multireaction sequences. Examine the starting materials and products, and determine which carbon–carbon bonds in each product could be made via a sequence that requires the fewest carbon–carbon bond-forming reactions. Do not concern yourself with the actual reactions. Consider only how many carbons are in each starting material and product and how the carbons of the starting material are interconnected in the product. Assume that no sources of carbon other than the starting material are used.

(a) $\underset{\text{O}}{\overset{\|}{\diagup\!\!\diagdown}} \longrightarrow \longrightarrow \longrightarrow$ product with NH$_2$ group

(b) CH$_3$OH $\longrightarrow \longrightarrow \longrightarrow$ product with OH group

(c) CH$_3$CH$_2$OH $\longrightarrow \longrightarrow \longrightarrow$ alkene product

## CHEMICAL PERSPECTIVES

### SYNTHESIS OF VERY LARGE AND COMPLEX MOLECULES

One of the most complex molecules ever synthesized in the laboratory is vitamin $B_{12}$, prepared by Robert B. Woodward of Harvard University and Albert Eschenmoser of the Eidgenössische Technische Hochschule in Zurich. Its synthesis required the effort of more than a hundred chemists working together over a period of 11 years. Woodward's group was also the first to synthesize chlorophyll a, the macromolecule responsible for the green color of plants. The description of the synthesis of chlorophyll a included the research of 17 coworkers and required an entire issue of a journal. The synthesis was considered so important that the journal's editors changed the cover of the journal to green (from its customary blue) to match the color of chlorophyll. (Check it at your library: *Tetrahedron, 46,* 7599, 1990.) Ironically, in view of this chromatic homage to the synthesis of chlorophyll, Woodward's favorite color was blue—he owned more than 200 ties, all the same shade of blue.

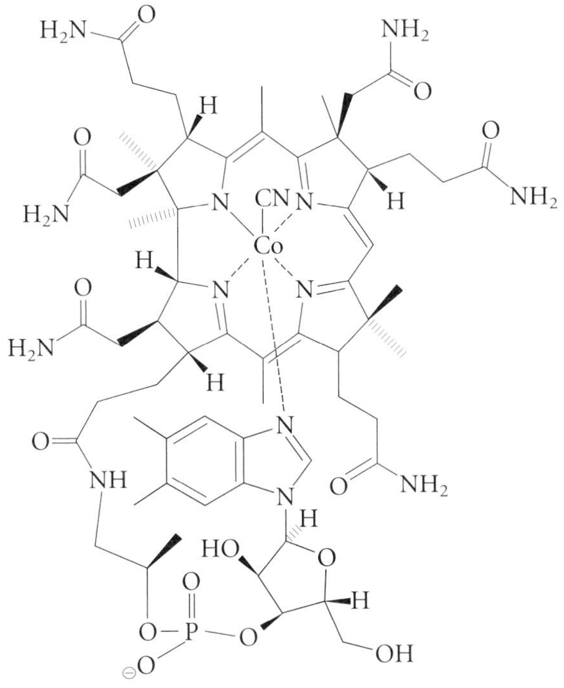

**Vitamin $B_{12}$**

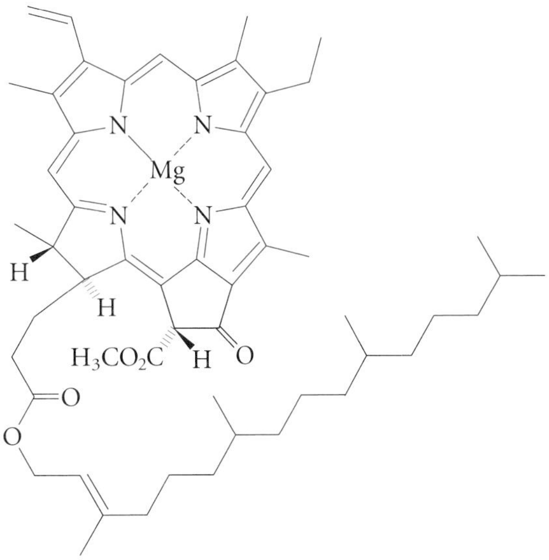

**Chlorophyll a**

## 15.3 Reactions Requiring Both Functional-Group Transformation and Skeletal Construction

The conversion of 2-propanol into 2-butanone might seem a rather trivial problem, hardly requiring the extensive retrosynthetic analysis we have outlined. However, let's continue this approach, adding one more level of complexity, to consider the overall transformation of 2-propanol into 2-butanol:

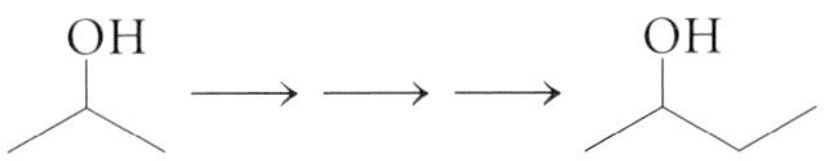

Knowing a sequence for transforming 2-propanol to 2-butanone, it is certainly tempting to use that synthesis, realizing that we need only reduce 2-butanone to arrive at our new objective, 2-butanol.

Let's see if we arrive at this same conclusion by retrosynthetic analysis. If we compare the ultimate product, 2-butanol, with the starting material, 2-propanol, we again ascertain that somewhere in the sequence a carbon–carbon bond must be formed. Returning to the compilation of carbon–carbon bond-forming reactions (Table 15.1), we find a transformation, the reaction of a Grignard reagent with a carbonyl group, that directly forms an alcohol.

The next step is to imagine what starting materials we might use for a Grignard synthesis of 2-butanol. A Grignard reaction forms a carbon–carbon bond to a carbinol carbon in the product. Therefore, only two sequences are possible, because there are only two such carbon–carbon bonds in 2-butanol:

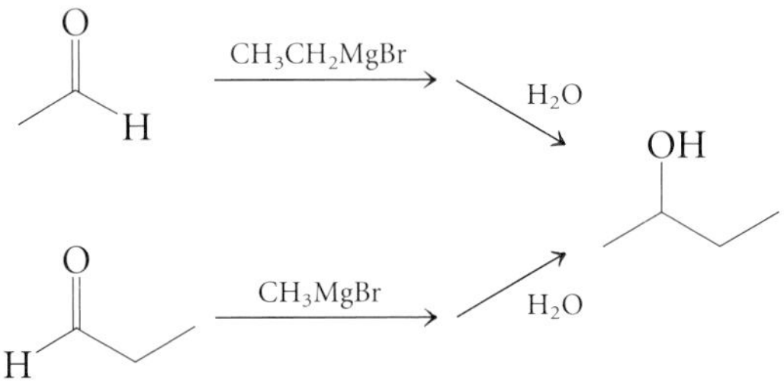

Of these two, the upper sequence is less convenient because it breaks 2-butanol into two two-carbon fragments, neither of which could be immediately derivable from the three-carbon starting material, 2-propanol. The lower sequence uses a three-carbon starting material, propanol. However, if we are to use the lower transformation, we must derive a connection between 2-propanol and propanal. (Rest assured that this does not look trivial even to practiced chemists.) Perhaps we are stuck, having followed a lead that took us down a blind alley.

We return to the starting point of the retrosynthetic analysis to see whether there is some other carbon–carbon bond-forming reaction that might directly produce 2-butanol. We draw a blank but realize that the formation of the required carbon–carbon bond need not be the last step in the sequence. Perhaps some other transformation (an oxidation, a reduction, or a functional-group transformation) can be used as the ultimate reaction.

The next logical step is to consider that the final reaction might be an oxidation (or reduction). When we examine the oxidation–reduction reactions in Table 15.2, we see that the reduction of a ketone produces a secondary alcohol. With this reaction, we can make use of the sequence pro-

753

15.4 Extending the
Retrosynthetic Approach:
Alternative Routes for
Synthesizing More Complex
Molecules

posed earlier for the synthesis of 2-butanone from 2-propanol, and then reduce the ketone functional group:

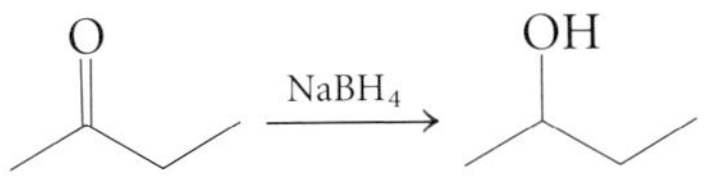

In summary, then, although a retrosynthetic analysis does not always
lead automatically to the shortest sequence of transformations connecting
starting material and product, it does provide alternative pathways. With
some practice, you can recognize fairly quickly when to abandon those sequences that would require too many steps or an unrealistically difficult
step.

### EXERCISE 15.3

Identify a sequence of reactions to transform propanal into 2-butanol. (You may
use any reagents, including organic ones, as long as the three carbons of propanal
are incorporated in the product.)

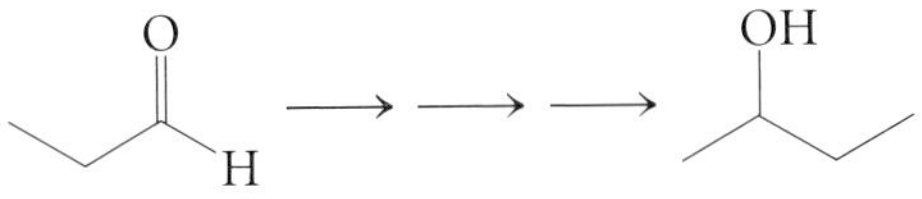

## 15.4

# Extending the Retrosynthetic Approach:
# Alternative Routes for Synthesizing
# More Complex Molecules

Often, more than one sequence of reactions connects a given starting material with a desired product. As an example, let's consider the transformation of 2-propanol into 3-methyl-2-butanone:

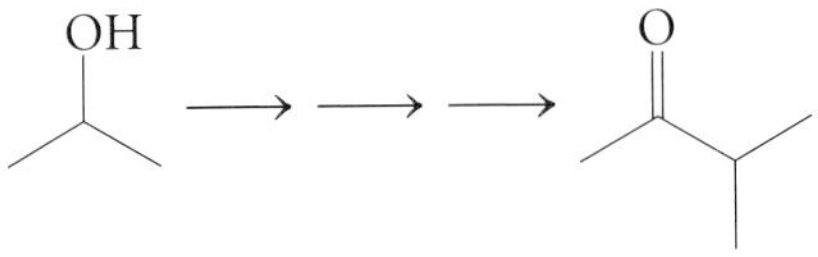

### Analyzing Individual Reactions
### in a Sequence

We begin by recognizing that there are two more carbon atoms in the
product than in the starting material. At some point, at least one carbon–carbon bond must be formed, and it is reasonable to consider whether
this might be accomplished as the last step in the sequence. Indeed, there
is such a reaction, exactly the same alkylation of a ketone that we used for
the synthesis of 2-butanone from acetone. The desired conversion of 2-bu-
tanone can be accomplished via formation of an enolate anion, followed by

alkylation with methyl iodide to form the desired product, 3-methyl-2-butanone.

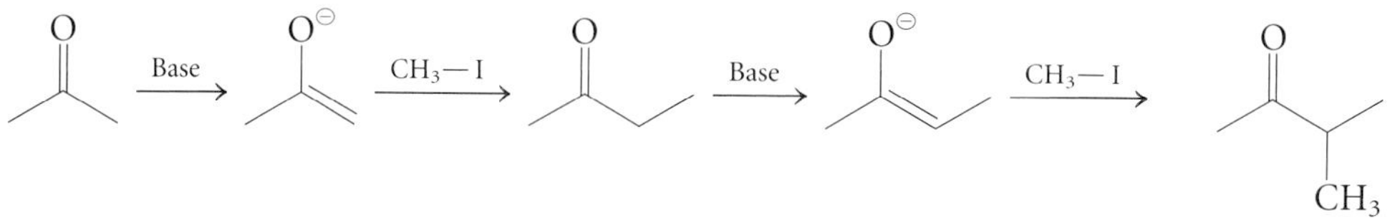

Acetone                    2-Butanone                    3-Methyl-2-butanone

Because 2-butanone is the product of the preceding problem, we now have an overall sequence that can be used to accomplish the desired transformation. However, there is one difficulty—the formation of an enolate from an unsymmetrical ketone generally leads to a mixture of regioisomers. Thus, this synthesis would produce not only 3-methyl-2-butanone, as desired, but also 3-pentanone:

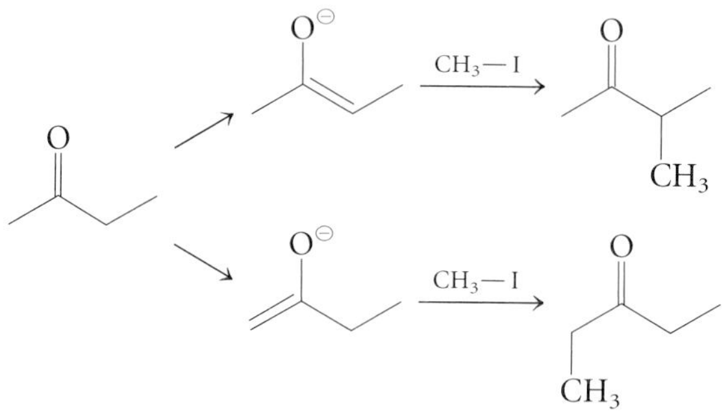

## Order of Chemical Transformations

Let us now return to the original objective, and search for a sequence that avoids forming a mixture of regioisomers. The list of carbon–carbon bond-forming reactions yielding products with ketone groups consists solely of the enolate alkylation reaction we have already examined. We move on, then, to consider other possible reactions for the last step—for example, an oxidation that would provide a route to the desired ketone. In fact, the oxidation of a secondary alcohol could be used to produce the desired 3-methyl-2-butanone:

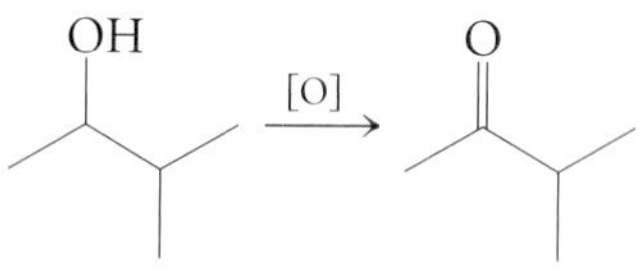

We then consider how we might arrive at this alcohol, 3-methyl-2-butanol. As before, we look first for a reaction that forms a carbon–carbon bond. From Table 15.1, we choose the reaction of a Grignard reagent (derived from the appropriate alkyl halide) with a carbonyl group to produce a secondary alcohol. The Grignard synthesis of alcohols always involves the formation of a carbon–carbon bond to the carbinol carbon. In this case, there are two different carbon–carbon bonds at the carbinol carbon, and thus two possible combinations of Grignard reagent and carbonyl com-

**755**

15.4 Extending the
Retrosynthetic Approach:
Alternative Routes for
Synthesizing More Complex
Molecules

pound can lead to this alcohol. The two combinations are shown in routes A and B:

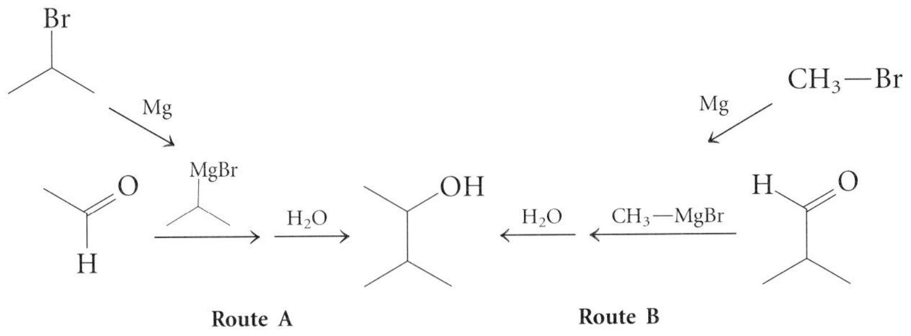

**Route A**          **Route B**

Route A starts with 2-bromopropane, which, after conversion into the Grignard reagent, reacts with acetaldehyde. The alternative sequence, B, is the reaction of 2-methylpropanal with the Grignard reagent derived from methyl bromide.

One of the two sequences, A or B, might be more efficient or shorter than the other. How do we choose between them? The overall objective is the conversion of 2-propanol into 3-methyl-2-butanone:

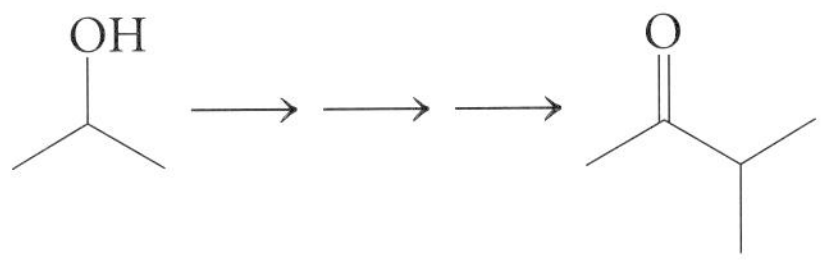

In our first analysis, we determined that at some point we needed to form a carbon–carbon bond, because the product has five carbons and the starting material only three. We can see, then, that a sequence that includes carbon–carbon bond formation between a three-carbon unit derived from isopropanol and another unspecified fragment with two carbons requires only one carbon–carbon bond-forming step. Conversely, two carbon–carbon bond-forming steps are needed if fragments containing only one carbon each are used.

Of the two alternatives, route A combines a three-carbon unit (2-bromopropane) with a two-carbon unit (acetaldehyde) to produce a five-carbon alcohol. Route B combines a four-carbon unit and a one-carbon unit to form the same product. Thus, an additional carbon–carbon bond-forming step would be required to prepare the four-carbon aldehyde from 2-propanol if route B were followed. The first approach, then, is to investigate route A.

We now have a sequence that leads from 2-bromopropane and acetaldehyde to our final objective, 3-methyl-2-butanone:

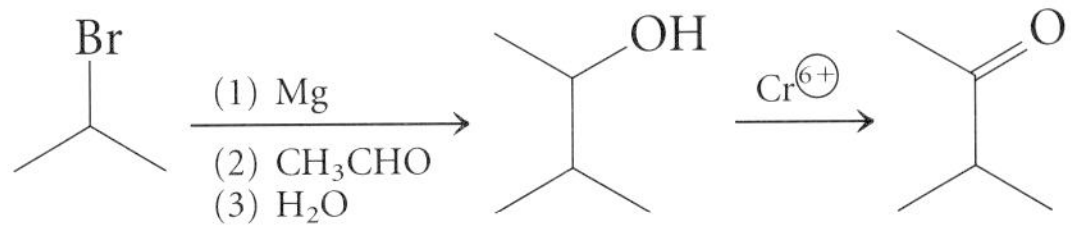

However, the synthesis is not yet complete, because we still need to prepare 2-bromopropane from 2-propanol. Clearly, this process does not require carbon–carbon bond formation and, by counting bonds to heteroatoms, we see that the oxidation levels of the carbon bonded to those heteroatoms in the two species are the same:

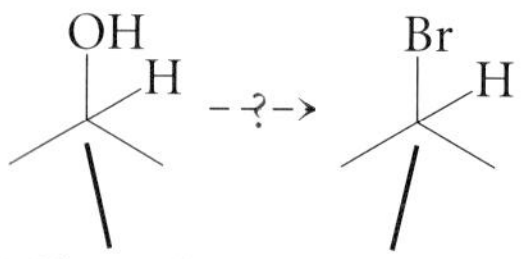

Oxidation level of carbon = 0

Because neither a carbon–carbon bond formation nor an oxidation–reduction is needed, we turn to Table 15.3, listing functional-group transformations, to see which one might be appropriate. We note that halides (both chlorides and bromides) can be conveniently prepared (using any of a variety of reagents) from the corresponding alcohols. Thus, it remains only to fill in the blank represented by the question mark on the dashed arrow with one such reagent—for example, HBr—to complete the sequence:

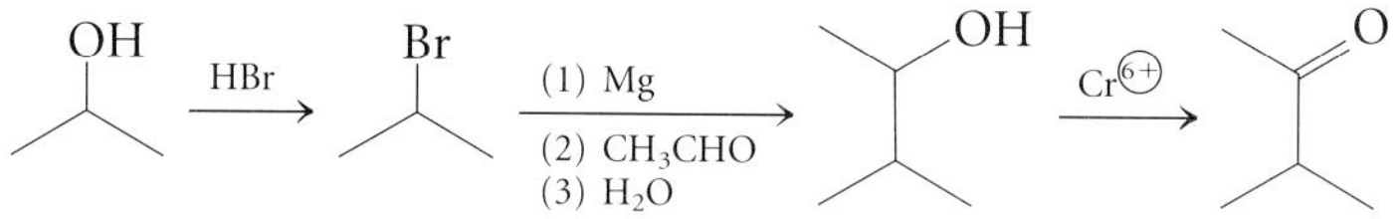

## EXERCISE 15.4

Devise a multistep reaction sequence for the transformation of ethanol into the following alkene:

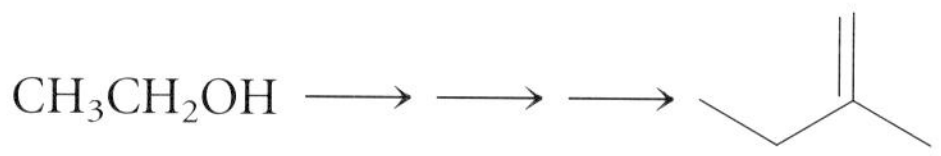

$$CH_3CH_2OH \longrightarrow \longrightarrow \longrightarrow$$

## 15.5

## Selecting the Best Synthetic Route

At this point, you might suspect that a "Christmas tree" depiction of possible pathways for a synthesis is an oversimplification because it implies that there is only one preferred route connecting starting material and ultimate product. Certainly, when all options are considered, it is often possible to end up with more than one workable route.

To examine further the principles of retrosynthetic analysis and the factors that govern the utility of a given synthetic pathway, let's return to route B, which also leads from an aldehyde to 3-methyl-2-butanol, although by a sequence judged to be less efficient overall than route A.

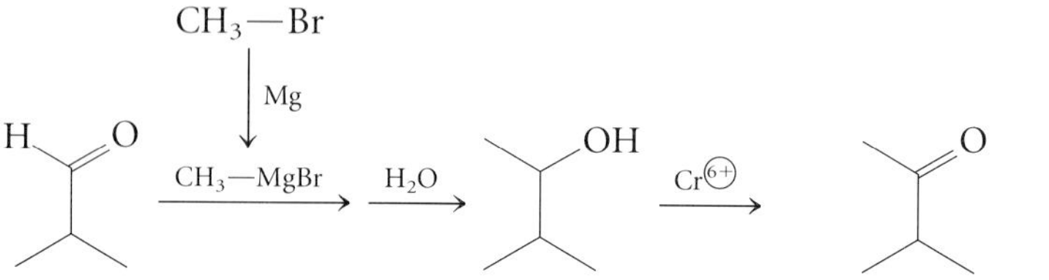

Route B

3-Methyl-2-butanone

To be able to use this sequence as part of the conversion of 2-propanol into the desired ketone, we must find reactions that can be used to convert 2-propanol efficiently into the aldehyde 2-methylpropanal.

We can, and should, view this simply as another problem in retrosynthetic analysis. Thus, we first consider the carbon content of 2-propanol and 2-methylpropanal to see whether we need to effect carbon–carbon bond formation at some point. Indeed, because 2-propanol has three carbons and 2-methylpropanal has four, we do need to form a carbon–carbon bond. From the list of reactions that make such bonds (Table 15.1), suppose we choose the enolate alkylation. In this case, the starting material is propanal, and the formation of its enolate anion(with a strong base such as a lithium dialkylamide), followed by reaction of the anion with an alkyl halide, results in the desired 2-methylpropanal:

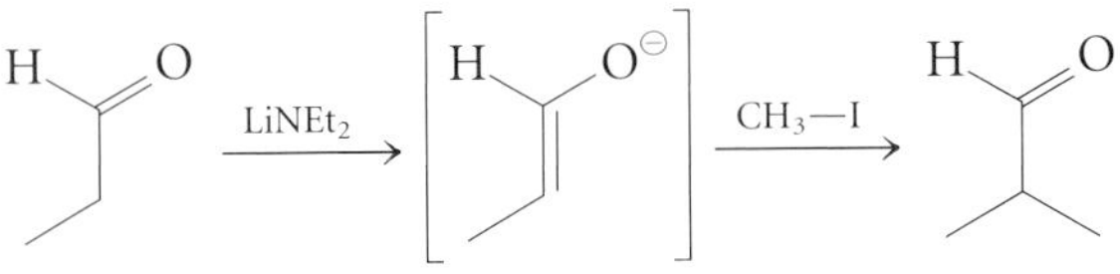

We have now simplified the overall problem, because the starting material for this step, propanal, has the same number of carbons as 2-propanol.

We can consider the conversion of 2-propanol into propanal, in turn, to be a separate synthesis and can conduct an investigation of possible routes as part of the retrosynthetic analysis. Although the number of carbons in these two species is the same, the oxidation levels of two carbons are different:

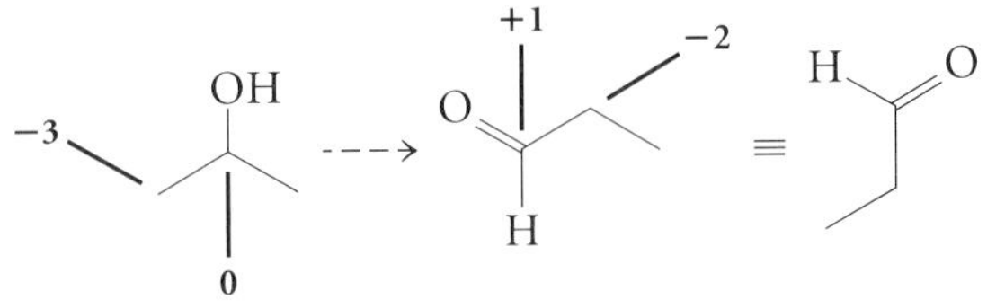

Furthermore, oxygen is bonded to a different carbon atom in the reactant and the product. Overall, we must reduce C-2 of 2-propanol and oxidize C-1 to create the aldehyde functional group. To find routes to accomplish this conversion, we turn to the table of oxidation–reduction reactions (Table 15.2). Because we need to accomplish both an oxidation and a reduction, it makes sense to pay special attention to those reactions that simultaneously effect both transformations. These are the dehydration of an alcohol and the reverse process, the hydration of an alkene to form an alcohol.

The latter process can be accomplished in two significantly different ways, which result in different regiochemical outcomes. Thus, simple hydration of an alkene follows a Markovnikov orientation and results in the hydroxyl group of the alcohol being on the more substituted carbon of the original alkene. Conversely, hydroboration–oxidation places the hydroxyl group on the less substituted carbon.

Can we use one or more of these dehydration or hydration processes to produce propanal from 2-propanol? Because none of them produces a carbonyl group, the answer is no. We are thus forced to examine other possible transformations that produce propanal from 2-propanol, keeping in mind that the major tasks to be accomplished are the oxidation of C-1 and

## CHEMICAL PERSPECTIVES

### SYNTHESIS OF VERY LARGE QUANTITIES OF MATERIALS

When industrial chemists consider various options, they must often take into account factors relating specifically to the scale of reactions. Although laboratory chemists generally perform reactions in flasks smaller than 20 liters, this size is much too small for the manufacture of bulk chemicals. To get some sense of the scale of industrial production, consider that 5 billion pounds of styrene was produced in 1995. Almost all of this amount was used in the manufacture of the plastic polystyrene (the material in polystyrene foam). A year's production amassed in one place would require a cubical container 150 meters on a side! (A football field is less than 100 meters long.)

the reduction of C-2. We might at this point recognize that propanal can be prepared by oxidation of the corresponding primary alcohol, 1-propanol. This compound, in turn, can be formed by hydroboration–oxidation of propene:

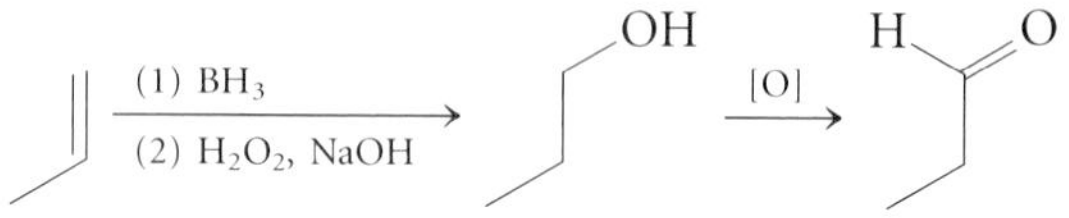


Once again we have a new problem to analyze, but one that should be relatively straightforward by this time. Indeed, we can prepare propene from 2-propanol by dehydration:

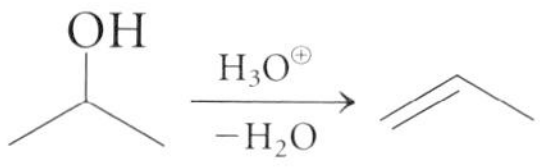


Thus, we have found a second sequence for the conversion of 2-propanol into 3-methyl-2-butanone that incorporates route B.

## EXERCISE 15.5

Suggest reagents that could be used to accomplish each of the following transformations:

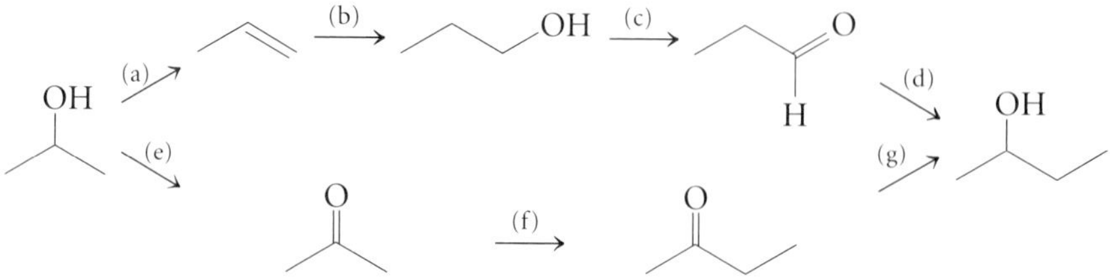


## EXERCISE 15.6

Compare and contrast the positive and negative features of the two routes proposed in Exercise 15.5 for the conversion of 2-propanol into 2-butanol.

## 15.6

# Criteria for Evaluating Synthetic Efficiency

Retrosynthetic analysis has led us to two separate routes from 2-propanol to 3-methyl-2-butanone:

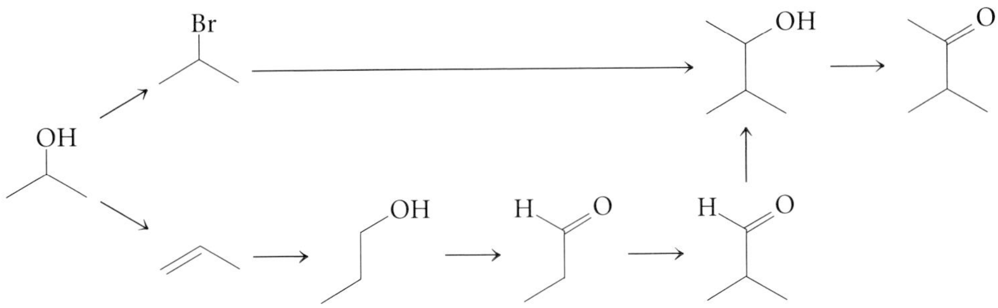

This is certainly an accomplishment; but before going into the laboratory to carry out such a conversion, we would need to decide which route is the "best." There is no simple answer because there are many factors to be considered. These are some important criteria for an efficient synthesis:

1. Number of steps
2. Yield of each step
3. Reaction conditions
4. Ease of purification of intermediates
5. Cost and availability of starting materials, reagents, and personnel time

Although these factors are often not independent, let's consider them separately. To simplify this analysis, we will assume in each case that the other criteria are fixed. For example, when analyzing the effect of the number of steps, we assume that all other factors (in this case, criteria 2 through 5) are constant.

### Linear Synthesis

If we assume that the yield in each step of a multistep synthesis is 90%, 75%, or 50%, we can easily calculate the effect of added steps on overall yield (Table 15.4, on page 760). With each additional step, the overall yield decreases. For example, after two steps, each with a yield of 90%, the overall yield drops to 81% ($0.90 \times 0.90 \times 100\%$). The decrease is even more dramatic if the yield per step is lower. With a 50% yield per step, the overall yield after five transformations is only 3%.

Therefore, when the yields in the individual steps are similar, the sequence with the lowest number of steps is preferable. Furthermore, a sequence of three steps, even if each step proceeds with 90% yield, results in an overall yield slightly lower than that attained in a single step with a 75% yield. We also see how quickly the effect of modest yields (such as 50% per step) can reduce the amount of material available from a synthesis.

**TABLE 15.4**

Overall Percent Yields for Multistep Syntheses

| Number of Steps | Yield per Step | | |
|---|---|---|---|
| | **90%** | **75%** | **50%** |
| 1 | 90 | 75 | 50 |
| 2 | 81 | 56 | 25 |
| 3 | 73 | 42 | 12 |
| 4 | 66 | 32 | 6 |
| 5 | 53 | 18 | 3 |
| 6 | 48 | 13 | 1.5 |

Because the objective of a synthesis is generally to prepare usable quantities of a product, we can also look at the effect of overall yield from this aspect. Assume that the objective is to prepare 10 grams of a product from a starting material whose molecular weight, for the sake of simplicity, is the same as the product's. A five-step sequence with a 50% yield per step requires 333 grams of starting material to begin the synthesis; a five-step sequence with a 90% yield per step requires only about 17 grams of starting material.

The number of steps, the overall yield, and the yield per step are clearly important, but another factor must also be considered. The type of synthesis we have dealt with so far is referred to as a **linear synthesis**—that is, one that effects sequential transformations, for example, a five-step sequence proceeding with 50% yield per step:

**Linear Synthesis**

$$A \xrightarrow{50\%} B \xrightarrow{50\%} C \xrightarrow{50\%} D \xrightarrow{50\%} E \xrightarrow{50\%} F$$

C: (50% × 50% = 25%)
D: (50% × 50% × 50% = 12.5%)
E: (50% × 50% × 50% × 50% = 6.2%)
F: (50% × 50% × 50% × 50% × 50% = 3%)

## ■ Convergent Synthesis

An alternative type of synthesis, called a **convergent synthesis,** is one in which two separate starting materials, A and D, are taken along separate routes to form intermediates C and F, which are combined to form the ultimate product, G:

**Convergent Synthesis**

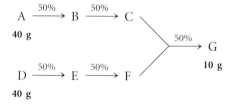

$$A \xrightarrow{50\%} B \xrightarrow{50\%} C$$
40 g
$$\xrightarrow{50\%} G$$
10 g
$$D \xrightarrow{50\%} E \xrightarrow{50\%} F$$
40 g

We again have a sequence with a total of five steps and assume that each step proceeds in 50% yield, but we cannot derive a simple overall yield because there are two branches. Nonetheless, we can examine the effect of this branching if we consider the amounts of starting material required, rather than the overall yield.

Assume for simplicity that half of the mass of the ultimate product, G, is moved along each sequence, A → C and D → F. To produce 10 grams of the ultimate product, G, requires 10 grams of C and 10 grams of F because the yield in the step that combines these two is only 50%. Furthermore, 40 grams of A is required to produce 10 grams of C, and 40 grams of D is required to produce 10 grams of F. Thus, overall, a total of 80 grams of starting materials is required to produce 10 grams of the ultimate product. This contrasts markedly with the 333 grams of starting material required by the linear synthesis, even though the number of steps and the yield per step for the linear and convergent syntheses are the same. Although it is not always possible to develop a convergent synthesis, it is clearly advantageous to do so when feasible.

## Logistical Factors

The overall yield, which determines the amount of starting material required for a synthesis, is not the only important consideration. Reactions that use simple, inexpensive reagents and solvents and do not require elaborate experimental precautions make a synthesis easier to carry out. Conversely, reactions that require very low or high temperatures, inert atmospheres (to prevent contamination by water or oxygen, or both), or unusual solvents should be avoided when simpler alternatives are available, even if those alternatives require some sacrifice in yield.

Another important factor is the number and nature of by-products. In a sense, this consideration is just a facet of chemical yield, but there are other ramifications of the production of by-products besides the simple reduction in the amount of the desired product. For example, reconsider the alkylation of 2-butanone, which can yield two products (resulting from alkylation at C-3 and C-1):

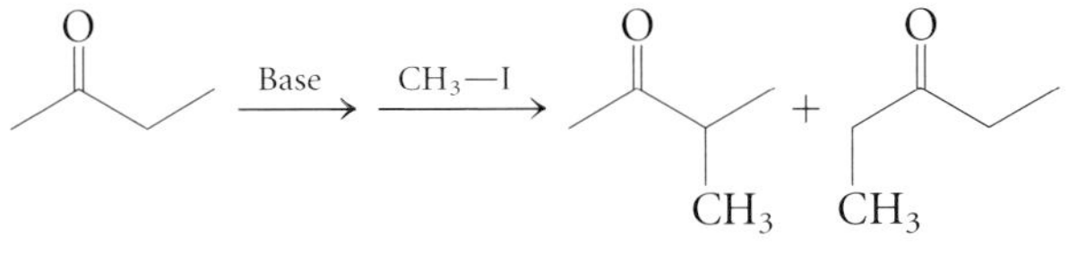

Before proceeding to the next step, these products must be separated and the desired compound (3-methyl-2-butanone) purified. The similarity between the two isomeric ketones makes this separation difficult. Separations like this often consume more time and effort than the chemical transformations themselves, and some of the desired product is frequently lost in the process.

How, then, do chemists select between various possible synthetic routes? In some cases, the choice is simplified, because most or all of the factors just described favor only one of several possible sequences. That, indeed, is the case for the two routes to 3-methyl-2-butanone we have developed. One requires only three transformations; the other requires six. Furthermore, the longer route includes a hydroboration–oxidation sequence, requiring use of the toxic and highly flammable reagent diborane. Finally, the hy-

### NAMING PHARMACEUTICAL AGENTS

Tagamet, a drug belonging to a group known as $H_2$ *blockers,* is used for the treatment of ulcers. Its success in the marketplace stimulated a search by many pharmaceutical companies for new $H_2$ blockers that would be more effective and have fewer side effects. At Glaxo, this research resulted in the discovery of a structurally similar compound marketed as Zantac (ranitidine). Sales of this drug accounted for more than half of Glaxo's gross revenues of $4.9 billion in 1994.

**Tagamet (cimetidine)**

**Zantac (ranitidine)**

Patent protection for Glaxo's manufacture of Zantac expired in 1995. As a result, many other manufacturers were able to start producing and selling Zantac, but they had to market the drug under its generic name, ranitidine. What's in a name? And how are names for pharmaceutical agents selected? The company that first prepares a successful drug is free to choose any proprietary name it wishes, and marketing specialists use their expertise to pick a name that is easy to pronounce, and perhaps even "catchy." An independent group helps decide on a generic name by providing a list of ten possibilities from which the originating company may select. The generic name that is least easy to pronounce and remember is chosen in the hope that, even after patent protection has expired, physicians and consumers will continue to opt for the familiar trade name under which the drug was first released, preserving market share for the original company.

droboration–oxidation sequence does not form the primary alcohol exclusively: rather, it gives a mixture of the dominant primary alcohol with a significant amount of the secondary alcohol (approximately 20%). We therefore have no difficulty in choosing the shorter sequence to accomplish the transformation of 2-propanol into 3-methyl-2-butanone. However, this example is rather unusual. In many cases, the alternative routes are not so markedly different. Indeed, even for practicing synthetic chemists, the choice of route is often difficult.

## EXERCISE 15.7

Calculate the net yield in the following synthesis, given the individual yields shown for each step. Calculate how much starting material is needed to prepare 10 grams of the product by this route. Categorize each step (a) through (f) as a carbon–carbon bond-forming reaction, an oxidation–reduction, or a functional-group transformation.

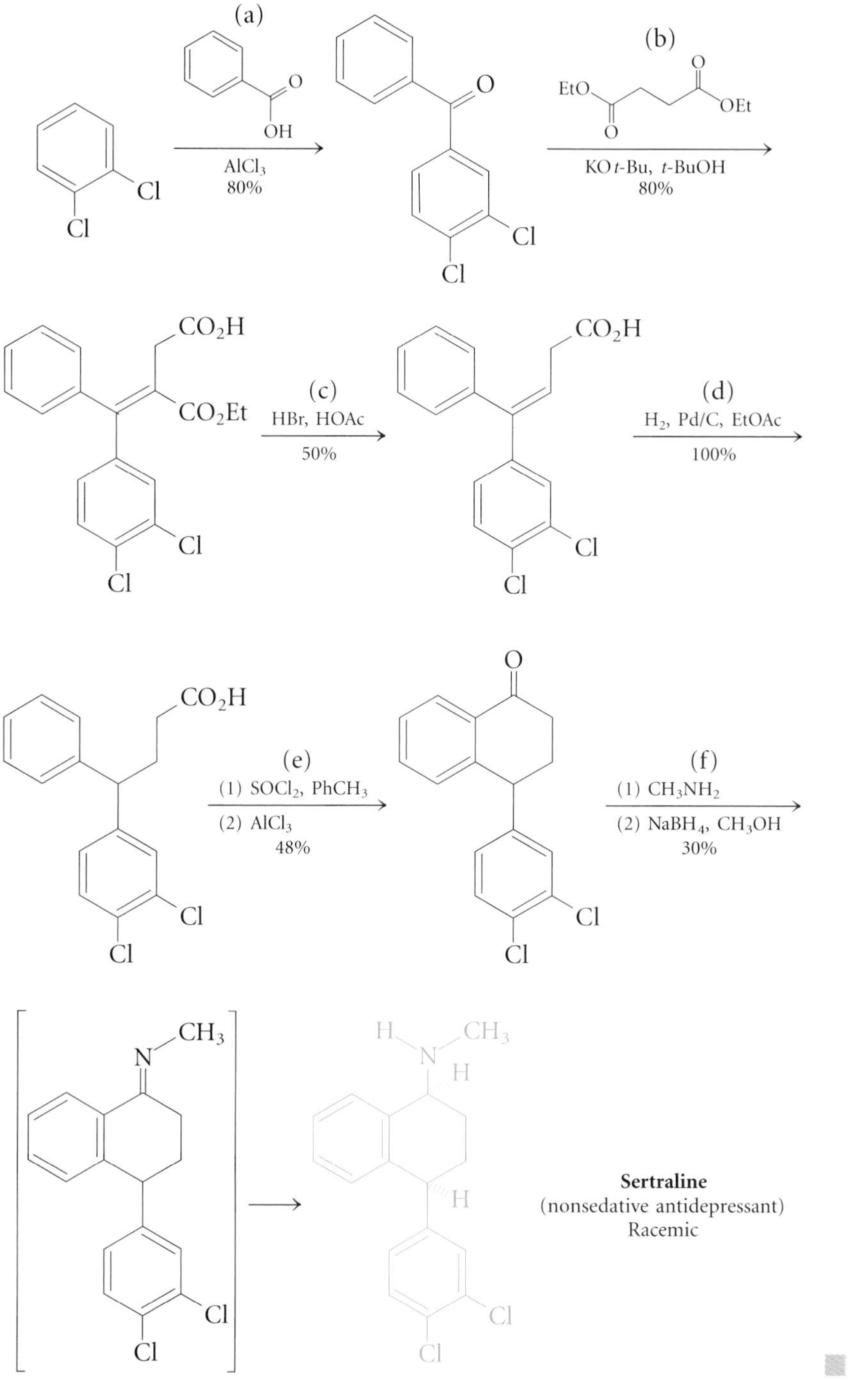


**Sertraline**
(nonsedative antidepressant)
Racemic

# Real-World Synthetic Objectives

## Multifunctional Compounds

The synthetic objectives considered so far have been simple molecules, generally having only a single functional group. However, the types of molecules of interest to practicing synthetic chemists are often complicated structures containing many different functional groups. For example, Figure 15.1 shows the ten best-selling prescription pharmaceuticals for 1995 and gives their generic names, trade names, and manufacturers. Some of these drugs are naturally occurring materials. For example, digoxin (Lanoxin) and levothyroxine (Synthoid) are obtained for commercial use by isolation from their natural sources. It is often more economical to do that than to prepare complex organic materials synthetically in the laboratory. Furthermore, naturally occurring materials can often be modified in relatively minor ways to provide highly active pharmaceutical agents. This type of modification is used to prepare amoxicillin (Amoxil), a highly effective antibiotic belonging to the penicillin class. Conjugated estrogen (Premarin) is prepared in a similar manner. The remaining six compounds—ranitidine (Zantac), diltiazem (Cardizem), enalapril (Vasotec), nifedipine (Procardia XL), fluoxetine (Prozac), and albuterol (Proventil)—are prepared by multistep syntheses. Designing syntheses for such complex molecules requires learning some "tricks of the trade."

## Functional-Group Compatibility

In devising synthetic sequences for the preparation of complex molecules, chemists must consider the issue of functional-group compatibility—that is, whether a reagent might react with a functional group other than the targeted one. Generally, it is desired that only one functional group interact with a given reagent.

Let's consider functional-group compatibility with respect to the possible conversion of a ketoester into a ketoalcohol:

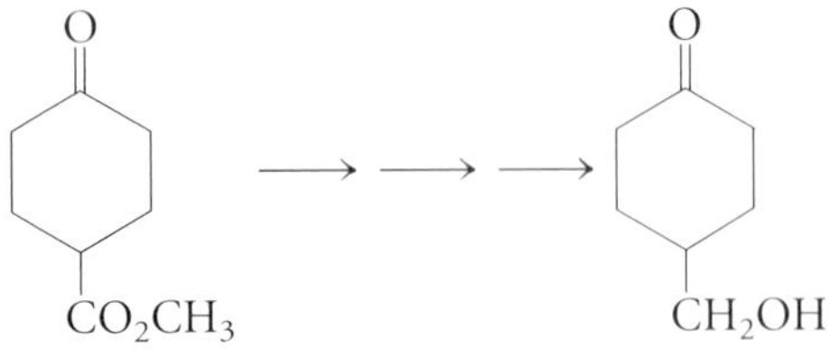

Retrosynthetic analysis reveals that the starting material and the product have the same number of carbons (not counting the methyl group of the ester, which is not connected by a carbon–carbon bond to the rest of the molecule). Therefore, the conversion does not require a carbon–carbon bond-forming reaction. Indeed, all that is needed is the reduction of the ester functional group in the starting material to a primary alcohol in the product, a reaction that can be accomplished with $LiAlH_4$. However, the starting material has two functional groups, a ketone and an ester; unfortunately, both are reduced by $LiAlH_4$. In fact, the resonance stabilization of

**1 Amoxicillin (Amoxil, Trimoxin)**
SmithKline Beecham, Squibb (antibiotic)

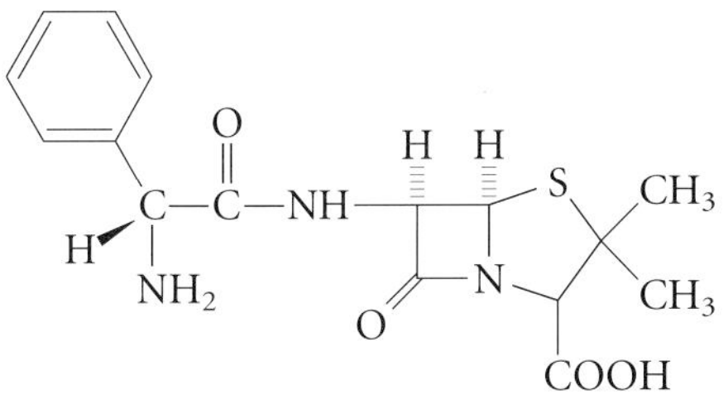

**2 Conjugated estrogen (Premarin)** Wyeth–Ayerst
(antimenopausal/estrogen replacement)

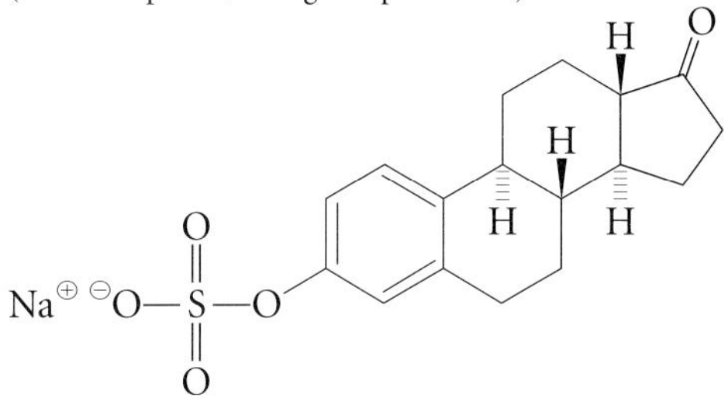

**3 Levothyroxine (Synthoid)**
Boots (for treatment of hypothyroidism)

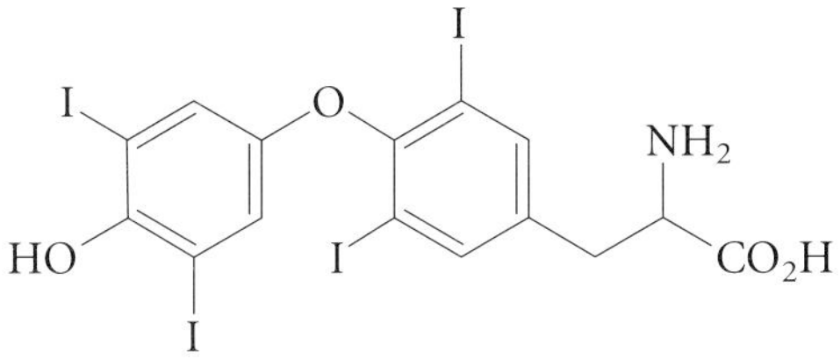

**4 Ranitidine (Zantac)**
Glaxo (antiulcerative)

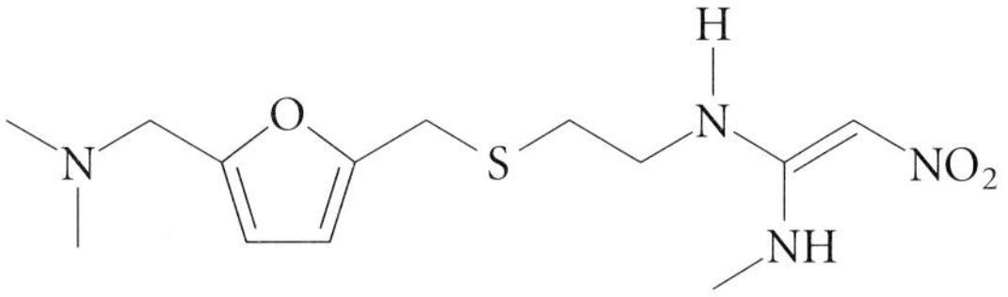

**5 Diltiazem (Cardizem)**
Marion Merrell Dow
(calcium channel blocker/
coronary vasodilator)

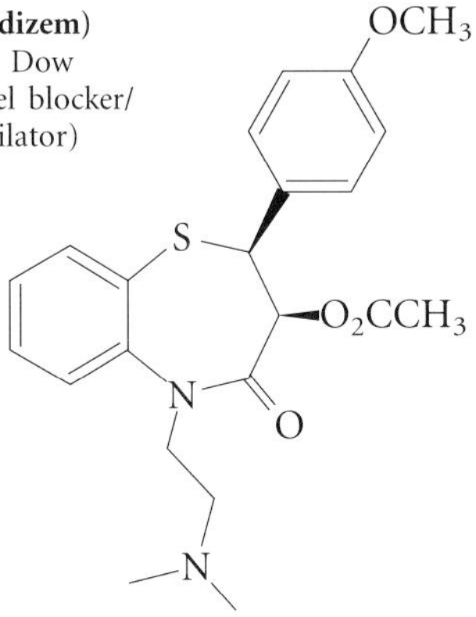

**6 Digoxin (Lanoxin)**
Burroughs Wellcome
(heartbeat regulator)

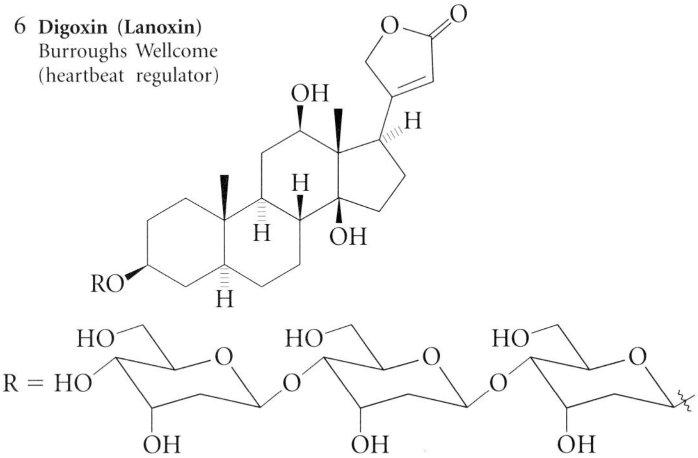

**7 Enalapril (Vasotec)**
Merck (antihypertensive)

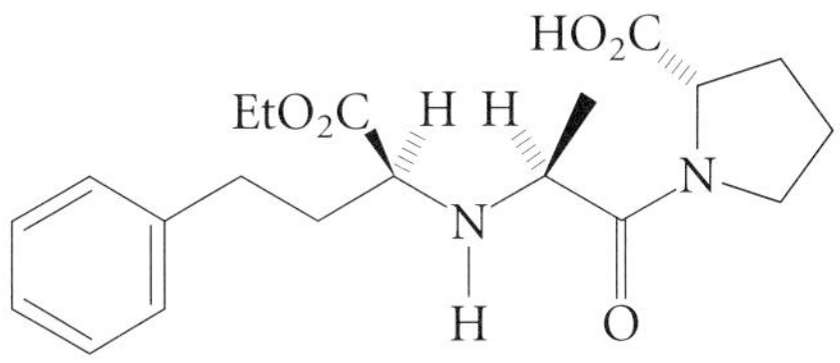

**8 Nifedipine (Procardia XL)**
Pfizer (antianginal; antihypertensive)

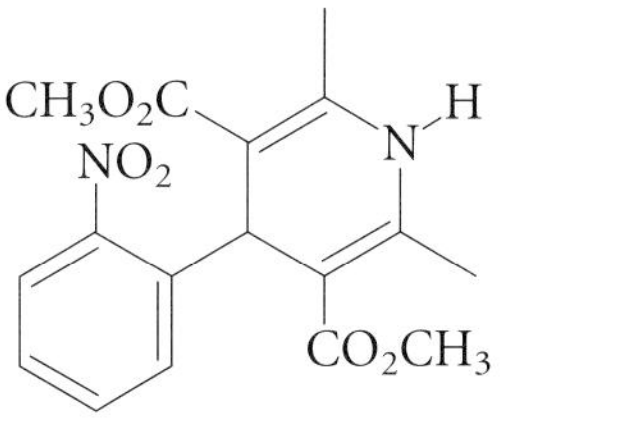

**9 Fluoxetine (Prozac)**
Lilly (antidepressant)

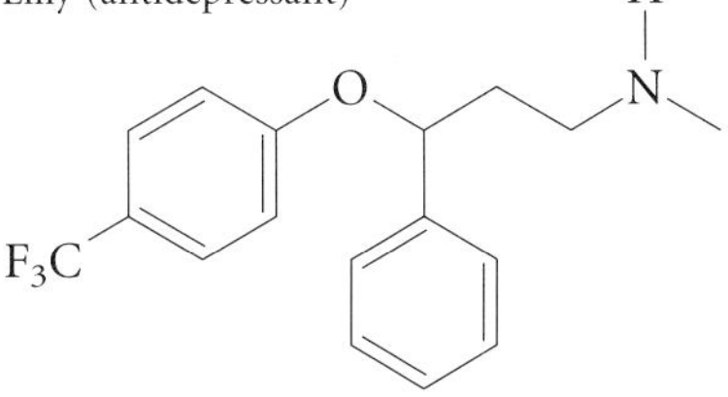

**10 Albuterol (Proventil)**
Schering (bronchodilator)

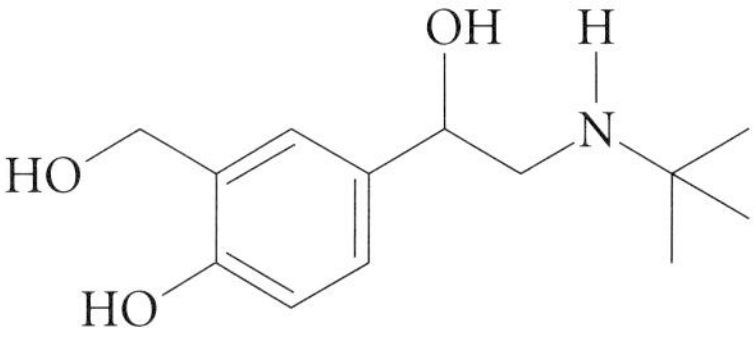

**FIGURE 15.1**

The "top ten" prescription drugs are organic molecules with several functional groups.

SEMISYNTHETIC PHARMACEUTICAL AGENTS

It is common in the pharmaceutical industry to make use of natural compounds for the production of valuable chemicals. For example, some antibiotics are produced by microbes and can be simply isolated and purified from the culture medium. However, some microbes do not produce products that have all of the necessary and desirable features (activity, stability, few side effects, for example). In such a case, it is often possible to "help" a microbe produce a desired product that is similar to one naturally produced. Feeding microbes appropriate structural subunits often results in the incorporation of these pieces into the final product, as, for example, in the industrial production of the antibiotic penicillin V. This antibiotic is prepared by feeding to a *Penicillium* mold 2-phenoxyethanol, which the mold oxidizes and uses for the amide side chain of penicillin V:

2-Phenoxyethanol

Penicillin V

an ester functional group makes it somewhat more resistant to reduction than a ketone. Reaction of the ketoester with LiAlH$_4$ results in the reduction of both functional groups, producing a diol:

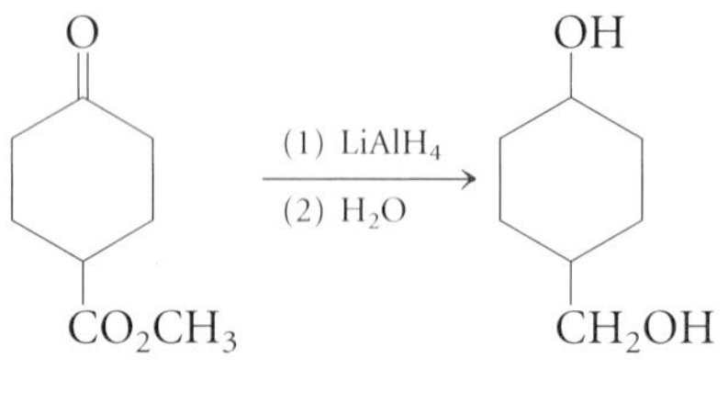

Note that if the problem were constituted differently, requiring the reduction of the ketone rather than the ester carbonyl, there would have been no difficulty—a milder reducing reagent, NaBH$_4$, could have been used to reduce only the ketone group:

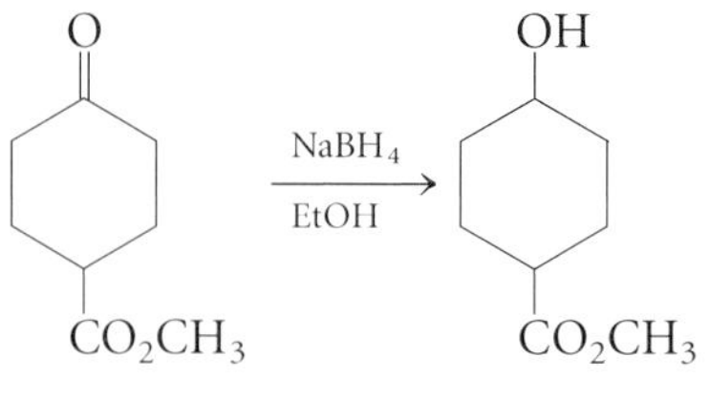

Analogous concerns apply to many other functional groups. In planning a synthesis, chemists must be aware of possible competing reactions that can be induced by a chosen reagent on other parts of a molecule. As we will see, one solution to effecting a transformation on only the less reactive functional group is first to convert the more reactive group to a different functional group that does not react with the chosen reagent—in other words, to use a protecting group.

## EXERCISE 15.8

For each of the following transformations, determine whether the indicated reagent is compatible with functional groups in starting material other than the one it is intended to affect. If not, predict the nature of the problem that would be encountered if the indicated transformation were attempted.

(a)

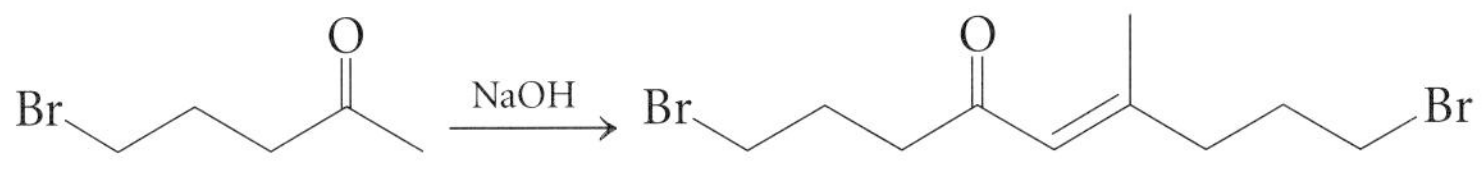

(b)

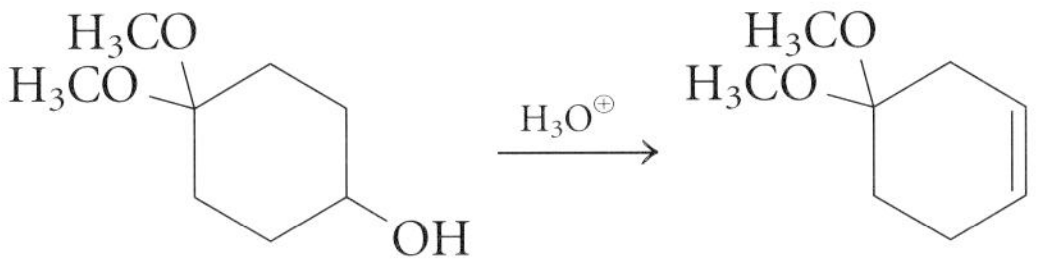

(c)

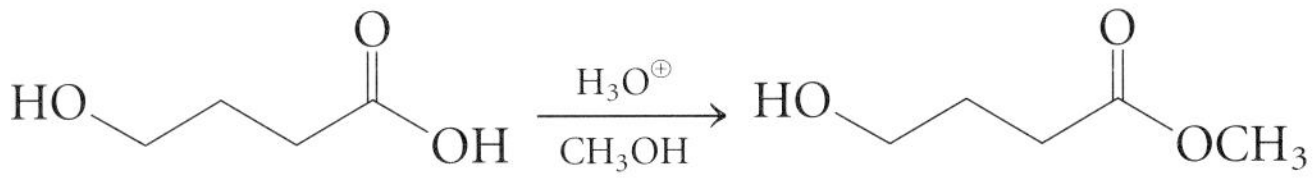

(d)

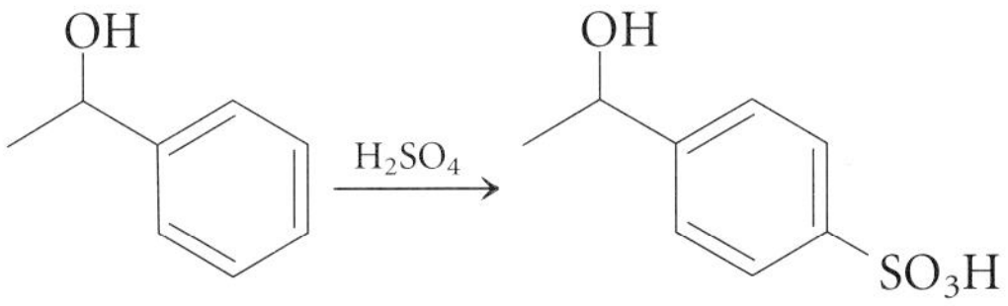

(e)

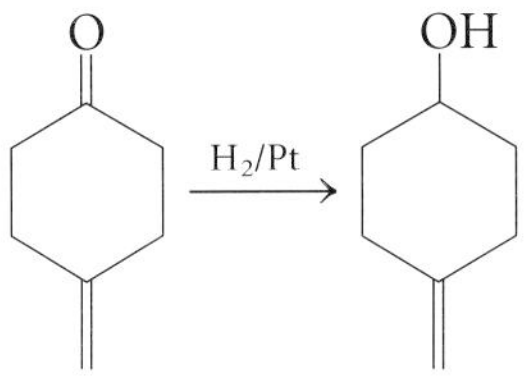

(f)

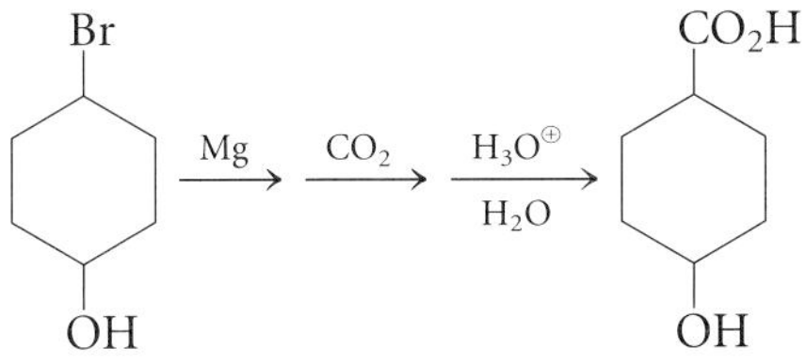

(g)

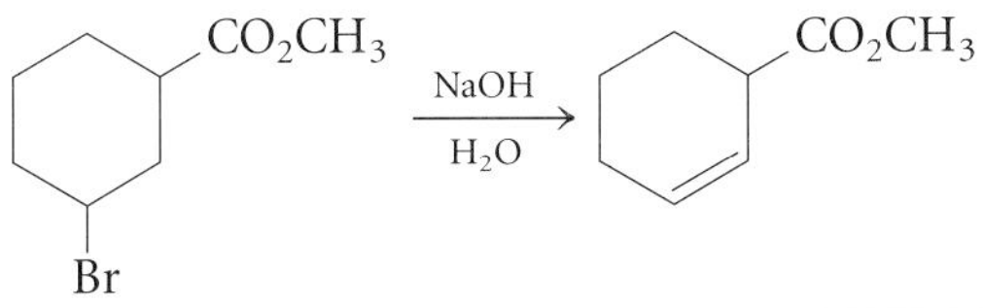

## 15.8

## Protecting Groups

The problem of selectively reducing an ester in the presence of a ketone is a difficult one, because there is no reduction reaction that leads directly to the desired ketoalcohol.

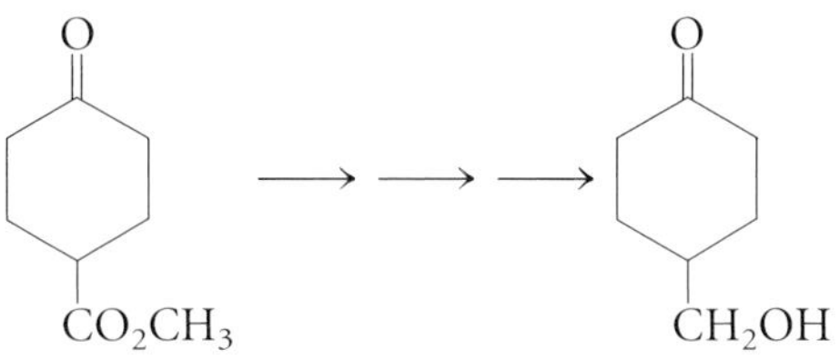

The only common reagent that will reduce an ester, namely $LiAlH_4$, will also reduce the ketone group.

At this point we return to a concept touched on earlier—that the reactivity of a functional group can be masked by a protecting group. Chapter 8 pointed out that acylation of thiols to form thiol esters protects the thiol group from alkylation and that the thiol could be regenerated by hydrolysis of the ester:

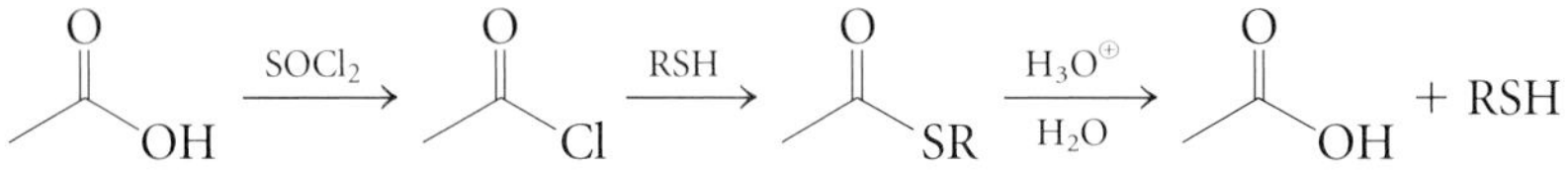

In Chapter 10, we saw that alkenes could be protected from reaction by masking them as vicinal dibromides, from which the alkene could be regenerated by debromination with zinc:

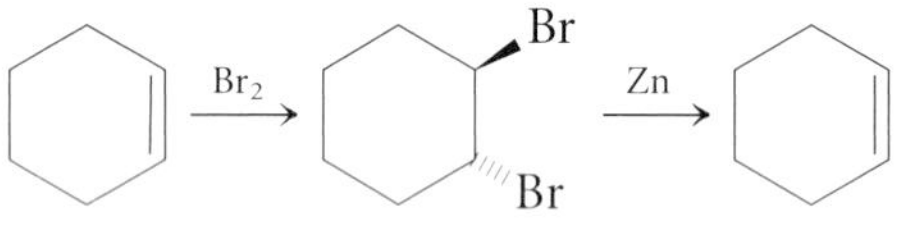

Let's apply this use of a protecting group to the ketone group of the keto-ester.

### ▨ Protection of Aldehydes and Ketones

Ketones are known to form ketals by treatment with alcohols under acidic conditions (Table 15.3), and the reaction can be reversed by treatment with water under acidic conditions:

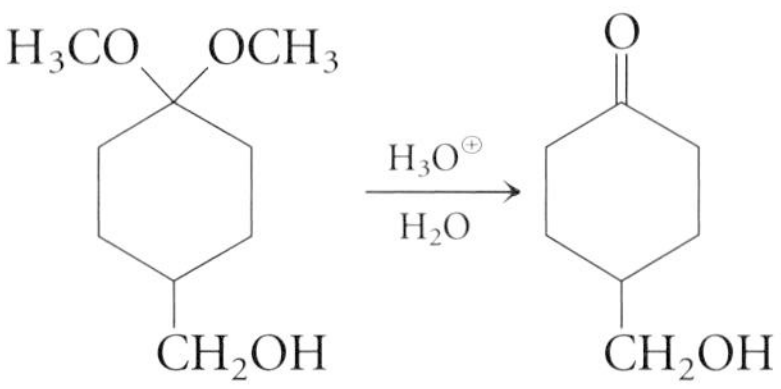

Furthermore, the ketal group (a 1,1-diether) is unaffected by the reagent $LiAlH_4$. Thus, an ester group can be reduced in the presence of a ketal:

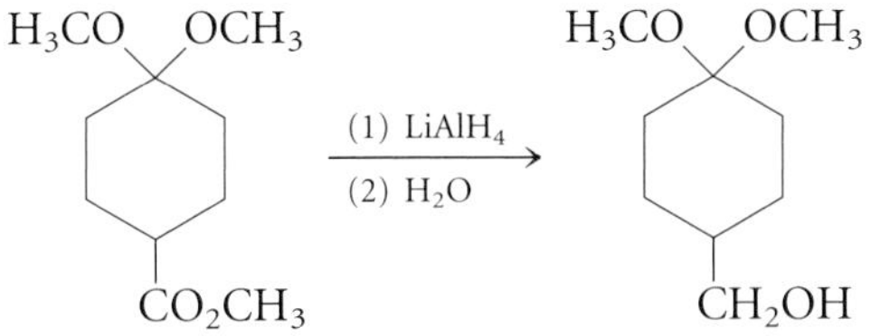

With these reactions at our disposal, we are ready to propose a scheme for carrying out the transformation of the ketoester to the ketoalcohol:

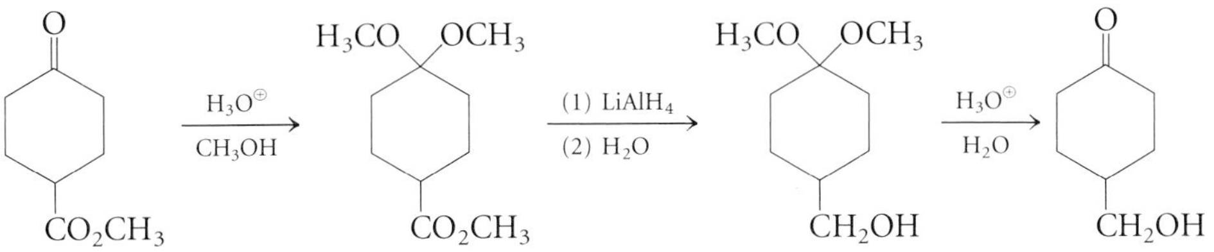

In the last step, the ketone function is restored by hydrolysis of the ketal under acid conditions. In this scheme, the ketal has functioned as a protecting group for the ketone.

## Requirements for the Use of Protecting Groups

Chemists have rather stringent requirements for the reactions employed to protect functionality, because the use of a protecting group requires two additional steps (protection and deprotection):

**1.** The yield of both reactions must be high.

**2.** The reactions must produce few by-products, because by-products can pose significant separation problems.

**3.** Reactions that protect and deprotect a functional group should require the use of only simple reagents and reaction conditions.

The formation of a ketal from a ketone and its subsequent hydrolysis fit all of these requirements.

Both reactions generally proceed with high chemical yields, no major by-products are formed, and the reaction conditions are relatively mild and affect few other functional groups.

A wide variety of protecting groups have been developed for all the common functional groups, but we will consider only one example for each of three other functional groups: alcohols, amines, and carboxylic acids.

Evaluate whether each of the transformations listed in Table 15.3 meets the three criteria listed above for the use of protecting groups. Select the transformations you think might serve as methods for protecting functional groups.

## Protection of Alcohols

There are many protecting groups for alcohols, possibly more than for any other functional group. Benzyl ethers are simple and quite useful examples that can be formed readily by the reaction of an alcohol with benzyl bromide under basic conditions (a Williamson ether synthesis):

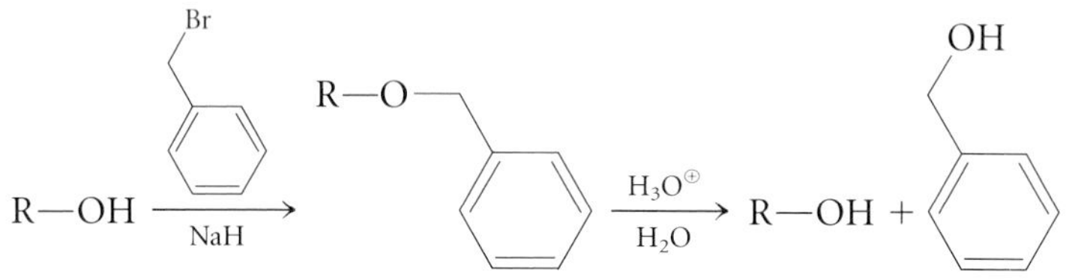

**A benzyl ether**

The normal acidity of an alcohol is masked in the ether, which is inert to almost all basic and mildly acidic reaction conditions. However, benzyl ethers are readily cleaved in strong acid by an $S_N1$ substitution at the benzylic carbon, as well as by reduction with $H_2$ and a metal catalyst such as Pd.

Write a detailed mechanism for the following benzyl ether hydrolysis:

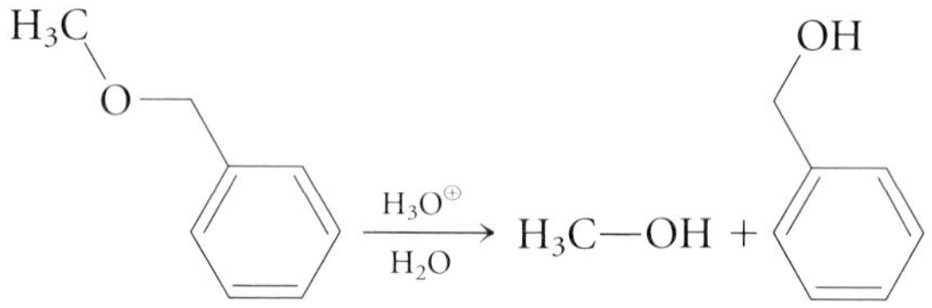

Consider carefully which carbon–oxygen bond is more likely to be cleaved under acidic conditions. Are there benzyl ethers with other R groups for which it is less clear which carbon–oxygen bond will be cleaved more rapidly?

## Protection of Carboxylates

The acidity of a carboxylic acid can be masked by converting it to a tertiary-butyl (*t*-butyl) ester:

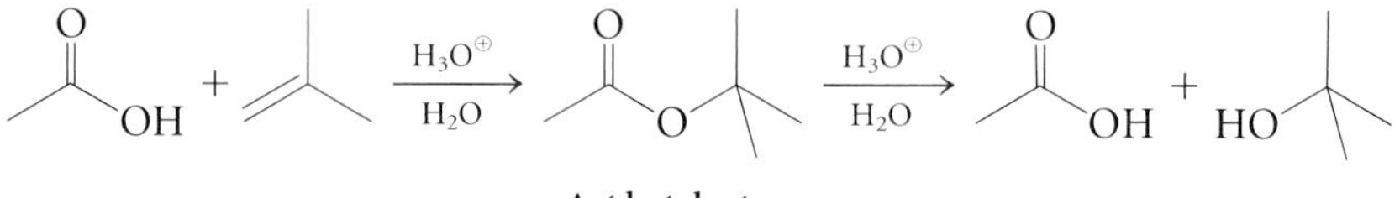

**A *t*-butyl ester**

In contrast to normal esterification by nucleophilic acyl substitution, the formation of *t*-butyl esters is an acid-catalyzed Markovnikov addition to the π bond of isobutylene (2-methylpropene).

### EXERCISE 15.11

Write a detailed mechanism, showing electron flow, for the formation of the following *t*-butyl ester:

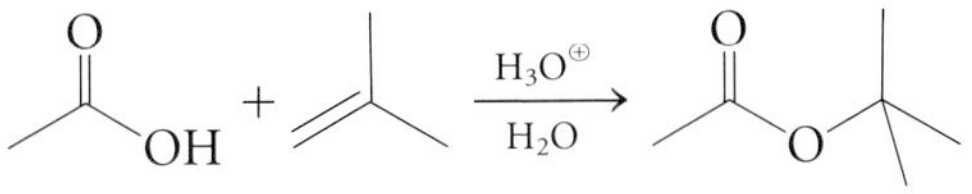

Hydrolysis of a *t*-butyl ester under basic conditions (by nucleophilic acyl substitution) is difficult, because of the additional steric interference due to the three methyl groups of the tertiary butyl group. However, the acid-catalyzed esterification can be reversed in the presence of water, and the treatment of *t*-butyl esters with relatively mild acid under aqueous conditions regenerates the carboxylic acid and *t*-butyl alcohol. Both the formation of *t*-butyl esters and their subsequent cleavage to reform the carboxylic acids take place under conditions significantly milder than those necessary to form and cleave esters (for example, methyl esters) via a nucleophilic acyl substitution pathway.

### EXERCISE 15.12

The cleavage of *t*-butyl esters can also be achieved with trifluoroacetic acid. This carboxylic acid is considerably stronger than acetic acid because of stabilization of the carboxylate anion by the three electron-withdrawing fluorine atoms. Under these conditions, however, isobutylene, not *t*-butyl alcohol, is formed. Explain.

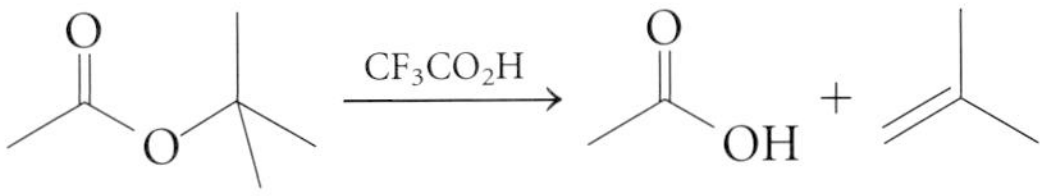

### Protection of Amines

Finally, we consider the protection of amines, a functional group that is considerably less acidic than either carboxylic acids or alcohols but has more significant nucleophilic character as a neutral species. The reaction of a primary amine with *t*-butyl chloroformate in the presence of a weak base (such as pyridine) forms a *urethane* (also called a *carbamate*).

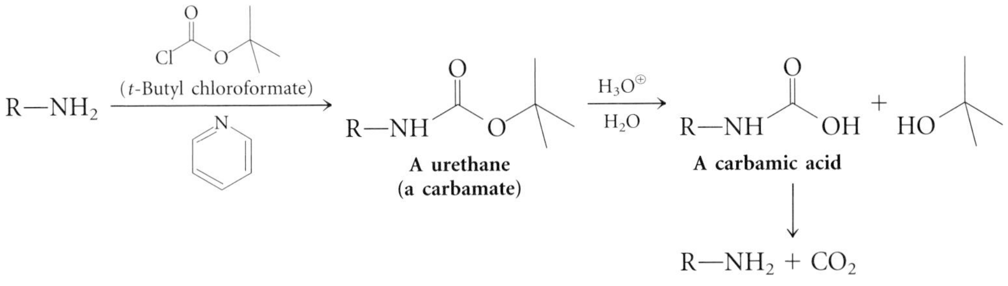

The urethane functional group resembles an ester on one side and a car-boxylic acid amide on the other. Indeed, the lone pair of electrons on the nitrogen in a urethane is delocalized into the carbonyl group in the same way as in an amide. The interaction of an electrophile with this lone pair requires disruption of resonance stabilization. As a result, urethanes are not sufficiently nucleophilic to react with most electrophiles. However, in the deprotection step, hydrolysis of the urethane formed from *t*-butyl chloro-formate resembles the hydrolysis of a *t*-butyl ester. That is, treatment with mild aqueous acid results in the formation of *t*-butyl alcohol and a nitro-gen-substituted carboxylic acid (a carbamic acid) that undergoes rapid de-carboxylation to form the amine.

### EXERCISE 15.13

Write significant resonance contributors for a urethane, an amide, and an ester. Which of these functional groups has the most charge on the carbonyl oxygen? Which has the least? Does the amide or the urethane have more charge on nitro-gen? Does the ester or the urethane have more charge on the ether oxygen?

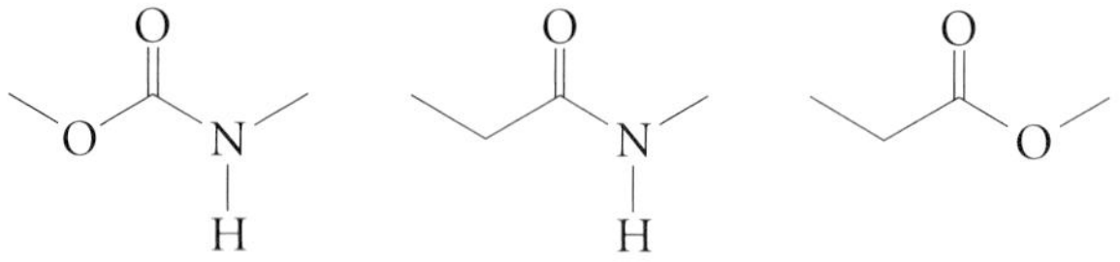

| A urethane | An amide | An ester |

### EXERCISE 15.14

Write a detailed mechanism for the decarboxylation of a carbamic acid under acidic conditions:

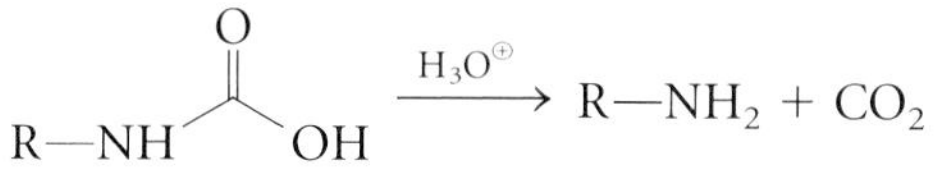

$$ R-NH \overset{O}{\underset{OH}{\|}} \xrightarrow{H_3O^{\oplus}} R-NH_2 + CO_2 $$

Explain why this decarboxylation has a lower activation energy than does that of a simple carboxylic acid.

### Use of an Alcohol Protecting Group

Let's consider how to use a protecting group to carry out the following conversion:

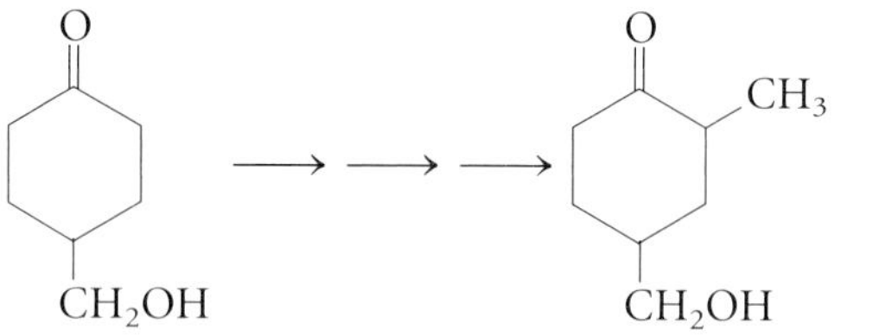

By comparing the starting material and product, we see that we need to form a carbon–carbon bond. After examining Table 15.1, we opt to use the conversion of a ketone into its enolate anion, followed by alkylation with an electrophile (in this case, methyl iodide). However, there is a potential problem with this approach. Although the two functional groups in the starting material are quite different (a ketone and a primary alcohol), both are alkylated when treated with base and an alkyl halide—the ketone undergoes enolate alkylation, and the alcohol forms an ether by the Williamson ether synthesis:

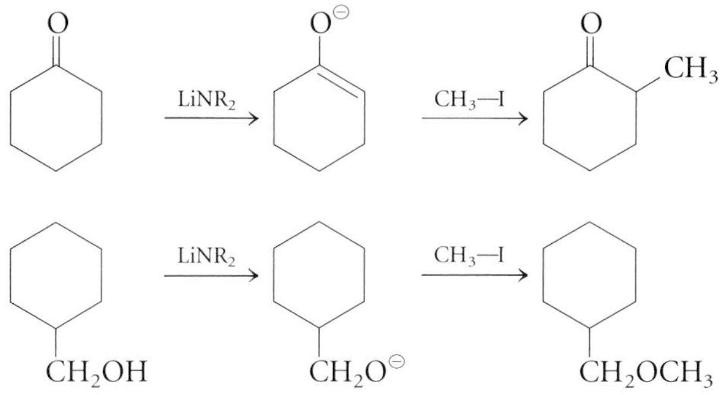

Furthermore, because the alcohol is more acidic than the ketone, deprotonation and alkylation of the ketone are not possible without converting the alcohol into its methyl ether.

Thus, the desired carbon–carbon bond-forming reaction can be effectively accomplished only if we temporarily mask the reactivity of the alcohol group. A satisfactory approach is first to convert the alcohol into its benzyl ether. The alkylation is then accomplished, and the benzyl ether group protecting the alcohol is removed.

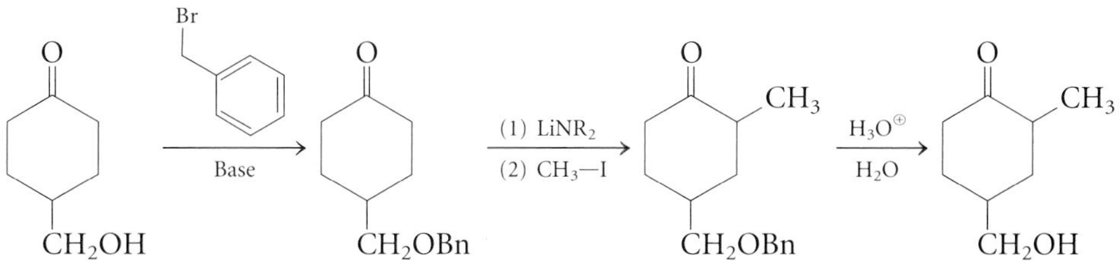

## 15.9

# Practical Examples of Multistep Syntheses

Having explored the thought processes required for retrosynthetic analysis, let's look at several commercially important, relatively short synthetic sequences. Rather than attempting to develop a retrosynthetic analysis, we will simply concentrate on how each transformation fits into the overall sequence and accomplishes the needed changes.

### Phenylpropionic Acid Analogs: Ibuprofen and Ketoprofen

Ibuprofen and ketoprofen are two anti-inflammatory analgesics that have a variety of uses. Originally sold by prescription only, ibuprofen is now available as an over-the-counter general pain reliever.

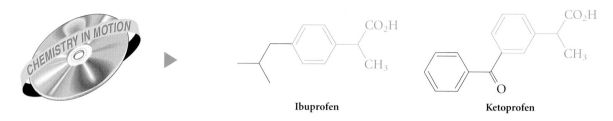

**Ibuprofen**

**Ketoprofen**

If we concentrate on the right side of these molecules, we see that each has an aromatic ring as a substituent on propanoic acid (a phenylpropionic acid unit). Indeed, many pharmaceutical agents have this same structural feature, which is modified by varying the functionality on the aromatic ring. Such "second-generation" drugs are often developed via a trial-and-error approach by pharmaceutical chemists who are trying to produce new agents that require lower dosages and have fewer side effects than existing products.

The syntheses of ibuprofen and ketoprofen address the construction of the phenylpropanoic acid unit in distinctly different ways (Figure 15.2). Note that in each case a sequence of five reactions is required.

We can view the problem of forming the phenylpropionic acid unit as the incorporation of a carboxylic acid and a methyl group on a benzylic carbon. Both syntheses depend on the displacement of a benzylic bromide by cyanide (the fourth step in the synthesis of ibuprofen and the second step in the synthesis of ketoprofen) to incorporate the required carboxylic acid functional group. In ibuprofen, the benzylic carbon is secondary and already bears the necessary methyl group. In contrast, the ketoprofen synthesis requires displacement by cyanide at a primary benzylic bromide site. Recall that an $S_N2$ substitution is subject to steric interference and that for secondary alkyl halides, E2 elimination often competes with $S_N2$ substitution (Chapter 8). With regard to this aspect, the ketoprofen synthesis is superior.

The ibuprofen synthesis begins with Friedel–Crafts acylation of isobutylbenzene. Reduction of the ketone and conversion of the resulting alcohol into the bromide precedes the critical cyanide displacement. Hydrolysis of the nitrile then yields the acid required in the product.

The ketoprofen synthesis begins by benzylic bromination of the methyl group of 3-methylbenzophenone. After the $S_N2$ displacement by cyanide ion, a methyl group must be introduced at a position $\alpha$ to the nitrile. A nitrile can be deprotonated at the $\alpha$ position to form an anion that can act as a nucleophile. There are then two acidic hydrogens that might be replaced by sequential reaction with an electrophile (for example, methyl iodide). If only one methyl group is required, as here, this is a potential problem. However, even if this double alkylation were not a problem, deprotonation would require the use of a strong base (such as lithium dialkyl amide), which would add substantially to the cost of the process.

An alternative is to convert the nitrile into a nitrile ester by treating it with a weak base and diethyl carbonate. Why is it possible to conduct a monoacylation, but not a monoalkylation? The answer lies in the fact that although there is still an acidic hydrogen between these two functional groups after the acylation, the anion formed by deprotonation is stabilized by resonance delocalization over the nitrile group, the benzophenone group, and an ester functional group, making this anion insufficiently reactive to undergo further reaction with diethyl carbonate. On the other hand, the

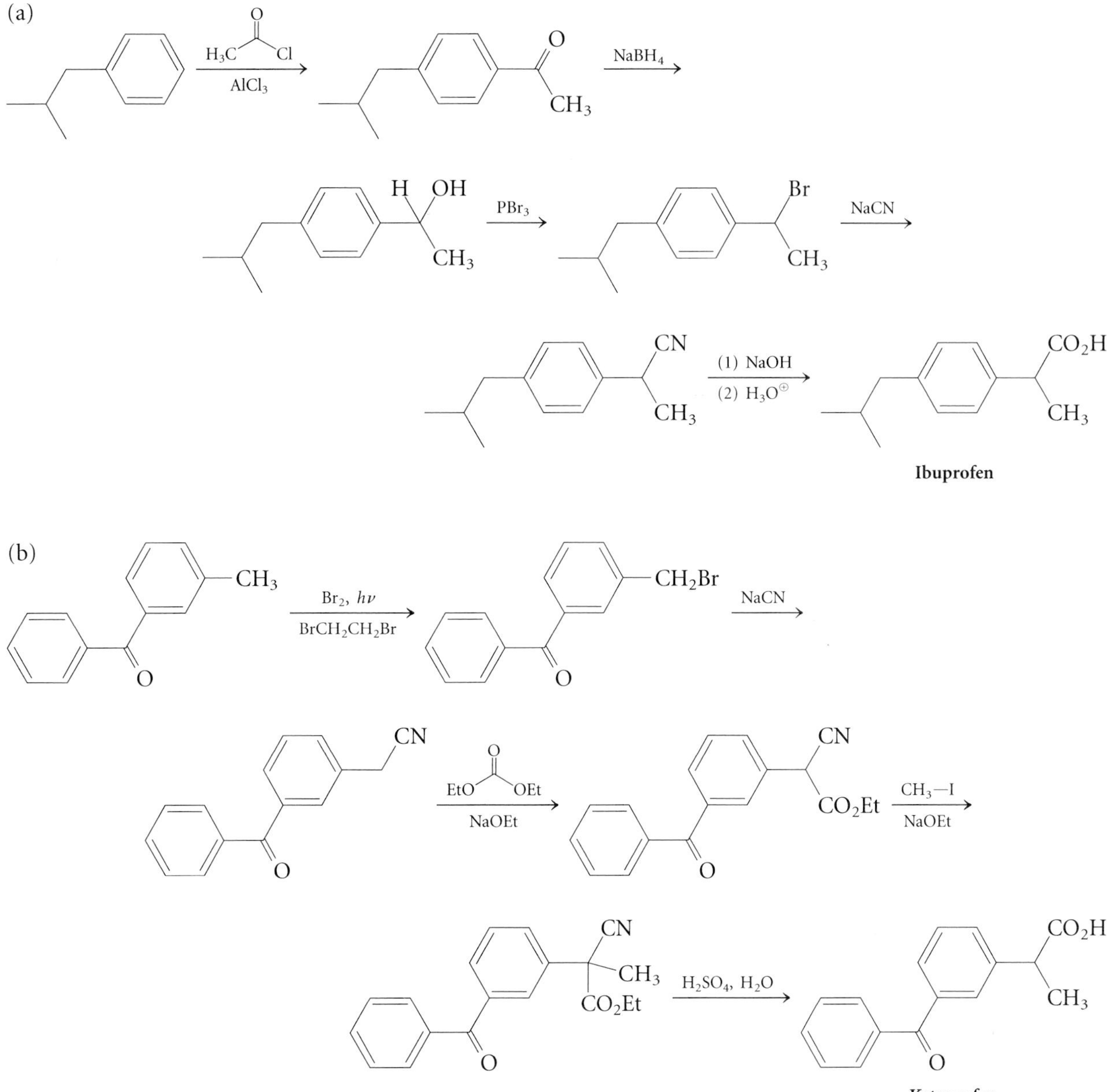

**FIGURE 15.2**

(a) The synthesis of ibuprofen begins with a Friedel–Crafts acylation of an alkylated benzene. Reduction of the ketone to an alcohol and then conversion to a bromide permit the introduction of an additional carbon atom as a nitrile by $S_N2$ displacement. Hydrolysis gives the carboxylic acid product. (b) The synthesis of ketoprofen begins with bromination at the benzylic position. After introduction of a nitrile by an $S_N2$ displacement, the remaining benzylic hydrogens are sufficiently acidic that an anion can be formed by deprotonation with alkoxide. Nucleophilic acyl substitution of diethyl carbonate introduces a —$CO_2Et$ group, making the remaining hydrogen very acidic. Deprotonation produces an anion that is alkylated by methyl iodide. The nitrile and ester are then hydrolyzed to give the substituted malonic acid ($\beta$-diacid), which readily loses carbon dioxide to form the monoacid product.

anion does react (but only once) with the more reactive electrophile, methyl iodide, as shown in Figure 15.2(b). This alkylation results in a quaternary center, and so no further alkylation is possible. The synthesis is finished by hydrolysis of both the nitrile and the ester groups to carboxylic acids. The resulting dicarboxylic acid is a malonic acid derivative, which undergoes spontaneous loss of carbon dioxide to form the monocarboxylic acid, ketoprofen.

---

### EXERCISE 15.15

Write a mechanism for the reaction of acetonitrile with diethyl carbonate in base:

$$H_3C-C\equiv N + \underset{\text{EtO}}{\overset{\text{O}}{\underset{|}{\text{C}}}}\text{OEt} \xrightarrow{\text{NaOEt}} \overset{N\equiv}{\underset{}{\text{C}}}-CH-\overset{\text{O}}{\underset{}{\text{C}}}-\text{OEt}$$

---

### Benzodiazepines: Valium

Valium (patented in 1962) was the first of a long series of psychoactive diazepines, a group of pharmaceutical agents used primarily as sedatives and antianxiety agents. The synthesis of Valium is shown in Figure 15.3

Let's examine the structure of Valium for features that lend themselves to a retrosynthetic analysis. The molecule contains a tertiary amide and an imine, as well as two aromatic rings. Because all are common functional groups, they can be formed by functional-group transformations listed in Table 15.3. Thus, the combination of an aminoketone and an aminocarboxylic acid should result in the formation of Valium:

**Valium**

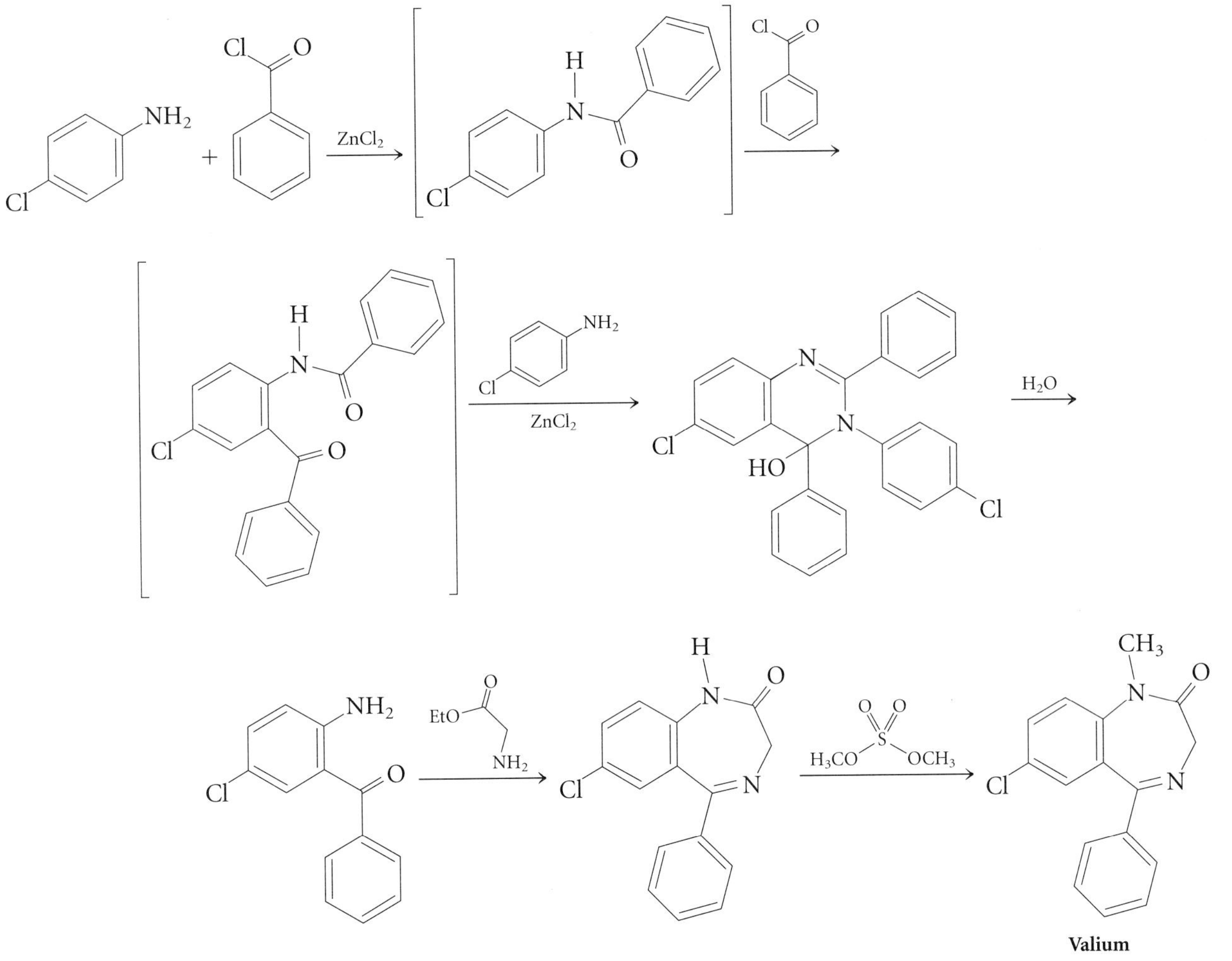

**FIGURE 15.3**

The first step in the synthesis of Valium is a double acylation of an aromatic amine—first by nucleophilic acyl substitution of an acid chloride by the amino group, and then by an electropohilic aromatic substitution on the ring. The resulting intermediate is trapped by a second equivalent of *p*-chloroaniline, leading to a six-member ring with two nitrogens. Hydrolysis reverses this cyclization and frees the amino group to give the simple electrophilic substitution product. A seven-member ring is then produced by reaction with an aminoester. This cyclization takes place by amide formation between the ring amino group and the ester group of the other reactant and by imidation of the ketone group by the primary amine. Methylation of the amide nitrogen by dimethyl sulfate completes the synthesis.

Indeed, this reaction is very close (bearing an extra *N*-methyl group) to the next-to-last transformation in the commercial preparation shown in Figure 15.3. In that sequence, an aminoester reacts with an aminoketone to form the seven-member ring in Valium. The final reaction converts the secondary amide into the tertiary amide by alkylation on nitrogen.

The reaction of dimethyl sulfate or dimethyl carbonate with a nucleophile leads to different products. Compare the mechanisms of the two reactions, and explain why they follow different pathways.

Dimethyl sulfate            Dimethyl carbonate

Referring to the discussion of the syntheses of ibuprofen and ketoprofen, explain why, in the synthesis of Valium, it is more desirable to carry out the alkylation of nitrogen with a methyl group after the amide is formed (as in the upper pathway here and in the commercial procedure in Figure 15.3) rather than before (as in the lower pathway).

For a cyclization to be the last step of the synthesis, we must have a route to prepare this acylated aniline:

The first step of the Valium synthesis (Figure 15.3) is the reaction of benzoyl chloride with *p*-chloroaniline. At first, this might seem to be a straightforward Friedel–Crafts acylation by which the required ketoamine can be prepared:

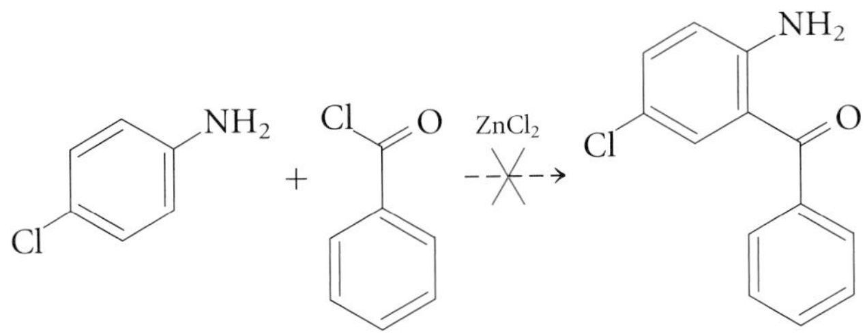

However, recall that amides are formed from the reaction of carboxylic acid chlorides with amines. This is precisely what happens, rather than a Friedel–Crafts acylation of the ring:

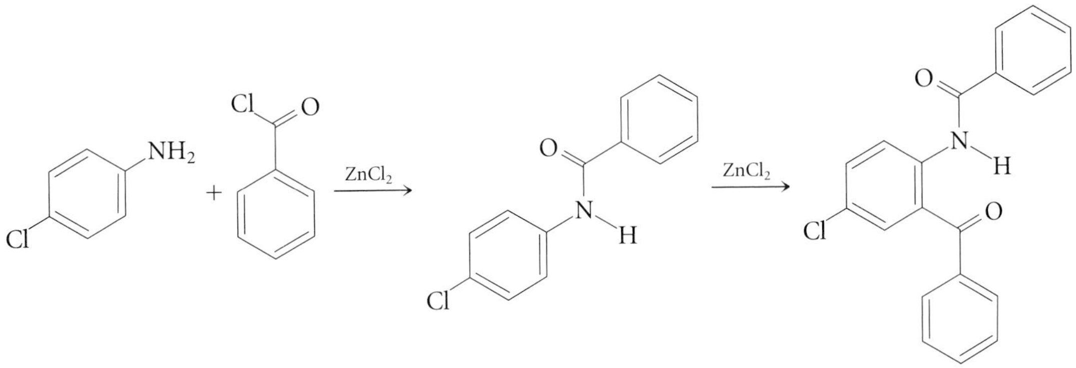

Once this occurs, the nitrogen is no longer nucleophilic, because its lone pair of electrons is delocalized into the carbonyl $\pi$ system. The aromatic ring then undergoes the desired Friedel–Crafts acylation, and, because the amide is more activating than chlorine, the reaction takes place at a position *ortho* to nitrogen. The reaction does not stop here. In the presence of zinc chloride, the substituted aniline adds to this intermediate at the keto group, producing an intermediate that attacks the carbonyl group of the amide. Although this sequence may seem complicated, it is in fact nothing more than a clever protection of the reactive nitrogen by the very reagent that is used to accomplish the desired reaction.

The concepts introduced in this chapter can be used to analyze syntheses that incorporate many of the reactions presented in Chapters 8 through 14. From this point on, you should begin to consider new reactions in terms of not only how they occur, but also what they accomplish. It will also be useful to begin to look at complex molecules from the point of view of how they might have been constructed from smaller ones.

## Summary

1. Synthetically useful reactions can be classified into three groups: carbon–carbon bond-forming reactions, oxidation–reduction reactions, and functional-group transformations.

**2.** An analysis that identifies the steps needed to transform a simple molecule into a relatively more complex one is best accomplished in a backward direction, proceeding from the ultimate product back to the starting material (retrosynthetic analysis).

**3.** A proposed pathway can be evaluated for synthetic efficiency by considering the number of steps, the yield of each step, the required reaction conditions, the ease of purification of intermediates, and the cost of reagents and personnel time.

**4.** A convergent synthesis, in which short separate routes combine to form a desired product, is generally preferred to a linear synthesis, which takes place as a sequential series of transformations.

**5.** A protecting group is a functional group that can be readily interconverted with another group but has significantly different reactivity toward common reagents. Protecting groups are used to control reactivity in synthetic intermediates when a desired conversion at one position may be incompatible with the presence of another, different group at another position. Ketals and acetals are used as protecting groups for ketones and aldehydes, benzyl ethers for alcohols, *t*-butyl esters for carboxylic acids, and urethanes for amines. Protecting groups are important in the construction of complex molecules that contain many functional groups.

## Review of Reactions

Benzyl Ether Hydrolysis

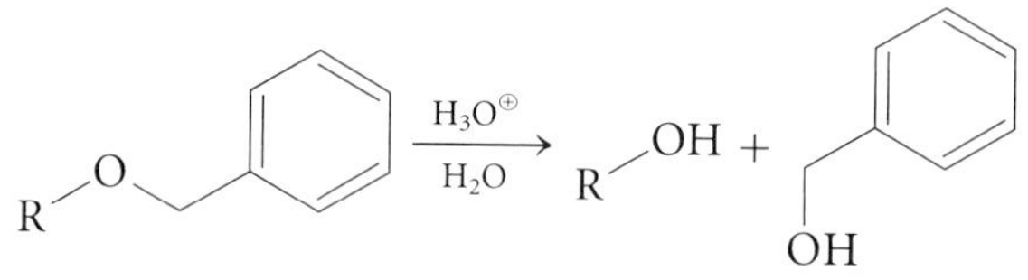

*t*-Butyl Ester Formation

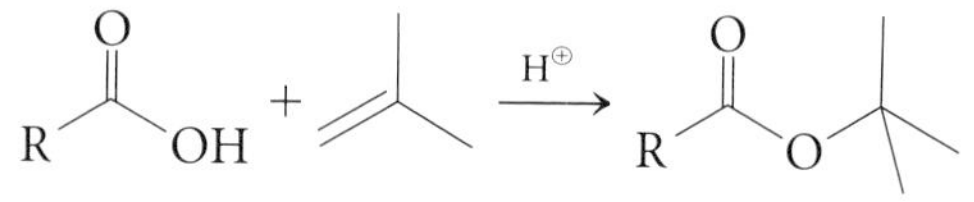

*t*-Butyl Ester Cleavage

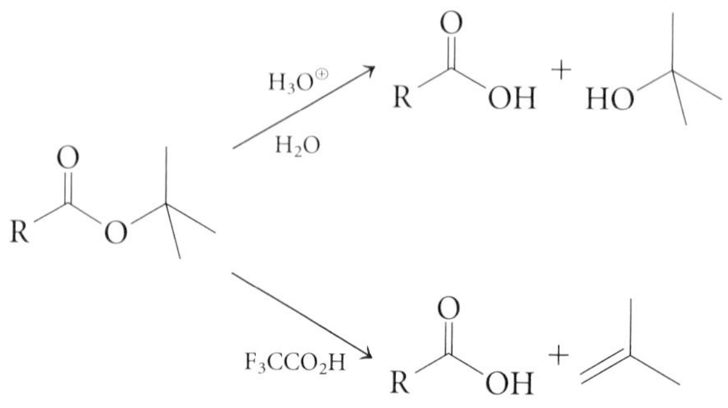

## Urethane Formation

## Urethane Hydrolysis

# Review Problems

**15.1** Classify each of the following transformations as a carbon–carbon bond formation, an oxidation–reduction, or a functional-group transformation.

**15.2** What reagents can be used to accomplish each transformation in Problem 15.1?

**15.3** Identify an efficient route for the conversion of methanol and ethanol into *t*-butyl alcohol. (You may use any inorganic reagents needed.)

**15.4** Devise a short and efficient route for the conversion of methanol and ethanol into 2-butanol:

**15.5** Each of the following syntheses requires only one reaction to accomplish the transformation. However, each starting material is bifunctional, and one group is to undergo the reaction while the other remains unchanged. This may require the use of a protecting group. Decide if a protecting group is needed, and, if so, which one can be used in each case. If a protecting group is needed, specify the reagents for the three steps necessary (protection, transformation, and deprotection).

(a)

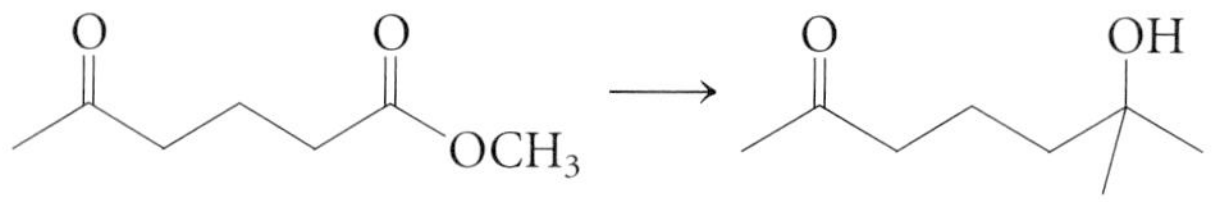

(b)

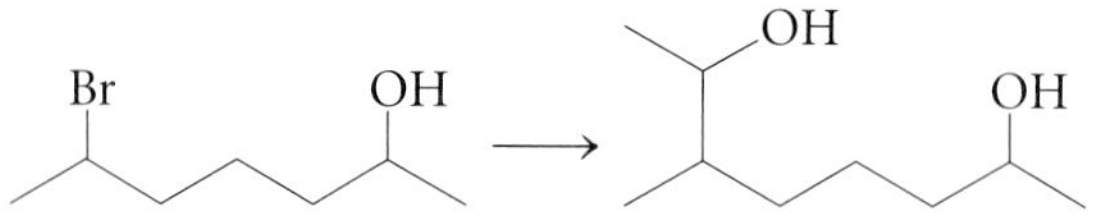

(c)

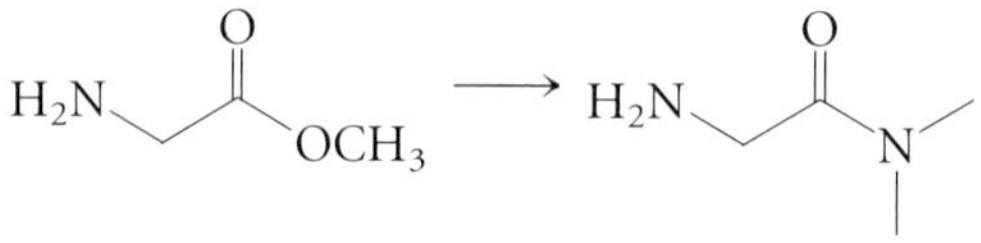

**15.6** Transforming a ketone into a ketal is the most common way of protecting this functional group. Review your understanding of the mechanisms of reactions of carbonyl groups by providing mechanisms for both the formation and the hydrolysis of the dimethyl ketal of acetone.

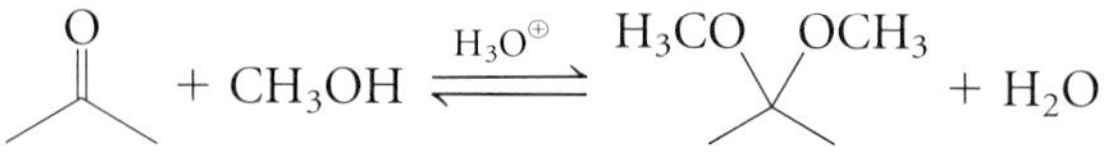

**15.7** Develop as short a synthesis as possible for *cis*-2-pentene, starting from methanol and ethanol as the only sources of carbon. (You may use any inorganic reagents needed.)

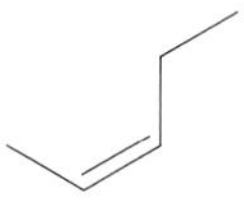

*cis*-2-Pentene

**15.8** What properties of the desired product in Problem 15.7 might lead to serious practical difficulties in the final step of its synthesis?

**15.9** How might the synthesis you developed in Problem 15.7 be modified in a simple way to prepare 2-methyl-2-pentene?

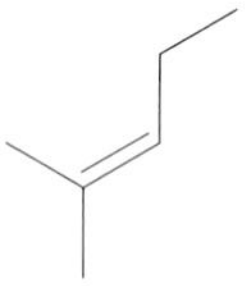

2-Methyl-2-pentene

**15.10** Develop a short synthesis of the following alcohol from starting materials that do not have rings. (*Hint:* Because a cyclic product is ultimately desired, examine Table 15.3 for reactions that make rings.)

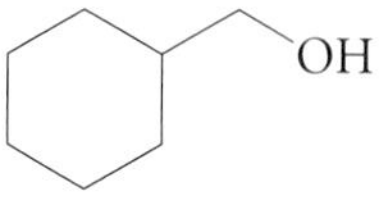

## Supplementary Problems

**15.11** A chemist plans to synthesize terfenadine by the following reaction:

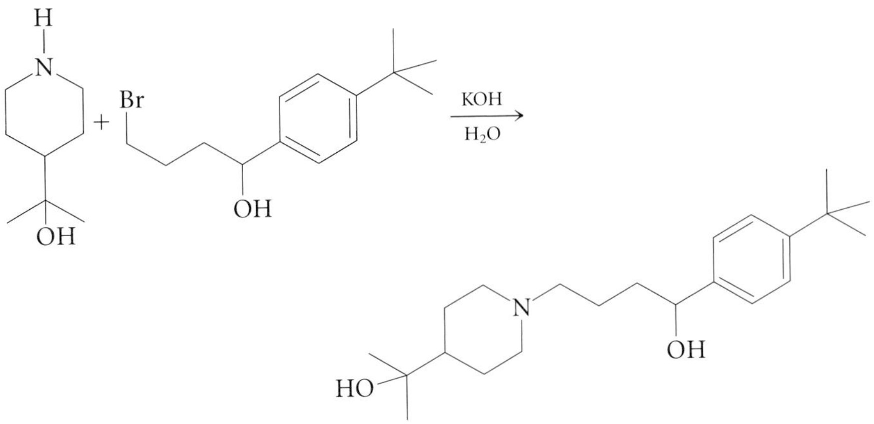


**Terfenadine**

What might go wrong with this procedure?

**15.12** The researcher mentioned in Problem 15.11 went off on spring break before trying the proposed synthesis. While catching some rays on Padre Island off the coast of Texas, the chemist devised an alternative synthesis:

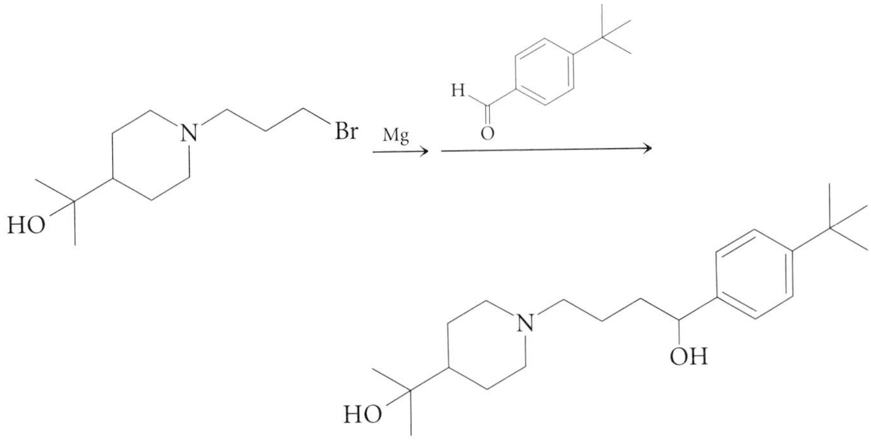


**Terfenadine**

(a) Would you advise going ahead with this reaction? If not, what might go wrong here?

(b) If you found something seriously wrong with the reaction, suggest a (relatively) simple solution so that the carbon–carbon bond-forming reaction could be used to construct the skeleton of terfenadine.

**15.13** Cyclic ethers such as tetrahydrofuran react with concentrated halogen acids (especially in the presence of Lewis acids) to form halohydrins in which one of the carbon–oxygen bonds of the starting ether has been replaced by a C—X bond.

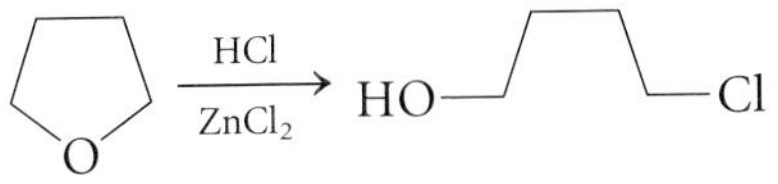

Do you think the following substituted tetrahydrofuran will yield the bromoalcohol? Why, or why not?

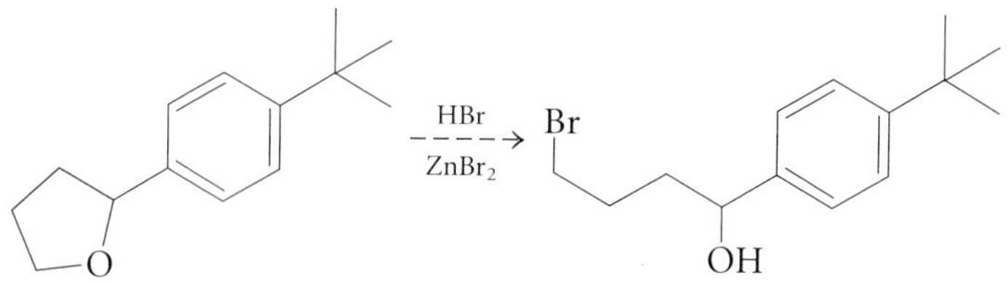

**15.14** The starting material for the synthesis of ibuprofen (Figure 15.2) is isobutylbenzene.

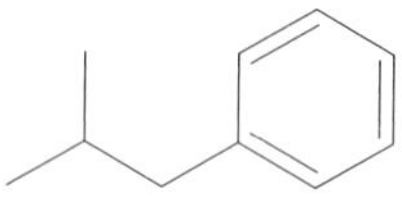

**Isobutylbenzene**

Develop a synthesis of this compound starting from benzene and using any other reagents necessary, keeping in mind what you learned about electrophilic aromatic substitution (especially alkylation) in Chapter 11.

**15.15** The starting material for the synthesis of ketoprofen (Figure 15.2) is 3-methylbenzophenone.

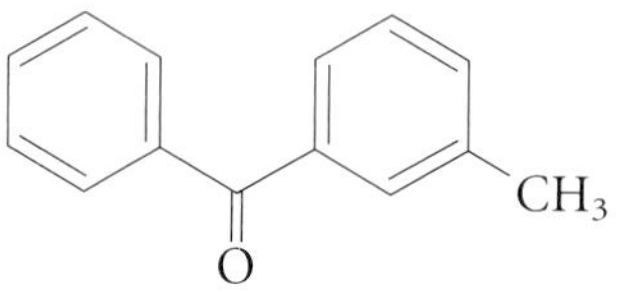

**3-Methylbenzophenone**

Develop a synthesis of this compound starting from benzene and using any other reagents necessary, keeping in mind what you learned about electrophilic aromatic substitution reactions (especially alkylation *and* acylation) in Chapter 11.

**15.16** Design a synthesis of the following compound, deriving all of the carbon atoms from reagents having no more than three carbons.

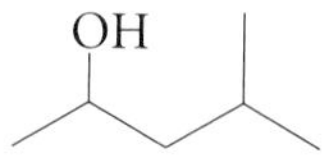

**15.17** Diol A can be synthesized from cyclopentenone using a Diels–Alder reaction in a sequence that requires only four steps:

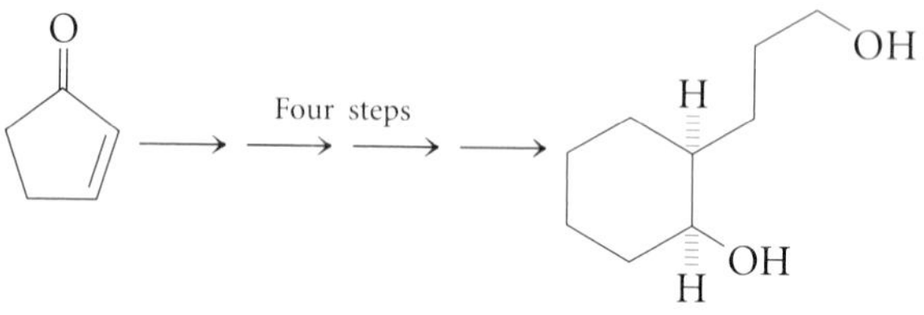

**A**

However, the synthesis is not an obvious one because of changes that occur in the skeleton of the five-member ring of the enone during the sequence. To uncover this route, start by comparing the number of carbons in cyclopentenone with the number of carbons in the diol. Use this difference to decide what the dienophile is likely to be, and then carry out the Diels–Alder reaction. Next, number the carbons of the diol (start with the carbinol carbon, work around the six-member ring, and then proceed out the side chain), and assign the same numbers to the corresponding carbon atoms of the Diels–Alder product. What are the three remaining steps in the synthesis?

**15.18** Upon treatment with acid, a diol undergoes pinacol rearrangement and forms a ketone by a 1,2-shift of a carbon substituent.

In a variant of this reaction, the diol is first reacted with methanesulfonyl chloride, converting one of the hydroxyl groups to a sulfonate ester. Upon warming in a protic polar solvent such as water, the sulfonate group is lost and the same 1,2-shift occurs as in the pinacol rearrangement. What diol could be used as the starting material in this variant of the pinacol rearrangement to prepare ketone B? Would it make a difference which of the two hydroxyl groups was converted to a sulfonate ester?

**B**

**15.19** When chemists actually carry out synthetic transformations like that in Problem 15.18, they must establish that the desired product was indeed formed and determine to what extent the starting material remains. Explain in qualitative terms how the $^1H$ and $^{13}C$ NMR spectra of the starting diol and the product ketone would differ. How would the infrared spectra differ?

**15.20** Red No. 2 is a dye once widely used to color food, especially maraschino cherries, until laboratory studies showed that it had mutagenic properties.

**Red No. 2**

The diazo linkage that joins the two naphthalene rings can be conveniently prepared by a reaction in which a diazo-substituted aromatic acts as an electrophile to effect an electrophilic aromatic substitution on another aromatic molecule.

In turn, the diazo intermediate can be prepared by reduction of a nitro aromatic and diazotization of the resulting amine. Because the two naphthalene rings of Red No. 2 are differently substituted, two combinations of reactants will form the diazo linkage of the dye. Taking into consideration the directing influence (*ortho, para,* or *meta*) of the substituents $SO_3Na$, $SO_3H$, $NO_2$, $NH_2$, and OH, develop a synthesis of Red No. 2 starting from naphthalene and 2-naphthol.

**Naphthalene**  **2-Naphthol**

(*Hint:* Consider first the two amino species that would be used for the diazo coupling reaction, and then how the necessary substituents could be introduced into the correct positions on the aromatic rings.)

**15.21** The following sequence includes all of the transformations used in a synthesis of the alkaloid retronecine, one of the pyrrolizidine alkaloids found in a variety of plants. For each of the lettered steps (a) through (j), provide the reagents required and indicate the general type of reaction involved.

**15.22** Suggest a synthesis for hexalure, a sex attractant for pink bollworm moths, using starting materials with fewer than nine carbon atoms.

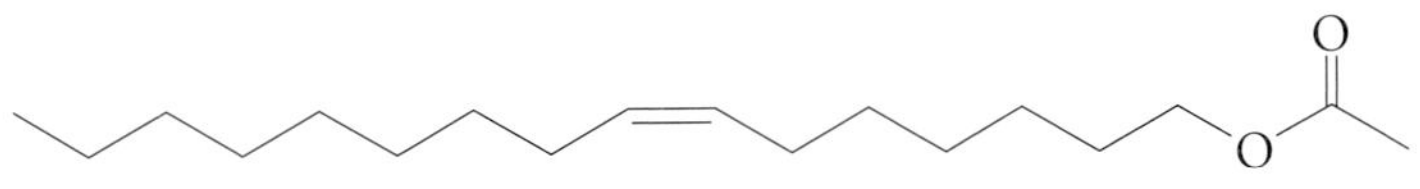

Hexalure

**15.23** A chemist was attempting to make an α,β-unsaturated ketone by treating α-chloroketone C with NaOH. However, the product obtained was a carboxylic acid.

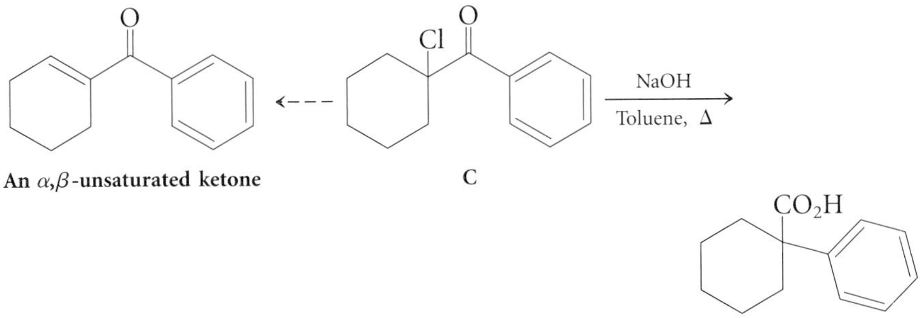

An α,β-unsaturated ketone       C

What type of reaction is involved in this transformation? Suggest a detailed, step-wise mechanism for this conversion. How might the reaction conditions be modified to obtain the desired α,β-unsaturated ketone?

**15.24** Suggest a synthesis of the herbicide (2,4-dichlorophenoxy)acetic acid (known as 2,4-D) starting from phenol and any other source of carbon and any required reagents.

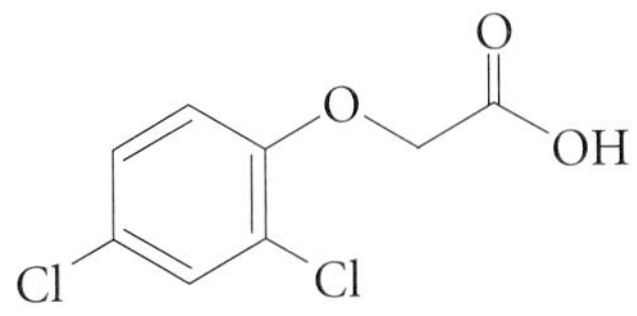

2,4-D

**15.25** Devise a synthesis of 9-ketodecenoic acid (the honeybee queen substance) starting from cycloheptanone and any other organic and inorganic compounds you need.

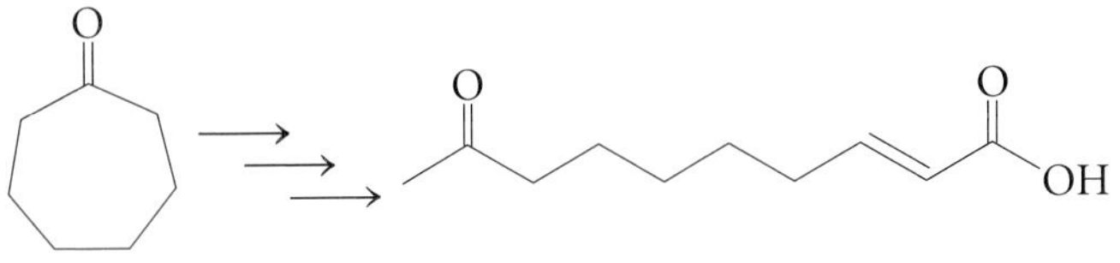

9-Ketodecenoic acid (*trans*)

**15.26** For each of the synthetic steps in Problem 15.25, decide what single spectroscopic technique you would use to determine that the desired conversion had occurred. For each step, explain briefly the differences you would expect to see between the spectrum of the reactant and that of the product.

# 16

# Polymeric Materials

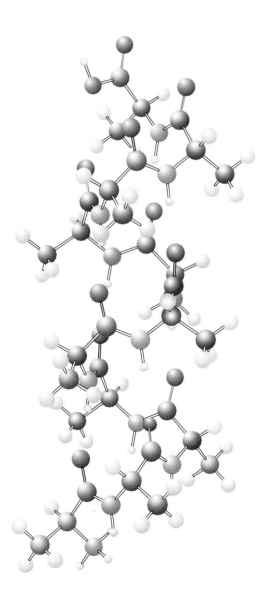

Infrared spectroscopy is a useful tool for the analysis of polymer structure. Helical polyalanine, shown as a ball-and-stick model, can be synthesized by growing the polymer on a metal surface with appropriate initiation sites. The carbonyl regions of the infrared spectra of both surface-grown and solution-grown polyalanine are shown. The carbonyl stretching frequencies for helical and β-pleated sheet arrangements of peptides are different. Only the helical polymer is formed on a metal surface; the polymer formed in solution is a mixture of helices and β-pleated sheets.

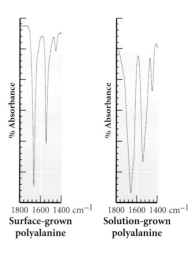

$U$ p to this point, we have studied the properties and transformations of relatively small molecules, typically with molecular weights less than 1000 and containing no more than 20 or 30 carbon atoms. However, the world is full of molecules that are tens and even hundreds of times larger. Examples range from the high-molecular-weight hydrocarbons in such materials as coal and crude petroleum to the highly specialized molecules such as enzymes and DNA that are important to living systems. When large molecules are composed of many repeating subunits, they are called **polymers,** a term derived from the Greek *polumeres* ("having many parts"). Johns Jakob Berzelius introduced this term in 1830, 22 years before the birth of Jacobus Henricus van't Hoff, who first described three-dimensional tetrahedral carbon. Berzelius was also the first to use the terms *catalyst, isomer,* and *protein,* all of which you will encounter in this chapter.

We will first examine, using a simple example, how a large molecule can be conceptually derived from repeating subunits. Then we will deal in some detail with the chemical and physical properties of polymers.

## 16.1

### Monomers and Polymers

#25    Polymers—
        Introduction

Consider the structure of poly(lactic acid) below, in which a subunit of the long, chainlike molecule is shown in blue. As you can see, this subunit is repeated along the chain. Such a repeating group is derived from a low-molecular-weight molecule called a **monomer** (here a simple hydroxyacid called lactic acid); the high-molecular-weight polymer is constructed of many monomers. Polymers are often named by adding the prefix *poly-* to the trivial name of the monomer from which the polymer is prepared. For this reason, the name of a polymer may not give an accurate picture of its structure. For example, the repeating group present in poly(lactic acid) is a carboxylic acid *ester,* even though the name ends in "acid." Quite logically, this macromolecule belongs to the class of polymers known as **polyesters.**

As you know from Chapter 12, esters can be prepared from alcohols and carboxylic acids. We can thus conceive of making this polyester from lactic acid by combining the hydroxy group of one molecule with the carboxylic acid group of another.

HO⟍⟍⟍OH  —H₂O⟶  Poly(lactic acid) structure

Lactic acid                                             Poly(lactic acid)

The process by which poly(lactic acid) is formed from lactic acid monomers is called **polymerization.** It begins with the reaction of two lactic acid molecules, the carboxylic acid functional group of one molecule reacting with the alcohol unit of the other. The product, a **dimer,** contains a newly formed ester functional group and also retains one alcohol and one carboxylic acid group.

The two functional groups present in the dimer of lactic acid are the same ones present in both monomers. Reaction of either the alcohol or the acid group of the dimer with a third molecule of lactic acid leads to a **trimer,** containing three repeating subunits and still retaining one alcohol and one acid functional group. In theory, this process can continue indefinitely, with the chain lengthening at either or both ends. This example illustrates a feature of all monomers: a pattern of functionality that makes it possible to form at least two bonds and thereby link each monomer to two others.

## EXERCISE 16.1

The dimer of lactic acid can form a cyclic structure in which a second ester linkage results from the reaction of the carboxylic acid group with the alcohol at the other end of the molecule, making a bis-lactone. Draw a structure for the bis-lactone (also known as a lactide). Write a reaction mechanism, assuming acid catalysis, for the conversion of the dimer into the lactide. (*Hint:* Recall Chapter 12.)

**Lactic acid dimer**

The concept of polymerization can be illustrated graphically in a general sense, by representing the monomer as a stick and representing the functional groups that link the monomers together as two balls. In lactic acid, the two functional groups are different and so are represented by red and blue balls:

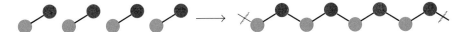

However, a monomer unit need not contain two *different* functional groups. For example, a polyester can be derived from the combination of a dicarboxylic acid and a diol. Here, each monomer unit has the same functional group at both ends, but a 1:1 pairing of the monomers is required to generate the polymer:

It is easy to visualize the monomer units required to form a polyester. The functional group in the polymer, an ester, is directly related to the two

791

functional groups from which it is made. However, as we will see, many polymers are formed by linking monomer units together with carbon–carbon bonds. It is often difficult, if not impossible, to distinguish between the carbon–carbon bonds present in the original monomer and those formed as a result of linking these units together. For example, ethylene undergoes polymerization to form long hydrocarbon chains in which the carbon–carbon $\sigma$ bonds are indistinguishable from those present in the monomer.

**Ethylene monomer units** → **Polyethylene**

## 16.2

### Linear and Branched Polymers

In the foregoing examples, each monomer unit is capable of attachment to another at either end. Thus, end-to-end linkage of these monomer units results in a **linear polymer.** Although each of the possible functional groups by which monomer units can link imparts different characteristics to the resulting polymer, linear polymers have some physical properties in common. For example, individual polymer chains associate with each other by electrostatic and van der Waals attractions. Because of the high molecular weights of polymers, the number of such attractions on a molar basis is quite large, and polymers often exist as solids or highly viscous liquids. This viscosity (or rigidity) decreases at higher temperatures. The decrease in viscosity results from progressively greater disruption of the attractive intermolecular interactions between the polymer molecules. These attractive forces can be large, even though each individual van der Waals or dipole–dipole interaction is weak, because there are a very great number of such interactions between these large molecules.

In contrast, other extremely large, multidimensional molecules have distinctly different properties from those of linear polymers. These polymers are said to be **branched,** or **cross-linked.** In such a polymer, chemical bonds interconnect chains, resulting in a complex network. One example is diamond, in which the smallest repeat unit is a single carbon that is connected to four others by carbon–carbon bonds. In fact, diamond is made under such extreme conditions of high temperature and pressure that scientists have been unable to define in detail how it is formed. Nonetheless, the bonding of each carbon to four partners in the diamond lattice results in a material that is connected in three dimensions by very strong, carbon–carbon bonds. As a consequence, diamond does not melt or soften as it is heated and is totally insoluble in all known solvents.

The structure of graphite is somewhat analogous, except that the carbons are $sp^2$-hybridized and thus each is linked to only three neighboring

**Diamond lattice**

carbons. Because of the planar arrangement of bonding to such a carbon, graphite is composed of sheets resembling fused polyaromatic arrays:

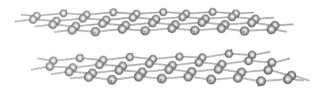

**Two sheets of graphite**

Because each sheet is planar, there is a fairly strong attraction between the carbon nuclei of one sheet and the $\pi$ electron cloud of the adjacent sheets, and graphite exhibits the same insolubility as does diamond. Although these van der Waals interactions are relatively strong, the attraction between the sheets changes relatively little as one sheet slides over another. Thus, the sheets can be moved with virtually no resistance, making graphite an excellent lubricant. Diamond and graphite are examples of three-dimensional and two-dimensional polymers, respectively, in which all of the monomer units are identical.

So far we have seen two distinct types of polymers: (1) linear polymers derived from monomer units that have only two possible attachment points, and (2) branched polymers, which are two- or three-dimensional arrays formed when each monomer can be attached to another by either three bonds, as in graphite, or four bonds, as in diamond. A three-dimensional network of chemical bonds in a polymer generally leads to a material that is harder and less flexible than the corresponding linear polymer with similar functional groups. Polymers can also be made from mixtures of different monomers, with one unit having three bonding sites and the other only two. Such a possibility can be represented schematically using blue balls to represent alcohol functional groups and red balls to represent carboxylic acids:

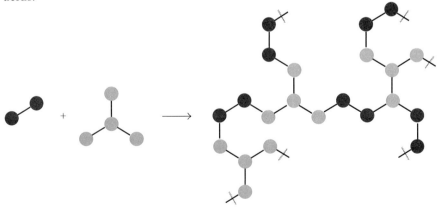

## 16.3

# Types of Polymerization

The chemical transformations that result in polymers can be divided into two major classes. The polyesters exemplify the type of material referred to as a **condensation polymer.** The reaction that forms such a polymer (in

**CHEM TV® II**

#26   Polymers Survey

this case, the reaction of an acid with an alcohol to produce a polyester) also produces a small molecular by-product (in this case, water). In contrast, in the conversion of ethylene into polyethylene, all atoms present in the monomer are retained in the polymeric product. Because the latter process consists of the addition of one molecule to another (recall electrophilic addition in Chapter 10), the resulting polymers are known as **addition polymers.**

These two types of polymerization—condensation and addition—often produce polymers whose structures and properties differ, and for this reason, they are usually treated as separate categories. However, it is sometimes possible, as with the nylons, to make quite similar polymers by either condensation or addition polymerization.

## 16.4

## Addition Polymerization

As discussed in Chapter 10, carbon–carbon $\pi$ bonds are susceptible to electrophilic attack in a process that results in the breaking of the $\pi$ bond and the simultaneous formation of a new $\sigma$ bond between the electron-deficient reagent (the electrophile) and a carbon of the original $\pi$ bond. The other carbon of the $\pi$ bond becomes a cation and is thus activated as an electrophile for reaction with a second equivalent of alkene. This interaction forms yet another carbon–carbon $\sigma$ bond, and repetition of the bond formation results in a carbon chain that continues to grow until all the alkene is consumed. Recall also that both radicals and cations are electron-deficient and therefore electrophilic: both can initiate addition polymerization.

### Radical Polymerization

The reaction of a radical with ethylene results in the formation of a C—X $\sigma$ bond at the expense of the C—C $\pi$ bond:

$$\Delta H^\circ = -20 \text{ kcal/mole}$$

Because the product of this initial reaction is itself a radical, it is capable of adding to yet another molecule of the alkene in a process that regenerates a carbon-centered radical, with a net change in bonding from one $\pi$ to one $\sigma$ carbon–carbon linkage. Thus, each step in this polymerization process is exothermic by approximately 20 kcal/mole, the difference in energy between a carbon–carbon $\pi$ and a carbon–carbon $\sigma$ bond. Each addition of a radical to ethylene lengthens the growing polymer chain by two carbons. The overall process is called **radical polymerization.**

The polyethylene made by radical polymerization can have a molecular weight ranging from 14,000 to 1,400,000 (corresponding to between 500 and 50,000 ethylene monomer units). Because the attractive van der Waals interactions holding these chains loosely together are much weaker than covalent linkages, the chains (like the sheets of graphite) can move relative to each other.

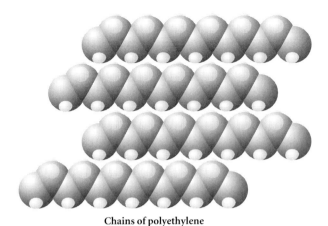

**Chains of polyethylene**

Thus, polyethylene and most other linear polymers are somewhat flexible. (Some linear polymers, such as the polymer used to make rubber bands, are especially flexible or "stretchy.") Flexibility increases with temperature, because the greater kinetic energy of the molecules permits the disruption of the attractive intermolecular interactions. It is thus possible to form polymers such as polyethylene into various shapes by softening them by heating and then molding them while they are relatively soft. Such materials are called **plastics** (a term derived from the Greek *plastikos,* meaning "fit to be molded").

Radical polymerization of ethylene need not go on indefinitely. As we have seen in the free-radical halogenation of alkanes in Chapter 7, processes that simultaneously consume two radicals terminate a radical chain reaction. This is also true of radical polymerization, which can be terminated by the reaction between the radical at the growing end of the polymer chain and another such radical (or the radical chain initiator).

### EXERCISE 16.2

How might the relative probabilities of the two termination reactions (combination of two growing chains versus combination of a chain with an initiator) change as ethylene is converted into polyethylene?

### EXERCISE 16.3

In the polymerization of ethylene, abstraction of a hydrogen atom from a position adjacent to the growing end of a chain by another carbon radical results in termination, with the formation of an alkene. Write a mechanism for this process. How many chains are terminated by this abstraction? Would you expect this reaction to be exothermic or endothermic?

## ▦  Ionic Polymerization

Ethylene and other alkenes also undergo polymerization by processes in which the growing end is either a carbocation (in **cationic polymerization**) or a carbanion (in **anionic polymerization**).

**Cationic Polymerization**

**Anionic Polymerization**

Cationic polymerization and radical polymerization are quite similar in that both take place through the interaction of an electron-deficient (and therefore electrophilic) species (either the initiator or a carbocation) with the $\pi$-electron density of an alkene. At first inspection, the reaction of a simple carbon–carbon $\pi$ bond with an anion, especially a carbanion, appears to be an unusual reaction, because it requires the interaction of an electron-rich species with a $\pi$ bond, which also has high electron density. (Recall from Chapter 11 that the characteristic addition reaction of alkenes is electrophilic addition.) However, keep in mind that, regardless of how ethylene is converted into a polymer, the thermodynamics of this conversion remains the same. Thus, all radical, cationic, and anionic polymerizations are exothermic by approximately 20 kcal/mole for each ethylene molecule incorporated into the polymer.

Anionic polymerization of ethylene is initiated by the addition of an alkyllithium reagent (such as butyllithium) to ethylene. Unlike the free-radical process, anionic (or cationic) polymerization of ethylene does not have termination steps because one carbanion does not react with another (nor do cations react with each other). As a result, the growing anionic chain is sometimes called a *living polymer,* meaning that the end of the polymer chain is chemically active until a quenching terminator is added. Polyethylene formed by anionic polymerization thus has much longer chains and a correspondingly higher molecular weight than polyethylene formed by radical polymerization. These longer chains intertwine to a greater extent than those of the lower-molecular-weight polyethylene; the polymer formed by anionic polymerization is therefore more rigid and has a higher density than the polymer formed by radical polymerization. Because the interactions between the various polymer chains in polyethylene are stronger than the interactions of each chain with solvent, polyethylene is a highly insoluble plastic.

*Polyethylenes.*  A variety of plastics are made by polymerization of substituted ethylenes. For example, **chloroethylene** forms the polymer **poly(vinyl chloride),** or **PVC:**

Vinyl chloride    Poly(vinyl chloride)
(PVC)

**Styrene** is converted into **polystyrene:**

Styrene          Polystyrene

Poly(vinyl chloride) is used in many applications, from plastic bags to water pipes, and polystyrene is the plastic in **Styrofoam.**

## EXERCISE 16.4

Explain why polystyrene polymerizes with a head-to-tail orientation (that is, with C-1 of one monomer attached to C-2 of its neighbors) rather than head-to-head. (*Hint:* Recall from Chapter 6 the factors that affect radical, carbocation, and carbanion stability.)

Head-to-tail          Head-to-head

Some other substituted ethylene monomers and the resulting linear polymers are shown in Figure 16.1. Each of these polymers has unique properties that depend on the functional group(s) present. For example, **Teflon** is an inert plastic because of the absence of carbon–hydrogen bonds and the high strength of the carbon–fluorine bond (110 kcal/mole). It is unreactive with all reagents except molten lithium, sodium, or potassium. Teflon is also very "slippery" because the surface of the polymer has only fluorine atoms exposed. Because of the high electronegativity of fluorine, these atoms do not participate in significant attractive interactions with other groups.

Teflon

Chlorotrifluoroethylene          Kel-F          Tetrafluoroethylene          Teflon

Methyl methacrylate          Plexiglas, Lucite          Methyl cyanoacrylate          "Instant Glue"          Acrylonitrile          Orlon

## FIGURE 16.1

Many common polymers are formed from vinyl monomers.

### SERENDIPITY IN SCIENCE

Chance discoveries have played major roles in the development of both science and technology. For example, R. J. Plunkett, a chemist at DuPont, was working with tetrafluoroethylene, taking samples for experiments from a metal cylinder in which the gas was stored. One day, to his surprise, no gas came from the cylinder when the valve was opened, even though he was sure that he had not used all of the gas. Indeed, upon (carefully) cutting open the cylinder with a metal saw, he discovered a white powder, the first sample of the polymer now known as Teflon. The best scientists are indeed those with inquiring minds, who always try to understand unexpected results.

The presence of the carbomethoxy group in **Plexiglas** imparts a high index of refraction, as well as optical clarity, to this material, which is therefore used as a lightweight replacement for glass. The carbomethoxy group in Plexiglas also contributes to the high solubility of this material in organic solvents, especially those with carbonyl groups, such as acetone and ethyl acetate. As a result of this high solubility, the smooth, transparent surface of Plexiglas is marred if it is splashed with any of these solvents. The surface layers of the plastic dissolve and then redeposit unevenly as the solvent evaporates.

### EXERCISE 16.5

(a) For one of the polymers in Figure 16.1, the monomer is not the shortest repeating structural subunit. Identify this polymer. (b) Is it possible to determine in which direction polymerization occurred by examining a middle section of a polymer? Why, or why not?

***Butadiene Polymers: Rubbers.*** Butadiene and other conjugated dienes also form polymers. For example, the polymer derived from **isoprene** (2-methylbutadiene) is synthetic rubber. The rubber tree and several other plants produce a similar natural polymer known as **latex rubber,** which differs from synthetic rubber in that essentially all of its double bonds have the *cis* configuration. **Chloroprene** (2-chlorobutadiene), a monomer structurally similar to isoprene, leads to a synthetic rubber called **neoprene.**

Isoprene    Synthetic rubber

Chloroprene    Neoprene

## NATURAL POLYENES

Because of the ease with which conjugated dienes undergo polymerization, they are relatively rare in plants. Notable exceptions of highly conjugated polyenes found in plants are vitamin A, which is essential for vision, and enanthotoxin from the water dropwort, the most poisonous plant in England.

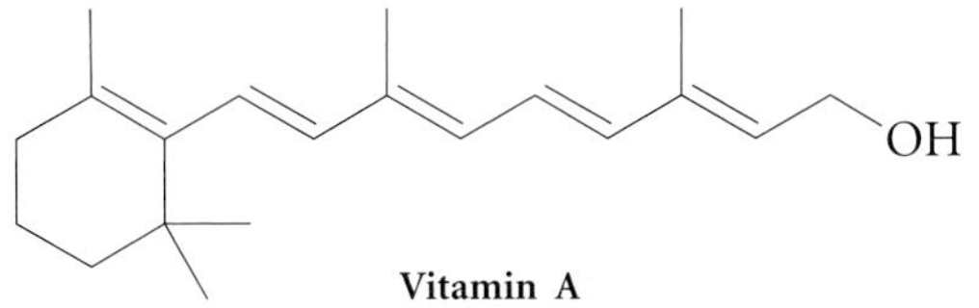

Vitamin A

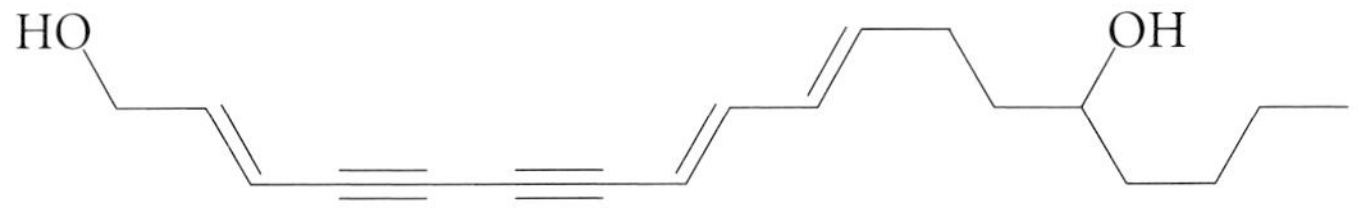

Enanthotoxin

Replacement of the methyl group of isoprene with a chlorine atom reduces the ability of neoprene to associate with hydrocarbons, and thus neoprene is more resistant to gasoline and oils than is synthetic rubber.

Although synthetic and naturally occurring rubbers are similar in many ways, some of their properties differ. The synthetic polymers consist of a very diverse mixture of structures, whereas the structure of the natural polymer is very consistent. Radical polymerization begins by formation of a bond between a radical initiator and one of the terminal atoms, resulting in a stabilized allylic radical:

Resonance-stabilized allylic radical

Polybutadiene

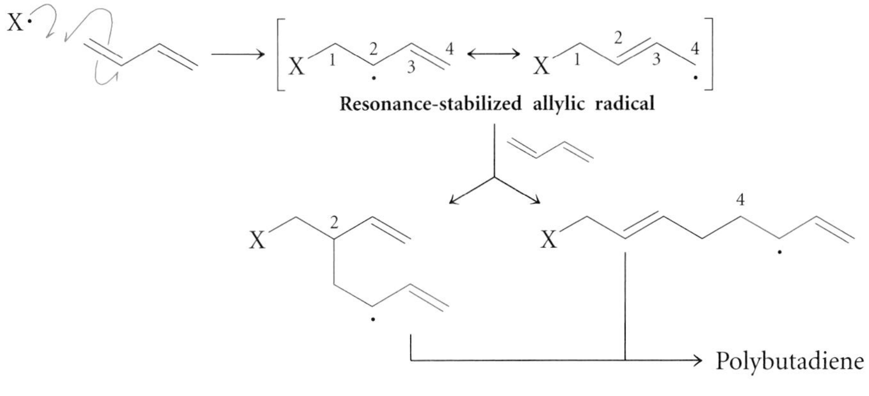

Two sites in an allylic radical bear radical character. Therefore, addition of this species to another butadiene molecule occurs with two different regiochemical outcomes: bond formation occurs either at C-2 or at C-3. Two dimers are formed, one linear and one branched. As *each* additional butadiene molecule adds to a growing chain, there are two possible outcomes: linear or branched addition. Furthermore, both *cis* and *trans* geometrical

NMR AND RUBBER BANDS

Typically NMR spectra are obtained using compounds dissolved in a solvent. However, such spectra can be run on solids. For example, the $^{13}C$ NMR spectrum below was obtained from small pieces of a latex rubber band ($D_2O$ was added to the tube to help stabilize the instrument). The spectrum shows five absorptions, one for each of the unique types of carbon atom present in the latex rubber. The signals are quite broad, in part because there are many different conformations in the solid, which are not free to interconvert as they would be in solution. When a rubber band is stretched, the polymer chains elongate as *gauche* conformations are converted into *anti* arrangements.

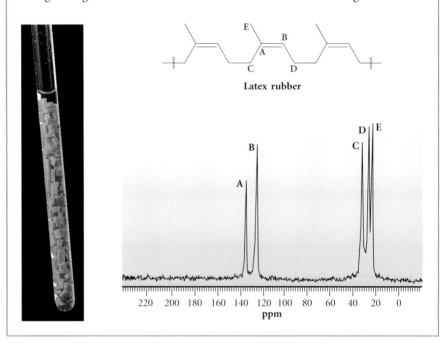

isomers will be formed, further expanding the number of different species formed. Thus, synthetic rubber is a quite complex mixture.

In contrast to synthetic rubber, latex rubber is produced in a living plant through catalysis by enzymes. These natural catalysts produce a polymer in which essentially all of the double bonds have the *cis* geometry. For this reason, there is less structural diversity among the polymer molecules found in natural rubber.

## Cross-Linking in Polymers

Most of the polymers described so far are derived from monomers with two bonding sites. As a result, the polymer chains are held together only by relatively weak van der Waals attractive interactions. In contrast, glass is a rigid polymeric material in which tetravalent silicon atoms provide a highly interconnected, three-dimensional covalent network:

**Glass**

**Cross-linking** is a process in which a bifunctional molecule (such as a diene) participates in polymerization and is incorporated into two separate polymer chains. Polymerization of a mixture of a simple alkene (such as styrene) with a diene (such as *p*-divinylbenzene) allows each of the two alkene units in the diene to be incorporated into a separate chain, thus linking the chains together. A molecule such as divinylbenzene can be viewed as a monomer capable of forming four bonds and thus of establishing cross-links between growing polymer chains.

The extent of cross-linking attained depends on the relative concentrations of styrene and divinylbenzene. But, even if divinylbenzene constitutes only a very small percentage of the mixture, the resulting plastic is much more rigid than polystyrene itself because of additional covalent bonds between the chains.

## EXERCISE 16.6

Draw a short section of polyisoprene consisting of three isoprene units (15 carbons) in which (a) all the double bonds are *trans* and (b) all the double bonds are *cis*. How would you expect the properties of either type of polymer to differ from one that was a random mixture of geometric and regiochemical isomers?

   Diene polymers such as latex or neoprene rubber can also be made more rigid by cross-linking with nonconjugated dienes. One such linking agent is **5-vinylnorbornene,** large quantities of which are produced each year by the Diels–Alder reaction of butadiene with cyclopentadiene:

Cyclopentadiene      Butadiene        5-Vinylnorbornene

The strained double bond of 5-vinylnorbornene is more reactive than the monosubstituted alkene substituent, and thus the former is incorporated into individual chains while the majority of the latter remains unreacted. When polymerization has consumed most of the available alkene, the less reactive double bonds participate in cross-linking of polymer chains. In this way, cross-linking is delayed until late in the polymerization, resulting in quite long chains. The inclusion of 5-vinylnorbornene in a monomer mixture results in a more rigid polymer by enhancing the degree of cross-linking between chains.

Alternatively, polymer chains can be linked by sulfur bridges. Transformation of the rather gummy and soft latex rubber into the much more rigid material that is used, for example, in automobile tires is accomplished by heating with sulfur in a process known as **vulcanization,** which was discovered by accident by Charles Firestone. (The term refers to Vulcan, the Roman god of fire, who was thought to be very strong.) The reaction almost certainly involves radicals, but the mechanism is not well understood. The bridges are depicted here as resulting from simple allylic substitution of sulfur for hydrogen.

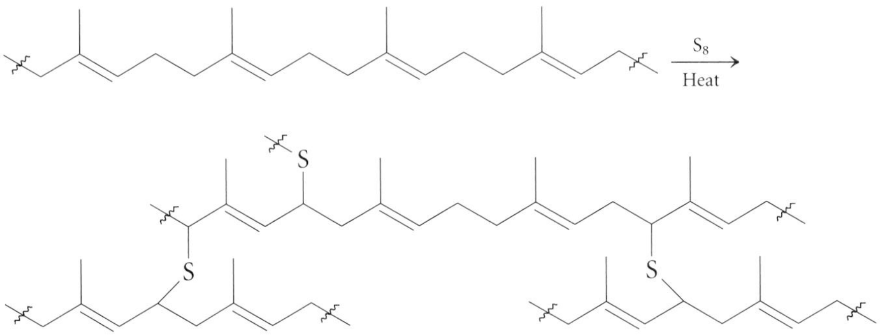

Neither the original polymer nor the cross-linked rubbers derived from isoprene have any functionality that produces color. The characteristic black color of the rubber is caused by the presence of **carbon black,** a material similar to graphite, which acts as a lubricant and imparts a greater lifetime to the rubber under conditions of repeated flexing. Large quantities of carbon black are produced in Eastern Europe. Unfortunately, the outdated technology in use in this region has resulted in the emission of substantial amounts of carbon black into the atmosphere, causing significant industrial pollution. Indeed, the proverbial "black sheep" can be found there in great abundance.

### ■ Heteroatom-Containing Addition Polymers

Substituted vinyl polymers have very low solubility in water because they lack any functional group that can form hydrogen bonds. Now we turn to polymers that incorporate oxygen—in the form of alcohol or ether functional groups. As discussed in Chapter 3, the attachment of heteroatoms to a carbon framework increases solvent–solute interactions, enhancing the solubility of the heteroatom-containing compound.

***Carbon-Linked Monomer Units: Polyols.*** The polyol known as **poly(vinyl alcohol),** or **PVA,** is highly water-soluble. Its name might seem to imply that this polymer is made from vinyl alcohol, but this enol is not present in significant amounts in equilibrium with its much more stable keto form, acetaldehyde.

Radical polymerization of vinyl acetate is used to form poly(vinyl acetate), a polymer with a hydrocarbon chain substituted with acetate esters. The esters are cleaved by acid- or base-catalyzed reaction with methanol, forming methyl acetate and PVA. The resulting polymer is much more soluble in water than are hydrocarbon polymers such as polyethylene and polystyrene. Aqueous solutions of polyols have a higher viscosity but a lower surface tension than pure water. (Viscosity is increased because the long polymer molecules are quite viscous, and the surface tension is reduced because the hydrogen bonding network of water is partially disrupted by the polymer.) Poly(vinyl alcohol) is therefore included in products ranging from hair sprays and styling gels to lubricants for molding rubber.

*Heteroatom-Linked Monomer Units.* The addition polymers discussed so far have carbon–carbon bonds linking the monomer units. However, there are also important classes of addition polymers in which the monomer units are linked by heteroatoms.

*Polyethers.* A very important class of addition polymers is formed by the addition polymerization of simple epoxides. The addition of a nucleophile such as hydroxide ion to ethylene oxide results in ring opening and yields the monoanion of ethylene glycol. (We studied a similar reaction of ethylene oxide with Grignard reagents in Chapter 8.) This ion can also serve as a nucleophile, reacting with another molecule of ethylene oxide.

This polymerization is quite similar to the anionic polymerization of ethylene. Each step in this ring-opening polymerization releases the ring strain of a three-member epoxide ring and is thus exothermic by approximately 25 kcal/mole.

The resulting polyether is called **poly(ethylene glycol),** or **PEG.** Synthetic PEGs are used in cosmetic creams, lotions, and deodorants and in antistatic agents. These plastics are marketed commercially as Carbowaxes and under other trade names. They have high water solubility because of hydrogen bonding to the ether oxygen atoms and have many of the same applications as poly(vinyl alcohol). Naturally occurring cyclic polyethers are important in the biological transport of cations across membranes.

## EXERCISE 16.7

Poly(vinyl alcohol) and poly(ethylene glycol) are similar in that both have oxygen-containing functional groups. Assuming nearly equal molecular weights, which of these polymers would have the greater solubility in water? In a hydrocarbon solvent? Which would be more viscous? Explain your reasoning.

*Polyacetals.* Another oxygen-containing polymer is **paraformaldehyde,** a polyacetal formed by the addition polymerization of formaldehyde, a reaction that takes place as an aqueous solution of formaldehyde is concentrated.

**Paraformaldehyde**

Note that the backbone of the polymer is formed of acetal linkages and that both ends are hemiacetals, which are active sites for further chain growth. Note, also, that all the atoms of the monomer are retained within the polyacetal, making it an addition polymer.

This addition polymerization is only slightly exothermic. As a result, paraformaldehyde undergoes depolymerization in water, reforming formaldehyde. Because formaldehyde is a strong antibacterial agent, aqueous solutions of paraformaldehyde are used as disinfectants, and the polymer is the active ingredient in some contraceptive creams.

Polyacetals made from formaldehyde, as well as polyacetals and polyketals made from other aldehydes and ketones, are strong plastics that are resistant to fatigue (breaking after repeated flexing) and have high electrical resistance, making them quite useful as components for computer hardware and automobile parts. To stabilize these materials toward hydrolysis, the hemiacetals (or hemiketals) at the ends are "capped" by reaction with either acetic anhydride (to form an ester) or ethylene oxide. Polyacetals (or polyketals) require more strongly acidic or basic conditions for their hydrolysis than does paraformaldehyde.

## EXERCISE 16.8

If you buy "formaldehyde" from a chemical supplier, it is supplied as a trioxane, a cyclic trimer. Write a clear, detailed reaction mechanism showing all steps in the conversion of formaldehyde into this trimer, catalyzed by a protic acid such as HCl. Can the reaction also be catalyzed by base? Why, or why not?

# 16.5

# Condensation Polymers

Unlike an addition reaction, in which all atoms of the reactants are incorporated within the product, a condensation reaction forms a more complex organic molecule from two less complex ones, with the expulsion of a small molecule. Some examples of condensation reactions include the acid-catalyzed esterification of a carboxylic acid discussed in Chapter 12 and the aldol and Claisen condensations discussed in Chapter 13; in all of these, water is formed as a by-product. When such a reaction is repeated many times with an appropriately functionalized monomer, a condensation polymer is formed.

## ▩ Polyesters

Because the interaction of dicarboxylic acids with diols to form polyesters produces one equivalent of water for each link formed in the polymer chain, polyester formation is a condensation polymerization. If esters are used in place of the carboxylic acids, the reaction is a transesterification. **Dacron** is a commercially important condensation polymer used as a fiber and also sold as a film called Mylar. It is formed by the reaction of dimethyl terephthalate with ethylene glycol, a transesterification in which methanol is produced as a by-product:

Dimethyl terephthalate     Ethylene glycol

Dacron, Mylar

**Polycarbonates** are condensation polymers that also result from transesterification. A diol often employed in this reaction is **bisphenol A** (the A in the name comes from acetone), produced by reaction between phenol and acetone in the presence of a Lewis acid (see page 806). Because phenols are better leaving groups than are aliphatic alcohols, diphenyl carbonate is more reactive in transesterification than is, for example, dimethyl carbonate. The polycarbonate formed from bisphenol A has many of the same properties as Plexiglas or polystyrene. All of these materials have high optical clarity, but the polycarbonate is much stronger and more rigid, with great impact resistance.

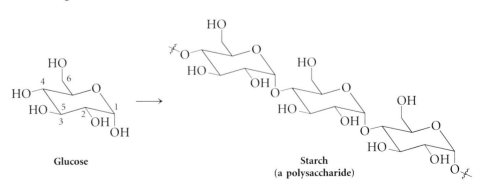

**Phenol**    **Acetone**                    **Bisphenol A**

**Polycarbonate**

### EXERCISE 16.9

Write a complete, detailed reaction mechanism for the conversion of phenol and acetone into bisphenol A in the presence of a protic acid. (Can you guess what bisphenol B is?)

### EXERCISE 16.10

Write a mechanism for formation of a polymer in the presence of hydroxide ion as base from (a) bisphenol A and diphenyl carbonate and (b) bisphenol A and dimethyl carbonate. Explain why the reaction with diphenyl carbonate is more efficient.

## Polysaccharides

#17    Di/Polysaccharides

Like polyesters, polysaccharides contain oxygen atoms within the polymer backbone. In naturally occurring **polysaccharides,** the monomer units are joined by acetal linkages. A hemiacetal group of one monomer condenses with an alcohol group of another, and water is eliminated. The monomer units of polysaccharides are **saccharides,** often called **sugars** or **carbohydrates.** The latter name arose from the fact that their molecular formulas can be written as $C_m(H_2O)_n$, suggesting structures that are hydrates of carbon. Polysaccharides have a variety of essential biological functions and are derived from simple sugars such as glucose. For example, starch is formed mainly from the reaction of the hydroxyl group at C-4 of one glucose molecule with the hemiacetal at C-1 of another:

**Glucose**                    **Starch**
                              **(a polysaccharide)**

Note that the cyclic form of glucose is a hemiacetal. In fact, this form is more stable than the corresponding open-chain hydroxyaldehyde.

As discussed in Chapter 12, the reaction of a hemiacetal with an alcohol produces an acetal and a molecule of water. In a similar reaction, two glucose molecules form a disaccharide. Thus, starch and related polysaccharides are condensation polymers.

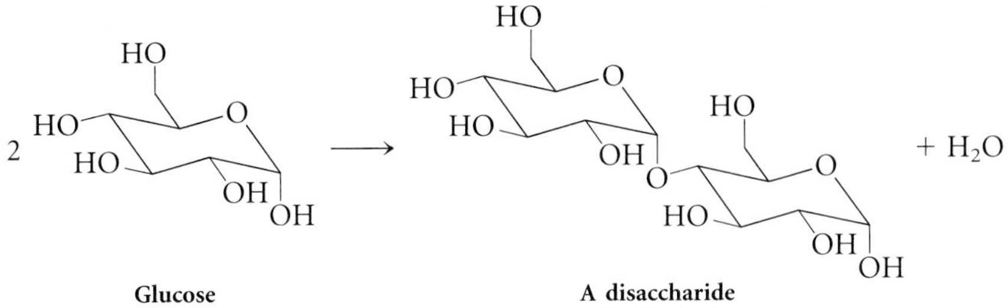

Glucose
(C₆H₁₂O₆)

A disaccharide
(C₁₂H₂₂O₁₁)

## EXERCISE 16.11

There are two possible pathways for the conversion of a hemiacetal such as glucose into an acetal by reaction with an alcohol and acid. Using the simple cyclic hemiacetal shown here, write a complete, detailed reaction mechanism for the formation of the cyclic acetal by a process that (a) cleaves the carbon–oxygen bond within the six-membered ring or (b) cleaves the carbon–oxygen bond of the hydroxyl group.

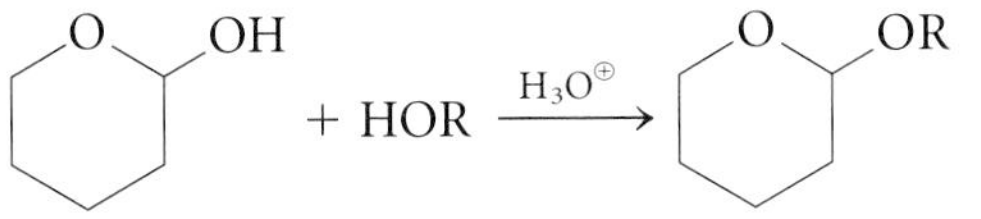

The presence of many oxygen atoms, as acetals and as hydroxyl groups, makes starch (also known as **amylose**) very water-soluble despite its high molecular weight. (It can comprise as many as 4000 glucose units.) In contrast, **amylopectin** is a water-insoluble starch consisting of as many as a million glucose units with occasional cross-links at C-6 hydroxyl groups. Its highly branched structure and very large size make it quite insoluble.

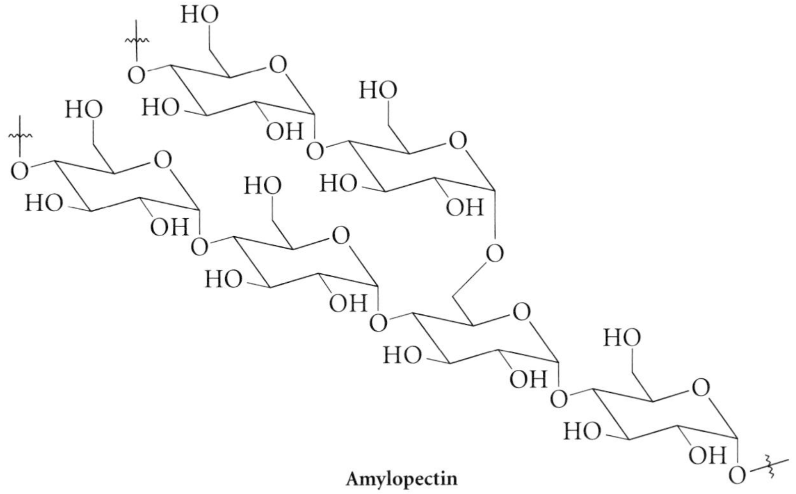

**Amylopectin**

Cellulose is a biological polymer that differs in structure from starch mainly in the stereochemistry of the linkage between glucose units.

**Cellulose**

Cellulose generally has 3000 to 5000 glucose units and is thus similar in size to amylose. However, unlike amylose, cellulose is water-insoluble. As a direct result of the stereochemistry at the linkage of one glucose unit to another, the polymer chains of cellulose fit together much better than do those of starch. Starch is used by plants as a storage medium for glucose and must be accessible to individual cells to meet the metabolic requirements of the plant, whereas cellulose is mainly a structural material that must withstand rain, humidity, and so forth. Thus, polymers with very different biological functions are produced from a common monomer, with the difference in properties resulting from a structural difference.

The reaction of cellulose with acetic anhydride converts many of the hydroxyl groups to acetate esters. Because of this change in functionality, the properties of cellulose acetate are quite different from those of cellulose. Besides possessing optical transparency, this modified natural polymer is more soluble in organic solvents than is cellulose and can be processed into thin sheets. Photographic film is one commercial application of cellulose acetate.

## Polyamides

A polymeric condensation product formed from a diacid and a diamine is called a **polyamide.** In a route parallel to that for polyester formation, adipic acid and 1,6-diaminohexane react (at high temperature) to form a polyamide known as **nylon 66.** (The numbers refer to the number of carbon atoms in the monomeric units.)

**Adipic acid**　　　　　**1,6-Diaminohexane**

**Nylon 66**

The designation 66 distinguishes this polymer from nylon 6, produced by the polymeric ring opening induced by the catalytic interaction of a nucleophile with caprolactam.

Cyclohexanone
oxime

Caprolactam

Nylon 6

This seven-member cyclic amide (a lactam) is prepared industrially by the Beckmann rearrangement of cyclohexanone oxime in sulfuric acid (as solvent). To isolate the product, the acid must be neutralized, often with ammonia. The by-product, ammonium sulfate, is sold as fertilizer.

The terms *condensation* and *addition* refer to the methods by which polymers are produced and therefore do not necessarily provide clues to the character of a polymer. Nylon 66 is produced by the condensation polymerization (with loss of water) of two different six-carbon monomers, one a diacid and the other a diamine. Nylon 6, on the other hand, is the result of the addition polymerization of a single six-carbon monomer (the cyclic amide). Both nylon 66 and nylon 6 are useful plastics that form very long-lasting and flexible fibers; the flexibility is due, in part, to the conformational freedom of the chains.

Condensation polymerization of terephthalic acid with *p*-phenylenediamine results in a polymer known as **PPTA** (short for *p*-phenyleneterephthalamide), which has very unusual properties.

*p*-Phenylenediamine        Terephthalic acid

PPTA

In contrast with nylon 6 and nylon 66, which have considerable conformational freedom, PPTA is quite rigid. As a result, this polymer can be formed into fibers that have great tensile strength and resist both compression and elongation. PPTA is five times stronger than steel on a per-weight basis. Fibers spun from this polymer are marketed as Kevlar and are used in applications such as bulletproof vests that require high strength and low weight.

## EXERCISE 16.12

Consider the reaction of acetic acid with methylamine to form *N*-methylacetamide. Could the reaction be either acid- or base-catalyzed? Explain.

Write mechanisms for the Beckmann rearrangement that forms caprolactam and for the ring-opening polymerization of this lactam to yield nylon 6.

## Polypeptides

The amide linkage in a polyamide resembles those found in polypeptides and proteins, which are natural polymeric materials derived from $\alpha$-amino acids. This linkage is referred to as a *peptide bond*, and thus the term *polypeptide* explicitly refers to the repeating amide linkage. In polypeptides (and proteins), the monomer units are linked by amide groups formed by a condensation reaction between the amino group of one $\alpha$-amino acid and the carboxylic acid group of another.

$\alpha$-**Amino acids**                    **Dimer**

**Polypeptide**

Polypeptides and proteins are naturally occurring polymers that have many different forms and a variety of functions. The distinction between the terms *polypeptide* and *protein* relates to molecular size. A polypeptide is a polymer containing no more than 100 amino acid subunits, and a protein is a larger molecule.

The synthetic (manufactured) polyamides were developed to mimic the properties found in silk and in animal hairs such as wool, both composed mainly of polypeptides. **Silk** is a polymer of the amino acids glycine and alanine. **Wool** is structurally more complex, having sulfur–sulfur bonds that link individual chains one to another and form a matrix somewhat like that of vulcanized rubber. These bonds in wool are the result of significant amounts of the sulfur-containing amino acid cysteine.

### MEDICAL APPLICATIONS OF POLYMERS

Biologically compatible polymers are an increasingly important type of specialty polymer. The polymer used in making contact lenses must be quite hydrophilic to permit easy lubrication of the eye. Therefore, a hydrogel in which free alcohol groups are attached to a poly(methyl methacrylate) is used. Polymers for making dental impressions, on the other hand, should not form local hydrous pockets but must be moderately hydrophilic in order to wet the oral tissue effectively. These polyethers will solidify, because they are capped with groups that can react by a ring-opening polymerization (like that discussed in this chapter for the polymerization of epoxides) to form cross-links.

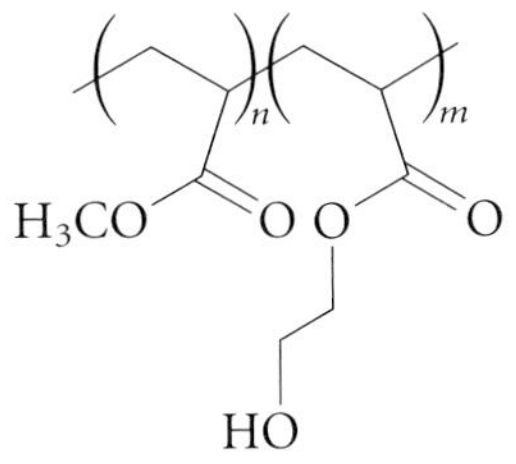

**Hydrogel**

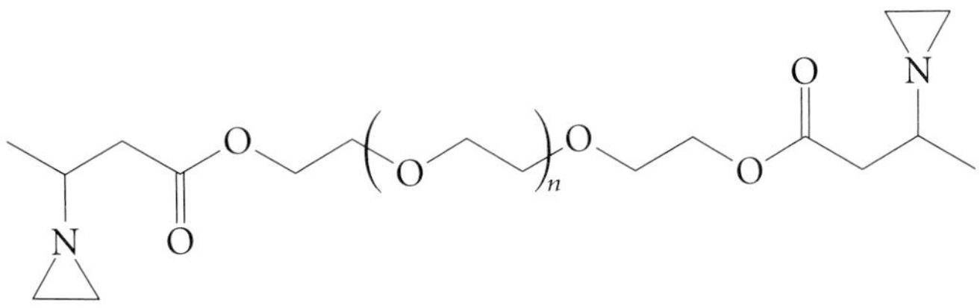

**Polymer for dental impressions**

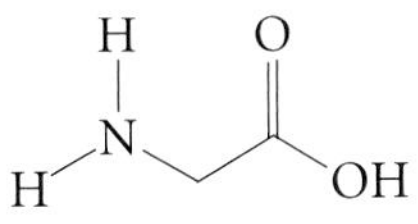

Glycine

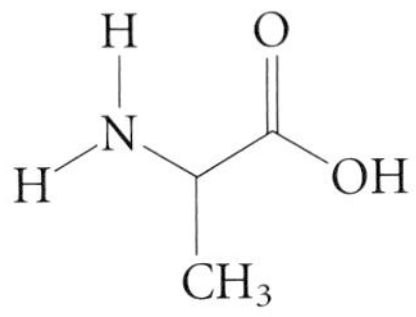

Alanine

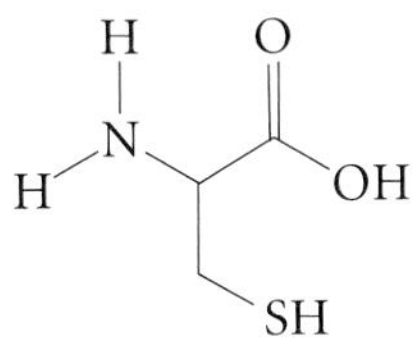

Cysteine

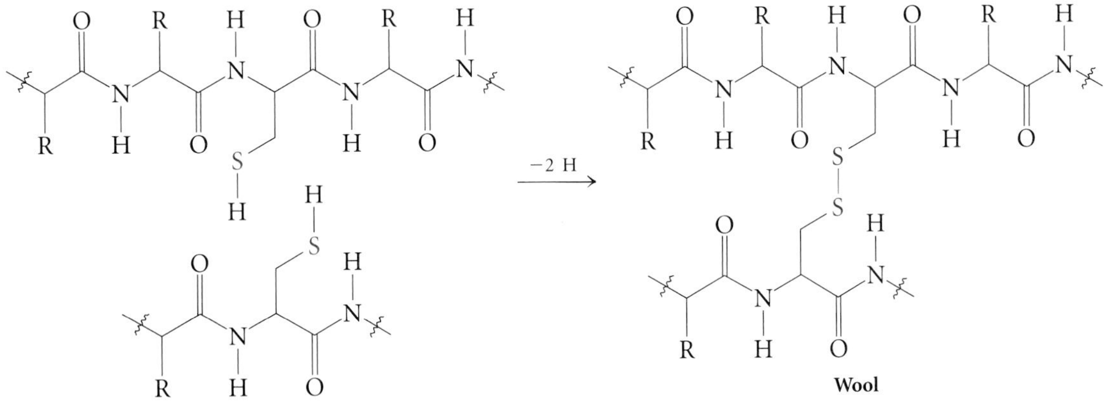

$-2$ H

**Wool**

Although a bond between two sulfurs may seem unusual, it is nonetheless very easily formed upon exposure of thiol functional groups to oxidants, and even molecular oxygen will effect this transformation. Sulfur–sulfur bonds play a crucial role in many biological processes.

**A urethane**

## Polyurethanes

Similar in structure to the polyamides are the **polyurethanes,** which have as one of the components in the chain a urethane (or carbamate) group. Urethanes are formed by the reaction of an isocyanate with an alcohol. (Recall from Chapter 14 that isocyanates are critical intermediates in rearrangements.) Polyurethanes are typically derived from low-molecular-weight polyesters (with terminal hydroxyl groups) and bis-isocyanates, which combine to form much longer chains linked by both ester and urethane groups. In Figure 16.2, blocks are used to represent each of the basic units, the polyester and the bis-urethane, allowing the functional groups participating in the polymerization to be seen more clearly.

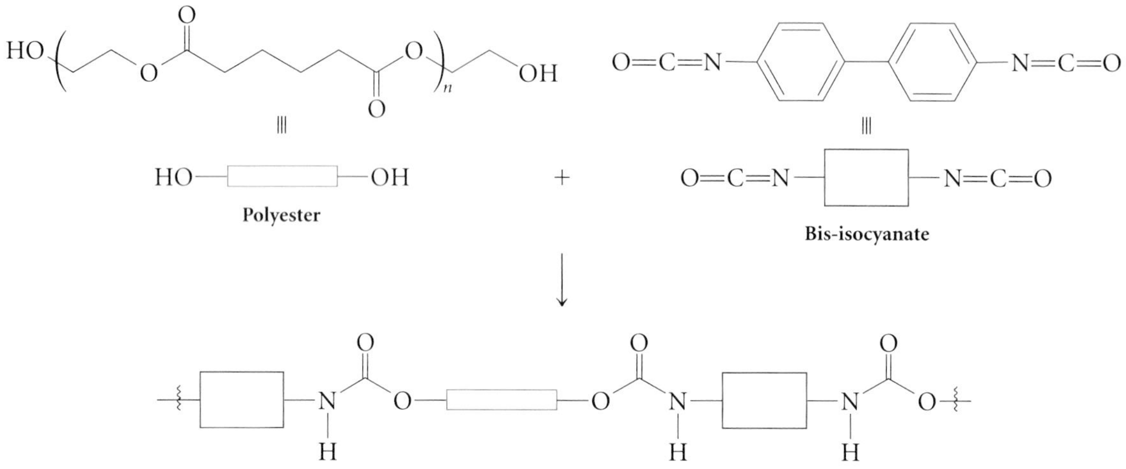

**FIGURE 16.2**

A polyester with an alcohol group at each terminus reacts with a bis-isocyanate to form a polyurethane. The addition step is mechanistically similar to the hydrolysis of an isocyanate in the Hofmann rearrangement.

When a compound with a low boiling point (for example, carbon dioxide or a volatile hydrocarbon such as methane or ethane) is dissolved under pressure in one of the starting materials and is then allowed to expand and vaporize as the polymerization proceeds, the polyurethane obtained has tiny "void" spaces sealed off by the surrounding polymer. The resulting polyurethane foam is a valuable lightweight material for building insulation and padding.

### EXERCISE 16.14

An alternative way of producing bubbles to yield polyurethane foam is to include a small amount of water along with the other components in the polymerization mixture. The water reacts with the isocyanate functional group to produce, ultimately, an amine and carbon dioxide, which is volatile. Write a complete, detailed mechanism for this conversion, assuming that a protic acid catalyzes the reaction. Do the same for the reaction in which hydroxide ion serves as the catalyst.

# Extensively Cross-Linked Polymers

Even in vulcanized rubber, the number of cross-linking bonds is usually small in comparison with the number of monomer units. On the other hand, monomers with three (or more) points of connection produce polymers that extend in two and three dimensions rather than having a linear arrangement as in simpler polymers such as polyethylene. These extensively cross-linked polymers are often very hard. Bakelite is one example; it is a resin (a highly viscous polymeric glass) derived from the reaction of phenol with formaldehyde (Figure 16.3).

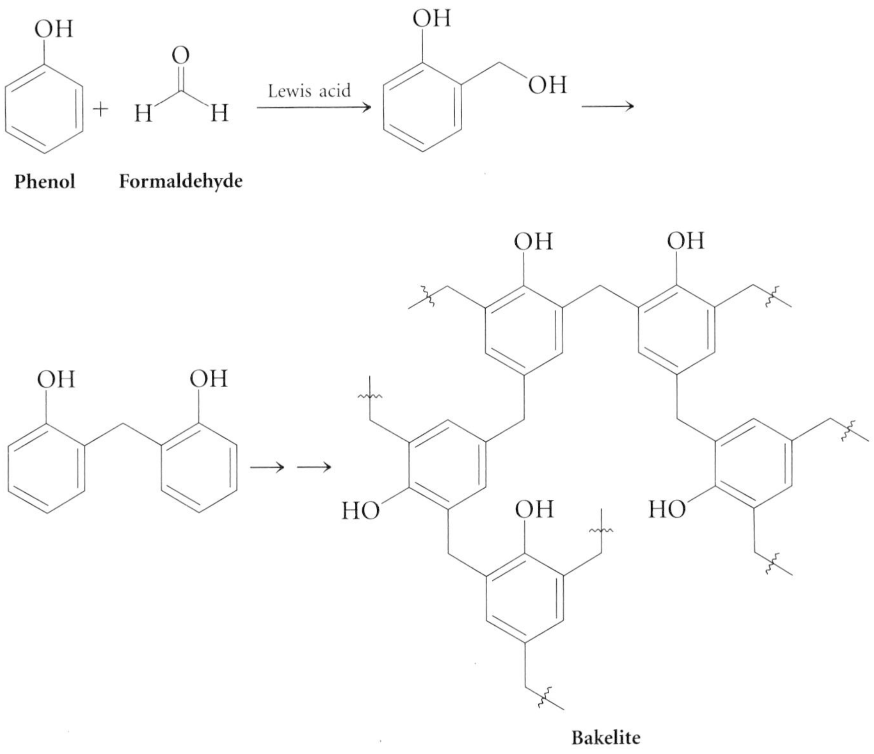

**FIGURE 16.3**

The activating effect of the phenolic —OH group directs electrophilic aromatic substitution to all three *ortho* and *para* positions. The benzylic alcohol unit alkylates another phenolic ring, at any or all of the *ortho* and *para* positions, to produce a highly cross-linked, very rigid polymer.

Each reaction in the formation of Bakelite can be viewed as an electrophilic aromatic substitution, taking place at both the *ortho* and *para* positions of phenol. These carbon–carbon bond-forming reactions are very similar to those that occur in the preparation of bisphenol A (see page 806). The resulting polymer is very highly branched. It is therefore quite rigid and does not soften significantly at elevated temperatures, which makes it a useful material for dishes, bowling balls, and the handles of cooking utensils.

Write a mechanism for the reaction of formaldehyde with phenol in the presence of a Lewis acid. Why does the reaction take place *ortho* and *para,* but not *meta,* to the hydroxyl group?

**Epoxy resins** constitute another important class of cross-linked polymers that have many applications as structural materials. Included in this class are the commonly used **epoxy glues.** Many laboratory bench tops are made of epoxy resin, and microelectronic chips are encapsulated in this material. Because such resins are expensive, a "filler" is added in the same way that sand and small stones are added to cement to make concrete. (For microchips, the filler is silicon dust.) The chemistry of epoxy resins is straightforward, but the structures are complex; for this reason, block diagrams are often used to represent structures such as bisphenol A:

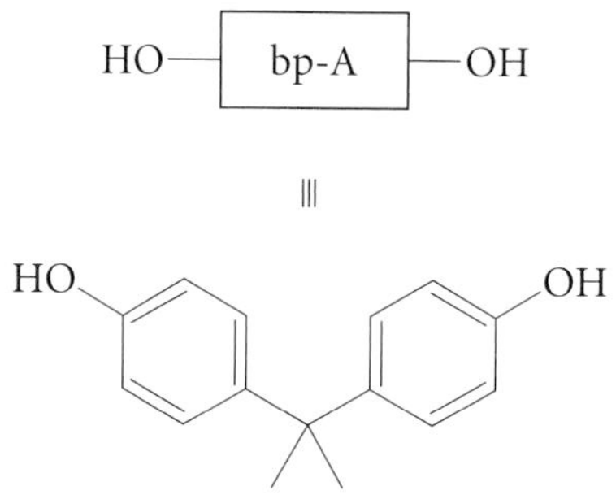

**Bisphenol A**

Reaction of bisphenol A with epichlorohydrin results in the formation of a bis-epoxide (Figure 16.4). This, in turn, reacts with additional bisphenol A in a one-to-one ratio, forming first a simple one-to-one adduct. This epoxy alcohol then reacts either with more bisphenol A to form a diol or with more bis-epoxide to form a larger bis-epoxide. These reactions continue, ultimately forming a mixture of short polymers composed of diols, epoxy alcohols, and bis-epoxides. This mixture is still fluid and is the clear, nearly odorless component of the two-part epoxy glues. The reaction of this complex mixture of linear polymers with a triamine, the fishy-smelling component, leads to the opening of the terminal epoxides and the formation of a three-dimensionally linked network that is much more rigid than the original resin.

The reaction of epichlorohydrin with an alcohol (or other nucleophile) can be viewed as a simple $S_N2$ displacement at the carbon bearing the chlorine atom. However, labeling studies indicate that the nucleophile is sometimes bonded to the carbon at the other end of the molecule. Suggest a mechanism that can account for these observations. Use methoxide ion as the nucleophile.

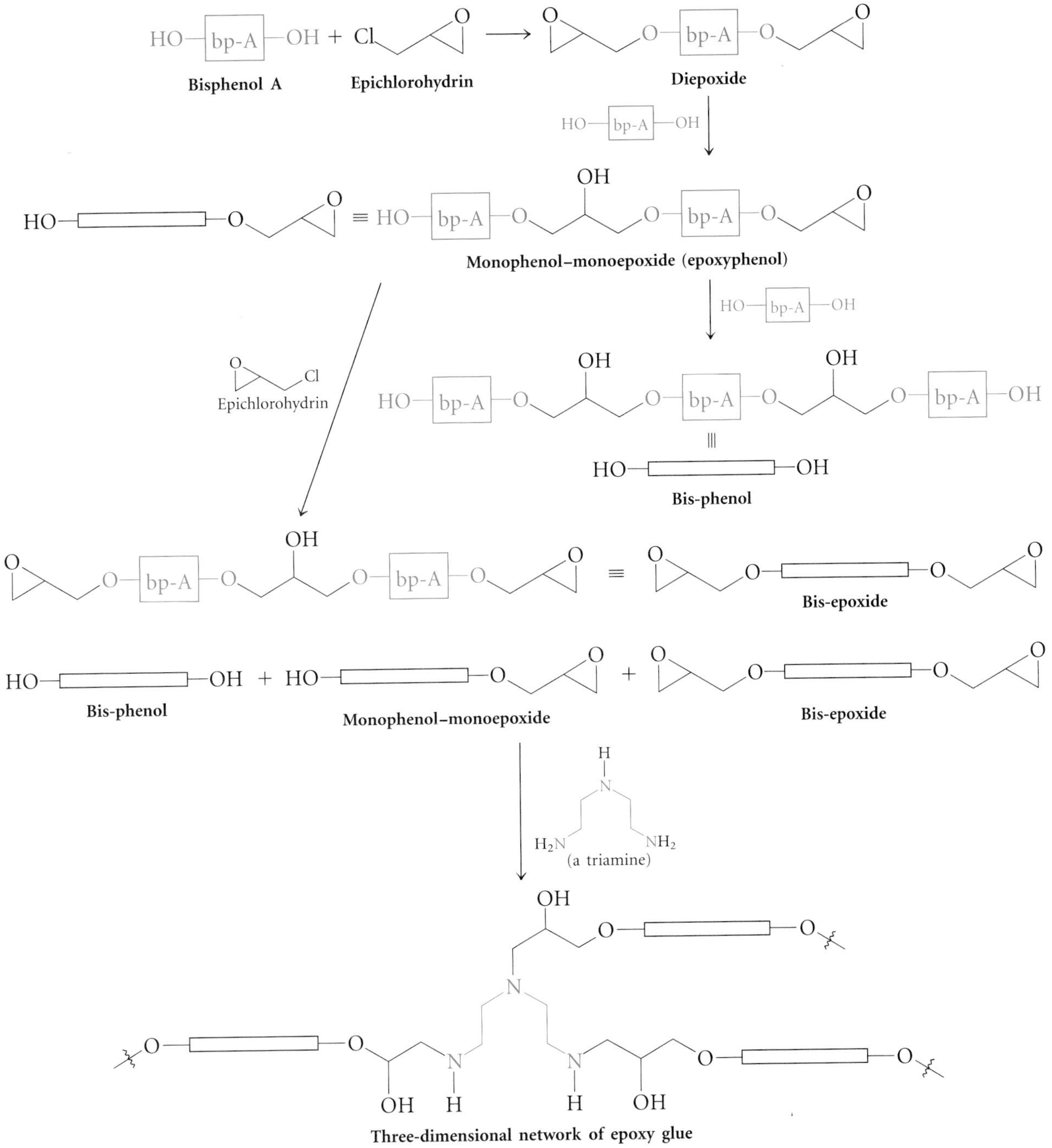

**FIGURE 16.4**

Reaction of bisphenol A with two equivalents of epichlorohydrin via two $S_N2$ displacements of chloride ion produces a diepoxide. Reaction of this diepoxide with additional bisphenol A at one of the epoxide groups produces an epoxyphenol; further reaction at the remaining epoxide group produces a longer molecule containing three bisphenol A subunits and two parts derived from epichlorohydrin. At the same time, some of the epoxyphenol reacts with additional epichlorohydrin, producing a bis-epoxide. In all, three types of products are formed: bis-phenol, monophenol–monoepoxide, and bis-epoxide.

# Three-Dimensional Structure of Polymers

The functional groups present in a polymer can dramatically affect its bulk properties. The contrast between the water insolubility of polyethylene and other hydrocarbon polymers and the water solubility of materials such as the poly(vinyl alcohols) and polysaccharides is dramatic. Indeed, such differences should be expected, based on the fact that various functional groups impart different properties to small molecules. Recall, however, that there is a large difference in water solubility between amylose and cellulose, even though both are approximately the same size and have the same functionality. Thus, there are further subtleties to be explored in connecting the structures and properties of polymers. The structures of amylose and cellulose differ in the stereochemistry of the linkage between the glucose subunits. This comparatively small stereochemical difference in the linkage translates into large differences in the three-dimensional shapes of the polymers. In this section, we examine some aspects of three-dimensional structure and how it affects the properties of a polymer.

## Polypropylene

*Stereochemical Designations.* Polypropylene is the simplest polymer whose bulk properties are influenced significantly by stereochemistry. When propylene is polymerized, the methyl groups along the polymer backbone can be oriented relative to each other in one of three ways: random, alternating, or all on one side. Simple radical, anionic, and cationic polymerizations yield a random orientation of the methyl groups, producing a stereochemistry referred to as **atactic**. In contrast, the use of Lewis acid complexes known as *Ziegler–Natta catalysts* as initiators produces a polymer in which almost all of the methyl groups are on the same side of the polymer backbone. This regular and constant stereochemistry is referred to as **isotactic**.

Isotactic stereochemistry

Although the stereochemistry of isotactic polypropylene is most easily seen with the carbon backbone in an extended, all-*anti* conformation, this arrangement leads to substantial steric interaction between the methyl groups. To avoid this interaction, isotactic polypropylene adopts a conformation in which some of the backbone carbon–carbon bonds are *gauche* (Figure 16.5).

A polymer chain with methyl groups alternating between front and back is known as **syndiotactic**.

Syndiotactic stereochemistry

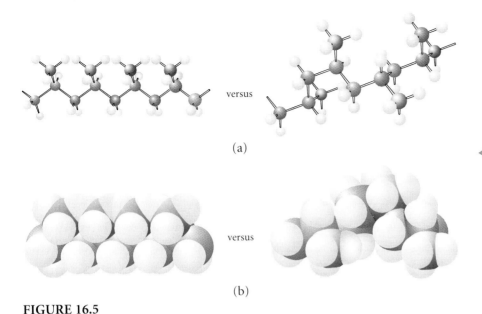

(a)

versus

(b)

**FIGURE 16.5**

Two conformations of isotactic polypropylene, shown as (a) ball-and-stick and (b) space-filling models. With the backbone in an extended all-*anti* conformation (left) (the backbone carbon atoms are shown in green), there is serious steric interaction between the appended methyl groups. These interactions are relieved in an arrangement in which some of the bonds of the backbone are in *gauche* conformations (right).

*Isotactic Polypropylene: Ziegler–Natta Polymerization.* Ziegler–Natta polymerization uses a **Ziegler–Natta catalyst,** an organometallic Lewis acid complex prepared by treatment of a trialkylaluminum with titanium trichloride. In addition to isotactic polypropylene, a high-density polyethylene with greater strength and heat resistance than was previously possible can be synthesized with a Ziegler–Natta catalyst. For this work, Karl Ziegler of the Max Planck Institute for Coal Research in Mülheim–Ruhr, Germany, and Giulio Natta of the Milan Polytechnic Institute received the Nobel prize in chemistry in 1963, only 10 years after the introduction of their catalyst. Van der Waals interactions between chains with a regular arrangement are stronger than those between chains with randomly oriented groups. As a result, isotactic polypropylene is harder and more rigid than the atactic polymer.

Let's first look at the chirality at each carbon in isotactic polypropylene. If the isotactic polypropylene formed by Ziegler–Natta polymerization is rotated end-to-end, an arrangement in which all of the methyl groups point to the back is obtained:

We could also arrive at almost the same arrangement by interchanging the hydrogen and methyl group (and thus the configuration) at each carbon. The result is almost the same because the two ends of the polymer chain are not identical: one bears the initiator as a substituent. Thus, although each chain of isotactic polypropylene is chiral, the difference between the enantiomers is insignificant for large polymer chains, and chains of opposite handedness are almost indistinguishable.

---

### EXERCISE 16.17

Reexamine the polymers in Figure 16.1 from the point of view of stereochemistry. Which ones can be formed with atactic, isotactic, or syndiotactic stereochemistry? Draw structures that illustrate the possible stereochemistries.

---

### Naturally Occurring Polypeptides

All naturally occurring polypeptides are derived from $\alpha$-amino acids. More than 20 $\alpha$-amino acids (differing in the alkyl substituent, R) are commonly found in natural polypeptides.

$\alpha$-Amino acid → Polypeptide

This diversity of building blocks allows for a virtually unlimited number of polypeptides differing in the sequence of the constituent amino acids. Nonetheless, these polypeptides have certain features in common that derive from the presence of regularly repeating amide groups.

***Properties of the Amide Bond.*** The three-dimensional properties of polypeptides derive from the unique properties of the amide functional group. As noted in Chapter 3, there is significant electron delocalization in amides, with donation of the nitrogen lone pair of electrons to the carbonyl group $\pi$ system:

Amide resonance contributors

Hybrid structure

The additional stability imparted by this delocalization has important implications for the chemistry of amides in general and specifically for the structure that this group imparts to polypeptides. One consequence of this delocalization is the planarity of the amide bond—that is, effective donation of lone-pair electron density from the nitrogen atom to the carbonyl $\pi$ system requires that the atoms involved be in the same plane. Experimental evidence is consistent with an energetic contribution from this delocalization of about 18 kcal/mole. Indeed, the amide group and the two flanking carbon atoms (six atoms in all) are essentially coplanar in virtually all amides.

There are two geometric isomers for the monosubstituted amides found in polypeptides. The arrangement with the carbon *syn* and the hydrogen *anti* to the carbonyl oxygen (known as the *syn* or *Z* isomer) is considerably more stable than the alternate *anti* (or *E*) arrangement. (The origin of this energetic difference has not yet been well defined.) As a consequence of the planar geometry, the amide group is a rigid linker for polypeptides.

A second important feature imparted to amides by electron delocalization is the polarization of the bonds involved. There are a significant partial negative charge on oxygen, a partial positive charge on nitrogen, and an increase in the polarization of the nitrogen–hydrogen bond, which leaves the hydrogen highly electron-deficient. The presence of both an electron-rich oxygen and an electron-deficient hydrogen in the amide functional group provides the opportunity for strong hydrogen bonding between amide units (Figure 16.6).

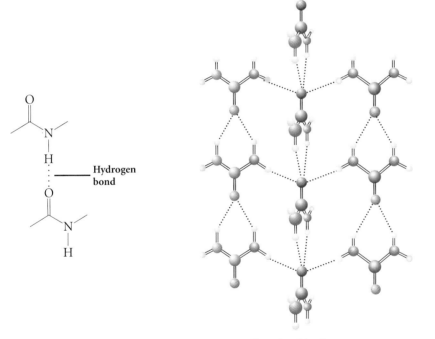

**Crystal packing for urea**

## FIGURE 16.6

Intermolecular hydrogen bonding is responsible for the regular order between amide groups in separate urea molecules packed as a solid crystal. Hydrogen bonds between the urea molecules are represented by dotted lines. Note that there are four hydrogen bonds to the oxygen atoms of the central urea molecules. Although each oxygen atom has only two lone pairs of electrons, each of these can associate with more than one hydrogen.

***Hydrogen Bonding in Amides.*** The crystal structure of urea illustrated in Figure 16.6 shows how hydrogen bonds can hold individual molecules together. Comparing the melting point of urea (133 °C) with that of acetone (−95 °C), a molecule of very similar molecular weight and functionality but without a hydrogen bonded to its heteroatom, reveals the importance of these hydrogen-bonding interactions.

Estimates for the magnitude of the energy of hydrogen bonds in various environments vary from less than 1 kcal to about 5 kcal per hydrogen bond. It is quite likely that this magnitude differs markedly with relatively subtle changes in distance and bond angle.

Note that the hydrogen bond illustrated at the left in Figure 16.6 is an arrangement in which the nitrogen–hydrogen and oxygen–carbon bond systems are colinear. This arrangement is most commonly found for hydrogen bonds in amides although not in the crystal structure of urea.

## EXERCISE 16.18

One way to view the hydrogen bond is as a proton partially bonded to two different electron-rich groups. These two partial bonds are generally quite different, one linkage being very strong and the other very weak, as indicated by the different bond lengths:

$$X—H\cdots\cdots\cdot X$$

and        versus    $X\cdots\cdots\cdot H\cdots\cdots\cdot X$

$$X\cdots\cdots\cdot H—X$$

Explain why this should be the case rather than there being two equal partial bonds between the hydrogen and its neighbors. Also, explain the planar arrangement of atoms described for urea in view of the simple concepts of bonding arrangements developed in Chapters 1, 2, and 3.

The amide groups in a polypeptide can interact through hydrogen bonding in either an intramolecular or an intermolecular sense. Let's consider the latter arrangement first, because it is a bit easier to see with a polypeptide in which the amino hydrogen atom and the carbonyl oxygen atom of each α-amino acid are on the same side of the chain.

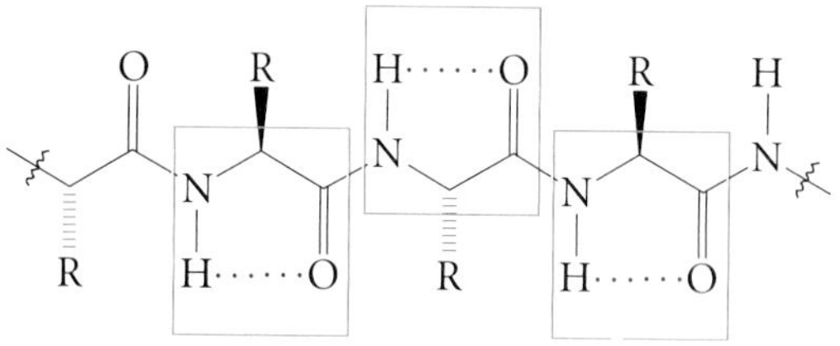

This arrangement is held in place by hydrogen bonding between the oxygen atom and the hydrogen of the same amino acid residue. This pair can further interact by hydrogen bonding with another pair in an adjacent polypeptide molecule. This is illustrated in Figure 16.7 for the interaction of three polypeptide chains.

The outer peptide chains have pairs of amino and carbonyl groups that can participate in further hydrogen bonding with other peptide chains. This

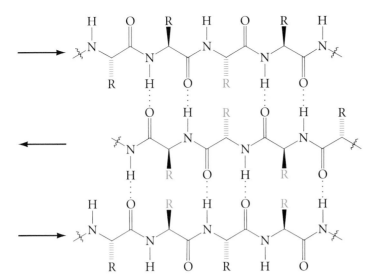

**FIGURE 16.7**

When adjacent polypeptide chains are oriented in opposite directions, they are properly positioned for intermolecular hydrogen bonding with another such chain on either side. (The arrows indicate the direction from the carboxyl to the amino terminus of the polypeptide chain. Here the chains are oriented in an antiparallel arrangement—that is, each runs in the direction opposite to that of its immediate neighbors.) The extended sheetlike structure resulting from this interaction is similar to that resulting from the urea crystal packing in Figure 16.6.

structure can be extended in a virtually unlimited fashion to form an essentially flat sheet of peptides. In order for the amide hydrogen and carbonyl groups from one chain to hydrogen bond with the carbonyl groups and hydrogens on a second chain, the chains must alternate direction (from carbonyl to amino group along the chain).

*The β-Pleated Sheet.* The stereochemistry of naturally occurring amino acids is almost universally *S*. As a consequence, the substituents (the R groups) nearest to each other on adjacent chains in the sheetlike arrangement of polypeptides are on the same side. If one substituent is a small alkyl group such as methyl and the other is hydrogen (or if both are hydrogen), there is essentially no steric interaction between the groups on adjacent chains. On the other hand, if both are large alkyl groups, there is a substantial repulsive interaction. This interaction results in a twisting of the peptide chains such that the alkyl groups rotate away from each other, reducing the steric interaction between them (Figure 16.8, on page 822). This deviation from a totally planar arrangement results in what is referred to as a pleating of the peptide sheets, and the arrangement is called a *β-pleated sheet*. This pleating is not without other consequences: in reducing the steric interaction between the alkyl groups, it brings the hydrogen atoms on the centers of chirality closer together. Furthermore, the twisting distorts the hydrogen bonds, further destabilizing the pleated-sheet arrangement. If all or most of the R groups are not hydrogen, the pleated-sheet arrangement is less stable than an arrangement in which the NH and C=O groups are intramolecularly hydrogen-bonded.

#20    β-Strands/Sheets

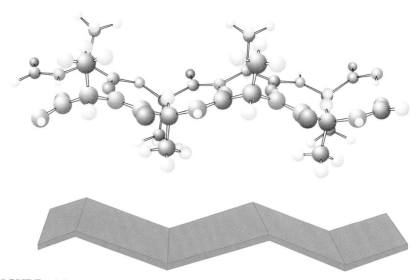

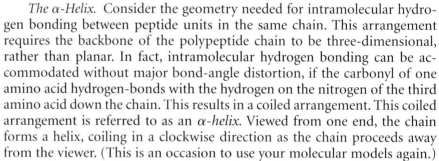

**FIGURE 16.8**

When the polymer backbone of the peptide is arranged in a sheetlike structure, the side-chain substituents (here, the methyl groups of alanine) on the
$\alpha$ carbons are directed toward the same region of space. The $\alpha$ carbons
rotate from the plane occupied by the amide bonds, reducing steric interaction and forming a $\beta$-pleated sheet.

#21    Alpha Helix

*The $\alpha$-Helix.*  Consider the geometry needed for intramolecular hydrogen bonding between peptide units in the same chain. This arrangement
requires the backbone of the polypeptide chain to be three-dimensional,
rather than planar. In fact, intramolecular hydrogen bonding can be accommodated without major bond-angle distortion, if the carbonyl of one
amino acid hydrogen-bonds with the hydrogen on the nitrogen of the third
amino acid down the chain. This results in a coiled arrangement. This coiled
arrangement is referred to as an *$\alpha$-helix.* Viewed from one end, the chain
forms a helix, coiling in a clockwise direction as the chain proceeds away
from the viewer. (This is an occasion to use your molecular models again.)
     At first, one might expect that a polypeptide chain could coil equally
well in either a left- or a right-handed fashion. However, the twisting of the
chain required to achieve intramolecular hydrogen bonding places the alkyl
and hydrogen substituents of the amino acid units in distinctly different
positions (Figure 16.9). In the $\alpha$-helix, the alkyl substituents of (*S*)-amino
acids are oriented more or less directly away from the helical structure; the
hydrogen atoms are pointed toward its interior. If the helix coils in the
opposite (counterclockwise) direction, producing a left-handed helix (a
$\beta$-helix), these positions are interchanged, and a very substantial steric interaction between the chain and the alkyl groups results.

**EXERCISE  16.19**

Draw three strands of peptides oriented in a parallel arrangement (rather than an
antiparallel one, as in Figure 16.7). Which arrangement, parallel or antiparallel, has
more hydrogen bonds?

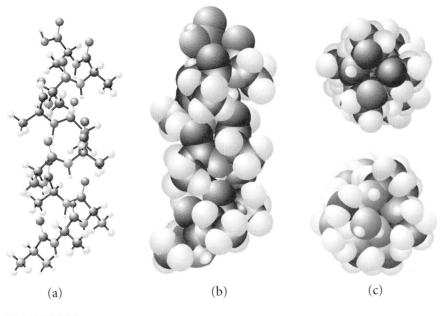

(a)           (b)           (c)

**FIGURE 16.9**

Coiling in an $\alpha$-helix takes place so as to direct the large groups (in this case, the methyl groups of alanine in polyalanine) bound to the $\alpha$ carbon toward the exterior of the helical coil. (a) Ball-and-stick and (b) space-filling models of an $\alpha$-helix. (c) Space-filling models viewed from the top down (showing carbonyl-group oxygen atoms) and from the bottom up (showing amide N—H groups).

*Primary, Secondary, Tertiary, and Quaternary Structure.* Many different molecules found in nature are based on the peptide bond. Some, whose molecular weights exceed 100,000, are composed of many hundreds of individual amino acid units. Molecules with 100 amino acids or fewer are known as **polypeptides;** larger species are called **proteins.** Each polypeptide or protein has a unique sequence of amino acids, referred to as its **primary structure.** The three-dimensional structure of large polypeptides is often quite complex, with both helical and $\beta$-pleated-sheet arrangements in different parts of the same chain. These specific arrangements, fixed in local regions, constitute the **secondary structure** of the polypeptide (Figure 16.10, page 824).

The schematic representation in Figure 16.10 also illustrates what is referred to as the **tertiary structure** of a protein—that is, how the $\beta$-pleated sheets and $\alpha$-helices are spatially dispersed. The protein shown has five distinct regions that contain $\beta$-pleated sheets. In these regions, the backbone chain of amino acids is twisted and folded around itself. In addition, two regions are helical. Together with the rest of the amino acid chain, these regions assume a complex three-dimensional arrangement. The precise tertiary structure of the protein results directly from the type and location of its individual constituent amino acids. Often, several large polypeptide or protein molecules join together to form a discrete complex, called a **quaternary structure,** which also ultimately derives its precise arrangement from the sequence of individual amino acids present.

CHEM TV® II

#22    Protein Organization

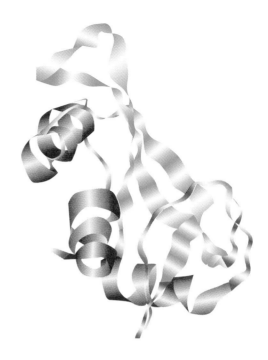

**FIGURE 16.10**

The structure of ribonuclease, a rather small protein responsible for the degradation of RNA. The primary structure consists of a unique sequence of 108 amino acids. Secondary structures are shown in red ($\alpha$-helices) and blue ($\beta$-pleated sheets). The tertiary structure is the spatial dispersal of the units of the secondary structures.

Each protein has its own unique tertiary structure, which is but one of an almost infinite number of possible arrangements. Assuming that each amino acid unit could adopt just three different possible conformations, the 108 amino acids of ribonuclease could be arranged into $3^{108} = 10^{51}$ different three-dimensional structures! Thus, statistically speaking, any one tertiary structure is highly unlikely and disfavored by entropy. Indeed, the entropy associated with ribonuclease is 285 cal $K^{-1}$mole$^{-1}$.

Recall that free energy is defined as

$$\Delta G° = \Delta H° - T\Delta S$$

For proteins to exist in tertiary structures, the enthalpy contribution to free energy ($\Delta H°$) must offset that of entropy. For ribonuclease, $\Delta H° = -95$ kcal/mole, in large part due to the many hydrogen bonds that literally hold the structure in its three-dimensional shape. At body temperature (37 °C = 310 K), this enthalpy contribution is enough to offset the entropy factor of 88.4 kcal/mole (285 cal $K^{-1}$mole$^{-1}$ × 310 K × 1 kcal/1000 cal). Thus, the folded structure of ribonuclease is favored by $95 - 88.4 = 6.6$ kcal/mole.

**EXERCISE 16.20**

Calculate the temperature at which the enthalpy and entropy contributions to the folding of ribonuclease are equal ($\Delta G° = 0$).

Linus Pauling (1901–1995) made many contributions to chemistry and the world during his highly productive career, ranging from fundamental studies of the chemical bond, for which he was awarded the 1954 Nobel prize in chemistry, to his efforts on behalf of peace which were acknowledged by the 1962 Nobel peace prize.

Pauling also made fundamental contributions to the understanding of the structures of large biopolymers. His early drawing of the hydrogen bonding leading to the $\alpha$-helix is shown here. (Make a copy of this drawing on transparency film. You can then roll the film into a coil that shows the helix.)

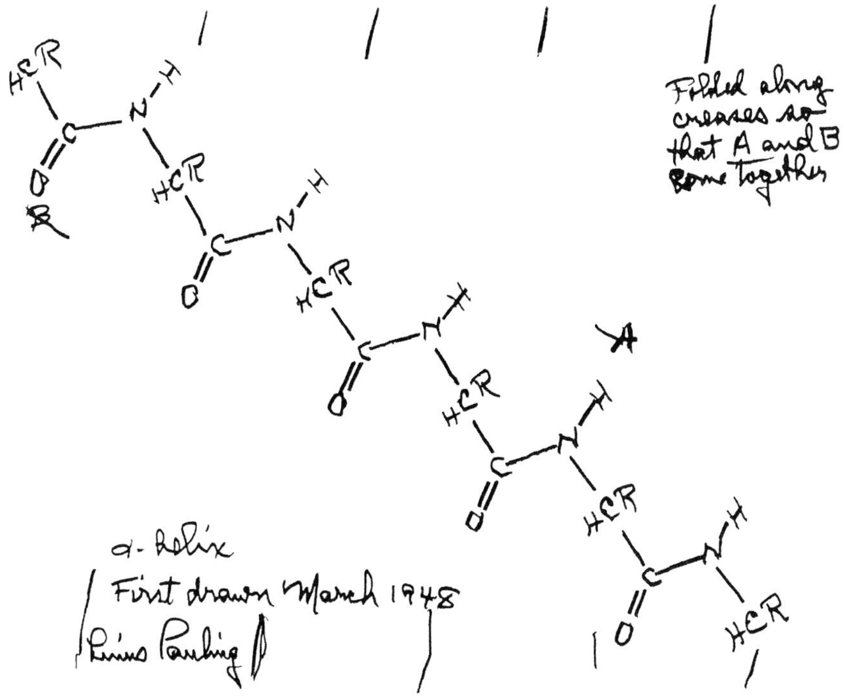

A major focus of research is to discover rules that may permit the prediction of a protein's tertiary structure from its primary sequence of amino acids. Such analyses are complicated by the observation that solvents are also important in determining three-dimensional structure. Solvents (such as water) that can form relatively strong hydrogen bonds to the amide groups can produce solvated polypeptide structures that are similar in energy to the $\alpha$-helix (in which all hydrogen bonds are internal). For example, in dioxane, a solvent that can serve only as a proton acceptor in hydrogen bonding, polyalanine is nearly completely coiled in a helical arrangement. Conversely, in water, the degree of coiling is only about 50%, the remainder being $\beta$-sheets. Although much smaller by comparison, solvent molecules constitute the environment in which large polymers exist. Thus, the shape of macromolecules varies with the solvent. The details of protein folding remain one of the most intriguing unanswered questions in biochemistry.

## Cellulose and Starch

Earlier in this chapter, it was pointed out that cellulose is much less soluble in water than is starch of comparable molecular weight. Now let's look at this difference from a molecular point of view. Recall that the structures of these naturally occurring polyacetals are quite similar, differing mainly in the stereochemistry at C-1:

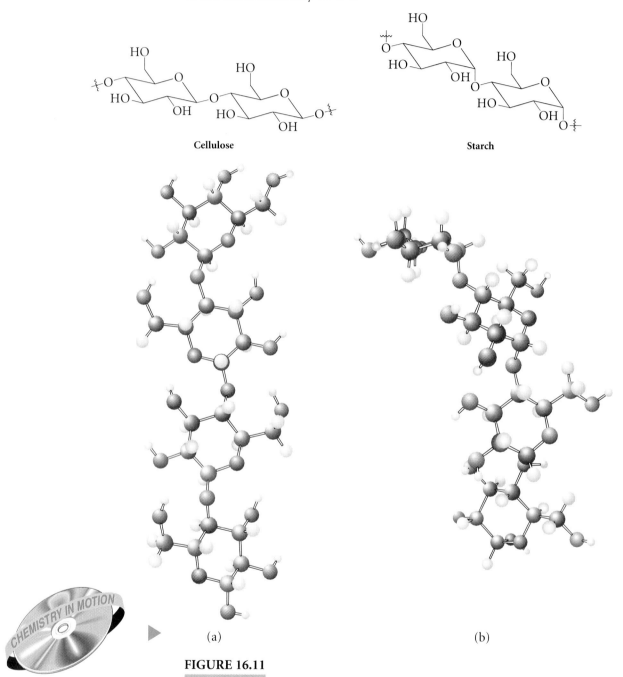

FIGURE 16.11

(a) Cellulose has hydrogen bonds between residues, but these hold the polymer chains in an almost flat line. (b) In contrast, a four-glucose subunit of starch shows the helical twist imparted by hydrogen bonds between its sugar residues.

This difference at C-1 has a dramatic effect on the ability of each polymer to participate in intramolecular hydrogen bonding between the hydroxyl groups. Because the linkage in starch has an axial oxygen substituent on one of the six-member rings, the individual rings can coil, forming a helix with hydrogen bonding between the hydroxyl groups on adjacent rings of the same polymer chain (Figure 16.11).

In contrast, the oxygen linking the units in cellulose is an equatorial substituent on both rings, and, as a result, hydrogen bonding between adjacent glucose units provides a linear structure without a twist. Individual chains of cellulose can readily participate in interchain hydrogen bonding, fixing the individual chains into a rigid, three-dimensional matrix, as shown in Figure 16.12. The helical twist of starch interferes with regular interchain hydrogen bonding, and thus the chains are held together more loosely.

Thus, relatively subtle differences in molecular structure result in quite dramatic differences in three-dimensional structure and in marked differences in the properties of starch and cellulose. And the differences between digestible starch and indigestible cellulose, which affect all of the inhabitants of the biosphere, are due to a difference in stereochemistry at one carbon atom in a molecular subunit. Similar examples are found in protein chemistry, where the substitution of an alkyl group for a hydrogen atom in one amino acid of a polypeptide leads to a change in arrangement from a sheet to a helix. These structural changes lead to marked chemical and physical differences that, in turn, affect function.

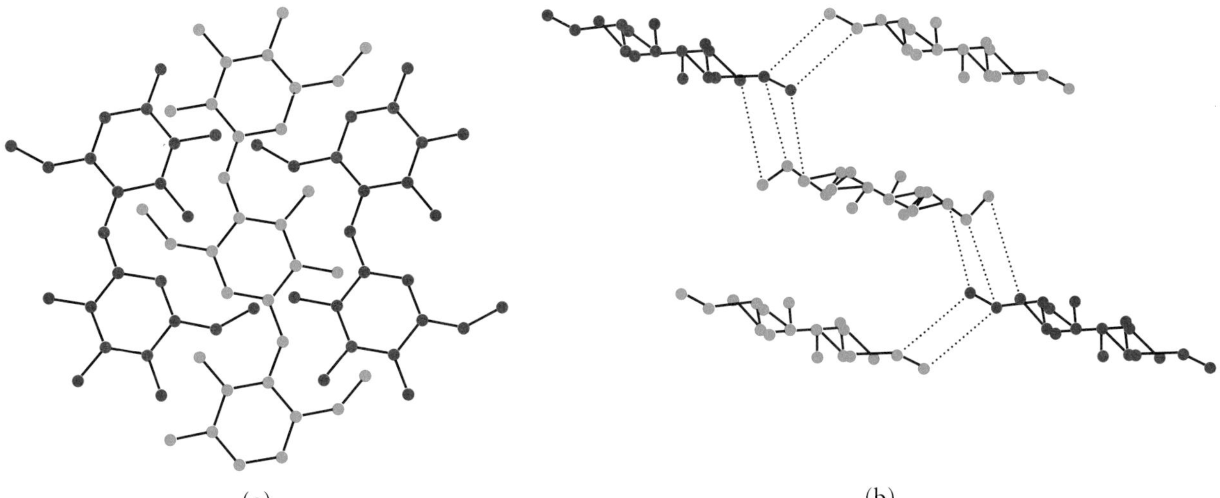

(a)                                                                                          (b)

**FIGURE 16.12**

Two views of the relative arrangement of cellulose chains as found in, for example, wood fibers. (a) In a "top" view, the six-member rings of the glucose units are clearly visible. (b) The chains are viewed end-on; the dotted lines represent the hydrogen bonding between glucose hydroxyl groups on adjacent chains that contribute to the strong, intermolecular attractive interactions that account for the strength and rigidity of cellulose. For clarity, only the carbon, oxygen, and hydroxyl hydrogen atoms are shown.

# Summary

**1.** The properties of a polymer are uniquely determined by the characteristics of its component functional groups and its three-dimensional structure.

**2.** Although the chemical characteristics of the functional group(s) in a polymer are similar to those of the monomer, the structure and physical properties are unique to the polymer.

**3.** The difference in structure between linear and branched polymers contributes to the macroscopic properties of hardness and strength, which determine the appropriate applications for these materials.

**4.** Monomers can have a variety of functional groups, including simple alkyl groups, aromatic substituents, esters, alcohols, amines, and halocarbons. These substituents may be appendages on the polymer backbone or may participate directly in the chemical bonding that links the monomer units together to form the polymer.

**5.** Linkages between monomers include carbon–carbon bonds in vinyl polymers, carbon–oxygen bonds in polyesters, polyacetals, and polyethers, and carbon–nitrogen bonds in polyamides and peptides.

**6.** Each functional group in a polymer, whether part of the backbone or a substituent on the backbone, contributes to the bulk properties of the polymer. Thus, for example, polyethylene is totally insoluble in most solvents, poly(methyl methacrylate) dissolves readily in polar organic solvents, and poly(vinyl alcohol) dissolves in water.

**7.** There are two broad classifications of polymers, addition and condensation, based on the type of reaction required for polymerization. Addition polymerization takes place through an intermediate cation, radical, or anion. The anionic intermediates are referred to as living polymers because the reactions are not self-terminating and will stop only when all of the starting materials have been consumed. Condensation polymers are formed by reactions in which monomers combine with the loss of a small by-product.

**8.** The properties of a polymer can vary significantly with stereochemistry and variations in the strength of the intermolecular attractive interactions. Covalent cross-linking can make a polymer very hard and rigid.

**9.** Hydrogen bonding plays an important role in the three-dimensional structure of nitrogen- and oxygen-containing polymers. Intermolecular hydrogen bonding produces a $\beta$-pleated sheet as a secondary structure in peptides, whereas intramolecular hydrogen bonding leads to a coiled, $\alpha$-helical secondary structure.

**10.** Hydrogen bonds contribute substantially to the secondary, tertiary, and quaternary structures of proteins. Intramolecular hydrogen bonding in starch reduces the number of intermolecular hydrogen bonds between individual starch chains. As a result, starch is water-soluble. In contrast, cellulose has little opportunity to participate in intramolecular hydrogen bonding. Therefore, the polymer chains are held quite tightly together by multiple contacts, and cellulose is not water-soluble.

### Radical Polymerization

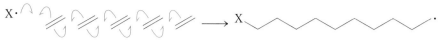

### Anionic Polymerization

### Cationic Polymerization

### Polyacetal Formation

# Review Problems

**16.1** Identify the repeating units in each of the following polymers:

(a)

(b)

(c)

(d)

(e)

(f)

**16.2** Identify a bond or functional group that could have been formed in synthesizing each polymer in Problem 16.1. Then, based on this analysis, write structure(s) for the monomeric unit(s) that could be used in each case.

**16.3** The polymer formed from formaldehyde, known as paraformaldehyde or polyacetal, is not very stable under either acidic or basic conditions. However, the stability can be increased dramatically by conversion of the hemiacetal end groups of the polymer into either —CH$_2$CH$_2$—OH groups by reaction with ethylene oxide to form the polymer Celcon or acetate esters by reaction with acetic anhydride to produce the polymer Deldrin.

Suggest a reason for this enhanced stability. One of these polymers is much more stable in the presence of a strong base than the other. Which is it, and why?

**16.4** Large quantities of ethylene are converted by oxidation with molecular oxygen into ethylene oxide, which is then treated with aqueous base to form ethylene glycol for use in making polyesters such as Dacron. Increasing the amount of water in the reaction mixture decreases the amounts of di- and tri(ethylene glycol) compared with ethylene glycol. Suggest a reason for this observation. When the reaction is conducted as a very dilute solution in water, almost all of the ethylene oxide is converted to ethylene glycol. Why might this not be a practical method for making ethylene glycol on a very large scale?

**16.5** The reaction of phenol with formaldehyde in the presence of acid produces a material called bisphenol F, which is analogous to bisphenol A. Write a structure for this material, and suggest a mechanism for its formation.

**Bisphenol A**

**16.6** Terephthalic acid (the *para* isomer of benzene dicarboxylic acid) is prepared industrially from *p*-dimethylbenzene (*p*-xylene). Suggest a reagent or sequence of reagents that you could use in the laboratory to carry out this transformation:

*p*-Xylene        Terephthalic acid

**16.7** Polyamides can be made from β-amino acids in much the same way as they are made from α-amino acids. Assuming a completely extended structure for these polymers as shown below, examine possible hydrogen-bonding interactions between chains. Would you expect these interactions to be stronger (on a per-weight basis) or weaker than those for $\alpha$-amino acids? How do the relations between the chains differ for polymers derived from $\alpha$- and $\beta$-amino acids?

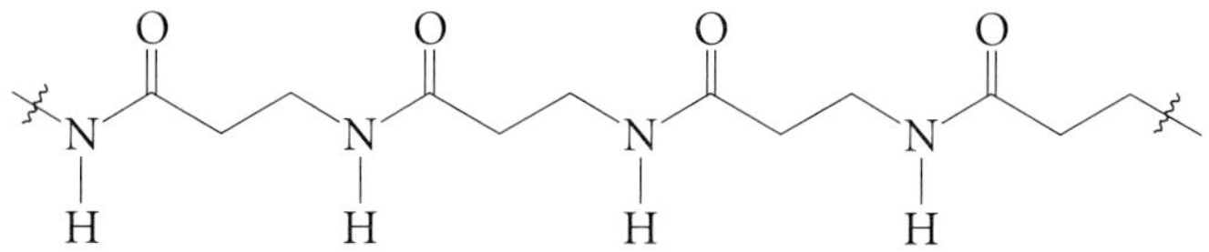

**16.8** Polypeptides can be formed from a derivative of an amino acid known as an *N*-carbonic anhydride. The polymerization is initiated by nucleophiles such as simple amines. Write a mechanism for the reaction of methylamine with the *N*-carbonic anhydride of alanine ($R = CH_3$). Next, write a mechanism for the reaction of the product with another equivalent of the *N*-carbonic anhydride, forming a dipeptide. Is this method of forming polypeptides a condensation or addition polymerization?

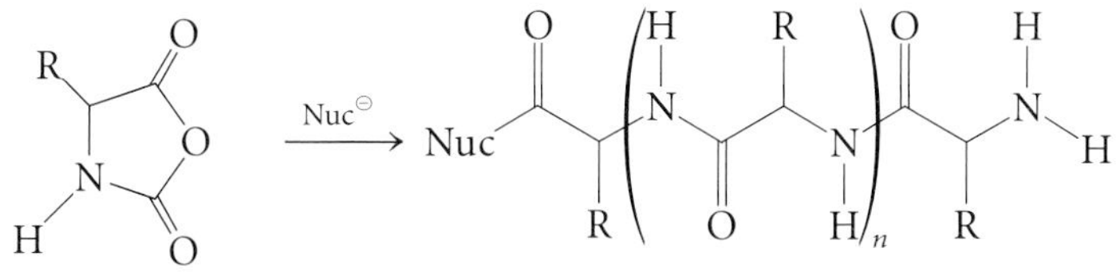

*N*-Carbonic anhydride

**16.9** The cross-linking diene 5-vinylnorbornene is produced by the Diels–Alder reaction of cyclopentadiene and butadiene. A small amount of diene A (bicyclo[4.3.0]nona-3,7-diene) is also obtained. Although compound A could have been produced by a Diels–Alder reaction of cyclopentadiene and butadiene in which cyclopentadiene served as the dienophile and butadiene served as the diene, it is known that compound A is produced during the reaction by a further transformation of 5-vinylnorbornene. Write a reaction mechanism for this transformation. (*Hint:* It involves a single step without the formation of intermediates.)

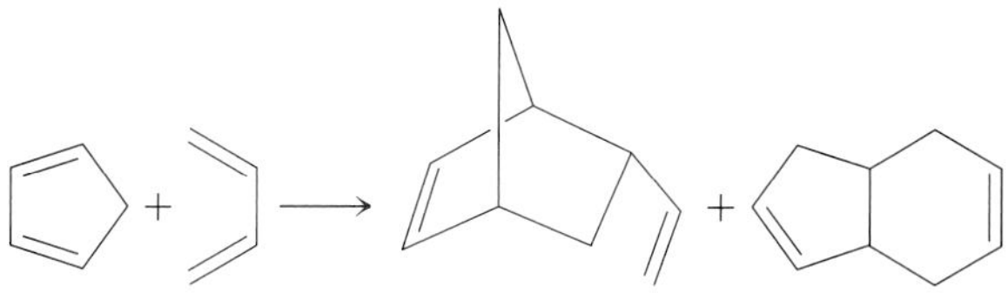

**5-Vinylnorbornene**          **A**

**16.10** Although both 5-vinylnorbornene and the diene A shown in Problem 16.9 can serve as cross-linking agents in radical polymerizations, use of the former results in superior polymers of styrene with higher molecular weights. (Indeed, diene A is removed by continuous fractional distillation and burned as fuel in the chemical plant where it is produced.) Production of polymers with higher molecular weights is a direct result of the difference in reactivities of the two double bonds in 5-vinylnorbornene. The less reactive double bond is also less reactive than styrene toward reaction with carbon radicals. Explain these differences in reactivity. Why does this difference in reactivity lead to polymers of higher molecular weight?

**16.11** Polystyrene undergoes oxidation with molecular oxygen considerably faster than does polyethylene, especially in the presence of light. Which carbon of polystyrene do you think is most easily oxidized?

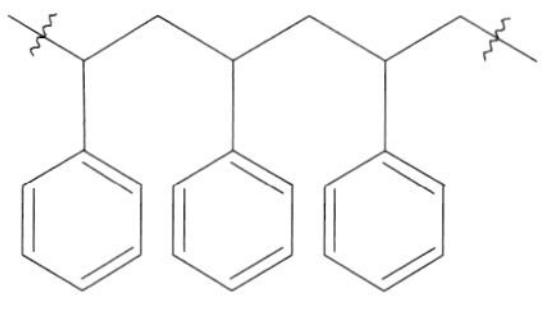

**Polystyrene**

## Supplementary Problems

**16.12**  Most polymerizations result in a polymer that is more dense than the starting monomer. Can you think of an explanation for this? This increase in density results in shrinkage during polymerization that can cause problems in specific applications. For example, computer chips are encapsulated in a polymer that protects the electronic circuitry from damage due to physical contact or exposure to oxygen and water. Epoxy resins are the polymer of choice in the microelectronic industry because of several desirable properties, including low to negligible shrinkage during the curing process. Can you think of a reason why the formation of epoxy resins results in less shrinkage than, for example, the formation of polystyrene from styrene?

**16.13**  Stretching of a polymer such as rubber results in (mainly) conformational changes that accommodate the increase in length and corresponding decrease in width of the material. In turn, these changes result in a temperature change. (Try it using a heavy rubber band and your upper lip, which is quite sensitive to small temperature variations.) When the rubber band is released, it returns to its relaxed state, and the temperature returns to near the original. First, determine whether the temperature of a rubber band increases or decreases upon stretching, and then rationalize this observation.

**16.14**  Ball-like conformations of long-chain polymer molecules are favored over extended conformations at high temperature; the reverse is true at lower temperatures. Straight-chain polymers such as poly(ethylene glycol) are added to motor oil as viscosity enhancers, which help maintain a constant viscosity over an extended temperature range. (Thin oils are desirable when an engine is cold, but may be too thin to provide adequate lubrication at the high operating temperatures of an internal combustion engine. An oil rated 10W-40 should have the viscosity of a 10-weight oil (thin) at cold temperatures and of a 40-weight oil at engine-operating temperatures.) Why does the conformational preference of straight-chain polymers change with temperature in the direction noted? How does the preference for ball-like conformations at high temperatures and extended conformations at low temperatures help maintain a constant viscosity over an extended temperature range?

**16.15**  The synthetic polymers added to motor oil are not as stable as the oil itself and break down into smaller fragments during engine operation. What effect does this have on the viscosity of the oil, at both low and high temperatures?

**16.16**  Five-, six-, and seven-member cyclic lactones are not stable at room temperature in the presence of trace impurities, even water. The polymers formed are polyesters that have not yet found wide commercial applications. Write a mechanism for the formation of the dimer of valerolactone in the presence of water, assuming that a small amount of acid is present.

Valerolactone                    Dimer

Polymer

# Naturally Occurring Oxygen-Containing Compounds

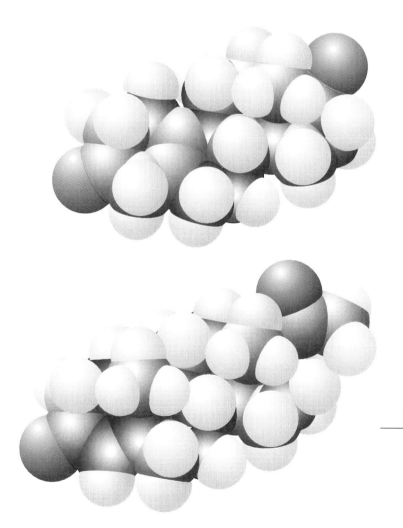

*The synthetic steroid norethyn-odrel (top) is sufficiently similar to the natural hormone proges-terone (bottom) that it mimics that hormone's action on the ovaries and suppresses ovulation.*

Typically, small molecules are used to illustrate the concepts relevant to various organic reactions in order to focus on the changes that occur in proceeding from starting material to product. This chapter will deal with large, biologically important molecules that contain oxygen. Although these large organic molecules often have many functional groups, the reactions they bring about and the changes they undergo are no different than those for small molecules with the same functional groups. However, to ensure that you are not intimidated by the sheer size of these molecules, we will first look closely at their structures and chemical properties and then relate these general features to the molecules' biochemical functions in living systems.

## 17.1

## Lipids

#18   Lipids

The **lipids** are a broad group of molecules whose common characteristic is their solubility in nonpolar solvents. The lipids contain only simple functional groups. Because they are composed mainly of carbon and hydrogen, they dissolve readily in hydrocarbon solvents.

### Fats and Waxes

**Fats** and **waxes,** the simplest lipids, are esters of long, straight-chain acids (known as **fatty acids**). The two classes are distinguished by the alcohol precursors for their ester groups. Fats originate from the triol **glycerol** and are frequently referred to as **triglycerides.** Waxes are derived from long, straight-chain alcohols.

$$
\begin{array}{lll}
\text{HO—CH}_2 & \text{CH}_3(\text{CH}_2)_n\text{COO—CH}_2 & \\
\text{HO—CH} & \text{CH}_3(\text{CH}_2)_m\text{COO—CH} & \text{CH}_3(\text{CH}_2)_n\text{COO(CH}_2)_m\text{CH}_3 \\
\text{HO—CH}_2 & \text{CH}_3(\text{CH}_2)_o\text{COO—CH}_2 & \\
\textbf{Glycerol} & \textbf{Fats} & \textbf{Waxes} \\
& (n, m, o \approx 8\text{–}18) & (n \approx 8\text{–}18, \ m \approx 16\text{–}36)
\end{array}
$$

Some common fatty acids are shown in Table 17.1, and the alcohols that are their ester partners in typical waxes are listed in Table 17.2. Both fats and waxes are found in nature as mixtures of many components. Thus, a sample of a fat contains many different triglycerides, each of which may be constituted from several different fatty acids. Furthermore, even if only two fatty acids are present as ester units, several different triglycerides can be formed. For example, in a fat sample from a biological source, an individual triglyceride composed primarily of esters of stearic and palmitic acids may have some molecules with three stearyl ester groups, some with two stearyl and one palmityl, and so forth. Each of these triglyceride molecules has slightly different properties. Waxes isolated from natural sources, such as the moisture-barrier coating of many insects and plants, also consist of mixtures of esters; in these compounds, the chain lengths of both the alcohol and the acid components vary.

## TABLE 17.1

Common Fatty Acids

| Formula | Name | Number of Carbons | Melting Point (°C) |
|---------|------|-------------------|--------------------|
|  $CO_2H$ | Caproic acid | 6 | $-3$ |
| 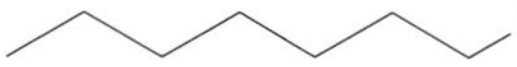 $CO_2H$ | Caprylic acid | 8 | 17 |
|  $CO_2H$ | Capric acid | 10 | 31 |
| 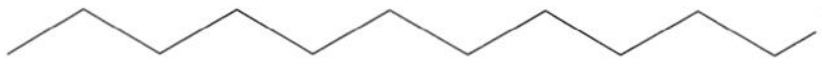 $CO_2H$ | Lauric acid | 12 | 44 |
|  $CO_2H$ | Myristic acid | 14 | 58 |
| 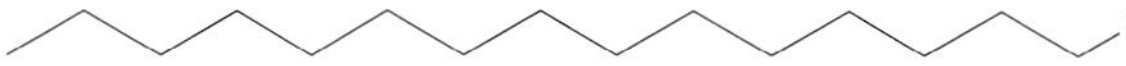 $CO_2H$ | Palmitic acid | 16 | 63 |
|  $CO_2H$ | Stearic acid | 18 | 70 |
| 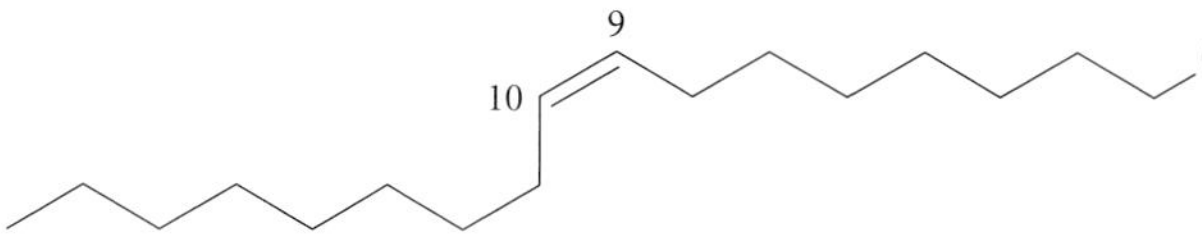 $CO_2H$ | Oleic acid | 18 | 4 |
| 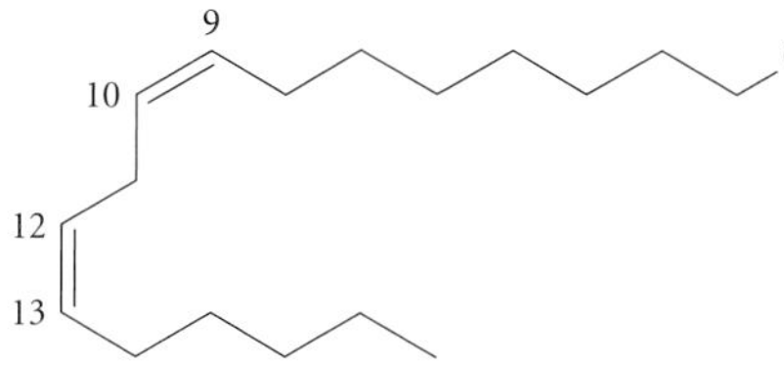 $CO_2H$ | Linoleic acid | 18 | $-12$ |
| 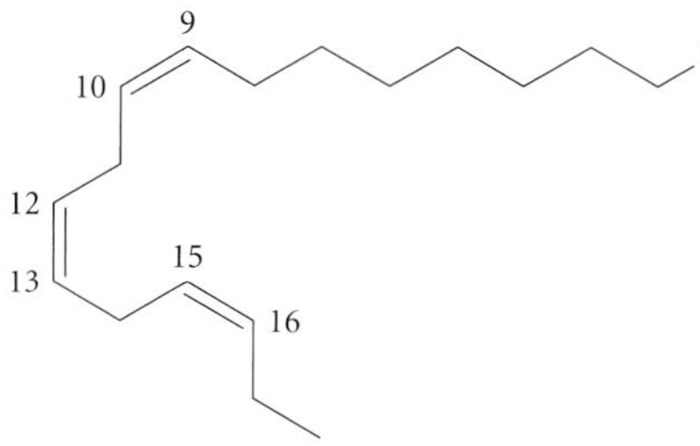 $CO_2H$ | Linolenic acid | 18 | $-11$ |

## TABLE 17.2

Common Long-Chain Alcohols

| Formula | Name | Number of Carbon Atoms | Melting Point (°C) |
|---------|------|------------------------|--------------------|
| $CH_3(CH_2)_{14}CH_2OH$ | Cetyl alcohol | 16 | 49 |
| $CH_3(CH_2)_{22}CH_2OH$ | Carnaubyl alcohol | 24 | — |
| $CH_3(CH_2)_{24}CH_2OH$ | Ceryl alcohol | 26 | 79 |
| $CH_3(CH_2)_{28}CH_2OH$ | Myricyl alcohol | 30 | 86 |

17-3

## EXERCISE 17.1

Consider a triglyceride in which all three acid units are different. In this case, C-2 of the glycerol unit is a chiral center. (Use molecular models if needed to understand this.) Determine the total number of possible isomers, both structural and stereochemical, and draw a structural formula such as the following for each isomer:

$$R_2COO \quad H$$

$$R_1COO \qquad OOCR_3$$

You may be surprised to note in Tables 17.1 and 17.2 that all of these common, naturally occurring fatty acids and alcohols have an even number of carbon atoms. This consistency follows from the biological synthesis of long alkane chains in which acetic acid, a two-carbon precursor, undergoes a Claisen condensation.

In long-chain fatty acids (and alcohols), the hydrocarbon portion of the molecule clearly dominates the structure. The waxes, then, can be viewed as structurally similar to the short chains of polyethylene. Not surprisingly, the macroscopic properties of waxes (as well as fats) are also reminiscent of polyethylene. Waxes are nearly completely insoluble in water but, as their classification as lipids implies, quite soluble in hydrocarbon solvents. Although their molecular weights are high enough that they are solids at room temperature, the fact that they are mixtures prevents them from being crystalline. Indeed, it is a difficult task to separate any single pure component from naturally occurring waxes (or fats).

The melting points for the fatty acids given in Table 17.1 increase gradually as the number of carbons in the chain grows. Although fats and waxes derived from these acids are not crystalline because they exist as mixtures, their **viscosity,** or "stiffness," nonetheless increases as the carbon chain lengthens. Exceptions to the trend in melting points are the three **unsaturated fatty acids** listed in Table 17.1: oleic acid, linoleic acid, and linolenic acid. Here, the term *unsaturated* has the same meaning as for hydrocarbons: unsaturated fatty acids have at least one double bond. The geometry and position of the double bond(s) are indicated in the systematic names for these acids. For example, in the systematic name for oleic acid, *cis*-$\Delta^9$-octadecenoic acid, $\Delta^9$ refers to the chain carbon at which the double bond starts.

Note that in the three 18-carbon unsaturated fatty acids, each additional double bond decreases the melting point; that is, the melting point decreases from oleic to linoleic to linolenic acid. The "kinks" in the structure of linolenic acid are revealed in the space-filling model in Figure 17.1. All of the double bonds in oleic, linoleic, and linolenic acid have the *cis* configuration, which interrupts the regular zig-zag arrangement of the extended conformation of the saturated fatty acids. A bend in the carbon chain inhibits close contact between the molecules, resulting in less crystallinity. The three unsaturated fatty acids thus impart a lower viscosity to the fats that contain them.

## CHEMICAL PERSPECTIVES

### VARIED USES OF ETHANOL

Ethyl alcohol, or ethanol, was probably the first organic chemical to be made routinely by humans. Fermentation of sugars and starches still accounts for most of the alcohol produced for human consumption. There are many industrial applications of ethanol, both as the alcohol itself and as a part of larger molecules. Most industrial ethanol is made by hydration of ethylene produced in the cracking of petroleum feedstock. However, the quantity of ethanol produced in this way has been dropping steadily—from 1.3 billion pounds in 1981 to only 546 million pounds in 1990. In part, this decline in the synthesis of ethanol from ethylene is due to the availability of ethanol from an alternative source—wine grapes, which have been widely overproduced, especially in Italy. The excess production is fermented and then distilled to produce ethanol for industrial use, as well as for mixing with gasoline to make gasohol.

Stearic acid

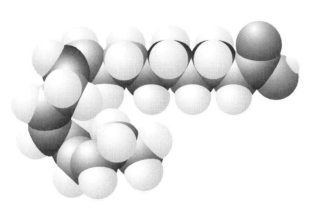

Linolenic acid

### FIGURE 17.1

The carbon chain of an unsaturated fatty acid can have "kinks" imparted by the presence of three *cis* double bonds, as can be seen in these space-filling models of the 18-carbon saturated stearic acid and unsaturated linolenic acid.

*Saponification.* Fatty acid esters can be readily hydrolyzed with aqueous sodium hydroxide in a process called **saponification,** because the carboxylic acid salts formed in this way are the constituents of soap. This reaction, known for thousands of years, originated in the treatment of animal fats with an alkaline solution obtained by soaking ashes in water.

$$
\begin{array}{l}
CH_3(CH_2)_nCOO-CH_2 \\
CH_3(CH_2)_mCOO-CH \\
CH_3(CH_2)_oCOO-CH_2
\end{array}
\xrightarrow[\text{H}_2\text{O}]{\text{NaOH}}
CH_3(CH_2)_xCO_2^{\ominus} {}^{\oplus}Na +
\begin{array}{l}
HO-CH_2 \\
HO-CH \\
HO-CH_2
\end{array}
$$

| Fats | Salts of fatty acids | Glycerol |

### EXERCISE 17.2

Hydrolysis of carboxylic acid derivatives takes place via nucleophilic acyl substitution by water under both acidic and basic conditions. Write a complete, detailed mechanism for the hydrolysis of a carboxylic acid ester in the presence of sodium hydroxide. Now do the same for the hydrolysis in the presence of hydrochloric acid. Only one of these processes is catalytic. Which is it? Explain why the other process is not catalytic.

### EXERCISE 17.3

In the presence of base, the monoesters of 1,2-diols (and 1,3-diols) rapidly equilibrate:

$$
CH_3COO \overset{OH}{\diagup} \quad \underset{\longleftarrow}{\overset{{}^{\ominus}OH}{\rightleftharpoons}} \quad HO \overset{OCOCH_3}{\diagup}
$$

Write a detailed mechanism for this interconversion. Would you expect this transformation to be faster or slower than the hydrolysis of a carboxylic acid ester by base? Explain.

*Micelles.* Salts of long-chain fatty acids have an unusual property that results from the presence, in a single molecule, of a part that is **hydrophobic** (the hydrocarbon unit) and one that is **hydrophilic** (the carboxylate anion). Because of the hydrophilic groups, the salts of long-chain fatty acids dissolve readily in water. In so doing, they form structures known as **micelles,** which are globular collections of many molecules in which most of the carboxylate anion ends are on the outer surface.

In a micelle, the carboxylate "heads" are located on the surface, where they can interact with the surrounding water, whereas the hydrocarbon "tails" are intertwined in the center. This central region is held together by van der Waals attractive interactions, like those that exist between polymer chains. However, the main driving force for forming micelles is the minimization of contact between the hydrocarbon portions of the fatty acid chains and water, which would disrupt hydrogen bonding between water molecules.

Micelles are roughly spherical simply because this shape has the minimum surface-to-volume ratio, and the smaller the surface area, the less disruption of hydrogen bonding between water molecules there is. Although the chains have approximately the same "diameter" along their length, the volume available to accommodate them decreases toward the center of the sphere.

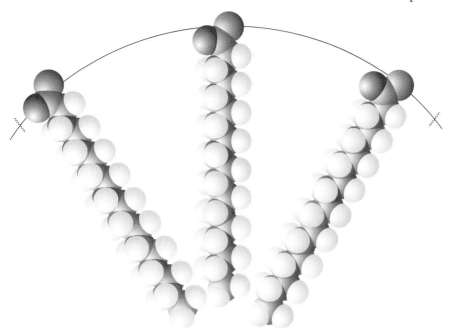

Thus, the spherical shape of a micelle makes it impossible for the hydrocarbon chains to adopt an all-*anti* conformation, and the chains instead become quite disordered in the interior of the micelle (as shown in Figure 17.2, on page 17-8). Because the central region of a micelle is structurally ill-defined, other hydrocarbon-like molecules can readily dissolve in this region. This is the basis for the cleansing action of soaps: the oily, water-insoluble materials that bind dirt to clothing and bodies dissolve in the interiors of micelles.

The calcium salts of long-chain fatty acids are not water-soluble and form a scummy film as they precipitate. Because many ground sources of water pass through limestone deposits and thus have high concentrations of calcium ions, soaps have been replaced for many cleaning purposes by detergents, which are similar in structure but are derivatives of sulfonic acids. The calcium (and other metal) salts of sulfonic acids are more soluble in water than are fatty acid salts.

$$SO_3^{\ominus}\ Na^{\oplus}$$

$CH_3(CH_2)_n$

**Sodium alkylbenzene sulfonate**

$$CH_3(CH_2)_n SO_3^{\ominus}\ Na^{\oplus}$$

**Sodium alkyl sulfonate**

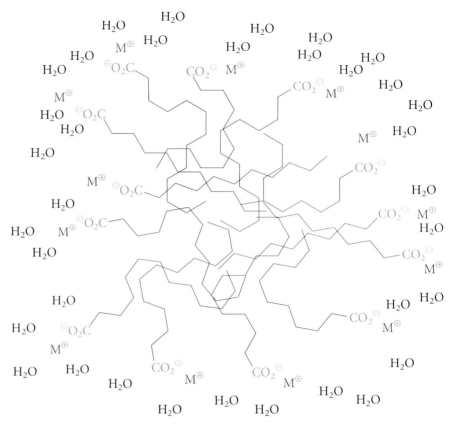

**FIGURE 17.2**

At concentrations above a critical value, long-chain fatty acid molecules associate to form a micelle. The hydrocarbon tails of these molecules are directed toward the core of the micelle, making the core hydrophobic. The polar carboxylate anion ends of the molecules are hydrophilic and are directed toward the external surface, producing a highly polar interface between the micelle and the aqueous solution. This interface contains both the carboxylate anion and its counterion, $M^{\oplus}$, as well as substantial quantities of solvating water molecules.

In the past, phosphates were also added to detergents to solubilize calcium salts. This practice is currently banned for most cleaning agents (dishwashing detergents being one exception), because phosphate serves as a source of phosphorus, resulting in the overgrowth of algae and other undesirable plants in streams, rivers, and lakes.

**Bilayer Membranes.** The fundamental structure of a micelle is the result of its component molecules having distinctly different hydrophobic and hydrophilic regions. An alternative to the roughly spherical structure of a micelle is the sandwich structure of a **bilayer.** Like the micelle, this is a self-organized, three-dimensional array. In living cells, such arrangements are found at many of the boundary regions known as **membranes.** A membrane forms the outer surface of a cell, separating the cell from the external aqueous medium. In addition, membranes form the boundaries of in-

## CHEMICAL PERSPECTIVES

### ORGANIC CHEMISTRY IN THE COMPUTER INDUSTRY

As scientists learn more about the earth's complex ecosystems, it becomes necessary to reevaluate the use of some chemicals. For example, chlorocarbons such as trichloroethane have many industrial uses because they have a good balance of solvent properties for both hydrocarbon-like and polar materials, are relatively nontoxic, and do not burn readily. They have been used as cleaning agents in both the dry-cleaning business and the computer industry, where the removal of all contaminants is essential for the production of microcomputer chips containing billions of circuits.

Unfortunately, some of these solvents inevitably escaped into the atmosphere, where they ultimately migrated to the upper layers and contributed to the destruction of the ozone layer. Recently, AT&T has experimented with replacing trichloroethane with other solvents as cleaning agents for chip manufacture. One satisfactory substitute is *n*-butyl butyrate. Although the *n*-butyl butyrate used by AT&T is synthetic, this ester is found naturally in many fruits, including cantaloupes, peaches, plums, and pineapples. Its smell is characteristic of ripe bananas.

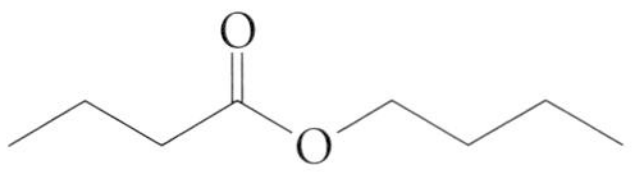

*n*-Butyl butyrate

ternal structures of the cell, such as the nucleus, mitochondria, and endoplasmic reticula.

The fundamental building blocks of membranes are **phospholipids,** which are similar in structure to the fatty acid esters used in making soaps. Phospholipids are dicarboxylate–monophosphate esters of glycerol (Figure 17.3). The phosphoric acid unit of a phospholipid is itself a diester, monoesterified on one side by one of the hydroxyl groups of glycerol and on the other by one of a number of alcohols that also contain an amino group.

$R_1 = $ Saturated fatty acid

$R_2 = $ Unsaturated fatty acid

$R_3 = -CH_2-CH_2-\overset{\oplus}{N}(CH_3)_3$    $\alpha$-Lecithins (phosphatidylcholines)

$= -CH_2-CH_2-NH_2$    Cephalines (phosphatidylethanolamines)

$= -CH_2-\underset{\underset{NH_2}{|}}{CH}-CO_2H$    Phosphatidylserines

**FIGURE 17.3**

Phospholipids are esters of glycerol, $HOCH_2CH(OH)CH_2OH$, in which two of the ester bonds are to fatty acids and the third is to a phosphoric acid derivative.

In aqueous solution near pH 7, the phosphoric acid is ionized and the amine protonated. The resulting zwitterion is a very polar species that imparts hydrophilic character to the head of the phospholipid (the upper end of the space-filling model of phosphatidylserine). In contrast, the remainder of a phospholipid molecule bears the long-chain hydrocarbon tails of the saturated and unsaturated fatty acids and is therefore quite hydrophobic. The phospholipid molecules are free to twist, rotate, and generally reorganize within the membrane. However, the strong interactions between the aminophosphate groups and the external aqueous environment prevent these groups from penetrating into the hydrocarbon interior of the membrane.

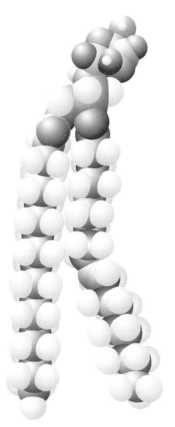

**Phosphatidylserine**

In a cell membrane, the phospholipids form a structure known as a **fluid mosaic,** which is basically a bilayer. This structure is formed by two layers of molecules whose hydrocarbon tails point toward each other. Figure 17.4 is a schematic rendering of a lipid bilayer formed from phospholipids. This lipid bilayer structure was first proposed in 1925 by two Dutch chemists, E. Gorter and F. Grendel. They based their structural suggestion on the observation that twice as much lipid is present in the total mass of a red blood cell as would be required to cover it with a single layer.

A lipid bilayer is held together by the association of the polar head groups of the phospholipids with water and by the energetic *dis*advantage

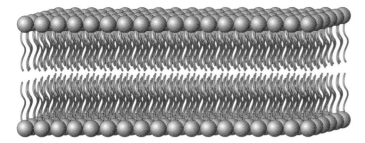

**FIGURE 17.4**

A lipid bilayer is formed when the hydrocarbon tails (represented in gray) of a phospholipid are directed toward each other, away from the surrounding aqueous solution, whereas the polar head groups (red spheres) are exposed to the aqueous solution.

of the hydrocarbon tails disrupting hydrogen bonding of the solvent. The van der Waals interactions between the hydrocarbon tails represent a relatively minor contribution. The three-dimensional structure of the bilayer is flexible and can be readily distorted and reshaped in much the same way as polyethylene. The name *fluid mosaic* is consonant with this lack of rigidity in the bilayer structure.

The bilayer structure of cell membranes is complicated by the presence of a number of other molecules that serve very important auxiliary functions (Figure 17.5). For example, there are protein molecules that have

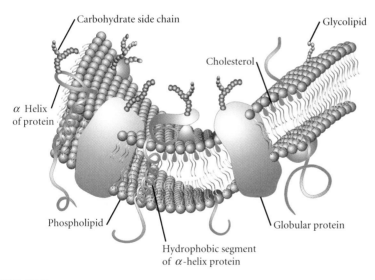

**FIGURE 17.5**

Large molecules forming "channels" that span the thickness of the lipid bilayer are active in the transport of neutral molecules or ions from one side of the membrane to the other. Often these channels are composed of helical peptides, sometimes containing a large number of ionized (charged) side chains. Such peptides are particularly effective at transporting ions and small drug molecules.

---

*CHEMICAL PERSPECTIVES*

BUBBLES

Foaming is associated with cleaning power in soaps and detergents. However, foaming is not connected with micelle formation, and thus foaming cleaners are not necessarily more effective in removing oily dirt. Nonetheless, people expect soaps to foam, and so chemists in the detergent industry often formulate products that foam in order to obtain better acceptance in the marketplace. In some circumstances, foaming can be harmful, for example, in dishwashers and clothes washers that lack the space to hold the foam.

Foaming can also denature protein molecules, which unfold their structures as they stretch from one side of a bubble to the other. For this reason, care must be taken not to shake protein preparations, and vials of insulin suspensions should be mixed by rolling between the hands rather than by shaking. Perhaps James Bond, in insisting that his vodka martinis be "shaken, not stirred," wanted to be sure than any enzymes present were denatured!

---

hydrophobic exteriors and internal regions containing several charged amino acid residues. Because of their hydrophobic exteriors, these proteins "dissolve" readily in the hydrophobic interior of the phospholipid bilayers. Collections of these protein molecules form tubes that connect one side of the membrane with the other and function as "channels" through which polar or charged molecules can move. Without these channels, it would not be possible for polar biological molecules and metal ions to pass through the membrane and into the interior of the cell. Small biomolecules having high polarity are "coated" with a very tightly held solvation shell of water molecules, which in turn are hydrogen bonded to the surrounding water molecules. In the absence of membrane channels, these water molecules would have to be shed, and the associated hydrogen bonds would have to be broken, to permit these polar molecules to pass into the inner, hydrocarbon region of the bilayer.

When a bilayer is extended to form a closed surface, the resulting structure is called a **vesicle,** shown in a cutaway view in Figure 17.6. A vesicle differs from a micelle in being a hollow structure having two hydrophilic surfaces, an outer and an inner; a micelle has only an outer hydrophilic surface.

Vesicles may be useful as vehicles for the delivery of pharmaceutical agents. Many therapeutic agents, though quite effective if delivered to a specific site of action, cannot reach the site before being destroyed by enzymes. For example, peptide hormones are known to control essential functions such as the reproductive cycle, and other peptides such as insulin play important regulatory roles as catalysts, controlling key biochemical pathways. It is not possible to correct a deficiency of these peptides by administering them orally, because enzymes in the stomach, called **peptidases,** rapidly cleave peptide bonds as part of the natural process of digestion. A peptide hormone "encapsulated" inside a vesicle might be protected by this coating until the vesicle, with its contents, has passed through the stomach wall.

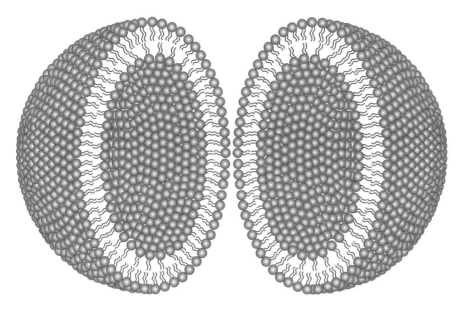

**FIGURE 17.6**

When an extended bilayer folds on itself, it forms a vesicle, a hollow spherical array (shown here cut in half). A vesicle contains an inner core of aqueous solution separated by the bilayer membrane from the aqueous solution in which the vesicle is suspended.

---

**EXERCISE 17.4**

It is important to understand how little material is required to form a bilayer membrane. Assume that a molecule is approximately cubic and measures 20 Å (1 Å = $10^{-7}$ mm) on a side.

(a) How many of these molecules would it take to build a cube that is 0.001 mm on a side (approximately the size of a human red blood cell)?

(b) How many molecules are on each surface of this cube?

(c) What fraction of the total volume occupied by the component molecules is accounted for by the surface layer?

---

### Terpenes

All the lipids described so far have been derived from linear carbon chains. We now turn to another class of lipids known as **terpenes,** which are derived from a branched five-carbon unit, **isopentenyl pyrophosphate.** (Terpenes are sometimes called **isoprenoids** because they formally consist of skeletal polymers of isoprene, which is 2-methyl-1,3-butadiene.)

**Isopentenyl pyrophosphate**          **Isoprene**

Most terpenes have a carbon count that is a multiple of five and have methyl group substituents arranged as in isoprene itself. An example is the terpene geraniol.

**Geraniol**

Not infrequently, a terpene may lack one or more of the expected number of carbons, but the basic arrangement of isoprene units is maintained. This head-to-tail arrangement, found in geraniol, links carbon atoms at opposite ends of the isoprene structural subunit. Head-to-head and tail-to-tail connections are exceptions to the rule.

**Head-to-tail**　　　　　**Head-to-head**　　　　　**Tail-to-tail**

A key distinction between the fats and waxes and the terpenes is that the molecular subunits of fats and waxes are joined by ester linkages, whereas the isoprene units of terpenes are joined by carbon–carbon bonds. For this reason, fats and waxes are often referred to as **saponifiable lipids,** and terpenes are called **nonsaponifiable lipids.** Another distinguishing feature is that terpenes often bear extensive oxygen substitution along the chain (as alcohols or carbonyl groups). Such oxygen functionality is usually absent in fats and waxes.

The terpenes are an extraordinarily diverse group of natural compounds. (There are more than 22,000 known structures.) This seems surprising given their common origin from isopentenyl pyrophosphate. The study of terpenes is made all the more complex because the functions of most of the terpenes present in plants are not known. Some of these compounds may at one time have served as defensive agents against predators (for example, insects) that no longer exist, and the plants synthesizing them have not yet evolved to the point where they no longer waste the energy of producing these now-unneeded chemical defense agents. It is also possible that some terpenes are simply chemical by-products of biochemical transformations in plants that are not as well tuned as those in mammals. Later we will discuss the importance of several terpenes (the steroids) whose biological functions in humans are well established.

*Classification of Terpenes.* Terpenes are subdivided into classes according to the number of isoprene units they contain (Figure 17.7).

Most terpenes that have been isolated have an even number of isoprene units. **Monoterpenes** are molecules with 10 carbons, accounting for two isoprene units, **diterpenes** have 20 carbons (four isoprene units), **triterpenes** have 30 carbons (six isoprene units), and **tetraterpenes** have 40 car-

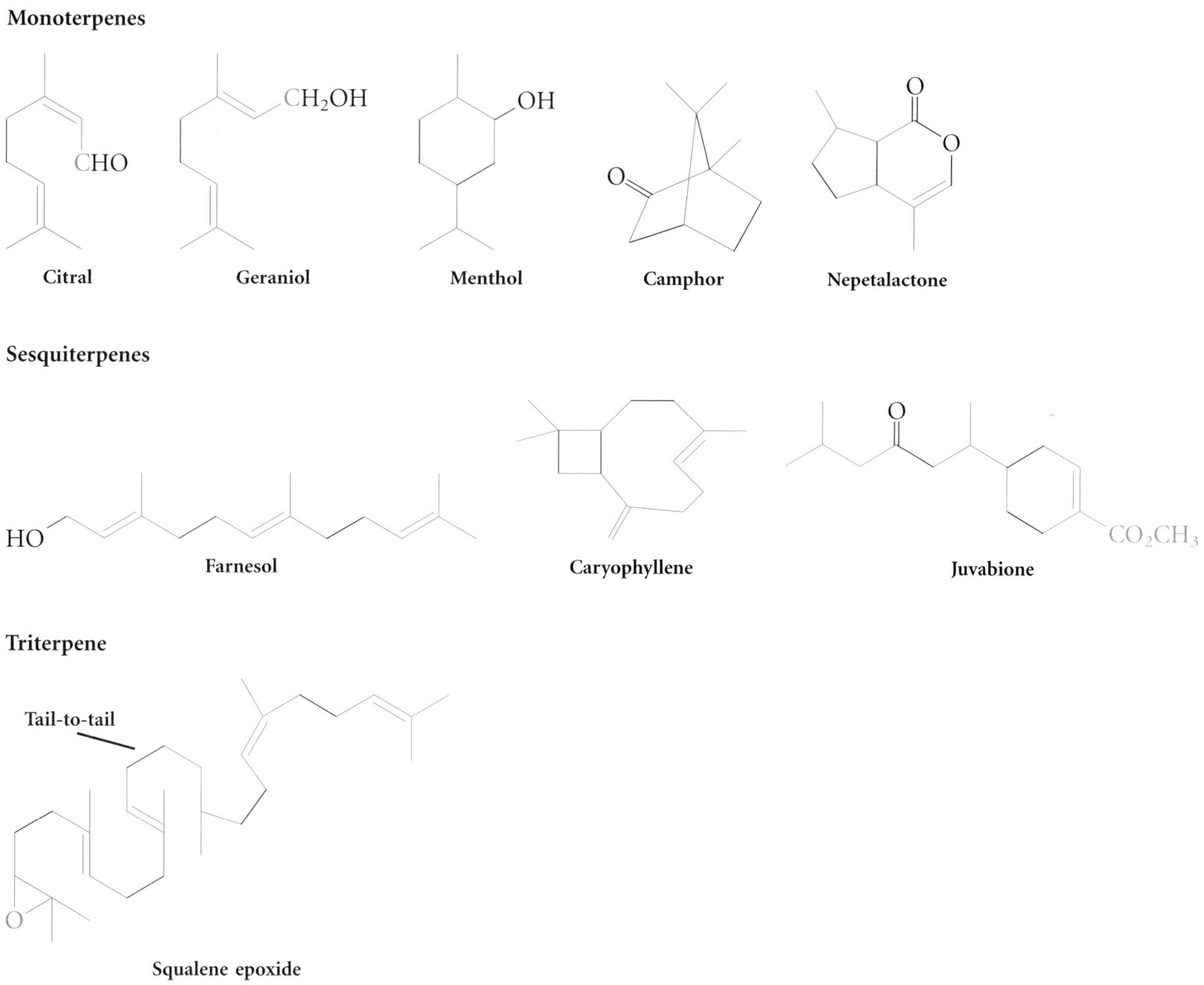

**Monoterpenes**

Citral  Geraniol  Menthol  Camphor  Nepetalactone

**Sesquiterpenes**

Farnesol  Caryophyllene  Juvabione

**Triterpene**

Tail-to-tail

Squalene epoxide

**FIGURE 17.7**

Representative isoprenoid terpenes containing 10 (monoterpene), 15 (sesquiterpene), and 30 (triterpene) carbon atoms. The isoprene units are indicated in blue. Squalene epoxide is made up of two 15-carbon terpenes linked in the middle by a tail-to-tail connection and is thus an exception to the head-to-tail linking of isoprene units.

bons (eight isoprene units). In addition, a significant number of terpenes with three isoprene units (15 carbons) have been isolated; these are the **sesquiterpenes.** A smaller group, the **sesterterpenes,** are terpenes composed of five isoprene units.

A number of biologically important compounds are terpenes—for example, lanosterol, a triterpene, is a biosynthetic precursor of cholesterol. Lanosterol cannot be broken into isoprene units because migration of methyl groups occurs in the cation of the isoprenoid structure that is its immediate biochemical precursor. These migrations are only some of the steps in the complex biosynthesis of cholesterol:

**17-15**

## NATURAL INSECTICIDES

Many oxygenated compounds that are toxic to insects have been isolated from plants. Indeed, some, such as the pyrethrins (from chrysanthemum flowers), are marketed as insecticides. Others, such as azadirachtin (from the seeds of the neem, a common tree in India and Burma), are more difficult to isolate from the plant. Because they are also too complicated to synthesize in quantities sufficient for large-scale use, they are quite expensive.

**Pyrethrin I**

**Azadirachtin**

**Lanosterol**

Methyl group migrations

**Cholesterol**

Substantial contributions toward working out this sequence were made by the American chemist Konrad Bloch, who received the Nobel prize in medicine in 1964, and by the British chemist John Cornforth, who received the Nobel prize in chemistry in 1975. Cholesterol is derived from lanosterol by a process that removes three methyl groups. Because cholesterol contains only 27 carbons, it is not a triterpene and is properly referred to as a **nortriterpene.** The prefix *nor-* indicates a substance whose structure closely resembles that indicated by the rest of the name but is lacking some small feature, usually a methyl group. Carbon degradation processes are encountered in the biosynthesis of other terpenes, and all classes of terpenes have norterpene members.

**Biological Activity of Terpenes.** Although the biological functions of most plant terpenes are not known, a number of terpenes and terpene-like compounds have pronounced biological effects in human beings and other mammals. For example, nepetalactone is the active constituent in catnip, a plant that affects many cats:

**Nepetalactone**

Many terpenes have pleasant odors and are used extensively in perfumery. In high concentrations, terpenes such as camphor (see Figure 17.7) have a strong, yet not unpleasant, odor. At one time, they were used as principal constituents of quack medicines and nostrums.

## EXERCISE 17.5

Identify the five-carbon fragments corresponding to the isoprene units in each of the following terpenes:

(a)

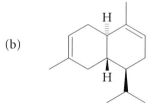

**Ambrosin**

(d)

**Citronellal**
(used as an insect repellent)

(b)

**β-Cadinene** (one of many sesquiterpenes from cedar trees)

(e)

**Grandisol** (sex attractant of the cotton boll weevil)

(c)

HO$_2$C ... CO$_2$CH$_3$

**Bixen** (used as a golden-yellow food coloring)

A number of terpenes, some with very complex structures, have been shown to have potent activity in inhibiting the division of cancerous cells. Most of them have proved too toxic to be of value in chemotherapy. A notable exception is taxol, used as a chemical agent for suppressing the growth of hard tumors:

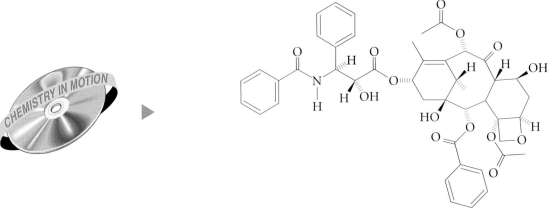

**Taxol**

Unfortunately, taxol is found in only very small amounts in the bark of the Pacific yew tree. Although this species is not rare, especially in the western United States and Canada, the bark of three trees is required to isolate sufficient taxol for the treatment of one patient, and harvesting of the bark usually kills the tree. Clinical use of taxol is thus limited by the available supply, and some scientists are concerned that demand for taxol may ultimately result in the extinction of the species.

## Steroids

A **steroid** is any of a large group of naturally occurring compounds that have a fused-ring system of three six-member rings and one five-member ring, as, for example, in cholesterol.

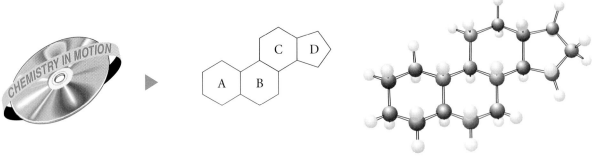

**Steroid structure**

In contrast to the functions of terpenes found in plants, the biological functions of many of the steroids found in animals have been established. For example, despite the negative press given to cholesterol because of its high concentration in the fatty deposits in arterial walls, this steroid is essential to life. Cholesterol is a constituent of cell membranes, or lipid bilayers, and is especially prevalent in the **myelin sheath** that surrounds and insulates nerve axons. Because the brain is essentially a collection of interconnected nerve cells, it is not surprising that cholesterol constitutes 10% of the dry weight of the brain.

The triterpene lanosterol and the nortriterpene cholesterol are formed from the sesquiterpene farnesol in a complex pathway that includes dimerization of farnesol, followed by a large number of other biochemical transformations. Through the common intermediate lanosterol, all of the steroidal triterpenes, which contain the basic skeleton of four fused rings, are derived from the linear sesquiterpene farnesol.

Many steroids have been shown to be chemical messengers, compounds produced in one organ and triggering a response in another. In humans, these hormones fall into two main classes: the corticosteroids regulate metabolic processes essential to life, and the sex hormones determine sexual characteristics and regulate sexual functions (Figure 17.8). The sex hormones are grouped into three categories: **estrogens** (female), **androgens** (male), and **progestins** (pregnancy). **Estradiol,** an estrogen produced by the ovaries, is responsible for secondary female sexual characteristics. It undergoes oxidation to **estrone** before being excreted in the urine. The corresponding male sex hormone, **testosterone,** is produced by the testes and is responsible for secondary male sexual characteristics. Reduction of testosterone produces **androsterone,** which, like estrone, is excreted in the urine.

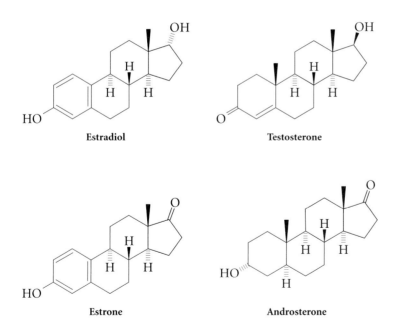

Estradiol

Testosterone

Estrone

Androsterone

**FIGURE 17.8**

Structures of several steroids that function as sex hormones. Like all steroids, they contain a 6-6-6-5 fused-ring system.

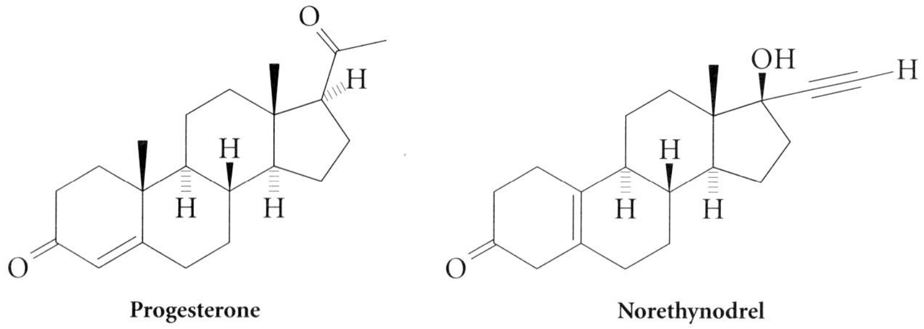

Progesterone                Norethynodrel

**FIGURE 17.9**

The synthetic steroid norethynodrel mimics the hormone activity of the
natural steroid progesterone, suppressing ovulation.

The progestins (Figure 17.9) are hormones specific to pregnancy. For
example, **progesterone** stimulates changes in the uterine lining necessary
for the implantation of an egg and simultaneously suppresses ovulation.
Synthetic steroids—for example, **norethynodrel** (Enovid)—have been developed as effective contraceptives, acting on the ovaries in the same way
as progesterone does. (Progesterone itself cannot be used as an oral contraceptive because it is readily degraded in the stomach.)

## 17.2

# Biosynthesis

Chemists have long been fascinated by the sequences by which biological
molecules are constructed in living systems. As a result, the biosynthetic
pathways are known for many natural products. Often such a synthesis follows closely the route a chemist would select for a laboratory synthesis, using reactions quite similar to those you have learned in the course. We will
examine the biosynthesis of several naturally occurring compounds (as well
as some degradation sequences) in this chapter. This will allow you to review your knowledge of organic reactions and at the same time see how nature goes about converting one molecule to another. We begin with the
biosynthesis of terpenes.

### Biosynthesis of Terpenes

In considering how terpenes are formed biologically, we will concentrate on the key carbon–carbon bond-forming processes. The two major
five-carbon building units are isopentenyl pyrophosphate and its isomer,
dimethylallyl pyrophosphate:

Isopentenyl pyrophosphate                Dimethylallyl pyrophosphate

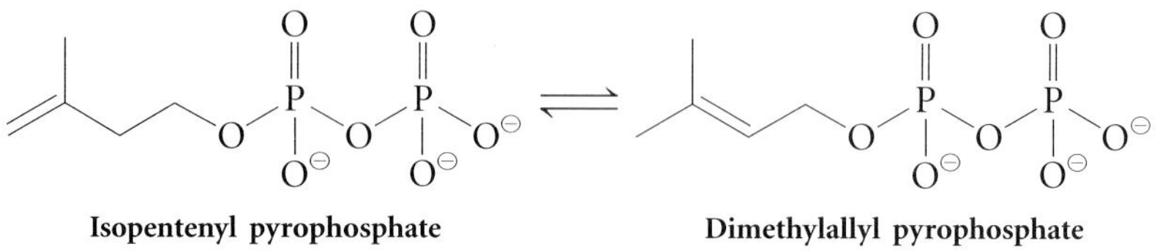

The pyrophosphate group is so named because phosphoric acid forms pyrophosphoric acid (as well as larger molecules) upon heating. Because water is lost, pyrophosphoric acid is an anhydride of phosphoric acid.

Phosphoric acid     Pyrophosphoric acid

The pyrophosphate group plays several diverse roles in biochemical reactions. One very important function is the storage of energy; the anhydride linkage reacts with water in an exothermic hydrolysis that is analogous to the hydrolysis of carboxylic acid anhydrides. In the biosynthesis of terpenes, the pyrophosphate group functions as a good leaving group.

Isopentenyl pyrophosphate and dimethylallyl pyrophosphate differ only in the position of the carbon–carbon double bond:

Isopentenyl pyrophosphate               Dimethylallyl pyrophosphate

These two isomers are easily interconverted in biological systems, and chemists can mimic these transformations in the laboratory, using processes that include cationic intermediates. Protonation of either isopentenyl or dimethylallyl pyrophosphate according to Markovnikov's rule leads to the same tertiary carbocation. In biological systems, an enzyme catalyzes the interconversion.

**EXERCISE 17.6**

Assuming that isopentenyl and dimethylallyl pyrophosphates are in equilibrium, which of these isomeric structures do you expect to be present in greater concentration? Explain.

### Phosphates as Leaving Groups

Isopentenyl and dimethylallyl pyrophosphates are bifunctional molecules, having a pyrophosphate unit as well as a carbon–carbon double bond. The reactivity of a phosphate or pyrophosphate group as a leaving group derives from the stability of the phosphate anion as the conjugate base of a strong acid.

Let's compare the relative rates of hydrolysis of three kinds of esters: those derived from a carboxylic acid, those derived from phosphoric acid, and those derived from sulfuric acid (Figure 17.10). These acids differ greatly in acidity as a direct consequence of the greater stability of the monoanion of sulfuric acid compared with the monoanions of phosphoric and acetic acids.

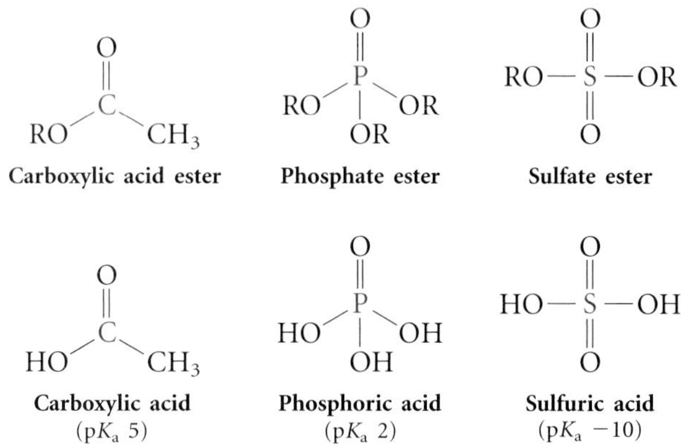

**FIGURE 17.10**

The structures of carboxylic, phosphoric, and sulfuric acids and their esters are analogous, although the stabilities of their monoanions vary.

The chemistry of the esters derived from these acids also varies quite dramatically. As you know, nucleophiles react with carboxylic acid esters in a nucleophilic acyl substitution that proceeds through a tetrahedral intermediate formed by attack at the carbonyl carbon (Figure 17.11). In contrast, sulfonate esters, such as tosylates, are derivatives of alcohols that react with nucleophiles at carbon rather than at sulfur. This contrasting behavior can be directly attributed to the greater ability of the sulfate and sulfonate anions to act as leaving groups compared with the less stable acetate anion. Because the acidity of phosphoric acid is midway between that of acetic acid and sulfuric acid, phosphate esters undergo both types of reactions, leading to products resulting from nucleophilic attacks at both phosphorus and carbon.

### EXERCISE 17.7

Recalling the factors that affect acidity, explain the qualitative differences in acidity (and therefore the trend in stability of the corresponding conjugate bases) of acetic acid, phosphoric acid, and sulfuric acid.

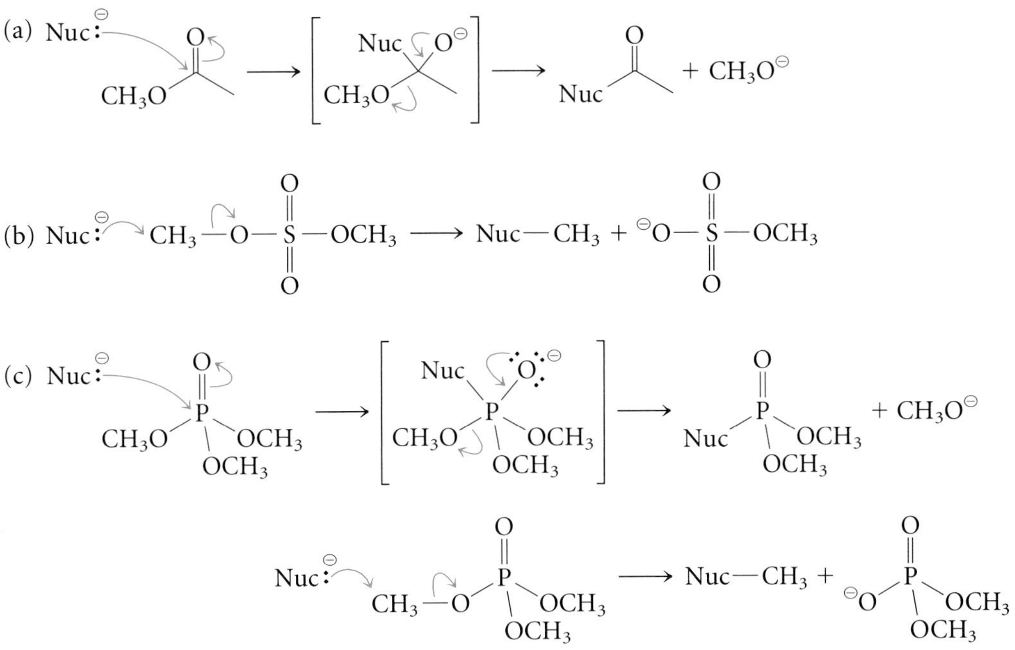

**FIGURE 17.11**

(a) Nucleophilic attack on a carboxylic acid ester takes place at the carbonyl carbon. (b) Nucleophilic attack on a sulfate ester takes place on the alkyl group. (c) Phosphate esters undergo nucleophilic attack at either phosphorus or an alkyl group carbon.

### Biosynthetic Pathways Involving Phosphates

In the biosynthesis of terpenes, the phosphate ester serves the same role as sulfonate or sulfate esters do for the laboratory chemist—they activate carbon toward substitution reactions. Dimethylallyl pyrophosphate is especially activated toward substitution at C-1 because that position is allylic. Thus, carbon–carbon bond formation results from the reaction of dimethylallyl pyrophosphate with the $\pi$ bond of isopentenyl pyrophosphate. The reaction is accompanied by simultaneous loss of a proton to base.

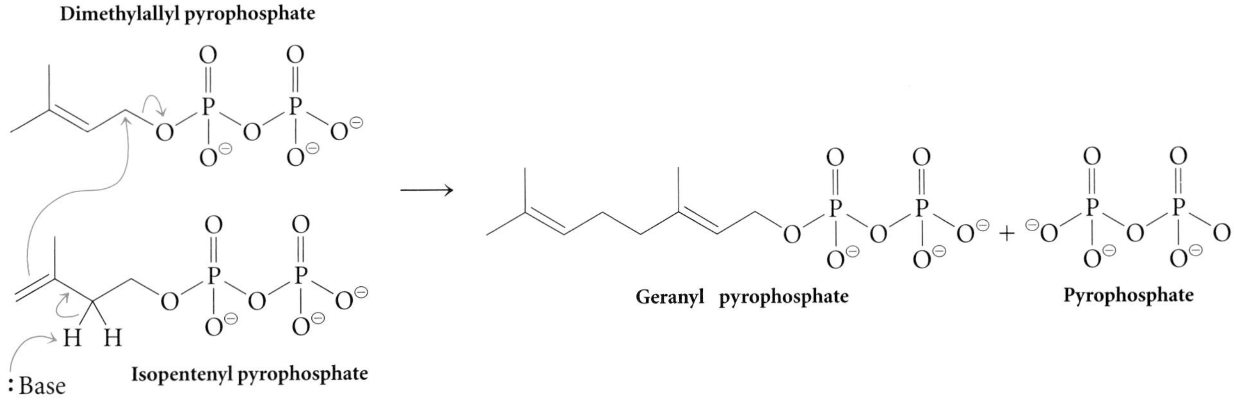

A carbon–carbon bond can then be formed between geranyl pyrophosphate and isopentenyl pyrophosphate, resulting in the construction of a 15-carbon terpene known as farnesyl pyrophosphate:

**Geranyl
pyrophosphate**

**Isopentenyl
pyrophosphate**

**: Base**

**Farnesyl  pyrophosphate**

+

**Pyrophosphate**

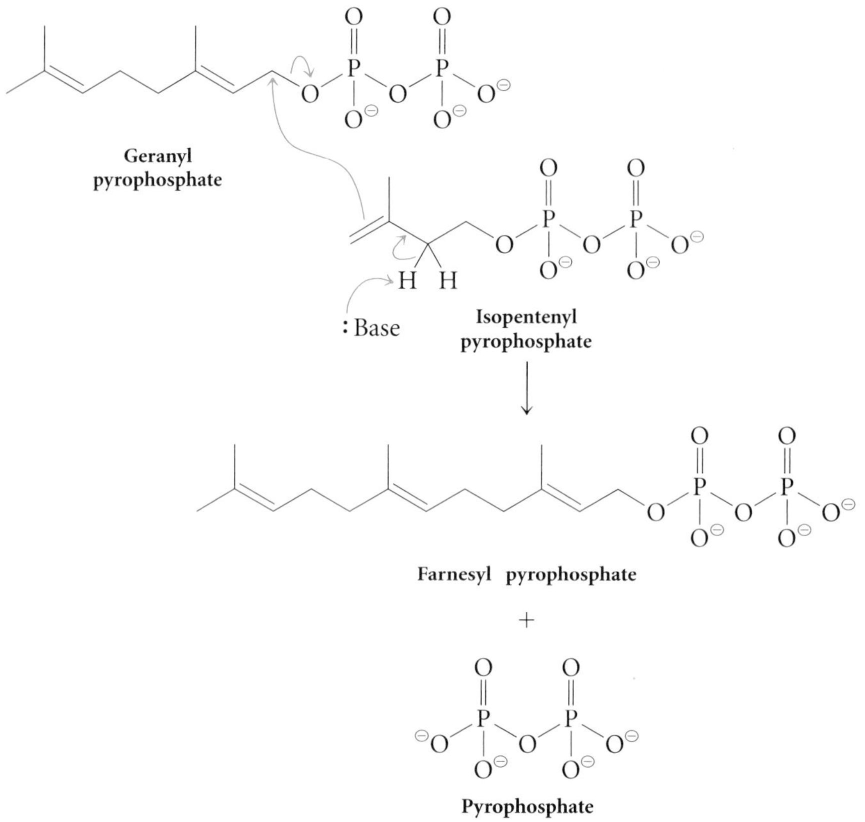

The alcohols corresponding to geranyl pyrophosphate and farnesyl pyrophosphate, **geraniol** and **farnesol,** are found in plants and contribute significantly to the scents of flowers. For example, geraniol has the odor of roses, and both this alcohol and its acetate ester are important materials in the perfume industry.

The carbon–carbon bond-forming process based on the allylic activation of a pyrophosphate group is also important in the formation of cyclic terpenes. For example, **nerol pyrophosphate** (a geometric isomer of geranyl pyrophosphate) undergoes cyclization with loss of the pyrophosphate group to form an intermediate cation. Then loss of a proton gives the monoterpene limonene.

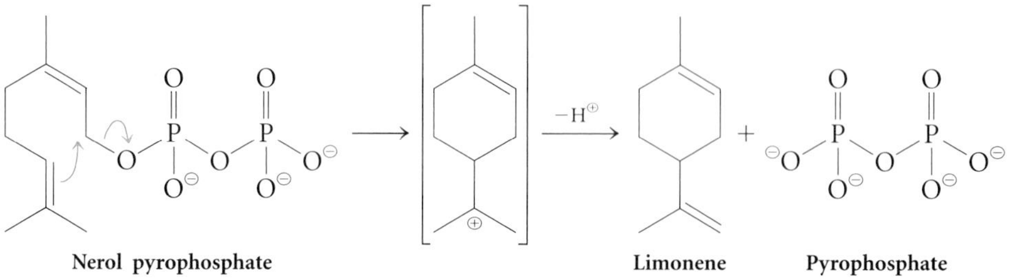

**Nerol  pyrophosphate**

$-H^{\oplus}$

**Limonene**

**Pyrophosphate**

## EXERCISE 17.8

Write a detailed, step-by-step mechanism for the formation of farnesyl pyrophosphate from geranyl pyrophosphate and isopentenyl pyrophosphate.

## EXERCISE 17.9

Limonene is a chiral molecule. Identify its center(s) of chirality. Generally, only one enantiomer of limonene is found in natural sources. However, limonene can be racemized (its enantiomers can be interconverted) by strong acid. Explain.

### Biosynthetic Pathways Involving Epoxides

Carbon–carbon bond formation in living systems is also initiated by other functional groups. For example, squalene monoepoxide undergoes opening of the three-member epoxide ring and cyclization to form a six-member ring, as shown in Figure 17.12. The intermediate cation undergoes further cyclization reactions in which all of the remaining carbon–carbon $\pi$ bonds are replaced by new carbon–carbon $\sigma$ bonds. In biological systems, this process takes place within a single enzyme, and no products that would result from any of the intermediate cations have been isolated.

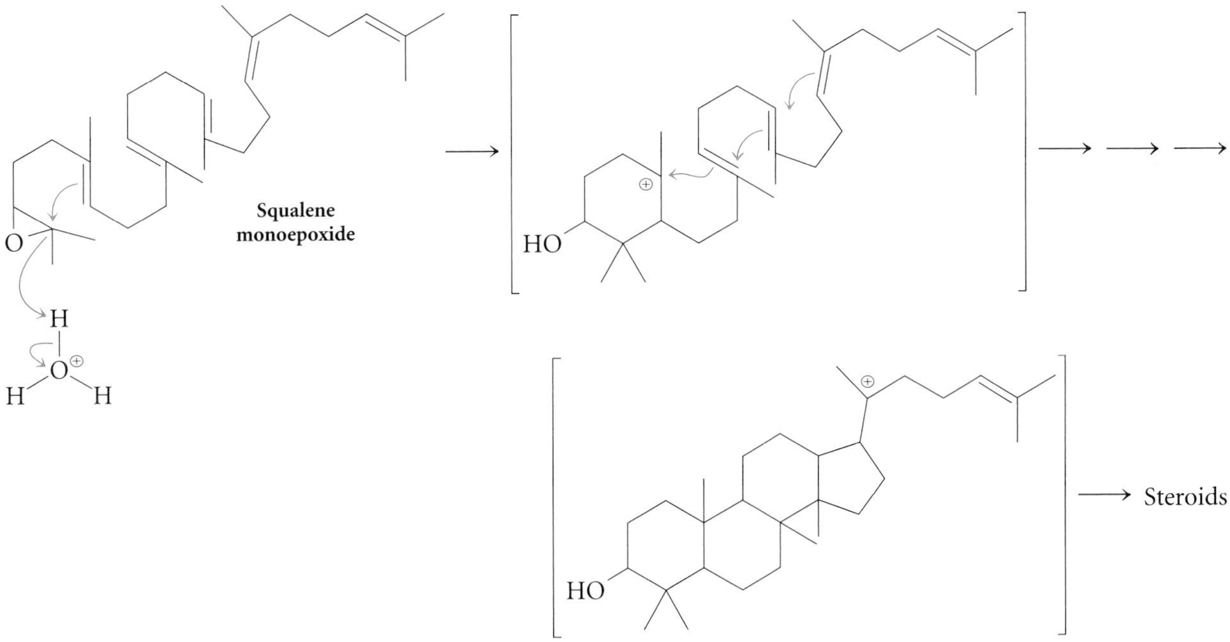

Squalene monoepoxide

**FIGURE 17.12**

The first cyclization of squalene monoepoxide involves the ring opening of a protonated epoxide and simultaneous carbon–carbon $\sigma$ bond formation using the electrons of a $\pi$ bond. The resulting tertiary carbocation undergoes a further cyclization, again forming a carbon–carbon $\sigma$ bond and breaking a $\pi$ bond. Two further cyclizations result in a carbon cation with the steroid ring system (three six- and one five-member ring).

Although it was once believed that the transformation of an acyclic molecule into a tetracyclic one takes place in a single step and includes only one transition state, it is now thought that it is more likely to happen in discrete stages through a number of cation intermediates. Many attempts to effect the same transformation in the absence of enzymes have led to products with, at most, three of the rings.

### EXERCISE 17.10

Recalling the thermodynamics of the polymerization of ethylene to polyethylene, predict $\Delta H°$ for the conversion of the monocyclic cation illustrated in Figure 17.12 to the tetracyclic cation, assuming no increase in ring strain.

## 17.3

## Carbohydrates

#16   Monosaccharides

The natural materials considered so far have been similar to hydrocarbons, in having a high proportion of carbon and hydrogen atoms, with relatively few heteroatoms. In contrast, the **carbohydrates** generally have one oxygen for each carbon, and their name implies that they are formally hydrates of carbon. For example, the formula of glucose, $C_6H_{12}O_6$, can alternatively be written as $C_6(H_2O)_6$. Glucose is a constituent of the important biopolymers starch and cellulose and is stable in the cyclic hemiacetal form:

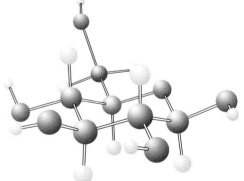

OH

HO

HO

O

OH

OH

**Glucose**
$C_6H_{12}O_6 = C_6(H_2O)_6$
Carbohydrates: $C_n(H_2O)_n$

It is appropriate to examine briefly the hemiacetal function in glucose because it (or the analogous hemiketal) is present in nearly all carbohydrates. Consider a simple, cyclic hemiacetal that undergoes ring opening to form a noncyclic hydroxyaldehyde:

O   OH        OH   CHO

⇌

**A hemiacetal**        **A hydroxyaldehyde**

These structures are quite analogous to those available for glucose, and, in fact, like the natural carbohydrate, this simpler system favors the cyclic form. This is generally the case for hydroxyaldehydes and hydroxyketones for which there is no undue strain in the cyclic hemiacetal (or hemiketal)—that is, when the cyclic form is either a five- or a six-member ring. In contrast, the combination of an aldehyde (or a ketone) with a second molecule containing a hydroxyl functional group is not favored:

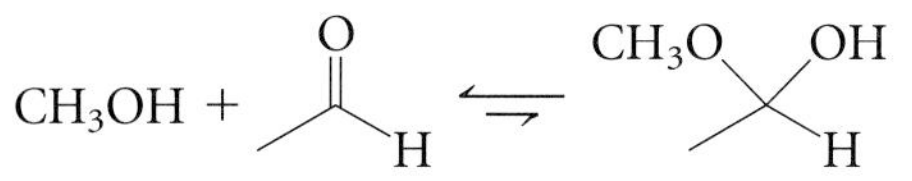

In this case, the separate molecules of alcohol and aldehyde are more stable than when they are combined into a hemiacetal. It is not clear why hemiacetal formation is favored to a greater extent for the cyclic system than for the reaction of separate alcohol and aldehyde molecules, although entropic factors are undoubtedly important. (Two molecules must be joined to form a noncyclic hemiacetal.)

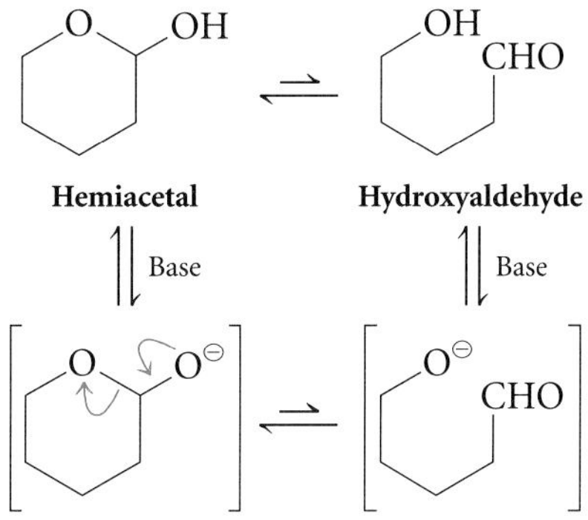

The conversion of a hemiacetal to a ring-opened hydroxyaldehyde in basic solution involves deprotonation to form an alkoxy anion. This anion then undergoes cleavage of the carbon–oxygen bond of the ring with simultaneous formation of the carbon–oxygen $\pi$ bond of the aldehyde. Protonation of the alkoxide produces the ring-opened hydroxyaldehyde.

## EXERCISE 17.11

The interconversion of an acyclic hydroxyaldehyde and a hemiacetal can also be effected by acid catalysis. Write a reaction mechanism for the acid-catalyzed conversion:

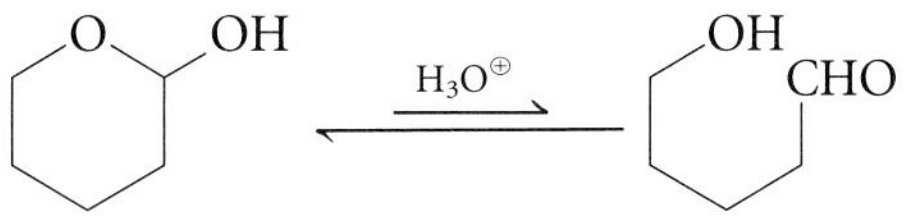

Carbohydrates with three, four, five, and six carbons are called **trioses, tetroses, pentoses,** and **hexoses,** respectively. Representatives of each of these classes are commonly found in nature. Note that all of the names of the simple carbohydrates end in *-ose* and that the number of carbon atoms is represented by a prefix derived from the Greek.

### Trioses

D-Glyceraldehyde and dihydroxyacetone are the simplest examples of trioses (Figure 17.13). The **stereochemical designator** D used in describing the terminal chiral centers in sugar molecules refers to an arrangement about a center of chirality that matches the three-dimensional arrangement in D-glyceraldehyde (that is, *R*). Thus, this compound serves as the three-dimensional reference point for all similarly functionalized molecules. Generally, only D enantiomers of carbohydrates occur naturally.

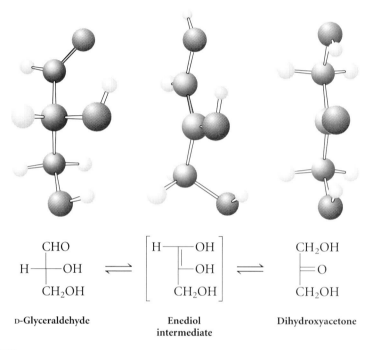

D-Glyceraldehyde      Enediol
intermediate      Dihydroxyacetone

**FIGURE 17.13**

The two trioses D-glyceraldehyde and dihydroxyacetone are interconverted through an intermediate enediol.

Glyceraldehyde, like other carbohydrates, can be depicted by a Fischer projection. Although Fischer projections do not adequately convey the three-dimensional shapes of molecules, they provide a means for comparing stereochemistry. (Recall that, in a Fischer projection, the vertical bonds to a center of chirality recede away from the viewer, and the horizontal bonds come toward the viewer.)

These two simple carbohydrates, D-glyceraldehyde and dihydroxyacetone, are also the simplest members of the two classes into which all carbohydrates are subdivided. Glyceraldehyde and other carbohydrates containing an aldehyde carbonyl group are referred to as **aldoses;** those like dihydroxyacetone, with a ketone, are called **ketoses.** Thus, these two sugars are also referred to as an **aldotriose** and a **ketotriose.** Upon enolization by treatment with acid or base, both of these trioses form the same enediol.

Although this enediol is less stable than either carbonyl form, it does serve as an intermediate for the conversion of one carbohydrate to the other. This process can be readily catalyzed by base or acid.

### EXERCISE 17.12

Using curved arrows to indicate the flow of electrons, write a detailed reaction mechanism for the acid-catalyzed conversion of glyceraldehyde into dihydroxyacetone. Do the same for the base-catalyzed conversion.

### Aldotetroses

Both D-erythrose and D-threose, common tetroses, contain two chiral centers (Figure 17.14). They differ only in the configuration at C-2 (counting from the aldehyde end of the molecule). Because there are two centers of chirality in these molecules, there are, in fact, four possible stereoisomers. Only the two shown in Figure 17.14 are found in nature. Here, as in almost all naturally occurring carbohydrates, the stereochemistry at the chiral center furthest from the aldehyde end has the D configuration; that is, the configuration is the same as in D-glyceraldehyde.

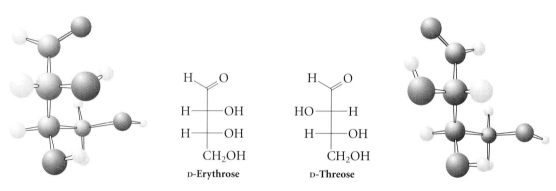

**D-Erythrose**      **D-Threose**

**FIGURE 17.14**

Two common tetroses are D-erythrose and D-threose, each of which has two centers of chirality.

### EXERCISE 17.13

Determine whether erythrose and threose can be interconverted by reversibly forming the enediol. Why is this possible, or not possible?

### EXERCISE 17.14

Tetroses can form cyclic hemiacetals. Draw a structure (with stereochemistry) for the cyclic hemiacetal of erythrose, and write a complete mechanism for its formation under acidic conditions.

## Aldopentoses

Because only D-carbohydrates generally occur naturally, it is possible to isolate only half the number of possible stereoisomers from natural sources. As a consequence, only four D-pentoses have been isolated, despite the fact that three chiral centers allow for eight ($2^3$) possible stereoisomers. These four pentoses, shown in Figure 17.15, are D-**ribose**, D-**arabinose**, D-**xylose**, and D-**lyxose**.

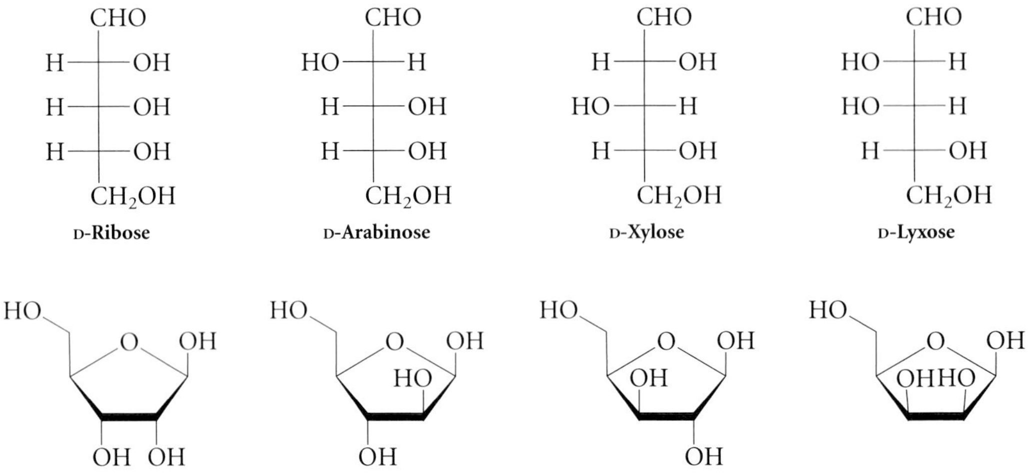

**FIGURE 17.15**

The four D-pentoses are represented here both as Fischer projections and as cyclic hemiacetal structures. (Only the $\beta$ anomers are shown.)

Of the four D-pentoses, only the first three are found to any extent in nature. Xylose is the second most prevalent simple sugar, after glucose. It is found primarily in a polymer that is very similar to cellulose. Arabinose is an exception to the general rule that naturally occurring carbohydrates are of the D-configuration; significant amounts of the L form also occur in nature. A fifth pentose of great importance is D-2-deoxyribose; it is a significant constituent of the biopolymer **deoxyribonucleic acid (DNA).** The prefix *deoxy-* means that an oxygen functional group (in this case, an OH group) bonded to carbon has been replaced by a hydrogen. Ribose is the corresponding carbohydrate subunit in the **ribonucleic acids (RNA).**

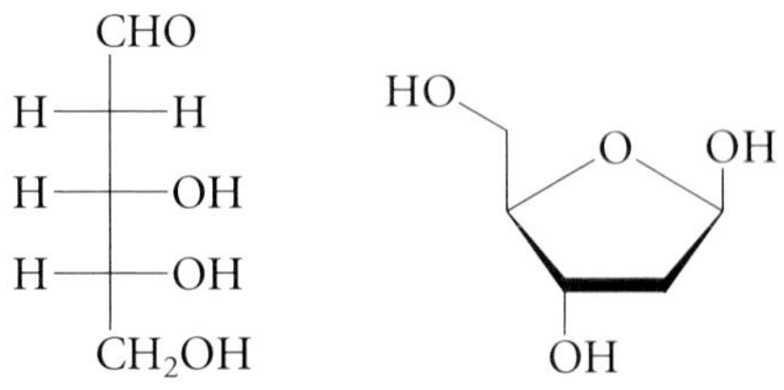

D-2-Deoxyribose

*Furanose and Pyranose Forms.* The common pentoses in Figure 17.15 are shown in both their acyclic forms as Fischer projections and their cyclic, hemiacetal forms. These five-member cyclic hemiacetals are known as **furanoses,** a name derived from that of the simple compound furan. Alternatively, the oxygen on C-5 can be bonded to C-1, forming a six-member cyclic hemiacetal, which also has α and β stereoisomers. Cyclic hemiacetals in the carbohydrate series having six-member rings are named **pyranoses,** after the simple parent pyran. Although ribose is present mainly (76%) in the pyranose form in aqueous solution, the free carbohydrate constitutes only a small fraction of the ribose found in nature. In the vast majority of cases, ribose is joined with other molecules and is in the β-furanose form.

*Anomers.* A cyclic hemiacetal (or hemiketal) of a carbohydrate has one more chiral center than is present in the noncyclic hydroxyaldehyde (or hydroxyketone). Thus, there are two diastereomeric forms of each cyclic acetal, and these are designated as the α and β **anomers.** The α and β anomers of the furanose and pyranose forms of D-ribose are shown in Figure 17.16. The β anomer in Figure 17.16 is defined for ribose, and for all carbohydrates, as the stereoisomer in which the hydroxyl group at C-1 is *cis* relative to the last carbon of the chain (in the case of ribose, C-5). In the pyranose forms of ribose, there is no stereochemistry at the last carbon, and α and β are assigned by analogy with the furanose forms. The α anomer has the opposite configuration at C-1, the hemiacetal carbon. That is, the α anomer is the stereoisomer in which the hydroxyl group at C-1 is *trans* to the last carbon of the chain.

**Furanose Forms**

Furan

**Pyranose Forms**

β-D-Ribose

α-D-Ribose

Pyran

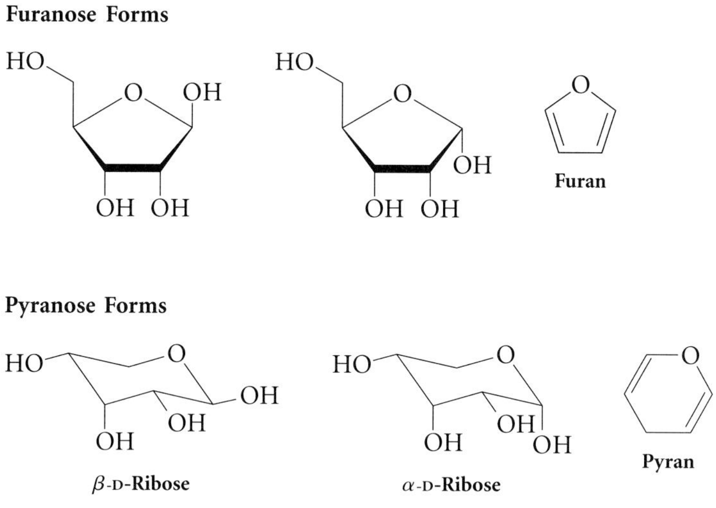

**FIGURE 17.16**

The α and β forms of cyclic hemiacetals differ in configuration at C-1. In a furanose form, a five-member ring is present; in a pyranose form, it is a six-member ring.

## Aldohexoses

Hexoses, with six carbons, have four stereocenters. As for the smaller, four- and five-carbon carbohydrates, the only common, naturally occurring forms of the hexoses have the D configuration, limiting the total number of naturally occurring stereoisomers to eight. Of the hexoses, only the three illustrated in Figure 17.17—glucose, mannose, and galactose—are commonly found in nature. Glucose is by far the most prevalent; it is also the only hexose in which (in the β form) all the hydroxyl substituents on the six-member ring are in equatorial positions. The other two hexoses found in nature, mannose and galactose, differ from glucose at only one chiral center. Thus, in their β forms, each of these sugars has one axial substituent on the ring; the other substituents are equatorial.

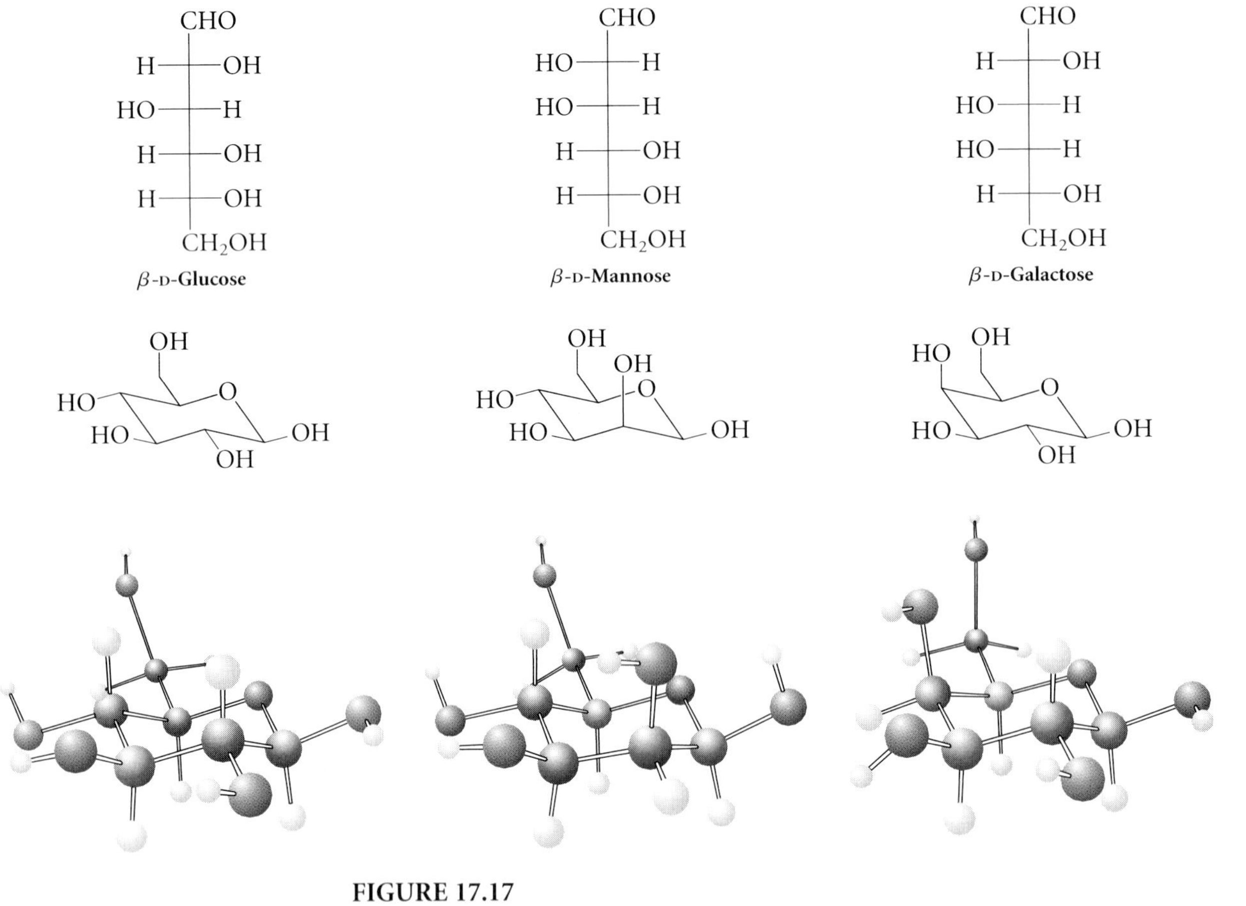

FIGURE 17.17

The three hexoses found commonly in nature are shown here as Fischer projections and as the β-anomers of the cyclic hemiacetals.

### EXERCISE 17.15

It is convenient to use the D,L notation to describe the stereochemistry of the last stereocenter in carbohydrates. Unfortunately, no convenient system exists for the specification of the stereochemistry at the other stereocenters. For practice, assign Cahn–Ingold–Prelog (R,S) designations to all the chiral centers of glucose, mannose, and galactose. (*Hint*: Keep in mind that interchanging the two horizontal substituents in a Fischer projection inverts the stereochemistry of that carbon. Thus, the problem is much simpler than it may at first appear.)

*Mutarotation.* The $\beta$ and $\alpha$ forms of glucose are chiral diastereomeric molecules, and each occurs in nature as only a single enantiomer. These cyclic forms differ in configuration only at C-1, and consequently are anomers. Each form has a unique specific rotation of plane-polarized light. The specific rotation of the $\alpha$ form of glucose is $+19°$, and that of the $\beta$ form is $+112°$. When each of these pure substances is dissolved in water, it undergoes a relatively slow process of uncatalyzed interconversion with its anomer, resulting in the same equilibrium mixture from either isomer. Because this process produces a change in rotation from that of either pure substance to that of the equilibrium mixture, the transformation is known as **mutarotation.**

### EXERCISE 17.16

Write a complete reaction mechanism for the mutarotation of glucose in acid solution. Do the same for base-catalyzed mutarotation.

The interconversion between the $\beta$ and $\alpha$ forms of glucose (and of all carbohydrates) is very rapid when catalyzed by either acid or base. Interconversion of these anomers can be accomplished by processes that cleave either the bond between C-1 and the oxygen in the ring or the bond between C-1 and the oxygen of the attached hydroxyl group. Although the $\beta$ form, with the hydroxyl group at C-1 in the equatorial position, is more stable than the $\alpha$ form, the difference in energy between the forms is very small—at equilibrium, the $\beta$ form is favored by a factor of only about 2:1.

$\beta$-D-**Glucose**      $\alpha$-D-**Glucose**

65 : 35

*The Anomeric Effect.* Glucose can be converted to a derivative known as a **glucoside** by replacing the C-1 hydroxyl group by an alkoxy group derived from an alcohol. In the simple case in which the alcohol is methanol, the two anomeric products are known as methyl glucosides. The mechanism for their formation is parallel to the conversion of a hemiacetal into an acetal.

**EXERCISE 17.17**

Write a detailed mechanism for the formation of β-D-methyl glucoside from β-D-glucose in the presence of acid. At what point in the sequence is the anomeric stereochemistry determined? Based on your mechanism, do you think there would be any difference in the stereochemistry of the product if the reaction started from α-D-glucose?

At equilibrium, the α anomer of methyl glucoside predominates by a factor of approximately 2:1. The origins of this anomalous preference for the axial position of the substituent on the six-member ring (as well as the relatively small preference for the β orientation of the hydroxyl group in glucose itself) are not well understood. This unusual favoring of the α, or axial, orientation (or, conversely, the disfavoring of the β anomer) is known as the **anomeric effect.** Other carbohydrates also form acetals upon reaction with an alcohol in the presence of acid, and these products are referred to as **glycosides.** The noncyclic C—O bond of the acetal is called the **glycosidic bond.**

## Ketoses

All the carbohydrates considered so far have the carbonyl group at the end of the chain and are thus aldehydes. Although most of the common sugars are aldehydes, the carbonyl group does occur at positions other than C-1; these sugars are known as *ketoses*. A commonly occurring ketose is fructose, which has the carbonyl group on the second carbon of the chain (Figure 17.18).

As you might expect, some of the chemistry of fructose differs from that of the aldoses in the same way that the reactions of ketones differ from those of aldehydes. For example, the aldoses can be oxidized under mild conditions to carboxylic acids, whereas oxidation of fructose and other ketoses requires more vigorous conditions and results in carbon–carbon bond cleavage. However, ketoses and aldoses also have many reactions in common. For example, fructose exists predominantly in a cyclic hemiketal form in solution. Both five- and six-member hemiketals are present, with the latter being favored 4:1 over the former. Although only the β forms are depicted in Figure 17.18, in solution fructose exists as α and β mixtures of both 2-pyranose and 2-furanose.

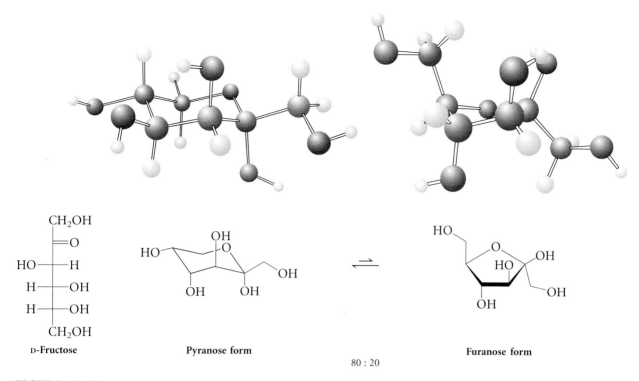

CH$_2$OH
$=$O
HO——H
H——OH
H——OH
CH$_2$OH

D-Fructose        Pyranose form        Furanose form

80 : 20

**FIGURE 17.18**

In solution, equilibrium is established between the six-member cyclic pyranose, the ring-opened form, and the five-member cyclic furanose. The pyranose form is favored over the other two.

**EXERCISE 17.18**

To establish the structure and stereochemistry of glucose, Emil Fischer, who won a Nobel prize in chemistry in 1902 for his characterization of the chemistry of carbohydrates, oxidized both C-1 and C-6 to carboxylic acids. Is the resulting diacid a chiral molecule? What would your answer be if the starting material were mannose? Galactose?

17.4

## Dimeric and Polymeric Carbohydrates

Carbohydrates constitute a very significant fraction of the mass of all biological materials. Indeed, if all the carbohydrates in plant and animal material in the world were converted to carbon dioxide (for example, by burning), the quantity of that gas in the atmosphere would increase by 50%. Most carbohydrates do not occur in nature in their free, monomeric forms. They are most commonly found in the polymers already described—cellulose, starch, and related materials. Carbohydrates are also found as dimers,

CHEM TV® II

#17   Di/Polysaccharides

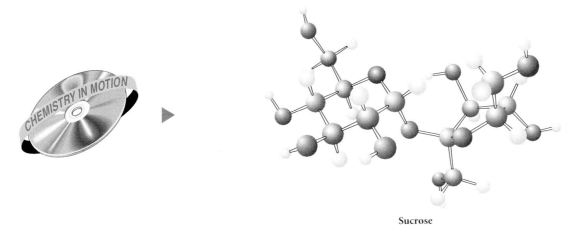

**Sucrose**     **Maltose**     **Lactose**

**FIGURE 17.19**

The three most common disaccharides.

known as **disaccharides,** of which the three most prevalent are **sucrose, maltose,** and **lactose** (Figure 17.19). Of the disaccharides, sucrose (ordinary table sugar) is the most common. It occurs naturally in many plants and is obtained commercially from both sugar cane and sugar beets.

**Sucrose**

Disaccharides result from acetal formation between two monosaccharides. The stereochemistry at the acetal linkage can be either $\beta$, as in lactose, or $\alpha$, as in maltose. In sucrose, the linking oxygen is attached to the carbonyl carbon in both the glucose and fructose subunits. This linkage is $\alpha$ with respect to glucose, but $\beta$ with respect to fructose. In addition, note that the linkage of the monosaccharide units in sucrose is between C-1 of glucose and C-2 of fructose. In maltose and lactose, the linkages are between C-1 and C-4. This type of arrangement, rather uncommon in nature, is referred to as a **bis-acetal,** in which one oxygen is shared between two acetal functional groups.

Hydrolysis (acid-catalyzed) of the acetal linkage of a disaccharide forms the individual component sugars. For example, the bis-acetal linkage be-

## CHEMICAL PERSPECTIVES

### GRITTY ICE CREAM

Because the two anomers of lactose are diastereomers, they have different physical properties. For example, the $\beta$ anomer has a sweeter taste than the $\alpha$, but is less soluble. During long-term storage of ice cream, the $\beta$ anomer crystallizes from the mixture, and these crystals impart a gritty texture to the ice cream.

tween the carbohydrate subunits in sucrose is readily cleaved under acidic conditions, producing a 1:1 mixture of glucose and fructose:

$$\xrightarrow[\text{H}_2\text{O}]{\text{H}_3\text{O}^{\oplus}} \text{Glucose} + \text{Fructose}$$

$$[\alpha]_\text{D} +52° \quad [\alpha]_\text{D} -92°$$
$$1:1$$
$$[\alpha]_\text{D} -20°$$

**Sucrose**

The specific rotation of sucrose differs from that of the 1:1 mixture of glucose and fructose. In fact, the optical rotation of this mixture is opposite in sign to that of sucrose (minus versus plus). For this reason, the 1:1 mixture of glucose and fructose obtained upon cleavage of sucrose is known as **invert sugar,** and the biological catalyst, an enzyme, that catalyzes this process is known as **invertase.** Enzymes are often named by adding the suffix *-ase* to a term descriptive of the transformation that is catalyzed by the enzyme.

Maltose is formed from two $\alpha$-linked glucose molecules, and lactose has a galactose $\beta$-linked to a glucose. In lactose, anomerization is possible at C-1 of the glucose subunit. In approximately 60% of the lactose molecules, the glucose subunit exists in the $\beta$ form, a proportion very similar to that found for glucose itself. Like sucrose, the disaccharides maltose and lactose can be cleaved both with aqueous acid and by enzymes. These enzymes are specialized catalysts that are generally active only for a particular type of linkage. For example, $\alpha$-glucosidase is capable of catalyzing the cleavage of only $\alpha$-glucosidic linkages. Correspondingly, lactose is cleaved by the specific enzyme **$\beta$-galactosidase,** an enzyme that is capable of selectively hydrolyzing a $\beta$ linkage between two sugars. This chemical specificity of distinct enzymes is very important for complex living systems, because many different chemical transformations must occur simultaneously without one interfering with another. In general, di- and polysaccharides having $\beta$ link-

FATTY ACIDS AND SUCROSE: AN UNNATURAL BUT TASTY COMBINATION

The relatively poor diet followed by many people has increased significantly the cost of health care in Western countries. Diets high in salt and fats have been linked to a variety of problems, the most significant of which are diseases of the circulatory system.

Research chemists at Procter & Gamble have sought a substitute for dietary fats that would lend the same taste and texture to foods as are provided by natural fats but would be indigestible. After 30 years (and at a cost of over $200 million dollars) an answer has been found in Olestra, a simple combination of sucrose and fatty acids obtained from oils such as soybean and cottonseed.

The ester linkages in Olestra are too hindered for digestive enzymes to cleave, and this octaester passes through the digestive system untouched, thus contributing neither calories nor fats to the body. Olestra passed all the tests required by the FDA and was approved for limited use in "savory snack foods" that are now on supermarket shelves.

As with everything that is consumed, moderation is in order. Olestra is a big "grease ball" that carries small amounts of the fat-soluble vitamins (A, D, E, and K) with it on its way through the intestinal tract.

**Olestra**

ages, such as cellulose, are not digested by human beings, who lack the enzymes that can cleave this $\beta$ linkage.

### EXERCISE 17.19

Write a mechanism for the acid-catalyzed cleavage of sucrose to form glucose and fructose.

## FOLK MEDICINE AND MODERN PHARMACEUTICALS

As science progressed into the twentieth century, there was a strong feeling of elitist superiority among scientists. For example, "home" remedies for diseases were dismissed as "unscientific" and unworthy of study. Fortunately, the situation changed as many chemists in the pharmaceutical industry realized that there were sound bases for at least some of these medicines.

Some earlier scientists had the insight to realize the importance of folk medicine. For example, in 1795, William Withering, a British physician, heard of a peasant woman who was famous for curing

chronic heart problems. Upon investigation, Withering discovered that one of the ingredients of the cure was the herb foxglove, and he began using this plant to treat patients with congestive heart failure. It was not until much later that chemists isolated a number of different steroidal glycosides (for example, digitoxin) from foxglove and demonstrated their dramatic effects as stimulants of heart action.

Digitoxin is an effective drug that is in use today; however, it must be administered with care because the difference between the effective and lethal doses is comparatively small. Indeed, extracts of this plant have been used as a poison on arrowheads by natives in various parts of the world. Both the ancient Egyptians and Romans used an extract from the sea onion containing similar steroidal glycosides as both a heart tonic and a rat poison. The therapeutic maintenance dose of digitoxin is 0.1 mg (or 1 mg/kg); the lethal dose (in cats) is 200 mg/kg.

**Digitoxin**

## Summary

**1.** The most simple oxygen-containing natural products are the fats and waxes. In fats, long-chain fatty acids are connected by ester functional groups to a glycerol core. In waxes, a simple ester linkage connects a fatty acid to a fatty alcohol. Despite the presence of these ester linkages, fats and waxes consist mainly of straight-chain hydrocarbon units. The esters undergo hydrolysis in strong base (saponification) to form salts of the free fatty acids.

**2.** Lipids can incorporate polar groups such as carboxylate, sulfonate, and phosphate esters. The presence of two sections of distinctly different polarity, one hydrocarbon-like (hydrophobic) and the other polar (hydrophilic), provides the driving force for the assembly of lipids into micelles, bilayers, and vesicles.

**3.** Biological membranes are formed of lipid bilayers. Some large protein molecules form tubes spanning the bilayer thickness to transport small polar molecules across the membrane, thereby eliciting specific biological responses in cells.

**4.** Terpenes are nonsaponifiable, naturally occurring molecules that are mainly hydrocarbon-like, although common oxygen-containing functional groups (such as carboxylic acids and esters, alcohols, epoxides, aldehydes, and ketones) also are present. Terpenes are constructed in living systems (by biosynthesis) beginning with isopentenyl pyrophosphate. This common source is the reason why terpenes generally contain multiples of five carbons and why their structures can often be subdivided into isoprene units. Many terpenes are biologically active, and the steroids (with four fused rings) play an important role as mammalian hormones, relaying information from one organ to another.

**5.** An important chemical function of pyrophosphate is as a leaving group. The chemical reactions of phosphate esters resemble those typically observed for both carboxylic acid esters and sulfate esters. Phosphate esters are important intermediates in biosynthetic pathways. Isopentenyl pyrophosphate and dimethylallyl pyrophosphate are the building blocks for terpene and steroid biosynthesis.

**6.** Biosynthetic pathways can be mimicked in the laboratory. Three of the four rings of the steroid skeleton can be formed by laboratory cyclization of the unsaturated triterpenoid, squalene epoxide, via ring opening of the epoxide.

**7.** Carbohydrates are an important class of natural products with a high degree of substitution by oxygen-containing functional groups. The presence of multiple functional groups makes carbohydrates much more water-soluble than fats, waxes, terpenes, or steroids. Carbohydrates are named by a prefix indicating the number of carbon atoms and the suffix -*ose*. In an aldose, an aldehyde is present at C-1; in a ketose, the carbonyl functionality is at another position of the carbon backbone. An aldose can be converted to a ketose through an enediol. The designation D or L relates the sugar's absolute configuration to the naturally occurring triose D-glyceraldehyde.

**8.** Because a carbohydrate contains either an aldehyde (or a ketone) in a molecule that bears a number of alcohol groups, it can undergo intramolecular cyclization to form a hemiacetal (or hemiketal). The cyclized structures can contain either five atoms (furanoses) or six atoms (pyranoses). In the ring-closed hemiacetal (or hemiketal) form, an additional center of asymmetry is produced, permitting the existence of $\alpha$- and $\beta$-anomers. Fischer projections are convenient tools for specifying stereochemistry at non-anomeric carbon atoms.

**9.** Glycosides are formed when the hemiacetal hydroxyl group is converted to an ester by reaction with either a simple alcohol or the hydroxyl group of another sugar. Individual carbohydrates can be linked together by acetal groups (called a glycosidic linkage) to form dimers (disaccharides) and polymers. The most common disaccharide is sucrose, composed of glucose and fructose. Starch and cellulose are polymers of glucose that are widely distributed throughout nature.

## Review of Reactions

Allylic Substitution of Pyrophosphate

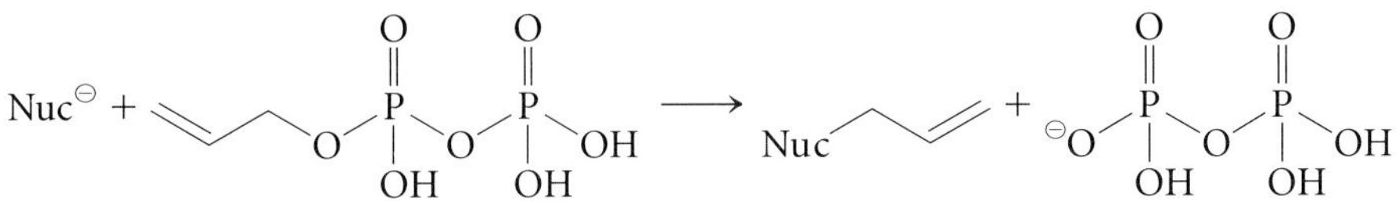

Cationic Cyclization

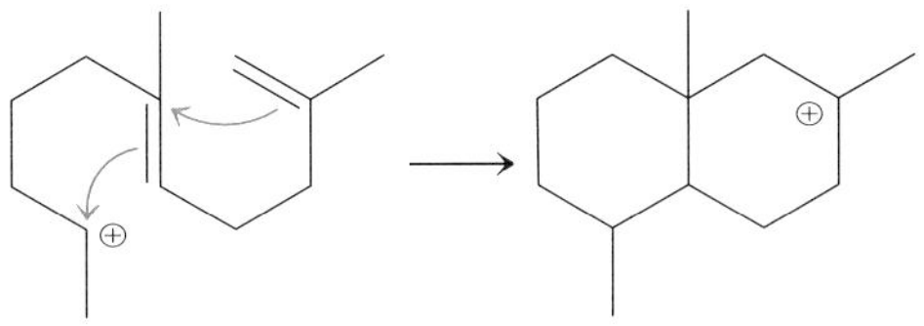

Anomerization

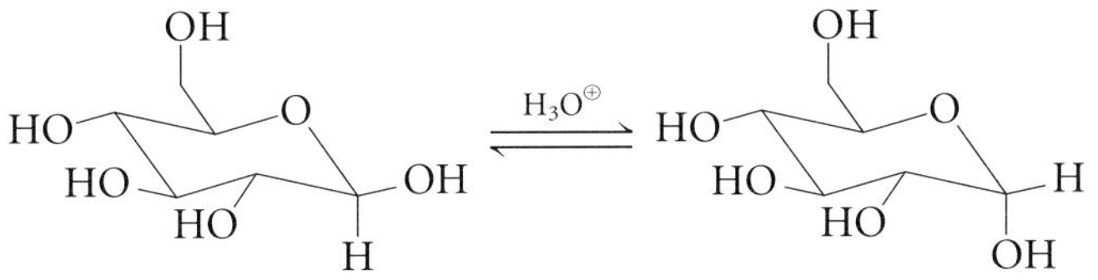

## Review Problems

**17.1** The melting points of *cis* unsaturated fatty acids are lower than those of the corresponding saturated acids. Would you expect the melting point of a *trans* unsaturated fatty acid to be closer to that of its *cis* isomer or that of the saturated fatty acid? Explain your reasoning.

**17.2** The phospholipids (illustrated in Figure 17.3), which are the fundamental building blocks of cell membranes, are mixed esters of glycerol and are chiral. Identify the center(s) of chirality in the following phospholipid, and draw three-dimensional representations of the possible stereoisomers.

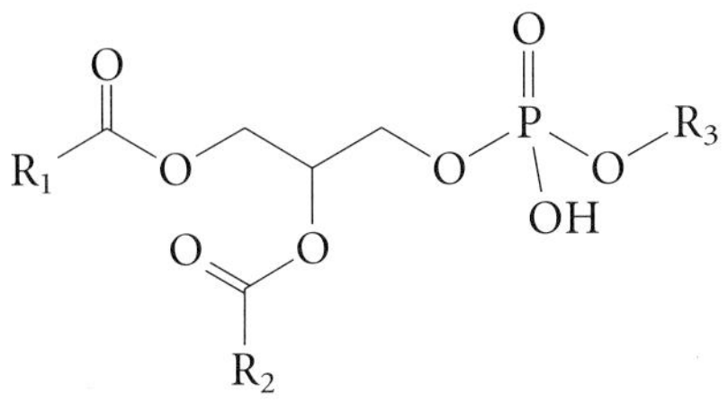

**17.3** Although terpenes are not derived biochemically from isoprene, this diene does polymerize under radical and cationic conditions to form polyisoprenoids. Predict whether such a polymerization will have a tendency to join monomer units in a head-to-head or a head-to-tail fashion. Explain your answer.

**Isoprene
units**
Head-to-head          **Isoprene
units**
Head-to-tail

**17.4** Recall that chair conformations of cyclohexyl ring systems are energetically favored and that *trans*-decalin is more stable than the *cis* isomer.

*trans*-**Decalin**          *cis*-**Decalin**

Note that the stereochemistry at the bridgehead carbons of *trans*-decalin is such that both substituent hydrogens are axial, and that in *cis*-decalin, one substituent hydrogen is axial and the other is equatorial. Now consider the stereochemistry at the bridgehead carbons of the following structure, which includes three of the four rings of a steroid system:

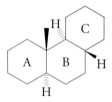

Decide whether the decalin systems formed of rings A and B and of rings B and C are fused *cis* or *trans*. Draw a three-dimensional representation of the structure that is based on either the *trans*- or the *cis*-decalin conformations.

**17.5** Many different types of organic cleaning agents are fundamentally similar to the salts of long-chain fatty acids. Two examples are as follows:

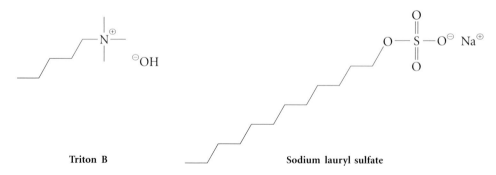

**Triton B**          **Sodium lauryl sulfate**

(You may find one or the other listed as an ingredient on your shampoo bottle.) Identify in each case the hydrophobic and the hydrophilic parts of the molecule.

**17.6** The three naturally occurring aldohexoses (glucose, mannose, and galactose) exist preferentially in solution as cyclic hemiacetal forms. You might expect glucose to be thermodynamically more stable than all other aldohexoses because (in the β form) all its substituents are in equatorial positions. Mannose and galactose should be only slightly less stable, because only one of the hydroxyl groups is axial (again in the β form). For practice, draw the other chair conformation of each of these sugars, noting the number of 1,3-diaxial interactions.

**17.7** The acid-catalyzed hydrolysis of sucrose (refer to Exercise 17.19) can be initiated by cleavage of any one of four different bonds. Indicate these bonds on the following abbreviated structure. Then write a detailed mechanism showing the electron flow for each bond-breaking step.

**17.8** Vitamin C has a variety of natural functions, including acting as an antioxidant. It resembles carbohydrates in that each carbon bears at least one oxygen, although not all are at the oxidation level found in, for example, glucose.

**Vitamin C**
**(ascorbic acid)**

(a) Identify which carbon(s) have higher and which have lower oxidation levels than a normal aldohexose.

(b) Note that vitamin C has an enediol functionality; it is this subunit that undergoes oxidation with oxidants such as molecular oxygen, preventing oxidation of other important biological molecules. Draw the two proton tautomers of vitamin C that result from isomerization of the enediol to ketoalcohols. Note the relation between carbonyl groups in these two isomers. Which of these isomers is more stable (based on the relation of the carbonyl groups to each other), and which carbon of the enediol system in vitamin C is more nucleophilic toward an electrophilic oxidant?

(c) Identify the center of chirality in ascorbic acid that corresponds to that used for the D,L classification system in carbohydrates. Does vitamin C have the D or the L configuration?

**17.9** Vitamin C is also called *ascorbic acid* because the first dissociation constant is 3.6. This value is significantly larger than the $pK_a$ values for simple alcohols, which are in the range 16–18. Suggest why the combination of functional groups present in vitamin C leads to a much stronger acid.

## Supplementary Problems

**17.10** Provide an IUPAC name for each of the following compounds. (Be sure to indicate the configuration of each center of chirality.)

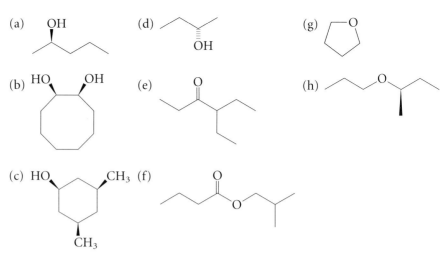

**17.11** Draw a structure that corresponds to each of the following names. (Be sure to represent correctly any centers of chirality present.) Then classify each in terms of the functional group present.

(a) (R)-2-hexanol

(b) 2,6-dimethyl-4-heptanone

(c) cyclohexyl propionate

(d) 2,2-dimethoxypropane

(e) *meso*-2,3-butanediol

(f) (S)-3-methylcyclohexanone

(g) methyl acetate

(h) cyclopentyl ether

**17.12** Refer to Figure 17.3 on page 17-9. Are phospholipids chiral molecules? Would they be chiral if both ester groups ($R_1$ and $R_2$) were the same?

**17.13** A vesicle is a hollow structure composed of molecules having one hydrophobic end and one hydrophilic end. The hydrophilic ends gather on both the outside and inside surfaces of the vesicle where they are in contact with water. Can there be exactly the same number of hydrophilic ends on the outside as on the inside of the vesicle? Why, or why not?

**17.14** The reduction of the carbonyl group of a carbohydrate produces a polyol that has some interesting properties and reactions. For example, reduction of mannose produces mannitol, which is used as a dietary supplement and dietetic sweetener. Mannitol undergoes double cyclization in concentrated hydrochloric acid to form the bicyclic diether known as dianhydromannitol. (Check the formulas of mannitol and dianhydromannitol to convince yourself that two equivalents of water are lost in this transformation.)

Compare the stereochemistry of the centers of chirality in mannitol with those in dianhydromannitol to determine if stereochemistry has been preserved or inverted at each center of chirality during the course of the reaction. Use this information to write a detailed mechanism for the formation of the cyclic bisether.

**17.15** Perillene, γ-decalactone, and rose furan are naturally occurring compounds that are used by thrips as defensive agents, especially against ants that "mine" these insects for the sweet nectar they produce. What class of oxygen-containing compounds does each of the three defensive agents resemble? If any are terpene-like, identify the isoprene subunits, and indicate if they are joined in a head-to-tail fashion.

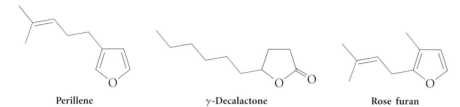

| Perillene | γ-Decalactone | Rose furan |

**17.16** A recent study has shown that *cis*-9,10-octadecenoamide is present in increased concentration in the spinal fluid of cats that have been deprived of sleep for 15 hours. (Figuring out how to keep cats awake for that long might be an interesting problem in its own right—cats typically sleep 75% of the time.) Furthermore, when injected into rats, this compound induced a rapid onset of sleep. To what class of natural materials does this primary amide belong?

**17.17** There are many ways in which *cis*-9,10-octadecenoamide could be prepared from smaller molecules. Keeping in mind the presence of the *cis* alkene linkage, devise a synthesis of this amide, starting from any materials having no more than nine carbon atoms.

**17.18** Draw all possible D-aldopentoses as Fischer diagrams and also as the more stable conformation of the six-member cyclic hemiacetal. Which aldopentoses are most stable in this cyclic form?

**17.19** Reduction of the aldehyde functional group of an aldohexose yields a polyol of six carbons, each carbon bearing one hydroxyl group. These alcohols are named as derivatives of the parent sugar—thus, the alcohol resulting from the reduction of mannose is called *mannitol*.

$$
\begin{array}{ccc}
\text{CHO} & & \text{CH}_2\text{OH} \\
\text{HO}\!\!-\!\!\text{H} & & \text{HO}\!\!-\!\!\text{H} \\
\text{HO}\!\!-\!\!\text{H} & \xrightarrow{\text{NaBH}_4} & \text{HO}\!\!-\!\!\text{H} \\
\text{H}\!\!-\!\!\text{OH} & & \text{H}\!\!-\!\!\text{OH} \\
\text{H}\!\!-\!\!\text{OH} & & \text{H}\!\!-\!\!\text{OH} \\
\text{CH}_2\text{OH} & & \text{CH}_2\text{OH} \\
\textbf{Mannose} & & \textbf{Mannitol}
\end{array}
$$

Draw Fischer diagrams of the alcohols that would result from reduction of each of the eight possible D-aldohexoses. Which of these are chiral molecules, and which are not? (*Note*: The reduction of the aldehyde to a primary alcohol removes the functional distinction between the ends of the molecule.)

**17.20** Reduction of the ketohexoses produces the same set of alcohols that was discussed in Problem 17.19. However, reduction of a ketose produces a new center of chirality, and typically a mixture of both possible diastereomers is produced. The reduction of fructose produces two products. Is either one the same as that obtained by the reduction of glucose?

**17.21** Upon treatment with warm, dilute nitric acid, both the aldehyde and primary alcohol functional groups of carbohydrates undergo oxidation to form dicarboxylic acids known as *aldaric acids*. Draw Fischer projections for the aldaric acids produced from the eight D-aldohexoses. How many of these are chiral? How many unique carbinol carbon signals would be expected in the $^{13}C$ NMR spectrum for each of these diacids?

**17.22** Ketals (or acetals) are used to mask temporarily the electrophilic reactivity of ketones (or aldehydes). Cyclic ketals can be prepared using diols such as 1,3-propanediol, and the formation of such ketals is less endothermic than the formation of ketals prepared from simple alcohols. Can you think of a reason for this difference? (*Hint:* Don't forget about entropy.)

**17.23** Formation of a cyclic ketal from 2,2-dimethyl-1,3-propanediol is even less endothermic than formation of a ketal from 1,3-propanediol (see Problem 17.22). What is the reason for this difference? (*Hint:* Consider the various possible conformations of 1,3-propanediol and of 2,2-dimethyl-1,3-propanediol.)

**17.24** Mixed anhydrides, in which the two carbonyl groups are derived from two different carboxylic acids, can be prepared. For example, benzoic acid can be reacted with trifluoroacetyl chloride to form such a mixed anhydride. Upon reaction with a nucleophile, a mixed anhydride invariably reacts preferentially at the carbonyl group derived from the *weaker* carboxylic acid.

(a) Write a detailed, stepwise reaction mechanism for the reaction of a mixed anhydride with a general nucleophile, $^{\ominus}$Nuc.

(b) Which symmetrical carboxylic anhydride, benzoic acid anhydride or trifluoroacetic acid anhydride, will be more reactive toward a nucleophile?

(c) Construct an energy diagram for the reaction of a mixed anhydride with a nucleophile that is consistent with the observation that the nucleophile effects substitution at the carbonyl group of the less acidic carboxylic acid.

**17.25** Devise a reasonable set of reactions that could be used to prepare 2-methylbutanoic acid from precursors that have no more than two carbons. (You may use any inorganic reagents needed.)

**2-Methylbutanoic acid**

**17.26** Upon treatment with NaNH$_2$, acetone is converted to isophorone:

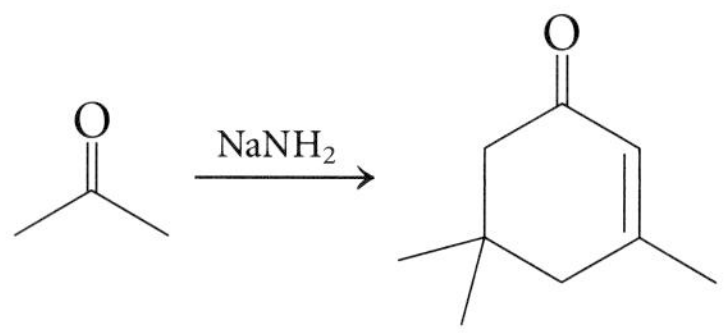

**Isophorone**

This ketone is thus readily available and has been used for the synthesis of several terpenes. Analyze the structure of isophorone in terms of isoprene units. Does isophorone have the normal head-to-tail arrangement? Devise a sequence of reactions that would account for the formation of isophorone from acetone. (There are many variants. Concern yourself only with the use of reasonable chemical transformations to construct the carbon skeleton and given functionality.)

**17.27** Frontalin is one of the pheromones used by the western pine bark beetle for communication.

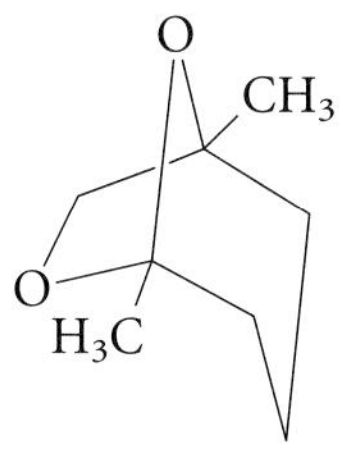

**Frontalin**

What functional groups are present in frontalin? Can you think of an acyclic compound from which frontalin could be prepared by forming only carbon–oxygen bonds? How many centers of chirality are present in this precursor, and how many in frontalin? How many reasonable diastereomers are there for frontalin?

**17.28** Maleic and fumaric acids are geometric isomers. Each has two carboxylic acid groups that are sequentially deprotonated as the pH is raised. For removal of the first proton, maleic acid is more acidic than fumaric acid; for removal of the second proton, fumaric acid is the stronger acid. Explain these differences.

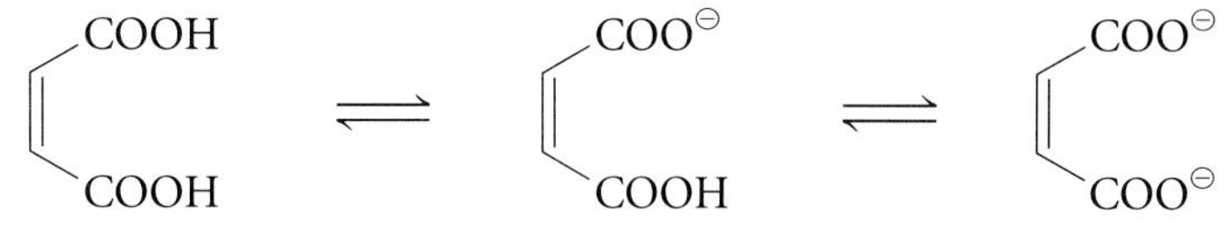

**Maleic acid**
(p$K_a$ 1.9)

(p$K_a$ 6.5)

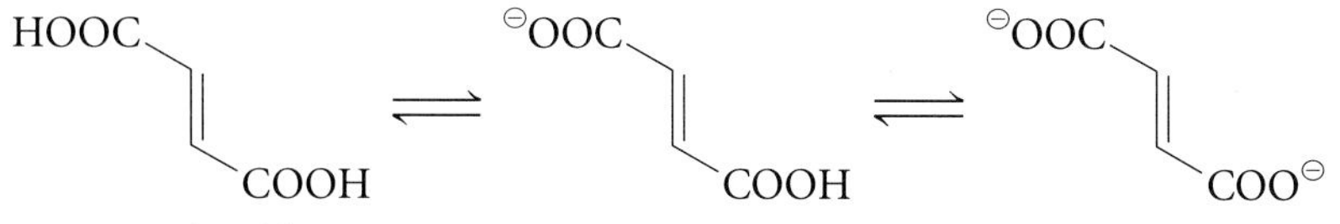

**Fumaric acid**
(p$K_a$ 3.0)

(p$K_a$ 4.5)

# Important New Terms

**aldose:** a sugar with an aldehyde at C-1 (17-28)

**$\alpha$ anomer:** a stereoisomer of the cyclic form of a carbohydrate in which the hydroxyl group at C-1 is *trans* to the last carbon of the chain (17-31)

**$\beta$ anomer:** a stereoisomer of the cyclic form of a carbohydrate in which the hydroxyl group at C-1 is *cis* to the last carbon of the chain (17-31)

**$\beta$-galactosidase:** an enzyme that catalyzes the hydrolysis of the $\beta$ linkage between galactose and another carbohydrate (17-37)

**bilayer:** a three-dimensional structure of two layers of lipids or surfactants in which the hydrocarbon tails point toward the interior and the polar groups are on the two surfaces, solvated by water (17-10)

**carbohydrate:** a polyhydroxylated aldehyde or ketone with the molecular formula $C_n(H_2O)_n$ (17-26)

**deoxyribonucleic acid (DNA):** the principal genetic information storage unit; a biopolymer composed of deoxyribonucleotide units linked through a sugar–phosphate backbone (17-30)

**disaccharide:** a dimer in which two carbohydrate units are covalently bonded (17-36)

**diterpene:** a terpene containing 20 carbons; derived from four isoprene units (17-14)

**fat:** a fatty acid ester of glycerol (17-2)

**fatty acid:** a long, straight-chain carboxylic acid containing an even number of carbon atoms (17-2)

**furanose:** a carbohydrate containing a five-member cyclic hemiacetal (17-31)

**glucoside:** a cyclic derivative of glucose in which the C-1 hemiacetal hydroxyl group of glucose has been replaced by an alkoxy group (17-33)

**glycoside:** a cyclic acetal derivative of a carbohydrate in which the hemiacetal (or hemiketal) hydroxyl group has been replaced by an alkoxy group (17-34)

**hexose:** a six-carbon sugar (17-27)

**hydrophilic:** having a preference for association with an aqueous environment (17-6)

**hydrophobic:** having a preference for association with a nonaqueous environment (17-6)

**invert sugar:** a 1:1 mixture of glucose and fructose obtained upon cleavage of sucrose (17-37)

**invertase:** an enzyme that catalyzes the cleavage of sucrose to a 1:1 mixture of glucose and fructose (17-37)

**isopentenyl pyrophosphate:** a branched five-carbon derivative of a 2-methylbutadiene that is the biochemical precursor of the terpenes (17-13)

**isoprenoids:** derivatized oligomers of isoprene (17-13)

**ketose:** a sugar with a ketone functional group, usually at C-2 (17-28)

**lipids:** a group of simple, naturally occurring molecules that are soluble in nonpolar solvents (17-2)

**membrane:** a lipid bilayer on the outer surface of a cell, separating it from the external aqueous medium (17-8)

**micelle:** a roughly spherical aggregation of many soap-like molecules with hydrophobic and hydrophilic portions, arranged with the polar "heads" at the surface surrounding a hydrocarbon-like core of "tails" (17-6)

**monoterpene:** a terpene containing 10 carbons; derived from two isoprene units (17-14)

**myelin sheath:** the lipid bilayer that surrounds and insulates nerve axons (17-19)

**nonsaponifiable lipid:** a lipid that cannot be hydrolyzed by aqueous base to form soap; a terpene (17-14)

**nor-:** prefix descriptor for a compound whose structure closely resembles that of another compound but is lacking some small feature, usually a methyl group (17-16)

**pentose:** a five-carbon sugar (17-27)

**phospholipid:** a dicarboxylate, monophosphate ester of glycerol (17-9)

**pyranose:** a carbohydrate containing a six-member cyclic hemiacetal (17-31)

**ribonucleic acid (RNA):** transcribes the genetic information stored in DNA and directs protein synthesis; a biopolymer composed of ribonucleotide units linked through a sugar–phosphate backbone (17-30)

**saponifiable lipid:** a lipid that can be hydrolyzed by aqueous base to fatty acids; a fat or wax (17-14)

**saponification:** the making of soap by hydrolysis of fatty acid esters with aqueous sodium hydroxide (17-6)

**sesquiterpene:** a terpene containing 15 carbons; derived from three isoprene units (17-15)

**sesterterpene:** a terpene containing 25 carbon atoms; derived from five isoprene units (17-15)

**steroid:** a naturally occurring, often oxygenated, tetracyclic compound with a fused-ring system of three six-member rings and one five-member ring (17-18)

**terpenes:** a family of relatively nonpolar natural compounds (lipids) derived from isopentenyl pyrophosphate and thus containing $5n$ carbon atoms (17-13)

**tetraterpene:** a terpene containing 40 carbon atoms; derived from eight isoprene units (17-14)

**tetrose:** a four-carbon sugar (17-27)

**triose:** a three-carbon sugar (17-27)

**triterpene:** a terpene containing 30 carbons; derived from six isoprene units (17-14)

**unsaturated fatty acid:** a straight-chain carboxylic acid containing at least one double bond along the chain (17-4)

**vesicle:** a bilayer extended to form a closed surface (17-12)

**wax:** a mixture of fatty acid esters of long, straight-chain alcohols (17-2)

# Naturally Occurring Nitrogen-Containing Compounds

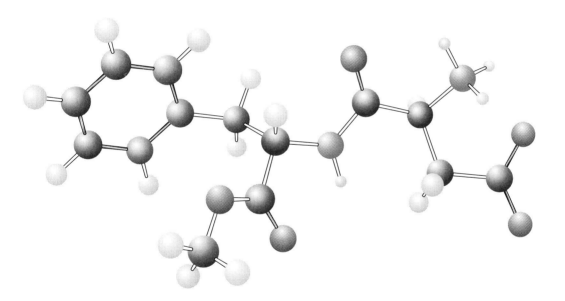

*Nutrasweet® is a simple dipeptide that has a sweet taste, similar to that of sucrose.*

*S*everal classes of natural compounds contain oxygen atoms in various functional groups. Nitrogen-containing functional groups also play an important role in biological chemistry, although the classification system for these materials is not as well defined. This chapter considers amino acids in detail, as well as polypeptides and proteins—polymeric materials with multiple amide functional groups. This chapter also examines the structure of nucleic acids—polymeric materials with sugar–phosphate backbones to which heterocyclic amines are attached, and it introduces alkaloids and aminocarbohydrates. The group that encompasses the alkaloids is an especially diverse one, including all materials that are isolated from plants and contain a basic nitrogen atom. Aminocarbohydrates, obtained by substituting nitrogen for one of the hydroxyl groups in a carbohydrate, are essential chemical elements of a variety of biological recognition systems.

We start by reviewing the methods for forming carbon–nitrogen bonds presented in earlier chapters. We then explore how these transformations are used in sequences that incorporate nitrogen into some biologically important organic molecules.

## 18.1

## Methods for Forming Carbon–Nitrogen Bonds: A Review

Carbon–nitrogen bonds can be formed by several methods, and now we will see that compounds containing carbon–nitrogen multiple bonds undergo some of the reactions of their carbonyl counterparts. We begin by reviewing bond formation via nucleophilic substitution at $sp^3$-hybridized carbon to form amines, via nucleophilic acyl substitution to form amides, and via nucleophilic addition to form imines. We will see that with imines C—C bond formation can take place by a pathway analogous to Grignard additions to carbonyl compounds. We will look at a condensation reaction that combines separate reactions described earlier. Finally, we will review nitrile reduction as a route to primary amines and the Beckmann rearrangement as a route to amides.

### Amines via Nucleophilic Substitution

Simple amines can be prepared by $S_N2$ substitution at a carbon that bears a leaving group. For example, methylamine can be readily converted to ethylmethylamine by an $S_N2$ reaction with ethyl iodide:

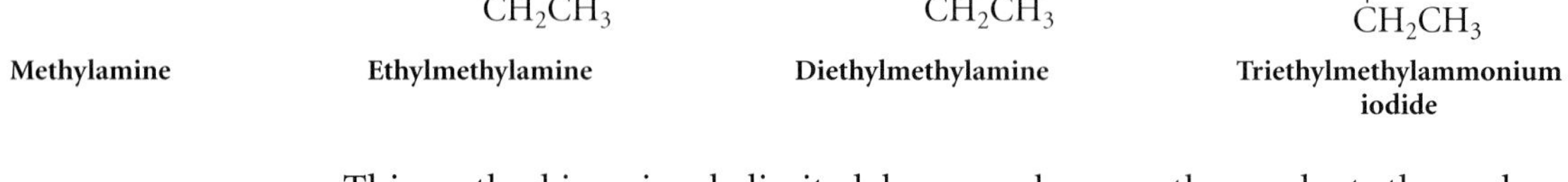

This method is seriously limited, however, because the products themselves are amines that undergo further reaction with the electrophile. In this example, the product reacts further to form diethylmethylamine and then even further to form triethylmethylammonium iodide.

One exception to the generalization that the direct alkylation of amines results in polyalkylation is the nucleophilic ring opening of an epoxide by an amine:

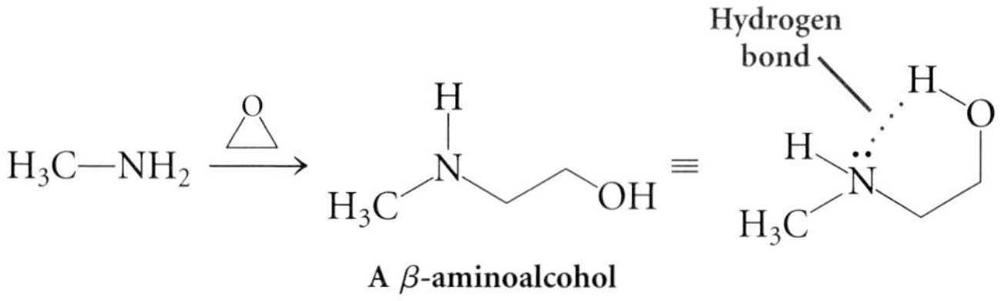

**A β-aminoalcohol**

The reaction is considerably slower than that of a comparably substituted alkyl bromide and, except for monosubstituted epoxides, requires elevated temperatures. The product, a β-aminoalcohol, is considerably less reactive than the starting amine, probably because of stabilization by intramolecular hydrogen bonding between the lone pair of electrons on the nitrogen and the hydrogen of the hydroxyl group. This hydrogen-bonding interaction also results in significantly lower basicity and nucleophilicity for the nitrogen in a β-aminoalcohol.

## Amides

*Synthesis of Amides.* The reaction of a primary or secondary amine (or ammonia) with a reactive carboxylic acid derivative results in the formation of an amide by nucleophilic acyl substitution. This reaction is of particular importance in the synthesis of peptides and will be discussed in more detail in Section 18.3.

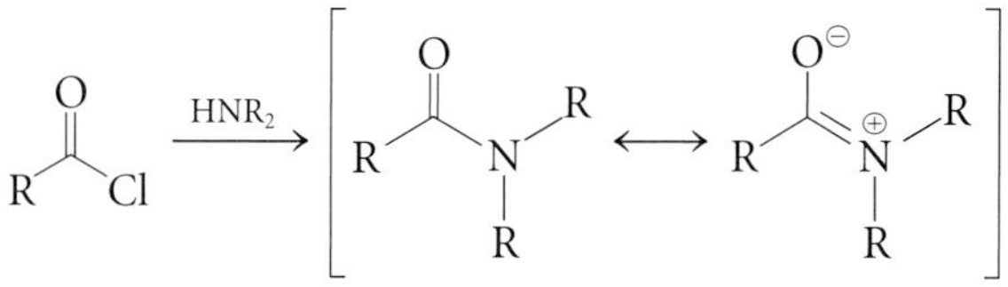

Because of resonance delocalization of the nitrogen lone pair, the nucleophilic character of the nitrogen in the product amide is substantially lower than that of the nitrogen in the starting amine, and thus, even with primary amides, only monoacylated products are obtained.

*Reduction of Amides to Amines.* Amides can be further transformed into amines by reduction. In the laboratory, this is most conveniently accomplished with LiAlH$_4$.

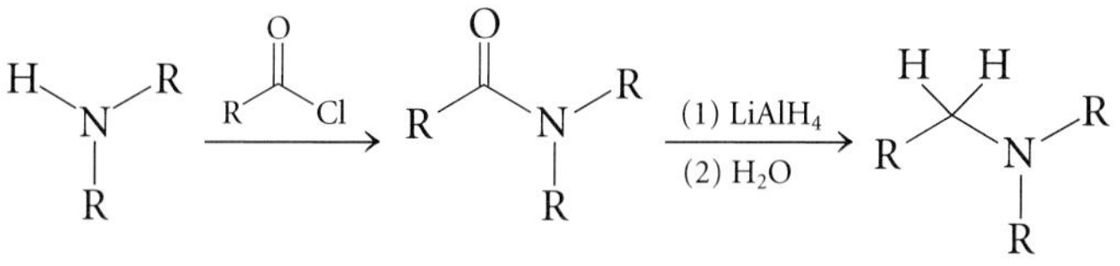

We can see that the overall process—amide formation followed by reduction—adds one substituent to the nitrogen. As a result of the reduction of the carbonyl group, the substituent has a methylene group adjacent to the nitrogen.

## Imines

Amines also react quite selectively with carbonyl compounds to yield products resulting from a one-to-one pairing of the reactants. The reaction of primary amines with ketones or aldehydes produces imines:

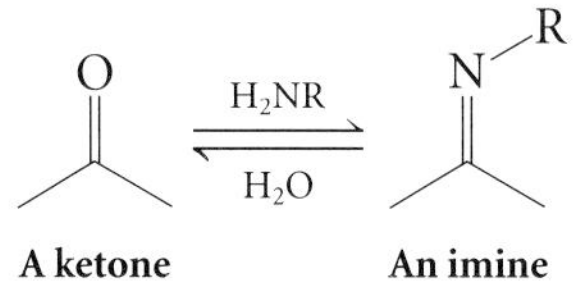

**A ketone**          **An imine**

The imine is not highly favored in the equilibrium, but the reaction can be driven toward the product imine by removal of water (for example, by azeotropic removal or with molecular sieves).

*Reduction of Imines to Amines.* An imine can be reduced to an amine, either with complex hydride reagents such as $NaBH_4$ or with molecular hydrogen and a metal catalyst:

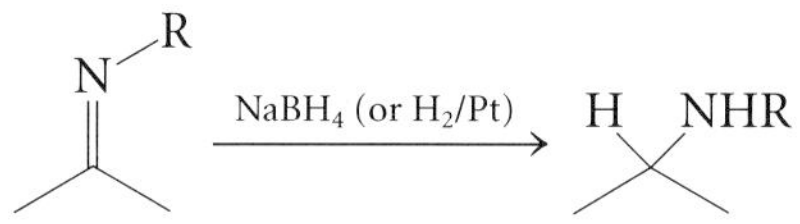

The amine product is produced by reduction at both carbon and nitrogen; in contrast, only the carbonyl carbon is reduced in the reduction of an amide. As a result, in the reduction of an imine originating from a primary amine, the product has one hydrogen and two carbon substituents on nitrogen (a secondary amine), with one alkyl substituent derived from the starting amine and the other from the carbonyl compound. The two-step sequence, in which reaction of an amine with a carbonyl compound is followed by reduction to form a more substituted amine, is called **reductive amination** and achieves a net alkylation of the amine.

Reductive amination can also be carried out with imines derived from ammonia. Generally, these imines are not sufficiently stable to allow their isolation. However, they can be trapped as they are formed by reduction with sodium cyanoborohydride, $NaBH_3CN$, a hydride reagent that is stable in protic solvents (such as water and methanol) at a pH at which imine formation proceeds at an appreciable rate. Reductive amination by this process works equally well with ammonia (R = H) to yield a primary amine and with a primary amine (R = alkyl) to yield a secondary amine.

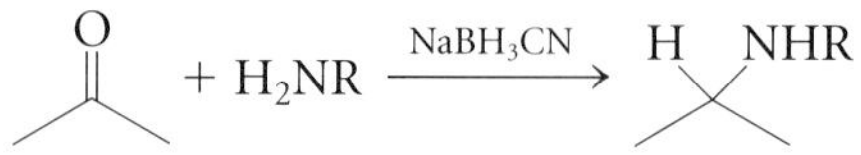

The reduction of an imine by $NaBH_4$ or $NaBH_3CN$ consists of the formal addition of both a hydride ion as a nucleophile to the carbon and a proton to the nitrogen atom.

Sodium cyanoborohydride is less active as a hydride donor than is sodium boro-hydride. As a result, NaBH$_3$CN can be used in aqueous solution under acidic pH conditions, where NaBH$_4$ is rapidly destroyed. Explain this difference in reactivity.

*Nucleophilic Addition to Imines.* A carbon nucleophile such as a Grignard or alkyllithium reagent can be added to the carbon of the imine functional group in a reaction analogous to the addition of such a nucle-ophile to a ketone or aldehyde.

This mode of reaction, resulting in the formation of a new carbon–carbon bond, is quite general, and a wide variety of Grignard reagents can be used. However, only secondary amines can be prepared by this method.

**EXERCISE 18.2**

Identify appropriate starting materials and reagents for the synthesis of each of the following amines.

## The Mannich Condensation

Carbon nucleophiles other than Grignard and alkyllithium reagents also add to imines. An important process in which such a reaction takes place is the Mannich condensation, which combines a simple aldehyde (often formaldehyde), a primary or secondary amine, and a ketone.

This process is a combination of three familiar reactions: activation of the α-carbon of a ketone as a nucleophile by enol formation, imine formation, and nucleophilic addition to a $C=X$ $\pi$ system. First, the ketone is tautomerized to its enolic form. Then, formaldehyde and a primary amine react to form an imine. In a final step, the imine reacts (as an electrophile) with the enol form of the ketone. Each of these steps is acid-catalyzed; generally, a small amount of a weak acid, such as acetic acid, is added to the reaction mixture as a catalyst.

The Mannich condensation is very useful in natural product synthesis. For example, the tropane ring system found in cocaine can be formed in a single reaction by a double Mannich condensation. The dialdehyde (compound A) is condensed with methylamine and dicarbomethoxyacetone (compound B) in the presence of an acid catalyst. This straightforward process provides the bicyclic product (compound C).

A          B                    C                          Cocaine

### EXERCISE 18.3

Write a detailed reaction mechanism for the formation of the bicyclic compound C in the synthesis of cocaine.

### EXERCISE 18.4

Suggest a sequence of reactions for converting compound C into cocaine. (*Hint:* Compound C contains a $\beta$-ketoester functional group.)

The nitrogen atom of an amine functional group can also be introduced into an organic substrate by $S_N2$ substitution with cyanide ion, followed by reduction of the resulting nitrile to form a primary amine.

$$R{-}CH_2{-}Br + K^{\oplus}\ {}^{\ominus}C{\equiv}N \xrightarrow{\text{DMSO}} R{-}CH_2{-}C{\equiv}N \xrightarrow[H^{\oplus}]{H_2/Pt}$$

$$R{-}CH_2{-}CH_2{-}\overset{\oplus}{N}H_3 \xrightarrow{Na_2CO_3} R{-}CH_2{-}CH_2{-}NH_2$$

**A primary amine**

This method works well only if the displacement is at a primary carbon. With more substituted alkyl halides, in addition to acting as a nucleophile, cyanide ion acts as a base, effecting elimination. The primary amines that can be produced are thus limited to those in which the nitrogen is attached to a chain of two methylene groups, $-CH_2-CH_2-$. Reduction of the nitrile can be accomplished with either $LiAlH_4$ or $H_2$ and a metal catalyst. Because amines deactivate metal catalysts and decrease the rate of catalytic hydrogenation, an acid such as acetic acid is included in the reaction mixture so that the amine product is converted into an ammonium ion. After reduction, neutralization of the acid with a base such as $Na_2CO_3$ frees the amine from the salt.

## The Beckmann Rearrangement of Oximes: Amides from Carbonyl Compounds

A very important industrial method of carbon–nitrogen bond formation is the Beckmann rearrangement, a reaction that converts an oxime to an amide by a rearrangement.

**An oxime**      **An amide**

Oximes are formed from ketones by reaction with hydroxylamine. This reaction is analogous to the formation of imines from ketones and amines. However, the oxime is more stable relative to the reactants than is the analogous imine, and equilibrium conditions favor formation of the oxime as product. Treatment of the oxime with concentrated acid initiates a series of steps that ultimately form the amide by migration of one of the original carbon substituents of the ketone to nitrogen. The sequence involves the migration of a carbon substituent from carbon to nitrogen with simultaneous loss of water.

Readdition of water to the carbocation and proton tautomerization (by protonation followed by deprotonation) gives the amide.

Large quantities (approximately 700 million kilograms) of caprolactam are produced each year from cyclohexanone by the Beckmann rearrangement.

**Caprolactam**

**Nylon 6**

### EXERCISE 18.5

The cationic intermediate in the Beckmann rearrangement has the following structure:

Are there other resonance contributors for this cation? Would you expect this cation to be bent, as shown, or linear? Explain.

### EXERCISE 18.6

Write a complete mechanism for the conversion of the oxime of cyclohexanone into caprolactam. Would you expect the intermediate carbocation formed by carbon migration in this reaction to be more or less stable than that formed in an acyclic system? Explain.

## 18.2

# Amino Acids: Structure and Properties

#19   Amino Acids and
       Peptide Bonds

**Amino acids** constitute an important class of nitrogen-containing, naturally occurring compounds. The biological function of amino acids is well established: they are the constituent monomer units from which the biopolymers based on the peptide bond are built (for example, polypeptides and proteins, including enzymes). Here, we will examine all the common amino acids found in nature.

### Structure of Amino Acids

For convenience, amino acids are divided into four classes based on the polarity and the acidity of the groups present in the side chains: (1) hydrophobic, (2) hydrophilic, (3) acidic, and (4) basic. The 20 amino acids commonly found in proteins in living systems are shown in Figure 18.1.

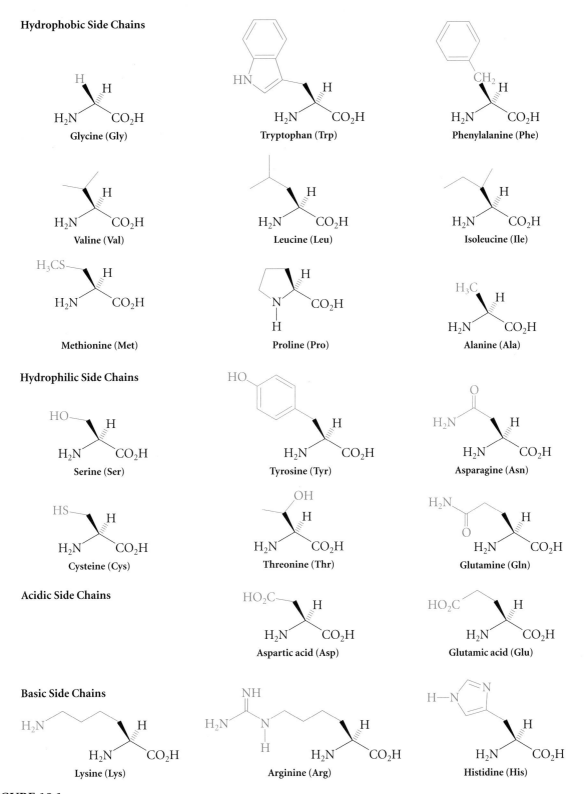

**FIGURE 18.1**

The 20 naturally occurring amino acids grouped into four classes based on polarity and acidity of the side chain. The names are not systematic; abbreviations are given in parentheses.

Each amino acid contains a unique side-chain, R, on a common substrate:

$$H_2N \overset{CO_2H}{\underset{R}{\longmapsto}} H$$

The names of the naturally occurring amino acids are arbitrary, but grouping them according to structure makes the names easier to remember. Glycine is the parent amino acid and the only one that is achiral. Phenylalanine derives both its name and its structure from alanine. Tyrosine is phenylalanine with a phenolic hydroxyl group added. Serine and cysteine differ from alanine in having a hydroxyl or a thiol group, respectively. Glutamine contains one more methylene group than asparagine. Just as glycine is unique in having no center of chirality, isoleucine and threonine are the only common amino acids that have two centers of chirality. Proline is unique among amino acids in that its side chain is joined in a ring containing the $\alpha$-amino nitrogen.

Because biochemists frequently work with biopolymers having hundreds or even more amino acid residues (units), they often use the three-letter abbreviations shown in Figure 18.1 to designate the individual amino acids, instead of writing their structures. These abbreviations are not used in this chapter because we are focusing specifically on the chemical properties imparted by the side chain of the amino acid.

Peptide chains prefer an extended, zig-zag arrangement.

However, this conformation is not possible if the chain includes a proline residue because of proline's five-member ring. Furthermore, a proline residue in a peptide chain does not have a hydrogen on nitrogen and can thus participate in hydrogen bonding only as a proton acceptor, not a donor. The hydrogen bonding that links one peptide chain to another is thus disrupted when proline is present. The interruption in the zig-zag arrangement induced by the presence of a proline residue also disrupts hydrogen bonding in the $\alpha$-helix.

## ▨ Properties of Amino Acids

Several other classification systems for amino acids that recognize their biological properties are commonly used. For example, amino acids are classified as either glucogenic or ketogenic, depending on how they are biodegraded. Alternatively, amino acids are grouped on the basis of the precur-

*CHEMICAL PERSPECTIVES*

NATURE'S SLEEPING PILL

Melatonin is a naturally occurring hormone (related to the amino acid tryptophan) produced by the pea-sized pineal gland found in the center of the brain. Production of melatonin is triggered by the absence of light, and output of the hormone drops dramatically when light falls on the retina. Melatonin production also varies with age, ranging from a high of 125 pg/L of blood at age 6 to less than 25 pg/L of blood in people over 80.

**Melatonin**

Melatonin has varied effects on many organisms. As days shorten during the fall, melatonin production increases, triggering birds to migrate. Conversely, the longer days of spring lead to a drop in melatonin production, a signal for dogs to shed their winter coats. For humans, melatonin regulates sleeping patterns. In part, you get sleepy at night because of the increase in melatonin production in the dark. However, the natural day–night cyclic production of melatonin is disrupted when travelers cross many time zones in a day. Melatonin has thus become a popular over-the-counter treatment for jet lag, taken on the night of arrival.

sors from which they are synthesized in living systems. In another scheme, amino acids are classified as essential or nonessential, depending on whether a particular living system must obtain them from dietary sources or can synthesize them internally. In human beings, eight amino acids are essential and must be obtained from the diet: isoleucine, leucine, lysine, methionine, phenylalanine, threonine, tryptophan, and valine.

*Hydrophobic and Hydrophilic Properties.* The amino acids with neutral side chains are divided into two classes: those with hydrocarbon-like (hydrophobic) side chains, and those with polar functional groups that impart hydrophilic character to the side chain. Other amino acids have functional groups that participate in important chemical reactions. For example, with reference to the structure of wool, cysteine residues in separate peptide chains can couple by forming sulfur–sulfur bonds, which serve as cross-links between the chains.

*Acidic and Basic Amino Acids.* Because of the polarity associated with acidic and basic functional groups, all of the amino acids with acidic or basic side chains are hydrophilic. Like the neutral amino acids, these amino acids have related structures. For example, **glutamic acid** has one more methylene group than **aspartic acid,** and both are related to the hydrophilic neutral amino acids **glutamine** and **asparagine. Arginine** can be viewed as a derivative of **lysine** in which the simple amino group of the latter has been converted into what is referred to as a **guanidine unit:**

$$-NHC{=}NH(NH)-$$

This functional group is much more basic than a simple amine; indeed, arginine is the most basic of the common amino acids. The acidity or basicity of the various functional groups present in amino acids is generally described in terms of the dissociation constant ($pK_a$) of each functionality in its acidic, protonated form. For example, the two dissociation constants of protonated alanine, $CH_3CH(NH_3{}^{\oplus})COOH$, are 2.35 and 9.87:

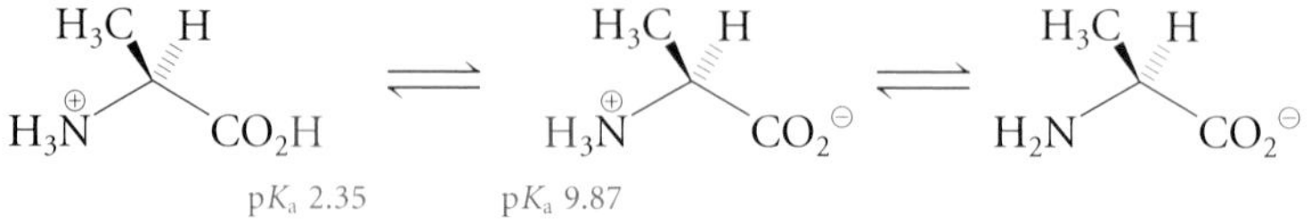

There are three dissociation constants for arginine: 2.18, 9.09, and 13.2. The first dissociation constant (the lowest number) is for the conversion of the carboxylic acid into its carboxylate anion. Both protonated alanine and protonated arginine are substantially more acidic than acetic acid ($pK_a$ 4.8).

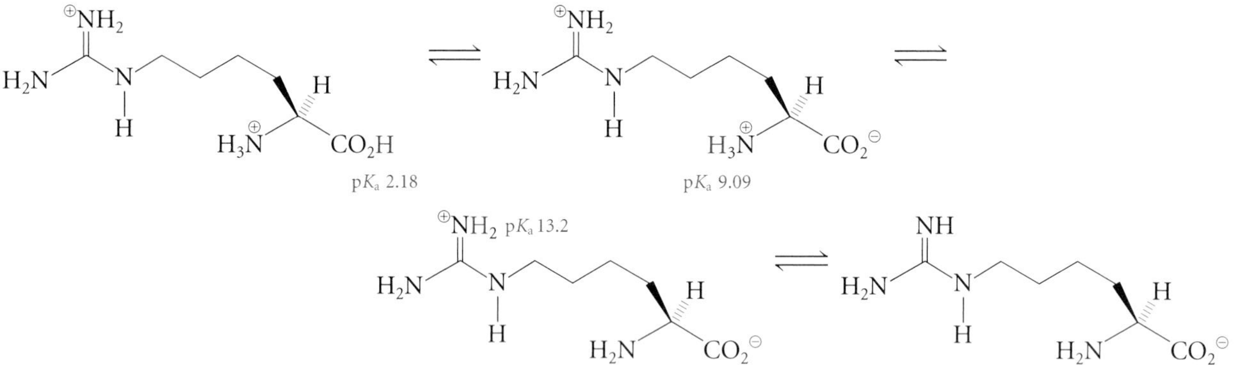

The second step for arginine is the loss of a proton from the $\alpha$-ammonium ion. The third deprotonation occurs at the guanidinium ion to form a neutral guanidine unit.

There are two tautomers for **histidine;** only one is shown in Figure 18.1. In solution, the two tautomers are in dynamic equilibrium:

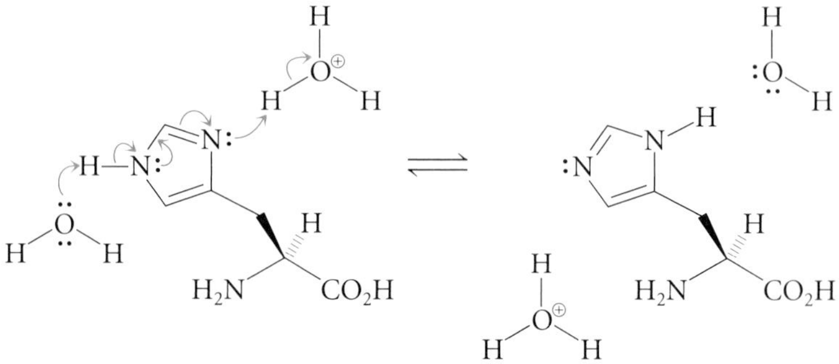

Histidine residues play an important role in enzymes because of the facility with which a proton can be gained by one ring nitrogen at the same time as another proton is lost from the other ring nitrogen.

---

**EXERCISE 18.7**

---

Explain why the carboxylic acid group of a protonated amino acid should be more acidic than a simple carboxylic acid such as acetic acid. Also explain why the guanidine unit is more basic than a simple amine ($pK_a$ 10–11 for the conjugate acids of amines).

---

*Zwitterionic Character of Amino Acids.* All amino acids discussed so far are **amphoteric:** they contain both an acidic carboxylic acid group and a basic $\alpha$-amino group. In aqueous solution at a specific pH (referred to as the **isoelectric point**) that varies with structure, an amino acid exists mainly as a zwitterion. In the zwitterionic form, the amino group has gained a proton and exists as a cation, and the carboxylic acid has lost a proton to form a carboxylate anion:

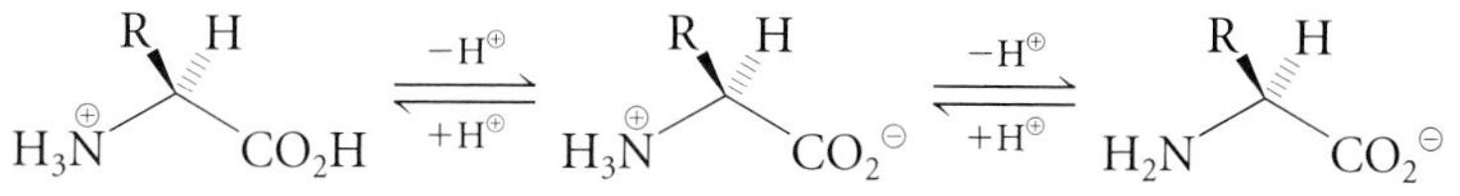

**A zwitterion**

When the acidity of the solution is increased (pH is lowered), the concentration of carboxylic acid increases as the carboxylate anion is protonated, and the equilibrium is displaced toward the cationic form of the amino acid, which has an ammonium ion and a carboxylic acid group. Conversely, when the solution becomes more basic (pH increases), the equilibrium is displaced toward the anionic form, in which there is a carboxylate anion and a neutral amine. In fact, in aqueous solution at any pH, the amount of the uncharged form of the amino acid, in which both the amine and the carboxylic acid are in the free state, is very small.

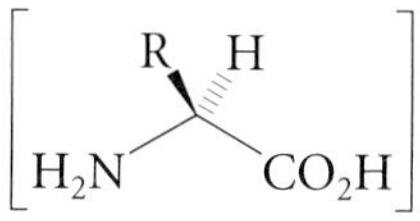

**Uncharged form
of an amino acid**

Amino acids can be separated by *electrophoresis,* a technique in which ions are drawn through a medium by an applied electric field. Only charged species migrate, and the greater the charge (on a weight-adjusted basis), the faster they move. Electrophoresis is a valuable analytical tool for studying biologically important compounds because many of them are ionic materials. Alternative techniques such as gas chromatography and liquid chromatography in which normal solid phases (for example, silica gel) are used are not appropriate because the high polarity of many biomolecules results in excessively strong interactions with the solid phase.

# Polypeptides: Structure, Function, and Synthesis

## Structure and Function of Polypeptides

The individual amino acids just discussed generally do not have important biological roles. On the other hand, when joined together as biopolymers, they have great significance. Smaller biopolymers, containing fewer than 100 amino acids, are referred to as *polypeptides,* and larger biopolymers are called *proteins.* The linking bond between two amino acids joined together to form an amide is referred to as a *peptide bond.* Polypeptides serve mainly as chemical messengers, or **hormones.** Proteins are used for a variety of purposes: as structural materials, as the critical recognition elements of the immune system, and, most importantly, as enzymes, the catalysts for biochemical transformations.

Although there are only 20 common amino acids, the number of possible ways to combine them, even in short polypeptides, is immense. For example, consider the peptide hormone **LHRH (luteinizing hormone release hormone),** which has 10 amino acid subunits (Figure 18.2). LHRH serves as a critical component in regulating human fertility; it communicates information from the hypothalamus gland, first to the pituitary gland, and then to the ovaries at the time that an ovum is to ripen and be released. If all we knew about this peptide was that it was a decapeptide, the number of possible sequences derived from the 20 naturally occurring amino acids would be $20^{10} = 1.0 \times 10^{13}$. Determination of which amino acids are present reduces the number of possible structures tremendously. However, if all ten are different, there are still 3,628,800 (10!) different possible sequences. It is this tremendous potential for structural diversity that allows amino acid polymers to have so many applications.

Except for the amino acid at either end, each amino acid of LHRH participates in two amide bonds, one from its carboxyl group to the amino group of the next amino acid (in one direction) and the other from its amino group to the carboxyl group of the next amino acid (in the other direction). The amino acid units at the two ends participate in only one amide bond each. Thus, at one end of the polypeptide, designated the **amino terminus,** or simply the **N terminus,** there is a free amino group. At the other end, the **carboxyl terminus,** or the **C terminus,** there is a carboxylic acid group. LHRH has the unusual amino acid pyroglutamic acid (pyroGlu) at the C terminus. The name for this cyclic amino acid derives from the ease with which it is formed when glutamic acid is heated above 100 °C:

Glutamic acid      Pyroglutamic acid

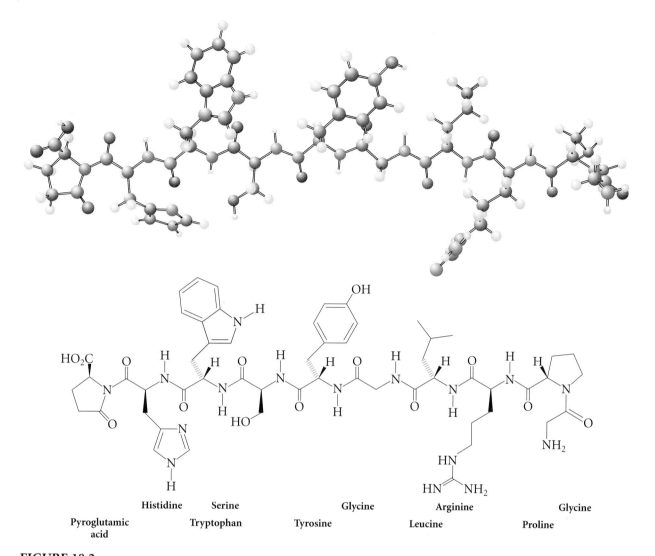

**Histidine**  **Serine**  **Glycine**  **Arginine**  **Glycine**

**Pyroglutamic**  **Tryptophan**  **Tyrosine**  **Leucine**  **Proline**
**acid**

**FIGURE 18.2**

The hormone LHRH is a relatively small polypeptide with 10 amino acid subunits.

At physiological pH, the terminal groups of polypeptides normally exist as an ammonium ion and a carboxylate group, respectively, just as in individual amino acids.

## Synthesis of Polypeptides

The laboratory synthesis of small to medium polypeptides is an important focus of research in the pharmaceutical industry, where the goal is to prepare useful quantities of peptide hormones, as well as analogs of these natural substances. In such analogs, unnatural amino acids (such as the enantiomeric D forms) substitute for those naturally present, either to modify the polypeptide's activity or to protect the polypeptide from

**18-15**

degradation in the body. The formation of amide linkages by nucleophilic acyl substitution is a key step in the synthesis of both naturally occurring and modified polypeptides.

Four features complicate the synthesis of polymers derived from amino acids:

**1.** Linking amino acids together requires the formation of an amide by a specific condensation reaction between the carboxylic acid of one amino acid and the amine of another. Because each amino acid contains both functional groups, protection of the amino group of one and the carboxyl group of the other is required. Otherwise, a complex mixture would result, consisting of two homogeneous and two heterogeneous dipeptides:

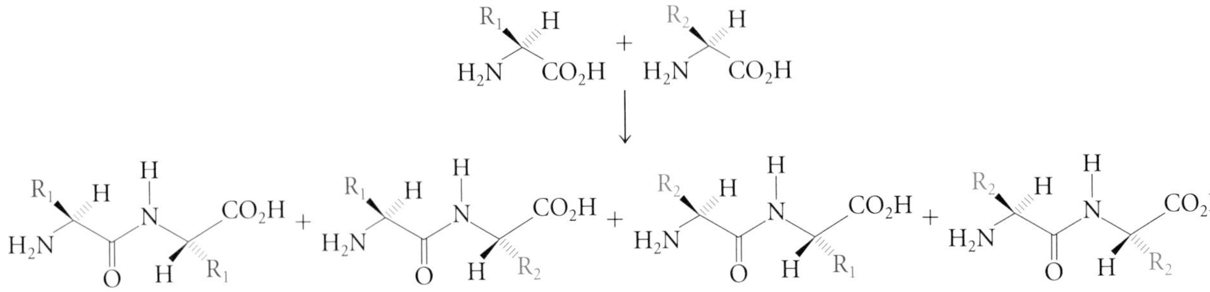

**Mixture of dipeptide products**

**2.** Amino acid derivatives are prone to epimerization of the $\alpha$ carbon by a deprotonation–reprotonation sequence:

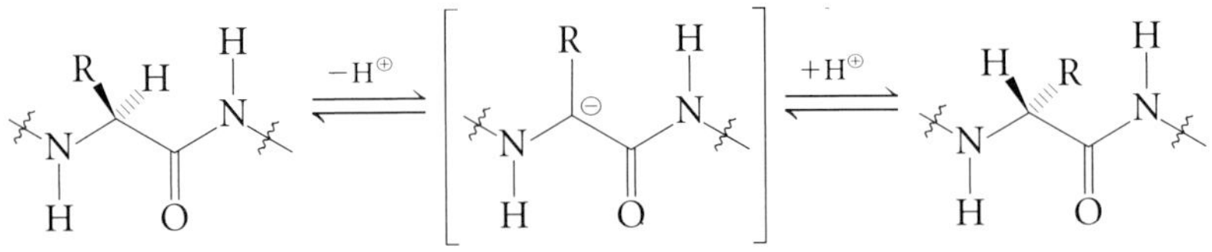

Even a small loss of stereochemical purity in each individual reaction of the multistep synthesis required to produce polypeptides will inevitably lead to a large number of diastereomeric impurities. Not only does this reduce the yield of the desired biologically active stereoisomer, but the polypeptides present as impurities are very similar chemically to the desired one, making isolation of that product extremely difficult.

**3.** The pure polypeptide product can be obtained only if each growing chain undergoes *every* condensation reaction in the sequence. A polypeptide lacking even one of the amino acids from the total sequence constitutes an impurity that is difficult to remove.

**4.** Many polypeptides are quite insoluble in solvents that are suitable for the condensation reactions used to form the amide linkages. For example, the simple polypeptide composed of six alanines linked together (hexaalanine) is almost completely insoluble in all solvents except water, and the nucleophilic reactivity of water precludes its use as a solvent for the reactions that form peptide bonds.

## EXERCISE 18.8

How many unique tripeptides can be formed by combining two different amino acids? Three different amino acids?

The stereochemical carbon $\alpha$ to the carboxyl carbon of an amino acid derivative epimerizes more rapidly when the amino group is part of an amide. Explain.

*The Merrifield Solid-Phase Synthesis.* The problem of the insolubility of amino acids was solved by the American chemist R. Bruce Merrifield. In the **Merrifield solid-phase synthesis,** the first amino acid of the polypeptide to be synthesized is chemically attached to porous beads of a polystyrene support through a $\sigma$ bond that is resistant to the reaction conditions required to form peptide bonds. Additional amino acids are added step by step, with appropriate protection and deprotection steps. When all the amino acids have been added, the complete polypeptide is released from the polystyrene support. In this way, the polypeptide is synthesized on a solid support and is released into solution only in the last step.

Merrifield announced in 1969 that he had synthesized the enzyme ribonuclease (124 amino acids) by an automated process involving 369 reactions and 11,931 separate procedures in a period of just six weeks. Merrifield was awarded the Nobel prize in 1984 for his contribution to the advancement of peptide chemistry. In the commercial production of polypeptides, computer-controlled machines are programmed to carry out all the individual reactions automatically, adding appropriate reagents at each point in the synthesis. Even quite complex polypeptides containing 100 amino acid residues can be prepared in this way.

*Polymer Support.* The polymer support for the Merrifield synthesis is prepared by chloromethylating polystyrene by a Friedel–Crafts alkylation. In this process, a chloromethyl group is added to the aromatic ring (mainly at the position *para* to the point of attachment to the polymer chain) by reaction with chloromethyl methyl ether:

**Polystyrene**  $\xrightarrow[\substack{Cl—H_2COCH_3 \\ \text{(chloromethyl} \\ \text{methyl ether)}}]{AlCl_3}$  **Chloromethylated polystyrene**

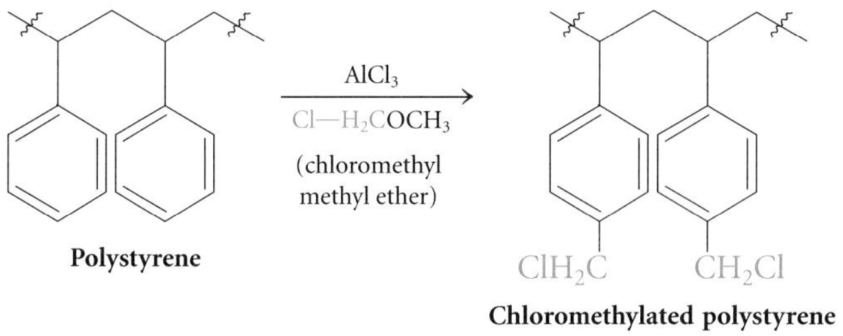

The first step almost certainly takes place via an $S_N1$ reaction involving the methoxymethyl cation formed by loss of chloride ion from chloromethyl methyl ether. Reaction of this cation with the aromatic ring produces a methoxymethylene ($—CH_2OCH_3$) substituent that is converted by a second $S_N1$ reaction with HCl and AlCl₃ into the chloromethylene group.

**EXERCISE 18.10**

Chloromethyl methyl ether is formed *in situ* from HCl, CH₃OH, and $H_2C=O$.

(a) Write a detailed and reasonable mechanism for this multistep conversion.

(b) This reaction also produces small amounts of bis-chloromethyl ether, whose potent carcinogenic activity was discovered, unfortunately, only through epidemiological studies of chemical plant workers. Write a detailed mechanism for the formation of this dangerous by-product.

*Use of Amino Protecting Groups.* Amino acid units cannot be added in a controlled fashion to a growing peptide chain if both the N terminus and the C terminus are free. Therefore, one of these ends must be protected. Let's envision the process of synthesis schematically. In Figure 18.3, a protecting group for the amino functionality is represented by P. The first amino acid unit (starting from the C terminus of the peptide) is attached to the polymer support by reaction of the carboxylic acid group with the benzylic chloride (an $S_N2$ reaction). In this displacement, a weak base such as triethylamine is used to convert the carboxylic acid into the carboxylate anion. Once the amino acid is attached to the polymer support via the carboxylate group, the protecting group on nitrogen is removed, exposing the free amino group for reaction with the carboxyl group of the next amino acid to be added.

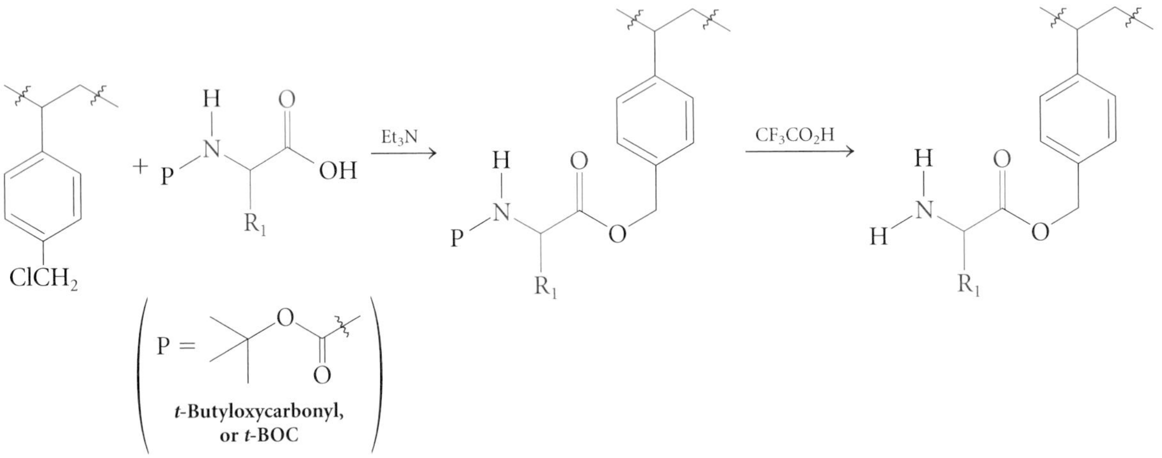

**FIGURE 18.3**

The amino group of the first amino acid of the polypeptide sequence is protected by a *t*-butyloxycarbonyl (*t*-BOC) group. The free carboxylate functionality effects $S_N2$ displacement of chloride ion from chloromethylated polystyrene, attaching the amino acid to the polymer support. Removal of the protecting group by treatment with trifluoroacetic acid produces an amino acid covalently attached to the support and bearing a free amino group.

What requirements must be met by such a protecting group? First, it must be stable to the reaction conditions used to form the amide bond during polypeptide synthesis. Second, conditions for removing the group from nitrogen must be sufficiently mild that the amide bonds already formed are not affected. The *t*-butyloxycarbonyl (*t*-BOC) group is used quite commonly as a protecting group in solid-phase synthesis. Deprotection of the amino group by removal of *t*-BOC is readily accomplished by treatment with trifluoroacetic acid.

The Merrifield synthesis requires two steps for the introduction of each amino acid unit: (1) formation of the peptide bond by reaction of the growing polypeptide with an *N*-protected amino acid, and (2) deprotection of the amino group in preparation for a repetition of the first step (Figure 18.4).

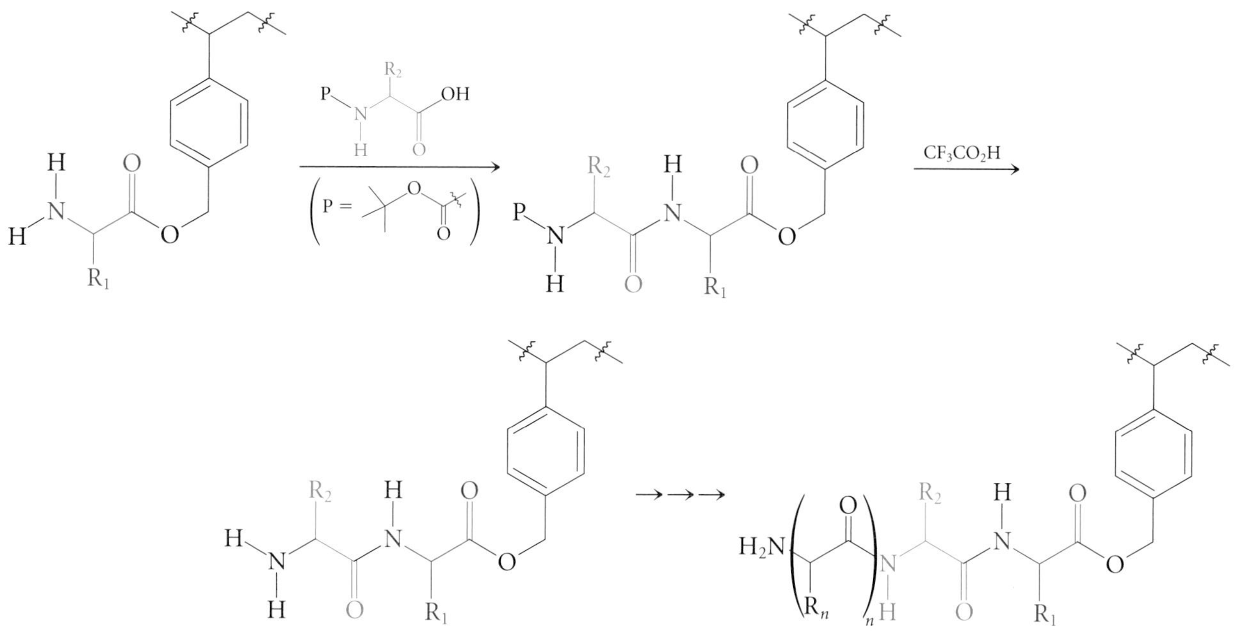

**FIGURE 18.4**

The free amino end of an amino acid (or polypeptide) attached to the polymer support reacts with the carboxylate end of an N-protected amino acid. A new amide bond is thus formed, extending the peptide chain. Deprotection of the N terminus by treatment with trifluoroacetic acid produces a covalently bound polypeptide with a free amino end that is ready for another amidation cycle.

## EXERCISE 18.11

Write a detailed reaction mechanism for the removal of the protecting group *t*-BOC from an amino group. (*Hint:* Remember that a tertiary carbocation is exceptionally stable.)

*Epimerization.* The Merrifield synthesis of polypeptides usually proceeds by attaching each additional amino acid to the N terminus of the growing peptide chain. This method is generally preferred to growth in the opposite direction (addition to the C terminus), because of a side reaction that results when the carboxylic acid end of a polypeptide is converted into a more reactive derivative. This process, illustrated in Figure 18.5 (on page 18-20) for a carboxylic acid chloride, is an intramolecular reaction that forms an **azlactone.** (This cyclization takes place spontaneously at room temperature, once the terminal carboxylic acid has been activated.) Although the azlactone can react with the free amine of a C-protected amino acid to form an amide, it is especially prone to epimerization because of the high acidity of the hydrogen $\alpha$ to nitrogen. Epimerization at the azlactone stage leads to the formation of diastereomeric contaminants.

18-19

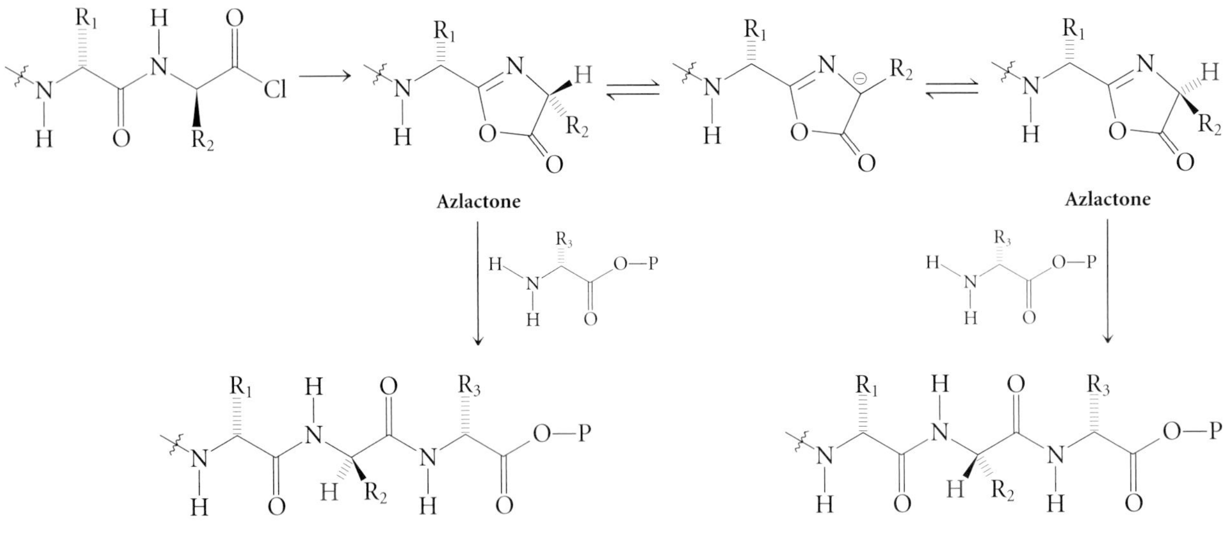

**FIGURE 18.5**

The azlactone group is formed when a free carboxylic acid end of a polypeptide is converted into an acid chloride (or other reactive acyl derivative).

---

**EXERCISE 18.12**

Write a detailed mechanism for the formation of an azlactone and its reaction with an amine to form an amide, as illustrated in Figure 18.5. (Assume that the azlactone is formed by nucleophilic acyl substitution.) What special feature of the anion of an azlactone makes it particularly stable?

---

*Activation of the Carboxylate Group.* The formation of an amide bond between two amino acids can be accomplished in a number of ways, but all require that the carboxylic acid be converted to an activated acyl derivative that is more susceptible to nucleophilic acyl substitution. However, highly activated acyl derivatives of carboxylic acids are more susceptible to racemization at the $\alpha$ carbon, because the anion formed by deprotonation is stabilized by resonance to a greater extent because of increased electron demand by the carbonyl group. Thus, choosing the activated acyl derivative involves balancing the speed of the desired acylation against the rate of the undesirable epimerization.

A particularly good choice for the acyl derivative is one formed from a **carbodiimide** and the carboxylic acid. This addition product, which is quite similar to an anhydride, is not isolated; rather it is generated *in situ*.

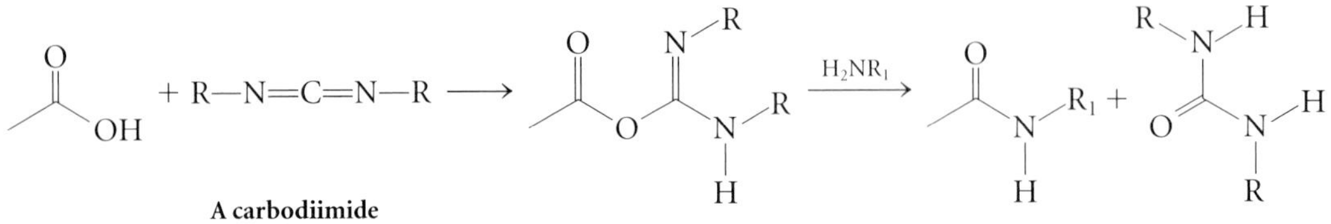

**A carbodiimide**
(R = cyclohexyl)

**An amide**     **A urea**

18-20

## ARTIFICIAL SWEETENERS

Although the science of chemistry has advanced greatly in the twentieth century, many important discoveries are still made by accident. For example, the artificial sweetener aspartame (NutraSweet®) is a dipeptide that corresponds to the last two amino acid units of the peptide gastrin, a potent natural stimulant of gastric secretions. In the course of research on gastrin, James Schlatter, a chemist working at Searle, prepared the methyl ester of the dipeptide of aspartic acid and phenylalanine in 1965 and by chance tasted some of the crystalline powder that had got-

ten on his hands. Because the dipeptide is composed of two natural amino acids, Schlatter and other chemists at Searle realized that it was likely to be nontoxic and might therefore replace "less natural" artificial sweeteners such as cyclamate and saccharin. Indeed, five years later, cyclamates were banned by the U.S. Food and Drug Administration. Aspartame is 160 times as sweet as sucrose; cyclamate is 30 times as sweet and saccharin is 500 times as sweet. Interestingly, neither of the constituent amino acids of aspartame is sweet; indeed, aspartic acid is bitter.

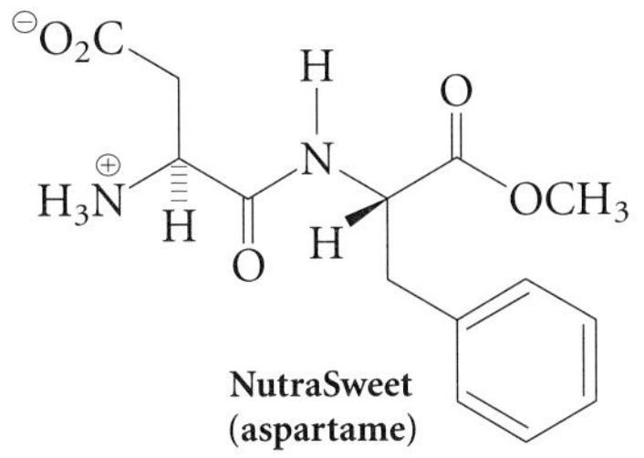

**NutraSweet**
**(aspartame)**

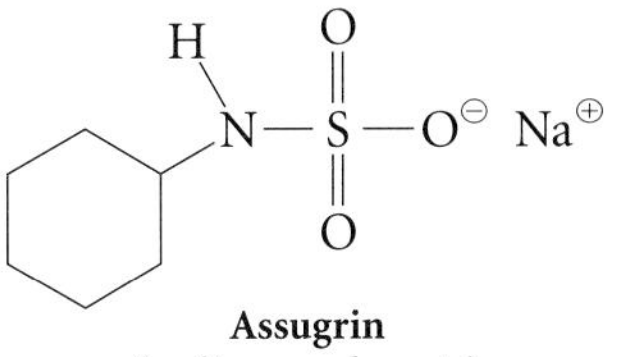

**Assugrin**
**(sodium cyclamate)**

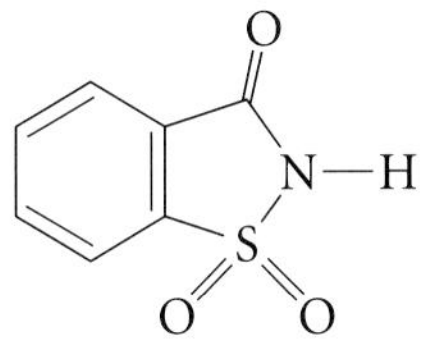

**Saccharin**

Nucleophilic acyl substitution takes place in the normal addition–elimination fashion and produces a urea (derived from the carbodiimide) in addition to the desired amide.

*Recovery of the Product.* The two-step process—nucleophilic displacement followed by deprotection—is repeated until all the peptide bonds have been formed. The product is then removed from the polymer support by cleavage of the benzylic carbon–oxygen bond connecting the polypeptide to the support. A strong acid, either HF or HBr, in trifluoroacetic acid effects an $S_N2$ displacement of the carboxylic acid (Figure 18.6).

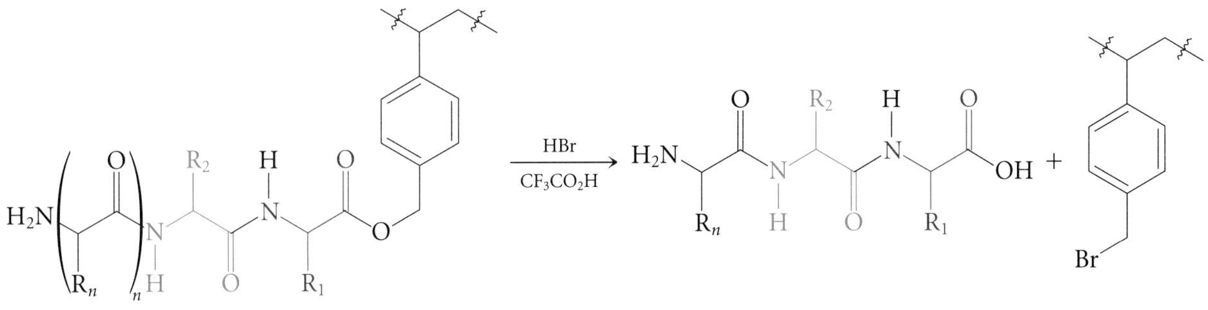

**FIGURE 18.6**

After the cycle shown in Figure 18.4 has been repeated sequentially to attach all of the desired amino acids, the polypeptide is removed from the polymer support by treatment with HBr.

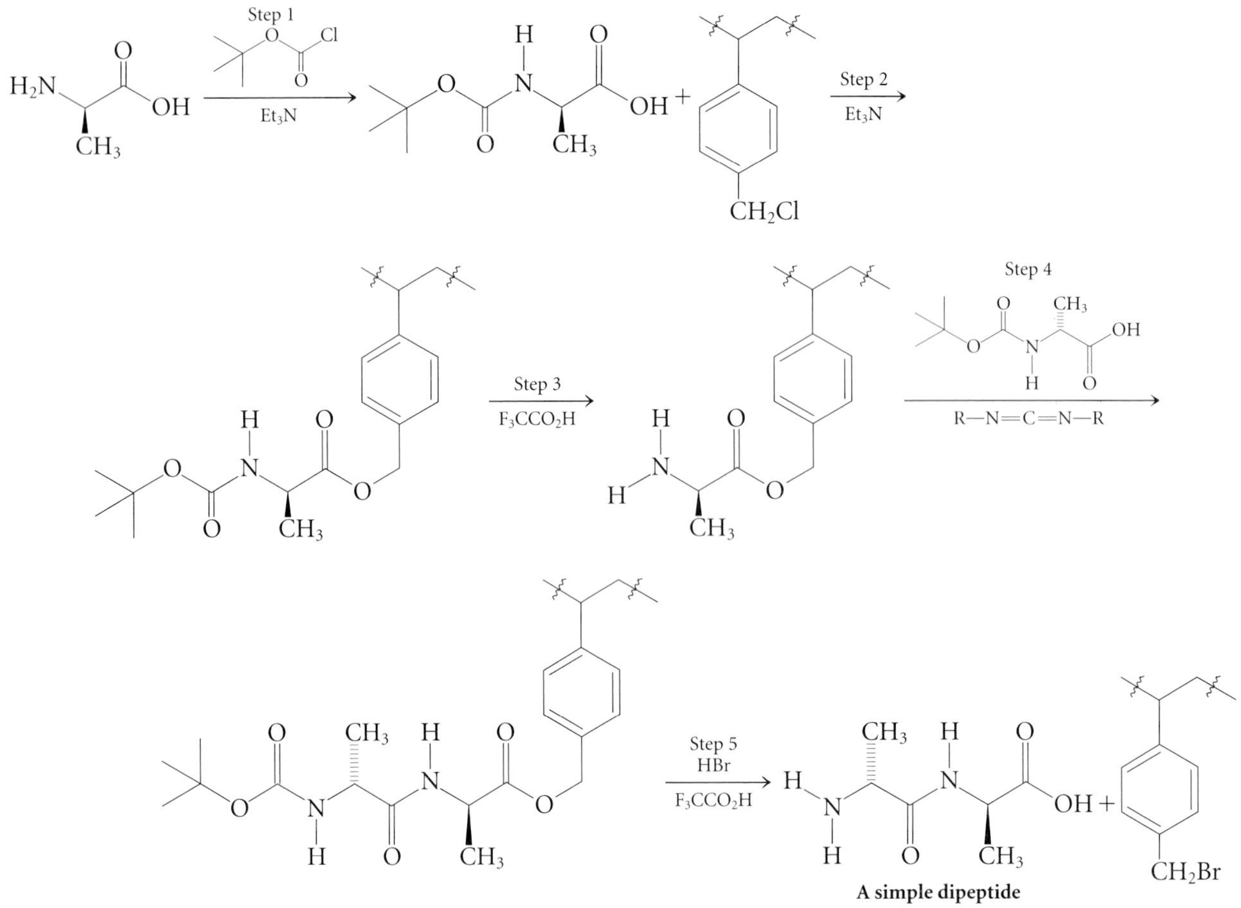

**FIGURE 18.7**

Combining the individual steps shown in Figures 18.3 through 18.6 leads to a sequence for the synthesis of a dipeptide of D-alanine. The steps are (1) protection of the amino group of the amino acid to be present as the N terminus of the dipeptide, (2) attachment to the polymer support, (3) deprotection of the amino group of the bound amino acid, (4) amidation of that amino group with the carboxylate end of a second N-protected amino acid, and (5) deprotection of the N terminus and removal of the dipeptide from the support.

*Synthesis of a Simple Dipeptide.* Let's consider a complete sequence for the synthesis of a simple dipeptide consisting of two D-alanine units, as shown in Figure 18.7. The presence of this dipeptide plays a pivotal role in the destruction of bacterial cell walls by certain antibiotics. The first step is the protection of the amino group of D-alanine by *t*-BOC. This N-protected amino acid is then attached to the polymer support by an $S_N2$ reaction. The protecting group is then removed by treatment with trifluoroacetic acid. A second N-protected amino acid—in this case, another D-alanine—is then added, along with a carbodiimide. This treatment results in the formation of a peptide bond between the N terminus of the bound amino acid and the C terminus of the second amino acid in solution. An excess of the amino acid in solution is used to ensure that conversion into product is as complete as possible. After each step of the sequence, the polymer support is washed with an appropriate solvent to remove by-products and excess reagents. The completed peptide chain is removed from the polymer support by cleavage of the link between the C terminus and the support with HBr in trifluoroacetic acid. A separate step to remove the N-protecting group from the last amino acid added is not required because the strongly acidic conditions used to cleave the bond to the support are more than sufficient for removing the *t*-BOC group as well.

### EXERCISE 18.13

Show in detail the sequence of steps and the reagents required to synthesize a tripeptide consisting of alanine, glycine, and valine, in that order.

## 18.4

# Alkaloids: Structure and Biological Activity

**Alkaloids** are naturally occurring materials that have a basic, nitrogen-containing functional group; they are found primarily in plants. Examples of structurally diverse alkaloids are shown in Figure 18.8 (on page 18-24).

Many plants contain a variety of alkaloids as complex mixtures. In most cases, the biological importance of these materials to the plants in which they are produced is not known. The only general explanation for the existence of this class of natural products (which comprises many tens of thousands of members) is that they serve to inhibit feeding on the plant material by predators. This supposition is based mainly on the observation that many of these compounds have a bitter taste (to humans). Some are extremely bitter indeed; for example, the slightly bitter taste of tonic water is the result of only a few thousandths of a percent of **quinine.**

Alkaloids have received more attention than any other single class of naturally occurring compounds, because many of them exhibit striking biological activity in insects and mammals, especially humans. For example, the alkaloid **nicotine** is a potent insecticide, and the physiological effects of nicotine, **mescaline,** and **heroin** on people are well known. Many alkaloids are poisonous, and some, such as **coniine,** the hemlock poison given to Socrates, are known specifically for this biological activity. Others have less drastic biological properties when given in small doses. For instance, **cocaine** is a frequently used illicit drug that causes artificial mood elevation and euphoria; **atropine** is used to dilate pupils for eye examinations; and **yohimbine,** a reputed aphrodisiac, has been used as an antidepressant.

Alkaloids have been isolated from plants for centuries—as both pure substances and mixtures. The use of these materials as medicinal agents has historically formed the basis for the cures of the medicine man or witch doctor in tribal societies and continues in today's homeopathy (more prevalent in Europe than the United States). Indeed, the use of plant extracts persists in the form of many home remedies used for a variety of ailments ranging from the common cold to cancer. In the past, physicians scoffed at

CHEM TV® II

#16    Heterocycles

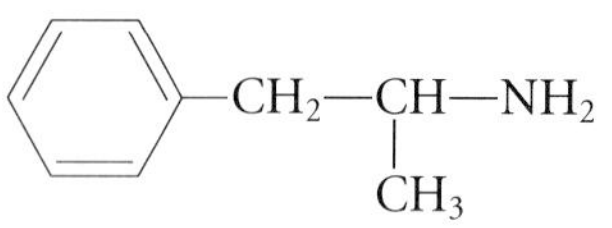

**Amphetamine**
(stimulant, LD$_{50}$ for rats = 180 mg/kg)

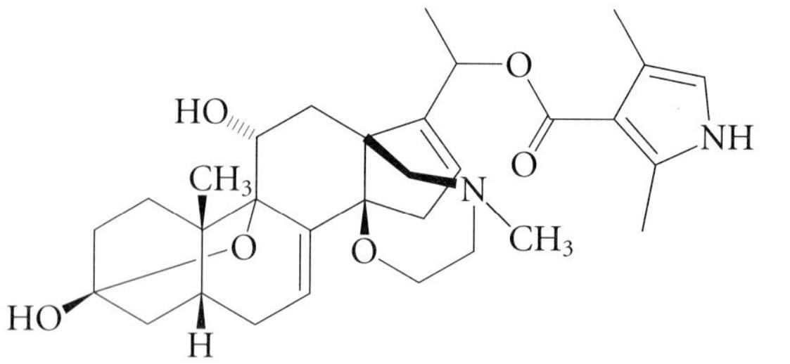

**Batrachotoxin**
(poison from frogs, LD$_{50}$ for mice = 2 $\mu$g/kg)

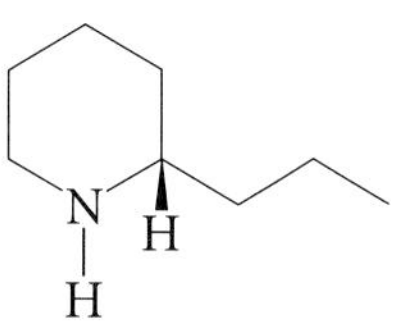

**Coniine**
(poison from hemlock,
LD$_{50}$ for humans = 10 mg/kg)

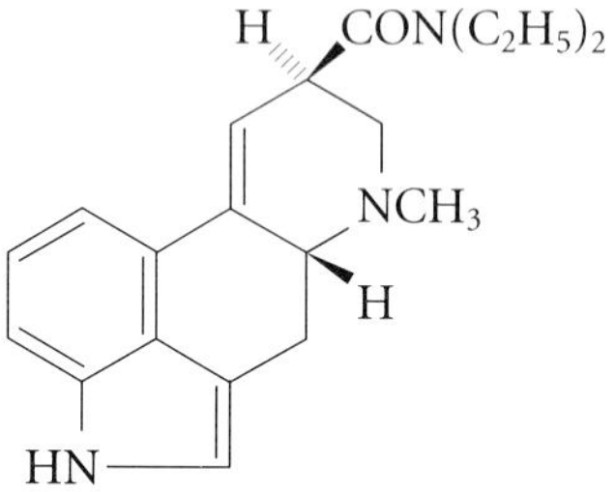

**Lysergic acid
diethylamide (LSD)**
(hallucinogen from ergot)

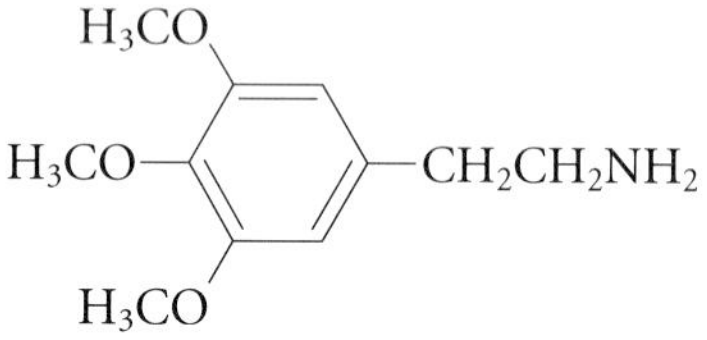

**Mescaline**
(hallucinogen from peyote cactus,
LD$_{50}$ for rats = 370 mg/kg)

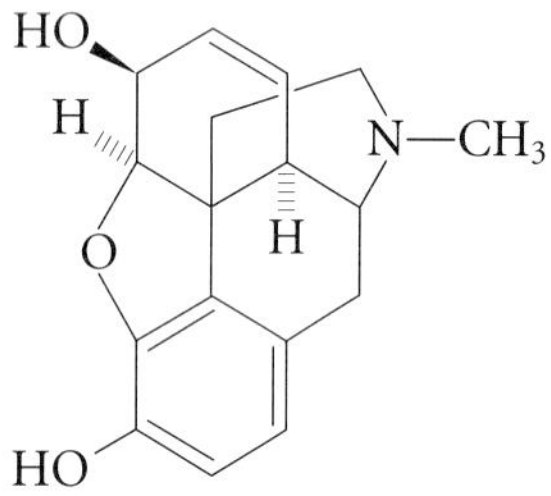

**Morphine**
(narcotic and analgesic
from opium poppy)

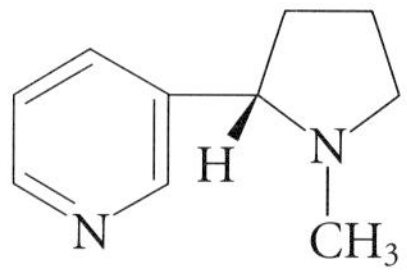

**Nicotine**
(from tobacco,
LD$_{50}$ for rats = 230 mg/kg)

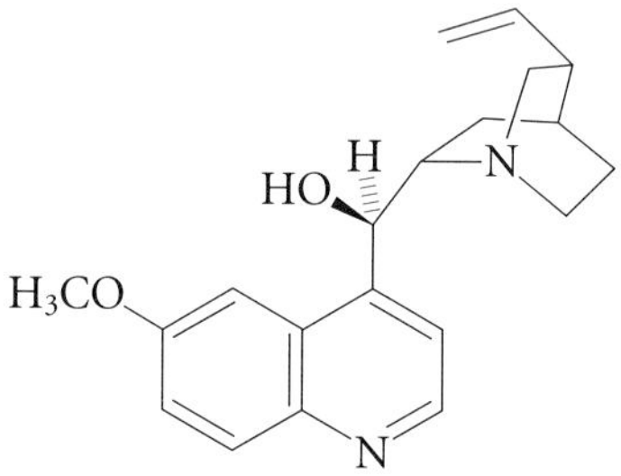

**Quinine**
(antimalarial
from cinchona tree)

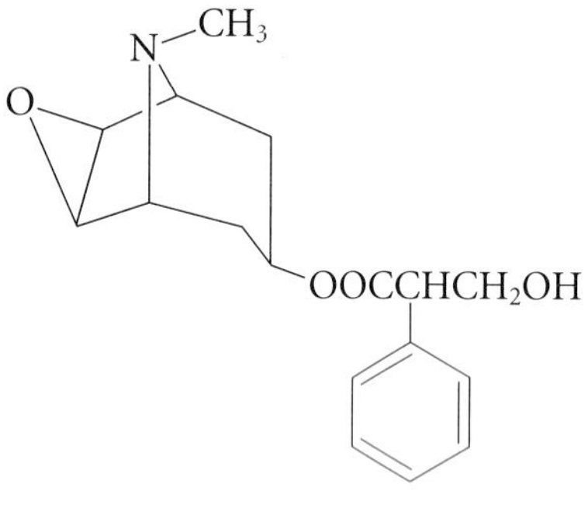

**Scopolamine**
(for treatment of motion
sickness, LD$_{50}$ for rats = 3.8 g/kg)

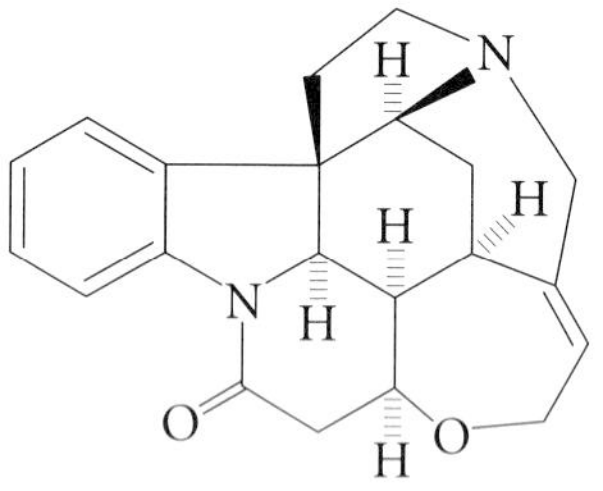

**Strychnine**
(poison from strychnos plant,
LD$_{50}$ for humans = 1 mg/kg)

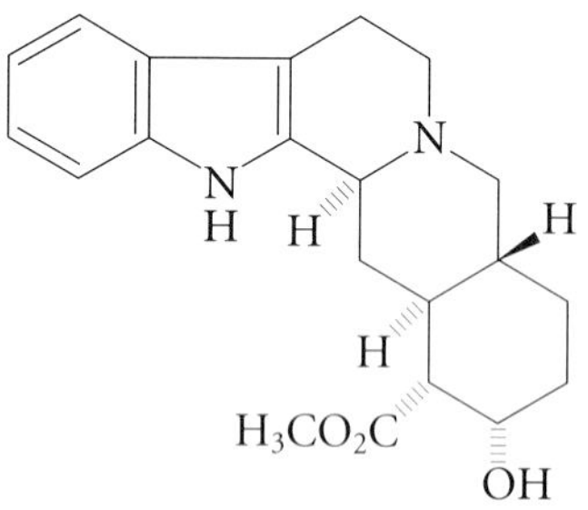

**Yohimbine**
(vasodilator and supposed aphrodisiac
from roots of *Rauwolfia serpentina*)

**FIGURE 18.8**

Alkaloids vary in structure, but all contain nitrogen, are basic, and have natural origins in plants. The LD$_{50}$ (lethal dose) is the amount required to kill 50% of a sample of the indicated animal.

18-24

## CHEMICAL PERSPECTIVES

### DOES NATURAL MEAN GOOD AND SYNTHETIC MEAN BAD?

There is a widespread belief that things that are "natural" must be good for us and that "unnatural chemicals" prepared in the laboratory are suspect, despite many examples to the contrary. Certainly, humanity would be much worse off without synthetic "wonder drugs" such as antibiotics. Nonetheless, this belief persists, and anxiety about the use of synthetic chemicals is frequently out of proportion with the possible negative consequences. For example, the synthetic compound Alar was developed and marketed as an inhibitor of the enzymes that cause apples to ripen and ultimately to rot. Alar is degraded in the body to 1,1-dimethylhydrazine, which, in turn, is oxidized to a nitrosamine. Every nitrosamine tested in laboratory animals has been shown to be carcinogenic, and this one is no exception. As a result, a campaign was mounted to ban Alar, and it succeeded, causing millions of dollars worth of apples treated with this compound to be destroyed. The action was based on only a limited knowledge of the facts: one ordinary mushroom contains ten times as much 1,1-dimethylhydrazine as two apples treated with Alar.

Alar        1,1-Dimethylhydrazine        Nitrosamine

the notion of such cures, but chemists in the pharmaceutical industry have been able to isolate many useful alkaloids (and other compounds) from plants. The destruction of large areas of natural vegetation in developing countries and the resulting extinction of large numbers of plant species have raised concern over the possible loss of valuable organic compounds.

The names, if not the structures, of many alkaloids are familiar because of their very potent ability to disrupt the central nervous system (CNS). A general structural feature of this group of psychoactive compounds is $\beta$-**phenethylamine** as a subunit.

$\beta$-Phenethylamine

### EXERCISE 18.14

Identify the β-phenethylamine subunit, if possible, for each alkaloid shown in Figure 18.8. Devise a synthesis of β-phenethylamine starting from benzene and any other reagents.

### NATURAL CHEMICAL DEFENSE SYSTEMS

Numerous toxic compounds are produced by animals, apparently as chemical defenses against predators. Frogs, especially those that live in the tropics, excrete alkaloids that can be fatal to mammals in very small doses. For example, bufotenine and bufotoxin have been isolated from the toad *Bufo marinus*. Both of these compounds increase the contractive power of weak heart muscles, and bufotoxin has been reported to cause terrifying hallucinations. The dried skin of this toad has been used in so-called zombie powder, leading to the suggestion (unsubstantiated) that these and perhaps other compounds present in the skin of the toad are responsible for inducing the trancelike state associated with zombies.

**Bufotenine**

**Bufotoxin**

A related alkaloid, homobatrachotoxin, has been isolated from birds of the genus *Pitohui*, endemic to the New Guinea subregion. The same compound (along with batrachotoxin) is found in Colombian poison-dart frogs of the genus *Phyllobates*. Only 2 μg of one of these alkaloids will kill a 1-kg mouse (proportionally less for mice of normal size). It is probable that frogs and birds synthesize these alkaloids by modification of closely related substances obtained by eating plants.

R = H in batrachotoxin
= CH₃ in homobatrachotoxin

## 18.5

# Structure of Nucleic Acids

Another important group of nitrogen-containing compounds are the **nucleic acids,** very large and quite complex molecules. This section describes the structure of these biopolymers, which serve to store and code information.

#23   Nucleic Acid Components

### Polynucleotide Backbone

A short section of a nucleic acid is depicted in Figure 18.9. The backbone is built of two different monomer units: a phosphate group and a ribose sugar unit (in the furanose form). The ribose bears a heterocyclic base. This polymer is known as a ribonucleic acid, or RNA. The structure of deoxyribonucleic acid, or DNA, is similar to that of RNA, but the sugar unit in the backbone of DNA is deoxyribose.

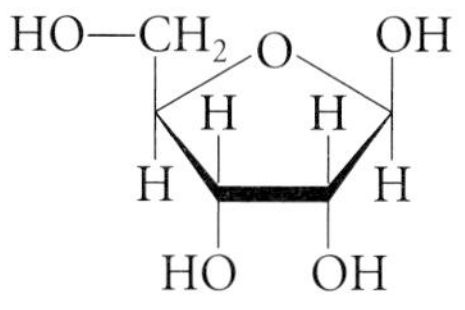

Ribose

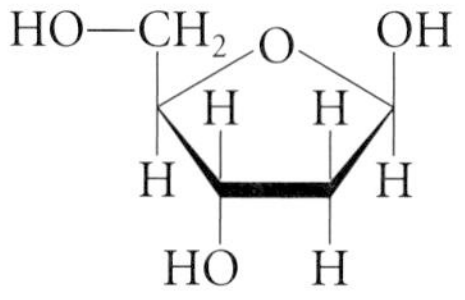

Deoxyribose

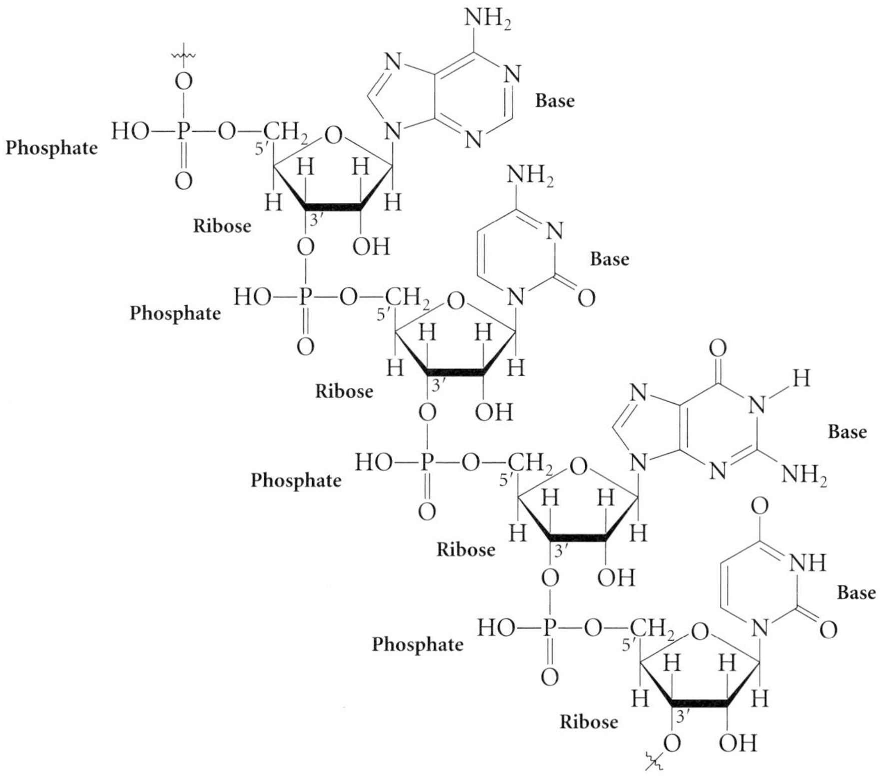

**FIGURE 18.9**

Short section of a nucleic acid. The backbone consists of ribose units connected through a phosphate diester linkage between the C-5′ hydroxyl group of one sugar and a C-3′ hydroxyl group of another sugar. C-1′ of each ribose unit is bonded to a purine or pyrimidine base.

### Nucleic Acid Bases

Let's consider the structures of the **nucleic acid bases** attached to the ribose (or deoxyribose) sugar units. These heterocyclic amines are divided into two classes: the **purine bases** (**adenine** and **guanine**) and the **pyrimidine bases** (**cytosine, uracil,** and **thymine**).

**Purine Bases**

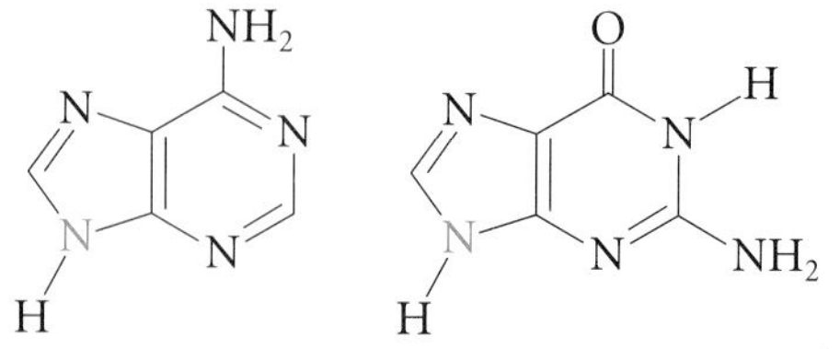

Adenine

Guanine

**Pyrimidine Bases**

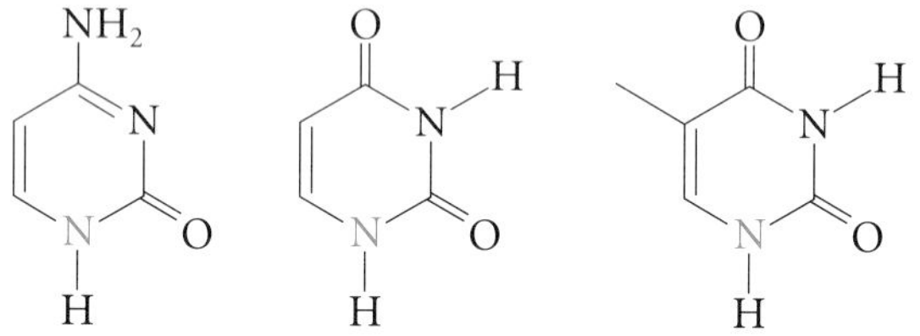

Cytosine

Uracil

Thymine

The class names, purine and pyrimidine, refer to the parent heterocyclic skeleton; the names of the individual bases are arbitrary or trivial (note that thymine is methyluracil). These bases are key units of nucleic acids; they are the chemical basis for the storage and coding of information. Their precise sequence in a nucleic acid chain ultimately determines the specific sequences of amino acids in polypeptides and enzymes.

These naturally occurring aromatic bases contain several heteroatoms, nitrogen and oxygen. The sequence of bases and hydrogen bonding between the bases serves as the key to information storage in the genetic code. One of the nitrogen atoms in the base (indicated in blue) links the base to the ribose unit.

Carbohydrates such as glucose can be converted into glycosides by reaction with alcohols, following the normal reaction pathway for the formation of acetals.

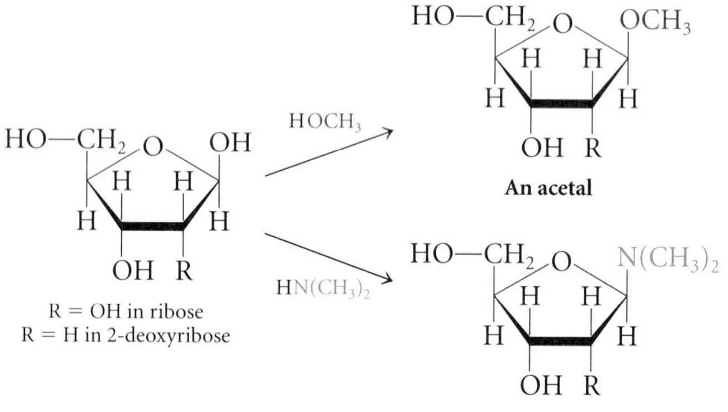

R = OH in ribose
R = H in 2-deoxyribose

**An acetal**

**An aminal**

Use of a secondary amine in place of the alcohol results in the formation of an aminal, whose structure is quite similar to that of an acetal. A similar reaction between a sugar unit and a base leads to a **nucleoside** (Figure 18.10), in which the base is a substituent on C-1 of a ribose (or deoxyribose) sugar unit.

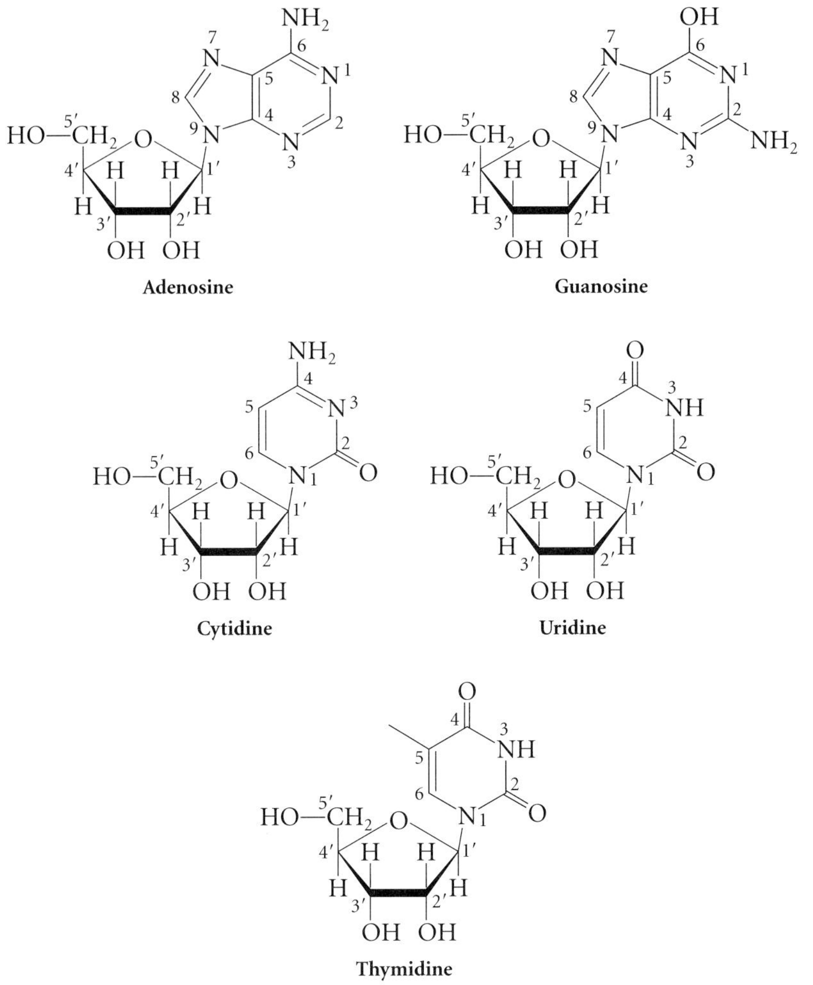

**FIGURE 18.10**

In a nucleoside, a heterocyclic base is attached to C-1 of ribose or deoxyribose. (The two units that constitute a nucleoside, the base and the carbohydrate fragments, have separate numbering systems. To distinguish them, the identifying numbers for atoms in the carbohydrate part have primes.)

### Attachment of Base to Sugar

Each base in a nucleic acid or a nucleoside has at least two nitrogen atoms. When the reaction between one of the bases and ribose is carried out under laboratory conditions, a mixture results (Figure 18.11, on page 18-30).

The five-member rings in adenine and guanine resemble the imidazole ring, present in the amino acid histidine (Section 18.2). In solution,

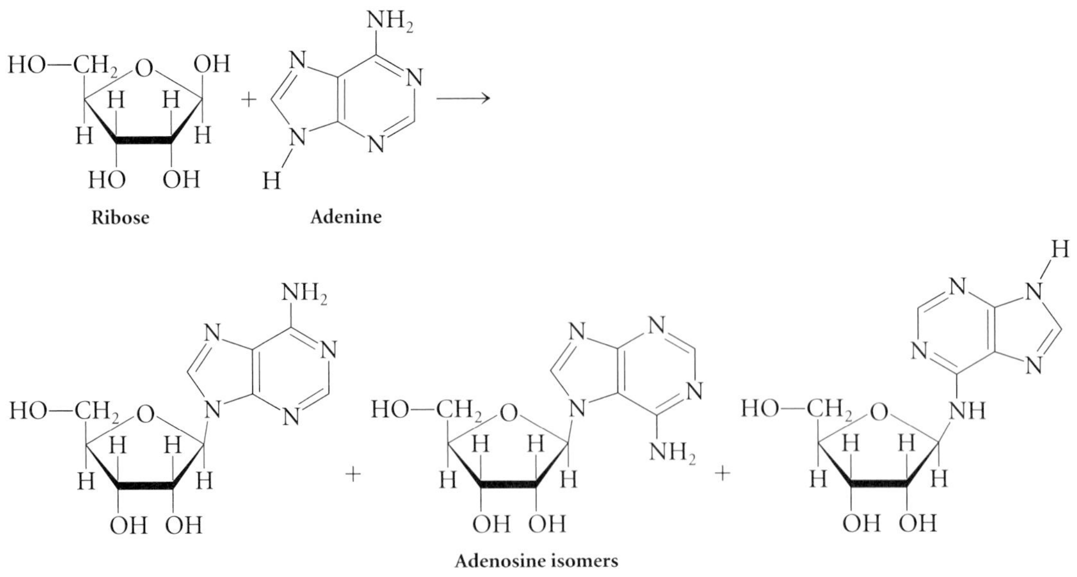

Ribose          Adenine

Adenosine isomers

## FIGURE 18.11

The nucleic acid bases exist as an equilibrium mixture of tautomers. Each different structure has nucleophilic character on different ring nitrogen atoms. Consequently, several isomeric products (differing in the site of the heterocyclic base at which the backbone sugar attaches) result from the reaction of ribose with a base, illustrated here with adenine.

adenine and guanine, like histidine, exist as mixtures of tautomers. Other possible proton tautomers for these heterocyclic bases come into play when hydrogen bonding of these bases is used for information storage.

**Tautomerization of Adenine**                **Tautomerization of Guanine**

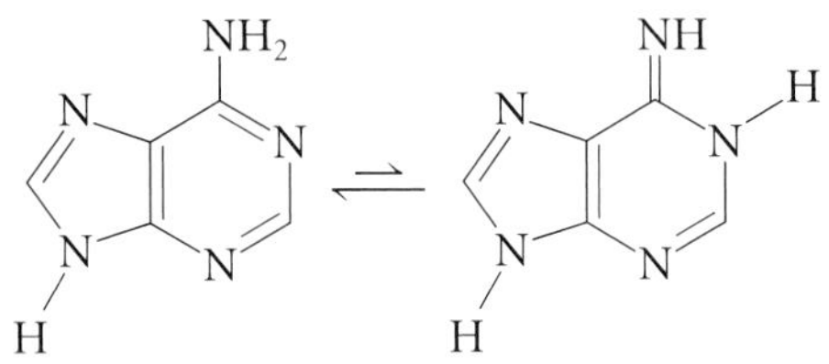

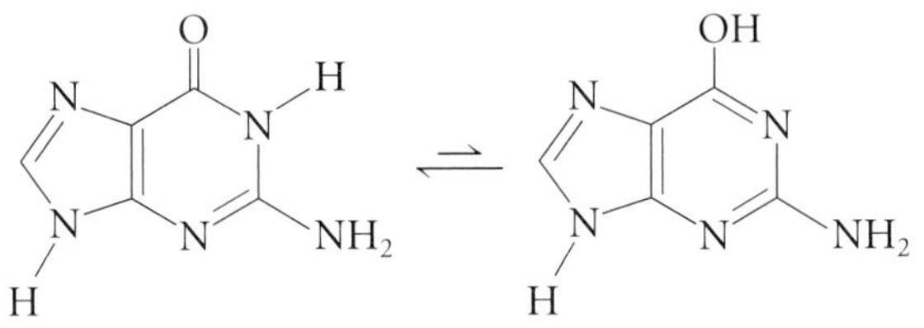

In contrast to the laboratory synthesis of nucleosides, the biochemical synthesis employs enzymes, which specifically catalyze the formation of only one of the possible isomeric nucleosides.

### EXERCISE 18.15

Write mechanisms for the interconversion of the two tautomers of adenine under both acidic and basic conditions.

Let's now turn to the polymer formed from the combination of individual nucleosides with phosphoric acid. Figure 18.12 shows a representation of the structure of a ribonucleic acid. This illustration allows us to view the nucleic acid as having been derived from a 1:1 mixture of two monomers—phosphoric acid and the nucleoside. Both monomers are polyfunctional and thus capable, at least in theory, of forming highly branched polymeric networks. Nonetheless, nucleic acids are formed only as shown in Figure 18.12, with only the C-3′ and C-5′ hydroxyl groups of ribose participating in the bonding.

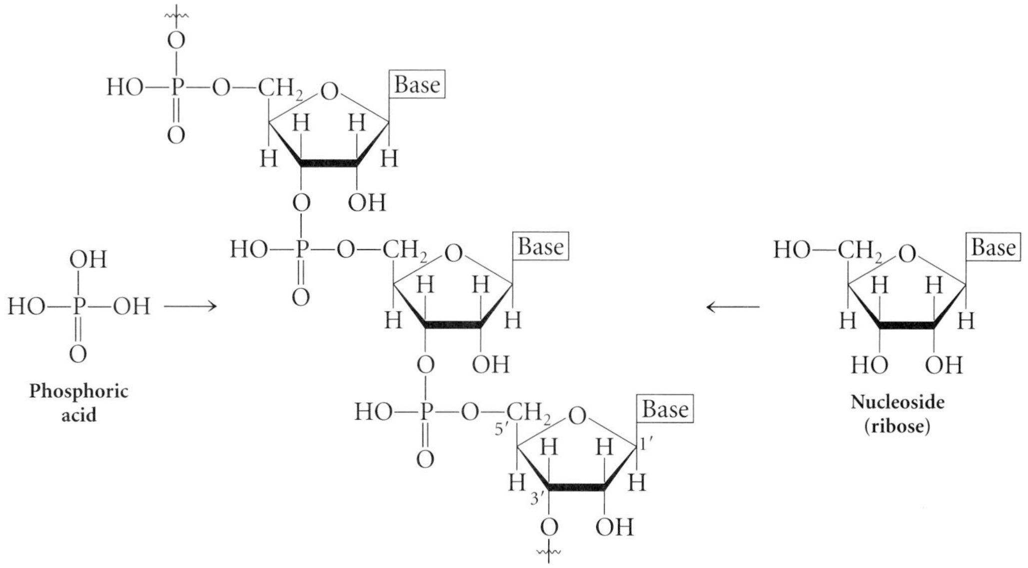

**FIGURE 18.12**

Representation of the structure of a ribonucleic acid. To emphasize the nucleoside–phosphate backbone, the heterocyclic bases are shown schematically.

Figure 18.12 represents the polymer as being composed of phosphate diesters in which the remaining oxygen is a free hydroxyl group. However, this group has a sufficiently high acidity ($pK_a$ 2) that it is essentially completely deprotonated under the near-neutral conditions found in living systems.

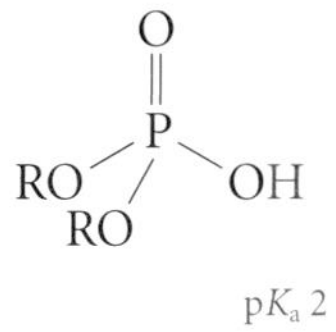

Thus, DNA and RNA exist as polyanions and are therefore **polyelectrolytes.** For this reason, electrophoresis is particularly valuable for the separation and identification of nucleic acids.

# Aminocarbohydrates: Structure and Function

Carbohydrates play a central role in energy storage. In addition, carbohydrates substituted with amino groups, called **aminocarbohydrates** or **aminosugars,** serve a crucial role in molecular recognition. For example, different aminocarbohydrate units distinguish the various blood groups. We will briefly discuss the chemical features that distinguish aminocarbohydrates from their all-oxygen analogs.

Many different aminocarbohydrates are found in nature. Figure 18.13 presents the structures of some of the more important ones, along with the structures of three antibiotics—**streptomycin, erythromycin,** and **kanamycin A**—which include at least one aminocarbohydrate. These antibiotics are resistant to biological degradation that occurs by cleavage of the acetal linkages, in part because of the presence of the basic nitrogen atom(s).

Aminocarbohydrates are also found in the exoskeletons of insects in the form of chitin, a polysaccharide formed from *N*-acetylglucosamine:

*N*-acetylglucosamine

As in cellulose, individual polymer chains of chitin are held together by hydrogen bonding. However, in chitin, hydrogen bonds are formed between amide units, forming an even tighter link between the chains.

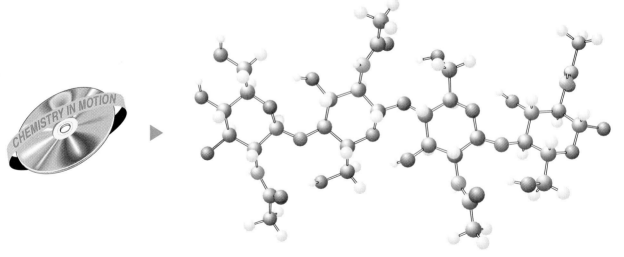

Segment of the structural polymer chitin

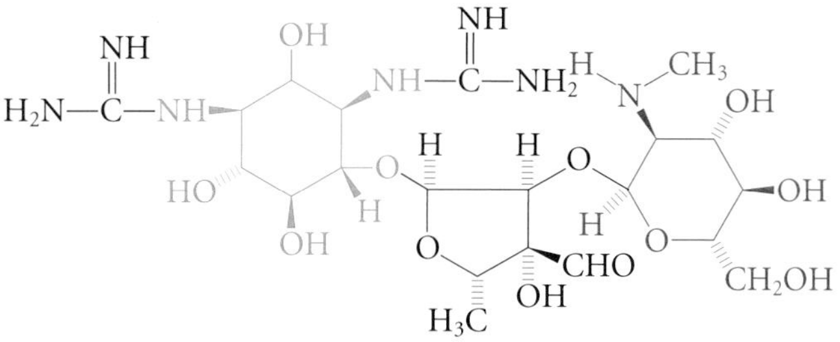

**Streptomycin**
**(antibiotic)**

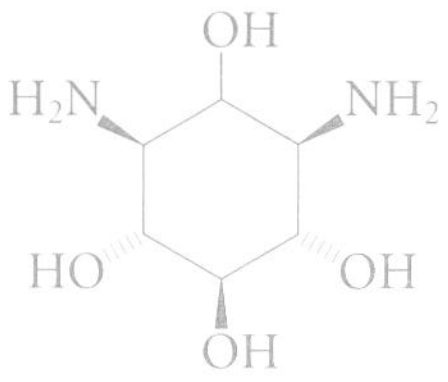

**Streptamine**

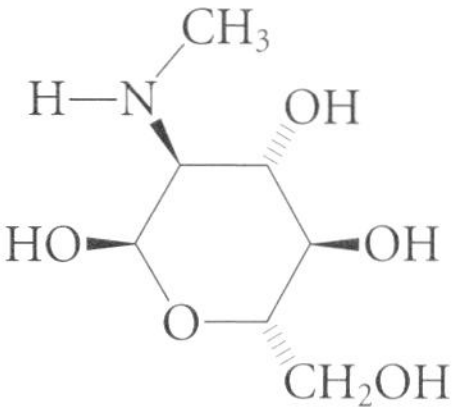

***N*-Methyl-α-L-glucosamine**

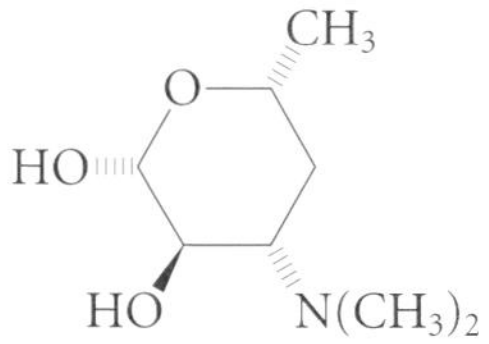

**Desosamine**

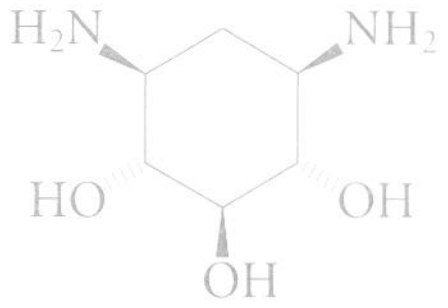

**2-Deoxystreptamine**

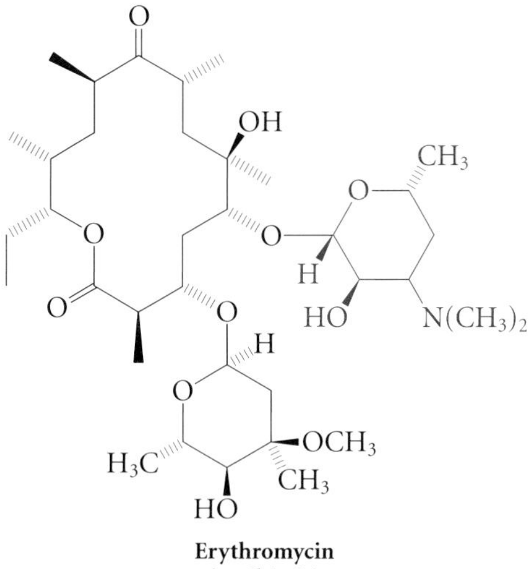

**Erythromycin**
**(antibiotic)**

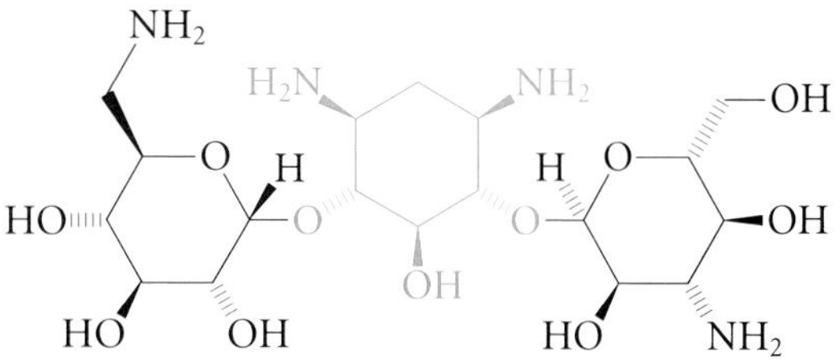

**Kanamycin A**
**(antibiotic)**

**FIGURE 18.13**

A number of potent antibiotics contain aminocarbohydrate subunits (red structures). Similar carbocyclic subunits in which each carbon bears a hydroxyl or amino group (blue structures) are also commonly found in antibiotics.

**18-33**

α-D-Glucose

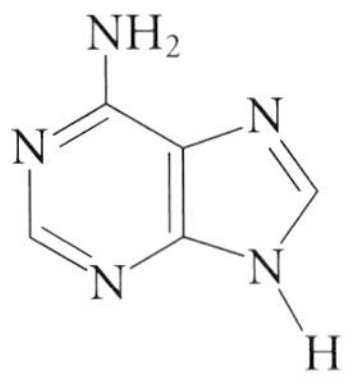

Adenine

## Abiotic Synthesis

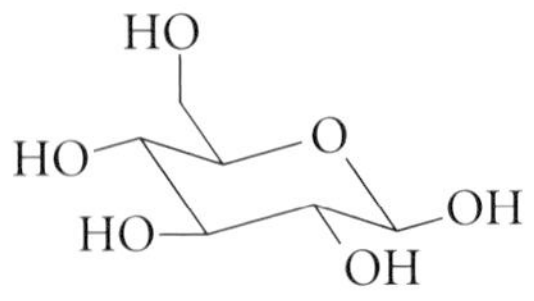

Many natural products are structurally complex. For example, carbohydrates such as glucose and nucleic acid bases such as adenine have several heteroatom-based functional groups, and the carbohydrates have numerous stereocenters as well. In part, this functional complexity of carbohydrates and the nucleic acids is necessary for their roles as components of complicated biochemical systems, because it establishes a basis for specific interactions between molecules. Were the molecules of natural systems less complicated, their interactions would be less specific.

Although the molecules of biochemical systems appear complex from a structural point of view, they constitute arrays that can be readily assembled conceptually from small building blocks such as ammonia, hydrogen cyanide, and formaldehyde. Thus, although nucleic acid bases are synthesized in living systems by a complex series of enzyme-mediated reactions, they are also formed (along with several of the amino acids) from mixtures of nitrogen, carbon monoxide, and carbon dioxide in the presence of an electrical discharge or intense short-wavelength irradiation. Both of these energy sources were likely to have been present on earth before life appeared. A synthesis of critical biochemicals in the absence of enzymes is often referred to as *prebiotic chemistry,* and some scientists attempt to draw conclusions relating to the origin of life and living systems from such experiments. However, such observations establish only that such molecules might have originated by nonenzymatic chemical transformations in primeval broths, not that current living systems have necessarily derived from them. The term **abiotic synthesis** is more appropriate, because it does not denote a time sequence.

### ▨ Adenine

Let's examine one possible abiotic synthesis of adenine (Figure 18.14). In fact, this abiotic synthesis closely resembles the biosynthesis of adenine that takes place in cells. Using the principles of retrosynthetic analysis, you can see that adenine has many carbon–nitrogen bonds but only two car-

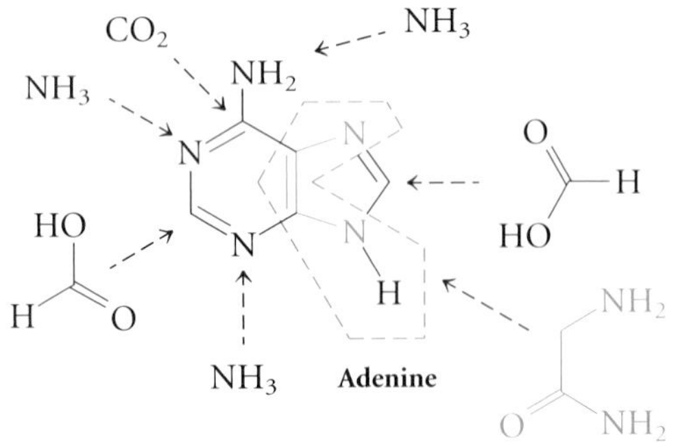

**FIGURE 18.14**

One possible abiotic synthesis of adenine uses simple, common compounds as the source of the atoms.

bon–carbon bonds, in a central three-carbon unit. An effective synthesis must therefore include oxidation–reduction and functional-group transformations. All the molecules shown in Figure 18.14 as "starting materials"—ammonia, carbon dioxide, formic acid, and the amide of the amino acid glycine—have been detected in interstellar space.

## EXERCISE 18.16

Identify which of the sources of carbon used as starting materials for the synthesis of adenine in Figure 18.14 must undergo a net reduction, which must be oxidized, and which remain at the same oxidation level.

The synthesis starts with the formation of the formamide of $\alpha$-aminoacetamide through condensation with formic acid (step 1):

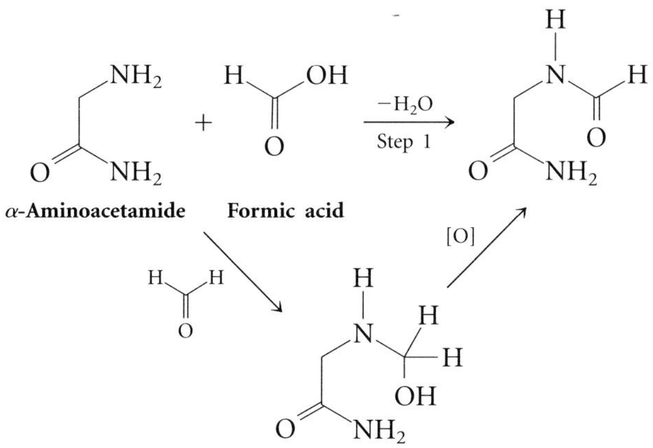

Amides can indeed be formed from amines and carboxylic acids at high temperature, although the severity of the reaction conditions limits the usefulness of the method for complex molecules. Alternatively, the same product can be formed by the addition of the amine to formaldehyde, followed by oxidation.

In step 2, condensation of ammonia with the acetamide carbonyl group forms a C=N bond similar to that present in a simple imine. Intramolecular condensation then forms the five-member ring (step 3). Dehydration (step 4) followed by tautomerization (step 5) establishes a six-electron $\pi$ system, the aromatic heterocyclic system known as imidazole.

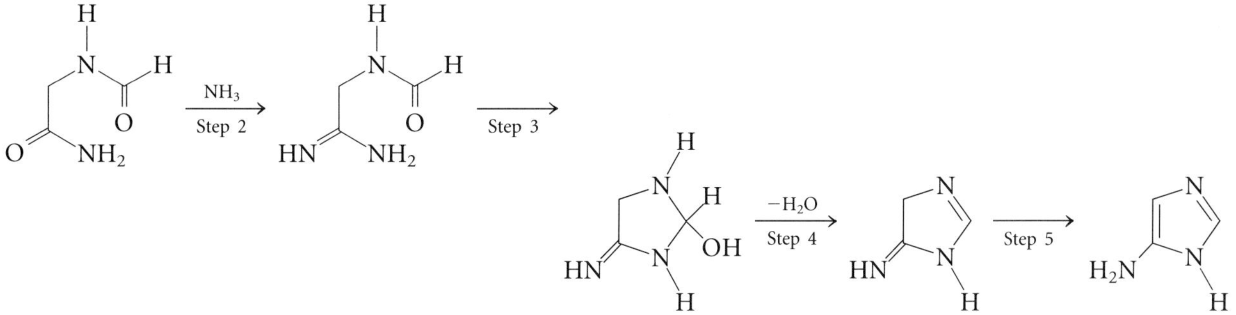

Aminoimidazole

**EXERCISE 18.17**

Write a step-by-step reaction mechanism for the condensation of the primary amide of glycine ($\alpha$-aminoacetamide) with formic acid:

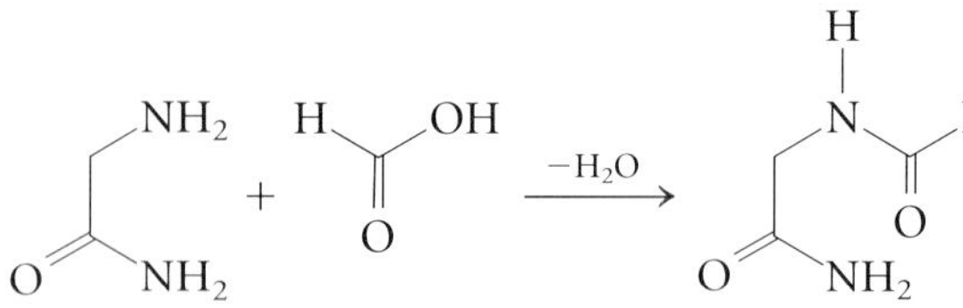

We can now see that the condensation reactions in this sequence are driven by the aromatic stabilization energy of the substituted imidazole. Looking ahead, we also realize that the final product (adenine) has an additional aromatic ring.

In the next stage of the synthesis, we begin to add the necessary atoms that constitute the six-member ring. Although the aminoimidazole is stabilized by aromaticity, it does have an enamine functional group. Recall that enamines are moderately active nucleophiles and can react with electrophiles. In this case, the electrophile is carbon dioxide, and the carbon–carbon bond formed in this reaction (step 6) completes the three-carbon central unit of adenine.

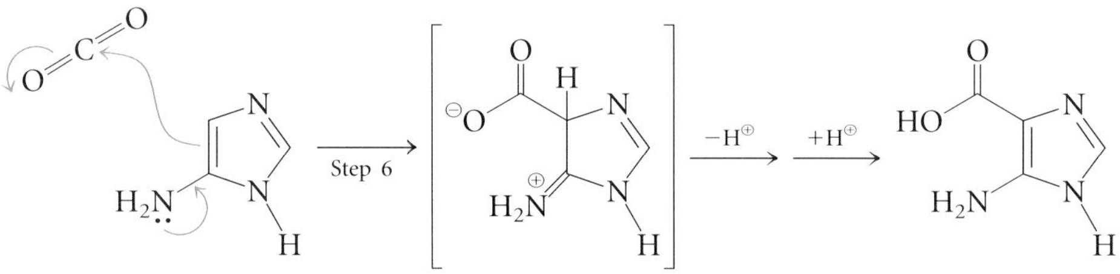

We are now close enough to the goal to see that what remains is the addition of one more carbon and two more nitrogen atoms. The next operation is the conversion of the just-formed carboxylic acid into the amide by reaction with ammonia (step 7), in a process analogous to the first step in the sequence. A second amide group is then formed by reaction of formic acid and the primary enamine group (step 8).

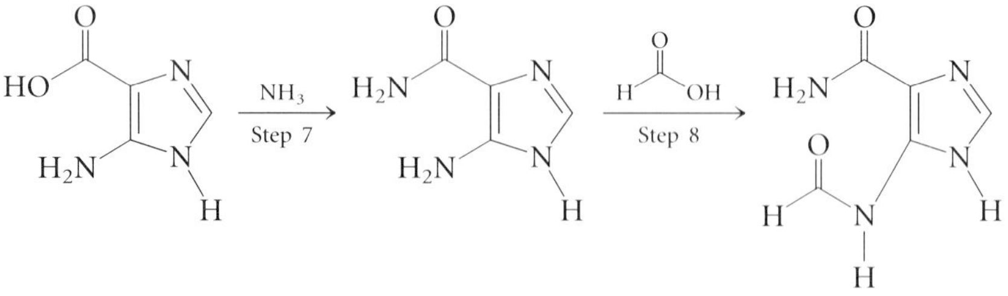

**EXERCISE 18.18**

Would you expect the primary enamine group that undergoes acylation in step 8 to be more or less reactive than a simple primary amine (such as methylamine) with respect to nucleophilic acyl substitution?

The six-member ring of adenine is formed via a condensation reaction analogous to steps 3 and 4, which form the imidazole ring. The final step is the reaction of ammonia with the remaining carbonyl group.

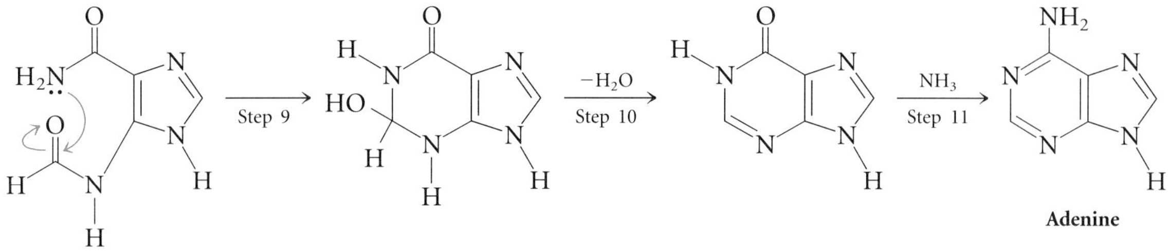

Adenine

## EXERCISE 18.19

Write a detailed mechanism for the transformation in step 11:

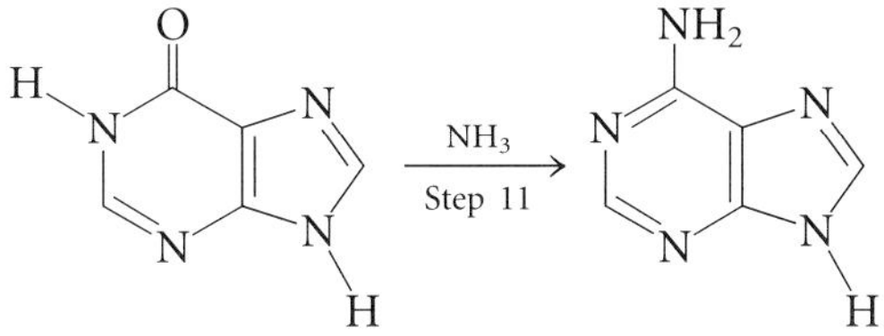

Adenine

There are 11 steps in this sequence, and, by the criterion of length, it is the most complicated synthesis we have examined so far. The individual steps of this synthesis can be grouped according to the type of transformation accomplished (Figure 18.15). From this perspective, the synthesis is quite straightforward, because it consists of a repeated use of two functional-group transformations—that is, the formation of amides (A) and imines (B)—a proton tautomerization (C), and the formation of a carbon–carbon bond (D).

Both functional-group transformations are accomplished by bond formation between nitrogen and a carbonyl carbon. Thus, although there is a

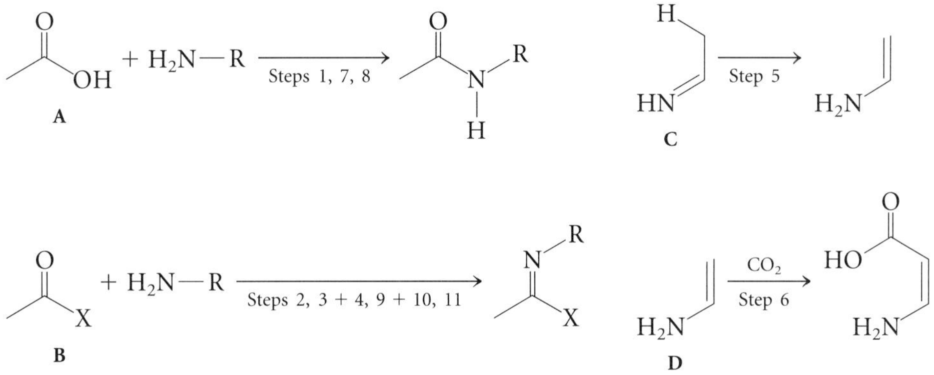

**FIGURE 18.15**

The adenine synthesis includes conversions of carboxylic acids to amides by reaction with a primary amine (A), iminations of an acid derivative (B), an imine–enamine tautomerization (C), and a nucleophilic carboxylation (D).

relatively complex array of functionalities in adenine, the molecule can be assembled using only a few elementary synthetic transformations. In a living system, it is important that relatively simple chemistry be required for the construction of complex molecules so as to maximize the overall chemical efficiency of the organism. From this perspective, adenine is a "natural."

### EXERCISE 18.20

Analyze the structure of cytosine in the way the structure of adenine is analyzed in Figure 18.14. How many fundamental synthetic operations (analogous to those summarized in Figure 18.15 for the synthesis of adenine) would be required to synthesize cytosine?

## Carbohydrates (Ribose)

Now let's consider the abiotic synthesis of ribose, the carbohydrate that, in combination with bases such as adenine, forms nucleosides and nucleic acids. Ribose (and other carbohydrates) can be formed by a multistep synthesis requiring only one reaction type. Ribose can be synthesized by the treatment of formaldehyde with calcium hydroxide:

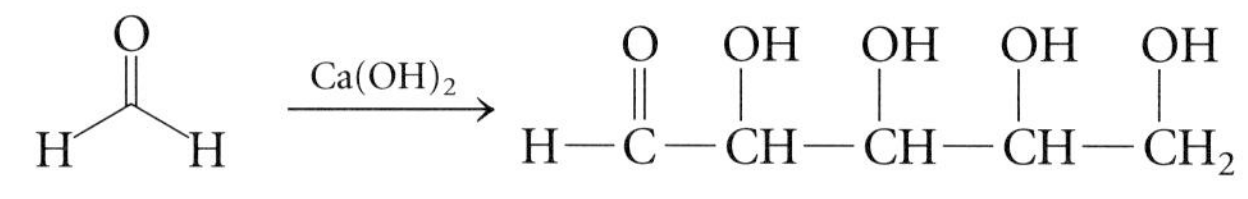

The carbon–carbon bonds are formed through nucleophilic addition to a carbonyl group of the anion derived by deprotonation of formaldehyde:

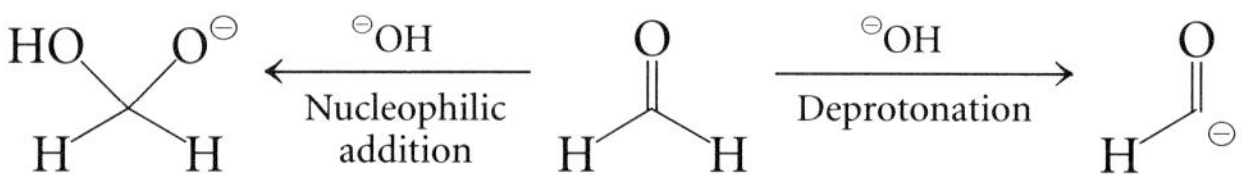

At first, the anion at the right may seem unusual because, in all earlier examples in which carbonyl groups were converted into anionic nucleophiles, the proton was removed from the $\alpha$ carbon rather than from the carbonyl carbon itself. But in formaldehyde and other aldehydes lacking $\alpha$ C—H bonds, the interaction with base cannot follow the normal course. As in the Cannizzaro reaction, hydroxide ion can also act as a nucleophile, adding to the carbonyl group. However, in this case, this process is reversible and nonproductive.

### EXERCISE 18.21

Would you expect the anion obtained by deprotonation of formaldehyde to be more or less stable than a vinyl anion? Explain your reasoning.

By default, then, hydroxide ion can function only as a strong base—removing the aldehydic proton and producing a relatively unstable but highly reactive anion. Nucleophilic addition of this formaldehyde anion to another molecule of formaldehyde produces the first and simplest of the carbohydrates.

### EXPLOSIVES

Many of the compounds used to make explosives contain nitrogen, although most of them do not occur naturally. Indeed, the majority of nonmilitary explosives are a mixture of $NH_4NO_3$ (ammonium nitrate) and diesel oil.

An explosion is not just a rapid release of energy and gases that push outward from the source. A true explosion requires the production of a shock wave that travels faster than the speed of sound. Indeed, modern explosives produce shock waves that travel at speeds close to 10 km/sec with pressures of $7 \times 10^4$ kg/cm$^3$. It is the power of this shock wave that is responsible for initiating the chemical decomposition of the bulk material and leads to essentially complete reaction before significant expansion has occurred.

The first explosive was black powder, first produced by the Chinese and reinvented in Europe in the fourteenth century. This mixture of $KNO_3$, charcoal (carbon), and sulfur is a relatively weak explosive because separate components must come together for the reaction to occur. On the other hand,

an explosive such as nitroglycerin (discovered in 1846 by Ascanio Sorero) has all the components required for reaction packaged in a single molecule.

Nitroglycerin is hard to handle and use by itself because it is a viscous liquid. Alfred Nobel (1833–1896) made what some have called the greatest contribution to civilization by mixing nitroglycerin with diatomaceous earth, forming dynamite, an easily handled yet still powerful explosive. However, Nobel was less than happy with his invention because the bulk of the diatomaceous earth contributed nothing to the explosion itself. He took advantage of the discovery by Frederick Abel in 1866 of a stable form of guncotton (nitrocellulose), cellulose treated with nitric acid, which converts most of the hydroxyl groups into the same nitrate esters found in nitroglycerin. Nobel mixed nitroglycerin with nitrocellulose, forming blasting gelatin. When squeezed through small holes, this mixture forms long cords that are cut into short segments to form cordite, the primary explosive for military projectiles.

This simple, two-carbon carbohydrate can be elaborated by an aldol reaction with itself to form a tetrose, which, in turn, can be further elaborated to a hexose.

**18-39**

The reaction of formaldehyde anion in which the carbonyl carbon acts as a nucleophile is certainly not the most likely possibility. After all, the vast majority of carbonyl transformations occur through reaction of the carbonyl carbon with the opposite electronic character, as an electrophile. Reactions like the one of formaldehyde anion, which require reactive species with an abnormal, or inverted, charge density, are commonly referred to as examples of *Umpolung,* a German word meaning, literally, "reversed polarity."

### EXERCISE 18.22

Recall that certain carbohydrates are more stable as cyclic hemiacetals. Use this knowledge to explain why the abiotic synthesis from formaldehyde tends to yield the pentose and hexose carbohydrates rather than shorter or longer chains.

## Summary

**1.** Naturally occurring nitrogen-containing molecules range in complexity from the relatively simple amino acids through polypeptides and alkaloids to the huge nucleic acids.

**2.** Carbon–nitrogen bonds can be formed by several possible routes: nucleophilic substitution of an alkyl halide, amidation of an acid derivative, reductive amination, nucleophilic addition to an imine, Mannich condensation, cyanide displacement followed by nitrile reduction, or oximation followed by a Beckmann rearrangement. Nucleophilic attack on an imine is similar to the attack of Grignard reagents on carbonyl groups, and the Mannich condensation is a combination of simpler reactions.

**3.** Amino acids exist as charged species and participate in acid–base equilibria involving a cation, the neutral zwitterion, and an anion. At a characteristic pH known as the isoelectric point, an amino acid exists almost completely in the zwitterionic form.

**4.** The side chains of the various $\alpha$-amino acids are the source of differences in their properties. Depending on the side chain, an amino acid can be neutral (hydrophobic or hydrophilic), acidic, or basic.

**5.** In polypeptides and proteins, amino acids are linked by amide bonds. The chemical synthesis of moderately sized polypeptides presents special challenges. The use of the Merrifield solid-phase synthesis solves most of these difficulties. This technique involves immobilization of an N-protected amino acid on a chloromethylated polystyrene, followed by deprotection to produce a bound amino acid with a free amino group. Treatment with an N-protected amino acid in the presence of a carboxylic acid activator (such as carbodiimide) produces a bound polypeptide. This sequence can be repeated many times, ultimately leading to a polypeptide that is cleaved from the polymer support by treatment with HBr or HF. The growth of a polypeptide from the C terminus is more problematic, because of epimerization of the azlactone intermediate.

**6.** Alkaloids are basic, nitrogen-containing compounds produced by plants. Many physiologically active alkaloids contain $\beta$-phenethylamine as a subunit.

**7.** The structures of the huge nucleic acids are built up from nucleosides (purine or pyrimidine aminals of ribose or deoxyribose) joined through a phosphate ester linkage between a C-3′ hydroxyl group of one nucleoside and a C-5′ hydroxyl group of another. The nucleic acids exist as anions at physiological pH because the phosphate acid backbone is deprotonated. They are therefore polyelectrolytes.

**8.** Abiotic synthesis of even such complex molecules as nucleic acid bases or sugars often employ simple combinations of the reactions covered earlier in this book. On the other hand, novel processes can also be involved. For example, formaldehyde can be deprotonated to form an acyl anion. This species reacts as a nucleophile in carbon–carbon bond formation, providing a route to carbohydrates.

# Review of Reactions

Mannich Condensation

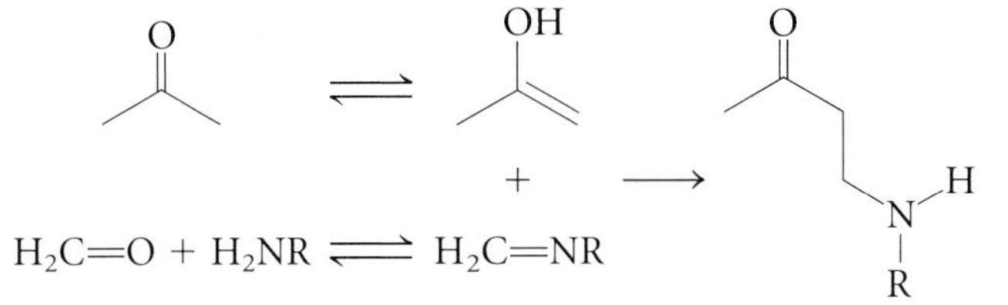

$H_2C=O + H_2NR \rightleftharpoons H_2C=NR$

Azlactone Synthesis

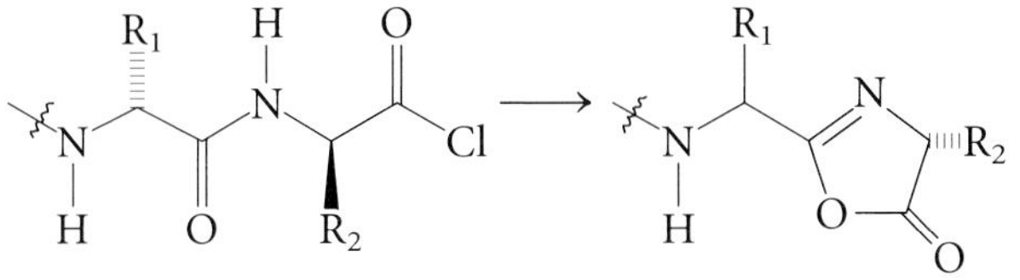

Formation of a Peptide Bond with Carbodiimide

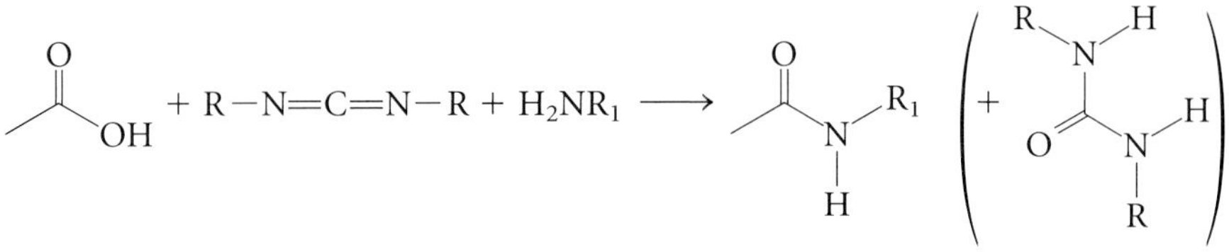

$+ R-N=C=N-R + H_2NR_1 \longrightarrow$

Aminol Formation

$\xrightarrow{R-NH_2}$

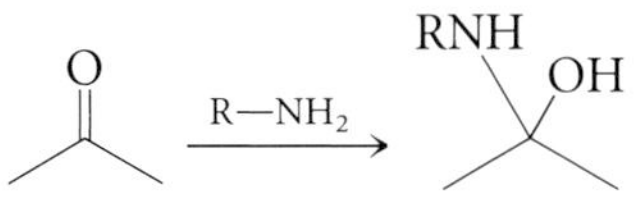

# Review Problems

**18.1** Each of the following aminoalcohols can be prepared by the reaction of an amine with an epoxide. Suggest appropriate starting materials that can produce each compound through nucleophilic ring opening of an epoxide by nitrogen.

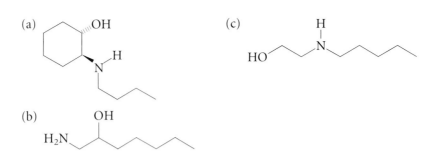

(a)

(b)

(c)

**18.2** Reduction of an amide by LiAlH₄ can be used to prepare each of the following amines. Indicate the starting material required for each product.

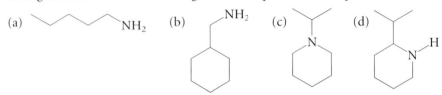

(a)      (b)      (c)      (d)

**18.3** Determine what starting materials (both amine and carbonyl compound) are required, and the sequence in which they must be combined, for the synthesis of each of the following amines by reductive amination.

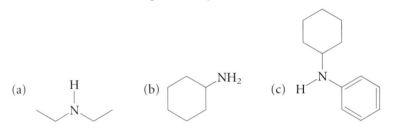

(a)      (b)      (c)

**18.4** Which imine and which Grignard reagent react to form each of the following amines?

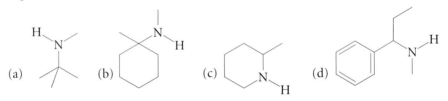

(a)      (b)      (c)      (d)

**18.5** Indicate the starting materials required for the synthesis of each of the following aminoketones by a Mannich condensation:

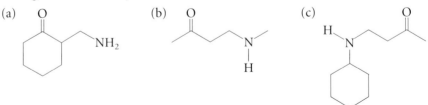

(a)      (b)      (c)

**18.6** Which of the following amines can be synthesized by reductive amination, Mannich condensation, reduction of a nitrile, or reduction of an amide?

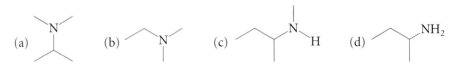

(a)      (b)      (c)      (d)

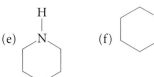 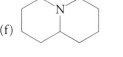 

(e) (f) (g) (h)

**18.7** Show the structure of the product that would be obtained from the Beckmann rearrangement of each of the following oximes. Be sure to consider the regiochemistry.

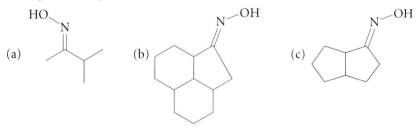

(a)  (b)  (c)

**18.8** Show the structure of the ketone that could be used as starting material for a Beckmann rearrangement that will produce each of the following amides or lactams:

(a)  (c)

(b)  (d)

**18.9** The synthesis of polypeptides by the sequential addition of one amino acid after another is a linear synthesis and, as such, has certain drawbacks. Determine the final yield of the complete and correct sequence of a decapeptide, if the yield in each coupling reaction is (a) 90%, (b) 95%, or (c) 99%.

**18.10** More reactive (toward nucleophilic acyl substitution) carboxylic acid derivatives undergo base-catalyzed epimerization more rapidly than do less reactive ones. Explain why a carboxylic acid chloride is more readily deprotonated at the α position than is a carboxylic acid ester.

**18.11** Acetal linkages formed from aminosugars are more resistant to acid-catalyzed cleavage than are typical carbohydrates. Suggest a rationale for this reduced reactivity, indicating which step(s) in the multistep process leading to hydrolysis is (are) most affected by the presence of the nitrogen.

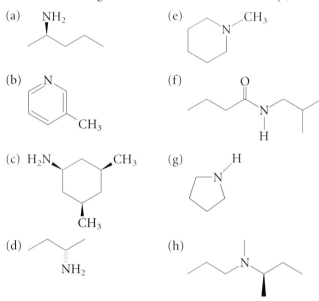

## Supplementary Problems

**18.12** Provide an IUPAC name for each of the following compounds. (Be sure to indicate the configuration of each center of chirality.)

(a)

(e)

(b)

(f)

(c)

(g)

(d)

(h)

**18.13** Draw a structure that corresponds to each of the following names. (Be sure to represent correctly any centers of chirality.) Then classify each in terms of the functional group present.

(a) di-(*R*)-2-butylamine

(e) *meso*-2,5-dimethylpyrrolidine

(b) 2,6-dimethyl-4-aminoheptane

(f) (*S*)-3-methylpiperidine

(c) cyclohexylamine

(g) 2-chloropyrrole

(d) 2,3-dimethoxypropanoamide

(h) *N*-methylpyridinium bromide

**18.14** What starting materials are required to synthesize the following amino-ketones using a Mannich condensation?

(a)

(b)

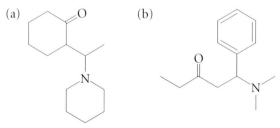

**18.15** Which of the following amines *cannot* be synthesized by reductive amination of a ketone or an aldehyde? Explain your answer.

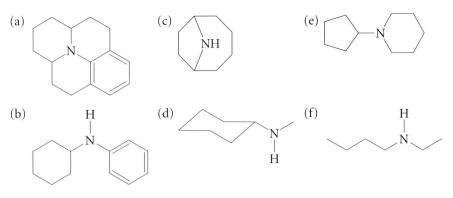

In those cases where reductive amination can be used to prepare the amine, show the starting materials necessary for the synthesis.

**18.16** NutraSweet® is a dipeptide composed of two different amino acids.

**NutraSweet**

What are they, and do they have the natural L configuration? Devise a synthesis of NutraSweet®, being careful to note that one of the amino acids has an acidic side chain with a carboxylic acid functional group.

**18.17** Alar can be prepared in a single step starting from 1,1-dimethylhydrazine and another organic material. What might the other starting material be?

**Alar**

**18.18** Figure 18.8 shows a number of alkaloids, several of which contain two nitrogen functional groups. For each of these, decide which nitrogen atom is more basic, and explain your reasoning.

**18.19** Each of the pyrimidine bases—cytosine, uracil, and thymine—can exist in a number of tautomeric forms. One tautomer is shown for cytosine and for uracil. Draw all other tautomeric forms.

**Cytosine**          **Uracil**

Indicate those features of the functional groups present in the tautomers shown that lend stability to these isomers. (It is not possible to compare these features quantitatively with those present in the other tautomers.)

**18.20** Write a complete mechanism for the reaction of acetic acid and methylamine to form *N*-methylacetamide in the presence of a carbodiimide:

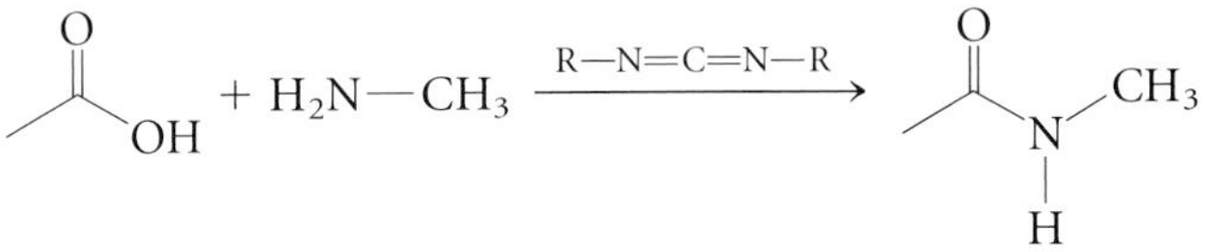

**N-Methylacetamide**

**18.21** Several of the methods discussed in Section 18.1 can be used to introduce the nitrogen atom into amphetamine, starting with a variety of different organic precursors.

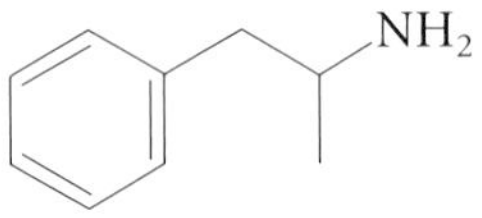

**Amphetamine**

Develop two different syntheses of amphetamine that do not involve carbon–carbon bond formation. (Don't try any of these at home, or anywhere else for that matter: the synthesis of amphetamine and its derivatives without license violates federal law, and federal prisons are *very* uncomfortable, we're told.)

**18.22** As in Problem 18.21, propose two syntheses of amphetamine that *do* involve carbon–carbon bond formation.

**18.23** Rank the following amines in order of basicity, and explain your reasoning.

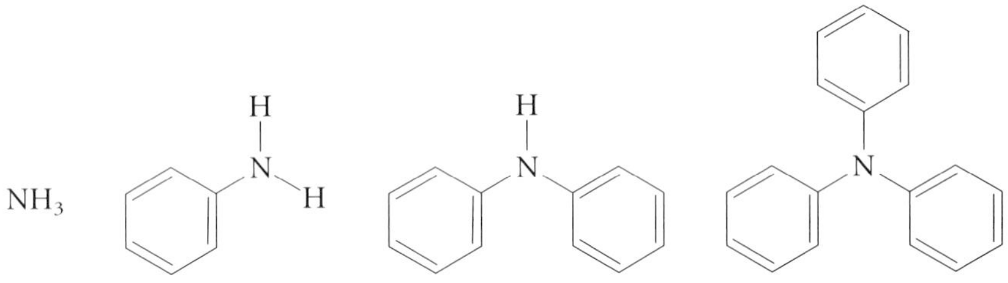

**18.24** The isoelectric point varies with the nature of the functional groups present in the side chain of an amino acid. For example, the three amino acids alanine, lysine, and aspartic acid have isoelectric points of 9.6, 2.7, and 6.0, not necessarily in that order. Decide which value belongs to which amino acid, and rationalize your choice.

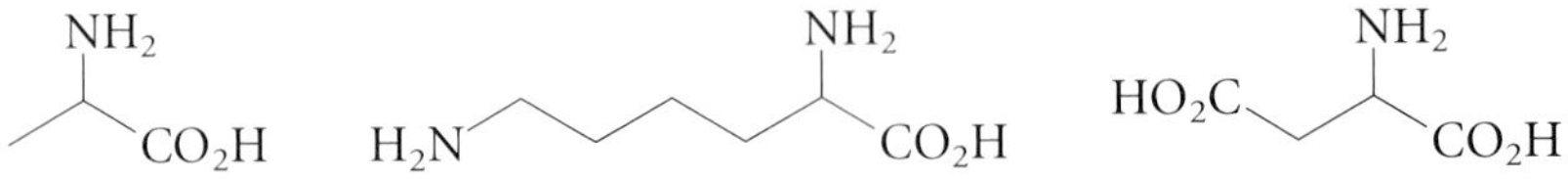

**18.25** Each of the three isomeric hydroxypyridines can exist in two tautomeric forms. Only for the first is the keto tautomer more stable. Explain what features lead to these differences.

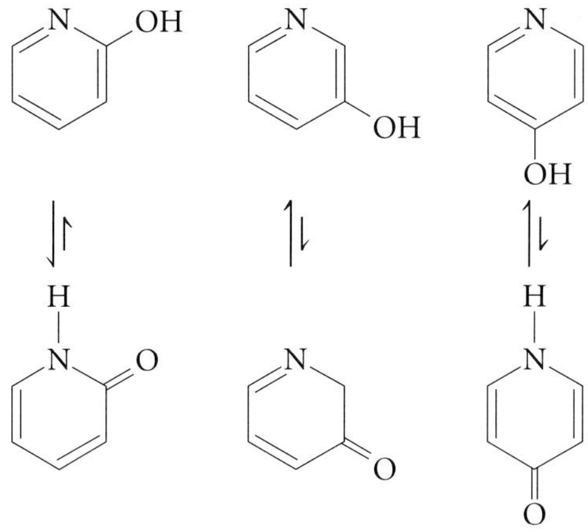

**18.26** Develop an appropriate sequence of reactions that could be used to carry out the following overall conversion. Be sure to include the structures of any intermediates that can be isolated, as well as the necessary reagents.

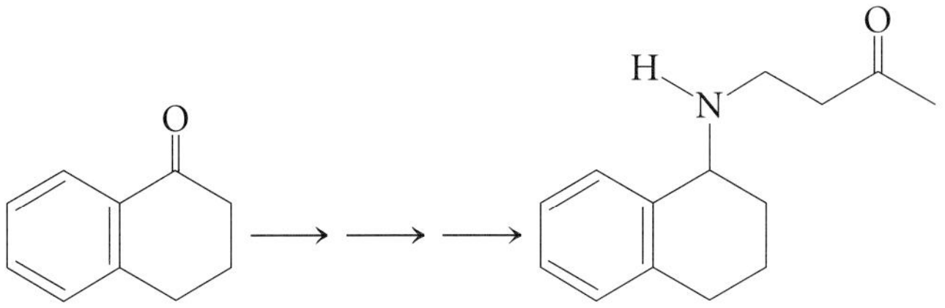

**18.27** Reaction of ammonia with formaldehyde produces a product, $C_6H_{12}N_4$, with very interesting spectral properties: only a single absorption is seen on the $^{13}C$ NMR spectrum; the $^1H$ NMR spectrum also exhibits a single signal; and the IR spectrum is devoid of N—H stretching absorptions. (*Hint:* There are no carbon–carbon bonds in this compound.) Draw a structure for this product, and account for its formation.

**18.28** Treatment of 2-chloropyridine with aqueous sodium hydroxide results in the formation of 2-pyridone. Devise a reasonable reaction mechanism for this transformation that is consistent with the observation that no products are obtained in which oxygen is introduced at any of the other carbons of the pyridine ring.

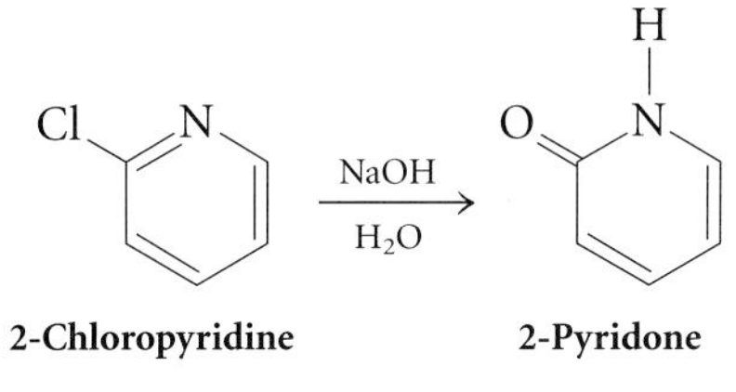

2-Chloropyridine      2-Pyridone

**18.29** The degree of aromatic stabilization decreases in the following order:

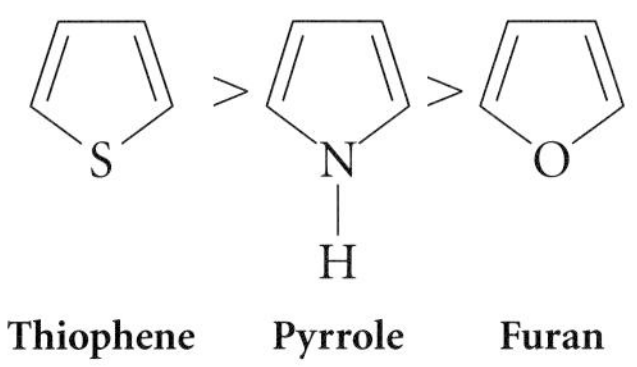

**Thiophene**    **Pyrrole**    **Furan**

Provide a rationalization for the greater aromatic character of thiophene compared with pyrrole and of pyrrole compared with furan.

# Important New Terms

**abiotic synthesis:** the preparation of a compound, often of biological relevance, without the use of biological agents such as enzymes or nucleic acids (18-34)

**alkaloids:** a diverse set of naturally occurring compounds that are isolated from plants and contain a basic, $sp^3$-hybridized nitrogen atom (18-23)

**aminocarbohydrates:** carbohydrates substituted with amino groups (18-32)

**aminosugars:** *see* **aminocarbohydrates**

**amphoteric:** containing both an acidic and a basic site—for example, $\alpha$-amino acids (18-13)

**azlactone:** an activated, cyclized derivative of the carboxyl terminus of a peptide or amino acid (18-19)

**$\beta$-phenethylamine:** an Ar—C—C—N subunit present in many psychoactive compounds (18-25)

**carbodiimide:** R—N=C=N—R; a reagent used to activate a carboxylic acid toward amide formation (18-20)

**guanidine unit:** —NHC=NH(NH)— (18-12)

**isoelectric point:** the pH in aqueous solution at which an amphoteric molecule exists as a neutral entity with an equal number of positive and negative charges; for a simple $\alpha$-amino acid, the pH at which it exists as a zwitterion (18-13)

**LHRH (luteinizing hormone release hormone):** a peptide hormone (18-14)

**Merrifield solid-phase synthesis:** a peptide synthesis carried out on a porous polystyrene support (18-17)

**nonessential:** descriptor of an amino acid that can be synthesized by the organism itself (18-1)

**nucleic acid:** a polymer composed of alternating sugar (ribose or deoxyribose) and phosphate units along a backbone and having one of several heterocyclic bases appended to the sugar unit joined through a phosphate ester linkage between a C-3′ OH group of one nucleoside and a C-5′ OH group of another; used to store and code information, which translates into sequences of amino acids in the peptides and enzymes critical to living systems (18-27)

**nucleic acid base:** one of the purine or pyrimidine bases found in DNA or RNA—adenine, guanine, cytosine, uracil, or thymine (18-28)

**nucleoside:** a component of RNA and DNA with a purine or pyrimidine base attached to C-1 of a ribose or deoxyribose sugar unit (18-29)

**polyelectrolyte:** a macromolecule in which the repeating unit bears a positive or negative charge (18-31)

**prebiotic chemistry:** the synthesis of critical biochemicals in the absence of chemicals formed by living systems, as might have occurred before life was present on earth (18-34)

**purine bases:** a family of bicyclic heteroaromatic molecules comprised of a five-member ring fused to a six-member ring and containing two nitrogens in a 1,3-relationship in each ring; two members of the family (adenine and guanine) are nucleic acid bases (18-5)

**pyrimidine bases:** a family of monocyclic heteroaromatic molecules comprised of a six-member ring containing two nitrogens in a 1,3-relationship; three members of the family (cytosine, uracil, and thymine) are nucleic acid bases (18-6)

**reductive amination:** a process by which a C=O group is converted to a $CH(NR_2)$ group (18-4)

***Umpolung:*** descriptor of a reagent with reversed polarity; a reaction employing an *Umpolung* reagent (18-40)

# Noncovalent Interactions and Molecular Recognition

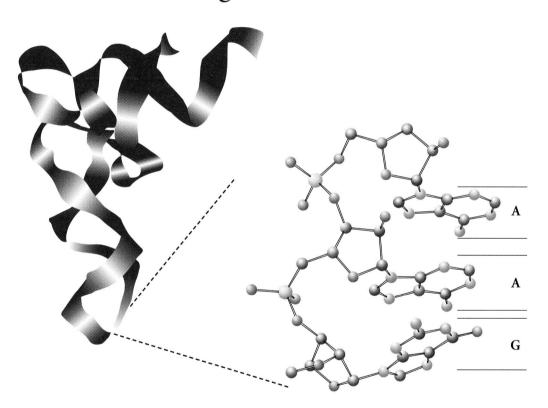

*A*

*A*

*G*

*Representation of the transfer RNA (from yeast) responsible for insertion of a phenylalanine residue in a growing peptide chain. The backbone of nucleoside bases is represented by the purple ribbon. The three bases (A-A-G) that form the anticodon for the specified amino acid are shown enlarged (without their hydrogen atoms, for clarity). Other atoms are not shown.*

$O$rganic chemistry includes many examples of both strong bonds, such as the covalent carbon–carbon bonds that hold molecules together, and weaker interactions, such as hydrogen bonds that form additional links within and between molecules. For example, hydrogen bonds are responsible for linking peptide chains together in a $\beta$-pleated sheet or for holding a single chain in an $\alpha$-helical secondary structure. These dramatically different shapes are fixed by the same basic force—the hydrogen bond. This chapter examines in detail how the hydrogen bond and other weak forces dramatically affect interactions between molecules. Hydrogen bonds provide one way by which molecules "recognize" those partners with which they form the strongest links.

Molecular recognition is defined as weak, reversible, selective binding between two chemical species. Selectivity is the key to molecular recognition, and, without such recognition, enzymes would not be able to accelerate specific reactions selectively. This chapter will consider some of the details of the relatively weak, noncovalent interactions that hold specific pairs of molecules together.

## Nonpolar (Hydrophobic) Interactions

Hydrogen is unique among the elements in that, when it is chemically combined, all of its electron density is used in bonding. Sharing its electron density with a neighboring atom leaves the hydrogen nucleus relatively exposed on the side opposite the bond. Thus, hydrogen atoms interact with the electron density surrounding other atoms in the same molecule or in other molecules by means of weak, but significant, electrostatic attractive interactions. When this type of attraction is to the electrons of another $\sigma$ bond, it is called a *van der Waals attractive interaction*. When the attraction is between a hydrogen atom participating in a polar covalent $\sigma$ bond and an atom with one or more accessible lone pairs of electrons, the term *hydrogen bond* is used. This latter interaction is generally stronger, because the concentration of electron density in a lone pair is greater than that surrounding a covalently bonded hydrogen atom.

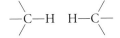

A van der Waals interaction

A hydrogen bond

### ▪ Influence of van der Waals Interactions on Physical Properties

Weak intermolecular interactions definitely influence the physical properties of organic compounds. The effect of these interactions can be illustrated by comparing the melting and boiling points of methane with those of mono-, di-, tri-, and tetrafluoromethane and of water (Table 19.1). Note that both the melting points and the boiling points increase as one and then two fluorine atoms replace hydrogen atoms of methane ($CH_2F_2$ does not crystallize) but then decrease for $CHF_3$ and $CF_4$, even though the molecular weight increases uniformly along the series. As hydrogen atoms are progressively replaced with fluorine atoms, the van der Waals attractive forces between the hydrogen atoms are replaced with stronger attractive interactions between hydrogen atoms and lone pairs of fluorine atoms. The strength of these attractive interactions reaches a maximum when the num-

**TABLE 19.1**

Melting and Boiling Points of Methane, Fluoromethanes, and Water

| Compound | Molecular Weight | Melting Point (°C) | Boiling Point (°C) |
|---|---|---|---|
| $CH_4$ | 16 | −182 | −164 |
| $CH_3F$ | 34 | −142 | −78 |
| $CH_2F_2$ | 52 | — | −52 |
| $CHF_3$ | 70 | −160 | −83 |
| $CF_4$ | 88 | −184 | −128 |
| $H_2O$ | 18 | 0 | 100 |

ber of hydrogen atoms and fluorine atoms is equal. As the number of fluorine atoms increases to three and then four, the repulsive interactions between the additional lone pairs of electrons dominate over the attractive forces due to intermolecular polar interactions. Note that water, although lighter than all four fluoromethanes, has higher melting and boiling points than any of them, a direct result of strong hydrogen bonding between a lone pair of electrons on the oxygen of one water molecule and a proton on the oxygen of another water molecule.

### Strength of van der Waals Interactions

Because the periphery of a typical organic molecule is often composed mostly of hydrogen atoms (molecules such as the fluorocarbons and carbohydrates being exceptions), the contribution of van der Waals interactions to intermolecular attractions does not differ greatly from one molecule to another. In contrast, the number and strength of hydrogen bonds can vary dramatically. For molecular recognition where the pairing of molecules depends on the strength of specific nonbonding interactions, the weak van der Waals forces play a much smaller role than do hydrogen bonds.

### Origin of van der Waals Interactions

Both van der Waals and dipole–dipole attractions have the same origin—the electrostatic attraction between unlike charges. In dipole–dipole attractions, two (or more) molecules have a net attractive interaction because they possess permanent dipoles. In van der Waals interactions, a net electrostatic attraction of the electrons of one molecule for the nuclei of nearby molecules results from induced polarization.

The electrons associated with hydrogen and carbon atoms in hydrocarbons are moderately polarizable. Thus, when a hydrogen atom participating in a C—H bond is brought near another hydrogen atom in a partially polarized C—H bond, the electron densities of the two C—H bonds repel each other. If we were able to take a photograph at any particular instant, we would find that the electron density surrounding one hydrogen would be closer to the other hydrogen (and farther from its attached carbon) than it is normally in an isolated molecule, whereas the electron density on the neighboring hydrogen would be shifted toward its carbon. These electron clouds do not behave totally independently of each other—rather, the temporary change in dipole resulting from the shift of electron density in one molecule results in an **induced dipole** in neighboring molecules. The distribution of electron density is not static—it is in a constant state of flux. Nonetheless, a net electrostatic attraction results.

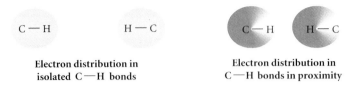

Electron distribution in          Electron distribution in
isolated C—H bonds               C—H bonds in proximity

## 19.2

## Polar Interactions

### Dipole–Dipole Interactions

When two atoms in a molecule share electrons unequally, a dipole results. The more electronegative atom is surrounded by electron density in excess of that required to balance its nuclear charge and therefore bears a partial negative charge. The less electronegative atom, lacking sufficient electron density to balance the nuclear charge, bears a partial positive charge. A dipole–dipole interaction is the electrostatic attraction between these regions of negative and positive charge in different molecules. For example, the triple bond of a nitrile is polarized toward the more electronegative nitrogen atom, resulting in a partial negative charge on nitrogen and a partial positive charge on carbon. Two molecules of nitrile can associate so that atoms with opposite charge are in proximity, resulting in a net attractive interaction.

$$H_3C-\overset{\delta+}{C}\equiv\overset{\delta-}{N}$$

$$\underset{\delta-}{N}\equiv\underset{\delta+}{C}-CH_3$$

*Effect of Dipole–Dipole Interactions on Boiling Point.* The strength of intermolecular attractions rises dramatically as the magnitudes of the dipoles in the individual bonds of two interacting molecules increase. As an example, compare the boiling points of nitriles with those of alkynes of similar molecular weights (Figure 19.1); the boiling point of the nitrile is consistently higher. This higher boiling point is the result of strong dipole–dipole interactions between pairs of nitrile molecules in the liquid phase, an association that must be broken as the molecules pass into the gaseous state. Because alkynes have only small dipole moments, dipole–dipole interactions do not significantly affect the boiling points of these molecules.

*Effect of Dipole–Dipole Interactions on Melting Point.* In contrast to their differing boiling points, the melting points of alkynes and nitriles, especially those of higher molecular weight, are very similar (Figure 19.2). The effects of dipole–dipole interactions differ somewhat for the liquid and solid states, because of differences in the patterns of association in the two states. In the liquid phase, each molecule is closely associated with only one other molecule or a small group of molecules, but in the solid phase, each molecule is surrounded by an array of nearest neighbors. Indeed, in the hexagonal close-packing arrangement, which is common for the solid phase

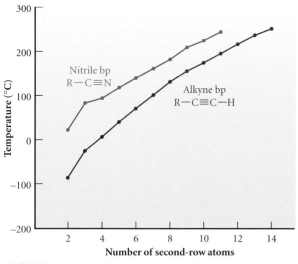

**FIGURE 19.1**

Boiling points of nitriles and alkynes as a function of molecular weight. The higher boiling points for the nitriles are a consequence of strong intermolecular dipole interactions involving the more highly polarized C≡N bond.

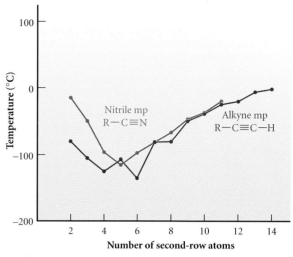

**FIGURE 19.2**

Melting points of nitriles and alkynes as a function of molecular weight. The melting points of nitriles and the related alkynes are nearly equal because attractive and repulsive dipole–dipole interactions nearly offset each other in the solid state.

of organic molecules (as well as for spheres such as marbles), each molecule is surrounded by six others in the same plane, plus three above and three below, for a total of 12 neighbors. (The central sphere is shown in a contrasting color for clarity.)

The consequence of these differences in association is that although dipole–dipole interactions are important in the liquid state, in a three-dimensional solid matrix in which each molecule is in close contact with many neighbors, attractive and repulsive interactions nearly cancel. Thus, for a covalent compound, the presence of dipole–dipole interactions leads to a marked increase in boiling point but has relatively little effect on melting point.

The change with molecular weight of the melting and boiling points of nitriles and alkynes can be represented by reasonably smooth curves (Figures 19.1 and 19.2). The reason for the zig-zag in the curve for the melting point of alkynes is not well understood but is observed quite often for molecules that are held together in the solid state mainly by van der Waals attractive interactions. The same behavior is observed, for example, with straight-chain alkanes. The abnormally high melting points of acetylene, propyne, 1-butyne, hydrogen cyanide, acetonitrile, and propionitrile (the points at the left in Figure 19.2) are due to the fact that each of these molecules has only one conformation, an effect that will be discussed shortly.

**Hexagonal close packing**

**Hexagonal close packing with the "lid" removed**

## Hydrogen Bonds

The large differences between both the boiling and melting points of water and methane are a clear indication that the contribution of hydrogen bonds to intermolecular attractions is much larger than that of van der Waals interactions.

**19-5**

X—H···:Y

Hydrogen
bond

**Nature of the Hydrogen Bond.** The hydrogen bond is a weak electrostatic interaction between a hydrogen atom and an electronegative atom bearing at least one lone pair of electrons—that is, a heteroatom (N, O, P, S) or a halogen (F, Cl, Br, I). Typically, the hydrogen atom is covalently bonded to one heteroatom and more weakly associated with a second heteroatom.

The hydrogen bond includes a total of three atoms: the hydrogen atom and two heteroatoms (typically), both of which are attracted to this single proton. The magnitude of this attractive interaction varies with the electronegativity of the two heteroatoms. The *more* electronegative is the heteroatom that is formally bonded to the hydrogen atom, the greater is the attraction of that hydrogen for additional electron density. Conversely, the *less* electronegative is the heteroatom donating the lone pair of electrons, the stronger is the hydrogen bond. With heteroatoms that are more electronegative than carbon, the hydrogen–heteroatom bond is more polarized than the carbon–hydrogen bond, and the hydrogen is more electron-deficient than if it were attached to carbon. However, even if both heteroatoms are of the same element, they are not equidistant from the hydrogen atom. For example, the distance between the $\sigma$-bonded hydrogen and oxygen atoms in alcohols is ~1 Å, whereas the distance for the hydrogen-bonded oxygen and hydrogen is 1.3–2 Å.

---

### EXERCISE 19.1

Suggest an order of boiling points for propane, ethylamine, and ethyl alcohol, all of which have approximately the same molecular weight. Check your answer (by consulting a table of boiling point data, perhaps in the *Handbook of Chemistry and Physics*). Then explain this order.

---

**Proton Transfer.** Might the hydrogen atom in a hydrogen bond be located symmetrically, midway between two heteroatoms? This arrangement is possible but is higher in energy than either of the arrangements in which the hydrogen atom is closer to one of the heteroatoms (at an optimal $\sigma$-bonding distance). In fact, an arrangement in which a hydrogen atom is equidistant from the two heteroatoms actually represents the transition state for the transfer of the proton from one heteroatom to the other. In simple systems, such a transfer is generally accompanied by a change in valency for both heteroatoms, as is the case when a proton is transferred from one water molecule to another. When this occurs, one of the heteroatoms is considerably more effective in attracting the proton—another argument for an unsymmetrical arrangement.

There are situations in which proton transfer results in an arrangement identical (or nearly so) in energy. For example, the intramolecular proton transfer in the enol of a $\beta$-diketone results in the same structure so long as the two ends have identical alkyl groups.

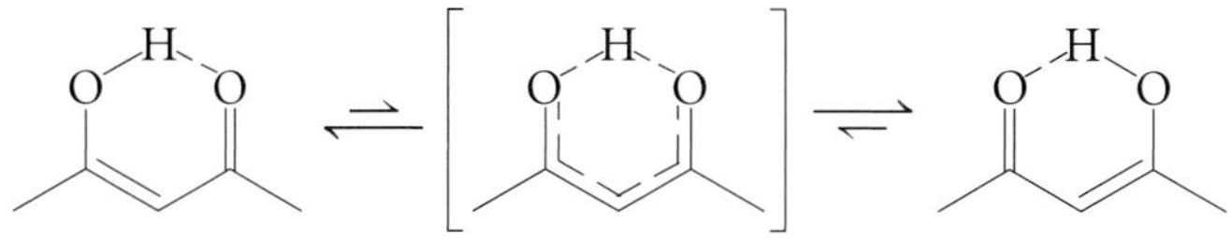

Similarly, the intermolecular transfer of three protons among three associated water molecules results in an identical arrangement:

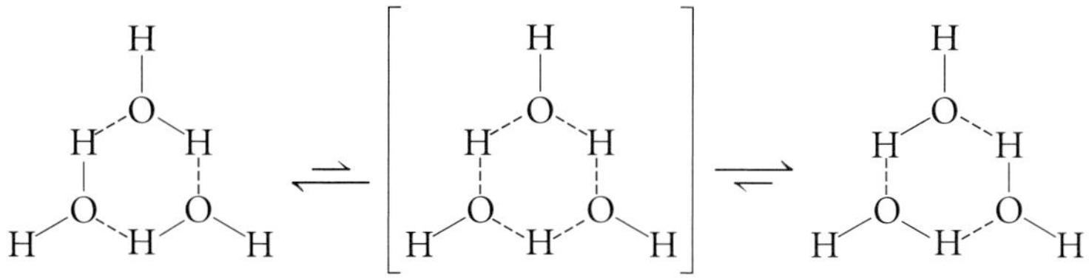

In both cases, the starting and ending states are identical, and the transition state is a structure in which each hydrogen atom is simultaneously and equally associated with two oxygen atoms. However, the total strength of these two bonds is weaker than the sum of a $\sigma$ bond between a hydrogen and one oxygen and a hydrogen bond between that hydrogen and a second oxygen.

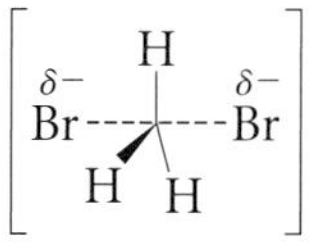

*Geometry of Hydrogen Bonds and Proton Transfer.* A symmetrical transition state also occurs in the $S_N2$ displacement of bromide from methyl bromide. Although organic chemists tend to focus on the carbon-based part of a reaction, the $S_N2$ reaction could be viewed alternatively as the transfer of the carbon from the leaving group to the nucleophile (for example, the methyl group being transferred from one bromine atom to another in the above transition state). Similarly, the symmetrical positioning of a hydrogen atom between two heteroatoms represents the transition state for the transfer of the proton from one heteroatom to the other. As in nucleophilic substitution, the arrangement of lowest energy for the three atoms taking part in a proton transfer is linear, with maximum separation of the two pairs of electrons associated with the proton. Furthermore, both unsymmetrical ground states (the hydrogen-bonded structures of the reactant and the product) leading to this transition state are lower in energy with this alignment. This linear relation exists in the hydrogen bonding in peptides, in both the $\beta$-pleated sheet and the $\alpha$-helical secondary structures (Figure 19.3, on page 19-8).

Although this linear arrangement is favored in both the ground state for hydrogen bonding and the transition state for proton transfer, other alignments are also possible. It is only necessary that the balance between attractive and repulsive interactions provide for net bonding. Although the alignments of hydrogen bonds in the enolic forms of $\beta$-diketones and $\beta$-ketoesters deviate substantially from the ideal, linear arrangement, they are nonetheless substantially stabilized by intramolecular hydrogen bonding:

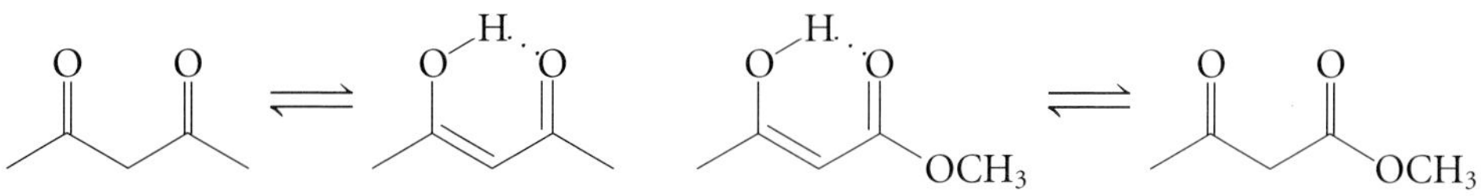

β-Pleated sheet                    α-Helix

**FIGURE 19.3**

Hydrogen bonds are responsible for the precise intermolecular interactions
in the β-pleated sheet and for the intramolecular interactions in the
α-helical structures of peptides.

**EXERCISE 19.2**

The percentage of the enol form present at equilibrium for three dicarbonyl compounds is as follows:

| | |
|---|---|
| acetylacetone ($CH_3COCH_2COCH_3$) | 76% |
| ethyl acetoacetate ($CH_3COCH_2COOCH_2CH_3$) | 8% |
| diethyl malonate ($CH_3CH_2OCOCH_2COOCH_2CH_3$) | 0.01% |

Explain these differences, keeping in mind the resonance stabilization of the ester
group.

***Hydrogen Bonding and Solubility.*** A simple way to examine the
strong influence of hydrogen bonding on intermolecular interactions is to
compare the solubilities of various compounds in water, a solvent that is
excellent at hydrogen bonding. For example, cyclopentane is virtually insoluble in water, whereas tetrahydrofuran is miscible with water. In this case,
water acts solely as a proton donor, because tetrahydrofuran does not have
a hydrogen bonded to a heteroatom.

The basic nitrogen atom of an amine serves as a proton acceptor, with
similar effects on solubility. Pyridine, a somewhat weaker base than a typ-

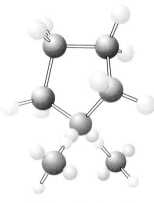

Hydrogen bonding of
tetrahydrofuran with
two molecules of water

### TABLE 19.2

Melting and Boiling Points of Isomeric Butanols

| Compound | Boiling Point (°C) | Melting Point (°C) |
|---|---|---|
| 1-Butanol (*n*-butyl alcohol) | 117 | −90 |
| 2-Methyl-1-propanol (*i*-butyl alcohol) | 108 | −108 |
| 2-Butanol (*s*-butyl alcohol) | 99 | −115 |
| 2-Methyl-2-propanol (*t*-butyl alcohol) | 82 | 26 |

ical aliphatic amine, is miscible with water, but the analogous hydrocarbon, benzene, dissolves in water only to the extent of 0.01%.

| | Cyclopentane | Tetrahydrofuran | Pyridine | Benzene |
|---|---|---|---|---|
| Water solubility: | 0.01% | ∞ | ∞ | 0.01% |

Hydrogen bonding also affects the boiling and melting points of isomeric alcohols (Table 19.2). Note that, as the degree of steric congestion about oxygen increases, and therefore the ability to participate in linear hydrogen bonding decreases, the boiling points of the isomeric butanols drop. A similar effect is seen in the melting points, except for that of *t*-butyl alcohol which is more than 100 °C higher than the melting points of the other three butanols.

*Role of Entropy in Phase Changes.* The dramatic difference between the melting point of *t*-butyl alcohol and those of its isomers is an example of the effect of *symmetry* and *rigidity* on the physical properties of molecules. An energetic factor that must also be considered in a comparison of the gaseous, liquid, and solid states is entropy—that is, the degree of randomness. In the gas state, not only are the separated molecules moving through space in a totally random fashion, but they can also tumble and spin freely relative to each other. In the liquid state, the degree of order is greater, and motion is restricted as a result of attractive interactions between neighboring molecules. In the crystalline state, the most rigid and regular of the three phases, the degree of order is quite high; the orientation of molecules with respect to one another is fixed, and molecules move only to a limited extent within the confines of the lattice structure.

Because long-range order is entropically unfavorable, the entropy contribution is always favorable for the transformation from a crystal to a liquid, and from a liquid to a gas. The symmetrical arrangement of molecules in a crystal is lost when the solid melts or evaporates, and this change in entropy is the same for all regular crystalline materials. However, for molecules that can adopt more than one conformation, the change from the crystalline to the liquid state is further favored by the presence of these additional arrangements in the liquid state. Of the four isomeric butanols, *t*-butyl alcohol has only a single conformation (even considering rotation about the C—O bond), whereas there are eight unique conformers for *n*-butyl alcohol. Thus, the favorable entropy change that accompanies melting of *n*-butyl alcohol is greater than that with melting of *t*-butyl alcohol.

### EXERCISE 19.3

Determine the number of unique conformations for *i*-butyl alcohol and *s*-butyl alcohol. (Do not fail to take into account enantiomeric conformations and conformational isomers due to rotation about the C—O bond, that is, the relative orientation of the OH hydrogen atom with respect to the rest of the structure.)

Molecular symmetry also influences the change in entropy upon melting of a crystalline compound. In a crystal, each molecule is held in a regular orientation relative to its neighbors, whereas in solution, molecules are free to orient in many different ways. However, molecular symmetry reduces the number of possible different orientations.

### EXERCISE 19.4

The effect of symmetry on the melting points of crystalline compounds is quite general. For each of the following pairs, predict, on the basis of symmetry, which molecule (A or B) has the higher melting point:

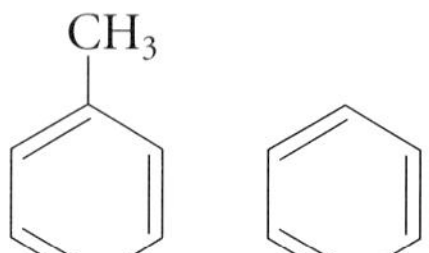

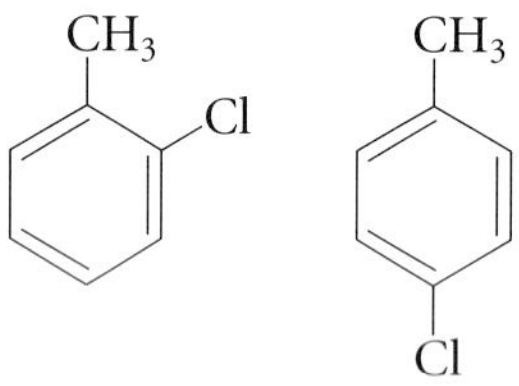

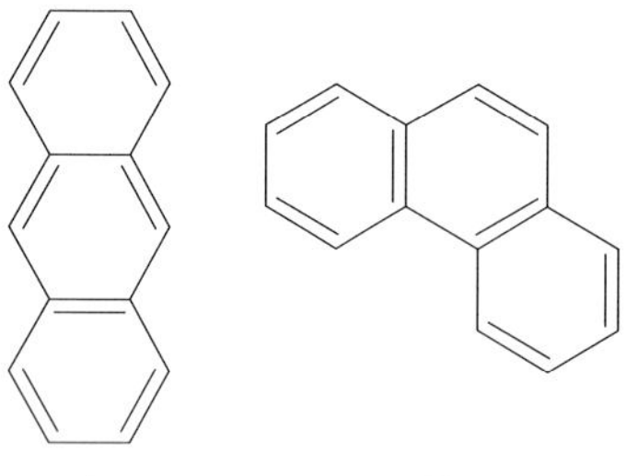

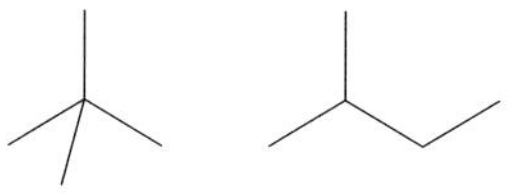

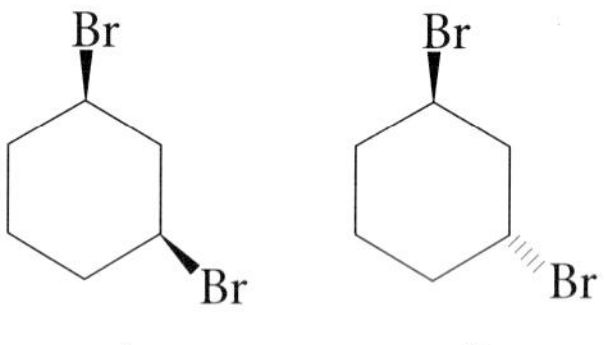

***Energetics of Hydrogen Bonding.*** Despite the large body of information available, a precise energy for a "typical" hydrogen bond cannot be given. Each hydrogen bond is only a relatively minor contributor to the total energy of a molecule and its arrangements, and it is not yet possible to factor out the contribution of hydrogen bonding from other factors such as entropy. Generally, a single hydrogen bond is believed to contribute 1–5 kcal/mole to attractive interactions between molecules, with the strongest contribution from the interaction between relatively basic heteroatoms and relatively acidic hydrogen atoms, as in the following example:

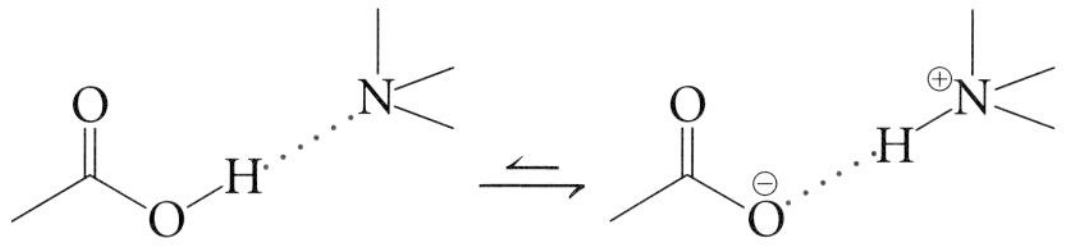

In the extreme, when these groups are sufficiently basic and sufficiently acidic, the proton is transferred in an acid–base reaction, becoming covalently bonded to the basic heteroatom.

***Multiple Hydrogen Bonds between Two Molecules.*** Hydrogen bonds are important contributors to molecular interactions, especially when the arrangement of atoms is such that two or more hydrogen bonds can simultaneously hold a pair of molecules together. Let's consider what happens when pairs of molecules are held in close contact by two or more hydrogen bonds, a situation that differs in several ways from the single hydrogen bonds considered so far. Perhaps the most obvious difference is that the strength of the association is greater, because the effects of the hydrogen bonds are additive. However, this is true only if the orientation of the three participating atoms (the two heteroatoms and the hydrogen) for each hydrogen bond is ideal. Therefore, in a system containing multiple hydrogen bonds, the relative positioning of the participating atoms is critical. Minor deviations from an ideal geometry can substantially decrease the strength of the bonding interactions and lead to a situation in which several nonideal hydrogen bonds are no stronger (or even are weaker) than a single bond with an optimal spatial orientation. Keep in mind that even the best of hydrogen bonds is weak. As a result, such bonds are in dynamic equilibrium, rapidly forming and breaking, under most conditions.

Multiple contacts between molecules are crucial in biological systems. For instance, they are often necessary to bind a substrate reversibly within an enzyme cavity. If the hydrogen bonding is too weak, the complex does not form, and no acceleration of the reaction occurs. On the other hand, if the hydrogen bonding is too strong, the enzyme (present in minor concentration) is tied up irreversibly and is not available for repeated cycles. Again, no acceleration occurs.

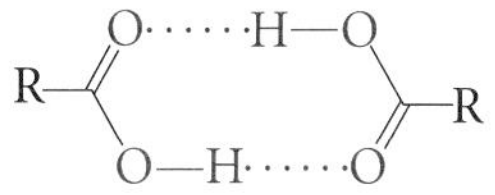

The carboxylic acid functional group can form two hydrogen bonds between a pair of molecules. Indeed, carboxylic acids exist as dimers in solution, except in solvents that are themselves very good at hydrogen bonding. Note that the dimer structure is symmetrical, with each carboxylic acid group donating one hydrogen and one lone pair to the two hydrogen bonds. The hydrogen atoms are located on a line between the pairs of oxygen atoms. The magnitude of this hydrogen-bonding interaction can be judged by comparing the boiling points of a straight-chain carboxylic acid with that of the methyl ester of the same molecular weight (the methyl ester of the carboxylic acid with one fewer carbon atom). This comparison is shown graphically in Figure 19.4 (on page 19-12). In each case, the acid has a considerably higher boiling point as a result of the two hydrogen bonds between a pair of molecules. These bonds are sufficiently strong that they are not broken in going from the liquid to the gaseous state. Thus, distillation of a carboxylic acid requires vaporization of the dimer, a complex twice as heavy as the carboxylic acid itself.

**Carboxylic acid dimer**

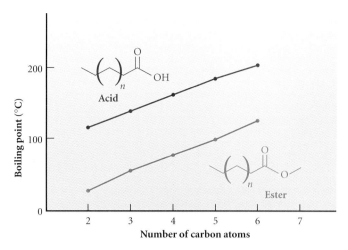

**FIGURE 19.4**

The boiling points of carboxylic acids are consistently higher than those of methyl esters of the same molecular weights because of strong, intermolecular hydrogen bonding in the acids.

### Metal–Heteroatom Bonds

Many small cations, especially alkali metal ions, are attracted to the lone-pair electron density on a heteroatom in much the same way as hydrogen atoms are. (Heteroatoms interacting with metal cations are referred to as **ligands.**) These interactions are one of the two major contributors to the water solubility of salts such as sodium chloride. The other major contributor is the interaction of the hydrogen atoms in water molecules with the negatively charged counterions (chloride ion, in the case of sodium chloride). Often these anions have four nonbonded lone pairs of electrons. Because both species in contact with water are ions, these interactions are generally stronger than simple hydrogen bonds with neutral heteroatoms.

Solvation of $Na^{\oplus}$ and $Cl^{\ominus}$ by $H_2O$ molecules

Interactions between metal ions and lone pairs are especially favorable when a single metal ion can simultaneously interact with lone pairs on several heteroatoms in a single molecule. Species capable of such interactions are called **ionophores.** When two heteroatoms are present, as in ethylenediamine, such ligands are referred to as being **bidentate** (meaning "having two teeth"). A ligand with three heteroatoms is **tridentate,** and one with even more heteroatoms, such as TRIS or EDTA, is **polydentate.**

## A TASTY BUT DEADLY FISH

Tetrodotoxin, a highly oxygenated compound resembling a carbohydrate, is found (mainly) in the liver and ovaries of many species of puffer fish. Its toxicity is based on its ability to carry ions across cell membranes. Tetrodotoxin destroys the ion imbalance required for nerve cells to fire and is therefore effective in blocking pain. Early hopes of medicinal applications were dimmed when it was discovered that the effective dose is just one-quarter of the lethal dose, which is 10 $\mu$g/kg (1 $\mu$g $= 10 \times 10^{-6}$ g) in mice.

Gourmets in Japan and other countries around the world consider the puffer fish a delicacy. For some, the ultimate experience is to sense a slight numbing effect on the lips caused by traces of tetrodotoxin in the fish. In Japan, preparation of the puffer fish can be done legally only by specially certified and licensed chefs.

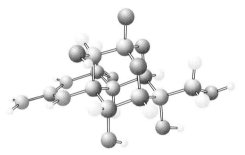

**Tetrodotoxin**

There are numerous naturally occurring ionophores (both cyclic and acyclic), and many of them play critical biological roles. For example, vitamin $B_{12}$, heme (the oxygen carrier in mammalian blood), and chlorophyll (a compound essential for photosynthesis) are tetradentate ligands that bind to different metal ions through coordination with nitrogen (Figure 19.5, on page 19-14).

Several naturally occurring polydentate ionophores have also been identified in which metal ions coordinate through oxygen atoms (Figure 19.6, on page 19-15). Some of these ionophores have antibacterial activity (and sometimes, unfortunately, toxic effects in mammals) as a result of their ability to encase metal ions within an exterior that is mostly hydrocarbon in nature. Wrapped in this hydrophobic shell, ions can permeate cell membranes, thus upsetting the normal ion balance within cells. For example, the antifungal compound amphotericin B is quite toxic but is used in low doses

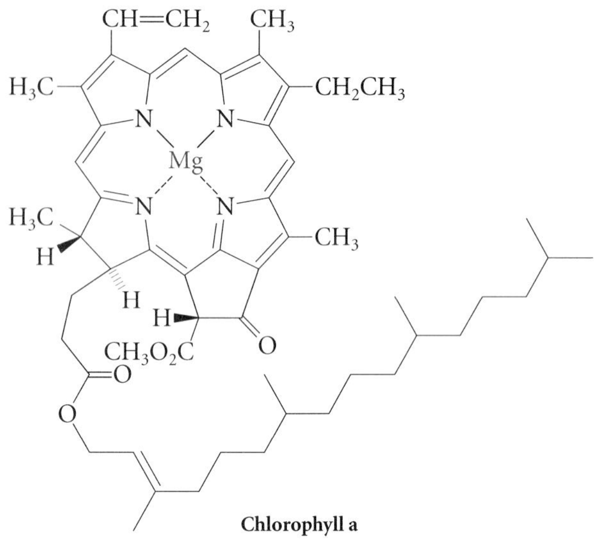

**Chlorophyll a**

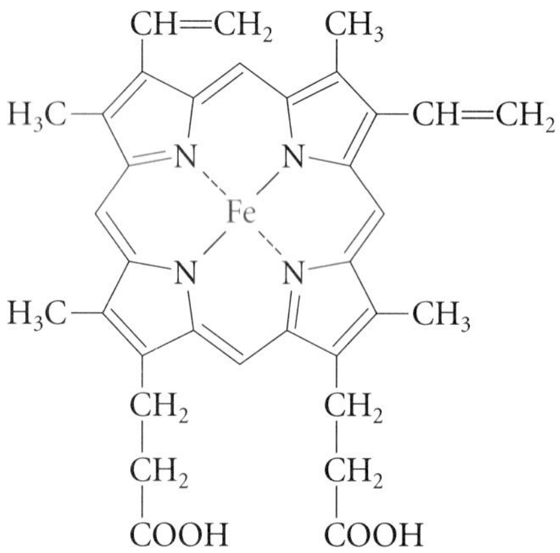

**Heme**

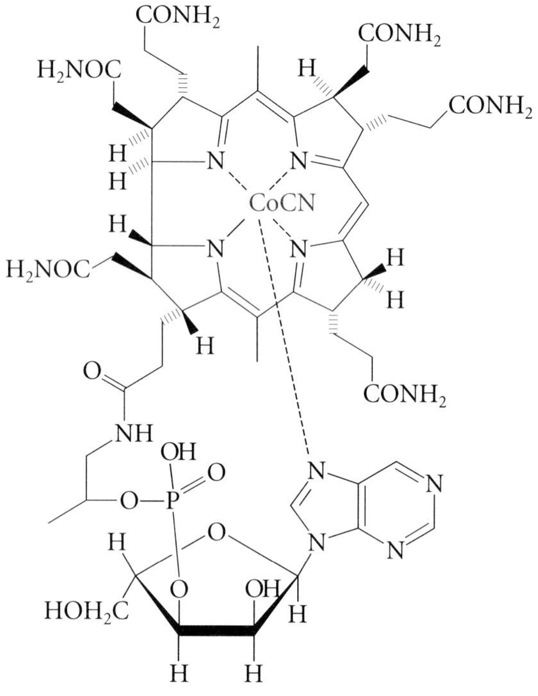

**Vitamin B$_{12}$**

## FIGURE 19.5

Chlorophyll a, heme, and vitamin B$_{12}$ are porphyrins, compounds in which four nitrogen atoms (present as four pyrrole nitrogen atoms in a large ring) coordinate to a metal ion. In vitamin B$_{12}$, the metal ion (cobalt) is also coordinated to a covalently attached nucleotide.

over extended periods for the treatment of leishmaniasis, a protozoan infection that affects 10–15 million people annually.

***Cyclic Ionophores.*** The naturally occurring ionophores displayed in Figure 19.6 are representative of a special class of cyclic ionophores. A particularly noteworthy characteristic of these cyclic ionophores is their usually high specificity for particular ions. A number of synthetic cyclic ionophores also exist, thanks to the pioneering research led by the American chemists Donald Cram and Charles Pedersen and the French scientist Jean-Marie Lehn. These researchers shared the Nobel prize in chemistry in 1987 for their work on these compounds.

The ion selectivity of a cyclic ionophore can be measured by determining the ability of the ionophore to "transport" a metal cation from an

#22   Crown Ethers

**Nigericin**

**Tetrodotoxin**

**Amphotericin B**

## FIGURE 19.6

Naturally occurring ionophores with many oxygen atoms assist in the transporting of metal ions across membranes. Because this transport significantly affects nerve transmission, some of these ionophores (such as tetrodotoxin) are deadly poisons, whereas others are medicinal agents.

aqueous medium to a hydrocarbon solvent. Essentially, the ionophore replaces the solvation interaction of the oxygen of water. However, to do so effectively, the ionophore must surround the cation and provide a sufficiently large number of heteroatoms to meet the ligand requirements of the metal. This ability, in turn, depends on the "fit" of the metal ion within the cavity in the center of the ligand. Cations that are too large cannot penetrate deeply enough to contact more than two or three of the heteroatoms. For cations that are too small, the distance from one side of the ionophore to the other is too large for the cation to interact effectively with all of the heteroatoms simultaneously. For example, the ionophore known as benzo[15-crown-5] ether has a cavity that is nearly the ideal size for a sodium ion, and, as a result, the interactions between this ion and the ionophore are quite strong. In contrast, both the lithium ion, which is smaller than the sodium ion, and the potassium ion, which is larger, bind much more poorly to this ionophore. This is an example of molecular recognition—between an ionophore and a specific ion.

**EXERCISE 19.5**

Suggest an experimental method for determining the relative ability of an ionophore to solubilize an inorganic salt in a nonpolar, organic solvent such as benzene.

*Entropy Effects on Ligand Binding.* Entropy also influences the magnitudes of interactions between a metal ion and otherwise comparable cyclic and noncyclic polydentate ligands. For example, compare the cyclic and noncyclic polyethers shown as space-filling representations in Figure 19.7. These two polyethers are very similar and, indeed, differ in formula by only two hydrogen atoms. However, their abilities to associate with a potassium ion differ by four orders of magnitude. This dramatic difference is due to the effect of entropy: the organization required to wrap the noncyclic ligand around the potassium ion deprives this ligand of significant conformational freedom. Indeed, there are 15 $\sigma$ bonds about which this molecule is free to rotate when it is not associated with a cation; thus, there are a large number of unique conformations. Interaction with the potassium ion deprives this ligand of all but one of these conformations, the one required for effective binding to the cation. The cyclic ligand has far less conformational freedom in its uncomplexed form. Therefore, association with the potassium ion (although still entropically unfavorable) is much less energetically costly for the cyclic ligand than for the acyclic ligand, because a

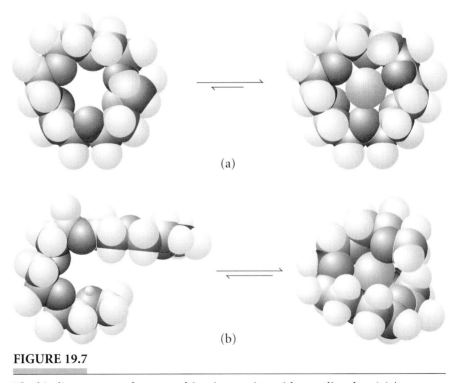

(a)

(b)

**FIGURE 19.7**

The binding constant for a metal ion interacting with a cyclic ether (a) is much higher than that for the same ion with the analogous open-chain ether (b) because of the entropy cost required to bring the open-chain ether into a conformation suitable for binding.

smaller change in entropy takes place with the former. A large fraction of the entropy cost associated with binding has been paid earlier in preparing the cyclic ligand, which has many fewer degrees of freedom than does the acyclic analog.

***Entropy Considerations in Cyclic Systems.*** Even such a conformationally "ideal" arrangement as chair cyclohexane can be relatively unstable from an entropic point of view when compared with its acyclic counterparts. For example, there is little difference in conformational enthalpy between the chair conformation of cyclohexane and the conformation of *n*-hexane in which all the carbon–carbon bonds are oriented in an *anti* fashion. On the other hand, there is only one stable conformation of cyclohexane (the chair) (Figure 19.8). For *n*-hexane, in addition to the all-*anti* conformation, there are four unique conformations in which there is only one *gauche* bond, and the remaining bonds are *anti*: two enantiomeric conformers in which the *gauche* bond is between C-2 and C-3, and two more in which the *gauche* bond is between C-3 and C-4. Each of these four conformers is about 0.9 kcal/mole less stable than the all-*anti* conformer. Thus, each is disfavored by 4:1 relative to the all-*anti* conformer. However, because there are four of these conformers, the total of their populations is equal to that for the all-*anti* form. For longer hydrocarbon chains, the number of *gauche* conformers is even greater, and for very long chains, the contribution of the all-*anti* conformer to the total equilibrium becomes quite small.

Thus, cyclic systems can be significantly less favorable from an entropic point of view than their acyclic counterparts. Furthermore, as the number of atoms in the cyclic arrangement increases, the influence of entropy (disfavoring the cyclic form) increases. Yet this entropic disadvantage inherent in cyclic systems is an advantage in molecular recognition, precisely because the cyclic system has already incorporated the entropic disadvantage associated with the interactions required for the recognition event.

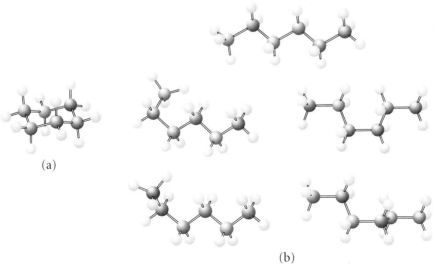

(a)

(b)

**FIGURE 19.8**

(a) There is only one energetically reasonable conformation for cyclohexane (the chair). (b) In contrast, for *n*-hexane, there are, in addition to the all-*anti* conformation, four unique conformations with one *gauche* bond.

#24   β-DNA Structure

# Genetic Coding

### Hydrogen Bonding in Biopolymers

Strong hydrogen bonds are formed to hydrogen atoms attached to nitrogen, as, for example, in the intermolecular bonding of peptides. Among many other biologically important pairs held together through hydrogen bonding involving nitrogen, the most important are the bases in DNA and RNA. These bases form the fundamental units for the storage and transmission of information in biological systems.

As the name of their class implies, these substances have one or more nitrogen atoms (although not all of the "bases" are strongly basic). These bases also have carbonyl groups that can serve as proton acceptors in hydrogen bonding. In addition, each NH group can function as a proton donor. Two sets of base pairings are illustrated in Figure 19.9, in which thymine is paired with adenine, and cytosine is paired with guanine. The distance between each pair of heteroatoms taking part in a hydrogen bond is nearly the same for all the hydrogen bonds; that is, each of the two hydrogen bonds joining thymine and adenine and of the three joining cytosine and guanine contributes substantially to the total interaction.

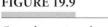

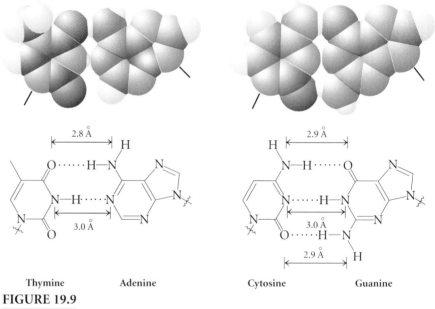

Thymine        Adenine          Cytosine          Guanine

**FIGURE 19.9**

Complementary base pairing between thymine and adenine is accomplished through two strong hydrogen bonds. Base pairing between cytosine and guanine takes place through three hydrogen bonds.

### Complementary Base Pairing

Hydrogen-bonding patterns such as those illustrated in Figure 19.9 form a simple system of recognition between molecules, and the appropriate pairings of guanine with cytosine (G-C) and adenine with thymine (A-T) are said to be **complementary.** Alternative base pairings—for example, between adenine and guanine or between adenine and cytosine—would have fewer hydrogen-bonding interactions than the preferred numbers shown in Figure 19.9. This specificity of the bases for their appropriate partners forms the essence of the duplication and reading of the genetic code, the system by which information is stored and used in living systems.

### EXERCISE 19.6

Draw representations of the hydrogen bonding for the pairing between adenine and guanine and that between adenine and cytosine, trying to maximize the number of hydrogen bonds. (It might be helpful to draw the structure of adenine on tracing paper so that you can readily overlay and move this base relative to cytosine and guanine.) Using your drawings, explain why adenine prefers to base-pair with thymine.

In the biopolymers DNA and RNA, these bases are attached to C-1 of a carbohydrate. For the bases in two polymer chains to associate through hydrogen bonding while the optimal bond angles in the sugar phosphate backbone are maintained, the two chains must twist about each other and be arranged in a head-to-tail fashion:

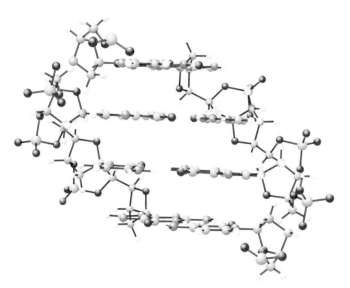

**Segment of DNA**

(The handedness of the twisting of the double helix is a result of the chirality of the carbohydrate. However, visualizing how the twist comes about is quite difficult without molecular models.) Imagine, then, that a sequence of bases in one DNA chain is G-A-C-T. For a second chain to associate effectively with the first, it must have the complementary sequence, C-T-G-A. This paired sequence represents information that, as we will see shortly, is translated into a specific sequence of amino acids in the biosynthesis of peptides and proteins, including enzymes.

Each base (guanine, adenine, cytosine, and uracil or thymine) is attached to either ribose (for RNA) or deoxyribose (for DNA). These subunits are called *nucleosides,* and the names of the bases are appropriately modified (for example, adenine forms the nucleoside adenosine). A **nucleotide** has a phosphoric acid ester appended to the C-5′ hydroxyl group.

**Base**

Adenine

**Nucleoside**

Adenosine

**Nucleotide**

Adenylic acid

## Replication

All living systems depend on the translation of the information contained in specific sequences of bases in DNA into peptide sequences. Thus, DNA represents stored information that, when read, provides the primary sequence of amino acids for all peptides, proteins, and enzymes. It is equally important that this information be in a form that can be reproduced and passed on from parent cell to "daughters" and from organism to offspring. Imagine that the two strands of a double helix unwind in a solution rich in the various nucleotides. Each separated strand can then associate by hydrogen bonding with free bases in solution, and the proper, complementary pairings will be favored. Enzyme-catalyzed formation of a phosphate ester bond between the nucleotides then links the bases together and completes the synthesis of a complementary strand for each original strand, in essence, duplicating the original double helix.

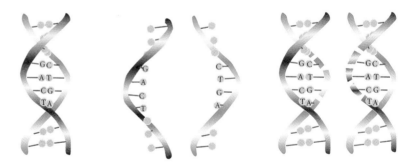

This process would be highly unfavorable energetically if the strands were to separate fully before associations with the nucleotides in the solution replaced the hydrogen bonds lost by unwinding of the original double helix. (DNA can be as long as $3 \times 10^9$ bases.) Instead, the complementary bases are replaced, and the phosphate bonds of the polymer are formed at the same time as the strands uncoil (Figure 19.10). In this fashion, the information stored in a specific sequence of bases within a DNA strand can be precisely duplicated in a process critical to the propagation (reproduction) of living systems.

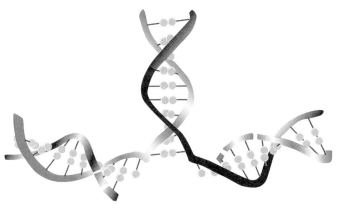

**FIGURE 19.10**

Only partial unwinding of the DNA double helix is needed for replication by complementary base pairing. As complementary bases associate with the unwinding sequence, polymerization produces strands that are bound to each strand of the original double helix. This specificity of sequence accomplishes information transfer through replication.

## Coding Requirements

The sequence of bases along a DNA strand determines the sequence of amino acids in a peptide. Because there are only four different bases, more than one base is needed to encode each of the 20 common amino acids. If a code unit consisted of two bases, there would be $4^2 = 16$ different permutations—also too few to encode 20 amino acids. In fact, each amino acid is encoded by a specific three-base sequence, because that allows $4^3 = 64$ possible combinations—more than enough to encode all the amino acids. There are several different codes for most amino acids, and therefore the genetic code is said to be **degenerate.** In many cases, the first two nucleotides are sufficient to identify the amino acid uniquely. For example, G-U-U, G-U-C, G-U-A, and G-U-G all code for valine. Because of this redundancy, 61 of the possible 64 permutations code for the amino acids. The remaining three codes (U-A-A, U-A-G, and U-G-A) are *stop* signals—indicators that the finished protein should be released.

## EXERCISE 19.7

Write the 16 possible pairs made from the bases C, T, A, and G.

## Reading the Genetic Code

In order for the genetic code for the synthesis of peptides to be read, the information from a segment of one strand of double-stranded DNA is encoded onto a complementary strand of RNA by a process called **transcription.** Recall that RNA is distinguished from DNA mainly by the presence of ribose rather than deoxyribose. There is another, more subtle difference: thymine is found only in DNA, and uracil is found only in RNA. These bases differ only by the presence of a methyl group in thymine; their ability to base-pair with adenine is essentially the same.

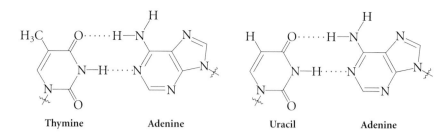

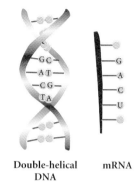

Double-helical    mRNA
DNA

The transcribed strand of RNA, known as **messenger RNA (mRNA),** is complementary to the short segment of DNA from which it is transcribed. Therefore, it contains the same sequence of bases as the other strand of the DNA segment (except for the substitution of uracil for thymine).

The sequence of the mRNA molecule carries the code for a single peptide. The mRNA moves out of the cell nucleus, where it was synthesized, to a ribosome—a body in the cell cytoplasm. The ribosome contains all the biological reagents needed to synthesize peptides—that is, to translate mRNA into a peptide. Critical to this process is **transfer RNA (tRNA),** a small nucleic acid molecule ranging in size from 73 to 95 nucleotides. Most of these nucleotides are the same in all tRNAs, except for two unique regions: one region that has a sequence of three bases (an **anticodon**) that complements the three-base codon in mRNA for a particular amino acid, and another region that provides a site for the attachment of a specific amino acid. The amino acid is chemically attached to a specific terminal nucleotide through a mixed anhydride (Figure 19.11) and is thus activated for nucleophilic acyl substitution.

A tRNA bearing its specified amino acid associates with a three-base codon of mRNA. With each incoming aminoacyl-tRNA, the amino acid at the other end of the tRNA becomes linked to the growing peptide chain by the formation of an amide (peptide) bond. In this short example, the G-C-U sequence in mRNA is complementary to the C-G-A sequence in the tRNA coding for alanine, and the next three bases of the mRNA, U-C-U, are

**FIGURE 19.11**

Peptide bond formation occurs between two amino acids (here alanine and serine), each tethered as an ester with the 3′ hydroxyl group at the end of tRNAs.

complementary to the set in tRNA for serine. The two amino acids bound to the tRNAs are brought into proximity and are then linked together by an enzyme in a process that is similar to the replication of DNA. A large number of coding sequences are required for the synthesis of a large peptide.

**EXERCISE 19.8**

A coding system using four bases per code sequence would provide far more unique codes than necessary. (How many?) What advantage might such a system offer in transcription of DNA? What disadvantage would there be for living systems?

## Misreading the Genetic Code

The system of information storage, transcription, and translation might seem foolproof, but it is not. Several chemical changes occur in the bases that can lead to the misreading of the code for a particular amino acid. For example, simple proton tautomerization of adenine converts it into a form that can form two hydrogen bonds to cytosine and thereby mimic guanine:

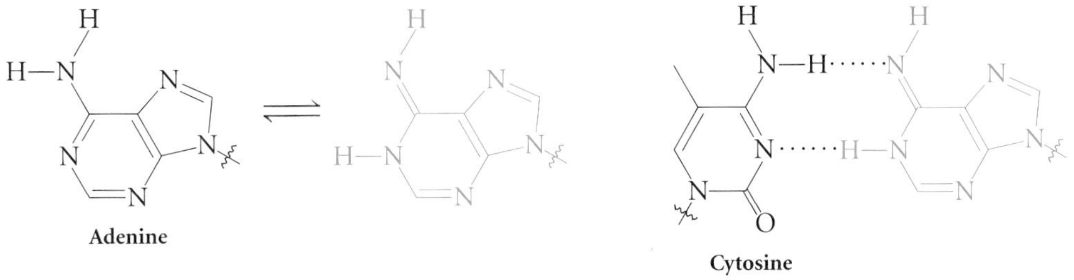

More permanent changes can also alter the hydrogen-bonding pattern of a base. Many chemicals in the diet are capable of altering the genetic code: nitrite ion, $NO_2^{\ominus}$, is perhaps the most notorious because of its widespread use as a meat preservative in the form of potassium (or sodium) nitrite. It is an effective preservative, functioning as an antibacterial and antifungal agent and as an antioxidant. However, primary amines react readily with nitrous acid, $HNO_2$, which is formed from nitrite ion in slightly acidic solution, to yield diazonium ions, which are excellent leaving groups. In aqueous systems, reaction of amines with nitrous acid can lead to net overall replacement of the amino nitrogen by a hydroxyl group. In this way, a cytosine residue can be converted into a uracil residue:

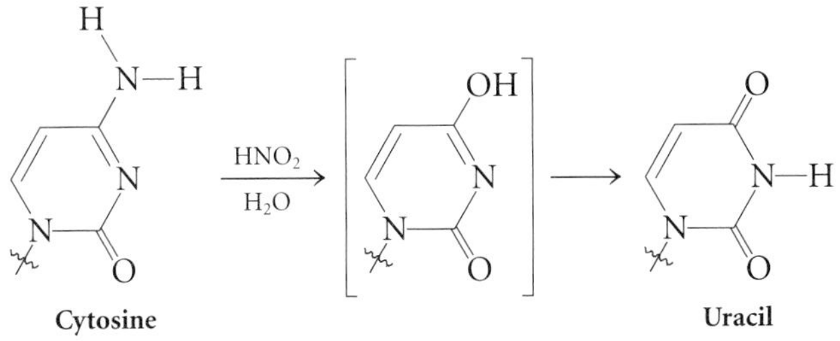

Were this transformation to change the sequence C-U-U to U-U-U, the specified amino acid would be phenylalanine instead of leucine.

In many cases, such a change has little consequence, especially if it alters a three-base sequence to one that codes for the same amino acid. However, in some cases, such an alteration leads to the substitution of one amino acid for another. The end result is the synthesis of a peptide with an "incorrect" amino acid residue. Even when such changes do affect the amino acid sequence in a peptide, it is rare that they have a major effect on the organism, because the modified peptide most likely will have little or no activity as a hormone or an enzyme.

### EXERCISE 19.9

Suggest a reasonable mechanism for the conversion of cytosine into uracil by a strong acid in water.

## 19.4

## Molecular Recognition of Chiral Molecules

Many naturally occurring molecules are chiral, almost always as a result of the presence of carbons with four different groups attached. In addition, almost all chiral molecules occur as single enantiomers, as is the case for the carbohydrates and the amino acids. In this section, we examine some of the underlying reasons for this handedness of nature, which is critical to the intimate associations at the transition states of most biochemical transformations.

### ▓ Necessity of Three-Point Contact
### for Chiral Recognition

For simplicity, let's represent a chiral molecule as a tetrahedron with four differently colored groups (with the priority scheme: red > green > blue > yellow):

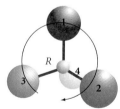

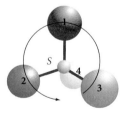

This depiction is just like that used to describe the Cahn–Ingold–Prelog rules for the assignment of the chirality specifiers $R$ and $S$. However, in the present context, such pictures represent chiral molecules in which the four regions differ in the types and strengths of possible weak interactions with other molecules. Thus, one group might be a large, hydrophobic hydrocarbon region, another a polar, hydrogen-bonding functional group such as a hydroxyl group, and so forth. Nonetheless, the $R,S$ notation is used to distinguish the enantiomers, as if the groups were substituents on a chiral center in the Cahn–Ingold–Prelog sense. By rotating these two enantiomeric representations so that three of the four groups of one enantiomer are oriented toward the same groups of its mirror image, we see that there is close proximity between the same groups (red to red, green to green, and blue to blue):

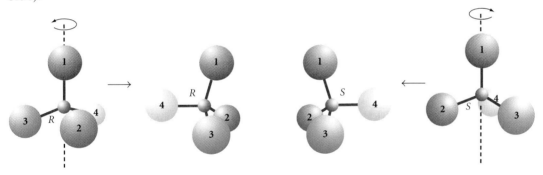

Now let's compare this pairing of an $R$ with an $S$ enantiomer with the pairings in which both partners are of the same configuration ($R$ with $R$, and $S$ with $S$). (Recall that interchanging any two substituents at a chiral center effects the conversion of one enantiomer into its mirror image.)

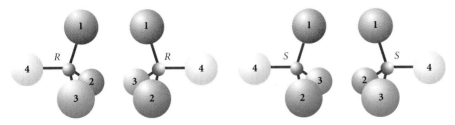

Note that only one interaction is between like groups (red and red), and the other two interactions are between unlike groups (green and blue, and blue and green). Clearly, then, the association of a chiral molecule with its enantiomer is not identical with that obtained by pairing the enantiomer with itself.

**19-25**

This three-point contact between enantiomers forms the basis for chiral recognition. Before we consider the consequences of these interactions, let's see why three contacts are required for chiral recognition by examining associations of pairs of chiral molecules in which only one or two of the four groups are in close contact. For example, consider two molecules oriented so that only the groups of lowest priority (4) are in proximity:

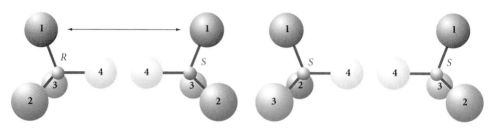

Note that if the chiral molecules touch at only one point, there will be little difference in energy between the unlike (*R* with *S*) and like (*S* with *S*) pairings. Indeed, to the extent that groups 1, 2, and 3 in one molecule are distant from those in the other, there is little or no difference in energy between these alternative arrangements. Even when two of the groups are in contact, the energy difference between the *R,S* and the *S,S* pairings is still small, because the green and blue groups (2 and 3) are directed away from each other in both pairs:

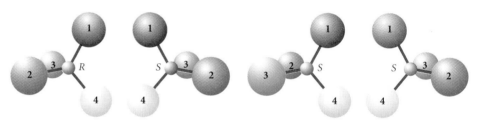

Thus, significant differentiation based on chirality can occur only when there is interaction between three groups, or three-point contact.

Simple physical analogies may help you understand and visualize the differences between matched and unmatched pairings. Hands and feet (as well as many other common objects) are chiral; that is, the object is not superimposable on its mirror image. A glove made for one hand does not fit the other. In fact, the misfit of a glove with the wrong hand is very similar to the mismatching of chiral stereocenters. If the palm of the glove is matched up with the palm of the wrong hand, then the thumb is on the wrong side. Conversely, turning the glove over so that the thumb is on the correct side puts the palm of the glove on the back of the hand. If the smallest finger and the thumb were the same shape and size, the hand would be symmetrical from left to right, and the glove could be worn with either face oriented toward the palm. Similarly, if there were no difference between the palm and the back side of the glove, then one shape would fit either hand equally well (as is the case for most children's mittens).

### Resolution of Enantiomers

The energetic consequences of chiral recognition are apparent in the separation of a racemic mixture into individual enantiomers. This process, known as *resolution*, consists of the interaction of the racemic mixture with a single enantiomer of another chiral compound, referred to as the *resolving agent*. There are two different possible pairings in a resolution: with a resolving agent of the *R* configuration, *R* can pair with *S* or *R*.

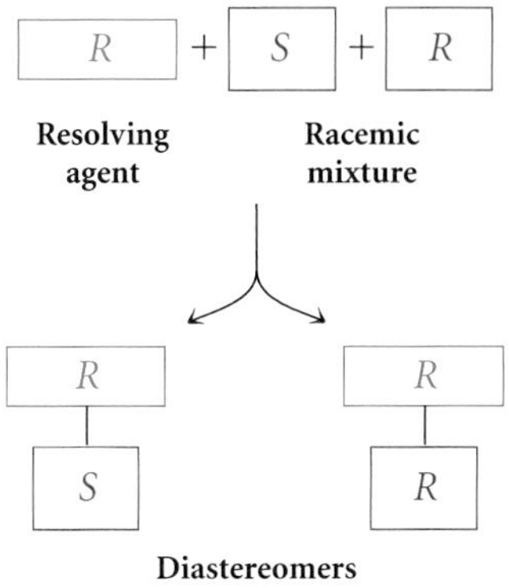

These alternative possible pairings are diastereomers and thus have different physical properties, such as solubility. However, there must be some chemical reason why the two different molecules combine in a single crystal. Indeed, most resolutions require either the combination of a racemic acid with a basic resolving agent, such as an amine, or the opposite combination of a racemic amine with the single enantiomer of a chiral acid. In either case, the acid–base reaction is a driving force for the formation of the crystalline salt.

In many cases, the acidic or basic resolving agent is one that is obtained from natural sources as a single enantiomer. As an example, consider the resolution of mandelic acid with the alkaloid cinchonine (pronounced "sing'ke-nen'''"), shown in Figure 19.12. This alkaloid, like the structurally related antimalarial alkaloid quinine, is obtained by extraction of the bark of the cinchona tree. Combining racemic mandelic acid with cinchonine results in the formation of two diastereomeric salts that differ dramatically in their melting points. Furthermore, the lower-melting diastereomer is twice as soluble in water (20 g/L versus 10 g/L). Therefore, these diastereomers are readily separated by recrystallization. After this resolution process, the individual enantiomers of mandelic acid and the resolving agent are easily separated by an acid–base extraction.

## EXERCISE 19.10

Suggest an appropriate sequence of extraction operations for the separation of mandelic acid and cinchonine from the salt.

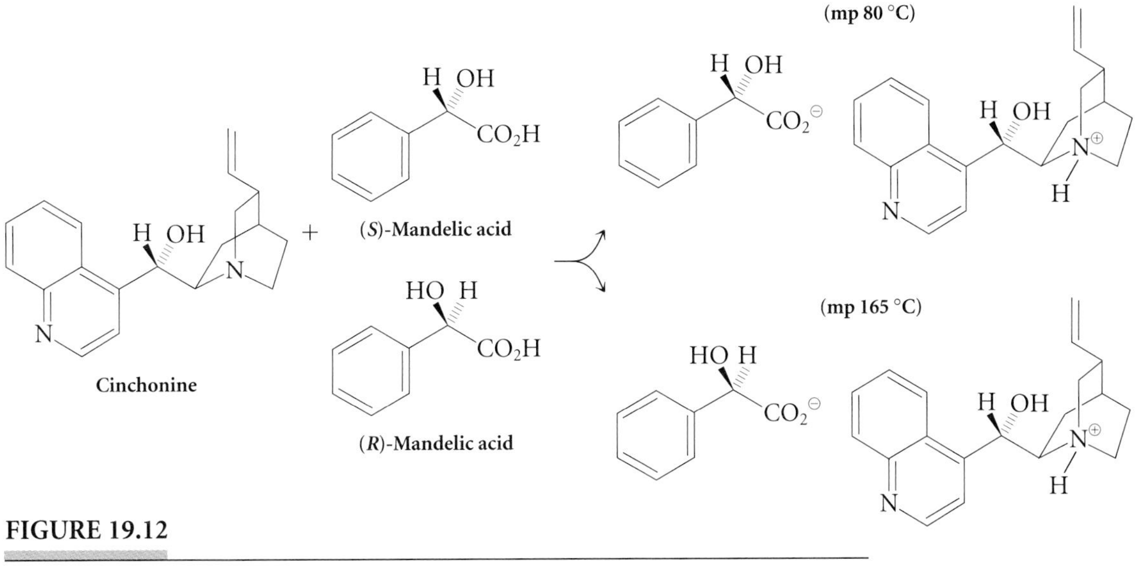

**FIGURE 19.12**

Resolution of racemic mandelic acid into separate enantiomers can be accomplished by separation of the diastereomeric salts formed by its acid–base reaction with cinchonine, a tertiary amine. Neutralization of each of the salts with sodium hydroxide will reform the neutral amine, which can be easily separated from the carboxylate ion of mandelic acid.

**19-27**

(−)-Carvone     (+)-Carvone

(R)-Thalidomide

(S)-Thalidomide

## Biological Significance of Chirality

Handedness has profound consequences in living systems. In most cases, the interactions that invoke a biological response take place at receptor sites in the body that are chiral and produced in only one enantiomeric form. Therefore, the two enantiomers of a biologically active material do not elicit the same response. Indeed, the differences are often dramatic. For example, both enantiomeric terpenes, (−)-carvone and (+)-carvone, have a distinctive odor, but the first is a major contributor to the odor of spearmint, and the second, to the odor of caraway seeds.

The spearmint and caraway plants produce only one enantiomer, whereas, interestingly, racemic carvone is found in gingergrass. Likewise, the naturally occurring L-enantiomers of leucine, phenylalanine, tyrosine, and tryptophan have a bitter taste, but the D-isomers are sweet. In this case, the two enantiomers must interact with different receptors, because the sensations of sweetness and bitterness result from stimulation of different areas of the tongue.

A tragic example of the interaction of enantiomers with different receptor sites was uncovered when the antidepressant thalidomide was sold as the racemic mixture in the early 1960s, even though only the R enantiomer has the desired activity. Because of the keen insight of Frances Kelsey, a researcher at the Food and Drug Administration (FDA), thalidomide was never approved for use in the United States. However, this prescription drug was already in use in Canada and European countries and, despite strong warnings against its use by pregnant women or even women likely to become pregnant, thalidomide was being prescribed for the treatment of "morning sickness." The antidepressant activity of thalidomide is due to one enantiomer; the other was found to be mutagenic and antiabortive. With this combination of activities, this enantiomer not only initiated genetic alterations that resulted in deformed fetuses, but also prevented the natural expulsion of fetuses damaged from other causes. The result of the use of thalidomide was the birth of many very seriously deformed infants, often having underdeveloped arms and legs. Curiously, the observation that Kelsey cited in withholding approval of thalidomide was that it caused abortions at high doses in rats.

## Summary

**1.** Weak interactions between molecules include van der Waals and dipole–dipole interactions and hydrogen bonds, which play a major role in holding molecules together. Such interactions significantly influence physical properties such as melting points, boiling points, and solubility.

**2.** Relatively weak, but significant, electrostatic interactions govern the association of small metal cations, such as those derived from the alkali earth metals, with lone pairs of electrons on electronegative atoms. Natural ionophores and synthetic ionophores bind metal ions most effectively when multiple ligand sites coordinate with the cation. Entropy considerations, including the effect of molecular symmetry, influence the equilibrium binding constant between multidentate ligands and metal ions.

3. Although a linear arrangement of the three atoms involved in a hydrogen bond is optimal, nonlinear hydrogen bonds often occur. Energies of hydrogen bonds range from about 1 to 5 kcal/mole.

4. Complementary base pairing is responsible for information storage and transcription in DNA and RNA. It is based on the specific recognition of cytosine by guanine and of adenine by thymine in DNA. Uracil fills the role of thymine in RNA, mRNA, and tRNA.

5. A three-letter code specifies each unique amino acid in building a peptide chain.

6. Chiral recognition results only from three-point contact. One- or two-point contact is inadequate to differentiate chiral centers. The resulting selective interactions based on subtle structural differences have wide-ranging consequences for the chemistry of living systems.

7. Chiral recognition is used in the laboratory to separate enantiomers. This process, called *resolution*, requires the use of a chiral resolving agent.

## Review Problems

**19.1** In each of the following pairs, indicate which substance (A or B) is expected to have the higher boiling point. Briefly explain your answer, indicating the key molecular difference in each case that leads to a difference in the degree of association of molecules in the liquid phase.

**19.2** Several heterocyclic systems exist as equilibrium mixtures of a hydroxyaromatic and a ketonic form. Suggest a reasonable explanation for the observation that the equilibrium between 2-hydroxypyridine and the ketonic form known as 2-pyridone favors the latter, whereas the corresponding equilibrium for hydroxybenzene (phenol) favors the hydroxyaromatic form. Write reasonable mechanisms for these interconversions as catalyzed by acid and by base.

2-Hydroxypyridine      2-Pyridone      Phenol

**19.3** Suggest an explanation for the observation that the reaction of 2-aminopyridine with strong base can lead to replacement of nitrogen by oxygen, whereas under similar reaction conditions, aniline is unreactive.

2-Aminopyridine          2-Pyridone

Aniline

Write a mechanism for the reaction leading from 2-aminopyridine to 2-pyridone.

**19.4** Dimethylsulfoxide (DMSO) and acetone are very similar in structure in that each has a polarized $\pi$ bond to oxygen. Although DMSO, $(CH_3)_2SO$, is only 20 mass units heavier than acetone (molecular weight of 68 versus 48), it has a very much higher boiling point (189 °C versus 57 °C). Can you suggest a reason for this difference? Dimethyl sulfone, $(CH_3)_2SO_2$, has a boiling point of 238 °C, which is 49 °C higher than that of DMSO, but a melting point substantially higher (109 °C versus 18 °C). What molecular feature of dimethyl sulfone contributes to its higher melting point?

**19.5** A common technique for extracting polar organic molecules from aqueous solution into an organic solvent is to add large quantities of sodium chloride to the water. Explain why adding salt decreases the solubility of organic compounds in water.

**19.6** How would you expect the boiling point of amides to change as the number of methyl groups (0, 1, and 2) on nitrogen increases? Check your prediction with experimental data (for example, in the *Handbook of Chemistry and Physics*), and then explain the observed order.

**19.7** Large peptide and protein molecules generally have both hydrophobic and hydrophilic side chains. When these polymers form three-dimensional globular structures, the amino acid residues with hydrophobic groups tend to be together, as do those with hydrophilic groups. Which would you expect to be on the outside and which on the inside of such a three-dimensional arrangement of a biologically important protein?

**19.8** In each of the following pairs, predict which isomer (A or B) is expected to have the higher solubility in water. In each case, provide a brief explanation for your choice.

(c)

O
‖
~~~^~~OH

A

O
‖
~CH~O~

B

(d)

O
‖
~~OH
|
NH₂

A

O
‖
HO~~~NH₂

B

which of the following amines is it possible to carry out a resolution into separate enantiomers by forming a salt with a chiral carboxylic acid?

A

B

C

NH₂

HO        OH

D

(all *cis*)

**19.10** One of the three isomers of hydroxybenzaldehyde is much less soluble in water and has a lower boiling point than the other two. Identify this isomer, and explain why its properties are different from those of the other two isomers.

CHO          HO     CHO                    CHO

OH                                    HO

## Supplementary Problems

**19.11** The most stable form (or conformation) of a molecule is often thought to be the one that is most prevalent. However, statistical factors (an entropic factor, $\Delta S°$) can often overcome a relatively modest enthalpic ($\Delta H°$) advantage. For example, the energy difference between *anti* and *gauche* butane is 0.9 kcal/mol, which by itself would lead to a ratio of 4:1. However, there are two different *gauche* conformers (enantiomers), and thus the observed ratio is only 2:1. Consider pentane, and determine how many different conformers there are with one *gauche* relationship. If each of these exists in a ratio of 1:4 with the single all-*anti* conformation, what is the ratio of all of these *gauche* conformers to the all-*anti* conformer? (Be sure that all conformations you consider are really unique and do not duplicate others.) Do the same analysis for decane.

**19.12** The bis-amine called bis-1,8-(dimethylamino)naphthalene is also known by the trivial name *super base*, because it is considerably more basic than a simple aromatic amine such as aniline.

(a) Explain why aniline is less basic than an aliphatic tertiary amine such as triethylamine.

(b) Examine the super base molecule for special features that provide enhanced basicity. (*Hint:* Ionic species often contribute more than neutral ones do to differences in both kinetic and thermodynamic properties.)

$(H_3C)_2N$      $N(CH_3)_2$

**Super base**

**19.13** The properties of nucleophilicity and basicity often shift in parallel with structural changes. Indeed, basicity can be viewed as nucleophilicity toward a proton, although it is important to remember that relative basicity is a thermodynamic property, whereas relative nucleophilicity refers to rates of reaction. Super base (see Problem 19.12) is useful not only because it is a strong base but also because it reacts very slowly, if at all, as a nucleophile. Consider the reaction of super base with methyl iodide, and develop an explanation for the low nucleophilicity of this strong base.

## Important New Terms

**anticodon:** a sequence of three bases that complements the three-base codon in mRNA and selects for a particular amino acid in protein synthesis (19-22)

**bidentate:** a compound with two ligating heteroatoms; from Latin, meaning "having two teeth" (19-12)

**complementary:** descriptive of a favorable hydrogen-bonding interaction between bases in DNA or RNA (19-19)

**induced dipole:** a shift of electron density within a molecule or bond that is induced by the environment (19-4)

**ionophore:** compound containing several heteroatoms arranged such that multiple, simultaneous contacts with a metal ion or other highly polar molecule are possible (19-12)

**ligand:** a heteroatom interacting with a metal cation (19-12)

**messenger RNA (mRNA):** a transcribed strand of RNA that is complementary to a segment of DNA; a sequence of complementary bases replicated from a DNA strand, except for the substitution of uracil for thymine (19-22)

**nucleotide:** a phosphate ester derivative of a nucleoside (19-20)

**polydentate:** having many ligating heteroatoms; from the Latin, meaning "with many teeth" (19-12)

**transcription:** the process of encoding information stored in DNA during RNA synthesis (19-21)

**transfer RNA (tRNA):** a small nucleic acid containing 73 to 95 nucleotides and having two unique regions, one consisting of an anticodon and another providing a site for the attachment of a specific amino acid (19-22)

**tridentate:** having three ligating heteroatoms; from the Latin, meaning "with three teeth" (19-12)

# Catalyzed Reactions

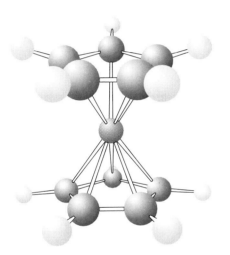

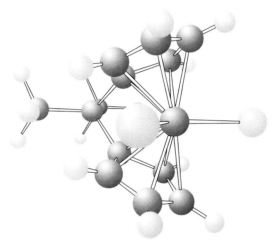

*Bonding between carbon and metals in organometallic compounds does not conform to conventional valence bond theory. The structure at the left is called ferrocene. It has a neutral iron atom sandwiched between two cyclopentadienyl rings (five-member rings composed of sp²-hybridized carbon atoms). The iron atom is equidistant from all ten carbon atoms. The structure at the right is a complex that catalyzes olefin polymerization. It has a zirconium atom sandwiched between two cyclopentadienyl rings that are connected to each other by a carbon atom. The zirconium atom also bears two chlorine ligands.*

$\mathcal{T}$ here is an incredibly wide variety of organic reactions, and many take place under widely varying conditions, with very different rates and efficiencies. For example, carboxylic acid derivatives have quite different reactivities. Carboxylic acid chlorides are sufficiently reactive to be hydrolyzed at room temperature within minutes (as long as they are somewhat soluble in water), whereas amides are stable toward hydrolysis in neutral aqueous solutions for years.

Such large differences in the reactivities of derivatives can be used to advantage in controlling the reaction rate of a functional group. Alternatively, the reactivity can be influenced by altering the concentrations of the reagents. For example, the rate of hydrolysis of an amide is pH-dependent and increases dramatically with increasing concentration of hydroxide ion. Other reactions, such as the dehydration of alcohols, are slow in the absence of hydronium ion and accelerate as the pH of the solution is decreased.

Living systems operate within a range of conditions narrowly defined by temperature, pH, and solvent. They do not have the flexibility chemists have in the laboratory. Each of the multitude of chemical transformations that take place in a living system must proceed at a reasonable rate under the same conditions of temperature, pH, and solvent. Regardless of the criteria used to define *life,* all living things are systems that change with time, and all are in a state of constant chemical flux. Most of the chemical reactions taking place in living systems are far from equilibrium, and all such systems depend on the compartmentalization of reactions into regions (such as cells) to achieve this vital nonequilibrium state. In fact, for a living system, to achieve equilibrium is equivalent to death.

As a comparison, consider an organic chemist who wishes to convert one compound into another using a series of reactions, each with its own reagent(s), solvent requirements, and reaction conditions. For example, 2-propanol can be converted into 2-butanol by the following series of reactions, carried out sequentially in separate reaction flasks:

OH
|

2-Propanol

$\xrightarrow[-H_2O]{H^{\oplus}}$

$\xrightarrow[\text{THF}]{B_2H_6}$ $\xrightarrow[\text{NaOH, H}_2O]{H_2O_2}$ OH $\xrightarrow[\text{Pyridine}]{CrO_3}$ O H $\xrightarrow{CH_3MgBr}$ $\xrightarrow{H_2O}$

OH
|

**2-Butanol**

Imagine the consequences of simply mixing the reagents and solvents for all of these reactions with 2-propanol in a single flask! Even the reagents involved in the two steps of hydroboration–oxidation are incompatible with each other.

There must be some feature of biological reactions such that, for example, oxidizing and reducing reagents do not react with each other but simply progress to a thermodynamic minimum. That is, the rate of each biological reaction must be controlled by nonthermodynamic factors. The special feature is, in fact, that virtually all biochemical transformations are greatly accelerated by the presence of biological catalysts, or enzymes. Without these catalysts, it would be impossible for all the molecules important to living systems to be present simultaneously and for their rates of reaction to be controlled as required by the metabolism of the organism.

To understand the reactions that take place in living systems, it is necessary to know the concepts of catalysis in general, as well as the specific features unique to enzymes. In this chapter, we will explore general concepts of reaction catalysis and then see how these ideas can be used to explain the catalysis achieved by enzymes. The importance of organization into supermolecular units (such as cells) is more appropriately covered in courses in biochemistry, microbiology, and molecular biology.

## 20.1

## General Concepts of Catalysis

A catalyst accelerates the rate of a chemical reaction without being consumed. Many catalytic cycles can be induced by a single catalyst molecule, and so stoichiometric quantities of the catalyst are not needed. Catalysis is accomplished in the laboratory by stabilizing the transition state of the rate-determining step of a chemical reaction. This might be done, for example, by providing electrostatic stabilization of partially charged centers that develop as the transition state is formed or by prearranging the reactants into a geometry appropriate for a desired bimolecular reaction, or by both methods.

## CHEMICAL PERSPECTIVES

### EFFECT OF pH ON ENZYME FUNCTION

Many biological reactions are catalyzed by enzymes, but each enzyme operates at a maximum rate only within a fairly narrow range of conditions. For example, although reaction rates generally increase with temperature, most enzymes denature at elevated temperatures, losing their catalytic properties. Acidity is also important. For example, digestive enzymes in the stomach (where the pH is ~1.5) function best under acidic conditions, whereas digestive enzymes in the small intestine (where the pH is ~8) prefer alkaline conditions. The enzyme that causes cut apples to turn brown on exposure to oxygen does not perform well below pH 3.5; for this reason, lemon juice applied to the surface of freshly cut apples keeps them from turning brown by increasing the acidity (lowering the pH).

## Transition-State Stabilization

There are two fundamentally distinct ways by which the rate of a chemical reaction can be increased. The first, and more common, is the use of a reaction pathway that is very similar to the uncatalyzed process except for some special feature, such as solvent polarity, that leads to a decrease in the energy of the transition state (or an increase in the energy of the starting materials). With such a catalyst, the activation energy ($\Delta G^{\ddagger}$) is lowered, and the reaction rate is increased (Figure 20.1). Even though the catalyzed re-

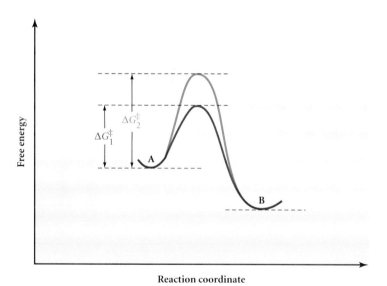

**FIGURE 20.1**

A chemical catalyst accelerates a reaction by reducing the activation energy, $\Delta G_1^{\ddagger}$, below the energy, $\Delta G_2^{\ddagger}$, needed in the absence of the catalyst.

action is accelerated, the nature and sequence of the bond-breaking and bond-making steps remain essentially unchanged. The second way to increase the reaction rate is to catalyze a reaction by making available an entirely different pathway—for example, one in which bond breaking and bond making are homolytic rather than heterolytic.

To understand catalysis, it is critical to have a firm grasp of transition-state theory and how it relates to reaction rates. The shape of an energy diagram for a simple, one-step, exothermic transformation is shown in Figure 20.2. From such a diagram, we can obtain direct information only for species that lie in energy wells, or local minima—in this case, the starting material, A, and the product, B. All other aspects of the diagram are inferred. For example, the energetic position of the transition state is inferred from the activation energies of the forward and reverse reactions, which, in turn, are derived from the rates of these transformations. Although we cannot get direct information on transition states, we can obtain much indirect evidence by examining such parameters as the effect of solvent polarity on the rate of reaction. For example, transition states more polar than the starting materials are favored to a greater extent by polar solvents. Solvation stabilizes the transition state, lowering its energy relative to that of the starting material. This results in a faster rate than for the same reaction carried out in a less polar or a nonpolar solvent.

$$\text{Rate} \propto e^{-(\Delta G^{\ddagger}/RT)}$$

$$\Delta G^{\circ} = \Delta G_{f}^{\ddagger} - \Delta G_{r}^{\ddagger}$$

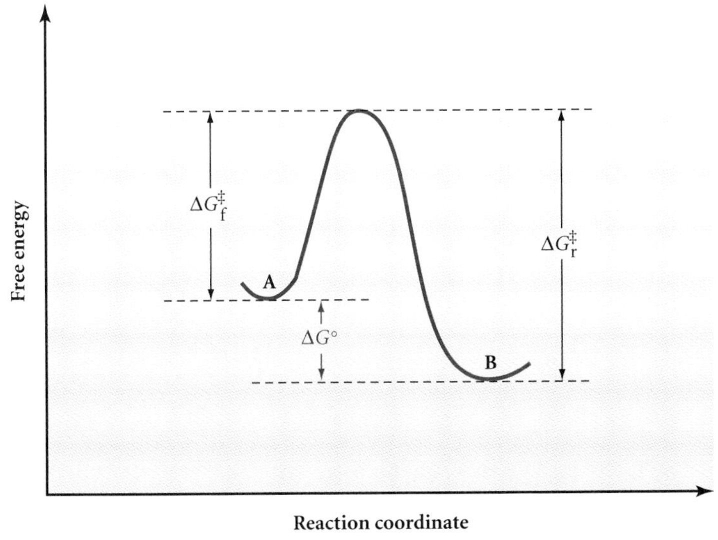

**FIGURE 20.2**

Assuming that entropy contributions can be neglected, the free energy for a concerted reaction (one that proceeds without intermediates) is equivalent to the difference in activation energies for the forward (A to B) and reverse (B to A) processes.

## Effect of Solvation on $S_N2$ Reactions

Figure 20.3 shows energy curves for a simple $S_N2$ reaction of a neutral amine (a nucleophile) with a neutral substrate (an alkyl halide) in water (a polar protic solvent) and in ethyl acetate (a less polar aprotic solvent):

$$H_3N + H_3C-I \longrightarrow \left[ H_3\overset{\delta+}{N} \cdots \overset{H}{\underset{H}{\overset{|}{C}}} \cdots \overset{\delta-}{I} \right] \longrightarrow H_3C-\overset{\oplus}{N}H_3 + I^{\ominus}$$

This reaction starts with an uncharged species but proceeds to charged products through a transition state in which partial charge is developed. Of the two solvents, water is much better at stabilizing ions, via hydrogen bonding. Thus, although water stabilizes the starting materials, its effect on the transition state is even greater. (The greatest stabilization is that for the product ions–species with full charges.) Because the energy of the transition state is lowered by solvation to a greater extent than is the energy of the reactants, the activation energy ($\Delta G^{\ddagger}$) is lower for the reaction in water than for that in ethyl acetate. For reactions that develop charged products from neutral starting materials, the rate of reaction is increased by polar (and especially polar protic) solvents. We can see the effect of polar solvents on the rates of reactions in which charged species combine to form neutral products by considering the reaction in Figure 20.3 in the reverse direction. In this case, the greater stabilization of the ions by a polar protic solvent such as water compared with the transition state leads to an increase in the activation energy ($\Delta G^{\ddagger}$) and a corresponding decrease in the rate of the reverse reaction.

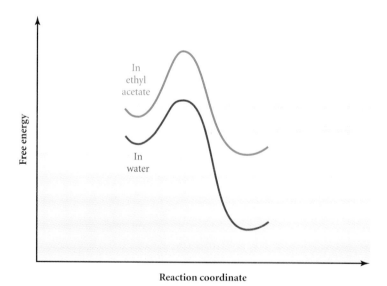

Reaction coordinate

**FIGURE 20.3**

Energy curves for the $S_N2$ reaction of ammonia with methyl iodide in ethyl acetate (blue) and in water (red).

A true catalyst does not affect the stability of either the starting material or the product. As a consequence, it does not significantly affect the equilibrium position. In the example in Figure 20.3, the product is stabilized by water to a greater extent than is either the starting material or the transition state. Although the rate of the reaction in the forward direction is accelerated by the presence of water, water is not a catalyst because it binds to the product through solvation. If this reaction were carried out with a "catalytic" amount (less than one equivalent) of water, it would be rapid at first but would then slow down dramatically, because water would bind tightly to the products and would not be available to stabilize the transition state for reaction of additional starting material. Nonetheless, when water is used as the reaction solvent, its effect is quite similar to that of a catalyst: the energy of the transition state is lowered relative to that of the starting materials, and the reaction's activation energy is decreased. We conclude that if a compound is to function as a catalyst, it cannot become more tightly bound to either starting material or product than it is to the transition state of a given reaction.

Let's consider the influence of solvation on the rates of reaction for the $S_N2$ displacement of bromide ion in methyl bromide by hydroxide ion (Figure 20.4):

$$M^{\oplus} \ {}^{\ominus}OH + H_3C—Br \longrightarrow \left[ \begin{array}{c} H \\ \delta{-} \quad | \quad \delta{-} \\ HO \text{-}\text{-}\text{-} \overset{|}{\underset{H}{\text{C}}} \text{-}\text{-}\text{-} Br \\ | \\ H \end{array} \right] \longrightarrow HO—CH_3 + Br^{\ominus} \ M^{\oplus}$$

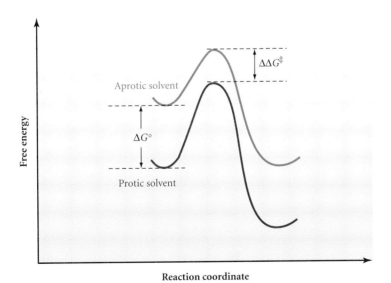

**FIGURE 20.4**

Energy diagram for the nucleophilic displacement of bromide by hydroxide in a protic solvent (red curve) and an aprotic solvent (blue curve). The rate is faster in the aprotic solvent.

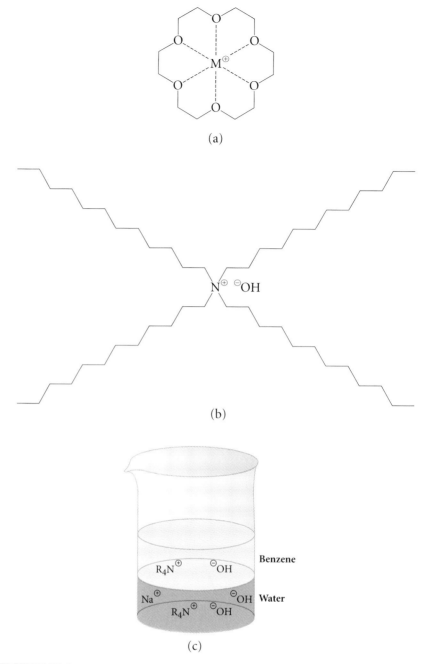

**FIGURE 20.5**

(a) A crown ether associates with a metal cation and decreases interaction between the cation and water. Transport of this cation complex from water into an organic solvent is more energetically favorable than if the cation were highly solvated by water. (b) The positively charged nitrogen atom of a tetraalkylammonium salt is similarly blocked from significant interaction with water by the hydrocarbon chains. (c) The movement of a reagent associated with a counterion from one phase to another, in which a reaction takes place, is called *phase-transfer catalysis.* For example, the counterion $R_4N^{\oplus}$ associates with the $^{\ominus}OH$ ion and transports it from water into a benzene layer.

This is a displacement reaction in which the concentrated negative charge in the reactant hydroxide ion is more widely dispersed in the transition state. Reactions of this type are slower in polar protic solvents, because the stabilization from solvation is greater for the starting material than for the transition state. Conversely, this type of reaction is greatly accelerated in an aprotic solvent, because the reactant hydroxide ion is less effectively stabilized in this medium than is the transition state. However, most metal hydroxides are poorly soluble in aprotic solvents, and, conversely, organic molecules such as methyl bromide dissolve to only a very limited extent in protic solvents such as water. There is, then, a dilemma here: the lowest activation energy for this substitution reaction is attained in a solvent in which only limited amounts of one of the reactants can dissolve.

Two methods are used in laboratories to solubilize such ions as hydroxide ion in relatively nonpolar solvents. Both techniques rely on decreasing the degree of solvation of the positively charged counterion by water (Figure 20.5). Macrocyclic ethers (crown ethers) can surround a small cation such as lithium or potassium with oxygen ligands, leaving the exterior of the complex very hydrocarbon-like and thereby increasing its solubility in lipophilic solvents. The second method exchanges a tetraalkylammonium cation for $Na^{\oplus}$ as a counterion for the hydroxide ion. The tetraalkylammonium cation resembles a crown ether complex in that a cation is surrounded by hydrocarbon chains with a high affinity for hydrocarbon-like solvents. These two methods differ in that the cation in the second case is covalently bonded to its lipophilic partner; whereas in the first case, the association is the result of electrostatic attraction between the cation and lone pairs of electrons on the ether's oxygens.

Tetraalkylammonium ions can be used as carriers to transport anions such as hydroxide ion between an aqueous solution and a hydrocarbon solvent such as benzene (Figure 20.5). The ammonium ion need not be used in a stoichiometric amount and, indeed, serves as a true catalyst for the reaction, increasing the rate by increasing the concentration of hydroxide ion without itself being consumed. Because the increase in rate is due to transfer of the reactive nucleophile from one phase to another phase, in which the reaction proceeds, this process is called **phase-transfer catalysis.**

### EXERCISE 20.1

How would the change from a nonpolar to a polar protic solvent affect the rate of an $S_N2$ reaction in which the nucleophile is neutral and the substrate is positively charged? Provide a specific example of a reaction of this type.

## 20.2

## Avoiding Charge Separation in Multistep Reactions

This section examines a slightly more complex reaction in which catalysis can operate. Figure 20.6 (on page 20-10) is a simple energy diagram describing the two-stage process that converts a carboxylic acid ester and water into the corresponding carboxylic acid and alcohol:

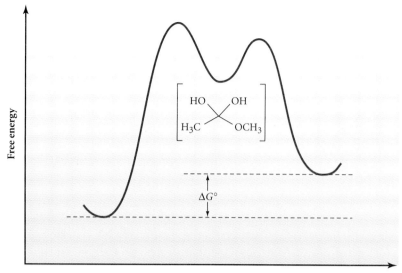

**Free energy** (vertical axis)

**Reaction coordinate** (horizontal axis)

$\Delta G°$

## FIGURE 20.6

As this energy diagram shows, transesterification is a two-step process that proceeds through a tetrahedral intermediate formed by nucleophilic addition in the initial step.

$$H_3C\text{—CO—OCH}_3 + H_2O \longrightarrow [\text{HO—C(OH)—OCH}_3] \longrightarrow H_3C\text{—CO—OH} + HOCH_3$$

Tetrahedral
intermediate

This reaction proceeds through a tetrahedral intermediate, represented here as a simple, uncharged species. However, let's consider what must take place in the bond-forming and bond-breaking steps that lead to this intermediate. As the oxygen atom of a water molecule approaches and begins to form a bond with the carbonyl carbon of the ester, it loses electron density, ultimately to become trivalent and therefore positively charged.

Simultaneously, the $\pi$ bond of the carbonyl group must be broken to an extent at least as great as that to which the new carbon–oxygen $\sigma$ bond is being formed, to prevent a buildup of more than eight electrons around carbon. Upon completion of these bond changes, the oxygen that was part of the carbonyl group is monovalent and thus bears a full negative charge. This tetrahedral intermediate has complete charge separation, unlike the intermediate shown in Figure 20.6. Such charge separation is highly unfavorable energetically and, indeed, the zwitterionic tetrahedral intermediate very rapidly undergoes proton transfer to form the neutral intermediate. (We will return to this proton transfer shortly.)

Now, let's see how the reaction differs when it is conducted in an acidic medium. Under these conditions, there is a significant concentration of the protonated form of the ester (B), in which the carbonyl oxygen bears a positive charge:

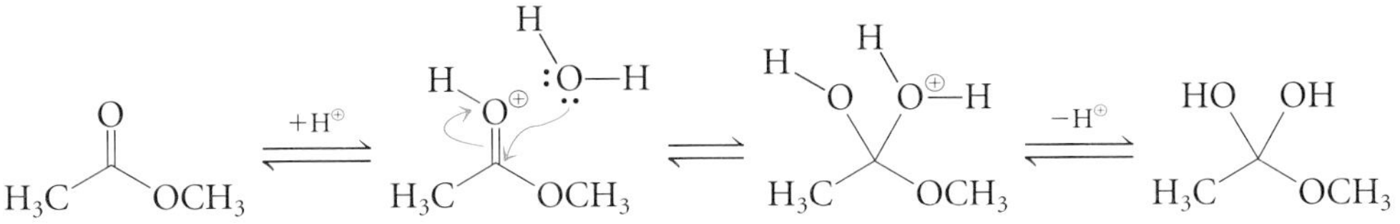

As a bond is formed between the carbonyl carbon and the oxygen of water, a positive charge develops on oxygen as it becomes trivalent. However, as the $\pi$ bond of the carbonyl system is broken, the positive charge on that oxygen decreases until it is formally neutral in the intermediate, C. Throughout this sequence, there is no charge separation—there is only a transfer of positive charge from the protonated ester's carbonyl oxygen to the oxygen of the attacking water molecule. This lack of charge separation is responsible for a dramatic energetic difference between the uncatalyzed and catalyzed processes. The transition state for oxygen–carbon bond formation in the catalyzed process is considerably more stable than that in the uncatalyzed reaction. Because the proton consumed at the beginning is ultimately released at the end, the reaction in acid constitutes an example of true catalysis.

Let's compare the energy curves for the catalyzed (c) and uncatalyzed (uc) processes (Figure 20.7, on page 20-12). The most important difference is that the activation energy for the catalyzed process (red curve) is considerably lower than that for the uncatalyzed reaction (blue curve). The protonated ester is more reactive toward nucleophilic addition, as evidenced by the lower activation energy required for the catalyzed pathway. Therefore, the catalyzed reaction is faster.

Besides showing the difference in activation energies, the energy diagram indicates that the catalyzed reaction has three transition states, whereas the uncatalyzed reaction has only two. In the uncatalyzed pathway, the first transition state corresponds to carbon–oxygen bond formation as the nucleophile bonds to the carbonyl carbon. The second transition state corresponds to the proton transfer required to neutralize the separated charges produced in the first step. The higher energy of the transition state leading to the zwitterionic intermediate is responsible for the slower rate in the absence of the acid catalyst. For the catalyzed pathway, the rate-limiting step (that with the highest activation energy) is preceded and followed

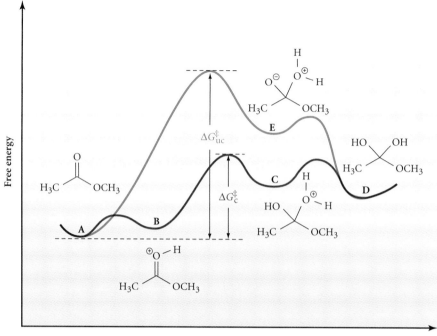

**FIGURE 20.7**

In the acid-catalyzed addition of water to methyl acetate, the neutral ester (A) is in equilibrium with B, which is protonated on the carbonyl oxygen. A cationic tetrahedral intermediate (C) is formed in this nucleophilic addition. Cation C equilibrates with its neutral form (D) in an acid–base reaction. In the uncatalyzed pathway, the neutral ester is transformed into a high-energy, zwitterionic intermediate (E), which is converted to the hydrate (D) by synchronous protonation and deprotonation on the two charged oxygen atoms.

by very rapid proton-transfer steps, neither of which results in an appreciable change in energy. We can quantitatively evaluate the energy change in the first step by comparing the acidities of the protonated ester and the hydronium ion:

$$\underset{(\text{p}K_a\ -6.5)}{\text{H}_3\text{C}-\text{C}(\overset{+}{\text{O}}-\text{H})-\text{OCH}_3} \qquad \underset{(\text{p}K_a\ -1.7)}{\text{H}-\overset{+}{\text{O}}(\text{H})-\text{H}}$$

This modest difference in acidity means that protonation of an ester by hydronium ion is endothermic by only 3.5 kcal/mole. The catalyzed and uncatalyzed pathways differ at almost all points along the two reaction profiles in Figure 20.7. Only the starting ester in water and the neutral tetrahedral intermediate (D) are common to both.

# EXERCISE 20.2

Using dashed lines to represent partial bonds that are undergoing either formation or cleavage, draw representations of the five transition states in the catalyzed and uncatalyzed reaction pathways of Figure 20.7.

## Intermolecular Proton Transfer

Let's return to the question of how the zwitterionic tetrahedral intermediate undergoes proton transfer to form the neutral tetrahedral intermediate. Bimolecular proton transfers between heteroatoms often occur with very small energies of activation. As a consequence, it is quite difficult to uncover the details of these reactions. However, some reasonable conclusions can be drawn on the basis of what is known about proton-transfer reactions and geometric constraints on hydrogen bonding. The various possibilities for transforming the tetrahedral zwitterionic intermediate into the neutral intermediate are shown in Figure 20.8: (1) intramolecular proton transfer through a four-member cyclic transition state (top reaction), (2) protonation of the alkoxide oxygen, followed by deprotonation at the oxonium ion (middle reaction), (3) deprotonation at the oxonium ion and protonation on the alkoxide oxygen (bottom reaction).

We can rule out intramolecular proton transfer, because of the very serious distortion in the cyclic transition state from the ideal, linear arrangement for proton transfer, and because this transition state would create bond-angle distortion (ring strain) in forming the four-member ring.

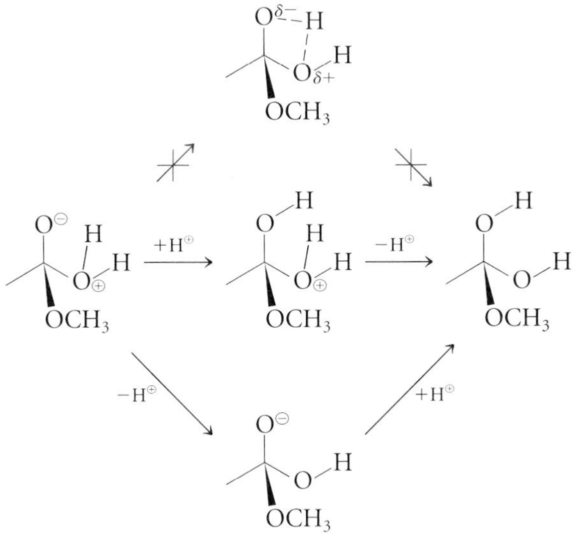

**FIGURE 20.8**

Three possible routes by which charge can be neutralized in the tetrahedral zwitterionic intermediate produced by hydration of methyl acetate under neutral conditions.

Instead, the neutral tetrahedral intermediate is formed via intermolecular proton transfer. The negatively charged oxygen gains a proton from the medium, and the proton on the positively charged oxygen is transferred to water (acting as a base):

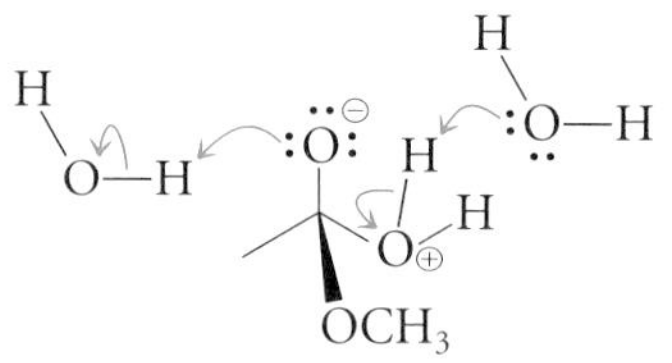

These two proton transfers might be simultaneous, but this is unlikely for two reasons: first, each bond-forming and bond-breaking process contributes its own activation energy to any reaction in which there are simultaneous, multiple transformations; second, in addition to this enthalpy factor, each proton transfer requires its own organization of molecules into the entropically unfavorable arrangement necessary for the transformation. Thus, for the two proton transfers—to and from the zwitterionic tetrahedral intermediate—to be simultaneous, three molecules (proton donor, proton acceptor, and intermediate) would have to come together simultaneously in the proper arrangement. Such three-body collisions are improbable.

By a process of elimination, then, we have arrived at two alternative multistep processes that differ only in the order in which the protons are added and removed. The two processes can be expected to have very similar activation energies and thus to contribute approximately equally to the overall process.

### Proton Transfer via Charge Relay

There is a final possibility for proton transfer—a pathway that includes bimolecular transfer but nonetheless has some of the features of the intramolecular transfer shown in Figure 20.8. The two-step sequence for intermolecular proton transfer in Figure 20.8 consists of the removal of a proton by one water molecule and the addition of a proton from another water molecule. Let's consider how both functions—removing and adding a proton—could be served by a single molecule of water:

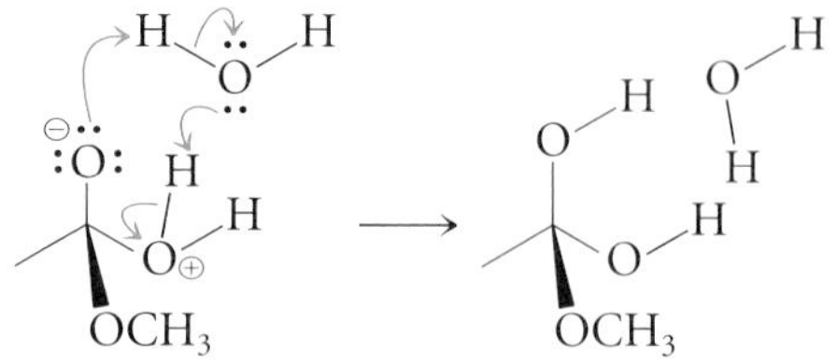

When a single water molecule serves both functions simultaneously, this molecule undergoes no net change. In this transition-state arrangement, a total of six atoms take part in the intermolecular proton transfer, instead of the four atoms participating in the intramolecular transfer shown in Figure 20.8, which was deemed too high in energy. Although the preferred linear arrangement of hydrogen between two oxygen atoms in a hydrogen bond is not attained in this transition state, the O—C—O and O—H—O angles are larger than those in the four-atom cyclic intermediate shown in the top reaction of Figure 20.8. However, this improvement in bond angles is achieved at a cost. There are now two bond-breaking and two bond-making steps taking place simultaneously, each adding to the energy of the transition state.

## EXERCISE 20.3

Write mechanisms for all four of the proton-addition and proton-removal steps shown in Figure 20.8. Be sure to use arrows to indicate the electron flow. Carefully examine the species present as starting materials and as products. (Be sure to include water and its product.) Determine whether you would expect any large changes in energy as each of these four reactions proceeds.

The expression **charge-relay mechanism** is used for this type of reaction, in which the function of the water molecule is to transfer a proton, and therefore the charge, from one charged atom in a molecule to an uncharged atom that is not spatially oriented correctly for direct transfer between these centers.

It might seem possible that water could relay charge at the same time as the bond between the oxygen of the attacking water molecule and the carbonyl carbon is formed, with one water molecule relaying charge and another acting as the nucleophile:

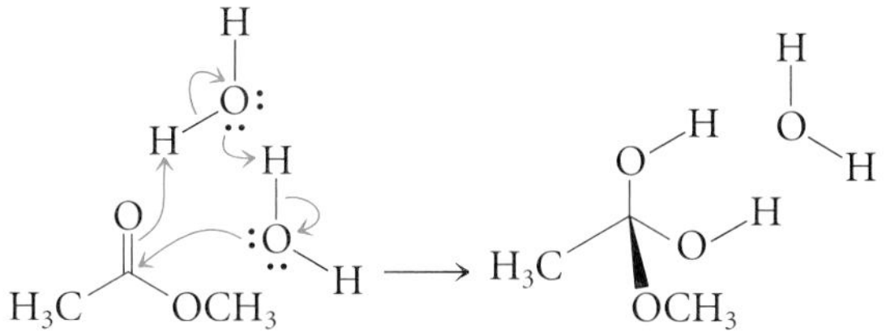

This mechanism has many favorable features, the most important of which is that no significant charge separation develops in the transition state. However, it is disfavored by entropy, because it requires the simultaneous interaction of three molecules—an ester molecule and two water molecules. The difficulty of organization in such cases can be surmounted by the intervention of enzymes, which are already highly organized. In essence, the cost of the entropic requirement of a highly organized transition state has already been paid in the construction of the enzyme. Because enzymes are catalysts and serve to transform thousands of molecules of starting material into product, the energetic cost of their synthesis is not a significant factor overall for the organism.

## 20.3

## Distinction between Catalysis and Induction

We have seen how an acidic medium can prevent charge separation in the transition state for the hydrolysis of an ester. Let's now consider whether an alkaline pH (increased hydroxide ion concentration) can also prevent the energetically unfavorable separation of charge. Indeed, with a sufficiently high concentration of sodium hydroxide, the negatively charged hydroxide ion becomes an effective nucleophile, forming a bond with the carbonyl carbon of the ester:

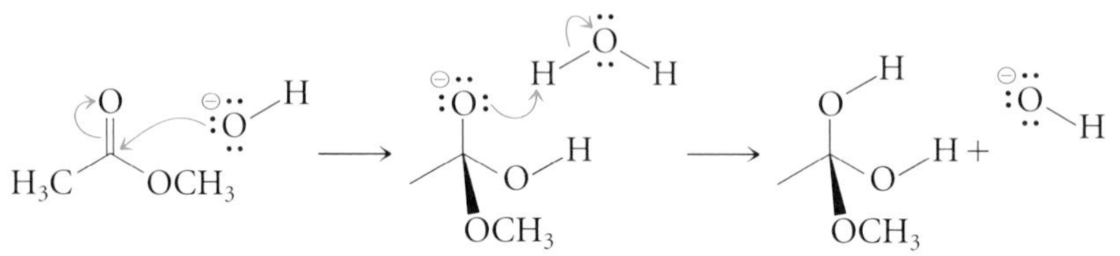

The nature of the charge in the starting materials and in this tetrahedral intermediate is essentially identical: in both cases, negative charge is localized on a single oxygen atom. Thus, nucleophilic addition of hydroxide ion consists only of the transfer of charge; no additional charge separation develops in the transition state. The neutral tetrahedral intermediate is then formed by transfer of a proton from water, reforming the hydroxide ion consumed in the first step. Again, the transition state for this step does not include the development of charge separation; it involves only transfer of negative charge from one oxygen to another. In this way, the ester is converted to the neutral tetrahedral intermediate in two steps, neither of which develops additional charge separation. In this sequence, hydroxide ion is a true catalyst. As a result, the formation of the tetrahedral intermediate is much more rapid when hydroxide ion, rather than water, is the nucleophile.

### EXERCISE 20.4

Draw an energy diagram for the following two steps, keeping in mind that the transfer of a proton from one heteroatom to another is generally a reaction with a very low energy of activation and a modest change in energy.

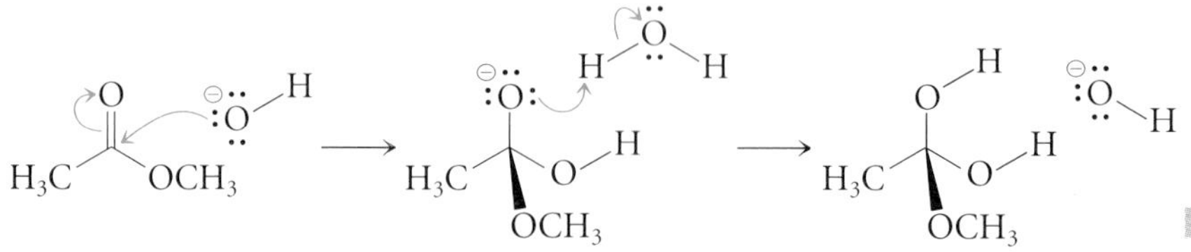

Let's continue to examine the ester hydrolysis reaction as it proceeds from the tetrahedral intermediate to the final product in both acidic and basic media. These sequences are affected by the pH in much the same way as the reactions leading to the tetrahedral intermediate.

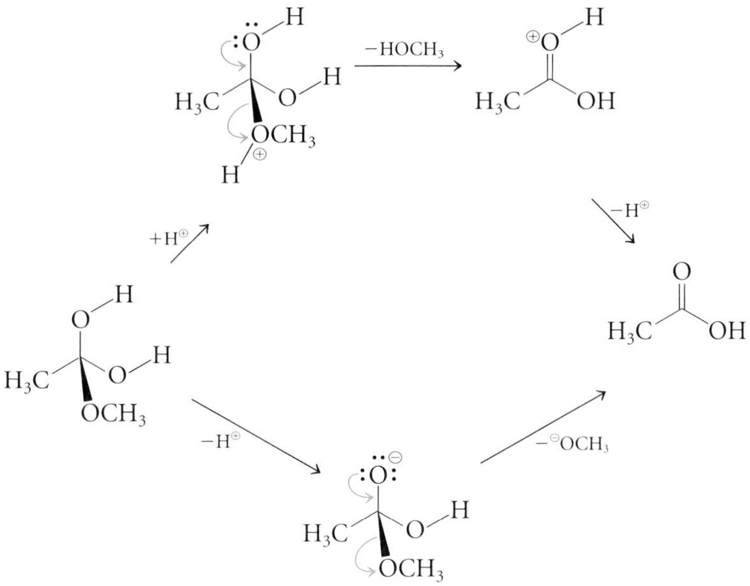

In the reaction in acid (top), a proton is gained in the first step and later lost. Therefore, this stage of the reaction is acid-catalyzed. In the reaction in base (bottom), base present in the medium removes a proton, reforming an anionic tetrahedral intermediate. Reformation of the carbonyl $\pi$ bond with loss of methoxide ion leads to the carboxylic acid. Because methanol is a weaker acid than acetic acid, these two species react to form methanol and the carboxylate ion. Indeed, the equilibrium favors acetate and methanol by some 11 orders of magnitude:

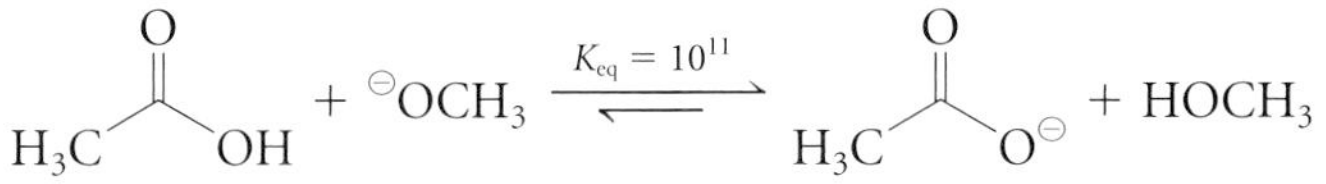

### EXERCISE 20.5

Verify the role of the acid as a catalyst in ester hydrolysis by showing how the reaction sequence leading from the neutral tetrahedral intermediate to the carboxylic acid might differ if the reaction were to take place at neutral pH.

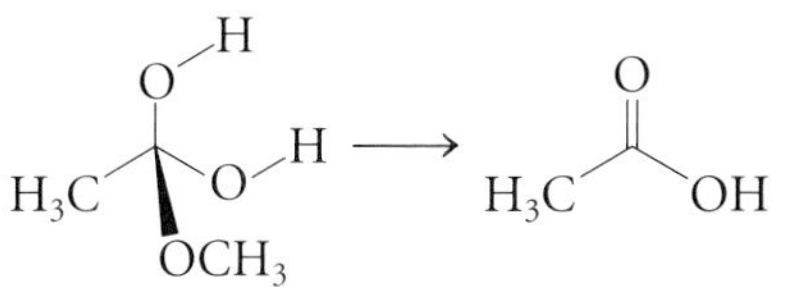

In the net conversion of methyl acetate and water into acetate ion and methanol, the hydroxide ion is not a catalyst because it is not regenerated. Thus, there is a fundamental difference between the hydrolysis of an ester in the presence of acid and the same hydrolysis in the presence of base. In the first case, the reaction is catalyzed; in the second, it is not. Chemists say that the second process is *base-induced,* to distinguish it from a truly catalytic sequence. Nonetheless, both hydrolysis reactions have many features in common, especially with regard to how they are accelerated. Many enzymatic processes resemble hydrolysis induced by base in that a separate molecule (called a *cofactor*) serves an essential role and is consumed in the process. Such reactions can still be correctly referred to as being catalyzed by the enzyme, because, unlike the cofactor, the enzyme remains unchanged at the end of the reaction.

The reaction conditions can be adjusted so that hydroxide ion does participate as a catalyst in the hydrolysis of an ester. Until the last (neutralization) step, hydrolysis under basic conditions is catalytic, because the methoxide ion generated can, and indeed does, react with water to form hydroxide ion. Recall that a buffered solution is one that contains a mixture of an acid and its conjugate base in rapid equilibrium. Small amounts of additional acid or base do not significantly affect the pH of such a solution, only the position of the equilibrium. If the hydrolysis of an ester is carried out in a buffered basic solution, the concentration of hydroxide ion remains relatively unchanged as long as the amount of buffer is significantly greater than the amount of acid generated by the hydrolysis. Under these conditions, the hydroxide ion is a true catalyst: it is the conjugate base of the buffer that is consumed, not the hydroxide ion.

The distinction between an acid-catalyzed and a base-induced hydrolysis of an ester is important for laboratory chemists. In the former case, the amount of acid can be adjusted so as to provide a practical rate of conversion. However, there must be at least one equivalent (referred to as a stoichiometric amount) of base present for the reaction to proceed to completion. In fact, even an equimolar amount of base is not sufficient, because, as the reaction nears completion, the concentration of the hydroxide ion and, correspondingly, the rate of the reaction is diminished. In contrast, the distinction between a catalyzed and an induced reaction is far less important in biological systems, where small amounts of material are converted in a large volume of buffered medium.

## 20.4

## Base Catalysis

Although the base that accelerates a reaction (for example, hydroxide ion) is frequently also the nucleophile, some other species can often act as a base to generate the active nucleophile by acid–base reaction with a protonated precursor. Consider the hydrolysis of an ester by sodium carbonate in methanol containing a small amount of water. Here, sodium carbonate, the base, generates methoxide ion (and ultimately hydroxide ion) as the nucleophile. (A mild base such as carbonate is often chosen when the molecule contains other functional groups that might also undergo a base-induced or base-catalyzed transformation.)

In the first step, methoxide ion effects transesterification, converting the cyclohexyl ester of acetic acid to methyl acetate. In the second step, methyl acetate is hydrolyzed to acetate ion by attack of hydroxide ion. Note that the first step does not consume base; it is thus base-catalyzed. In the second step, the acetic acid initially produced reacts irreversibly with base, making this step base-induced, not base-catalyzed.

Because sodium carbonate is only slightly soluble in methanol, the effective pH of the solution is only mildly basic. Nonetheless, methoxide and hydroxide ions are produced in sufficient quantity for them to react as nucleophiles. In essence, the reaction contains a sufficient quantity of base (some that is not even in solution) to effect the desired hydrolysis without the high pH that would result if an equivalent of methoxide or hydroxide ion were present at the start.

Because sodium carbonate is not as basic as sodium hydroxide, the presence of a full equivalent of sodium carbonate does not yield as basic a solution as would be obtained with sodium methoxide. Yet, the equilibrium between sodium carbonate and sodium methoxide is quite rapid, and even though it favors the former by some eight orders of magnitude, sufficient methoxide is present for the reaction to proceed.

The base-catalyzed hydrolysis requires a balance between the concentration of the nucleophile and the pH of the solution. Clearly, when the pH is very low, the rate of the reaction is retarded to a point at which it is too slow to be usable. Conversely, at higher pH, the rate of reaction increases, but so do those of other undesirable base-catalyzed and base-induced reactions.

**20.5**

# Comparison of Intermolecular and Intramolecular Reactions

Many organic reactions are bimolecular—that is, two reactants come together in the rate-determining transition state. For example, the hydrolysis of an ester, whether carried out under acidic or basic conditions, has a transition state in which a bond is formed between the carbonyl carbon and a nucleophilic oxygen of either water or hydroxide ion. Several energy barriers must be overcome in reaching such a transition state. First, two separate molecules must be brought together, making the process entropically unfavorable. Furthermore, these two reacting species cannot be brought together in a random fashion but must be oriented in an arrangement favorable for bond formation, thus adding additional entropic cost to the formation of the transition state. In addition, each reactant molecule is originally surrounded by (and stabilized by) solvent molecules. For a reaction to occur, some of the associated solvent molecules must be removed so the two species can approach to within the distance necessary for bond formation in the transition state.

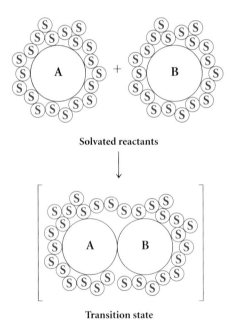

Solvated reactants

Transition state

   Both entropy and solvation are far less important when the two reacting partners are already connected by bonds that are not affected by the reaction. For example, acid-catalyzed esterification of carboxylic acids by alcohols (equation 1) is much slower than intramolecular ring closure of hydroxyacids to form lactones (equations 2 and 3). (Recall that a lactone is a cyclic ester.)

$$RCO_2H + HOR \xrightarrow{\ H^{\oplus}\ } RCO_2R \tag{1}$$

$$\tag{2}$$

$$\tag{3}$$

In reaction 1, the two separate reagents must become associated to arrive at the transition state. In contrast, the reactions that form lactones (2 and 3) are unimolecular and have less entropically unfavorable transition states. As a result, they have higher rates. Not only are the two reactive functional groups (—OH and —$CO_2H$) already associated by covalent bonds, but less desolvation is required to reach the transition state.

Two intramolecular examples of ester (lactone) formation are shown: one in which the hydroxyl group is attached at the end of a flexible alkyl chain (reaction 2) and the other in which the hydroxyl group is attached to a more rigid, aromatic spacer (reaction 3). The rates of formation of the lactones are expected to differ. The reacting groups are held more closely together in reaction 3 than in reaction 2, and less additional organization is required to reach the transition state for lactone formation. Furthermore, as a result of the proximity of these two reacting groups in the starting hydroxyacid, fewer intervening solvent molecules must be removed from the region between them for bond formation to occur. The effects of entropy and solvation are even more important in enzyme-catalyzed reactions.

## 20.6

# Transition-Metal Catalysis

### Catalytic Addition of Hydrogen to Alkenes

Transition metals such as palladium and platinum catalyze the addition of molecular hydrogen to an alkene, a reaction that otherwise does not proceed at a measurable rate.

#18   Catalytic Hydrogenation of Alkenes

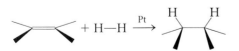

To see why the metal catalyst is required, let's examine conceivable mechanisms that follow homolytic or heterolytic pathways.

## CHEMICAL PERSPECTIVES

The first light bulbs invented by Thomas A. Edison worked very much in the same way as the sun—as black-body radiators. On the other hand, there are sources of light that do not depend on thermal excitation for the emission of light. One example is the cold light sticks marketed under the trade name Cyalum. The chemistry involved in these light sticks appears simple, yet it demonstrates the profound effect of electronic factors on concerted electrocyclic reaction pathways.

Within each stick is a solution of diphenyloxalate and a dye, plus a thin glass vial holding a separate solution of hydrogen peroxide and sodium hydroxide.

**Diphenyloxalate**

When the vial is broken, the two separate solutions mix, and the anion of hydrogen peroxide begins its reaction with the oxalate ester via nucleophilic acyl substitution. The proton of the peroxyl group is then removed by hydroxide ion, and a second, now intramolecular, acyl substitution ensues.

The resulting peroxyanhydride contains a four-member ring and is very much less stable than two carbon dioxide molecules. Nevertheless, because the transition state for such an electrocyclic conversion involves four electrons in bond making and bond breaking, the conversion cannot occur by the usual pathway (that is, to produce two molecules of carbon dioxide in the ground state). Instead, the symmetry of the electronic states involved requires that one of the product carbon dioxide molecules be in the first excited state (that is, with an electron in an antibonding orbital). When the electron in the excited state drops down to the ground state (to a bonding orbital), a photon in the ultraviolet region of the spectrum is released. This transition is invisible to the human eye. However, in the presence of a dye, the excited carbon dioxide transfers some of its excitation energy to the dye molecule and in so doing raises an electron in that molecule to an antibonding orbital. (The remainder of the energy is released as heat.) When this electron drops to a bonding orbital, a photon of visible light is released.

Several animals possess light-emitting chemical systems. For example, fireflies produce luciferin (and an enzyme, luciferase). Luciferase catalyzes the reaction of luciferin with ATP and molecular oxygen, with the emission of light. Firefly luciferin is oxidized to a four-member cyclic perester, which loses carbon dioxide spontaneously. The resulting heterocyclic compound is produced in an excited state, and light is emitted when the excited electron drops to a bonding orbital.

**Firefly luciferin**

The reaction has been used to test for the presence of ATP. Because there are light detectors that can count single photons, concentrations of ATP as low as $10^{-11}$ M can be detected. Intense research into light emission by fireflies in the 1950s and 1960s led to a dramatic decline in the population of these insects—it takes 17,000 fireflies to produce 10 mg of luciferin.

First, we can rule out the concerted pathway:

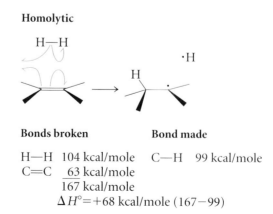

A simple rule usually applies to concerted electrocyclic processes between reactants in the ground state: they take place most readily when the number of electrons involved is the same as that in an aromatic system, as defined by Hückel's rule. For example, the Diels–Alder reaction, requiring a combination of four electrons in a diene with two electrons in a dienophile, proceeds by a concerted pathway because the transition state involves moving $4n + 2$ electrons. On the other hand, the direct addition of a molecule of hydrogen to an alkene, with only four electrons (two from the $\sigma$ bond in $H_2$ and two from the $\pi$ bond of the alkene) cannot be a concerted process because such a reaction involves only $4n$ electrons.

There are two possible multistep processes for accomplishing the bond making and bond breaking required for the addition of $H_2$ to an alkene. However, simple bond cleavage by either the homolytic or the heterolytic mode is energetically prohibitive, because each route requires the input of at least as much energy as is needed for the dissociation of the hydrogen–hydrogen bond: 104 kcal/mole.

$$H\cdot + H\cdot \qquad \Delta H° = 104 \text{ kcal/mole}$$

$$H{-}H$$

$$H^{\oplus} + H^{\ominus} \qquad \Delta H° = 104 \text{ kcal/mole} + ?$$

In fact, heterolytic rupture of the hydrogen–hydrogen bond requires even more energy because additional energy is needed for charge separation.

Alternatively, we can consider the possible interaction of a molecule of hydrogen with an alkene by a homolytic or heterolytic process:

**Homolytic**

| Bonds broken | Bond made |
|---|---|
| H—H  104 kcal/mole | C—H  99 kcal/mole |
| C=C   63 kcal/mole | |
| 167 kcal/mole | |

$$\Delta H° = +68 \text{ kcal/mole} \ (167 - 99)$$

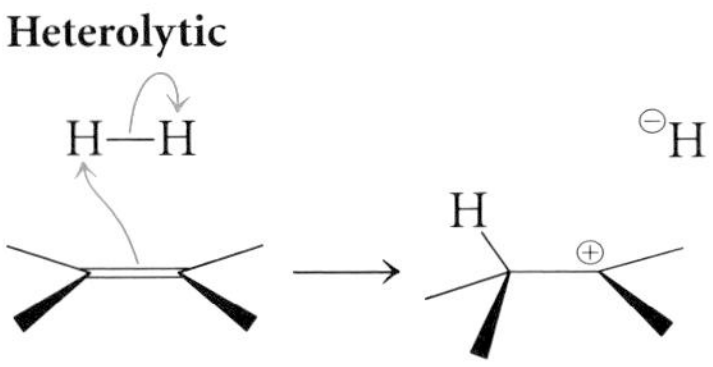

Again, we can arrive at a minimum energy requirement for both these possibilities by evaluating the bond breaking and bond making that take place in the homolytic pathway. Two bonds are broken (the hydrogen–hydrogen bond and the carbon–carbon $\pi$ bond), and a single carbon–hydrogen bond is formed. The energy released upon formation of the carbon–hydrogen bond does not nearly compensate for that consumed in the bond-breaking processes, and the homolytic pathway requires at least 68 kcal/mole, a prohibitively large value. (Recall that the activation energy for an endothermic process must be equal to or greater than $\Delta H°$.) The endothermicity of the heterolytic pathway is even greater, as a result of charge separation. Thus, there appears to be no direct way by which an alkene and molecular hydrogen can react to form an alkane.

Indeed, the hydrogenation of alkenes requires a metal catalyst. The mechanisms of such reactions are imperfectly understood, but some clues as to the intermediate species involved can be obtained from studies on the reactions of soluble transition metal complexes. Molecular hydrogen is absorbed by a metal such as platinum in a process in which the hydrogen–hydrogen bond is broken and then replaced by two new platinum–hydrogen bonds.

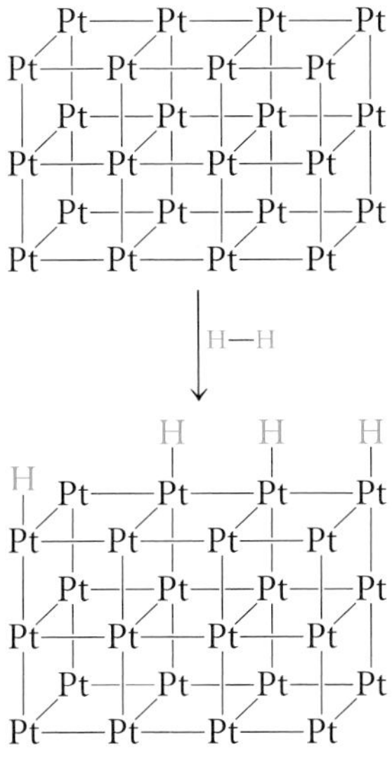

The energy released in this bond making more than compensates for the energy required to cleave the hydrogen molecule. Although the hydrogen atoms are shown here as bonded only to the metal atoms on the top surface, hydrogen atoms in fact migrate freely throughout the metal. (Hydrogen is the only element whose atoms are small enough to do this.) This process, like others that will be considered shortly, is facilitated greatly by the ability of a transition metal to expand and contract its valence shell and number of associated ligands. This characteristic of transition metals results from the presence of unfilled valence shells, so that the addition or subtraction of an electron does not greatly alter the energy. In contrast, elements in the second row of the periodic table have normal valence requirements that result in filled electron shells. With atoms such as carbon and nitrogen, addition or subtraction of electron density results in an electronic state of much higher energy.

The next step consists of the interaction of the $\pi$ electrons of the alkene with a platinum atom not bonded to a hydrogen:

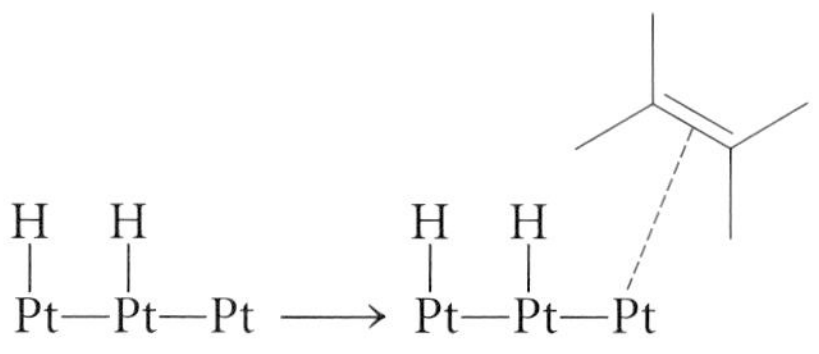

Empty orbitals of the metal are available to overlap with the $\pi$ bonding orbital of the alkene, producing a molecular orbital of lower energy than the overlapping orbitals. By bonding through this orbital, the alkene and platinum form a $\pi$ complex. With both the hydrogen atoms and the alkene associated with the platinum surface, hydrogen is transferred from platinum to carbon, with simultaneous formation of a platinum–carbon bond.

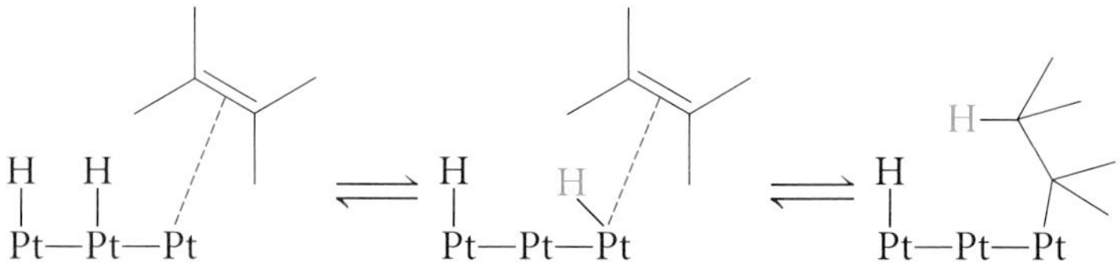

This step is neither highly exothermic nor endothermic, and it has a low energy of activation. The energy consumed in the cleavage of the platinum–hydrogen bond and the carbon–carbon $\pi$ bond is compensated for by the energy released in the formation of the platinum–carbon and carbon–hydrogen bonds.

The hydrogen atoms associated with the platinum are very mobile and can readily move from one platinum atom to another on the metal surface. The next step is the migration of a hydrogen to the platinum already bonded to the alkyl fragment.

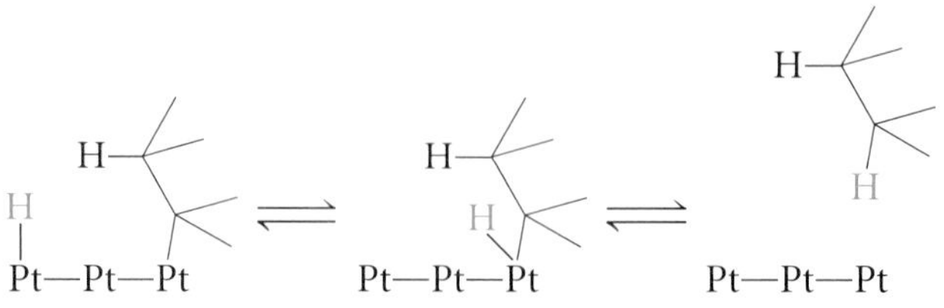

In the final step, both the platinum–hydrogen and the platinum–carbon bonds are cleaved in a process that simultaneously forms the second carbon–hydrogen bond in the product. This step is known as a **deinsertion reaction,** because the reverse process, an insertion reaction, can be viewed simply as the interjection of platinum between a hydrogen and a carbon of an alkane.

Unlike the starting alkene, the product alkane does not have readily accessible electron density for interaction with the platinum surface and thus returns to the solution, exposing the metal surface for another cycle of hydrogenation. Although the overall reduction reaction is exothermic, all of the steps in the addition of hydrogen to an alkene in the presence of a noble metal catalyst are reversible. The reverse reaction is favored by entropy (two molecules formed from one), and, because the contribution of entropy to a free-energy change varies with temperature ($\Delta G° = \Delta H° - T\Delta S°$), the alkene and hydrogen are favored at high temperature.

Reverse hydrogenation (oxidation) is an important industrial reaction. Alkanes burn much more rapidly than do alkenes, and too rapid a rate of oxidation results in knocking (pinging) in the cylinder of an internal combustion engine. Because of the rapid oxidation, pressure builds up before the piston reaches the top of its stroke in the cylinder. When the crude oil feedstocks obtained from oil wells are too rich in alkanes, high-temperature treatment with noble metal catalysts is used to effect partial dehydrogenation. This results in the conversion of some of the alkanes into alkenes, as well as production of aromatics from cyclohexanes and formation of molecular hydrogen, a valuable by-product.

---

**EXERCISE 20.6**

---

Explain why the mechanism for catalytic hydrogenation presented in this section results in *cis* addition of hydrogen.

---

### Catalysis of Olefin Polymerization

A Ziegler–Natta catalyst (a mixture of aluminum and titanium Lewis acids) provides a catalytic system for olefin polymerization that produces unique properties in the resulting polymers. Among these properties are regular stereochemistry and significantly longer chains than can be obtained by free-radical polymerization. Details of how this catalytic system functions are obscured by the complexity of the mixture of species present, many of which are in dynamic equilibrium with each other. Identifying the active catalyst (or catalysts) is an exceedingly difficult task. As a result, attempts to alter the properties of the catalyst system are hampered by lack of detailed knowledge of its structure.

Alternative catalysts for olefin polymerization have recently emerged from a number of chemical companies. These catalysts, of striking similarity in structure and function, are unique in that they are well-defined, single structures that can be studied in detail in the laboratory. When the structure of an active catalyst is known, structural modification can be correlated with performance as an olefin polymerization catalyst. In one example of this new breed of catalyst, the transition metal zirconium is complexed to two five-member rings in which all the carbon atoms are $sp^2$-hybridized:

**Organozirconium catalyst
for olefin polymerization**

The bonding between the metal atom and the carbon atoms of the two five-member rings cannot be portrayed by the classical approach to chemical bonds. The bonding in such organometallic catalysts is rather complex, and its study occupies a full-semester course in the curriculum of chemistry majors.

To this point, our focus has been primarily on bonding by second-row elements, for which the number of bonds can easily be predicted from the octet rule. For example, fluorine, with electron configuration $1s^2\ 2s^2\ 2p^5$ and seven electrons in the valence shell, forms one bond to attain a stable octet of electrons in the outer shell. Similarly, oxygen forms two bonds, nitrogen three, and carbon four. The situation for transition metals such as zirconium is considerably different. These elements have much larger numbers of electrons, and the energies of the orbitals in the outer shells lie very close to one another. Moreover, the energies of the different orbitals differ such that the electron configurations of the transition series vary in a complex manner. For example, the electron configuration for zirconium is $1s^2\ 2s^2\ 2p^6\ 3s^2\ 3p^6\ 3d^{10}\ 4s^2\ 4p^6\ 4d^2\ 5s^2$. Note that the $5s$ shell is filled before the $4d$ shell. What rules can be formulated to describe the type and number of bonds Zr can form? Will the bonding be ionic or covalent? Are all of the fourth- and fifth-level electrons of Zr (12 total) available to form covalent bonds with neighboring atoms? Should we look at the number of electrons that can be accommodated in the $4d$ subshell? In the $5p$ subshell? The $5d$ subshell? A further consideration with transition elements is that the symmetry properties of higher atomic orbitals are more complex than those of the $2s$ and $2p$ orbitals. As a consequence of these factors, the number and type of bonds formed by the transition series elements far exceed anything we have encountered, and there are no simple rules to describe the bonding for such elements.

Despite difficulties in the theoretical treatment of bonding in transition metals, a wealth of experimental information exists on the fascinating compounds of these metals. The conventions employed to convey structural information about these compounds are different from those used by organic chemists. In structures of transition metal complexes with carbon, lines drawn between carbon and other second-row elements denote "classical" bonds, whereas dashed lines between the metal and other elements imply a spatial relationship. For example, the neutral organoiron compound ferrocene is depicted here with a total of 10 dashed lines between carbons and iron in order to show that the iron is equidistant from each carbon atom of the two cyclopentadienyl rings:

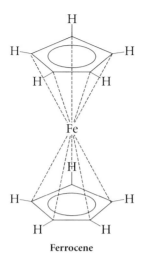

**Ferrocene**

The mechanism of how these organometallic catalysts effect olefin polymerization is well understood and has similarities with catalytic hydrogenation. In the first step, an alkene associates with the catalyst through a $\pi$ complex. Next, the complexed olefin inserts itself between the metal and its chlorine ligand, forming new C—Zr and C—Cl bonds.

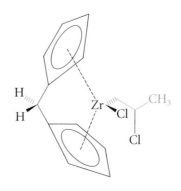

Note that from the viewpoint of the metal, these two steps have effected the replacement of a chlorine by a carbon substituent. A second molecule of olefin now complexes with the metal. Insertion of the second olefin into the C—Zr bond forms a new C—Zr bond and, importantly, a new C—C $\sigma$ bond.

These two steps, olefin complexation followed by insertion, represent the propagation steps in this polymerization.

Polymers formed by these organometallic catalysts are more regular and have less branching than those formed using Ziegler–Natta catalysis, and both of these methods yield less branching than does free-radical polymerization. Individual polymer strands can be made 10 to 100 times longer, improving polymer strength and reducing the amount of catalyst required. Further, a relatively minor change in the catalyst results in the polymerization of propylene to form either syndiotactic or isotactic polypropylene. However, these catalysts are quite expensive, costing more than $1000 per kilogram. Nonetheless, the advantages of the catalysts in polymer production have led major producers of polyolefins to convert to organometallic catalysis.

## 20.7

## Catalysis by Enzymes

Chemical reactions in living systems are much more varied and complex than any conversions yet devised and carried out in the laboratory. For the many reactions of nature to take place simultaneously in water near pH 7 requires that each one be accelerated by a separate catalyst specific for that process. This is the role of enzymes, large polypeptide molecules that have molecular weights as high as 100,000, and sometimes even higher. Their size allows for a virtually unlimited number of possible permutations and variations. Indeed, for a peptide consisting of 63 amino acid units, there are $20^{63} = 10^{82}$ possible permutations of the 20 common amino acids, a number larger than the estimated number of all elementary particles in the universe.

### Enzyme–Substrate Binding

Large polypeptides adopt a folded and twisted three-dimensional structure that is uniquely determined by the sequence of amino acids present. Nonbonding intermolecular interactions control, in part, whether a peptide chain forms an $\alpha$-helix or a $\beta$-pleated sheet along any particular segment and how these segments fold back on each other so as to maximize van der Waals, hydrogen-bonding, and electrostatic interactions between the amino acid side chains. These attractive interactions often result in a large, three-dimensional mass, and although much of the volume is filled by the backbone and residues of the polypeptide, there are often hydrophobic and hydrophilic pockets on the surface and within cavities. The three-dimensional structure of an enzyme is chiral. Only one enantiomeric form is present, because the constituent amino acids are generally all L-enantiomers. The pockets and holes are chiral, and it is in these sites that catalysis occurs in most enzymes. The relatively small part of an enzyme in which catalysis occurs is referred to as the **active site.**

The binding of a molecule at an active site often requires a close matching between the shapes and electrostatic properties of the reactant and pocket. As a result, the molecule undergoing reaction, called the **substrate,** often binds rather tightly in the active site before reaction. This process is represented schematically as follows:

$$E + S \longrightarrow E \cdot S \longrightarrow E \cdot P \longrightarrow E + P$$

Here E represents the enzyme, S the substrate, P the product, and $E \cdot S$ and $E \cdot P$ the enzyme bound to the substrate and the product, respectively. In many cases, the first step, forming the $E \cdot S$ complex, is very fast and is limited only by the rate of diffusion of the substrate through the solution to the active site. The overall reaction rate is determined by the rate of the second step, which includes the actual bond changes that convert the substrate into the product.

The binding between the enzyme and the product is not very strong. If it were, the reaction rate would decrease in direct proportion to the amount of product, and the catalysis would all but stop when the concentration of the product equaled that of the enzyme. If this does happen, the enzyme is said to be product-inhibited. However, relatively weak, favorable interactions between the enzyme and the product can serve a useful function in complex biological systems. In such cases, the product serves to regulate its own rate of formation via a feedback process. An important and developing area of pharmaceutical research is based on the concept of moderating the rate of biological transformations that overproduce product. Structural mimics of the product or the starting material can act as inhibitors by binding tightly to the enzyme (without themselves undergoing reaction) and blocking the natural substrate.

### ▨ Catalysis by the Enzyme Chymotrypsin

Let's consider the mechanism of catalysis by a specific enzyme, chymotrypsin, whose structure has been determined by single-crystal, x-ray crystallographic techniques. This enzyme catalyzes the hydrolysis of peptide bonds. Such enzymes are referred to as **proteases,** emphasizing their role in cleaving proteins. Recall from Section 20.2 that the hydrolysis of esters (and, indeed, other carboxylic acid derivatives such as amides) at or near pH 7 is extremely slow because the addition of water as a nucleophile to the carbonyl carbon results in charge separation in the transition state for the reaction. Conversely, at high concentrations of hydroxide ion, the pH is too basic for many other functional groups to be stable. However, in Section 20.4 we saw that a balance between reaction rate and pH can be established in base-catalyzed reactions.

When catalyzed by chymotrypsin, amide hydrolysis is achieved with a carboxylate ion acting as a base. Because the $pK_a$ of a typical carboxylic acid is $\sim$5, the equilibrium favors the carboxylate anion at pH 7. Furthermore, because a carboxylate is present in the enzyme cavity quite near the active site, its effective concentration is much higher than if a base such as sodium acetate were merely added to the solution.

Let's look specifically at the hydrolysis of the amide bond to a phenylalanine residue in a peptide chain. The entire process leading from E · S to E · P is shown in Figure 20.9. In this sequence, the substrate becomes temporarily attached through an ester linkage to a serine residue in the enzyme—therefore, this process is called **covalent catalysis.**

Chymotrypsin-catalyzed hydrolysis of peptides proceeds in four steps:

**1.** Nucleophilic attack by the hydroxyl group of serine on the peptide amide linkage to form a tetrahedral intermediate

**2.** Loss of an amine from the tetrahedral intermediate to yield an ester with the serine hydroxyl group

**3.** Nucleophilic addition of water to the serine ester to form a second tetrahedral intermediate

**4.** Cleavage of the second tetrahedral intermediate to free the carboxylic acid group and regenerate serine

In step 1, the hydroxyl oxygen of serine (part of the covalent structure of chymotrypsin) serves as a nucleophile, adding to the carbonyl carbon of the amide group. Simultaneously, the proton of the same hydroxyl group is transferred to one nitrogen of the imidazole of a nearby histidine residue, and the imidazole loses a proton to the carboxylate ion of an aspartate residue. The imidazole group serves to relay a proton from serine to the carboxylate ion and, in so doing, undergoes tautomerization. The histidine residue does not directly take part in the net reaction; it functions as a catalyst. No charge separation develops as a result of these proton transfers. This first step results only in the transfer of negative charge from the carboxylate ion to the tetrahedral intermediate.

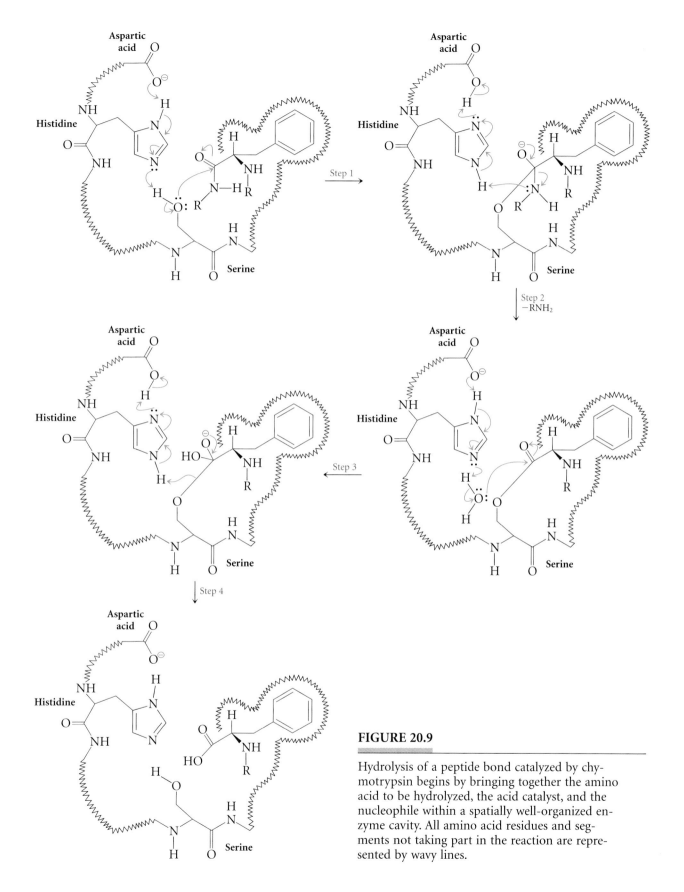

**FIGURE 20.9**

Hydrolysis of a peptide bond catalyzed by chymotrypsin begins by bringing together the amino acid to be hydrolyzed, the acid catalyst, and the nucleophile within a spatially well-organized enzyme cavity. All amino acid residues and segments not taking part in the reaction are represented by wavy lines.

In step 2, the $\pi$ bond of the carbonyl group is reformed from the tetra-hedral intermediate, with loss of the nitrogen substituent RNH$_2$ and re-generation of the general base (the aspartate carboxylate). A second series of proton transfers, in which the histidine residue again serves as catalyst, results in the simultaneous protonation of the nitrogen in the tetrahedral intermediate and deprotonation of oxygen in the aspartate residue. The leaving group is effectively the free amine, not its anion. At this stage, the resulting ester is covalently bonded to the enzyme through the serine hy-droxyl group.

In step 3, a water molecule acts as a nucleophile, adding to the ester carbonyl group formed in step 2, to form a new tetrahedral intermediate. Again, stepwise proton transfer from water to the aspartate carboxylate an-ion ensures that this transformation results only in charge transfer, not charge separation. In this proton transfer, a tautomer of histidine is formed via deprotonation by an enzyme-bound aspartate carboxylate anion, per-mitting reprotonation of histidine by a water molecule present near the ac-tive site. This activates the water molecule as a nucleophile, permitting at-tack on the covalently attached ester and producing a second tetrahedral intermediate.

In step 4, collapse of the second tetrahedral intermediate results in for-mation of the carbonyl $\pi$ bond and rupture of the carbon–oxygen bond linking the substrate to the serine residue. Reformation of the carbon–oxy-gen $\pi$ bond, with expulsion of the serine hydroxyl group, produces the free hydrolyzed carboxylic acid, as the remote carboxylic acid (aspartate) re-protonates the histidine tautomer. The product acid now has no covalent bonds to the enzyme and diffuses into the solution. At this point, the en-zyme has been restored to its initial condition, with a free serine residue and a histidine in a tautomeric form ready to activate another nucleophile by general base. Overall, the rate of this catalytic process is accelerated be-cause charge separation does not develop in these pathways.

## Stabilization of Transition States by Enzymes

The rate acceleration of amide hydrolysis by chymotrypsin is achieved through a reaction pathway quite different from the one followed (very much more slowly) without the enzyme. This is not the only way in which enzymes provide for catalysis. Many enzymes operate in the simple way ex-plored in Section 20.1, by stabilizing the transition state through favorable interactions with amino acid residues present in the active site. For exam-ple, zinc and other metal cations that are present near the active site in many enzymes effect the hydrolysis of carboxylic and phosphoric acid derivatives. These metal ions effectively lower the energy of the transition state through favorable electrostatic interactions with developing negative charge.

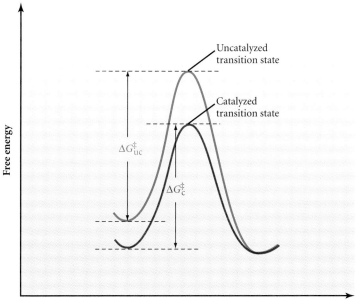

**FIGURE 20.10**

A reaction is accelerated by complexation with a reagent, if the transition state is stabilized to a greater extent than the reactants.

Thus, the activation energy for the reaction is lowered as a direct result of the stabilization of the transition state (Figure 20.10).

Note that in the energy diagram in Figure 20.10, the starting material in the catalyzed reaction is also stabilized to a certain extent. Indeed, this is often the case for enzymatic reactions in which the E · S complex is more stable than the separate substrate and enzyme. However, it is important that this stabilization of the complex be less than that of the transition state. Otherwise, the activation energy would be raised instead of lowered.

**EXERCISE 20.7**

Which amino acid residues in an enzyme could provide positive charge in the vicinity of the developing negative charge on the carbonyl oxygen of an amide or ester substrate?

### Enzymes and Chiral Recognition

Chymotrypsin is an effective catalyst for the hydrolysis of only certain peptide bonds—namely, those in which the carboxyl group is that of phenylalanine. Even minor structural variations in this amino acid residue result in a vastly lower rate or even no reaction at all. Because enzymes are chiral, it should not be surprising that even the seemingly subtle change from L- to D-phenylalanine has a dramatic effect (Figure 20.11, on page 20-36). Indeed, the active site of chymotrypsin is a pocket that has regions of the proper size and shape for three of the groups attached to the center of chi-

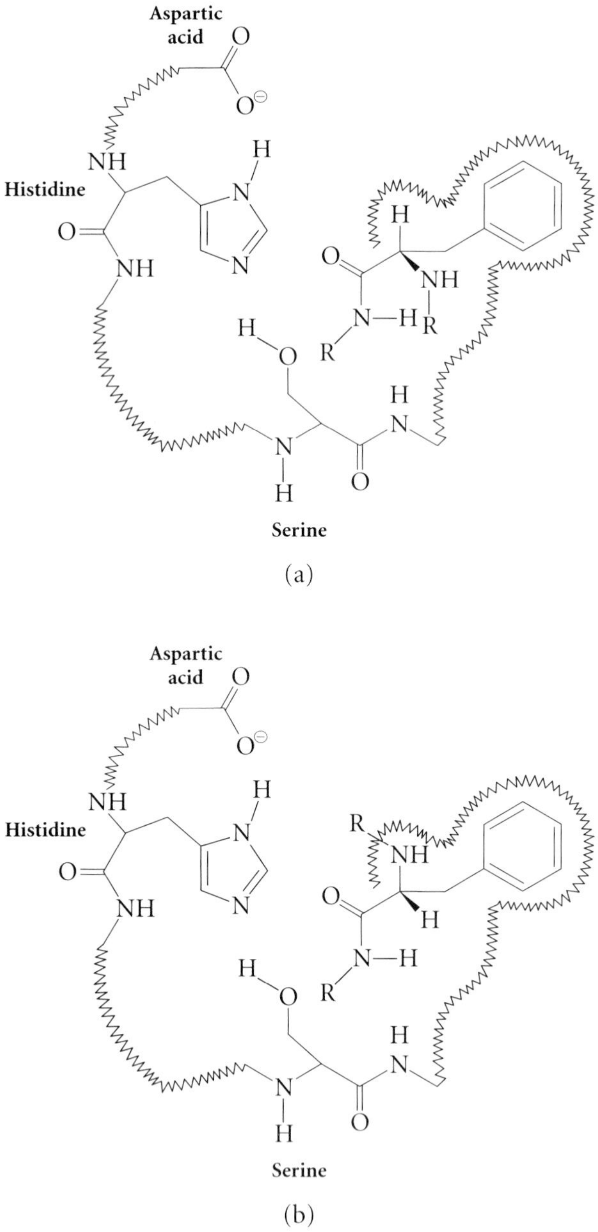

(a)

(b)

**FIGURE 20.11**

The "fit" of an amino acid residue into the chiral cavity at the active site of chymotrypsin is enantiospecific. The cavity matches well with (a) the L-enantiomer but not at all with (b) the D-enantiomer.

rality of phenylalanine: a hydrogen, a nitrogen, and a phenyl group. The enzyme is thus matched to this particular substrate, L-phenylalanine, with what is generally referred to as a *lock-and-key fit*. Just as only a key of just the right shape and size fits a given lock, only a limited number of quite similar substrates fit in the active site of a given enzyme and undergo catalyzed reaction.

**TABLE 20.1**

Relative Rates of Hydrolysis of Various Amides
with Chymotrypsin Catalysis

| Substrate | Relative Rate of Hydrolysis |
| --- | --- |
| | $3 \times 10^4$ |
| | 0 |
| | 0 |
| | 1 |
| | 7 |

An increase in the size of any of the substituents of a substrate means
that it can no longer fit into the active site, often preventing any reaction
from taking place. You can see from Table 20.1 that replacement of a hydrogen by a methyl group, as well as the interchange of two of the substituents (which produces the enantiomeric substrate), results in no reaction. In both cases, a relatively large substituent would have to fit in the
small pocket that normally accommodates hydrogen in order for the enzyme to accept the substrate (Figure 20.12, on page 20-38).

Furthermore, attractive interactions with an enzyme's amino acid
residues hold the substrate relatively tightly in the active site. Changing the
nature of the substrate results in a dramatic decrease in the rate acceleration induced by the enzyme. For example, without the phenyl group (that
is, with alanine rather than phenylalanine), the rate of the chymotrypsin-
catalyzed hydrolysis drops by more than four orders of magnitude, and
without the nitrogen, it drops by more than three. It is clear, then, that chymotrypsin is a selective enzyme that is an effective catalyst for the hydrolysis of only certain peptide bonds. This specificity is the result of the very

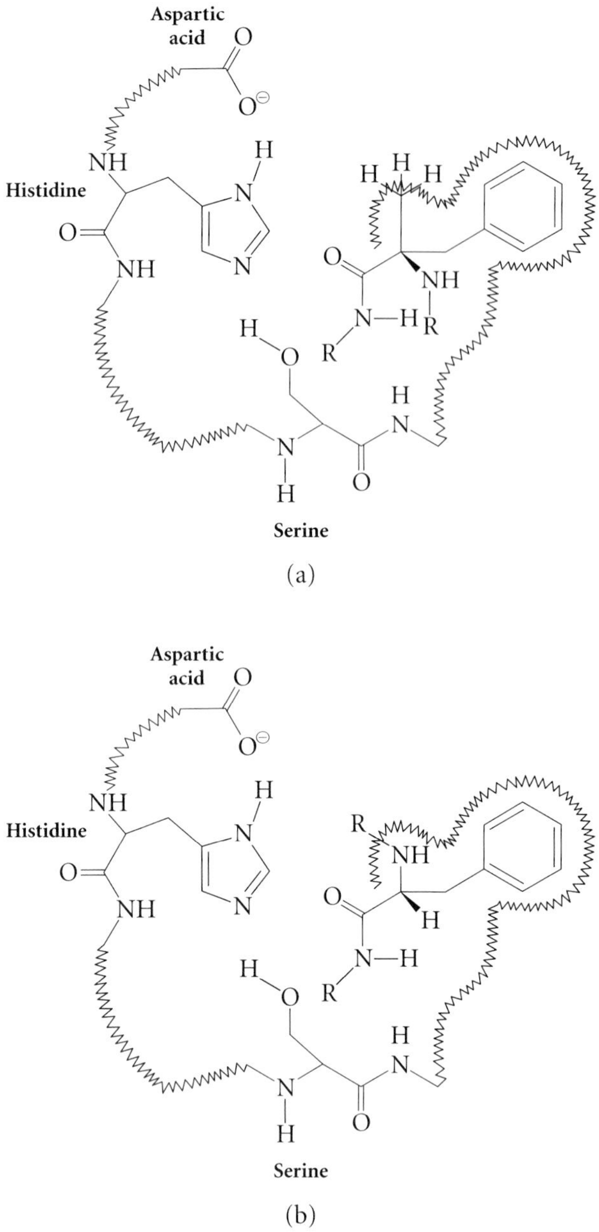

(a)

(b)

**FIGURE 20.12**

A bulky substituent at the $\alpha$ position of a polypeptide to be hydrolyzed interferes with binding to the enzyme. Lower binding equilibrium constants are observed with (a) an $\alpha$-alkylated substrate and (b) the "wrong" enantiomer.

well-defined structure of the active site. The existence of a very large number of different enzymes, each with its own set of specific substrates (and reactions), is possible only because of the structural complexity allowed by the large size of enzyme molecules.

Many enzymes are just as selective as chymotrypsin, but some others accept a wide range of substrates. For example, oxidizing enzymes in the liver are responsible for increasing the water solubility of unwanted materials by the addition of oxygen substituents. It is logical for these enzymes to be relatively nonselective, because their function is to remove an undefined variety of unwanted, and possibly even toxic, substances. Unfortunately, these liver oxidases also operate on certain external contaminants, with undesirable consequences. Benzene and polycyclic aromatic hydrocarbons such as benzo[a]pyrene are oxidized to polyepoxides that are sometimes further transformed into diols:

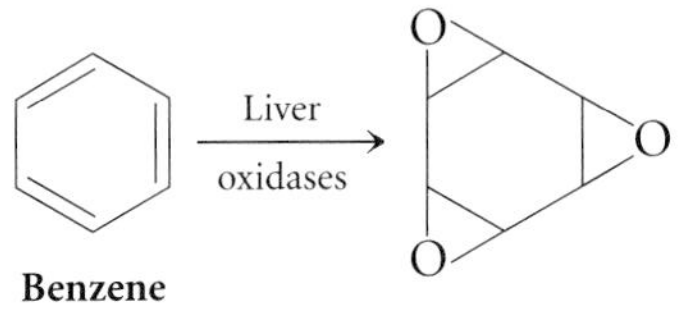

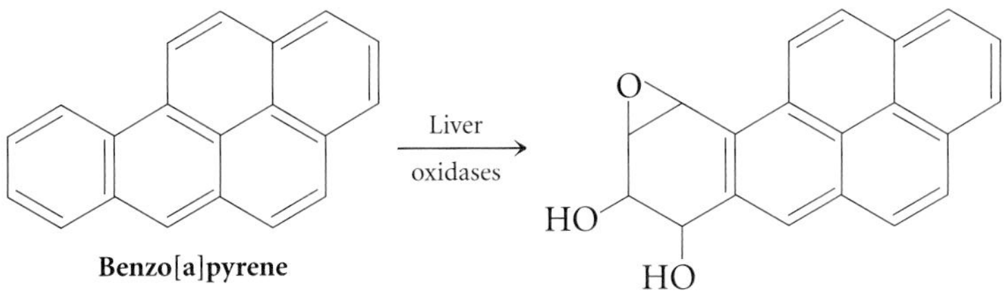

In general, these oxygen functional groups are quite reactive toward nucleophilic addition and substitution. It is believed that some of the product polyepoxides and glycols react with nucleotides on both strands of DNA, linking the double helix permanently together by strong chemical bonds. This association dramatically alters cell replication and can lead to the cell becoming cancerous. For example, exposure to benzo[a]pyrene leads to cancer, apparently after this polycyclic aromatic hydrocarbon is oxidized to the epoxydiol that chemically attaches to DNA. (Benzo[a]pyrene has been identified as one of the hydrocarbons in tobacco smoke.)

## Artificial Enzymes: Catalytic Antibodies

An exciting field of organic chemistry deals with the design and construction of artificial enzymes. These are molecules that have some of the features of natural enzymes and are able to catalyze organic reactions. The process of developing artificial enzymes is complicated, because few of the required design criteria are known. Therefore, it is necessary to fine-tune the structure in an almost trial-and-error fashion. Because such catalysts must have at least some of the structural complexity of enzymes, they are difficult to construct by planned sequences in the laboratory. Although some artificial enzymes have shown high levels of catalytic activity, the substrates often have been chosen to match the catalyst.

Researchers working at the interface between organic chemistry and molecular biology have devised a way of inducing animal and bacterial cells

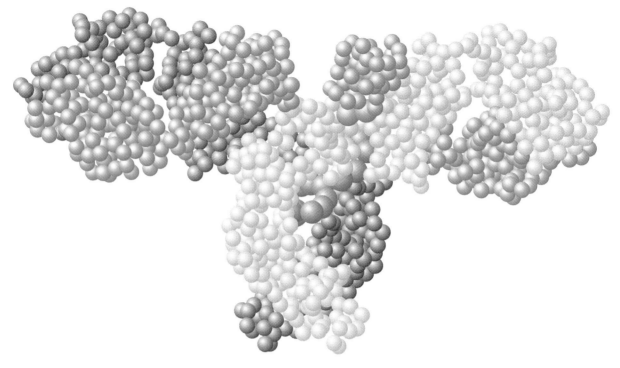

**FIGURE 20.13**

Large sections of all catalytic antibodies appear to be the same. The variable region found within the antigen-binding fragment is responsible for the chemical selectivity of the antibody.

to produce catalysts for selected reactions. This process takes advantage of the natural defense system based on **antibodies,** moderately sized peptide complexes that are responsible for alerting the immune system to the presence of foreign substances. Antibodies possess many different regions: some regions are the same for all antibodies produced by a given cell, and others, called **variable regions,** are produced with nearly unlimited diversity (Figure 20.13). When a foreign substance (called an *antigen*) by chance binds into a cavity of a variable region of an antibody, the immune system is activated to produce more of this specific antibody and to destroy the "captured" molecule.

Chemists have developed methods for stimulating cells to produce antibodies that bind a specific molecule. Complementary techniques have also been perfected for the isolation of these specific antibodies. When the molecules used to stimulate antibody production closely resemble the transition state of a particular reaction, antibodies that bind this transition-state mimic are isolated. In some cases, these antibodies also bind the substrate(s) for the reaction and, because the antibodies have been selected to match the transition-state mimic, binding to the true transition state is stronger than to the starting material. As we have seen, a species that interacts more strongly with the transition state than with the starting material is a catalyst for the reaction. These naturally produced catalysts are called, logically enough, **catalytic antibodies.** An example of a **transition-state analog** is

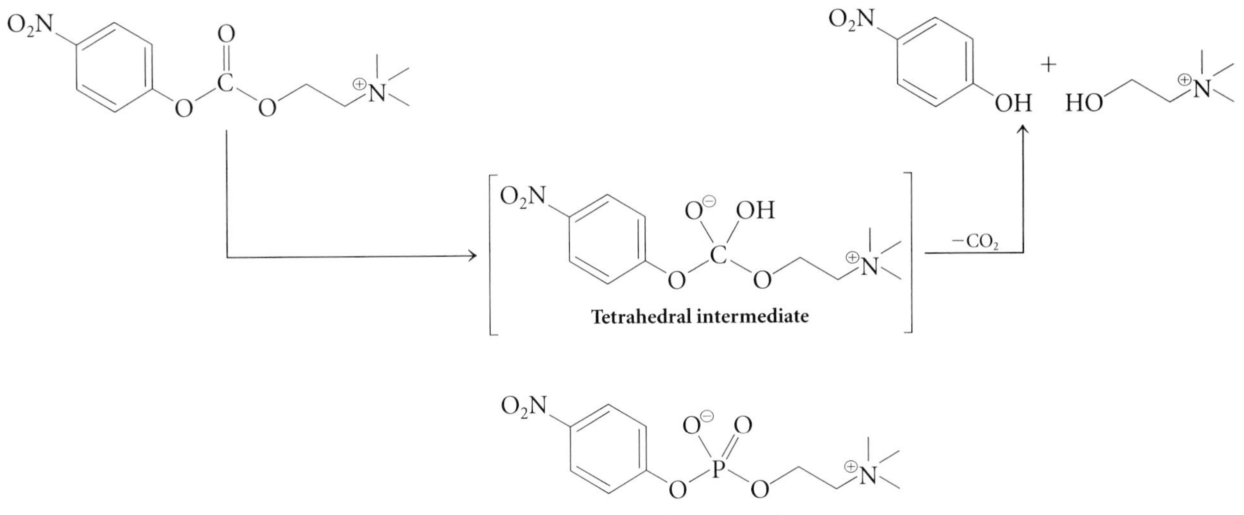

**FIGURE 20.14**

The transition-state analog shown here is a phosphate ester that is structurally analogous to the tetrahedral intermediate formed in the base-induced hydrolysis of the carbonate ester. Both the charge separation and the geometry are similar in the phosphate ester and the tetrahedral intermediate. In response to exposure to the phosphate ester, antibodies catalyze the hydrolysis reaction.

shown in Figure 20.14, along with the reaction that is catalyzed by the antibodies that bind to this mimic.

The phosphate ester exists at neutral pH mainly as the anion. It provides a charge distribution and three-dimensional shape that map closely to those of the tetrahedral intermediate in hydrolysis of the carbonate ester. Because the first step in the hydrolysis is endothermic, the transition state resembles the tetrahedral intermediate and thus the phosphate ester. Presumably, the antibody has amino acid residues in close proximity to the phosphate group that stabilize the negative charge through electrostatic interaction in much the same fashion as do enzymes with metal ions.

The development of a catalytic antibody for a particular reaction requires that an appropriate mimic for the transition state be available. Although this may not be possible for all reactions, catalytic antibodies capable of accelerating a large number of useful transformations are likely to be developed. One especially promising use for catalytic antibodies is the synthesis of single enantiomers of pharmaceutical agents. Only one enantiomer of a racemate generally has the desired effect, and, in some cases, administration of both enantiomers (as the racemate) can have disastrous consequences.

### EXERCISE 20.8

What is the energetic consequence if an antibody mimics the product of a reaction? Would such an antibody accelerate the rate of the reaction? Would it be a catalyst?

# Summary

**1.** The rates of chemical reactions are accelerated by catalysts—species that affect the activation energy of a reaction without themselves being consumed in the process.

**2.** Catalysis is accomplished by stabilizing the transition state of the rate-determining step in a chemical transformation, or by providing an alternative and lower-energy pathway.

**3.** When a greater separation of charge develops in the transition state of a given reaction than in the reactant, polar solvents or local environments (for example, inside a polar enzyme cavity) preferentially stabilize the transition state and enhance the rate of reaction. When a reaction takes place through a transition state having a lower degree of charge separation than in the reactant, nonpolar (hydrophobic) environments enhance reactivity.

**4.** A catalyst cannot become more tightly bound to either starting material or product than it is to the transition state of a given reaction. However, it can act through a charge relay mechanism.

**5.** Acid and base catalysts accelerate the rates of reactions by causing the reactants to bear substantial amounts of localized charge. This results in charge being more highly dispersed in the transition state than in the reactant. For example, an acid accelerates a nucleophilic acyl substitution by protonation of the carboxyl oxygen, producing a cationic reactant; a base accelerates the same transformations by producing a highly activated anionic nucleophile. These catalysts thereby minimize the separation of charge along the reaction pathway.

**6.** An intramolecular reaction usually takes place at a faster rate than does the corresponding intermolecular reaction, unless the intramolecular pathway includes an especially strained transition state. Thus, preassociation of a reagent with a substrate (forming a "unimolecular" reactant) usually accelerates a given reaction.

**7.** In phase-transfer catalysis, reversible association of a reagent with a suitable complex enables the reagent to move to a site allowing enhanced reactivity.

**8.** When a reagent (for example, an acid or a base) is fully regenerated after completing an accelerated cycle, it is said to act as a catalyst. When a reagent is consumed, the reaction is said to be induced by the reagent, to distinguish it from a truly catalytic sequence.

**9.** Interaction of the $\pi$ bond of alkenes with zero-valent transition-metal surfaces accounts for the utility of such metals as hydrogenation catalysts. Metals such as platinum provide a reactive surface for the catalyzed addition of hydrogen to alkenes, a process that brings together two species in an exothermic reaction that would not proceed without this catalytic association and activation.

**10.** The structural complexity of enzymes is responsible for both the catalytic activity and the substrate specificity necessary for the biochemical transformations critical to living systems.

**11.** Enzymes bind both a specific substrate and the required reagent within a common cavity (or active site), preorganizing these reactants and reducing the extent of desolvation required in order to approach the transition state. The equilibrium constant for binding of a given substrate to an enzyme is controlled by the size of the enzyme cavity and by the charge distribution within that active site.

**12.** Enzymatic binding is sufficiently specific to permit chiral recognition: the binding constant for one enantiomer is much higher than for the other. Both steric and electrostatic factors influence the arrangement of the reagent, substrate, and acid- or base-catalysts within the active site.

**13.** Chymotrypsin catalyzes hydrolysis reactions by covalent catalysis, a process in which an intermediate is reversibly covalently bonded to a residue present within the enzyme cavity. The preferential stabilization of a given transition state within the active site accounts for the enhanced reactivity induced by the enzyme.

**14.** Tight binding of a product molecule within the active site inhibits the catalyzed reaction. Weak, reversible binding of a product molecule is a mechanism for feedback inhibition.

**15.** Catalytic antibodies bind with molecules (mimics or antigens) that structurally and electrostatically resemble the transition state of the reaction being accelerated. When a living organism produces an immune response to the antigen, a catalytic antibody capable of binding preferentially to a transition state of a desired reaction is produced. Catalytic antibodies can thus provide selectivity, as well as rate acceleration, for some organic transformations.

## Review Problems

**20.1** Each of the following transformations requires the presence of either acid or base, as indicated above the reaction arrow. Indicate whether each reaction is acid- (or base-) induced or catalyzed. That is, must the acid or base be present in stoichiometric amounts, or will a catalytic amount suffice?

(a)

(b)

(c)

(d)

**20.2** A solvent can alter the activation energy and therefore change the rate of a reaction. How would you expect the rate of the following reaction to differ in a polar protic solvent such as water compared with a nonpolar aprotic solvent such as dimethyl ether? Explain your answer.

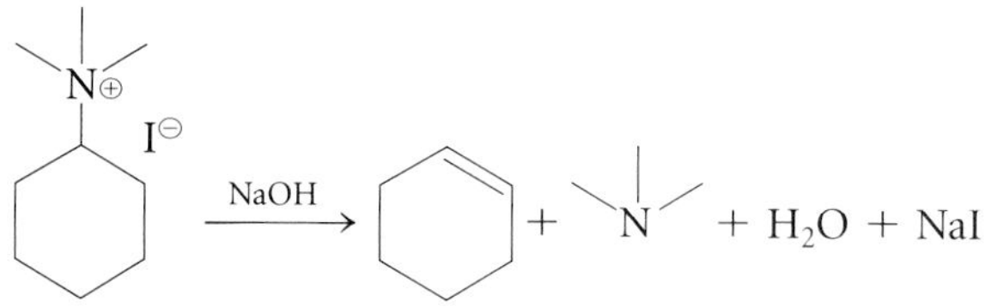

**20.3** Isomerization of alkenes is sometimes detected in catalytic reduction when $D_2$ is used instead of $H_2$. In such cases, deuterium is found in unexpected places, as shown in this example:

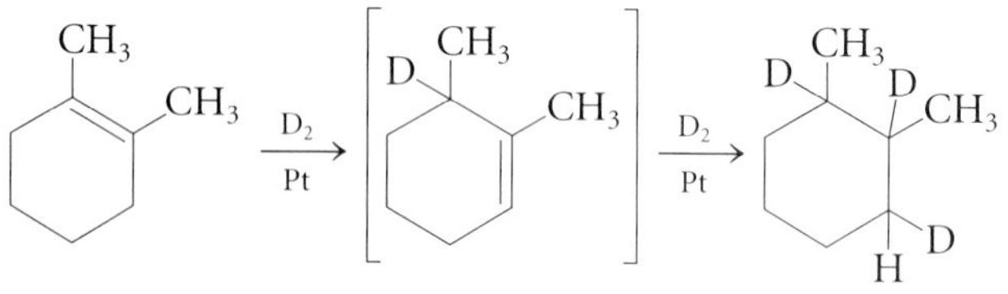

This anomalous behavior can be explained on the basis of the known observation that all steps in the catalytic hydrogenation of an alkene to an alkane are reversible. However, the sequence need not be carried through to completion before reversal to effect alkene isomerization. What is the first point in the mechanistic sequence detailed in Section 20.6 (pages 20-24 through 20-26) in which reversal to alkene could lead to isomerization? Write a mechanistic sequence for this isomerization leading to the formation of product A.

**20.4** Phospholipase C is an important zinc-containing enzyme that is responsible for disassembling cell-wall material by catalyzing the hydrolysis of phospholipids. In the reaction, a phosphate ester functionality is cleaved, resulting in the loss of the ionic head that is critical to the formation of the lipid bilayer structure of the cell:

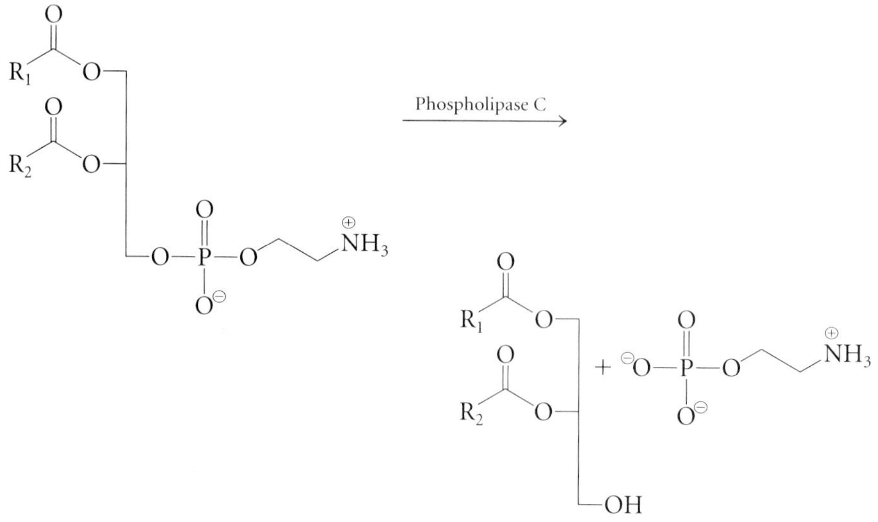

Write a simple mechanism for this hydrolysis that is parallel to that for nucleophilic acyl substitution, using hydroxide as a base. Then show how zinc ions, properly positioned, could stabilize the transition state leading to the intermediate in this substitution reaction.

**20.5** Polyunsaturated fatty acid triglycerides derived from plant sources have been highly recommended as dietary substitutes for fats derived from animal sources (for example, butter). However, these plant-derived fats are typically liquids at room temperature, and to make an acceptable substitute for butter (margarine), some of the double bonds are removed by catalytic reduction. In this process, some of the naturally occurring *cis*-alkenes are isomerized to *trans*-alkenes in which the double bond is not necessarily between the same two carbon atoms:

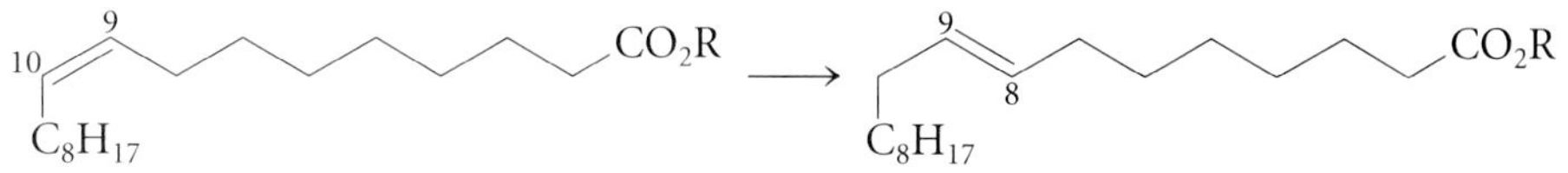

Examine the sequence involved in catalytic hydrogenation (pages 20-24 through 20-26), and determine at what point and how the isomerization occurs. (*Hint*: Recall from Problem 20.3 that the steps in catalytic hydrogenation are reversible.)

**20.6** The following reactions are useful for preparing both laboratory- and industrial-scale quantities of materials. In each case, the presence of an acid or a base is required. Determine whether the acid or base is involved as a catalyst or is required in stoichiometric quantities because it is consumed.

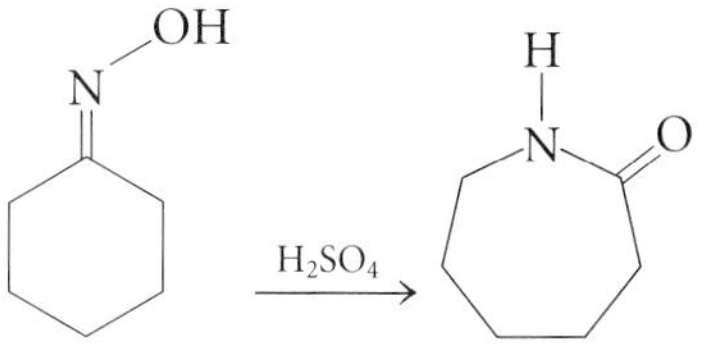

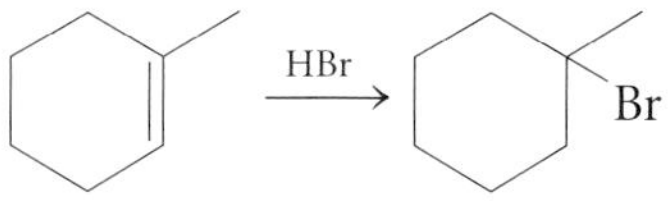

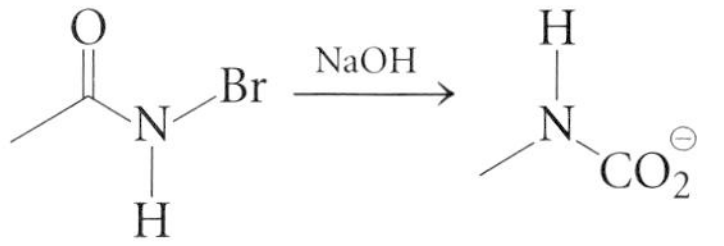

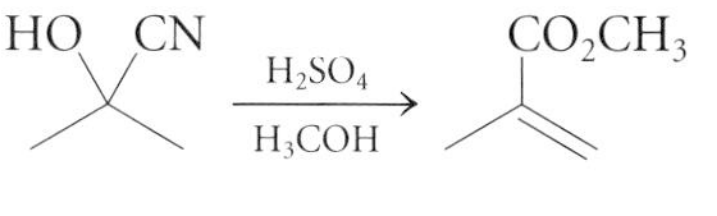

**20.7** The addition of water to an alkene to form an alcohol is catalyzed by acids such as $H_2SO_4$.

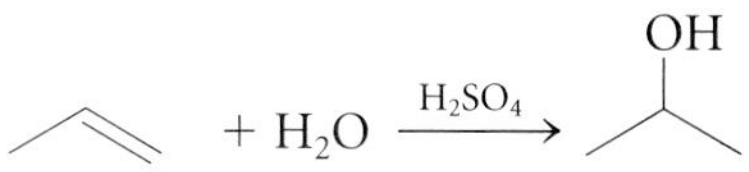

(a) Write a detailed mechanism for this reaction, being sure to include the structures of all intermediates.

(b) Construct an energy diagram for this process, paying particular attention to the step that is rate-limiting.

## Supplementary Problems

**20.8** The reaction of an enol ether with water in the presence of acid proceeds initially through steps that are essentially identical to those for the hydration of an alkene (see Problem 20.7). However, with an enol ether, further reaction ensues, ultimately producing a ketone.

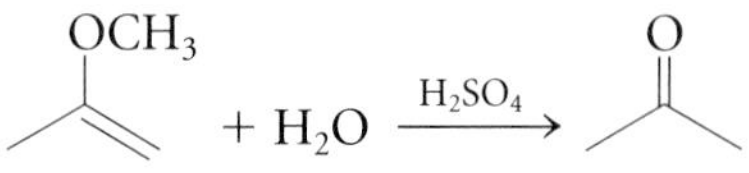

(a) Write a detailed reaction mechanism for the conversion of an enol ether (and water) to a ketone in the presence of acid. Is the acid catalytic or stoichiometric in this reaction?

(b) As in Problem 20.7, construct an energy diagram for this conversion, again being sure to place the transition state of the rate-limiting step higher than all others.

**20.9** The structures of phenolic ethers such as anisole are similar to those of enol ethers. Nonetheless, anisole reacts only very slowly with water in the presence of strong acid to form phenol. In the presence of very strong Lewis acids such as $BCl_3$, anisole is converted to phenol and methyl chloride:

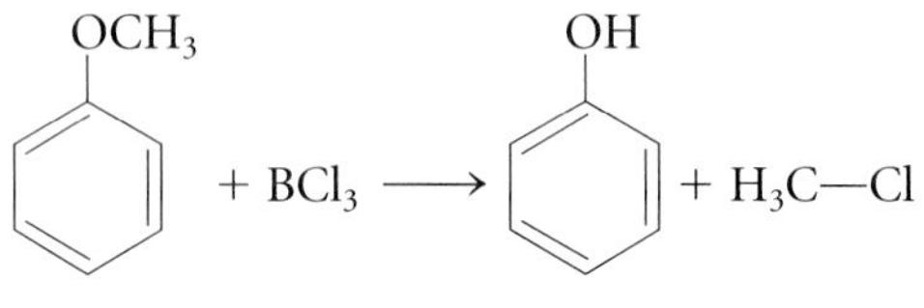

The mechanism for this reaction differs substantially from that for the hydrolysis of simple enol ethers (Problem 20.8). Devise a reaction mechanism for the conversion of anisole to phenol. Does $BCl_3$ function as a catalyst for this conversion? Why do simple enol ethers and phenolic ethers differ so markedly in reactivity toward acid and water?

**20.10** Two reactions form carbon–carbon bonds to aromatic rings by electrophilic aromatic substitution: Friedel–Crafts alkylation and Friedel–Crafts acylation. Both reactions require the presence of a strong Lewis acid such as $AlCl_3$, but they differ in one important feature: alkylation requires only a catalytic amount of the Lewis acid, whereas acylation requires a full equivalent. Explain this key difference between these apparently quite similar reactions.

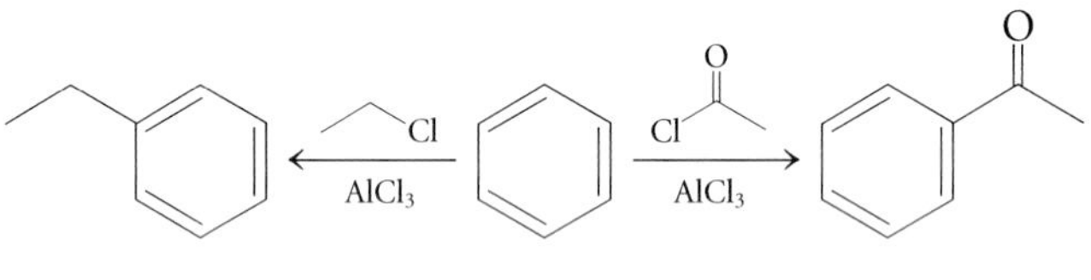

**20.11** The Diels–Alder reaction of an unsaturated ester with a diene is catalyzed by Lewis acids such as $AlCl_3$. Explain the role of the Lewis acid in increasing the rate of this Diels–Alder reaction:

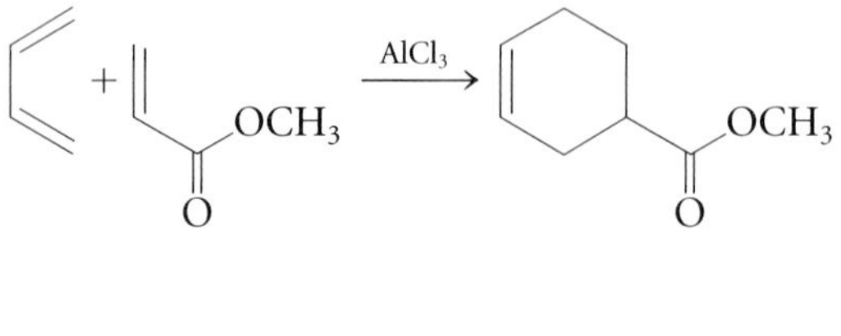

Why is the acid catalytic in this reaction, unlike the Friedel–Crafts acylation, which also involves interaction of a Lewis acid with one of the starting materials?

**20.12** The reaction of chlorobenzene with NaOH at elevated temperatures leads to a substitution reaction, producing phenol. Is this reaction base-catalyzed by hydroxide? Explain the reasoning behind your answer.

**20.13** The hydrolysis of a β-ketoester with NaOH in water to form a β-ketocarboxylate is not base-catalyzed. However, over an extended period of time, the conversion of a β-ketoester to a ketone requires only a catalytic amount of NaOH. Explain why there is a difference in the roles of NaOH in the first step (simple hydrolysis) and in the overall reaction (hydrolysis combined with decarboxylation).

**active site:** the relatively small portion of an enzyme where catalysis actually occurs (20-31)

**antibodies:** moderately sized peptide complexes that are responsible for alerting the immune system to the presence of foreign substances (20-40)

**catalytic antibody:** a protein expressed by the immune system of some organism in response to an injected transition-state analog of a desired reaction (20-40)

**charge-relay mechanism:** a reaction in which some molecule, often water, transfers a proton (and, therefore, charge) from one position to another in the same molecule or to a position in a different molecule when direct transfer is impossible because of the spatial orientation and the distance separating the two sites (20-15)

**covalent catalysis:** an accelerated chemical reaction in which the substrate becomes temporarily bonded through a covalent linkage to an active site on a catalyst (20-32)

**deinsertion reaction:** a reaction in which an atom, often a transition metal, covalently associated with two groups, is removed as the two groups become covalently bonded to each other (20-27)

**lock-and-key fit:** the highly specific, tight association between a substrate and the active site of an enzyme (20-36)

**phase-transfer catalysis:** the process by which a catalyst provides enhanced solubility in organic solvents to a reagent through reversible binding, providing for greatly increased concentrations of the reagent in a nonaqueous phase (20-9)

**protease:** an enzyme that catalyzes the hydrolysis of peptide bonds (20-32)

**substrate:** a molecule undergoing reaction under the influence of one or more external reagents (20-31)

**transition-state analog:** a stable species that closely mimics the geometry and charge distribution of the transition state of a reaction (20-40)

**variable region:** the portion of an antibody that is unique for a particular foreign invader and is responsible for antibody specificity (20-40)

# Cofactors for
# Biological Reactions

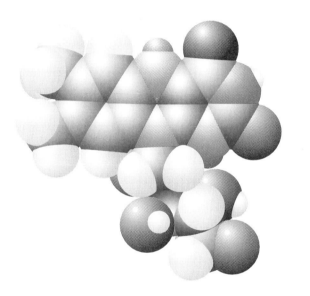

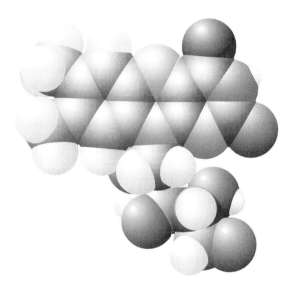

*The reduced (FADH₂, left) and oxidized (FAD, right) forms of the flavin cofactor differ in overall molecular shape and size by only two hydrogen atoms, shown in green.*

*I*n this chapter, we will examine why structurally complex reagents—or cofactors—are required for seemingly simple chemical reactions in living organisms. To function within a biological system, a cofactor must have the ability to recognize a target molecule within the complex mixture present in biological fluids. It must also be able to be recycled, so that it can be used again by the same organism. We will look at the rather complex structures and functions of specific reagents that take part in many biochemical transformations. For example, the complex redox reagents (pyridoxamine phosphate, nicotinamide adenine dinucleotide, and flavin adenine dinucleotide) act to open up a pathway between the oxidized and reduced forms of an organic compound. We will also see how thiol esters are formed in biological systems to activate carboxylic acids for nucleophilic acyl substitution.

Finally, we will examine mechanisms for decarboxylating $\alpha$-ketoacids (by interaction with thiamine pyrophosphate) and for transferring one carbon atom from a reagent to a growing carbon chain (by interaction with tetrahydrofolic acid). In considering the reactivity of thiamine pyrophosphate in the presence of lipoic acid, we will encounter in a biological context an example of *Umpolung*, or the reversal of normal chemical reactivity. The cofactor thiamine pyrophosphate functions by inducing in an $\alpha$-ketoacid the chemical reactivity more typical of a $\beta$-ketoacid. Analyzing the centers of reactivity of the various carbon atoms in a molecule provides a deeper understanding of the phenomenon of *Umpolung*, both in biological systems and in analogous reactions in the laboratory.

21.1

## Molecular Recognition

Biological reagents generally have higher molecular weights and more functional groups than do laboratory oxidation and reduction reagents. Why are the reagents in biological transformations so complex? Simply because they must function within the diverse mixture of materials present in living systems. Biological molecules must be able to pass through or be excluded by membranes at different times. Even after a reagent reaches the site where a reaction is to take place (inside or outside a given cell), it must interact only with a specific reaction partner. The reagent must be able to distinguish a target reactant from all other molecules, even those having similar functional groups.

An enzyme matches a substrate with a reagent (cofactor) by providing a defined environment that brings the substrate and the reagent together in a well-defined geometry (Figure 21.1). An enzyme has an irregular cavity into which the specific reagent and its target substrate can bind in a way that enables a chemical reaction to proceed rapidly. Only those reactions that are accelerated by the close association of a substrate with a reagent within the enzyme take place with high chemical efficiency. This process of molecular recognition is critical to the operation of complex living systems, which requires that oxidation of some species occur simultaneously with reduction of other species.

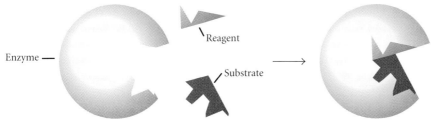

FIGURE 21.1

The shape of an enzyme cavity matches closely with the shapes of the substrate and any required biochemical reagent (cofactor).

## 21.2

## Recycling of Biological Reagents

Despite the structural complexity of each biological reagent, we can simply focus on the part of the molecule, "the working end," that effects the transformation. However, keep in mind that the rest of the molecule is critical for molecular recognition and for delivery of the reagent to the site where reaction is to take place.

In addition to the greater structural complexity of biological reagents, there is another significant distinction between laboratory reagents and their biochemical counterparts. Most reactions carried out in the laboratory go essentially to completion, whereas those in biological systems take place under near-equilibrium conditions. Many laboratory reagents (such as lithium aluminum hydride) are highly reactive, and the reactions in which they take part are quite exothermic and, thus, essentially irreversible. For the laboratory chemist, this exothermicity is highly desirable, because it ensures that the reaction can be carried to completion within a convenient period of time. The reagent itself is fully consumed and is converted into a form from which it is typically difficult to derive the original reagent. For example, the reduction of carbonyl groups with sodium borohydride ultimately yields the borate ion, and reforming the reagent (borohydride) from the product (borate) can be accomplished only with great difficulty. In the laboratory, this by-product is discarded rather than reused.

$$\text{O} \xrightarrow[\text{EtOH}]{\text{NaBH}_4} \xrightarrow[\text{H}_2\text{O}]{\text{H}_3\text{O}^\oplus} \text{H} \quad \text{OH} + \text{B(OH)}_3$$

It would be highly inefficient for living systems to construct reagents having the complexity required for molecular recognition and then use them only once. Instead, biological reagents are "recycled," often in a place within the cell or organism that is quite distant from the site where they are used. For recycling to be effective and efficient, the energy difference between the active and "spent" forms of the reagents must not be large. As a consequence, these reagents do not provide a large driving force for biochemical transformations. Often, these reactions have equilibrium constants near 1. Clearly, this would not be desirable in a laboratory, where the goal is to obtain the complete conversion of a starting material into a product. In biological systems, reactions are driven toward product by the constant influx of starting materials, produced by other reactions, and are pulled toward product by the further transformation of the product into yet another material. This driving of biochemical sequences is an application of Le Chatelier's principle. The dynamic nature of living systems is the key feature that distinguishes biological chemistry from chemistry that takes place in a closed reaction flask, where change stops either when the starting material has been consumed (for essentially irreversible processes) or when equilibrium is reached.

The following representative multistep sequence starts with compound A and ultimately arrives at compound D through intermediates B and C:

$$A \rightleftharpoons B \rightleftharpoons C \rightleftharpoons D$$

The thermodynamics of the conversion of A to D does not require that each step (A → B, B → C, and C → D) be exothermic; it is only necessary that the overall process be energetically favorable.

**EXERCISE 21.1**

From the energy diagram, determine which steps are endothermic and which are exothermic. Is the conversion of B into C faster or slower than the reformation of A from B? Is the conversion of C into D faster or slower than the reformation of B from C?

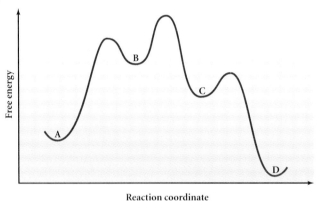

Reaction coordinate

## 21.3

# Cofactors: Chemical Reagents for Biological Transformations

Before we turn to the roles of several cofactors in biological systems, it is worthwhile to reiterate some important points. **Cofactors** are the biochemical equivalents of laboratory reagents, except that cofactors are always recycled. Cofactors are complex molecules that function in combination with enzymes. Enzymes act as catalysts to control the rate of biological transformations and, as such, are themselves unchanged as a result of a reaction. Cofactors, on the other hand, act as reagents to accomplish net chemical conversions and are transformed. This transformation creates a need for a second reaction to reform the original cofactor.

We look first at three common cofactors that function as biochemical reducing agents: **pyridoxamine phosphate, nicotinamide adenine dinucleotide (NADH),** and **flavin adenine dinucleotide (FADH₂).** Their structures are shown in Figure 21.2. In their reduced forms, these cofactors act formally as hydride transfer agents. Despite their much greater structural complexity, these cofactors behave chemically like the complex metal hydrides.

Each cofactor shown in Figure 21.2 is an active reducing agent derived from a simpler molecule, called its **precursor.** Pyridoxamine phosphate is a derivative of **pyridoxol (vitamin B₆).** NADH and NADPH are biosyn-

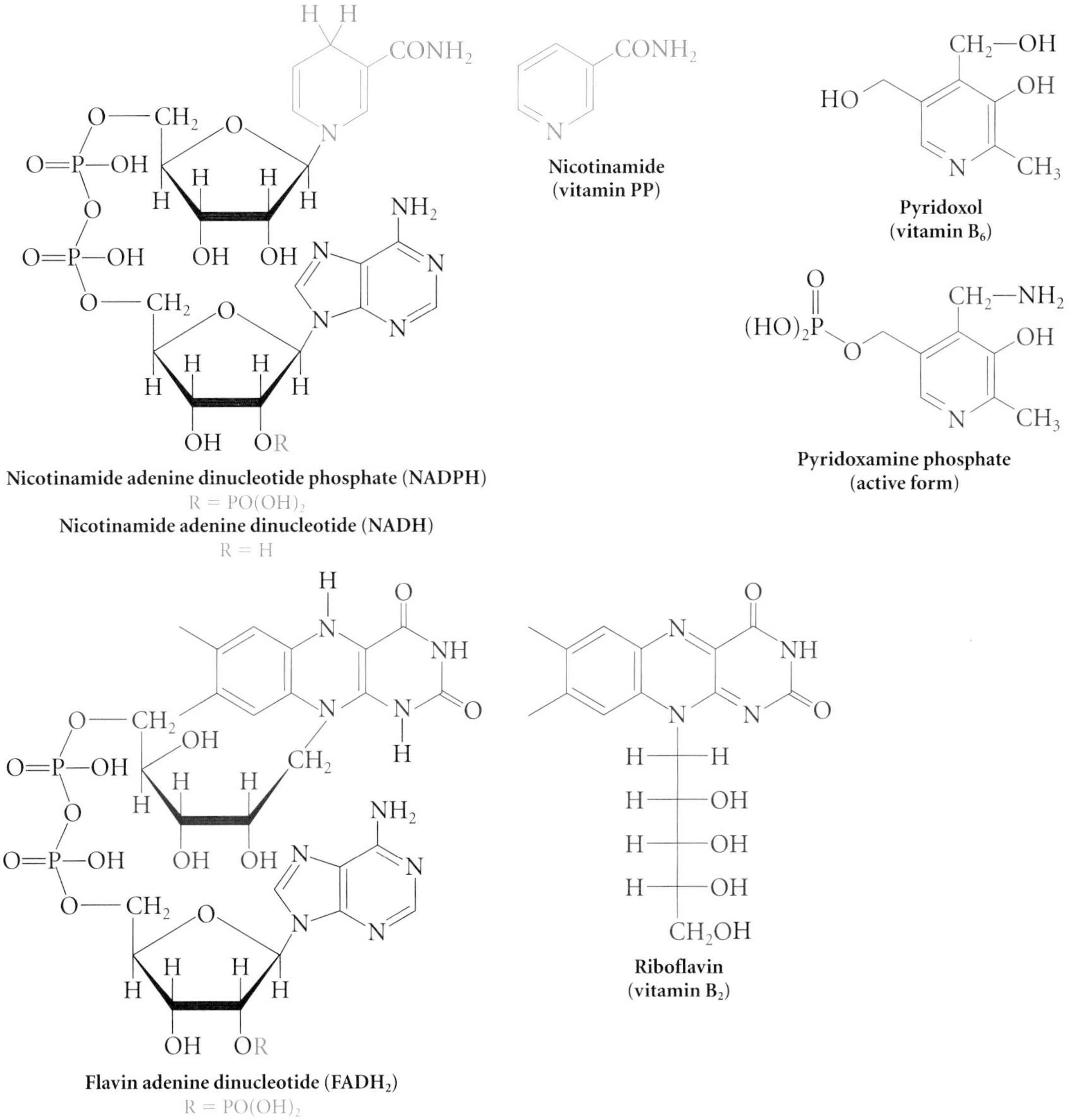

**FIGURE 21.2**

The cofactors NADH, FADH₂, and pyridoxamine phosphate are derivatives of the vitamins shown with them. These cofactors function as biological reducing agents.

thetically related to **nicotinamide** (**vitamin PP,** one of the B vitamins). FADH$_2$ is a derivative of **riboflavin** (**vitamin B$_2$**). Many precursor molecules, although essential to living systems, cannot be synthesized within an organism—they must be obtained from the diet. Molecules required to sustain life but that cannot be synthesized by an organism are referred to as **vitamins.** Because vitamins are recycled in most biochemical transformations, they are usually needed in only small amounts.

## 21.4

## Cofactors for Redox Reactions

### Pyridoxamine Phosphate

*Reductive Amination of α-Ketoacids as a Route to α-Amino Acids.* Amino acids, essential components of proteins and polypeptides, must be either ingested or synthesized in the body. Pyridoxamine phosphate, formed by the reductive amination of pyridoxal phosphate, serves as both a reducing agent and a source of nitrogen for the production of α-amino acids (Figure 21.3). The starting material for the α-amino acid is the corresponding α-ketoacid. This reaction requires not only the α-ketoacid and pyridoxamine phosphate, but also an enzyme that functions as a catalyst. The enzyme allows for molecular recognition of the cofactor and the substrate (in this case, pyridoxamine phosphate and the α-ketoacid) and provides a defined environment with a lower activation energy for the reaction.

**FIGURE 21.3**

Production of α-amino acids from α-ketoacids proceeds via a coupled reaction, in which an α-ketoacid is reductively aminated (to an α-amino acid) as pyridoxamine phosphate is oxidatively deaminated (to pyridoxal phosphate).

The key transformations in this reductive amination are shown in Figure 21.4:

**1.** Reaction of the amino group of pyridoxamine phosphate with the carbonyl group of the α-ketoacid to form a **ketimine** (an imine of a ketone)

**2.** Tautomerization of the ketimine to a quinoid intermediate

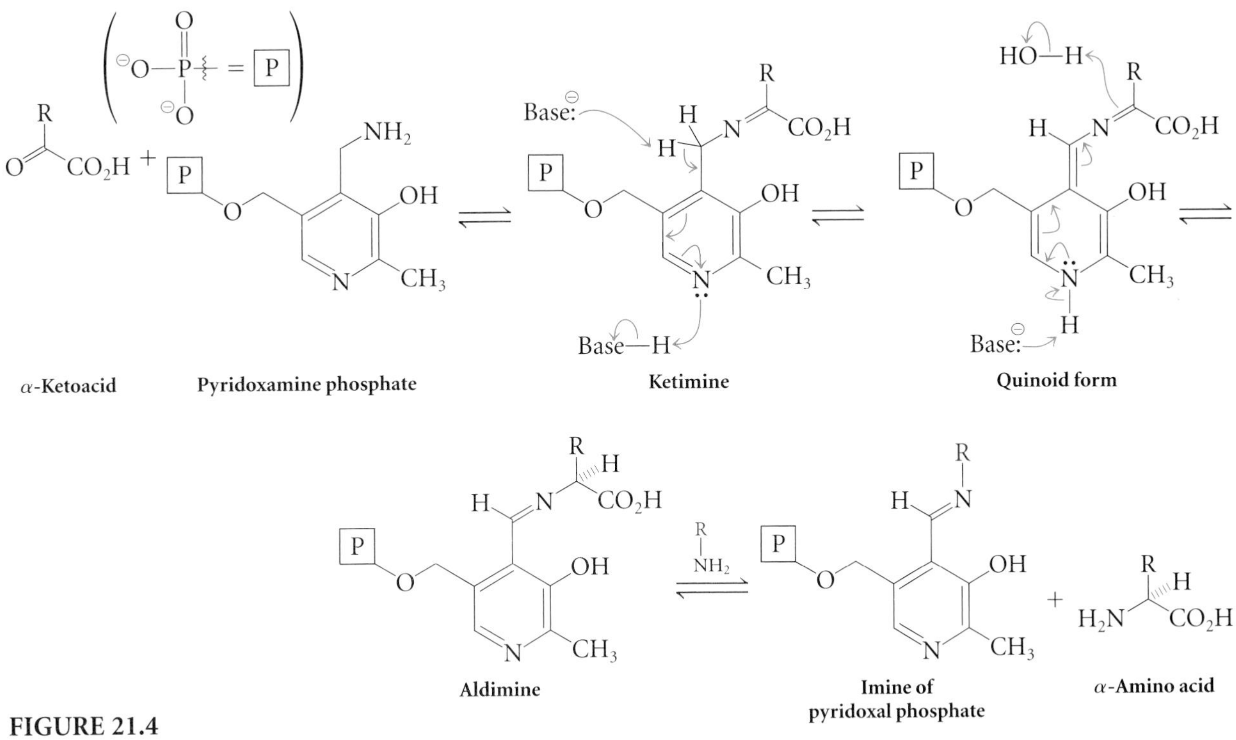

**FIGURE 21.4**

α-Ketoacid     Pyridoxamine phosphate     Ketimine     Quinoid form

Aldimine     Imine of pyridoxal phosphate     α-Amino acid

In reductive amination of an $\alpha$-ketoacid, the transformation of the $\alpha$-ketoacid into an $\alpha$-amino acid is coupled with the transformation of pyridoxamine phosphate into an imine of pyridoxal phosphate.

**3.** Tautomerization of the quinoid intermediate to an **aldimine** (an imine of an aldehyde)

**4.** Displacement of the $\alpha$-amino acid from the aldimine by the side-chain amino group of a lysine unit of the enzyme

Note that in this process pyridoxamine phosphate is transformed into the imine of pyridoxal phosphate, and that pyridoxamine phosphate is ultimately regenerated via coupling with an oxidative deamination reaction:

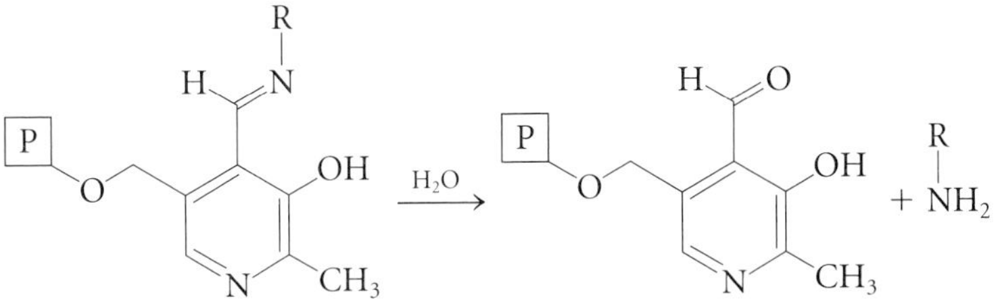

The first step of the reductive amination sequence begins with a nucleophilic attack of the amino group of pyridoxamine phosphate on the ketone group of the $\alpha$-ketoacid, producing a ketimine. The next two steps, the transformation of the ketimine into the quinoid form and the transformation of the latter into the aldimine, are simple tautomerizations. Deprotonation at the position adjacent to the imine nitrogen and reprotonation on the ring nitrogen at the other end of the conjugated $\pi$ system

21-7

produce a quinoid intermediate. The overall result of these proton transfers is a shift of a proton from the $sp^3$-hybridized carbon that is attached to the imine nitrogen in pyridoxamine phosphate to the imine carbon, converting the ketimine into an aldimine. The $sp^3$-hybridized carbon attached to nitrogen in the aldimine was the ketone carbon in the starting $\alpha$-ketoacid. The ketimine exists in equilibrium with the quinoid form and the aldimine. Because the deprotonation of the amine and the reprotonation on the imine carbon take place within a chiral pocket of an enzyme, the aldimine is produced exclusively as the L-enantiomer. Reaction of the aldimine with the amino group of a lysine residue present on the enzyme near the site of reaction releases the free $\alpha$-amino acid.

### EXERCISE 21.2

Recalling that imines bearing an $\alpha$ C—H bond can tautomerize to form enamines, consider the ketimine in Figure 21.4 and assume that R = $CH_3$. Write a complete mechanism for an imine–enamine tautomerization, showing the structures of all relevant intermediates. Explain why this tautomerization is less likely than the ketimine–aldimine tautomerization shown in the figure.

The ketimine–aldimine tautomerization in Figure 21.4 results in the loss of a proton from one carbon and its replacement on another carbon. This overall exchange can take place either in acid (by the addition of one proton, followed by the loss of the other) or in base (by the reverse sequence—the removal of one proton, followed by the addition of the other). Either sequence results in a charged intermediate with substantially higher energy than the neutral reactant—namely, a cation in acid or an anion in base. Such charged intermediates are avoided when the addition and removal of the protons occur simultaneously. This can be accomplished only if an acidic site is near the atom to be protonated, and a basic site is near the proton to be removed. The enzyme provides an environment in which acidic and basic sites are precisely positioned, so that the ring nitrogen of pyridine can be protonated at the same time as the C—H proton is removed. Like the imidazole ring of histidine, the pyridine nitrogen accomplishes the transfer of a proton to a more remote site. In addition, the aromatic ring of pyridine serves alternately as an electron sink and an electron source. The $\alpha$-amino acid is released from the aldimine in the last step by transamination. In this process, an imine is formed at the primary amino group of a lysine present in the enzyme catalyzing the overall transformation.

***Oxidative Deamination and the Amino Acid Pool.*** All of the steps illustrated in Figure 21.4 are readily reversible, because they are neither highly exothermic nor highly endothermic. The pyridoxamine phosphate consumed in the overall production of one amino acid is regenerated by the biodegradation of another amino acid to form a ketoacid by oxidative deamination (Figure 21.5). In this way, nitrogen is obtained from one or more amino acids brought into the living system in the diet. The unneeded $\alpha$-ketoacids that are produced are ultimately degraded to carbon dioxide, in a process that also releases energy. At the same time, required $\alpha$-amino acids are produced by reductive amination of the corresponding $\alpha$-ketoacids. Thus, the random mixture of $\alpha$-amino acids taken in by an organism is converted into the specific amino acid mixture needed for a given biological function.

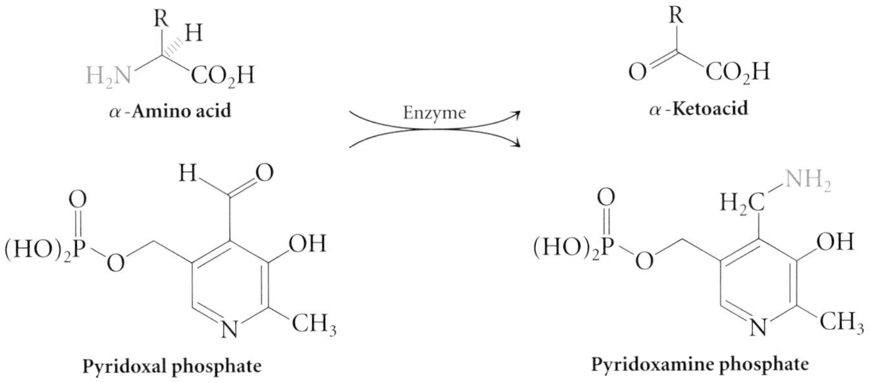

R H
$H_2N$ $CO_2H$
**α-Amino acid**

R
O $CO_2H$
**α-Ketoacid**

Enzyme

H O
$(HO)_2P$ O OH
N $CH_3$
**Pyridoxal phosphate**

NH$_2$
O $H_2C$
$(HO)_2P$ O OH
N $CH_3$
**Pyridoxamine phosphate**

**FIGURE 21.5**

In an oxidative deamination, a CH(NH$_2$) group is converted into a C=O group. Here, pyridoxal phosphate is reductively aminated as an α-amino acid is oxidatively deaminated. These reactions constitute the inverse of the reactions in Figure 21.4.

This process of degrading amino acids to provide a source of nitrogen and energy and then rebuilding the carbon framework, followed by reductive amination, may at first seem wasteful. However, the advantage is that the amount of each amino acid available to a living system is controlled by the rate of production of the required ketoacid and not by the ratio of amino acids ingested in the diet.

## EXERCISE 21.3

Write a complete mechanism for the last step of the reaction shown in Figure 21.4 in which the aldimine is converted into a different imine as the α-amino acid is released. (For simplicity, abbreviate the structures—for example, use RCH$_2$NH$_2$ to represent pyridoxamine phosphate.)

## NADPH

The process of photosynthesis forms carbohydrates and involves reduction of the carbon atom of carbon dioxide. Conversely, the process of respiration involves oxidation. Overall, photosynthesis is endothermic and thus requires the input of chemical potential energy. Some of this chemical potential is recovered in respiration.

$$CO_2 + H_2O + [H] \underset{\text{Respiration}}{\overset{\text{Photosynthesis}}{\rightleftharpoons}} C_6H_{12}O_6$$

The same analysis applies to the synthesis of fatty acids from acetic acid. Chemical potential energy is consumed in the synthesis, a portion of which is recovered when fatty acids are degraded to acetate.

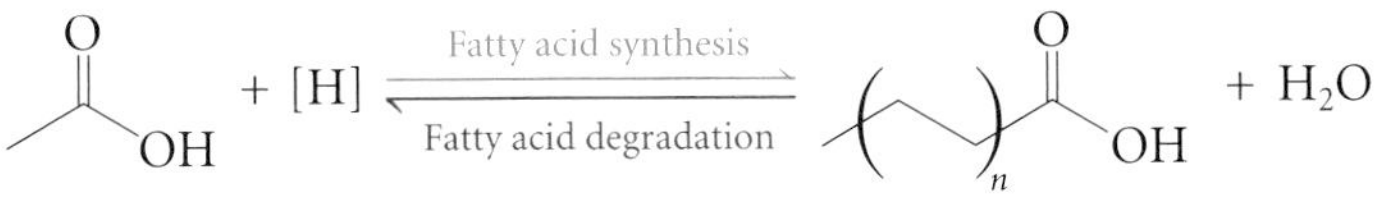

O
OH + [H] $\underset{\text{Fatty acid degradation}}{\overset{\text{Fatty acid synthesis}}{\rightleftharpoons}}$ O
OH$_n$ + H$_2$O

***Hydride Reduction of β-Ketoacids.*** In the reduction of a carbonyl group, NADPH functions very much like the laboratory reagents lithium aluminum hydride and sodium borohydride. All three compounds transfer to a carbonyl carbon a hydride equivalent—a proton plus the two electrons needed to form a carbon–hydrogen bond. In Figure 21.6 (on page 21-10), NADPH effects the reduction of a β-ketoester to a β-hydroxyester, one of

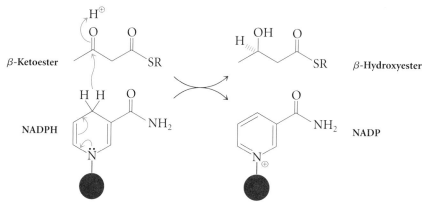

**FIGURE 21.6**

Nicotinamide adenine dinucleotide phosphate (NADPH) is a biological reducing agent that acts as a formal source of hydride. The part of the NADPH
molecule not directly participating in the reduction is represented here by a
red circle, to focus attention on the part of the molecule responsible for the
reduction. (The complete structure of NADPH is shown in Figure 21.2.)

several steps in the biosynthesis of straight-chain fatty acids. The presence
of an electron pair on the nitrogen of the dihydropyridine enhances the
electron density of the ring, making NADPH more active as a hydride transfer agent.

The formal loss of hydride ion ($H^{\ominus}$) leaves the cofactor NADP with a
positive charge. Although the oxidized form of NADP contains an aromatic
pyridine ring, it is quite electron-deficient, because the ring nitrogen is positively charged. There is little change in energy in this reaction, and the
NADP formed can (and does) serve as a hydride acceptor in other reactions
that result in the oxidation of another substrate. In this way, the reagent
pair NADPH and NADP can be recycled, with one acting as an oxidizing
agent and the other as a reducing agent.

### EXERCISE 21.4

NADP is a biologically important pyridinium ion. For the simple pyridinium ion
shown in the margin, draw alternative resonance structures that do not place positive charge on nitrogen. Consider the reaction of a pyridinium ion with a nucleophile to form a new bond. Which atoms of the pyridinium ring are most electrophilic and therefore most easily attacked by the nucleophile?

*Oxidation of β-Hydroxyacids.* It might seem reasonable that the oxidative degradation of fatty acids would take place via the exact reverse of
the reactions illustrated in Figure 21.6, but this is not the case. There are
two important differences. First, the oxidizing agent in fatty acid degradation is not NADP but a dephosphorylated analog known as NAD. Second,
the oxidation specifically involves the *R* configuration of the hydroxythiol
ester, whereas the biosynthesis produces the *S* configuration of the alcohol
(Figure 21.7). Because the absolute configuration at the center of chirality
differs, the forward and reverse reactions must take place at different enzymatic sites.

NADP

NADPH

O O
SR

NADH

NAD

HO H O
S SR

H OH O
R SR

Fatty acids

**Fatty Acid Degradation**

## FIGURE 21.7

A key step of fatty acid synthesis is the conversion of a $\beta$-ketoester into a
$\beta$-hydroxyester. Equally important in fatty acid degradation is the oxidation
of the hydroxyl group of a $\beta$-hydroxyester to the corresponding $\beta$-ketoester.
The former conversion, in the upper half of the cycle, is effected by NADPH.
The latter conversion, in the lower half of the cycle, is induced by NAD.
Each reaction is accelerated by a different enzyme.

## EXERCISE 21.5

NMR spectroscopy can be used to follow the course of biological transformations
by "feeding" to an animal samples of biochemical intermediates that have been enriched in carbon-13. As the material is consumed and transformed, changes in the
chemical shift of the carbon can be followed by NMR spectroscopy. How might
this technique be used to establish that the L-enantiomer of a $\beta$-hydroxyester is an
intermediate in the synthesis of fatty acids, whereas the D-enantiomer is part of the
degradation sequence? Which carbon in the $\beta$-hydroxyester should be replaced by
the isotope to obtain the greatest differences in shifts?

## FADH$_2$

Reductions by both FADH$_2$ (alkene to alkane) and NADPH (ketone to
alcohol) provide a means of "storing" energy, which is later released when
the fatty acids are consumed in the reverse, oxidation reactions.

***Mechanisms of FADH$_2$ Reduction.*** The reducing part of flavin adenine dinucleotide (FADH$_2$) is the flavin subunit shown in red in Figure 21.2
(here the remainder of the cofactor is represented as a red circle). This unit
serves as the source of hydrogen when the cofactor effects a biological reduction. In some reactions, FADH$_2$ functions as a source of hydride in a
manner analogous to that for NADPH. In other reactions, FADH$_2$ serves as
a source of electrons. When FADH$_2$ acts as an electron source, the reactions
resemble dissolving-metal reductions in which an alkali metal such as
lithium or sodium provides a source of electrons and hydrogen is supplied
in a second step in the form of one or more protons. The mechanism in-

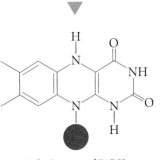

**Reducing part of FADH$_2$**

volving $FADH_2$ is sometimes called **single-electron transfer,** because the individual steps transfer one electron at a time. The reduction of a substrate by $FADH_2$ results in the oxidation of the cofactor to FAD. Just as occurs with the NADH–NAD pair, FAD later functions as an oxidizing agent, once again being reduced to $FADH_2$.

### Electron-Transfer Reduction of an α,β-Unsaturated Thiol Ester.

A biological reduction by $FADH_2$ is illustrated in Figure 21.8. An unsaturated thiol ester is reduced to a saturated one, a transformation that, like the reduction of β-ketoesters, is part of fatty acid biosynthesis. Overall, the mechanism of this reduction includes the transfer of two electrons and two protons to the unsaturated ester (Figure 21.9).

In the first step, an electron is transferred from $FADH_2$ to the unsaturated ester. This electron occupies the lowest-lying antibonding orbital, resulting in a radical anion, for which one resonance contributor is shown in Figure 21.9. (Recall that the $\pi$ bond of the neutral molecule is weakened upon forming a radical anion.) Protonation of this radical anion on the β carbon in the second step yields the most stable radical possible, one that is resonance-stabilized by conjugation with the carbonyl group (as in enolate anions derived from aldehydes, ketones, and esters). The third step of the reduction is the transfer of a second electron to the radical to produce the enolate anion of the thiol ester. Protonation of this enolate anion on carbon gives the saturated product, a thiol ester.

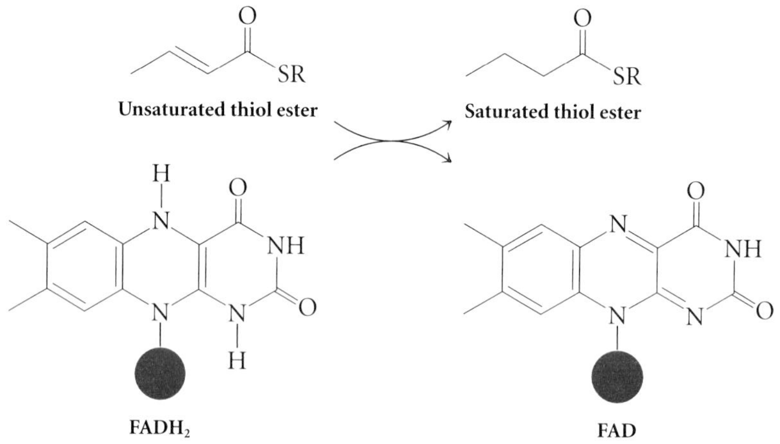

**FIGURE 21.8**

The reduced form of flavin adenine dinucleotide ($FADH_2$) is a reducing agent that effects the reduction of a carbon–carbon double bond in an α,β-unsaturated thiol ester, forming the saturated thiol ester.

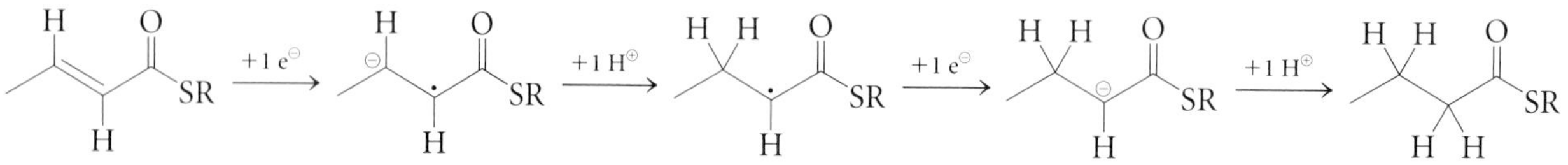

**FIGURE 21.9**

Two electron-transfer steps effect reduction of an α,β-unsaturated thiol ester by $FADH_2$.

## EXERCISE 21.6

There are many additional resonance contributors for the radical anion shown in Figure 21.9. These can be subdivided into four groups according to which atom (carbonyl oxygen, carbonyl carbon, $\alpha$ carbon, or $\beta$ carbon) bears the negative charge. Draw one resonance structure from each group. Which group is likely to contribute the most to the hybrid structure, based solely on the position of the negative charge?

## EXERCISE 21.7

For the reduction shown in Figure 21.9, write an alternative mechanism in which FADH$_2$ acts as a hydride donor in a Michael sense. How could this alternative mechanism be differentiated from that shown in Figure 21.9?

## 21.5

# Acetyl CoA: Cofactor for Acyl Transfer

### The Role of Thiol Esters

Thiol esters play an essential role in carboxylic acid chemistry in living systems. For example, thiol esters are the substrates for each of the biological redox reactions discussed in Section 21.4. In many circumstances, the thiol ester serves as a carboxylic acid derivative that is more reactive toward nucleophilic attack (for example, toward hydrolysis) than is the corresponding oxygen ester. The enhanced reactivity of thiol esters is the result of less extensive donation of sulfur's lone pair of electrons into the carbonyl $\pi$ system compared to that from oxygen.

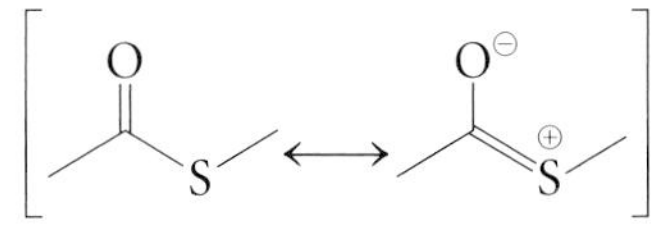

Comparing the two zwitterionic resonance structures at the right, we see positive charge on sulfur or oxygen and a $\pi$ bond between carbon and sulfur or between carbon and oxygen. Although the more electronegative oxygen is less stable than sulfur when positively charged, the delocalization of lone-pair electron density from sulfur into the carbonyl $\pi$ system requires overlap between a $3p$ orbital on sulfur and the $\pi$ system of the carbonyl group, which is composed of $2p$ orbitals. Because of the large difference in size between these orbitals, such overlap plays only a minor role in thiol esters. Addition of a nucleophile to the carbonyl carbon of an ester results in the breaking of the carbon–oxygen or carbon–sulfur $\pi$ bond and thus also prevents any stabilization resulting from delocalization. Because thiol esters are less stabilized by delocalization, they undergo addition of a nucleophile to the carbonyl carbon more rapidly than do carboxylic acid esters.

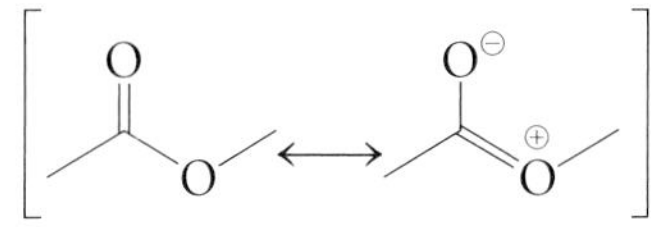

Thiol ester

Ester

### Activation of Carboxylic Acids (as Thiol Esters) toward Nucleophilic Attack

The most important thiol compound in living systems is **coenzyme A (CoA),** shown in Figure 21.10 (on page 21-14). As you might expect, coenzyme A often participates in reactions that require the activation of a carboxylic acid. The relatively complex structure of this cofactor contains some familiar pieces. For example, in NADPH, FADH$_2$, and coenzyme A, the active part of the molecule is attached through a diphosphate unit to an ade-

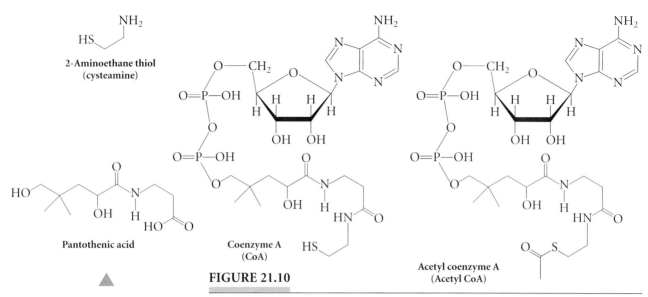

**FIGURE 21.10**

The cofactor coenzyme A is constructed of three pieces: adenosine diphosphate, pantothenic acid (vitamin B₅), and 2-aminoethane thiol. Acetylation of CoA on the terminal thiol group produces a thiol ester.

nine nucleotide. Critical to the function of coenzyme A is the presence of a thiol group, derived from the very simple molecule 2-aminoethane thiol (known as **cysteamine**), which is linked through an amide bond to the vitamin **pantothenic acid.**

### EXERCISE 21.8

Write mechanisms for hydrolysis of acetyl CoA under both acidic and basic conditions. Explain why the thiol ester rather than the amide is hydrolyzed under these conditions.

## 21.6

## Tetrahydrofolic Acid: A One-Carbon Transfer Cofactor for Methylation of Nucleic Acids

The synthesis of many biochemically important compounds involves steps that add one carbon atom, often as a methyl group. For example, as we will see in this section, tetrahydrofolic acid transfers one carbon atom from serine to deoxyuridylic acid, forming deoxythymidylic acid.

Tetrahydrofolic acid serves as a carbon carrier, transferring one carbon obtained from biodegradation to a biosynthetic intermediate.

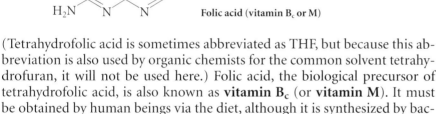

**Tetrahydrofolic acid**

**Folic acid (vitamin B_c or M)**

(Tetrahydrofolic acid is sometimes abbreviated as THF, but because this abbreviation is also used by organic chemists for the common solvent tetrahydrofuran, it will not be used here.) Folic acid, the biological precursor of tetrahydrofolic acid, is also known as **vitamin $B_c$** (or **vitamin M**). It must be obtained by human beings via the diet, although it is synthesized by bacteria.

### EXERCISE 21.9

(a) Identify the atoms that are reduced when tetrahydrofolic acid is formed from folic acid. How many equivalents of hydrogen would be required if this process were carried out by catalytic hydrogenation?

(b) Note that there are other differences between the structures of folic acid and tetrahydrofolic acid, besides the numbers of hydrogens. What special feature(s) of each of these structures provides unique stabilization?

The carbon transferred by tetrahydrofolic acid is derived from serine by transfer of a formaldehyde equivalent, a process that produces glycine. This carbon is then available for transfer to another molecule, for example, to **deoxyuridylic acid,** forming **deoxythymidylic acid,** one of the monomer units of DNA (Figure 21.11, on page 21-16).

The methylene carbon of methylene-tetrahydrofolic acid is functionally similar to that in an acetal, in that it is attached to two heteroatoms. Indeed, this functionality, known as an **aminal,** undergoes reactions quite similar to those of an acetal. An aminal is formed by the reaction of two equivalents of an amine (or, as in this case, a diamine) with an aldehyde. Aminals react rapidly with water under acidic conditions to regenerate the original carbonyl compound (Figure 21.12, on page 21-16).

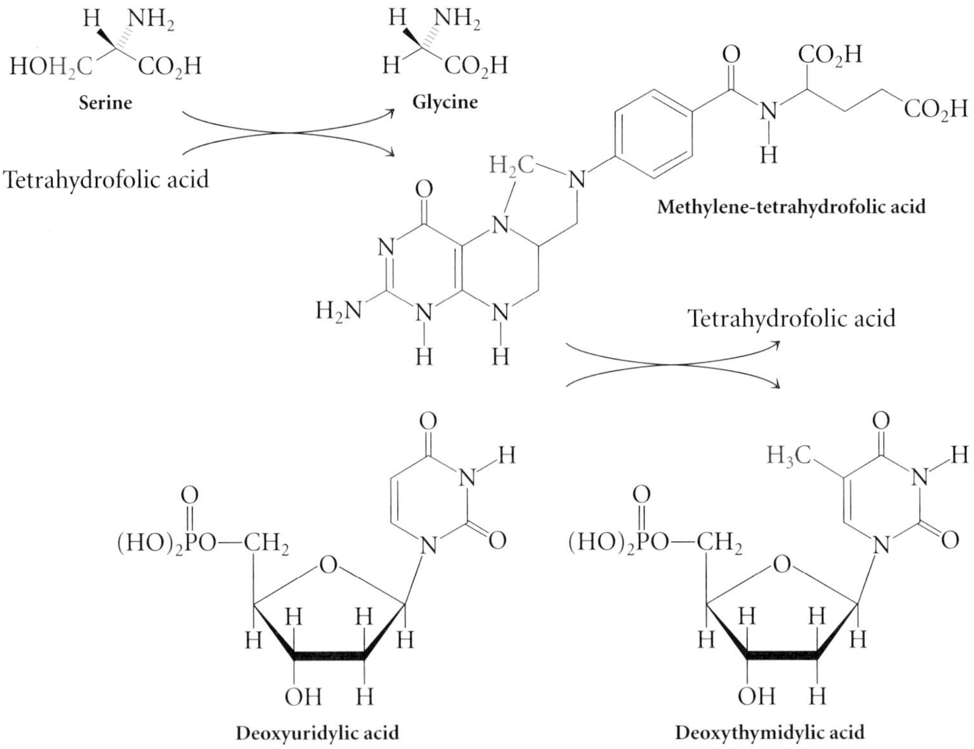

**FIGURE 21.11**

Serine transfers a single carbon atom as a methylene group to tetrahydrofolic acid. Methylene-tetrahydrofolic acid then relays the methylene unit to deoxyuridylic acid, producing deoxythymidylic acid, which is incorporated directly into DNA.

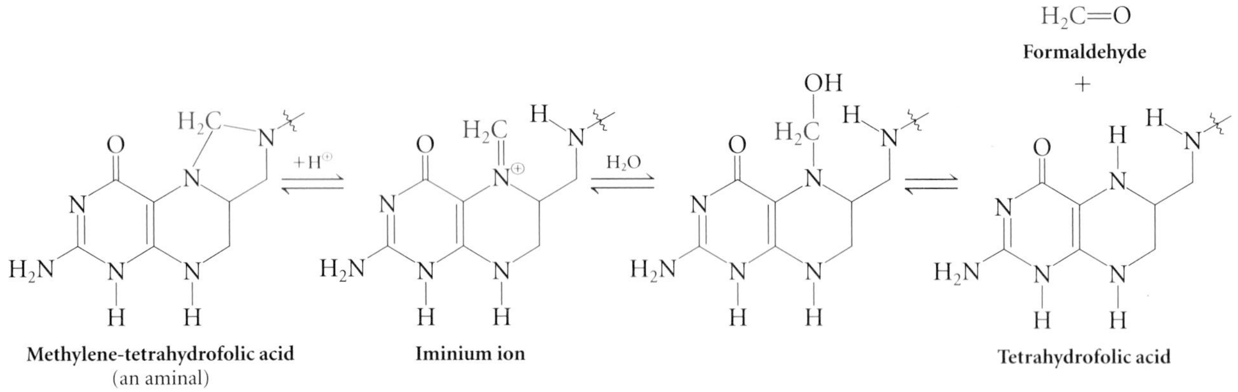

**FIGURE 21.12**

In the equilibrium between methylene-tetrahydrofolic acid, tetrahydrofolic acid, and formaldehyde, the aminal exists in equilibrium with the iminium ion. Hydrolysis of the iminium ion releases formaldehyde, forming tetrahydrofolic acid.

**EXERCISE 21.10**

Write a mechanism for the formation of an aminal from formaldehyde and ethylenediamine, $H_2N-CH_2CH_2-NH_2$.

Write a clear, step-by-step mechanism for the sequence shown in Figure 21.12. Which will react more rapidly with water—aminals or acetals? Explain your answer.

## Transfer of a Methylene Fragment

How might one carbon be transferred from serine to tetrahydrofolic acid to form methylene-tetrahydrofolic acid? (The actual details of the biological sequence are not yet known.) Overall, two bonds to the hydroxymethylene carbon in serine must be replaced by two new bonds to nitrogen atoms in tetrahydrofolic acid. This sequence is outlined in Figure 21.13):

1. Dehydration of serine

2. Nucleophilic addition of folic acid to the $\alpha,\beta$-unsaturated serine derivative

3. Elimination of $\alpha$-amino acid anion equivalent

4. Intramolecular nucleophilic substitution by nitrogen to form a cyclic aminal

Dehydration of serine is facilitated because the hydroxyl group is $\beta$ to a carbonyl group (recall the aldol condensation), and the reaction produces an $\alpha,\beta$-unsaturated acid. Nucleophilic attack on this unsaturated acid by the cyclic amino group of tetrahydrofolic acid forms a carbon–nitrogen bond. The carbon–carbon bond can then be cleaved. The lone pair of electrons on nitrogen assists in the formal loss of a carbonyl-stabilized $\alpha$-amino acid anion, producing an iminium ion. Nucleophilic attack by the second amino group of tetrahydrofolic acid on the iminium carbon forms the cyclic aminal, completing the sequence. A comparison of this product with the starting material (tetrahydrofolic acid) reveals that a methylene unit has been transferred to folic acid.

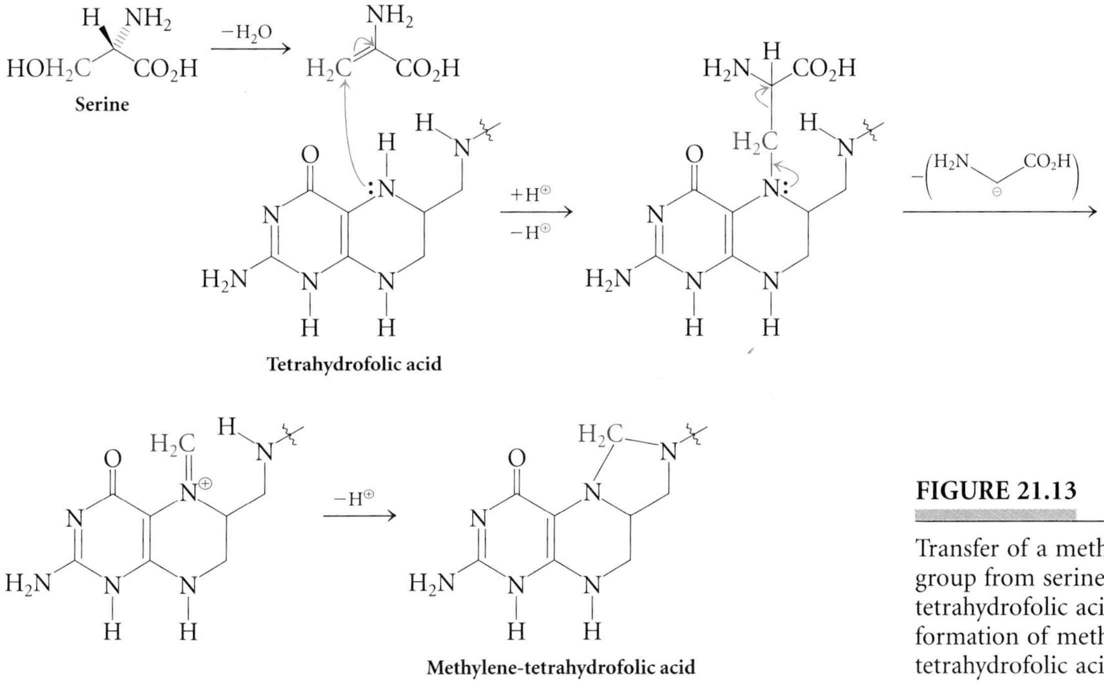

**FIGURE 21.13**

Transfer of a methylene group from serine to tetrahydrofolic acid and formation of methylene-tetrahydrofolic acid.

## Reductive Methylation of Deoxyuridylic Acid to Form Deoxythymidylic Acid

The sequence of steps in Figure 21.13 can readily be reversed, resulting in the transfer of the methylene carbon to another site. With deoxyuridylic acid, nucleophilic addition to the intermediate iminium ion introduces the carbon–carbon bond necessary to form deoxythymidylic acid (Figure 21.14). After deprotonation, the tetrahydrofolic acid residue is lost. Formal addition of a hydride equivalent (for example, from NADH) completes the carbon transfer and the formation of deoxythymidylic acid.

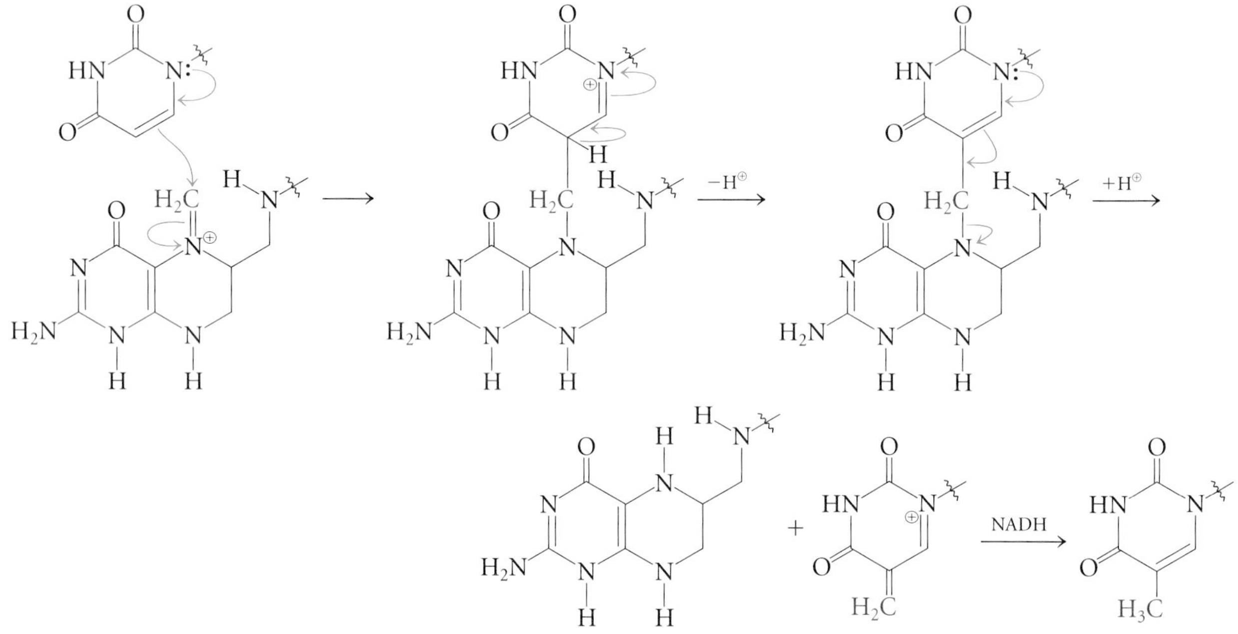

**FIGURE 21.14**

Reductive methylation of deoxyuridylic acid to deoxythymidylic acid (partial structure). Intermolecular transfer of the methylene group is followed by reduction with NADH.

### EXERCISE 21.12

A route for forming methylene-tetrahydrofolic acid is shown in Figure 21.13. An alternative route for forming the same molecule requires an $S_N2$ displacement of $H_2N$—$^{\ominus}CH$—$CO_2H$ by the second nitrogen atom:

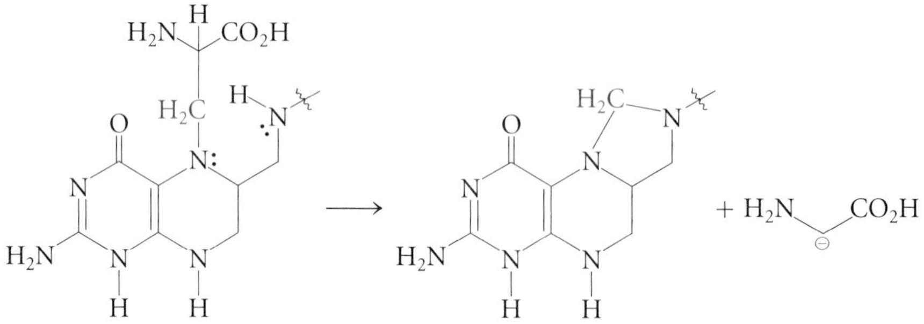

Write a detailed reaction mechanism for this $S_N2$ pathway. Which factors favor the route shown in Figure 21.13? Which favor this pathway?

21-19

21.7 Thiamine Pyrophosphate
and Lipoic Acid: Cofactors
for the Decarboxylation
of α-Ketoacids

## 21.7

# Thiamine Pyrophosphate and Lipoic Acid: Cofactors for the Decarboxylation of α-Ketoacids

Oxidation deamination of amino acids produces α-ketoacids as intermediates, Futher degradation of these α-ketoacids involves oxidative decarboxylation to the corresponding acids with one fewer carbon atom.

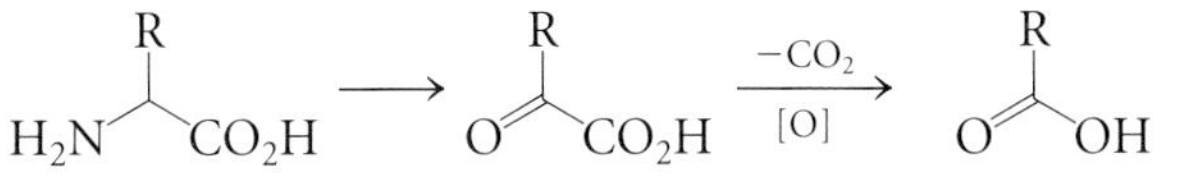

### Mechanisms of Decarboxylation

Oxidative decarboxylation of α-ketoacids is reminiscent of the decarboxylation of β-ketoacids. However, there is an important difference in the stability of the products. The difference is due to the fact that the two carbonyl groups are separated by a carbon atom in β-ketoacids, whereas the two carbonyl groups are adjacent in α-ketoacids.

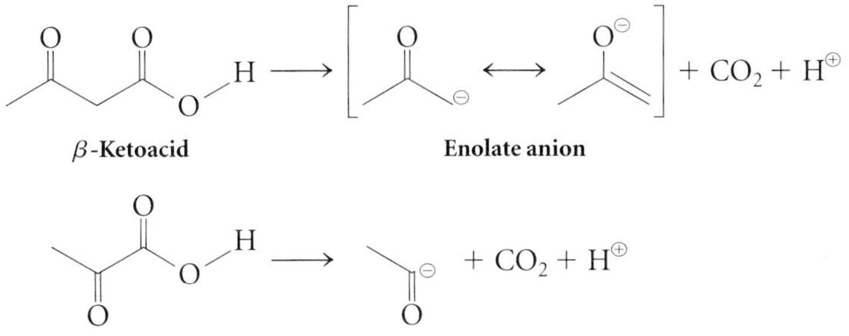

The loss of carbon dioxide from the salt of a β-ketoacid results in the formation of a resonance-stabilized enolate anion. However, decarboxylation of an α-ketoacid by a parallel route leads to a very unstable acyl anion. Because the resulting anion would be unstable, α-ketoacids do not readily undergo decarboxylation in biological systems. Instead, decarboxylation of α-ketoacids requires that the ketoacids first be modified so that the anion resulting from the loss of carbon dioxide is stabilized.

### EXERCISE 21.13

Compare and contrast the structures and stabilities of an acyl anion and an enolate anion, explaining why the latter is more stable.

### Structure and Function of the Cofactors Thiamine Pyrophosphate and Lipoic Acid

To understand how the ketoacid decarboxylation is accomplished biochemically, let's look at the structures of the cofactors involved—thiamine pyrophosphate and lipoic acid.

**Thiamine pyrophosphate** is derived from **vitamin B$_1$** (also known as **thiamine**) by phosphorylation of the hydroxyl group:

For convenience, the pyrophosphate unit is represented here in the acid form, but keep in mind that, in water near pH 7, the fully deprotonated form dominates.

The chemical reactivity of thiamine derives from the relatively high acidity of the single hydrogen on the thiazole ring (p$K_a$ 17.6) as compared with other hydrogens bonded to carbon. Although anions derived from the deprotonation of $sp^2$-hybridized carbons are uncommon, this example is a special case in which stability is imparted to the vinyl carbanion by the adjacent positively charged nitrogen:

**Thiamine**

In contrast to the complex cofactor structures presented so far, **lipoic acid** has a simple structure. Lipoic acid serves as an oxidizing agent and in the process is reduced to an acyclic dithiol:

**Lipoic acid**

### Decarboxylation of α-Ketoacids in Biological Systems

Thiamine and lipoic acid, in combination, effect the oxidative decarboxylation of an α-ketoacid, a process very difficult to accomplish directly. The sequence of reactions shown in Figure 21.15 brings about decarboxylation and regenerates thiamine through the following steps:

**1.** Nucleophilic attack of thiamine on the α-ketoacid to form A, an adduct having a C=N bond β to the carboxylic acid function

**2.** Decarboxylation of intermediate A to yield an activated enol, B

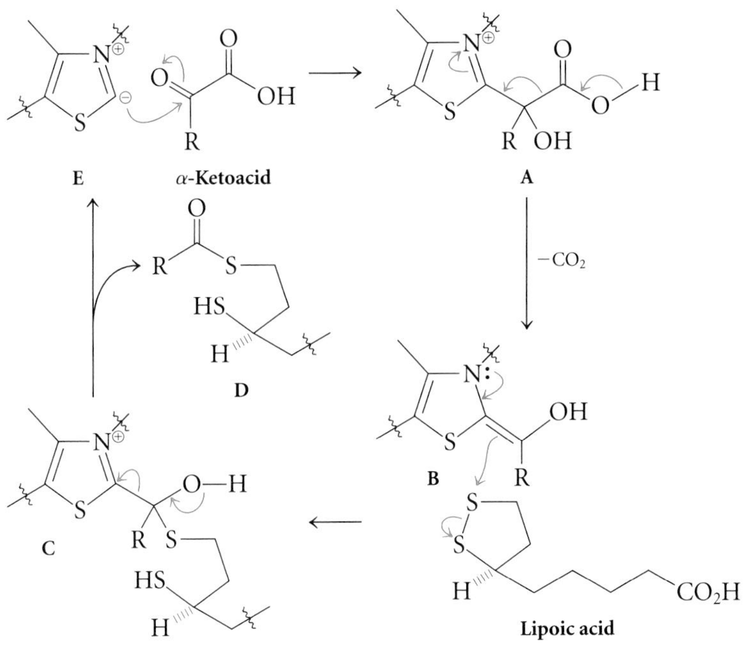

**21-21**

21.7 Thiamine Pyrophosphate
and Lipoic Acid: Cofactors
for the Decarboxylation
of $\alpha$-Ketoacids

**FIGURE 21.15**

Decarboxylation of an $\alpha$-ketoacid by thiamine and lipoic acid.

**3.** Nucleophilic attack by enol B on lipoic acid to cleave the disulfide bond, forming tetrahedral intermediate C

**4.** Loss by C of a thiol ester, D, to regenerate thiamine

The reaction sequence begins with deprotonation of thiamine pyrophosphate to form an ylide, E. Nucleophilic attack by the ylide carbon atom on the carbonyl group of the $\alpha$-ketoacid forms adduct A. The $\alpha$-ketone group is converted into an alcohol, and a carbon–heteroatom $\pi$ bond is formed at a position $\beta$ to the carboxylic acid carbon. The arrangement of functional groups is thus analogous to that of a $\beta$-ketoacid.

Decarboxylation of adduct A is now easy. The thiamine nitrogen is already positively charged in A, and loss of carbon dioxide and a proton results in a neutral intermediate, B. Thus, although the presence of the iminium structure acts as an electron sink in the same way as the ketone group in a $\beta$-ketoacid, this process differs from the decarboxylation of a $\beta$-ketoacid, which consists of the intramolecular transfer of the proton from the carboxylic acid to the ketone oxygen through a cyclic transition state.

The carbon residue of the starting $\alpha$-ketoacid is covalently bonded to thiamine pyrophosphate in intermediate B. The remaining steps are necessary to disconnect these two pieces, forming the decarboxylated acid and regenerating thiamine pyrophosphate. In the next step, intermediate B acts as a nucleophile, cleaving the disulfide bond in the oxidized form of lipoic acid and simultaneously forming a new carbon–sulfur bond. This reaction employs the electron pair of the ring nitrogen (an enamine) to displace a thiol from the disulfide by an $S_N2$ reaction. Comparing the resulting intermediate C with adduct A, we see that one of the sulfur atoms of lipoic acid has replaced the carboxyl group of the original $\alpha$-ketoacid. Regeneration of the carbonyl group and thiamine pyrophosphate produces compound D, and hydrolysis of D reforms lipoic acid and the decarboxylated acid.

Show how the decomposition of intermediate C in Figure 21.15 is similar to the nucleophilic substitution of a simple thiol ester.

## 21.8

# Mimicking Biological Activation with Reverse-Polarity Reagents

By decarboxylating an $\alpha$-ketoacid, thiamine induces chemistry normally observed only with $\beta$-ketoacids. This cofactor exemplifies the use of reagents to invert normal chemical reactivity, a methodology that is very important for the design of new and unusual chemical reactions. One reverse-polarity (*Umpolung*) reagent, the formyl anion, permits reversal of the typically characteristic electrophilic reactivity of a carbonyl carbon. In like fashion, thiamine can be viewed as a nucleophile at one stage in the oxidative decarboxylation of an $\alpha$-ketoacid and as a leaving group at a later stage. These two functions have opposite electronic requirements: nucleophiles readily donate electron density to form a new bond, and leaving groups readily accept the electron density of a bond undergoing cleavage.

### Analysis of Carbon Reactivity

Carbonyl carbons typically react as electrophiles; that is, because they are electron-deficient, they typically react with nucleophiles. Conversely, the $\alpha$ carbon can be readily converted into an enolate anion, which is a reactive nucleophile. This reactivity is represented in a pictorial fashion by a plus or minus enclosed within parentheses, $(+)$ or $(-)$, to distinguish them from formal charges. Pluses (for electrophilic sites that react with nucleophiles) or minuses (for nucleophilic sites that react with electrophiles) are assigned to the relevant atoms of acetone and methyl vinyl ketone, an $\alpha,\beta$-unsaturated ketone. Be aware that these symbols represent reactivity, not charges, although the pattern of electrophilic and nucleophilic reactivity often parallels the pattern of partial charges resulting from the presence of electron-withdrawing or electron-donating groups.

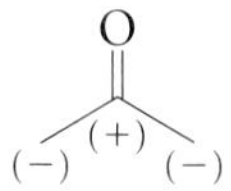

It is also useful to consider that the $\beta$ carbon of a saturated ketone is a site of potential electrophilic reactivity, as illustrated here for 2-butanone:

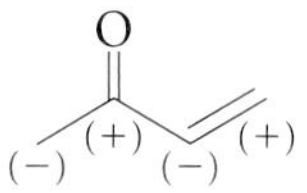

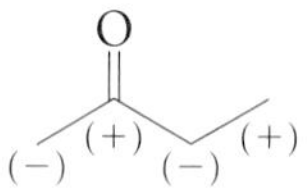

Acetone     Methyl vinyl ketone

2-Butanone

Just as the $\alpha$ carbon of a neutral ketone is a potential rather than a real nucleophile, the $\beta$ carbon is a potential electrophile because conversion of a saturated ketone into an $\alpha,\beta$-unsaturated ketone results in electrophilic reactivity at this site. It is important to note that reactivity alternates in a regular fashion along the carbon chain. This alternation is quite reasonable, because the strong electron-withdrawing nature of the carbonyl group imparts electrophilic reactivity to the carbonyl carbon and stabilizes a carbanion at the $\alpha$ carbon.

## EXERCISE 21.15

Justify the reactivities shown for methyl vinyl ketone by providing an example of a chemical reaction likely to occur at each of the carbon atoms:

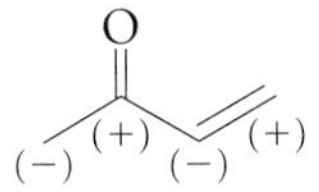

When more than one functionality is present, color is useful to distinguish the reactivities imparted by each functional group to adjacent atoms, as in the $\beta$- and $\alpha$-ketoesters in the margin. The sequence of reactivities shown in red is that imparted by the ester, whereas that shown in blue represents the effect of the ketone. In the $\beta$-ketoester, the reactivity imparted by both carbonyl groups to each carbon atom of the chain is the same. As a consequence, the alternation of reactivities imparted by each individual carbonyl group is preserved in the $\beta$-ketoester. Recall that the carbon between the two carbonyl groups is especially acidic ($pK_a = 11$ for a $\beta$-ketoester). Indeed, the superimposition of the reactivities provides a sense of the additional stability imparted to the anion at this position. On the other hand, the reactivities imparted to each carbon of an $\alpha$-ketoester by the two carbonyl groups do not match. We will see the consequence of this mismatch shortly.

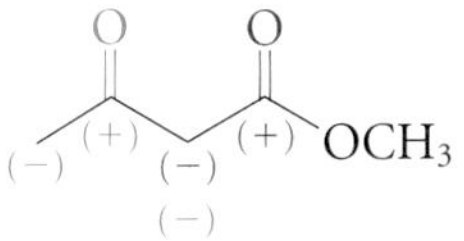

### Analysis of Carbon Reactivity in $\beta$-Ketoesters

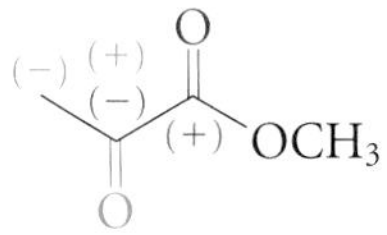

A consideration of potential reactivity of a target molecule can help in retrosynthetic analysis because the characteristic reactivity at a specific site does not generally change from starting material to product as long as the fundamental nature of the functional group remains unaltered. For example, let's analyze the transformation of two molecules of methyl acetate into methyl acetoacetate, a reaction known as the Claisen condensation.

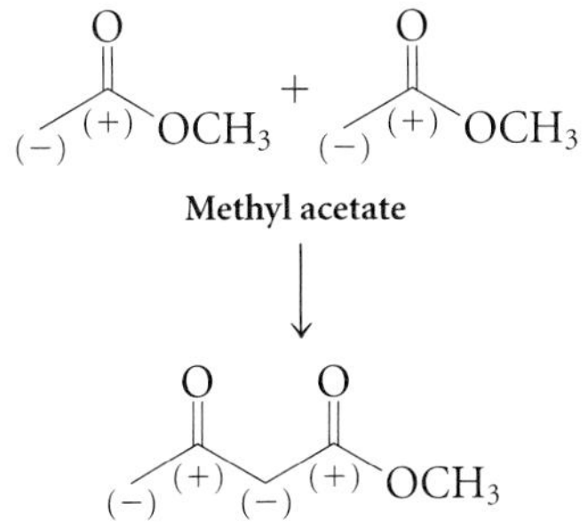

Recall that the Claisen condensation consists of the interaction of the $\alpha$ carbon of one ester molecule as an enolate anion in a nucleophilic acyl substitution at the carbonyl group of a second ester molecule. The reactivity of each carbon of the starting materials remains the same in the product $\beta$-ketoester. However, imagine that we do not know about this specific reaction but have, nonetheless, the task of constructing methyl acetoacetate from starting materials with fewer carbon atoms. Such a synthesis requires carbon–carbon bond formation. This kind of reaction should come late in a proposed synthetic sequence (that is, early in the retrosynthetic analysis). Rather than consulting a list of carbon–carbon bond-forming reactions, let's consider, in turn, the formation of each carbon–carbon bond by the combination of a nucleophile with an electrophile in such a way as to reflect the reactivity pattern of the target molecule. Our analysis is simplified because there are only three carbon–carbon bonds in methyl acetoacetate. The three possible nucleophile–electrophile combinations are illustrated in Figure 21.16 (on page 21-24). Note that, in each set of fragments, a pair of electrons is available on an atom that we have determined to have potential nucleophilic character in the product. Because the reactivities alternate along the chain, the positively charged carbon atom in the derived fragment is naturally one that has electrophilic character in the product $\beta$-ketoester.

$\beta$-Ketoester

$\alpha$-Ketoester

Methyl acetate

Methyl acetoacetate

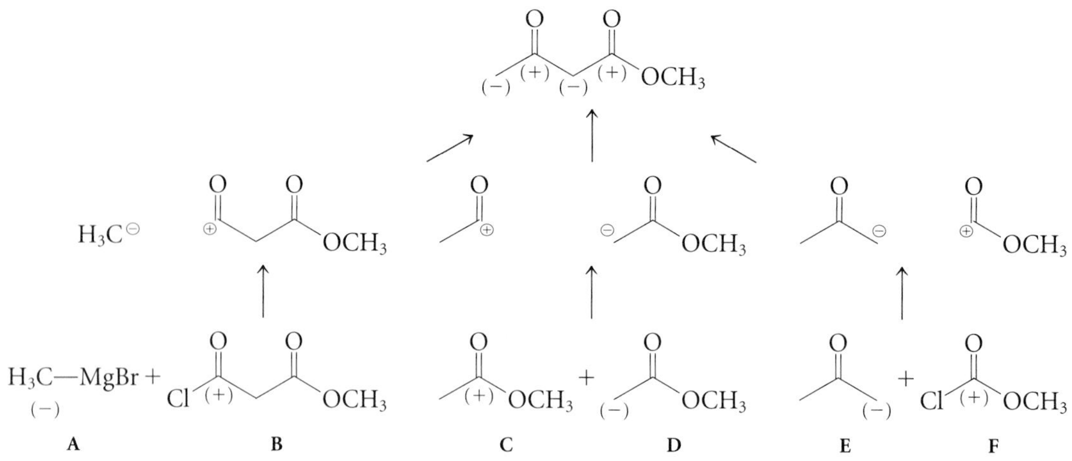

Because the characteristic reactivities of the atoms of a β-ketoester alternate along the carbon chain in a regular fashion, it is possible to construct this species by three separate routes, each combining an electrophilic and a nucleophilic starting material.

**FIGURE 21.16**

The analysis to this point has led us to ionic fragments, requiring the implausible simultaneous formation of both a carbanion and a carbocation. However, in each case, we can envision stable reactants whose reactivity will enable the bond to be formed by a reaction in which specific atoms act as nucleophiles or electrophiles, as appropriate. For example, we can envision the $CH_3^{\ominus}$ fragment shown as A in Figure 21.16 as being derived from an organometallic reagent, such as methyllithium or methylmagnesium bromide. Each of these reagents has substantial negative charge on the methyl carbon as a result of the highly polarized carbon–metal bond. The reaction of a Grignard reagent with B, the half ester, half acid chloride of malonic acid, might well be expected to result in methyl acetoacetate by nucleophilic acyl substitution on the more reactive of the two carboxylic acid derivatives (the acid chloride). This reaction represents a chemically reasonable combination of reagents. Note that in the product β-ketoester, the added methyl group is adjacent to a carbonyl group and thus has (potential) nucleophilic character. The carbonyl group of the acid chloride remains in the product and has electrophilic character both before and after the reaction.

In addition, our analysis has uncovered two other possibilities involving enolate anions of an ester (center reaction in Figure 21.16) and of a ketone (right-hand reaction). The center reaction in Figure 21.16 is a "rediscovery" of the Claisen condensation as a combination of functional groups that should readily form a β-ketoester. Similarly, nucleophilic acyl substitution on methyl chloroformate by the enolate anion produced by deprotonation of acetone (the reaction at the right) should also produce the desired product. Although we have concentrated on the chemistry of only one of these syntheses of a β-ketoester, all three indeed work in the laboratory.

In this example, the chemical reactivity of each carbon atom in the product matches that of the corresponding atom in the proposed starting materials in all three combinations. This is possible because the two functional groups in the product are positioned in such a way that the alternating reactivities imparted by each group match up. This matching is a consequence of the alternation in reactivity imparted by carbonyl groups and of the positioning of the two functional groups of the product so that their reactivities correspond. Using this kind of analysis of electron demand is one way to apply knowledge of organic chemistry to design new synthetic transformations.

## EXERCISE 21.16

The three routes for constructing a new carbon–carbon bond in the synthesis of methyl acetoacetate have not been evaluated from a practical point of view. Suppose that, instead of using methyl chloroformate as F in the reaction at the right in Figure 21.16, we chose to react the enolate anion of acetone (E) with dimethyl carbonate ($CH_3OCO_2CH_3$). What problem(s) might be encountered with this proposed combination of reagents? Might there be similar difficulties with methyl chloroformate? (*Hint:* Consider the relative acidities and electrophilic reactivities of each of the starting materials and the product.)

## Analysis of Carbon Reactivity in α-Ketoesters

Let's undertake a similar analysis for an α-ketoester (Figure 21.17). Here, the relation of the carbonyl groups is such that they impart opposite senses of reactivity to each carbon in the chain. Because the potential reactivities imparted by the two carbonyl groups in the product do not match (the signs are opposite for the ketone carbon), an analysis parallel to that used in Figure 21.16 does not lead to an unambiguous, straightforward reaction between a nucleophile and an electrophile. We cannot find a unique pairing of starting materials that leads naturally to the product. Each of the two possible nucleophile–electrophile combinations shown in Figure 21.17 seems equally inappropriate, because both require the use of an unstable, acyl anion as a nucleophile.

How, then, can we form an α-ketoester if the potential reactivities imparted by functional groups do not match? This analysis suggests that at some point we must either (1) use a reaction in which a functional group reacts in an "unnatural" fashion or (2) convert functional groups so that the natural reactivity is reversed. An example of the first option is the use of the acyl anion of formaldehyde as a nucleophile. However, such anions are not particularly stable and are restricted to aldehydes lacking hydrogen atoms on the α carbon. As a consequence, such anions are only rarely used in carbon–carbon bond-forming reactions. The second possibility, reversal of reactivity, or *Umpolung,* is discussed in the next section.

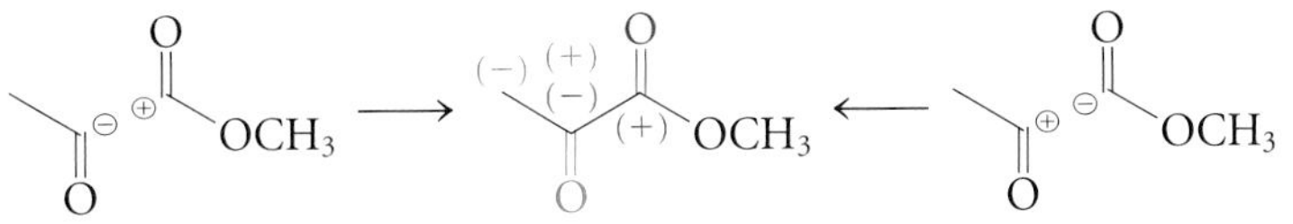

**FIGURE 21.17**

With the reactivities expected from an analysis focusing on the ester (shown in red), the fragments shown at the right are needed. With the reactivities expected from an analysis focusing on the ketone group (shown in blue), the fragments shown at the left are required.

### Reversal of Reactivity

We have already seen several examples in which the characteristic reactivity of a carbon in a reagent can be reversed by the transformation of one functional group into another. The conversion of an alkyl halide into an organometallic reagent such as a Grignard reagent is a familiar example of such a reversal of reactivity.

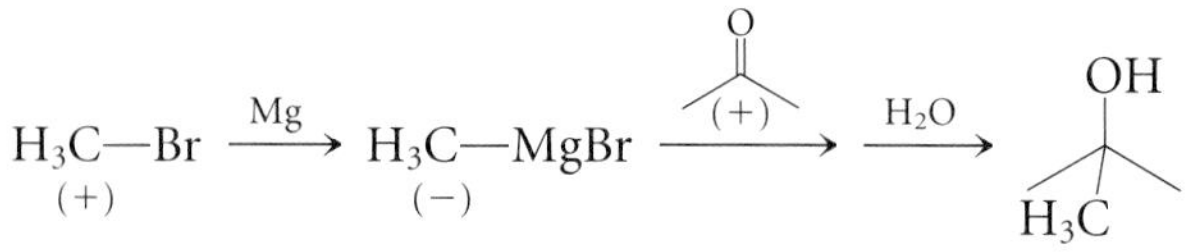

Because of the polarization of the carbon–halogen bond, the carbon of an alkyl bromide characteristically reacts as an electrophile. However, in the derived organomagnesium reagent, the replacement of the more electronegative halogen by the less electronegative metal results in a polarization of the carbon–metal bond that is the reverse of that in an alkyl halide. Thus, the carbon of a Grignard reagent reacts as a nucleophile.

Similarly, the conversion of an alkyl halide into a phosphonium salt and then into an ylide (a Wittig reagent) also produces a nucleophilic carbon, which reacts with acetone.

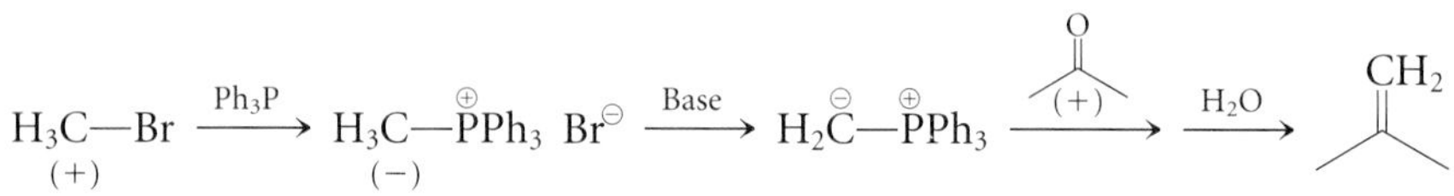

In both cases, because of the nature of the transformation, the reactivities of the carbons in the product leave no clue as to the original, electrophilic reactivity of the carbon in the reagent (the added carbon), because the original functionality, an alkyl halide, is not present in the product.

Chemists have formulated a number of reactions that temporarily mask the normal reactivity of a functional group while imparting the opposite reactivity. For example, consider the conversion of an aldehyde into a dithioacetal (Figure 21.18): the normal electrophilic activity of the carbonyl

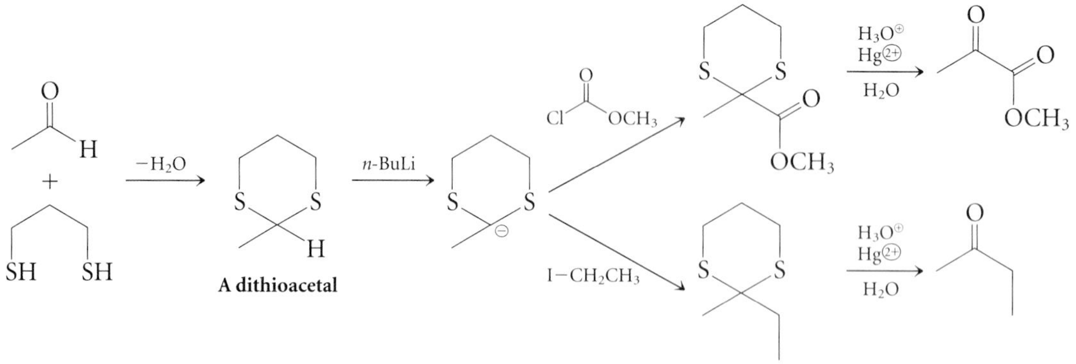

**FIGURE 21.18**

Reversal of reactivity is achieved by formation of a dithioacetal from an aldehyde. Deprotonation of the dithioacetal produces an anion that can function as a nucleophile for reaction with an alkyl iodide by $S_N2$ displacement (left) and with an acid chloride by nucleophilic acyl substitution on the acid chloride (right) to form a new carbon–carbon bond. Hydrolysis of the dithioacetal group forms the carbonyl group of the ketone product.

carbon is replaced by potential nucleophilic activity in the anion of a dithioacetal. Formation of the dithioacetal proceeds by a mechanism essentially identical with that for the formation of the corresponding oxygen acetal. In the dithioacetal, the normal electrophilic reactivity of the carbonyl carbon of the aldehyde is no longer present. Moreover, the proton on the carbon bearing the two sulfur atoms is moderately acidic and can be removed by a strong base such as a lithium dialkylamide.

The anion formed from the dithioacetal reacts as a nucleophile with many electrophilic carbon species, such as alkyl halides and acid chlorides. Once the carbon–carbon bond is formed, the product dithioketal is hydrolyzed to regenerate the carbonyl group. The overall process forms a carbon–carbon bond to the carbonyl carbon in a reaction employing the carbonyl carbon as a nucleophile. As shown in Figure 21.18, this sequence makes it possible to convert an aldehyde into a ketone or an $\alpha$-ketoester. The dithioacetal anion functions just like an acyl anion. Many such acyl anion equivalents have been developed by synthetic chemists.

**Dithioacetal anion**          **Acyl anion**

### EXERCISE 21.17

Write a step-by-step mechanism for the formation of a dithioacetal, shown as the first step of Figure 21.18.

### EXERCISE 21.18

Show how you could use the dithioacetal anion shown above to prepare 2,3-hexanedione and 2,5-hexanedione.

## Summary

1. Cofactors are important reagents that are essential components of several key biochemical transformations. Although much of the chemistry of cofactors is analogous to that observed for laboratory reagents, cofactors are structurally more complex than most laboratory reagents.

2. The structural complexity of cofactors is necessary because they must have parts that enable them to recognize their target molecules and that make them recognizable to enzymes.

3. Biological reactions involving cofactors differ from laboratory reactions in that they are close to thermoneutral ($\Delta G°$ near 0) in order that the "spent" cofactor can be converted back into its active form for reuse.

4. Cofactors often react by pathways similar to those for reactions involving simpler reagents. Pyridoxamine pyrophosphate accomplishes reductive amination by a route similar to straightforward nucleophilic addition and reduction. NADH and NADPH react by pathways quite similar to those followed by the hydride donors $LiAlH_4$ and $NaBH_4$. $FADH_2$ is a biological reducing agent that can function either as a hydride donor or by single-electron transfer, which is a pathway similar to that of a dissolving-metal reduction. Acetyl CoA activates carboxylic acid derivatives by forming a thiol ester, a functional-group transformation analogous to the activation of carboxylic acids as acid chlorides or anhydrides.

5. The transfer of a single carbon in the course of biodegradation and biosynthesis can be accomplished through the reversible binding of a meth-

ylene group to tetrahydrofolic acid. In an intermediate step, the one-carbon unit is transferred from serine by reversible aminal formation.

**6.** The cofactor thiamine pyrophosphate is responsible for the removal of carbon dioxide in the biodegradation of $\alpha$-ketoacids. The special function of thiamine pyrophosphate is to invert the normal reactivity of an $\alpha$-ketoacid to mimic the decarboxylation reaction usually observed with a $\beta$-ketoacid. Thiamine's function in this decarboxylation thus places it in the special class of reverse-polarity reagents.

**7.** Lipoic acid, in its oxidized, disulfide form, reacts in combination with thiamine to effect the overall oxidative decarboxylation of an $\alpha$-keto-acid, yielding a simple carboxylic acid with one fewer carbon.

**8.** The reversal of reactivity imparted by thiamine in the decarboxylation of $\alpha$-ketoacids has analogies in laboratory reactions. A dithioacetal is a simple reverse-polarity reagent that inverts the normal chemical reactivity of the carbonyl carbon atom of an aldehyde. Deprotonation of a dithioacetal produces an acyl anion equivalent, which can be trapped by alkylation reactions. The ketone can be regenerated by hydrolysis.

## Review Problems

**21.1** Classify each of the following reactions as (1) a highly exothermic reaction, (2) a reaction having an equilibrium constant near 1, or (3) an endothermic reaction favoring starting materials.

(a) [structure] + NaBH$_4$ $\longrightarrow$ [structure] + B(OH)$_3$

(b) [structure] + CH$_3$OH $\longrightarrow$ [structure] + H$_2$O

(c) [structure] + CH$_3$OH $\longrightarrow$ [structure] + H$_2$O

(d) [structure] + Cl$_2$ $\longrightarrow$ [structure] + HCl

(e) [structure] $\xrightarrow{H_3O^{\oplus}}$ [structure]

**21.2** Hydroquinone is an important antioxidant that readily traps oxidizing agents, especially radicals, by undergoing oxidation to form a quinone.
(a) Use a radical (R·) to remove a hydrogen atom from one hydroxyl group of hydroquinone, and show, with resonance structures, how this radical is highly delocalized and therefore stabilized.
(b) Remove the other hydroxyl hydrogen from hydroquinone, and use arrows (remember, half-headed) to show how benzoquinone is produced.

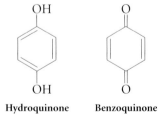

**Hydroquinone**   **Benzoquinone**

**21.3** Butylated hydroxytoluene (BHT) is an important antioxidant in food; it has been approved by the U.S. Food and Drug Administration for human consumption. Compare the structure of this phenol with that of hydroquinone, noting the similarities and the differences.

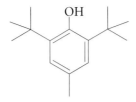

**Butylated hydroxytoluene (BHT)**

**21.4** The oxidation of FADH$_2$ to FAD can take place through radical intermediates. Use the ideas you developed in solving Problem 21.2 to write a mechanism for this oxidation, assuming that two radicals, R·, are responsible for sequentially removing the two hydrogen atoms (shown in red).

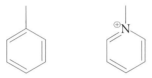

**21.5** Compare the structures of toluene and *N*-methylpyridinium ion, paying particular attention to the arrangement of electrons. Draw resonance structures for the ion that do not have positive charge on nitrogen, noting which carbons bear partial positive charge in the hybrid structure.

**Toluene**     ***N*-Methylpyridinium ion**

**21.6** Thiol esters are more reactive toward nucleophiles than their oxygen analogs, because of reduced orbital overlap between the sulfur atom and the carbonyl $\pi$ system. Thione esters differ from thiol esters in the position of the sulfur and oxygen atoms. How would you expect thione esters to differ from thiol esters in stability and reactivity toward nucleophiles? From "normal" carboxylic acid esters?

**Thiol ester**     **Thione ester**

**21.7** In some ways, the reactions of tetrahydrofolic acid in transferring one carbon resemble the steps in the Mannich condensation. Review that reaction, and then compare it with the steps in Figure 21.13. Which steps are comparable?

**21.8** Analyze the patterns of reactivity imparted by each functional group in the following bifunctional molecules. In which cases could the molecule be formed by a connection of nucleophilic and electrophilic carbons whose reactivity is consistent with that imparted by the functional groups?

(a)   (c)

(b)   (d)

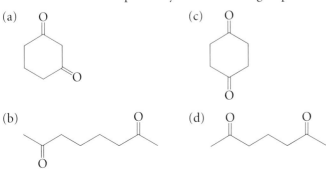

## Supplementary Problems

**21.9** A number of functional groups can be reduced by sequential transfers of two electrons and two protons. One such reduction is the conversion of an $\alpha,\beta$-unsaturated thiol ester to a saturated thiol ester (Section 21.5). For example, simple ketones can be reduced to alcohols by treatment with metallic sodium in liquid ammonia with *t*-butyl alcohol as the source of the protons:

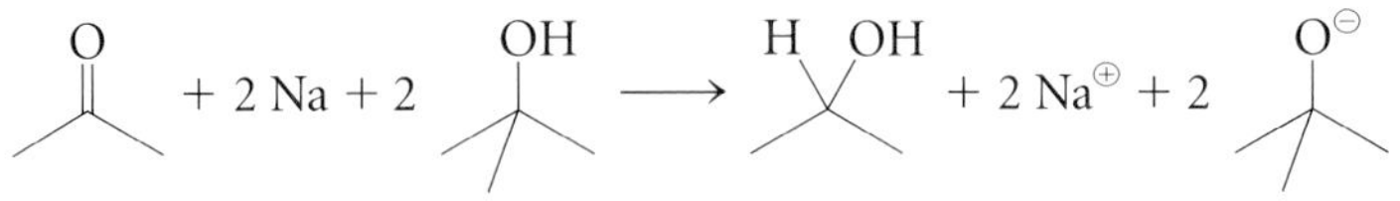

Write a detailed mechanism for this transformation. (Do not concern yourself with the details of how the electron is transferred from the metal—simply write $+1\ e^{\ominus}$ over the appropriate arrow.)

**21.10** Aromatic compounds can be reduced using the same conditions as in Problem 21.9 in a reaction known as the *Birch reduction.* Provide a step-by-step mechanism (again the transfer of an electron is simplified as $+1\ e^{\ominus}$). (*Hint:* The product is the nonconjugated rather than the conjugated diene, but do not concern yourself with why the former isomer is favored. It might be helpful to first balance the reaction. Keep in mind the need for protons and electrons. The protons come from *t*-butyl alcohol, and one electron is available from each sodium atom.)

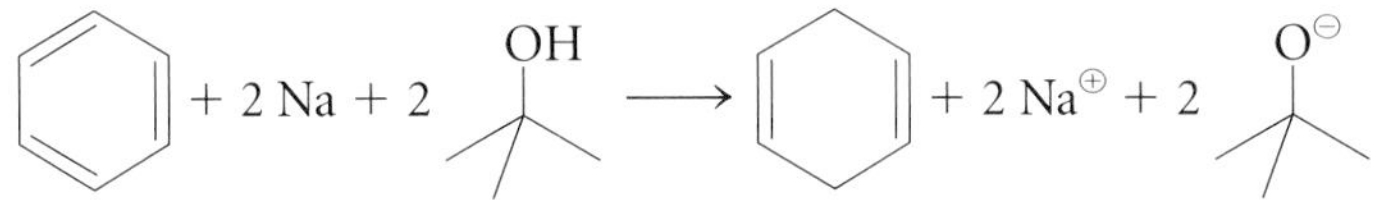

**21.11** Some oxidation reactions occur by stepwise removal of hydrogen atoms. For example, quinones like tetracyanoquinone can be used to form aromatic systems from more reduced cyclohexyl compounds.

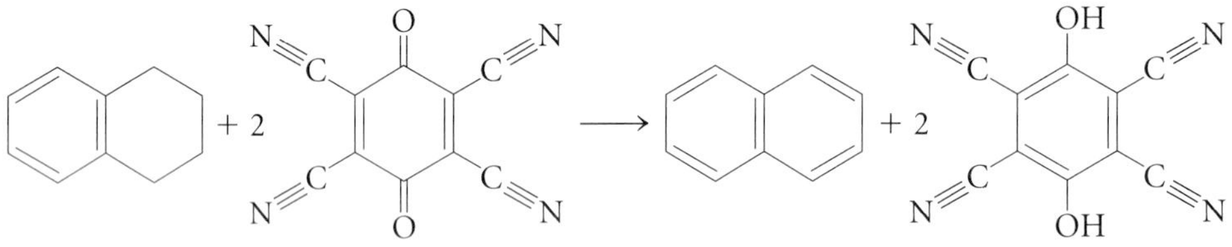

Tetracyanoquinone

Write a detailed mechanism for this transformation. (Show the abstraction of hydrogen atoms by simply writing $-H\cdot$ over the arrow, as appropriate. This exercise requires only that you decide in which order the hydrogen atoms are abstracted.)

**21.12** Expand your analysis of the reaction in Problem 21.11 to consider how benzoquinone might serve as a hydrogen atom acceptor. Use R—H as a hydrogen atom donor. Be sure to include the appropriate half-headed curved arrows.

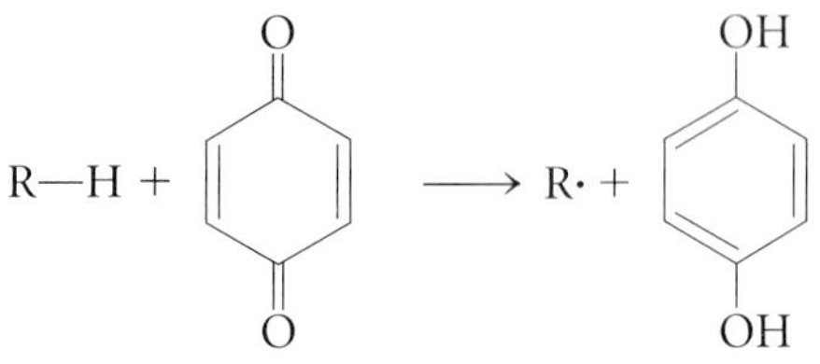

Benzoquinone         Hydroquinone

**21.13** The cyano substituents of tetracyanoquinone (used in the reaction in Problem 21.11) make this quinone a significantly more powerful oxidizing agent than the parent benzoquinone. Can you explain this difference in reactivity?

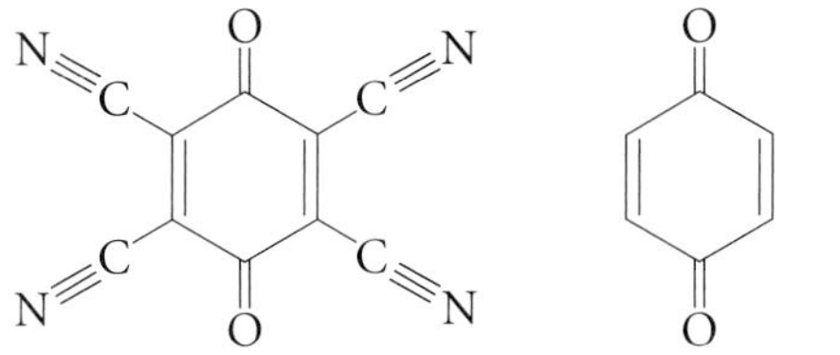

**Tetracyanoquinone**
(more reactive)

**Benzoquinone**
(less reactive)

**21.14** Both $\beta$-ketoacids and $\beta$-dicarboxylic acids undergo loss of carbon dioxide, to form simple ketones and simple carboxylic acids, respectively:

Faster → + $CO_2$

Slower → + $CO_2$

The decarboxylation of $\beta$-ketoacids is significantly faster than that of $\beta$-diacids. Write detailed mechanisms for all steps of both reactions, and account for the difference in reaction rate.

**21.15** Decarboxylation of $\beta$-ketoacids is normally relatively rapid (see Problem 21.14), and solutions of these acids often "bubble" at room temperature as $CO_2$ is released. On the other hand, there are exceptions. Explain why decarboxylation of the following $\beta$-ketoacid is unusually slow:

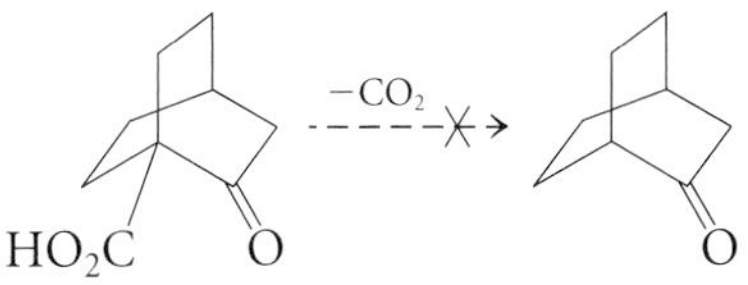

**21.16** Treatment of pyridine with potassium hydroxide in water at elevated temperature results in the formation of 2-pyridone. Does this transformation involve a change in oxidation level? If so, what might be reduced? Account for this transformation with a detailed mechanism.

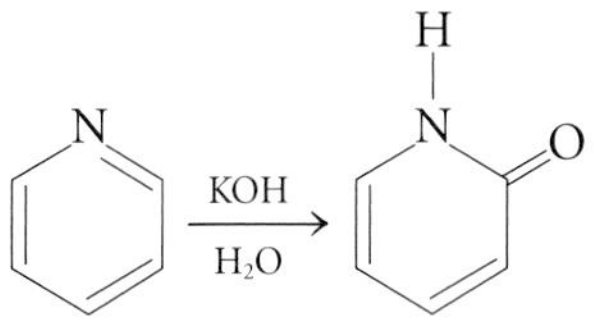

**Pyridine**   **2-Pyridone**

**21.17** The presence of amino acids can be determined by a simple test. Reaction of the primary amino group of an amino acid with ninhydrin hydrate results in the formation of a product that has a deep purple color. Account for the formation of this product with a detailed reaction mechanism. Why is this test specific for $\alpha$-amino acids? (Simple primary amines do not give a purple color.)

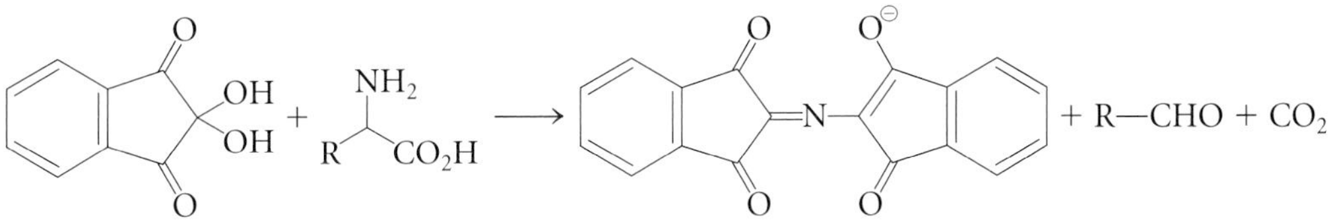

+ R—CHO + $CO_2$

**Ninhydrin hydrate**          **Purple product**

**aldimine:** an imine of an aldehyde (21-7)

**aminal:** a functional group with one hydrogen, an alkyl group, and two amino groups on one carbon atom [$RCH(NR'_2)_2$]; produced in the reaction of a secondary amine with an aldehyde (21-15)

**coenzyme A (CoA):** a complex thiol that, as a thiol ester derivative, accelerates nucleophilic acyl substitution in several biochemical transformations (21-13)

**cofactor:** a recyclable biological reagent (21-4)

**flavin adenine dinucleotide (FADH$_2$):** a cofactor used for the electron-transfer reduction of, for example, an $\alpha,\beta$-unsaturated thiol ester in fatty acid synthesis (21-5)

**ketimine:** an imine of a ketone (21-6)

**lipoic acid:** a dithiol cofactor important in the oxidative degradation of $\alpha$-ketoacids (21-20)

**nicotinamide adenine dinucleotide (NADH):** a cofactor that effects the reduction of $\alpha$-ketoacids in fatty acid biosynthesis (21-5)

**pyridoxamine phosphate:** the product of reductive amination of pyridoxal phosphate; a cofactor that serves as both a reducing agent and a source of nitrogen for the production of $\alpha$-amino acids (21-5)

**single-electron transfer:** a chemical reaction in which the key step consists of the transfer of one electron (21-12)

**tetrahydrofolic acid:** a cofactor that effects the methylation of nucleic acids by transferring one carbon from serine (21-15)

**thiamine pyrophosphate:** a cofactor important in the degradation of amino acids (21-20)

**vitamin:** a molecule that is required to sustain life but cannot be synthesized by an animal and must therefore be obtained from the diet (21-6)

# Energy Storage in Organic Molecules

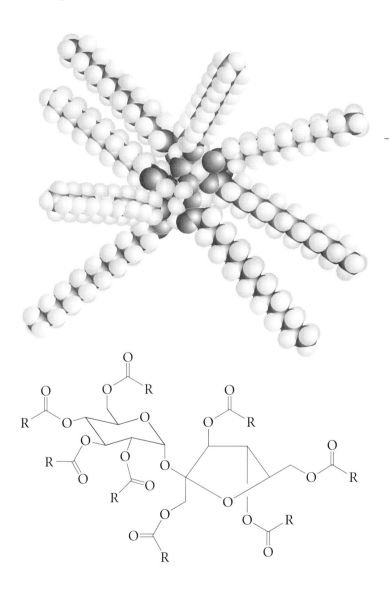

Olestra® is a semisynthetic, nondigestible, and therefore noncaloric, fat mimic marketed by Procter & Gamble. It is constructed of a sucrose core esterified with fatty acids (represented here as the octapalmitic acid ester with R = $C_{14}H_{29}$: Digestion of Olestra® is inhibited by steric hindrance to hydrolysis of the ester linkages. If 1 mole of this compound were converted to carbon dioxide and water by the pathways covered in this chapter, 8,176 kcal of chemical potential energy would be stored in various cofactors, and 12,284 kcal of heat would be released (enough to raise the temperature of a large bathtubful of water by 100 °C).

*T*his chapter deals specifically with the chemical properties of lipids and carbohydrates, especially the way these properties relate to the storage of chemical energy. We will see how these compounds are broken into smaller and smaller pieces, ultimately forming carbon dioxide, water, and heat. In addition, some of these reactions are coupled to the conversion of a diphosphate into a triphosphate, a process that itself is endothermic by approximately 7 kcal/mole. Thus, the degradation of lipids and carbohydrates serves two functions: (1) producing thermal energy, and (2) supplying stored, chemical potential energy. Throughout the chapter, a close correspondence between biochemical transformations and laboratory equivalents will be evident.

## 22.1

## Reaction Energetics

### Kinetics and Thermodynamics

We begin by reviewing some basic concepts of reaction kinetics and thermodynamics so that we can apply these concepts to the more complicated, multistep processes of biochemical transformations. Because we will be dealing with relatively small energy differences, we will include the effects of entropy; that is, discussions and calculations will be based on free energy, $G°$, rather than enthalpy, $H°$. (Recall that $G° = H° - TS°$.)

Consider an energy diagram relating two species, A and B, in which A is higher in energy (and therefore less stable) than B (Figure 22.1). The state of equilibrium is defined as that condition in which the rate of conversion of A to B is precisely balanced by the rate of conversion of B to A. In this example, the activation energy for the forward reaction (A → B), $\Delta G_f^{\ddagger}$, is less than that for the backward reaction (B → A), $\Delta G_r^{\ddagger}$. Because the rate is proportional to $\exp(-\Delta G^{\ddagger}/RT)$, the intrinsic rate of conversion of A into B is higher than the rate of conversion of B into A. To maintain equilibrium, the concentration of species A must be lower than that of B. Indeed, because

$$K = \frac{[B]}{[A]}$$

and

$$\Delta G° = -RT \ln K$$

the ratio of concentrations of A and B can be expressed as $\exp -(\Delta G°/RT)$ in which $\Delta G° = \Delta G_f^{\ddagger} - \Delta G_r^{\ddagger}$, and in which $\Delta G°$ is always identical to the difference in the free energies of activation for the forward and backward reactions.

Although in some simple organic transformations, a single starting material is converted into a single product, most transformations involve at least two starting materials and result in more than one product. For example, the oxidation of an alcohol to a ketone requires an oxidizing agent

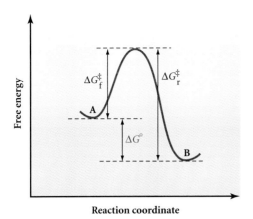

$$A \rightleftharpoons B$$

$$\text{Rate} \propto e^{-(\Delta G^{\ddagger}/RT)}$$

$$\Delta G_f^{\ddagger} - \Delta G_r^{\ddagger} = \Delta G^{\circ}$$

$$\frac{[B]}{[A]} = e^{-(\Delta G^{\circ}/RT)}$$

**FIGURE 22.1**

In all reactions, the rate is controlled by the magnitude of the activation energy, $\Delta G^{\ddagger}$. For a reaction in which the number of reactant and product molecules is the same and entropy effects ($\Delta S^{\circ}$) are assumed to be negligible, the enthalpy of reaction ($\Delta H^{\circ}$) is the same as the free energy change ($\Delta G^{\circ}$). It is also equal to the difference in activation energies for the forward and reverse reactions. The equilibrium constant, which describes the ratio of the concentrations of product B to reactant A at steady state, is dependent on the magnitude of the free-energy difference, $\Delta G^{\circ}$ ($\Delta G^{\circ} = \Delta H^{\circ} - T\Delta S^{\circ}$).

such as $Cr^{6+}$, which itself is ultimately transformed in the reaction into $Cr^{3+}$:

It is thus impossible to speak of the thermodynamics of the transformation of an alcohol into a ketone solely on the basis of the stabilities of the two organic species, the reactant and the product. Indeed, an alcohol and a ketone cannot be directly related in terms of energy, because they are not isomeric—the ketone lacks two of the hydrogen atoms present in the alcohol. At best, we can compare the relative stabilities of the reactants (the alcohol and the oxidizing agent) and the products (the ketone and the reduced form of the oxidizing agent).

We cannot speak of the thermodynamics of the A-to-B conversion as being independent of the C-to-D conversion; the energetics of the reaction require a coupling of the two transformations. In Figure 22.2 (on page 22-4), these additional species have been added to the energy diagram. The concentrations of the two products (B and D) and the two starting materials (A and C) are related by the same equation used for the simpler case (Figure 22.1). Thus, a reaction profile provides important information about kinetics and thermodynamics. Because biological systems accelerate specific chemical reactions, we next consider how a reaction rate can be influenced by the addition of a catalyst and compare the efficiencies of a multistep and a single-step conversion.

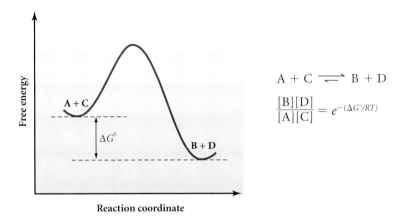

**FIGURE 22.2**

For a bimolecular reaction between A and C yielding B and D, the equilibrium constant can be expressed as a ratio of the product of the concentrations of the products to the product of the concentrations of the reactants. As in the unimolecular reaction of Figure 22.1, the position of the equilibrium is dependent on the magnitude of the difference in free energy of the reactants and products, $\Delta G°$.

## Catalysis

What is the effect of a catalyst on the energetics of a reaction? Recall that, by definition, a *catalyst* is a species that increases the rate of a reaction but does not itself undergo a net chemical transformation. Because the catalyst is not changed, the difference in energy between starting materials and products is the same whether the catalyst is present or absent (Figure 22.3).

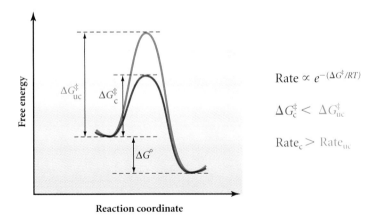

**FIGURE 22.3**

A catalyst influences the magnitude of the activation energies for the forward and reverse reactions but does not affect the potential energies of the reactants or products. A catalyst therefore influences the rate at which an equilibrium is established but not its position. In lowering the activation energy barrier for the forward reaction, a catalyst accelerates the rate of formation of product.

Therefore, a catalyst cannot change the position of an equilibrium. Nonetheless, the catalyst does influence the reaction by providing a pathway with a transition state of lower energy. As a result, the activation energy for a catalyzed process is lower than that for the same transformation in the absence of the catalyst. It is important to remember that only the *rate* of the reaction is affected by the catalyst, not the position of the equilibrium.

Most biochemical reactions are catalyzed by enzymes, which accelerate the rates of these reactions. Because an enzyme, like all catalysts, does not undergo a net transformation, it is not consumed in a reaction and therefore need not be present in stoichiometric amounts. Nonetheless, the rate of the reaction is proportional to the concentration of the catalyst (as well as of each reactant), and the rate of reaction decreases as the amount of the catalyst is reduced.

Enzymes are important not only because they can accelerate biochemical reactions, but also because the rate at which product is produced is controlled directly by changes in the concentration of the enzyme. The effect of the enzyme, as for all catalysts, is to provide a transition state with a lower energy. This has an identical effect on the energies of activation (and therefore the rates) of both the forward and reverse reactions. This is an important point to keep in mind because some of the pathways for the synthesis of complex biological molecules are the same as those used to degrade these compounds.

## Multistep Transformations

The second important point relating to reaction energetics concerns the consequence on reaction rates of multiple steps. Figure 22.4 shows two possible scenarios for a two-step reaction that takes place through the formation of an intermediate (I) that is higher in energy than either the starting material (A) or the product (B). Although there are two discrete steps in each of these reactions, and therefore two activation energies ($\Delta G_1^{\ddagger}$ and

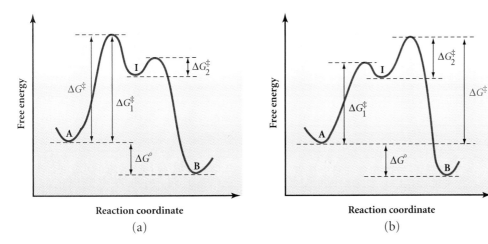

**FIGURE 22.4**

In a two-step sequence taking place through an intermediate, I, either (a) the first step or (b) the second step can be rate-determining.

## POLYMERS IN BATTERIES

The energy used in many areas of daily life is stored in various materials in the form of reduction potential. This energy is released as substantial quantities of heat upon combustion with molecular oxygen. Both natural gas (methane) and gasoline (a complex mixture of organic hydrocarbons) release relatively large quantities of energy on burning.

$$CH_4 + 2\,O_2 \rightarrow CO_2 + 2\,H_2O$$

$$\Delta H° = -213 \text{ kcal/mole}$$

There are other ways in which organic molecules can participate in an energy storage system. For ex-ample, lithium batteries are composed of three layers: (1) an oxidized form of polyaniline, (2) an intervening polymer layer rich in water, and (3) a layer of lithium metal. When the first layer is connected with the third by an external circuit, electrons flow from lithium through the circuit to the polyaniline layer, effecting oxidation of the lithium and reduction of the polymer. Concurrently, lithium ions and the anions associated with the polymer migrate into the intervening water-rich layer. Electric current will flow until all of the lithium has been oxidized and all of the polymer reduced. The process can be reversed by applying a potential in the opposite direction. Then the lithium ions are reduced to the metal, and the polyaniline layer is oxidized.

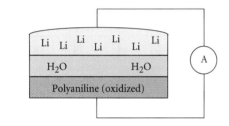

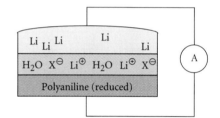

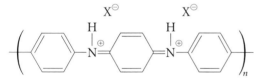

Oxidized form of polyaniline

Reduced form of polyaniline

$\Delta G_2^{\ddagger}$), the rate of each reaction is determined solely by the amount of energy required ($\Delta G^{\ddagger}$) to proceed from the starting material to the highest point of the energy curve. This energy of activation is equal to $\Delta G_1^{\ddagger}$ in the first diagram but is equal to neither $\Delta G_1^{\ddagger}$ nor $\Delta G_2^{\ddagger}$ in the second diagram. In both examples, the equilibrium positions of the overall reactions are determined solely by $\Delta G°$ and are not influenced by the shapes of the energy curves or by the presence or absence of relatively high-energy intermediates.

One or more endothermic steps are often included in multistep biochemical sequences, because the process is favored overall because of one or more exothermic steps. However, the last step must be exothermic if the sequence is to yield significant quantities of the ultimate product.

## 22.2

## Complex Reaction Cycles

Chemical reactions are often written as in equation 1, with the required reagent, C, placed above the arrow:

$$A \xrightarrow{C} B \tag{1}$$

However, when the reagent is consumed, a more complete representation is that in equation 2, explicitly showing both the reagent, C, and the product, D, derived from it:

$$A + C \longrightarrow B + D \tag{2}$$

In depicting biochemical reactions, it is often more convenient to use the method shown in equation 3. Here the use of the two intersecting curved arrows indicates that the transformation of C into D is intimately coupled with (and required for) the transformation of A into B:

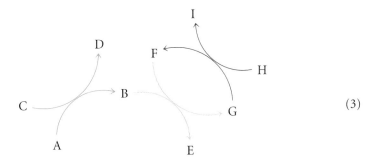

$$(3)$$

The representation in equation 3 is particularly convenient for showing reactions in which several transformations are coupled together. Recall that the curved arrows indicating coupled reactions have a different meaning from those used to indicate electron flow in reaction mechanisms. The intersecting curved arrows shown here are used to emphasize the link between coupled reactions. For example, the net conversion of A to E takes place in two steps. The first step simultaneously converts C to D as A is converted to B. The second step converts F to G, as B is converted to E. Species G is converted back to F, for use in another cycle, as H is transformed into I. With such complex sequences, it is important to keep in mind that it is the overall energetics of all the transformations taken together that determine the position of the equilibrium. For example, a highly exothermic conversion of G and H into F and I can compensate for the endothermicity of the two preceding steps.

Many biosynthetic sequences are used reversibly either to prepare or to degrade biological materials. Clearly, the forward and reverse reactions cannot both be exothermic. The endothermic process must be coupled with one or more other transformations that are sufficiently exothermic to compensate for its endothermicity.

## 22.3

# Energy in Living Organisms

The ultimate source of energy for almost all living systems on earth is the sun (some life has been found near thermal vents deep in the oceans beyond the reach of sunlight). Through photosynthesis, plants use light from the sun to fix the carbon from carbon dioxide into a vast array of carbon compounds. A considerable part of the light energy received by a plant is stored in the form of carbohydrates and other carbon compounds. This energy can be released if these compounds are oxidized, either by direct oxidation with molecular oxygen (burning) or by the metabolic processes of other living organisms.

Being unable to carry out photosynthesis, members of the animal kingdom derive energy from the sun indirectly, by ingesting plants or other animals that have eaten plants. The energy ingested in the form of food can be utilized immediately for growth or for fueling metabolic processes through respiration. Respiration is a process by which energy is released as various organic compounds are oxidized to carbon dioxide, in which carbon is at its highest possible oxidation level. Alternatively, the chemical energy in food can be stored as carbohydrate (glycogen) or fats (long-chain fatty acid esters of glycerol). A large number of different reactions are involved in these processes, but they can all be analyzed and understood in terms of basic organic reactions.

## 22.4

# Energy Transfer via Phosphoric Acid Anhydrides

Three phosphorylated species are critical in biochemical energy transfer. Each of these molecules is based on the adenine–ribose unit present in the cofactors NAD and NADP, and each contains one, two, or three phosphoric acid units attached to C-5' of ribose. These three phosphorylated species are **adenosine monophosphate (AMP), adenosine diphosphate (ADP),** and **adenosine triphosphate (ATP).** In the latter two, the additional phosphate groups are linked by anhydride bonds that are effective storers of chemical potential energy.

**Adenosine monophosphate (AMP)**

**Adenosine diphosphate (ADP)**

**Adenosine triphosphate (ATP)**

The **phosphoric acid anhydride** found in the di- and triphosphates is quite similar to a carboxylic acid anhydride in that it is hydrolytically unstable relative to two separate acid units. In effect, this means that the hydrolysis of both types of anhydrides is exothermic:

Phosphoric acid anhydride

Carboxylic acid anhydride

Thus, the hydrolysis of the phosphoric acid anhydride units in ADP and ATP should also be exothermic. Indeed, the hydrolysis of ATP to ADP (and phosphate) is exothermic by 7.0 kcal/mole, and the hydrolysis of both anhydride bonds in converting ATP to AMP (and phosphate) produces 16.3 kcal/mole of energy:

$$\text{ATP} + \text{H}_2\text{O} \rightleftharpoons \text{ADP} + \qquad \Delta G° = -7.0 \text{ kcal/mole} \qquad (4)$$

$$\text{ATP} + 2\ \text{H}_2\text{O} \rightleftharpoons \text{AMP} + 2 \qquad \Delta G° = -16.3 \text{ kcal/mole} \qquad (5)$$

For this reason, these phosphoric acid anhydride units are known as **high-energy phosphate bonds.**

The exact amount of energy released upon hydrolysis of ATP to either ADP or AMP depends on the experimental conditions. Such factors as the acidity of the solution and the presence of cations (especially divalent ones such as magnesium) have small, but very real, effects on the equilibrium position. For example, the phosphoric acid anhydride units of ATP, ADP, and AMP are essentially completely deprotonated at pH 7. With monovalent cations such as sodium and potassium as counterions for the phosphate units, the energy of the phosphoric acid anhydride units is increased by repulsion between the negatively charged oxygen anions. This is one reason why the $\Delta G°$ for the hydrolysis of both anhydride linkages (ATP $\rightarrow$ AMP) is more than twice that for the hydrolysis of one (16.3 versus 7.0 kcal/mole). On the other hand, in the presence of $\text{Mg}^{2+}$, the anhydride linkage is slightly stabilized by the bridge formed by the metal cation between the oxygen anion of one phosphate unit and that of another, reducing the value of $\Delta G°$. In no case, however, do these factors reverse the energetics: the process is always exothermic in the direction of cleavage of the phosphoric acid anhydride units. We will see throughout this chapter that the conversion of ATP into ADP is a source of chemical energy used to compensate for the endothermicity of other transformations to which it is coupled.

Pyrophosphate
chelated with magnesium ion

## EXERCISE 22.1

The interconversion of various derivatives of phosphoric acid is believed to take place by means of a nucleophilic addition–elimination mechanism.

(a) Write such a mechanism for the hydrolysis of trimethyl phosphate in aqueous acid, forming first dimethyl hydrogen phosphate and then methyl dihydrogen phosphate:

| Trimethyl phosphate | Dimethyl hydrogen phosphate | Methyl dihydrogen phosphate |

(b) Do the same for hydrolysis in aqueous sodium hydroxide. Be careful to consider the state of protonation of each intermediate and product as it would exist in an alkaline medium.

## 22.5

## Energy Transfer through Redox Reactions

A second form of potential energy is stored when NAD (or NADP) is reduced to NADH (or NADPH). Nature frequently uses oxidation–reduction couples such as NAD–NADH to store potential energy. Green plants effect the conversion of carbon dioxide and water to glucose (and other carbohydrates) and molecular oxygen under the influence of light.

$$6 \ CO_2 + 6 \ H_2O \underset{\text{Respiration}}{\overset{\text{Photosynthesis}}{\rightleftharpoons}} C_6H_{12}O_6 + 6 \ O_2 \qquad \Delta G° = +686 \ \text{kcal/mole}$$

In the reaction as written, the carbon of carbon dioxide, which is at the highest possible oxidation level (+4), is reduced to an oxidation level of 0 in glucose by the process of **photosynthesis.**

The formation of glucose (the forward reaction), a complex transformation proceeding by many steps, is very endothermic and achieves the storage of solar energy in carbon–hydrogen and carbon–carbon bonds. The reverse reaction (oxidation of glucose to carbon dioxide by bond cleavage, a process called **respiration**) is exothermic and constitutes a major source of both chemical and simple thermal energy for living systems. We can think of carbon dioxide reduction (photosynthesis) as a reaction that leads to a high-energy product (a sugar), which later releases its energy when the reverse process of oxidation (respiration) is carried out.

One of the ways in which nature temporarily stores energy is in the reduced form of various oxidation–reduction couples—for instance, in NADH. An oxidation–reduction couple such as NAD–NADH is also used to transfer energy, accepting energy when the reduced molecule is formed and releasing energy when that form is used to reduce other compounds. This transfer of energy is noted by the free-energy change that accompanies the coupled reaction.

The conversion of a ketone into an alcohol by reduction with NADPH can be represented as coupled oxidation–reduction reactions. Typically, the oxidation of NADPH to NADP coupled with the reduction of a ketone to an alcohol is, overall, exothermic. The exothermicity is not the result of the conversion either of the ketone to the alcohol or of NADPH to NADP. The exothermicity is the result of both transformations and requires that all four species take part. The reverse reaction, the reduction of NADP with the accompanying oxidation of the alcohol to the ketone, must be endothermic. Contrast this endothermic reaction with the exothermic conversion of an alcohol into a ketone, effected in the laboratory by $Cr^{6+}$. The energetics of the interconversion of a ketone and an alcohol depend not only on these two species, but also on the reagent(s) used in the redox reaction.

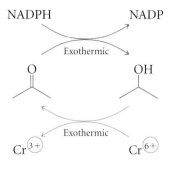

In biological systems, molecular oxygen, $O_2$, is the ultimate oxidizing agent. Couples such as NADPH–NADP enable a system to release energy in a controlled fashion by the stepwise oxidation of carbon compounds. The reaction of molecular oxygen as a biochemical oxidant is very complicated, especially in its early stages. Although many of the mechanistic features of this process have been established, a detailed description of these reactions is more appropriate to a course in biochemistry. However, the process can be represented as a very simple scheme in which NADH is oxidized by molecular oxygen to NAD in a reaction that is exothermic by 52.4 kcal/mole. Similarly, the oxidation of $FADH_2$ by molecular oxygen releases 36.2 kcal/mole of energy.

$$NADH + H^\oplus + \tfrac{1}{2} O_2 \longrightarrow NAD + H_2O \qquad \Delta G° = -52.4 \text{ kcal/mole} \quad (6)$$

$$FADH_2 + \tfrac{1}{2} O_2 \longrightarrow FAD + H_2O \qquad \Delta G° = -36.2 \text{ kcal/mole} \quad (7)$$

## 22.6

### Energy Storage in Fatty Acid Biosynthesis

In living organisms, fats are valuable as storers of energy. We can apply what we know about simple biochemical energetics to see how the synthesis of fatty acids achieves the storage of potential energy. The direct oxidation of fatty acids is an exothermic process; for example, the burning of palmitic acid in air releases a large quantity of energy:

$$\underset{\textbf{Palmitic acid}}{CH_3(CH_2)_{14}CO_2H} + 23 O_2 \longrightarrow 16 CO_2 + 16 H_2O \qquad \Delta G° = -2386 \text{ kcal/mole}$$
$$(8)$$

It follows that the reverse reaction, the conversion of carbon dioxide into a fatty acid, must be endothermic. Let's examine in some detail how this process occurs in nature.

**Fatty acid biosynthesis** is a process that builds long-chain fatty acids from acetic acid. The formation of carbon–carbon bonds is an intrinsic part of the process, as are the reductions required to replace most of the oxygen atoms of the acetic acid building blocks with hydrogen atoms.

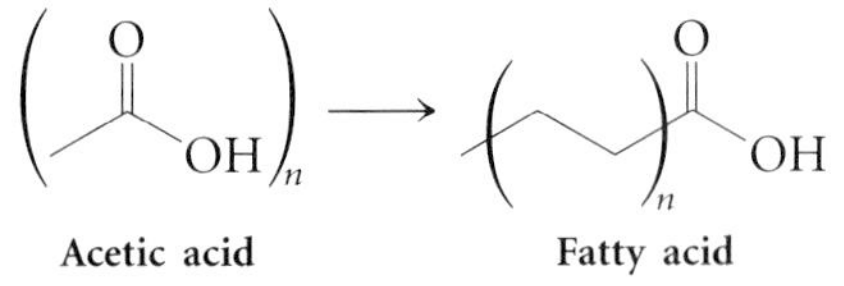

Acetic acid        Fatty acid

## Carbon–Carbon Bond Formation

Carbon–carbon bond formation in fatty acid biosynthesis is very similar to the Claisen condensation. However, there are several differences between the biosynthetic pathway for carbon–carbon bond formation and the laboratory-based condensation reaction, shown here.

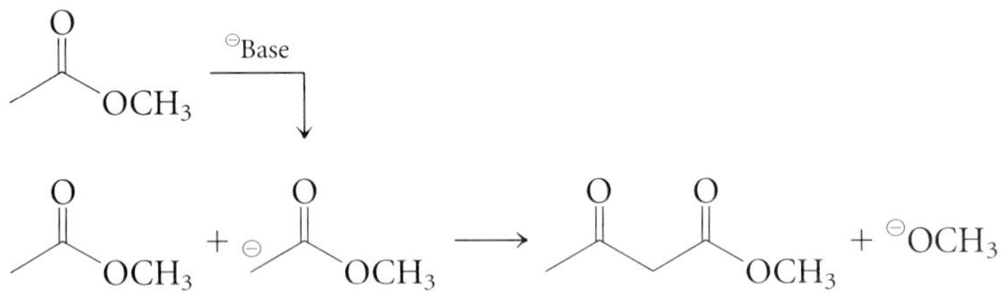

One difference is that in biological systems, thiol esters are involved instead of oxygen esters, presumably because thiol esters are significantly more reactive toward nucleophilic acyl substitution than are their oxygen analogs.

---

### EXERCISE 22.2

Explain the chemical rationale for the higher reactivity of a thiol ester compared with a carboxylic acid ester toward nucleophilic acyl substitution.

---

The reaction of a thiol with a carboxylic acid to produce a thiol ester and water, as illustrated in Figure 22.5, is endothermic and must therefore be coupled with an exothermic reaction in order to take place. The reaction of water with ATP to produce AMP (and the pyrophosphate ion) results in the cleavage of a high-energy bond. Because the latter process is more exothermic than the former is endothermic, the combination of the two processes to produce a thiol ester and AMP is energetically favorable.

The second difference between the biosynthetic pathway and the classic Claisen condensation is that the nucleophile in the nucleophilic acyl substitution is a derivative of malonic acid, formed from a thiol ester of acetic acid by carboxylation at the $\alpha$ carbon (Figure 22.6). (Recall that the Claisen condensation involves a carboxylic acid ester.) This carboxylation is not well understood, and so we won't deal with it here in detail. However, it is known that biotin (vitamin H) participates as a temporary carrier of carbon dioxide and that the overall process consumes one molecule of ATP. The condensation of this nucleophile (perhaps in its enol form) with the coenzyme A ester of acetic acid is illustrated in Figure 22.6. This reaction initially forms a $\beta$-carboxy-$\beta$-ketoester. Loss of carbon dioxide from this tricarbonyl compound is a highly exothermic process that helps make the overall process more exothermic than it otherwise would be.

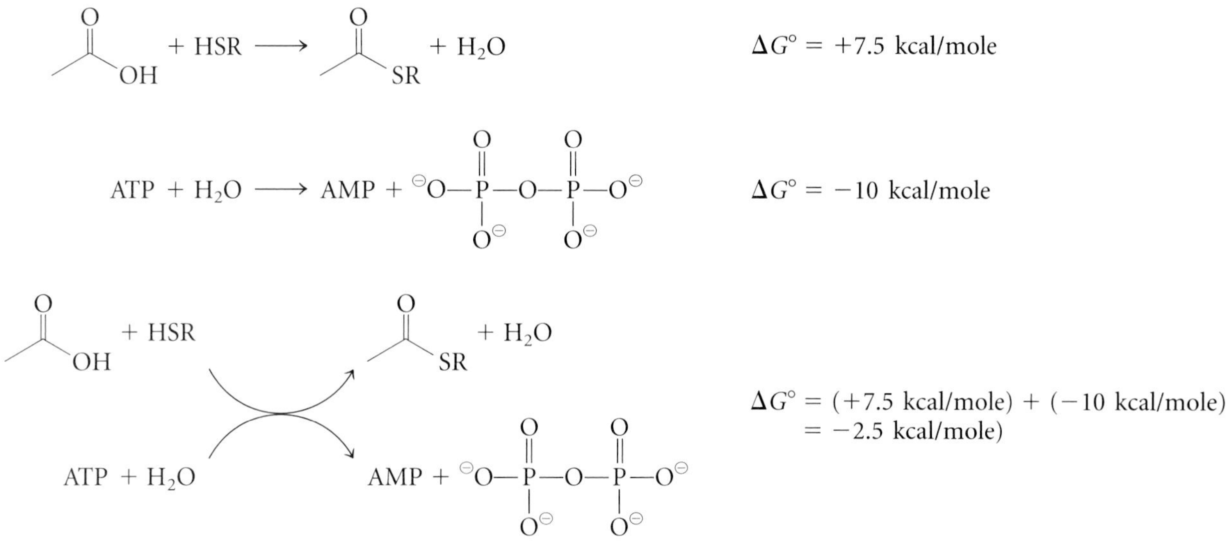
+ HSR
OH
SR
$+ H_2O$
ΔG° = +7.5 kcal/mole
$ATP + H_2O$
AMP +
ΔG° = −10 kcal/mole
ΔG° = (+7.5 kcal/mole) + (−10 kcal/mole)
= −2.5 kcal/mole)

**FIGURE 22.5**

The endothermicity of the formation of a thiol ester from a carboxylic acid is compensated for by the exothermicity of the hydrolysis of ATP to AMP, when these reactions are coupled.

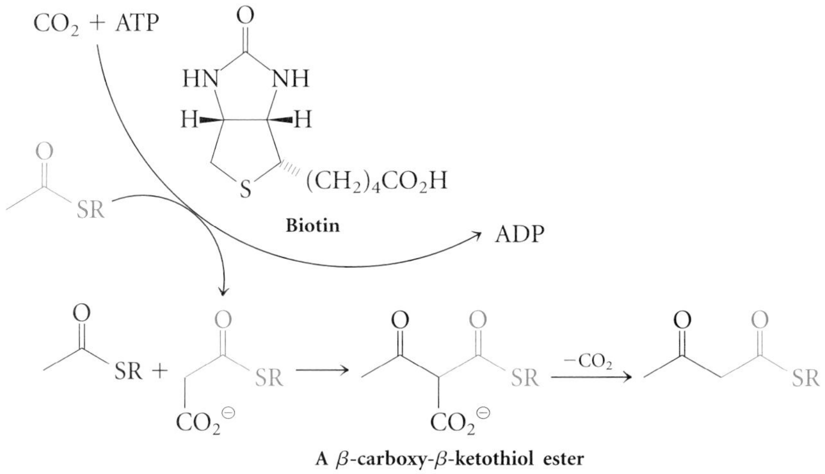
$CO_2$ + ATP
Biotin
$(CH_2)_4CO_2H$
ADP
$-CO_2$
A β-carboxy-β-ketothiol ester

**FIGURE 22.6**

Carboxylation of a thiol ester by biotin installs a second carboxylic acid functional group onto the thiol ester. By analogy with a malonic ester, this compound is activated toward the formation of either an enol or an enolate anion, which initiates the Claisen condensation. Once the new carbon–carbon bond has been formed via the Claisen condensation, the carboxylate group is no longer needed and is lost as carbon dioxide. This last decarboxylation step is assisted by the presence of two β carbonyl groups.

We can think of carbon dioxide and biotin as catalysts: overall, two molecules of acetic acid thiol ester and one molecule of ATP are consumed. This process can be repeated to build more complex chains, with two carbons being added in each cycle.

## EXERCISE 22.3

Write a mechanism for the formation of the intermediate β-carboxy-β-ketothiol ester shown in Figure 22.6. What is the chemical function of the carboxylation step?

## EXERCISE 22.4

Explain why the decarboxylation of the intermediate β-carboxy-β-ketothiol ester in Figure 22.6 is relatively fast. Write a mechanism for that conversion.

## EXERCISE 22.5

Refresh your memory of the mechanism for nucleophilic acyl substitution by writing a detailed mechanism for the following reaction, induced by strong base. Why can this reaction not be catalyzed by strong base? (*Hint:* Reexamine your answer to Exercise 22.3.)

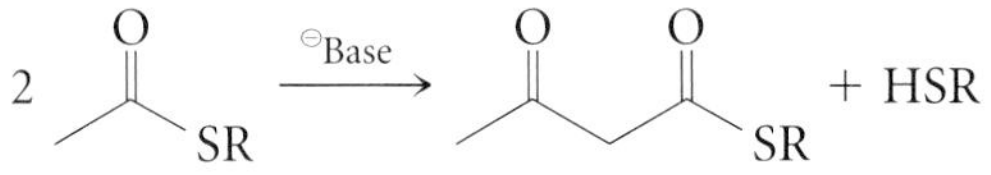

$$2 \quad \overset{O}{\underset{SR}{\|}} \quad \xrightarrow{\ominus \text{Base}} \quad \overset{O}{\underset{SR}{\|}} \quad + \text{ HSR}$$

## Reduction

The sequence described in the preceding section accomplishes the formation of a carbon–carbon bond between two acetic acid molecules. To proceed to a fatty acid, the β-ketothiol ester must be reduced, replacing the oxygen of the ketone group with two hydrogen atoms. This is accomplished by the three steps shown in equations 9–11.

$$\overset{O \quad O}{\underset{SR}{\|\quad\|}} + \text{NADPH} \longrightarrow \overset{OH \quad O}{\underset{SR}{\|}} + \text{NADP} \qquad (9)$$

$$\overset{OH \quad O}{\underset{SR}{\|}} \xrightarrow{-H_2O} \overset{O}{\underset{SR}{\|}} \qquad (10)$$

$$\overset{O}{\underset{SR}{\|}} + \text{NADPH} \longrightarrow \overset{O}{\underset{SR}{\|}} + \text{NADP} \qquad (11)$$

First, the keto group of the β-ketothiol ester is reduced by NADPH, forming a β-hydroxythiol ester (equation 9). Next, the hydroxyl group is lost by dehydration in a process that does not require a cofactor but is nonetheless enzyme-catalyzed (equation 10) and stereospecific. Last, further reduction of the unsaturated ester is effected by NADPH (equation 11) in a process similar to a metal hydride reduction or a Michael addition.

Overall, then, the chemical potential energy of one ATP and two NADPH molecules has been stored temporarily in the thiol ester of butyric acid, by means of the formation of a carbon–carbon bond and the chemical reactions presented in equations 9–11.

## EXERCISE 22.6

What are some alternative reactions for the reduction of a carbonyl to a methylene group?

## EXERCISE 22.7

Loss of water from a $\beta$-hydroxythiol ester is highly favored in the direction of the conjugated regioisomer. Explain this result, assuming that the reaction is induced by base. Now do the same assuming that the reaction is catalyzed by acid.

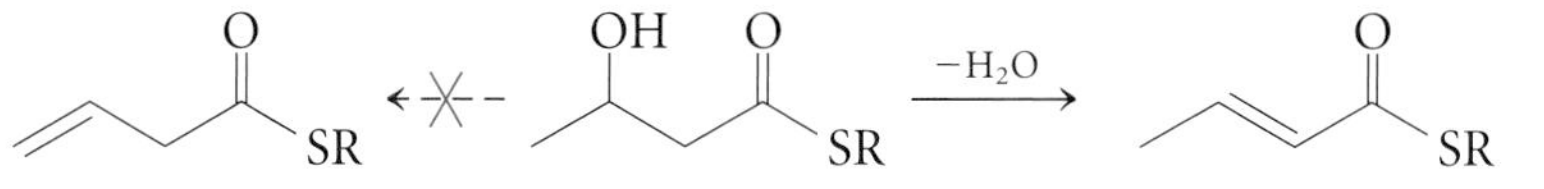

A $\beta$-hydroxythiol
ester

Conjugated regioisomer
(favored)

## EXERCISE 22.8

Assign oxidation levels to the $\alpha$ and $\beta$ carbons of the hydroxythiol ester in equation 10. Is the dehydration a redox reaction? That is, is there a net change in oxidation level in the molecule?

## Synthesis of Longer Chains

Clearly, butanoic acid is not a long-chain fatty acid. However, repetition of carbon–carbon bond formation and the steps in equations 9–11 can be used to build longer chains. The carboxylated thiol acetic acid ester shown in Figure 22.6 is used again, but this time it is condensed with the thiol butanoic acid ester produced in equation 11 rather than with an acetate ester. The steps (shown in Figure 22.7) are exactly analogous to those illustrated in Figure 22.6, except that one of the two components to be joined

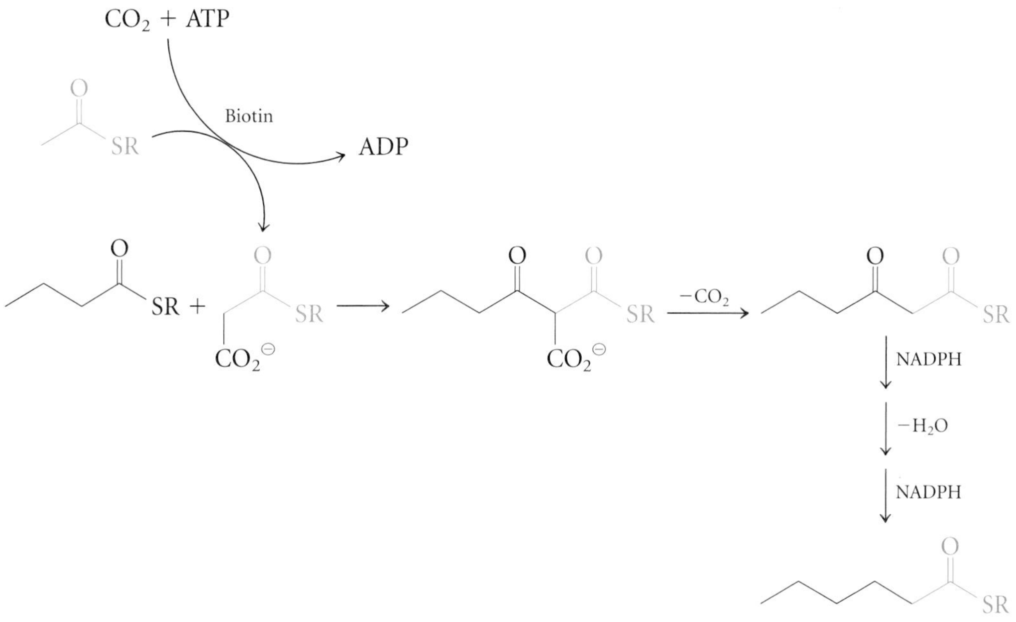

**FIGURE 22.7**

The thiol ester of butanoic acid (produced as shown in equations 9–11) is elaborated to a six-carbon thiol ester by the same reactions by which it was formed. The equivalent of a Claisen condensation with carboxylated thiol acetate forms the carbon–carbon bond. Reduction is then effected by repetition of the reactions in equations 9–11.

by the new carbon–carbon bond has four carbons. The resulting $\beta$-keto-thiol ester, then, has six carbons. In accord with the steps shown in equations 9–11, the keto group of this product is reduced, the resulting alcohol dehydrated, and the unsaturated ester reduced to form a six-carbon fatty acid chain (as a thiol ester).

Each cycle of the sequence adds two carbons (from the malonic acid thiol ester unit) to the growing straight-chain thiol ester. As noted earlier, the third carbon of the malonic acid derivative is lost as carbon dioxide, thus providing a driving force for this step in the sequence. The original thiol ester of acetic acid gains two carbons to become a butanoic acid thiol ester, to which two more carbons are added, forming the thiol ester of hexanoic acid. Each time this process is repeated, two carbons are added, which explains why fatty acids have an even number of carbon atoms. The process of fatty acid biosynthesis is regulated by the "rate-limiting" step of conversion of the thiol ester of acetic acid into the malonic acid thiol ester. This step consumes ATP, and, in its absence, the biosynthesis of fatty acids stops.

### EXERCISE 22.9

Suppose that acetic acid molecules in which the carboxylate carbon is isotopically labeled with $^{13}C$ are available to an organism conducting fatty acid biosynthesis.

(a) Indicate the positions of $^{13}C$ atoms in octanoic acid if it were formed in this way.

(b) How might $CH_3{}^{13}CO_2H$ be prepared in the laboratory starting from $^{13}CH_3OH$? From $^{13}CH_3CO_2H$?

## 22.7

## Energy Release in Fatty Acid Degradation

We have just seen how fatty acids are built progressively from acetic acid, with each two-carbon chain-lengthening step consuming one ATP and two NADPH molecules. A significant fraction of the energy stored in the fatty acids can be released later, when biological demands require it, by their degradation. In the process of fatty acid degradation, except for minor details, the steps in fatty acid biosynthesis are reversed. When an organism has a low demand for energy, fatty acid degradation stops, and the fatty acids accumulate as lipids. Unfortunately, this accumulation of lipids has well-known consequences in human beings!

The following steps in fatty acid degradation (oxidation) are illustrated in Figure 22.8, beginning with the six-carbon hexanoic acid thiol ester:

**1.** Loss of two hydrogen atoms from the fatty acid thiol ester to form an $\alpha,\beta$-unsaturated thiol ester

**2.** Hydration of the $\alpha,\beta$-unsaturated thiol ester to a $\beta$-hydroxythiol ester

**3.** Oxidation of the $\beta$-hydroxythiol ester to a $\beta$-ketothiol ester

**4.** Retro-Claisen condensation to split off a two-carbon unit, shortening the carbon chain by two to form butanoic acid thiol ester

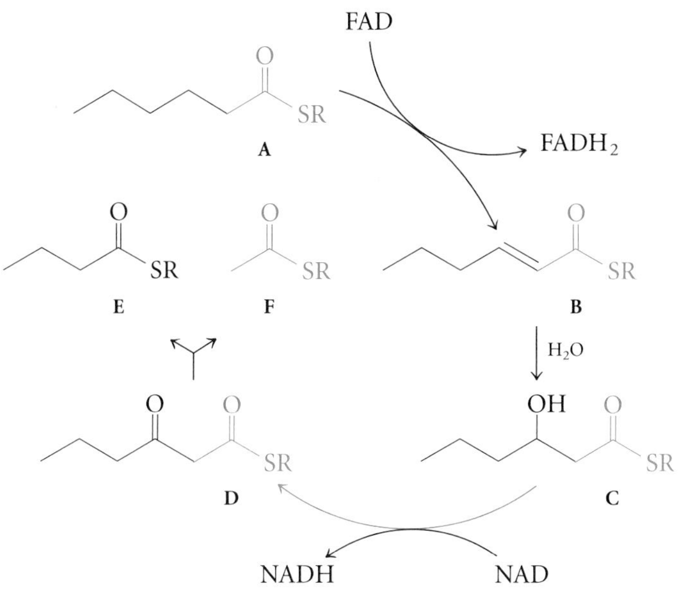

**FIGURE 22.8**

Most of the steps by which a fatty acid is degraded are the reverse of the
reactions for its construction. The FAD-mediated oxidation of a fatty acid
thiol ester (A) to the $\alpha,\beta$-unsaturated thiol ester (B), followed by hydrolysis, produces a $\beta$-hydroxythiol ester (C). Oxidation of the hydroxyl group,
coupled with the reduction of NAD, leads to a $\beta$-ketothiol ester (D), from
which a retro-Claisen condensation produces two shorter thiol esters (E
and F). Overall, the two-carbon acetate unit shown in blue is lost from
the starting fatty acid.

In the first step of fatty acid degradation, two hydrogen atoms are lost
to form an $\alpha,\beta$-unsaturated thiol ester. This oxidation is accomplished with
FAD, not with NADP (which would exactly reverse the fatty acid biosynthesis step). The reasons for this difference in pathway are not clear. It is
likely that this oxidation includes radical intermediates. In contrast, the corresponding reduction in the final stage of fatty acid biosynthesis, which is
effected by NADPH, is usually represented as the transfer of a hydride ion.

In the next step, the double bond is hydrated by the addition of water.
As in fatty acid synthesis, this step requires an enzyme (as catalyst) but no
cofactor. An enzyme directs the hydration. Only one stereoisomer of this
alcohol is formed; its configuration is opposite that produced in the biosynthesis by the reduction of a $\beta$-ketothiol ester (equation 9). Oxidation of the
alcohol to the $\beta$-ketothiol ester is effected by the consumption of NAD, producing NADH. Finally, the $\beta$-ketothiol ester undergoes a reverse (retro-)
Claisen condensation. This last reaction cleaves the carbon–carbon bond
between the C-2 and C-3 of the six-carbon chain, producing a two-carbon
fragment (a thiol ester of acetic acid) and the thiol ester of butanoic acid.

By the repetition of these degradation steps, saturated fatty acids are
broken down into thiol esters of acetic acid:

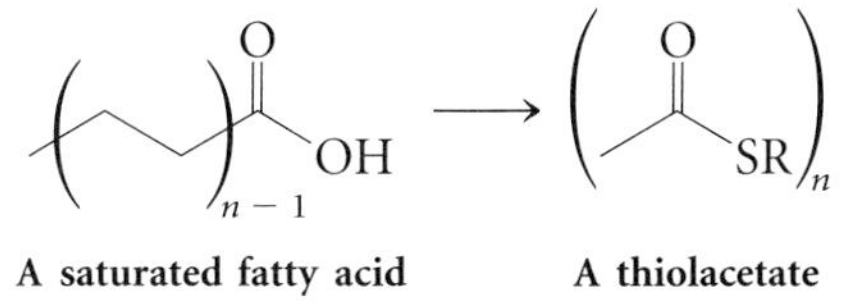

**A saturated fatty acid**          **A thiolacetate**

For example, the 18-carbon fatty acid stearic acid is degraded ultimately to nine acetate units via this pathway. This process constitutes an oxidation, because half the carbons in the acetate products are at the carboxylic acid oxidation level, compared with only one in the original fatty acid.

Overall, two biochemical reducing agents, $FADH_2$ and NADH, are produced in this sequence. These acquire some of the chemical potential energy stored in the fatty acid. In the simplest energetic sense, either of these molecules can reduce molecular oxygen in an exothermic process. Furthermore, either of these reducing agents provides sufficient driving force for the reduction of a ketone. Thus, fatty acids serve as a storage medium for reducing equivalents in a structurally quite simple form. Because fatty acids contain only $CH_3$, $CH_2$, and $CO_2H$ groups, they are chemically quite stable in the absence of the enzymes that initiate the beginning stages of fatty acid degradation.

### EXERCISE 22.10

What intermediate would be formed if the conversion of compound A into compound B in Figure 22.8 were accomplished by single-electron transfer, as for $FADH_2$ reductions? Show the electron flow needed to convert this intermediate into the observed product.

## 22.8

## The Krebs Cycle: Release and Transfer of Energy from Acetate

### Oxidative Decarboxylation: Overview of the Krebs Cycle

The acetate units produced in the degradation of fatty acids are further oxidized in a fascinating sequence of steps known as the **tricarboxylic acid (TCA) cycle** (or citric acid cycle, or **Krebs cycle**). Its discoverer, Sir Hans Adolf Krebs, shared the Nobel prize for physiology and medicine in 1953 with Fritz Albert Lipmann, who discovered coenzyme A. This sequence of reactions results ultimately in the oxidative decarboxylation of acetic acid to two molecules of carbon dioxide and produces additional equivalents of reduction potential, stored in NADH and $FADH_2$. It thus serves to release stored energy by providing a pathway for the controlled oxidation of acetate, and temporarily stores some of the energy released in that process in NADH and $FADH_2$. The energy in NADH and $FADH_2$ can then be transferred by coupling with other reactions.

In order for the energy recovery from acetate to be maximized, both carbons must be oxidized to their highest level (+4, in carbon dioxide). This requires cleavage of the carbon–carbon bond in acetate—that is, decarboxylation—which creates a dilemma. First, recall that acetic acid cannot readily lose carbon dioxide chemically (equation 12) in the same way as a $\beta$-ketoacid can (equation 13). Also, decarboxylation of $\alpha$-ketoacids is not a favored process; it fails (equation 14) under conditions in which carbon dioxide is lost easily from a $\beta$-ketoacid.

$$\text{(12)}$$

$$\text{(13)}$$

$$\text{(14)}$$

The ketone group in a $\beta$-ketoacid provides a means by which the negative charge produced in the loss of carbon dioxide is stabilized as an enolate anion (equation 13). The corresponding $\alpha$-ketoacid, even though it has a ketone group, cannot generate a stable anion upon loss of carbon dioxide (equation 14). Instead, it requires a sequence of several steps involving thiamine for decarboxylation, as in the biochemical degradation of amino acids.

The situation with simple carboxylic acids is similar to that for $\alpha$-ketoacids. Direct loss of carbon dioxide would lead to the highly unstable, charge-intensive methyl anion (equation 12), a process that does not take place under normal reaction conditions. Therefore, energy release through decarboxylation of acetic acid requires the formation of a derivative with the functionality properly positioned to provide anion stabilization after loss of carbon dioxide.

Decarboxylation is an integral part of the TCA cycle, which is outlined on Figure 22.9 (on page 22-20). Acetate feeds into the TCA cycle (as a thiol ester) by reaction with a key starting material, oxaloacetic acid. As this cycle proceeds, two of the carbons of oxaloacetic acid are lost as carbon dioxide, and two carbons are added in the form of another acetate unit. Thus, oxaloacetic acid is not consumed as acetate is degraded to two molecules of carbon dioxide. Several oxidation steps in the cycle take place with the simultaneous conversions of NAD to NADH and of FAD to $FADH_2$. There are also other similarities with the steps in fatty acid degradation. Indeed, the cycle begins with the reaction of a thiol ester of acetic acid with a ketone group of oxaloacetic acid in an aldol-like reaction, which is similar to the Claisen condensation that initiates fatty acid biosynthesis. The product, citric acid, has a new carbon–carbon bond and a tertiary alcohol group.

Both the TCA cycle and fatty acid degradation take place completely within mitochondria, small structures within the cell that are responsible for the production of ATP from ADP. The NADH produced in the TCA cycle does not leave the mitochondria but is oxidized (indirectly) by molecular oxygen, producing one molecule of NAD and three molecules of ATP. Thus, the chemical potential energy produced in this cycle is available to the cell only as ATP, which moves freely into the body of the cell, and not as the reduction potential of NADH.

## Prochiral Centers in Citric Acid

Citric acid is formed by the nucleophilic addition of the $\alpha$ carbon of thiolacetate to the $\alpha$-keto group of oxaloacetic acid, followed by hydrolysis of the thiol ester:

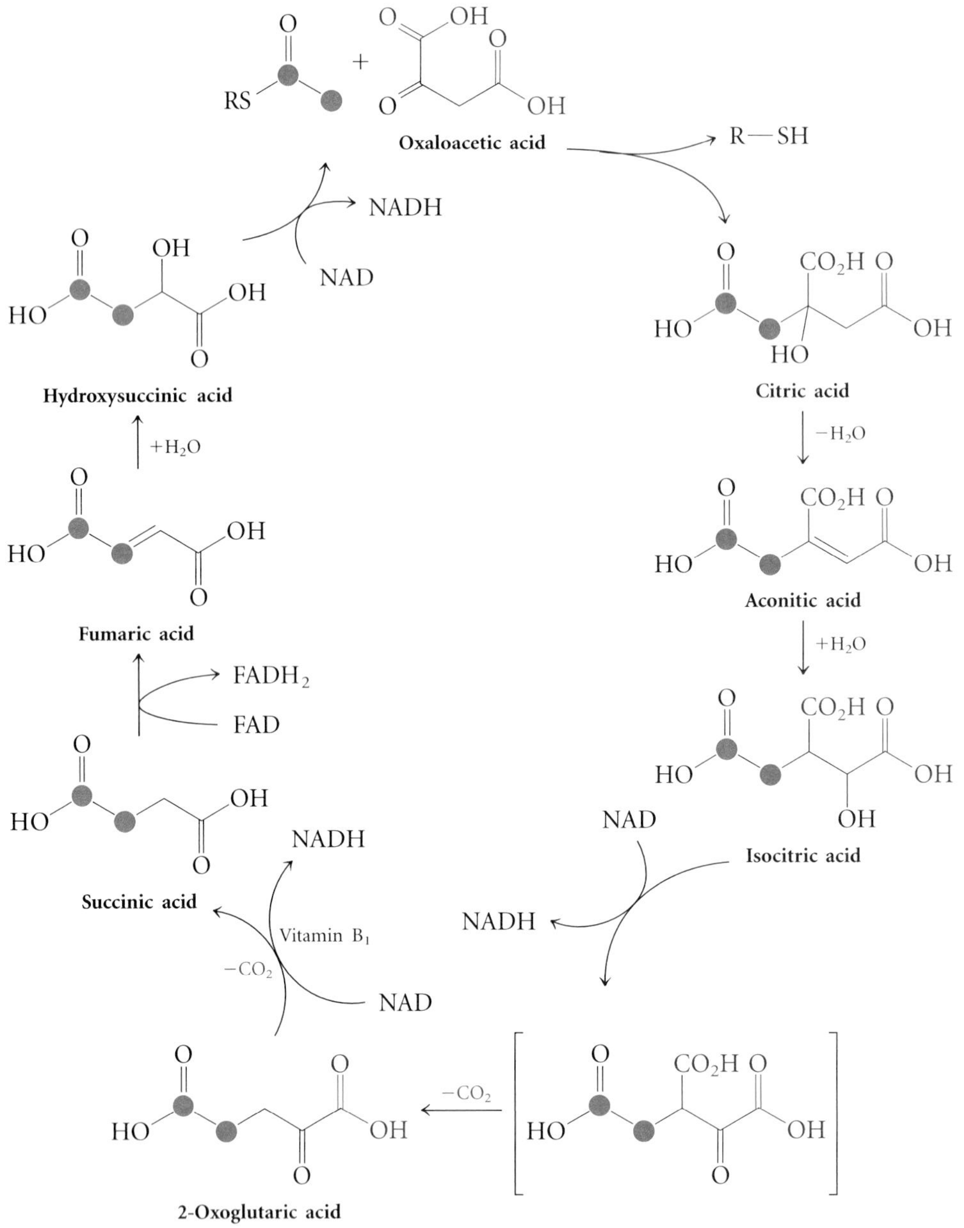

**FIGURE 22.9**

In the tricarboxylic acid (TCA) cycle, citric acid undergoes sequential dehydration, rehydration, and oxidation. These steps introduce a ketone carbonyl group $\beta$ to a carboxylic acid (bottom right), which undergoes ready decarboxylation to produce 2-oxoglutaric acid. Decarboxylation of 2-oxoglutaric acid (an $\alpha$-ketoacid) takes place with the production of succinic acid. The succinic acid is oxidized to fumaric acid, which undergoes hydration followed by oxidation to form oxaloacetic acid. Condensation of a thiol ester with this $\alpha$-ketoester reforms citric acid and permits the cycle to be repeated. One complete cycle converts an acetate unit (shown entering the cycle at top left) to two equivalents of carbon dioxide (shown in red, leaving the cycle at bottom and bottom left). The stored energy is released as heat and transferred to the NADH–NAD and FADH$_2$–FAD couples. The carbons of the thiol ester are shown as green circles so that their progress through the cycle can be readily seen.

$H_2O$

Oxaloacetic acid       Citric acid       + R—SH

## EXERCISE 22.11

Write a step-by-step mechanism for the formation of citric acid from oxaloacetic acid.

On first examining the structure of citric acid, you might think that C-4 is identical to C-2, because it appears that the left- and right-hand sides of the molecule are the same. However, this is not the case. A tetrahedral carbon atom with three of its four substituents different (like C-3 here) is a **prochiral center.**

(3S)-Monomethyl ester       Citric acid       (3R)-Monomethyl ester

Although the citric acid molecule is superimposable on its mirror image and therefore achiral by definition, C-3 becomes a center of chirality as a result of any transformation that changes one of the two identical groups. The handedness of this new center of chirality depends on which of the identical groups has been changed (and what change has been effected); thus, from this point of view, these two groups are really not identical. Transformation of one group leads to the $R$ configuration at the new center, so the group is called **pro-$R$.** Correspondingly, the other group is termed **pro-$S$.**

For example, if the C-1 carboxylic acid group of citric acid is converted to a methyl ester, the two groups attached to C-3 are no longer identical, and C-3 becomes a center of chirality. Thus, we can distinguish between these two groups on the basis of the configuration of the product formed by selective transformation of one of them. Such groups are called *enantiotopic*, because they lie on opposite sides of the plane of symmetry of an achiral molecule. When the carbon at C-1 is esterified, C-2 has a higher priority than C-4, and the resulting ester has the $S$ configuration. Conversely, when the carboxylic acid at C-5 is converted to a methyl ester, C-3 has the $R$ configuration. As noted earlier, enzymes almost always control processes that generate stereochemical centers in such a way that only one configuration is formed with a given enzyme.

You can visualize this process of chiral recognition with the help of Figure 22.10 (on page 22-22). Geometric shapes are used to represent the structure of citric acid, exemplifying the general case of a prochiral center for which three of the four groups are different. Imagine that the four groups correspond to two spheres, a cube, and a prism, and the structure is placed on a board containing three holes (square, round, and triangular), such that the sphere labeled 5 (to correspond to C-5 of citric acid) is aligned with a

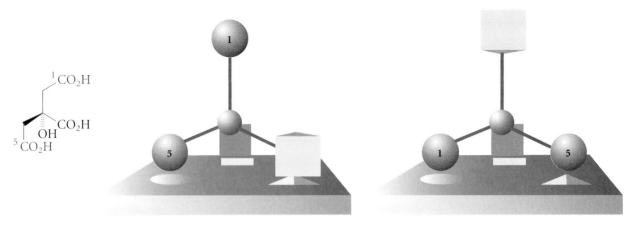

**FIGURE 22.10**

A prochiral molecule can interact with a chiral surface (for example, at the active site of an enzyme) in different ways, depending on which of the two identical (prochiral) substituents approaches the surface.

round hole on the surface. Both the cube and the prism also align with their corresponding geometric holes (Figure 22.10, left). If we keep the cube in the square hole but rotate the structure so that sphere 1 aligns with the round hole in the board, we have no choice but to attempt to match sphere 5 with the triangular hole (Figure 22.10, right). This analogy contains no real chemical details, but it serves to describe how an enzyme is capable of recognizing the details of the stereochemistry of molecules. Like a board having holes of various geometric shapes, the pocket of an enzyme has cavities that differ from each other in polarity, size, and shape. Each cavity is tuned to a part of a substrate molecule, enabling recognition of that particular part. However, as a result of this precise matching, many enzymes serve as catalysts for only a limited number of substrates, and some are effective for only one specific molecule.

In the next two steps of the TCA cycle, citric acid is isomerized to **isocitric acid** by dehydration and then rehydration. Note that in this conversion, a hydroxyl group is moved from the center of the molecule to C-4.

Citric acid $\xrightarrow{-H_2O}$ Aconitic acid $\xrightarrow{+H_2O}$ Isocitric acid

In the first step, an enzyme discriminates between the pro-*R* and the pro-*S* groups, and dehydration occurs only with the loss of hydrogen from C-4. This selective process has no stereochemical consequence, because there are no centers of chirality in the product alkene. However, the consequence of removal of a hydrogen atom specifically from C-4 and introduction of a hydroxyl group at this carbon is that neither carbon atom of the acetate unit introduced in the first step is lost as carbon dioxide in subsequent reactions.

*CHEMICAL PERSPECTIVES*

FLUOROACETIC ACID: A SIMPLE BUT DEADLY MOLECULE

Fluoroacetic acid is very toxic, as was first observed in animals that ate the leaves of the South African plant *Gifblaar* (*Dichapetalum cymosum* Hook), which contains this unusual natural compound. The toxic effect of fluoroacetic acid results from its incorporation into the TCA cycle to form fluorocitric acid. This compound binds tightly to the enzyme responsible for the isomerization of citric acid to isocitric acid and blocks it from performing this critical role. As a result, the level of citric acid increases dramatically and, deprived of the chemical potential energy and reduction potential produced in the TCA cycle, the animals often die. Only about 2–5 mg per kilogram of body weight (less than 0.5 g) is required to kill an average-sized man.

Fluoroacetic acid                Fluorocitric acid

## Energy Transfer

The first three steps of the TCA cycle set up the system so that the remaining steps can release energy by decarboxylation and reduction to NADH and FADH$_2$. Isocitric acid is oxidized by NAD (producing NADH) to a molecule that has both an $\alpha$-ketoacid and a $\beta$-ketoacid functional group.

Isocitric acid                                    2-Oxoglutaric acid

This ketotriacid readily undergoes decarboxylation. The carbon lost in this step was present in oxaloacetic acid and is not derived from the added acetic acid residue. As an $\alpha$-ketoacid, the resulting 2-oxoglutaric acid undergoes decarboxylation. Vitamin B$_1$ (thiamine) serves as cofactor, and another molecule of NAD is reduced to NADH in the process.

2-Oxoglutaric acid                                Succinic acid

The resulting symmetrical diacid, succinic acid, is oxidized by FAD to an unsaturated diacid, fumaric acid, in a fashion analogous to fatty acid degradation. As in fatty acid degradation, the double bond of fumaric acid undergoes hydration, and the resulting hydroxyl group formed is oxidized to a ketone with NAD.

Succinic acid          Fumaric acid

Hydroxysuccinic acid          Oxaloacetic acid

This product ketodiacid is oxaloacetic acid, and the cycle has returned to its starting point. In this cycle, C-5 and C-6 of citric acid are lost in the two decarboxylation steps.

Oxaloacetic acid          Citric acid

2 $CO_2$ (C-5 and C-6)

Each time the cycle is completed, two carbon atoms are lost as carbon dioxide, and two different ones are added from acetate.

The net conversion accomplished in the TCA cycle, in which one molecule of acetate is consumed, is the production of two molecules of carbon dioxide, three of NADH, and one each of $FADH_2$ and coenzyme A (written as R—SH).

$$2\ CO_2 + R-SH + \underbrace{3\ NADH + FADH_2}$$

$$\downarrow 2\ O_2$$

$$3\ NAD + FAD + 4\ H_2O$$

$$\Delta G° = -193\ \text{kcal/mole}$$

If all of the reduction potential in the NADH and FADH$_2$ formed in the cycle were consumed in the reduction of molecular oxygen to water, a total of 193 kcal/mole would be released as a consequence of this oxidation of a thiol ester of acetic acid to carbon dioxide. This is an enormous amount of energy to be released at once: it is thus important to the organism that it be released in carefully controlled fashion.

### EXERCISE 22.12

All steps in the TCA cycle are catalyzed by enzymes. Therefore, the outcome may not correspond to what would happen under laboratory conditions. Under acid-catalyzed conditions, which isomer—citric acid or isocitric acid—will be formed more rapidly by hydration of the intermediate alkene? Write a mechanism for this hydration reaction.

### EXERCISE 22.13

Draw the transition state for the following decarboxylation. Indicate the electron flow that accompanies bond-breaking and bond-making.

2-Oxoglutaric acid

## 22.9

# Maximizing the Efficiency of Energy Release by Controlling Heat Release

The energy stored in fatty acids can be released as heat by direct oxidation with molecular oxygen to form carbon dioxide and water. In addition, some of this energy can be stored in other chemical bonds. The efficiency of the transfer of chemical energy from fatty acids to other molecules depends on how much energy is lost as heat during the degradation process. An organism needs to be able to control the amount of stored energy that will be released as heat, so that energy will be available to carry out chemical transformations.

Let's examine the oxidative degradation of palmitic acid in more detail. The energy released by the simple combustion of palmitic acid is 2386 kcal/mole:

$$CH_3(CH_2)_{14}CO_2H + 23\ O_2 \longrightarrow 16\ CO_2 + 16\ H_2O \qquad \Delta G° = -2386\ \text{kcal/mole}$$

**Palmitic acid**

The biological breakdown of palmitic acid into 8 thiol acetic acid units produces 2 NADPH molecules for each acetate (a total of 16 NADPH molecules). Degradation through the TCA cycle of the 8 acetic acid equivalents produced from palmitic acid produces 24 NADH and 8 FADH$_2$ molecules.

## ENERGY CONSERVATION

The process of making rum from molasses begins with fermentation of sugars into ethanol. However, the microorganisms responsible for this reduction process cannot survive alcohol concentrations above 13% (26 proof). Therefore, to obtain 80-proof rum, the crude mixture of alcohol, water, and other volatile compounds is separated by the use of large distillation columns. At the Bacardi production facility outside San Juan, Puerto Rico, 70% of the energy required to run the entire operation, including the stills that produce 100,000 gallons of rum daily, is obtained by converting organic waste materials into methane gas, using microbes. (The microbes carry out this reduction efficiently when deprived of oxygen.)

$$8 \text{ NADP} + 24 \text{ NAD} + 8 \text{ FAD} + 2386 \text{ kcal/mole}$$

In total, then, the biological degradation of palmitic acid to 16 carbon dioxide molecules produces 40 NADH (or NADPH) molecules and 8 $FADH_2$ molecules.

If all of these reduced cofactors were reoxidized by molecular oxygen, 2386 kcal of energy would be released. This value is precisely the same as that obtained when palmitic acid is converted by direct oxidation with molecular oxygen into carbon dioxide and water. This is exactly as expected, because thermodynamics does not depend on reaction pathway. Were such a complete oxidation to be the process followed biologically, all of the energy stored in the fatty acid would be released as heat, and none would be available for other biochemical transformations. Furthermore, the heat released would be dangerous to the organism. Instead, the potential energy is much more efficiently utilized by being broken into smaller amounts that can be incorporated into subsequent biochemical transformations.

It is often advantageous to have a source of chemical energy, or driving force, that can be incorporated into a reaction that would otherwise be unfavorable. By coupling chemical reactions with oxidation–reduction agents, biological systems effect the oxidation of NADH (as well as of NADPH and $FADH_2$) and the simultaneous conversion of ADP to ATP.

$$\frac{1}{2} O_2 + NADH + 3 ADP \xrightarrow{\text{Enzymes}} NAD + 2 H_2O + 3 ATP \qquad \Delta G° = -31.4 \text{ kcal/mole} \qquad (15)$$

$$\frac{1}{2} O_2 + FADH_2 + 2 ADP \xrightarrow{\text{Enzymes}} FAD + H_2O + 2 ATP \qquad \Delta G° = -22.2 \text{ kcal/mole} \qquad (16)$$

Unlike the direct oxidation of NADH (and FADH$_2$) with molecular oxygen (which releases all of the stored energy as heat), the biological process stores some of this energy in the high-energy phosphate bond of ATP.

Recall that the simple hydrolysis of ATP to ADP releases 7.0 kcal/mole of energy:

$$ATP + H_2O \rightleftharpoons ADP + \qquad \Delta G° = -7.0 \text{ kcal/mole} \quad (17)$$

Thus, when one NADH molecule is consumed to produce one NAD molecule and three ATP molecules, 21.0 (3 × 7.0) kcal/mole of energy is stored in the ATP. Similarly, the exothermicity of the combination of molecular oxygen with FADH$_2$ is coupled with the conversion of two ADP molecules to two ATP molecules. Here, direct oxidation of FADH$_2$ would release 36.2 kcal/mole (equation 7), but when the process is mediated biologically, 14.0 kcal/mole (2 × 7.0, equation 4) is stored in the two molecules of ATP, and only 22.2 kcal/mole is released as heat (equation 16). Once again, the process is still exothermic, but when FADH$_2$ is converted into FAD, not all of the energy stored in FADH$_2$ is released as thermal energy.

We can now look at the overall energetics of the conversion of palmitic acid to carbon dioxide from the perspective of how much ATP can be produced. Let's assume that, instead of being directly oxidized, NADH and FADH$_2$ are oxidized in processes biologically coupled with the conversion of ADP into ATP. In this case, the 16 molecules of NADPH plus the 24 molecules of NADH produce 120 molecules of ATP, and the 8 molecules of FADH$_2$ produce 16 molecules of ATP, for a total of 136 ATP molecules.

$$40 \text{ NAD(P)H} + 8 \text{ FADH}_2 \xrightarrow{\begin{array}{c} 24 \text{ O}_2 \quad 48 \text{ H}_2\text{O} \\ \end{array}} 40 \text{ NAD(P)} + 8 \text{ FAD} \qquad \Delta G° = -952 \text{ kcal/mole} \qquad (18)$$
$$136 \text{ ADP} \quad 136 \text{ ATP}$$

Because each molecule of ATP represents 7.0 kcal/mole of stored chemical energy, the total energy produced from one mole of palmitic acid is 952 kcal. This amount represents 40% of the total of 2386 kcal/mole released on direct oxidation. The efficiency of the conversion of the stored energy of palmitic acid into ATP is impressive. Analogous efficiencies for energy conversion outside the biological realm are rare. For instance, in the conversion of thermal energy into electricity in power plants, 40% efficiency is the best yet obtained.

**22-27**

22.10

# Energy Release from Carbohydrates through Glycolysis

Carbohydrates provide another major store of energy for living systems. Let's compare the biological sequence for releasing the energy stored in carbohydrates to that for fatty acid degradation. Like the oxidative degradation of fatty acids, the breakdown of carbohydrates, called **glycolysis,** also produces acetic acid (as a thiol ester), which, as we have seen, enters the TCA cycle, a major pathway for energy release. Like fatty acid degradation and the TCA cycle, the transformation of the carbohydrate glucose to acetic acid takes place through reactions analogous to standard organic reactions.

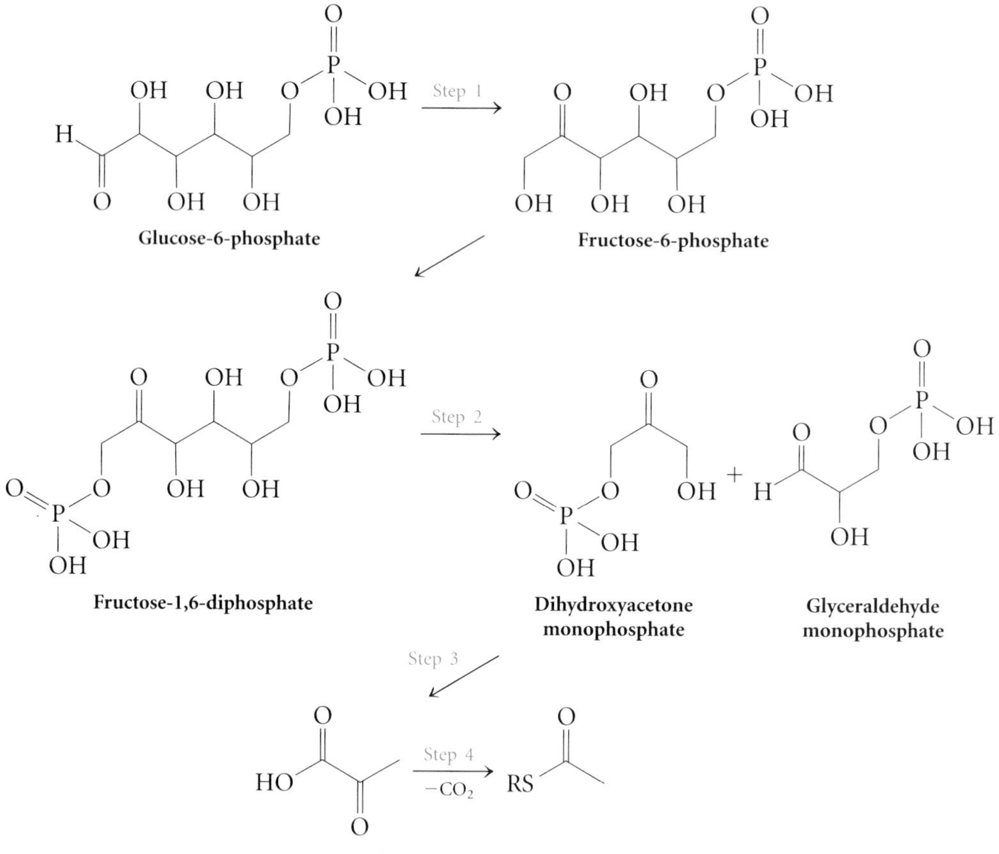

**FIGURE 22.11**

The degradation of carbohydrates begins with the isomerization of glucose-6-phosphate (an aldose) to fructose-6-phosphate (a ketose) in step 1. Phosphorylation of the C-1 alcohol produces fructose-1,6-diphosphate, which reacts by a retro-aldol reaction in step 2 to form two three-carbon fragments, dihydroxyacetone monophosphate and glyceraldehyde monophosphate. After the oxidation levels of the three carbons of dihydroxyacetone monophosphate are adjusted to produce pyruvic acid in step 3, decarboxylation can take place in step 4 by routes similar to those for other $\alpha$-ketoacids.

Glycolysis takes place in a sequence of four steps, shown in Figure 22.11:

**1.** Isomerization of glucose to fructose

**2.** Cleavage of fructose into three-carbon fragments through a retro-aldol reaction

**3.** Functionalization of the three-carbon fragments (forming pyruvic acid) to set the stage for decarboxylation

**4.** Decarboxylation of pyruvic acid to form a thiol ester of acetic acid

There are many steps in this biological process, but not much stored chemical energy is released. However, glucose (and other carbohydrates) ultimately provide significant amounts of energy because the acetic acid produced by glycolysis enters the tricarboxylic acid cycle and is further degraded to carbon dioxide. In the process, considerable stored potential energy is transferred to the reducing species NADPH and $FADH_2$. Let's look more closely at the first three steps in glycolysis.

### Isomerization of Glucose to Fructose

As glucose enters the cell's mitochondria, it is phosphorylated by ATP to form glucose-6-phosphate:

Glucose-6-phosphate

Thus, glycolysis starts by consuming chemical potential energy. The phosphorylated derivative next undergoes isomerization to fructose-6-phosphate:

Glucose-6-phosphate          Fructose-6-phosphate

You might expect the ketose (fructose) to be more stable than the aldose (glucose), but this is not the case, and the isomerization is endothermic by 0.4 kcal/mole. However, recall that both glucose and fructose exist in solution in cyclic forms, without free carbonyl groups. It is the cyclic hemiacetal form of glucose that is more stable than the hemiketal form of fructose. (Anomerization of these sugars takes place through the open-chain forms, and so these species are quite accessible as chemical intermediates.)

The isomerization of glucose to fructose is accomplished through a sequence of tautomerizations analogous to the reversible conversion of a carbonyl group to an enol. The open-chain form of glucose has an aldehyde functional group at C-1, from which an enol can be formed by the shift of a proton from C-2 to the aldehyde oxygen. The resulting enol bears two hydroxyl groups on the carbons of the double bond; it is therefore referred to as an *enediol*. Dihydroxyacetone with a ketone at C-2 is now accessible via tautomerization.

We can examine the chemistry of enediols in more detail by considering the interconversion of glyceraldehyde and dihydroxyacetone. This is a simple system that nonetheless includes the functional transformation that takes place in the conversion of glucose into fructose.

Glyceraldehyde          Enediol          Dihydroxyacetone

Simple enols are tautomers of ketones, and the same carbon bears the hydroxyl in the enol and the carbonyl oxygen in the ketone. However, in the enediol intermediate, each carbon of the double bond (C-1 and C-2) has a hydroxyl group. Thus, it is not possible to determine from the structure of the enediol whether it derived from an aldehyde carbonyl group at C-1 (glyceraldehyde) or from a ketone carbonyl group at C-2 (dihydroxyacetone); that is, this intermediate can serve as a precursor for either carbonyl compound. Protonation at C-2 forms glyceraldehyde after the loss of a proton from the C-1 hydroxyl group, whereas protonation at C-1 forms dihydroxyacetone after deprotonation of the C-2 hydroxyl group. We will encounter this isomerization again, later in the glycolysis sequence. Although the details of the sequence of proton removals and replacements in this isomerization are not known with certainty, a species analogous to the enediol must be formed as an intermediate.

### EXERCISE 22.14

Write mechanisms for the transformation of glucose into fructose under both acidic and basic conditions. Suggest a reason why the carbonyl group of fructose might not be as stable as that of a simple ketone such as acetone.

### Cleavage of Fructose into Three-Carbon Fragments

After glucose has been converted into fructose, glycolysis proceeds with cleavage of the bond between C-3 and C-4 via the equivalent of a retro-aldol reaction. Let's review the aldol reaction, a transformation similar to the carbonyl addition reactions in fatty acid biosynthesis and in the TCA cycle. The aldol reaction of acetone is

The aldol reaction produces a carbon–carbon bond as the nucleophilic carbon of an enolate anion or an enol reacts with an electrophilic carbonyl carbon. The reverse of this transformation, the retro-aldol reaction, is closely analogous to the loss of carbon dioxide from a $\beta$-ketoacid.

**Retro-aldol Reaction**     **Decarboxylation of a $\beta$-Ketoacid**

However, unlike the decarboxylation, the retro-aldol reaction does not have the large energetic driving force from formation of the very stable carbon dioxide molecule. Indeed, the equilibrium slightly favors the aldol product.

Drawing fructose as an open-chain ketone rather than as a cyclic hemiketal shows clearly that this molecule is a $\beta$-hydroxyketone. Because fructose is a $\beta$-hydroxyketone, it can undergo a retro-aldol reaction.

In the biological process of glycolysis, fructose-6-phosphate is converted into fructose-1,6-diphosphate (consuming yet another equivalent of ATP):

Fructose

Fructose-6-phosphate     Fructose-1,6-diphosphate

This diphosphate of fructose then undergoes a retro-aldol reaction, producing two three-carbon fragments by cleavage of the central carbon–carbon bond of fructose:

Fructose-1,6-diphosphate     A zinc enolate     Glyceraldehyde monophosphate

Dihydroxyacetone monophosphate

The enzyme that catalyzes this retro-aldol reaction, aldolase, has a zinc cation located near the carbonyl oxygen of fructose bound in the active site. The retro-aldol reaction initially produces glyceraldehyde monophosphate and a zinc enolate, which upon protonation yields dihydroxyacetone monophosphate. Each of these monophosphates represents half of fructose diphosphate.

## EXERCISE 22.15

It is conceivable that glucose (in its straight-chain aldehyde form) could undergo a retro-aldol reaction in the presence of base, although this reaction is not known to occur naturally. What products would be formed in this retro-aldol reaction? Write a detailed mechanism accounting for the bond-breaking and bond-making in this proposed transformation.

Dihydroxyacetone monophosphate is next isomerized to glyceraldehyde monophosphate, converting an achiral molecule into a chiral one:

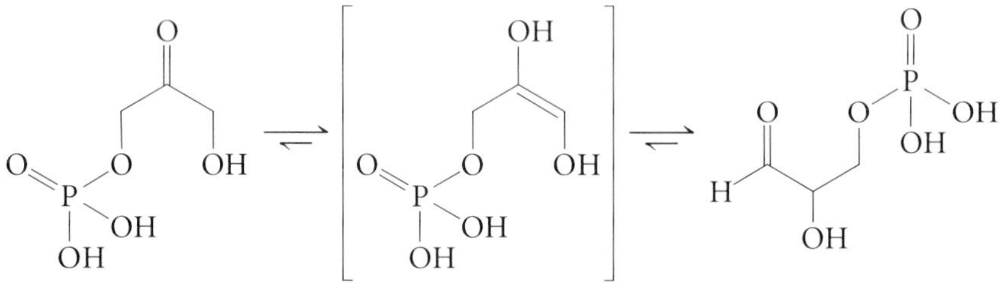

**Dihydroxyacetone monophosphate** **Glyceraldehyde monophosphate**

When effected by the enzyme isomerase, this transformation forms only the D-enantiomer. Addition of a proton to one face of the enediol results in the observed D configuration:

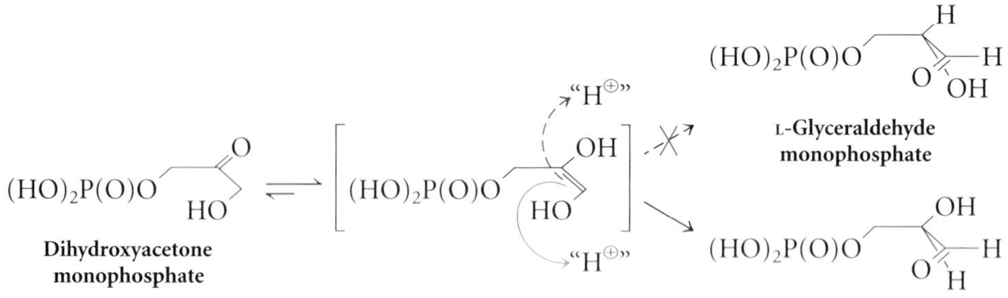

**Dihydroxyacetone monophosphate** **L-Glyceraldehyde monophosphate** **D-Glyceraldehyde monophosphate**

The alternative process, in which a proton is added from the other face to produce L-glyceraldehyde, is not observed in biological systems, where the enzyme isomerase catalyzes the protonation. Once again, we see that enzymes are capable of creating a stereochemical distinction. Because the addition of the proton results in the formation of a center of chirality, the $sp^2$-hybridized carbon atom in the precursor alkene is prochiral, in the same sense that C-3 of citric acid is prochiral. Similar stereochemical control is also observed in both fatty acid biosynthesis and the TCA cycle.

Knowledge of simple carbonyl-group chemistry tells us that it is reasonable for the ketone to be the more stable product derivable from the enediol, although the energy difference is not large (1.8 kcal/mole). Although the equilibrium actually favors dihydroxyacetone monophosphate over glyceraldehyde monophosphate, further conversion of this aldehyde (to a carboxylic acid) "pulls" the equilibrium as the aldehyde is consumed.

It is interesting to consider the reverse of the reaction that cleaves fructose diphosphate, because both the forward and the reverse reactions have the same transition state. The reverse carbon–carbon bond-forming reaction, in which fructose-1,6-diphosphate is built from dihydroxyacetone monophosphate and glyceraldehyde monophosphate, produces two centers of chirality:

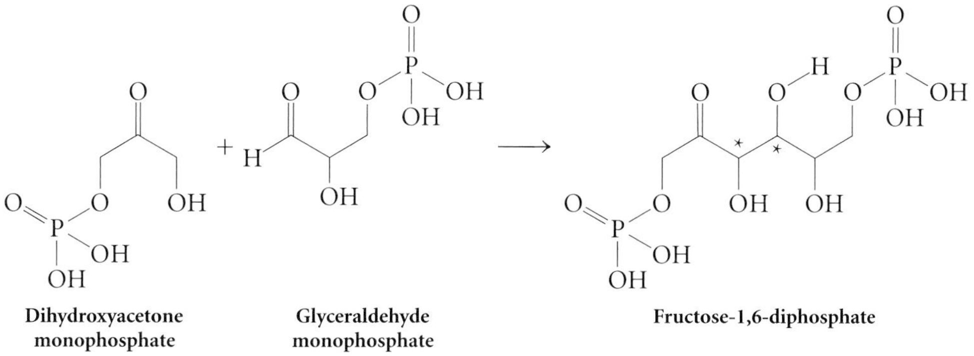

Dihydroxyacetone
monophosphate

Glyceraldehyde
monophosphate

Fructose-1,6-diphosphate

This process can be accomplished in the laboratory with simple glyceraldehyde (that is, lacking a phosphate group). Barium hydroxide acts as a base to catalyze the interconversion of glyceraldehyde and dihydroxyacetone. Then the enolate anion of dihydroxyacetone undergoes an aldol reaction with glyceraldehyde to produce the carbon skeleton of fructose:

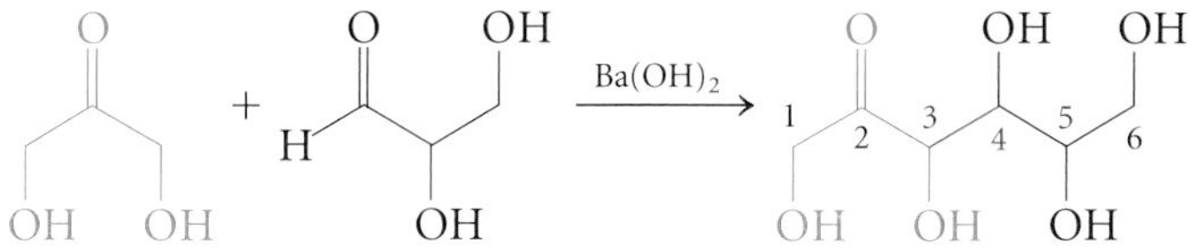

Of the six carbons of fructose, three are centers of chirality (C-3, C-4, and C-5). Only one of these stereocenters (C-5) is present in the starting materials, coming from C-2 of glyceraldehyde. Thus, the carbons linked in this process, C-3 and C-4, are new stereocenters formed in the reaction. Because each of them can have either the $R$ or the $S$ configuration, four stereoisomers are possible, $R,R$; $R,S$; $S,R$; and $S,S$.

## EXERCISE 22.16

Draw three-dimensional representations with hatched and wedge bonds and Fischer representations of the four possible stereoisomers that result from the aldol reaction of dihydroxyacetone with D-glyceraldehyde.

In the laboratory, only two of the four possible stereoisomers are formed to any appreciable amount—one is fructose. The presence of the divalent

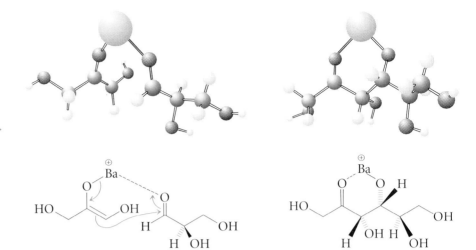

**FIGURE 22.12**

Reaction of the barium enolate of dihydroxyacetone with D-glyceraldehyde
proceeds through a cyclic transition state with the barium ion linking the
two fragments joined in the reaction.

cation barium in the reaction is responsible for this selectivity, functioning
to bridge the enolate oxygen and the carbonyl group of the reacting aldehyde, as illustrated in Figure 22.12. In general, the energetic difference between competing cyclic transition states is greater than that for noncyclic
transition states that accomplish the same transformation. The noncyclic
transition state offers more conformational options whereby groups can
avoid steric interactions. Because six atoms participate directly in the cyclic
process, the transition state can be conformationally defined roughly as a
chair. As a result of this conformation being favored, only one of the four
stereoisomers is formed.

The sequence of steps to this point in glycolysis converts glucose
through fructose diphosphate to two molecules of glyceraldehyde monophosphate. Although glucose constitutes a major source of chemical potential energy, this part of the glycolysis sequence actually consumes two
molecules of ATP to form the two phosphate bonds present in the two molecules of glyceraldehyde monophosphate. Even with the added driving force
of the ATP-to-ADP conversion, the transformation of glucose into two molecules of glyceraldehyde monophosphate is exothermic by only 1.3 kcal/
mole:

$\Delta G° = -1.3$ kcal/mole

## Conversion of the Three-Carbon Fragments into Acetic Acid Derivatives

Glyceraldehyde monophosphate is further transformed into carbon dioxide, acetic acid, and stored chemical potential energy in the form of NADH and ATP. The conversion of glyceraldehyde into pyruvic acid, an important intermediate in this pathway, requires changes in the oxidation levels of all three carbons.

***Oxidation of Glyceraldehyde at C-1.*** First, inorganic phosphate is added to the carbonyl carbon of glyceraldehyde monophosphate in the same way as other nucleophilic species, such as water and cyanide, react to make hydrates and cyanohydrins.

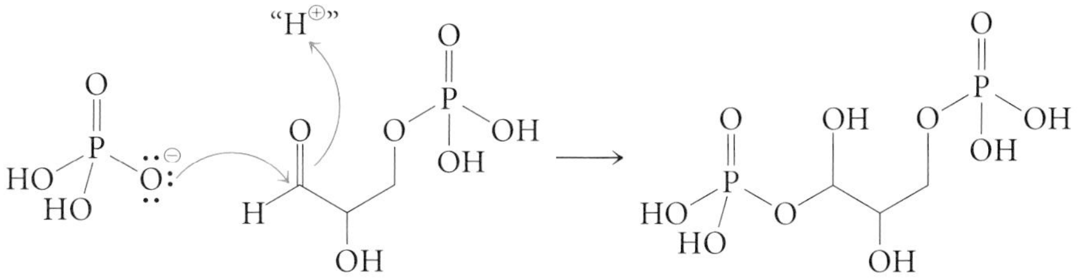

The hemiacetal derivative formed by this addition undergoes oxidation by NAD to form an interesting **mixed anhydride** between phosphoric acid and a substituted carboxylic acid:

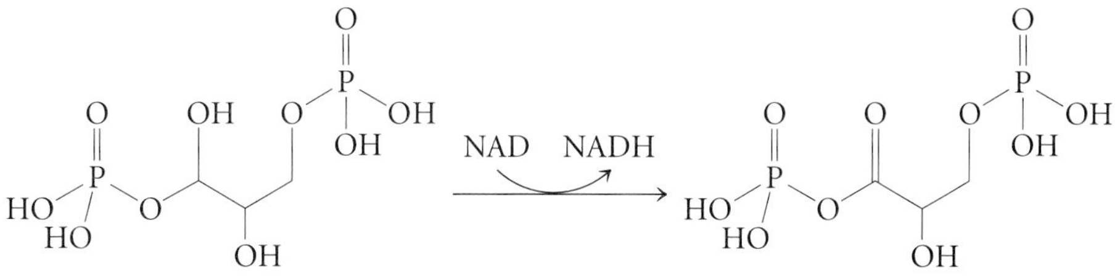

**A mixed anhydride**

The reaction of this mixed anhydride with ADP transfers a phosphate group, forming ATP and releasing the carboxylic acid:

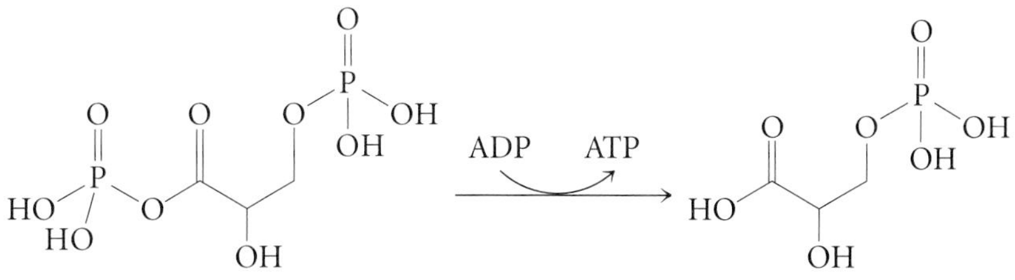

**A mixed anhydride**

This sequence thus achieves oxidation of the aldehyde of D-glyceraldehyde monophosphate, while leaving the oxidation state of the secondary alcohol unchanged.

***Adjusting the Oxidation Levels of C-2 and C-3.*** The next reaction is yet another isomerization, one that transfers the phosphate group from the hydroxyl at C-3 of glyceraldehyde to the hydroxyl at C-2.

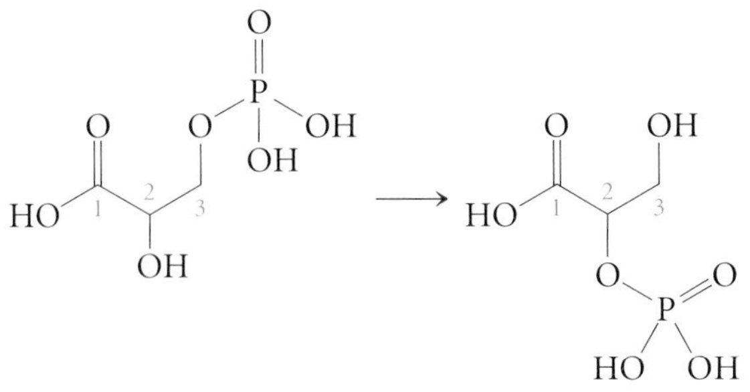

Let's consider how this conversion takes place. Isomerizations in which an ester group is transferred to an adjacent hydroxyl group are very common not only with phosphate esters, but also with carboxylic acid esters. It may be easier to visualize this process with a carboxylic acid ester group, as in the isomerization of a monoacetate ester of *cis*-1,2-cyclohexane diol:

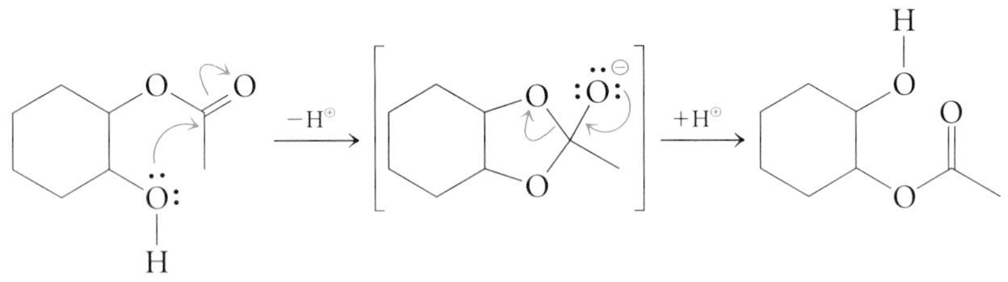

This transformation is similar to other nucleophilic acyl substitution reactions of carboxylic acid derivatives. It proceeds through a tetrahedral intermediate that, lacking a carbonyl group, is less stable than either the starting material or the product. Reactions that transfer carboxyl (or phosphoryl) groups intramolecularly often take place very rapidly when they proceed through five- or six-member cyclic transition states or intermediates. The transfer of the phosphate group from the C-3 to the C-2 hydroxyl of glyceraldehyde proceeds through an analogous cyclic intermediate and is very fast. Enzymatic catalysis accelerates the reaction even further.

In glycolysis, the next step in the sequence is the loss of water from the monophosphate ester to yield an enol phosphate carboxylic acid. In this step, the oxidation level of the terminal carbon in the product decreases, whereas that of the carbon bearing the phosphate group increases.

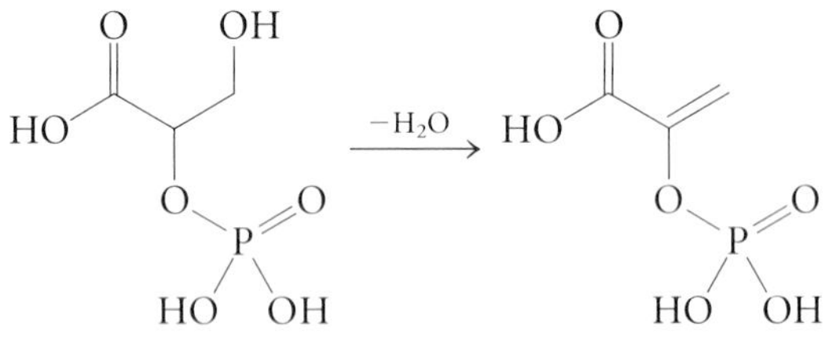

This transformation is similar to a sequence in the TCA cycle in which the oxidation level of a carbon atom bonded to a hydroxyl group is transferred to an adjacent carbon by dehydration, followed by rehydration. This sequence is illustrated by the conversion of 1-propanol to 2-propanol:

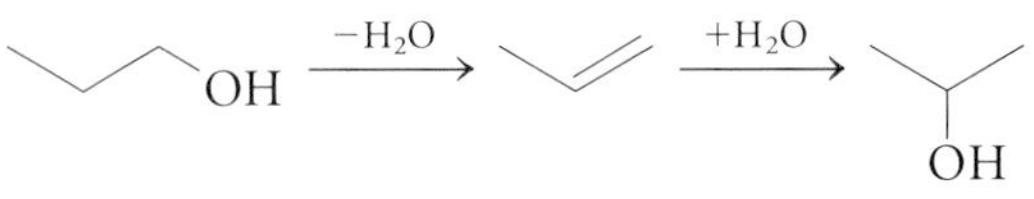

THE CHEMISTRY OF ACHES AND PAINS

Yogurt is made by treating milk with the organism *Lactobacillus bulgaricus,* which, among other things, produces lactic acid through the sequence of glycolysis. The reactions are similar to those described for the production of pyruvic acid. In animals, lactic acid is not normally produced by glycolysis to any significant extent because an excess of oxidation potential is available in the form of NAD.

However, upon extended exertion, the concentration of NADH builds up in muscle tissue, as glycolysis is called upon to provide a rapid increase in ATP as a source of energy. Reduction of pyruvic acid to lactic acid accompanies this process. It is the lactic acid that causes tired muscles to ache. Lactic acid contributes the gamey flavor attributed to meat from animals that have been killed on the run by hunters.

Pyruvic acid          L-Lactic acid

Although the overall conversion does not result in net oxidation or reduction, the oxidation level of C-1 decreases and that of C-2 increases. This oxidation and reduction occurs in two stages. In the intermediate alkene, 1-propene, the oxidation level of C-1 has decreased, whereas that of C-2 has increased. The same change happens again as the alkene is rehydrated to produce 2-propanol.

The enol phosphate formed in glycolysis by dehydration can be viewed, as its name implies, as a derivative of a carbonyl compound. Indeed, acid-catalyzed hydrolysis of this functional group releases pyruvic acid as well as inorganic phosphate:

Pyruvic acid

This simple process releases a large amount of thermal energy. Enolic derivatives of ketones are considerably less stable than the ketones themselves, because of the greater strength of the $\pi$ bond of a carbonyl group compared with that of an alkene (93 versus 63 kcal/mole). In biological systems, this energy is not released fully as heat; some is captured by the simultaneous conversion of ADP to ATP.

***Entry into the TCA Cycle via Decarboxylation of Pyruvic Acid.*** Pyruvic acid is degraded by decarboxylation (with the use of thiamine) to produce acetic acid. After conversion into a thiol ester, acetic acid enters the tricarboxylic acid cycle and is converted into two molecules of carbon dioxide. Thus, respiration, which overall constitutes the complete oxidative degradation of a hexose to carbon dioxide, is now complete.

**22-37**

The net transformation accomplished in glycolysis is summarized in Figure 22.13.

Conversion of each molecule of glucose to fructose-1,6-diphosphate consumes 2 ATPs. Degradation of fructose-1,6-diphosphate produces 2 pyruvate molecules, 2 ATPs, and 4 NADHs. This latter reduction potential is equivalent to 12 ATPs (equation 15), and thus the conversion of glucose to pyruvate produces net chemical potential energy equivalent to 12 ATPs.

Decarboxylation of the 2 pyruvate molecules produces 2 acetates and 2 NADHs. The 2 acetates enter the citric acid cycle, producing 4 molecules of $CO_2$, 6 NADHs, and 2 $FADH_2$s. Again, counting each NADH as the equivalent of 3 ATPs and each $FADH_2$ as 2 ATPs gives a total of 28 ATPs from conversion of 2 pyruvates to 2 $CO_2$ molecules. Adding these 28 ATPs to the 12 produced by conversion of glucose to pyruvate gives a total of 40 ATPs generated by respiration in the conversion of glucose into carbon dioxide. Because each ATP represents 7.0 kcal/mole of chemical potential energy, each mole of glucose is converted into 280 kcal/mole ($40 \times 7.0$) of potential energy. This value is 38% of that obtained when glucose is burned in air to form carbon dioxide and water, an efficiency almost the same as that for fatty acid degradation.

$$C_6H_{12}O_6 + 6\ O_2 + 40\ ADP \xrightarrow{\text{Respiration}} 6\ CO_2 + 6\ H_2O + 40\ ATP$$

$$\Delta G° = -280\ \text{kcal/mole}$$

$$C_6H_{12}O_6 + 6\ O_2 \xrightarrow{\text{Combustion}} 6\ CO_2 + 6\ H_2O$$

$$\Delta G° = -686\ \text{kcal/mole}$$

### EXERCISE 22.17

Representing ADP by the abbreviated structure shown below, write a reasonable reaction mechanism for the conversion of the enol phosphate into pyruvic acid concurrently with the transfer of the phosphate group to produce ATP.

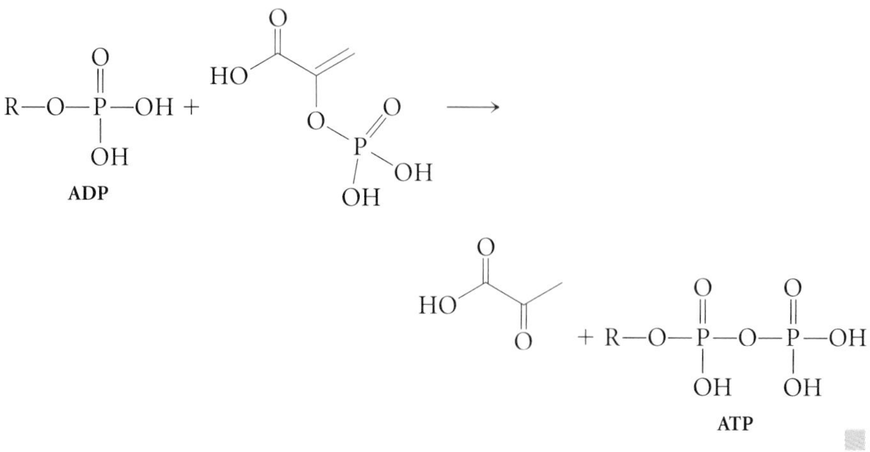

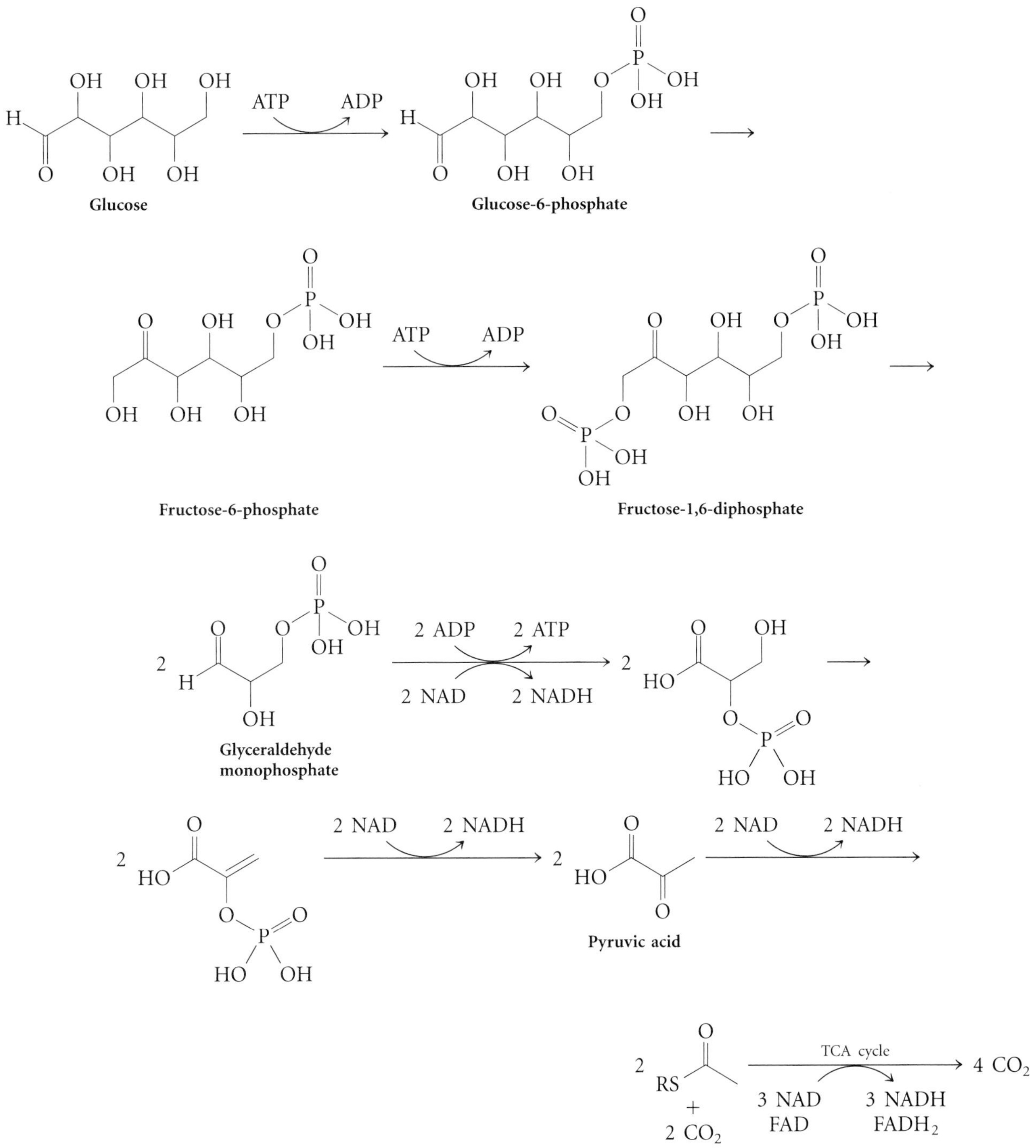

**FIGURE 22.13**

Metabolic breakdown of sugars (glycolysis) in animals takes place in several steps: phosphorylation of glucose; isomerization to fructose-6-phosphate; a second phosphorylation to fructose-1,6-diphosphate; a retro-aldol reaction producing glyceraldehyde monophosphate; dehydration and intramolecular transesterification of the phosphate ester to an enol phosphate; hydrolysis to pyruvic acid; decarboxylation to an acetate, which is converted into a thiol ester and degraded in the TCA cycle to carbon dioxide.

**22-39**

22.11

# Biological Reactions in Energy Storage and Utilization

In this chapter, we have considered a number of biochemical transformations that are involved in the storage and retrieval of energy. These biologically important processes are summarized in Table 22.1. They form an essential part of the chemistry of living organisms, both plants and animals. Nonetheless, each of the reactions involved in these processes belongs to one of the basic categories, including carbon–carbon bond formation, oxidation–reduction, and functional-group transformation.

## TABLE 22.1

Biological Reactions Involved in Energy Storage and Utilization

| Biological Process | Type of Reaction | Example |
|---|---|---|
| **Fatty acid biosynthesis** | Claisen-like condensation | |
| | Reduction, dehydration, and reduction | |
| **Fatty acid degradation** | Oxidation (alkane to alkene) | |
| | Hydration | |
| | Oxidation (alcohol to ketone) | |
| | Retro-Claisen-like condensation | |
| **Tricarboxylic acid cycle** | Aldol-like reaction | |

**TABLE 22.1**

*(continued)*

| Biological Process | Type of Reaction | Example |
|---|---|---|
| **Tricarboxylic acid cycle** | Dehydration | 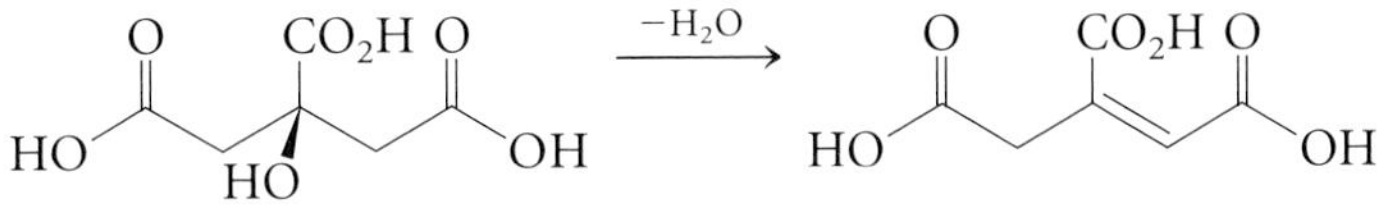 Citric acid $\xrightarrow{-H_2O}$ Aconitic acid |
| | Hydration of an alkene | 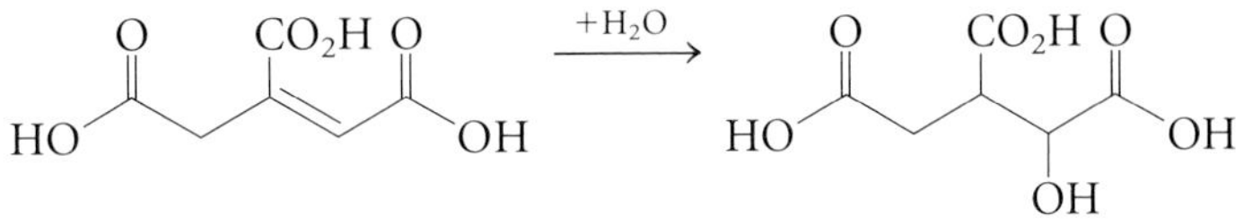 Aconitic acid $\xrightarrow{+H_2O}$ Isocitric acid |
| | Oxidation of an alcohol to a ketone | 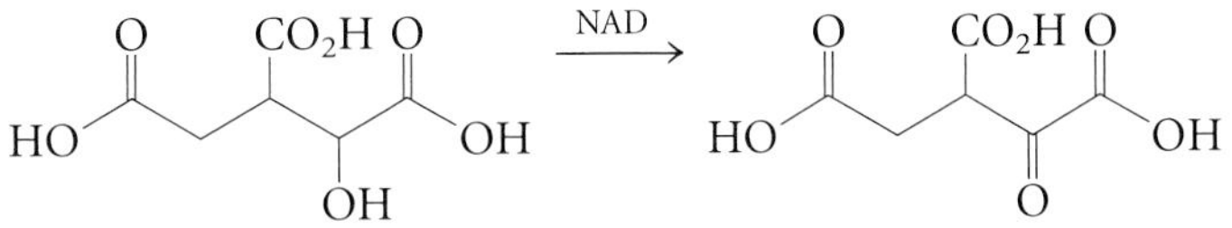 Isocitric acid $\xrightarrow{NAD}$ |
| | Decarboxylation of a $\beta$-diacid | 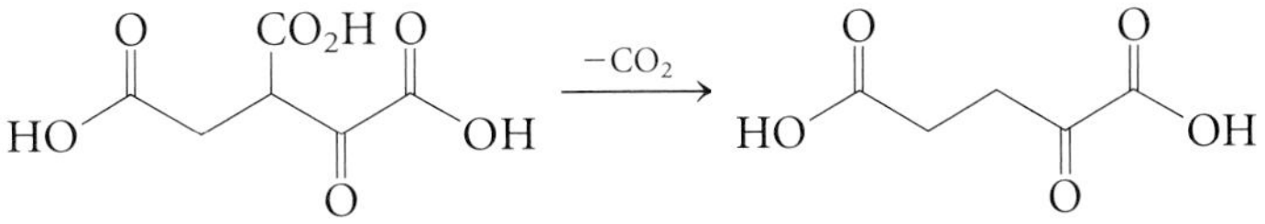 $\xrightarrow{-CO_2}$ 2-Oxoglutaric acid |
| | Oxidative decarboxylation of an $\alpha$-ketoacid | 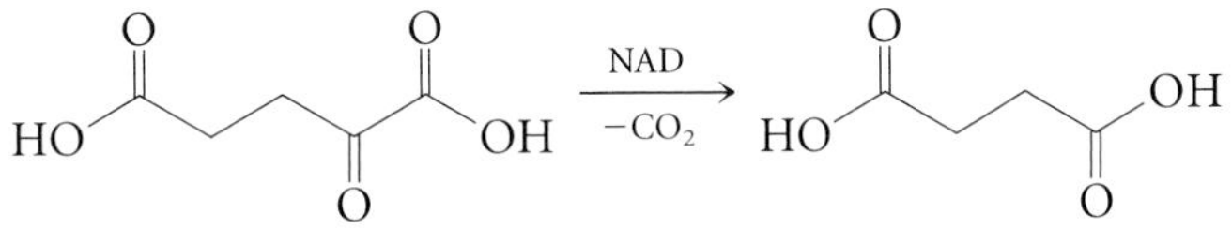 2-Oxoglutaric acid $\xrightarrow[-CO_2]{NAD}$ Succinic acid |
| | Oxidation of an alkane to an alkene | 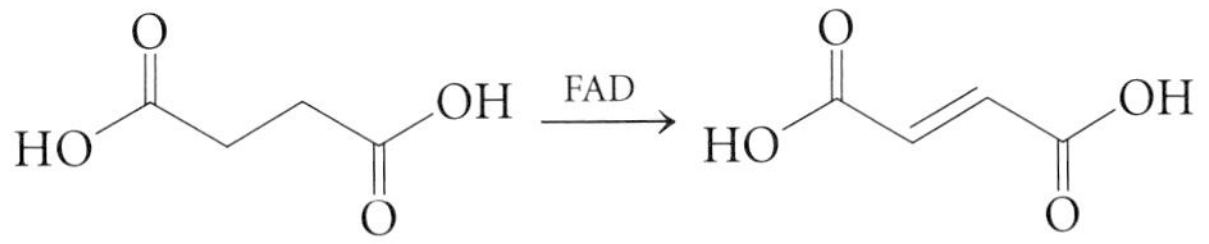 Succinic acid $\xrightarrow{FAD}$ Fumaric acid |
| | Hydration of an alkene | 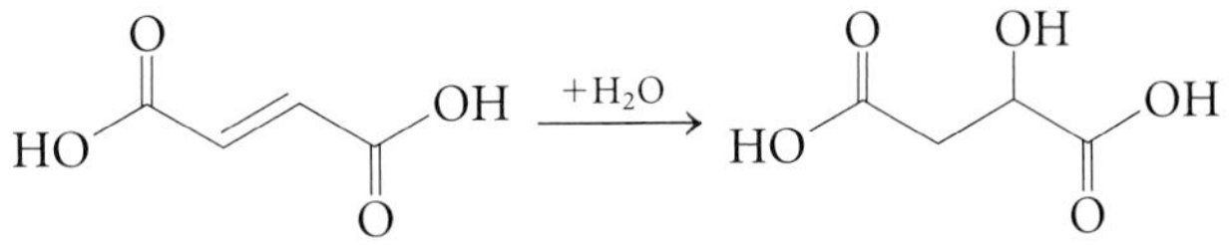 Fumaric acid $\xrightarrow{+H_2O}$ 2-Oxoglutaric acid |
| | Oxidation of an alcohol to a ketone | 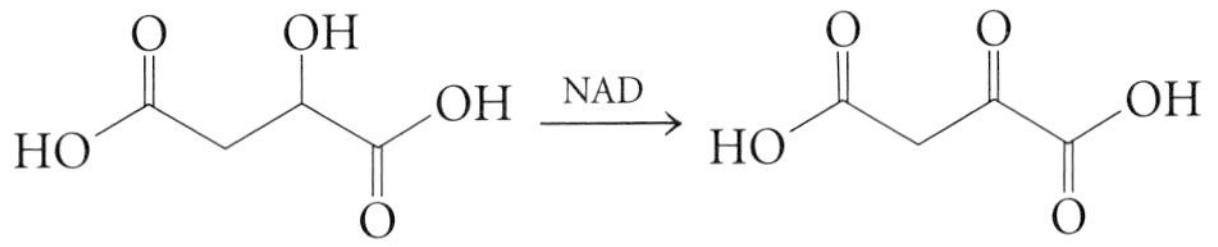 2-Oxoglutaric acid $\xrightarrow{NAD}$ Oxaloacetic acid |

*(continued)*

**TABLE 22.1**

*(continued)*

| Biological Process | Type of Reaction | Example |
|---|---|---|
| **Glycolysis** | Retro-aldol reaction | 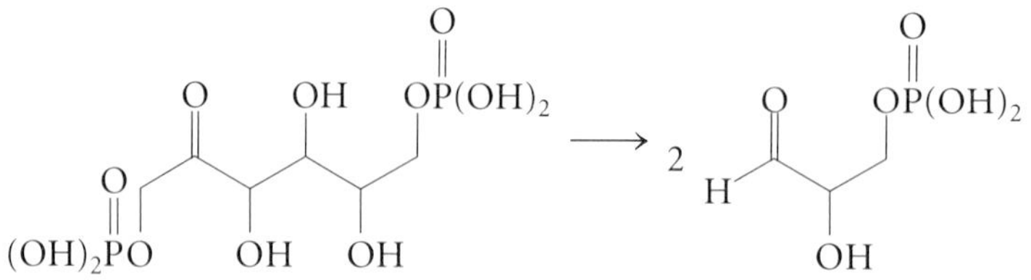 Fructose-1,6-diphosphate → 2 Glyceraldehyde monophosphate |
| | Oxidation of an aldehyde to a carboxylic acid | 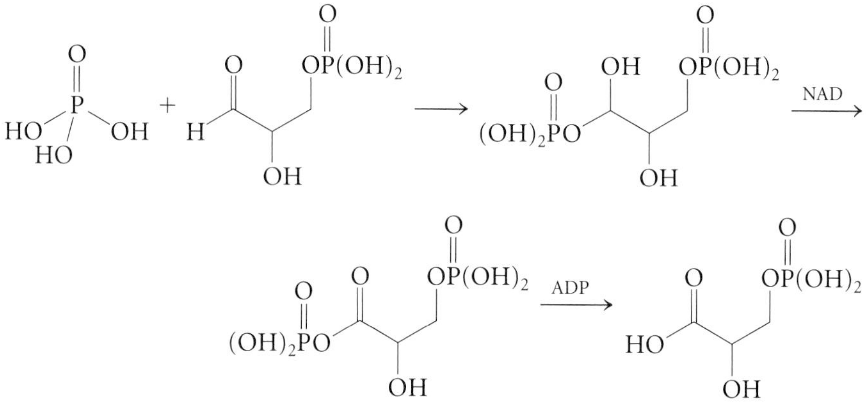 |
| | Intramolecular transesterification | 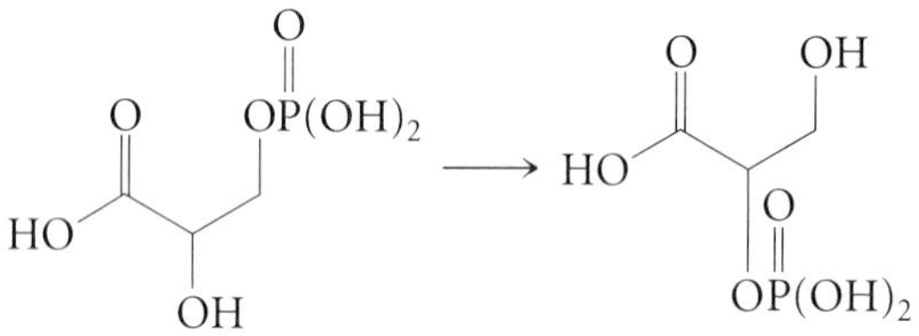 |
| | Dehydration | 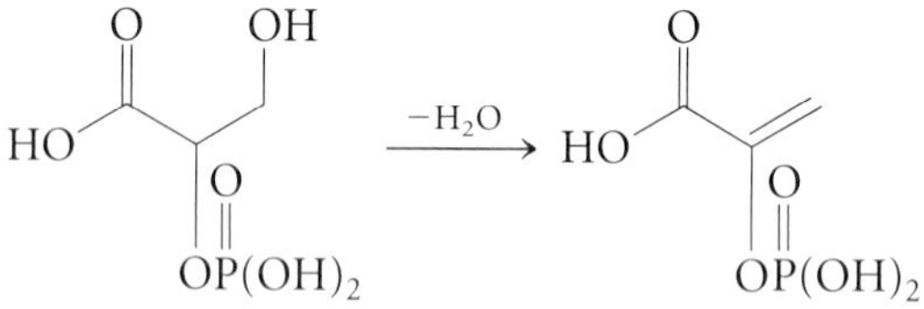 |
| | Hydrolysis of an enol derivative to a ketone | 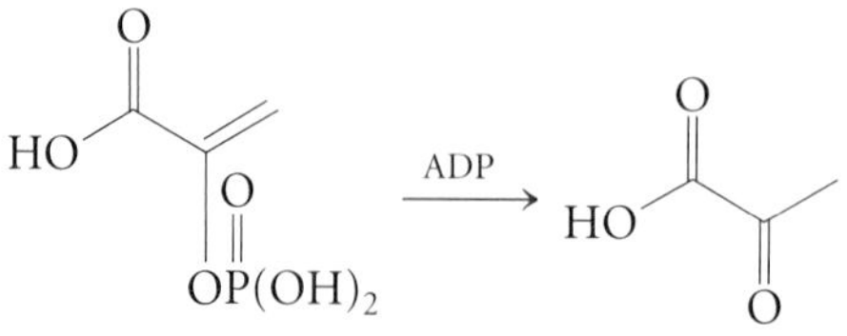 Pyruvic acid |

**1.** In the energetics of reactions in biological systems, a key and recurring theme is the coupling of several reactions in which one or more endothermic steps are driven by an exothermic partner.

**2.** Biological transformations are closely analogous to the elementary reactions of organic chemistry. In some cases, the chemistry seems complex, because the structures themselves are large and have a variety of functional groups. However, usually only one of these groups undergoes reaction.

**3.** Among the most common energy-releasing reactions in biological systems are the hydrolyses of ATP and ADP, the oxidation of NADPH to NADP, the oxidation of $FADH_2$ to FAD, and the decarboxylation of $\alpha$- and $\beta$-ketoesters and thiol esters. When coupled with highly exothermic steps, the reverse reactions constitute energy storage steps.

**4.** Energy stored in biomolecules, especially in fats and carbohydrates, is released both as heat and as chemical potential energy. The latter can be called upon to drive an otherwise unfavorable reaction toward product in a controlled fashion.

**5.** The breakdown of both fats and carbohydrates quite efficiently converts the heats of these reactions into stored chemical energy, with approximately 40% of available energy being ultimately stored in ATP.

**6.** Fatty acid biosynthesis begins with the conversion of acetic acid to acetyl CoA, a thiol ester. Carboxylation of acetyl CoA, catalyzed by biotin, introduces a carboxylic acid equivalent into the thiol ester, activating it toward a reaction analogous to the Claisen condensation. The resulting $\beta$-ketothiol ester is reduced, dehydrated, and reduced again to provide a thiol ester whose chain has been extended by two carbons. Repetition of this sequence, followed by hydrolysis, produces fatty acids with even numbers of carbon atoms.

**7.** Fatty acid degradation is accomplished by the reverse sequence, initiated by FAD-mediated oxidation of a fatty acid thiol ester to an $\alpha,\beta$-unsaturated thiol ester. Hydrolysis, followed by oxidation, produces a $\beta$-ketothiol ester, from which two shorter thiol esters are formed in a retro-Claisen-like condensation.

**8.** The tricarboxylic acid (TCA) or Krebs cycle is a series of biochemical transformations that are responsible for the oxidative degradation of acetate to carbon dioxide.

**9.** In the TCA cycle, citric acid is dehydrated, rehydrated, oxidized, and decarboxylated to produce 2-oxoglutaric acid. Decarboxylation of this $\alpha$-ketoacid leads to succinic acid, which is oxidized to fumaric acid. Upon hydration and oxidation, oxaloacetic acid is formed. Condensation with acetyl CoA reforms citric acid and permits the cycle to be repeated. One complete cycle incorporates an acetate unit and releases two equivalents of carbon dioxide, while producing stored chemical energy.

**10.** Some of the energy stored in acetic acid is converted by the TCA cycle into chemical potential energy in NADH and $FADH_2$, in a form that can be used later in other complex chemical transformations.

**11.** The breakdown of carbohydrates takes place in stages. First, glucose (a hexose) is split into two three-carbon fragments that undergo oxidative loss of carbon dioxide to form acetic acid. This sequence begins with the isomerization of glucose-6-phosphate (an aldose) to fructose-6-phosphate (a ketose). Phosphorylation is needed for molecular recognition by an enzyme, which induces a retro-aldol reaction to form two three-carbon fragments (dihydroxyacetone monophosphate and glyceraldehyde monophosphate). In the second stage of carbohydrate breakdown, oxidation of glyceraldehyde phosphate produces pyruvic acid, which is decarboxylated to acetic acid. As a thiol ester derivative (acetyl CoA), the acetic acid enters the TCA cycle and is oxidatively degraded to carbon dioxide. Most of the original chemical potential energy of the carbohydrate is released in this last stage.

**12.** Reactions in which a center of chirality is formed are controlled by enzymes so that only one absolute configuration is produced. Molecules with mirror symmetry are transformed into those having chiral centers: prochiral centers are involved in both the TCA cycle and glycolysis. This control, which is a natural consequence of the chirality of the enzymes themselves, has additional significance in that enantiomers are treated as unique molecules in living systems.

## Review Problems

**22.1** For some of the following reactions, the reagent above the arrow must be present in a stoichiometric amount; in others, the reagent is a catalyst and is not consumed. Indicate for each reaction whether the reagent is a catalyst.

(a)

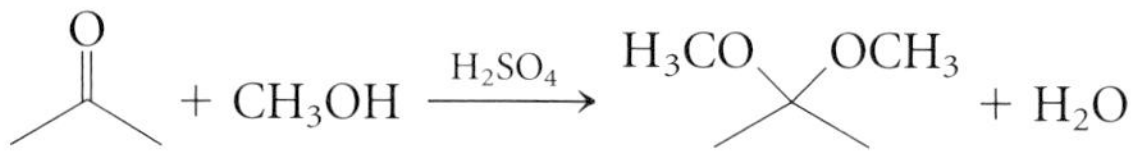

(b)

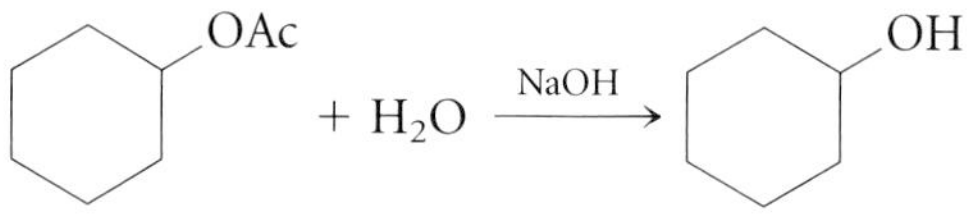

(c)

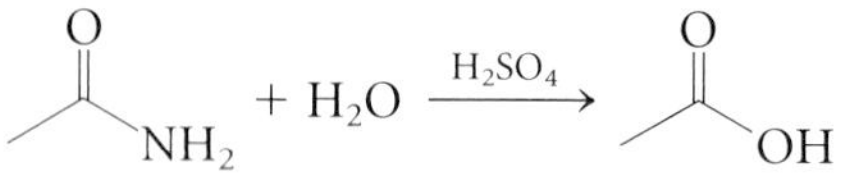

(d)

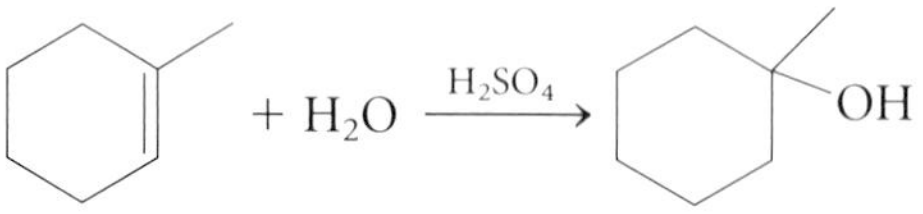

(e) $CH_3Br \xrightarrow{\text{NaOH}} CH_3OH$

**22.2** Some of the reactions in Problem 22.1 take place by a pathway that includes a single transition state with no intermediates. Others are multistep reactions with intermediates (and more than one transition state). Classify the reactions as either single step (SS) or multistep (MS), and draw an energy diagram for each. For multistep reactions, indicate which transition state is of highest energy and therefore formed in the rate-limiting step of the overall reaction.

**22.3** For each of the following energy diagrams, identify the transition state of highest energy. Decide if the equilibrium between the intermediate, I, and the reactant, A, favors I or the starting material. Do the same for the conversion of I into the product, B. Identify the transition state of second-highest energy. Would the overall reaction proceed more rapidly if this transition state were lower in energy?

(a)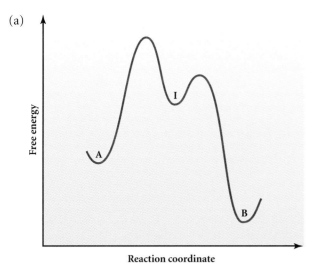

(b)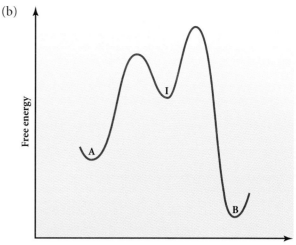

**22.4** The rate of nucleophilic acyl substitution is usually limited by the addition of the nucleophile to form a tetrahedral intermediate, rather than by the collapse of this species and expulsion of the leaving group to form product. Typically, the energy profile for acyl substitution resembles the first energy diagram in Problem 22.3, in which the first step has the higher activation energy. Under this condition, does the stability of the leaving group directly affect the rate of reaction? Would your answer be different if the reaction profile resembled the second energy diagram in Problem 22.3?

**22.5** Anhydrides react with water to form carboxylic acids in an exothermic process. Anhydrides can therefore be viewed as a chemical means for the storage of thermal energy. Explain why an anhydride is a less stable and more reactive carbonyl unit toward nucleophiles than the acid. Will each of the following anhydrides react more or less exothermically with water than does acetic acid anhydride?

(a) $F_3C$—C(=O)—O—C(=O)—$CF_3$  (b) $H$—C(=O)—O—C(=O)—$H$  (c) $H_3CO$—C(=O)—O—C(=O)—$OCH_3$

**22.6** Consider a propanoic thiol ester taking part in the biosynthesis of a fatty acid. What saturated fatty acid would be produced from one acetate unit and one propanoic thiol ester unit if the key nucleophilic acyl substitution took place on the acetate unit? On the propionate unit?

**22.7** Apply the sequence detailed in this chapter for fatty acid biosynthesis to prepare the following fatty acid:

Is there any step at which the normal course of the sequence cannot occur because of the presence of the extra methyl group? Explain your answer.

**22.8** Unsaturated fatty acids have *cis* double bonds between C-9 and C-10, C-12 and C-13, and C-15 and C-16. Review the structures of these fatty acids. Decide whether each double bond is in the correct regiochemical position to have resulted from a synthesis in which the $\alpha,\beta$-unsaturated ester intermediate did not undergo reduction of the double bond. (Note that, even if the position were correct, the geometry would not be, because the intermediate in the synthesis has *trans* geometry.)

**22-45**

**22.9** At one stage of fatty acid degradation, an $\alpha,\beta$-unsaturated ester is hydrated to produce a $\beta$-hydroxyester. This process represents formal addition of water across the double bond and can, in principle, occur with the opposite regiochemistry to that observed—that is, with the introduction of the hydroxyl group at the $\alpha$ carbon.

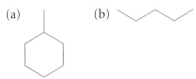

Write mechanisms for this hydration under both basic and acidic conditions, and explain why the formation of the $\beta$-hydroxyester is faster than that of the $\alpha$-regioisomer.

**22.10** In the TCA cycle, two seemingly identical hydrogens on a methylene group are in fact unique. Replacement of one hydrogen leads to one enantiomer, whereas replacement of the other hydrogen results in the other enantiomer—for example, in the formation of (R)- and (S)-2-butanol from butane:

Each of the following molecules has several methylene groups. For each such $CH_2$ group, determine if replacement of one of the hydrogens generates a new center of chirality. If so, are enantiomers or diastereomers generated?

(a)          (b)

**22.11** Ascorbic acid contains a stable enediol functionality. Draw structures for two alternative hydroxycarbonyl forms (tautomers),and write a mechanism for the acid-catalyzed interconversion of these two ketonic forms proceeding through vitamin C.

**Vitamin C**
**(ascorbic acid)**

**22.12** The formula of vitamin C ($C_6H_8O_6$) is close to that of hexose ($C_6H_{12}O_6$). Is vitamin C at a higher or lower oxidation state than the carbohydrates? How many centers of chirality are in vitamin C? (Refer to the structure in Problem 22.11.) Identify the end of the carbon chain having the lower oxidation level, and classify vitamin C as belonging to either the D or the L series.

**22.13** Suppose 1.0 mole of acetic anhydride was dissolved in sufficient water to produce 1 L of solution and underwent reaction to form acetic acid. How much would the temperature of the water increase as a result of the release of heat from the reaction? (Assume that the heat capacity of the final solution is the same as that of pure water, 1 cal/g °C, but don't forget that you have added mass from the acetic acid.)

**22.14** Which of the intermediates involved in the TCA cycle (Figure 22.9) have centers of chirality?

**22.15** Only one of the intermediates in the TCA cycle has two centers of chirality. Identify this compound, and draw three-dimensional representations with hatched and wedge bonds of all four stereoisomers.

**22.16** Write a step-by-step mechanism for the following interconversion, assuming that the reaction is acid-catalyzed.

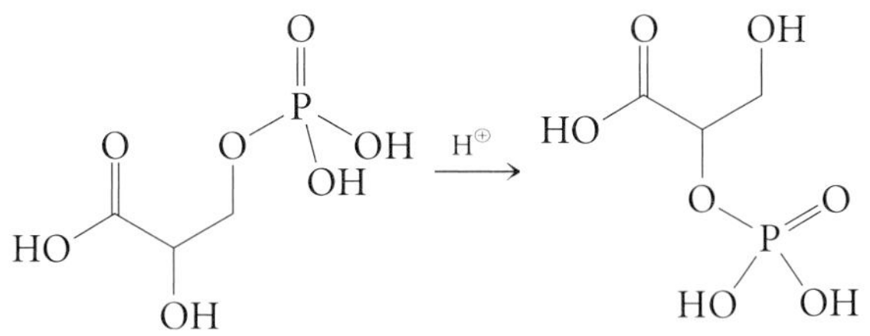

**22.17** Fluorocitric acid binds tightly to one of the enzymes involved in the TCA cycle, preventing this enzyme from serving its normal role as a catalyst.

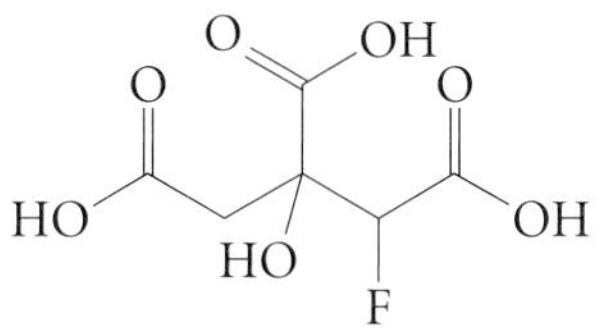

Fluorocitric acid

Consider all the functional groups that are near the fluorine substituent. What effect might the fluorine have on biomolecular interactions of these groups?

**22.18** (a) Write a mechanism for the acid-catalyzed conversion of citric acid to isocitric acid. Draw an energy diagram for the conversion that includes all intermediates involved. Which step is rate-limiting for the overall reaction?

(b) Would fluorocitric acid undergo conversion to the intermediate aconitic acid bearing a fluorine substituent at a rate faster or slower than that for citric acid? Explain your reasoning.

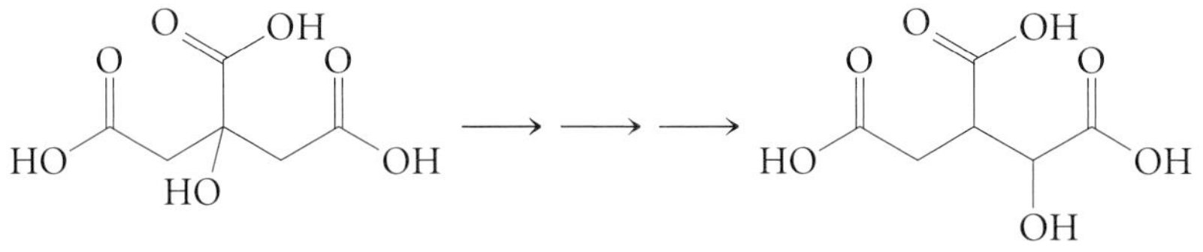

Citric acid

Isocitric acid

## Important New Terms

**adenosine diphosphate (ADP):** a diphosphate ester of adenine; a principal energy-storage molecule in biological systems (22-8)

**adenosine monophosphate (AMP):** a monophosphate ester of adenine (22-8)

**adenosine triphosphate (ATP):** a triphosphate ester of adenine; a principal energy-storage molecule in biological systems (22-8)

**citric acid cycle:** *see* **tricarboxylic acid cycle**

**glycolysis:** the breakdown of carbohydrates, in which a retro-aldol-like reaction is a key step (22-28)

**high-energy phosphate bonds:** phosphoric acid anhydride units that are critical in biological energy storage (22-9)

**Krebs cycle:** *see* **tricarboxylic acid cycle**

**mixed anhydride:** an anhydride with two different carboxylic acid subunits; often an anhydride derived from a carboxylic acid and a phosphoric acid (22-35)

**phosphoric acid anhydride:** the condensation product obtained by dehydration of two equivalents of phosphoric acid (22-9)

**photosynthesis:** the complex biological process by which carbon dioxide is converted to carbohydrates in a series of reactions initiated by the absorption of light energy (22-10)

**respiration:** the biological oxidation of glucose to carbon dioxide (22-10)

**tricarboxylic acid (TCA) cycle:** a sequence of reactions by which acetate units in the form of acetyl CoA are degraded to carbon dioxide; some of the energy released during this oxidation is stored as chemical reduction potential in the cofactors $FADH_2$, NADH, and NADPH (22-18)

# Molecular Basis for Drug Action

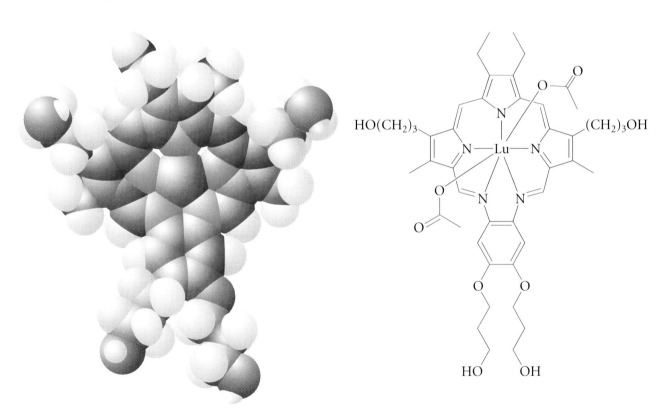

Macrocyclic ligands designed and synthesized by Jonathan Sessler (University of Texas at Austin) have five nitrogen atoms that can coordinate to a single metal atom, here an atom of the lanthanide element lutetium. (Two acetate groups also bonded to lutetium above and below the overall plane of the molecule are hidden in the space-filling representation in order to reveal the "star" of five nitrogen atoms.) Compounds like this are currently being investigated in clinical trials for photodynamic therapy of cancer.

$O$ne important area for the practice of organic chemistry is the design and synthesis of new drugs in the pharmaceutical industry. Development of an effective pharmaceutical agent is a very complex endeavor. Despite tremendous advances in the understanding of the chemical basis of disease states, the success rate for the progression from initial drug concept to final product is low. Only about 1 of every 50 compounds tested in pharmaceutical laboratories shows promising biological activity. Structural modification of this class of compounds produces an agent worthy of further study only 1 time in 100. Of these agents, only 1 in 10 passes successfully through clinical trials with human patients and becomes marketable. There are many reasons for this low success rate. Effective pharmaceutical agents must not only have structures that can elicit the desired response, but also possess many other features. They must have carefully balanced solubility characteristics (fat versus water solubility) that enables them to reach the target organ or tissue; they must be at least somewhat resistant to chemical degradation in the body; and, perhaps most important, they must not interfere with any other aspects of the complex biochemical dance that is essential to life.

Organic chemistry has played a major role in the development of new drugs. Knowledge of disease states and their causes and of the details of biochemistry is expanding at a tremendous rate. Using this knowledge, organic chemists are playing an ever-increasing role in the design and development of new and effective pharmaceutical agents.

The design and synthesis of a medicinally active molecule bring into play many of the basic concepts of bonding and structure that are part of a first-year course in chemistry. These basic features can be extended to the relatively large collections of atoms present in organic substances. Stereochemistry, both configurational and conformational, accounts for some of the exceptional diversity found among organic compounds. Detailed analysis and classification of the various reactions common to organic chemistry is based on their mechanisms. A fundamental knowledge of structure and reactivity of functional groups can be applied to predict some of the macroscopic properties of polymers. Various naturally occurring organic molecules, including many biochemicals, are of critical importance to living systems in eliciting specific molecular recognition, which enables various chemical species to perform a particular role without taking part in other transformations. Finally, catalysts are often critical both in laboratory transformations and in biologically important processes involving enzymes.

This chapter will apply such basic knowledge to understanding the molecular basis for the action of a variety of drugs, or pharmaceutical agents. The use of chemicals for "healing" is generally referred to as **chemotherapy,** although that term is often used specifically to refer to drug treatment of cancer.

## 23.1

## Chemical Basis of Disease States

In a general sense, disease states typically result from one of the following conditions:

1. Deficiency of a critical biochemical

## *CHEMICAL PERSPECTIVES*

### STEREOCHEMISTRY AND PHARMACEUTICALS

Approximately 1850 different compounds are marketed worldwide as pharmaceutical agents. The chart reveals that most of these are prepared in the laboratory. (*Semisynthetic* means that a compound from nature is modified in the laboratory.)

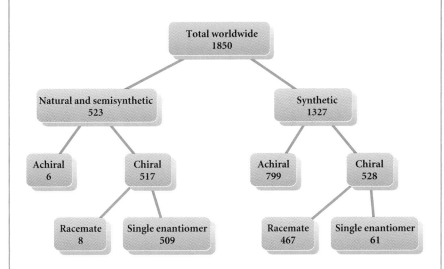

Note that almost all of the natural and semisynthetic drugs are chiral, and that most of these are marketed as single enantiomers. On the other hand, most synthetic drugs are achiral, and most synthetic chiral drugs are sold as racemates. It is almost invariably the case that only one enantiomer of a racemic mixture has the desired effect. Because the other enantiomer can have undesirable effects, the trend in the pharmaceutical industry is to market single enantiomers.

**2.** Surfeit of a critical biochemical

**3.** Invasion of an alien species that produces substances toxic to the host or consumes materials required by the host

**4.** Uncontrolled and overly rapid growth of part of the organism

Of these conditions, the first is generally the easiest to control because all that is required is a supplement from an external source to make up for the deficiency. For example, diabetics produce insufficient insulin, a peptide hormone with 51 amino acid residues that triggers the cleavage of glycogen to glucose, initiating the process of energy release through glycolysis. The major symptoms of diabetes can be relieved by administration of insulin. Because the enzyme pepsin, which cleaves peptide bonds, is present in the stomach, insulin cannot be taken orally. However, subcutaneous injection transfers insulin rapidly into the bloodstream, which distributes it to various body tissues.

Many other disease states result from a deficiency of some essential biochemical. Indeed, the importance of vitamins was first discovered because of the diseases that resulted from dietary deficiencies. Deficiency diseases can be caused by lack of organic compounds—for example, scurvy (vitamin C), rickets (vitamin D), and beriberi (vitamin $B_6$, or thiamine)—or of inorganic materials. For example, goiter is caused by too little iodide ion. Dietary deficiencies are generally easy to correct, but other deficiency diseases, such as those caused by genetic faults that interfere with protein synthesis, are much more difficult to treat. For example, sickle-cell anemia is caused by alterations in the amino acid sequence of hemoglobin, reducing its ability to carry oxygen. The abnormal structure of sickle-cell hemoglobin also causes abnormal clotting of blood within body tissues.

The undesirable effects of a surfeit of essential biochemicals can be seen in the overproduction of acid in the stomach. Not only is this painful, but it can lead to the destruction of the lining of the digestive tract, producing ulcers. A number of agents are used to treat this condition, including cimetidine (Tagamet). Cimetidine blocks acid production by mimicking the structure of the natural material histamine, which is responsible for triggering acid production.

However, not all areas of the body can be reached as readily as the stomach. For example, transport across the membrane surrounding a nerve cell is quite selective, and generally only small, nonpolar molecules can pass into the interior of these cells.

Diseases can also be caused by the attack of parasites, bacteria, and viruses. Although these organisms have characteristics that are common to all living things, they also have differences derived from the precise structure of certain characteristic biomolecules. These differences are exploited by the immune system present in animals, including human beings, to provide a natural defense against alien entities. The immune system relies on subtle differences in the exterior composition of cell walls and a large collection of antibodies produced by the immune system. Owing to different permutations of amino acids in the part of an antibody referred to as the *variable region,* which typically contains about 63 amino acids, approximately $10^{12}$ different antibodies can be produced. Some of these antibodies are capable of tight association with a specific alien material—for example, specific binding to one carbohydrate on the surface of an invading cell. Production of the antibody that binds specifically to the invading cells is then increased, and special cells (called *T cells*) consume the antibodies along with their attached alien material.

An immunogenic "fingerprint" differentiates all cellular organisms, even within the same species. For example, the difference in the carbohydrates present on the surface of blood cells is the chemical basis for the classification of different human **blood types** as A, B, O, and AB. Figure 23.1 shows the oligosaccharides responsible for the classification of types O, A, and B. Although these structures are complex, they differ *only* in the region shown in blue. When blood of a different type is administered to a patient, antibodies associate with these foreign sequences of carbohydrates and, as a result, the alien cells are destroyed. Each person lacks antibodies to his or her own blood cells, because those antibodies, which would trigger the person's immune system against itself, are removed in the early stages of fetal development.

**Type O**

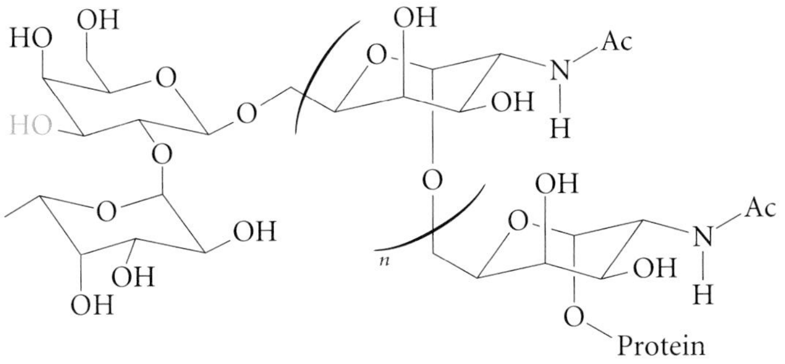

**Type A**

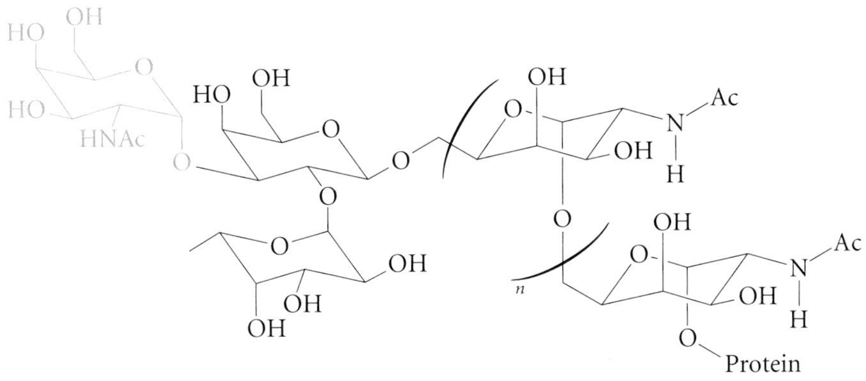

**Type B**

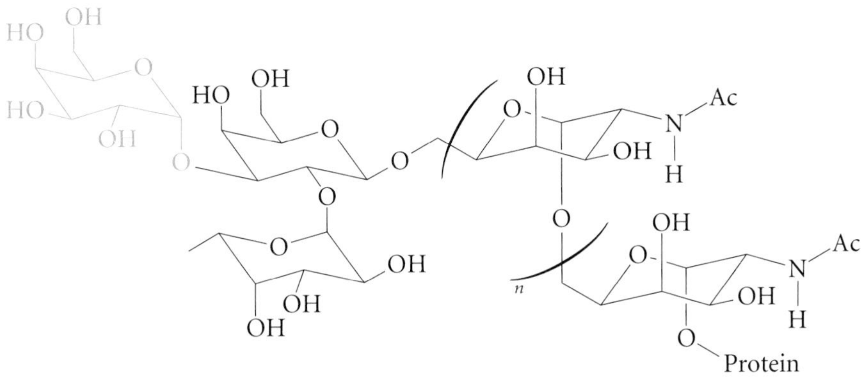

**FIGURE 23.1**

Different human blood types have slightly different saccharides, shown here,
on the surface of blood cells. These small differences are sufficient to trigger
the immune system when, for example, type B blood is given to a person
who has type A. Type O blood lacks the sugar substituent and thus does not
trigger the immune system. Type O blood can be given to people having any
blood type (the universal donor).

The differences between the cells of different individuals within the same species are sufficient to cause rejection of blood transfusions and tissue grafts. There are even greater differences between the cells of different species, such as humans and bacteria, and some of these differences form the basis for the action of antibacterial drugs. For example, bacteria synthesize tetrahydrofolic acid, the cofactor that transfers single carbon units, but mammals do not and must obtain this essential compound in the diet. A drug that interferes with the synthesis of tetrahydrofolic acid can therefore kill bacteria (which are incapable of incorporating folate from outside the cell) while not affecting human beings, who derive this cofactor from the diet. Cephalosporin and penicillin antibiotics function in a different fashion, by interfering with the construction of bacterial cell walls. The selective, deleterious effect of these chemicals on bacteria is possible because of a fundamental difference between the cell walls of bacteria and those of animals.

Cancer is a very difficult disease to treat—not only because of the uncontrolled growth of certain cells, but also because these cells can metastasize and invade tissues distant from the point of origin. Uncontrolled growth is characteristic of transformed cells that have become cancerous, and the enhanced rate of division of cancer cells compared with normal cells is exploited by chemotherapeutic drugs.

## 23.2

## Intact Biological Systems as Chemical Factories

Equilibrium can be equated with death: with only a few exceptions (such as viruses), living systems are in a constant state of flux. New raw materials constantly enter a system, and products and by-products are excreted. It is difficult to imagine how this dynamic state could be maintained without compartmentalization. Indeed, the basic building block of living systems, the cell, is itself divided into smaller units such as the nucleus and ribosomes. Cell membranes are constructed of lipid bilayers containing special substructural features that facilitate the flow of inorganic ions and small organic molecules into and out of the cell. The membranes serve a dual role: they isolate the contents of the cell and its substructures, and they permit selective transport. Many antibacterial agents are effective because they interfere with one or the other of these two critical functions of membranes.

It is one thing to destroy foreign organisms in a laboratory environment, or *in vitro* (literally, "in glass"), and quite another to do so *in vivo* (in a living organism) without simultaneously destroying the host. There are many simple ways to destroy the delicate balance intrinsic to life. All cells, for example, are sensitive to such environmental features as temperature and pH. No known life forms survive temperatures higher than 150 °C or treatment with concentrated sulfuric acid (even cockroaches succumb after some time), which denatures critical biopolymers (peptides, enzymes, nucleic acids). As we will see, to destroy an invader organism, such as a virus, bacterium, or cancer cell, a pharmaceutical agent not only must be effective at the assigned task, but also should cause only minimal damage to the host cells.

## 23.3

# H$_2$ Blockers: Modern Antacids

For many years, the standard treatment for the overproduction of acid in the stomach was the administration of antacids such as sodium bicarbonate or a mixture of aluminum and magnesium hydroxides. These treatments are still used: check the ingredient statement on a commercial antacid such as Maalox or Alka-Seltzer. However, excessive bicarbonate can make the stomach alkaline, triggering the production of even more acid. The metal hydroxides constitute a better treatment, because they are not soluble in water and do not increase the pH above neutrality. In either case, however, because of their mode of action, antacids address only the symptoms of excessive acidity, not the cause.

A major breakthrough in the treatment of hyperacidity was the discovery that **histamine,** a potent vasodilator, also stimulates the secretion of pepsin and acid into the stomach. Histamine is released by the body in response to injury. It has a variety of actions, including the contraction of smooth muscle in the bronchi and gut and the relaxation of other muscles, such as those in the walls of fine blood vessels. Histamine is also responsible for the nasal congestion associated with common colds and with allergic response to pollen. Certain synthetic compounds such as **brompheniramine (Dimetapp)** and **terfenadine (Seldane)** act as **antihistamines,** interfering with the natural action of histamine by competing with it for binding to receptor sites where it exerts its effect.

Histamine

Brompheniramine
(Dimetapp, Dimetane)

Terfenadine (Seldane)

These traditional antihistamines have no effect on the secretion of acid by the stomach, which is due to binding of histamine to different receptors (H$_2$) from those involved in allergic reactions (H$_1$). The discovery that histamine is involved in the release of stomach acid spurred the search for new compounds, structurally resembling antihistamines, that would block acid secretion. Chemists at Smith-Kline-Beecham in England conducted a successful search for such a compound and ultimately produced cimetidine:

**Cimetidine (Tagamet)**

Cimetidine is marketed by Smith-Kline-Beecham as Tagamet, and by other pharmaceutical companies under different names. Annual sales of these drugs for the treatment of ulcers exceeds $2 billion worldwide.

### EXERCISE 23.1

In some cases, the structural analogy between a substance such as an antihistamine and the natural substance with which it competes for binding is clear; in other cases, the relation is less obvious. Compare the structures of histamine, bromphen-iramine, terfenadine, and cimetidine. Identify the structural features that the three synthetic compounds have in common with histamine.

The process of discovering a new drug is challenging and fascinating because of the many requirements that must be met by a pharmaceutical agent. It must have the desired effect (in this case, binding to, and thus blocking, the receptor site for histamine without triggering the histamine response). It must also be able to move through the body to the target organ. (Preferably this should be through the stomach wall so that the drug can be administered orally. Ability to use the oral route also requires that the drug not be deactivated or degraded in the stomach.) The drug must not elicit other responses by taking part in other biochemical transformations. Finally, it must be excreted, either in its original form or after degradation in the liver, within a reasonable period of time. Some of these requirements can be addressed at the molecular level. For example, transport throughout the body requires some water solubility. Other requirements can be dealt with at the macroscopic level. For example, the drug can be coated with a substance that is resistant to the action of stomach acid and pepsin but is ultimately degraded in the intestines. The macroscopic aspect of drug development is called **formulation** and is just as important as the molecular aspects. Unfortunately, finding the right balance among all the necessary features and desired properties of a drug still requires substantial trial and error.

The discovery of cimetidine is an example of the modern approach to the development of pharmaceutical agents, which requires a reasonable level of understanding of the biochemical processes of a disease state. Using this knowledge, skilled synthetic chemists can prepare molecules designed to elicit a desired response. However, a number of very useful drugs were developed and used for many years before such knowledge was available. Indeed, most "miracle drugs" were uncovered by chance. The next three sections deal with several biologically active and useful therapeutic agents that were originally uncovered by chance, although many details of their modes of action are now known.

## 23.4

## Neurologically Active Drugs: β-Phenethylamines

Morphine has been used for many centuries as a **psychoactive drug** to provide relief from pain and to achieve a state of euphoria.

Morphine          Heroin

Heroin is hydrolyzed inside brain cells to form morphine, thus producing an identical physiological effect. However, heroin is preferred by drug abusers, because it is less polar than morphine and crosses from the blood into the brain more readily, resulting in a more rapid and more elevated "high." More recently, similar drug-induced neurological activity has been used medicinally to relieve severe pain—for example, in terminally ill cancer patients. Morphine, the major active component of the opium poppy, rightly derives its name from the Greek god of dreams, Morpheus. The veil that morphine hangs upon the mind is unsurpassed in clouding the sensation of both physical and mental pain. Most unfortunately, the pain-abating character of morphine is accompanied by a high potential for developing physical and psychological dependence.

Morphine is an "old" compound—it was the first of the naturally occurring bases to be isolated in pure form. A German apothecary named Friedrich Sertürner reported his discovery in 1805, but it was not until 1835 that the French scientist Jean Baptiste André Dumas coined the term *alkaloid* for this group of compounds. Although morphine was the first alkaloid isolated, its structure was not established until 1925. Because of its long history, it is surprising that morphine's mode of action was not described in detail years ago. However, only when details of how morphine affects the brain were uncovered did it become clear why this discovery had taken such a long time. Morphine mimics the natural action of peptides called **endorphins** (a name derived from *endo*genous mo*rphine*) that are normally present in the brain in extraordinarily low concentrations. Endorphins serve as natural pain relievers that function to change or remove the perception of nerve signals. It is believed that their concentration is increased under conditions of high stress—for example, when long-distance runners push themselves to their limits and beyond (described as "hitting the wall").

Two pentapeptide endorphins known specifically as the **enkephalins,** Met-enkephalin and Leu-enkephalin, are shown here. These peptides are distinguished by the presence of either a methionine (Met) or a leucine (Leu) residue at the carboxyl terminus. (Note that these residues are quite similar in molecular shape.)

**Met-enkephalin**

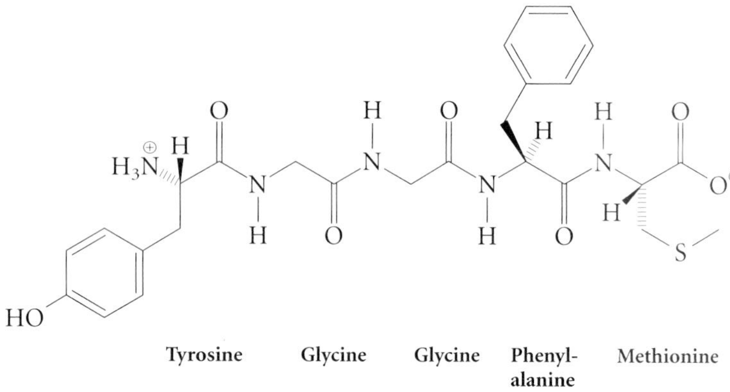
H
O
H
O
H
O
H
H
$H_3N^{\oplus}$
N
N
N
N
$O^{\ominus}$
H
H
O
H
O
S
HO
Tyrosine
Glycine
Glycine
Phenyl-
alanine
Methionine

**Leu-enkephalin**

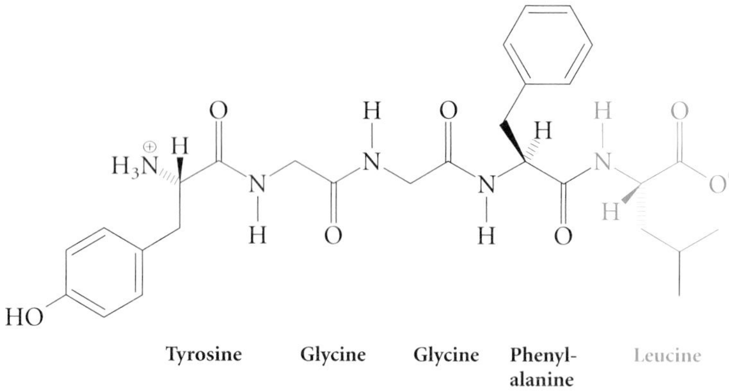
H
O
H
O
H
O
H
H
$H_3N^{\oplus}$
N
N
N
N
$O^{\ominus}$
H
H
O
H
O
HO
Tyrosine
Glycine
Glycine
Phenyl-
alanine
Leucine

The endorphins are believed to be naturally produced painkillers, selectively intervening with the perception of pain but not with nerve signals from other senses. It is thought that morphine has the same effect, because its three-dimensional shape mimics that of the enkephalins and produces the same response. Figure 23.2 shows a representation of Met-enkephalin in a conformation similar to that of morphine. Through a detailed study of the biological effects of many different peptides resembling the natural

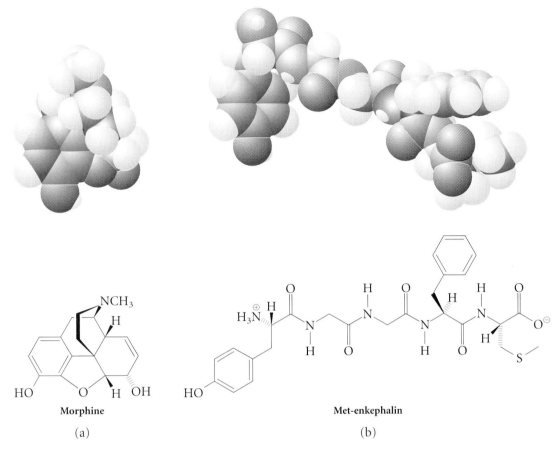

**FIGURE 23.2**

Three-dimensional representations of (a) morphine and (b) Met-enkephalin. The left-hand region of each molecule is believed to be the part that conforms to a common binding site in the brain.

enkephalins, it has been possible to demonstrate that the phenolic ring and amino group of tyrosine, as well as the phenyl ring of phenylalanine, are essential for pain-relieving activity.

Many compounds that affect the brain have a common structural feature, the β-phenethylamine subunit. Some important neurologically active compounds are presented in Figure 23.3 (on page 23-12). An important feature common to all psychoactive drugs is the ability to cross the blood–brain barrier. The blood vessels in the brain are surrounded by tissue that prevents the passage of charged ions and large molecules into brain tissue. Nutrients enter the brain by special transport mechanisms. Small neutral molecules, particularly those that are fat-soluble such as the β-phenethylamine derivatives shown in Figure 23.3, pass freely into brain tissue.

## EXERCISE 23.2

Identify the β-phenethylamine subunit present in mescaline, mesembrine, levodopa, and lysergide (shown in Figure 23.3).

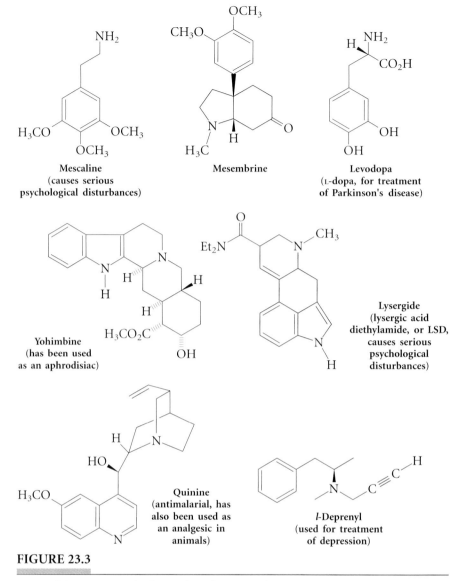

**FIGURE 23.3**

Most of the compounds shown here have two features in common: (1) They have a $\beta$-phenethylamine structural subunit. (Exceptions are yohimbine and quinine, both of which have an amino group $\beta$ to an aromatic ring.) (2) They affect the central nervous system of human beings and other animals.

## 23.5

# Antibiotics

As the science of chemistry progressed in the nineteenth and twentieth centuries, the idea took hold that chemicals could be found that would adversely affect invading bacteria but not the host. The German bacteriologist Paul Ehrlich (1854–1915) was the originator and most ardent advocate of the search for a "magic bullet," a chemical compound that would kill some cells but not others.

## Blocking Synthesis of Tetrahydrofolic Acid

Starting as a young medical student, Ehrlich was fascinated by the well-known observation that some dyes interact selectively with cells and cellular structures. For example, the dye methylene blue binds rather tightly to nerve cells, permitting researchers to highlight nerve paths.

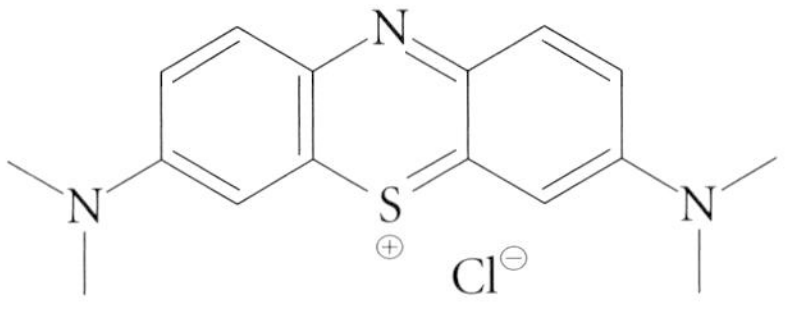

Methylene blue

Ehrlich concluded that compounds other than dyes might selectively interact with a parasite in the presence of its host. He also explored the possibility that altering the structure of a toxic substance could change its relative toxicity, or the *chemotherapeutic index* (the ratio of toxic dose for the host to that for the invading organism). Ehrlich's initial studies led to the synthesis of atoxyl, which he prepared by heating the extremely toxic metal arsenic with aniline in air:

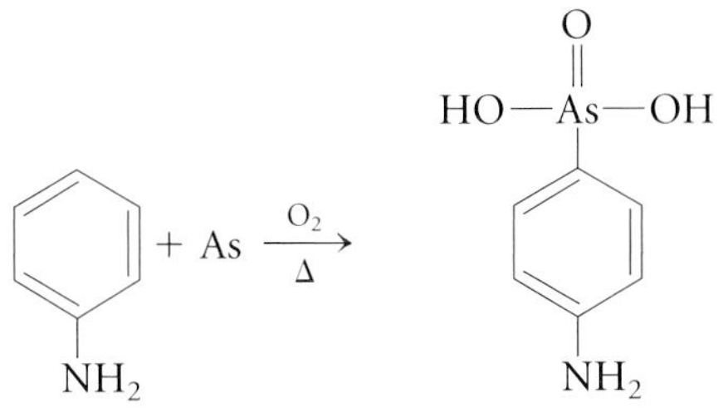

p-Aminophenylarsonic acid
(atoxyl)

The chemotherapeutic index of atoxyl was only about 10, and its long-term use for the treatment of syphilis gave evidence of serious toxicity to the host. Further modification of the structure led, in 1912, to arsphenamine, known as salvarsan, the first really effective treatment for syphilis. Although salvarsan is toxic to human beings, its effect on the spirochete that causes syphilis is much greater.

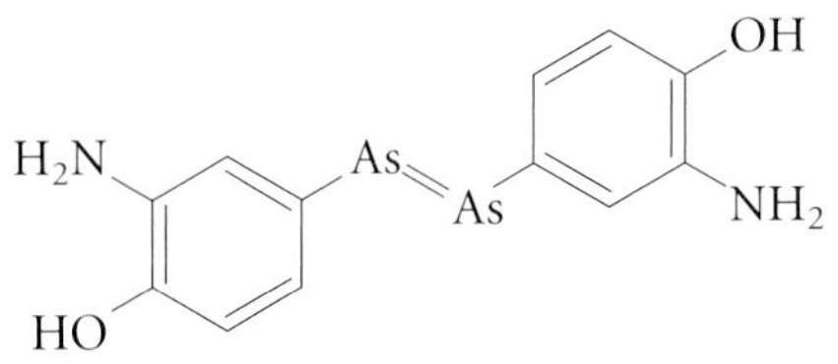

Salvarsan

Ehrlich continued his studies, concentrating on dyes produced by chemical companies. These dyes were the first chemicals produced on a large scale, and the new chemical industry established to produce them was located in Basel, Switzerland, on the Rhine River. (This location at the border ensured that the industrial waste deposited into the Rhine would flow away from Switzerland into France and Germany.) Then, as now, most dyes were produced by the coupling reaction (diazo coupling) between an amine and an aryl diazonium salt.

**Diazo Coupling**

An azo dye

The extended conjugation in the product is responsible for the strong absorption of light in the visible region of the spectrum.

The structural similarity between azo dyes and salvarsan is clear, with the arsenic in salvarsan taking the place of nitrogen in the dye. The rise of the dye industry was fortuitous for Ehrlich. Not only was there a ready supply of new chemicals, but Ehrlich was particularly interested in studying dyes because he hoped to find compounds that would selectively bind to bacteria. Any preference in binding was easily observed with these highly colored materials. The first effective antibacterial agent developed from this investigation was Prontosil, prepared in 1932.

Prontosil                  Sulfanilamide

Chemists then began a study of the relation between structure and activity to ascertain what molecular features were necessary for antibacterial properties. It was soon discovered that just the right-hand part of Prontosil, in the form of *p*-aminobenzenesulfonamide (sulfanilamide), has antibacterial activity, and the era of sulfa drugs was born.

## EXERCISE 23.3

There are two ways in which Prontosil can be prepared by using the reaction of a diazonium salt with an aromatic ring. Show both ways, and suggest which is the better method.

How can such a simple molecule as sulfanilamide kill bacteria? And why isn't it toxic to human beings? The answer to both questions is that bacteria must synthesize a precursor of tetrahydrofolic acid from *p*-aminobenzoic acid and a pteridine pyrophosphate (Figure 23.4), whereas mammals, lacking the ability to synthesize tetrahydrofolic acid, obtain this cofactor in the diet. Sulfanilamide (and other sulfa drugs) are sufficiently similar in structure and reactivity to *p*-aminobenzoic acid that they become incorporated into the biosynthetic pathway for tetrahydrofolic acid that proceeds

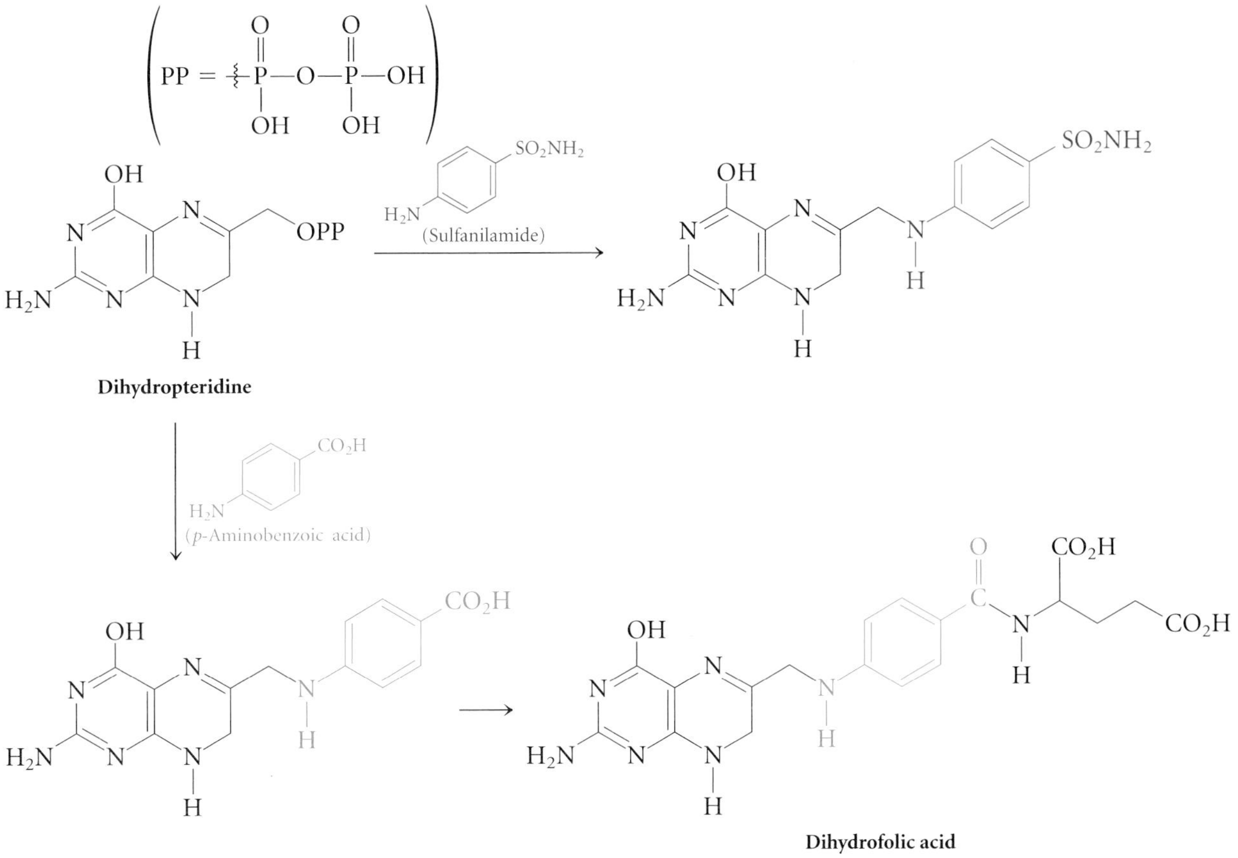

**FIGURE 23.4**

The essential cofactor dihydrofolic acid is synthesized in bacteria by the combination of a dihydropteridine with *p*-aminobenzoic acid. In the presence of sulfanilamide, a bacterium incorporates the dihydropteridine into a structure similar to the natural cofactor.

through dihydrofolic acid. Apparently, the modified cofactor produced in this way consumes pteridine, reducing the amount of the "real" cofactor, and results in the death of the bacterium.

Although it is easy to recognize the structural similarity between sulfanilamide and *p*-aminobenzoic acid, unraveling the details of how the antibiotic interferes with bacteria was quite complex. In addition to the hypothesis just summarized (for which there is substantial experimental verification), there are other possible ways in which sulfanilamide could interfere with the synthesis of tetrahydrofolic acid. For example, the antibiotic could bind tightly to the enzyme responsible for the synthesis of *p*-aminobenzoic acid or to the enzyme responsible for its incorporation into tetrahydrofolate. Alternatively, the bogus tetrahydrofolate itself could block the enzyme responsible for its synthesis (as well as that of tetrahydrofolic acid). Unraveling such details is an important activity that chemists carry out in developing an understanding of the molecular basis for drug action. This information, in turn, is an invaluable resource for chemists who design and synthesize new pharmaceutical agents.

## Disruption of Membrane Structure and Interference with Ion Balance across Membranes

It has been known for some time that surfactants (surface-active agents) have an adverse effect on bacteria. For example, many common household cleaning agents contain tetraalkylammonium salts. (Your knowledge of organic chemistry should help you to understand the function of various ingredients in commercial products.) These salts appear to have antibacterial activity because they are absorbed into and therefore disrupt the bacterial cell membrane.

Physical disruption of the cell membrane is not the only mechanism for the action of antibiotics. One class of antibiotics, the ionophores, destroy bacteria by disrupting the normal ionic balance across cell membranes. Some ionophores are effective antibacterial agents because they encapsulate metal ions in a hydrophobic exterior, and thus carry them through a membrane. Other ionophores function by becoming part of the cell membrane and forming new channels through which ions can move relatively freely. Recall that the unique chemical property of ionophores is their ability to act as multidentate ligands for metal ions. Simple ionophores, such as the cyclic polyether 18-crown-6, bind tightly to alkali metal cations; the polyether can then carry the cation, with its associated anion(s), from water into an organic solvent.

Several naturally occurring antibiotics bind $Na^{\oplus}$, $K^{\oplus}$, and $Li^{\oplus}$ within a substantially hydrocarbon-like exterior. These encapsulated ions can then be transported through the lipid bilayer of cell membranes, producing a dramatic change from the natural ionic balance. Valinomycin is a naturally occurring, cyclic dodecadepsipeptide that has both L- and D-amino acid residues. (Depsipeptides are those that have ester linkages in addition to amide, or peptide, bonds.)

**Valinomycin**

Valinomycin carries ions across membranes, disrupting the normal balance and producing catastrophic effects on the cell. There are many antibacterial agents with apparently the same mode of action as valinomycin. However, like valinomycin, most of them are also quite toxic to mammals and have a low chemotherapeutic index. As a result, they are not of clinical importance.

Gramicidins are linear pentadecapeptides produced by *Bacillus brevis*, a common bacterium found in soil. The major peptide (88%) is gramicidin A (Figure 23.5). Gramicidin A contains D- and L-amino acids in an alter-

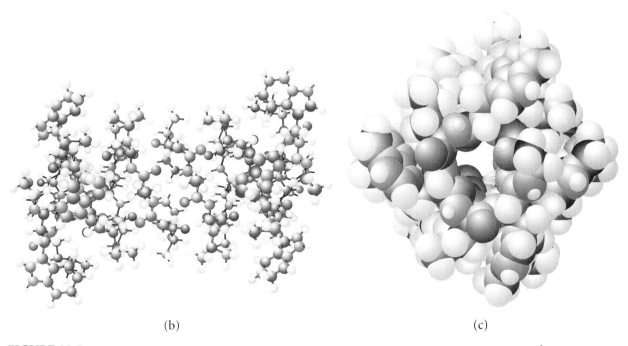

Gramicidin A

(a)

(b)                                              (c)

**FIGURE 23.5**

(a) Gramicidin A is a linear pentadecapeptide with amino acids of L or D
configuration alternating. (b) Because of the presence of a number of amino
acids with the unnatural D configuration, gramicidin A forms an unusual
secondary structure, a left-handed helix, shown in an edge-on view. This helix is considerably larger than that normally formed from peptides composed entirely of L-amino acids, and it has 6.3 instead of 3.4 amino acids per
turn. (c) The end-on view of the helix shows the cylindrical space available
for bonding to and transporting ions.

nating sequence, except for the presence of glycine, an achiral residue, at the second unit from the amino end. The substitution of a single D-amino acid in a linear peptide chain prevents the formation of the $\alpha$-helix for a length of several residues on either side. However, the alternation of configuration found in gramicidin allows the molecule to form a left-handed helix with 6.3 amino acid residues per turn. This helix is markedly different from the $\alpha$-helix with 3.4 residues per turn. The gramicidin helix also differs from the normal helix in that it has a larger central hole, measuring about 4 Å. Thus, it is possible for ions as large as $Cs^{\oplus}$ to be contained inside and move freely from one end of this "tube" to the other. It is believed that two of these helices join together to form a conduit for ions across the cell membrane.

The polyene antibiotics also appear to function by creating additional ion channels through cell membranes. Amphotericin B is produced by a soil bacterium, *Streptomyces nodosus,* apparently to protect itself from attack by fungi.

**Amphotericin B**

Not all bacteria are harmful to their hosts, and, indeed, some are beneficial. Bacteria in the intestines are a major source of folic acid for mammals and aid in the constant battle with fungus infections. The cell membranes of fungi more closely resemble those of mammals than do the cell membranes of bacteria. Therefore, developing a selective antifungal agent is generally much more difficult than finding an antibiotic. Outbreaks of infections by the yeast *Candida albicans* can occur in the mouth (oral thrush) and the vagina, especially when the beneficial bacteria that control these yeasts are destroyed.

### ▨ Disruption of Bacterial Cell Walls

*Structure of Bacterial Cell Walls.* Because bacteria must survive in different media in order to be transferred from one person to another (for example, in tap water and "salty" blood), the bacterial cell wall must be able to resist the swelling that occurs when the ionic strength inside the bacterium is several orders of magnitude greater than that outside. In this situation, water migrates into the cell faster than it leaves (osmosis), resulting in an increase in pressure inside the bacterium by as much as 25 atmospheres. An ordinary lipid bilayer membrane is very weak and does not remain intact under even much lower pressure differentials. In a bacterium, the cell membrane has a protective outside wall composed of peptidoglycans (Figure 23.6), long chains of carbohydrates cross-linked by short peptides (7–12 amino acids). Interestingly and very importantly, several "un-

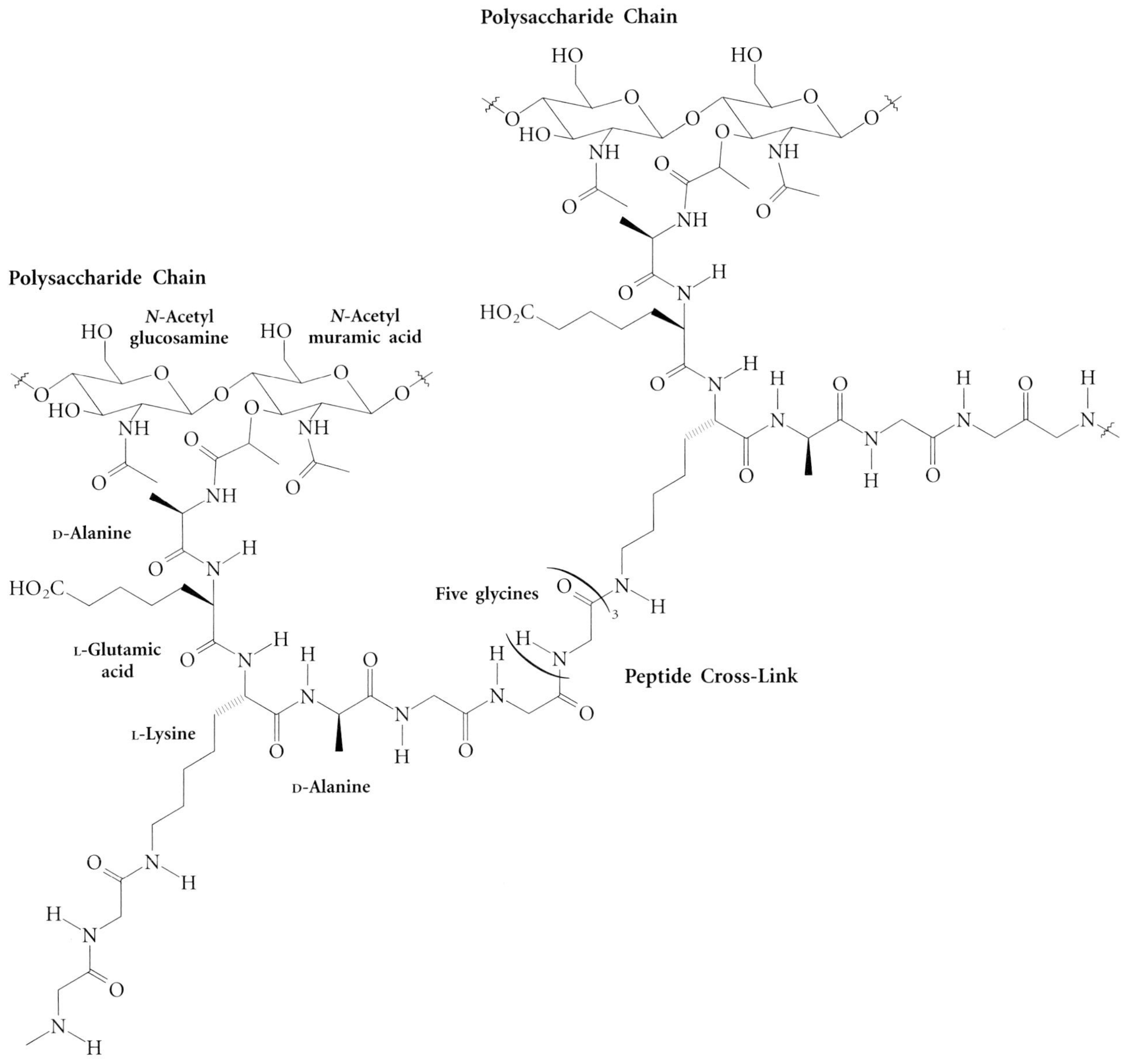

**FIGURE 23.6**

The cell walls of bacteria are composed of polysaccharide chains that are cross-linked by short peptide chains to form a rigid polymer matrix. The penicillin antibiotics greatly weaken bacterial cell walls by interfering with the construction of these cross-links at the point containing D-alanine. Without complete cross-linking, the cell wall cannot withstand the large internal pressure developed when the bacterium is exposed to water of low ionic strength. The cell wall then ruptures, and the bacterium dies.

natural" D-amino acids are incorporated into these peptides, providing the opportunity for a selective chemical action involving bacteria because these D-amino acids are not present in animals.

Although a relatively rigid protective wall is essential to bacterial survival, the wall must be able to expand as the cell divides. To provide for this expansion, bacteria have enzymes that can cleave peptide bonds, thereby removing the cross-links, and other enzymes that can repair this damage to

**23-19**

the outer wall by forming new peptide bonds. As the cell divides, there is a net addition of a new material as the wall is unstitched and then restitched until there is enough for two cells. Clearly, interfering with the ability to remove the cross-links prevents bacteria from multiplying, and blocking the formation of the cross-links results in rupture of the cell wall. This process is illustrated in Figure 23.7 for *Staphylococcus aureus* by the reaction at a critical stage of cross-linking. The carbohydrate chain of the bacterial cell wall has a tripeptide side chain that terminates in two D-alanines. The terminal residue is removed from the branch, and the rest of the branch is covalently bonded to the enzyme transpeptidase at the C-terminal D-alanine. This D-alanine is then linked by a peptide bond to the amino group of the terminal glycine of a neighboring chain.

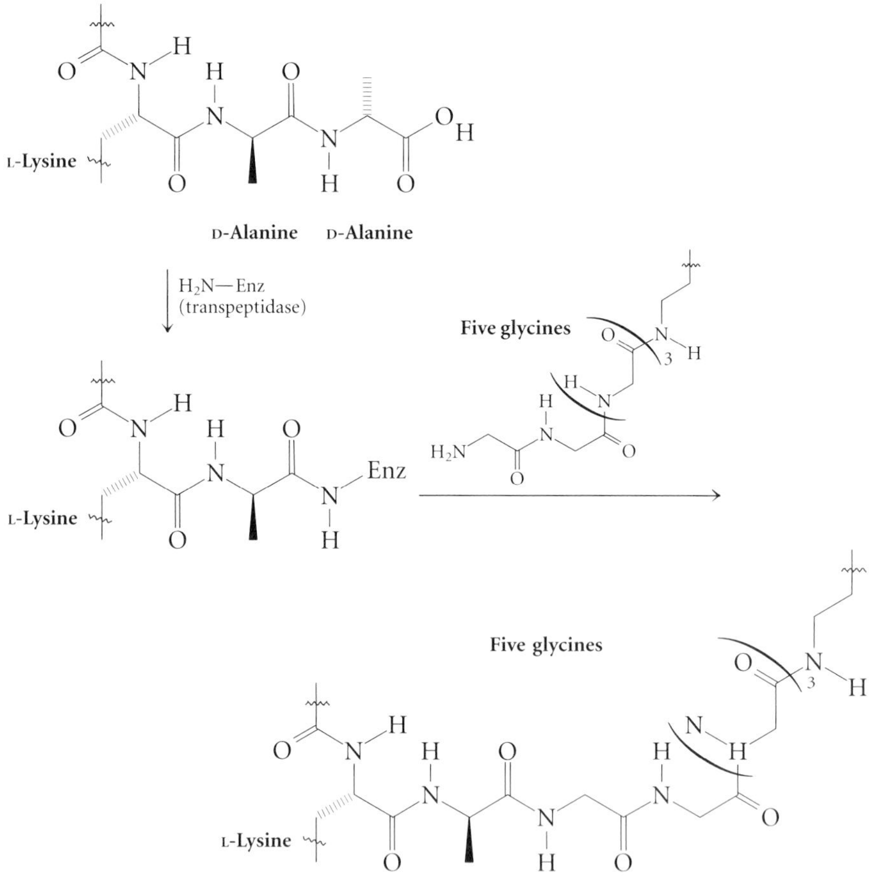

**FIGURE 23.7**

The construction of the cross-link between polysaccharide chains in bacterial cell walls requires the formation of peptide bonds. A short (tripeptide) chain attached to one polysaccharide chain is first covalently bound by an amide bond to the enzyme transpeptidase. By transamidation, the carboxyl functional group is transferred to the amino group of a terminal glycine of a pentapeptide attached to another polysaccharide chain. The formation of this amide bond covalently links one polysaccharide to another.

*CHEMICAL PERSPECTIVES*

TETRAHYDROFOLIC ACID: A COFACTOR
SUPPLIED BY BACTERIA

Taking sulfa drugs for infections kills not only harmful bacteria, but also beneficial bacteria that are part of the "fauna and flora" of the intestinal system. These bacteria are normally a major supplier of tetrahydrofolic acid. Thus, taking a sulfa drug orally for an extended period (2–3 months), as in, for example, the treatment of chronic acne, can lead to a deficiency in this essential cofactor. For this reason, doctors often recommend that sulfa drug therapy be accompanied by eating yogurt with active culture.

*Discovery and Structure of Penicillin Antibiotics.* The discovery by Alexander Fleming in 1928 of the antibacterial properties of a *Penicillium* fungus dramatically altered the course of history, and his observations constitute one of the most amazing cases of serendipity in science. Fleming, a British bacteriologist, was making more or less routine studies for contribution to a textbook on bacteriology. He was growing various bacteria known to be pathogenic (that is, capable of causing disease), such as *Staphylococcus aureus,* at different temperatures. In another laboratory, other scientists were studying molds, and it is suspected that the contamination of Fleming's Petri dish came from this source. He left the dish at room temperature and departed for a week's vacation; during that time, the fungi grew rapidly, producing penicillin and retarding the growth of the bacteria. Had Fleming put this particular dish in an incubator at 37 °C, the course of history might well have been quite different. At that temperature, the *Staphylococcus* bacteria grow much more rapidly than the mold, and once a bacterial colony is mature, it is unaffected by penicillin. That Fleming did not discard this dish marks him as a true scientist. Rather than discarding the contaminated specimen and trying to complete his routine studies, he was able to recognize the significance of the unexpected result, which led to the development of modern antibiotics.

Isolation and purification of the active component was a daunting task, but led ultimately (after 13 years) to the accumulation of sufficient material for use in clinical trials. Initially, this work was carried out by H.W. Florey and his group at Oxford University. Later, large-scale culture and purification was carried out in the United States. (Florey and Fleming shared the 1945 Nobel prize for medicine for their independent contributions to the development of penicillin.) It became apparent that the structures of the penicillins isolated by the two groups were different. Both were effective antibiotics, but the penicillin G produced in the United States (Figure 23.8, page 23-22) became the most widely used form, and it is the standard against which the effectiveness of other penicillins is measured. Many modifications of this basic structure have been made, all affecting the side-chain group attached to nitrogen. These modifications are the origin of such familiar antibiotics as ampicillin and amoxicillin. Interestingly, the simple penicillin structure lacking a substituent on nitrogen has no antibacterial activity.

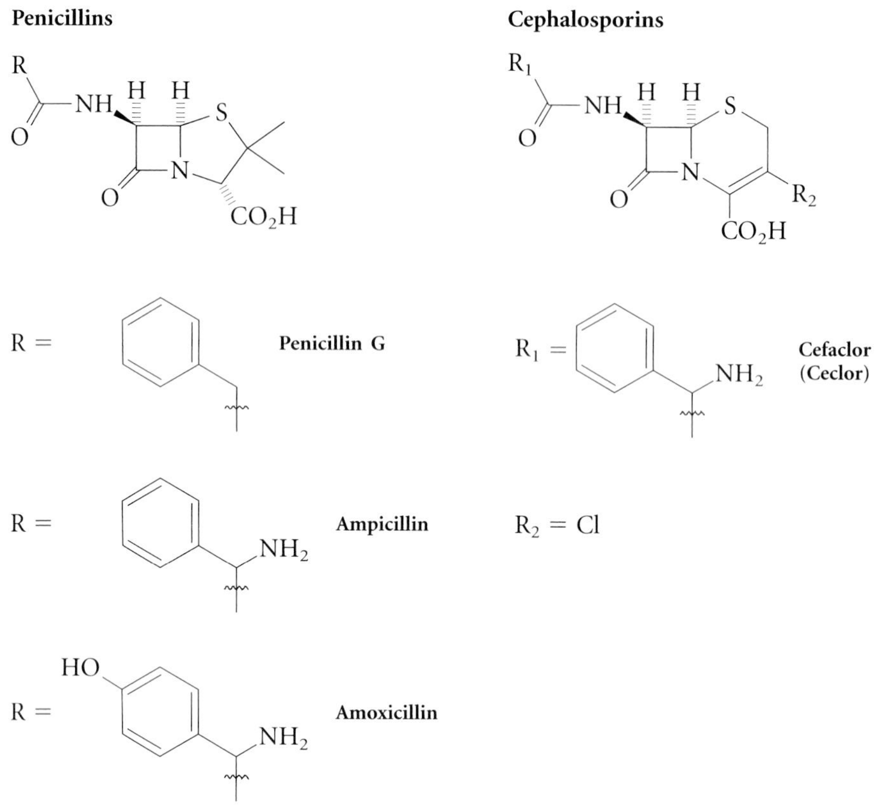

**Penicillins**

**Cephalosporins**

R =  Penicillin G

$R_1$ =  Cefaclor (Ceclor)

R =  Ampicillin

$R_2$ = Cl

R =  Amoxicillin

**FIGURE 23.8**

All penicillins and cephalosporins have a four-member lactam ring fused at nitrogen to another ring that contains sulfur. Shown here are a few of the many $\beta$-lactam antibacterial agents.

*Mode of Action of Penicillin Antibiotics.* All penicillin antibiotics, as well as the structurally related cephalosporins, function in similar ways. They have in common a four-member cyclic lactam ($\beta$-lactam), with an amino group attached to the $\alpha$ carbon. They are thus $\alpha$-amino acid derivatives and are of the L configuration. They also have a carboxylic acid functional group with an $\alpha$-amine substituent, but of the D configuration. Three-dimensional representations of the penicillin nucleus and the D-alanine–D-alanine branch that is critical to cross-linking in bacterial cell walls are shown in Figure 23.9. Because of the close similarity in shape of the two fragments shown in Figure 23.9, the bacterial enzyme responsible for cross-linking incorporates the antibiotic into the cell wall instead of the carbohydrate's peptide side chain. This prevents the formation of the necessary cross-link. It has been suggested that it requires only one such "nick" in the stitching of the cell wall to lead to the rupture of a bacterium. This is certainly consistent with the observation that $\beta$-lactam antibiotics are effective at very low concentrations. For example, penicillin G can kill bacteria at a concentration of 1:50,000,000, which corresponds to only 20 $\mu$g per liter. Figure 23.10 shows the dramatic change in the cell wall of *Staphylococcus* bacteria after treatment with penicillin.

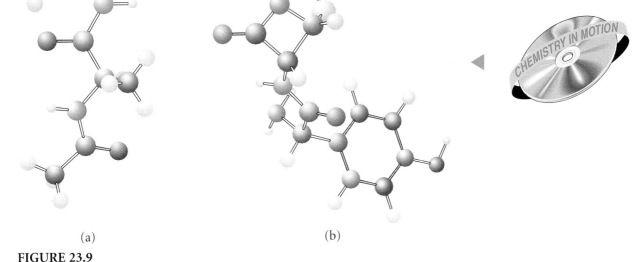

(a)                                                     (b)

**FIGURE 23.9**

These ball-and-stick models show the overall structural similarity between (a) the dipeptide D-alanine–D-alanine and (b) a representative penicillin antibiotic. Because of this similarity in structure, penicillins become attached to and thus block the action of the transpeptidase enzyme that is critical to the construction of cross-links in bacterial cell walls.

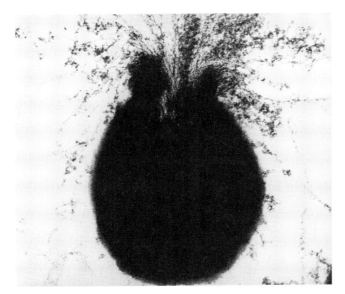

**FIGURE 23.10**

This photomicrograph shows a *Staphylococcus aureus* bacterium literally exploding as a result of cell-membrane defects caused by exposure to penicillin.

It may well be that this mechanism of action for the penicillins is only one of several ways in which the synthesis of bacterial cell walls can be impeded. For example, instead of becoming incorporated into the cell wall, the antibiotic might become covalently attached to the enzyme, thus permanently blocking the reaction site and killing the enzyme. This mode of inactivation of enzymes is referred to as *suicide inhibition,* because the enzyme undergoes an irreversible transformation that blocks its catalytic function.

*Stability of Antibiotics in Vivo.* The strain in the four-member ring of penicillin and cephalosporin antibiotics makes these cyclic amides unusually reactive. Indeed, the 25 kcal/mole of strain energy exceeds the resonance-stabilization energy (18 kcal/mole) of an amide. As a result, some of these antibiotics are rapidly hydrolyzed in the stomach's acidic conditions and cannot be administered orally.

Bacterial mutation sometimes produces new enzymes that catalyze the cleavage of the $\beta$-lactam ring of antibiotics. These enzymes, referred to as *β-lactamases,* destroy antibiotics before they can disrupt bacterial cell membranes. This effect can be partially overcome through the administration of a $\beta$-lactamase inhibitor, a compound that binds strongly at the active site of the enzyme, blocking its normal function. One such compound is cloxacillin, which is mixed with ampicillin to protect the ampicillin from enzymatic cleavage.

**Cloxacillin**

Many criteria determine the suitability of a pharmaceutical agent for the treatment of a disease state. For antibiotics, especially those that must be injected to avoid degradation in the stomach, resistance to enzymatic degradation in the blood is a very important characteristic. In fact, the cephalosporin called ceftriaxone is the leading injectable antibiotic precisely because it need be administered only once each day. (It is effective against

*CHEMICAL PERSPECTIVES*

**A PHARMACEUTICAL AGENT FOR
TICKS, FLEAS, AND LICE**

Carbomycin is a macrolide aminosugar isolated from fermentation broths
of the mold *Streptomyces halstedii.* Carbomycin, marketed under the name
Magnamycin, is a potent antibiotic effective against many of the same bacteria as penicillin. However, in addition, carbomycin is active against
*Rickettsiae,* a group of parasitic bacteria carried by ticks, fleas, and lice. These
bacteria are responsible for such diseases as typhus and Rocky Mountain
spotted fever.

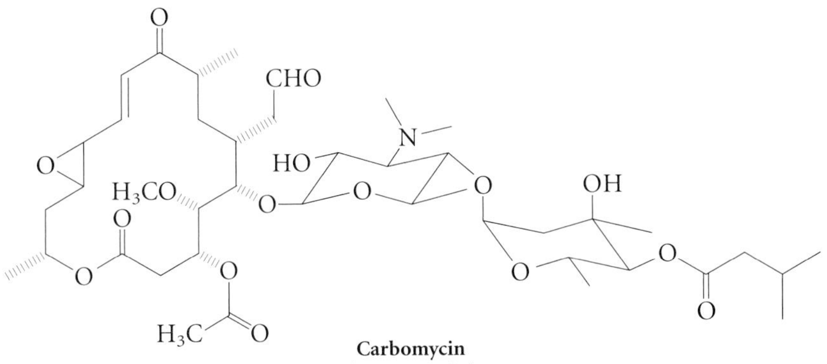

**Carbomycin
(Magnamycin)**

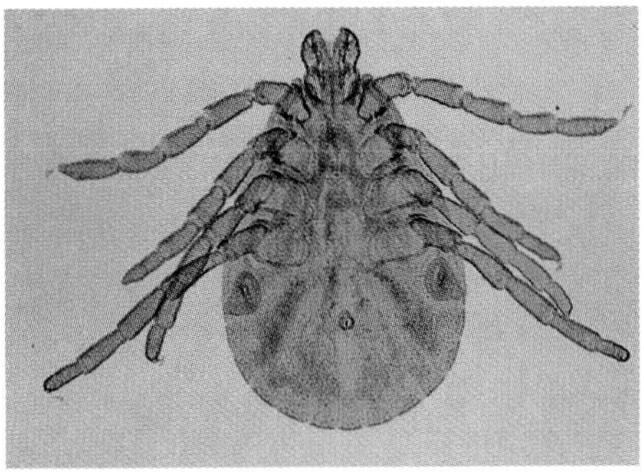

Lyme disease and drug-resistant gonorrhea, as well as many other bacterial infections.)

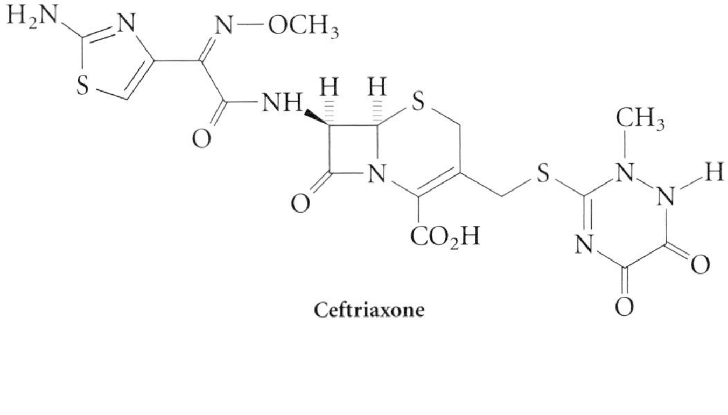

**Ceftriaxone**

## 23.6

# Antiviral Agents

Drugs are widely and effectively used for the treatment of bacterial diseases, but viral infections are, for the most part, controlled through the administration of preventive vaccines. The difference in the approach to combating bacterial and viral diseases is rooted in the nature of the causative agents. Bacteria are living entities (with chemistry that is far from being in equilibrium), but viruses are quite simple and inert packages of biological materials. Replication of a virus requires the reproductive machinery of a host cell, and so stopping viral reproduction seemingly requires destruction of the host cell as well. We first examine the structure of viruses and then see how a virus tricks the host cell into making copies of it. Then, we will examine the mode of action of some antiviral agents currently in use.

### Viral Structure

Key to the structure (and the action) of all viruses is a central core of genetic material. Some viruses contain DNA; others have RNA. This difference is the basis for a broad division of viruses into two categories. This genetic material represents the code necessary to induce the host cell to produce anywhere from 1 to 240 proteins. This genetic core is surrounded by a coating of protein material known as a **capsid,** and together, the core and the protein coat form a **virion** (also called a nucleocapsid).

Virions occur in a variety of shapes, primarily dictated by the structure of the protein outer core. For example, the tobacco mosaic virus is composed of an inner core of RNA twisted into an α-helix, surrounded by a collection of some 2000 identical protein molecules, arranged as wedges in an ascending "staircase" about the core. Other viruses, such as the AIDS virus, have a capsid containing several different proteins. A virion is tiny, typically only 200–400 Å on a side. These tiny particles can pass easily through filters that stop the smallest bacteria.

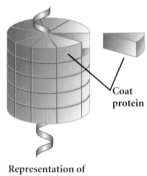

Coat
protein

Representation of
tobacco mosaic virus

### Viral Attack on Cells

Let's follow the invasion of a healthy cell by a virus. The sequence is well defined and only slightly modified depending on whether the virus's genetic material is DNA or RNA.

Critical to the invasion of the virus is its attachment to the surface of the target cell. This process involves specific chemical recognition and varies with the surface coating of the host cell. This recognition is the responsibility of the protein coat of the virus particle, and, in certain cases, can be blocked by chemical agents. Once attached to the cell surface, the virus "blends" with the cell membrane and is then transported to the interior of the cell, where the protein coat is removed, revealing the virus's genetic material. When the core of the virus is DNA, this DNA is replicated and incorporated into the DNA in the host cell's nucleus. This DNA is reproduced without hindrance by the host cell, producing additional DNA and, just as important, the mRNA (usually single-stranded) that will carry the genetic code to the ribosomes. In the ribosomes, tRNA reads the code transmitted by mRNA and transfers the individual amino acids required for the synthesis of proteins. In a cell infected by a virus, the cell's ribosomes synthesize the proteins that make up the outer coat of the virus.

If the attacking virus contains RNA, the mechanism of viral replication is yet more complex. The overall scheme for replication of RNA viruses is shown in Figure 23.11. The viral RNA is accompanied by an enzyme, re-

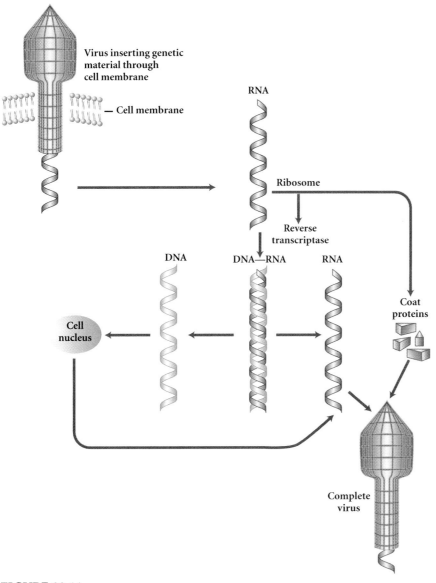

**FIGURE 23.11**

Replication of an RNA virus begins by insertion of the viral RNA through the host cell's membrane. The genetic code of this RNA is then read by ribosomes, forming the coat proteins of the virus, and a reverse transcriptase enzyme that translates the virus RNA into DNA. The DNA is incorporated into the host cell's nuclear genetic material, programming the host cell to produce additional viral RNA. Viral particles are then assembled from viral RNA and coat proteins. The completed virus escapes through the cell membrane, invading new host cells.

verse transcriptase, that translates the genetic code of the viral RNA into a strand of DNA. Normally, this process does not occur in cells; instead, the genetic code is transferred only in the direction from DNA to mRNA. For this reason, these viruses are frequently referred to as **retroviruses.** The DNA produced from RNA serves the same role as the DNA inserted by those viruses whose genetic core consists of DNA.

Another highly significant aspect of viral infection that complicates treatment is the existence, in certain viruses, of a built-in time clock. Although incorporated into the cell, the genetic material of many viruses is not reproduced until a so-called third-party event occurs. This trigger can be the effect of another virus or a bacterium on the immune system. In the interval before activation, the virus may have installed itself into the genetic code of a large number of host cells. When reproduction of the virus is activated, the massive release of viral particles and their invasion of many more host cells can overwhelm the natural response of the host organism's immune system. In the case of the AIDS virus, the virus invades and destroys the T cells of the immune system, disabling the body's own natural defense mechanism.

### Viral Replication: The "Coup de Grace"

The virus has put into place the mechanism for its own reproduction. The chemical machinery of cells is a blind transcriber—it follows (almost perfectly) the coded instructions delivered from the cell nucleus by mRNA. Thus, if an mRNA delivers an incorrect three-base code sequence to a ribosome, a tRNA dutifully delivers the wrong amino acid to the growing peptide chain. And so it is with the mRNA generated from the viral genetic material—except that the nature of the material produced has been dictated by the viral DNA, and protein for the outer coat of new viral particles is produced by the hard-working ribosomes. Once the protein coat is assembled around the nucleic acid core, a new copy of the virus is ready to emerge and seek new host cells, to continue the reproduction of viruses in a chain reaction.

Why does the virus kill the host cell? In many cases, it doesn't. But for some viruses, destruction of the cell is the mechanism by which the daughter virus particles are released. Even when this is not the case, the reproductive machinery of the cell has been significantly diverted to the production of materials (DNA or RNA, and proteins) for the new virus particles. Moreover, there are many other consequences of viral infection that can be devastating to the host organism.

### Antiviral Therapy: Looking for a "Magic Bullet"

Because a virus uses the host cell for its own reproduction, it might seem at first that there would be no means that could interfere with this process that would not also kill the host. However, once again, seeking a "magic bullet" for viruses (as Ehrlich did for bacteria) entails finding unique aspects of the infective process that enable the pharmaceutical agent to target the invader specifically. The root of the problem with viral infection is that, once DNA has been integrated into the genetic material contained in the nucleus of the host cell, it is difficult to expunge this foreign code.

However, there are a number of features of the viral cycle that offer potential for controlling viral infections: initiation of the attack by recognition and penetration of the host cell membrane, shedding of the protein "coat" that surrounds the nucleic acid core, replication of the nucleic acid core and the protein coat, and reassembly of the virus by wrapping of the protein around the nucleic acid core. In the case of retroviruses, the reverse transcription of RNA to DNA presents another opportunity for dealing with the virus, especially as this process does not occur in normal cells. Increasing knowledge of details of the viral cycle and the development of new antiviral drugs are proceeding hand in hand, although progress is disappointingly slow.

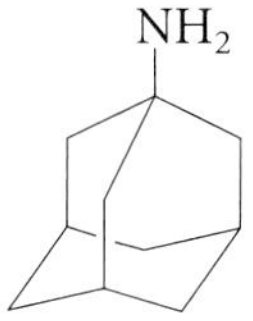

*Blocking Viral Attack: Amantadine.* A limited group of compounds can prevent a virus particle's critical association with and/or transmission through the membrane of the host cell. One of the first antiviral compounds discovered was amantadine (1-adamantanamine), which operates via this mechanism. This relatively simple amine (administered as its hydrochloride salt) blocks association of a variety of viruses, including influenza and rubella, with the target host cells. Unfortunately, because amantadine can cross the blood–brain barrier, it also has serious adverse effects on the central nervous system. Significant alteration in the interpretation of sensory signals has been noted, and some volunteers even reported that their worlds turned upside-down instantaneously. (Fortunately, their views corrected once treatment stopped.)

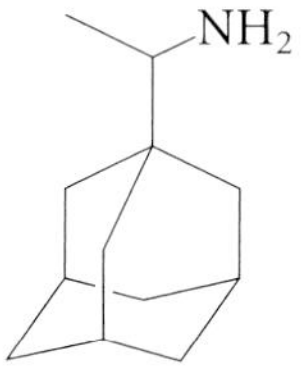

Rimantadine, an analog of amantadine, is even more effective against the influenza A virus and appears to have fewer neurological side effects. However, the incidence of hallucinations, nightmares, and vomiting has so far prevented approval of this drug for use in the United States.

*Blocking DNA Replication: Acyclovir.* Acyclovir is a nucleoside analog that has a DNA base (guanine) attached to an abnormal (and short) mimic of the carbohydrate portion of a nucleoside:

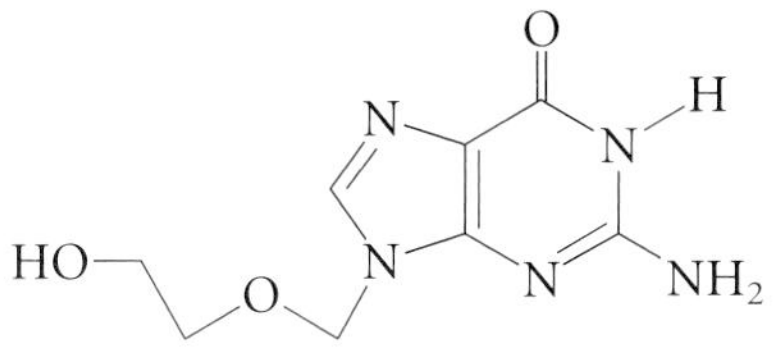

**1-Adamantanamine**

**Rimantadine**

**Acyclovir**
**9-[(2-hydroxyethoxy)methyl]-guanine**

A variety of mechanisms combine to make acyclovir an effective antiviral for the treatment of herpes viruses, which are responsible for cold sores, genital herpes, and herpes simplex encephalitis. In infected cells, acyclovir assumes the role of a nucleoside, being converted to a triphosphate derivative, which is then incorporated into a growing DNA chain. Because acyclovir lacks the free sugar hydroxyl group characteristic of the nucleosides, further elongation of the DNA chain is blocked. Incorporation of acyclovir into growing DNA chains also takes place in normal cells, but the process in normal cells is much slower because acyclovir is phosphorylated by the viral enzymes much faster than by normal cellular enzymes. In addition, the phosphorylated acyclovir has a much higher affinity for viral DNA polymerase than for normal cellular DNA polymerase. Thus, viral DNA replication incorporates acyclovir to a far greater extent than does cellular DNA replication. In a final blow to the virus, the DNA incorporating acyclovir binds irreversibly to the viral DNA polymerase, ending its utility.

*Interfering with Reverse Transcription: AZT.* Viruses containing RNA are critically dependent on the enzyme reverse transcriptase to wreak havoc on host cells. Because host cells do not possess reverse transcriptase, a drug that targets this enzyme selectively would seem an ideal way to take out a retrovirus. Like acyclovir, AZT (azidothymidine) is phosphorylated in the body to form a nucleoside mimic. This AZT triphosphate binds very tightly to HIV reverse transcriptase and blocks its critical role of producing DNA from viral RNA. Moreover, when the AZT triphosphate bound to the viral enzyme is incorporated into DNA, the absence of a hydroxyl group at the 3′ position blocks further elongation of the DNA. AZT remains one of the most useful antiviral agents for the treatment of HIV (human immunodeficiency virus).

AZT (3′-azido-2′,3′-dideoxythymidine)

*Protease Inhibitors.* The HIV virus is unusual in that its genetic material directs the host cell's ribosomes to synthesize a protease enzyme. Although this enzyme may have additional functions, it is required to form the proteins that make up the capsid of the virion particle. These proteins are formed by cleavage of a large polyprotein produced by the host cell at the direction of the viral genes. Intense research has been directed at developing inhibitors of this rather specialized enzyme. However, a candidate has not yet been found that meets the rather stringent requirements for a protease inhibitor. For example, protease inhibitors are typically peptides or peptide-like materials and are therefore open to degradation by the host. In addition, enzyme inhibition must be selective and not block cellular proteases required by the host.

A typical HIV-protease inhibitor

### Identification of Disease-Causing Agents

Identification of the causative agent of a disease state can range from being relatively easy to being extremely difficult. For most bacterial infections, a sample from the infected host can be distributed on an appropriate growing medium, and, after an appropriate incubation period, bacterial colonies can be seen and examined. If samples of the culture are injected into a healthy host and infection ensues, the bacterium responsible for the disease has been identified.

Viral infections are much more difficult to assay, in part because viruses require a host for replication and do not multiply *in vitro*. In addition, many viral diseases have an induction period that extends from initial exposure until the appearance of specific disease symptoms in the host. This is the case with the AIDS virus, which, having orchestrated insertion of its genetic code into the host cell nucleus, can lay dormant for many years.

However, most difficult to characterize are the causative agents in diseases caused by prions (short for proteinaceous infectious particles). Indeed, the nature of prions is not yet entirely clear.

### Diseases Caused by Prions

Three diseases—all of which affect the brain and are fatal—are believed to be caused by prions: scrapie, a disease that causes goats and sheep to scrape off their wool; kuru, a disease found among cannibals of Papua, New Guinea; and bovine spongiform encephalopathy (BSF, or much more commonly, mad cow disease). Scrapie has been studied to a much greater extent than the other two diseases, and much is known about its causative agent. Scrapie agent is a hydrophobic glycoprotein with 208 amino acid residues. Digestion in the host by cellular proteases breaks down scrapie agent into smaller proteins, which then aggregate into insoluble masses because of their hydrophobic exterior. These deposits, known as *amyloid plaques*, amass in the brain and interfere with its normal functions.

But how do prions reproduce if they contain no genetic material? Several possible answers to this question are being considered—and some go strongly against the grain of current biological theory. For example, prions may act as reverse transcriptase enzymes that translate their own amino acid sequence into genetic code, which is then transcribed by the host cell. Although genetic information flow has always been thought of as a one-way street, this dogma was modified in 1970 when the reverse transcriptase enzymes in RNA viruses were identified independently by Howard Temin and David Baltimore. Yet the concept of a protein coding for DNA is a much greater imaginative leap from known biochemical pathways than the rather minor shift from DNA-to-RNA translation to RNA-to-DNA translation. It might also be possible that prions do contain genetic material that, to this point, has escaped detection.

Perhaps the simplest explanation is that the genetic material for production of the prion exists in the healthy host and that expression of this gene is triggered by some event, as yet unidentified. Indeed, the genetic code for the protein sequence in the scrapie agent has been found in both healthy and infected mice. The explanation could be even simpler. It is possible that the scrapie agent is produced by a minor genetic misread that otherwise produces a useful protein for the host. The modification in the agent causes it to become significantly less soluble, and association of the agent with normal protein causes both to become insoluble. Although intriguing, this explanation does not fully account for the apparent transmission of kuru from victim to feaster during cannibalistic rituals in the Pacific islands.

### Protein Misfolding and Alzheimer's Disease

Alzheimer's disease is a degenerative brain condition that eventually leads to complete loss of mental function in many patients. Although no causative agent has been found, the deposits found in the brain tissue of Alzheimer's patients bear striking resemblance to the amyloid plaque found in the brains of sheep affected with scrapie. Thus, it is possible that there is a prion agent that triggers the onset of Alzheimer's disease.

An alternative and intriguing explanation has also been offered. In the normal course of protein synthesis by ribosomes, the growing polypeptide chain emerges without the complete folding pattern that represents its secondary and tertiary structures. These secondary and tertiary structures are assumed gradually and must be complete before the protein can assume its role (for example, as a structural protein or an enzyme). If the protein were to fail to achieve its final shape, it might be left with an exterior too hydrophobic for it to remain soluble in water. It would then precipitate, and could also aggregate with other misfolded or only partially folded proteins, forming the insoluble plaques that characterize Alzheimer's disease.

## 23.8

# Cancer Chemotherapy

### Characteristics of Cancer

Cancer is a group of diseases characterized by the rapid growth of transformed tissue and by migration of cancerous cells to parts of the body distant from the point of origin. The difficulties associated with the treatment of cancer can be gauged by the fear with which it is regarded. Nonetheless, some progress in the control and cure of cancer has been made.

A tumor is the result of the uncontrolled growth of cells. A tumor may be benign or, if it has the ability to metastasize (spread to form tumors in other tissues), malignant. Cancer is the term used to describe the large number of diseases that share these characteristics of uncontrolled and life-threatening growth and the ability to metastasize. Cancers of the blood (such as leukemias) do not involve tumor formation. As with other cancers, there is uncontrolled growth, but in this case the overproduction of single cells (leukocytes or erythrocytes) becomes life-threatening.

The clinical manifestations of cancer reflect differences between malignant and normal cells. The rapid growth characteristic of cancer is reflected in the elevated metabolic rate of cancer cells. Because all cancers at some stage require rapid and abnormal growth, it is possible to target cancer cells selectively with chemical agents that disrupt or inhibit cell division. This is the mode of action of many effective chemotherapeutic agents.

Cancer cells exhibit other distinctive characteristics. For instance, cultures of normal cells anchor themselves to the surface of the culture dish, grow to form a single-cell layer, and then stop growing. In contrast, cancer cells grow even when floating freely. The growth of cancer cells does not stop when the surface of the culture dish is covered, but continues as cells pile up on one another. These differences are attributed to differences in the membranes of normal and cancer cells, and certain anticancer drugs appear to act by disrupting cell membrane structure and processes.

Other differences between cancer and normal cells have been noted with regard to the cell cycle. A cell cycle is the period from the birth of a new cell to the time that the new cell divides. This period encompasses four phases: $G_1$, S, $G_2$, and M. In $G_1$ (gap 1), which occurs after cell division, growth of cytoplasmic material occurs but no DNA is synthesized. All DNA synthesis occurs in the next stage, S (for synthesis). In $G_2$, the structures required for mitosis (such as spindle structures) are synthesized, in preparation for the M (mitosis) phase, during which the two DNA copies separate. Under certain conditions, normal cells can pass to a rest state $G_0$ from the $G_1$ state; under the same conditions, cancer cells continue to divide. Cancer cells will stop dividing if deprived of essential nutrients. Moreover, unlike normal cells, cancer cells will stop growing at any point in the cell cycle, not just at the $G_1$ phase, and unlike normal cells, they cannot resume growth from this resting state; instead, they die. Therefore, depriving cells of essential nutrients is expected to be more devastating to the cancer cell than to the normal cell.

## Causes of Cancer

The causes of cancer are not well understood. Although the causes of some cancers have been pinpointed, in general, the cause is not known. In two relatively well-defined cases where viruses and chemical agents have been shown to be causative, modification of genetic material (DNA) appears to be responsible for alteration of normal cell behavior.

In 1911, the American pathologist Peyton Rous was the first to show that cancer could be induced in chickens by a virus. For this discovery, Rous (1879–1970) was awarded the 1966 Nobel prize in medicine. His seminal contribution led others to search for other viruses that cause cancer. Several are now known, including Epstein–Barr virus, which is responsible for Burkitt's lymphoma and nasopharyngeal cancer, and the hepatitis B virus (a DNA virus), which causes liver cancer (mainly among Asians). Therapy for these cancers can be focused on the causative virus and prevention of viral infection.

A wide variety of chemicals have been shown to induce tumors in laboratory mice that have been selectively bred to be hypersensitive to cancer induction. Extrapolation of these findings to "normal" mice and then to humans is tenuous, at best. Nonetheless, such studies serve as notice to eliminate unnecessary exposure to these compounds.

In some cases, the link between specific chemicals (and groups of compounds, such as "tobacco smoke") and cancer in humans is well established. Many of these compounds are found in the environment, such as, for example, benzo[a]pyrene (found in coal tar and shale oil) and aflatoxin $B_1$ (which, along with other aflatoxins, is produced by a fungus that thrives on peanuts).

**Aflatoxin $B_1$**

**Benzo[a]pyrene**

Induction of cancer in humans by chemical agents often has an induction period of years or even decades. Indeed, the discovery of the carcinogenic effects of aflatoxins was made quite difficult by a typical 20-year delay between exposure and development of cancer. Thus, although prevention is the best remedy, the difficulties in identifying carcinogens can compound the problems inherent in removing environmental carcinogens, both natural and synthetic. As with other disease states, the focus of medical efforts is on both prevention and cure.

## ▨ Treatment of Cancer: Chemotherapeutic Agents

Numerous methods have been used to treat cancer. Currently, surgery, radiation therapy, chemotherapy, and immunologic therapy are used in varying combinations. Their success is usually gauged by comparing survival rates.

Chemical agents for cancer treatment, of which there are many, are called **antineoplastics.** The selectivity of many antineoplastic agents is based on the fact that cancer cells grow more rapidly than normal cells. Some normal cells, including those of hair follicles, the intestinal lining, and bone marrow (critical to the immune system), also exhibit rapid growth. Thus, most chemotherapeutic agents currently available have serious side effects, because they affect such rapidly growing normal cells. The side effects include hair loss, destruction of the intestinal lining, and immunosuppression.

To see how chemotherapeutic agents might operate, let's look at a hypothetical drug that interferes with DNA synthesis. In the life cycle of the cell, DNA synthesis takes place only during a limited period, the S phase. If a drug that interferes with DNA synthesis is taken, a cell in the S phase cannot complete this part of the cell cycle; thus, the cell dies or becomes susceptible to attack by normal body defenses. Cells not in the S phase are unaffected. Thus, during the time the chemotherapeutic agent is in the body, the fraction of rapidly dividing (cancer) cells destroyed will be larger than the fraction of normal cells destroyed, because relatively few normal cells will have passed through the S phase. A similar analysis can be made for agents that interfere with other parts of the cell cycle.

## ▨ Drugs Affecting Nucleic Acid Synthesis

*Antimetabolites.* All living systems use nucleic acids for the storage of genetic information. Therefore, compounds that interfere with the synthesis of these vital materials are potentially toxic to all life forms. Such compounds are called **antimetabolites.** Recall that the sequence for the synthesis of adenine is parallel to its biosynthesis and includes stages that sequentially add carbon and nitrogen atoms to glycine. Two representative chemotherapeutic agents, methotrexate and 5-fluorouracil, inhibit the biosynthesis of the purine-based nucleosides, but in different ways.

*Methotrexate.* Methotrexate acts by inhibiting the synthesis of tetrahydrofolic acid, which is responsible for the addition of single carbon atoms in the biosynthesis of nucleic acid bases and a number of other important biochemicals. For example, this cofactor is the source of the two carbons of adenosine shown here in green:

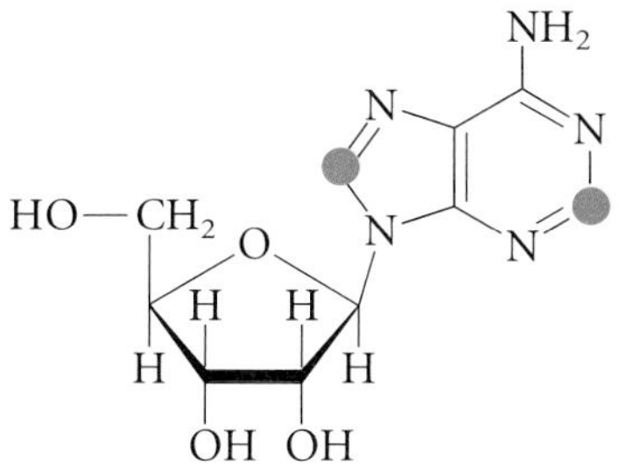

**Adenosine**

Methotrexate is structurally quite similar to dihydro- and tetrahydrofolic
acid (Figure 23.12). By irreversible binding, methotrexate inhibits the enzyme dihydrofolate reductase, a catalyst for the formation of tetrahydrofolic acid.

(a)

**Methotrexate**

(b)

**Dihydrofolic acid**

Dehydrofolate
reductase

**Tetrahydrofolic acid**

**Leucovorin calcium**

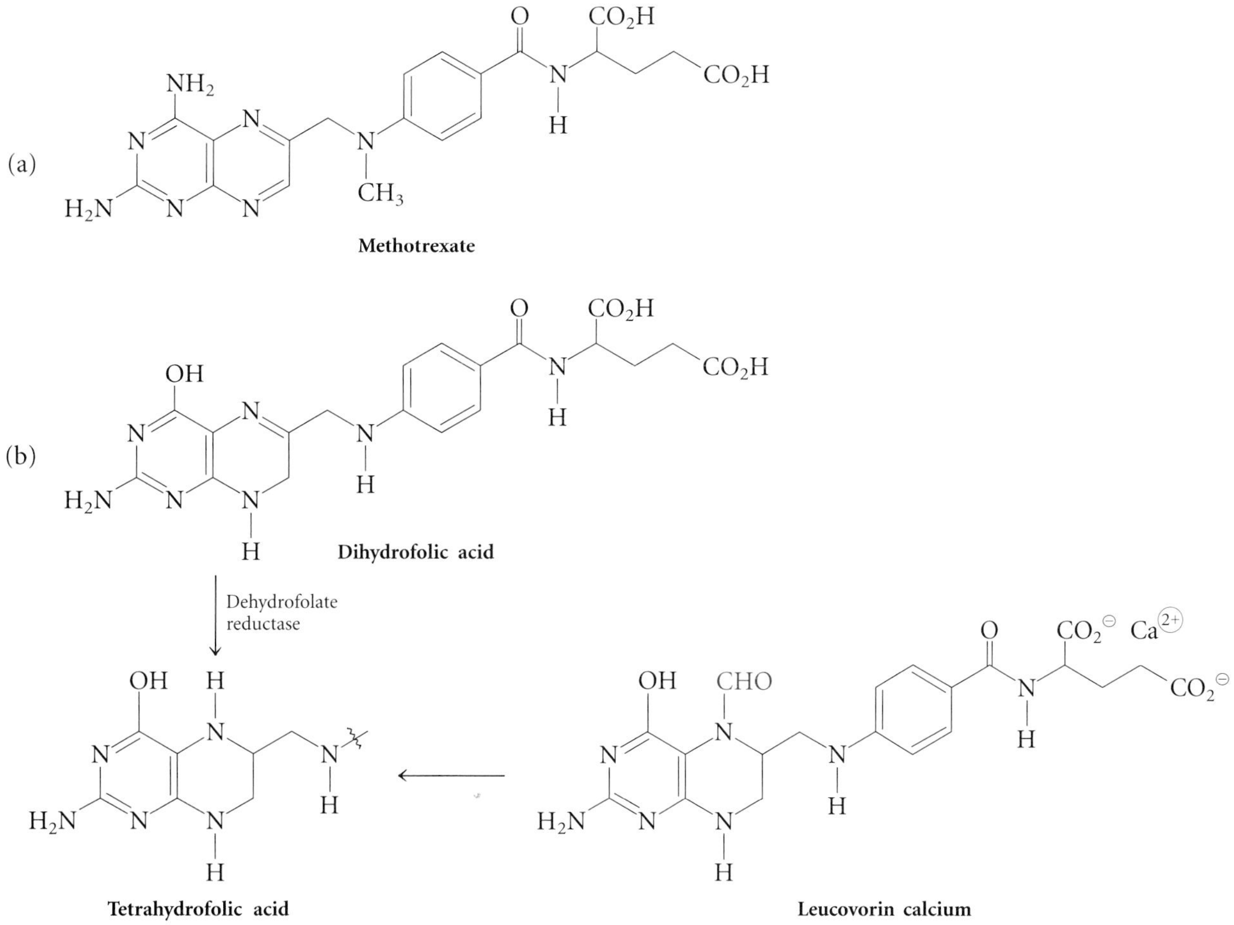

**FIGURE 23.12**

(a) Methotrexate closely resembles (b) dihydrofolic acid. After methotrexate
treatment for cancer, leucovorin is administered as a source of tetrahydrofolic acid to aid in the recovery of normal cells.

CANCER AND THE DELANEY AMENDMENT

The Delaney amendment prohibits the addition to food of any substance that has been shown to cause cancer in *any* animal assay. Although applauded by many as protection from chemical carcinogens, this legislation failed to address many common naturally occurring chemical agents known to cause cancer (at least in animal tests). Examples include 1,1-dimethylhydrazine from common table mushrooms, benzene in gasoline (stand upwind when you fill your gas tank), and aflatoxins from fungus-infected peanuts. Indeed, it is next to impossible to produce fungus-free peanuts, and virtually all peanut butter, even the "organic" variety, contains aflatoxins—after all, the fungus is "organic."

Although methotrexate has a pronounced effect on cancer cells, it also is destructive to normally dividing cells, which also require tetrahydrofolate. To counteract this effect on normal cells, leucovorin calcium is typically administered after methotrexate. This synthetic drug is degraded in the body to tetrahydrofolic acid by loss of the aldehyde group (shown in red in Figure 23.12) and helps normal cells recover from the effect of the antineoplastic agent.

*5-Fluorouracil (5-FU).* Another agent that interferes with DNA synthesis is 5-fluorouracil. This derivative of the natural base uracil inhibits the enzyme that converts uracil to thymine by methylation. (Recall that fluorine and hydrogen are not too dissimilar in size.)

5-Fluorouracil

Degradation of 5-fluorouracil is rapid in the body, but leads to the production of fluorocitric acid, which is a potent inhibitor of the TCA cycle. (The antimetabolite properties of 5-FU have also been put to use in preparing "drug cocktails" for the treatment of AIDS.)

*Hadacidin.* Hadacidin inhibits DNA replication by interfering with the introduction of one of the nitrogen atoms of adenosine.

**Hadacidin**

In the biosynthesis of adenosine, aspartic acid serves as the source of the C-6 amino group:

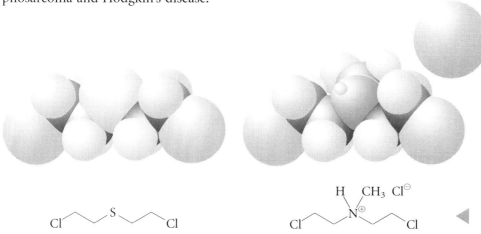

Hadacidin is sufficiently similar to aspartic acid in shape and electronic properties that it binds tightly to the enzyme responsible for catalyzing this amino transfer reaction.

## EXERCISE 23.4

Suggest plausible reaction mechanisms for both of the steps shown for the synthesis of adenosine via transfer of nitrogen from aspartic acid.

*DNA Cross-Linkers.* DNA replication can also be inhibited by bifunctional alkylating agents that link together the two strands of DNA in the double helix. Separation of the strands thus becomes impossible, and replication fails.

Most of the bis-alkylating agents used today as anticancer agents are structurally related to the mustard gas that was introduced as a chemical weapon in World War I. (The names for mustard gas and the nitrogen mustards are derived from their mustard-like odor.) Individuals who had been heavily gassed during the war suffered damage to bone marrow and lymphoid tissues. This observation led to the hypothesis that such compounds might be useful in treatment of cancers of lymphoid tissue such as lymphosarcoma and Hodgkin's disease.

Chlorambucil and cyclophosphamide are two current-generation nitrogen mustards. The latter finds wide use in treating Hodgkin's disease and breast, ovarian, and lung cancers. Cyclophosphamide itself is not active, but is converted by enzymes in the liver to the antineoplastic agent.

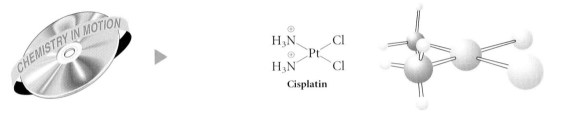

**Chlorambucil**          **Cyclophosphamide**

Only one compound containing a heavy metal, cisplatin, is used as an antineoplastic agent. Cisplatin (*cis*-dichlorodiammineplatinum) is a planar structure in which the chlorine substituents are rather loosely held by the metal and are easily replaced by stronger bonds to amino groups.

**Cisplatin**

Reaction of cisplatin with amino groups of guanine residues forms bridges across the DNA double helix and, as with the nitrogen mustards, inhibits the DNA replication essential for cell division.

**Cross-linking of DNA strands by**
**guanine—NH—Pt—HN—guanine bridge**

***DNA Binding Agents.*** A group of anticancer drugs inhibit DNA replication by initially forming noncovalent bonds to the double helix. DNA binding agents are characterized by extended aromatic systems as, for example, in the anthracyclines daunorubicin and doxorubicin produced by *Streptomyces* species. Many of the DNA binding agents are derived from natural sources.

**Daunorubicin**          **Doxorubicin**

*CHEMICAL PERSPECTIVES*

ANTICANCER COMPOUNDS FROM MOLDS

Many unusual compounds that inhibit the growth of cancer cells have been isolated from plants and other natural sources. For example, dynemicin A (produced by a mold isolated from a Texas soil sample) contains an unusual ene–diyne conjugated system (one alkene and two alkynes) that undergoes cyclization, forming a biradical:

Dynemicin A

This process is energetically favorable because of the formation of the aromatic ring, yet the product biradical is quite reactive. Each radical carbon center reacts with one of the chains of double-stranded DNA, linking the strands and thus preventing replication of the genetic material, a process that requires that the two strands separate. Because cancer cells divide much more frequently than do normal cells, they are affected to a greater extent. Dynemicin A is also toxic to normal cells, and so cannot be used for the treatment of cancer. Chemists are currently synthesizing other molecules containing the ene–diyne system. It is hoped that these will be less toxic to normal cells and yet retain the anticancer activity of the natural compound.

Both daunorubicin and doxorubicin have quite a broad spectrum of activity. They appear to exert their anticancer activity not only by binding to the DNA helix but also by reacting directly with cell membranes. Although daunorubicin and doxorubicin differ only in that the latter has an extra hydroxyl group (shown in red), their biological action is surprisingly different. Doxorubicin is active primarily against a wide variety of hard tumors such as breast and lung cancer, and daunorubicin is most useful in cancer of the lymph system and in leukemia. Indeed, dramatic alteration of biological activity often results from quite minor changes in molecular structure. Recall that heroin provides a "quicker rush" than does morphine solely because it passes the blood–brain barrier more rapidly; once within brain cells, it is hydrolyzed, forming morphine.

*Modes of DNA Binding.* The manner in which antitumor agents are bound to DNA and the way in which they damage cells are areas of active investigation, and understanding of these processes is constantly evolving. Existing experimental evidence has led to some reasonable hypotheses.

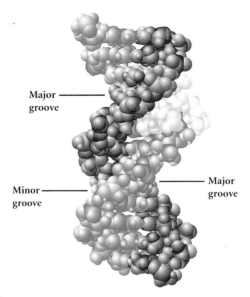

## FIGURE 23.13

A model for doxorubicin in the major groove of DNA. The two DNA chains are shown with red and blue atoms for clarity, and the atoms of the intercalating anthracycline in yellow.

Because the two strands of the double helix of DNA are not rotated by quite 180° relative to each other, there is a wide and a narrow groove, referred to as the major and minor grooves, respectively. Certain antitumor agents, such as anthramycin, bind in the narrow (minor) groove of the DNA double helix. A model of how doxorubicin might bind to DNA is shown in Figure 23.13.

Other aromatic compounds, especially those that are hydrophobic, small, and relatively flat, can slip between the stacked layers of base pairs in the DNA double helix, as illustrated in Figure 23.14 for benzopyrene in a short segment of DNA. This mode of binding, where a foreign agent is sand-

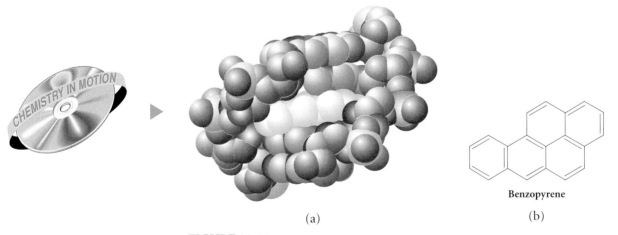

**Benzopyrene**

(a)           (b)

## FIGURE 23.14

Space-filling model (with hydrogen atoms omitted) showing benzopyrene intercalated into DNA.

wiched between the nucleic acid bases that form the steps of the spiral stair-case structure of the DNA helix, is called **intercalation.** This mode of binding requires that the DNA open to make room for the inserted molecule. We will shortly see how advantage can be taken of this mode of binding to cross-link the two strands of DNA.

*Action of Binding Agents.* The mode of action of many of these DNA binding agents is not well understood, but general mechanisms for blocking replication may involve one or more of the following:

**1.** Tight noncovalent binding that inhibits unfolding and/or replication

**2.** Covalent binding with the same consequences as noncovalent binding

**3.** Binding followed by reduction of the drug to form a radical species

Many DNA-bound anticancer compounds generate radicals. These radicals may damage DNA directly, or they may be involved in the generation of superoxide ($O_2^{\ominus}$) or hydroxy ($\cdot OH$) radicals. In any event, DNA can be damaged in such a way that faithful replication is impaired.

## Photochemotherapy

It is often said that "there is nothing new under the sun." This quote seems apropos as we consider how the "magic bullet" concept of Ehrlich and the ancient practice of photochemotherapy are combined in an exciting new approach to cancer treatment.

Photochemotherapy was first used by the Egyptians in 2000 B.C. to help reduce the effects of leukoderma (vitiligo), a disorder characterized by patches of skin that lack pigmentation. A similar regimen was used in India in 1200 B.C. In both cases, plant extracts were administered, either orally as teas or by direct application to the skin. Irradiation of the patient with sunlight (not hard to find in either country) then darkened the skin, completing the treatment. This phenomenon was reinvestigated in 1948 by El Mofty, an Egyptian physician.

In the vitiligo treatment, both plant extracts and sunlight were required to bring about the desired effect. This suggested that it might be possible to find other therapeutic agents that could be activated by light. The active components of the plants used by the Egyptians (*Ammi majus*) and the Indians (*Psoralea corylifolia*) are 5-methoxy- and 8-methoxypsoralen:

5-Methoxypsoralen      8-Methoxypsoralen

Investigation of these compounds showed that they had a number of photobiological properties. In addition to the phototoxic effect used in the treatment of vitiligo, these compounds showed antiproliferative, mutagenic, and carcinogenic effects when their administration was followed by exposure to ultraviolet light. Psoralens undergo many different reactions on irradiation with short-wavelength light (UV-A), including degradation of the compound itself into fragments, and dimerization reactions. Within the cell, psoralens bind to DNA, and there, upon irradiation, undergo a 2 + 2 photocyclization with a nucleoside base, illustrated here with thymidine and 8-methoxypsoralen (Figure 23.15). For the cell, the consequences of such treatment are similar to the results of exposure to the nitrogen mustards discussed earlier.

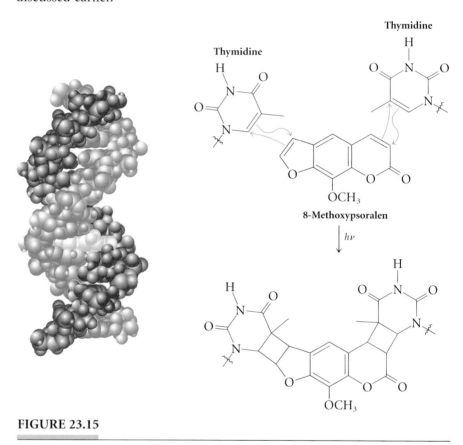

**FIGURE 23.15**

By successive photochemical reaction at both ends of the molecule, a psoralen can form a bridge linking the strands of the DNA double helix.

A rational approach to photochemotherapy was proposed by J. A. Parrish and coworkers at Harvard University, and in recent years, there has been intensive research in this area in an effort to target cancer cells selectively for destruction. We noted earlier that Ehrlich was fascinated by the ability of numerous dyes to differentiate between various cells. A number of dyes—in particular, acridines, xanthenes, psoralens, and porphyrins—bind selectively to cancer cells. However, unlike the "magic bullets" envisioned by Ehrlich, most of these dyes lack the necessary reactivity by themselves to effect damage to the tumor. Upon irradiation, some of them can produce tissue destruction, either through direct photochemical reaction or through sensitizing the production of superoxide and hydroxyl radicals, powerful oxidants that can destroy virtually any tissue.

Porphyrin dyes—in particular, hematoporphyrins and their derivatives—have superior tumor-localizing properties. Hematoporphyrins (Hp) are macrocyclic, highly colored compounds obtained from blood and structurally related to chlorophyll.

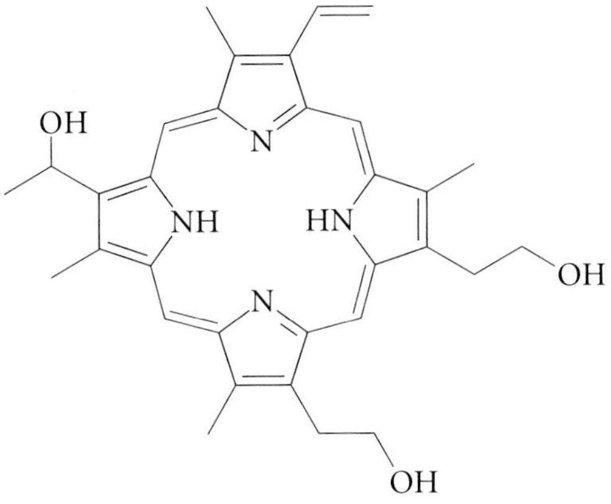

**A hematoporphyrin**

A mixture of related hematoporphyrins, known as photofrin, is used for photodynamic therapy. Photofrin is injected intravenously and is selectively absorbed in tumor cells. The drug is cleared more slowly from cancer cells than from normal cells, and after 1 to 2 days, the differential in the photofrin concentration in tumor and normal cells is at a maximum. Irradiation with visible light results in excitation of the porphyrin followed by energy transfer to molecular oxygen to form singlet oxygen:

$$Hp \xrightarrow{h\nu} Hp^* + :\overset{..}{\underset{..}{O}} — \overset{..}{\underset{..}{O}}: \longrightarrow Hp + :\underset{..}{O}=\underset{..}{O}:$$

**Triplet**              **Singlet**
**oxygen**             **oxygen**

Singlet oxygen is a reactive oxidant that has several modes of reaction; we will deal with only one here.

"Normal" molecular oxygen is unusual in that the most stable electronic state is the triplet, in which there is only one bond between the atoms. The excited state is the singlet, with both a $\sigma$ and a $\pi$ bond between the atoms. Singlet oxygen reacts with alkenes, forming hydroperoxides:

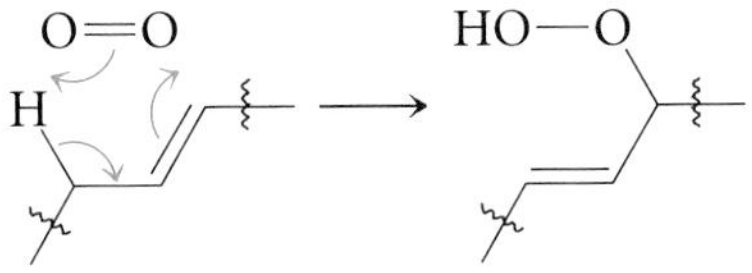

Because hematoporphyrins are localized mainly in the cytoplasmic membrane, oxidation affects the proteins, sterols, and double bonds of the phospholipids making up the cell membranes. As these components are oxidized, the membrane structure is destroyed, and the cell dies. Usually there is no damage at the nuclear level, because porphyrins are not bound there.

Clinical investigations have shown that photochemotherapy can cure small tumors (less than 2 cm) and can be useful for the treatment of early cancers of skin, bladder, vagina, mouth, and lungs, all tissues located in body cavities and therefore accessible to light. Similar techniques involving treatment with psoralen derivatives, followed by ultraviolet irradiation of the blood, have been used against T-cell lymphoma. The potential for this type of therapy seems immense; a multitude of compounds could be designed by varying both binding and photochemical characteristics.

# Organic Chemistry: Retrospective and Prospects

This chapter's exploration of drug interactions has provided you with an opportunity to see how the concepts studied in a course on organic chemistry can enhance your understanding of life processes. Over a period of almost two centuries, the science of organic chemistry has been furthered by the efforts of many individuals who sought to understand and explain the phenomena they observed in the laboratory. Their efforts resulted in the development of the concepts of isomerism, stereochemistry, functional group chemistry, reaction mechanisms, thermodynamics, and kinetics—to name a few. These concepts have been and will continue to be applied to the design and synthesis of new materials, organic, polymeric, and organometallic, and to arriving at a deeper understanding of the complex biochemical and biological dance of living things. In turn, this scientific knowledge will provide the basic understanding from which chemists and other scientists will continue to provide materials that will greatly enhance the quality of life.

## Summary

   **1.** The design of medicinal agents has progressed dramatically from the early development of salvarsan to the design of effective $H_2$-blockers for the treatment of hyperacidity. Many of the concepts explored in this course are important for understanding the mode of action of modern pharmaceutical agents.

   **2.** Many disease states are caused by chemical imbalance due to over- or underproduction of critical biochemicals or deficiencies resulting from an inadequate diet. Many of these diseases can be effectively treated through chemical means. A deficiency can be corrected by administering the needed chemical, and an enzyme inhibitor can be used to reduce the amount of a substance that is overproduced by a natural pathway.

   **3.** A biological response can sometimes result from administering compounds that resemble, even if vaguely, the natural material. For example, the alkaloid morphine relieves pain because its shape mimics that of the endorphins, naturally occurring painkillers. Many biologically active agents have in common with morphine the $\beta$-phenethylamine unit.

   **4.** Disease states resulting from invasion by another living entity (such as a bacterium) also can be treated with chemicals. Effective antibacterial agents have been developed that are destructive to the microbe while being relatively harmless to human beings and other animals.

**5.** The selective toxicity of modern antibiotics results from distinct differences between bacteria and animals at the molecular level. Bacterial cell walls are substantially stronger than those of mammalian cells, because bacteria must be protected from the harsh environment. Penicillin antibiotics destroy bacterial cells by interfering with the construction of the cell walls at the critical stage of cross-linking, when much of the strength is imparted. Bacteria must manufacture tetrahydrofolic acid, a critical cofactor responsible for the transfer of one carbon in biosynthesis and biodegradation. Sulfa drugs are effective antibiotics because they block the synthesis of tetrahydrofolic acid.

**6.** Developing effective antiviral agents is considerably more difficult than developing antibiotics, because viruses use the host cell's machinery to replicate.

**7.** Antiviral agents operate by interfering at various stages in the viral infective cycle. Amantadine prevents association with and/or penetration of the host cell membrane. Acyclovir inactivates the replication of viral DNA by the host cell. Acyclovir is phosphorylated rapidly by viral enzymes and has such a high affinity for viral DNA polymerase that it inactivates it. When incorporated into growing viral DNA, it terminates chain growth. AZT binds so tightly to reverse transcriptase that transcription of viral RNA to DNA is effectively inhibited. AZT is also incorporated into growing chains of viral DNA and stops chain elongation.

**8.** Cancer chemotherapy focuses on exploiting the differences between cancer cells and normal cells. One of the chief differences is the much faster metabolic rate of cancer cells.

**9.** A number of chemotherapeutic agents interfere with DNA replication, an integral part of cell replication. Antimetabolites, DNA cross-linkers, and DNA binding agents can all adversely affect DNA replication. Because cancer cells reproduce faster than normal cells, more of their time is spent in DNA replication, when they are susceptible to drug attack.

**10.** Photochemotherapy extends the opportunity to target cancer cells selectively by activating compounds that bind selectively to them.

## Review Problems

**23.1** Many molecules with a β-phenethylamine subunit can have pronounced effects on the mind. For review, devise two different routes to synthesize β-phenethylamine, starting with benzene and any other needed organic and inorganic reagents.

**23.2** The reaction of an aryl diazonium ion with another aryl ring bearing an electron-donating substituent G produces a diazo linkage between the benzene rings. Write a detailed reaction mechanism for this transformation, keeping in mind what you have learned about electrophilic aromatic substitution.

**23.3** Write a detailed reaction mechanism for the conversion of aniline into its diazonium ion with nitrous acid.

**23.4** Although aromatic diazonium ions prepared from aryl amines are sufficiently stable to be isolated, the treatment of aliphatic amines with nitrous acid generally leads to a number of products. Explain this dramatic difference in reactivity between aryl and aliphatic diazonium ions.

**23.5** Examine the structure of valinomycin to determine if there are any sequences of amino acids that repeat. (Ignore stereochemistry at this point.)

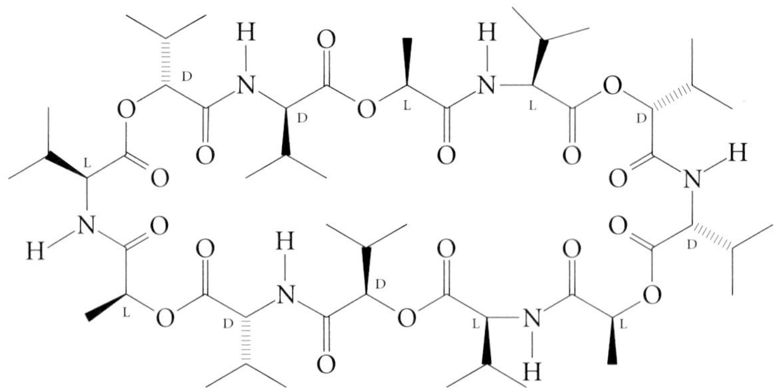

**Valinomycin**

(a) What are the largest such segments, and how many are there?
(b) Calculate the number of different stereoisomers possible from the presence
of 12 amino acid residues, all chiral. Can any of them be *meso* isomers?

**23.6** Lefetamine is a synthetic material that has some medicinal properties similar to those of morphine. These result from binding of lefetamine to the morphine receptors. Suggest what portions of lefetamine correspond to structural
units of morphine. What is the stereochemical configuration of the center of chirality in lefetamine? Do you think that the enantiomer would have the same
medicinal properties? Explain your answer.

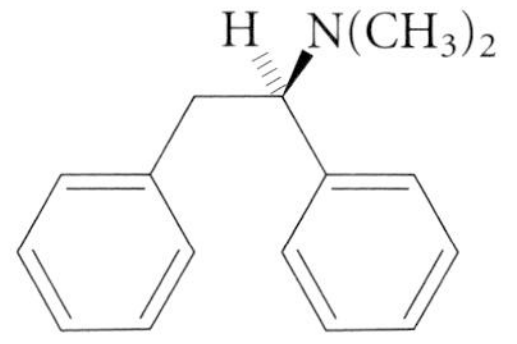

**Lefetamine**

**23.7** Suggest a synthesis of lefetamine (see Problem 23.6), starting from
phenylacetic acid and any other organic and inorganic materials you need.
(Phenylacetic acid has a disagreeable odor that resembles that of a person who
has not bathed in a *very* long time.)

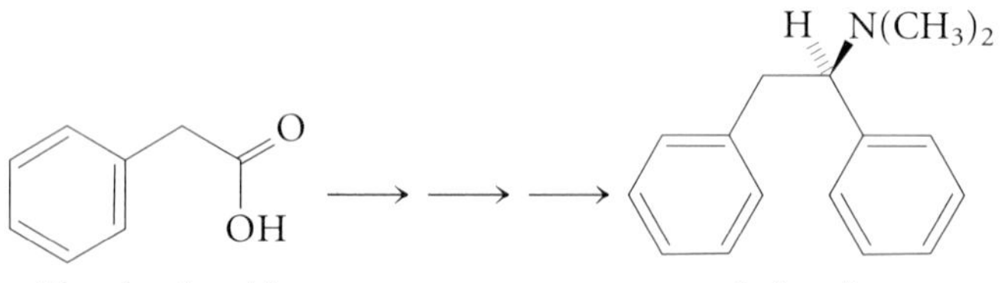

**Phenylacetic acid**                                    **Lefetamine**

**23.8** Polynoxylin is the trade name for a polymer formed from urea and
formaldehyde. This polymer has been used as a topical antibacterial agent and
probably functions by the slow release of formaldehyde, a potent antibacterial
agent. Suggest mechanisms for the release of formaldehyde from Polynoxylin under both acidic and basic conditions.

**Polynoxylin** → $H^{\oplus}$ (or $^{\ominus}OH$) → **Formaldehyde**

**23.9** When *p*-methylbenzyltrimethylammonium hydroxide is heated in toluene at reflux, a cyclic compound (known as a paracyclophane) is formed, in addition to a polymer. Suggest a mechanism for the formation of the paracyclophane that proceeds through an intermediate diradical.

*p*-**Methylbenzyltrimethylammonium hydroxide**

$\xrightarrow[\text{Toluene}]{\Delta}$

**A paracyclophane**
(17–19% yield)

**antihistamine:** a compound that inhibits vasodilation by competing with histamine for binding at a physiologically active site (23-3)

**antimetabolite:** a compound that interferes with the synthesis of nucleic acids (23-8)

**antineoplastic:** a chemical agent for treating cancer (23-8)

**blood type:** one of four classifications of human blood cells differentiated by the identity of carbohydrate residues present on the cell surface (23-1)

**capsid:** the coating of protein surrounding the genetic core of a virus (23-6)

**chemotherapeutic index:** the ratio of a drug's toxicity to the host to that for the invading organism (23-1)

**endorphin:** a natural pentapeptide found in the brain in extraordinarily low concentration that induces euphoria or blocks pain; the name is derived from *endo*genous mo*rphine* (23-4)

**enkephalin:** a pentapeptide endorphin (23-4)

**formulation:** the final preparation of a pharmaceutical in a form acceptable for delivery to the target organ or organism (23-3)

**in vitro:** describing a reaction conducted in a laboratory environment; from the Latin for "in glass" (23-2)

**in vivo:** describing a reaction conducted within a living organism; from the Latin for "in life" (23-2)

**intercalation:** the sandwiching of a foreign agent between nucleic acid bases in DNA (23-8)

**psychoactive drugs:** pharmaceutical agents used to achieve a state of euphoria or to block intense pain (23-4)

**retrovirus:** a virus in which viral RNA is accompanied by an enzyme, reverse transcriptase, that translates the genetic code of the RNA into a strand of DNA (23-6)

**suicide inhibition:** an irreversible transformation in which the catalytic function of an enzyme is blocked by irreversible covalent bond formation (23-5)

**virion:** an intact virus consisting of a core of genetic material surrounded by coat proteins (23-7)

# Appendix:
# Summary of
# Synthetic Methods

The following table summarizes the major classes of reactions covered in Chapters 8 through 14 and organizes them according to the type of bond formed. Practicing organic chemists use these reactions, along with some others, in various combinations to make new compounds and to prepare naturally-occurring compounds in the laboratory.

| Bond Formed | Type of Reaction | Example |
|---|---|---|
| C—H | Catalytic hydrogenation of an alkene (or alkyne) | |
| | Hydrolysis of a Grignard reagent | $R-MgBr \xrightarrow{H_2O} R-H$ |
| | Clemmensen reduction | |
| | Wolff–Kishner reduction | |
| | Decarboxylation of a $\beta$-ketoacid | |
| | Catalytic hydrogenation of an alkyne | |
| | Dissolving metal reduction of an alkyne | |
| C—C | S$_N$2 displacement by cyanide | $R-Br \xrightarrow{{}^{\ominus}C{\equiv}N} R-C{\equiv}N$ |
| | S$_N$2 displacement by acetylide anion | |
| | Grignard addition | |
| | | |
| | | |
| | | |

| Bond Formed | Type of Reaction | Example |
|---|---|---|
| C—C | Friedel–Crafts acylation | benzene + acyl chloride $\xrightarrow{\text{AlCl}_3}$ aryl ketone |
| | Friedel–Crafts alkylation | benzene + Cl—R $\xrightarrow{\text{AlCl}_3}$ alkylbenzene (R) |
| | Diels–Alder reaction | diene + dienophile $\xrightarrow{\Delta}$ cyclohexene |
| | Conjugate addition to an $\alpha,\beta$-unsaturated carbonyl group | enone + $R_2CuLi \longrightarrow$ R-added ketone |
| | Michael reaction | enone + $(RO_2C)_2CH^{\ominus} \longrightarrow (RO_2C)_2CH$—ketone |
| | Aldol reaction | acetone $\xrightarrow{\text{Base}}$ $\beta$-hydroxy ketone (OH) |
| | Alkylation of ketone enolate anion | cyclohexanone $\xrightarrow{\text{LiN}(i\text{-Pr})_2}$ $\xrightarrow{\text{H}_3\text{CBr}}$ 2-methylcyclohexanone ($CH_3$) |
| | Acetoacetic ester synthesis | acetoacetic ester $\xrightarrow{\ominus\text{OEt}}$ $\xrightarrow{\text{R—Br}}$ |
| | | alkylated acetoacetic ester (R) $\xrightarrow[\text{H}_2\text{O}]{\text{H}_3\text{O}^{\oplus}}$ ketone (R) |
| | Malonic ester synthesis | malonic ester (RO, OR) $\xrightarrow{\ominus\text{OEt}}$ $\xrightarrow{\text{R—Br}}$ |
| | | alkylated malonic ester (RO, OR, R) $\xrightarrow[\text{H}_2\text{O}]{\text{H}_3\text{O}^{\oplus}}$ carboxylic acid (R, OH) |

(continued)

| Bond Formed | Type of Reaction | Example |
|---|---|---|
| C—C | Claisen condensation | |
| | Cope rearrangement | |
| | Claisen rearrangement | |
| C=C | Dehydrohalogenation | |
| | Dehydration | |
| | Hofmann elimination | |
| | Catalytic hydrogenation of an alkyne | |
| | Dissolving metal reduction of an alkyne | |
| | Reductive elimination of a vicinal dihalide | |
| | Wittig reaction | |

| Bond Formed | Type of Reaction | Example |
|---|---|---|
| C=C | Aldol condensation | |
| C≡C | Dehydrohalogenation | |
| | $S_N2$ displacement by an acetylide anion | $R-C{\equiv}C^{\ominus} + R-Br \longrightarrow R-C{\equiv}C-R$ |
| C—X | Free-radical halogenation | $R-H \xrightarrow[h\nu]{X_2} R-X$ |
| | Addition of H—X | |
| | Addition of $X_2$ | |
| | Conversion of an alcohol to an alkyl halide | $R-OH \xrightarrow{PX_3,\ POX_3,\ or\ HX} R-X$ |
| | Electrophilic aromatic substitution | |
| | α-Halogenation of a ketone | |
| | Hell–Volhard–Zelinski reaction | |
| | Chlorination of an alkene | |

(continued)

| Bond Formed | Type of Reaction | Example |
|---|---|---|
| X X | Bromination of an alkene | |
| C—OH | Hydrolysis of an alkyl halide | $R-X \xrightarrow{\ominus OH} R-OH$ |
| | Hydration of an alkene (Markovnikov regiochemistry) | |
| | Oxymercuration–demercuration (Markovnikov regiochemistry) | |
| | Hydroboration–oxidation (anti-Markovnikov regiochemistry) | |
| | Grignard reaction of an aldehyde or ketone | |
| | Grignard reaction of an ester | |
| | Metal hydride reduction of an aldehyde or ketone | |
| | Metal hydride reduction of an ester | |
| | Aldol reaction | |
| | Cannizzaro reaction | |

| Bond Formed | Type of Reaction | Example |
|---|---|---|
| C—OH | Cyanohydrin formation | |
| HO, OH (structure) | *cis*-Hydroxylation | |
| | Nucleophilic opening of an epoxide | |
| R—C≡N | S$_N$2 displacement by cyanide | R—Br + $^{\ominus}$C≡N ⟶ R—C≡N |
| | Cyanohydrin formation | |
| | Dehydration of an amide | |
| R—NH$_2$ | Aminolysis of an alkyl halide | R—X $\xrightarrow{\text{NH}_3}$ R—NH$_2$ |
| | Gabriel synthesis | |
| | Reduction of an aromatic nitro compound | |
| | Reductive amination of a ketone | |
| | Lithium aluminum hydride reduction of an amide | |

*(continued)*

| Bond Formed | Type of Reaction | Example |
|---|---|---|
| R—NH$_2$ | Hofmann rearrangement | |
| R—O—R′ | Williamson ether synthesis | R—O$^{\ominus}$ + Br—R′ ⟶ R—O—R′ |
| | Peracid oxidation of an alkene | |
| | Oxidation of a primary alcohol | |
| | Ozonolysis of an alkene | |
| | Hydrolysis of an acetal | |
| | Hydrolysis of a ketal | |
| | Hydrolysis of a terminal alkyne | |
| | Chromate oxidation of a secondary alcohol | |
| | Friedel–Crafts acylation | |
| | Claisen condensation | |

| Bond Formed | Type of Reaction | Example |
|---|---|---|

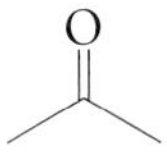

**Bond Formed:** (ketone carbonyl, C=O)

Pinacol rearrangement

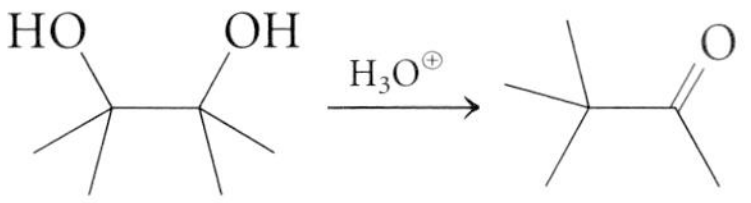

$$\text{(diol)} \xrightarrow{H_3O^{\oplus}} \text{(ketone)}$$

Claisen rearrangement

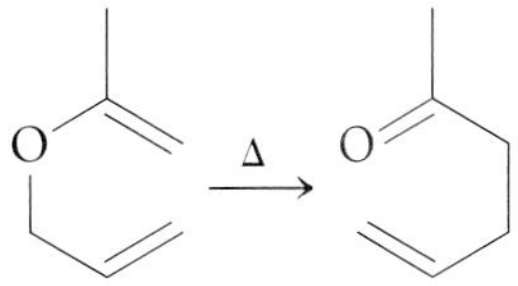

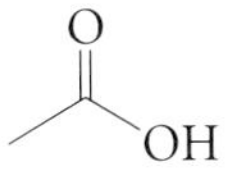

$$\text{(allyl vinyl ether)} \xrightarrow{\Delta} \text{(carbonyl)}$$

**Bond Formed:** (carboxylic acid, C=O and OH)

Hydrolysis of a carboxylic acid derivative

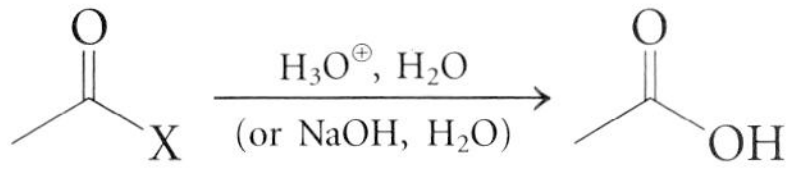

$$\text{R-C(=O)-X} \xrightarrow[\text{(or NaOH, H}_2\text{O)}]{H_3O^{\oplus}, H_2O} \text{R-C(=O)-OH}$$

X = Cl, OR, OAc, NR$_2$, SR

Hydrolysis of a nitrile

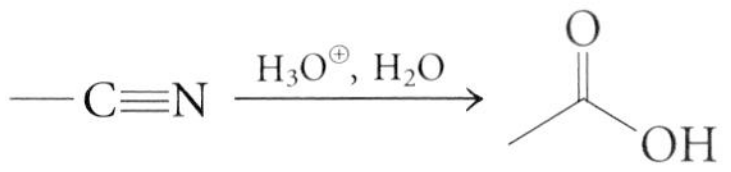

$$-C\equiv N \xrightarrow{H_3O^{\oplus}, H_2O} \text{R-C(=O)-OH}$$

Oxidation of a primary alcohol

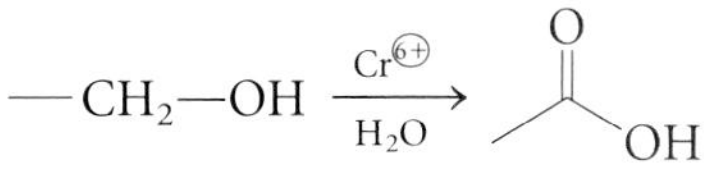

$$-CH_2-OH \xrightarrow[H_2O]{Cr^{6+}} \text{R-C(=O)-OH}$$

Oxidation of an aldehyde

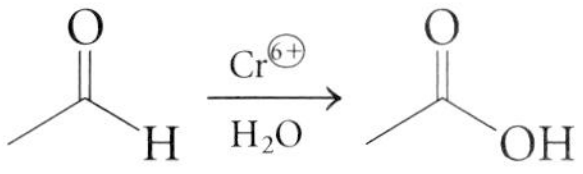

$$\text{R-C(=O)-H} \xrightarrow[H_2O]{Cr^{6+}} \text{R-C(=O)-OH}$$

Permanganate oxidation of an alkyl side chain of an arene

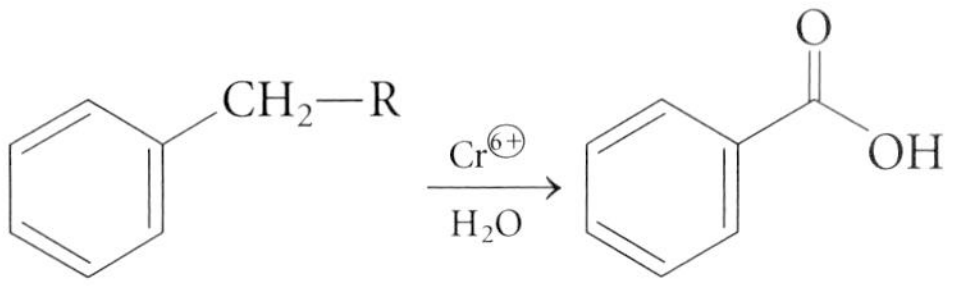

$$\text{Ar-CH}_2\text{-R} \xrightarrow[H_2O]{Cr^{6+}} \text{Ar-C(=O)-OH}$$

Iodoform reaction

$$\text{R-C(=O)-CH}_3 \xrightarrow{I_2, \text{NaOH}} \xrightarrow{H_3O^{\oplus}} \text{R-C(=O)-OH}$$

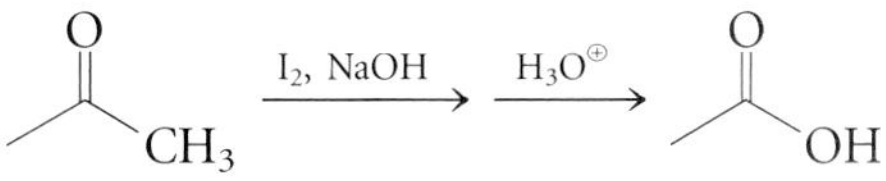

Carboxylation of a Grignard reagent

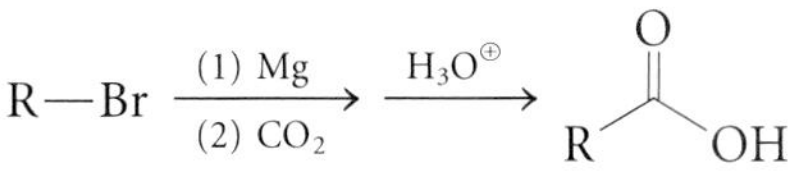

$$R-Br \xrightarrow[(2)\ CO_2]{(1)\ Mg} \xrightarrow{H_3O^{\oplus}} \text{R-C(=O)-OH}$$

Benzilic acid rearrangement

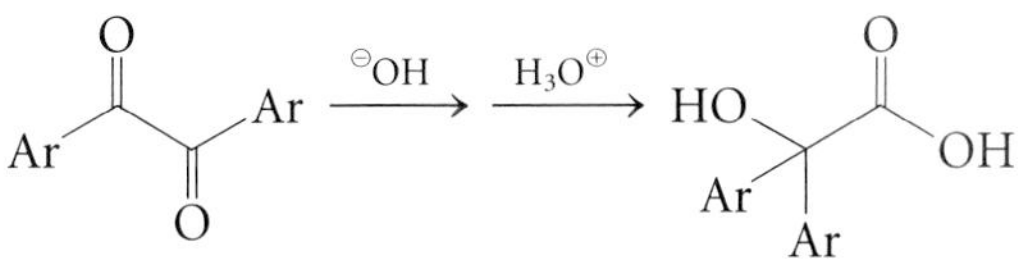

$$\text{Ar-C(=O)-C(=O)-Ar} \xrightarrow{^{\ominus}OH} \xrightarrow{H_3O^{\oplus}} \text{HO-C(Ar)(Ar)-C(=O)-OH}$$

*(continued)*

A-9

| Bond Formed | Type of Reaction | Example |
|---|---|---|
| | Treatment of an acid with thionyl chloride | |
| | Acid dehydration | |
| | Amidation of a carboxylic acid derivative | <br>X = Cl, OR, OAc |
| | Beckmann rearrangement | |
| | Esterification of a carboxylic acid | |
| | Transesterification | |
| | Baeyer–Villiger oxidation | |
| | Ketal (acetal) formation | |
| | Imine formation | |
| | Enamine formation | |

# Glossary

**absolute configuration:** the three-dimensional structure of a molecule that has one or more centers of chirality (5.6)

**absolute stereochemistry:** the unambiguous specification of all spatial positions about a center of chirality (5.6)

**absorption spectroscopy:** the measurement of the dependence of the intensity of absorbed light on wavelength for light in the visible and ultraviolet regions (4.3)

**acetal:** a functional group bearing an alkyl group, a hydrogen atom, and two alkoxy groups on one carbon atom [$RCH(OR)_2$]; produced in the acid-catalyzed alcoholysis of an aldehyde or a hemiacetal (12.3)

**acetoacetic ester:** $CH_3(CO)CH_2CO_2R$; an $\alpha$-acetylated derivative of an ester (13.5)

**acetoacetic ester synthesis:** a method for preparing an $\alpha$-mono- or dialkylated derivative of a methyl ketone by sequentially alkylating an acetoacetic ester anion, hydrolyzing the alkylated ester, and decarboxylating the resulting $\beta$-ketoacid (13.5)

**acetyl CoA:** a thiol ester of coenzyme A (CoA) and acetic acid; a critical intermediate in fatty acid biosynthesis and degradation, in the citric acid cycle, and in glycolysis (21.5)

**acetylene:** *see* **ethyne**

**acetylide anion:** *see* **alkynide anion**

**achiral:** descriptor for a molecule in which at least one conformation has a mirror plane of symmetry; lacking "handedness" (5.5)

**acid chloride:** $RCOCl$; a functional group in which a carbonyl carbon bears an alkyl or aryl group and a chlorine atom (3.9)

**acid-catalyzed reaction:** a reaction that is accelerated in the presence of an acid but in which the acid is not consumed in forming the product (10.1)

**acid-induced reaction:** a reaction in which acid is required and is not regenerated at the end of the sequence (20.3)

**activation energy ($\Delta H^{\ddagger}$ or $E_{act}$):** the energy difference between a ground-state reactant and the transition state (5.1)

**activation energy barrier:** *see* **activation energy**

**active electrophile:** a form of an electrophilic reagent that is more active than normal and is often prepared by interaction of an electrophilic reagent with a Lewis or Brønsted acid (11.2)

**active site:** the relatively small portion of an enzyme where catalysis actually occurs (20.7)

**acyclic:** lacking rings (1.3)

**acyclovir:** a nucleoside analog that blocks DNA synthesis (23.6)

**acyl anion equivalent:** a reagent that provides a nucleophilic equivalent of $RC{=}O^{\ominus}$ (21.8)

**acylation:** the replacement of a hydrogen by an acyl group (11.2)

**acylium ion:** $RC{\equiv}O^{\oplus}$; a resonance-stabilized cation in which positive charge is distributed between carbon and oxygen (11.2)

**1,2-addition:** a mode of addition in which two groups are bonded to adjacent carbons in the product (10.1)

**1,4-addition:** a mode of addition in which two groups are bonded to the ends of a four-atom system in the product; *see also* **conjugate addition** (10.1)

**addition polymer:** a macromolecule produced in a polymerization in which all atoms present in the monomer are retained in the polymeric product (16.3, 16.4)

**addition reaction:** a chemical conversion in which two reactant molecules combine to form a product containing all the atoms of both reactants (7.1)

**adenine:** $C_5H_5N_5$; a biologically important heteroaromatic base (3.11)

**adenosine diphosphate (ADP):** a diphosphate ester of adenine; a principal energy-storage molecule in biological systems (22.4)

**adenosine monophosphate (AMP):** a monophosphate ester of adenine (22.4)

**adenosine triphosphate (ATP):** a triphosphate ester of adenine; a principal energy-storage molecule in biological systems (22.4)

**ADP:** *see* **adenosine diphosphate**

**adsorption:** association with a solid surface, often reversible (4.2)

**aflatoxin:** a carcinogenic compound produced by a fungus that thrives on peanuts (23.8)

**alcohol:** a compound bearing the OH functional group (3.5)

**alcoholysis:** a reaction in which an alcohol displaces a leaving group or is added across a multiple bond (12.3)

**aldehyde:** RCHO; a functional group in which a carbonyl carbon bears a hydrogen and an alkyl or aryl group (3.8)

**aldimine:** an imine of an aldehyde (21.4)

**aldol:** a $\beta$-hydroxyalcohol; a molecule containing both an aldehyde and an alcohol functional group (13.3)

**aldol condensation:** the production of a more complex $\alpha,\beta$-unsaturated aldehyde (or ketone), with the elimination of water, upon treatment of two equivalents of an aldehyde (or ketone) with acid or base (7.1, 13.3)

**aldol reaction:** formation of a $\beta$-hydroxyaldehyde (or ketone) from two molecules of an aldehyde (or ketone) (13.3)

**aldolase:** an enzyme that catalyzes a retro-aldol reaction, such as the degradation of fructose diphosphate in glycolysis (22.10)

**aldose:** a sugar with an aldehyde functional group at C-1 (17.3)

**aliphatic hydrocarbons:** a family of compounds containing hydrogen and carbon atoms, but no aromatic rings (2.3)

**alkanes:** a family of saturated hydrocarbons with the empirical formula $C_nH_{2n+2}$ for acyclic members (1.3)

**alkenes:** a family of unsaturated hydrocarbons containing one or more double bonds; compounds with the empirical formula $C_nH_{2n}$ for acyclic members with one double bond (2.1)

**alkoxide:** an anion obtained by deprotonation of the OH group of an alcohol (8.3)

**alkyl group:** a fragment derived from an alkane by removal of one hydrogen (1.5)

**alkyl halide:** R—X (X = F, Cl, Br, I); a compound in which carbon is bonded to a halogen atom (3.12)

**alkylborane:** a functional group in which carbon is attached to a trivalent boron atom (10.4)

**alkynes:** a family of hydrocarbons containing a triple bond; compounds with an empirical formula $C_nH_{2n-2}$ for acyclic members with one triple bond (2.4)

**alkynide anion:** $RC\equiv C^\ominus$; an anion formed by deprotonation of a terminal alkyne (8.4)

**allene:** an unsaturated hydrocarbon containing two double bonds emanating in opposite directions from a common *sp*-hybridized carbon atom (2.4)

**allyl cation:** a resonance-stabilized carbocation in which the vacant *p* orbital is adjacent to a $\pi$ bond (3.6)

**allyl group:** —$CH_2CH=CH_2$; an alkyl substituent in which the point of attachment is adjacent to a double bond (2.3)

**aluminate:** a species containing an oxygen–aluminum bond (12.2)

**Alzheimer's disease:** a degenerative brain disease characterized by loss of memory and general brain function; its cause is not yet known (23.6)

**amantadine:** an antiviral agent that prevents association of the virus particle with the host cell membrane (23.6)

**ambiphilicity:** the tendency of a molecule XH to act as both an acid and a base (3.2)

**amide:** $RCONR_2$; a functional group in which a carbonyl carbon bears an alkyl or aryl group and an amino group (3.9)

**aminal:** $RCH(NR'_2)_2$; a functional group with one hydrogen, an alkyl group, and two amino groups on one carbon atom; produced in the reaction of a secondary amine with an aldehyde (21.6)

**amine:** an alkyl or aryl derivative of ammonia (3.1)

**$\alpha$-amino acid:** a compound in which an amino group and a carboxylic acid are attached to the same carbon atom (16.5)

**amino group:** an $NH_2$ substituent (3.1)

**ammonia:** $NH_3$; the simplest compound containing $sp^3$-hybridized nitrogen (3.1)

**AMP:** *see* **adenosine monophosphate**

**amylopectin:** a highly branched, water-insoluble starch (16.5)

**amylose:** *see* **starch**

**androgen:** a male hormone (17.1)

**angle strain:** the destabilization caused by deformation from normal bonding angles for atoms in a cyclic compound (1.4)

**anhydride:** $RCO_2COR$; a functional group in which two carbonyl carbons bearing alkyl or aryl groups are linked through an oxygen atom (3.9)

**aniline:** $C_6H_5NH_2$; amino-substituted benzene (3.11)

**anion:** a negatively charged ion (3.2)

**anionic polymerization:** the formation of a polymer by a process in which the growing end is a carbanion (16.4)

**annulation:** formation of a ring on an existing ring (13.3)

**anomer:** a stereoisomer of a cyclic hemiacetal, usually a carbohydrate; anomers differ in configuration at the hemiacetal carbon (17.3)

**$\alpha$ anomer:** a stereoisomer of the cyclic form of a carbohydrate in which the hydroxyl group at C-1 is *trans* to the last carbon of the chain (axial in six-member cyclic carbohydrates) (17.3)

**$\beta$ anomer:** a stereoisomer of the cyclic form of a carbohydrate in which the hydroxyl group at C-1 is *cis* to the last carbon of the chain (equatorial in six-member cyclic carbohydrates) (17.3)

**anomeric effect:** the unusual favoring of the $\alpha$, or axial, orientation (or, conversely, the disfavoring of the $\beta$ anomer) in an anomeric equilibrium (17.3)

**anomerization:** the interconversion between the $\alpha$ and $\beta$ anomers of a carbohydrate (17.3)

***anti* addition:** the formation of an addition product by delivery of an electrophile and a nucleophile to opposite faces of a double (or triple) bond (10.2)

***anti* conformer:** a conformational isomer in which two large groups on adjacent atoms are separated by a 180° dihedral angle (5.2)

**antiaromatic hydrocarbon:** a planar, conjugated, cyclic, unsaturated hydrocarbon composed of $sp^2$-hybridized carbon atoms, lacking the chemical stability of a Hückel aromatic; most such systems contain $4n$ $\pi$ electrons and are said to be conjugatively destabilized, although the choice of an appropriate model with which to compare them is not absolutely clear (2.3)

**antibodies:** moderately sized peptide complexes that are responsible for alerting the immune system to the presence of foreign substances (20.7)

**antibonding molecular orbital:** a molecular orbital that, when occupied by electrons, destabilizes a molecule relative to the separated atoms (2.1)

**anticodon:** a sequence of three bases that complements the three-base codon in mRNA and selects for a particular amino acid for protein synthesis (19.3)

**antifungal agent:** a pharmaceutical that selectively attacks and destroys fungi (23.5)

**antihistamine:** a compound that inhibits vasodilation by competing with histamine for binding at a physiologically active site (23.3)

**anti-Markovnikov regiochemistry:** regiochemistry opposite to that predicted by Markovnikov's rule, occurring via an addition in which a proton is delivered to the more-substituted carbon and the nucleophile to the less-substituted carbon of an alkene (10.3)

**antimetabolite:** a compound that interferes with the synthesis of nucleic acids (23.8)

**antineoplastic:** a chemical agent for treating cancer (23.8)

***anti*-periplanar:** a geometric relationship in which the bonds to substituents on adjacent atoms of a $\sigma$ bond are coplanar, with a dihedral angle of 180°; the preferred geometry for an E2 elimination (9.3)

**applied field ($H_{app}$):** the external magnetic field applied to a sample in a nuclear magnetic resonance spectrometer (4.3)

**aprotic solvent:** a solvent molecule lacking a polar heteroatom–hydrogen bond (3.5)

**arene:** an aromatic hydrocarbon or derivative (2.3)

**aromatic hydrocarbons:** a family of planar, $sp^2$-hybridized, conjugated, cyclic, unsaturated hydrocarbons with unusual chemical stability; according to Hückel's rule, such compounds contain $4n + 2$ electrons in their $\pi$ systems (2.3)

**aromaticity:** the special stability afforded by a planar cyclic array of $p$ orbitals containing a Hückel number ($4n + 2$) of electrons (2.3)

**Arrhenius equation:** $k_{obs} = Ae^{-\Delta H^{\ddagger}/RT}$; mathematical correlation of the rate of a reaction with its activation energy (5.2)

**arrow notation:** use of half- or full-headed curved arrows to indicate electron motion in a reaction mechanism (7.2)

**arrow pushing:** the use of curved arrows to describe the movement of electrons as a reaction proceeds (7.2)

**aryl group:** an arene fragment lacking one substituent from a ring carbon (2.3)

**aryl halide:** a functional group in which a halogen is attached to an arene ring (9.8)

**-ase:** suffix descriptor for an enzyme; a biological catalyst for a specific transformation (17.3)

**atactic:** stereochemical designator for a polymer with random orientation of groups at centers of chirality (16.7)

**atomic orbitals:** the probability surfaces within which an electron associated with an atom is likely to be found (1.2)

**ATP:** *see* **adenosine triphosphate**

**average bond energy:** the typical energy of a specific type of bond; obtained from heats of formation (for example, the average bond energy of a C—H bond is obtained as 1/4 of the heat required to convert methane to carbon and hydrogen—that is, for $CH_4 \rightarrow C + 4$ H, $\Delta H°/4 = 99$ kcal/mole; the average bond energy of a C—C bond was obtained by measuring the heat of formation of ethane and subtracting the bond energies of six C—H bonds) (3.3)

**axial:** descriptor for a group pointing roughly orthogonally from the pseudoplane of a chair conformation (5.4)

**azo dyes:** highly colored compounds containing the —N=N— linkage; among the first synthetic colorfast agents (11.3)

**AZT (azidothymidine):** a nucleoside analog that binds tightly to the enzyme reverse transcriptase and thus blocks replication of RNA viruses, especially HIV (23.6)

**back-side attack:** approach of a nucleophilic reagent from the side opposite that from which the leaving group is displaced (7.4)

**Baeyer–Villiger oxidation:** the transformation of a ketone into an ester by reaction with a peracid; the net change is the insertion of an oxygen atom between the carbonyl carbon and an adjacent carbon atom (14.3)

**ball-and-stick model:** three-dimensional representation of a molecule in which bonds are sticks and atoms are spheres of different colors and sizes (1.2)

**base pairing:** simple system of molecular recognition based on optimal hydrogen-bonding patterns between nucleic acid bases; for example, guanine with cytosine (G–C) and adenine with thymine (A–T) (19.3)

**base peak:** the most intense peak in a mass spectrum (4.3)

**base-induced reaction:** a chemical conversion in which a base that is required for the reaction is consumed as product is formed (12.5)

**baseline separation:** an efficient chromatographic separation of two compounds in which the peaks detected as representative of elution of the component molecules do not overlap; that is, the detector response returns to the baseline between peaks (4.2)

**Beckmann rearrangement:** the acid-catalyzed reaction through which the oxime of a ketone is converted to an amide in which one of the carbon substituents originally on the carbonyl carbon has migrated to nitrogen (7.1, 14.2)

**benzenoid ring:** a six-member ring in a polycyclic aromatic compound that retains three formal double bonds (11.5)

**benzilic acid rearrangement:** the anionic skeletal rearrangement of an $\alpha$-diketone to an $\alpha$-hydroxyacid, induced by treatment with aqueous hydroxide (14.1)

**benzopyrene:** a cancer-causing agent that binds to DNA (23.8)

**benzyl cation:** a resonance-stabilized carbocation in which the vacant $p$ orbital is adjacent to an aryl ring (3.6)

**benzyl ether:** a protecting group for an alcohol (15.8)

**benzyne:** $C_6H_4$; a highly reactive cyclic compound related to benzene but having two hydrogen atoms removed from adjacent ring positions (9.4)

**bidentate:** descriptor for a compound with two ligating heteroatoms; from the Latin meaning "two teeth" (19.2)

**bilayer:** a three-dimensional structure of two layers of lipids or surfactants in which the hydrocarbon tails point toward the interior and the polar groups are on the two surfaces, solvated by water (or another polar protic solvent); a bilayer is held together by the van der Waals attractive interactions between the hydrocarbon tails (17.1)

**bimolecular reaction:** a reaction that requires a collision between two reactants in the rate-determining step (6.9)

**biochemical energy storage:** the storage of energy in chemical bonds, often as anhydrides of phosphoric acid and reduced forms of redox cofactors (22.3)

**biochemical reducing agent:** a cofactor that acts as a reducing agent, either by transferring a hydride equivalent or by providing electrons (21.3)

**biotin:** vitamin H; a biological carrier of carbon dioxide (22.7)

**biradical:** a chemical species bearing two noninteracting radical centers (2.1, 2.3)

**bis-acetal:** a functional group in which one oxygen is shared by two acetal functional groups (17.3)

**blood type:** one of four classifications of human blood cells differentiated by the identity of carbohydrate residues present on the cell surface (23.1)

**boat conformation:** the eclipsed conformation of cyclohexane or an analogous six-atom cyclic compound in which the spatial placement of C-1 and C-4 roughly resembles the bow and stern of a boat (5.4)

**bond alternation:** a repeating sequence of short and long (single and double) bonds in an extended $\pi$ system (2.3)

**bond angle:** the angle formed by two bonds intersecting at an atom; typically about 109.5° at an $sp^3$-hybridized atom, 120° at an $sp^2$-hybridized atom, and 180° at an $sp$-hybridized atom (1.2)

**bond dissociation energy:** the quantity of heat consumed when a covalent bond is homolytically cleaved (3.6)

**bond length:** the equilibrium distance between two covalently bonded atoms (1.2)

**bonding molecular orbital:** a molecular orbital that, when occupied by electrons, stabilizes a molecule relative to the separated atoms (2.1)

**borate:** a species containing one or more oxygen–boron bonds (12.2)

**bovine spongiform encephalopathy (BSF):** a disease of cattle thought to be caused by a prion (also called *mad cow disease*) (23.7)

**branched polymer:** a macromolecule in which chemical bonds interconnect chains, forming a complex, three-dimensional network (16.2)

**bridgehead atom:** an atom that is common to both rings in a bicyclic (or multicyclic) compound (5.4)

**bromonium ion:** a three-member cyclic intermediate in which bromine bears a formal positive charge; formed by the addition of $Br^{\oplus}$ (or a source of this species) to an alkene (10.2)

**Brønsted acid:** a proton ($H^{\oplus}$) donor (3.2)

**Brønsted base:** a proton ($H^{\oplus}$) acceptor (3.2)

**$t$-butyl ester:** a protecting group for a carboxylic acid (15.8)

**$n$-butyl group:** $—(CH_2)_3CH_3$; an unbranched four-carbon alkyl group attached through the primary carbon (1.5)

**s-butyl group:** —CH(CH₃)CH₂CH₃; a four-carbon alkyl chain attached through the secondary carbon (1.5)

**t-butyl group:** —C(CH₃)₃, a four-carbon alkyl group attached through the tertiary carbon (1.5)

**Cahn–Ingold–Prelog rules:** rules used in specifying absolute stereochemistry (5.6)

**calorimeter:** a device with which the heat released or consumed in a chemical reaction can be accurately measured (1.6)

**Cannizzaro reaction:** the conversion of an aldehyde lacking α hydrogens to equal amounts of the corresponding carboxylic acid and alcohol upon treatment with sodium hydroxide or potassium hydroxide (12.3)

**capsid:** the coating of protein surrounding the genetic core of a virus (23.6)

**carbamate:** *see* **urethane**

**carbanion:** a negatively charged trivalent carbon bearing an unshared electron pair (6.4)

**carbene:** a neutral reactive intermediate in which a carbon atom bears two σ bonds and two unshared electrons and contains only six electrons in its outer shell (6.4)

**carbocation:** a positively charged trivalent carbon atom containing only six electrons in its outer shell (3.6)

**carbocation stability:** $3° > 2° > 1°$ (3.6)

**carbohydrate:** a polyhydroxylated aldehyde or ketone with the molecular formula $C_m(H_2O)_n$ (16.5, 17.3)

**carbon–carbon bond-forming reaction:** a chemical transformation in which two previously unconnected carbon atoms become covalently bonded (15.1)

**carbonium ion:** *see* **carbocation**

**carbonyl group:** C=O; a functional group containing a carbon–oxygen double bond (3.8)

**carbowax:** a synthetic poly(ethylene glycol) (16.4)

**carboxylic acid:** $RCO_2H$; a functional group in which a carbonyl carbon bears an alkyl or aryl group and an OH group (3.9)

**carboxylic acid anhydride:** *see* **anhydride**

**carcinogens:** cancer-inducing agents (2.3)

**catalyst:** a reagent that facilitates a reaction without itself ultimately forming chemical bonds in the product or appearing in the stoichiometric equation describing the reaction; a catalyst is recovered unchanged after a reaction, but is needed for the reaction to proceed at a reasonable rate (2.1, 7.1)

**catalytic antibody:** a protein expressed by the immune system of an organism in response to the injection of a transition-state analog of a desired reaction (20.7)

**catalytic cycle:** the complete sequence of steps by which a chemical transformation is accelerated in the presence of a catalyst (20.7)

**catalytic hydrogenation:** a reaction catalyzed by a heterogeneous catalyst (usually a noble metal) in which hydrogen is added across one or more multiple bonds (2.1, 7.1)

**cation:** a positively charged ion (3.5)

**cationic polymerization:** the formation of a polymer by a process in which the growing end is a carbocation (16.4)

**cellulose:** a water-soluble biopolymer that contains 3000–5000 glucose units connected exclusively by β linkages (16.5)

**cellulose acetate:** an optically transparent polymer obtained by treating cellulose with acetic anhydride, which converts many of the polysaccharide hydroxyl groups to acetate esters (16.5)

**center of chirality:** a tetrahedral atom (usually carbon) bearing four different groups (5.5)

**chain reaction:** a chemical conversion in which one of the products is a reactive species that initiates another cycle; a reaction that, after initiation, repeats a cycle of propagation steps until one of the reactants is consumed (7.6)

**chair conformation:** the staggered conformation of cyclohexane or an analogous six-atom cyclic compound roughly resembling the back, seat, and footrest of a chair (5.4)

**charge-relay mechanism:** a reaction in which some molecule, often water, transfers a proton (and, therefore, charge) from one position to another in the same molecule or in another molecule, where direct transfer is impossible because of the spatial orientation of and the distance separating the two sites (20.2)

**charge separation:** the development of centers of positive and negative charge upon interaction of neutral reagents (20.2)

**chemical bond:** an energetically favorable interaction between two atoms induced by a pair of electrons mutually attracted to both nuclei or by electrostatic attraction between two ions (1.2)

**chemical shift:** the magnitude of the change of the observed resonance energy for a given nucleus relative to that observed for a standard (usually, tetramethylsilane); the position on an NMR spectrum at which a given nucleus absorbs (4.3)

**chemotherapeutic index:** the ratio of a drug's toxicity to the host to that for the invading organism (23.1)

**chiral:** lacking a mirror plane through any conformation (5.5)

**chiral center:** *see* **center of chirality**

**chiral molecule:** a molecule that lacks an internal plane of symmetry and is not superimposable on its mirror image; the most common indicator of chirality is the presence of a carbon atom bonded to four different groups (5.5)

**chiral recognition:** specific, reversible interaction between two chiral molecules based on three-point contact, the interaction being different for the different diastereomeric pairings (19.4)

**chirality:** "handedness"; a property of an object or a molecule that makes it not superimposable on its mirror image (5.5)

**chlorambucil:** a nitrogen mustard used in cancer chemotherapy (23.8)

**chloronium ion:** a three-member cyclic intermediate in which chlorine bears a formal positive charge; formed by the addition of $Cl^{\oplus}$ (or a source of this species) to an alkene (10.2)

**chlorosulfite ester:** an intermediate in the conversion of an alcohol to an alkyl halide with thionyl chloride (8.3)

**chromate oxidation:** oxidation with $Cr^{\textcircled{6}}$, often of alcohols to aldehydes, ketones, or carboxylic acids, which is accompanied by a color change of the inorganic reagent from red-orange to green ($Cr^{\textcircled{3}}$) (9.9)

**chromatogram:** a plot of a detector's response as a function either of the volume of effluent flowing through the column or of time (4.2)

**chromatographic resolution:** the degree of separation of a mixture of compounds (4.2)

**chromatographic separation:** the isolation of individual components of a mixture through a chromatographic technique (4.2)

**chromatography:** the technique by which components of a mixture are partitioned between two different phases, thus attaining separation because of a difference in solubility of the component molecules in each phase (4.1)

*cis* **isomer:** a geometric isomer in which the largest groups are on the same side of a double bond or ring (1.5, 2.1)

**cisplatin:** $(H_3N)_2PtCl_2$; a DNA cross-linking agent used in cancer chemotherapy (23.8)

**citric acid cycle:** *see* **tricarboxylic acid cycle**

**Claisen condensation:** reaction producing a $\beta$-ketoester upon treatment of an ester with base (13.4)

**Claisen rearrangement:** a pericyclic reaction in which allyl vinyl ether is converted to a rearranged $\beta,\gamma$-enone; sometimes called an oxa-Cope rearrangement (14.3)

**Clemmensen reduction:** the reduction of a ketone to a methylene group by treatment with zinc in HCl (11.2)

**codon:** a three-base sequence on mRNA that specifies the amino acid to be used in protein synthesis and is complementary to the anticodon on tRNA (19.3)

**coenzyme A (CoA):** a complex thiol that, as a thiol ester derivative, accelerates nucleophilic acyl substitution in several biochemical transformations (21.5)

**cofactor:** a recyclable biological reagent (21.3)

**column chromatography:** liquid chromatography conducted with an open chromatography column through which the eluent flows in response to gravity (4.2)

**combustion:** burning in air (1.6)

**complementary:** descriptor for a favorable hydrogen-bonding interaction between bases in DNA and/or RNA (19.3)

**complex metal hydride:** a reagent in which hydride is bonded to boron or aluminum and which is soluble in organic solvents, providing the equivalent of the hydride ion in nucleophilic reactions; the most common examples are $NaBH_4$, $LiAlH_4$, and $NaBH_3(CN)$ (12.2)

**complex metal hydride reduction:** the use of a complex metal hydride to convert an aldehyde to the corresponding primary alcohol, a ketone to a secondary alcohol, an ester to a primary alcohol, an imine to an amine, or an amide to an amine (12.2)

**concerted reaction:** a reaction that proceeds directly from reactant to product through a single transition state and without intermediates (6.1)

**condensation polymer:** a macromolecule produced in a polymerization in which a small molecule is formed as a by-product (16.5)

**condensation reaction:** a chemical conversion in which two molecules combine to form a more complex product, with the loss of a small molecule, usually water or an alcohol (7.1)

**configurational isomers:** stereoisomers that can be interconverted only by the breaking and reforming of a covalent bond (5.1)

**conformational analysis:** an energetic description of a conformational interconversion that relates the relative atomic positions to the changes in potential energy during rotation about a $\sigma$ bond (5.2)

**conformational anchor:** a substituent (usually large) that so strongly prefers the equatorial position that it blocks conformational flipping of the six-member ring to which it is attached (5.4)

**conformational isomers:** stereoisomers that can be interconverted by rotation about a $\sigma$ bond (1.3, 5.2)

**conformational lock:** *see* **conformational anchor**

**conformer:** a conformational isomer (5.2)

**conjugate acid:** a species obtained by the addition of a proton to a Brønsted base (3.2, 6.8)

**conjugate addition:** the addition of a reagent across a four-carbon conjugated $\pi$ system, producing a 1,4 adduct and a double bond between C-2 and C-3 (10.2)

**conjugate base:** a species obtained by the removal of a proton from a Brønsted acid (6.8)

**conjugated diene:** a diene with an array of $p$ orbitals on adjacent atoms, that is, one in which the double bonds comprising the $\pi$ system interact directly without interruption by an intervening $sp^3$-hybridized atom (2.2)

**conjugation:** formation of a series of alternating single and double bonds along a carbon chain with adjacent *p* orbitals (2.2)

**connectivity:** representation of the attachments of atoms in a molecule (1.2)

**constitutional isomers:** compounds with the same molecular weight but with their atoms connected in different sequences (2.1, 5.1)

**convergent synthesis:** a branched synthesis in which two or more synthetic intermediates react with each other (15.6)

**Cope rearrangement:** a [3,3] sigmatropic shift by which a new carbon–carbon $\sigma$ bond is formed between C-1 and C-6 in a substituted 1,5-hexadiene at the same time as the bond between C-3 and C-4 is broken, with both $\pi$ bonds shifting to take up new positions between different carbon atoms (14.1)

**coupling:** the interaction of the magnetic spin of a nucleus with that of one or more neighboring nuclei, causing a nuclear magnetic resonance spectroscopy (NMR) signal to be split into a characteristic pattern reflecting the number of magnetically active neighboring nuclei (4.3)

**coupling constant:** the magnitude of splitting of an NMR signal by one or more magnetically active neighboring nuclei (4.3)

**covalent catalysis:** an accelerated chemical reaction in which the substrate becomes temporarily bound through a covalent linkage to an active site on the catalyst (20.7)

**crossed aldol condensation:** an aldol condensation between two different carbonyl compounds (13.3)

**crossed Claisen condensation:** a Claisen condensation between two different esters (13.4)

**cross-linking:** the covalent interconnections between polymer chains that result in creation of a three-dimensional network; the process in which a bifunctional molecule is incorporated in two separate polymer chains (16.2, 16.6)

**cumulated diene:** a diene in which the two orthogonal double bonds share a common carbon atom (2.2)

**cyanide ion:** $^{\ominus}C{\equiv}N$ (8.4)

**cyano group:** $R{-}C{\equiv}N$; a functional group with a carbon–nitrogen triple bond (also called a *nitrile group*) (3.4)

**cyclic:** containing one or more rings (1.4)

**cyclic bromonium ion:** *see* **bromonium ion**

**cyclic chloronium ion:** *see* **chloronium ion**

**cyclic halonium ion:** *see* **halonium ion**

**cycloaddition reaction:** a pericyclic reaction that combines two separate $\pi$ systems into a cyclic product (14.1)

**cycloalkanes:** saturated hydrocarbons containing one or more rings; in the empirical formula for a cycloalkane two

hydrogens per ring are subtracted from that for an acyclic alkane ($C_nH_{2n+2}$) (1.4)

**cyclophosphamide:** a nitrogen mustard used in cancer chemotherapy (23.8)

**cyclopropane:** $C_3H_6$; the simplest cycloalkane (1.4)

**cycloreversion:** a pericyclic reaction in which a cyclic molecule fragments into two or more smaller $\pi$ systems (14.1)

**cytosine:** $C_4H_5N_3O$; a biologically important heteroaromatic base (3.11)

**d:** relative stereochemical designator for a molecule with positive (dextrorotatory) rotation; from the Greek for "right-rotating" (5.8)

**D:** absolute stereochemical descriptor that relates substituent disposition at a given center of chirality to that in natural D-glyceraldehyde (5.8)

**d,l:** indicator for an optically inactive racemic modification (5.8)

**D,L:** absolute stereochemical descriptors that relate substituent disposition at a center of chirality to that in D- and L-glyceraldehyde; refers to a racemic mixture when used together as D,L (17.3)

**Dacron:** a commercial polyester produced by linking dimethyl terephthalate with ethylene glycol (16.5)

**daunorubicin:** a DNA binding agent used in cancer chemotherapy (23.8)

**decalin:** bicyclo[4.4.0]decane; two fused six-carbon rings (5.4)

**decarboxylation:** the loss of $CO_2$, usually from a carboxylic acid (13.5)

**degenerate rearrangement:** a skeletal rearrangement in which the breaking and forming of bonds leads to a product that is chemically identical to the reactant (14.1)

**dehalogenation:** the formal loss of $X_2$ from a dihalide (9.6)

**dehydration:** the formal loss of water, usually from an alcohol (3.5, 7.1, 9.5)

**dehydrobromination:** the loss of HBr from an alkyl bromide (9.3)

**dehydrohalogenation:** the formal loss of HX from an alkyl halide (7.1)

**deinsertion reaction:** a reaction in which an atom, often a transition metal covalently associated with two groups, is removed as the two groups become covalently bonded to each other; the opposite of an insertion reaction (20.6)

**delocalization:** the spreading of $\pi$ electron density over an entire $\pi$ system (2.3)

**deoxy-:** prefix indicating that an oxygen-containing functional group (often OH) has been replaced by a C—H bond (17.3)

**deoxyribonucleic acid:** *see* **DNA**

**deoxyribose:** a sugar unit found in the backbone of DNA (17.3)

**depsipeptide:** a compound with both ester and peptide linkages (23.5)

**detector:** a device that produces a signal in response to the presence of a substance of interest (4.2)

**dextrorotatory:** *see* **d**

**diastereomers:** stereoisomers that are not mirror images of one another (5.8)

**1,3-diaxial interaction:** the steric interaction between axial substituents bonded to carbon atoms in a six-member ring, resulting in destabilization (5.4)

**diazo coupling:** the connection of two aromatic rings through an azo linkage, usually via electrophilic attack on one ring by an aryl diazonium salt (11.3, 23.5)

**diazonium salt:** $[Ar-N\equiv N]^{\oplus}\ X^{\ominus}$; prepared by treatment of a primary aniline with nitrous acid, $HNO_2$ (11.3)

**diazotization:** the conversion of a primary aniline to a diazonium salt (11.3)

**β-dicarbonyl compound:** a functional group containing two carbonyl groups attached to a common atom (13.5)

**Dieckmann condensation:** an intramolecular variant of the Claisen condensation (13.4)

**Diels–Alder reaction:** the concerted cyclization of a conjugated diene and an alkene (called a dienophile) to produce a cyclohexene; the most frequently encountered $[4 + 2]$ cycloaddition (6.6)

**dienes:** compounds containing two double bonds (2.2)

**dienophile:** in a Diels–Alder reaction, the alkene that reacts with a diene (6.6)

**digonal:** descriptor for a carbon atom with only two bonds (6.4)

**dihedral angle:** the angle formed by two intersecting planes (5.1)

**dimer:** a molecule containing most or all of the atoms of two molecules of a starting material (10.3)

**dimethylallyl pyrophosphate:** an isomer of isopentenyl pyrophosphate that is involved in terpene biosynthesis and has the formula $(CH_3)_2CH-CHCH_2OPO_3PO_3^{\textcircled{3}\ominus}$ (17.2)

**dipolar aprotic solvent:** *see* **polar protic solvent**

**dipole–dipole interaction:** the intermolecular attraction or repulsion deriving from the electrostatic forces between bond dipoles in two interacting molecules (19.1)

**dipole moment:** the vector pointing from the center of positive charge to the center of negative charge in a molecule (3.2)

**directive effect:** a substituent effect that influences the regiochemistry of a reaction (11.3)

**disaccharide:** a dimer in which two carbohydrate units are joined through an acetal or ketal linkage (17.4)

**disease state:** unnatural condition of an organism caused by under- or overproduction of a critical biochemical, invasion by an alien living species that produces substances that are toxic to the host, or too rapid growth of part of the organism (23.1)

**disproportionation:** a reaction in which a species of intermediate oxidation level is converted to equal amounts of a more oxidized and a more reduced product (12.2)

**diterpene:** a terpene containing 20 carbon atoms; derived from four isoprene units (17.1)

**dithiane:** a dithioacetal or dithioketal (21.8)

**dithioacetal:** $RCH(SR)_2$; a functional group that has one hydrogen, an alkyl group, and two sulfide groups on one carbon atom and is produced in the reaction of a thiol with an aldehyde (21.8)

**dithioketal:** $R_2C(SR)_2$; a functional group that has two alkyl groups and two sulfide groups on one carbon atom and is produced in the reaction of a thiol with a ketone (21.8)

**DNA:** deoxyribonucleic acid; the principal genetic information storage unit, which is found in cell nuclei; a biopolymer composed of deoxyribonucleotide units linked through a sugar–phosphate backbone (17.3)

**DNA cross-linkers:** compounds that covalently bond to both strands of the double helix of DNA, linking them together and blocking replication (23.7)

**double bond:** a $\sigma$ and a $\pi$ bond between $sp^2$-hybridized atoms (2.1)

**doublet:** a two-line multiplet on an NMR spectrum (4.3)

**downfield:** descriptor for the chemical shift of a nucleus that resonates at a higher $\delta$ value than a reference nucleus; shifted to a lower frequency, or deshielded; left-hand portion of an NMR spectrum (4.3)

**doxorubicin:** a DNA binding agent used in cancer chemotherapy (23.7)

**E1 reaction:** a unimolecular heterolytic elimination reaction in which breaking of the C—LG $\sigma$ bond, with the formation of a carbocation, is the rate-determining step (9.4)

**E1cB reaction:** a unimolecular heterolytic elimination reaction in which loss of a leaving group from the deprotonated form (anionic conjugate base) of the neutral substrate is the rate-determining step (9.2)

**E2 reaction:** a bimolecular concerted elimination reaction in which bonds to both the proton and leaving group are broken in the rate-determining step (9.3)

**early transition state:** a reactant-like transition state (6.3)

**eclipsed conformation:** a spatial arrangement in which each $\sigma$ bond at one carbon atom is coplanar with a $\sigma$ bond on an adjacent atom (dihedral angle = 0°); when viewed in a Newman projection, the conformation has bonds aligned on adjacent atoms (5.2)

**effective collision:** a collision between two reactants with the correct orientation and with sufficient energy to overcome the activation energy barrier (6.9)

**effective field ($H_{eff}$):** the net magnetic field at a nucleus of interest in NMR; differs from the applied field by the tiny local magnetic field ($H_{loc}$) induced by the electron cloud surrounding the nucleus (4.3)

***E*-isomer:** a geometric isomer in which the groups of highest priority are on opposite sides of a double bond; from the German *entgegen*, "opposite" (2.1)

**electrocyclic reaction:** a concerted, pericyclic, intramolecular ring-forming reaction (14.1)

**electromagnetic radiation:** particles (called photons) or waves traveling at the speed of light; includes infrared, visible, ultraviolet, and x-rays (4.3)

**electron acceptor:** a group that withdraws electron density from an attached atom (11.4)

**electron configuration:** an atomic orbital description of the electrons associated with a given atom (1.2)

**electron donor:** a group that releases electron density to an attached atom (11.4)

**electronegativity:** the tendency of an atom to attract electrons, thus polarizing a covalent bond (1.2)

**electronic effect:** the perturbation of molecular properties by shifts in electron density due to a substituent (11.4)

**electrophile:** an electron-deficient reagent that attacks centers of electron density; from the Greek *electros*, "electron," and *philos*, "loving" (3.7)

**electrophilic addition:** a chemical reaction in which a C=C $\pi$ bond is replaced by two $\sigma$ bonds upon reaction with an electrophilic reagent, often a proton, followed by addition of a nucleophile (7.5, 10.1)

**electrophilic aromatic substitution:** *see* **electrophilic substitution**

**electrophilic substitution:** the replacement of a substituent (usually hydrogen) on an aromatic ring upon interaction of the $\pi$ system with an active electrophile (11.1)

**electrophilicity:** the tendency of an atom, ion, or group of atoms to accept electron density from a carbon center (3.7)

**electrophoresis:** the migration of a charged molecule under the influence of an electric field; used to separate charged organic species, often proteins, nucleic acids, and other polyelectrolytes (4.2)

**electrostatic attraction:** the favorable interaction between two species of opposite charge (1.2)

**electrostatic repulsion:** the unfavorable interaction between two species of like charge (1.2)

**elimination:** a chemical reaction in which two groups on adjacent atoms are lost as a double bond is formed (7.1, 9.1)

**eluent:** the mobile phase in liquid chromatography (4.2)

**elution:** the motion of solute and solvent through the stationary phase in a chromatography column (4.2)

**elution time:** the time required for a given compound to pass through a chromatography column (4.2)

**empirical formula:** quantitative description of the relative proportion of elements present in a compound in smallest whole numbers (1.3)

**enantiomeric excess:** the predominance of one enantiomer over the other (5.7)

**enantiomers:** stereoisomers related to each other as non-superimposable mirror images; stereoisomers with opposite configuration at each center of chirality (5.5)

**enantiotopic:** descriptor for identical groups that lie on opposite sides of the plane of symmetry of an achiral molecule (22.8)

**endergonic reaction:** a chemical transformation for which free-energy input is needed; a reaction in which the free-energy content of the products is higher than that of the reactants (*see also* **endothermic reaction**) (6.1)

**endorphin:** a natural pentapeptide that is found in the brain in extraordinarily low concentration and induces euphoria or blocks pain; the name is derived from *en*do*genous* mo*rphine* (23.4)

**endothermic:** requiring input of energy (3.6)

**endothermic reaction:** a chemical conversion with a positive enthalpy change (*see also* **endergonic reaction**) (6.1)

**enediol:** a functional group bearing two hydroxyl groups on the carbons of a double bond (22.10)

**energy barrier:** the amount of energy required to reach the most unfavorable point along the path followed in the conversion of one species to another (5.1)

**energy diagram:** a graphic representation of the change in free energy (or enthalpy) encountered during the course of a reaction (6.1)

**energy of activation:** *see* **activation energy**

**enkephalin:** a pentapeptide endorphin (23.4)

**enol:** a functional group in which a hydroxyl group is attached to an alkenyl carbon (6.2)

**enolate anion:** a resonance-stabilized anionic intermediate obtained by removal of a proton from the $\alpha$ position of a carbonyl compound or the OH group of an enol (6.2)

**enolization:** keto to enol tautomerization; conversion of a ketone or aldehyde to its enol form (13.1)

***entgegen:*** *see* ***E*-isomer**

**enthalpy:** heat of reaction (5.1)

**enthalpy change ($\Delta H°$):** heat of reaction; the difference between the bond energies of the reactants and those of the products (5.1)

**entropy:** disorder; free motion (5.1)

**entropy change ($\Delta S°$):** the difference in disorder between reactants and products (5.1)

**enyne:** an organic compound containing a double bond and a triple bond (2.4)

**enzyme:** a protein that functions as a biological catalyst (17.4)

**enzyme catalysis:** the acceleration of a chemical reaction by reversible association of a substrate with an enzyme active site (20.7)

**epoxidation:** the preparation of an epoxide from an alkene (10.4)

**epoxide:** a three-member cyclic compound containing oxygen (10.4)

**epoxy resin:** a structurally rigid material obtained by cross-linking a diol with epichlorohydrin (16.6)

**equatorial:** descriptor for a group lying roughly parallel with the pseudoplane of a chair conformation (5.4)

**equilibrium:** the state in which the forward rate of an ideally reversible reaction is equal to the reverse rate (5.2)

**equilibrium constant:** $K = [C][D]/[A][B]$; a measure of the equilibrium position of the reaction $A + B = C + D$; the ratio of the forward and reverse rate constants of a reversible reaction at equilibrium (6.7)

**ester:** $RCO_2R$; a functional group in which a carbonyl carbon bears an OR group (3.9)

**ester enolate anion:** a resonance-stabilized anionic species obtained by removal of a proton from the $\alpha$ position of an ester (13.1)

**estrogen:** a female hormone (17.1)

**ethane:** $C_2H_6$; the simplest saturated hydrocarbon containing a C—C bond (1.3)

**ethene:** $C_2H_4$; the simplest unsaturated hydrocarbon containing a double bond between $sp^2$-hybridized carbon atoms (also called *ethylene*) (2.1)

**ether:** a functional group in which two alkyl or aryl groups are attached to an $sp^3$-hybridized oxygen atom (3.5)

**ethylene glycol:** $HOCH_2CH_2OH$ (16.4)

**ethylene oxide:** $C_2H_4O$; the simplest epoxide (8.4)

**ethyne:** $HC\equiv CH$; the simplest alkyne containing a triple bond (also called *acetylene*) (2.4)

**excited state:** an electronic configuration with a higher energy content than the ground state; often produced by absorption of a photon, which promotes an electron from a bonding or nonbonding molecular orbital to an anti-bonding molecular orbital (4.3)

**exergonic reaction:** a reaction in which free energy is released, that is, in which the total free energy content of the products is lower than that of the reactants (6.1)

**exothermic reaction:** a chemical conversion with a negative enthalpy change (6.1)

**extraction:** the selective partitioning of a compound between two immiscible liquids, often a nonpolar organic phase and an aqueous or alcoholic phase (4.2)

**FAD:** flavin adenine dinucleotide; a cofactor used for the electron-transfer oxidation of, for example, a saturated to an $\alpha,\beta$-unsaturated thiol ester in fatty acid degradation (21.4)

**FADH$_2$:** flavin adenine dinucleotide (reduced form); a cofactor used for the electron-transfer reduction of, for example, an $\alpha,\beta$-unsaturated thiol ester in fatty acid synthesis (21.4)

**fat:** a fatty acid ester of glycerol (17.1)

**fatty acid:** a long, straight-chain carboxylic acid containing an even number of carbon atoms (17.1)

**fatty acid biosynthesis:** the biosynthetic pathway by which acetate (as acetyl CoA) is converted to a long-chain, unbranched carboxylic acid through a series of Claisen-like condensations and reductions (21.4)

**fatty acid degradation:** the biosynthetic pathway by which long-chain, unbranched carboxylic acids are converted to acetyl CoA through a series of retro-Claisen-like condensations (22.7)

**feedback:** a process by which a product serves to regulate its own rate of formation; often accomplished through partial product inhibition of enzyme catalysis (20.7)

**fingerprint region:** the region in an infrared spectrum (400 cm$^{-1}$ to about 1100 cm$^{-1}$) that usually exhibits a series of complex, low-energy bands that are characteristic of a specific molecule (rather than a functional group) (4.3)

**Fischer projection:** a type of notation used to indicate absolute configuration, in which the intersection of two lines indicates the position of a chiral carbon, with horizontal lines indicating substituents directed toward the observer and vertical lines indicating substituents directed away from the observer (5.8)

**flagpole hydrogens:** the two hydrogens that are located in a 1,4-relationship on a boat cyclohexane and point at each other (5.4)

**flame ionization detector:** a gas chromatography detector that senses the presence of ions that are generated as the effluent from the column is burned in a hydrogen flame (4.2)

**fluid mosaic:** a term used to indicate the mobile nature of lipid bilayers (17.1)

**5-fluorouracil (5-FU):** an antimetabolite used in cancer chemotherapy (23.8)

**formal charge:** a construct used to describe electron distribution in a molecule by comparing the number of valence electrons in a neutral atom with the sum of the number of unshared electrons plus half the number of shared electrons available to that atom; difference between the number of electrons accessed by an atom in a molecule and in its elemental state (1.2)

**formulation:** the final preparation of a pharmaceutical in a form acceptable for delivery to the target organ or organism (23.3)

**formyl anion:** $HC{=}O^{\ominus}$; an anion produced by deprotonation of formaldehyde (21.7)

**fragmentation pattern:** the specific set of ion fragments obtained by bombarding a neutral molecule with high-energy electrons in a mass spectrometer (4.3)

**free energy:** a state property of a system, with contributions from both enthalpy ($H°$) and entropy ($S°$); a measure of the potential energy of a molecule or group of molecules (6.1)

**free rotation:** the motion attained when orbital overlap is unaffected by rotation about the internuclear axis of a $\sigma$ bond (1.2)

**free-energy change ($\Delta G°$):** a measure of the potential energy change during a chemical reaction; includes enthalpy ($\Delta H°$) and entropy ($\Delta S°$) components; $\Delta G° = \Delta H° - T\Delta S° = -RT \ln K$ (6.1, 6.6)

**free-radical halogenation:** a homolytic substitution of halogen for hydrogen, often in an alkane (7.6)

**Friedel–Crafts acylation:** the reaction of a carboxylic acid chloride with an aromatic compound in the presence of a Lewis acid, resulting in the replacement of a hydrogen by an acyl substituent (11.2)

**Friedel–Crafts alkylation:** the reaction of an alkyl halide with an aromatic compound in the presence of a Lewis acid, resulting in the replacement of a hydrogen by an alkyl substituent (11.2)

**full-headed curved arrow:** used to indicate the movement of an electron pair (3.6)

**functional group:** a site at which a molecule undergoes characteristic and selective chemical reactions (2.1)

**functional-group compatibility:** a characteristic of a reagent or reaction that is sufficiently chemically selective so that only the desired functional group (of the several present in the molecule) interacts with the reagent (15.7)

**functional-group transformation:** a chemical reaction in which one functional group is changed into another (15.1)

**furan:** $C_4H_4O$; a five-atom cyclic heteroaromatic molecule containing oxygen (3.11)

**furanose:** a carbohydrate containing a five-member cyclic hemiacetal (17.3)

**Gabriel synthesis:** the synthesis of a primary amine by alkylation of phthalimide anion, followed by treatment of the resulting $N$-alkylphthalimide with hydrazine (8.3)

**β-galactosidase:** an enzyme that catalyzes the hydrolysis of the $\beta$ linkage between galactose and another carbohydrate (17.3)

**gas chromatography:** a chromatographic technique in which a vaporized sample is carried by a gaseous mobile phase over a stationary phase (usually either a solid or a solid coated with a nonvolatile liquid) (4.2)

***gauche* conformer:** a conformational isomer in which two large groups on adjacent atoms are separated by a 60° dihedral angle (5.2)

**gel electrophoresis:** a separation technique that uses an electric field to induce movement of polyelectrolytes through a gel (*see also* **electrophoresis**) (4.2)

**geminal diol:** a functional group bearing two —OH substituents on the same carbon atom (*see also* **hydrate**) (12.3)

**genetic code:** a system of information storage, transcription, and translation based on complementary base pairing in DNA; the sequence of base pairings in DNA–RNA transcription (19.3)

**geometric isomerization:** a chemical conversion in which geometric isomers are interconverted (7.1)

**geometric isomers:** isomers with the same connectivity along the backbone, but different spatial disposition of one or more groups around a bond with restricted rotation; *cis–trans* isomers (2.1, 5.1)

**glass:** a polymer based on a three-dimensional network of tetrahedrally arranged silicon atoms linked by oxygen (16.4)

**glucoside:** a cyclic derivative of glucose in which the C-1 hemiacetal hydroxyl group of glucose has been replaced by an alkoxy group (17.3)

**D-glyceraldehyde:** ($2R$)-propanal-2,3-diol; a triose carbohydrate that serves as a reference compound for stereochemical designation of sugars (17.3)

**glycerol:** 1,2,3-propanetriol (17.1)

**glycol:** a 1,2- or 1,3-diol (16.4)

**glycolysis:** the breakdown of carbohydrates, in which a retro-aldol-like reaction is a key step (22.10)

**α-glucosidase:** an enzyme that catalyzes the cleavage of $\alpha$-glucosidic linkages (17.3)

**glycoside:** a cyclic acetal derivative of a carbohydrate in which the hemiacetal (or hemiketal) hydroxyl group has been replaced by an alkoxy group (17.3)

**Grignard reagent:** a reagent in which carbon is directly bonded to magnesium (8.4)

**ground state:** the most stable, lowest-energy electron configuration (4.3)

**guanine:** $C_5H_5N_5O$; a biologically important heteroaromatic base (3.11)

**H$_2$ blocker:** a compound that interferes with binding to certain receptor sites, for example, those that stimulate acid production in the stomach (23.3)

**hadacidin:** an antimetabolite that interferes with the biosynthesis of adenosine; used in cancer chemotherapy (23.8)

**half-chair:** a high-energy conformation that represents the transition state in converting from a chair to a boat conformation and has all but one atom of the ring in the same plane (5.4)

**half-headed curved arrow:** used to indicate the movement of a single electron (3.6)

**haloform reaction:** the conversion of a methyl ketone to the corresponding carboxylic acid and haloform ($CHX_3$) upon treatment with aqueous base and dihalogen (13.1)

**halogenation:** the formal addition of dihalogen to an alkene (10.2)

**halonium ion:** a three-member cyclic intermediate in which a halogen (usually bromine or chlorine) bears a formal positive charge; formed by the reaction of $X^{\oplus}$ with an alkene (10.2)

**Hammond postulate:** an assertion that a transition state most closely resembles the stable species that lies closest to it in energy (6.3)

**hard:** descriptor for a charge-intensive reagent; often applied to nucleophiles, electrophiles, acids, and bases (13.3)

**hatched line:** in a line structure, a graphic representation indicating a group or atom positioned away from the observer (1.2)

**heat of combustion:** the heat released when one mole of a compound is completely oxidized to $CO_2$ and $H_2O$ (1.6)

**heat of formation:** a theoretical description of the energy that would be released if a molecule were formed from its component elemental atoms in their standard states (1.6)

**heat of hydrogenation:** the heat released when one mole of an unsaturated compound is completely hydrogenated to a saturated compound (2.1, 2.3)

**heat of reaction:** the energy difference between a reactant and a product (5.1)

**$\alpha$-helix:** a right-handed spiraling structure imposed by intramolecular hydrogen bonding between groups along a single peptide chain (16.7)

**$\beta$-helix:** a left-handed spiraling structure imposed by intramolecular hydrogen bonding between groups along a single peptide chain; not found with naturally occurring $\alpha$-amino acids (16.7)

**Hell–Volhard–Zelinski reaction:** a method for monobromination $\alpha$ to a carboxyl group by treatment of a carboxylic acid bearing $\alpha$ hydrogen atoms with bromine in the presence of phosphorus tribromide (13.1)

**hematoporphyrins:** porphyrins used in photochemotherapy (23.8)

**hemiacetal:** RCH(OR)(OH); a functional group having an alkyl group, a hydrogen atom, an alkoxy group, and a hydroxy group on one carbon atom; the product of the nucleophilic addition of an alcohol to an aldehyde (12.6)

**hemiketal:** RRC(OR)(OH); a functional group having two alkyl groups, an alkoxy group, and a hydroxy group on one carbon atom; the product of the nucleophilic addition of an alcohol to a ketone (12.3)

**heteroaromatic molecule:** an aromatic molecule containing a ring heteroatom (3.12)

**heteroatom:** any atom besides carbon and hydrogen (3.1)

**heterocycle:** a cyclic molecule in which the ring contains one or more heteroatoms (3.11)

**heterocyclic aromatic:** *see* **heteroaromatic molecule**

**heterolysis:** *see* **heterolytic cleavage**

**heterolytic cleavage:** the cleavage of a bond in which both electrons are shifted to one atom of the bond (3.6, 7.2)

**hexose:** a six-carbon sugar (17.3)

**high-energy phosphate bonds:** phosphoric acid anhydride units critical in biological energy storage (22.4)

**high-performance liquid chromatography:** *see* **high-pressure liquid chromatography**

**high-pressure liquid chromatography (HPLC):** liquid chromatography in which the mobile phase is driven through a sealed chromatography column by a mechanical pump (4.2)

**Hofmann elimination:** a kinetically controlled elimination reaction in which the less substituted alkene is formed preferentially (9.3)

**Hofmann rearrangement:** the conversion of an amide to an amine containing one fewer carbon upon treatment with bromine in aqueous base (14.2)

**HOMO:** highest occupied molecular orbital (4.3)

**homolysis:** the cleavage of a bond in which one electron is shifted to each of the atoms of the bond; synonymous with homolytic cleavage (7.2)

**homolytic cleavage:** the cleavage of a bond with one electron shifted to each of the atoms of the bond (3.6)

**homolytic substitution:** *see* **free-radical halogenation**

**hormone:** a compound that controls essential biological functions and plays an important regulatory role in controlling key biochemical pathways (17.1)

**Hückel's rule:** an empirical generalization that any planar, cyclic, conjugated system containing $(4n + 2)\pi$ electrons (in which $n$ is an integer) experiences unusual aromatic stabilization, whereas those containing $(4n)\pi$ electrons do not (2.3)

**Hund's rule:** when possible, electrons singly occupy orbitals of identical energy (1.2)

**hybrid orbitals:** the orbitals formed by mixing hydrogenic (*s, p, d,* etc.) atomic orbitals (1.2, 2.1, 2.4)

**hybridization effect:** the influence of mixing *s* and *p* orbitals; the greater the fraction of *s* character (50% in an *sp* hybrid; 33% in an *sp*²-hybrid; 25% in an *sp*³-hybrid) of the hybrid orbital, the more electronegative is the atom (6.7)

**hydrate:** the product of nucleophilic addition of water to an aldehyde or ketone (12.3)

**hydration:** the addition of water to a multiple bond (7.1, 10.1)

**hydrazone:** $R_2C{=}NNH_2$; a condensation product of hydrazine ($H_2NNH_2$) with an aldehyde or ketone; often a highly colored solid used as a diagnostic test for the presence of a carbonyl group (12.4)

**hydroboration:** the addition of a carbon–boron and a carbon–hydrogen bond to an alkene (10.4)

**hydroboration–oxidation:** a reaction sequence used to achieve anti-Markovnikov hydration of an alkene; initiated by concerted *syn*-addition of borane, which is followed by oxidation with basic hydroperoxide (10.4)

**hydrocarbons:** compounds that contain only carbon and hydrogen (1.3)

**hydrogen bond:** the weak association of a hydrogen atom attached to one electronegative heteroatom with a nonbonding electron pair on a second electronegative atom in the same or another molecule (X—H · · · Y, where X and Y are electronegative heteroatoms) (3.2, 3.5)

**hydrogen peroxide:** $H_2O_2$ (10.4)

**hydrogenation:** the addition of $H_2$ (2.1, 2.3, 2.4)

**hydrogenic atomic orbitals:** the atomic orbitals calculated precisely for hydrogen, including spherical *s* orbitals, propeller-shaped *p* orbitals, dumbbell-shaped *d* orbitals, etc. (1.2)

**hydrohalogenation:** the formal addition of HX (X = halide) to a multiple bond (10.1)

**hydrolysis:** a reaction in which water displaces a leaving group or is added across a multiple bond (7.1, 7.5)

**hydronium ion:** $H_3O^{\oplus}$ (10.1)

**hydrophilic:** having a preference for association with an aqueous environment; a property of polar molecules (17.1)

**hydrophobic:** having a preference for association with a nonaqueous environment; a property of nonpolar molecules (17.1)

**hyperconjugation:** an orbital description of the stabilizing effect derived from interaction of an aligned $\sigma$ bond with an adjacent *p* orbital (2.1)

**imide:** RCONHCOR′; a functional group in which two carbonyl carbons bearing an alkyl or aryl group are linked through a nitrogen atom (3.9)

**imine:** a family of compounds whose members contain a C=N double bond (3.3, 12.4)

**imine–enamine tautomerization:** the process by which a proton is shifted from the $\alpha$ carbon of an imine to the nitrogen or from the N—H group of an enamine to the adjacent alkenyl carbon; a 1,3-shift of a proton in an imine or enamine (12.4)

**immune system:** the biological system that produces antibodies to a foreign substance and effects its destruction or excretion (20.7)

***in vitro:*** describing a reaction conducted in a laboratory environment; from the Latin for "in glass" (23.2)

***in vivo:*** describing a reaction conducted within a living organism; from the Latin for "in life" (23.2)

**index of hydrogen deficiency:** half the difference between the number of hydrogen atoms in a hydrocarbon and the number expected for a straight-chain alkane ($2n + 2$); indicative of the number of multiple bonds and/or rings present (2.1)

**induced dipole:** the shift of electron density within a molecule or bond induced by the environment (19.1)

**inductive effect:** polarization through a series of $\sigma$ bonds, causing a shift of electron density from or to a charged or polar site (6.8)

**inert gas:** an atom that does not readily enter into chemical bonds with other atoms because the valence electron shell is filled; found at the far right column of the periodic table (1.2)

**infrared spectroscopy:** a technique that measures the absorption of light energies of between about 4000 and 400 cm$^{-1}$ (4.3)

**inhibitor:** a species that blocks a catalyzed chemical reaction by binding to the catalyst (without itself undergoing reaction), thus blocking the substrate (20.7)

**initiation step:** the first step of a radical chain reaction in which the number of radicals produced is greater than the number of radicals present in the reactants (7.6)

**initiator:** a substance with an easily broken covalent bond that fragments to radicals that can induce a radical chain reaction (7.6)

**insertion reaction:** a reaction in which an atom, often a metal, becomes bonded to two atoms that were themselves originally covalently bonded (8.4, 20.6)

**integration:** the measurement of the relative area under each peak on a spectrum (4.2)

**integration curve:** an indicator of the relative area under each peak of a spectrum or chromatogram (4.3)

**intercalation:** the sandwiching of a foreign agent between nucleic acid bases in DNA (23.8)

**intermediate:** *see* **reactive intermediate**

**intermolecular hydrogen bond:** a hydrogen bond connecting electronegative atoms in separate molecules (3.2)

**intermolecular proton transfer:** the movement of $H^{\ominus}$ from a bonded position in one molecule to a bonded position in another molecule (19.2)

**intramolecular hydrogen bond:** a hydrogen bond connecting electronegative atoms within the same molecule (3.2)

**intramolecular proton transfer:** the movement of $H^{\oplus}$ from a bonded position in one molecule to another position in the same molecule (19.2)

**inversion of configuration:** the reversal of configuration at a center of chirality, attained by forming a new bond on the opposite face from the site where a bond is broken (7.4)

**invert sugar:** a 1:1 mixture of glucose and fructose, obtained upon cleavage of sucrose (17.4)

**invertase:** an enzyme that catalyzes the cleavage of sucrose to a 1:1 mixture of glucose and fructose (17.4)

**iodoform test:** a chemical test for the presence of an $R(CO)CH_3$ functionality by treatment with aqueous base and iodine; a positive result is the formation of a yellow precipitate of $CHI_3$ (13.1)

**ion channels:** compounds, often proteins, with hydrophobic exteriors and hydrophilic internal regions, sometimes containing several charged amino acid residues, that dissolve readily in the interior of a phospholipid bilayer, thus spanning the bilayer, through which polar or charged molecules can move from one side of the membrane to the other (17.1)

**ion pairing:** the electrostatic association between oppositely charged ions (6.4)

**ionic bond:** an attractive electrostatic association between two oppositely charged ions (1.2)

**ionic polymerization:** the formation of a polymer by a process in which the growing end is an ion (16.4)

**ionophore:** molecule containing several heteroatoms arranged so that multiple, simultaneous contacts with a metal ion or other highly polar species are possible (19.2)

**irreversible reaction:** an exothermic reaction in which the activation energy for the reverse reaction is sufficiently large that the reaction proceeds only in the forward direction under practical conditions (6.1)

**isobutyl group:** $—C(CH_3)_3$; a branched four-carbon alkyl group attached through a tertiary carbon (1.5)

**isocyanate:** $RN{=}C{=}O$; an intermediate in the Hofmann rearrangement (14.2)

**isoelectronic:** descriptor for two atoms with the same electron configuration (6.4)

**isolated diene:** a diene in which the double bonds do not interact directly with each other because of one or more intervening $sp^3$-hybridized atoms (2.2)

**isomerase:** an enzyme that catalyzes isomerization (22.10)

**isomerization:** a chemical conversion in which compounds with the same molecular formula, but different structures, are interconverted (7.1)

**isomers:** different structural arrangements of the same atoms (1.3)

**isopentenyl pyrophosphate:** a branched five-carbon derivative of 2-methylbutadiene that is the biochemical precursor of the terpenes and has the formula $CH_2{=}C(CH_3)CH_2CH_2OPO_3PO_3^{3\ominus}$ (17.1)

**isoprene:** 2-methylbutadiene (16.4)

**isoprenoids:** derivatized skeletal oligomers of isoprene (17.1)

**isopropyl group:** $—CH_2CH(CH_3)_2$; a branched four-carbon alkyl group attached through a primary carbon (1.5)

**isotactic:** stereochemical designator for a polymer in which all groups at centers of chirality along the chain point in the same direction (16.7)

**isotopic labeling:** the replacement of an isotope of highest natural abundance with another isotope at a specific position in a molecule; for example, replacement of $^1H$ by $^2H$ (D) or of $^{12}C$ by $^{13}C$ (8.4)

**IUPAC rules:** a set of rules for naming organic compounds in which a root word describes the number of backbone carbon atoms, a suffix defines the functional group, and a prefix gives the position of each substituent (1.5)

**$K_a$:** the acid-dissociation equilibrium constant; $K_a = K[H_2O] = [A^{\ominus}][H_3O^{\oplus}]/[HA]$ (6.7)

**Kekulé structures:** cyclic six-carbon structures, devised by August Kekulé, that depict benzene as having localized double bonds (2.3)

**ketal:** $R_2C(OR)_2$; a functional group having two alkyl groups and two alkoxy groups on one carbon atom, produced in the acid-catalyzed alcoholysis of a ketone or a hemiketal (12.3)

**ketimine:** an imine of a ketone (21.4)

**$\alpha$-ketoacid:** a carboxylic acid bearing a ketone functional group at the $\alpha$ position (21.7)

**$\beta$-ketoacid:** a carboxylic acid bearing a ketone group at the $\beta$ position (13.5)

**keto–enol tautomerization:** the process by which a proton is shifted from the $\alpha$ carbon of a ketone to the carbonyl oxygen or from the OH group of an enol to the remote alkenyl carbon; a 1,3-shift of a proton in an aldehyde or ketone (6.6)

**ketone:** $R_2CO$; a functional group in which a carbonyl carbon bears alkyl and/or aryl groups (3.8)

**ketose:** a sugar with a ketone functional group, usually at C-2 (17.3)

**ketyl:** a radical anion obtained when an electron is added to the carbonyl group of a ketone (6.4)

**kinetic control:** descriptor for a chemical reaction for which the reverse reaction takes place slowly or not at all, so that the relative concentration of products directly correlates with the relative rates of their formation rather than their relative stabilities (6.6)

**kinetics:** the factors influencing the rate at which a reaction proceeds (6.5)

**Krebs cycle:** *see* **tricarboxylic acid cycle**

**kuru:** a disease found among cannibals of Papua New Guinea and thought to be caused by a prion (23.7)

*l***:** relative stereochemical designator for a molecule with a negative (levorotatory) specific rotation; from the Greek for "left-rotating" (5.8)

L: absolute stereochemical descriptor that relates substituent disposition at a given center of chirality to that in natural L-glyceraldehyde (5.8)

**lactam:** a cyclic amide (16.5)

**lactone:** a cyclic ester (14.3)

**late transition state:** a transition state that is product-like (6.3)

**Le Chatelier's principle:** observation that the position of an equilibrium A + B $\rightleftharpoons$ C + D can be shifted to the right, either by increasing the concentration of A and/or B or by decreasing the concentration of C and/or D (7.1)

**leaving group:** a group displaced from a reactant in a substitution or elimination reaction (7.5)

**leukoderma (vitiligo):** a disease characterized by lack of skin pigmentation (23.7)

**levorotatory:** *see l*

**Lewis acid:** an electron-pair acceptor (3.2)

**Lewis base:** an electron-pair donor (3.2)

**Lewis dot structure:** a representation in which electrons available to a given atom are shown either as a nonbonding lone pair (by a pair of dots) or as a shared bonding pair (by a pair of dots between two atoms) (1.2)

**linear polymer:** a macromolecule in which the monomer units are attached end-to-end (16.2)

**linear synthesis:** a sequence of transformations in which the product of one reaction is the reactant in the next (15.6)

**lipid bilayer:** *see* **bilayer**

**lipids:** a group of simple, naturally occurring molecules that are soluble in nonpolar solvents and are composed mostly of carbon, hydrogen, and oxygen atoms (17.1)

**lipoic acid:** a dithiol cofactor important in the oxidative degradation of $\alpha$-ketoacids (21.7)

**lipophilic:** hydrophobic (17.1)

**liquid chromatography:** a chromatographic technique in which a solid or liquid sample is carried by a liquid mobile phase over a stationary phase (usually a solid composed of small particles around which the liquid phase can flow) (4.2)

**lithium dialkylcuprate:** $R_2Cu^{\ominus}$ $Li^{\oplus}$; an alkylating agent for alkyl halides (8.4)

**living polymer:** a macromolecule in which the ends of the chain are chemically reactive but will not react with each other (16.4)

**lock-and-key fit:** the highly specific, tight association between a substrate and an active site in an enzyme (20.7)

**lone pair:** two nonbonding electrons of opposite spin accommodated in an atomic or hybrid atomic orbital (3.1)

**Lucas reagent:** a mixture of Brønsted and Lewis acids that induces the conversion of an alcohol to the corresponding alkyl chloride (3.6)

**Lucas test:** a chemical means for distinguishing tertiary, secondary, and primary alcohols by the rate of formation of the corresponding alkyl chloride from an alcohol upon treatment with the Lucas reagent (3.6)

**LUMO:** lowest unoccupied molecular orbital (4.3)

**macromolecule:** *see* **polymer**

**magnetic resonance imaging (MRI):** the creation of a three-dimensional map of water concentration in an object; often used in medical applications for visualizing organs or anomalous growths (4.3)

**malonic ester:** $CH_2(CO_2R)_2$; a diester in which both ester groups are bonded to the same carbon atom (13.5)

**malonic ester synthesis:** a method for preparing mono- and dialkylated carboxylic acids by sequential alkylation of a malonic ester anion, hydrolysis of the alkylated diester, and decarboxylation of the resulting $\beta$-diacid (13.5)

**Markovnikov's rule:** an empirical prediction that the regiochemistry of the addition of HX to an unsymmetrical alkene places the proton on the less substituted carbon atom of the multiple bond (7.5, 10.1)

**mass spectroscopy:** a technique that determines the mass of ions formed when molecules are bombarded with high-energy electrons (4.3)

**mechanism:** *see* **reaction mechanism**

**mechanistic organic chemistry:** the subarea of organic chemistry that focuses on the study of how reactions take place (7.3)

**membrane:** a lipid bilayer, found on the outer surface of a cell and separating it from the external aqueous medium (17.1)

**mercaptan:** *see* **thiol**

***meso* compound:** an optically inactive molecule that contains a mirror plane or center of symmetry interrelating centers of chirality in the molecule (5.8)

**messenger RNA (mRNA):** a transcribed strand of RNA

complementary to a segment of DNA; a replicated sequence of complementary bases of a DNA strand, except in the substitution of uracil for thymine (19.3)

**mesylate:** $ROSO_2CH_3$; a methanesulfonate ester (8.3)

*meta* **director:** a functional group that directs electrophilic aromatic substitution to the *meta* position (11.4)

*meta* **substitution:** a reaction that results in substituents that are 1,3 to each other on an aromatic ring (2.3)

**methane:** $CH_4$; the simplest hydrocarbon, which consists of a carbon atom surrounded by four hydrogens (1.3)

**methotrexate:** an antimetabolite used in cancer chemotherapy (23.8)

**micelle:** a roughly spherical aggregation of many soap-like molecules with hydrophobic and hydrophilic portions, arranged with polar or ionic heads at the surface and surrounding a hydrocarbon-like core; *see also* **bilayer** (17.1)

**Michael addition:** a reaction in which a resonance-stabilized carbanion reacts with an $\alpha,\beta$-enone in a conjugate addition (13.3)

**microscopic reversibility:** the principle dictating that the pathway followed in the forward and reverse directions of a given reaction must be the same (6.5, 7.4)

**(−):** levorotatory; *see* ***l*** (5.8)

**mirror image:** a reflected projection of an object (5.5)

**mirror plane:** a plane through an object such that each part of the object on one side of the plane is reflected by an identical part on the opposite side (5.5)

**mixed anhydride:** an anhydride with two different carboxylic acid subunits; often one derived from a carboxylic acid and a phosphoric acid (22.10)

**mobile phase:** the flowing medium used in chromatography to carry a mixture through the stationary phase; flow can be induced by gravity, pressure, or capillary action (4.2)

**mobility:** a measure of the ease with which a given compound can move (for example, through a chromatography column) (4.2)

**molecular formula:** description of the number of each type of atom present in a molecule; *see also* **empirical formula** (1.3)

**molecular ion:** in mass spectrometry, an unfragmented (parent) ion formed by loss of an electron from a molecule and having the same mass as that molecule (4.3)

**molecular orbital:** a calculated probability surface within which an electron is likely to be found in a molecule; a molecular orbital can be constructed by the mathematical combination of atomic orbitals (2.1)

**molecular recognition:** selective, weak, reversible binding between two reagents; the energetically favorable association of two molecules through hydrogen bonding and/or van der Waals interactions (19.4, 21.1)

**molozonide:** a five-member ring containing three oxygen atoms; produced by the direct addition of $O_3$ to an alkene (10.4)

**monomer:** the chemical precursor of a polymer (16.1)

**monoterpene:** a terpene containing ten carbons and derived from two isoprene units (17.1)

**multiplet:** a pattern obtained by splitting the NMR signal for a magnetically active nucleus into several lines (4.3)

**multiplicity:** the number of peaks into which an NMR signal is split (4.3)

**mustard gas:** $ClCH_2SCH_2Cl$; a highly toxic DNA cross-linking agent (23.7)

**mutarotation:** a process that produces a change in the optical rotation of a solution of two or more equilibrating species to that of the equilibrium mixture; a change in optical rotation that takes place when a pure sugar anomer is dissolved (17.3)

**myelin sheath:** the lipid bilayer that surrounds and insulates nerve axons (17.1)

**n,$\pi^\star$ transition:** an electronic transition of an electron from one of the nonbonded, lone pairs of electrons to a $\pi^\star$ (antibonding) orbital (4.3)

**NADH:** *see* **nicotinamide adenine dinucleotide**

**NADPH:** a phosphorylated derivative of NADH, with many of the same functions (21.4)

**neoprene:** poly(2-chlorobutadiene) (16.4)

**Newman projection:** a representation used to indicate stereochemical relationships between groups bonded to adjacent carbon atoms; conformational representation in which a triad juncture inscribed within a circle represents dihedral angles between $\sigma$ bonds on one carbon and those attached to the adjacent atom (5.2)

**nicotinamide adenine dinucleotide (NADH):** a biological reducing agent that provides a hydride equivalent; a cofactor that effects the reduction of $\alpha$-ketoacids in fatty acid biosynthesis (12.3, 21.4)

**nitration:** the replacement of H by an $NO_2$ group (11.2)

**nitrile:** RC≡N; a functional group in which an *sp*-hybridized nitrogen atom is triply bonded to carbon (also called a *cyano group*) (3.4)

**nitrilium cation:** R—C≡N$^\oplus$ —R; a resonance-stabilized alkylated nitrile cation, encountered as an intermediate in the Beckmann rearrangement (14.2)

**nitrogen inversion:** the rapid redisposition of the nonbonding lone pair of an amine to the opposite side of the molecule, converting the starting amine to its mirror image (5.10)

**nitrogen mustard:** a nitrogen analog of mustard gas used in cancer chemotherapy (23.8)

**nodal surface:** a position in an atomic or molecular orbital at which electron density is zero (1.2)

**nonpolar covalent bond:** a chemical bond characterized by the absence of appreciable partial charge separation because of nearly equal sharing of the electrons comprising the bond by the two bonded atoms (1.2)

**nonsaponifiable lipid:** a lipid that cannot be hydrolyzed by aqueous base to soaps; a terpene (17.1)

*nor-:* prefix describing a substance whose structure closely resembles that of another but is lacking some small feature, usually a methyl group (17.1)

**normal alkane:** a straight-chain alkane (1.3)

**normal phase chromatography:** a liquid chromatographic technique in which less polar compounds elute first through a polar stationary phase, often unmodified silica gel or alumina (4.2)

**nuclear magnetic resonance (NMR) spectroscopy:** a spectroscopic technique for measuring the amount of energy needed to bring a nucleus (most commonly $^1$H or $^{13}$C in organic molecules) into resonance when a molecule is placed in a strong magnetic field and is irradiated with radio-frequency waves (4.3)

**nucleic acid:** a polymer that has alternating sugar (ribose or deoxyribose) and phosphate units along its backbone, with one of several heterocyclic bases joined to the sugar unit through a phosphate ester linkage between a C-3′ OH group of one nucleoside and a C-5′ OH group of another; used to store and code information that translates into sequences of amino acids in the peptides and enzymes critical to living systems (18.5)

**nucleic acid base:** a purine or pyrimidine base found in DNA or RNA; the bases are adenine, guanine, cytosine, uracil, and thymine (18.5)

**nucleocapsid:** *see* **virion**

**nucleophile:** an electron-rich reagent that attacks centers of positive charge; from the Greek *nucleo,* "nucleus," and *philos,* "loving" (3.7)

**nucleophilic acyl substitution:** *see* **nucleophilic substitution**

**nucleophilic addition:** an addition reaction initiated by attack by an electron-rich reagent (a nucleophile) on a carbonyl compound or derivative (12.1)

**nucleophilic substitution:** a chemical conversion in which a leaving group is displaced by an electron-rich (nucleophilic) reagent (7.4, 7.5, 12.5)

**nucleophilicity:** the tendency of an atom, ion, or group of atoms to release electron density to form a bond with a carbon atom (3.7)

**nucleoside:** a component of RNA and DNA having a purine or pyrimidine base attached to C-1 of a ribose or deoxyribose sugar unit (18.5)

**nucleotide:** a phosphate ester derivative of a nucleoside (18.5)

**nylon 6:** a polyamide formed in the ring-opening polymerization of caprolactam (16.5)

**nylon 66:** a polyamide formed in the cross reaction between adipic acid and 1,6-diaminohexane (16.5)

**olefin:** an alkene (2.1)

**optical isomers:** isomers that differ in the three-dimensional relationship of substituents about one or more atoms (5.1)

**optical purity:** a measure of the excess of one enantiomer over the other in a mixture, as determined by comparison of the optical rotation of the sample with that of a sample presumed to be a single enantiomer (5.7)

**optically active:** descriptor for a sample that rotates the plane of polarized light and thus contains an excess of one of a pair of enantiomers (5.7)

**optically inactive:** descriptor for a sample that does not rotate the plane of polarized light (5.7)

**orbital:** the probability surface describing the volume in which an electron is likely to be found (1.2)

**orbital overlap:** the spatial intersection of atomic or hybrid atomic orbitals required for the formation of a chemical bond (1.2)

**orbital phasing:** the relative wave property of electrons in orbitals that results in either favorable or unfavorable interaction; like phasing results in bonding, and unlike phasing results in antibonding interactions (2.1)

**organic chemistry:** the chemistry of carbon compounds (1.1)

**organic synthesis:** a subarea of organic chemistry that focuses on the construction of new or existing molecules (for example, natural products) (7.3)

**organocuprate:** a reagent in which carbon is directly bonded to copper (8.4)

**organolithium:** a reagent in which carbon is directly bonded to lithium (8.4)

**organomagnesium compound:** *see* **Grignard reagent**

**organometallic compound:** a reagent in which carbon is directly bonded to a metal atom (8.4)

*ortho, para* **director:** a functional group that directs electrophilic aromatic substitution to the *ortho* and *para* positions (11.4)

*ortho* **substitution:** a reaction that results in substituents at the 1 and 2 positions of an aromatic ring (2.3)

**osmosis:** the migration of a gas or liquid (usually water) across a membrane, from the side containing the lower concentration of the substance to the side containing the higher concentration (23.5)

**oxidation:** a chemical transformation resulting in the loss of electrons and hydrogen atoms and/or the addition of oxygen atoms or other electronegative heteroatoms (3.3)

**oxidation–reduction reaction:** a chemical transformation in which the oxidation level of a reactant and its reaction partner are equivalently changed, with one substrate gaining electrons and the other losing them; also, a reaction in which a substrate undergoes both oxidation at one atom and reduction at another (15.1)

**oxidative decarboxylation:** the conversion of a functionalized carboxylic acid (often an $\alpha$-ketoacid or $\alpha$-amino acid) to a carboxylic acid with one fewer carbon by the loss of $CO_2$ (21.7)

**oxidative degradation:** the cleavage of a carbon skeleton (often at a C=C double bond) with the introduction of new carbon–oxygen bonds (10.4)

**oxidizing agent:** an agent that effects an oxidation (3.3)

**oxime:** $R_2C$=NOH; a condensation product of hydroxylamine ($NH_2OH$) with an aldehyde or ketone; often, a highly colored solid used to test for the presence of a carbonyl group (12.4)

**oxirane:** *see* **epoxide**

**oxonium ion:** a cation produced when oxygen bears three $\sigma$ bonds (3.6)

**oxymercuration–demercuration:** a reaction sequence used to achieve Markovnikov hydration of an alkene without accompanying skeletal rearrangements; initiated by treating the alkene with mercuric acetate in aqueous acid, followed by $NaBH_4$ (10.2)

**ozonation:** the addition of $O_3$ to an alkene (10.4)

**ozone:** $O_3$; an electrophilic allotrope of oxygen that exists in a zwitterionic form in which the central oxygen bears formal positive charge (10.4)

**ozonide:** a five-member ring containing three oxygen atoms; produced by rearrangement of a molozonide in the addition of $O_3$ to an alkene (10.4)

**ozonolysis:** a sequence in which a C=C double bond is oxidatively converted to two carbonyl groups through treatment with $O_3$, followed by Zn in acetic acid (10.4)

**paper chromatography:** a chromatographic technique in which a mixture of compounds is separated by passing the liquid phase by capillary action through a sheet of chromatographic paper (4.2)

**Pauli exclusion principle:** a theoretical statement that each electron must be unique (that is, each must have a distinct set of principal, secondary, azimuthal, and spin quantum numbers) and that no more than two electrons, which must have opposite spins, can occupy the same orbital (1.2)

**pentose:** a five-carbon sugar (17.3)

**peptidase:** an enzyme that catalyzes the cleavage of peptide bonds (17.1)

**peptide:** a polyamide composed of 2–10 $\alpha$-amino acid residues (the term is sometimes used interchangeably with *polypeptide*) (16.5, 16.7)

**peptide bond:** an amide linkage (16.5)

**peptide mimic:** a pharmaceutical agent that is structurally and functionally similar to a small peptide hormone and that mimics the physiological function of the natural compound (23.3)

**peptidoglycan:** a constituent of a bacterial cell wall composed of long chains of carbohydrates cross-linked by short peptides (23.5)

**peracid:** $RCO_3H$; an oxygenated relative of a carboxylic acid (10.4)

**pericyclic reaction:** a concerted chemical conversion taking place through a transition state that can be described as a cyclic array of interacting orbitals (14.1)

**pericyclic rearrangement:** a skeletal rearrangement proceeding through a concerted, pericyclic transition state (14.1)

**periodic table:** an orderly arrangement of the elements grouped according to their atomic number and electron configuration (1.2)

**periplanar:** *see anti*-**periplanar** and *syn*-**periplanar**

**peroxide:** ROOR; a functional group containing an oxygen–oxygen $\sigma$ bond (10.3)

**pharmaceuticals:** biologically active compounds that are sold by drug companies and may be synthetic, semi-synthetic, or obtained from natural sources (15.7)

**phase-transfer catalyst:** a compound that provides enhanced solubility in organic solvents to a reagent through reversible binding, providing for greatly increased concentrations of the reagent in a nonaqueous phase (20.1)

**phenol:** $C_6H_5OH$; an OH-substituted benzene (3.11)

**phenyl anion:** an unstable anion formed by deprotonation of benzene (9.8)

**phenyl group:** a $C_6H_5$ fragment with one fewer hydrogen than benzene (2.3)

**phenylhydrazone:** $R_2C$=NNHPh; a condensation product of phenylhydrazine ($PhNHNH_2$) with an aldehyde or ketone; often, a highly colored solid that is diagnostic for the presence of a carbonyl group (12.4)

**phosphate ester:** $(RO)_3P$=O; an ester of phosphoric acid (17.2)

**phosphine:** $PR_3$; a functional group containing trivalent phosphorus (8.3)

**phospholipid:** a dicarboxylate, monophosphate ester of glycerol (17.1)

**phosphonium salt:** a tetravalent phosphorus cation ($^{\oplus}PR_4$) obtained by protonation or alkylation of a phosphine (8.3)

**phosphonium ylide:** $R_3P^{\oplus}$—$(CR_2)^{\ominus}$; an $\alpha$-deprotonated phosphonium salt (8.3)

**phosphorane:** *see* **phosphonium ylide**

**phosphoric acid anhydride:** the condensation product that is obtained by dehydration of two equivalents of phosphoric acid and has the formula —OP(O)(OH)—O—P(O)(OH)—O— (22.4)

**phosphoric acid derivatives:** a family of compounds containing the —PO(OR)$_3$ group (12.6)

**photochemotherapy:** a cancer treatment that uses light to activate an otherwise nontoxic agent that localizes in cancer tissue (23.8)

**photoexcitation:** the process by which a photon ($hv$) is absorbed by a molecule, causing the promotion of one of the electrons from a bonding to an antibonding orbital (4.3)

**photofrin:** a mixture of hematoporphyrins used in photochemotherapy (23.8)

**photosynthesis:** the complex biological process by which carbon dioxide is converted to carbohydrates in a series of reactions initiated by the absorption of light energy (22.5)

**physical organic chemistry:** a subarea of organic chemistry that relates structure to reactivity and reaction mechanisms (7.3)

**pi ($\pi$) bond:** a covalent bond in which electron density is symmetrically arranged above and below the axis connecting the two bonded atoms; results from the sideways overlap of $p$ orbitals (2.1)

**$\pi,\pi^\star$ transition:** an electronic transition involving the promotion of an electron in a $\pi$ (bonding) orbital to a $\pi^\star$ (antibonding) orbital (4.3)

**pinacol rearrangement:** the acid-catalyzed conversion of a 1,2-diol to a ketone with migration of a carbon–carbon bond (7.1)

**p$K_a$:** the negative logarithm of $K_a$; a larger positive value denotes a weaker acid (6.6)

**plane of symmetry:** a plane surface that bisects a molecule such that half of the molecule is the mirror image of the other half (5.5)

**plane-polarized light:** light in which the electromagnetic vectors of all photons are aligned in a single plane; obtained by passing ordinary light through a polarizer (5.7)

**plastics:** polymers that can be heated and molded while relatively soft; from the Greek *plastikos*, "fit to be molded" (16.4)

**$\beta$-pleated sheet:** a folded sheet-like structure imposed by intermolecular hydrogen bonding between peptide chains (16.7)

**pleating:** the deviation from a planar arrangement in the structure of intermolecularly associated peptide chains to avoid steric interaction of the alkyl groups at the $\alpha$ position (16.7)

**Plexiglas:** —(CH$_2$C(CH$_3$)(CO$_2$CH$_3$))$_n$—; poly(methyl methacrylate) (16.4)

**(+):** dextrorotatory; *see d*

**polar covalent bond:** a chemical bond characterized by appreciable partial charge separation because of unequal sharing of the electrons comprising the bond between two bonded atoms (1.2, 3.1)

**polar protic solvent:** a solvent that has an acidic proton on a heteroatom (3.5)

**polarimeter:** an instrument used to measure optical rotation (5.7)

**polarizability:** a measure of the ease with which the electron distribution in a molecule can shift in response to a change in electric field; the ability of an atom to accommodate a change in electron density (3.12)

**polarization:** a partial charge separation induced by a difference in electronegativity between carbon and a heteroatom (1.2)

**polarized light:** *see* **plane-polarized light**

**polyacetal:** —(CHRO)$_n$— (16.4)

**polyamide:** polymer in which the repeat units are joined by an amide linkage (16.5)

**polycarbonate:** —(ROCO$_2$)$_n$— (16.5)

**polycyclic aromatic hydrocarbon:** an aromatic compound containing fused rings (2.3)

**polydentate:** descriptor for a compound with many ligating heteroatoms; from the Latin meaning "many teeth" (19.2)

**polyelectrolyte:** a high-molecular-weight molecule that readily ionizes to form a multiply charged species when dissolved in water or other polar solvents; a macromolecule in which the repeat unit bears a positive or negative charge (4.2, 18.5)

**polyene:** an unsaturated hydrocarbon or derivative containing more than two double bonds (2.2)

**polyene antibiotic:** a long-chain, multiply unsaturated compound that functions by creating additional ion channels through cell membranes (23.5)

**polyester:** a polymer in which the repeat units are joined by an ester linkage (16.1, 16.5)

**polyether:** a polymer in which the repeat units are joined by an ether linkage (16.4)

**poly(ethylene glycol):** —(CH$_2$CH$_2$O)$_n$—; condensation polymer from ethylene glycol (16.4)

**polymer:** a large molecule composed of many repeating subunits; from the Greek *polumeres*, "having many parts" (16.1)

**polymerization:** the process of linking monomer units into a polymeric matrix (16.1)

**polynucleic acid:** *see* **nucleic acid**

**polypeptide:** a polyamide derived from $\alpha$-amino acids, specifically composed of 10–100 $\alpha$-amino acids (the term is sometimes used interchangeably with *peptide*) (16.5, 16.7)

**polysaccharide:** a polyacetal formed by condensation of a hemiacetal group of one sugar unit with an alcohol group of another sugar unit, taking place with the loss of water (16.5)

**polystyrene:** $—(CH_2CH(Ph))_n—$ (16.4)

**polyurethane:** $—(OCONH)_n—$; polymer in which the repeat units are joined by a urethane (carbamate) linkage (16.5)

**poly(vinyl alcohol):** $—(CH_2CH(OH))_n—$ (16.4)

**poly(vinyl chloride):** $—(CH_2CH(Cl))_n—$ (16.4)

**positional isomerization:** a chemical conversion in which the position of a functional group is altered (7.1)

**positional isomers:** isomers in which the sequence along the chain differs in the position of one or more functional groups (2.1)

**potential energy surface:** a plot of the changes in potential energy taking place as a reaction proceeds (6.1)

**potential energy well:** an energy minimum along a potential energy diagram representing a molecule or intermediate with a real-time existence (5.2)

**prebiotic chemistry:** the synthesis of critical biochemicals in the absence of living systems, as might have occurred before life was present on earth (18.7)

**primary alcohol:** $RCH_2OH$; an alcohol in which the OH group is attached to a primary carbon atom (3.5)

**primary amine:** $RNH_2$; an amine in which nitrogen is attached to one carbon substituent (3.1)

**primary carbon:** a carbon atom chemically bonded to only one other carbon atom (1.5)

**primary structure of a peptide or protein:** the sequence of amino acid units along a peptide or protein chain (16.7)

**prion (proteinaceous infectious particle):** an as yet not well characterized protein thought to be responsible for scrapie, kuru, and bovine spongiform encephalopathy (23.7)

**prochiral:** descriptor for an achiral center that can become a center of chirality, either by replacement of one of two identical groups or by addition to a $\pi$ system (10.1)

**product inhibition:** the binding of a product to a catalyst, inhibiting further catalytic cycles (20.7)

**propagation steps:** the principal product-forming sequence in a free-radical chain reaction in which a reactant radical is converted to product and a different radical; the number of product radicals in a propagation step is equal to the number of reactant radicals (7.6)

***n*-propyl group:** $—(CH_2)_2CH_3$; an unbranched three-carbon alkyl group attached through the primary carbon (1.5)

**protease:** an enzyme that catalyzes the hydrolysis of peptide bonds (20.7)

**protease inhibitor:** a compound that blocks the enzyme protease; especially important in therapy for HIV (23.6)

**protecting group:** a functional group that masks the characteristic reactivity of another group into which it can later be converted (8.3, 10.2)

**protein:** a poly($\alpha$-amino acid) composed of more than 100 $\alpha$-amino acids (16.7)

**protic solvent:** a solvent molecule incorporating a polar X—H bond (3.5)

**proton decoupling:** a technique used routinely to simplify $^{13}C$ NMR spectra by irradiating the sample with radio-frequencies, either at a specific region or over the entire chemical shift range at which protons absorb; results in saturation of the populations in the high spin state and loss of coupling to the irradiated nuclei (4.3)

**proton transfer:** the movement of $H^{\oplus}$ from an acidic to a basic site (3.2)

**protonated alcohol:** a cationic species produced upon association of a proton with a nonbonding lone pair of the oxygen atom of an alcohol (3.6)

**protonation:** the covalent attachment of a proton ($H^{\oplus}$) to an atom either bearing a nonbonding lone pair of electrons or involved in a $\pi$ bond (7.5)

**psoralen:** a naturally occurring compound used in phototherapy for leukoderma (23.8)

**psychoactive drugs:** pharmaceutical agents used to achieve a state of euphoria or to block intense pain (23.4)

**puckered:** descriptor for a nonplanar cycloalkane that has fewer eclipsing C—H interactions and lower torsional strain than its planar analog (5.3)

**purines:** a family of bicyclic heteroaromatic molecules having a five-member ring fused to a six-member ring and containing two nitrogens in a 1,3 relationship in each ring; two members of the family (adenine and guanine) are nucleic acid bases (18.5)

**pyramidal:** descriptor for a spatial arrangement in which a central atom and three attached groups are located at the corners of a pyramid (3.1)

**pyranose:** a carbohydrate containing a six-member cyclic hemiacetal (17.3)

**pyridine:** $C_5H_5N$; a six-member cyclic, nitrogen-containing, heteroaromatic molecule (3.11)

**pyridoxamine phosphate:** the product of reductive amination of pyridoxal phosphate; a cofactor that serves as both a reducing agent and a source of nitrogen for the production of $\alpha$-amino acids (21.4)

**pyrimidines:** a family of monocyclic heteroaromatic molecules having a six-member ring containing two nitrogens in a 1,3 relationship; three members of the family (cytosine, uracil, and thymine) are nucleic acid bases (18.6)

**pyrophosphate group:** the monoanhydride of phosphoric acid (17.2)

**pyrrole:** $C_4H_4NH$; a five-member, cyclic, nitrogen-containing, heteroaromatic molecule (3.11)

**quartet:** a four-line multiplet on an NMR spectrum (4.3)

**quaternary ammonium ion:** a positively charged ion in which nitrogen is attached to four carbon substituents (3.1)

**quaternary structure of a peptide or protein:** clusters formed as several large polypeptide or protein units join together to form a functional object (16.7)

**quinoid form:** a six-member ring resembling quinone, with one or more exocyclic double bonds (21.4)

**R:** absolute stereochemical designator employed in the Cahn–Ingold–Prelog rules to describe the stereoisomer for which a clockwise rotation is required to move from the highest-priority to the next-to-lowest-priority group attached to a chiral tetravalent atom when the substituent of lowest priority is directed away from the observer (5.8)

**racemate:** *see* **racemic mixture**

**racemic mixture:** an optically inactive mixture composed of equal amounts of enantiomers (5.7)

**racemic modification:** *see* **racemic mixture**

**racemization:** the loss of optical activity when one enantiomer is converted to a 50:50 mixture of enantiomers (7.5)

**radical:** a chemical species bearing a single unpaired electron on one atom; a chemical species with an odd number of electrons (2.1)

**radical anion:** a reactive intermediate with one more electron than needed to achieve the electron configuration of a stable neutral molecule (6.4)

**radical cation:** a reactive intermediate lacking one electron from the complement needed for a stable neutral molecule (6.4)

**radical chain reaction:** a chain reaction in which a free radical is produced in the initiation and propagation steps and consumed in the termination steps (7.7)

**radical hydrobromination:** an anti-Markovnikov hydrobromination of an alkene initiated by peroxide decomposition and taking place through the radical addition of a bromine atom (10.3)

**radical polymerization:** polymerization initiated by a radical and in which the chain-carrying step is a radical (16.4)

**radical stability:** $3° > 2° > 1°$ (3.6)

**rate-determining step:** the slowest step in a multistep re-

action; the step whose transition state lies at the highest energy (6.3, 7.5)

**rate-limiting step:** *see* **rate-determining step**

**reaction coordinate:** the variation of a specific structural feature (such as bond length or angle) that measures how far a reaction has proceeded (6.1)

**reaction mechanism:** the sequence of bond-making and bond-breaking by which a reactant is converted to a product; a detailed description of the electron flow, including the identity of any intermediate(s) formed, that takes place during a chemical reaction (7.1, 7.3)

**reaction profile:** *see* **energy diagram**

**reactive intermediate:** a metastable species with a high energy relative to a reactant and a product; encountered at an energy minimum (in a potential energy well) along a reaction coordinate (6.1, 6.4)

**rearrangement reaction:** a chemical conversion in which the molecular skeleton is altered so that the sequence in which atoms are attached is changed (7.1)

**recycling of biological reagents:** a natural process whereby cofactors are reused (21.2)

**redox reaction:** a reaction involving oxidation or reduction (7.1)

**redox reagent:** a reagent that can induce oxidation or reduction (3.3)

**reducing agent:** an agent that effects a reduction (3.3)

**reduction:** a chemical transformation induced by the addition of electrons or hydrogen atoms and/or the removal of oxygen or other electronegative atoms (3.3)

**reductive amination:** a process by which a $C\!=\!O$ group is converted to a $CH(NR_2)$ group through reduction of an intermediate imine (12.4, 18.1)

**Reformatsky reaction:** a Claisen-like condensation of a preformed zinc ester enolate with a ketone or an aldehyde (13.4)

**refractive index:** the ratio of the speed of light in a vacuum to that in a material; the path of light is bent upon passing from one medium to another of different refractive index (4.2)

**refractive index detector:** a device that produces an electrical signal in response to the difference in refractive index of a solvent with and without a solute; often used in conjunction with high-pressure liquid chromatography (HPLC) (4.2)

**regiochemistry:** the orientation of a chemical reaction on an unsymmetrical substrate (7.6)

**regiocontrol:** the formation of one regioisomer to a greater extent than others in a chemical reaction (7.6)

**regioselective:** descriptor for a reaction in which there is a clear preference for one of two or more possible regioisomers (7.6)

**relative stereochemistry:** the specification of the stereo-chemical relationship between two molecules (5.6)

**repeat unit:** a segment that is encountered again and again along a polymer chain (16.1)

**resin:** a highly viscous polymeric glass (16.6)

**resolution:** the separation of a racemic mixture into two pure enantiomers; often accomplished by forming and then separating diastereomers, followed by regeneration of the original reactant (5.8)

**resonance:** in NMR, the condition in which the applied radio-frequency energy matches the energy difference between the parallel and antiparallel spin states of a nucleus, so that the energy is absorbed, causing the spin to "flip" from the lower-energy parallel state to the higher-energy antiparallel state (4.3)

**resonance contributors:** *see* **resonance structures**

**resonance effect:** stabilization by delocalization of $\pi$ electrons; the donation or withdrawal of electron density by overlap with a neighboring $\pi$ system (6.8)

**resonance hybrid:** an energetically weighted composite of contributing resonance structures (2.3)

**resonance structures:** valence bond representations of possible distributions of electrons in a molecule, differing only in positions of electrons and *not* in positions of atoms (2.3)

**respiration:** the biological oxidation of glucose to carbon dioxide (22.5)

**restricted rotation:** the inhibition of rotation about a $\sigma$ bond (3.3)

**retention time:** the interval required for a compound to elute from a chromatography column; influenced by the magnitude of noncovalent interactions between the compounds being separated and the stationary phase (4.2)

**retro-aldol reaction:** the reverse of an aldol reaction, by which a $\beta$-hydroxycarbonyl compound is cleaved to two carbonyl derivatives (22.10)

**retro-Diels–Alder reaction:** the concerted fragmentation of cyclohexene (or a derivative) to butadiene (or a derivative) and an alkene (or a derivative); the reverse of a Diels–Alder reaction (6.6)

**retrosynthetic analysis:** a method for planning an organic synthesis by working backward, step-by-step, from the product to the possible starting materials (7.3, 15.1, 15.2)

**retrovirus:** a virus in which viral RNA is accompanied by an enzyme, reverse transcriptase, that translates the genetic code of the RNA into a strand of DNA (23.6)

**reverse polarity reagent:** a reagent with a functional group that has greater reactivity (as a nucleophile or electrophile) than that of another functional group from which it is derived (21.8)

**reverse transcriptase:** an enzyme that translates RNA into DNA (23.6)

**reverse-phase chromatography:** a liquid chromatographic technique in which more polar compounds elute first through a nonpolar stationary phase, often silica gel coated with a long-chain alkylsilane (4.2)

**reversible reaction:** a reaction that can proceed backward or forward with similar ease (6.1)

**$R_f$ value:** the ratio of the distance migrated by a substance compared with the solvent front (4.2)

**ribonucleic acid:** *see* **RNA**

**ribose:** a sugar unit found in the backbone of RNA (17.3)

**ribosome:** a body in the cell cytoplasm containing all biological reagents needed to synthesize peptides (19.3)

**rigidity:** stiffness (19.2)

**ring annulation:** *see* **annulation**

**ring flip:** conformational interconversion of one ring conformation to another of the same type; often used to describe chair-to-chair or boat-to-boat interconversions (5.4)

**ring strain:** the destabilization caused by angle strain and eclipsing interactions in a cyclic compound (1.4)

**ring-opening polymerization:** a polymerization reaction in which the driving force for bond formation between repeat units is supplied by relief of ring strain in a monomer (16.4)

**RNA:** ribonucleic acid; a biopolymer found in cell nuclei and composed of ribonucleotide units linked through a sugar–phosphate backbone; RNA transcribes the genetic information stored in DNA and directs protein synthesis (17.3)

**Robinson ring annulation:** the use of an intramolecular aldol reaction to construct a six-member ring fused to another ring (13.3)

**rubber:** naturally occurring poly(2-methylbutadiene) (16.4)

**S:** absolute stereochemical designator employed in the Cahn–Ingold–Prelog rules to describe the stereoisomer for which a counterclockwise rotation is required to move from the highest-priority to the next-to-lowest-priority group attached to a chiral tetravalent atom when the substituent of lowest priority is directed away from the observer (5.8)

**saccharide:** *see* **carbohydrate**

**saponifiable lipid:** a lipid that can be hydrolyzed by aqueous base to fatty acids; a fat or wax (17.1)

**saponification:** the making of soaps by hydrolysis of fatty acid esters with aqueous hydroxide; one of the oldest known chemical reactions (17.1)

**saturation:** the condition of a compound that contains only $sp^3$-hybridized atoms; consequently, a description of a molecule lacking multiple bonds (1.3, 2.1)

**sawhorse representation:** a graphic method of representing three-dimensional structures using solid wedges and hatched lines (1.2)

**Schiff base:** $R_2C=NR$; an *N*-alkylated imine (12.4)

**scrapie:** a disease, possibly caused by a prion, that causes goats and sheep to scrape off their wool (23.6)

**secondary alcohol:** $R_2CHOH$; an alcohol in which the OH group is attached to a secondary carbon atom (3.5)

**secondary amine:** $R_2NH$; an amine in which nitrogen is attached to two carbon substituents (3.1)

**secondary carbon:** a carbon atom chemically bonded to only two other carbon atoms (1.5)

**secondary structure of a peptide or protein:** a complex three-dimensional structure describing local organization of chain segments such as $\alpha$-helices and $\beta$-pleated sheets (16.7)

**selectivity:** preference for reaction with one reagent over another at one site rather than another; the formation of one product in preference to other possible products (7.6, 8.2)

**self-exchange reaction:** a substitution reaction in which the incoming group and leaving group are identical (7.4)

**semicarbazone:** $R_2C=NNHC(O)NH_2$; condensation product of an aldehyde or ketone with semicarbazide ($H_2NNHC(O)NH_2$); often a highly colored solid that is diagnostic for the presence of a carbonyl group (12.4)

**semisynthetic:** descriptor for a naturally occurring (or cultured) material that is chemically altered, sometimes in a relatively minor way, in the laboratory (15.7)

**sesquiterpene:** a terpene containing 15 carbons and derived from three isoprene units (17.1)

**sesterterpene:** terpene containing 25 carbon atoms and derived from five isoprene units (17.1)

**sex hormones:** steroids that determine sexual characteristics and regulate sexual functions (17.1)

**shielding:** the shift of a nuclear magnetic resonance (NMR) signal from that expected from the applied field, caused by donation of electron density to the nucleus of interest (4.3)

**side-chain oxidation:** the conversion of an alkyl or acyl side chain on an aromatic ring to a $-CO_2H$ group upon treatment with hot aqueous $KMnO_4$ (11.3)

**sigma ($\sigma$) bond:** a covalent chemical bond in which electron density is arranged symmetrically along the axis connecting the two bonding atoms; results from direct overlap of hybrid orbitals having some *s* character (1.2)

**sigmatropic rearrangement:** a skeletal rearrangement accomplished through the shift of a $\sigma$ bond to the opposite end of a $\pi$ system, for example, as in the Cope rearrangement; involves the migration of a group from one end of a $\pi$ system to the other (14.1)

**sigmatropic shift:** a pericyclic reaction in which a $\sigma$-bonded substituent migrates from one end of a $\pi$ system to the other (14.1)

**silk:** a protein containing high fractions of glycine and alanine (16.5)

**Simmons–Smith reaction:** formation of a cyclopropane through stereospecific carbene addition to an alkene by treatment of a vicinal dihalide with a zinc–copper couple in the presence of an alkene (10.4)

**single-electron transfer:** a chemical reaction in which the key step consists of the transfer of one electron (21.4)

**singlet:** a molecule in which all electrons are paired, generally with two electrons of opposite spin paired in each molecular orbital (6.4)

**$S_N1$ reaction:** a stepwise, unimolecular, nucleophilic substitution that proceeds through an intermediate carbocation (7.5)

**$S_N2$ reaction:** a concerted, bimolecular, nucleophilic substitution that takes place via back-side attack by a nucleophile and leads to a substitution product with inverted configuration at the substituted carbon (7.4)

**soap:** a mixture of salts of long-chain fatty acids, obtained by base hydrolysis of fats (17.1)

**soft:** descriptor for a charge-diffuse reagent; often applied to nucleophiles, electrophiles, acids, and bases (13.3)

**solid wedges:** in a line structure, a graphic representation indicating a group or atom positioned nearer to the observer (1.2)

**solvation:** the association of solvent molecules about a solute (3.2, 3.13)

**solvent front:** the furthest point reached by the solvent in chromatography (4.2)

**solvolysis reaction:** a reaction in which the solvent displaces a leaving group or is added across a multiple bond (7.5)

**sp-hybrid orbitals:** a pair of hybrid orbitals that are formed by mathematically mixing one *s* and one *p* atomic orbital and are dispersed in a linear array separated by 180° (2.4)

**$sp^2$-hybrid orbitals:** a set of three hybrid orbitals that are formed by mathematically mixing one *s* and two *p* atomic orbitals and are dispersed in a plane and separated by 120° (2.1)

**$sp^3$-hybrid orbitals:** a set of four hybrid orbitals that are formed by mathematically mixing one *s* and three *p* atomic orbitals and are tetrahedrally dispersed (separated by 109.5°) around the atom's nucleus (1.2)

**space-filling model:** a three-dimensional representation indicating van der Waals radii of all component atoms oriented in space (1.2)

**specific rotation [$\alpha$]:** the extent to which a given molecule (on a weight basis) rotates a plane of polarized light; the observed rotation is the product of the specific rota-

**tion,** the concentration in the sample compartment, and the path length of the sample cell; $\alpha = [\alpha] \times l \times c$, where $l$ = path length (in dm) and $c$ = concentration (in g/mL) (5.7)

**spectroscopy:** a set of techniques that measure the response of a molecule to the input of energy (4.1)

**spectrum:** a display of peak intensity detected for a given spectroscopic method as a function of incident energy (4.3)

**staggered conformation:** a spatial arrangement in which each $\sigma$ bond on one carbon atom is fixed at a 60° dihedral angle from a $\sigma$ bond on an adjacent atom; when viewed end-on in a Newman projection, the conformation in which the bonds on one atom exactly bisect those on the adjacent atom (5.2)

**starch:** a water-soluble biopolymer containing as many as 4000 glucose units connected by $\alpha$ linkages (16.5)

**stationary phase:** an immobile medium (usually a solid or highly viscous liquid) through which a mixture passes in chromatography (4.2)

**stereoelectronic control:** a requirement for precise orbital alignment for a proposed reaction (9.3)

**stereoisomers:** isomers that differ only in the position of atoms in space (2.1, 5.1)

**stereorandom:** descriptor for a reaction that occurs without any stereochemical preference (10.1)

**steric effect:** the destabilization resulting from van der Waals repulsion between groups that are too close to each other (5.2)

**steric strain:** *see* **steric effect**

**steroid:** a member of a group of naturally occurring, often oxygenated, tetracyclic compounds that have a fused-ring system of three six-member rings and one five-member ring; as a class, steroids often have important hormonal functions (17.1)

**structural isomers:** isomers in which the carbon backbones differ (1.3)

**substituent effect:** the altered reactivity induced by the presence of a substituent group on a reactant, often affecting rate, stereochemistry, or regiochemistry (or all three) of a reaction (11.4)

**substitution reaction:** a chemical conversion in which one atom or group of atoms in a molecule is replaced by another (3.6, 7.1)

**substrate:** a molecule undergoing reaction under the influence of one or more external reagents (20.7)

**sugar:** *see* **carbohydrate**

**suicide inhibition:** an irreversible transformation in which the catalytic function of an enzyme is blocked by irreversible covalent bond formation (23.5)

**sulfate ester:** $(RO)_2SO_2$; an ester of sulfuric acid (17.2)

**sulfonamide:** $RSO_2NR_2$; a family of compounds containing the $-SO_2NR_2$ group (3.10, 12.6)

**sulfonate ester:** $ROSO_2R$; an alkylated derivative of a sulfonic acid (8.3)

**sulfonation:** the replacement of H by an $-SO_3H$ group (11.2)

**sulfonic acid:** $RSO_3H$; a functional group in which an $-SO_3H$ group is attached to an alkyl or aryl group (3.10, 12.6)

**sulfanilamide:** a *p*-aminobenzenesulfonamide antibiotic; a sulfonamide derivative of aniline (23.5)

**superimposable:** descriptor for the relationship of two molecules for which a conformation exists such that each of the four substituents at a center of chirality can be superimposed (5.5)

***syn* addition:** the formation of product by delivery of an electrophile and a nucleophile to the same face of a multiple bond (10.4)

***syn* eclipsed conformer:** a conformational isomer in which two large groups on adjacent carbons are at a dihedral angle of 0° (5.2)

**syndiotactic:** stereochemical designator for a polymer in which the alkyl groups at centers of asymmetry along the chain point alternately in one direction and in the opposite direction (16.7)

***syn*-periplanar:** descriptor for the geometric relationship in which the bonds to substituents on adjacent atoms of a $\sigma$ bond are coplanar, with a dihedral angle of 180°; a possible geometry for an E2 elimination, although it is less preferred than the *anti*-periplanar alignment (9.3)

**synthetic:** prepared in the laboratory (15.7)

**synthetic efficiency:** the evaluation of the utility of a proposed synthesis, depending on the number of steps, the yield of each step, the ease and safety of the reaction conditions, the ease of purification of intermediates, and the cost of starting materials, reagents, and personnel time (15.6)

**tautomerization:** a transformation from one structure to another in which the only changes are the position of attachment of a hydrogen atom and the position of $\pi$ bond(s); typically, a 1,3 (or 1,5) shift of a proton to or from a heteroatom in a three-atom system containing a double bond; catalyzed by acid or base (6.2)

**tautomers:** constitutional isomers that differ only in the position of an acidic hydrogen along a three-atom segment containing a heteroatom and a double bond (6.2)

**Teflon:** $-(CF_2)_n-$ (16.4)

**termination step:** a reaction that stops a chain reaction by consuming a reactive intermediate without producing another or by converting two reactive intermediates into one stable product (7.6)

**termolecular reaction:** a rare type of reaction that requires a collision between three reactants in the rate-determining step (6.9)

**terpene biosynthesis:** the condensation of isopentenyl pyrophosphate (17.2)

**terpenes:** a family of relatively nonpolar natural products (lipids) derived biochemically from isopentenyl pyrophosphate and thus containing $5n$ ($n$ = integer) carbon atoms (17.1)

**tertiary alcohol:** $R_3COH$; an alcohol in which the OH group is attached to a tertiary carbon atom (3.5)

**tertiary amine:** $R_3N$; an amine in which nitrogen is attached to three carbon substituents (3.1)

**tertiary carbon:** a carbon atom chemically bonded to only three other carbon atoms (1.5)

**tertiary structure of a peptide or protein:** the three-dimensional spatial dispersal of $\beta$-pleated sheets and $\alpha$-helices; describes protein folding (16.7)

**tetrahedral carbon:** an $sp^3$-hybridized carbon atom bearing four substituents directed at 109.5° from each other (1.2)

**tetrahedral intermediate:** an intermediate in nucleophilic addition and nucleophilic acyl substitution obtained upon covalent bond formation between an attacking nucleophile and a carbonyl carbon (12.5)

**tetrahydrofolic acid:** a cofactor that effects the methylation of nucleic acids by transferring one carbon from serine (21.6)

**tetraterpene:** a terpene containing 40 carbon atoms and derived from eight isoprene units (17.1)

**tetravalent intermediate:** *see* **tetrahedral intermediate**

**tetrose:** a four-carbon sugar (17.3)

**thermal conductivity detector:** a gas chromatography detector that measures the difference in thermal conductivity between the carrier gas alone and that observed as a sample elutes from the column (4.2)

**thermodynamic control:** descriptor for a chemical reaction for which the relative stabilities of the possible products, rather than the activation energies for their formation, define the course, so that the reaction preferentially forms the most stable product (6.8)

**thermodynamics:** a description of the relative energies of the reactants and products and the equilibrium established between them (6.5)

**thermoneutral reaction:** a conversion in which the reactants and products have the same energy content (6.1)

**thiamine pyrophosphate:** a cofactor important in the degradation of amino acids (21.7)

**thin-layer chromatography (TLC):** chromatographic technique in which a mixture of compounds is separated by elution of the liquid phase by capillary action through a flat solid support such as a sheet of glass, plastic, or aluminum foil coated with a thin layer of silica gel or alumina (4.2)

**thioacetal:** *see* **dithioacetal**

**thioether:** a functional group in which two alkyl or aryl groups are attached to an $sp^3$-hybridized sulfur atom (3.10)

**thioketal:** *see* **dithioketal**

**thiol:** a compound having the SH functional group (3.10)

**thiol ester:** RCOSR; a functional group in which a carbonyl carbon bears an alkyl or aryl group and an SR group (3.10)

**thionyl chloride:** $SOCl_2$; an effective reagent for the conversion of alcohols to alkyl chlorides or carboxylic acids to acid chlorides (8.3)

**thiophene:** $C_4H_4S$; a five-member cyclic, sulfur-containing, heteroaromatic molecule (3.11)

**three-point contact:** an association between pairs of molecules in which three of four groups are in close contact (19.4)

**thymine:** $C_5H_6N_2O_2$; a biologically important heteroaromatic base (3.11)

**torsional strain:** destabilization of an eclipsed conformation relative to a staggered conformation (5.2)

**tosylate:** $ROSO_2\text{-}p\text{-}C_6H_4(CH_3)$; a $p$-toluenesulfonate ester (8.3)

***trans* isomer:** a geometric isomer in which the largest groups are on opposite sides of a double bond or ring (1.5, 2.1)

**transcription:** the reading of the encoded information stored in DNA during RNA synthesis (19.3)

**transesterification:** the interconversion of one carboxylic acid ester into another (12.5)

**transfer RNA (tRNA):** a small nucleic acid containing 70 to 80 nucleotides with two unique regions, one containing an anticodon and another providing a site for the attachment of a specific amino acid (19.3)

**transition metal:** a metallic element with an incomplete inner electron shell; broadly found in the center of the third, fourth, and fifth rows of the periodic table and existing in multiple stable valence states (20.6)

**transition state:** the highest-energy arrangement of atoms along a reaction pathway (5.1)

**transition-state analog:** a stable species that closely mimics the geometry and charge distribution of the transition state of a reaction (20.7)

**transition-state stabilization:** a decrease in the energy of the highest-energy arrangement along a reaction coordinate; attained by altering the environment or the identity of the interacting reagents (20.1)

**transition-state theory:** a theory that asserts that the rate of a reaction varies exponentially with the energy required to reach the transition state (6.9)

**transmetallation:** an exchange of metals between an organometallic compound and either a metal or a different organometallic compound (8.4)

**tricarboxylic acid (TCA) cycle:** a sequence of reactions by which acetate units in the form of acetyl CoA are degraded to carbon dioxide; some of the energy released during this oxidation is stored as chemical reduction potential in the cofactors $FADH_2$, NADH, and NADPH (22.8)

**tridentate:** descriptor for a molecule having three ligating heteroatoms; from the Latin, meaning "three teeth" (19.2)

**triglyceride:** *see* **fat**

**trigonal:** descriptor for an atom with three $\sigma$ bonds (6.4)

**trimer:** a compound containing most or all of the atoms of three molecules of a starting material (10.3)

**triose:** a three-carbon sugar (17.3)

**triple bond:** a $\sigma$ and two $\pi$ bonds between adjacent $sp$-hybridized atoms (2.4)

**triplet:** a molecule in which not all electrons are spin-paired, with two electrons of the same spin being accommodated in two different orbitals; a three-line multiplet in an NMR spectrum (4.3, 6.4)

**triterpenes:** terpenes containing 30 carbons and derived from six isoprene units (17.1)

**twist-boat conformation:** a distorted boat conformation in which the steric interaction of the flagpole hydrogens has been relieved by conformational twisting (5.4)

**ultraviolet (UV) spectroscopy:** a technique that measures a molecule's tendency to absorb light of wavelengths of 200–400 nm (a region of energy just higher than that detectable by the human eye) (4.3)

***Umpolung:*** descriptor for a reagent with reversed polarity (18.7)

**unimolecular reaction:** a reaction involving only a single species in the rate-determining step (6.9)

**unsaturated fatty acid:** a long, straight-chain carboxylic acid containing at least one double bond along the chain (17.1)

**unsaturation:** the condition of a molecule that has some non-$sp^3$-hybridized atoms, consequently, containing one or more multiple bonds (2.1)

**upfield:** the chemical shift of a nucleus that resonates at a lower $\delta$ value than a reference nucleus; that is, for most uncharged molecules, at a higher frequency and thus closer to that of tetramethylsilane; the right-hand portion of an NMR spectrum (4.3)

**uracil:** $C_4H_4N_2O_2$; a biologically important heteroaromatic base (3.11)

**urethane:** a carbonyl group bonded on one side to the nitrogen of an amine and on the other side to the oxygen of an alcohol; a protecting group for an amine (15.8, 16.5)

**valence electrons:** electrons occupying an incompletely filled quantum level (1.2)

**valence shell:** the outermost atomic shell that typically contains electrons (1.2)

**van der Waals attraction:** the energetically favorable force resulting from the interaction of the bonded electrons of one molecule and the nuclei of another (1.2)

**van der Waals radius:** the effective size of an atom (1.2)

**van der Waals repulsion:** the energetically unfavorable force resulting from the interaction of the bonded electrons of one molecule and those of another or the interaction of the nuclei of one molecule with those of another; repulsive intermolecular dipole–dipole interaction (1.2)

**variable region:** portion of an antibody that is unique for a particular foreign invader and is responsible for antibody specificity (20.7)

**vesicle:** a bilayer extended to form a closed surface (17.1)

**vinyl group:** an alkene fragment lacking one substituent from the double bond (2.3)

**vinyl halide:** a functional group in which a halogen is attached to an alkenyl carbon (9.7)

**virion:** an intact virus consisting of a core of genetic material surrounded by coat proteins (23.6)

**viscosity:** resistance to flow; "stiffness" (17.1)

**visible spectroscopy:** a technique that measures a molecule's tendency to absorb light of wavelengths of about 400–800 nm (the region of energy detectable by the human eye) (4.3)

**vitamin:** a molecule that is required to sustain life but cannot be synthesized by a host animal and must therefore be obtained via the diet (21.6)

**vitiligo:** *see* **leukoderma**

**vulcanization:** the cross-linking of a polymer by heating it with sulfur (16.4)

**Wagner–Meerwein rearrangement:** cationic rearrangement in which a carbon substituent participates in a 1,2-shift (14.1)

**water:** $H_2O$; the simplest compound containing $sp^3$-hybridized oxygen (3.5)

**wavelength:** the distance from peak to peak of a wave (4.3)

**wax:** a mixture of fatty acid esters of long, straight-chain alcohols (17.1)

**wedge:** in a line structure, a graphic representation indicating a group or atom positioned nearer to the observer (1.2)

**Williamson ether synthesis:** the reaction of an alkoxide ion with an alkyl halide or tosylate to produce an ether (8.3)

**Wolff–Kishner reduction:** the reduction of a ketone to a methylene group by treatment with basic hydrazine $NH_2NH_2$ (11.2)

**wool:** a structurally complex, naturally occurring protein heavily cross-linked by sulfur–sulfur bonds (16.5)

**ylide:** a zwitterion bearing opposite charges on adjacent atoms (8.3)

**Zaitsev's rule:** an empirical prediction of preferential formation of the thermodynamically more stable, more highly substituted regioisomer in an elimination reaction (9.3)

**Ziegler–Natta catalyst:** an organometallic polymerization initiator that produces isotactic polypropylene (16.7)

**Z-isomer:** a geometric isomer in which the groups of highest priority are on the same side of a double bond; from the German *zusammen,* "together" (2.1)

***zusammen:*** *see* **Z-isomer**

**zwitterions:** neutral species that contain equal numbers of local charge (plus and minus) centers (2.3)

# Index

**I-15**

**I-18**

from benzyne, 465, 475
by diazo substitution, 556
2-Phenoxyethanol, 766
Phenyl acetate, 566
Phenyl anion, 464
Phenyl cation, 464
Phenyl group, 75
Phenylacetaldehyde, 635
N-Phenylacetamide, 566
Phenylacetonitrile, 415
Phenylalanine, 635, 18-9, 18-10, 18-11
in chymotrypsin hydrolysis, 20-35
N-Phenylbenzamide, 716
4-Phenylbut-3-en-2-one, 676
trans-4-Phenylbut-3-en-2-one, 675
p-Phenylenediamine, 809
p-Phenyleneterephthalamide, 809
α-Phenylethyl alcohol, 496, 497
α-Phenethylamine, 262
β-Phenethylamines, 18-25, 23-9–23-12
2-Phenylhexane, 548, 549
Phenylhydrazines, 617
Phenylhydrazones, 617
Phenyllithium, 417
3-Phenyl-1-propene, 498
Phenylpropionic acid analogs, 773–776
Pheromones, 595
Phosphate ($P_i$), 155, 190
biosynthetic pathways involving, 17-23–17-25
Phosphate ester, 17-22
nucleophilic attack on, 17-23
Phosphatidylserines, 17-9, 17-10
Phosphines, substitution of halogen by, 411–412
Phosphite ester, 402
Phosphocreatine, 155, 190
Phospholipids, 17-9
Phosphonium salt, 411, 412, 423
Phosphonium ylide, 411
Phosphorane, 411, 412, 423, 640
Phosphoric acid, 17-21, 17-22
Phosphoric acid anhydride, 22-9
Phosphoric acid derivatives, 633–634
Phosphorus trihalides, in formation of alkyl halides from alcohols, 402
Photochemical trans–cis isomerization, 227–229, 270
Photochemotherapy, 23-41–23-44
Photoexcitation, 203
Photofrin, 23-43
Photosynthesis, 21-9, 22-8, 22-10
Phthalimide, 409
Physical organic chemistry, 343
Physical properties, using to establish structure, 156–158
Pi ($\pi$) bonds, 46
energetics of rotation about, 225
Pi ($\pi$) molecular orbitals, 49–50
Pinacol, 705, 706
Pinacol rearrangement, 335–336, 705–707, 724

Pinacolone, 705, 706
α-Pinene, 704
p$K_a$
definition of, 304
values of, 99, 305–306
Planar conformation, 29, 238
Plastics, 795
Platinum, as transition metal catalyst, 20-21, 20-25–20-26
β-Pleated sheet, 821–822
hydrogen bonds and, 19-8
nonbonding intermolecular interactions and, 20-31
Plexiglas, properties of, 797, 798
Plunkett, R. J., 798
Polar covalent bonding, 17
in amines, 92–100
Polar interactions, 19-4–19-17
dipole–dipole interactions, 19-4–19-5
hydrogen bonds, 19-5–19-12
metal–heteroatom bonds, 19-12–19-17
Polarimetry, 257–258
Polyacetals, 804
formation reaction, 829
Polyacrylamide, 164
Polyalanine, 789
Polyamides, 808–810
Polyaniline, 22-6
Polycarbonates, 805–806
Polycyclic aromatic compounds, electrophilic attack on, 575–577
Polycyclic aromatic hydrocarbons, 72
Polydentate, 19-12
Polyelectrolytes, 164, 18-31
Polyene antibiotics, 23-18
Polyenes, 63, 204
natural, 799
Polyesters, 790, 805–806
Polyethers, 803–804
cyclic versus noncyclic, 19-16
Polyethylene, 792, 794–795, 796, 17-4
Poly(ethylene glycol) (PEG), 803–804
Poly(lactic acid), 790
Polymer chitin, 18-32
Polymer support, for Merrifield solid-phase synthesis, 18-17
Polymerization, 790
addition, 794–804
anionic, 796, 829
cationic, 533, 796, 829
free-radical, 512–513, 533, 794–795, 829
ionic, 796–800
olefin, 20-27–20-30
types of, 793–794
Ziegler–Natta, 817–818
Polymers, 509, 790
addition, 794
in batteries, 22-6
branched, 792–793

condensation, 793–794, 805–812
cross-linked, 792–793, 800–802
extensively cross-linked, 813–815
heteroatom-containing, 802–804
linear, 792–793
medical applications of, 811
stereochemical designations for, 816–817
sulfur bridges in, 802
three-dimensional structure of, 816–827
Poly(methyl methacrylate), 811
Polynucleotide backbone of nucleic acids, 18-27
Polyols, 802–803
Polypeptides, 810–811, 823, 18-14
geometric isomers of, 819
α-helix in, 822–823
naturally occurring, 818–825
β-pleated sheet in, 821–822
primary structure of, 823
secondary structure of, 823
structure and function of, 18-14–18-15
synthesis of, 18-15–18-23
tertiary structure of, 823, 824, 825
Polypropylene, 816–818
Polysaccharides, 806–808, 17-37
in bacterial cell walls, 23-19, 23-20
Polystyrene, 509, 512, 513, 797
large-scale production of, 758
for Merrifield solid-phase synthesis, 18-17
Polyurethanes, 812
Poly(vinyl acetate), 803
Poly(vinyl alcohol) (PVA), 802–803
Poly(vinyl)chloride (PVC), 796, 797
Porphyrin dyes, 23-43
Porphyrins, 19-14
Positional isomerism, 336, 358
Potassium t-butoxide, 449, 450
Potassium nitrite, alterations in genetic code by, 19-23
Potassium permanganate, 520, 521
PPTA, 809
Prebiotic chemistry, 18-34
Precalciferol, 710
Precor, 292
Precursor, 21-5
Premarin, 764, 765
Priestley, Joseph, 397
Primary alcohols, 109
Primary amines, 90
Primary carbon, 34
Primary structure of a polypeptide, 823
Principal quantum number ($n$), 8
Principle of microscopic reversibility, 348
Prions, 23-31–23-32
Alzheimer's disease and, 23-32
diseases caused by, 23-31–23-32
Procardia XL, 764, 765

## Average Bond Energies (kcal/mole)

*Example:* $CH_4 \rightarrow C + 4\,H$; $\Delta H°/4 = 99$ kcal/mole

| | | | | |
|---|---|---|---|---|
| C—H 99 | C—C 83 | C=C 146 | C≡C 200 | |
| N—H 93 | C—N 73 | C=N 147 | C≡N 213 | |
| O—H 111 | C—O 86 | C=O 179 | C≡O 257 | O=C=O 225 (each) |
| H—H 104 | N—N 39 | N=N 100 | N≡N 226 | |
| | O—O 35 | $^3$(O=O) 119 | | |
| H—F 136 | C—F 108 | F—F 38 | | |
| H—Cl 103 | C—Cl 81 | Cl—Cl 58 | | |
| H—Br 87 | C—Br 68 | Br—Br 46 | | |
| H—I 71 | C—I 51 | I—I 36 | | |

## Bond-Dissociation Energies (kcal/mole)

| Bond | X | | | | | | | |
|---|---|---|---|---|---|---|---|---|
| | H | F | Cl | Br | I | OH | NH$_2$ | CH$_3$ |
| Ph—X | 111 | 126 | 96 | 81 | 65 | 111 | 102 | 101 |
| CH$_3$—X | 105 | 108 | 85 | 70 | 57 | 92 | 85 | 90 |
| CH$_3$CH$_2$—X | 100 | 108 | 80 | 68 | 53 | 94 | 84 | 88 |
| (CH$_3$)$_2$CH—X | 96 | 107 | 81 | 68 | 54 | 94 | 84 | 86 |
| (CH$_3$)$_3$C—X | 93 | — | 82 | 68 | 51 | 93 | 82 | 84 |
| PhCH$_2$—X | 88 | — | 72 | 58 | 48 | 81 | — | 75 |
| H$_2$C=CHCH$_2$—X | 86 | — | 68 | 54 | 41 | 78 | — | 74 |
| H—X | 104 | 136 | 103 | 87 | 71 | 119 | 107 | 105 |
| X—X | 104 | 38 | 59 | 46 | 36 | 51 | 66 | 90 |